APPLICATION OF ACCELERATORS IN RESEARCH AND INDUSTRY

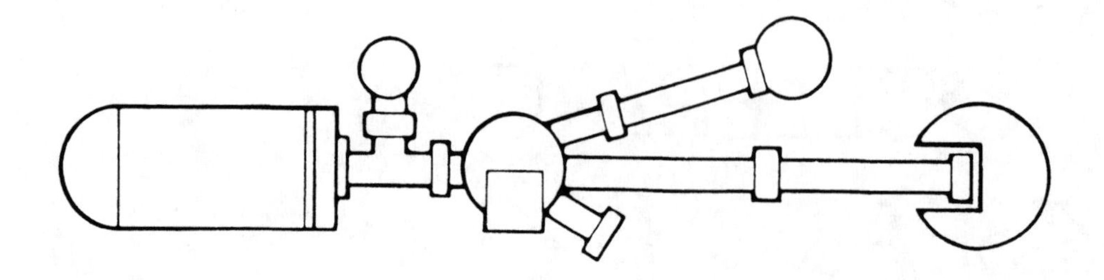

CONFERENCE ON THE APPLICATION OF ACCELERATORS IN RESEARCH AND INDUSTRY

Physics Department • University of North Texas
Denton, Texas 76203-305370

SUPPORTED BY:

U. S. Department of Energy

National Science Foundation

University of North Texas

and

Educational Institute for Superconductivity (EIS)
[formerly the International Industrial Symposium on the Super Collider (IISSC)]

APPLICATION OF ACCELERATORS IN RESEARCH AND INDUSTRY

Proceedings of the Fifteenth International Conference

Denton, Texas November 1998

PART ONE

EDITORS

J. L. Duggan
University of North Texas, Denton

I. L. Morgan
International Isotopes Inc.

American Institute of Physics

AIP CONFERENCE PROCEEDINGS 475

Woodbury, New York

Editors:

Jerome L. Duggan
University of North Texas
Department of Physics
P.O. Box 305370
Denton, TX 76203-5370
U.S.A.

E-mail: jduggan@unt.edu

I. Lon Morgan
International Isotopes, Inc.
3100 Jim Christal Road
Denton, TX 76207

E-mail: ilmorgan@intiso.com

L.C. Catalog Card No. 99-62167
ISSN 0094-243X

ISBN 1-56396-825-8 (Set)
1-56396-890-8 (Part One)
1-56396-891-6 (Part Two)

DOE CONF- 981122

Printed in the United States of America

CONTENTS

PART ONE

Section I. Atomic and Molecular Physics

*see footnote

*Note added in proof: For "A Galilean Invariant Theory of Atomic Collisions: Spectra of Ejected Electrons" by S. Y. Ovchinnikov and J. H. Macek see pp. 1147–1150.

Section II. Nuclear Physics and Radioactive Ion Beam Facilities and Experiments

Section III. Positron Sources, Experiments and Theory

Section IV. Clusters, Fullerenes, Biomolecules

PART TWO

Section VI. Synchrotron Storage Rings, X-Ray Tomography Facilities and Experiments

Section VII. Accelerator Mass Spectroscopy

Section VIII. Neutron Techniques for Nondestructive Analysis, Neutron Sources, FNAA

Section IX. Ion Implantation: Materials Modification, Semiconductors, Facilities, Codes

Section XI. Accelerator Technology, Facilities, New Accelerators, Injectors, Ion Sources, Beam Monitors, Power Supplies, Targets, Codes

Section XII. Radiation Processing, Electron Accelerators, Techniques

Section XIII. Medical Applications: Proton and Neutron Isotope Production, PET, New Facilities, BNCT, Beam Line Modeling

Section XIV. Radiography

Section XV. Teaching, Training

Section XVI. IBIC

Section XVII. Safety

Appendix

Preface

The Fifteenth International Conference on the Application of Accelerators in Research and Industry was held on the campus of the University of North Texas (UNT) November 4-7, 1998. The major sponsors of the conference were The U.S. Department of Energy, The National Science Foundation and UNT. The conference is also a topical conference of the American Physical Society sponsored through the Division of Nuclear Physics. An industrial exhibit show composed of 47 companies that distribute components of interest to accelerator users was held in parallel to the conference for the first two days. Approximately 900 accelerator scientists attended the conference from 47 countries. There were 485 invited papers in 82 sessions and 192 poster papers. Of the invited papers 26 were invited poster papers. Participants who gave invited poster papers were allowed five minutes to deliver an oral presentation during a regular session. In that oral presentation the speaker overviewed what you would see if you visited the poster which was shown at the regular poster session which was held from 6:00 PM until 10:00 PM Friday evening, November 6th in the UNT Environmental Educational Science and Technology building. A social was held during the poster session, which seemed to be well received.

As was mentioned above, there were 82 four-hour sessions in the main conference. Most of the sessions had six to eight invited speakers. The conference was opened with three plenary speakers. Lee Grodzins of MIT, who gave a comprehensive overview of the current automated techniques that, are being used to search for drugs and explosives in luggage and cargo. This was followed by Ted Litherland of Isotrace, Toronto, who gave an overview of Accelerator Mass Spectrometry. This discussion included an overview of the early machines and the most modern completely automated machines that are used today for rapid sample analysis in the pharmaceutical industry. The final plenary talk was by Pat Richard of Kansas State University, who gave a comprehensive discussion of Ion-Atom and Ion-Electron Collision Dynamics. This talk was particularly appropriate since one third of the audience was composed of scientists who do Atomic Physics or use techniques that have been developed from Accelerator Based Atomic Physics.

As has been the case for most of the previous conferences in this series, Accelerator Based Atomic Physics had the most sessions. There were 18 sessions that dealt directly with atomic physics. There were also two well attended sessions on synchrotrons that contained many papers on basic Atomic Physics. In addition, there were 47 poster papers, which fell under the general heading of Atomic Physics. There were eight sessions and 43 poster papers that dealt generally with the subject of Accelerator Technology. These sessions covered topics such as new accelerators, beam handling systems, ion sources, detectors, spectrometers, magnets, etc. Radioactive Beams and Nuclear Physics were the topics for eight sessions, which were very well attended. There were 13 sessions and 56 posters that dealt generally with ion beam analysis. These sessions covered such topics as Rutherford Backscattering Analysis, Particle Induced X-Ray Emission, Elastic Recoil Detection and Nuclear Reaction Analysis, Accelerator Mass Spectrometry and Activation Analysis. Additional sessions and posters covered such topics as; Radiation Processing, Tomography, Single Event Upsets, Ion Implantation, Targetry for Experiments, Production of Radioisotopes for Medicine, BNCT, Radiation Safety, Detectors and Spectrometers, Energy Loss, Clusters, and Free Electron Lasers. Finally, there were two sessions on Pulsed Accelerator Applications and Non-Destructive Testing. These sessions dealt primarily with the use of neutrons for Non-Destructive Analysis.

The conference banquet was held Saturday night, November 7th, in the Silver Eagle Suites of the UNT University Union building. The banquet was followed by a concert from the UNT award winning One O'clock Lab Band.

The editors would like to thank the major sponsoring agencies, namely, The United States Department of Energy, The National Science Foundation, and UNT for their continuing support of this conference series. Thanks are also due to the 47 industrial sponsors. They not only helped financially but also provided two complete days of industrial exhibits that add greatly to the total conference experience that each participant enjoyed. We are also indebted to the program and advisory committees for the excellent slate of invited speakers whose presentations were given at the conference. Our gratitude also goes out to the 94 session chairpersons and co-chairpersons who not only helped to organize many of the sessions but also conducted the sessions at the conference. The editors now have about 500 referees that can be called on to help with the refereeing of the papers. We wish to especially thank the individuals that helped us referee papers for these proceedings. With faxes, email, and overnight mail we accomplished this task rapidly with the outstanding contribution of these referees.

We wish to thank the administrative staff, students and professors at the University of North Texas for the monumental effort of helping us put the conference together and following through with the final publication of these proceedings. Without any question, the single person who does more work on this conference than anyone else is Barbie Stippec. As most of you know, Barbie has been with this conference almost 14 years. By now she knows this community very well. For almost each of the participants she has done something beyond the normal call of duty. She is extremely well organized, computerized, e-mailized, faxized and quite able to find and communicate with the participants no matter where they are. Barbie makes the conference work and for that we owe her a great debt. We also wish to thank Barbie's staff, Sara Wright and Dominic Tran, for doing an outstanding job. We wish to also thank the following: Terry Garland and Sameer Datar for setting up the industrial show; Alan Bigelow and Baonian Guo for planning, setting up and conducting the poster display at the conference; Dana Necsoiu, Mohit Nigam and Songling Li for coordinating the Physics graduate students in the operation of the media equipment for the sessions; and Dan Scheffer who helped with the Poster Session but most of all, maintained the conference web site. In addition, we extend our thanks to all of these students and our other graduate students for helping with the many other tasks associated with a conference of this magnitude. Our thanks is also extended to an employee of the Oak Ridge National Laboratory, Linda Saddiq, who has traveled to Denton many times, at our invitation, to help with the conference registration. We are indebted to Jonathan Reynolds of the UNT Center for Media Production, who took all of the pictures that appear in the front of these proceedings.

Finally, we wish to thank the UNT administration and faculty for the support that has been given this conference series since the first time it was held at UNT in 1974. It is not easy to bring 900 visitors onto a campus during the fall semester and provide classrooms, etc. for the meeting. Without the total support of the University, this would not be possible.

The next conference in this series will take place, as usual, on the campus of UNT, November 2000.

J.L. DUGGAN
I.L. MORGAN

Editors

Dedication

These Proceedings are dedicated to two "very much alive" accelerator scientists who have contributed not only to methodology of ion beam techniques but also to the understanding of these techniques. Each of these individuals have been very important to the accelerator community. It is therefore an honor to dedicate the proceedings to these two outstanding individuals.

Klaus Bethge

Born in Berlin in 1931, Klaus Bethge began studying physics at the Technische Universität Berlin and subsequently earned his diploma and doctoral degrees from the Universität Heidelberg. Early on, ion beams became a life-long interest for Bethge. As assistant professor at the Universität Heidelberg, he researched the development of ion sources, among other projects. Bethge also spent two years at the University of Pennsylvania at Philadelphia where he used the tandem accelerator for nuclear physics studies. Promoted to associate professor after returning to Heidelberg, he continued his work in nuclear physics and ion source physics.

In 1973, Bethge left Heidelberg for the Institut für Kernphysik at the J.W. Goethe Universität, Frankfurt am Main, where he remains today. Ion beams continued to be an important research interest. He broadened this research to include applications of ion beams and accelerators to materials science. Students and staff were gathered together over time to build a center that investigates many areas: ion beam analysis including NRA, CPAA, RBS and channeling, low- and high-energy ion implantation, ion-beam deposition, and sputtering. Besides scientific studies, Bethge and his group made significant contributions in technological and industrial applications of these methods. As a result, numerous collaborations with research laboratories and industries evolved. Bethge is a prolific author, citing over 250 publications, several physics textbooks, and editor of monographs on accelerator physics and applications. He has supervised more than 100 diploma and 60 doctoral theses as well.

Klaus Bethge has lead the field in promoting the usefulness and capabilities of ion beams and accelerators. He has given many invited talks in academia and at scientific conferences. As an active member of the European Physical Society, he founded a divisional conference in 1980 on Nuclear Physics Methods in Materials Research, held in Darmstadt, Germany. Firm in his opinion that applications of accelerator-based techniques and the exchange of information between accelerator developers and users be advanced from a European viewpoint, Bethge founded the European Conference on Accelerators in Applied Research and Technology (ECAART) in 1989 in Frankfurt am Main. Today, ECAART is a successful biennial conference, which draws scientists from around the world.

Klaus Bethge remains a leader in the field of ion beams and accelerators. In recognition of his many achievements, these proceedings are in part dedicated to Klaus Bethge.

Horst Baumann and Friedel Rauch
Institut für Kernphysik
J. W. Goethe-Universität
Frankfurt am Main, Germany

Thomas A. Tombrello, Jr.

One of Tom Tombrello's hallmarks is his boyish delight in new discovery; another is his disregard for traditional boundaries between research disciplines. Thanks in part to these characteristics, Tom has enjoyed a highly productive career grounded in nuclear and ion beam physics, and marked by innovative approaches to problems in fields ranging from astrophysics to geochemistry to materials science. His teaching is similarly innovative. For his many contributions and cutting-edge research, Tom stands out as a leader in the accelerator community.

Tom earned his Ph.D. in nuclear physics from Rice University in 1961 at the age of 24, after which he went to the California Institute of Technology for two years as a Research Fellow. He spent a year at Yale as an assistant professor before returning to Caltech, where he has remained for almost the entire time since. He took a two-year leave of absence in 1987-89 to serve as a vice president and Director of Research for Schlumberger-Doll, Ltd., one of the world's leading oil well service companies. Today he holds the William R. Kenan, Jr. Professorship at Caltech, and is also the recently appointed Chairman of the Institute's Division of Physics, Mathematics, and Astronomy.

As alluded to above, Tom's research interests have covered a broad spectrum, now documented in approximately 300 publications. At Rice and in his early days at Caltech's Kellogg Radiation Laboratory, he used accelerators to study a variety of problems in nuclear physics, many of which had astrophysical implications. When given the opportunity to analyze lunar samples returned by the Apollo missions in the early 1970's, he and student David Leich invented the technique of hydrogen depth profiling using the $^{1}H(^{19}F,\alpha\gamma)^{16}O$ resonant nuclear reaction, and suggested resonant reactions for the analysis of several other elements. These are now among the standard tools for measuring shallow depth profiles of elements such as hydrogen, fluorine, and nitrogen in materials.

The lunar sample analysis also opened up another avenue for Tom's research. In seeking to understand the influence of the space radiation environment on the lunar surface, he became interested in radiation damage in materials. This led to extensive studies of track formation and sputtering by MeV ions in dielectric materials, for example, and investigation of such effects as isotopic fractionation in collisional sputtering processes. In collaboration with Mark Shapiro, he has done a great deal of work with molecular dynamics simulations of sputtering, most recently on sputtering by clusters.

In addition to the important fundamental work on radiation effects in materials, Tom has also pursued many applications ranging from astrophysical to geological to technological. Examples include his work on MeV ion implantation in various semiconducting materials, and on the enhancement of thin-film adhesion by particle irradiation. There is much more. Not surprisingly, the boundaries of Tom's work encompass more than just accelerator-based science. One very interesting project was his development of a network of remotely-operated detectors that monitored radon levels in ground water across Southern California to forecast seismic activity. The breadth of Tom's interests is even reflected in the name of his research group at Caltech, the Basic Applied Physics Group, well known for its numerous Brown Bag preprints.

Tom's many honors and distinguished accomplishments include being named an Alfred P. Sloan Foundation Research Fellow in physics in 1971-73, and a Humboldt Fellow in 1983-84. He is a Fellow of the American Physical Society. He was granted an honorary Ph.D. by Sweden's Uppsala University in 1997, and was selected as a Distinguished Alumnus of Rice University in 1998.

Not only has Tom contributed to the scientific and technological communities through his research, but he has also given his time and expertise as a consultant and member of numerous committees and advisory boards for industry, government, and academia. To name a few, he has served on various committees for the National Academy of Sciences and the American Physical Society, and has advised and consulted for the Los Alamos, Lawrence Berkeley, Lawrence Livermore, and Argonne National Laboratories. He served on the Vice President's Space Policy Advisory Board in 1992. He has also served as an associate editor or editorial advisor for several journals including Radiation Effects and Nuclear Instruments and Methods B.

This dedication would be incomplete without mentioning Tom's contributions as a teacher and mentor. To date he has been the major professor for 33 Ph.D. students, along the way challenging them and infusing them with his enthusiasm, resourcefulness, and innovation in attacking problems. He has been recognized for his undergraduate instruction several times by the Caltech student body. Several years ago he developed Physics 11, a yearlong research course populated by a competitively selected handful of freshmen who are put to work immediately on independent research problems. Students in the course have produced a number of very well regarded publications; we even had the pleasure of having a student present his work at this conference. Additionally, many of these students have gone on to win top awards at Caltech, Goldwater scholarships, and top-rank fellowships to graduate schools. Recently, Tom has started a summer research program in the same vein for outstanding high school students from across the country.

For his many contributions, the breadth and depth of his expertise, and his leadership in accelerator-based research, it is a great pleasure indeed to dedicate these proceedings to Tom Tombrello, a distinguished colleague, mentor, and friend of the accelerator community.

Duncan L. Weathers
Department of Physics
University of North Texas
Denton, Texas 76203-5370, USA

Mark H. Shapiro
Chair, Department of Physics
California State University, Fullerton
Fullerton, California 92834-6866, USA

ORGANIZATIONAL COMMITTEE

ADVISORY COMMITTEE

Marshall CLELAND,	Ion Beam Applications, S.A.
Geoff DEARNALEY,	Southwest Research Institute
Norman HACKERMAN,	The Robert A. Welch Foundation
Robert L. (Cotton) HANCE,	Motorola Incorporated
Homer B. HUPF,	International Isotopes, Inc.
Chris E. KUYATT,	National Institute of Standards & Technology
Gregory A. NORTON,	National Electrostatics Corporation
Robert D. RATHMALL,	Eaton Semiconductor Equipment

INDUSTRIAL SESSION — *Chairman, Ira Lon MORGAN, International Isotopes Inc.*

Gerald D. ALTON,	Oak Ridge National Laboratory
Joe E. BEAVER,	International Isotopes Incorporated
Frank CHMARA,	Peabody Scientific
Cary N. DAVIDS,	Argonne National Laboratory
Barney L. DOYLE,	Sandia National Laboratories
Robert W. HAMM,	AccSys Technology, Inc.
George M. KLODY,	National Elecrostatics Corporation
Kenneth H. PURSER,	University of Toronto
James F. ZIEGLER,	IBM-Research

RESEARCH SESSION — *Chairman, Jerome L. Duggan, University of North Texas*

Frank T. AVIGNONE,	University of South Carolina
Klaus H. BETHGE,	J. W. Goethe-Universität
Bert BRIJS,	IMEC
H. Ken CARTER,	Oak Ridge Associated Universities
Wei-Ken CHU,	University of Houston
Sheldon DATZ,	Oak Ridge National Laboratory
David L. EDERER,	Tulane University
K. O. GROENVELD,	J. W. Goethe Universität
Lester D. HULETT,	Oak Ridge National Laboratory
Keith W. JONES,	Brookhaven National Laboratory
Andreas M. KOEHLER,	Harvard University
James M. LAMBERT,	Georgetown University
Joseph MACEK	University of Nebraska
Carl J. MAGGIORE,	Los Alamos National Laboratory
Klas G. MALMQVIST,	Lund University & Institute of Technology
Floyd D. McDANIEL,	University of North Texas
Themis PARADELLIS,	NCSR 'Demokritos'
David J. PEGG,	University of Tennessee
Friedel RAUCH,	J.W. Goethe Universität
John F. READING,	Texas A&M University
Patrick RICHARD,	Kansas State University
William S. RODNEY,	Georgetown University
Claus ROLFS,	RUHR Universität
M. Eugene RUDD,	University of Nebraska
Emile A. SCHWEIKERT,	Texas A&M University
Ivan A. SELLIN,	Oak Ridge National Laboratory
Stephen M. SHAFROTH,	University of North Carolina
Soey H. SIE,	CSIRO: Exploration & Mining
Joseph R. TESMER,	Los Alamos National Laboratory
Vlado VALKOVIC,	Analysis and Control Technologies Ltd.
S. L. VARGHESE,	University of South Alabama
George VOURVOPOULOS,	Western Kentucky University
Richard L. WALTER,	Duke University
Isao YAMADA,	Kyoto University

SESSION CHAIRPERSONS

Gerald D. Alton, *Oak Ridge National Laboratory*
Bill R. Appleton, *Oak Ridge National Laboratory*
James H. Arps, *Southwest Research Institute*
Frank T. Avignone, III, *University of South Carolina*

Paul Bergstrom, *Lawrence Livermore National Laboratory*
Klaus H. Bethge, *University of Frankfurt*
Alan W. Bigelow, *University of North Texas*
Bert Brijs, *IMEC vzw MAP/ARS*
Joachim Burgdoerfer, *University of Tennessee*

Wei-Kan Chu, *University of Houston*
William T. Chu, *Lawrence Berkeley National Laboratory*
Sam J. Cipolla, *Creighton University*
Marshall R. Cleland, *Ion Beam Applications, s.a.*
Benjamin C. Craft, *Louisiana State University*

William E. Dance, *Consultant, Neutron Radiology*
Geoff Dearnaley, *Southwest Research Institute*
Barney L. Doyle, *Sandia National Laboratory*
William M. Dunn, *International Isotopes Incorporated*

Alan K. Edwards, *University of Georgia*
Mohamed El Bouanani, *University of North Texas*

James M. Feagin, *California State University @ Fullerton*
Richard J. Fortner, *Lawrence Livermore National Laboratory*

Kenneth S. Grabowski, *Naval Research Laboratory*
Richard Gregory, *Motorola Inc.*
Noel A. Guardala, *Naval Surface Warfare Center*

Marianne E. Hamm, *AccSys Technology*
Robert W. Hamm, *AccSys Technology*
R. L. "Cotton" Hance, *Motorola Inc.*
Yale D. Harker, *Idaho National Engineering and Environmental Laboratory*
J. Frank Harmon, *Idaho State University*
Richard D. Hichwa, *University of Iowa*
Homer B. Hupf, *International Isotopes Incorporated*

Daryush Ila, *Alabama A & M University*

Amitabh Jain, *Texas Instruments Incorporated*
William D. James, *Texas A & M University*
Keith W. Jones, *Brookhaven National Laboratory*

Joseph Keenan, *Texas Instruments Incorporated*
William E. Kieser, *University of Toronto*
David L. Knies, *Naval Research Laboratory*
Andreas M. Koehler, *Harvard University*

James M. Lambert, *Georgetown University, Retired*
Richard C. Lanza, *Massachusetts Institute of Technology*
Gregory Lapicki, *East Carolina University*
JiaRui Liu, *University of Houston*

Hans J. Maier, *University of Munich*
Carl J. Maggiore, *Los Alamos National Laboratory*
Daniel K. Marble, *Tarleton State University*
Floyd D. McDaniel, *University of North Texas*
Rahul Mehta, *University of Central Arkansas*
Marcus H. Mendenhall, *Vanderbilt University*
James K. Milam, *CTI Incorporated*
Gary E. Mitchell, *North Carolina State University*

David W. Nigg, *Idaho State University*

Themis Paradellis, *NCSR 'Demokritos'*
David J. Pegg, *University of Tennessee*
Randolph S. Peterson, *University of the South*
Darden Powers, *Baylor University*
Jack L. Price, *Naval Surface Warfare Center*
Kenneth H. Purser, *Southern Cross Corporation*

Carroll A. Quarles, *Texas Christian University*

Robert D. Rathmell, *Eaton Semiconductor Equipment*
John F. Reading, *Texas A & M University*
Patrick Richard, *Kansas State University*
William S. Rodney, *Georgetown University*

David J. Schyler, *Brookhaven National Laboratory*
Emile A. Schweikert, *Texas A&M University*
Ivan A. Sellin, *University of Tennessee*
Stephen M. Shafroth, *University of North Carolina @ Chapel Hill*
Soey H. Sie, *CSIRO Exploration and Mining*
Susan B. Sinnott, *University of Kentucky*
Janet M. Sisterson, *Harvard University*
Michael S. Smith, *Oak Ridge National Laboratory*
Bertram Somieski, *Oak Ridge National Laboratory*

William L. Talbert, *Amparo Corporation*
James R. Tesmer, *Los Alamos National Laboratory*
Larry H. Toburen, *East Carolina University*
Thomas A. Tombrello, *California Institute of Technology*

Vlado Valkovic, *Analysis and Control Technologies, Ltd.*
S. L. Varghese, *University of South Alabama*
George Vourvopoulos, *Western Kentucky University*

Richard L. Walter, *Duke University*
H. R. J. Walters, *The Queen's University*
Duncan L. Weathers, *University of North Texas*
Alex Weiss, *University of Texas @ Arlington*
Stephen A. Wender, *Los Alamos National Laboratory*
Colm T. Whelan, *University of Cambridge*
Gary D. White, *Northwestern State University of Louisiana*

Isao Yamada, *Kyoto University*

Chuck Yarling, *Ion Beam Press*

FIFTEENTH INTERNATIONAL CONFERENCE ON THE APPLICATION OF ACCELERATORS IN RESEARCH AND INDUSTRY
******CAARI'98******
4-7 NOVEMBER 1998

INDUSTRIAL SPONSORS

A & C T, Prilesje 4, 10 000 Zagreb, Croatia
Accelsoft Inc., P O Box 2813, Del Mar CA 92014
Accsys Technology Inc., 1177 Quarry Lane Suite 208, Pleasanton CA 94566
Alcatel: See THT Sales Company
Alphatech International Ltd & Dehnel Consulting Ltd., 1512 Falls Street, Nelson, B.C. Canada VIL 1J4 (P O Box 201, Nelson, B.C. Canada V1L 5P9)
Amptek Inc., 6 De Angelo Drive, Bedford MA 01730
Bioscan Inc., 4590 MacArthur Boulevard N.W., Washington DC 10007
Ceramaseal: See THT Sales Company
Charles Evans & Associates, 301 Chesapeake Drive, Redwood City CA 94063
Continental Electronics, P O Box 2700879, Dallas TX 75227
Copley Controls Corp., 410 University Avenue, Westwood MA 02090
CTI PET Systems, Inc., 810 Innovation Drive, Knoxville TN 37932
Dexter Magnetics, 48460 Kato Road, Fremont CA 94538
EEV, Inc., 4 Westchester Plaza, Elmsford NY 10523
EG&G ORTEC, 100 Midland Road, Oak Ridge TN 37830
Eurisys Mesures Inc., P O Box 6258, Oak Ridge TN 37831-3848
European Accelerator Consortium, University of Surrey Ion Beam Center, Guildford, GU2 5XH, England
FusionStar Daimler-Benz Aerospace, Space Infrastructure Division, Center Trauen, Eugen-Saenger Str-52, 29328 Fassberg, Germany
GMW Associates & Danfysik A/S, P O Box 2578, Redwood City CA 94064
High Energy Systems Division of American Science & Engineering, 3300 Keller St., Bldg 101, Santa Clara CA 95054
High Voltage Engineering Europa B.V., P O Box 99, 3800 AB Amersfoort, The Netherlands
International Isotopes Incorporated, 3100 Jim Christal Road, Denton TX 76207
Ion Beam Applications S.A., Chemin du Cyclotron, 3, B-1348 Louvain-la-Neuve Belgium
Iso-Tex Diagnostics, P O Box 909, Friendswood TX 77546
Isotopes Products Laboratories, 1811 N. Keystone Street, Burbank CA 91504
JP Accelerator Works Inc., 2245 47th Street, Los Alamos NM 87544
Kurt J. Lesker Company, 1515 Worthington Avenue, Clairton PA 15025
Linac Systems, 2167 N. Highway 77, Waxahachie TX 75165
Magnet Sales & Manufacturing Company, 11248 Playa Court, Culver City CA 90230
Maxwell Technologies Inc., 4949 Greencraig Lane, San Diego CA 92123
MEV Technology 5150 Shadow Estates, San Jose CA 95135
MF Physics Corporation, 5074 List Drive, Colorado Springs CO 80919
Motorola Inc., 3501 Ed Bluestein Blvd., MD K-10, Austin TX 78721
National Electrostatics Corp., P O Box 620310, 7540 Graber Road, Middleton WI 53562-0310
Newton Scientific Inc., 7 Red Coach Lane, Winchester MA 01890
Nor-Cal Products Inc: See THT Sales Company
North Star Research Corporation, 4421 McLeod Road, N.W., Suite A., Albuquerque NM 87109-2217
Nuclear Research Corporation, 125 Titus Avenue, Warrington PA 18976
Oxford Microbeams Ltd., 3 Lawrence Road, Oxford OX4 3ER United Kingdom
Peabody Scientific, P O Box 2009, Peabody MA 01960
Physical Electronics: See THT Sales Company
Potentials Inc., 1704 Hydro Drive, Austin TX 78728
Princeton Scientific, P O Box 143, Princeton NJ 08542
SODERN, 20, Avenue Descartes – B.P. 23, 94451 Limeil-Brevannes Cedex France
Spectrum Technologies, 305 Oak Ridge Turnpike, Oak Ridge TN 37830
Thomson Components and Tubes Corporation, 40G Commerce Way, Totowa NJ 07511
THT Sales Company, 13438 Floyd Circle, Dallas TX 75243
U.S. Department of Energy,The Office of Medical, Industrial, and Research Isotope Supply, 19901 Germantown Road, Germantown MD 20874
Varian Vacuum Products, 2340 Glen Ridge Drive, Highland Village TX 75067
VAT Inc., 500 West Cummings Park, Woburn MA 01801
Whistle Soft Inc., 168 Dos Brazos, Los Alamos NM 87544

In Appreciation
Oxygen Depth Pro

SECTION I

ATOMIC AND MOLECULAR PHYSICS

Molecular Effects in K-Shell Ionization by Charged Particles

L.H. Toburen, A.M. Williams, C.R. Moreau, J.L. Shinpaugh, and E. Justiniano

Department of Physics, East Carolina University, Greenville, North Carolina 27858

Measurements of atomic inner-shell ionization cross sections for low-Z elements are complicated by small fluorescence yields and because elements do not occur naturally in atomic form. These conditions favor the measurement of Auger-electron yields to determine vacancy formation and they require the use of molecular gas targets. Because molecular structure has little influence on the energy levels of inner shells it is expected to have little effect on K-shell ionization probabilities. On the other hand, Auger-electron transitions involve molecular orbitals which introduce molecular effects in the spectra and could potentially influence yields. Variations in measured atomic K-shell ionization cross sections for different molecular targets have generally been found to be small (10-15%) and explained by simple geometric effects. Some recent measurements have, however, exhibited molecular effects as large as a factor of three which are beyond simple explanations. This has revived our interest in trying to further quantify the extent of molecular effects in K-shell ionization. We have employed a new method using well tested theoretical benchmarks to derive atomic cross sections from molecular targets to study K-shell ionization of carbon and fluorine in a number of fluorocarbon molecules excited by MeV protons and lithium ions. The atomic cross sections derived from these measurements are not found to exhibit effects of molecular structure within the experimental uncertainties of approximately 10% for protons and 25% for lithium ions.

Introduction

The study of inner-shell ionization by charged particle impact has been the subject of study for more than three decades. An accurate determination of the cross sections for inner shell vacancy formation can provide definitive tests of collision theory and the data are useful in a number of applied research areas from trace element analysis to plasma physics. The development of efficient low-energy electron detectors and high-resolution x-ray detectors in the 1960's enabled accurate measurement of Auger-electrons and x-rays. The remaining challenge was to associate measured yields with the probability that an inner shell vacancy had been formed. This association requires an accurate knowledge of the fluorescence and Auger yields and an assessment of the effects of the atomic matrix on the production and/or decay of the inner-shell vacancies.

For low-Z elements, the small and uncertain nature of fluorescence yields can be a limiting factor in the accuracy of inner-shell vacancy determinations made using x-ray yield measurements. In addition, soft x-rays are easily absorbed in the sample and detector windows leading to large changes in window absorption with small fluctuations in the x-ray energy. These factors have led to application of Auger-electron spectroscopy as a primary tool for the study of inner-shell vacancy production in low-Z elements where, for second-row elements, the Auger yield is nearly unity.

The short range of electrons in solids and the need to perform Auger-electron energy analysis and detection in a vacuum presents a number of experimental challenges for the determination of absolute Auger electron yields. The use of differentially pumped gas targets to reduce interactions of electrons in the target prior to detection makes accurate determination of the target density difficult. In addition, the target element is generally a constituent of a molecule which can lead to possible effects of the chemical environment on the Auger decay process. Measurements of target densities based on pressure measurement is complicated by the diffusion kinetics of differentially pumped targets, non-ideal-gas properties of some molecular species at temperatures and pressures common to experiments, and possible variations in the sensitivity of pressure-measuring devices to different molecular species. Still, by careful attention to experimental detail, absolute K-shell ionization cross sections have been published with quoted uncertainties of order 25% and smaller uncertainties in relative yields.

Effects of valence electron structure on the probability of inner shell ionization are expected to be small, e.g., atomic K-shell binding energies show little variation with the structure of the molecule in which they reside[1]. Density effects, the possibility that the proximity of additional atoms in a molecule might increase the probability of a second atom being excited in the same collision, are also expected to be small. Simple calculations, based on interatomic separation, K-shell radii, and use of the Massy criterion to estimate interaction distances needed to transfer sufficient energy to ionize an inner shell, suggest that the K-shell

CP475, *Applications of Accelerators in Research and Industry*,
edited by J. L. Duggan and I. L. Morgan

ionization cross sections might increase by as much as a couple of percent for each additional "like" atom in the molecule owing to this increased "density".[2] Such an increase has not been observed, suggesting this calculation based on geometric considerations is likely an overestimation. *Spectra* of Auger electrons are, however, markedly influenced by the molecular structure;[1] Auger transition energies depend on the outer shell molecular binding energies. Still, if all emitted Auger electrons are detected, i.e., one determines the total intensity of the Auger distribution resulting from an initial inner-shell vacancy, this yield should be representative of the probability of vacancy formation. Fluorescence yields are known to vary for transitions to different electronic configurations, however the variations are generally small, or represent a small fraction of the total decay probability. Thus for low-Z elements ($Z<10$), variations of more than a few percent from an overall Auger yield of unity is not expected.

Measurements of inner-shell vacancy production for atoms in different molecules have not generally shown a statistically significant effect of the molecular environment. Although some of the earliest work on carbon x-ray yields from electron and charged particle impact suggested sizable variations in yields with the molecular environment,[3,4] the uncertainties in the measured yields were also large. Subsequent x-ray yield measurements by Bissinger et al.[2] using improved experimental methods to obtain relative uncertainties of about 2% showed only small variations among a wide range of molecular targets. They measured x-ray yields using a thin window proportional counter and used nuclear Rutherford scattering at 90 degrees to normalize x-ray yields to target density. Their data exhibited a definite trend, with the carbon x-ray yield per atom decreasing by about 10% in going from hydrocarbons to CO and CO_2 targets. They were able to successfully model this trend based on the number of valence electrons with *p* character available to fill the initial vacancy.

By far the most extensive data exploring effects of molecular structure on inner-shell ionization are based on Auger electron yields where uncertainties due to small fluorescence yields are negligible. In general there is excellent agreement between investigators and little indication of any sizable effects of target molecular structure. Ionization of carbon by protons in CH_4, C_2H_2, C_2H_4, C_2H_6, CO, and CO_2 by Toburen;[5] in CH_4 and CO_2 by Stolterfoht;[6] and in CH_4, C_2H_2, C_2H_6, CF_4, C_2F_6, CO, CO_2, CCl_4, and $(CH_3)_2NH$ by McElroy et al.[7] gave cross sections that were all within experimental error. There were some indications of the trends observed by Bissinger et al.,[2] but the experimental errors were too large (10-20%) for definitive conclusions. Measurements of C K-shell ionization by protons in CH_4, CF_4, and CCl_4 by Matthews and Hopkins[8] resulted in Auger electron yields from CCl_4 significantly smaller than the other two molecules and, although not statistically significant, the data of McElroy et al.[7] also gave smaller values for CCl_4. Matthews and Hopkins suggest that in large symmetric molecules, such as CCl_4 and SF_6, when inner-shell ionization occurs in the central atom the exiting Auger electron can be inelastically scattered from the electrons of the outer atoms. This scattering might reduce the yield of Auger electrons detected in the Auger spectrum. This model was applied by Matthews and Hopkins and found in excellent agreement with the decreased yields found in CCl_4 relative to CH_4.

Measurements of K-shell ionization in other low-Z elements are more limited. Chaturverdi et al.[9] published cross sections based on Auger-electron yields for K-shell ionization of fluorine by fast protons in CF_4, C_2F_6, BF_3, SF_6, and TeF_6; variations were within experimental uncertainties. Measurements by Toburen et al.[10] for ionization of BF_3, SF_6 and TeF_6 by 0.3 - 2 MeV protons also gave fluorine cross sections which agreed among molecules to within experimental uncertainty of about 20%. More recently, however, Ghebremedhin et al.[11] measured cross sections for K-shell ionization of fluorine in different fluorocarbons by He^+ impact and found greater than a factor of two differences in the fluorine cross sections between $C_2H_2F_2$ and the molecules C_4F_8 and C_2F_6; the latter being smaller than the former. Their estimate of contributions from inelastic electron scattering were unable to account for such large variations and the trend is opposite to that expected based the increased density effect of molecular atoms. The possibility of large molecular effects in certain fluorocarbon-based K-shell ionization cross sections has encouraged us to revisit the subject of inner-shell ionization. We have focused on K-shell ionization of carbon and fluorine in a number of different molecules. Our objective was to study a wider range of molecules than previously studied, to use proton impact to minimize effects of multiple ionization and projectile electrons, and to calibrate the system using well known and tested theory for absolute ionization cross sections to eliminate the need for direct determination of target density. The latter can be especially difficult for large fluorocarbon molecules with low vapor pressures.

Experimental methods

Our experimental approach was to measure the spectra for electrons ejected by fast proton impact on a crossed beam molecular target and to use the well known Rutherford energy-loss cross sections for establishing the absolute cross sections. Rutherford cross sections relate to collisions with free electrons, but have been shown to be a reliable estimate for energy-loss involving ejected electron energies much greater than their initial binding energy and less than the kinematic maximum energy that can be transferred in a collision between a fast proton a free electron.[12] The ratio of measured singly differential cross sections for electron ejection $\sigma(\varepsilon)$, where ε is the ejected electron energy, to the corresponding Rutherford cross section (per target electron) should be equal to the number of electrons in the target atom or molecule. This relationship has been found true for a wide range of low-Z atoms and molecules;[12] an example, for 1.0 MeV protons on neon, is shown in Figure 1. This feature is used to normalize measured yields to absolute cross sections in the current measurements.

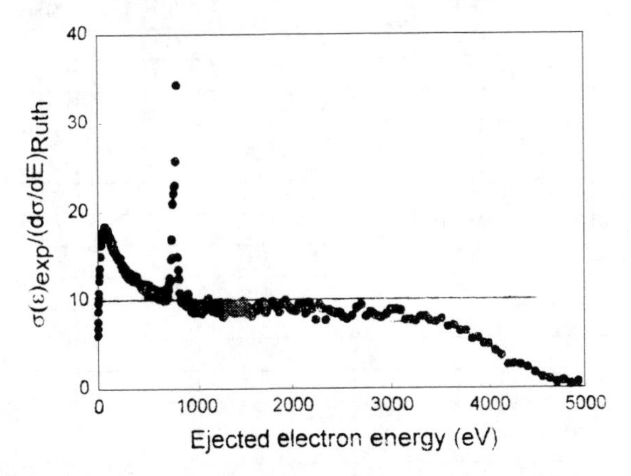

Figure 1. Ratio of the measured singly-differential ionization cross section to the Rutherford energy loss cross section for ionization of neon by 2 MeV protons. The data is from Toburen.[12]

In our measurements, doubly differential cross sections, differential in ejected electron energy and emission angle, were measured for a series of angles from 10 to 140 degrees. These yields were integrated with respect to emission angle for comparison to the Rutherford energy-loss cross sections. Measurements were made for 2 MeV proton impact on a number of fluorocarbon gases using the apparatus previously used in studies at the Pacific Northwest National laboratory[13] and currently installed at the East Carolina University 2-MeV tandem Van de Graaff accelerator. Electron energy spectra were obtained using a cylindrical-mirror electrostatic energy analyzer that could be positioned to measure electrons ejected at angles of 10 to 140 degrees with respect to the proton beam. The target was a crossed beam of target molecules generated by passing the gas through a collimated holes structure at a constant driving pressure. The doubly differential electron yields were integrated with respect to emission angle to obtain singly differential yields and converted to absolute cross sections by comparison to the corresponding Rutherford energy-loss cross sections at ejected electron energies of approximately 1 keV. The K-shell ionization cross sections were obtained by integrating the Auger electron spectra after background subtraction. The continuum background subtraction was performed by fitting the continuum above and below the Auger spectrum by a 2nd order polynomial to provide an estimate of the continuum in the region of the Auger electron contribution. The integrated Auger-electron yields thus obtained (per molecule) were divided by the number of appropriate atoms per molecule to obtain atomic cross sections. The results are shown in Table 1 along with previous measurements where several molecular targets have been studied. In the case of the molecules CF_4, CH_3F, CHF_3, C_2F_6, and C_4F_8, our proton measurements were repeated after some revisions to the apparatus. The reproducibility was within about ±7%; the data in Table 1 are an average of these two measurements. Some preliminary measurements were also made of the Auger electron yields for Li ion impact at 0.42 MeV/u. These data are shown in Table 1, scaled by Z^2, for comparison to similarly scaled He^+ data.

Discussion

The atomic K-shell ionization cross sections obtained in the ECU measurements for proton impact show very little variation from molecule to molecule and these variations are well within the estimated experimental precision of 10%. Data for carbon K-shell ionization by lithium ion impact show greater variation from molecule to molecule than the proton data. The source of these variations is not obvious and might simply be a result of the increased difficulty in background subtraction (Auger-to-continuum ratios decrease with decreasing ion velocity), or instabilities in the system during the prolonged data accumulation time required for the weak beam currents available for Li^{2+}. Our conclusion is that these variations are a result of the former because the results from the same electron spectrum for the analysis of the fluorine Auger yields do not exhibit similar variations; fluorine spectra occur at somewhat higher energies where the continuum background is smaller. The largest variation in the lithium ion data for fluorine is an increase of about 50% in the yield for C_2H_6 over C_4F_8. This is in contrast to the work of Ghebremedhin et al.[11] whose largest cross section was for $C_2H_2F_2$, about 130% larger that C_4F_8, and his results for C_2F_6 and C_4F_8 are within experimental error at this ion velocity. In conclusion, there does not appear to be any consistent evidence of molecular effects in the K-shell ionization data greater than experimental uncertainty, with the exception of the fluorine cross sections of Ghebremedhin et al.[11]

Table 1. Cross Sections for K-shell Ionization of Carbon and Fluorine in Various Molecules by Fast Charged Particles. Energies are given in MeV/u and Cross Sections Are Given in Units of 10^{-18} cm^2

	Carbon Cross Sections/z^2_{ion}					Fluorine Cross sections/z^2_{ion}			
Ion	**$Li^{2+}/9$**	**$He^{+}/4$**	**H^{+}**	**H^{+}**	**H^{+}**	**H^{+}**	**$He^{+}/4$**	**H^{+}**	**$Li^{2+}/9$**
Author	**ECU**	**McElroy[7]**	**Toburen[5]**	**ECU**	**Bissinger[2]**	**Chaturvedi[9]**	**Ghebremedhin[11]**	**ECU**	**ECU**
Energy (MeV/u)	**0.42**	**0.5**	**1.0**	**2.0**	**2.0**	**1.5**	**0.5**	**2.0**	**0.42**
Target									
CH_4		1.20 (0.10)	1.08 (0.07)		1.0[c]				
C_2H_2		1.16 (0.07)	1.00 (0.06)		0.83				
C_2H_4			1.06 (0.04)		0.88				
C_2H_6		1.24 (0.10)	1.10 (0.07)		0.95				
C_4H_8					0.90				
CH_3F	0.50			0.57				0.116	0.047
CH_2F_2								0.125	
CHF_3	.69			0.54				0.119	0.052
CF_4	.394	0.93 (0.15)		0.47	0.85	0.161 (0.008)		0.125	0.050
$C_2H_2F_2$							0.070 (0.013)	0.122	
C_2F_6	.460	0.99 (0.10)		0.47	0.76	0.156 (0.008)	0.041 (0.007)	0.131	0.078
C_4F_8	.584			0.52			0.032 (0.006)	0.121	0.054 0.058 (Li^{+})
BF_3						0.151 (0.008)			
SF_6						0.144 (0.007)			
TeF_6						0.166 (0.008)			
CO		1.17 (0.12)	0.98 (0.07)		0.84				
CO_2		1.00 (0.15)	0.96 (0.05)		0.88				
O_2									
CCl_4		0.95 (0.15)							
$(CH_3)_2NH$		1.35 (0.09)							

[c]Estimated relative uncertaintites are ±2%.

The authors thank the USDOE/PNL for the loan of research equipment and one of us (JLS) wishes to acknowledge financial support from the Research Corporation and East Carolina University

References

1. Seigbahn, K., Nordling, C., Johansson, G., Hedman, J., Hedén, P. F., Hamrin, K., Gelius, U., Bergmark, T., Werme, L. O., Manne, R., and Baer, Y., *ESCA Applied to Free Molecules*, Amsterdam, North-Holland Publishing Co., 1969.
2. Bissinger, G., Joyce, J. M., Tanis, J. A., and Varghese, S. L., *Phys. Rev. Lett.* **44**, 241-244 (1980).
3. Harrison, K. G., Tawara, H., and DeHeer, F. J., *Chem. Phys. Lett.* **14**, 285-290 (1972).
4. Harrison, K. G., Tawara, H., and DeHeer, F. J., *Physica* **66**, 16-32 (1973).
5. Toburen, L. H., *Phys. Rev. A* **5**, 2482-2487 (1972).
6. Stolterfoht, N., "Cross Sections for Inner Shell Ionization of Gaseous Molecules by 50- 500 keV Protons," in Proceeding of the International Conference on Inner Shell Ionization Phenomena and Future Applications, ed. by RW Fink, ST Manson, JM Palms, and P Venugopala, CONF-720404 (USAEC, Oak Ridge, TN, 1973) pp 979-988.
7. McElroy, R. D., Jr., Ariyasinghe, W. M., and Powers, D., *Phys. Rev. A* **36**, 3674-3681 (1987).
8. Matthews, D. L., and Hopkins, F., *Phys. Rev Lett.* **40** 1326-1329 (1978).
9. Chaturvedi, R. P., Lynch, D.J., Toburen, L. H., and Wilson, W. E., *Phys. Lett.* **61A**, 101-103 (1977).
10. Toburen, L. H., Wilson, W. E., and Porter, L. E., *J Chem Phys.* **67**, 4212-4221 (1977).
11. Ghebremedhin, A., Ariyasinghe, M. W., and Powers, D., *Phys. Rev. A* **53**, 1537-1544 (1996).
12. Toburen, L. H., "Atomic and Molecular Physics in the Gas Phase," in *Physical and Chemical Mechanisms in Molecular Radiation Biology*, ed. by WA Glass and MN Varma, New York, Plenum Press, 1991. Pp 51-94
13. Toburen, L. H., DuBois, R. D., Reinhold, C. O., Schultz, D. R., and Olson, R. E., *Phys. Rev. A* **42**, 5338-5347 (1990).

Description of ionization in ion-atom collisions from low to intermediate energies

B. Pons[1], L.F. Errea[2], C. Harel[1], C. Illescas[2], H. Jouin[1], L. Mendez[2] and A. Riera[2]

[1] *Laboratoire des Collisions Atomiques, CPTMB (UPRES 5468 CNRS), Université de Bordeaux-I, 351 Cours de la Libération, F-33405 Talence (France)*
[2] *Departamento de Quimica, Universidad Autonoma de Madrid, Cantoblanco, E-28049 Madrid (Spain)*

A careful analysis of the ionization mechanism at low and intermediate energies is still lacking. It is generally agreed upon that this study should be performed in the frame of a close-coupling formalism. Using a Classical Trajectory Monte Carlo study of the ionization mechanism at intermediate energies as a guideline, we recently proposed a new scheme for molecular close-coupling calculations, in terms of a "triple-center" basis of adiabatic molecular orbitals, augmented with a set of gaussian pseudostates centred at a point between the nuclei. The construction of this new basis and its application to close-coupling calculations for ion-atom collisions are presented.

INTRODUCTION

In fast ion-atom collisions (where the velocity v of the impinging ion is much bigger than the velocity v_e of the active target electron), ionization mechanisms are rather well understood because of the ability of perturbative approaches to describe experimental results [1]. In the low ($v<v_e$) and intermediate ($v \approx v_e$) velocity ranges, they still remain misunderstood: similar experiments have drawn opposite conclusions (in particular on the existence of the so-called saddle-point mechanism, see [2,3] and references therein) and theoretical treatments are difficult because they have to accurately take into account the strong couplings giving rise to charge exchange, excitation and ionization. Consequently, perturbative approaches are no longer valid and close-coupling expansions are more adequate. In practice, most of these expansions are in terms of atomic basis, including L^2-square integrable wavefunctions of positive energy, usually called pseudostates, to represent the ionization channel. Intrinsic properties and failures of atomic expansions can be found in [4-10] and references therein.

In this work, we focus on an alternative to atomic expansions, which is the *molecular close-coupling* formalism. This method usually employs a basis of adiabatic molecular wavefunctions (called OEDMs for single electron systems), modified by a common translation factor (CTF) [11]. We have recently shown that this approach can accurately reproduce excitation and electron-loss (charge exchange plus ionization) cross sections at intermediate energies [11,12,13]. At first, this latter feature appeared somewhat as a paradox, because of the lack of a representation of the molecular continuum in the basis. We showed [13] that there is no paradox, since a bound-state molecular basis is able to describe unbound states of the moving atoms, hence ionization. We found also that such a description of ionization is incomplete; the expansion should include pseudostates to properly separate the capture and ionization fluxes.

To obtain some quantitative requirements on pseudostates, Illescas *et al.* undertook a classical trajectory Monte Carlo study of He^{2+}+H(1s) collisional mechanisms at intermediate energies [14,15]. The evolution of ionizing densities during the collision pointed out the usefulness of wavefunctions centred on a third center. The specific form of these functions are then determined by a fitting procedure of the classical ionizing densities. In this work, we introduce the third-centred pseudostates in the context of a molecular expansion; by analogy to the well-known three-center atomic orbitals (3CAO) calculations [9,10], we call the method "triple-center" (3CMO).

DESCRIPTION OF IONIZATION BY MOLECULAR EXPANSIONS

We consider the one electron problem B^{Q+}+A, and an impact parameter approach in which the nuclei follow rectilinear trajectories with constant velocity **v** and impact parameter **b**; the electronic state of the colliding system is a solution of:

$$\left(H - i\left.\frac{\partial}{\partial t}\right|_o\right)\Psi = 0 \qquad (1)$$

where H is the fixed nuclei electronic Hamiltonian and the derivate $\partial/\partial t|_O$ is taken by keeping the electronic

CP475, *Applications of Accelerators in Research and Industry*, edited by J. L. Duggan and I. L. Morgan

coordinates fixed with respect to an origin O situated at distances pR and qR of nuclei A and B, respectively (p+q=1).

In molecular close-coupling expansions,

$$\Psi(\mathbf{r},t) = e^{iU(\mathbf{r},t)} \sum_k a_k(t)\chi_k(\mathbf{r},R)e^{-i\int E_k(t')dt'} \quad (2)$$

where e^{iU} is a CTF that accounts for the momentum transfer problem [11], R the internuclear distance at time t and $E_k = \langle \chi_k | H | \chi_k \rangle$.

Atoms A and B are moving with respect to the origin O. Accordingly, in the region of asymptotic orthogonality, a traveling basis function $e^{iU}\chi_k$ such that its energy is negative with respect to moving atom A may be taken to represent elastic or excitation channels. If its energy with respect to moving atom B is negative, it may be taken to represent exchange channels. When both energies are positive, it describes ionization.

In [13], we identified the description of ionization by the molecular expansion (2). First, a part of the electron cloud does not succeed in following the fast rotation of the internuclear axis at short R; this part is taken into account through a superposition of σ and π orbitals, involving mainly rotational transitions. As time elapses (at larger R), this flux is left behind the nuclei, in the internuclear region, and is promoted up to the highest states included in the expansion through a relay-race mechanism, involving mainly radial transitions. This whole scheme corresponds to the account of genuine ionization when the nuclear velocity is high enough that the energy of the traveling orbitals involved is locally positive with respect to both moving atomic reference frames. Nevertheless, the molecular basis are always limited and the promotion of ionizing flux ends in an unphysical trapping by bound states (both at short and large R) as soon as no more orbitals of positive atomic energies are available. Accordingly, pseudostates are required to improve the molecular basis at all internuclear distances.

PSEUDOSTATES

In [13], we noted that the OEDMs involved in the ionization mechanism at large R present a maximum electronic density between both nuclei, as well as positive energies with respect to both moving atomic reference frames. In their classical treatment of He^{2+}+H(1s) collisions, Illescas *et al.* [14,15] also stressed that a sizeable part of the ionizing density is found in the internuclear region. We thus choose the functional form of the pseudostates such that these qualitative properties are fulfilled:

$$G(\mathbf{r}) = x^{n_1}\left[z+(p-g)R\right]^{n_2} e^{-\gamma_n r_G^2} \quad (3)$$

with :

$$r_G^2 = x^2 + y^2 + \left[z+(p-g)R^2\right]. \quad (4)$$

These gaussian pseudostates are centred at distance gR of the target. {x,y,z} are the electronic coordinates in the rotating molecular frame with z along the internuclear axis and x-axis element of the collisional plane. The parameters γ_n are in geometrical series, which can be different for each (n_1,n_2) symmetry :

$$\gamma_n = \gamma_0\beta^n. \quad (5)$$

The third-centre basis is thus defined by the parameters g, γ_0, β, n_{max} such that $0 \le n \le n_{max}$ and L_{max} such that $0 \le n_1+n_2 \le L_{max}$.

Further, quantitative requirements on pseudostates are obtained by least-square fits of the classical ionizing densities, along representative nuclear trajectories. The fixed-laboratoty coordinate {X,Y,Z} space is divided in N^3 cubic cells of dimension δ with its origin on the target such that $-N/3 \le X,Y,Z \le 2N/3$. The simplest procedure is to define a classical counterpart of the electronic wavefunction :

$$\Psi^{clas}(\mathbf{r_i},v,b,t) = \sqrt{\rho(\mathbf{r_i},v,b,t)} \quad (6)$$

where $\rho(\mathbf{r}_i,v,b,t)$ is the classical ionizing density at the point $\mathbf{r_i}$ in the ith cell. Then we fit Ψ_{clas} with a real molecular counterpart, written in terms of OEDMs plus pseudostates, without translation factor :

$$\Psi^{mol}(\mathbf{r},t) = \sum_k c_k(t)\chi_k(\mathbf{r},t) \quad (7)$$

where the parameters $\{c_k\}$ are obtained by minimizing :

$$\kappa = \sum_{i=1}^{N^3}\left[\Psi_i^{clas} - \Psi_i^{mol}\right]^2 \quad (8)$$

and the degree of approximation of the classical density by the close-coupling expansion can be gauged from the overlap :

$$S(v,b,t) = \frac{\langle \Psi^{clas} | \Psi^{mol} \rangle}{\Omega^{clas}\Omega^{mol}} = \frac{\langle \Psi^{clas} | \Psi^{clas} \rangle + \langle \Psi^{mol} | \Psi^{mol} \rangle - \kappa}{2\Omega^{clas}\Omega^{mol}} \quad (9)$$

where $\Omega^{clas,mol} = \sqrt{\langle \Psi^{clas,mol} | \Psi^{clas,mol} \rangle}$. The closer S is to unity, the more complete and adequate the augmented molecular basis to represent ionization.

" TRIPLE-CENTRE " MOLECULAR CALCULATIONS

First implementations of our 3CMO method concern H^+, He^{2+} and Li^{3+}+H(1s) collisions. Capture and excitation states are represented by OEDMs and the CTF used in expansion (2) is of the form :

$$U(\mathbf{r},t) = f(\mathbf{r},R)\mathbf{v}.\mathbf{r} - \frac{1}{2}f^2(\mathbf{r},R)v^2t$$
$$f(\mathbf{r},R) = \frac{1}{2}\left(g_\alpha(\mu) + d\right) \tag{10}$$
$$g_\alpha(\mu) = \alpha^{\alpha/2}\mu\left(\mu^2 - 1 + \alpha^2\right)^{\alpha/2}$$

where $\mu=(r_A-r_B)/R$ is the prolate spheroidal coordinate and $d=1-2p$. The parameter $\alpha=1.25$ was choosen by checking the stability of the calculated cross sections with respect to its variation (see for example [12,16,17]).

For He^{2+}+H(1s) collisions, we performed least-squares fits of the classical ionizing densities, using molecular basis of increasing dimension, including or not pseudostates [17]. Drawings of these classical densities are reported in figure 1 for the nuclear trajectory defined by v=2 a.u. and b=2 a.u.. We found that molecular expansions in terms of OEDM orbitals are sufficiently complete to accurately represent the classical flux at small t ($S\approx1$). As in the molecular framework, ionization classically appears at small t as the result of the inertia of the electron cloud to follow the fast rotation of the internuclear axis. As time elapses, the description of the classical ionizing cloud by OEDM expansions worsens, and this happens the sooner the smaller the OEDM expansion is. In molecular calculations, it dynamically appears through the unphysical trapping of population by bound states. Addition of a mid-centred gaussian set {G} to these limited expansions improves the representation of the classical density at all times so that our 3CMO basis are able to represent the ionizing density.

The 3CMO basis used for H^+, He^{2+} and Li^{3+}+H(1s) collisions are described in Table 1. The ionization cross sections obtained are compared in Figure 2 with the experimental data of Shah *et al.* [18-21], the two-center atomic results of Toshima [6,22] and Kuang and Lin [7,23], the classical results of Illescas *et al.* [14,15] and the three-center atomic results of Winter [9] and McLaughin et al. [10]. The overall agreement of our results with other theoretical predictions and experiment is very good. Excitation and capture cross sections are not reported here

TABLE 1. Description of the 3CMO basis used for bare ion I^{Q+}+H(1s) collisions.

Q	OEDM set *		Gaussian set **				
	n_c	n_e	g	γ_0	β	n_{max}	L_{max}
1	3	3	0.5	0.006	2.5	5	5
2	5	3	0.5	0.00375	2.5	5	5
3	5	2	0.5	0.03	2	4	5

* Capture and excitation manifolds are represented up to $n=n_c$ and n_e respectively, including all nlm states with $m\leq2$.

** For (n1=0,n2) symmetries, the geometrical series have been restricted to their lowest terms.

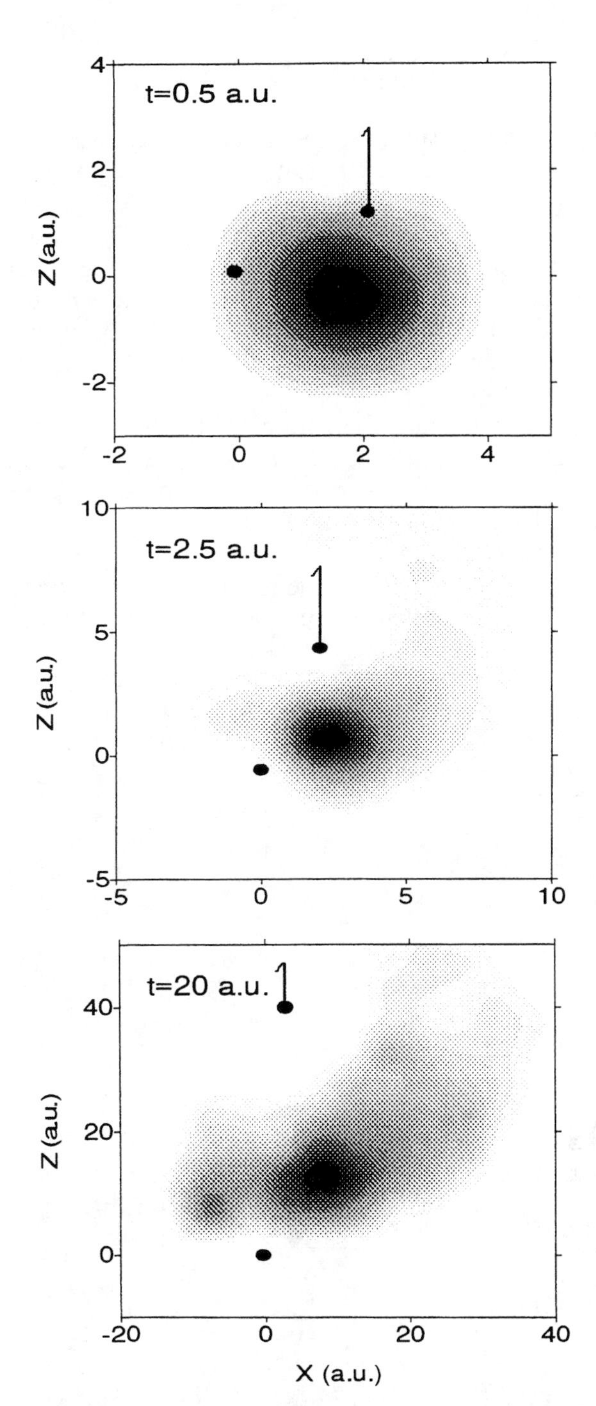

FIGURE 1. Classical ionizing densities lying in a slab $-\varepsilon\leq Y\leq\varepsilon$ about the collision plane at times t=0.5 a.u. (ε=0.5 a.u .), 2.5 a.u. (ε=0.5 a.u.) and 20 a.u. (ε=1 a.u.) for He^{2+}+H(1s) collision and a nuclear trajectory defined by v=2 a.u. and b=2 a.u. *The black points indicate the nuclei positions.*

for sake of conciseness ; for H^+ and He^{2+}+H(1s) cases, they can be found in [17] where we also present convergence checks of the cross sections with respect to all the parameters of the 3CMO calculations (number of OEDMs introduced, gaussian and CTF parameters). Variations of collisional results have been found to be less than 10%.

PERSPECTIVES

Using careful analysis of the ionization mechanism from both molecular and classical standpoints, we have proposed a new scheme for molecular close-coupling calculations, in terms of a "triple-center" basis of adiabatic molecular orbitals, augmented with a set of gaussian pseudostates centred at a point between the nuclei. Such expansions provide accurate results for capture, excitation and ionization in ion-atom collisions from low to intermediate energies.

With the development of cold target recoil ion momentum spectroscopy, kinematically complete ionization experiments have become possible [24,25]. We plan to look at how our 3CMO treatment describes the experimental features and is able to provide ejected electron momemtum distributions. Recently, Illescas and Riera [26] stressed that, even at low velocities, ionization takes place within a small range $0<t<10$ a.u.; later, most of the electrons move freely and their momentum **p** and position **r** are related by **p=r/t**. It means that two-center effects, although responsible for the shape of momentum distributions at small and intermediate impact velocities, have a limited range. Accordingly, (large enough) monocentric atomic expansions would be able to describe the main features of the ionization mechanism. Further work along these lines is in progress.

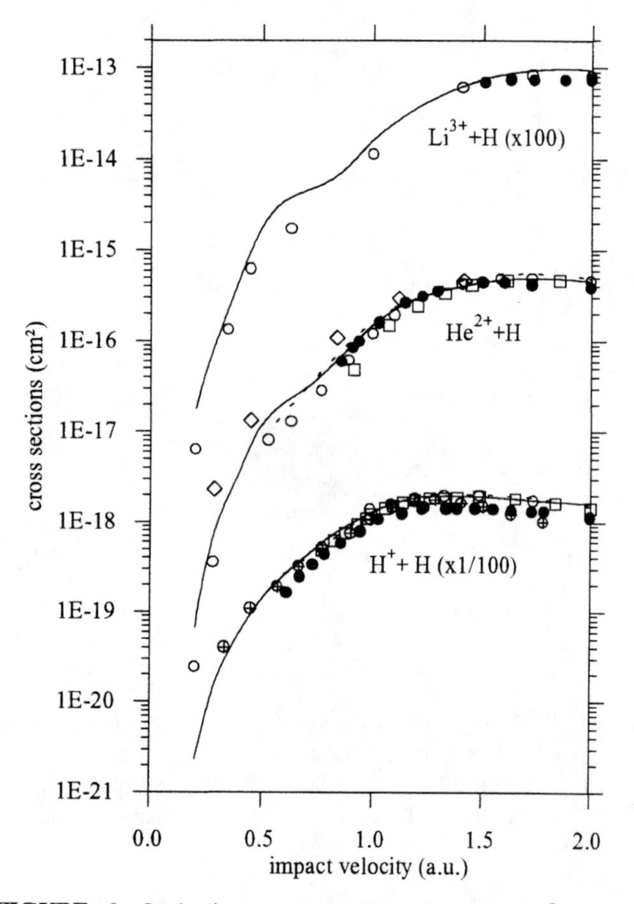

FIGURE 2. Ionization cross sections for H^+, He^{2+} and Li^{3+}+H(1s) collisions: 3CMO results, - - - classical results of Illescas *et al.* [14,15], ● experiment by Shah et al. [18-21], ❍ 2CAO results of Toshima [6,22], 3CAO results of Winter [9], 3CAO results of McLaughin et al. [10], ❑ 2CAO results of Kuang and Lin [7,23]. *For sake of clarity, the cross sections for H^+ and Li^{3+} +H(1s) collisions are multiplied by 1/100 and 100 respectively.*

REFERENCES

[1] Stolterfoht, N., DuBois, R.D., and Rivarola, R.D., *Electron Emission in Heavy Ion-Atom Collisions*, Berlin: Springer (1997)

[2] Bernardi, G., and Meckbach, W., *Phys. Rev. A* **51** 1709 (1995)

[3] Irby, V.D., *Phys. Rev. A* **51** 1713 (1995)

[4] Hall, K.A., Reading, J.F., and Ford, A.L., *J. Phys.B : At. Mol. Opt. Phys.* **29** 6123 (1996)

[5] Henne, A., Ludde, H.J., and Dreizler, R.M., *J. Phys.B : At. Mol. Opt. Phys.* **30** L565 (1997)

[6] Toshima, N., *Phys. Rev. A* **50** 3940 (1994)

[7] Kuang, J., and Lin, C.D., *J. Phys.B : At. Mol. Opt. Phys.* **29** 1207 (1996)

[8] Toshima, N., *J. Phys.B : At. Mol. Opt. Phys.* **30** L131 (1997)

[9] Winter, T.G., *Phys. Rev. A* **37** 4656 (1988)

[10] McLaughin, B.M., Winter, T.G., and McCann, J.F., *J. Phys.B : At. Mol. Opt. Phys.* **30** 1043 (1997)

[11] Errea, L.F., *et al.*, *J. Phys.B : At. Mol. Opt. Phys.* **27** 3603 (1994)

[12] Errea, L.F., *et al.*, *Phys. Rev. A* **46** 5617 (1992)

[13] Harel., C., *et al.*, *Phys. Rev. A* **55** 287 (1997)

[14] Illescas, C., Rabadan, I., and Riera, A., *J. Phys.B : At. Mol. Opt. Phys.* **30** 1765 (1997)

[15] Illescas, C., Rabadan, I., and Riera, A., *Phys. Rev. A* **57** 1809 (1998)

[16] Errea, L.F., *et al.*, *Physica Scripta* **T62** 27 (1996)

[17] Errea, L.F., *et al.*, *J. Phys.B : At. Mol. Opt. Phys.* **31** 3199 (1998)

[18] Shah, M.B., and Gilbody, H.B., *J. Phys.B : At. Mol. Opt. Phys.* **15** 413 (1982)

[19] Shah, M.B., and Gilbody, H.B., *J. Phys.B : At. Mol. Opt. Phys.* **14** 2361 (1981)

[20] Shah, M.B., Elliot, D.S., and Gilbody, H.B., *J. Phys.B : At. Mol. Opt. Phys.* **20** 2481 (1987)

[21] Shah, M.B., *et al.*, *J. Phys.B : At. Mol. Opt. Phys.* **21** 2455 (1988)

[22] Toshima, N., *Phys. Lett.* **175** 133 (1993)

[23] Kuang, J., and Lin, C.D., *J. Phys.B : At. Mol. Opt. Phys.* **30** 101 (1997)

[24] Kravis, S.D., *et al.*, *Phys. Rev. A* **54** 1394 (1996)

[25] Dorner, R., *et al.*, *Phys. Rev. Lett.* **77** 4520 (1996)

[26] Illescas, C., and Riera, A., *Phys. Rev. Lett* **80** 3029 (1998)

An application of the classical trajectory Monte Carlo method to the study of collisions between highly charged ions and surfaces

J. A. PEREZ and R. E. OLSON

Department of Physics, University of Missouri - Rolla, Rolla, MO 65401

We have developed a classical trajectory Monte Carlo code for use in the study of collisions between highly charged ions and systems with multiple targets, such as surfaces. We have simulated a collision between the bare ions C^{6+}, Kr^{36+}, Ne^{10+}, Ar^{18+}, and Xe^{54+}, and a configuration of approximately 400 individual atoms. The projectile has an initial energy of 0.25 keV/u with the velocity perpendicular to the surface. To simulate a simplified surface, the target atoms are held in a simple cubic lattice arrangement by the use of Morse potentials between target nuclei. Each target nucleus has one electron with a binding energy of 12 eV initially localized about it. Initial conditions of the electrons are restricted to represent the 2p electrons of LiF anions. The forces between all particles are calculated at each step in the simulation and the trajectory of every particle is followed. Results for the critical radius of capture, and the principal numbers are shown. Details of the capture of the first three electrons by Ar^{18+} as it approaches the surface are given.

INTRODUCTION

In the last decade there has been growing interest in research projects dealing with collisions of highly charged ions (HCIs) and surfaces (for a recent review see (1)). Initially most of the experimental and theoretical work dealt with metallic surfaces (2-5). With the development of high current ion sources and the subsequent interest by the semiconductor and information industries in possible applications including nanostructure technology, interest in HCI collisions with semiconductor and insulating surfaces has increased. For metallic surfaces the "classical over-the-barrier model" (CBM) first used for electron capture in ion-atom collisions (6,7), then modified for use in electron capture in ion-surface collisions (4,8), adequately describes the charge exchange process before the HCI gets close to the surface. Recently Hägg, Reinhold, and Burgdörfer have modified the CBM to describe collisions of HCIs with insulating surfaces (9), as have Ducrée, Casali, and Thumm (10), with reasonable results.

As the HCI approaches the surface it begins capturing electrons at relatively large distances above the surface and into states with high principal quantum numbers (n »1). The correspondence principle states we should be able to describe this classically. Because of the complex dynamical processes that occur during the collision, and the fact that the above surface collision can be explained classically, it would seem reasonable to model the collision using the classical trajectory Monte Carlo (CTMC) method (11,12). One distinguishing property of the CTMC method is that it includes interactions between all particles and allows the trajectories of each individual particle to be followed. Therefore, the use of the CTMC method to simulate collisions between HCIs and surfaces will lead to greater insight into the complicated dynamical processes that occur.

In collisions between HCIs and insulating surfaces, the experimental and theoretical focus has been on LiF (9,10,13). The electrons most easily removed from LiF are the 2p electrons from the F^- ions. The electron affinity of the free anion is about 3.4 eV, and the work function of LiF is 12 eV. The band gap of LiF is 14 eV meaning the 2p electrons must be lifted into the positive energy continuum. Because there are no conduction bands to allow for resonant loss from highly excited projectile states, LiF provides an opportunity to investigate the role conduction bands play in the neutralization process of HCIs.

In this work we have developed a multiple target classical trajectory Monte Carlo method to simulate a collision between a HCI and a collection of atoms that are arranged in a simple cubic lattice arrangement. We set up the configuration of atoms so as to mimic some of the characteristics of LiF using very simple interaction potentials. Unless stated otherwise atomic units are used throughout this paper.

SIMULATION

At the beginning of the simulation, at t=0, the bare projectile, with charge Z, is located at a distance $R_z < 0$ away from the configuration of atoms (see Fig. 1). The N individual target nuclei, each with ne electrons associated

CP475, *Applications of Accelerators in Research and Industry*, edited by J. L. Duggan and I. L. Morgan

with them and a charge of $q=1$, are positioned in a cubic lattice arrangement with a lattice spacing of $a = 7.6$. We limit ourselves to $ne = 1$ in this work. The first layer of atoms is located in the $z = 0$ plane, with the origin of the coordinate system located at the center of the first layer. The locations of the electrons and target nuclei are indicated by $\vec{r}_i$ and $\vec{r}_i'$ respectively, where $i = 1,\ldots,N$. The initial conditions imposed on the electrons are generated from a microcanonical ensemble. The electrons are given an initial binding energy of W_b=12eV, and the eccentricity of the orbit is restricted to represent the 2p electrons of F^- ions.

The projectiles distance from the origin is denoted by R. Initially the projectile is given a speed, v_z, directed in the positive z direction perpendicular to the layers of atoms. The lateral position of the projectile is randomly chosen over an area that includes nearest neighbor ions.

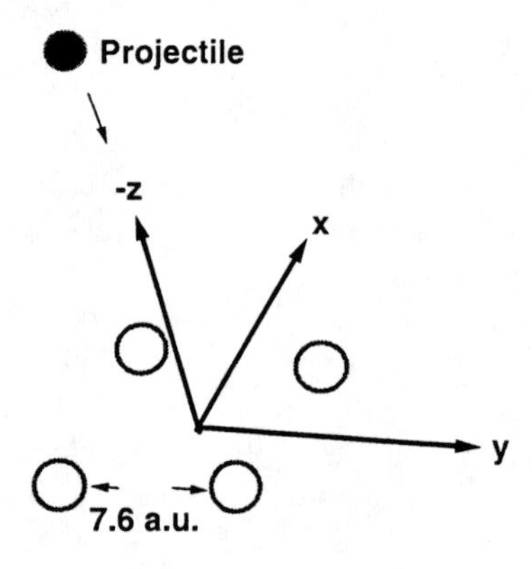

Figure 1. Projectile approaches the center of the coordinate system from the negative z direction. For our applications we allow the projectile to have a lateral position that lies within a circle centered at the origin and encircles the four atoms closest to the origin.

In the CTMC method, the forces acting on individual particles are calculated using Hamilton's equations of motion. The total Hamiltonian for our particular system can be written as

$$H = \sum_{i=1}^{nbody} \frac{p_i^2}{2m_i} + V_{tt}(\vec{r}') + V_{pt}(\vec{R},\vec{r}') + V_{pe}(\vec{R},\vec{r}) + V_{te}(\vec{r}',\vec{r}) + V_{ee}(\vec{r}). \tag{1}$$

The first term in equation (1) is the total kinetic energy of all the particles. The second term designates the interactions between the target nuclei that hold the atoms in a lattice configuration. This is done with a Morse potential of the form

$$V_{morse}(\vec{r}') = D_e(1 - e^{-B(r' - r_e)})^2 \tag{2}$$

where r_e is the equilibrium distance between nuclei, B is a curvature constant, and D_e is the value of the potential when the nuclei are at the equilibrium distance with respect to one another. For this work we have used r_e=7.6, B=1.69, and D_e=0.385. The value for D_e represents the approximate value for the cohesive energy for an ion pair in LiF (14), and the value of r_e is the lattice spacing between F^- ions in LiF.

Once any electron has a positive energy relative to its parent nucleus, it is considered to have been removed from this nucleus, and a coulomb interaction is included between it and all target nuclei and all other electrons, including those still bound. These interactions are turned on "slowly" depending upon the positive energy of the electron relative to its parent nucleus. So, including the contributions from N target nuclei, the second term in equation (1) becomes

$$V_{tt}(\vec{r}',\varepsilon) = \sum_{i=j+1}^{N}\sum_{j=1}^{N} \frac{1}{\left|(\vec{r}_j' - \vec{r}_i')\right|}(1-\exp(-\varepsilon_j))\Theta(\varepsilon_j) + \sum_{i=j+1}^{N}\sum_{j=1}^{N} D_e\left(1-\exp\left[\left(r_e - \Delta r_{ij}'\right)\right]\right)^2 \Theta\left(-\varepsilon_j\right) \tag{3}$$

where $\Delta r_{ij}' = \left|\vec{r}_j' - \vec{r}_i'\right|$, Θ is the heaviside step function and ε_j is the energy of the jth electron relative to its host nucleus. The factor $[1-\exp(-\varepsilon)]$ is used so that the potential is not turned on too rapidly. In this work the number of targets, N, is between 100 and 400 depending on the charge of the bare projectile ion.

The third and fourth terms in equation (1) are the coulomb potentials between the projectile and the target nuclei and associated electrons,

$$V_{pt}(\vec{R},\vec{r}') = \sum_{j=1}^{N} \frac{Z}{\left|\vec{R} - \vec{r}_j'\right|} \tag{4}$$

and

$$V_{pe}(\vec{R},\vec{r}) = \sum_{k=1}^{N} \frac{-Z}{\left|\vec{R} - \vec{r}_k\right|}. \tag{5}$$

The fifth term donates the interactions between target nuclei and the electrons. Because of stability problems initially only interactions between electrons and their host target nucleus are used unless the electron is removed from the parent nucleus, whereupon interactions are turned on "slowly" between the electron and the other target nuclei. This interaction potential can be written as

$$V_{te}(\vec{r},\vec{r}',\varepsilon)=\sum_{i=1}^{N}\frac{-1}{\left|(\vec{r}_i'-\vec{r}_i)\right|}\Theta(-\varepsilon_i)$$
$$+\sum_{i=1}^{N}\sum_{\substack{j=1\\ j\neq i}}^{N}\frac{-1}{\left|(\vec{r}_i'-\vec{r}_j)\right|}\left(1-\exp(-\varepsilon_j)\right)\Theta(\varepsilon_j). \tag{6}$$

The last term in equation (1) consist of the interactions between the electrons. For simplicity, these interactions are not turned on unless an electron has been removed from its parent nucleus. Then the coulomb potential is "slowly" turned on between the electron that has been removed and all other electrons,

$$V_{ee}(\vec{r},\varepsilon)=\sum_{i=j+1}^{N}\sum_{j=1}^{N}\frac{1}{\left|(\vec{r}_i-\vec{r}_j)\right|}\left(1-\exp(-\varepsilon_j)\right)\Theta(\varepsilon_j). \tag{7}$$

In this work we assume an electron has been captured by the projectile if it has a positive energy relative to its parent nucleus and a negative energy relative to the projectile.

NUMERICAL RESULTS

Figure 2 shows the critical distances for capture and the principal numbers associated with the states that these electrons are captured into. The principal numbers are calculated using

$$n=\frac{Z}{\sqrt{2U}}, \tag{8}$$

where U is the electron energy with respect to the projectile.

Becker and Mackellar have shown that the classical principal number, n, is related to the principal quantum number, n' (15), by the expression

$$\left[\left(n'-\frac{1}{2}\right)(n'-1)n'\right]^{1/3}<n\leq\left[\left(n'+\frac{1}{2}\right)(n'+1)n'\right]^{1/3} \tag{9}$$

which can be approximated by $\left(n'-\frac{1}{2}\right)<n\leq\left(n'+\frac{1}{2}\right)$ for large n .

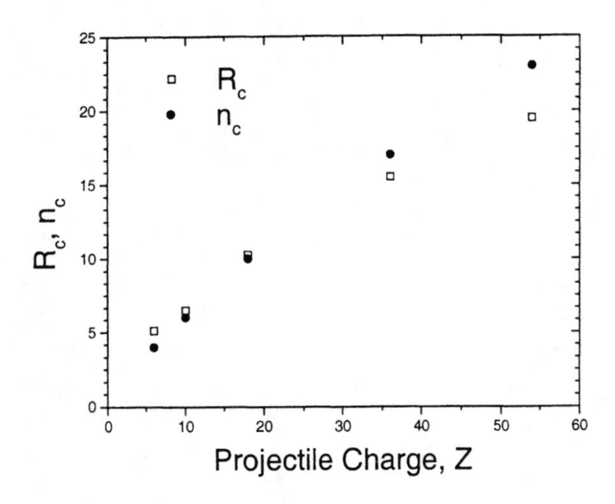

Figure 2. The distance above the surface where the first electron is captured is called the critical distance, R_c. The principal number associated with the first electron capture (equation (8)) is n_c.

In Fig. 3 we have plotted the distribution of distances above the surface where the first, second,and third electrons are captured for Ar^{18+}. Once electron capture begins it proceeds rapidly. At this speed, $v_z=0.1$, the projectile captures the third electron approximately 0.7fs after the first. The capture rate obtained from Fig. 3 will be lower than the actual rate for LiF because we have not included the image potential of the electron or projectile into our simulation.

The distribution of states populated by the first electron captured are shown in Fig. 4 for several projectiles. As the charge increases more than one state begins to be populated.

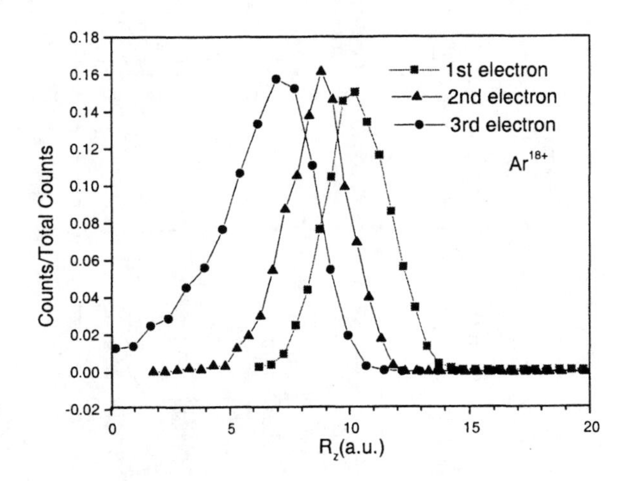

Figure 3. The capture of the first, second and third electrons vs projectile distance above the surface, R_z.

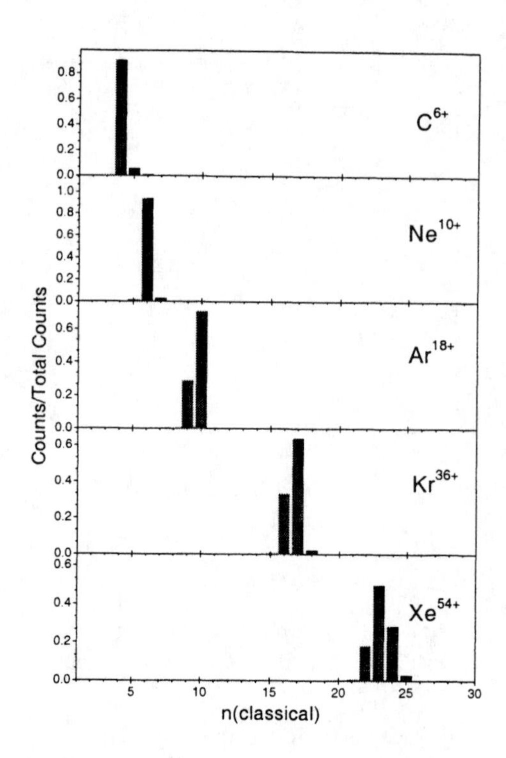

Figure 4. Projectile states populated by the the first captured electron.

As the projectile begins capturing more than one electron they interact with one another and the there is a spread in the distribution of states populated (Fig. 5). Lower states also begin to be populated.

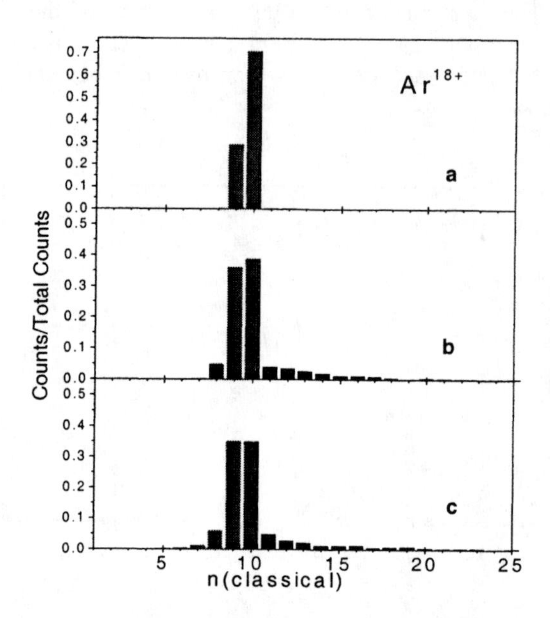

Figure 5. States populated by (a) the first (b) second and (c) the third electron captured by Ar^{18+}.

CONCLUSIONS

We have developed a multiple target classical trajectory Monte Carlo code and have used it to simulate a collision between a highly charged ion and a configuration of approximately 400 atoms. We have shown that this method can be used to investigate certain aspects of collisions between Highly Charged Ions and surfaces.

ACKNOWLEDGMENTS

This work was supported by the Office of Fusion Energy Sciences, Department of Energy.

REFERENCES

1. Arnau A., Aumayr F., Echenique P. M., Grether M., Heiland W., Limburg J., Morgenstern R., Roncin P., Schippers S., Schuch R., Stolterfoht N., Varga P., Zouros T. J. M., Winter H. P., *Surface Science Reports* **27,** 113-240 (1997).
2. Hagstrum H. D.and Becker G., *Phys. Rev. B* **8**, 107 (1973).
3. Burgdörfer J., Lerner P., and Meyer F. W., *Phys. Rev. A* **44**, 5674 (1991).
4. Aumayr F., Kurz H., Schneider D., Briere M. A., McDonald J. W., Cunningham C. E., and Winter H. P., *Phys. Rev. Lett.* **71**, 1943 (1993).
5. Limburg J., Schippers S., Hoekstra R., Morgenstern R., Kurz H., Aumyr F., and Winter H. P., *Phys. Rev. Lett.* **75**, 217 (1995).
6. Ryufuku H, Sasaki K., and Watanabe T., *Phys. Rev. A* **21**, 745 (1980).
7. Bárány A. , Astner G., Cederquist H., Danared H., Huldt S., Hvelplund P., Johnson A., Knudsen A., Liljeby L., and Rensfeld K. G., *Nucl. Instrum. Methods B* **9**, 397 (1985).
8. Burgdörfer J. Meyer F., *Phys. Rev. A* **47**, R20 (1993).
9. Hägg L., Reinhold C. O., and Burgdörfer J., *Phys. Rev.* A **55**, 2097 (1997).
10. Ducree J. J Casali F., Thumm U., *Phys. Rev. A* **57**, 338 (1998).
11. Abrines R. and Percival J. C., *Proc. Phys. Soc.* **88**, 861 (1966).
12. Olson R. E., in *Physics of Electronic and Atomic Collisions*, edited by H. B. Gilbody *et al.*, North-Holland, Amsterdam, 1987, p. 271.
13. Limburg J., Schippers S., Hoekstra R., Morgenstern R., Kurz H., Aumayr F., and Winter H. P., *Phys. Rev. Lett.* **75**, 217 (1995).
14. Ashcroft N. W. and Mermin N. D., *Solid State Physics*, Orlando, Harcourt Brace College Publishers, 1976, p. 406.
15. Becker R. L. and Mackellar A. D., *J. Phys. B* **17**, 3923 (1984).

RECOIL-ION MOMENTUM SPECTROSCOPY: IONIZATION IN ION-ATOM COLLISIONS

[1]S.F.C. O'Rourke, [2]W. Schmitt, [3]J. Kendrick, [4]Kh Khayyat, [5]R. Moshammer, [5]J.Ullrich, [4]R. Dörner and [4]H. Schmidt-Böcking

[1]*School of Mathematics and Physics, The Queen's University of Belfast, Belfast BT7 1NN, UK*
[2]*Gesellschaft für Schwerionenforschung D-64220, Darmstadt, Germany*
[3]*Gonville and Caius College, University of Cambridge, Cambridge, UK*
[4]*Institut für Kernphysik, August-Euler-Strasse 6, D-60486 Frankfurt, Germany*
[5]*Universität Freiburg, Hermann-Herder Strasse 3, D-79104, Freiburg, Germany*

The rapid development in the field of recoil-ion momentum spectroscopy now allows one to perform complete momentum experiments of all the reaction products for single target ionization. This experimental technique can address distinct features characterizing ionization processes in ion-atom collisions and can serve as a most stringent test for theory. The quantum mechanical models used most commomly in the description of ion-atom ionization will be discussed. In particular we will survey some new theoretical results which will illustrate the suitability of the continuum-distorted-wave (CDW) and continuum-distorted-wave eikonal-initial-state (CDW-EIS) models for both high and low Z projectiles in fast ion-atom collisions. The influence of the post collision interaction (PCI) effects in these collisions will also examined.

INTRODUCTION

Due to the recent advances in recoil-ion momentum spectroscopy (1) the study of final-state momentum distributions of the collision products in single target ionization by energetic ion-impact has received a great deal of attention recently. Two quantum mechanical models which have proved to be very popular in the study of ionization of helium by ion-impact are the continuum-distorted-wave (CDW) and the continuum-distorted-wave eikonal-initial-state (CDW-EIS) models (2,3). Both approximations were originally developed to describe the ionization of hydrogen-like target systems (4,5).

In this paper we report on some recent experiments for both high (6) and low (7) Z projectiles in singly ionizing collisions with helium, which illustrate the influence of the post collision interaction (PCI) effects on the longitudinal momentum distributions. Both the CDW and CDW-EIS approximations model the PCI effects as they take account of the long-range interaction of the projectile and target Coulomb field. In the next section we give a brief description of the essential equations for the longitudinal momentum distributions. In the third section we compare theory with experiment. Finally in the last section of this paper we summarize our results.

CALCULATIONAL DETAILS

In this section we consider the final-state longitudinal momentum distributions of the ejected electron and recoil-ion for the single ionization of helium by light or heavy swift ion impact. We closely follow the notation of Rodriguez et al (8). For fast collisions in the laboratory frame the longitudinal momentum conservation requires that

$$p_{R\parallel} = p_{P\parallel} - p_{e\parallel} = \frac{\Delta\varepsilon}{v} - k_e \cos\theta_e, \quad (1)$$

where $p_{R\parallel}$ and $p_{e\parallel}$ are respectively the longitudinal momenta of the recoiling ion and ionized electron and $p_{P\parallel}$ is the longitudinal projectile momentum transfer. The resonance defect of the reaction is given by $\Delta\varepsilon = \varepsilon_e - \varepsilon_i$ where ε_i and ε_e are the electron energy in the initial-bound and final-continuum states respectively. Here k_e is the momentum of the ionized electron and θ_e is the polar emission angle of k_e with respect to v the collision velocity. Equation (1) is valid provided that the energy loss of the projectile is small compared to its initial energy and if the projectile scattering angle is small. Both these conditions are generally fulfilled in swift ion-atom collisions. By means of equation (1) we can show that the longitudinal momentum distributions are related to the doubly differential cross sections (DDCS) in electron energy and emission angle.

For a given $p_{e\parallel}$ the longitudinal electron momentum distributions $d\sigma / dp_{e\parallel}$ can be obtained by integrating over the DDCS

$$\frac{d\sigma}{dp_{e\parallel}} = \int_{p_{e\parallel}^2/2}^{\infty} \frac{1}{k_e} \frac{d^2\sigma}{d\varepsilon_e d(\cos\theta_e)} d\varepsilon_e. \quad (2)$$

CP475, *Applications of Accelerators in Research and Industry*,
edited by J. L. Duggan and I. L. Morgan

Similarily for a given $p_{R\parallel}$ the longitudinal recoil-ion momentum distribution $d\sigma / dp_{R\parallel}$ is given by

$$\frac{d\sigma}{dp_{R\parallel}} = \int_{\varepsilon_e^-}^{\varepsilon_e^+} \frac{1}{k_e} \frac{d^2\sigma}{d\varepsilon_e d(\cos\theta_e)} d\varepsilon_e. \quad (3)$$

where the integration limits $\varepsilon_e^{\pm} = \frac{1}{2}(k_e^{\pm})^2$ with

$$k_e^{\pm} = v\cos\theta_e \pm \sqrt{v^2\cos^2\theta_e + 2(p_{R\parallel}v - |\varepsilon_i|)} \quad (4)$$

are determined from equation (1). Here k_e^+ corresponds to $\theta_e = 0$ and $|k_e^-|$ corresponds to $\theta_e = \pi$.

The theory outlined above is quite general and is independent of the theoretical approximations used to evaluate the transition matrix from which the doubly differential cross section (DDCS) is calculated. In the non-perturbative regime both the CDW and CDW-EIS models have proved to be successful in describing ionization collisions at high energies. The reasons for the suitability of these models are that; (a) the ionized electron sees the Coulomb field from both the target and projectile ion, (b) the wave functions satisfy the correct asymptotic boundary conditions. In the case of the CDW-EIS model both the initial and final states are normalised. The problem concerning the lack of normalization in the CDW model only arises in the case of low impact velocity. In our work we have adopted the independent electron model to treat the two electron helium target atom. The initial helium state is described by the Hartree-Fock wave function (9) while the final state is given by the hydogenic wave function with an effective target charge of 1.34 a.u..

COMPARISON BETWEEN THEORY AND EXPERIMENT

In this section only a brief report of recent theoretical and experimental results are presented. The examples selected illustrate the role that the PCI effect plays in both low and high Z projectiles for fast collisions with helium atoms. In figures 1(a) and (b) the cross sections differential in longitudinal momenta of the electron and recoil-ion are presented for the single ionization of 3.6 MeV/u Au^{24+} on helium (6).

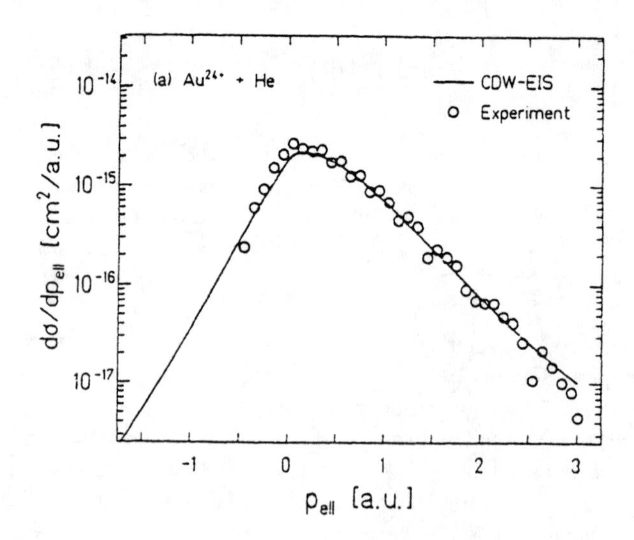

Figure 1(a). Calculated longitudinal momentum distributions of the ejected electron in single ionization of He by 3.6 MeV/u Au^{24+} ions. Experimental data are from Schmitt et al (6). Theoretical results; present CDW-EIS results.

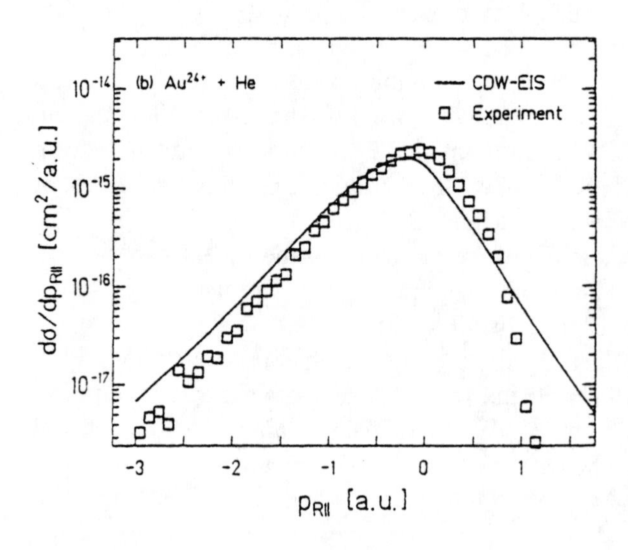

Figure 1(b). Same as in figure 1(a) but for the longitudinal momentum distribution of the recoil-ion.

As in previous measurements (10,11) of 3.6 MeV/u Ni^{24+} and Se^{28+} projectiles, the electron is found to be emitted dominantly into the forward direction, while the recoil-ion is dominantly scattered into the backward direction. This forward-backward asymmetry of the electron and recoil-ion is interpreted as being due to the PCI effect. This effect is caused by the long-range potential of the outgoing highly charged projectile with the low-energy electron in essence 'pulling' it behind. This finding is supported by the current CDW-EIS theory. The fact that the recoil-ion and electron are emitted back to back corresponds to the small momentum transferred by the projectile during the collision. Thus as pointed out by Moshammer et al (11) the action of the fast heavy projectile reveals similiarties to photoionization, where the electron perfectly balances the recoil-ion momentum to the extent of the neglibly small momentum transferred by the absorbed photon. We also performed the CDW-EIS calculation with a reversed projectile charge Au^{24-}. The result was reversed emission direction for both the electron and recoil-ion. Thus the emission characteristic of the electron and recoil-ion reflect the strength and sign of the projectile potential.

In figures 2(a) and (b) we show the longitudinal momentum distributions for anti-proton and proton impact at 945 keV impinging on helium. These results differ from the findings in figures 1(a) and (b) for highly charged ions. Here for the case of low Z projectiles experiment (7) predicts forward emission for the electrons for both the protons and anti-protons. These results are supported qualitatively by the CDW calculations (7). In figure 2(b) it can be seen that the recoil-ion has an almost symmetric distribution about the zero momentum which means that the recoil-ion behaves as a spectator during the collision process. Thus the recoil-ion momentum distribution is mainly given by its momentum in the initial state which is set free in the collision since the electron is knocked out by the projectile. This result is again supported qualitatively by the CDW calculation (7) within experimental error. Thus from figures 2(a) and (b) we may conclude that for fast low Z projectiles the PCI effects which depend on the sign of the projectile charge are very small in contrast to the results for fast high Z projectiles.

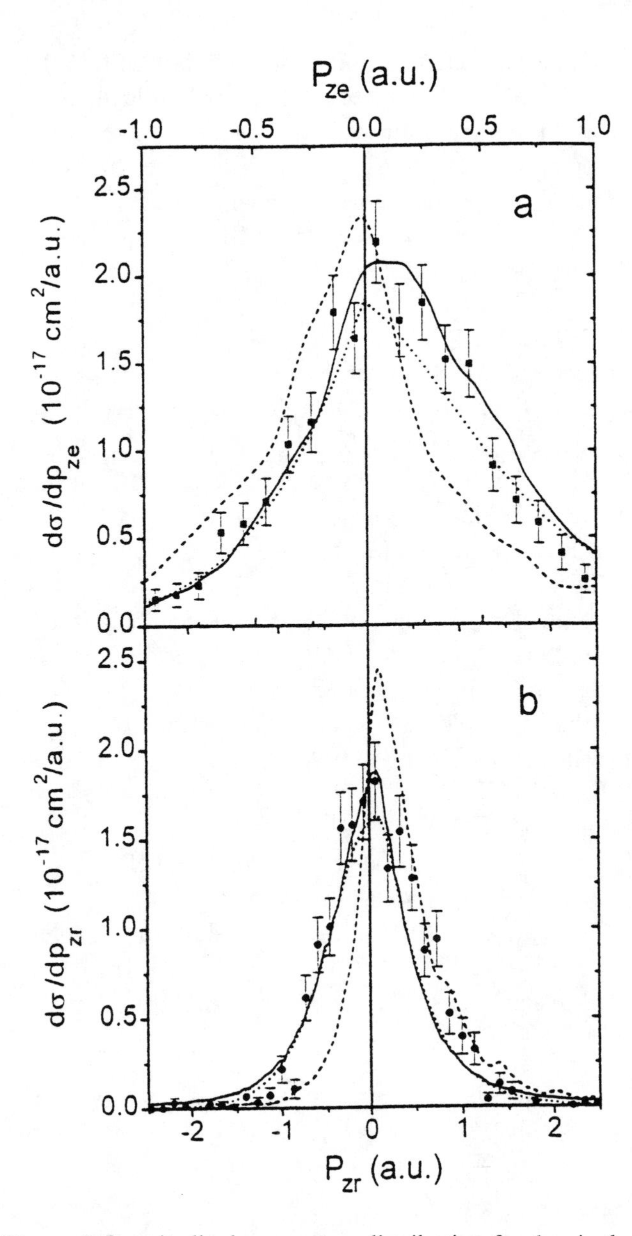

Figure 2. Longitudinal momentum distribution for the single ionization of helium by 945 keV antiproton (data points) in comparison with proton collision (solid line). a: represents electron momentum data; b: represents recoil-ion data. The theoretical caculations represent the antiproton collision where the dotted line is the CDW result and the dashed line is the CTMC result. Here p_{ze} and p_{zr} are equivalent to the notation $p_{e||}$ and $p_{R||}$ of figure 1.

SUMMARY

We have studied the influence of PCI effects in single ionization of helium with fast low and high Z projectiles. It has been shown that in the case of fast highly charged ion collisions eg 3.6 MeV/u Au^{24+} (6) on helium the longitudinal momentum of the electron is almost completely balanced by the backscattered recoil-ion. It was also found that for these collisions the projectile charge asymmetries had a major role to play in the collision dynamics. These findings are in contrast to the case of low Z projectiles considered here (eg 945 keV antiproton and proton impact) where it has been shown that the electron is the particle which compensates the projectile momentum loss while the recoil-ion behaves primarily as a spectator.

The data presented for this case also showed that the sign of the low Z projectile does not influence the collision dynamics above 1 MeV. At the present time it remains a challenge to the experimentalists to observe projectile charge asymmetries at low antiproton energies where the PCI is expected to influence the collision process.

REFERENCES

1.Ullrich, J., Moshammer, R., Dörner, R., Jagutzki, O., Mergel, V., Schmidt-Böcking, H. and Speilberger, L., J. Phys. B. **30** pp. 2917-2974 (1997).
2. O'Rourke, S.F.C. and Crothers, D.S.F., J. Phys. B. **30** pp 2443-2454 (1997).
3. O'Rourke, S.F.C., Moshammer, R. and Ullrich. J., J. Phys. B. **30** pp 5281-5291 (1997).
4. Belkic, Dz., J. Phys. B. **11** pp 3529-3552 (1978).
5. Crothers, D.S.F. and McCann J.F., J. Phys. B. **16** pp. 3229-3552 (1983).
6. Schmitt, W., Moshammer, R. and Ullrich, J., private communication (1998).
7. Khayyat, Kh. et al, Phys. Rev. Lett. submitted for publication (1998).
8. Rodriguez, V.D., Wang, Y.D. and Lin, C.D., J. Phys. B. **28** pp. L471-476 (1995).
9. Clementi, E. and Roetti, R., At. Data. Nucl. Tables **14** pp. 177- (1974).
10. Moshammer, R. et al Phys. Rev. Lett. **73** pp. 3371-3374 (1994).
11. Moshammer, R. et al Phys. Rev. A **56** pp. 1351-1363 (1997).

A generalized description of time-dependent ion-atom collisions

B. Fricke, J. Anton, K. Schulze, D. Geschke, W.-D. Sepp, S. Fritzsche

Universität Kassel, Fachbereich Physik, D-34109 Kassel / Germany

During the last years we were able to describe electron transfer and excitation in many electron ion-atom collisions as a time-dependent process. As basis function we used relativistic molecular wave functions from static molecular calculations. This is correct for the adiabatic collisions, i.e. slow collisions. In case of high energetic ion-atom collisions the diabatic description would be much more appropriate. In this paper we describe a generalized ansatz by using explicit translational factors. This description not only changes the coupling matrix elements in the coupled channel calculations but allows an automatic adaption of the basis to the energy of the electrons in the collision. The inclusion of centre of mass centred basis functions in addition allows an interpretation of the saddle-point electrons emitted in such collisions.

INTRODUCTION

In the semi-classical approximation of a time dependent ion-atom scattering problem the electronic motion is described by the time dependent, many particle Schrödinger- (Dirac-) equation

$$\left[\hat{H}_{el}(\vec{r}_i,t) - i\hbar\frac{\partial}{\partial t}\right]\Psi(\vec{r}_i,t) = 0\,, \qquad (1)$$

where $\vec{r}_i$ are the electronic coordinates in the molecular frame. i runs over all N electrons in the scattering system. This equation has to be solved for the initial condition

$$\lim_{t\to-\infty}\Psi(\vec{r}_i,t) = \Psi^0(\vec{r}_i,t)\cdot e^{\frac{i}{\hbar}\left(\sum\limits_{j=1}^{N_P} m\vec{v}_P\cdot\vec{r}_j + \sum\limits_{j=1}^{N_T} m\vec{v}_T\cdot\vec{r}_j\right)} \qquad (2)$$

where $\Psi^0(\vec{r}_i,t)$ is the channel function. N_P and N_T are the numbers of electrons attracted to the projectile and target in the incoming channel. $\vec{V}_P$ and $\vec{V}_T$ denote the projectile and target velocity respectively. The exponential function result from the fact that the electrons move in the asymptotic region with the velocity of the nuclei along a straight line.

We approximate the many-particle wave function Ψ in eqn. (1) by a Slater-Determinant

$$\Psi(\vec{r}_1,...,\vec{r}_N,t) = \frac{1}{\sqrt{N!}}\begin{vmatrix}\psi_1(\vec{r}_1,t) & \cdots & \psi_1(\vec{r}_N,t)\\ \cdots & \cdots & \cdots \\ \psi_N(\vec{r}_1,t) & \cdots & \psi_N(\vec{r}_N,t)\end{vmatrix} \qquad (3)$$

which is built from time dependent, single particle wave functions $\psi_i(\vec{r}_i,t)$. Inserting this ansatz in equation (1) we yield the time dependent single particle Dirac-Fock equations which have (in Slater approximation of the exchange potential) the form

$$\underbrace{[\hat{t} + \hat{V}^N(\vec{R}(t)) + \hat{V}^C(\rho(t)) + \hat{V}_\alpha^{Ex}(\rho(t))]}_{\hat{h}^{TDHF}(t)}\psi_i(\vec{r},t) = i\hbar\frac{\partial}{\partial t}\psi_i(\vec{r},t) \qquad i=1,...,N\,. \qquad (4)$$

Here $\hat{t}$ is the operator of the kinetic energy, $\hat{V}^N$ the operator of the electron-nucleus interaction, $\hat{V}^C$ the Coulomb part and $\hat{V}_\alpha^{Ex}$ the exchange part of the electron-electron interaction.

Following the initial condition (2) these equations (4) have to be solved for the N starting conditions

$$\lim_{t\to-\infty}\psi_i(\vec{r},t) = \psi_i^0(\vec{r})\cdot e^{\frac{i}{\hbar}m\vec{V}_{K'}\cdot\vec{r}}\,, \qquad (5)$$

where the functions $\psi_i^0(\vec{r})$ are the initially occupied, single-particle atomic states of both the projectile and target. The exponential functions are the well known electronic translation factors.

In order to solve the equation (4) we expand the time dependent, single particle wave functions in a set of basis functions $\phi_j(\vec{r},\vec{R}(t))$

$$\psi_i(\vec{r},t) = \sum_{j=1}^{M} a_{ji}(t)\,\phi_j(\vec{r},\vec{R}(t))\,e^{-\frac{i}{\hbar}\int^t \varepsilon_j dt'}\,, \qquad (6)$$

where $\vec{R}$ is the internuclear coordinate in the molecular frame. Inserting (6) into (4) one gets the matrix equation

$$i\hbar\,\underline{\underline{S}}\,\dot{\underline{\underline{a}}} = \underline{\underline{H}}\,\underline{\underline{a}} \qquad (7)$$

Here is $\underline{\underline{S}}$ the overlap matrix, $\underline{\underline{H}}$ the Fock-matrix and $\underline{\underline{a}}$ are the single particle probability amplitudes.

MO-PICTURE

As a first approximation we use as basis functions $\phi_j(\vec{r},\vec{R}(t))$ in eqn. (6) the molecular static orbitals

CP475, *Applications of Accelerators in Research and Industry*,
edited by J. L. Duggan and I. L. Morgan

(MO), which we get by solving the molecular static Dirac-Fock-Slater equations [1]

$$\underbrace{[\hat{t} + \hat{V}^N(\vec{R}) + \hat{V}^C(\rho(R)) + \hat{V}_\alpha^{Ex}(\rho(R))]}_{\hat{h}^{MO}(R)} \phi_j(\vec{r}, R) = \epsilon \phi_j(\vec{r}, R) . \quad (8)$$

In this case equation (7) reduces to the well known coupled channel equations :

$$\dot{a}_{li} = \sum_{j=1}^{M} -a_{ji} < \phi_l | \frac{\partial}{\partial t} | \phi_j > e^{-\frac{i}{\hbar}\int(\varepsilon_j - \varepsilon_l)dt'} . \quad (9)$$

In order to actually solve these equations (9) we use the LCAO-methode e.g. we expand the MO's in a set of atomic functions $\varphi_\nu(\vec{\xi}(\vec{r}, \vec{R}), \vec{R})$

$$\phi_j(\vec{r}, \vec{R}) = \sum_{\nu=1}^{S} d_{j\nu}(\vec{R}) \, \varphi_\nu(\vec{\xi}(\vec{r}, \vec{R}), \vec{R}) \quad (10)$$

which in our case leads to are the numerical solutions of the atomic DFS-equation.

R-DEPENDENT ATOMIC BASIS FUNCTIONS

To describe the molecular single particle functions $\phi_j(\vec{r}, \vec{R})$ in eqn. (10) we use the separated atoms basis functions only. This basis is good for large internuclear distances but not good for intermediate and bad for small internuclear distances. In order to improve the atomic basis $\varphi_\nu(\vec{\xi}(\vec{r}, \vec{R}), \vec{R})$ we have developed a R-dependent atomic basis. In addition to the target and projectile atomic basis functions we add a thrid basis which is centred on the centre of mass (charge). For all three basis sets we use by the numerical solution of the atomic Dirac-Fock-Slater-equation potential the monopol part of the whole potential of the scattering system at each centre for each distance separately as external potential. Details can be found in [2]. With this R-dependent atomic basis set we are able to describe the molecular system both at very large internuclear distances with the basis functions centred on the projectile and target and at very small internuclear distances with the basis functions centred on the centre of mass (charge). At the intermediate internuclear distance this basis too is a good basis to the solution of the static molecular DFS-equations. So we are able to get an accurate calculation of the eigenvalues and coupling matrix elements, which can be introduced in eqn. (8) and (9) to solve the time dependent ion-atom problem in a even better approximation.

EXPLICIT TRANSLATION FACTORS

As shown above we have to solve the matrix-equation (7) for the N initial conditions (5). The electronic translation factors result from the fact, that the atomic functions $\psi_i^0 = \lim_{R\to\infty} \varphi_\nu(\vec{\xi}(\vec{r}, \vec{R}), \vec{R})$ describe the atomic system in the atomic centre of mass frame at each of the three centres and not in the molecular frame. In order to describe the atomic system in the molecular frame we have to Galilei-transform the functions ψ_i^0 from the atomic to the molecular frame.

Therefore instead of eqn. (10) our LCAO-Ansatz with electronic translation factors have now the form

$$\begin{aligned} \phi_j(\vec{r}, \vec{R}) &= \sum_{\nu=1}^{S_P} d_{j\nu}^P(\vec{R}) \, \varphi_\nu^P(\vec{\xi}(\vec{r}, \vec{R}), \vec{R}) \cdot \exp\left\{\frac{i}{\hbar} m \vec{V}_P \cdot \vec{r}\right\} \\ &+ \sum_{\nu=1}^{S_T} d_{j\nu}^T(\vec{R}) \, \varphi_\nu^T(\vec{\xi}(\vec{r}, \vec{R}), \vec{R}) \cdot \exp\left\{\frac{i}{\hbar} m \vec{V}_T \cdot \vec{r}\right\} \\ &+ \sum_{\nu=1}^{S_C} d_{j\nu}^C(\vec{R}) \, \varphi_\nu^C(\vec{\xi}(\vec{r}), \vec{R}) \\ &\equiv \sum_{K',\nu} d_{j\nu}^{K'}(\vec{R}) \, \hat{S}_{K'} \left|\varphi_\nu^{K'}\right\rangle \end{aligned} \quad (11)$$

In other words, our new basis functions $\phi_j(\vec{r}, \vec{R})$ are a superposition of projectile respectively target atomic basis functions with the translation factors of target atomic basis functions plus the center of mass basis functions without translation factors, since these functions are already in the centre of mass frame. As a general nomenclature we chose the expansion coefficients $d_{j\nu}^{K'}$ so that the overlap-Matrix $\underline{\underline{S}}$ and the Fock-Matrix $\underline{\underline{H}}$ are orthogonal in the Basis $\phi_j(\vec{r}, \vec{R})$ i.e. the expansion coefficients $d_{j\nu}^{K'}$ are the solutions of the matrix equation

$$\underline{\underline{h}}^{QMO} \, \underline{\underline{d}} = \underline{\underline{\varepsilon}} \, \underline{\underline{S}} \, \underline{\underline{d}} .$$

The quasi-molecular operator $\hat{h}^{QMO}$

$$\hat{h}^{QMO}(\vec{R}) = \hat{t} + \hat{V}^N(R) + \hat{V}^C(\rho(\vec{R})) + \hat{V}_\alpha^{Ex}(\rho(\vec{R}))$$

is constructed very similar to the static Dirac-Fock-Slater-Operator $\hat{h}^{MO}$ in eqn. (8). The electronic density is

$$\rho(\vec{r}, \vec{R}) = \sum_{j}^{occ} \phi_j^\dagger(\vec{r}, \vec{R}) \phi_j(\vec{r}, \vec{R}) .$$

Inserting the ansatz (11) in the matrix-equation (7) we get for the matrix elements

$$S_{lj} = \delta_{lj}$$

$$H_{lj} = e^{-\frac{i}{\hbar}\int^{t}(\varepsilon_j-\varepsilon_l)dt'} \sum_{K,K'}\sum_{\mu,\nu} \left(d^{K}_{l\mu}\right)^* d^{K'}_{\nu j} \langle \varphi^{K}_{\mu} | \hat{S}^{+}_{K}\hat{S}_{K'}$$

$$\left(i\hbar\frac{\partial}{\partial t} + \varepsilon_j - \hat{h}^{TDHF} + \left[\hat{t},\hat{S}_{K'}\right] + \left[i\hbar\frac{\partial}{\partial t},\hat{S}_{K'}\right]\right)|\varphi^{K'}_{\nu}\rangle$$

In the non relativistic case the commutators have the form

$$\left[\hat{t},\hat{S}_{K'}\right] = i\hbar\vec{V}_{K'}\left(\vec{\nabla}_r\right)_{\vec{R}} + \frac{mV^2_{K'}}{2}$$
$$\left[i\hbar\frac{\partial}{\partial t},\hat{S}_{K'}\right] = m\vec{V}(R)\left(\vec{r}\cdot\vec{\nabla}_R\right)\vec{V}_{K'}(R) \quad (12)$$

where V is the relative velocity of the nuclei in the molecular frame $\vec{V} = \vec{V}_T - \vec{V}_P$. If we neglect the potential coupling

$$< \phi_l|\varepsilon_j - \hat{h}^{TDHF} - \frac{mV^2_{K'}}{2}|\phi_j> \approx 0 \quad (13)$$

we get for the time dependent single particle amplitudes a_{ji} the modified coupled-channel equations

$$\dot{a}_{li} = \sum_{j=1}^{M} -a_{ji} < \phi_l|\left(\frac{\partial}{\partial t}\right)_{\vec{\xi}_{K'}}$$
$$+ m\vec{V}(R)\left(\vec{r}\cdot\vec{\nabla}_R\right)\vec{V}_{K'}(R)|\phi_j> e^{-\frac{i}{\hbar}\int(\varepsilon_j-\varepsilon_l)dt'} \quad (14)$$

where we have used the relation

$$i\hbar\frac{\partial}{\partial t} + i\hbar\vec{V}_{K'}\left(\vec{\nabla}_r\right)_{\vec{R}} \equiv i\hbar\vec{V}(R)\left(\vec{\nabla}_R\right)_{\vec{r}}$$
$$+ i\hbar\vec{V}_{K'}(R)\left(\vec{\nabla}_r\right)_{\vec{R}} = i\hbar\vec{V}(R)\left(\vec{\nabla}_r\right)_{\vec{\xi}} \equiv i\hbar\left(\frac{\partial}{\partial t}\right)_{\vec{\xi}_{K'}}.$$

The matrix element $< \phi_l|\frac{\partial}{\partial t}|\phi_j >$ from eqn. (9) can now compared with the analogue matrix element in eqn. (14) has the form:

$$\left\langle\phi_l\left|\frac{\partial}{\partial t}\right|\phi_j\right\rangle = \dot{R}\sum_{K,K'}\sum_{\mu,\nu}\left(d^{K}_{l\mu}\right)^*\frac{\partial d^{K'}_{\nu j}}{\partial R}\left\langle\varphi^{K}_{\mu}\left|\hat{S}^{+}_{K}\hat{S}_{K'}\right|\varphi^{K'}_{\nu}\right\rangle$$
$$+ i\frac{\dot{R}}{\hbar}\sum_{K,K'}\sum_{\mu,\nu}\left(d^{K}_{l\mu}\right)^* d^{K'}_{\nu j}\left\langle\varphi^{K}_{\mu}\left|\hat{S}^{+}_{K}\hat{S}_{K'}\frac{\partial R^{K'}}{\partial R}\hat{p}^{K'}_{z}\right|\varphi^{K'}_{\nu}\right\rangle$$
$$+ \dot{R}\sum_{K,K'}\sum_{\mu,\nu}\left(d^{K}_{l\mu}\right)^* d^{K'}_{\nu j}\left\langle\varphi^{K}_{\mu}\left|\hat{S}^{+}_{K}\hat{S}_{K'}\left(\frac{\partial}{\partial R}\right)_{\xi_{K'}}\right|\varphi^{K'}_{\nu}\right\rangle$$
$$- i\frac{\dot{\theta}}{\hbar}\sum_{K,K'}\sum_{\mu,\nu}\left(d^{K}_{l\mu}\right)^* d^{K'}_{\nu j}\left\langle\varphi^{K}_{\mu}\left|\hat{S}^{+}_{K}\hat{S}_{K'}\hat{j}^{K'}_{y}\right|\varphi^{K'}_{\nu}\right\rangle$$
$$- i\frac{\dot{\theta}}{\hbar}\sum_{K,K'}\sum_{\mu,\nu}\left(d^{K}_{l\mu}\right)^* d^{K'}_{\nu j}R_{K'}\left\langle\varphi^{K}_{\mu}\left|\hat{S}^{+}_{K}\hat{S}_{K'}\hat{p}^{K'}_{x}\right|\varphi^{K'}_{\nu}\right\rangle$$

$$= < \phi_l|\left(\frac{\partial}{\partial t}\right)_{\xi_{K'}}|\phi_j> - M_2 - M_5\,.$$

The new part in the coupling matrix element can be written as:

$$< \phi_l|m\vec{V}(R)\left(\vec{r}\cdot\vec{\nabla}_R\right)\vec{V}_{K'}(R)|\phi_j> = -\frac{m}{\mathcal{M}^2}\frac{\partial W(R)}{\partial R}$$
$$\times\left[\sum_{K,K'}\sum_{\mu,\nu}\left(d^{K}_{l\mu}\right)^* d^{K'}_{\nu j}\left\langle\varphi^{K}_{\mu}\left|\pm\hat{S}^{+}_{K}\hat{S}_{K'}M_{K'}z\right|\varphi^{K'}_{\nu}\right\rangle\cos 2\theta\right.$$
$$\left.+\sum_{K,K'}\sum_{\mu,\nu}\left(d^{K}_{l\mu}\right)^* d^{K'}_{\nu j}\left\langle\varphi^{K}_{\mu}\left|\pm\hat{S}^{+}_{K}\hat{S}_{K'}M_{K'}x\right|\varphi^{K'}_{\nu}\right\rangle\sin 2\theta\right]$$

Here $W(R)$ is the potential which is used to calculate the classical trajectory of the nuclei $R(t)$, $M_{K'}$ is the mass of projectile or target and $\mathcal{M}$ is the reduced mass of nuclei.

FIRST APPLICATION FOR THE SYSTEM p-He

In order to apply this new concept we have started to calculate the simplest many-electron problem, the system p on He with 2 electrons. As discussed in the first part of this paper we use the atomic basis functions $1s$ to $2p$ at the position of the p and the $1s$ to $2p$ functions at the site of He with the two electrons in the $1s$ level. These functions are calculated in the field of the nucleus plus the monopole part of the other. In addition we use wave functions at the center of mass which are generated in a R-dependent manner in the monopole potential of the two nuclei p and He. Using the explicit translational factors we get correlation diagrams where we show one example in Fig. 1. In this case the wave functions and coupling matrix elements are calculated in the center of mass for an impact energy of 15 keV.

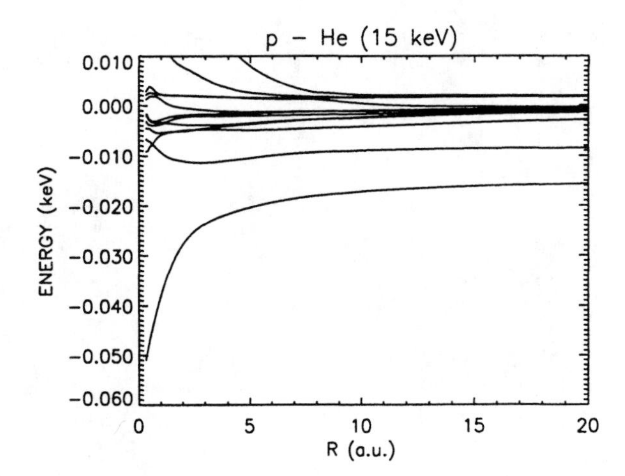

FIGURE 1. Correlation diagram of the system p-He at 15 keV impact energy

Solving the coupled channel equations (14) we get probabilities of the distribution of the two electrons in the final channels. Fig. 2 shows as example the probability to find one resp. two electrons in the wave functions of the monopole basis at the center of mass which are called 'saddle point electrons'. Using these two curves we are able to calculate the single electron ionisation to be $1.44 \cdot 10^{-17}$ cm^2 and $1.38 \cdot 10^{-19}$ cm^2 for a double ionisation. The ratio is about 1%. Comparison with experiment [3] shows that this is in the right order of magnitude. In a future paper we will be discussing the detailed results of this very extensive calculations.

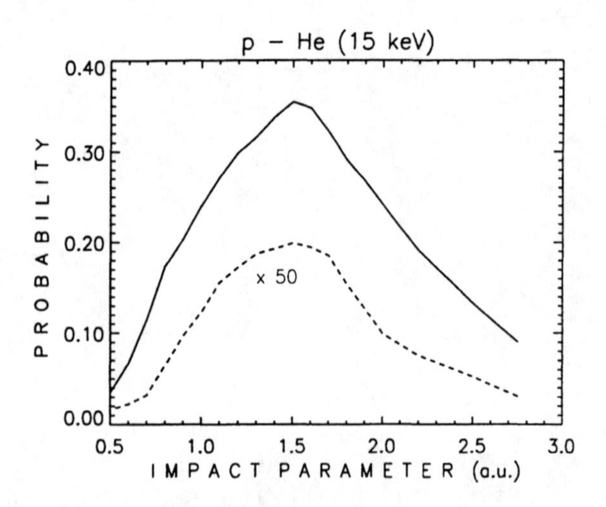

FIGURE 2. Probability for single (full line) and double ionisation (dashed line) as function of the impact parameter

REFERENCES

1. Sepp W.-D., Kolb D., Sengler W., Hartung H. and Fricke B., Phys. Rev. A **33** 3679-3687 (1986)
2. Anton J., Schulze K., Kürpick P., Sepp W.-D. and Fricke B., Hyperfine Interactions **108**, 89-94 (1997)
3. Shah M. B., McCallion P. and Gilbody H. B., J. Phys. B**22** , 3037-3045 (1989)

M-SHELL IONIZATION CROSS SECTIONS FOR 75-300 KEV PROTONS ON TM USING THICK AND THIN TARGETS

Kevin Welsh and Sam J. Cipolla

Physics Department, Creighton University, Omaha NE 68178

Low energy protons were used to excite M x-rays from elemental targets of Thulium. The x-rays were measured with a high-resolution Si(Li) detector equipped with an ultra-thin window. Separate measurements were taken with a thick target and a thin target. The accuracy and precision of the results using each target will be contrasted and x-ray production cross sections will be compared with the ECPSSR theory (1).

INTRODUCTION

There have been relatively few comparisons between experimental M-shell cross sections from proton impact and predictions from ECPSSR theory. Most studies of M-shell x-rays have involved the use of heavy elements (Z>79) or high proton energies (2-4,13). As the proton energy increases, the agreement between ECPSSR theory and experiment improves. However, at lower proton energies there is a consistent disagreement between theory and experiment, with theory underpredicting experimental cross sections. This disagreement has been observed in both heavy and rare-earth elements.

In the present work, M-subshell x-ray cross sections have been measured for 75 - 300 keV protons incident on the rare-earth element Thulium, using both thick and thin targets. The cross sections measured from each target are compared to each other as well as to the ECPSSR M-shell cross sections (1).

EXPERIMENTAL PROCEDURE

Thulium targets were mounted on a vertical target ladder in the collision chamber. For the thick target, the ladder was moved for each proton energy in order to minimize target damage. The thin target was mounted onto an attachment on the target ladder that did not allow for adjustment of the target with respect to the beam. The thin target was prepared by evaporation onto a carbon backing (5). The thickness of the backing and thin target was determined by alpha gauging using a Po-210 source to be 45.7 ± 1.2 μg and 55.84± 5.7 μg/cm^2, respectively.

A 350kV Cockcroft-Walton accelerator was used to accelerate protons onto the Thulium targets. The targets were situated at 45° with respect to the beam. The resulting x-rays were detected using a high resolution 30mm^2 x 3mm Si(Li) detector with an ultra-thin window, which was oriented at 90° to the beam line. A doubly - aluminized Mylar absorber was used to minimize dead time (<2%). For the thin target, the charge was collected in a Faraday cup located behind the ladder and from the target itself. The charge delivered to the thick target was collected directly. In both cases, a current integrator was used to measure the collected charge. The setups are shown in Fig. 1.

Typical beam currents were 2 - 6 nA for the thick target and 0.2 - 0.6 nA for the thin target. The efficiency of the Si(Li) detector was determined using the standard efficiency model (6). X-ray peaks for both targets were fitted to a sum of gaussian functions on a linear background, using a PC program (XSPEC).

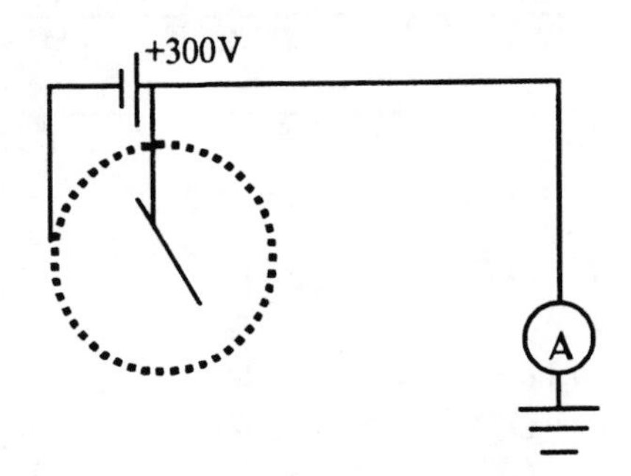

FIGURE 1a. Current collection for the thick target. The dashed line represents an electron suppression cage.

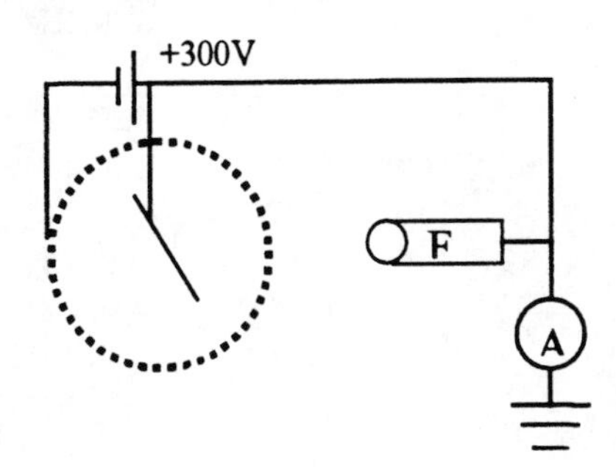

FIGURE 1b. Current collection for the thin target. Both target and Faraday cup are connected to the current integrator.

CP475, *Applications of Accelerators in Research and Industry*, edited by J. L. Duggan and I. L. Morgan

DATA ANALYSIS AND RESULTS

Thick Target

X-ray production cross sections, $\sigma_x(E)$, for each identified transition were obtained through the Merzbacher-Lewis equation (7):

$$\sigma_x(E) = \frac{1}{\varepsilon N}\left[S(E)\frac{dY}{dE} + \frac{\mu}{\rho}Y(E)\right] \quad (1)$$

where $Y(E) = N_x/N_p$ is the x-ray yield (x-rays/proton), S(E) is the target stopping power for protons (8), μ/ρ is the target mass absorption coefficient for x-rays (9), N is the target atom density, ε is the detector efficiency, and dY/dE was determined from an analytical fit to the x-ray yields. The fit and subsequent calculations were done in an Excel workbook (10).

Table 1 lists the x-ray production cross sections for the major M-shell transitions: M1O2,3, M2N4, M3N4,5(γ), M4N6(β), M5N6,7(α) and M5N3,M4N2 ($\zeta_{1,2}$). The peak cross sections were calculated from eq. (1), and the subshell cross sections were determined as:

$$\sigma_{Mi} = \sigma_x \frac{\Gamma_{Mi}}{\Gamma_x} \quad (2)$$

where Γ_x and Γ_{Mi} are individual x-ray transition rates and total transition rates, respectively (11).

Figure 2 shows the ratio of experimental to ECPSSR values for $\sigma_{M\text{-}Total}$ as a function of proton energy. As Fig. 2 shows, the theory tends to under-estimate cross sections at low proton energies. There is increasing agreement as the proton energy increases, with theory eventually over-estimating cross sections. This is a trend that has been observed by others in various elements (15).

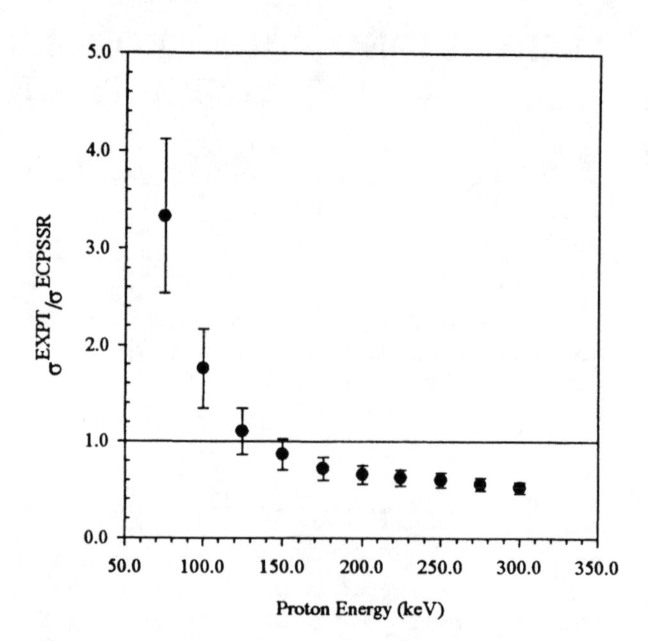

FIGURE 2. Total M-shell cross section ratio of experiment to ECPSSR theory for protons incident on a Thulium thick target.

Thin Target

Thin target X-ray production cross sections were obtained through the standard thin target formula:

$$\sigma_x(E) = \frac{Y}{N\varepsilon\rho t}F(E_0, \Delta E) \quad (3)$$

where Y, N and ε were defined earlier, ρt is the thickness of the target, and $F(E_0, \Delta E)$ is a correction factor accounting for x-ray self absorption and proton energy loss in the target (14). The same transitions used in the thick target analysis were again used in the thin target

TABLE 1. Thulium M-subshell X-Ray Production Cross Sections (barns) for the Thick Target*

Energy (keV)	M1O2,3	M2N4	M3N4,5 γ	M4N6 β	M5N6,7 α	M5N3,M4N2 ζ1,2
75		3.36-2 (13)	7.68-2 (15)	1.79 (33)	4.81 (31)	6.16-1 (14)
100	6.00-3 (19)	1.01-1 (12)	2.24-1 (14)	4.38 (29)	10.2 (32)	1.56 (14)
125	10.2-3 (18)	1.94-1 (12)	4.38-1 (13)	7.10 (27)	15.4 (30)	2.76 (13)
150	15.9-3 (15)	3.17-1 (12)	7.65-1 (13)	12.1 (23)	20.8 (28)	4.33 (13)
175	20.0-3 (14)	4.90-1 (12)	1.13 (12)	16.9 (21)	26.6 (25)	6.04 (12)
200	24.1-3 (19)	6.36-1 (12)	1.62 (12)	24.2 (19)	32.6 (23)	8.08 (12)
225	28.0-3 (14)	8.83-1 (12)	2.25 (12)	34.4 (17)	37.1 (22)	9.87 (12)
250	37.7-3 (14)	1.07 (12)	2.90 (12)	42.6 (16)	44.1 (20)	12.4 (12)
275	42.9-3 (13)	1.33 (12)	3.47 (11)	44.0 (16)	54.0 (18)	15.0 (12)
300	39.7-3 (14)	1.65 (11)	4.17 (11)	45.1 (17)	63.5 (17)	17.6 (11)

*N-n represents $N\text{x}10^{-n}$, number in parentheses represent % uncertainties.

TABLE 2. Thulium M-subshell X-Ray Production Cross Sections (barns) for the Thin Target*

Energy (keV)	M1O2,3	M2N4	M3N4,5 γ	M4N6 β	M5N6,7 α	M5N3,M4N2 ζ1,2
75		2.69-2 (14)	4.29-2 (15)	1.18 (15)	3.44 (15)	2.53-1 (12)
100	2.54-3 (67)	2.07-1 (14)	2.69-1 (15)	8.73 (15)	21.4 (15)	1.83 (12)
125	1.38-2 (69)	6.13-1 (14)	5.61-1 (15)	27.2 (15)	35.8 (15)	4.07 (12)
150	4.49-2 (69)	5.16-1 (14)	5.55-1 (15)	16.7 (15)	44.4 (15)	4.21 (12)
175	6.13-2 (23)	7.52-1 (14)	1.04 (13)	32.7 (16)	66.4 (13)	7.11 (12)
200	4.63-2 (28)	7.93-1 (13)	1.29 (13)	31.1 (16)	64.3 (13)	7.06 (12)
225	6.88-2 (28)	8.02-1 (13)	1.31 (13)	26.6 (16)	73.1 (13)	7.39 (12)
250	10.0-2 (15)	8.01-1 (17)	1.44 (13)	40.4 (16)	78.0 (13)	7.93 (12)
275	10.2-2 (15)	8.83-1 (17)	1.54 (13)	52.9 (16)	75.8 (13)	8.63 (12)
300	8.27-2 (15)	1.07-1 (17)	1.83 (13)	35.9 (16)	105 (13)	9.20 (12)

*N-n represents $N \times 10^{-n}$; numbers in parentheses represent % uncertainties.

analysis. Table 2 lists the thin target cross sections for these transitions. The M5N3, M4N2 peak cross sections were not corrected with the correction factor because they were fit as one peak. The correction factors are calculated based on individual subshells.

Figure 3 shows the ratio of experimental to ECPSSR values for $\sigma_{M\text{-total}}$ as a function of proton energy. The results provide numerous paradoxes that have proven to be difficult to explain. The expected trend of theory underpredicting experimental cross sections at low energy and coming into agreement with increasing energy is seen in the thin target results. Additionally, the thin target cross sections are more in line with theory than the thick target values, as seen in a comparison of σ_{Thick} to σ_{Thin} total cross sections in Figure 4.

DISCUSSION AND CONCLUSION

The thick target analysis gives results that are consistent with previously observed results with other elements (3,4,10,12). The thin target results are not so simple to analyze.

There are a couple of possibilities to consider in the thin target analysis. One involves the assumption that the target atom density is the same for a thin target as for a thick target. The second is that the target thickness remains unchanged during the measurements.

The standard configuration of a particle detector to measure the back-scattered protons was not employed due to geometry restrictions in the collision chamber. Planned modifications will allow this setup in the future so that

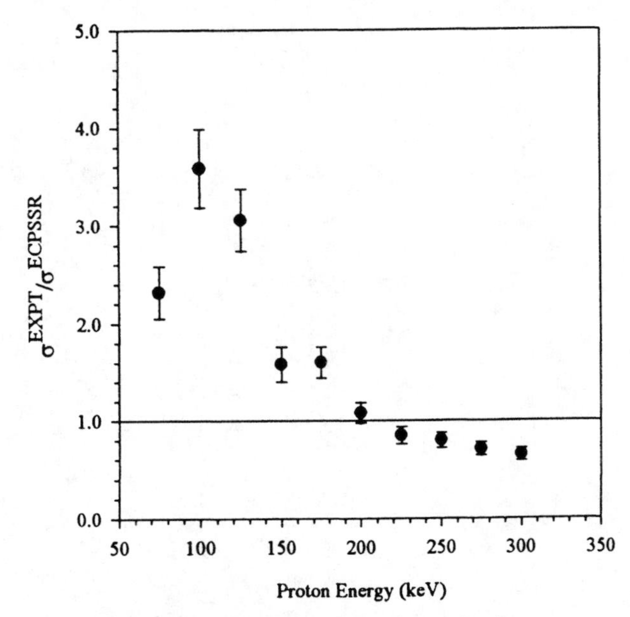

FIGURE 3. Total M-shell cross section ratio of experiment to ECPSSR theory for protons incident on a thin target of Thulium.

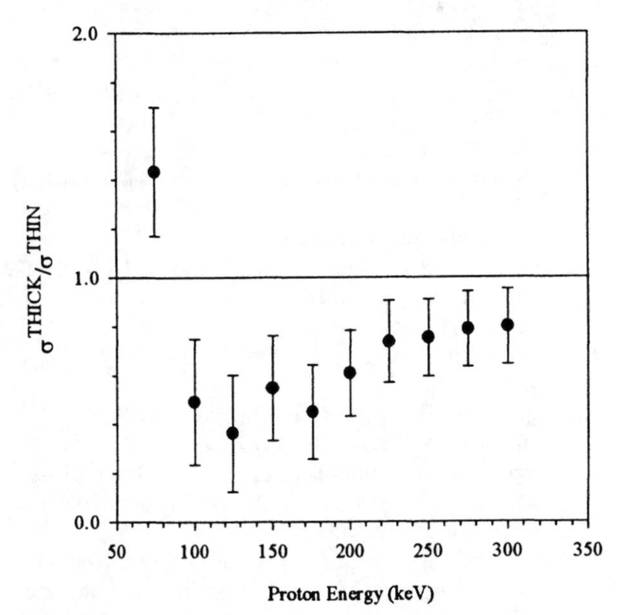

FIGURE 4. A plot of the ratio of Thick/Thin target cross sections for protons on Thulium.

target thickness can be measured during the x-ray measurement.

The experimental results for both targets show a larger cross section than predicted by ECPSSR theory at the lower energies. In both cases, experiment and theory converge as the proton energy increases. In comparison to each other, the thick and thin target cross sections follow a consistent trend with proton energy of theory under-predicting experiment at lower energies and over-predicting as the energies increase. Additionally, thick and thin target cross sections come into better agreement with each other as the proton energy increases. Except for 75 keV protons, the thin target cross sections tend to be higher.

Generally, more external parameters need to be employed using thick targets, such as stopping powers and absorption coefficients, which may account for the observed differences.

ACKNOWLEDGEMENTS

This work was conducted at the University of Nebraska at Lincoln and is supported by the Creighton University Graduate School and the Research Corporation.

REFERENCES

1. Brandt, W. and Lapicki, G., *Phys. Review* **A23**, 1717 (1981); Lapicki, G. , Private Communication.
2. Mehta , R., Duggan, J.L., Price, J.L., McDaniel, F.D. and Lapicki, G., *Phys. Review* **A26**, 1883 (1982).
3. Gresset, J.D., Marble, D.K., McDaniel, F.D., Duggan, J.L., Culwell, J.F. and Lapciki, G., *Nucl. Insr. & Meth* **B40/41**, 116 (1989).
4. Mehta , R., Duggan, J.L., Price, J.L., Kocur, P.M., McDaniel, F.D. and Lapicki, G., *Phys. Review* **A 28**, 3217 (1983).
5. Micromatter Corp, Deer Harbor, WA.
6. Cipolla, S.J. and Gallagher, W.J., *Nucl. Instr. & Meth.* **122**, 405 (1974); Cipolla, S.J. and Watson, S., *Nucl. Instr. & Meth.* **B10/11**, 946 (1985).
7. Merzbacher, E., and Lewis, H.W., *Handbook of Physics*, **34**, 166 (1958).
8. Ziegler, J., "Transport of Ions in Matter,"(TRIM) ver. 91.07, IBM, Yorktown Hts., Juy 26, 1991.
9. Berger, M.J., and Hubbell, J.H., "XCOM: Photon Cross Section on a Personal Computer," NBS Report NBSIR 87-3595, Gaithersburg, Md., 1987.
10. Teeter, P.J., *M-Shell X-Ray Production Cross Sections for Lead and Tungsten,* Masters Thesis, Creighton University (1997)
11. Bhalla, C.P., *J. Phys.* **B3**, 916 (1970).
12. McClure, J.D., *M-Shell X-Ray Cross Sections for Low Energy Protons on Holmium & Hafnium* Masters Research Report, Creighton University (1997).
13. Sarkar, M. , Mommsen, H., Sarter, W., Schurkes, P., *J. Phys.*, **B14**, 3163 (1981).
14. Pajek, M., Kobzev, A.P., Sandrik, R., Iklhamov, R.A.. and Khusmurodov, S.H., *Nucl. Instr. & Meth.* **B42**, 346 (1989).
15. Biendowski, A., Braziewicz, J., Czyzewski, T., Glowacka, L., Jaskola, M., Lapicki, G., and Pajek, M., *Nucl. Instr & Meth.* **B49**, 19 (1990).

Systematic Study of the L-Subshell Ionization Cross Sections of Ho, Er, and Tm by MeV α-Particle Bombardments

B. A. Shehadeh* and A. B. Hallak**

*Department of Physics and Astronomy, Iowa State University of Science and Technology, Ames, IA 50011.

**Physics Department, UAE University, P.O.Box17551, Al-Ain, United Arab Emirates.

The experimental L-subshell ionization cross sections, induced by MeV α particle bombardments, of Ho, Er, and Tm are compared with the ECPSSR theory. The general tendencies show systematic discrepancies for the three subshells. To account for the effects of induced intrashell transitions, the coupled state model is incorporated within the ECPSSR calculations, thus yielding the ECPSSR-IS. Systematic discrepancies between the predictions of the ECPSSR-IS and the experimental measurements are significantly reduced. Yet, velocity independent discrepancies remain, attributed primarily to the overestimation of the electronic binding energies. The ECPSSR-IS is recalculated at binding energies correspond to the united atom, thus yielding The ECPSSR-IS-UA. The later model is capable of describing the experimental ionization cross sections for the L_2 and the L_3 subshells over all energies, but a significant discrepancy remains for the L_1 subshell. The reasons for the L1-subshell discrepancy are discussed and proved.

1. INTRODUCTION

In a previous work we measured the L-subshell ionization cross sections induced by α-particle bombardments of Ho, Er, and Tm. The experimental procedure and the calculations are presented in details elsewhere[1]. Studying the ratios $\sigma_{L_i}^{EXP}/\sigma_{L_i}^{ECPSSR}$ versus the reduced velocity parameter ξ_{L_i} of α-particle impact for L_1, L_2, and L_3 subshells, systematic disagreements is observed especially for the $\xi_{L_i} < 0.3$ domain as shown in figure (1).

Neglecting the multi-vacancy effect[1], the systematic discrepancies are attributed to more effective physical phenomenon deduced from the general tendency of the ratios $\sigma_{L_i}^{EXP}/\sigma_{L_i}^{ECPSSR}$ for the three elements. These phenomenon are the induced intra-shell transitions among the subshells due to the Coulomb field of the α particle and the binding energy of the electron which must not exceed the united atom binding energy.

In the present paper, we study these effects by incorporating them within the ECPSSR formalism and test the validity of the modified models for describing the experimental data.

2. THE INTRASHELL TRANSITION EFFECT

The systematic discrepancies between the experimental ionization cross sections and the predictions of the ECPSSR theory, that have been observed at low-velocity range, can be reduced after taking into account the coupling between the subshells. The ionization process can not be treated independently for individual L subshells, because of the induced intrashell transitions (IS) among the subshells due to the Coulomb field of the projectile. This effect occurs mainly after the ionization process.

Using the coupled state model of Sarkadi and Mukoyama[2–5], the IS effect is incorporated within the ECPSSR via the multiplicative correction factor $C_{L_i}^{IS}(\xi_{L_i})$, given by[5]

$$C_{L_i}^{IS}(\xi_{L_i}) = 1 + n\frac{Z_{2L}R_\infty}{I_{L_i}}u_{L_i}(\xi_{Li}), \tag{1}$$

where $Z_{2L} = Z_2 - 4.15$ is the effective charge of the target atom, I_{L_i} is the L_i-subshell electron binding energy, R_∞ is Rydberg constant, $u_{L_i}(\xi_{L_i})$ is a universal function for specific L_i subshell as a function of ξ_{L_i}. Here, the function $u_{L_i}(\xi_{L_i})$ is redefined[6] to be compatible with the selected original definition of ξ_{L_i} given by Basbas *et al.*[7] as

$$\xi_{L_i} = \frac{2Z_{2L}}{n^2 I_{L_i}}\sqrt{\frac{m}{M_1}R_\infty E}, \tag{2}$$

where M_1 and m are the target atom and the electeron masses, respectively. E is the energy of the projectile. The resulting model, ECPSSR-IS, is compared with the experimental data as shown in figure (2). The main features obtained from fig.(2) indicate that the systematic disagreements, previously observed for the

CP475, *Applications of Accelerators in Research and Industry*, edited by J. L. Duggan and I. L. Morgan

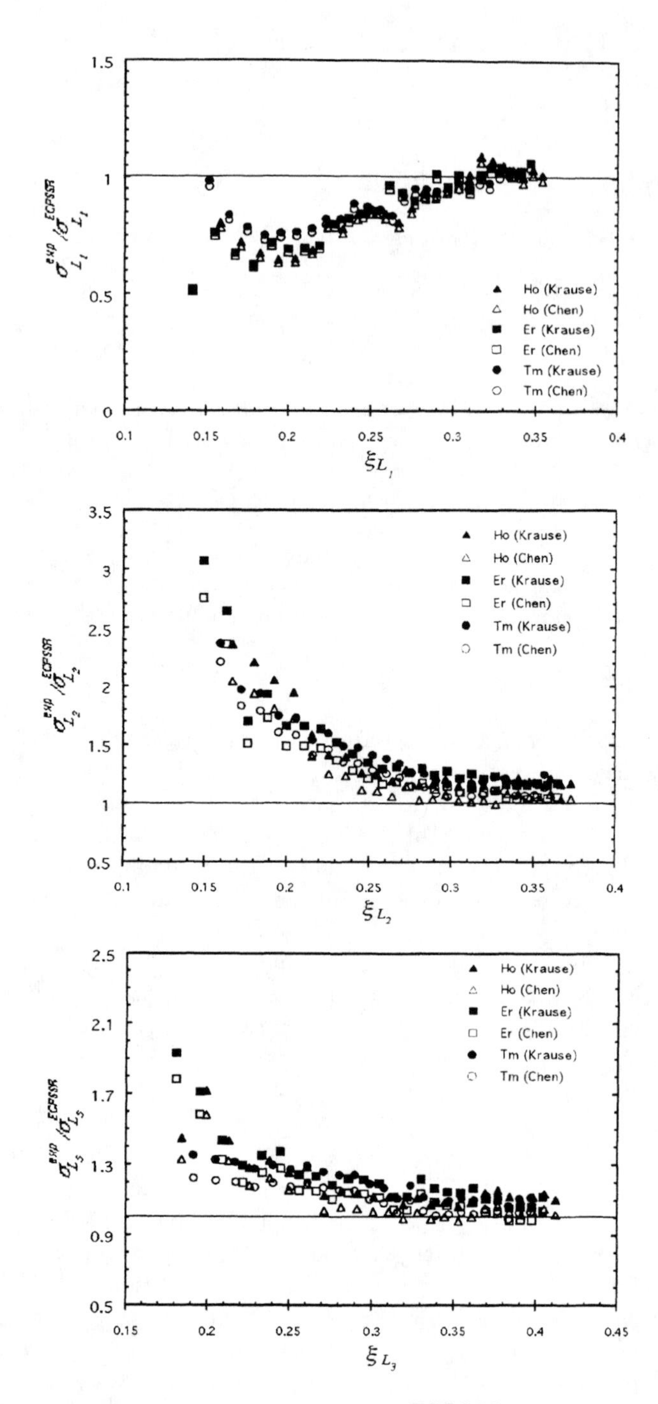

FIGURE 1. The ratio $\sigma_{L_i}^{\exp}/\sigma_{L_i}^{ECPSSR}$ versus the reduced velocity parameter ξ_{L_i} for L_1, L_2, and L_3-subshell ionization induced by α-particle bombardments. Each element is marked by a different symbol (see legend). Dark and open symbols indicate that the decay parameters of Krause and of Chen *et al.* are adopted, respectively, in the calculations of the ionization cross section.

ECPSSR model, are appreciably reduced for the L_2, and the L_3 subshells. Yet, the velocity-independent discrepancies remain, being even increased, especially for the L_1 subshell.

For the L_1 subshell, the average deviation when $\xi_{L_1} < 0.27$ is reduced to 12.4% and 14.4% between the ECPSSR-IS predictions and the experimental values calculated using the decay parameters of Chen *et al.*[8] and Krause[9], respectively. Whereas the average deviation when $\xi_{L_1} > 0.27$ is raised to 12% and 14% between the ECPSSR-IS predictions and the experimental values calculated using the decay parameters of Chen *et al.* and Krause, respectively.

For the L_2 subshell, the ECPSSR-IS underestimates the experimental values up to a factor of 1.86 instead of 3 in the ECPSSR. When $\xi_{L_2} < 0.28$ the average deviation is reduced to 26.6% and 34% for the experimental values calculated using the decay parameters of Chen *et al.* and Krause, respectively. When $\xi_{L_2} > 0.28$, the average deviation is raised to 17% (using the decay parameters of Chen *et al.*) and 25% (using the decay parameters of Krause).

Similar tendency is observed for the L_3 subshell. However, even at low velocities (when $\xi_{L_3} < 0.30$) the deviation increases to 27.7% (for the data calculated using the decay parameters of Chen *et al.*) and 33% (for the data calculated using the decay parameters of Krause), which corresponds to 10% increment with respect to the ECPSSR. Additionally, when $\xi_{L_3} > 0.30$, the average deviation increases to 11.35% (for the data calculated using the decay parameters of Chen *et al.*) and 17.7% (for the data calculated using the decay parameters of Krause).

3. THE BINDING CORRECTION

Following the suggestion of Vigilante *et al.*[10], that the ECPSSR overestimates the binding effect, and at low velocities, one must "saturate" the binding correction to a value that corresponds to the binding energy of the united atom (UA). Sarkadi and Mukoyama[5] account for this effect by using the function $g_{L_i}(\xi_{L_i})$ only to interpolate between the separated and united atom binding energies instead of using its absolute value, thus[5]

$$\epsilon_{L_i} = 1 + \frac{I_i(Z_1 + Z_2) - I_i(Z_2)}{I_i(Z_2)} \tag{3}$$

At low velocities $g_{L_i}(\xi_{L_i}) \approx 1$[5], and thereby $\epsilon_{L_i} I_{L_i}(Z_2) \approx I_{L_i}(Z_1 + Z_2)$. Whereas at high velocities $g_{L_i}(\xi_{L_i}) \approx 0$[5], and thus $\epsilon_{L_i} \approx 1$.

The cross section is recalculated using the recipe of Sarkadi and Mukoyama[5], by taking the function $g_{L_i}(\xi_{L_i}) = 1$ through our range of the reduced velocity parameter, since we always have $\xi_{L_i} < 1$ which means, according to Basbas *et al.*[7], that the collision is

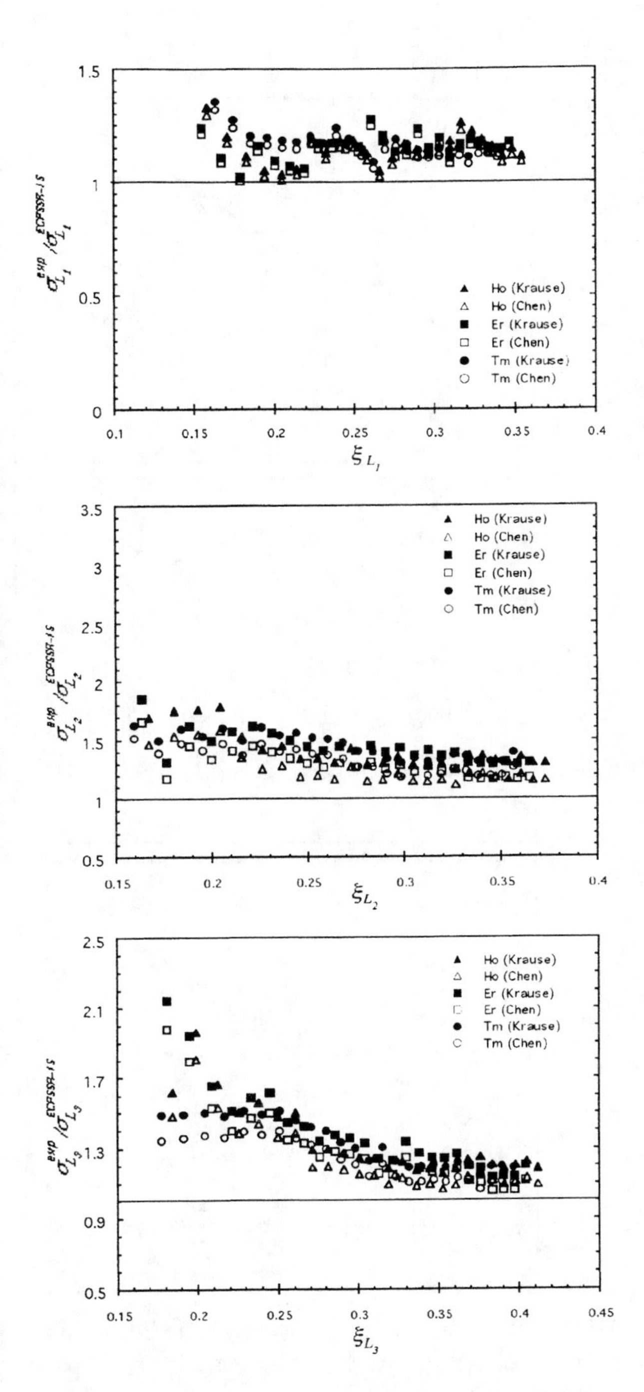

FIGURE 2. The ratio $\sigma_{L_i}^{\exp}/\sigma_{L_i}^{ECPSSR-IS}$ versus the reduced velocity parameter ξ_{L_i} for L_1, L_2, and L_3-subshell ionization induced by α-particle bombardments. Symbols and legends are decriped in figure (1).

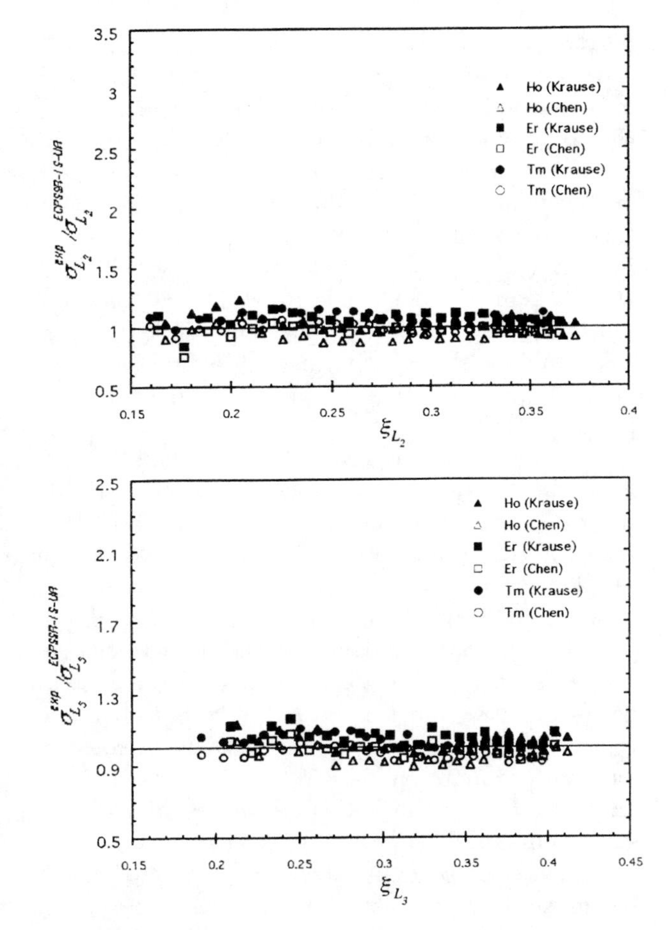

FIGURE 3. The ratio $\sigma_{L_i}^{\exp}/\sigma_{L_i}^{ECPSSR-IS-UA}$ versus the reduced velocity parameter ξ_{L_i} for L_2, and L_3-subshell ionization induced by α-particle bombardments. Symbols and legends are decriped in figure (1).

slow. The binding energies are taken from Bearden and Murr[11]. This modification, yielding the ECPSSR-IS-UA model, causes significant changes in the cross section for the L_2 and the L_3 subshells as shown in figure (3). The agreement is excellent through the whole range of the reduced velocity parameter. The average deviation for the L_2 subshell is reduced to 5.2% (using the decay parameters of Chen *et al.*) and 7.3% (using the decay parameters of Krause). Whereas the average deviation for the L_3 subshell is reduced to 3.8% (using the decay parameters of Chen *et al.*) and 4.3% (using the decay parameters of Krause).

For the L_1 subshell, the ECPSSR-IS-UA model does not give a pragmatic description for the experimental data. To verify the problem, we refer to the plot of the ionization cross section versus the bombardment energy, shown in figure (4), for the three elements, Ho, Er, and Tm, respectively. When $E \leq 3.2$ MeV, the ECPSSR-IS-UA model poorly describes the experimental data for the three elements. Moreover, a large departure is observed at energies higher than 3.2 MeV.

We attribute this phenomena to the fact that the calculated binding energy of the electron in the L_1 subshell should account for the screening effect, which reduces the effective charge of the projectile. This effect is especially significant for fast collisions in which the response time of the bound electron is much shorter than the collision time. Therefore, the electron does not experience the charge of the α particle to be +2. For slow collisions, however, the electronic response time is longer or comparable to the collision time, therefore, the electron becomes more sensitive to the actual Coulomb field of the projectile.

Accordingly, we re-calculate the ECPSSR-IS-UA model at a saturated binding energy corresponding to $I_{L_i}(Z_2+1)$ value. The modified ECPSSR-IS-UA model is represented by the dashed curves in figure (4). For the $E > 4$ MeV domain, good agreement can be achieved when the binding energy correction in the united atom approach considers the effective charge of the projectile to be between +1 and +2. As shown in fig. (4), the two versions of the ECPSSR-IS-UA model coincide when $E < 4$ MeV. This behavior is consistent with our presumption.

Yet, we have significant discrepancies around the L_1 subshell bump. The presence of the bump in the energy dependence of the L_1-subshell ionization cross section by charged particles, is one of the most characteristic features of that subshell. This bump is related to the extra node of the $2s_{\frac{1}{2}}$ radial wave function. Although it successfully predicts it, the theory exhibits a smoother bump than that demonstrated by the experimental data points. This indicates that the non-relativistic wave function, adopted within the original ECPSSR formalism, does not adequately describe the actual behavior of the L_1-subshell electron.

Attempts to overcome this problem are made by Sarkadi and Mukoyama[4,5]. They show that the use of the Dirac-Hartree-Slater (DHS) wave function instead of the screened hydrogen-like (SH) wave function produces changes in the L_1-subshell ionization cross section up to 60% for the lowest measured velocities, improving the agreement between theory and experiment[5,12]. Another attempt is carried out by DeCesare *et al.*[13], in which they use the RPWBA-BC of Chen and Crasemann[8] for protons to calculate the ECPSSR-IS-UA-DHS for α particles. This can be implemented by virtue of the scaling law after recalculating the proton RPWBA-BC at approximate integration limits of the momentum transfer, and hence incorporating more refined Coulomb deflection corrections. This yields some improvements in the agreement between the theory and the experimental data[13].

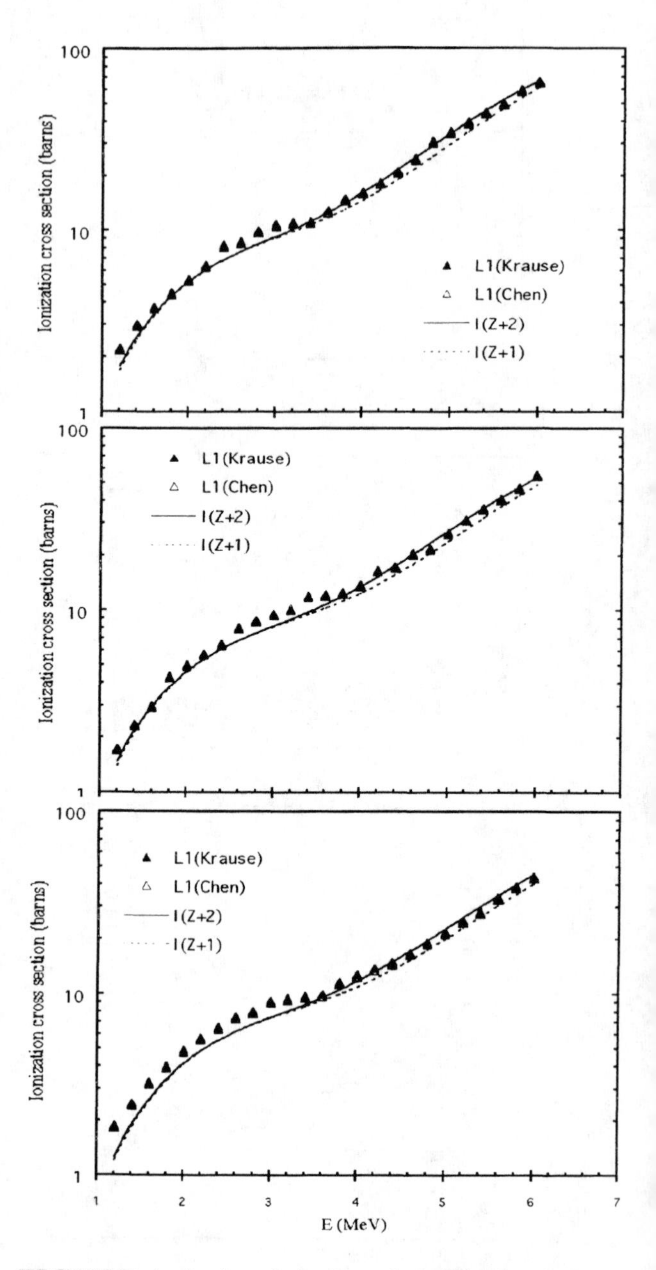

FIGURE 4. A plot of the L_1-subshell ionization cross sections of Ho, Er, and Tm induced by by α-particle bombardment, versus the bombardment energy. The solid and the dashed curves represent the ECPSSR-IS-UA models, calculated at saturated binding energies equal to $I(Z_1+2)$ and $I(Z_1+1)$, respectively.

4. CONCLUSION

The ECPSSR-IS-UA is capable of giving a full description of the L_2 and the L_3 subshell ionization cross sections of Ho, Er, and Tm induced by the α-particle bombardments. The discrepancies remain for the L_1 subshell through the entire range of the bombardment energy. The discrepancy at high bombardment energy is attributed to the screening effect, which becomes significant for fast collisions. Therefore the effective charge of the α particle, experienced by the electron at fast collision, should be reduced when incorporated in the binding calculations.

The discrepancy around the bump of the L_1-subshell ionization cross section as a function of the bombardment energy suggests that the screened hydrogenic wave functions in the original ECPSSR formalism does not adequately describe the actual behavior of the electrons in the L_1 subshell. There are evidences that the DHS wave functions can be a realistic alternative when incorporated within the theory, yielding the ECPSSR-IS-UA-DHS model.

Furthermore, the experimental ionization cross sections, calculated using the atomic decay parameters of Chen *et al.*[8], have better agreements with various versions of the theory than those calculated using the atomic decay parameters of Krause[9]. This implies that the theoretical atomic decay parameters of Chen *et al.*[8] are more realistic.

Accordingly, a full description of the L_1-subshell ionization cross section can be achieved when the ECPSSR-IS-UA is recalculated using the DHS wave functions, taking into account the precise effective charge of the projectile within the united-atom calculation scheme.

ACKNOWLEDGEMENTS

The authors are thankful to the KFUPM for their help and support to accomplish this work. Particular thanks to the Physics department and the Energy Resources Laboratory of the KFUPM Research Institute. One of the authors (Shehadeh) is thankful to Iowa State University for its support. Special thanks to S. Jun and J. M. Hill for their assistance in editing and revising the paper.

REFERENCES

1. Shehadeh B. A., and A. B. Hallak, under preparation.
2. Sarkadi, L., *J. Phys. B: At. Mol. Phys.*, **19**, 2519(1986).
3. Sarkadi, L., *J. Phys. B: At. Mol. Phys.*, **19**, L755(1986).
4. Sarkadi, L. and T. Mukoyama, *Phys. Rev.*, **A37**, 4540(1988).
5. Sarkadi, L. and T. Mukoyama, *Nucl. Inst. Meth.*, **B61**, 167(1991).
6. Shehadeh, B. A., *Measurements of L-Subshell Ionization Cross Sections by Energetic α Particles for Ho, Er, and Tm.* Dissertation, Dhahran Saudi Arabia, University of Petroleum and Minerals, 1994.
7. Basbas, G., W. Brandt, and R. Laubert, *Phys. Rev.* **A7**, 983(1973).
8. Chen, M., and B. Crasemann, *Atom. Data. Nucl. Data Tables*, **33**, 218(1985).
9. Krause, M. O. *J. Phys. Chem. Ref. Data,* **8**, 307(1979).
10. Vigilante, M., P. Cuzzocrea, N. DeCesare, F. Murolo, E. Perillo, and G. Spadaccini, *Nucl. Inst. Meth.*, **B51**, 232(1990).
11. Bearden, J. A. and A. F. Burr, *Rev. Mod. Phys.*, **39**, 125(1967).
12. Semaniak, J., J. Braziewicz, M. Pajek, A. Kobzev, and T. Trautmann, *Nucl. Inst. Meth.*, **B75**, 63(1993).
13. DeSesare, N., F. Murolo, E. Perillo, G.Spadaccini, and M. Vigilante, *Nucl. Inst. Meth.*, **B84**, 295(1994).

MULTIPLE IONIZATION IN M-, N- AND O-SHELL IN COLLISIONS OF O, SI AND S IONS WITH HEAVY ATOMS

M. Pajek [1], D. Banaś [1], J. Braziewicz [1], U. Majewska [1], J. Semaniak [1], T. Czyżewski [2] M. Jaskóła [2], W. Kretschmer [3], T. Mukoyama [4], D. Trautmann [5] and G. Lapicki [6]

1) Institute of Physics, Pedagogical University, 25-405 Kielce, Poland
2) Soltan Institute for Nuclear Studies, 05-400 Otwock-Świerk, Poland
3) Physikalisches Institut, Universität Erlangen-Nürnberg, D-91058 Erlangen, Germany
4) Institute for Chemical Research, Kyoto University, Kyoto 606, Japan
5) Institute of Physics, University of Basel, CH-4056 Basel, Switzerland
6) Department of Physics, East Carolina University, Greenville, NC 27858, USA

Multiple ionization in M-, N- and O-shells in solid Au, Bi, Th and U targets was studied for O^{q+} , Si^{q+} and S^{q+} ions of energies 0.4-2 MeV/amu under charge equilibration condition. From L-x-rays, detected by Si(Li) detector, the ionization probabilities for M- and N-shells were derived. This was done by fitting the measured spectra using a developed model, which describes x-ray energy shifts and line broadening in terms of the ionization probabilities. Derived ionization probabilities are compared with the predictions of the geometrical model (GM). For M- and N-shell the data agree quite well with calculations. The ionization probabilities for O-shell, estimated from line intensity ratios, are in strong disagreement with theoretical predictions, indicating an influence of solid state effects. An importance of multiple ionization effects on derivation of L-subshell ionization cross sections is evidenced and discussed.

INTRODUCTION

In collisions of heavy, energetic ions with atoms, due to the interaction of strong Coulomb field of projectile with target electrons, more than one electron can be ionized. Such multiple ionization, by reducing a nuclear charge screening, leads to occurring of the satellite transitions in emitted x-ray spectra. Consequently, a structure of satellite transitions contains information on a degree of multiple ionization in different atomic shells and by applying high resolution x-ray spectroscopy the ionization probabilities can be measured.

In the present paper we demonstrate that almost the same information on the ionization probabilities in outer shells can be obtained by using conventional Si(Li) detectors. This can be done by including the multiple ionization effects in the fitting procedure to resolve the measured x-ray spectra. In fact, as we showed earlier (1), the *shifts* and *widths* of x-ray transitions modified by the multiple ionization, can be expressed in terms of the ionization probabilities, allowing their unique and accurate estimation in fitting x-ray spectra. The developed method of x-ray spectra analysis was applied to study the multiple ionization effects in M-, N- and O-shells in collisions of O, Si and S ions with selected heavy atoms (Au, Bi, Th and U). By analyzing measured L_γ x-rays, the ionization probabilities for M- and N-shell were derived, which were interpreted as the ionization probabilities at the zero impact parameter. The ionization probabilities for the O-shell were derived from the measured ratios $L_{\gamma1}/L_{\gamma6}$, which indicated a reduction of the degree of ionization in O-shell, as compared to the theoretical predictions. This finding indicates an importance of the solid state effect for weakly bound electrons in O-shell.

Theoretical ionization probabilities can be calculated by using the geometrical model developed by Sulik et al. (2,3). In calculating the ionization probabilities the question of the projectile nuclear charge screening by its electrons has to be addressed, especially for outer shells with weakly bound electrons. Moreover, for collisions in solid targets, in which projectile charge equilibration takes place, a screening can be further modified. Both these effects can thus influence the theoretical predictions of the ionization probabilities.

EXPERIMENT AND METHOD

The measurements were performed at the Institute of Physics of the University Erlangen-Nürnberg, Germany. Thin solid targets of Au, Bi, Th, and U were bombarded by $^{16}O^{q+}$, $^{28}Si^{q+}$ and $^{32}S^{q+}$ ions of energy 0.4 - 2.0 MeV/amu from the EN tandem accelerator. The ion beam charge states q ranged between 3+ and 6+. Target thickness were 10-30$\mu g/cm^2$, thus allowing projectile charge equilibration in the targets. This was checked in separate measurements by comparing the X-ray yields for different target thickness. The x-rays were detected with a carefully energy

CP475, *Applications of Accelerators in Research and Industry*, edited by J. L. Duggan and I. L. Morgan

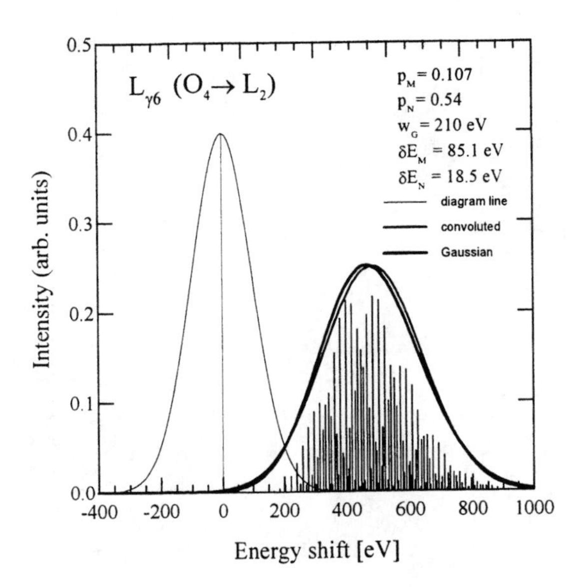

FIGURE 1. Simulated modification of x-ray spectrum of $L_{\gamma6}$ (L_2-O_4) transition (diagram line) in Au due to the multiple ionization. A convolution of the satellite transitions with the energy spread in Si(Li) detector is compared with predicted Gaussian shape described by Eqs. 1 and 2. The parameters used in calculations are shown in the figure.

calibrated Si(Li) detector of 180 eV resolution for 5.9 keV line.

Multiple ionization of outer M-, N-, and O-shells, that results in appearance of satellite structure, gives rise to the energy shift of individual L_γ-x-ray peaks, with respect to their diagram energies, as well as their broadening. By assuming a binomial distribution of intensities of satellite transitions and their Gaussian energy spread in a Si(Li) detector, the resulting x-ray peaks preserve Gaussian shape. It is evidenced in Fig. 1, where the simulation of the x-ray satellite structure for $L_{\gamma6}$ (L_2-O_4) transition, as expected to

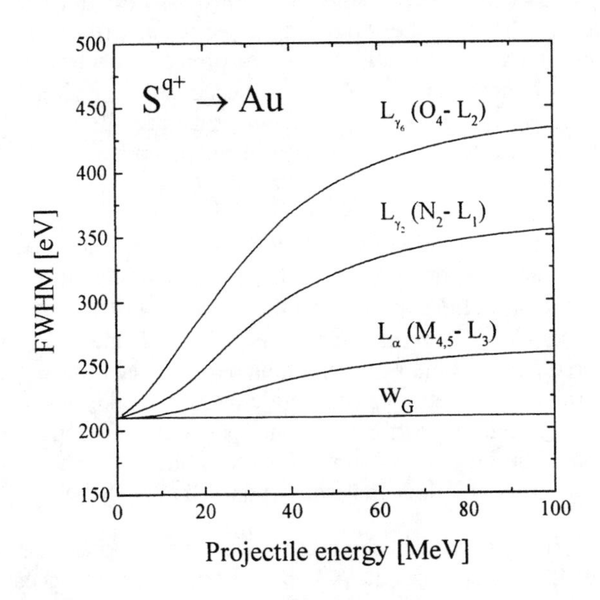

FIGURE 2. Expected dependence of a width of different x-ray lines on the projectile energy (Eq. 2) for S ions on Au. The ionization probabilities were calculated according to the geometrical model (3).

be measured by a Si(Li) detector, is shown. As we demonstrated earlier (1), the x-ray energy shifts ΔE and corresponding widths w can be expressed in terms of the multiple ionization probabilities p_i as follows:

$$\Delta E = \sum_i n_i p_i \delta E_i \tag{1}$$

$$w = \left[w_G^2 + 8 \ln 2 \sum_i n_i p_i (1 - p_i)(\delta E_i)^2 \right]^{1/2} \tag{2}$$

where n_i is the number of electrons in the i = M, N, O shells. Here, a Gaussian shape of the detector response function with a width w_G has been assumed. The energy shifts per spectator vacancy in outer M, N and O shells, δE_i, have been calculated for individual L_γ transitions by performing Dirac-Fock calculations (4). We note here that an important ingredient of the present approach is the introduction of a dependence of x-ray peak width on the ionization probability. As it is demonstrated in Fig. 2 this results in a substantial increase, up to a factor of two, of L_γ peak widths for ion impact energies of interest.

Performing a fit of measured L_γ x-ray spectra according to the outlined procedure we were able to derive ionization probabilities for M- and N-shell. We found that this method is not sensitive enough to estimate the ionization probabilities for the O-shell, mainly due to very the small contribution of the O-shell to the observed x-ray shifts. Final uncertainties of the measured ionization probabilities are below 20%, and they are attributed mainly to the uncertainties of the x-ray energy calibration (± 5 eV), calculated energy shifts per vacancy δE_i (± 1 eV) and our assumption of the Gaussian form of the x-ray peaks.

RESULTS

The multiple ionization probabilities per electron in M- and N-shell have been derived from L_γ-x-ray spectra of Au, Bi, Th, and U measured for O, Si, and S bombardment, over the energy range between 0.4 - 2.0 MeV. A typical measured and analyzed L_γ spectrum excited by 41.6-MeV $^{32}S^{q+}$ ions on Au is shown in Figure 3. The energy dependence of the ionization probabilities for M- and N-shell measured for Au target bombarded by sulfur ions is shown in Fig. 4. The ionization probabilities for M-shell, generally, do not exceed 0.1, while for N-shell they range between 0.15 and 0.6 for O and S ions, respectively.

The vacancies created initially in collision can be further redistributed within the same shell via Coster -Kronig transitions, as well as between different shells via Auger or radiative transitions. The time scales for these processes (5)

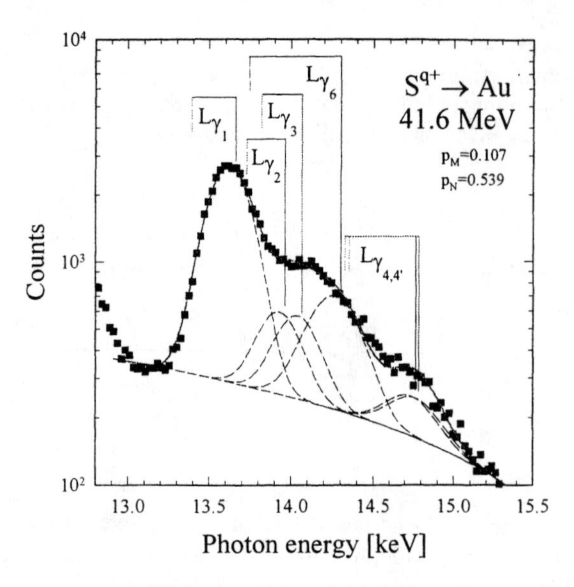

FIGURE 3. Typical L_γ x-ray spectrum measured for 41.6 MeV S ions on Au. Fitted ionization probabilities are also shown.

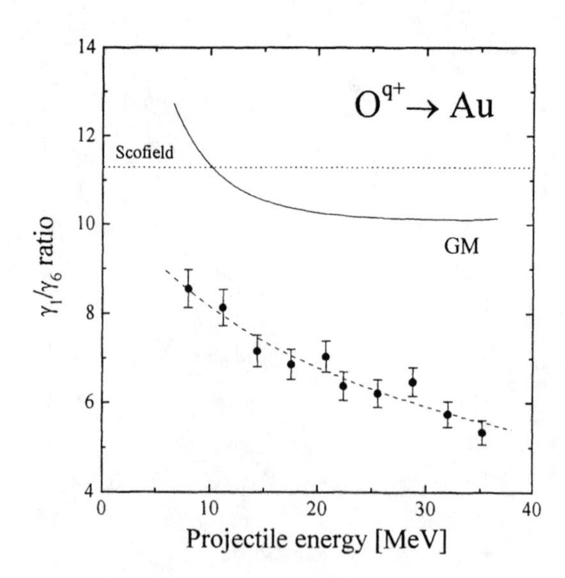

FIGURE 5. Measured intensity ratio $L_{\gamma1}/L_{\gamma6}$ in comparison with the ratio expected from the geometrical model (solid line) and the ratio of emission rates according to Scofield (9). Dashed curve through the experimental points is to guide the eyes.

show that during L_γ-x-ray emission most of the vacancies are in outer subshells. This is so because after fast Coster-Kronig transitions, the Auger rates are strongly suppressed, because they are dominated by Auger transitions from outer subshells of the next shell, which are also strongly ionized. According to the performed calculations (4) the energy shift per spectator vacancy in M-shell is practically constant for all M- subshells and changes within 10% in N-shell. Thus the measured ionization probabilities reflect a number of primarily created vacancies. Since the adiabatic radius for L-shell ionization is much smaller then for outer M- and N-shells, these data can be interpreted as the ionization probabilities in near central collisions, i.e. at the zero impact parameter.

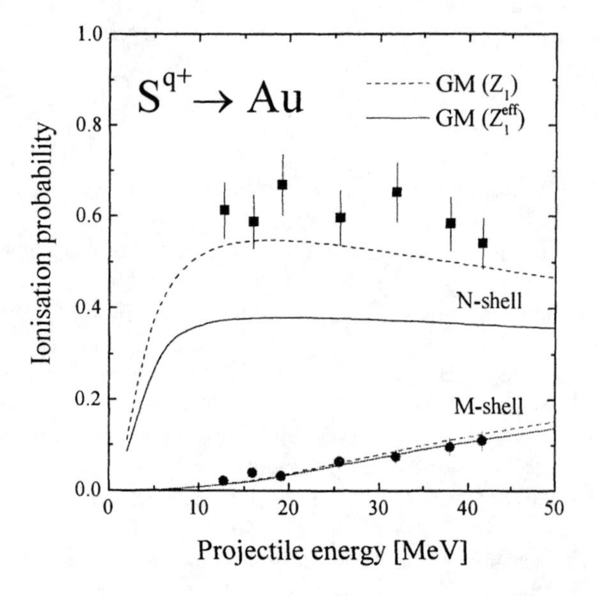

FIGURE 4. Derived ionization probabilities for M- and N-shell in comparison with the predictions of the geometrical model using the projectile nuclear charge (dashed line) and effective projectile charge (solid line).

The measured data are compared to the ionization probabilities for M- and N-shell at the zero impact parameter predicted by, the so called, geometrical model of Sulik et al. (2,3). The geometrical model (GM), developed on the basis of the binary encounter approximation (BEA), can be used to calculate the ionization probability adopting as a projectile charge the nuclear charge Z_1, as well as a screened, effective projectile charge Z_1^{eff} (6). The screening of the projectile during the collision has been calculated according to the procedure proposed by Toburen at al. (7). The effective projectile charges Z_1^{eff} for a charge equilibrated projectiles were calculated by weighting screened charges for a given projectile charge state $Z_1^{eff}(q)$ by the equilibrium charge state fractions F(q) (8). For the studied energies the effective charge of the projectile seen by M-shell electrons is close to the projectile nuclear charge Z_1, whereas for N-shell it is substantially smaller. Both theoretical approaches describe satisfactorily the data measured for M-shell, however, the experimental results for N-shell agree much better with the GM model using the projectile nuclear charge. Both models predict high multiple ionization probability for O-shell and thus a substantial reduction of L-O transition intensities. On contrary, in the experiment an enhancement of the intensity of $L_{\gamma6}$ (L_2-O_4) transition (see Fig. 5), as compared to the predictions of the geometrical model, has been found. This probably indicates a solid state effect for weakly bound O-shell electrons, being close, and partly in, the conduction band. In this case fast transitions from the conduction band can strongly influence vacancy population in O-shell.

The application in the present work of described method of L-x-ray spectra analysis has important consequences for determination of L_i-subshell cross sections. Usually they are derived from $L_{\alpha1,2}$, $L_{\gamma1}$ and $L_{\gamma2,3,6}$ x-ray production cross sections using the fluorescence and Coster-Kronig yields and x-ray emission rates for single vacancy configuration.

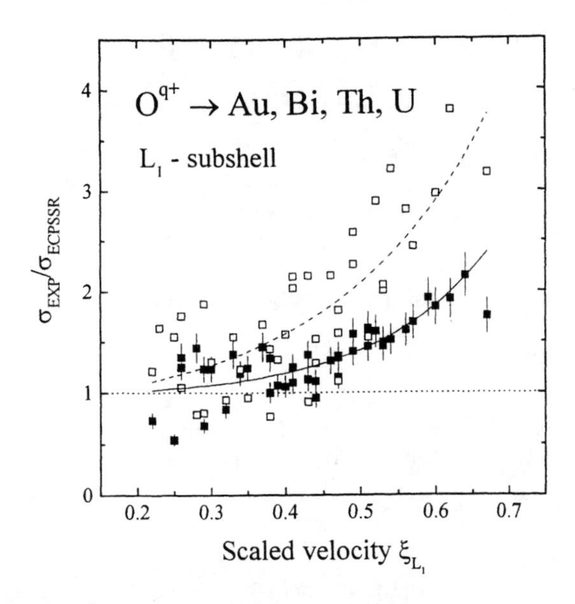

FIGURE 6. Measured L_1-subshell ionization cross sections for Au, Bi , Th and U targets bombarded by O ions normalized to the predictions of the ECPSSR theory (10). The results are plotted versus scaled velocity. The data marked by closed symbols were obtained by applying the fitting method of L-x-ray spectra described in the text, while the open symbols were obtained using standard fitting procedure.

In the present method, a more realistic description of x-ray lines modified by the multiple ionization effects (energy shifts and line broadening) allows unique resolution L_γ x-ray structure by reducing the number of parameters to be fitted. This is especially important for determination of L_1-subshell ionization cross sections using resolved $L_{\gamma 2,3,6}$ lines, instead correcting for a contribution of the $L_{\gamma 6}$ transition. A result of applying the present method to derive L_1-subshell ionization cross sections for different heavy targets (Au, Bi, Th, and U) bombarded by oxygen ions, as compared to the predictions of the ECPSSR theory (10), is shown in Fig. 6. In the figure one finds almost a factor of two reduction of discrepancies with theoretical prediction, as well as a substantial reduction of a spread of the data.

In conclusion, we have developed the method of analysis of x-ray spectra modified by the multiple ionization effects, which allows the accurate determination of the ionization probabilities for M- N- and O-shells. By using this method more realistic determination of L-subshell ionization cross sections is achieved, as well as the investigation of solid state effects is possible. The last question, however, needs further experimental and theoretical studies.

ACKNOWLEDGEMENTS

This work as supported by the Federal Ministry for Research and Technology, Germany, under Contract No. N-88-94 and by the Polish State Committee for Scientific Research under Grant No. 2PO3BO6514. We wish to express our appreciation to the staff of the EN tandem accelerator in Erlangen for their kind collaboration during the measurements.

REFERENCES

1. M. Pajek, D. Banaś, J. Braziewicz, T. Czyżewski, M. Jaskóła, W. Kretschmer and J. Semaniak in *Book of Abstracts, 20th ICPEAC*, ed. F. Aumayr, G. Betz and HP. Winter, Vienna, 1997, p. 158.
2. B. Sulik, G. Hock and D. Berényi, *J. Phys. B*17, 3239 (1984).
3. B. Sulik, I. Kádár, S. Ricz, D. Varga, J. Végh, G. Hock and D. Berényi, *Nucl. Instr. and Meth. B*28, 509 (1987).
4. T. Mukoyama et al. – to be published
5. O. Keski-Rahkonen and M.O. Krause, *At. Data and Nucl. Data Tables*, **14**, 139 (1974).
6. G. Hock , B. Sulik, J. Végh , I. Kádár, S. Ricz and D. Varga, *Nucl. Instr. and Meth. A*240, 475 (1985).
7. L.H. Toburen, N. Stolterfoht, P. Ziem, and D. Schneider, *Phys. Rev. B*24, 1741 (1981).
8. K. Shima, N. Kuno, M. Yamanouchi and H. Tawara, At. *Data and Nucl. Data Tables*, **51**, 173 (1992).
9. J. H. Scofield, *At. Data. Nucl. Data Tables* **14**, 121 (1974).
10. W. Brandt and G. Lapicki, *Phys. Rev. A*20, 465 (1979); *Phys. Rev. A*23, 1717 (1981).

M X-RAY PRODUCTION IN Yb, Hf, W, AND Pb FROM IMPACT BY 75-300 keV PROTONS

Sam J. Cipolla, Paul J. Teeter, and Jeffrey D. McClure

Physics Department, Creighton University, Omaha NE 68178

M subshell and total x-ray production cross sections have been measured for 75-300 keV protons impacting pure elemental thick targets. The results are compared with the ECPSSR theory and other measurements.

INTRODUCTION

M x-rays from inner-shell ionization of heavy atoms can be measured effectively with current high resolution Si(Li) detectors equipped with ultra-thin or no windows. Although M x-rays are not commonly used in applications, they are of interest in testing ionization models, such as ECPSSR(1). The ECPSSR (Perturbed Stationary State with Relativistic, Coulomb deflection, and Energy loss corrections) theory has application over a wide projectile range. It has been developed mostly to describe direct ionization of K shells by simple charged ions, and much work recently involves testing ECPSSR to describe L-shell ionization. M-shell ionization has been studied the least due to the complexity of the energy spectrum and interference from contaminant low-energy x-rays. These obstacles can be overcome with the use of modern Si(Li) detectors and appropriate spectrum fitting techniques.

EXPERIMENTAL METHODS

A high-resolution $30mm^2$x3mm Si(Li) detector equipped with an ultra-thin boron nitride window was used to measure M x-ray spectra for protons striking thick foil targets of Yb, Hf, W, and Pb. A double layer of 6-μm aluminized Mylar absorbers was also used to cut down detection of very soft x-rays. The detector was calibrated for efficiency using a model-based formula presented earlier(2). The protons were obtained from a 350-kV Cockcroft-Walton accelerator equipped with beam optics to steer and focus the beam, a mass analyzer magnet, and a biased 1.5-mm beam collimator. The targets were arranged on vertical ladder so that a different spot was used in each measurement. A biased screen surrounded the ladder to suppress secondary electrons. Beam currents were kept low (< 2% dead time) to minimize pile-up.

X-ray peaks were fitted to a sum of gaussian functions plus a fixed linear background. The gaussian widths were fitted to FWHM = $2.355(A + B(E_x)^{1/2})$, where A and B are the fit parameters, and the x-ray energy E_x is determined from $E_x = a + bX + cX^2$, where X, a, b, and c are fitted parameters, X being the peak channel number.

X-ray production cross sections, $\sigma_x(E)$, for each transition were obtained from(3),

$$\sigma_x(E)=\frac{1}{N\epsilon}[S(E)\frac{dY}{dE}+\frac{\mu}{\rho}Y(E)] \qquad (1)$$

where $Y(E) = N_x/N_p$ is the x-ray yield (x-rays/proton), S(E) is the target stopping power for protons(4), μ/ρ is the target mass absorption coefficient(5), N is the target atom density, ε is the detection efficiency(3), and dY/dE was determined analytically from a fit to the yields according to $Y(E) = A(E-C)^B$, where A, B, and C are the fit parameters.

RESULTS

Table 1 presents the x-ray production cross sections, σ_x, measured for the dominant transitions. Of these transitions, those used to obtain the subshell cross sections were: M_1N_3, M_2N_4, $M_3N_{4,5}(\gamma)$, $M_4N_6(\beta)$, $M_5N_{6,7}(\alpha)$. The Pb target turned out to be a composite with Sb, which resulted in Sb L x-rays interfering with Pb M_1 x-rays so that no M_1 subshell analysis was possible. Subshell x-ray production cross sections, σ_{Mi}, were obtained from σ_x according to,

$$\sigma_{Mi}=\sigma_x\frac{\Gamma_{Mi}}{\Gamma_x} \qquad (2)$$

where Γ_x and Γ_{Mi} are the x-ray transition rates for a transition (peak) to the M_i(i=1,2,3,4,5) subshell and the total rate for the subshell, respectively(6). Figures 1-5 display ratios of σ_{Mi} to ECPSSR values(7) as a function of the reduced projectile velocity,

$$\xi_s=\frac{2v_1}{\theta_s v_{2s}} \qquad (3)$$

CP475, *Applications of Accelerators in Research and Industry*,
edited by J. L. Duggan and I. L. Morgan

TABLE 1. M X-Ray Production Cross Sections (barns).*

	E(keV)	M_ζ	M_α	M_β	M_γ	M_2N_4	M_1N_3	$M_1O_{2,3}$	
Γ_{Mi}/Γ_x=		0.0563	0.950	0.929	0.658	0.679	0.401	0.112	Yb
	75	6.81-1(11)	3.65(11)	4.43-1(20)	3.43-2(20)	1.38-2(29)	2.86-3(31)	3.77-4(15)	
	100	2.08(11)	11.0(11)	1.03(18)	1.16-1(14)	4.19-2(19)	8.86-3(20)	1.80-3(13)	
	125	4.08(11)	21.0(11)	1.90(16)	2.24-1(13)	7.64-2(17)	1.93-2(20)	4.03-3(13)	
	150	6.73(10)	33.4(10)	3.69(17)	3.58-1(12)	8.32-2(13)	2.47-2(16)	7.38-3(12)	
	175	10.3(10)	47.6(10)	6.48(16)	6.43-1(12)	1.32-1(13)	4.36-2(16)	1.10-2(12)	
	200	14.1(10)	66.2(10)	8.97(15)	7.67-1(11)	1.87-1(12)	3.71-2(13)	1.52-2(11)	
	225	18.0(10)	82.8(10)	12.6(14)	1.00(11)	2.69-1(12)	7.67-2(13)	1.77-2(11)	
	250	22.2(10)	100(10)	25.4(14)	1.41(11)	2.25-1(11)	6.13-2(12)	2.39-2(11)	
	275	28.8(10)	129(10)	18.1(12)	1.73(11)	4.50-1(11)	1.25-1(12)	2.85-2(11)	
	300	34.4(10)	152(10)	27.0(12)	2.29(11)	4.42-1(11)	9.32-2(11)	3.43-2(11)	
Γ_{Mi}/Γ_x=		0.0524	0.948	0.932	0.659	0.677	0.385	0.119	Hf
	75	4.52-1(20)	2.60(18)	4.70-1(28)	4.66-2(12)	9.9-3(26)	2.32-3(24)	3.90-4(51)	
	100	1.24(19)	7.18(18)	1.39(25)	1.40-1(11)	3.32-2(22)	7.11-3(21)	1.33-3(48)	
	125	2.42(18)	14.0(17)	3.05(21)	2.84-1(11)	7.22-2(20)	1.53-2(18)	3.14-3(42)	
	150	4.10(16)	23.4(16)	5.28(20)	4.94-1(10)	1.34-1(18)	2.75-2(15)	6.62-3(34)	
	175	6.00(15)	34.6(15)	7.48(19)	7.47-1(9)	2.12-1(17)	4.02-2(15)	1.07-2(31)	
	200	8.65(14)	47.5(14)	9.78(19)	1.17(9)	3.03-1(16)	6.12-2(13)	1.49-2(31)	
	225	11.1(14)	61.8(14)	15.7(16)	1.54(10)	4.71-1(14)	7.58-2(13)	2.10-2(29)	
	250	13.3(14)	77.4(14)	17.8(17)	1.87(10)	5.42-1(15)	9.10-2(13)	2.50-2(31)	
	275	16.5(13)	95.5(14)	23.4(16)	2.52(10)	7.13-1(14)	1.13-1(13)	3.02-2(32)	
	300	20.1(13)	116(13)	28.5(15)	3.12(10)	8.84-1(14)	1.34-1(13)	3.60-2(32)	
Γ_{Mi}/Γ_x=		0.0429	0.957	0.935	0.743	0.674	0.369		W
	75	3.11-1(9)	2.32(9)	8.26-1(11)	3.26-2(9)	7.44-3(9)	1.80-3(26)		
	100	9.15-1(9)	6.86(8)	2.58(11)	9.8-2(8)	2.54-2(8)	7.26-3(25)		
	125	1.74(8)	12.8(8)	4.85(10)	1.85-1(8)	5.03-2(8)	1.46-2(23)		
	150	2.78(8)	20.4(8)	8.02(10)	3.05-1(8)	8.42-2(8)	2.55-2(19)		
	175	4.07(8)	29.6(8)	11.8(9)	4.79-1(8)	1.29-1(8)	3.92-2(17)		
	200	5.55(8)	39.7(8)	16.4(9)	6.43-1(8)	1.85-1(8)	4.95-2(17)		
	225	6.95(8)	50.7(8)	20.4(9)	8.44-1(9)	2.34-1(8)	6.46-2(15)		
	250	8.91(8)	63.9(8)	26.1(9)	1.09(9)	3.02-1(9)	7.15-2(16)		
	275	10.9(8)	77.6(9)	33.1(9)	1.39(9)	3.91-1(9)	8.88-2(15)		
	300	12.8(9)	91.7(8)	38.6(9)	1.72(9)	4.54-1(9)	1.02-1(15)		
Γ_{Mi}/Γ_x=		0.0337	0.913	0.943	0.600	0.601			Pb
	75	1.33-2(18)	2.16-1(31)	9.52-2(20)	5.14-3(15)	3.22-4(28)			
	100	7.87-2(18)	1.31(32)	6.06-1(21)	2.70-2(15)	2.69-3(28)			
	125	2.05-1(17)	3.45(31)	1.67(20)	6.87-2(14)	8.35-3(26)			
	150	4.07-1(16)	6.81(31)	3.40(20)	1.35-1(13)	1.95-2(24)			
	175	6.9-1(16)	11.4(30)	5.85(19)	2.27-1(12)	3.56-2(22)			
	200	1.05(15	18.7(27)	8.96(19)	3.3-1(12)	5.59-2(21)			
	225	1.47(14)	24.2(28)	12.7(19)	4.62-1(11)	8.14-2(21)			
	250	1.95(14)	32.1(28)	17.1(18)	6.09-1(11)	1.12-1(21)			
	275	2.53(13)	41.6(27)	22.5(18)	8.07-1(11)	1.52-1(20)			
	300	3.10(14)	51.5(27)	28.2(18)	1.01(11)	1.98-1(19)			

*N-n represents $N\times10^{-n}$; () are percent uncertainties.

where s is the subshell, v_1 is the projectile velocity, θ_s is the shell binding energy, and v_{2s} is the electron orbital velocity.

DISCUSSION

A perusal of fig. 1-5 shows that ECPSSR theory tends to under-estimate our experimental M_1, M_2, and M_3 cross sections for all four elements, the more so as ξ_s decreases. The M2 shell results for Pb are an apparent exception in that the ratio $\sigma_{expt}/\sigma_{ECPSSR}$ is constant at value slightly higher than unity. Regarding the M_4 shell, ECPSSR under-estimates our results for W and Pb, while the it begins to over-estimate the cross sections for Yb and Hf at around $\xi_s \geq 0.5$. The M_5-shell comparison shows that ECPSSR under-estimates the Pb cross sections and over-estimates the rest up to around $\xi_s \geq 0.6$.

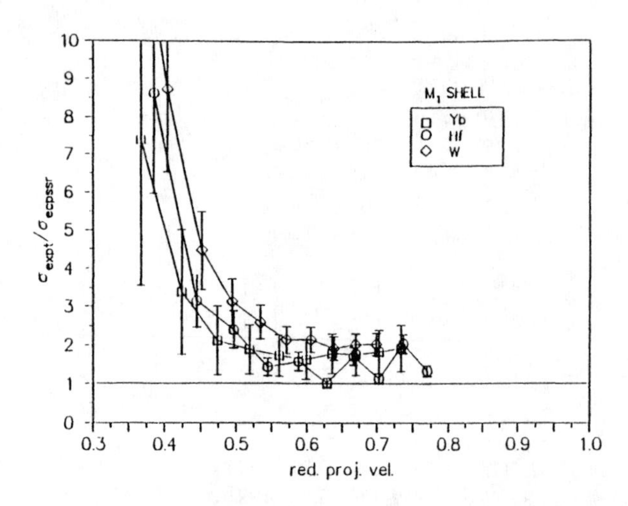

FIGURE 1. Comparison of M_1 shell x-ray production cross sections with ECPSSR calculations(7).

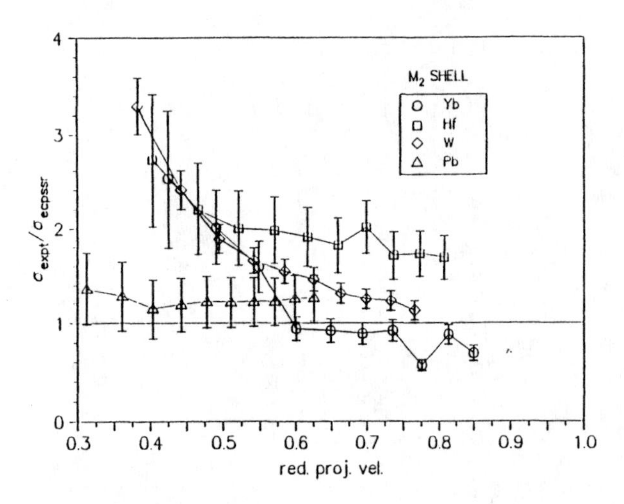

FIGURE 2. Comparison of M_2 shell x-ray production cross sections with ECPSSR calculations(7).

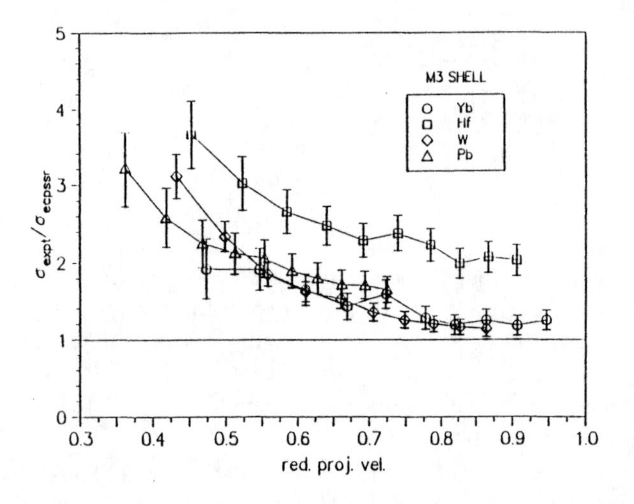

FIGURE 3. Comparison of M_3 shell x-ray production cross sections with ECPSSR calcualtions(7).

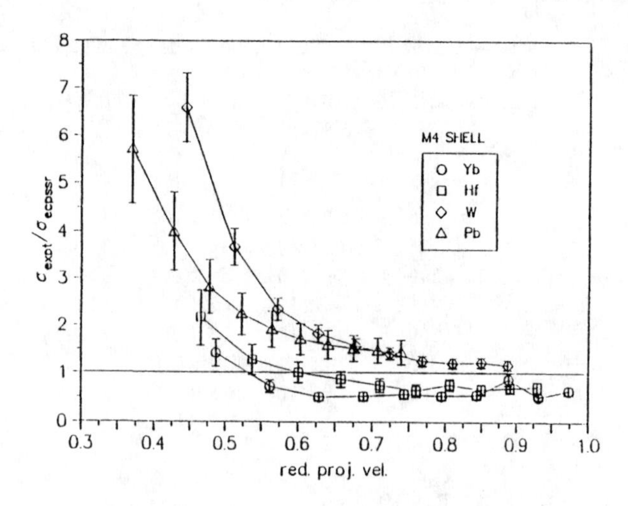

FIGURE 4. Comparison of M_4 shell x-ray production cross sections with ECPSSR calculations(7).

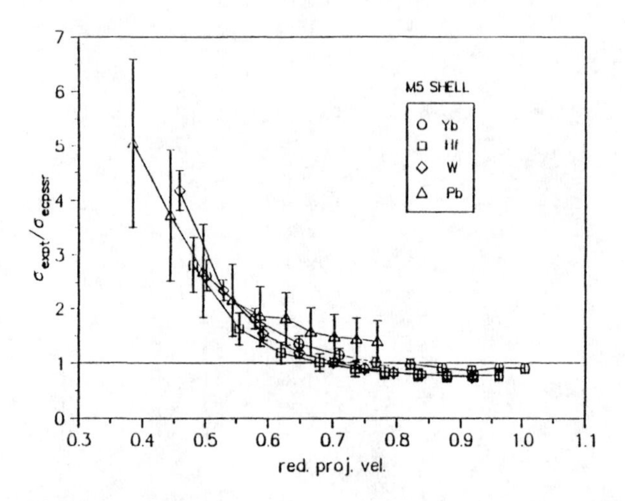

FIGURE 5. Comparison of M_5 shell x-ray production cross sections with ECPSSR calculations(7).

There are few measurements by others in this energy region. Needham and Sartwell(8) measured total M-shell ionization cross sections for 60-200 keV protons on Hf and W. Mehta et al.(9) measured the total M x-ray production cross sections for 200 keV protons on Yb; Mehta et al.(10) and de Castro Faria et al.(11) for 300 keV protons on Pb. Gressett, et al.(12) measured M shell x-ray production cross sections for 70-200 keV protons on Pb. Bienkowski et al.(13) reported total M x-ray production cross sections for 100-600 keV protons on W. Figures 6-8 compare a sampling from these references with our work.

Figures 6-8 also show that ECPSSR tends to underestimate the total x-ray production cross section as the target atomic number increases for these elements. At 300 keV (fig. 8), these results are very similar to those found by Pajek et al.(14) for 3 MeV protons.

We also calculated ECPSSR x-ray production cross sections using the computer program ISICS(15), which uses a united-atom approach for the binding-energy effect and

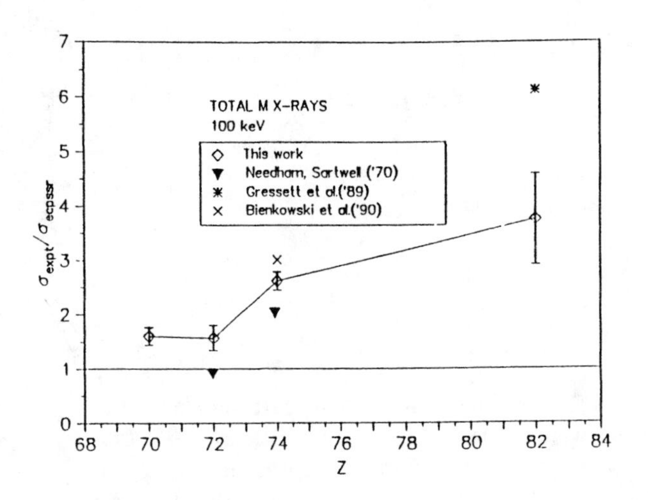

FIGURE 6. Comparison of the total M x-ray production cross section with ECPSSR as a function of target atomic number for 100 keV proton bombardment.

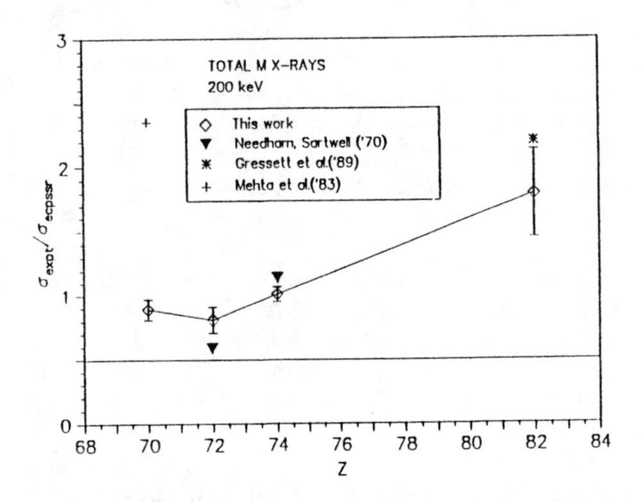

FIGURE 7. Comparison of the total M x-ray production cross section with ECPSSR as a function of target atomic number for 200 keV proton bombardment.

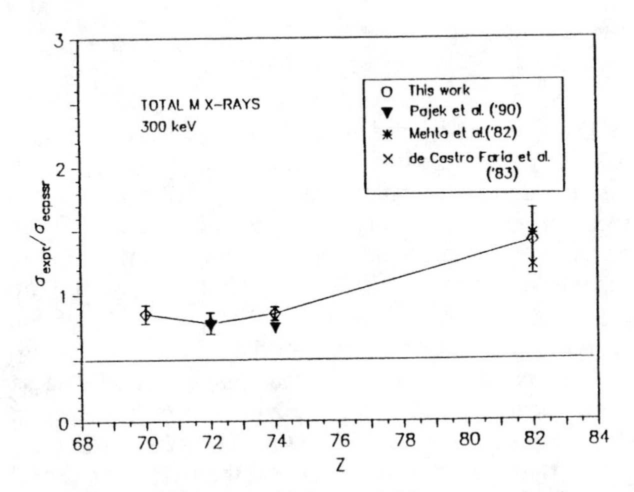

FIGURE 8. Comparison of the total M x-ray production cross section with ECPSSR as a function of target atomic number for 300 keV proton bombardment.

has no relativistic correction, in order to test for these effects. But the cross sections were insignificantly different from those used in the analyses presented here(7).

CONCLUSION

The M sub-shell x-ray production cross sections for 75-300 keV protons on thick targets of Yb, Hf, W, and Pb measured in this work show a variation when compared with ECPSSR theory. Our results are consistent with the few total x-ray production cross sections measured by others and showed a similar trend versus target atomic number or reduced projectile velocity. The ECPSSR M x-ray cross sections were also seen to be relatively insensitive to the relativistic correction by comparison with calculations from the program ISICS.

ACKNOWLEDGEMENTS

This work was supported by a Cottrell College Science grant from the Research Corporation and was conducted at the University of Nebraska at Lincoln.

REFERENCES

1. Brandt, W., and Lapicki, G., *Phys. Rev.* A **23**, 1717 (1981).
2. Gallagher, W. and Cipolla, S., *Nucl. Instr. Meth.* **122**, 405-414 (1974); Cipolla, S. and Watson, S., *Nucl. Instr. Meth.* **B10/11**, 946-949 (1985).
3. Merzbacher, E., and Lewis, H., *Handbuch der Physik*, vol. **34**, Berlin: Springer Press, 1958, p. 166.
4. Ziegler, J., "Transport of Ions in Matter,"(TRIM) ver. 91.07, IBM, Yorktown Hts., July 26, 1991.
5. Berger, M.J. and Hubbell, J.H., "XCOM: Photon Cross Section on a Personal Computer," NBS Report NBSIR 87-3595, Gaithersburg, Md., 1987.
6. Bhalla, C. P., *J. Phys.* **B3**, 916 (1970).
7. Lapicki, G., private communication.
8. Needham, P. B. and Sartwell, B. D., *Phys. Rev.* **A2**, 1686 (1970).
9. Mehta, R., Duggan, J. L., Price, J. L., McDaniel, F. D., and Lapicki, G., *Phys. Rev.* **A26**, 1883 (1982).
10. Mehta, R., Duggan, J. L., Price, J. L., Kocur, F. D., McDaniel, F. D., and Lapicki, G., *Phys. Rev.* **A28**, 3217 (1983).
11. de Castro Faria, N. V., Freire, F. L., de Pinho, A. G., and da Silveira, E. F., *Phys. Rev.* **A28**, 2770 (1983).
12. Gressett, J., Marble, D., McDaniel, F. D., Duggan, J. L., Culwell, J., and Lapicki, G., *Nucl. Instr. & Meth.* **B40/41**, 116 (1989).
13. Bienkowski, A., Braziewicz, J., Czyzewski, T., Glowacka, L., Jaskola, M., Lapicki, G., and Pajek, M., *Nucl. Instr. & Meth.* **B49**, 19-23 (1990).
14. Pajek, M., Kobzev, A. P., Sandrik, R., Skrypnik, A. V., Ilkhamov, R. A., Khusmurodov, S. H., and Lapicki, G., *Phys. Rev.* **A42**, 261-272 (1990).

L-Shell Ionization Probabilities for Near-Central Collisions of MeV Protons with Low-Z Atoms

M. Kavčič, M. Budnar, Ž. Šmit

J. Stefan Institute, Ljubljana, Slovenia

High resolution K_α X-ray spectra of Ca, Cr and Fe induced by 0.7 - 1.5 MeV protons and of Ti induced by 0.7 - 4 MeV protons were measured by a flat crystal spectrometer. The experiment was mainly performed at the J. Stefan Institute. Ti spectra induced by 2, 3 and 4 MeV protons were measured at the Ruđer Bošković Institute in Zagreb. From the relative yields of the KL^1 satellites, which are the result of multiple inner-shell ionization, the average L-shell ionization probabilities in near-central collisions were determined. The effects of the rearrangement of the inner-shell holes prior to the K X-ray emission and the changes of the fluorescence yields due to the multiple ionization were taken into account in the evaluation of the ionization probabilities. It was shown that the rearrangement of the L-shell holes significantly changes the initial inner-shell holes distribution. The change of the K-shell fluorescence yield due to additional L shell hole was estimated to be negligible. The inner shell ionization induced by proton collisions with light atoms is mainly result of the direct Coulomb interaction between the inner shell electron and the moving projectile. The direct Coulomb ionization probabilities in the near-central collisions were determined from the relative yields of the KL^1 satellites after subtracting the shake contribution. The values obtained were compared with semiclassical approximation calculations exploiting relativistic hydrogenic and Hartree-Fock wave functions. The importance of a realistic atomic description using Hartree-Fock wave functions was demonstrated.

INTRODUCTION

In a collision of fast ion with low or mid-Z atom an inner shell electron can easily be removed from the atom. The probability to remove an inner-shell electron strongly depends on the type of the projectile and its energy. In the collision of proton with low-Z atom the dominant ionization mechanism is the direct ionization. The Coulomb interaction between the inner shell electron and the moving projectile perturbs the atomic potential and may result in the ejection of inner shell electrons. In the case of high energy projectiles, several inner shell vacancies can be created in the collision. These additional spectator vacancies changes the electron binding energies resulting in energy shift of the X-rays emitted in radiative transitions of such multiply ionized states. The resulting X-ray lines in the spectrum are characterized as satellites as they are shifted towards higher energies compared to the energies of the diagram lines in singly ionized atoms. Using a high resolution crystal diffraction spectrometer, the satellite lines can be resolved from the diagram lines.

From the KL^1 satellite lines intensities we have determined the L shell ionization probability arising from direct Coulomb ionization in proton collisions with low Z elements, being aware that important processes like the rearrangement of vacancies prior to the photon emission are taking place during the ionization-decay process. The data obtained in the present experiment were compared with the semiclassical approximation calculations (SCA) using hydrogenic [1,2,3,4] and Hartree-Fock (HF) wave functions [5]. In the comparison of calculated and measured data [6] we showed the importance of using realistic HF wave functions for the atomic description. The ionization probability was studied as a function of the reduced projectile velocity, which was defined as the ratio of the projectile, and the L-shell electron velocity. The differences between theory and experiment were clearly velocity dependent

EXPERIMENT

The K_α X-ray spectra of Ca, Ti, Cr and Fe ionized with 0.7 - 1.5 MeV protons were measured at the J. Stefan Institute, Ljubljana, with the use of high resolution crystal spectrometer. Protons were accelerated by the 2 MV Van de Graaff accelerator. Further measurements were performed at the Ruđer Bošković Institute in Zagreb. With their 6 MV Tandem accelerator, Ti K_α spectra with proton energies up to 4 MeV were obtained.

The protons impinged the thick metal target perpendicularly. The outcoming X-rays were diffracted in the first order on a flat LiF crystal. The <200> reflecting planes were used for Ca and Ti, and the <220> for Cr and

CP475, *Applications of Accelerators in Research and Industry*, edited by J. L. Duggan and I. L. Morgan

Fe. The reflected X-rays were detected by the position sensitive detector (PSD), which was of proportional type. The lateral resolution of the position sensitive detector was 0.4 mm.

The intrinsic resolution of the spectrometer determined mainly by the spatial resolution of the position sensitive detector was 2.2 eV at 4 keV. The distance between the target and the PSD was 866 mm which rendered a very good angular resolution. The experimental energy resolution depended mainly on the dimensions of the beam spot on the target which was 0.5-0.7 mm wide and 10 mm high. This size implied an energy resolution of 3.5 - 6.5 eV which was sufficient for resolving the KL^1 satellites from the K_α diagram lines clearly.

DATA ANALYSIS

In the independent particle framework, the probability that the proton with an impact parameter b produces m vacancies in the K shell, and n vacancies in the L shell, can be expressed by the binomial distribution. To obtain the cross section, the probability is integrated over the impact parameter. Since the L shell ionization probability is nearly constant over the impact parameter region where the K shell ionization may occur, we obtain

$$\sigma_{K^m L^n} = 2\pi \binom{8}{n} (p_L(0))^n (1 - p_L(0))^{8-n} \times \int_0^\infty \binom{2}{m} (p_K(b))^m (1 - p_K(b))^{2-m}\, b db, \tag{1}$$

where p_K is the K shell ionization probability per electron, and p_L is the mean ionization probability per electron for the three L subshells. Considering that the primary vacancy yield $I_{KL}{}^n$ is proportional to the cross section $\sigma_{KL}{}^n$, one obtains

$$\frac{I_{KL^n}}{I_{KL^0}} = \frac{\sigma_{KL^n}}{\sigma_{KL^0}} = \frac{\binom{8}{n} [p_L(0)]^n}{[1 - p_L(0)]^n}. \tag{2}$$

It is evident that the KL^n satellite yield relative to the K_α diagram line gives us the L shell ionization probability for the near-central collisions (zero impact parameter) within the independent particle framework.

In order to obtain the relative satellite intensities we have fitted the measured spectra using the following model. The $K_{\alpha 1}$ and $K_{\alpha 2}$ lines were analyzed using two Voigt functions. The Lorentzian widths of the $K_{\alpha 1}$ and $K_{\alpha 2}$ lines were assumed to be equal, but let free through the fit. The KL^1 satellite was described with a single Voigt function, but the widths of both components were left free. With this model we were able to successfully fit the measured spectra (Fig. 1).

The relative yields of the KL^1 satellite lines as follow from the measurements reflect the situation in the L shell at the moment of the K X-ray emission. In order to find the ionization probabilities, we have to take into account the rearrangement of inner shell vacancies prior to the K X-ray emission. This process alters the primary vacancy distribution produced in the collision. In the case of low Z elements where we can not resolve contributions from the different L subshells, the rearrangement between the subshells do not need to be considered.

The original vacancy distribution of the L shell is essentially altered by the vacancy promotion to higher shells (mainly M). For low Z elements, the energy shifts of the KM^1 satellite lines are comparable to the K_α natural line-width, so they can not be resolved from the diagram line. The vacancy promotion into higher shells actually

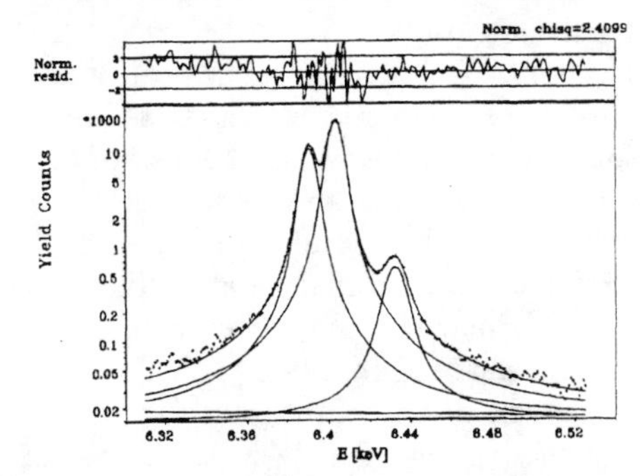

FIGURE 1. Fe K_α spectrum induced by 1.0 MeV protons, decomposed by the fitting method as explained in the text.

redistributes some of the KL^1 satellite line into the diagram one. The second effect that needs to be considered concerns the partial fluorescence yield for the K_α transition since it may vary due to the presence of additional vacancies in the L shell which is involved in the transition. On account of both effects, the yield of the KL^1 satellite line relative to the K_α diagram line can be expressed as

$$I_1^X = \frac{\sigma^X_{KL^1}}{\sigma^X_{KL^0}} = \frac{[\sigma_{KL^1}(1 - R(L^1))]\omega_{K_\alpha}(L^1)}{[\sigma_{KL^0} + \sigma_{KL^1} R(L^1)]\omega_{K_\alpha}(L^0)}, \tag{3}$$

where I_1^X is the relative yield of the KL^1 satellite line extracted from the fitting procedure and corrected for the absorption in the target and on the way to detector, R is the rearrangement factor, and $\omega_{K\alpha}$ is the partial fluorescence yield for the K_α transition. The rearrangement factor R and the partial fluorescence yield for the K_α transition are defined as

$$R = \frac{\Gamma_L^A + \Gamma_L^R}{\Gamma_L + \Gamma_K}, \qquad \omega_{K_\alpha} = \frac{\Gamma^R_{K\alpha}}{\Gamma_K}. \tag{4}$$

where Γ_i is the total atomic level width of the particular shell, while Γ_i^R and Γ_i^A are the partial widths for the radiative and Auger transition processes of the shell, respectively. The influence of Coster Kronig transitions is neglected since these transitions redistribute the vacancies between different L subshells, but they do not change the total number of L shell vacancies produced in the collision.

The relative vacancy yield (Eq. 2) of the KL^1 satellite produced in the near-central proton collision with low Z atoms can be expressed as

$$\frac{I_{KL^1}}{I_{KL^0}} = \frac{\sigma_{KL^1}}{\sigma_{KL^0}} = \frac{8p_L(0)}{(1-p_L(0))}$$
$$= I_1^X \frac{\omega_{K_\alpha}(L^0)}{\left[1-R(L^1)\left(1+I_1^X\frac{\omega_{K_\alpha}(L^0)}{\omega_{K_\alpha}(L^1)}\right)\right]\omega_{K_\alpha}(L^1)} . \tag{5}$$

For the estimation of $\omega_{K\alpha}(L^1)$ and $R(L^1)$ in Eq. 5, it is necessary first to estimate the level width of the K shell which is influenced by an additional L shell vacancy. The respective tabulated values have to be modified by a suitable correction. We used the statistical method of Larkins [7], where the width of a certain transition is in the first approximation proportional to the number of electrons available for the transition. It has been shown that this approach gives very realistic results for the K shell level widths [8]. For the radiative and nonradiative transition widths in a singly ionized atom we used the values of Scofield, Kostroun and McGuirre [9,10,11,12].

The rearrangement influences the primary vacancy yield to a much higher extent than the variations of the fluorescence yield, which can in first approximation be neglected. The original vacancy distribution can be changed significantly prior to the K X-ray emission. This strong rearrangement process originates mainly in the competitive LMM Auger transitions which have the level width comparable to that for the K radiative transition. The values of the rearrangement factor for doubly ionized states of Ca, Ti, Cr and Fe are collected in Table 1 together with the ratio of the K_α partial fluorescence yields. It is apparent that the rearrangement of the L shell vacancies prior to the K X-ray emission is extremely important. Its influence can not be neglected especially for the case of elements of low-Z elements.

TABLE 1. Rearrangement factor R and the ratio of the K_α partial fluorescence yields for the doubly (KL^1) and singly ionized atoms. The level widths for the doubly ionized atoms were calculated according to the statistical procedure [7].

Element	Z	$\omega_{K\alpha}(L^1)/\omega_{K\alpha}(L^0)$	R
Ca	20	1.010	0.260
Ti	22	1.011	0.271
Cr	24	1.009	0.281
Fe	26	1.013	0.331

Besides the direct Coulomb ionization, the shake process may also contribute to the satellite production. The KL^1 satellite yield is therefore composed of the direct ionization and shake contributions. Since the L shell ionization probabilities are rather small for proton impact (~1% for low Z elements), and Eq. 4 relates to the L shell ionization probability independently of its mechanism, we can write the probability obtained from the KL^1 satellite yield as a sum of probabilities

$$p_L = p_L^{DI} + p_L^S , \tag{6}$$

where DI and S refer to the direct ionization and shake process, respectively. In order to obtain the direct ionization probabilities, we subtracted the shake probabilities from our experimental values. We used the probabilities of Mukoyama and Taniguchi [13] that were calculated with the Hartree-Fock-Slater wave functions within the sudden approximation model.

TABLE 2. L shell ionization probabilities for the direct Coulomb ionization of Ca, Ti, Cr, and Fe induced in the near-central proton collision.

E_p	$p_{L(Ca)}(0)$	$p_{L(Ti)}(0)$	$p_{L(Cr)}(0)$	$p_{L(Fe)}(0)$
0.7	1.30±0.18	0.88±0.13		
0.8	1.27±0.18	0.92±0.14	0.66±0.11	0.43±0.09
0.9	1.29±0.18	0.95±0.14	0.68±0.11	0.47±0.10
1.0	1.31±0.18	0.95±0.14	0.73±0.12	0.53±0.11
1.1	1.27±0.18	0.96±0.14	0.75±0.12	0.56±0.12
1.2	1.27±0.18	0.99±0.15	0.79±0.13	0.60±0.12
1.3	1.27±0.18	0.96±0.14	0.81±0.13	0.64±0.13
1.4	1.25±0.18	0.99±0.15	0.82±0.13	0.66±0.14
1.5	1.23±0.17	1.01±0.15	0.83±0.13	0.69±0.14
2.0		0.86±0.26		
3.0		0.70±0.11		
4.0		0.51±0.08		

RESULTS AND DISCUSSION

Following the procedure as outlined above, the L shell ionization probabilities for the direct Coulomb ionization in central collisions were evaluated from the measured KL^1 satellite yields. The elements of interest were Ca, Cr, and Fe ionized with 0.7 - 1.5 MeV protons, and Ti ionized with 0.7 - 4 MeV protons. The ionization probabilities are collected in Table 2. The uncertainties of the deduced probabilities are mainly result of the uncertainties of the level widths used for the calculation of the rearrangement factors. Especially the level widths for LMM and some KLL Auger transitions have large uncertainties that yield the total uncertainty of the rearrangement factor up to 40%. The uncertainty contribution originating from the fitting procedure is generally less important.

For proton induced ionization of low-Z elements, the predominant ionization mechanism is the direct Coulomb ionization. As we are interested in the ionization probability at zero impact parameter, the respective theoretical values have to be obtained in the semiclassical approximation (SCA). This is the first order perturbation theory which allows us to calculate the impact parameter dependence of the single electron ionization probability.

The experimental values were compared to the SCA calculations of Trautmann and Rösel [1,2], Šmit [3,4] and Halabuka [5]. All these calculations employ classic hyperbolic trajectories, but differ in the choice of atomic wave functions. The codes of Trautmann and Rösel and that of Šmit employ screened hydrogenic Dirac wave functions, but in different approximations. In the calculation of Halabuka [5], the concise relativistic Hartree-Fock wave functions were employed.

The energy dependence of the experimental and theoretical values for Ti target are shown in Fig. 3, and the reduced velocity dependence of the ratios between the experimental and theoretical values are given in Fig. 2.

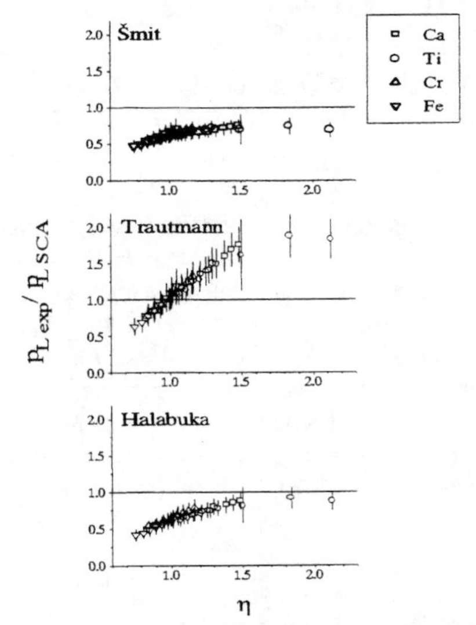

FIGURE 2. The ratio of the experimental and theoretical SCA ionization probabilities as a function of the reduced projectile velocity. The three models differ in the use of wave functions and employ in turn two different types of hydrogenic Dirac ([1,2] and [3,4], respectively), and the relativistic Hartree-Fock wave functions [5].

The probabilities calculated by the three methods have similar energy dependence, though they differ significantly in size. The values of Trautmann and Rösel [1,2] are generally by a factor of 1.5-2 lower than the values of Šmit [3,4] and Halabuka [5]. The latter two differ by a few percent at low collision velocities, and by 30% at high collision velocities. However, the differences at high velocities may be up to 10% higher since the calculation [3,4] involved the partial waves up to $l=2$ only.

The results of Figs. 2,3 suggest that the calculation [5] employing relativistic Hartree-Fock wave functions yields data which are closest to the experiment. For the method of [3,4], which uses a simple yet effective approximation for the states in atomic potential, the differences between theory and experiment are by 30% larger. This indicates that the proper choice of wave functions is essential for the calculation of ionization probabilities at near-zero impact parameters.

CONCLUSION

High resolution measurements have been performed for the Ca, Ti, Cr, and Fe K_α lines induced by protons. The energies of the proton beams were 0.7-1.5 MeV, for Ti 0.7-4.0 MeV. The relative KL satellite yields were used to determine the L-shell ionization probabilities in near-central collisions. The importance of the rearrangement process due to the L Auger transitions has been demonstrated for low Z elements (R=0.26 for Ca, 0.33 for Fe). The direct Coulomb ionization probabilities were determined from the satellite yields after subtracting the shake contribution

The obtained experimental results were compared with the SCA calculations. The calculated values depend strongly on the use of wave functions. The comparison supports the use of relativistic Hartree-Fock wave functions. In the low energy region ($\eta<1.5$) current SCA calculations fail to reproduce the measured values regardless of the type of wave functions used.

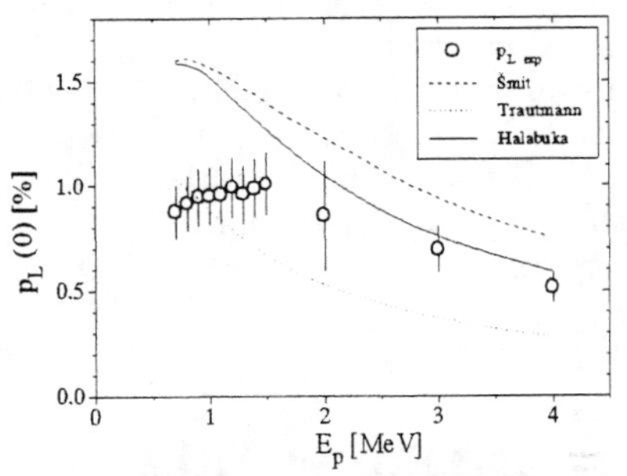

FIGURE 3. L shell ionization probabilities for the direct Coulomb ionization of Ti as a function of the proton energy. The experimental data are compared with the SCA calculations of Trautmann and Rösel [1,2] (dotted line), Šmit [3,4] (dashed line) and Halabuka [5] (solid line).

This work was mainly supported by the Slovenian Ministry of Science and Technology (Project No. J1-7473).

REFERENCES

1. D. Trautmann and F. Rösel, Nucl. Instrum. Methods **169**, 259 (1980).
2. D. Trautmann and F. Rösel, Nucl. Instrum. Methods **214**, 21 (1983).
3. Ž. Šmit and I. Orlić, Phys. Rev. A **50**, 1301 (1994).
4. Ž. Šmit, Phys. Rev. A **53**, 4145 (1996).
5. Z. Halabuka, W. Perger and D. Trautmann, Z. Phys. D **29**, 151 (1994).
6. M. Kavčič, Ž. Šmit, M. Budnar, Z. Halabuka, Phys. Rev. A **6**, 4675 (1997)
7. F. P. Larkins, J. Phys. B: J. Phys. B **4**, L29 (1971).
8. D. F. Anagnostopoulos, J. Phys. B **28**, 47 (1995).
9. J. H. Scofield, Phys. Rev. **179**, 9 (1969).
10. V. O. Kostroun, M. H. Chen and B. Crasemann, Phys. Rev. A **3**, 533 (1969).
11. E. J. McGuirre, Phys. Rev. A **3**, 587 (1971).
12. E. J. McGuirre, Phys. Rev. A **3**, 1801 (1971).
13. T. Mukoyama and K. Taniguchi, Phys. Rev. A **36**, 693 (1987)

Collisional deexcitation of metastable ions: A new technique to separate radiative and nonradiative contributions

H. T. Schmidt, S. H. Schwartz, A. Fardi, K. Haghighat, A. Langereis, H. Cederquist, L. Liljeby[*], J. C. Levin[**], and I. A. Sellin[**]

Department of Physics, Stockholm University, S-104 05 Stockholm, Sweden, [*]Manne Siegbahn Laboratory, Stockholm University, S-104 05 Stockholm, Sweden, [**]Department of Physics and Astronomy, University of Tennessee, Knoxville, TN 37996-1200, USA

A method is presented to measure separate cross sections for collisional deexcitation of metastable ions via radiative and non-radiative processes. The principle of the experiment is to first determine the total collisional deexcitation cross section in an attenuation measurement. After this the non-radiative part is determined separately in a measurement where the ionized target atom (unique to the non-radiative process) is detected. Here, we discuss recently published results on deexcitation of metastable $He^+(2s)$ ions colliding with Ar at 1.65 keV/amu (1,2) as well as preliminary results for Xe and H_2 targets. In all cases we find that the dominating contribution from radiative deexcitation agrees with the result of a semi-classical calculation of the $2s$-$2p$ mixing driven by the induced dipole field of the target atom or molecule modified by taking competing processes, for which we measured the cross sections, into account. Further, the observed time-of-flight spectra with the molecular target H_2 are discussed and evidence of low-energy dissociation is presented.

INTRODUCTION

When atoms, ions or molecules in a gas or plasma environment populate long-lived metastable states, their deexcitation rate to the ground state may be dominated by collision-induced rather than spontaneous deexcitation. In this work we are concerned with the deexcitation of the metastable $2s$ level of He^+ and other H-like ions. Our objective is to determine separately radiative,

$$P^{(Z_P-1)+}(2s)+T \rightarrow P^{(Z_P-1)+}(1s)+T+h\nu$$

and nonradiative,

$$P^{(Z_P-1)+}(2s)+T \rightarrow P^{(Z_P-1)+}(1s)+T^{+}+e^{-}$$

deexcitation cross sections for H-like projectile ions, $P^{(Z_P-1)+}$, with nuclear charge Z_P colliding with different target atoms or molecules, T.

The mechanism for radiative collisional deexcitation can be described as follows: The target atom is polarized in the electric field of the approaching projectile ion. This induced electric dipole field of the target atom in turn acts on the projectile to Stark mix the $2s$ and $2p$ states. After the collision there is therefore an impact-parameter dependent probability that the ion is found in the $2p_{1/2}$ state resulting in decay to the ground state on a 100 ps time scale (for He^+). For thermal velocities the dominating mechanism for *nonradiative* deexcitation is autoionization of the quasi-molecular collision complex, whereas at the keV range collisions considered here the dominating mechanism is expected to be electron capture to a doubly excited state of the projectile followed by autoionization at large internuclear distance.

In contrast to the situation for deexcitation of metastable neutral atoms in thermal collisions, the literature concerning collisional deexcitation of metastable ions is sparse. In the 1970s three measurements of the *total* collisional deexcitation cross section for $He^+(2s)$ ions colliding with various targets at collision energies ranging from near-thermal (3,4) to keV (5) energies were performed. In the present work, we emphasize a new experimental method to separately determine the cross sections for radiative and nonradiative deexcitation. Theoretically this problem was first considered by Lamb in connection with radiative level shift measurements for He^+

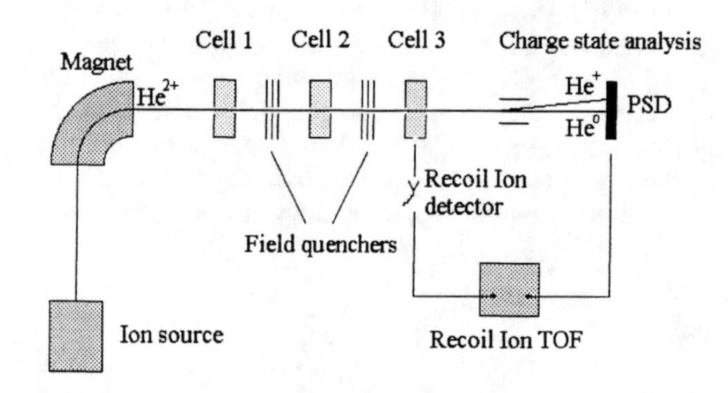

FIGURE 1. Schematic of the experimental arrangement.

CP475, *Applications of Accelerators in Research and Industry*, edited by J. L. Duggan and I. L. Morgan

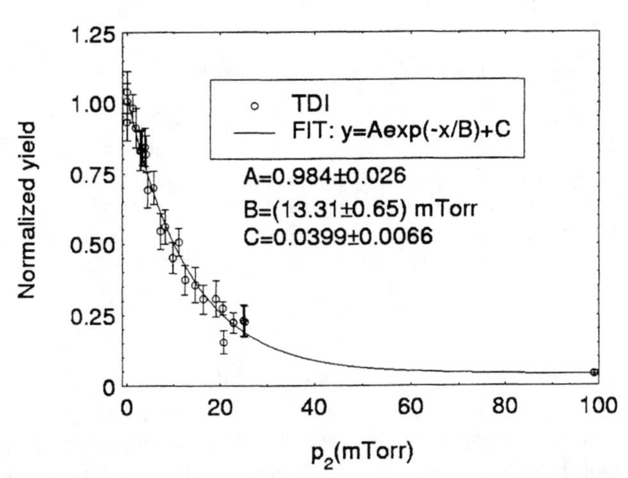

FIGURE 2. Effective cross section for transfer double ionization as a function of the pressure, p_2, of hydrogen in cell 2. The point at p_2=100 mTorr is the result of a measurement with the field quenchers on to establish the ground state contribution to the observed effective cross section. The full curve is a fit curve to extract the total deexcitation cross section.

ions (6). It was found that the radiative deexcitation cross section accounts for 68 % of the total deexcitation cross section, $\sigma_{de}^{tot} = 4.4 \cdot 10^{-15}\,\text{cm}^2$, at thermal collision energies (23 meV). In the work of Prior and Wang (3) a semi-classical approach is used to derive the cross section for radiative deexcitation. This model is modified by taking into account competing electron capture mechanisms at small impact parameters (2).

In the following section, the experimental method is described and results for 1.65 keV/amu $He^+(2s)$ on argon (1,2), xenon, and molecular hydrogen are presented and discussed on the basis of the modified semi-classical model. In the final section it is demonstrated how the time-of-flight spectra of recoil-ions from a molecular target (H_2) can yield information on the dynamics of fragmenting molecular ions.

EXPERIMENTAL METHOD AND RESULTS

The experimental method is described in detail elsewhere (2), and only a brief description is given here. The basic idea is to perform two series of measurements, one of which is devoted to a determination of the total collisional deexcitation cross section, σ_{de}^{tot}, while the aim of the other measurement series is to determine separately the cross section for non-radiative collisional deexcitation, σ_{de}^{NR}. The radiative deexcitation cross section, σ_{de}^{R}, is finally obtained by subtraction.

The experimental setup, which is shown in Fig. 1, consists of three consecutive movable gas cells C1, C2, and C3 connected to a Baratron for accurate pressure determination. In C1 the H-like ions are formed by electron capture to a bare ion from a suitable target gas. In the case of 6.6 keV He^+ a Kr target gave a metastable fraction of about 10 %. After C1 the H-like ions are separated from the bare ions and threaded through C2 and C3. For the measurement of the total deexcitation cross section the target gas of interest is let into C2. The collisional deexcitation taking place in C2 then leads to a reduced metastable fraction for the beam entering C3. In C3 a target gas is chosen for which a certain monitor process has a much higher cross section for metastable- than for ground-state ions. By measuring the projectile- and recoil-ion charge states in coincidence by means of a position-sensitive projectile detector and a time-of-flight recoil-ion spectrometer the monitor process events are identified and their rate is recorded as a function of the pressure of the gas in C2. The properly normalized coincidence rate as a function of p_2 yields a decay curve, where the decay constant gives the total deexcitation cross section when corrected for effects of beam loss due to electron capture to metastable- and ground-state ions in C2 (2). For He^+ projectiles, Xe in C3 provides two good monitor processes for metastable ions surviving the passage of C2: Single-ionization (He^+-Xe^+ coincidence in C3) and transfer-double ionization (He^0-Xe^{3+} coincidence in C3). In Fig. 2 is shown the normalized yield of transfer double ionization events in C3 when the pressure of the target gas in C2 (in this particular case H_2) is varied. From this curve (and a similar curve for the other monitor process, single ionization) the total deexcitation cross section in H_2, $\sigma_{de}^{tot}(H_2)=(7.1\pm0.9)\cdot10^{-16}$ cm^2, is deduced. The nonradiative deexcitation cross section is then determined separately by letting the target gas of interest into C3 (C2 now empty) and measuring the yield of coincidence events in C3 where the projectile keeps its charge state and recoil ions are produced. To correct for the much weaker single-ionization signal from the ground-state beam component this is repeated with a pure ground-state beam obtained by efficient electric-field quenching of the metastable component. From the yields with and without the electric-field quenchers on and the metastable fraction $F_0=(10\pm2)$ % (2,7), the nonradiative deexcitation cross section is found. The total, radiative, and nonradiative cross sections and the radiative branching ratio found for $He^+(2s)$ colliding with Ar (1,2) and Xe and H_2 (preliminary results) are presented in the first four columns of Table 1. It is remarkable that in spite of a variation of a factor of two in total cross sections, the branching ratios are similar. Furthermore, they all agree with the value calculated by Lamb for thermal $He^+(2s)$-He collisions.

Prior and Wang (3) applied the sudden approximation (8) and assumed straight-line trajectories to calculate the probability for a $He^+(2s)$ ion to perform a transition to the $2p$ state when being exposed to the electric field of the induced dipole of the target atom as a function of the impact parameter. At small impact parameters, b, the sudden approximation is not valid and in fact yields a transition probability larger than one. For small b the real transition probability will oscillate rapidly between zero

$He^+(2s)$ + ..	$\sigma_{de}^{tot}(exp)$	BR(%)	$\sigma_{de}^{NR}(exp)$	$\sigma_{de}^{R}(exp)$	$\sigma_{de}^{R}(SEM)$	$\sigma_{de}^{R}(SCC)$	$\alpha_T(Å^3)$
Ar	7.6±1.2	70±8	2.28±0.51	5.4±1.3	5.6	9.6	1.64
Xe	13.7±1.6	68±8	4.44±0.93	9.3±1.9	8.5	15.0	4.02
H_2	7.1±0.9	75±8	1.82±0.49	5.3±1.0	5.5	6.7	0.806

TABLE 1. The measured deexcitation cross sections for 6.6 keV $^4He^+(2s)$ on Ar, Xe, and H_2 are given (in units of 10^{-16} cm^2) together with the results of the semiclassical calculation with and without the correction for the competing processes. Experimental results are followed by (exp), results of the semi-classical calculation without corrections by (SCC), and the semi-empirically corrected results by (SEM). BR is the radiative branching ratio and α_T is the target polarizability. The experimental results for Xe and H_2 targets should be considered as preliminary.

and unity as a function of b. In the calculation of the cross section the transition probability is approximated by one half for the range of b where the sudden approximation yields a probability larger than one (2,3). When the found transition probability is integrated over b the following expression for the radiative deexcitation cross section for a H-like projectile ion with nuclear charge Z_P is found,

$$\sigma_{de}^{R}(SCC) = \frac{5}{3}\pi\sqrt{\frac{2(Z_P - 1)\alpha_T\hbar}{Z_P v_P m_e}}\,. \quad (1)$$

Here α_T is the target polarizability, m_e is the electron mass, and v_P is the projectile velocity. To account for the influence of the competing processes nonradiative deexcitation and electron capture, *half* the sum of their measured cross sections is subtracted from the calculated $\sigma_{de}^{R}(SCC)$ since these processes will most probably take place for impact parameters where the $2s$-$2p$ transition leading to radiative deexcitation will have a probability of one half (on the average) (2). The result of this semi-empirical model is thus simply,

$$\sigma_{de}^{R}(SEM) = \sigma_{de}^{R}(SCC) - (\sigma_{de}^{NR} + \sigma_{10}^{2s})/2\,. \quad (2)$$

Here, σ_{10}^{2s} is the measured total electron capture cross section for metastable ions. The cross sections with and without the correction for competing processes are given in Table 1 for the three cases corresponding to the measurements.

We see from Table 1 that the calculated radiative deexcitation cross sections compare favorably with the experimental results for all the three cases considered when the effect of the competing processes is taken into account. It is somewhat surprising to find that the simple description seems valid even for the molecular target H_2 as discussed in the following section.

DISSOCIATIVE IONIZATION OF H_2 MOLECULES

When the target gas is a molecule rather than a noble gas atom more degrees of freedom are available and not only ionization, but also dissociation is possible. It is a potential source of error in our experiment that a collisional deexcitation event, where the excitation energy is spent dissociating the target molecule into two neutral fragments will be counted as a radiative deexcitation event since no recoil-ion is produced. On the other hand the cross section for this process is probably very small since no neutral H_2 state is available 40 eV above the ground state to allow for an effective dissociative excitation process. With the present experimental setup we cannot investigate this problem, but we can consider events where both ionization and dissociation takes place by analyzing the time-of-flight recoil-ion spectra for electron capture and single ionization events. In Fig. 3 the time-of-flight spectrum for recoil-ions from $He^+(1s)+H_2 \rightarrow He+...$ collisions is shown. The dominating peak (off-scale in Fig. 3) is caused by H_2^+ recoil ions, whereas the triple peak is due to protons emitted from dissociating H_2^+ ions. The triple structure can be explained in the following way: The immediate result of the capture process is always the formation of a H_2^+ ion and if this ion is not in the $1s\sigma_g$ electronic ground state of the molecular ion (cf. Fig. 4) it will immediately dissociate along the antibonding potential energy curve releasing of the order of 10 eV kinetic energy. Due to the limited acceptance of the time-of-flight spectrometer only protons emitted at small angles relative to the spectrometer-axis will be detected. Therefore two peaks related to protons emitted from electronically excited H_2^+ towards and away from the detector are observed. The middle peak of the triplet we ascribe to formation of H_2^+ ions in the electronic ground state but with energy slightly above the dissociation

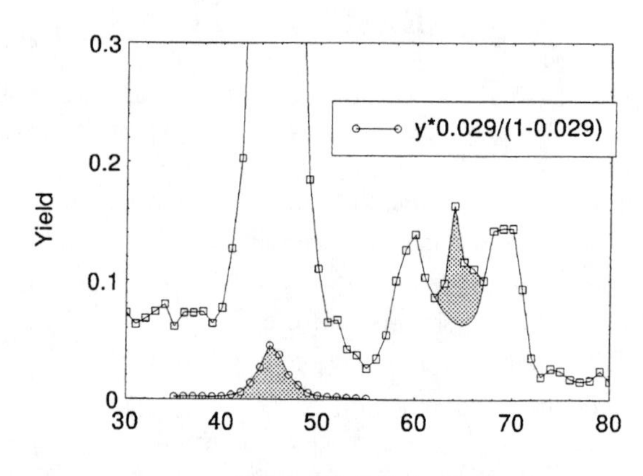

FIGURE 3. Time-of-flight spectrum for recoil ions in coincidence with neutralized projectiles from $He^+(1s)$-H_2 collisions.

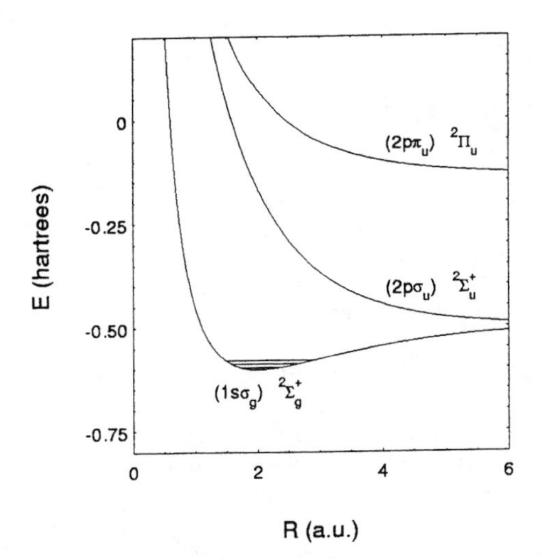

FIGURE 4. The lowest potential energy curves of the hydrogen molecular ion. Formation of H_2^+ in the electronic ground state gives rise to a H_2^+ recoil ion if the ion is found in one of the vibrational eigenstates. There is, however, a 2.9% probability for 'proton shake-off' giving rise to a slow proton recoil ion (the middle peak of the triplet in Fig. 3).

limit leading to release of a proton with essentially zero kinetic energy. In the velocity range considered here, the collision time is much shorter than a characteristic period of the vibrational motion, and thus a vertical transition from H_2 to H_2^+ is expected. The probability to find the H_2^+ ion in a specific vibrational level after the collision is therefore expected to be given by the Franck-Condon nuclear wave function overlap factors (9). If we sum the Franck-Condon factors for all the vibrational levels of the H_2^+ ground state, we get 0.971. That is, in 2.9% of the collisions forming H_2^+ in $1s\sigma_g$ a low energy proton is emitted. In Fig. 3, we have scaled down the H_2^+ peak to the expected intensity of slow protons formed in this manner to directly compare this with the observed intensity of slow protons (the middle peak of the triplet). Since there is no clear experimental separation between the fast and slow protons we can only make a qualitative comparison. We expect the middle peak to have the same intensity as the reduced H_2^+ peak. To illustrate the comparison we have shaded the reduced H_2^+ peak and an arbitrarily shaped but equally large area to represent the slow protons. The comparison indicates that the interpretation is correct.

CONCLUSION AND OUTLOOK

Separate radiative and nonradiative collisional deexcitation cross sections have been measured for metastable 1.65 keV/amu $He^+(2s)$ ions colliding with Ar, Xe, and H_2. A semi-classical calculation of the $2s$-$2p$ mixing driven by the induced dipole field of the target with corrections for competing processes is performed and shown to reproduce the experimental results for radiative deexcitation for all the three targets considered (the experimental results for Xe and H_2 targets are preliminary). In the case of a molecular hydrogen target time-of-flight spectra were recorded, and the H^+ peak were seen to be split in three. This is explained by dissociative ionization via different electronic levels of the H_2^+ ion.

For future measurements with He^+ as projectile, we want to extend the variation of target polarizability beyond the range from 0.806 Å^3 (H_2) to 4.02 Å^3 (Xe) considered so far to further test the semi-classical approach. Helium, which has a polarizability of 0.206 Å^3, has already been used as a target and necessary supplementary measurements are scheduled for December 1998. Further we will consider molecular targets with permanent electric-dipole moments. All the experimental efforts on deexcitation of metastable H-like ions have so far been limited to He^+ projectiles. We are considering to extend our method to multiply charged H-like ions. From Eq.(1) we see that the radiative deexcitation cross section is only weakly depending on the nuclear charge of the H-like projectile ion. In contrast the nonradiative deexcitation process, which for the relevant velocity range is expected to be dominantly electron capture to doubly excited states followed by autoionization, is expected to be roughly proportional to the projectile charge according to the classical over-barrier model (10). Thus for higher charges we expect to see an increase in the total deexcitation cross section and a shift in dominance from the radiative to the nonradiative part.

ACKNOWLEDGMENTS

This work is supported by the Swedish Natural Science Research Council (F-GF 08801-315, F-AA/FU 08801-320), by the Danish Natural Science Research Council (9600989), by the US National Science Foundation (PHY-9417924), and by the Commission of the European Union (ERBFMBICT961754).

REFERENCES

1. H. T. Schmidt, S. H. Schwartz, H. Cederquist, L. Liljeby, J. C. Levin, and I. A. Sellin, Phys. Rev. A, **57**, R4082 (1998)
2. H. T. Schmidt, S. H. Schwartz, A. Fardi, K. Haghighat, H. Cederquist, L. Liljeby, A. Langereis, J. C. Levin, and I. A. Sellin, Phys. Rev. A, **58**, 2887 (1998)
3. M. H. Prior and E. C. Wang, Phys. Rev. A. **9**, 2383 (1974)
4. C. A. Kocher, J. E. Clendenin, and R. Novick, Phys. Rev. Lett., **29**, 615 (1972)
5. M. B. Shah and H. B. Gilbody, J. Phys. B. **9**, 2685 (1976)
6. W.E. Lamb, Jr. and M.S. Skinner, Phys. Rev. **78**, 539 (1950); E. Lipworth and R. Novick, Phys. Rev. **108**, 1434 (1957); and W.E. Lamb, Jr., private communication (1997)
7. M. B. Shah and H. B. Gilbody, J. Phys. B. **9**, 1933 (1976); M. B. Shah and H. B. Gilbody, J. Phys. B. **7**, 256 (1974)
8. L. I. Schiff, *Quantum Mechanics*, 3rd edition (McGraw-Hill, New York, 1968), p. 292; W. Pauli, *Handbuch der Physik*, vol. **24**, Pt. **1**, 2nd ed. (Springer, Berlin, 1933), p. 164
9. G. H. Dunn, J. Chem. Phys. **44**, 2592 (1966)
10. A. Bárány, *et al.*, Nucl. Instr. Methods B. **9**, 397 (1985)

Techniques for Enhancing the Performance of High Charge State ECR Ion Sources

Z. Q. Xie

Nuclear Science Division
Lawrence Berkeley National Laboratory, Berkeley, California, 94720

Electron Cyclotron Resonance ion source (ECRIS), which produces singly to highly charged ions, is widely used in heavy ion accelerators and is finding applications in industry. It has progressed significantly in recent years thanks to a few techniques, such as multiple-frequency plasma heating, higher mirror magnetic fields and a better cold electron donor. These techniques greatly enhance the production of highly charged ions. More than 1 emA of He^{2+} and O^{6+}, hundreds of eμA of O^{7+}, Ne^{8+}, Ar^{12+}, more than 100 eμA of intermediate heavy ions with charge states up to Ne^{9+}, Ar^{13+}, Ca^{13+}, Fe^{13+}, Co^{14+} and Kr^{18+}, tens of eμA of heavy ions with charge states up to Xe^{28+}, Au^{35+}, Bi^{34+} and U^{34+} were produced at cw mode operation. At an intensity of about 1 eμA, the charge states for the heavy ions increased up to Xe^{36+}, Au^{46+}, Bi^{47+} and U^{48+}. More than an order of magnitude enhancement of fully stripped argon ions was achieved ($I \geq 0.1$ eμA). Higher charge state ions up to Kr^{35+}, Xe^{46+} and U^{64+} at low intensities were produced for the first time from an ECRIS.

INTRODUCTION

Because of its capabilities of reliably producing highly charged ion beams with high intensities and good ionization efficiencies, ECRIS has become nowadays the choice of ion source for heavy ion accelerators worldwide. Applications of ECRIS are also found in atomic physics research, production of radioactive ion beams and industry ion implantation.

The basic mechanism in an ECRIS is to couple microwave energy through electron cyclotron resonance heating (ECRH) into a plasma confined in a magnetic bottle to produce singly to highly charged ions. The magnetic bottle in a high charge state ECRIS is a minimum-B field configuration. This minimum-B field is a combination of an axial solenoid mirror field and typically a radial sextupole field which can provide many closed and nested ECR surfaces as schematic shown in Fig. 3. Typically microwaves of a single-frequency are used to drive the ECR plasma in which only one ECR surface is utilized. The electrons, traveling in a varying magnetic field in space, gain energy through a thin heating zone that occurs in a vicinity where the electron cyclotron frequency is about the same as the microwave frequency, i.e., $\omega_c = eB/m_e \approx \omega_f$. The primary cold electrons in an ECR plasma come from the electron impact ionization in which stepwise ionization is the dominant process for the production of highly charged ions. There is evidence that there are two groups of electrons in a high charge state ECRIS. The dominant electrons are the "cold" electrons with energy of tens to hundreds of eV and diffusion confined with lifetimes in the order of hundred microseconds. The other group are the hot electrons, due to ECR heating, with energy of above a keV and are magnetically confined with lifetimes of a few to tens of milliseconds (1). A high density of hot electrons in an ECRIS is very essential to the production of highly charged ions because of their high ionization potentials and dramatically decreased ionization cross sections. As long as the plasma in a given ECRIS is stable, the production of highly charged ions typically requires the source to operate at as low a neutral pressure as possible to reduce the charge exchange loss, and high microwave power to increase the hot electron density.

The basic architecture of ECRIS was established by Dr. Geller and his coworkers in the early 70's (2). Since then there are no major breakthroughs in the source basic concepts or theoretical understanding of the detailed physics processes involved in an ECR plasma. However, tremendous progress on source performance has been achieved, especially in the last few years with a few techniques. These techniques include multiple-frequency plasma heating, better additional cold electron donors and higher magnetic mirror fields (3-6). With these techniques, ECRIS performance has been greatly enhanced both in ion charge state and beam intensity. More than 1 emA of He^{2+} and O^{6+}, hundreds of eμA of O^{7+}, Ne^{8+}, Ar^{12+}, more than 100 eμA of intermediate heavy ions with charge states up to Ne^{9+}, Ar^{13+}, Ca^{13+}, Fe^{13+}, Co^{14+} and Kr^{18+}, tens of eμA of heavy ions with charge states up to Xe^{28+}, Au^{35+}, Bi^{34+} and U^{34+} were produced at cw mode operation. At an intensity of about 1 eμA, the charge states for the heavy ions increased up to Xe^{36+}, Au^{46+}, Bi^{47+} and U^{48+}. More than an order of magnitude enhancement of fully stripped argon ions was achieved ($I \geq 0.1$ eμA). Higher charge state ions up to Kr^{35+}, Xe^{46+} and U^{64+} at low intensities were produced for the first time from an ECRIS. Figure 1 graphically shows a summary of the cw performance of high charge state ECRISs at various intensity levels as a function of atomic number. This article will present a brief review and discussion of these techniques and future ECRIS developments.

CP475, *Applications of Accelerators in Research and Industry*,
edited by J. L. Duggan and I. L. Morgan

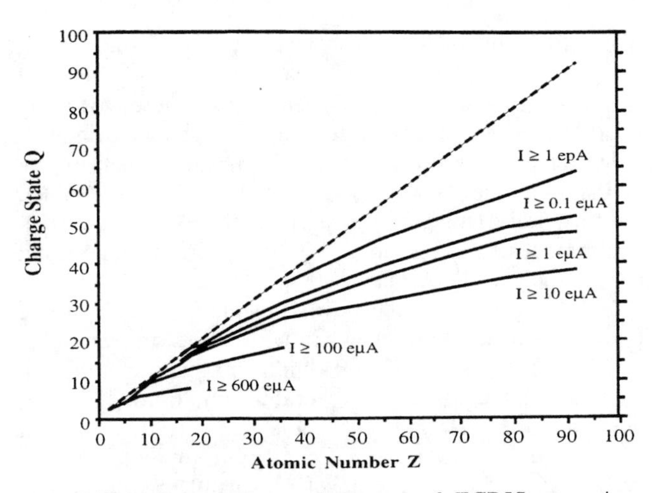

FIGURE 1. Present cw performance of ECRIS at various intensity levels as a function of atomic number up to uranium.

ADDITIONAL COLD ELECTRON DONOR

It was experimentally discovered that ECR plasma needs an additional supply of cold electrons, besides the primary cold electrons coming from ionization, to replace the electrons escaped from the plasma for the production of highly charged ions. With the additional cold electrons the ECR plasma stability is significantly improved, which enables the source to operate at lower neutral pressures and with more microwave power launched into the ECR plasma. A lower operating neutral pressure and high microwave power lead to a higher density of hot electrons that is the key to the production of highly charged ions. During the development of ECRIS, extra cold electrons have been provided to ECR plasmas with various methods. These include microwave-driven first stages, an electron gun, bias probes, plasma cathode and plasma chamber surface coatings with high yield of secondary electrons, such as SiO_2, ThO_2 and Al_2O_3 (7). Among all of these methods, surface coating is the most effective and economic technique to provide the needed additional cold electrons. Besides the desired high yield of secondary electrons, a good surface coating for ECRIS should be resistant to the constant plasma etching and with low material memory for easy daily operation. Although its maximum secondary electron coefficient of 9 is not the highest, Al_2O_3 has all of the desired characteristics and has been demonstrated as the best surface coating for ECRISs. In addition, plasma potential measurements have shown that Al_2O_3 coating substantially reduces the positive average ECR plasma potential, from a few ten of volts to about ten volts, and it is almost independent of the microwave power. A lower plasma potential reduces the ion sputtering and improves the plasma stability. Greatly enhanced production of highest charge state ions with an Al_2O_3 chamber surface coating has been reported by various ECR groups worldwide. Figure 2 shows a comparison of the performance of the LBNL AECR ion source on high charge state bismuth ion beams with and without an aluminum oxide coating.

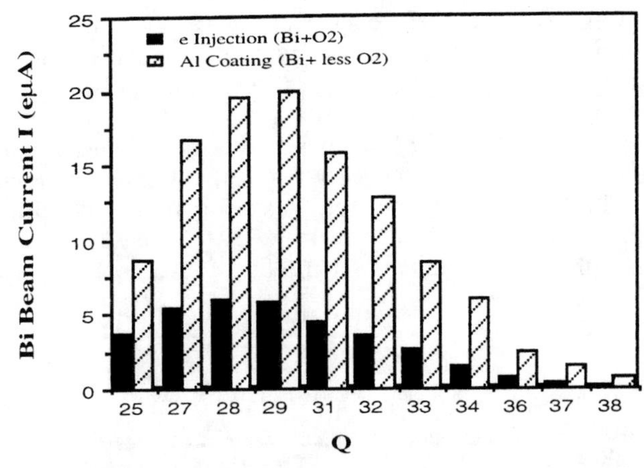

FIGURE 2. Optimized performances on bismuth ion beams for the cases with an aluminum oxide coating (Al Coating) on the plasma chamber surface and without the aluminum oxide coating but with cold electrons injected by an electron gun (e Injection). In these tests the LBNL AECR ion source operated at a single-frequency of 14 GHz.

MULTIPLE-FREQUENCY PLASMA HEATING

As mentioned in the Introduction the minimum-B field configuration in an ECRIS can support many closed, egg-shaped and well-separated ECR surfaces, as graphically shown in Fig. 3, if the incoming microwaves with various well-separated frequencies. Well-separated heating zones have presumably better heating efficiency. A high charge state ECRIS typically operates with a plasma below the so called cut-off density for the operating frequency. Therefore microwaves with certain frequency spacings, depending on the minimum-B field configuration, can be simultaneously launched into and absorbed by ECR plasma. So if two or more different frequencies are used, two or more well-separated and nested ECR surfaces will exist in the ECR plasma. The electrons then can be heated four times or more for one pass from one mirror point to the other while the electrons are heated only twice in the case of single-frequency heating. Multiple-frequency heating can couple more microwave power with better efficiency into the plasma which leads to a higher density of the hot electrons.

Two-frequency plasma heating has been successfully tested with the LBNL AECR, the upgraded AECR (AECR-U) and ECR-II (a similar version of the AECR-U) built at Argonne National Laboratory (3,8,9). All of these ECRISs have a good Al_2O_3 chamber surface coating. Since the multiple-frequency plasma heating can generate a higher density of hot electrons, a good plasma chamber surface coating is a necessity to withstand and reduce the ion sputtering. Tests with two-frequency plasma heating have shown that plasma stability improves and more total microwave power can be launched into the plasma. With the improved plasma stability, the source can operate at lower neutral input which indicates a lower neutral pressure since the mechanical pumping of the source system remains constant. The lower neutral pressure and higher microwave power lead to a higher "temperature" plasma with increased density of the hot electrons as evidenced by the enhanced production of the highly charged ions. Figure 4 shows the optimized uranium charge state distribution with two-frequency (14 and 10 GHz) plasma heating in comparison to the case of single-frequency (14 GHz) plasma heating in the LBNL AECR ion source. Besides a shift in the peak charge state from 33+ (Curve 1) to 36+ (Curve 2), there are also great enhancements for the higher charge state ions from 35+ to 43+.

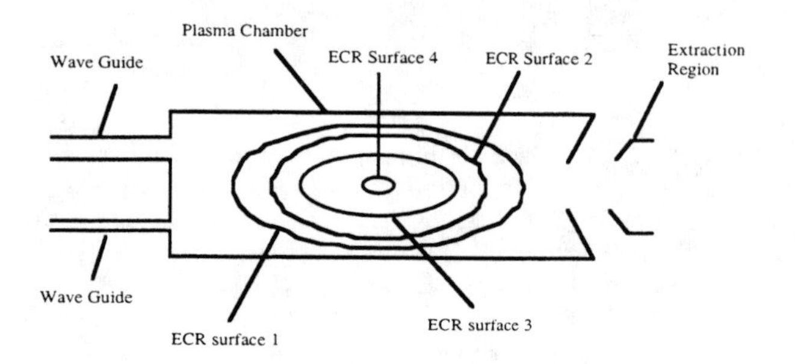

FIGURE 3. Schematic view of 4 nested ECR surfaces in a high charge state ECR ion source for 4 well-separated frequency waves. Wave guides for 2 frequencies are shown.

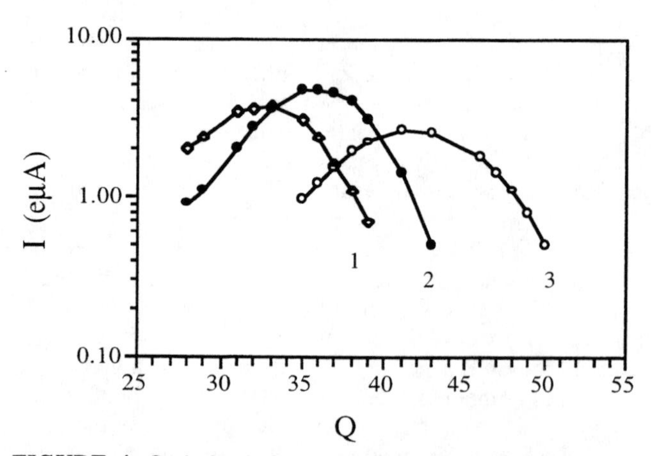

FIGURE 4. Optimized charge state distributions for uranium produced with the LBNL AECR and AECR-U ion sources. Curve 1 indicates the case of a single-frequency (14 GHz) heating and Curve 2 is the case of two-frequency (14+10 GHz) heating, both in the AECR. Curve 3 shows the higher charge state uranium ion beams produced by the AECR-U (14+10 GHz) with higher magnetic mirror fields and higher microwave power.

HIGHER MAGNETIC MIRROR FIELDS

The typical magnetic field configurations in the early ECRISs were designed with maximum mirror ratios of about 2 to 3. When a new source was designed to operate at higher frequency, the magnetic field was scaled up accordingly to ensure a closed ECR surface. The greatly enhanced performance of the Grenoble MINIMAFIOS operating at 16 GHz as compared to its performance at 10 GHz with lower magnetic field led Dr. Geller and his colleagues to propose a so called "frequency scaling law" for ECRIS (10). This scaling law states that operating at higher frequency will increase the plasma density according to $n_c \propto f^2$, the relationship between the critical density of an ideal plasma and the plasma frequency, thereby leading to an enhanced high charge state ECRIS performance. However the relationship of ECRIS plasma density to the operating frequency was not directly demonstrated. Guided by this empirical scaling, a number of ECRISs with operating frequencies up to 30 GHz were designed and built in the late 80s and early 90s.

At the same time the developments of the Grenoble CAPRICE operating at 10 GHz with higher mirror fields and later the NSCL SCECR, built with superconducting magnets and designed to operate from 6 to 30 GHz, operating at 6 GHz but with higher mirror ratios up to 6, have demonstrated significantly improved source performance (5,6). These developments clearly show that the nominal magnetic field configuration is not yet optimum for its designed operating frequency. Higher magnetic field can support a higher plasma power density which is a product of the plasma density and energy. ECRISs operating at higher magnetic fields can produce either a higher density plasma or a hotter plasma with somewhat lower density but the same power density. There is no evidence that ECRISs operate at or above the critical plasma density during the production of highly charged ions.

In recent years, new or upgraded ECRISs all have been built with higher magnetic mirror fields and have demonstrated improved source performance. The LBNL AECR-U, which has produced many record high charge state ion beams, is an outstanding example among the high field ECRISs. The LBNL AECR was upgraded (LBNL AECR-U) in 1996 by increasing the magnetic fields to further enhance the source performance (8). With the modified solenoid magnets and at no increase in ac power, the maximum axial fields of AECR-U increased from 1.0 to 1.7 Tesla at the injection side and from 0.7 to 1.1 Tesla at the extraction region. While the center field remains at about 0.4 Tesla, the mirror ratios increased from 2.4 to 4.2 at the injection side and from 1.8 to 2.8 at the extraction region. A new set of NdFeB permanent sextupole magnets raised the maximum radial field from 0.62 to 0.85 Tesla at the inner surface of the plasma chamber, which was made from aluminum. The LBNL AECR-U is one of the few ECRISs that has combined the multiple-frequency plasma heating, good cold electron donors (Al_2O_3 surface coating and a bias probe), higher magnetic fields and radial pumping that provides a better control of the neutral background in the plasma chamber. After the magnetic field configuration was optimized to match the two-frequency plasma heating (14+10 GHz), the AECR-U demonstrated significantly enhanced production of highly charged ions with maximum total microwave power of 2.1 kW from both of the 14 and 10 GHz klystrons (1.5 kW and 0.6 kW, respectively). It further shifted the peak charge state of uranium from 36+ up to 41+ with greatly enhanced production of highly charged ions as indicated by Curves 3 in Fig. 4. So far up to 360 eμA of O^{7+}, more than 100 eμA of intermediate heavy ions for charge states up to Ar^{13+}, Ca^{13+}, Co^{13+} and Kr^{18+}, tens of eμA of heavy ions with charge states to Kr^{26+}, Xe^{28+}, Au^{35+}, Bi^{34+} and U^{34+} were produced from the AECR-U. At an intensity of about 1 eμA, the charge states for the heavy ions increase up to Xe^{36+}, Au^{46+}, Bi^{47+} and U^{48+}. The production of 1 eμA of the heaviest natural element with more than half its electrons removed represents a milestone in ECRIS development. Besides the improvement on the heavy ions, more than an order of magnitude enhancement for fully stripped argon ions ($I \geq 0.1$ eμA) was achieved. Hydrogen-like krypton ions at an intensity of about 1 epA were produced for the first time from an ECR ion source.

Higher charge state heavy ion beams at low intensities produced with the AECR-U have been accelerated and extracted from the 88-Inch Cyclotron at LBNL. The extracted beam intensities from the cyclotron were 1×10^7 pps of xenon 41+, a few hundred pps of xenon 46+, 3×10^4 pps of $^{238}U^{55+}$ and about 10 pps of $^{238}U^{64+}$. Uranium 64+ ion beam and its total energy of 2.06 GeV are the highest charge state ion beam and the highest beam energy ever produced by the 88-Inch Cyclotron.

Recently a high magnetic field ECRIS built with superconducting magnets (SERSE) for INFN, Italy, is under commissioning in Grenoble, France (11). It is presently the highest magnetic field ECRIS in operation. The maximum peak fields of this ion source are 2.7 and 1.5 Tesla on axis at the injection and extraction regions and 1.4 Tesla radial field at the inner surface of the stainless steel plasma chamber. Its overall magnetic strengths are about 50% higher than the LBNL AECR-U. In the initial tests it operates at a single-frequency of 14 GHz with about 2 kW microwave power provided by two transmitters. Test results with gaseous ion beams are very encouraging, showing its performance on highly charged argon ions is very comparable to the LBNL AECR-U, as shown in Fig. 5. The SERSE's results again demonstrate the effect of high magnet fields on ECRISs for the production of highly charged ions.

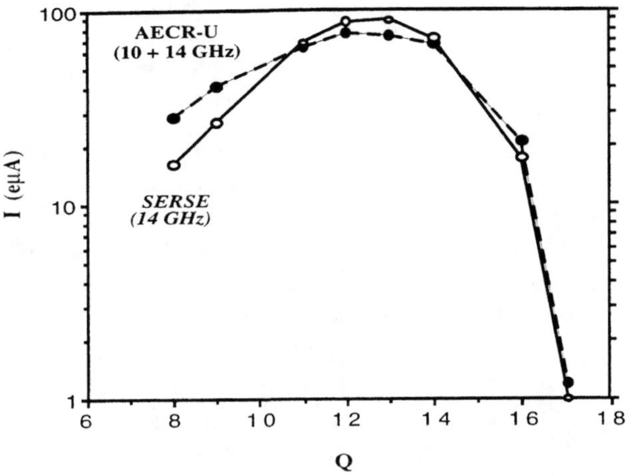

FIGURE 5. Argon charge state distributions with the LBNL AECR-U and SERSE. Both source tunings were optimized for Ar^{16+}.

FUTURE ECRIS DEVELOPMENT

The greatly enhanced capabilities of cyclotrons, synchrotrons and heavy-ion linacs with the application of ECRISs have demonstrated that continued development of ECRIS technology is essential. Recent ECRIS progress has shown that further enhanced production of high charge state ions remains possible with higher magnetic fields above 3 Tesla. New superconducting ECRIS with magnetic fields up to 4 Tesla and maximum mirror of 10 is under construction at LBNL (12). Shown in Table 1 are the design parameters for the new superconducting LBNL 3rd Generation ECRIS. Such high magnetic mirror fields combined with multiple-frequency heating and better techniques to provide the extra cold electrons should further greatly enhance the ECRIS performance.

Besides the techniques reviewed in this paper, other possible techniques such as improving ion beam extraction and transport efficiencies can also contribute significantly to the beam currents and charge states available for the use of highly charged ions.

TABLE 1. Design parameters of the LBNL 3rd Generation ECR

Parameter	Value
Mirror field on axis (T)	4, 3
Central field (T)	0.4
Radial field on I.D. of plasma chamber (T)	2.4
Mirror-mirror spacing (cm)	50
I.D. of plasma chamber (cm)	15
Proposed operating frequency (GHz)	10+14+18 or 14+18+28 ?

ACKNOWLEDGMENTS

The author would like to thank Dr. D. Clark for his help in preparing the manuscript presented here.

This work was supported by the Director, Office of Energy Research, Division of Nuclear Physics of the Office of High Energy and Nuclear Physics of the U.S. Department of Energy under Contract DE AC03-76SF00098.

REFERENCES

1. Melin, G. *et al.*, Rev. Sci. Instrum. **61**, 236 (1990).
2. Geller, R., IEEE Trans. Nucl. Sci. NS-**23**, 904 (1976).
3. Xie, Z. Q. and Lyneis, C. M., Rev. Sci. Instrum. **66**, 4281 (1995).
4. Nakagawa, T., Jpn. J. Appl. Phys. **30**, 930 (1991).
5. Jacquot, B. and Geller, R., Proceedings of the International Conference on ECR Ion Sources and their applications, E. Lansing, NSCL Report #MSUCP-47, 1987, p. 254.
6. Antaya, T. A. and Gammino, S., Rev. Sci. Instrum. **65**, 1723 (1994).
7. Xie, Z. Q., Rev. Sci. Instrum. **69**, 625 (1998).
8. Xie, Z. Q. and Lyneis, C. M., Proceedings of the 13th International Workshop on ECR Ion Sources, College Station, Texas, Feb 1997, p. 16.
9. Schlapp, M., *et al.*, Rev. Sci. Instrum. **69**, 631 (1998).
10. Geller, R., *et al.*, Proceedings of the International Conference on ECR Ion Sources and their applications, East Lansing, NSCL Report #MSUCP-47, 1987, p. 1.
11. Ludwig, p., *et al.*, Rev. Sci. Instrum. **69**, 653 (1998).
12. Lyneis, C. M., Xie, Z. Q. and Taylor, C. E., Rev. Sci. Instrum. **69**, 682 (1998).

X-ray and Auger Transitions from Highly-ionized Ca Ions

K. R. Karim,* L. Logan* and C. P. Bhalla**

* Department of Physics, Illinois State University, Normal, IL 61790-4560, USA

** Department of Physics, Kansas State University, Manhattan, KS 66506, USA

We have calculated radiative and Auger transition rates, x-ray wavelengths, lifetimes, and fluorescence yields of variously ionized calcium atoms with electronic configurations $1s2s^m2p^n$, $(m = 0-2,\ n = 1-6)$ with all allowed values of the orbital angular momentum quantum number l. The calculation was performed in the intermediate coupling scheme using the Hartree-Fock atomic model. The results from the present calculation have been compared with those available in the literature. Excellent agreement is found in calculated transition energies among the various calculations. The agreement in transition rates is generally good for prominent lines. The radiative and Auger transition rates decrease as electrons are removed from the outer shells. The rate of decrease of Auger transition rates, however, is faster than that of radiative transition rates. The fluorescence yields are found to increase with ionization.

1. Introduction

Calcium ions in various charge states are found in the laboratory and astrophysical plasmas. Such ions are also produced in ion-surface interactions. Theoretical values of x-ray and Auger transition rates, energies, and fluorescence yields of calcium atoms at various stages of ionization and excitation are needed to analyze the experimental spectra. There has been several calculations of variously ionized atoms by Chen et al [1], Nilsen et al [2], Karim et al [3], Safronova et al [4], and Bhalla et al [5]. We have recently reported [6] on a calculation of Auger and radiative de-excitation rates of muli-excited argon atoms with configurations $1s^02s^m2p^n, m = 0-2, n = 1-6$. In the present paper we report on a similar extensive calculation of transition rates, energies, lifetimes, and fluorescence yields of calcium atoms with configurations $1s^12s^m2p^n$, $m = 0-2$, $n = 1-6$. The data from the present calculation are compared with those available in the literature.

The radiative and Auger transition rates can be calculated using time-dependent perturbation theory,

$$\Gamma(i \rightarrow f) = \frac{2\pi}{\hbar} \mid < \psi_f \mid O \mid \psi_i > \mid^2 \ \rho(\epsilon), \quad (1)$$

where, ψ_i and ψ_f are, respectively, the antisymmetrized many-electron wave functions of the initial and final states, $\rho(\epsilon)$ is the density of final state, and O is is the appropriate operator. The transition probability for spontaneous emission of a photon of angular frequency ω is

$$\Gamma_x(i \rightarrow f) = \frac{4\omega^3}{3\hbar c^3} \ \frac{1}{2J+1} \ \mid < \gamma' J' \ \| D \ \| \gamma J > \mid^2, \quad (2)$$

where $i \equiv |\gamma J>$ and $f \equiv |\gamma' J'>$ represent , respectively, the initial and final states of the system, D is the electric-dipole operator, and $< \gamma' J' \| D \| \gamma J >$ is the reduced matrix element. In case of Auger transition, the operator O in Eq. (1) represents the electron-electron electrostatic interaction $O = \sum 1/r_{ij}$ and the matrix element can be expressed as a weighted sum of radial Slater integrals [6].

The line fluorescence yield $\omega(i \rightarrow f)$ is the probability that the excited state ψ_i will decay by the radiative transition $\psi_i \rightarrow \psi_f$ with the emission of a photon:

$$\omega(i \rightarrow f) = \frac{\Gamma_x(i \rightarrow f)}{\Gamma_x(i) + \Gamma_a(i)} = \frac{\Gamma_x(i \rightarrow f)}{\Gamma_t(i)}. \quad (3)$$

Here $\Gamma_x(i \rightarrow f)$ is the radiative transition rate for the initial state ψ_i decaying to a final state ψ_f and $\Gamma_a(i)$, $\Gamma_x(i)$, $\Gamma_t(i)$ are respectively, the total Auger transition rate, the total x-ray transition rate, and the total transition rate for the state ψ_i: $\Gamma_x(i) = \sum_f \Gamma_x(i \rightarrow f)$, $\Gamma_a(i) = \sum_{f'} \Gamma_a(i \rightarrow f')$, and $\Gamma_t(i) = \Gamma_x(i) + \Gamma_a(i)$. The lifetime of the excited state ψ_i may be defined as $\tau(i) = 1/\Gamma_t(i)$.

The Hartree-Fock atomic model was used to generate the single-particle wave functions. The appropriate radial matrix elements were calculated using these wave functions. A matrix representation of the many-electron hamiltonian was constructed using the Hartree-Fock basis functions including

CP475, *Applications of Accelerators in Research and Industry*,
edited by J. L. Duggan and I. L. Morgan

spin-orbit interactions. The hamiltonian matrices of the same total angular momentum and parity were diagonalized to obtain the atomic state functions ψ. These were then used to calculate the transition rates. The details of the theory and the numerical procedure are described elsewhere [6].

2. Results and Discusion

Table 1 contains the x-ray wavelengths λ, Auger transition rates Γ_a, radiative transition rates Γ_x, lifetimes τ, and line florescence yields ω for all possible states of variously ionized calcium atoms with configurations $1s^1 2s^0 2p^n$, $1s^0 2s^1 2p^n$, $1s^0 2s^2 2p^n$, $n = 1 - 6$. The transitions for which the values of fluorescence yields are less than 0.20 are not included in the table. Each state is a mixture of many basis states and only the dominant component is used for an approximate identification of the initial and final state.

Table 1. Radiative and Auger transition rates, lifetimes, and fluorescence yields of calcium atoms with electron configuration $1s^1 2s^m 2p^n$, $(m = 0-2,\ n = 2-6)$. The wavelengths (in Angstrom), total Auger transition rates (in units of $10^{13} s^{-1}$), partial x-ray transition rates (in units of $10^{13} s^{-1}$), lifetime (in units of $10^{-15} s$), and fluorescence yields are represented, respectively, by λ, Γ_a, Γ_x, τ, and ω.

Transition	λ	Γ_a	Γ_x	τ	$\omega \times 100$
$1s2p^2\ ^4P_{1/2}$ - $1s^2 2p\ ^2P_{1/2}$	3.2214	.10	.17	355.00	59.46
$1s2p^2\ ^4P_{5/2}$ - $1s^2 2p\ ^2P_{3/2}$	3.2212	.40	.21	165.00	34.10
$1s2p^2\ ^2D_{5/2}$ - $1s^2 2p\ ^2P_{3/2}$	3.2064	15.60	8.18	4.21	34.45
$1s2p^2\ ^2P_{1/2}$ - $1s^2 2p\ ^2P_{3/2}$	3.2050	.27	5.80	3.92	22.74
$1s2p^2\ ^2D_{3/2}$ - $1s^2 2p\ ^2P_{1/2}$	3.2033	14.20	10.10	4.08	41.31
$1s2p^2\ ^2P_{1/2}$ - $1s^2 2p\ ^2P_{1/2}$	3.2008	.27	19.40	3.92	76.24
$1s2p^2\ ^2P_{3/2}$ - $1s^2 2p\ ^2P_{3/2}$	3.2005	1.66	22.90	3.87	88.62
$1s2p^2\ ^2S_{1/2}$ - $1s^2 2p\ ^2P_{3/2}$	3.1925	6.39	8.55	6.43	54.98
$1s2p^3\ ^3D_3$ - $1s^2 2p^2\ ^3P_2$	3.2337	22.30	7.51	3.31	24.87
$1s2p^3\ ^1P_1$ - $1s^2 2p^2\ ^1S_0$	3.2324	16.20	10.70	2.44	26.19
$1s2p^3\ ^3D_2$ - $1s^2 2p^2\ ^3P_1$	3.2322	21.80	7.44	3.35	24.92
$1s2p^3\ ^3D_1$ - $1s^2 2p^2\ ^3P_0$	3.2304	21.70	6.87	3.31	22.73
$1s2p^3\ ^3S_1$ - $1s^2 2p^2\ ^3P_2$	3.2285	3.02	10.30	3.08	31.62
$1s2p^3\ ^1D_2$ - $1s^2 2p^2\ ^1D_2$	3.2278	21.10	21.20	2.36	50.03
$1s2p^3\ ^3S_1$ - $1s^2 2p^2\ ^3P_1$	3.2262	3.02	15.20	3.08	46.71
$1s2p^3\ ^3P_2$ - $1s^2 2p^2\ ^3P_2$	3.2249	18.00	8.50	3.41	28.99
$1s2p^3\ ^3P_1$ - $1s^2 2p^2\ ^3P_2$	3.2238	14.50	11.10	3.82	42.27
$1s2p^4\ ^4P_{5/2}$ - $1s^2 2p^3\ ^4S_{3/2}$	3.2564	25.10	7.48	3.04	22.75
$1s2p^4\ ^2P_{1/2}$ - $1s^2 2p^3\ ^2P_{1/2}$	3.2561	25.40	11.20	1.79	20.05
$1s2p^4\ ^4P_{3/2}$ - $1s^2 2p^3\ ^4S_{3/2}$	3.2538	25.00	7.65	3.05	23.33
$1s2p^4\ ^2D_{3/2}$ - $1s^2 2p^3\ ^2D_{3/2}$	3.2528	36.00	15.70	1.84	28.84
$1s2p^4\ ^4P_{1/2}$ - $1s^2 2p^3\ ^4S_{3/2}$	3.2526	25.00	7.65	3.05	23.33
$1s2p^4\ ^2P_{3/2}$ - $1s^2 2p^3\ ^2D_{5/2}$	3.2489	27.60	18.30	1.78	32.53
$1s2p^4\ ^2S_{1/2}$ - $1s^2 2p^3\ ^2P_{3/2}$	3.2481	30.10	14.90	2.13	31.63
$1s2p^5\ ^3P_0$ - $1s^2 2p^4\ ^3P_1$	3.2767	46.20	14.70	1.64	24.11
$1s2p^5\ ^1P_1$ - $1s^2 2p^4\ ^1D_2$	3.2745	46.20	24.40	1.32	32.25
$1s2s2p\ ^4P_{1/2}$ - $1s^2 2s\ ^2S_{1/2}$	3.2246	.00	.03	2620.00	90.65
$1s2s2p\ ^4P_{3/2}$ - $1s^2 2s\ ^2S_{1/2}$	3.2231	.01	.10	874.00	89.58
$1s2s2p\ ^2P_{1/2}$ - $1s^2 2s\ ^2S_{1/2}$	3.1999	1.88	12.90	6.76	87.20
$1s2s2p\ ^2P_{3/2}$ - $1s^2 2s\ ^2S_{1/2}$	3.1976	.27	15.90	6.18	98.42
$1s2s2p\ ^2P_{1/2}$ - $1s^2 2s\ ^2S_{1/2}$	3.1879	8.46	3.88	8.10	31.43
$1s2s2p^2\ ^5P_2$ - $1s^2 2s2p\ ^3P_2$	3.2562	.02	.03	1730.00	59.51

Table 1. (Continued.)

Transition	λ	Γ_a	Γ_x	τ	$\omega \times 100$
$1s2s2p^2\ ^5P_1$ - $1s^22s2p\ ^3P_1$	3.2551	.05	.05	825.00	39.05
$1s2s2p^2\ ^5P_3$ - $1s^22s2p\ ^3P_2$	3.2538	.12	.05	586.00	31.90
$1s2s2p^2\ ^3D_3$ - $1s^22s2p\ ^3P_2$	3.2265	17.00	8.14	3.98	32.41
$1s2s2p^2\ ^3P_0$ - $1s^22s2p\ ^3P_1$	3.2258	2.68	22.40	3.99	89.38
$1s2s2p^2\ ^3D_2$ - $1s^22s2p\ ^3P_1$	3.2246	13.70	11.10	3.97	43.91
$1s2s2p^2\ ^3P_2$ - $1s^22s2p\ ^3P_2$	3.2241	5.41	18.70	3.97	74.16
$1s2s2p^2\ ^3D_1$ - $1s^22s2p\ ^3P_0$	3.2239	7.42	12.90	4.00	51.73
$1s2s2p^2\ ^3D_1$ - $1s^22s2p\ ^3P_1$	3.2233	11.60	9.23	3.97	36.66
$1s2s2p^2\ ^1P_1$ - $1s^22s2p\ ^1P_1$	3.2206	5.95	25.10	3.22	80.82
$1s2s2p^2\ ^1S_0$ - $1s^22s2p\ ^1P_1$	3.2186	19.90	7.39	3.66	27.05
$1s2s2p^2\ ^3S_1$ - $1s^22s2p\ ^3P_2$	3.2138	9.54	4.37	5.94	25.94
$1s2s2p^3\ ^4D_{7/2}$ - $1s^22s2p^2\ ^4P_{5/2}$	3.2533	22.90	7.71	3.26	25.14
$1s2s2p^3\ ^4D_{5/2}$ - $1s^22s2p^2\ ^4P_{3/2}$	3.2518	22.60	6.92	3.29	22.76
$1s2s2p^3\ ^4S_{3/2}$ - $1s^22s2p^2\ ^4P_{5/2}$	3.2515	4.17	11.40	3.04	34.58
$1s2s2p^3\ ^2D_{3/2}$ - $1s^22s2p^2\ ^2D_{3/2}$	3.2513	30.40	16.40	1.96	32.10
$1s2s2p^3\ ^2D_{5/2}$ - $1s^22s2p^2\ ^2D_{5/2}$	3.2510	29.90	19.00	1.96	37.24
$1s2s2p^3\ ^4D_{3/2}$ - $1s^22s2p^2\ ^4P_{1/2}$	3.2504	22.00	6.35	3.28	20.83
$1s2s2p^3\ ^4D_{1/2}$ - $1s^22s2p^2\ ^4P_{1/2}$	3.2502	22.80	7.15	3.26	23.31
$1s2s2p^3\ ^4S_{3/2}$ - $1s^22s2p^2\ ^4P_{3/2}$	3.2494	4.17	13.10	3.04	39.75
$1s2s2p^4\ ^3P_0$ - $1s^22s2p^3\ ^3P_1$	3.2782	32.40	15.30	1.64	25.09
$1s2s2p^4\ ^5P_3$ - $1s^22s2p^3\ ^5S_2$	3.2752	25.30	7.37	3.06	22.56
$1s2s2p^4\ ^1S_0$ - $1s^22s2p^3\ ^1P_1$	3.2744	53.10	13.30	1.50	19.95
$1s2s2p^4\ ^5P_2$ - $1s^22s2p^3\ ^5S_2$	3.2730	25.20	7.42	3.06	22.71
$1s2s2p^4\ ^3P_2$ - $1s^22s2p^3\ ^3D_2$	3.2729	34.50	11.90	1.69	20.18
$1s2s2p^4\ ^3D_1$ - $1s^22s2p^3\ ^3D_1$	3.2718	37.30	15.20	1.78	27.12
$1s2s2p^4\ ^5P_1$ - $1s^22s2p^3\ ^5S_2$	3.2717	25.20	7.43	3.06	22.75
$1s2s2p^5\ ^2P_{3/2}$ - $1s^22s2p^4\ ^2D_{5/2}$	3.2968	61.30	17.80	1.15	20.50
$1s2s2p^5\ ^2P_{1/2}$ - $1s^22s2p^4\ ^2D_{3/2}$	3.2950	55.90	22.50	1.20	27.00
$1s2s^22p\ ^1P_1$ - $1s^22s^2\ ^1S_0$	3.2189	8.77	15.80	4.07	64.17
$1s2s^22p^2\ ^2D_{3/2}$ - $1s^22s^22p\ ^2P_{1/2}$	3.2465	29.30	9.28	2.58	23.93
$1s2s^22p^2\ ^2P_{1/2}$ - $1s^22s^22p\ ^2P_{1/2}$	3.2444	11.80	17.20	2.92	50.22
$1s2s^22p^2\ ^2P_{3/2}$ - $1s^22s^22p\ ^2P_{3/2}$	3.2441	13.60	20.30	2.86	58.13
$1s2s^22p^2\ ^2S_{1/2}$ - $1s^22s^22p\ ^2P_{3/2}$	3.2358	22.90	7.65	3.21	24.56
$1s2s^22p^3\ ^3S_1$ - $1s^22s^22p^2\ ^3P_2$	3.2710	17.00	9.87	2.30	22.69
$1s2s^22p^3\ ^1S_2$ - $1s^22s^22p^2\ ^1D_2$	3.2702	36.50	18.60	1.81	33.59
$1s2s^22p^3\ ^3D_1$ - $1s^22s^22p^2\ ^3P_1$	3.2689	17.00	13.30	2.30	30.59
$1s2s^22p^3\ ^3P_1$ - $1s^22s^22p^2\ ^3P_2$	3.2660	34.60	9.03	2.25	20.33
$1s2s^22p^3\ ^1P_1$ - $1s^22s^22p^2\ ^1D_2$	3.2619	31.00	12.40	1.89	23.44
$1s2s^22p^4\ ^2P_{3/2}$ - $1s^22s^22p^3\ ^2D_{5/2}$	3.2903	44.40	16.00	1.44	23.04
$1s2s^22p^4\ ^2S_{1/2}$ - $1s^22s^22p^3\ ^2P_{3/2}$	3.2892	50.20	13.10	1.53	20.04
$1s2s^22p^5\ ^1P_1$ - $1s^22s^22p^4\ ^1D_2$	3.3145	63.30	21.70	1.12	24.30

We have compared the results from the present calculation with those available in the literature. Good agreements are found among various calculations for prominent radiative and Auger lines. In Table 2 we list Auger and x-ray rates for a few selected transitions from the present calculation and compare those with the Z-expansion calculations of Vainshtein and Safronova [7] and Dirac-Fock calculations of Chen et al [1]. The Auger rates of Vainshtein and Safronova are generally larger (typically 15%) than our values while those reported by Chen et al [1] are usually smaller (typically 8%).

Table 2. Comparison of the total Auger and the total x-ray transition rates (in units of $10^{13}s^{-1}$) of variously ionized calcium atoms. $\Gamma_a^{[P]}$, $\Gamma_a^{[V]}$, and $\Gamma_a^{[C]}$ represent, respectively, the total Auger transition rates for the state $|i>$ from the present Hartree-Fock calculation, the Z-expansion calculation of Vainshtein and Safronova [Ref. 7], and the Dirac-Fock calculations of Chen et al [Ref. 1]. Similarly, $\Gamma_x^{[P]}$, $\Gamma_x^{[V]}$, and $\Gamma_x^{[C]}$ represent, respectively, the x-ray transition rates from the present calculation, the calculation of Vainshtein and Safronova, and that of Chen et al.

$\|i>$	$\Gamma_a^{[P]}$	$\Gamma_a^{[V]}$	$\Gamma_a^{[C]}$	$\Gamma_x^{[P]}$	$\Gamma_x^{[V]}$	$\Gamma_x^{[C]}$
$1s2s2p(^1P)\ ^2P_{1/2}$	8.46	10.5	-	3.88	3.51	-
$1s2p^2\ ^2P_{3/2}$	1.66	2.04	1.59	24.14	22.78	22.66
$1s2p^2\ ^2D_{3/2}$	14.20	16.50	13.71	10.26	14.26	10.16
$1s2p^4\ ^2D_{3/2}$	36.00	-	33.30	18.49	-	17.00
$1s2p^4\ ^2D_{5/2}$	38.50	-	36.50	15.22	-	13.80
$1s2p^4\ ^2S_{1/2}$	30.10	-	25.20	14.90	-	14.00
$1s2s2p^3\ ^4D_{7/2}$	22.90	-	21.90	7.71	-	7.05
$1s2s2p^3\ ^4D_{5/2}$	22.60	-	21.40	7.78	-	7.04
$1s2s2p^3\ ^4S_{3/2}$	4.17	-	5.44	28.75	-	26.20
$1s2s2p^3\ ^4D_{3/2}$	22.00	-	20.70	8.41	-	7.85
$1s2s^22p^2\ ^2D_{3/2}$	29.30	-	30.70	9.45	-	9.03
$1s2s^22p^2\ ^2D_{5/2}$	31.40	-	33.20	7.48	-	7.00

As the number of spectator electrons increase the x-ray and Auger rates both increase. This is expected since more electrons can participate in the decay process. However, the Auger transition rates increase at a faster rate than the x-ray transition rates. This is because the Auger transitions are caused by electron-electron interactions. The line fluorescence yields, which is the ratio of x-ray transition rate and the total rate, decreases as the number of spectator electrons increase.

Acknowledgment

This work was supported by the University Research Grant Office of Illinois State University, and in part by the Division of Chemical Sciences, Office of Basic Energy Sciences, Office of Energy Research, U.S. Department of Energy.

References

1. Chen, M. H., Crasemann, B. and Mark, H., Phys. Rev. A **27** 544 (1983); Chen, M. H., Crasemann, B. and Mark, H., Phys. Rev. A **26** 1441 (1982); Chen, M. H., Atomic Data Nuclear Data Tables **34** 301 (1986); Chen, M. H., Phys. Rev. A. **31** 1449 (1985); Chen, M. H.and Crasemann, B., Atomic Data Nuclear Data Tables **37** 419 (1987); Chen, M. H. and Crasemann, B., Phys. Rev. A **35** 4579 (1987), Chen, M. H.and Crasemann, B., Atomic Data Nuclear Data Tables **38** 381 (1988); Chen, M. H., Reed, K. J., McWilliams, D. M., Guo, D. S., Barlow, L., Lee, M. and Walker, V., Atomic Data Nuclear Data Tables **65** 2 89 (1997).
2. Nilsen, J., Atomic Data Nuclear Data Tables **38** 339 (1988); Nilsen, J., Safronova U. I. and Safronova, M. S., Physica Scripta **51** 589 (1995).
3. Karim, K. R. and Bhalla, C. P., J. Quant. Spectrosc. Radiat. Transfer **51** 557 (1994); Phys. Rev. A **45** 3932 (1992); Phys. Rev. A **43**615 (1991).
4. Safronova, U. I., Shlyaptseva, A. S. and Golovkin, I. E., Physica Scripta **52** 277 (1995); Safronova, U. I. and Shlyaptseva, A. S., Physica Scripta **52** 277 (1995); Safronova, U. I., Safronova, M. S., Burch, R. and Vainshtein, L. A., Physica Scripta **51** 471 (1995).
5. Bhalla, C. P., Folland N. O. and Hein M. A., Phys. Rev. A **8**649 (1973); Bhalla, C. P., Phys. Rev. A **8**2877 (1973); Bhalla, C. P. and Tunnell, T. W., J. Quant. Spectrosc. Radiat. Transfer **32**141 (1984).
6. Karim, K. R., Grabbe, S. R. and Bhalla, C. P., J. Phys. B **29**4007 (1996).
7. Vainshtein, L. A. and Safronova, U. I., Atomic Data Nuclear Data Tables **21** 49 (1978).

Highly Charged Ion Trapping and Cooling

L. Gruber*, B. R. Beck*, J. Steiger*, D. Schneider*, J. P. Holder+, D. A. Church+

*Lawrence Livermore National Laboratory, Livermore, CA 94550
+Physics Department, Texas A&M University, College Station, TX 77843-4242

Abstract. In the past few years a cryogenic Penning trap (RETRAP) has been operational at the Electron Beam Ion Trap (EBIT) facility at Lawrence Livermore National Laboratory. The combination of RETRAP and EBIT provides a unique possibility of producing and re-trapping highly charged ions and cooling them to very low temperatures. Due to the high Coulomb potentials in such an ensemble of cold highly charged ions the Coulomb coupling parameter (the ratio of Coulomb potential to the thermal energy) can easily reach values of 172 and more. To study such systems is not only of interest in astrophysics to simulate White Dwarf star interiors but opens up new possibilities in a variety of areas (e.g. laser spectroscopy, cold highly charged ion beams).

INTRODUCTION

Cold highly charged ion (HCI) plasmas are of interest in spectroscopy, astrophysics and ion beam physics. Fine and hyperfine transitions can potentially be measured to an unprecedented precision. In ion beam physics the reduction of phase space can lead to the development of high brilliance ions beams (1) and in astrophysics these kind of plasmas can simulate the interior of White Dwarf stars (2). The Coulomb coupling parameter Γ, a parameter classifying how strongly coupled the plasma is, is the ratio of the Coulomb potential energy to the thermal energy of the ions. One-component singly-charged ion crystals have been observed (3). Theory predicts (4) a fluid-solid transition at $\Gamma = 172$. In the following, efforts to create strongly coupled plasmas with HCIs are described.

EXPERIMENTAL SETUP

EBIT and Beam Transport

An Electron Beam Ion Trap (EBIT) produces ions by successive electron impact ionization (5). These ions are confined axially by static electric fields and radially by the space charge of the electron beam. The ions can be extracted by ramping up the middle drift tube potential (6) (Fig. 1). The extracted ion bunch is guided with static electric and magnetic ion optical elements (Fig. 2) to RETRAP, a cryogenic Penning trap. A 90° bending magnet in the beamline selects only ions with the same mass to charge ratio for transport to RETRAP. A typical ion bunch is about 5 μs long and contains $4 \cdot 10^4$ Xe^{44+} ions at an energy of 2 keV/u.

At RETRAP the ions are decelerated upon entering a tube biased at approximately the extraction potential. The tube potential is then rapidly (10 ns) switched to ground thereby reducing the ion kinetic energy to $\approx$70 eV/u. These ions can then be trapped by pulsing a capture electrodes of RETRAP.

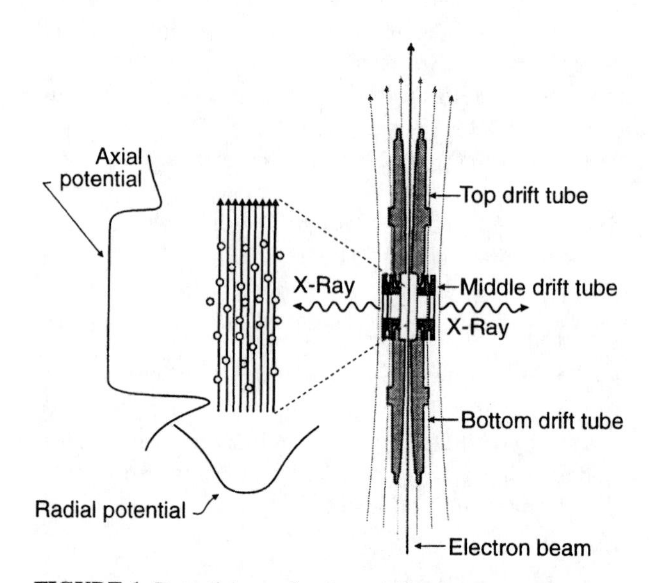

FIGURE 1. Potentials on the three drift tubes create the axial trap in EBIT and the electron beam (indicated by the arrow in the center) creates the radial potential. The potential of the bottom drift tube is higher than the top drift tube potential to assure that the ions escape through the top drift tube when the middle drift tube potential is ramped up.

CP475, *Applications of Accelerators in Research and Industry*,
edited by J. L. Duggan and I. L. Morgan

RETRAP

RETRAP (7) is a set of cryogenic Penning traps with a magnetic induction B set to 4.4 T. One of these traps has been used for the experiments described below. This trap has hyperbolic electrodes for electric potential harmonicity and holes in the ring electrode for laser beams and fluorescence light detection. For ion capture and release additional electrodes have been installed above and below the endcap electrodes (Fig. 3).

An electrostatic potential V_0 applied between the ring and the endcap electrodes produces a potential $V(\rho,z)$ inside the trap which can be described as

$$V(\rho,z) = V_0 \frac{z^2 - \rho^2/2}{2d^2} \tag{1}$$

where z is the axial (parallel to the magnetic field) and ρ the radial coordinate. The characteristic length d depends on the half length z_0 and the radius ρ_0 of the trap and is given by $d^2 = \frac{1}{2} \cdot (z_0^2 + \frac{\rho_0^2}{2})$.

The motion of an ion with charge q and mass m in such a potential is a superposition of three independent harmonic oscillations: an axial oscillation along the magnetic field lines with the frequency ω_z,

$$\omega_z = \sqrt{\frac{4qV_0}{md^2}} \tag{2}$$

and two radial oscillations with the frequencies $\omega_\pm$

$$\omega_\pm = \frac{\omega_c}{2} \pm \sqrt{\frac{\omega_c^2}{4} - \frac{\omega_z^2}{2}}. \tag{3}$$

Here $\omega_c = \frac{qB}{m}$ is the cyclotron frequency. Equations (2) and (3) are valid only for a single ion moving in the trap. If there are more ions in the trap the potential $V(\rho,z)$ is altered by the space charge of the ions and all frequencies shift by small amounts. Also, when ions are hot and oscillate with large amplitudes, anharmonicities in the trap potential cause broadening and shifts in the resonances.

To detect the axial motion of the ions the trap endcap electrodes are connected with an inductor designed to form a tuned circuit with high quality factor Q and the resonance frequency ω_0. The axial motion is detected by tuning the oscillation frequency ω_z of the ions to the resonance frequency ω_0 by setting the potential V_0. An amplifier coupled capacitively to the inductor picks up the

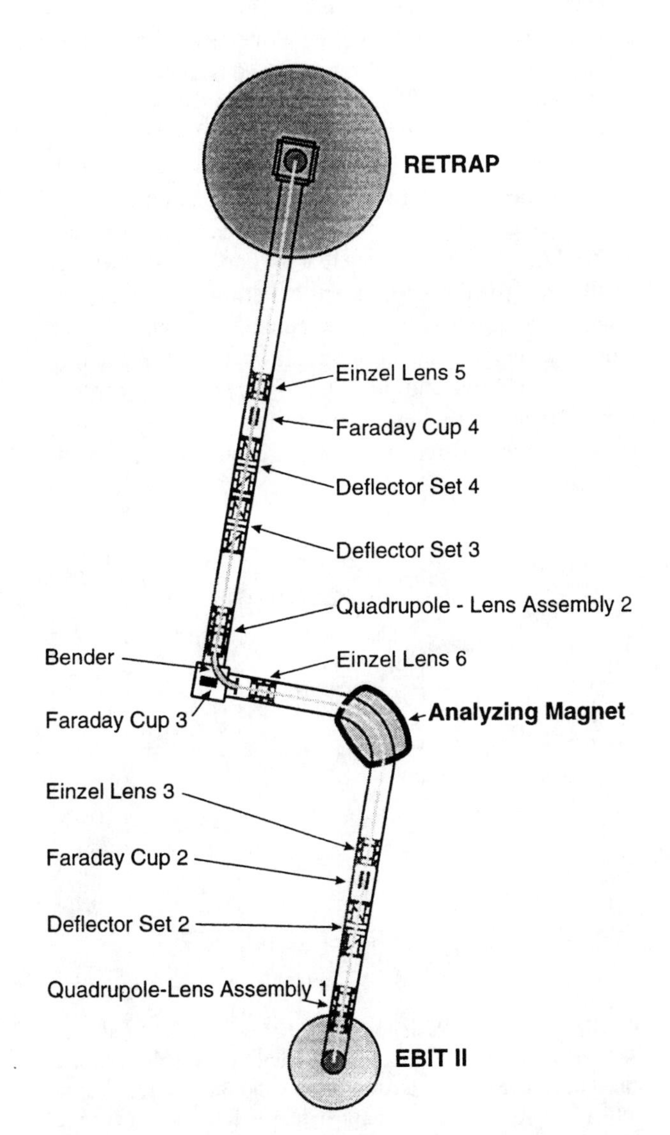

FIGURE 2. Top view of the beamline connecting EBIT II with RETRAP. The path of the ions is out of the paper plane from EBIT II and into the paper plane at RETRAP.

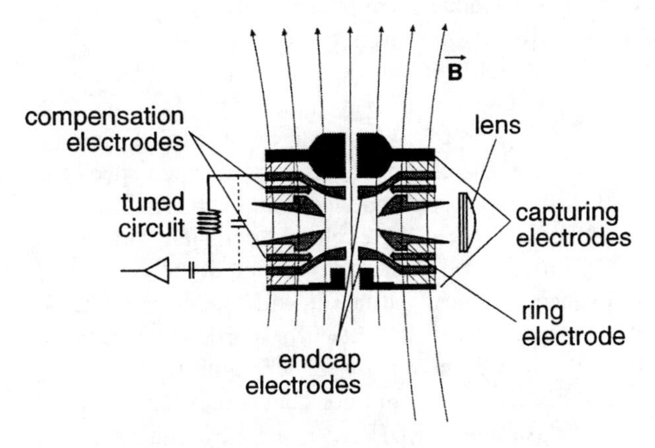

FIGURE 3. Schematic of the hyperbolic trap used in the described experiments. The lens increases the solid angle for fluorescence light detection.

increased noise signal V_S produced by the ions confined in the trap:

$$V_S^2 = K N Q^2 q^2 \frac{U}{m} \quad (4)$$

where K is a constant for a specific trap geometry, U is the average kinetic energy per ion, q is the charge state, m is the mass and N is the number of the ions on resonance.

This technique probes the trap content nondestructively. Provided the number of ions in the trap does not change, the change of temperature over time can be measured. If the temperature is kept constant the ion number can be determined. Assuming all the ions are in thermal equilibrium the relative numbers of different ion species and charge states in the trap can be determined as well.

COOLING TECHNIQUES

Singly charged ions with strong electric dipole transitions near the visible have been laser cooled to temperatures below 10 mK (8). Direct laser cooling is not feasible for HCIs since there are no electric dipole transitions accessible with laser light. Different techniques are needed to cool the HCIs to temperatures below 1 K. One possible scheme is to merge the HCIs with a laser cooled Be^+ cloud and cool the HCIs by Coulomb collisions (sympathetic cooling) to the temperature of the Be^+ ions. Different cooling techniques are described below.

An ion oscillating in the trap induces a current in the tuned circuit. If the ion oscillation frequency matches the tuned circuit frequency, the energy of the ion is dissipated in the ohmic resistance of the circuit. The result is an exponential decrease in the ion energy with a time constant τ_R of

$$\tau_R = \frac{4 m z_0^2}{Q \beta^2 q^2 \omega_0 L} \quad (5)$$

where L is the inductance of the circuit and β is a coupling constant ($\beta \approx 0.8$ for this trap geometry) (9).

For RETRAP ($Q = 400, \omega_0 = 2\pi \cdot 2.5$ MHz, $L =$ 200 mH and $z_0 = 5 \cdot 10^{-3}$ m) the calculated cooling time of Be^{2+} is $\tau_R = 18$ s. It should be noted that only the axial degree of freedom of this particular ion species is cooled in this process and the radial degrees of freedom and other ion species are cooled by collisions among the Be^+. This, as well as the fact that not all the ions are on resonance with the tuned circuit, will increase the time for the resistive cooling. Since τ_R is proportional to m/q^2 the time constant drops rapidly for ions with higher charge states. For Xe^{44+} this time constant is only 0.56 s.

Laser cooling in RETRAP is done by shining a laser beam into the trap through the holes in the ring, perpendicular to the magnetic field. Therefore the laser cools the two radial degrees of freedom of Be^+, when tuned to a wavelength just above that of the resonance transition. If the laser wavelength is tuned below the resonance, heating occurs.

Resistive cooling becomes more effective for higher charges [Eq. (5)], but the cooling limit is given by the temperature of the tuned circuit (4.2 K for RETRAP) (9).

RESULTS

Resistive cooling and laser cooling

Due to the nature of the beryllium ion source and the applied injection technique, most often Be^+ and Be^{2+} were caught in the trap. Figure 4 shows a typical cooling curve for Be^{2+} (the number of Be^{2+} ions is constant on that time scale; therefore the plotted signal is proportional to the Be^{2+} temperature). Two different regions of the data were fitted with an exponential decay: the first region is 0 s – 2200 s and the second is 2300 s – 2800 s. The two different exponential decays fit the data well. Cooling the axial motion of the Be^{2+} reduces the temperature of the two radial degrees of freedom as well as all degrees of freedom of the Be^+. Note that the cooling laser was on at all times for this experiment but at high temperatures it has a negligible effect on the cooling. At about 2300 s a drastic change in the cooling rate occurs: laser cooling becomes effective and the drop in temperature becomes faster. In this phase the cooling is being done by the two radial degrees of freedom of the Be^+ but still the axial

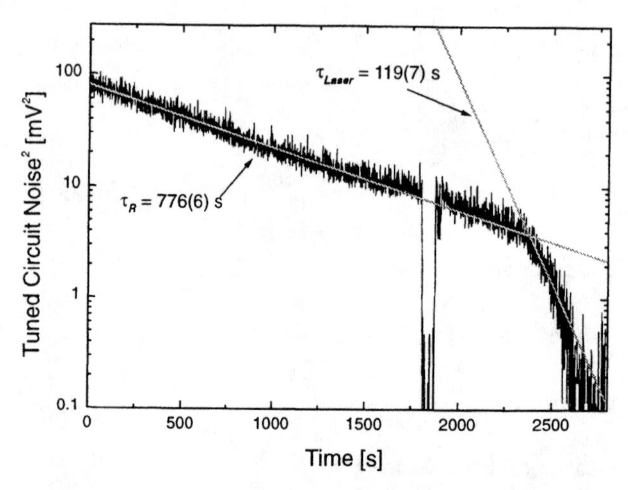

FIGURE 4. Tuned circuit noise signal of Be^{2+} ions stored with Be^+ ions. At first Be^{2+} is cooled mainly resistively with the tuned circuit. Be^+ is cooled mostly by the Be^{2+} via Coulomb collisions. As soon as the temperature is low enough for the laser cooling to be effective (at about 2300 s), laser cooling of Be^+ becomes faster and the Be^{2+} is now cooled mainly by the Be^+, indicated by the kink in the noise signal.

signal of the Be^{2+} is observed. This is a first indication of sympathetic cooling of the monitored Be^{2+} by the laser cooled Be^{+}.

The dip in the data at about 1800 s is due to tuning the ions off resonance to find the background level for background subtraction. The background is due to the Johnson noise across the input resistor of the amplifier.

Sympathetic cooling of Xe^{q+}

Sympathetic cooling has already been seen in Fig. 4 involving Be^{+} and Be^{2+}. Sympathetic cooling of Xe^{q+} (q>30) by Be^{2+} is shown in Fig. 5. Be^{2+} ions were loaded into the trap and soon thereafter Xe^{44+} ions were captured in the same trap. After 10 s the potential of the ring electrode was swept through the resonance of the Be^{2+} and the HCIs to probe the charge distribution. The maximum axial energy E_{max} that an ion of charge q can have in a potential well of depth U is given by:

$$E_{max} = qU = \frac{1}{2}k_B T. \qquad (6)$$

Since U is the same for both Be^{2+} and Xe^{44+}, the Be^{2+} temperature T can be a factor of 20 lower than the Xe^{44+} temperature. Here k_B is the Boltzmann constant. Both species will equilibrate collisionally. Ions with large axial amplitude (i.e. hot ions) have a lower oscillation frequency and need a deeper potential well to be on resonance with the tuned circuit. This causes the detected signal to be broadened towards higher voltages V_0. If the Be^{2+} ions are in the trap as well then the distribution starts to show a structure which corresponds to the individual charge states of the HCIs and the signal of the Be^{2+} ions can be seen as well. This is a strong indication of sympathetic cooling of HCIs by Be^{2+}.

The next step is to catch the HCIs into a laser cooled beryllium plasma (T < 2 K). Two different methods of detecting the ion cloud have been chosen and applied and a third method is being developed. A photomultiplier tube (PMT) detects scattered light through the ring electrode perpendicular to the cooling beam and is opposite a cryogenically cooled CCD camera that takes side view images of the fluorescing cloud. The PMT provides the time resolution whereas the CCD is an integrating detector that gives spatial information about the cloud.

Once the beryllium ions are cold, HCIs can be caught in the same fashion as explained above. This involves ramping and pulsing of certain electrodes. This can be a heat source and if so, an increase in the PMT signal should be observed. In Fig. 6 a PMT signal is displayed while two HCI loading sequences were done; the first sequence with no HCI beam coming from EBIT (short duration of higher scatter rate), then a second with loading of EBIT ions into the cold beryllium plasma. For the latter case the Be^{+} ions heat up more than in the first case, since the additional energy of the HCI is transferred to the Be^{+}.

The CCD camera gives additional information for the capture of HCIs into the cold beryllium plasma. In

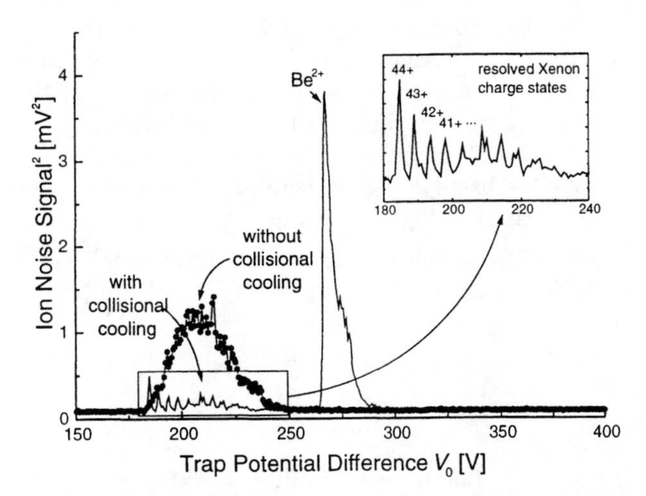

FIGURE 5. The tuned circuit signal indicates sympathetic cooling of Xe^{q+} ions. Without the presence of Be^{2+} a broad peak due to a wide energy distribution is visible when hot HCIs are captured (plot with symbols). When HCIs are merged into Be^{2+}, a set of distinct peaks appear, revealing the charge state distribution of the colder xenon ions.

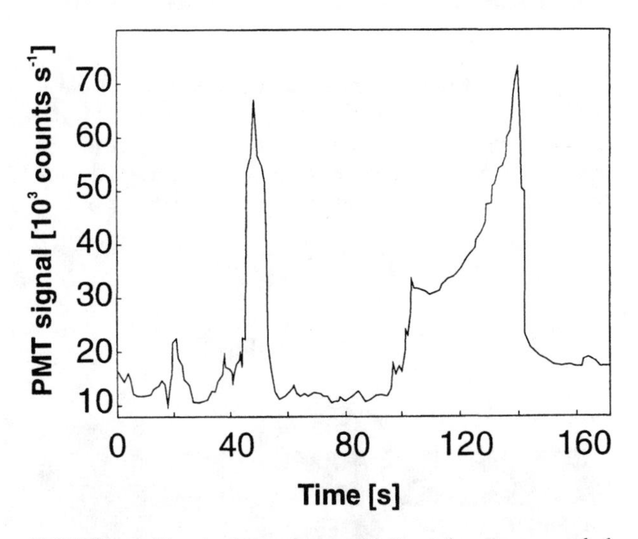

FIGURE 6. Signal of the photomultiplier tube. Laser cooled Be^{+} was in the trap. At the first peak (50 s) the electrodes were pulsed for xenon capture, but no xenon was admitted. In the second peak (100 s – 140 s) xenon was captured into the trap. It can be seen that the Be^{+} ions were heated more (PMT shows a high scatter rate for a longer time) in the case when xenon ions were injected into the pre-cooled ions in the trap.

Fig. 7 a) a side view of a cold Be^+ cloud can be seen. The two spots can be explained in the following way: As mentioned before Be^+ and Be^{2+} ions are typically caught in the same trap but only Be^+ is visible since there are no transitions excited for Be^{2+}. As the ions cool they start to centrifugally separate due to different mass to charge ratios (13). Be^+ forms an annulus around the Be^{2+}. Since the laser beam has a diameter close to the extent of the Be^{2+} cloud (i.e. the gap between the spots in Fig. 7 a) only two small sections of the Be^+ annulus can be observed (Fig. 8).

If HCIs ($m/q \approx 3$) are introduced into the trap, they are expected to concentrate in the center. Therefore they would displace some of the Be^{2+} and make the gap between the two spots even bigger. This is what can be seen in Fig. 7 b).

A third method of visualizing the HCIs in the trap would be to project them onto a CCD chip located in the fringe field of the magnet below the trap. The detection efficiency of a HCI impacted on a CCD is very high compared to a low charge state ion (e.g. Be^+ or Be^{2+}). This opens the possibility to check for HCIs in the trap, since the CCD would be almost 'blind' for beryllium ions

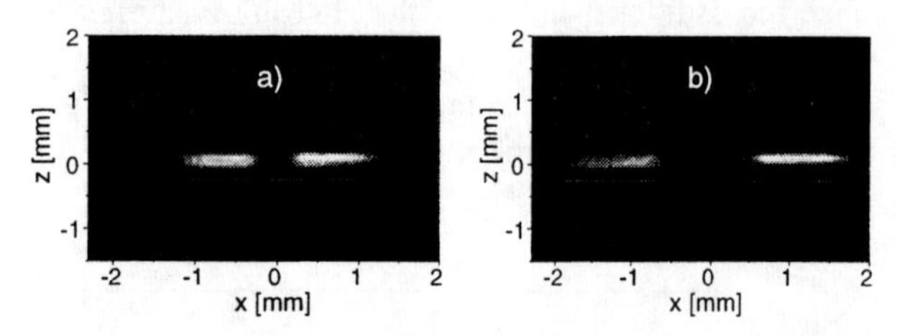

FIGURE 7. CCD side view pictures of the cold Be^+ cloud. In a) there were only Be^+ and Be^{2+} in the trap. In b) highly charged xenon ions were added. It can be seen that the gap between the two light spots increases. This indicates a radial displacement of the Be^{2+} ions by the additional xenon ions.

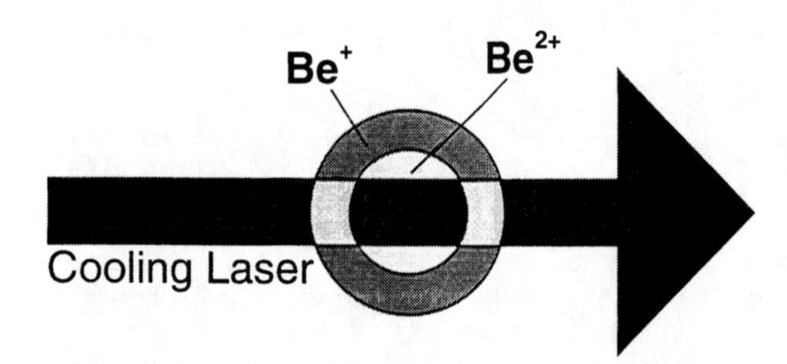

FIGURE 8. Schematic top view of the cold Be^+-Be^{2+} cloud. Only the Be^+ cloud overlapping with the laser beam can scatter light (indicated by the lighter areas in the Be^+ cloud) producing two spots in the side view images.

but would see every single HCI. The development of this method is in progress.

In HCIs, fine and hyperfine structure transitions of certain ions can be accessible with a laser (10), which would make an additional detection method feasible. The HCI cloud might be imaged directly with a CCD camera.

Coulomb coupling parameter Γ

To characterize the plasma the density and temperature has to be determined. For cases where only Be^+ was caught in the trap, the aspect ratio of the ellipsoidal cloud can be measured from the side view images from which the plasma density can be determined (11). Since the volume can be calculated from the same image, the number of trapped cold Be^+ ions can be determined (typical: $4 \cdot 10^5$). With this the number of Be^{2+} and Xe^{q+} ions can be extracted from the tuned circuit signal, which gives the relative numbers of ions assuming that the ions are in thermal equilibrium. The number of xenon ions was determined to be ≈ 400. This is in good agreement with the number of counted ions after release from the trap when there was no beryllium in the trap.

Assuming the fluid model is applicable for this plasma a minimum density for the ions in the trap (11) can be obtained ($3 \cdot 10^7$ cm^{-3} for Xe^{44+}).

The temperature has been estimated in different ways: The cooling laser has been swept over the resonance and the measured FWHM of the fluorescence signal was about 300 MHz. This indicates a temperature of 1.7 K. Another indication of the temperature is the numerical simulation of a mixture of Be^+ and Be^{2+} ions in a Penning trap. The mixture starts separating at around 0.25 K (12). And a third possibility is to use a model developed by O'Neil (13), predicting a separation at temperatures below 0.5 K based on a characteristic scaling length.

If a temperature of 1.7 K and a minimal density of $3 \cdot 10^7$ cm^{-3} is assumed the Coulomb coupling parameter Γ for Xe^{44+} can be calculated:

$$\Gamma = \frac{q^2}{4\pi\varepsilon_0 a_0} \frac{1}{k_B T} \tag{7}$$

with $a_0 = (\frac{3}{4\pi n})^{1/3}$. For an infinitely large one-component plasma theory predicts a crystallization at $\Gamma = 172$. This means that the plasma in RETRAP should have crystallized. Therefore the effects of an ordered structure can expected to be seen when the cloud of strongly coupled HCIs is projected onto the CCD chip below the trap.

CONCLUSION

With the described setup it was possible to trap several hundred HCIs in RETRAP and cool them to temperatures below 2 K. Assuming the fluid model is valid the minimum density of these ions is estimated to $3 \cdot 10^7$ cm^{-3}. This gives a Coulomb coupling parameter exceeding 1000, indicating a highly charged ion crystal in the trap.

ACKNOWLEDGMENTS

We would like to thank J. W. McDonald for the excellent preparation of the ion beam from EBIT and D. Nelson and E. Magee for the technical support.

This work was performed under the auspices of the U.S. Dept. of Energy by the Lawrence Livermore National Laboratories under contract #W-7405-ENG-48 and supported by the Texas Advanced Research Program.

REFERENCES

1. Moore, R. B., *Hyp. Int.* **81**, 45–70 (1993)
2. Van Horn, H. M., *Astr. J.* **151**, 227, 1968;
   Van Horn, H. M., *Strongly Coupled Plasma Physics*, Rochester, NY, University of Rochester Press, 1990, 3–19;
3. Diedrich, F., Peik, E., Chen, J. M., Quint, W., Walther, H., *Phys. Rev. Lett.* **59**, 2931–2934 (1987);
   Wineland, D. J., Bergquist, J. C., Itano, W. M., Bollinger, J. J., Manney, C. H., *Phys. Rev. Lett.* **59**, 2935–2938 (1987);
4. Brush, S. G., Sahlin, H. L., Teller, E., *J. Chem. Phys.* **45**, 2102–2118 (1966);
   Slattery, W. L., Doolen, G. D., DeWitt, H. E., *Phys. Rev. A* **26**, 2255–2258 (1982)
5. Marrs, R. E., Levine, M. A., Knapp, D. A., Henderson, J. R., *Phys. Rev. Lett.* **60**, 1715–1718 (1988);
   Levine, M. A., Marrs, R. E., Henderson, J. R., Knapp, D. A., Schneider, M. B., *Phys. Scr.* **T22**, 157-163 (1988)
6. Schneider, D., DeWitt, D. E., Clark, M. W., Schuch, R., Cocke, C. L., Schmieder, R., Reed, K. J., Chen, M. H., Marrs, R. E., Levine, E., Fortner, R., *Phys. Rev. A* **42**, 3889-3895 (1990)
7. Schneider, D., Church, D. A., Weinberg, G. M., Steiger, J., Beck, B. R., McDonald, J. W, Magee E., Knapp, D., *Rev.Sci.Instrum* **65**, 3472-3478 (1994)
8. Bergquist, J. C.,Itano, W. M., Wineland, D. J., *Phys. Rev. A* **36**, 428–430 (1987)
9. Walls, F. L., Dehmelt, H. G., *Phys. Rev. Lett.* **21**, 127–131 (1968)
10. Klaft, I., Borneis, S., Engel, T., Fricke, B.,Grieser, R., Huber, G., Kühl, T., Marx, D., Neumann, R., Schröder, S., Seelig, P., Völker, L. *Phys. Rev. Lett.* **73**, 2425–2427 (1994);
    Creso López-Urrutia, J.R., Beiersdorfer, P., Savin, D., Widmann, K., *Phys. Rev. Lett.* **77**, 826–829 (1996);
    Seelig, P. et al. *to be published*;
11. Brewer, L. R., Prestage, J. D., Bollinger, J. J., Itano, W. M., Larson, D. J., Wineland, D. J., *Phys. Rev. A* **38**, 859–873 (1988)
12. DeWitt, H. E., Pollock, R., *priv. comm.*, (1998)
13. O'Neil, T. M., *Phys. Fluids* **24**, 1447-1451 (1981)

CHARGE EXCHANGE AT VERY LOW COLLISION ENERGIES

I. Ben-Itzhak*, E. Wells, K.D. Carnes, and B.D. Esry†

James R. Macdonald Laboratory, Department of Physics,
Kansas State University, Manhattan, KS 66506-2604
†Institute for Theoretical Atomic and Molecular Physics
Harvard-Smithsonian Center for Astrophysics, 60 Garden Street Cambridge, MA 02138

Two of the simplest collision systems one can study are H^+ + H(1s) and H^+ + D(1s). Electron transfer is resonant in the first and nearly resonant in the latter because of the 3.7 meV gap between H(1s) and D(1s). Once the collision velocity becomes small enough quantum effects become more pronounced. However, these very low energies, of a few meV, are inaccessible using standard collision techniques. *We hereby suggest a method in which a dissociating HD^+ molecular ion is the "accelerator" used to measure electron transfer in the H^+ + D(1s) collision system down to a few meV.* When a HD molecule is ionized quickly about 1% of the $HD^+(1s\sigma)$ is in the vibrational continuum. During the dissociation, the electron initially centered on the D core can make a transition to the H core when the $2p\sigma$ and the $1s\sigma$ potential energy curves associated with the two dissociation limits get close to each other. It is important to note that during molecular dissociation the "avoided crossing" is crossed only once in contrast to twice during a full collision.

Slow proton collisions with atomic hydrogen provide the best testing ground for electron transfer studies because of the simplicity of the colliding system. Electron transfer in slow H^+ + H(1s) collisions is a resonant process involving the two lowest electronic states of the transient H_2^+ molecular ion formed during the collision. The transfer between the $1s\sigma$ and $2p\sigma$ states occurs at large internuclear separation where the two states merge together and their coupling is strong. The cross sections for this process were calculated, for example, by Hunter and Kuriyan [1] from 0.1 meV to 10 eV collision energy. Experimental determination of electron transfer cross sections are straight forward at keV energies and higher. However, the task becomes increasingly harder as the collision energy gets smaller. (See review by Gilbody [2]). Using a beam overtaking technique (i.e., neutralizing part of the beam and creating a velocity difference between the neutral and charged beam), Belyaev *et al* [3] managed to measure those cross sections down to about 5 eV. The experimental value is about 40% (2.7σ) above theory as shown in Fig. 1.

One would expect theory to be more precise than experiment for such collision systems and thus associate the difference to systematic experimental uncertainties. We are aware of no other measurements for slower collisions, which leaves the energy range below 1 eV for theorists only. Note that more structure appears as the collision energy is reduced.

*Corresponding author eMail: ibi@phys.ksu.edu.

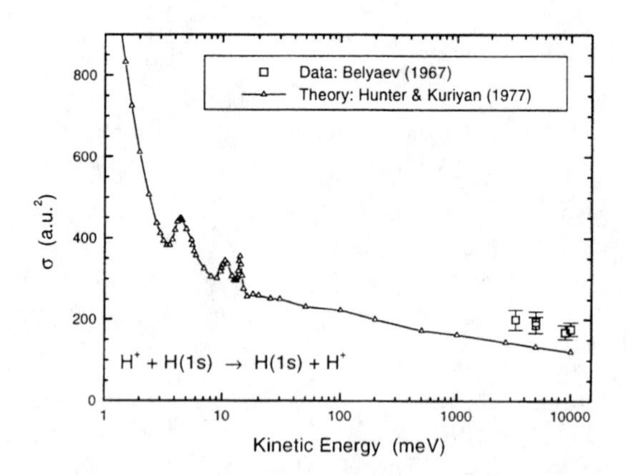

FIGURE 1. Electron transfer cross section in slow H^+ + H(1s) collisions.

The H^+ + D(1s) and D^+ + H(1s) collision systems are similar to the one above in their simplicity. In contrast to the H^+ + H(1s) collision system, however, electron transfer in the heteronuclear systems is only a near resonant process involving the same two lowest electronic states of the transient HD^+ molecular ion formed during the collision. The transfer between the $1s\sigma$ and $2p\sigma$ states occurs near the avoided crossing at around 12 a.u. of internuclear separation. The energy gap of 3.7 meV between the ground $1s\sigma$ state and the first excited $2p\sigma$ molecular state, shown in Fig. 2, is due to the difference in nuclear mass between H and

CP475, *Applications of Accelerators in Research and Industry*,
edited by J. L. Duggan and I. L. Morgan

D and signals the breakdown of the Born-Oppenheimer approximation.

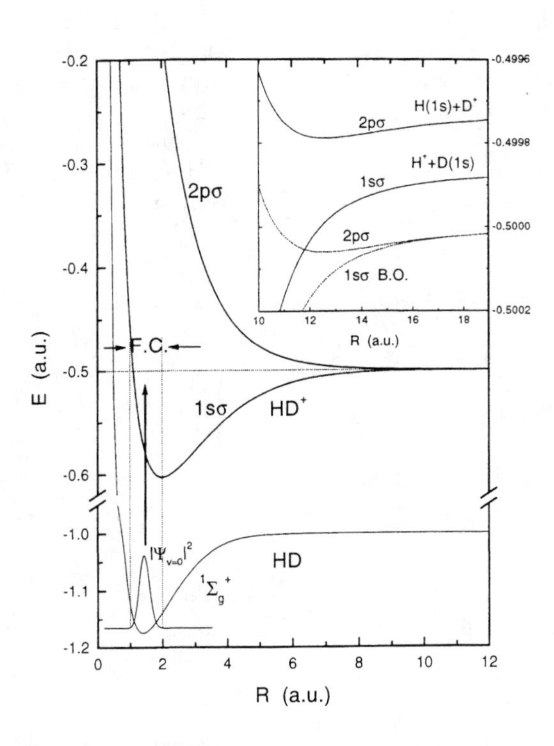

FIGURE 2. Potential energy curves of HD^+ (from Chen 1991 [4]) and HD (H_2 from Kołos *et al* [5]). Note that the Born-Oppenheimer approximation is valid near the minimum and the difference between HD and H_2 is in the vibrational part.

The cross sections for electron transfer in these collision systems were calculated by Hunter and Kuriyan [6] and also by Davis and Thorson [7]. The two calculations for $H^+ + D(1s)$ collisions are compared to each other in Fig. 3. Both calculations show more structure at lower energies that both groups attribute mostly to the interference between electron transfer on the way in and out, i.e. the two times the avoided crossing is traversed during the collision. According to Davis and Thorson [7], the finer structure might be due to the finer grid they used in their calculations. Surprisingly, the agreement between the two calculations is not as good as one would expect at the lower energy end. In spite of this disappointing disagreement, we failed to find any recent and more precise calculation. Theory is in good agreement with merged beam measurements of Newman *et al* [8], which extend down to 0.1 eV. However, the experimental precision deteriorates as the collision energy is reduced, both for the cross section measurement and for determining the collision energy. It seems important to extend the experimental work to lower collision energies in order to revive interest in this simple collision system, especially if one could reach collision energies of the order of the energy gap. In contrast to the simplicity of this collision system from the theorist's view point, conducting an experiment with sufficient energy resolution down to a few meV provides the experimentalist with quite a challenge, as can be seen from the data presented. *In this paper we suggest a new approach to bypass these experimental difficulties.*

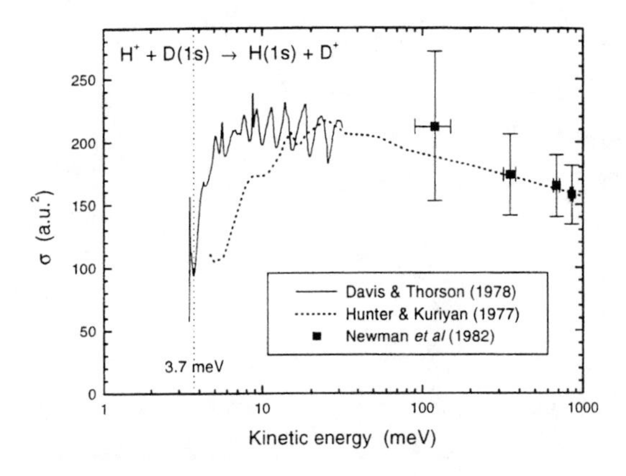

FIGURE 3. Electron transfer cross section in slow $H^+ + D(1s)$ collisions.

We suggest using ground state dissociation of HD^+ as a means to measure electron transfer from the $1s\sigma$ to the $2p\sigma$ state at large internuclear separation. Ground state dissociation of HD^+ is the process in which first a vertical ionization (fixed internuclear distance, R) of HD populates the vibrational continuum of the HD^+ ground state (i.e. the $1s\sigma$ state which is associated with H^++D($1s$) at the separated atom limit). This is then followed by a dissociation into either $H^+ + D(1s)$ if no transition occurs, or $H(1s) + D^+$ if the electron is transferred to the $2p\sigma$ state. It has been shown that this dissociation channel can be experimentally distinguished from the dissociation of the excited electronic states of HD^+ because the fragments have much lower dissociation energies [9,10]. For example, if the $2p\sigma$ state is populated directly by vertical ionization then the fragments have about 15 eV of kinetic energy while those associated with the $1s\sigma$ ground state have energies ranging from zero to less then 0.5 eV, thus enabling the separation of these dissociation channels [9,10].

Many different schemes can be used to ionize the HD molecule to the $1s\sigma$ state keeping the internuclear distance fixed. We have chosen to use fast proton impact ionization, which populates the vibrational continuum of the HD^+ ground state while producing only a small amount of fast fragments (from the dissociation of excited states of HD^+ and its double ionization [10,11]). Fast electron impact or photo-ionization would also be reasonable choices. In Fig. 4 we show the calculated kinetic energy release distribution for ground state dissociation of HD^+ [12]. The calculations are just the Franck-Condon factor between the HD(v=0) wave function and the $HD^+(1s\sigma)$ vibrational continuum wave functions. We have used many discrete vibrational states to represent the continuum as discussed in [10,12,13]. This distribution peaks at zero kinetic energy, i.e. the dissociation threshold, and falls off approximately exponentially with a FWHM of about 250 meV. It amounts to about 1% of the $HD(^1\Sigma^+)$ to $HD^+(1s\sigma)$ transitions.

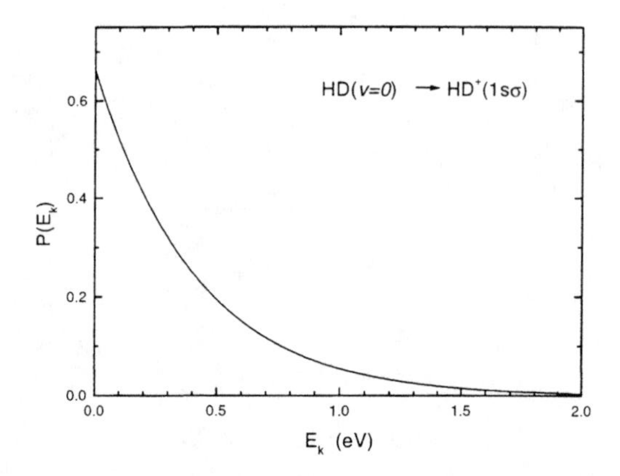

FIGURE 4. Computed kinetic energy release distribution for HD^+ ground state dissociation.

Using a fast proton beam to prepare the HD^+ in the dissociating $1s\sigma$ state of interest, the whole energy range from zero to about 0.5 eV in CM will be measured "simultaneously". Thus, instead of tuning the collision energy prior to the interaction, as done in most collision experiments, we will determine it afterwards by measuring the charged fragment energy. Clearly the energy resolution of such a measurement is the key issue in this method. The fragment ions are extracted toward a micro channel plate detector using a weak electric field, and time-of-flight is used to identify the H^+ and D^+ fragments. The deviation of each ion from the time-of-flight expected if it had no initial velocity toward the detector ($v_x = 0$) is used to determined this component of velocity. The other velocity components, perpendicular to the extraction field, are evaluated using the impact position of each fragment on the detector. A 2D resistive anode decodes this position information with resolution of about 0.15 mm. The main limiting factors in the energy resolution are: (1) thermal motion of the HD target molecules (about 25 meV at room temperature); and (2) extended target length along the beam direction. To reduce these energy-broadening effects an effusive gas HD target jet is used after pre-cooling the HD gas to about 10-30 K in a small cell mounted on a cold head of a cryo-pump . A schematic view of the experimental setup is shown in Fig. 5. This cooling is also expected to reduce the water contaminant in the target, thus improving the quality of the H^+ fragment data (see Ref. [13] for details). To further reduce the energy broadening caused by the target length along the beam direction, space focusing by the extraction field is used as discussed by Mergel *et al* [14]. The fragment energy resolution also depends on the strength of the extraction field, and it can be improved by lowering the field strength. However, while using a weaker extraction field only part of the dissociating vibrational continuum shown in Fig. 4 will be detected simultaneously. The best expected energy resolution is about 2 meV.

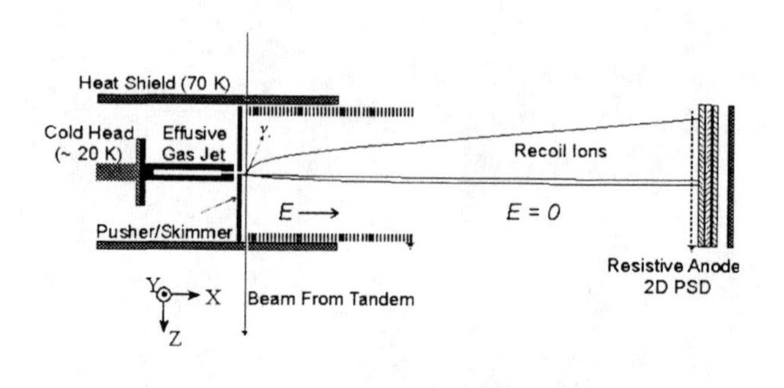

FIGURE 5. Schematic view of experimental set-up.

The measured yields of H^+ and D^+ fragments as a function of the dissociation energy in the CM frame of reference provide a direct measure of the electron transfer cross section because each slow D^+ detected indicates an electron transfer from the initial $1s\sigma$ to the final $2p\sigma$ state. Note that this dissociation process which provides an experimental probe for very slow collisions is not identical to the $H^+ + D(1s)$ collisions calculated by Hunter and Kuriyan [6], and by Davis and Thorson [7]. In molecular dissociation the avoided crossing is encountered only once, and it can be viewed as a "single pass" collision in contrast to the full collision. As a result, the interference between electron transfer on the way in and out is removed, and the calculations discussed above can not be directly compared to the measurement. In "single pass" collisions the transition probability is proportional to the square of the transition amplitude, to which it can be compared directly if available. We have used instead the Meyerhof formula [15] to evaluate the total electron transfer probability and found it to be in agreement with our measurement of the ratio $D^+/[H^++D^+]$ integrated over all dissociation energies [12]. We show in Fig. 6 the predictions of this model calculation for the H^+ and D^+ yields as a function of the dissociation energy, E_k. Note that when E_k is below the energy gap only H^+ is expected. While at high dissociation energies, $E_k \gg E_{gap}$, the yields approach the same value. The prediction of the behavior of these yields just above the energy gap, when electron transfer between the two states starts happening, is in question because this model is not expected to be valid unless $E_k \gg E_{gap}$. New theoretical calculations of the HD^+ ground state dissociation process focusing on the energy region near the gap are under way.

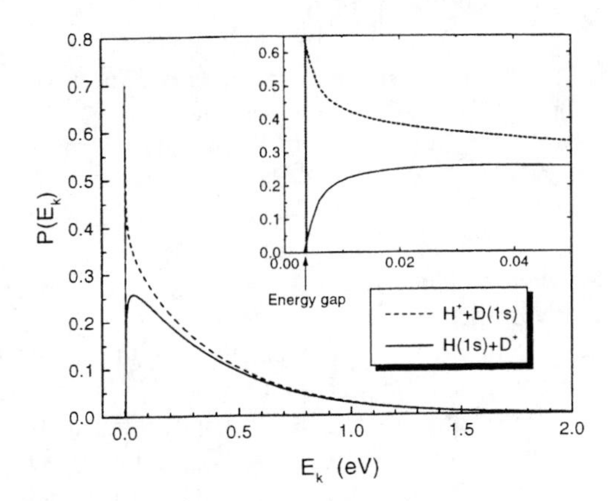

FIGURE 6. The yields of H^+ and D^+ as a function of dissociation energy as predicted by the Meyerhof formula (see text).

If one looks at the magnified view of the potential-energy curves around the energy gap shown in Fig. 7, one can see that the $2p\sigma$ state can sustain some vibrational states (see Carrington *et al* [16]). When the coupling to the $1s\sigma$ vibrational continuum is included, such states might become Feshbach resonances appearing in the $H^+ + D(1s)$ cross section. It is clearly questionable whether one can achieve the needed energy resolution to see these resonances. However, one can hope to use the new theoretical calculations of HD^+ ground state dissociation to estimate what energy resolution is needed before these resonances smear out.

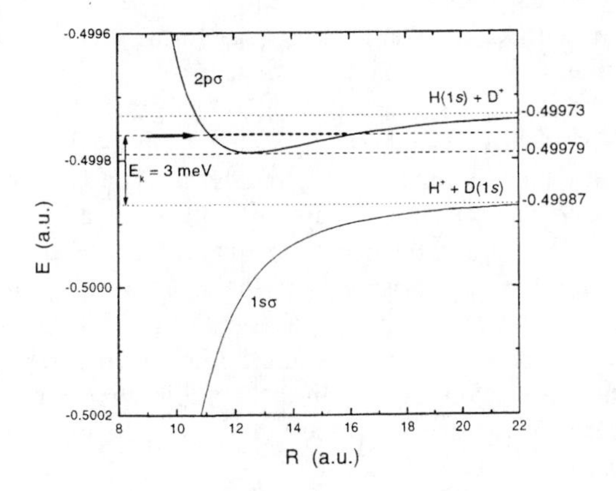

FIGURE 7. Magnified view of the potential energy curves of HD^+ around the energy gap.

In summary, the use of ground state dissociation of HD^+ has been suggested as a method to study $H^+ + D(1s)$ "single pass" collisions down to collision energies of a few meV. As for the energy resolution one can achieve, this is still to be determined.

ACKNOWLEDGMENT

This work was supported by the Division of Chemical Sciences, Office of Basic Energy Sciences, Office of Energy Research, U.S. Department of Energy. BDE is supported by the National Science Foundation through the Institute for Theoretical Atomic and Molecular Physics.

---

[1] Hunter, G., and Kuriyan, M., *Proc. Roy. Soc. Lond. A* **353**, 575 (1977).
[2] Gilbody, H.B., *Advances in Atomic, Moelcular, and Optical Physics* **33**, 149 (1994).
[3] Belyaev, V.A., Brezhnev, B.G., and Erastov, E.M., *Soviet Phys. JETP* **25**, 777 (1967).
[4] Chen Z, PhD thesis, Kansas State University 1991 (unpublished), and private communication.
[5] Kołos, W., Szalewicz, K., and Monkhorst, H.J., *J. Chem. Phys.* **84**, 3278 (1986).
[6] Hunter, G., and Kuriyan, M., *Proc. Roy. Soc. Lond. A* **358**, 321 (1977); and references within.
[7] Davis, J.P., and Thorson, W.R., *Can. J. Phys.* **56**, 996 (1978).
[8] Newman, J.H., Cogan, J.D., Ziegler, D.L., Nitz, D.E., Rundel, R.D., Smith, K.A., and Stebbings, R.F., *Phys. Rev. A* **25**, 2976 (1982).
[9] Edwards, A.K., Wood, R.M., Beard, A.S., and Ezell, R.L., *Phys. Rev A* **42**, 1367 (1990).
[10] Ben-Itzhak, I., Krishnamurthi, Vidhya, Carnes, K.D., Aliabadi, H., Knudsen, H., Mikkelsen, U., and Esry, B.D., *J. Phys. B* **29**, L21 (1996).
[11] Ben-Itzhak, I., Krishnamurthi, Vidhya, Carnes, K.D., Alibadi, H., Knudsen, H., Mikkelsen, U., *Nucl. Instr. and Meth. B* **99**, 104 (1995).
[12] Ben-Itzhak, I., Wells, E., Carnes, K.D., Krishnamurthi, Vidhya, Weaver, O.L., and Esry, B.D., submitted for publication (1998).
[13] Ben-Itzhak, I., Wells, E., Krishnamurthi, Vidhya, Carnes, K.D., Aliabadi, H., Mikkelsen, U., Weaver, O.L., and Esry, B.D., *Nucl. Instr. and Meth. B* **129**, 117 (1997).
[14] Mergel, V., *et al*, *Phys. Rev. Lett.* **79**, 387 (1997).
[15] Meyerhof, W.E., *Phys. Rev. Lett.* **31**, 1341 (1973).
[16] Carrington, A., Leach, C.A., Marr, A.J. Moss, R.E., *J. Chem. Phys.* **98**, 5290 (1993).

Charge state fractions of lead ions formed in one-electron capture collisions by fast H^+ and He^{2+} ions

P. C. E. McCartney, M. B. Shah, J. Geddes and H. B. Gilbody

Department of Pure and Applied Physics, The Queen's University of Belfast, Belfast BT7 1NN, UK

A crossed beam coincidence counting technique incorporating time of flight analysis of target charged products has been used to measure the charge state fractions of slow multiply charged Pb^{q+} ions formed in one-electron capture by H^+ and He^{2+} ions in single collisions with ground state Pb atoms within the energy range 50-600keV amu^{-1}. An attempt has been made to describe the measured fractions of Pb^{q+} ions for $q = 1 - 8$ in terms of a simple model based on an independent electron description.

INTRODUCTION

The processes leading to the multiple ionization of heavy metal atoms by fast ions are of much intrinsic interest. Additionally, the role of such processes are relevant to the understanding of impurity control and plasma diagnostics in high temperature fusion devices (1). In previous work in this laboratory (2, 3, 4, 5) we have used a crossed beam technique together with time of flight analysis and coincidence counting of the collision products to investigate electron capture, transfer ionization and pure ionization. In the case of capture processes with H^+ and He^{2+} incident on ground state $3p^6 3d^6 4s^2\ ^5D_4$ Fe, $3p^6 3d^{10} 4s^2\ ^2S_{1/2}$ Cu and $3d^{10} 4s^2 4p\ ^2P_{1/2}$ Ga, the formation of Fe^{q+}, Cu^{q+} and Ga^{q+} has been measured for $q = 1 - 4$, $q = 1 - 5$ and $q = 1 - 6$, respectively, within the energy range 35 – 720 keV amu^{-1}. The relevant processes are

$$H^+ + X \rightarrow H + X^{q+} + (q-1)e \qquad (1)$$

with cross section ${}_{10}\sigma_{0q}$ and

$$He^{2+} + X \rightarrow He^+ + X^{q+} + (q-1)e \qquad (2)$$

with cross section ${}_{20}\sigma_{1q}$. Here $q = 1$ corresponds to a simple charge transfer while $q > 1$ corresponds to transfer ionization where electron capture takes place simultaneously with multiple ionization of the target. The heavy metal atom beams were provided by a specially developed oven source (6).

The experimental results for Fe and Cu have been described satisfactorily in terms of an independent electron model in which electron removal takes place primarily from the 3s and 3d subshells and the simple charge transfer process ($q = 1$) was dominant. In the case of gallium where the 4p subshell was also involved, electron capture from the inner 3d subshell followed by autoionization led to emission of an additional electron from the target ($q = 2$) and simple charge transfer (q=1) was found to be greatly reduced in magnitude. Inclusion of autoionization allowed the independent electron model to give satisfactory agreement with the experimental data.

In the present work we have carried out similar studies of electron capture and transfer ionization for ground state $5d^{10} 6s^2 6p^2\ ^3P_0$ Pb atom. The electron subshell structure of Pb is significantly different to that for Ga, Cu and Fe hence it is of interest to determine the extent to which the independent model can be used to describe Pb.

EXPERIMENTAL APPROACH

The apparatus and measurement procedure was similar to that used previously in this laboratory (7,8) and only the essential features need be outlined here.

A primary beam of momentum analysed H^+ or He^{2+} ions of the required energy was arranged to intersect (at right angles in a high-vacuum region) a thermal energy beam of ground-state Pb atoms effusing from our specially developed oven source (6) housed in a separate differentially pumped region. The slow Pb^+ ions formed in the crossed-beam region were extracted with high efficiency by a transverse electric field (applied between two high-transparency grids) strong enough to ensure complete collection, and after a further stage of acceleration, were separately counted by particle multipliers. Pb^{q+} ions in any particular charge state q were selectively identified and distinguished from background gas product ions by time-of-flight analysis. The Pb^{q+} ions arising from one electron capture collisions could be identified by counting them in coincidence with the fast H atoms or the He^+ ions arising from the same events which were recorded (after charge analysis by electrostatic deflection) by a second particle multiplier located beyond the beam intersection region. In this way Pb^{q+} ions arising from electron capture processes could be distinguished from pure ionization events.

Analysis of the Pb^{q+} - fast atom/ion coincidence spectra in the way described previously (3) allowed determination of the separate relative cross sections, ${}_{10}\sigma_{0q}$,

CP475, *Applications of Accelerators in Research and Industry*, edited by J. L. Duggan and I. L. Morgan

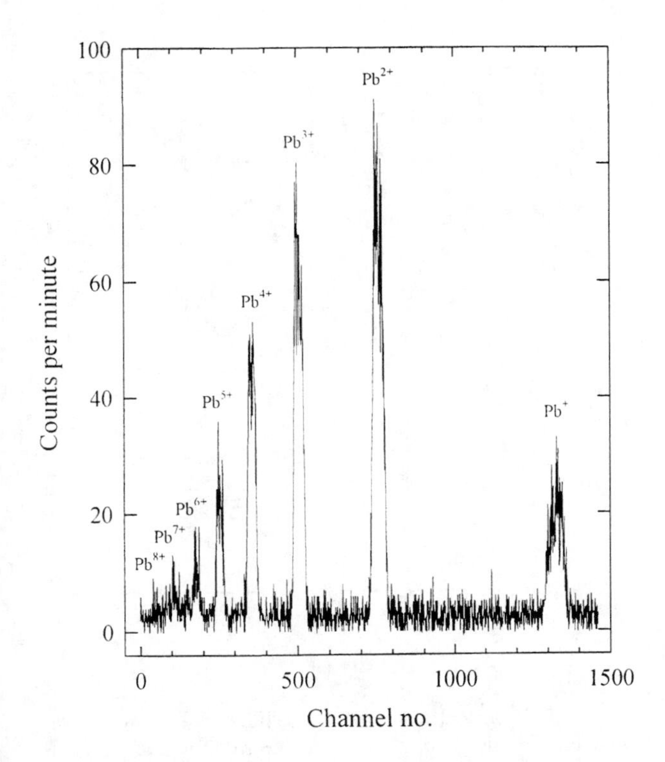

FIGURE 1. Fast He^{+}/slow Pb^{q+} ion time-of-flight coincidence spectrum resulting from one-electron capture by 290 keV amu^{-1} He^{2+} ions in collisions with Pb atoms. Adjacent channel separation is 2 ns.

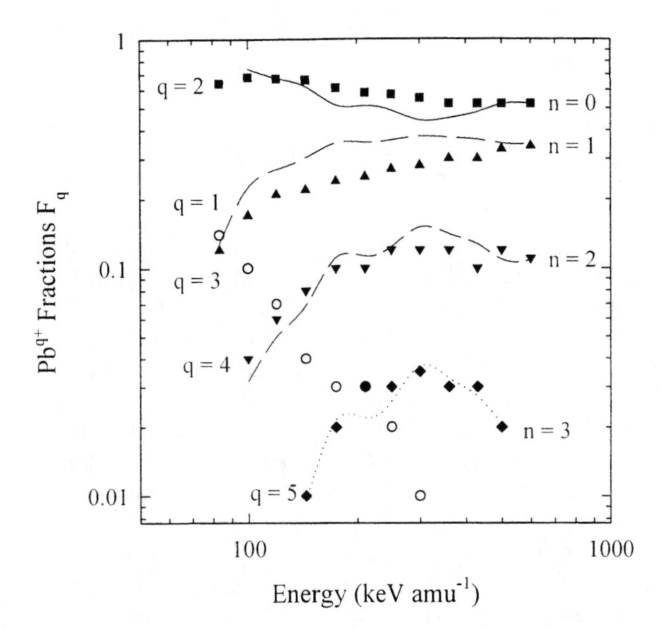

FIGURE 2. Measured charge-state fractions F_q (shown as data points) for Pb^{q+} ions formed by one-electron capture in single collisions between H^{+} ions and ground-state Pb atoms. Curves show calculated probabilities P_n where $n = q - 2$ (see text) based on binomial distributions.

for one-electron capture by 83 - 600 keV amu^{-1} H^{+} ions where $q = 1 - 5$ and for, ${}_{20}\sigma_{1q}$ one-electron capture by 50 – 360 keV amu^{-1} He^{2+} ions where $q = 1 - 8$ (see Fig. 1).

RESULTS AND DISCUSSION

Figure 2 shows the energy dependence of our measured fractions F_q of Pb^{q+} ions formed by one-electron capture and transfer ionization in H^{+}- Pb collisions for $q = 1$ to 5. The fraction F_1 arising from simple charge transfer decreases rapidly over the energy range considered. The fractions F_2 arising from transfer ionization exceeds F_1 over the entire energy range 83 - 600 keV while the other fractions F_3, F_4 and F_5 (also arising from transfer ionization) exceed F_1 above 90, 120 and 210 keV amu^{-1} respectively. At very low energies below the present energy range simple charge transfer ($q = 1$) will involve the capture of weakly bound outer shell electrons. In the present energy range capture to more tightly bound 6s, 5d and even 5p electrons would be expected to be significant. The creation of an electron vacancy in an inner shell of the Pb atom can lead to Pb^{2+} formation through autoionization. This process can lead to a reduction in the value of F_1 and enhancement of F_2 which is in accord with experimental observation. Similar enhancement of higher q values is possible. Evidence for enhancement of multiple ionization processes in Pb by autoionization has been observed in our earlier work on electron impact ionization (9).

In our previous studies of Fe and Cu (10, 3) we successfully described our observed values of F_q in terms of an independent electron model (11). The probability of one electron capture and simultaneous ionization was expressed as a product of a one electron capture probability P_c and an ionization probability P_n for the removal of n electrons from the target where $n = (q\text{-}1) \geq 0$.. The cross section ${}_{10}\sigma_{0q}$ may be expressed as

$$ {}_{10}\sigma_{0q} = 2\pi \int_{0}^{\infty} b P_c(b) P_n(b)\, db \tag{3} $$

where b is the impact parameter. Assuming $P_n(b)$ to be constant considered over the relatively small range of impact parameters where electron capture occurs then

$$ P_n = {}_{10}\sigma_{0q} / \sigma_{10} \tag{4} $$

where the total electron capture cross section $\sigma_{10} = \sum_q {}_{10}\sigma_{0q}$ Measured values of F_q can then be identified with P_n through equation 4.

Values of P_n can be expressed in terms of a binomial distribution

$$P_n = \left(\frac{N!}{n!(N-n)!}\right) p^n (1-p)^{(N-n)} \qquad (5)$$

where p and N represent the single electron ionization probability and the total number of available electrons respectively.

In the application of this procedure to Ga (5) where autoionization following one electron capture increases the charge state, we assumed $n = (q - 2)$ and were able to fit the experimental data for all q. A similar procedure has been applied in the case of Pb with electrons in the 5d, 6s and 6p subshells (N = 13) available for removal by ionization following one electron capture. Here it is also assumed that the time for autoionization is longer than the interaction time of the fast projectile ion. In our energy range the values of P_n have been fitted to the experimental value F_q using the weighted least squares method. The values of p are shown in Fig 3. In this case, unlike our

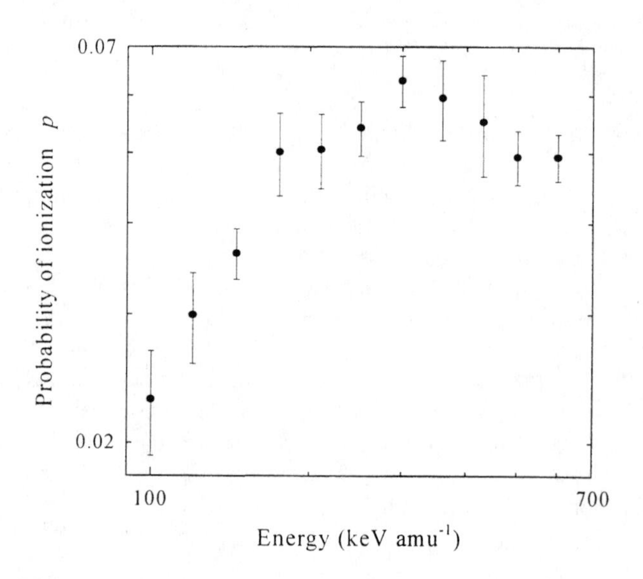

FIGURE 3. Ionization probabilities, p for a single electron derived from binomial fits to measured charge-state distributions arising from one-electron capture by H^+ ions in collisions with Pb atoms.

earlier work with Fe, Cu and Ga, the independent electron model gives a good fit to the experimental data only at the higher q values. It is believed that the allowance for the various possible autoionization pathways has been over-simplified in this treatment. Increasing the pool of available electrons to include the 5p subshell did not improve the fit to the experimental data.

In the case of one-electron capture in He^{2+}- Pb collisions (Fig. 4) the simple charge-transfer fraction, F_1 decreases very rapidly with increasing energy and at

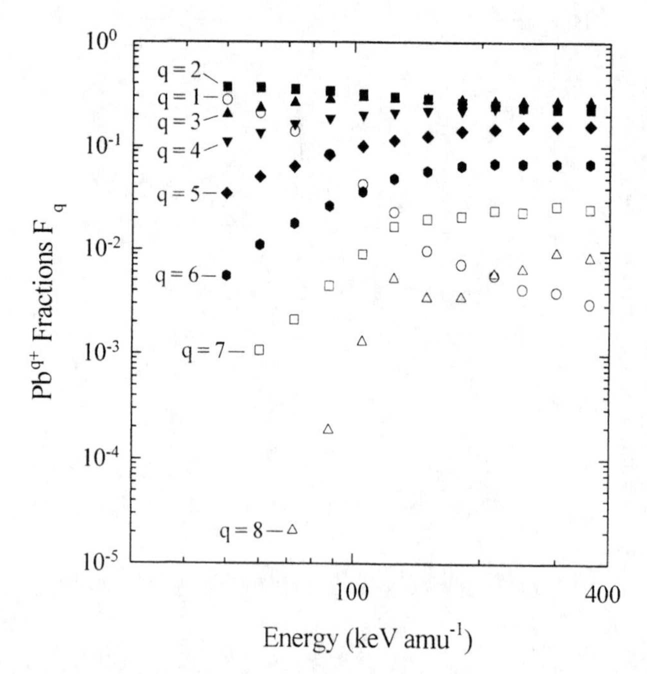

FIGURE 4. Measured charge-state fractions F_q (shown as data points) for Pb^{q+} ions formed by one-electron capture in single collisions between He^{2+} ions and ground-state Pb atoms.

energies exceeding 250 keV amu^{-1} is smaller than fractions F_q for q = 2 to 8 representing transfer ionization contributions. F_2, although dominant at the lowest energies, is exceeded by both F_3 and F_4 at our high energy limit. The measured fractions were again compared with our prediction for P_n based on a binomial fit and allowing for autoionization. However the fits in this case were very unsatisfactory. The results for the fits for He^{2+} provided a strong indication that even deeper inner shell electrons are being captured, leading to a cascading effect involving multiple Auger emissions.

ACKNOWLEDGEMENTS

This work is part of a large programme supported by the U.K Engineering and Physical Sciences Council. One of us (P.C.E McC) is also indebted to the Department of Education, Northern Ireland for the award of a Research Studentship.

REFERENCES

1. Janev R.K., 1993 Summary Report of IAEA Technical Committee Meeting on Atomic and Molecular Data for Fusion Reactor Technology, Cadarache, France *IAEA Report* INDC(NDS)-277, Vienna.

2. Patton C.J., Bolorizahdeh M.A., Shah M.B., Geddes J. and Gilbody H.B. *J.Phys. B: At. Mol. Opt. Phys.* **27**, 3695-3706 (1994).
3. Shah M.B., Patton C.J., Bolorizahdeh M.A., Geddes J. and Gilbody H.B. *J.Phys. B: At. Mol. Opt. Phys.* **28**, 1821-1833 (1995).
4. Patton C.J., Shah M.B., Bolorizahdeh M.A., Geddes J. and Gilbody H.B. *J.Phys. B: At. Mol. Opt. Phys.* **28**, 3889-3899 (1995).
5. Lozhkin K.O., Patton C.J., McCartney P.C.E., Sant'anna M., Shah M.B., Geddes J. and Gilbody H.B. *J.Phys. B: At. Mol. Opt. Phys.* **30**, 1785-1798 (1997).
6. Shah M.B., Bolorizahdeh M.A., Patton C.J. and Gilbody H.B. 1996b *Meas. Sci. Technol.* **7** 709
7. Shah M.B. and Gilbody H.B. *J.Phys. B: At. Mol. Opt. Phys.* **14**, 2361-2377 (1981).
8. Shah M.B., McCallion P., Itoh Y. and Gilbody H.B. *J.Phys. B: At. Mol. Opt. Phys.* **25**, 3693-3708 (1992).
9. McCartney P.C.E., Shah M. B., Geddes J. and Gilbody H. B. *J.Phys. B: At. Mol. Opt. Phys* (Accepted 1998).
10. Shah M.B., Patton C.J., Geddes J. and Gilbody H.B. *Nuclear Instruments & Methods in Physics Research.* **B98**, 280-283 (1995)
11. McGuire J H 1991 *Advances in Atomic, Molecular and Optical Physics* **29**, ed D R Bates and B Bederson (New York: Academic) p217

Electron Capture From Hydrocarbon Molecules by Proton Projectiles in the 60–120 keV Energy Range

J.M. Sanders and S.L. Varghese

Dept. of Physics, University of South of Alabama, Mobile, AL 36688

Cross sections for electron capture in collisions of 60- to 120-keV protons with molecular hydrogen, methane (CH_4), ethane (C_2H_6), and propane (C_3H_8) have been measured. The cross sections were found to scale roughly linearly with the number of carbon atoms, but the increase was not as large as the simple Bragg additivity rule would predict. Cross section ratio of ethane to methane increases with projectile energy, while the cross section ratio of propane to methane is relatively insensitive to the projectile energy.

INTRODUCTION

Cross sections for electron capture in collisions involving molecular targets are often related to cross sections involving atomic targets by assuming that the molecular cross section is the sum of the cross sections of its constituent atoms. This assumption is called the Bragg additivity rule. Over the years, several investigations of the validity of the additivity rule have been made (1,2,3,4,5). These previous experiments have concentrated on relatively high-energy collisions of bare projectiles with hydrocarbon gases. H^+ projectiles colliding with a variety of hydrocarbon gases were studied for energies from 800 to 3000 keV (2,3). In these experiments, significant deviations from additivity were observed, namely, the cross sections did not increase with the number of carbon atoms as strongly as the Bragg rule would predict. To explain the apparent reduction in the capture cross section, it was proposed that the projectile could lose its captured electron in collisions with other atoms on its way out of the molecule. This proposed mechanism was called the "exit effect."

Dillingham *et al.* (5) extended the investigation to F^{9+} and F^{8+} ions colliding with hydrocarbon gases for energies from 500 keV/u to 1500 keV/u. In this case, the Bragg rule was found to be valid. However, unlike the case with H^+ projectiles, for F^{9+} and F^{8+} projectiles the loss cross sections are much smaller than the capture cross sections, so the exit effect is not expected to play a significant role.

The exit effect depends strongly on the relative size of the loss and capture cross sections. For fast ($E > 300$ keV) protons, loss cross sections are at least two orders of magnitude greater than capture cross sections (8). Thus for high energy protons, the exit effect causes significant deviations from the additivity rule. At the other extreme, for fast ($E > 500$ keV/u) bare and hydrogen-like fluorine ions, capture cross sections are generally an order of magnitude greater than loss cross sections, and the exit effect is not significant (5).

In the present experiment, we investigate capture cross sections for 60- to 120-keV H^+ colliding with H_2, methane (CH_4), ethane (C_2H_6), and propane (C_3H_8). In these collisions, loss is of the same order of magnitude as capture. Also in this energy range, the capture cross section from hydrogen is not negligible compared to the cross section for capture from carbon.

EXPERIMENT

Proton beams with energies ranging from 60 to 120 keV were produced by a 150 kV Cockroft-Walton accelerator. After passing through water-cooled slits to collimate the beam and to control the projectile ion flux, the beam was momentum-analyzed by a 15-degree analyzing magnet. The protons then passed through a differentially-pumped target gas cell. A region 19 cm long with entrance and exit apertures 4 mm in diameter was pumped by a diffusion pump, and a gas cell 3.12 cm long with entrance and exit apertures 1.8 mm in diameter was mounted in the center of this region. The flow of target gas into the cell was controlled by a needle valve, and the pressure was measured by a capacitance manometer. In order to insure single collision conditions, target gas pressures were chosen such that the H^0 beam fraction did not exceed 10%. An additional indication that double collisions were not significant was the fact that the H^0 fraction varied linearly with the target gas cell pressure over the range of pressures employed in the experiment.

Upon leaving the target gas cell, the projectile beam passed between the plates of an electrostatic analyzer which separated the H^+ and H^0 projectiles. Both the H^+ and H^0 projectiles were detected with surface barrier detectors, and the projectile flux was controlled to insure that the projectile rate did not exceed 1000 s^{-1}. After amplification, the signals from the two detectors were pulse-height-analyzed with multichannel analyzers.

The cross sections were obtained using the growth method (6). For each proton energy and target gas, the H^0 fraction of the beam emerging from the gas cell was

CP475, *Applications of Accelerators in Research and Industry*,
edited by J. L. Duggan and I. L. Morgan

measured for several target gas pressures. A generalized linear least squares fit of the H^0 fraction versus pressure yielded a straight line the slope of which was proportional to the cross section. Relative cross sections can be obtained by comparing these slopes, and absolute cross sections are given by

$$\sigma = \frac{d\phi}{dP}\frac{k_B T}{l} \tag{1}$$

where ϕ is the H^0 fraction, P is the pressure, k_B is the Boltzmann constant, T is the temperature, l is the effective gas cell length. The effective length of the gas cell is somewhat larger than the physical length due to streaming of the target gas out of the entrance and exit apertures. A usual estimate of the effect of this end correction is that the length is increased by an amount equal to the diameter of the apertures. Thus the effective length for the gas cell would be 3.48 cm. Our cross sections, calculated using this effective gas cell length, are found to be in excellent agreement with those of Stier and Barnett (7) for H^+ on H_2 and of Toburen *et al.* (8) for H^+ on CH_4 and C_2H_6.

The major contribution to the uncertainty in the reported cross sections is the uncertainty in the effective gas cell length. We have taken this uncertainty to be equal to the end correction itself (0.36 cm or 10%). However, the effective gas cell length affects only the absolute normalization of the cross sections, and does not affect the relative cross sections. For the relative cross sections, the major contributions to the uncertainty are the variation in the gas pressure during data taking and the statistical uncertainty in the counting statistics for the charge fractions. In taking the pressure-dependence data, sufficient counts were accumulated in both the H^+ and H^0 channels to assure statistical uncertainties of 5% or better. Variation of the gas pressure during data accumulation was less than 5%.

RESULTS AND ANALYSIS

Cross sections for single electron capture for H^+ projectiles colliding with H_2, CH_4, C_2H_6, and C_3H_8 are shown in Fig. (1). Also shown in the figure is the data of Stier and Barnett (7) for H^+ on H_2. Ratios of the cross sections for single electron capture from C_2H_6 to that of CH_4 are shown in Fig. (2). In addition, similar ratios for C_3H_8 are shown. Note that the cross section ratio for C_2H_6 increases slightly with increasing energy, in contrast the cross section ratio for C_3H_8 seems to be independent of the projectile energy.

If capture from hydrogen were completely negligible and additivity were valid, one would expect the ratio C_2H_6/CH_4 to equal 2 and the ratio C_3H_8/CH_4 to equal 3. However, at the energy range of these measurements, the capture cross section from hydrogen is not negligible.

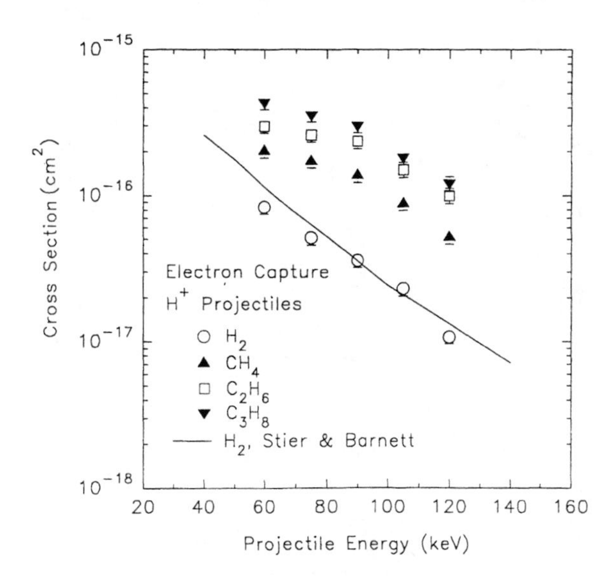

FIGURE 1: Single electron capture cross sections for H^+ colliding with H_2, CH_4, C_2H_6, and C_3H_8. The solid line represents the H_2 data of Stier and Barnett (7).

We therefore must first account for the differing numbers of hydrogen atoms on the various molecules. For a molecule C_mH_n, the cross section for capture from H is subtracted by using

$$\sigma_{C_m} = \sigma_{C_m H_n} - \frac{n}{2}\sigma_{H_2}. \tag{2}$$

In Fig. (3), the ratios obtained from these "corrected" cross sections are shown. In comparing Figs. (2) and (3), it will be noted that the general trends remain. In particular the C_2H_6 ratios are still linear and approach a value of 2 at high energies, while the C_3H_8 ratios show no particular energy dependence and are well below the value of 3 predicted by the additivity rule. The additivity rule does not appear to be valid for proton projectiles at these collision energies. Since the loss cross sections are large for these energies, the exit effect is likely to play a major role.

The exit effect can be estimated using the model of Bissinger *et al.* (2). In this model, it is assumed that capture from carbon atoms is much more probable than capture from hydrogen, so that capture can be considered to be all due to capture from carbon (at the present energies, this is a somewhat tenuous assumption, since $\sigma_c^H/\sigma_c^C \simeq 0.25$). Then, further assuming that, on average, the proton captures an electron near the middle of the molecule, the fraction f^0 of the original neutrals that survive on exiting the molecule is given by

$$f^0 = \frac{\sigma_c + \sigma_l e^{-(\sigma_c+\sigma_l)x}}{\sigma_c + \sigma_l} \tag{3}$$

where σ_c and σ_l are the capture and loss cross sections per atom and x is the areal density of atoms in the

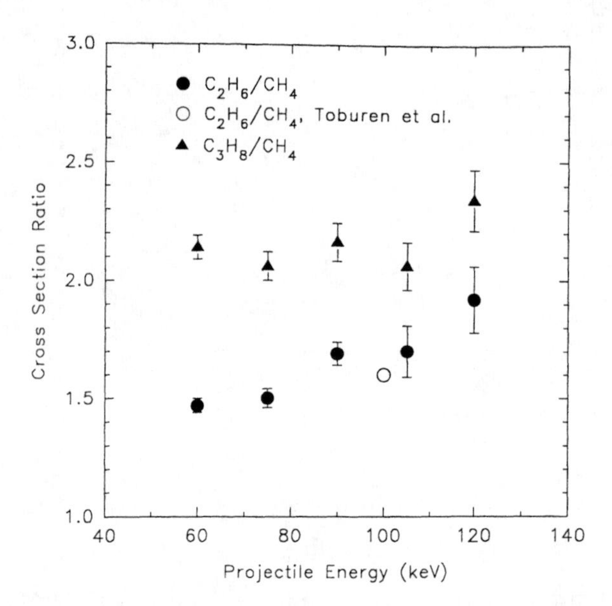

FIGURE 2: Single electron capture cross sections for C_2H_6 and C_3H_8 divided by the cross section for CH_4. The open circle is the ratio computed from the C_2H_6 and CH_4 data of Toburen *et al.* (8).

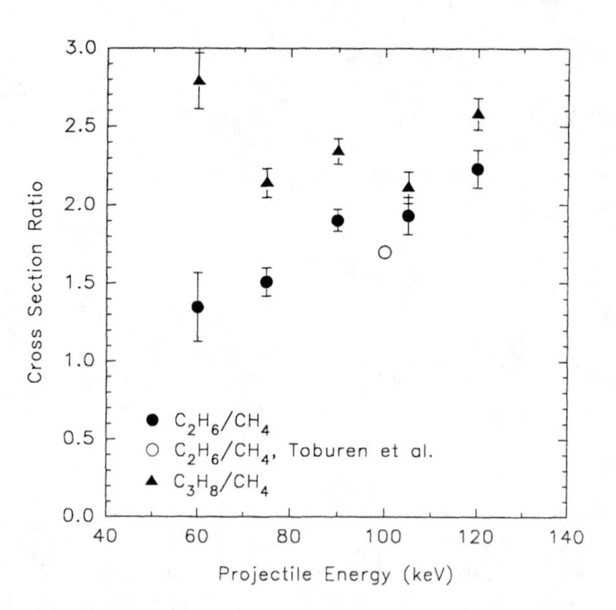

FIGURE 3: Cross section ratios as in Fig. (2). Contributions due to hydrogen atoms have been subtracted using Eq. (2).

remaining part of the molecule. The argument of the exponential in Eq. (3) for a molecule C_mH_n is approximated by

$$(\sigma_c + \sigma_l)x = \frac{1}{2}(m-1)(\sigma_c^C + \sigma_l^C)x_C + \frac{1}{2}n(\sigma_c^H + \sigma_l^H)x_H \quad (4)$$

where x_C and x_H are the areal densities for carbon and hydrogen, respectively, and $\sigma_c^{C,H}$ and $\sigma_l^{C,H}$ are the cross sections for capture and loss, respectively, for the individual atoms. Following Ref. (2), we take $x_C = 2.1 \times 10^{15}$ cm^{-2} for all hydrocarbons, while $x_H = 4.2 \times 10^{15}$ cm^{-2} for CH_4 and for all other hydrocarbons $x_H = 3.56 \times 10^{15}$ cm^{-2}. The atomic cross sections $\sigma_c^{C,H}$ and $\sigma_l^{C,H}$ were estimated from the data of Ref. (8) at 100 keV. Finally, the ratio of capture cross sections is expected to be

$$R_{C_mH_n} \equiv \frac{\sigma(C_mH_n)}{\sigma(CH_4)} = \frac{m\, f^0(C_mH_n)}{f^0(CH_4)}. \quad (5)$$

Using the values given above, we compute $R_{C_2H_6} = 1.38$ compared to an experimental ratio at 105 keV of 1.70 ± 0.11, and $R_{C_3H_8} = 1.61$ compared to an experimental value of 2.06 ± 0.10 at 105 keV. The model provides a general estimate of the exit effect, but it tends to overestimate its magnitude for the collision systems considered here.

In summary, we have measured electron capture cross sections for 60- to 120-keV protons colliding with H_2, CH_4, C_2H_6, and C_3H_8. It was found that the Bragg additivity rule was not obeyed by these cross sections and that a model calculation of the exit effect provided a rough agreement with the measured cross section ratios.

ACKNOWLEDGEMENTS

We thank Mr. Mark Byrne and Mr. George Soosai for their assistance in constructing apparatus and taking data. This work was supported in part by grants from Research Corporation (SLV) and the University of South Alabama Research Council (JMS).

REFERENCES

1. Varghese, S.L., Bissinger, G., Joyce, J.M., and Laubert, R., *Nucl. Instrum. Methods* **170**, 269-273 (1980).
2. Bissinger, G., Joyce, J.M., Lapicki, G., Laubert, R., and Varghese, S.L., *Phys. Rev. Lett.* **49**, 318-322 (1982).
3. Varghese, S.L., Bissinger, G., Joyce, J.M., and Laubert, R., *Phys. Rev. A* **31**, 2202-2209 (1985).
4. Varghese, S.L., *Nucl. Instrum. Methods B* **24/25**, 115-118 (1987).
5. Dillingham, T.R., Doughty, B.M, Hall, J.M., Tipping, T.N., Sanders, J.M., and Shinpaugh, J.L., *Nucl. Instrum. Methods B* **40/41**, 40-43 (1989).
6. McDaniel, E.W., Mitchell, J.B.A., and Rudd, M.E., *Atomic Collisions: Heavy Particle Projectiles*, New York: John Wiley & Sons, 1993.
7. Stier, P.M. and Barnett, C.F., *Phys. Rev.* **103**, 896-907 (1956).
8. Toburen, L.H., Nakai, M.Y., and Langley, R.A., *Phys. Rev.* **171**, 114-122 (1968).

Electron capture and ionization cross sections from ion collisions with oriented elliptic Rydberg atoms

K. R. Cornelius and R. E. Olson

Department of Physics, University of Missouri-Rolla, Rolla, MO 65401

The classical trajectory Monte Carlo method has been used to calculate the total electron capture and ionization cross sections for collisions between multiply charged ions and a Rydberg target. The Rydberg targets are prepared in an oriented, elliptic state prior to simulating the system. The electron capture cross sections show a strong dependence on eccentricity as well as target alignment over various reduced collisions speeds. Electron ionization cross sections are less effected by alignment and show very little dependence on eccentricity at the higher collision speeds.

INTRODUCTION

The use of Rydberg targets in collisions involving multiply charged ions has lead to a greater understanding of collision dynamics. This has allowed the development of experiments which are directly comparable to collision theory. The development of aligned circular and elliptic Rydberg states has allowed further comparisons with theory by providing even more detail information about a collision [1,2]. Specific orbits and alignments allows the collision system to be dissected into various aspects and studied more closely. One of the most straightforward calculations that can be compared to experiment is that of the total cross section.

The ions chosen for our calculations were a proton (H^+), alpha particle (He^{++}) and a fully striped carbon atom (C^{6+}). The collision speeds are in terms of the reduced speed of the projectile with respect to the electron, given by

$$v^* = \frac{v_p}{v_e} \quad \text{or} \quad v^* = n \cdot v_p \,. \tag{1}$$

It is important to note that an electron in a Rydberg atom has an rms speed equal to n^{-1} in atomic units (Note: atomic units will be used for the remainder of the paper unless otherwise stated).

METHOD

For our calculations, we have used the classical trajectory Monte Carlo method to simulate the collision system. This method is well documented in the literature and has provided excellent agreement between theory and experiments in many cases [3,4,5,6]. This method is most applicable for collision speeds in the classical regime, meaning all the involved particles are non-relativistic and can be treated using classical mechanics.

In this article, we present calculations for the total electron capture and ionization cross sections for multiply charged ions incident on an elliptic Rydberg hydrogen atom in an $n = 25$ state. The electron of the Rydberg atom is placed in a classically defined, elliptical orbit by specifying an eccentricity, ε, common to all Kepler orbits. In a method similar to Cornelius *et al*, our target Rydberg atom is configured into an $\Phi = 0^\circ$ (L_z parallel to beam) or $\Phi = 90^\circ$ (L_z perpendicular to beam) alignment relative to the incident beam [7]. Thus, a $\Phi = 0^\circ$ alignment provides the ions with an entire view of the orbit of the electron and a $\Phi = 90^\circ$ alignment provides the ions with an edge only

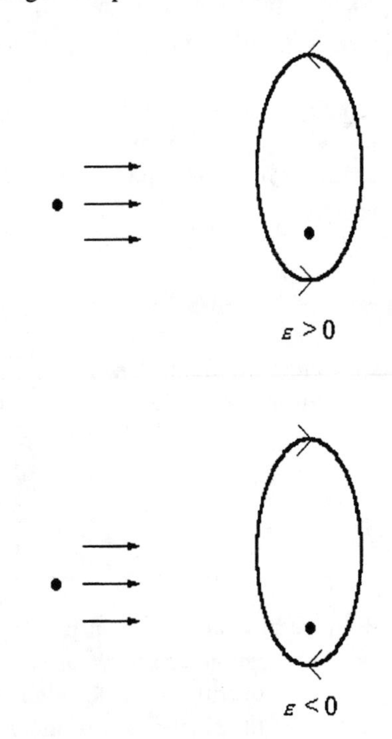

Figure 1. Schematic representation of the definition of positive and negative eccentricity relative to the incident ion beam

CP475, *Applications of Accelerators in Research and Industry*, edited by J. L. Duggan and I. L. Morgan

view of the orbit. We also define a positive and negative eccentricity such that the momentum vector of the electron at perihelion in an $\Phi = 90^o$ alignment is either parallel ($\varepsilon > 0$) or anti-parallel ($\varepsilon < 0$) to the incident ion beam [2] (see figure 1). Thus, the reactions have the form

$$A^{+q} + H(n=25,\varepsilon) \rightarrow A^{+q-1} + H^{+} \tag{2}$$

and

$$A^{+q} + H(n=25,\varepsilon) \rightarrow A^{+q} + H^{+} + e^{-}, \tag{3}$$

in which equations (2) and (3) represent electron capture and ionization processes, respectively.

Once the alignment and eccentricity of the target is chosen and the speed of the ion set, the system is propagated by numerically solving the classical Hamiltonian for a full three-body coulomb system. At the end of the trajectory, the system is tested to see if a capture or ionization event has occurred before calculating the next trajectory. Multiple trajectories are required before a reliable base of statistics has been gathered. Once all the trajectories have been completed, the cross sections are calculated as follows:

$$\sigma = \left(\frac{N_i}{N}\right)\pi b^2{}_{\max} \tag{4}$$

where N_i is the total number of electron capture or ionization events, N is the total number of trajectories and b_{max} is the largest value of the impact parameter for which a given event could occur. The number of trajectories for each calculation was altered in order to produce 1500 capture or ionization events. This condition was chosen to keep statistical errors less than 5%.

RESULTS

The results for electron capture are shown in figures 2 and 3. We have chosen to plot the unitless reduced cross section defined as

$$\sigma^* = \frac{\sigma}{\pi a_o^2 n^4} \tag{5}$$

with a_o the Bohr radius and n the principal quantum number. The reduced electron capture cross section, σ_{cap}*, is shown plotted on a logarithmic scale. Figure 2 shows the symmetric nature of the planer view of the orbit. Since the momentum vector of the electron is perpendicular to the ion beam, the direction of propagation of the electron around the orbit has no effect on the total cross section. Thus, the total cross section would be identical for $\pm\varepsilon$ of the same value. At certain collision speeds, there are some eccentricities that provide a total cross section that is larger or smaller than the circular orbit.

The reduced electron capture cross sections for $\Phi = 90^o$ are quite different compared to $\Phi = 0^o$. The most noticeable difference is the asymmetry about $\varepsilon = 0$. This asymmetry is due primarily to direct velocity matching between the electron and the projectile [8]. If the electron's velocity vector is in the same direction as

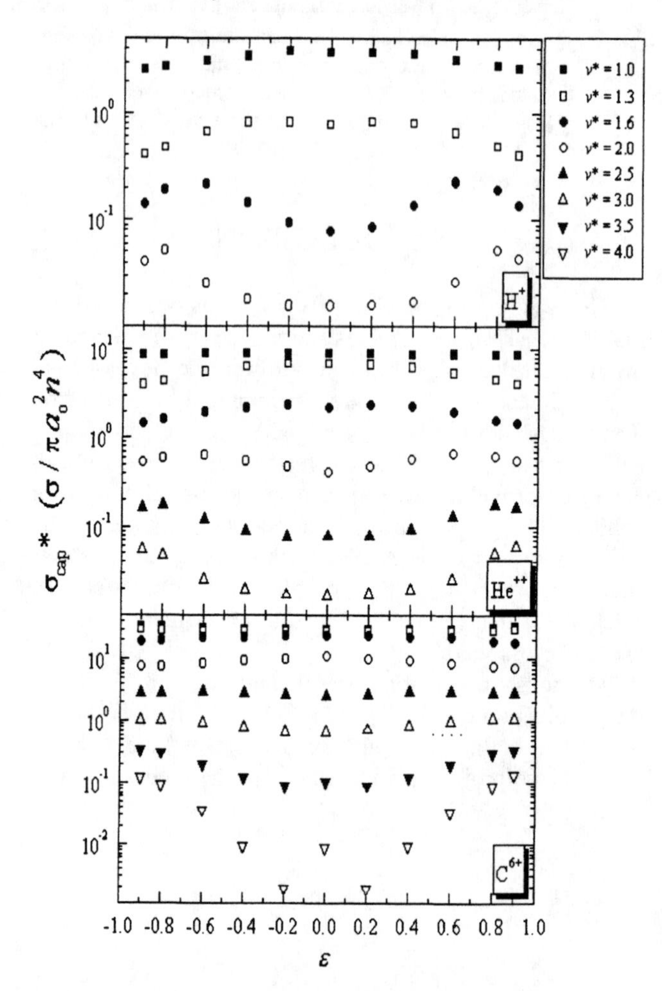

Figure 2. Total reduced electron capture cross sections for an $\Phi = 0^o$ alignment. Shown are σ_{cap}* for H^+, He^{++} and C^{6+} for various speeds (v*) incident on H(n=25) in elliptical states with eccentricity (ε). The projectile speeds for a given symbol are given in the legend.

the projectiles, electron capture is much more likely since the electron does not need to be accelerated as much in the projectiles direction.

Another important feature of figure 3 is the peak in the cross sections for a given projectile speed. This peak is linked to the eccentricity where the electron has a velocity closest to the projectile at perihelion. This leads to a larger cross section due to direct velocity matching. As v*

increases, values of ε closer to plus unity are required before direct velocity matching can occur. This results in the peak shifting toward unity. The peak will actually shift to negative values of ε for collision speeds less than $v^* = 1.0$. This occurs as direct velocity matching takes place at aphelion when the electron is parallel to the incident ion and at its slowest speed during the orbit [5].

Figures 4 and 5 represent the ionization cross sections for the $\Phi = 0^\circ$ and $\Phi = 90^\circ$ alignments. Figure 4 is the total reduced ionization cross section, σ_{ion}^*, for $\Phi = 0^\circ$. As with the capture cross sections, the ionization cross sections are symmetric about $\varepsilon = 0$. The ionization cross sections are shown for only those velocities where the cross section has peaked and begins to decrease as v^* increases.

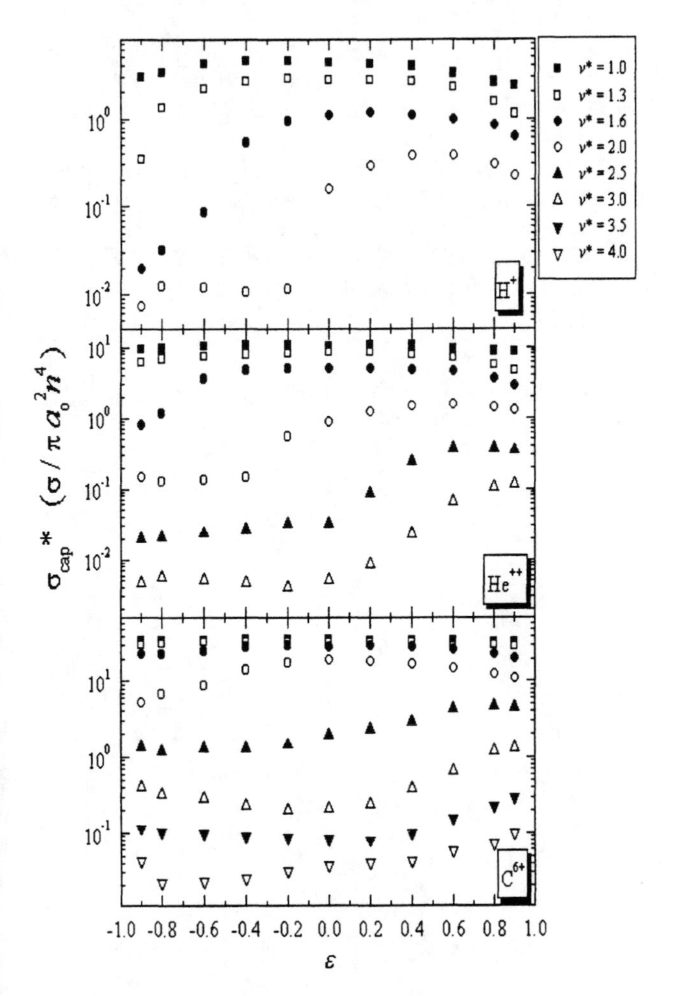

Figure 3. Same as figure 2 except for $\Phi = 90^\circ$.

The ionization cross sections symmetry is more uniform at each velocity than for the capture cross sections. In addition, the total ionization cross sections decrease at a much slower rate for all three projectiles, dropping only one power of ten between $1.0 \leq v^* \leq 6.0$, as opposed to several orders of magnitude for electron capture over the same region.

The total ionization cross sections for $\Phi = 90^\circ$ (figure 5) are not quite as asymmetric as it was for electron capture. A small asymmetry can be seen, but the nearly symmetric nature of the ionization cross sections has to do with the projectile speed being large enough such that ionization is not greatly effected by orientation. Almost all of the ionization's that occur at these high speeds are from hard two-body collisions between the projectile and the electron [7]. This leads to the total ionization cross sections of the two orientations being very close to the same values for the same projectile speed, which is what is observed in figures 4 and 5.

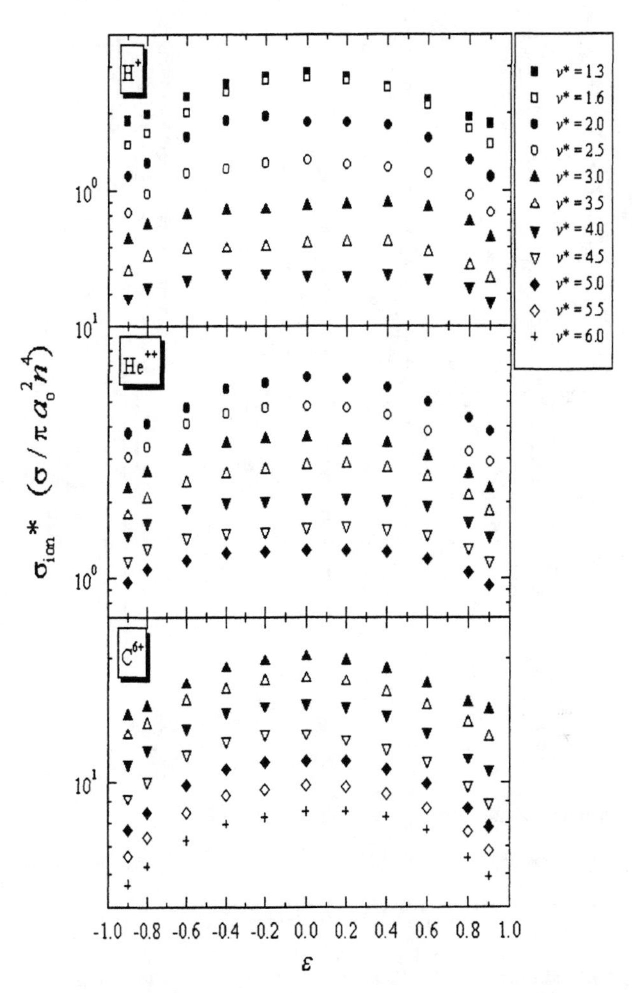

Figure 4. Total reduced electron ionization cross sections for an $\Phi = 0^\circ$ alignment. Shown are σ_{ion}^* for H^+, He^{++} and C^{6+} for various speeds (v^*) incident on H(n=25) in elliptical states with eccentricity (ε). The projectile speeds for a given symbol are given in the legend.

The final thing to point out about the total ionization cross sections is that both orientations approach a $1/E$ scaling dependence as v^* increases, differing only by a multiplicative constant, where E is the energy of the projectile. The ionization cross sections also appear to be more scaleable than the total capture cross sections, not only as a function of v^*, but also for $\varepsilon \neq 0$. Ionization cross sections do not exhibit the fluctuations in the eccentricity, but remain in a monotonic descending fashion on both sides of $\varepsilon = 0$. Only for the slowest speeds do any fluctuations appear.

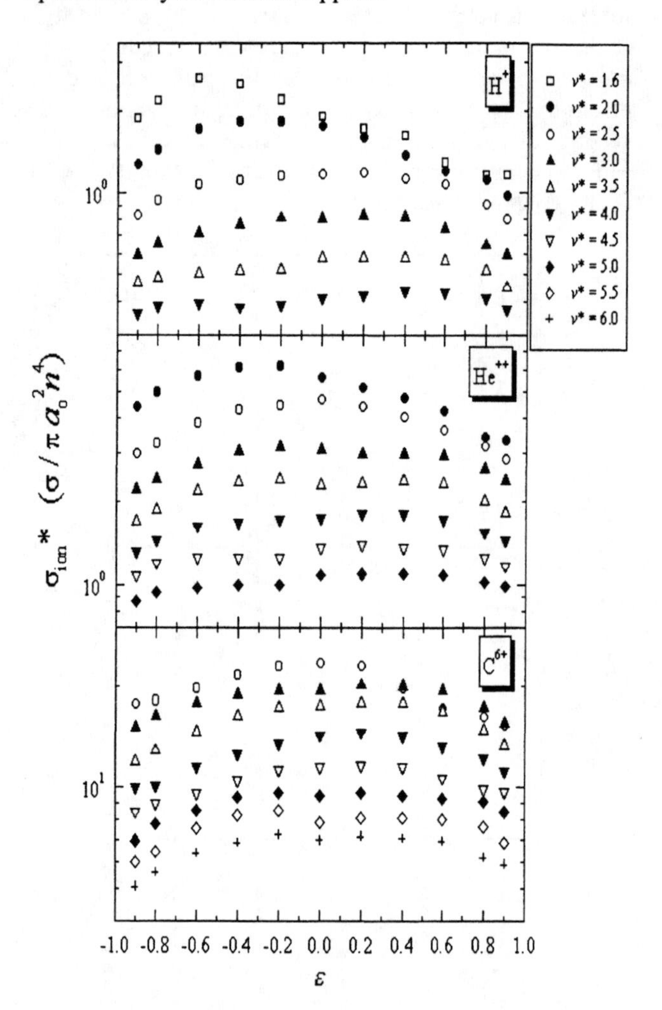

Figure 5. Same as figure 4 except for $\Phi = 90^{\circ}$.

CONCLUSION

In summary, the total electron capture and ionization cross sections from a Rydberg target have been presented for collisions with various charged particles. Electron capture cross sections displayed symmetric and anti-symmetric properties depending on the orbital alignment and the speed of the projectile which is a clear indication of direct and indirect velocity matching mechanisms. Ionization cross sections are less effected by orbital alignment or collision speed due to hard binary collisions between the electron and projectiles and appear to be more scaleable as a function of v^* and ε.

ACKNOWLEDGMENTS

This work was supported by the Department of Energy, Office of Fusion Energy Sciences.

REFERENCES

1. Day J C, Ehrenreich T, Hansen S B, Horsdal-Pedersen E, Mogensen K S and Taulbjerg K, *Phys. Rev. Lett.* **72** 1612 (1994)
2. Ehrenreich T, Day J C, Hansen S B, Horsdal-Pedersen E, MacAdam K B and Mogensen K S, *J. Phys. B : At. Mol. And Opt. Phys.* **27** L383 (1994)
3. Percival I C and Richards D *Adv. At. Mol. Phys.* **11** 1 (1975)
4. Olson R E and Salop A *Phys. Rev. A* **16** 531 (1977)
5. Bradenbrink, H Reihl, Wörmann Th, Roller-Lutz Z and Lutz H O, *J. Phys. B : At. Mol. And Opt. Phys.* **27** L391 (1994)
6. Wang J, McGuire J H and Olson R E, *AIP Conference Proceedings 360*, 619 (1995)
7. Cornelius K R, Wang J and Olson R E (accepted for publication- *J. Phys. B.:At. Mol. and Opt. Phys.* 1998)
8. Shakeshaft R and Spruch L *Rev. Mod. Phys.* **51** 369 (1979)

ABSOLUTE ELECTRON CAPTURE CROSS SECTIONS OF Kr^+ WITH Ar

H. Martínez, P. G. Reyes* and J. M. Hernández

Instituto de Física, Laboratorio de Cuernavaca, UNAM, Apartado Postal 48-3, 62151, Cuernavaca, Morelos, México. *Facultad de Ciencias, UNAM, Apartado Postal 70-542, 04510, México, D. F.*

Absolute differential and total cross sections for single electron capture were measured for Kr^+ ions on Ar in the energy range of 0.3 to 5.0 keV. The laboratory angular scan for the distributions ranged from -4.0 to 4.0 degrees. The absolute total cross section shows an oscillatory behavior. The total cross section is compared with other available measurements and with a semiempirical model. The results are in good agreement with previous measurements and the semiempirical model.

I.- INTRODUCTION

Most experiments concerned with the measurement of single electron capture cross sections for Kr^+ [1] have been performed at energies below 350 eV. On the other hand, in recent papers [2, 3], we report the single electron capture processes of Kr^+ ions in collisions with He and Xe. These investigations have shown that several excited states of the reactants are involved in the single electron capture processes. In particular, the Kr^+-Ar collisional system was chosen because:

(1) There has been a continuous interest in energy-transfer reactions between rare gas metastable atoms and rare gas ions because of their application to the pumping of rare gas ion lasers [4, 5].

(2) Data on total cross sections for the Kr^+-Ar system, exhibiting a well defined oscillatory behavior have been published [1], but not in the low keV region.

We report absolute measurements of the differential and total cross sections of single electron capture for Kr^+ collisions with Ar atoms. In addition, we present the results of a calculation based on the Olson [6] theoretical analyses for single electron capture. The energy range of the present study is from 0.3 to 5.0 keV. The laboratory angular scan for the distributions ranged from -4.0 to 4.0 degrees. This energy range includes the low-energy region where a quasimolecular picture of the collision is appropriate, as well as the region where molecular level crossings associated with the evolution of potential states of the collision system play a critical role in the dynamics of the collision.

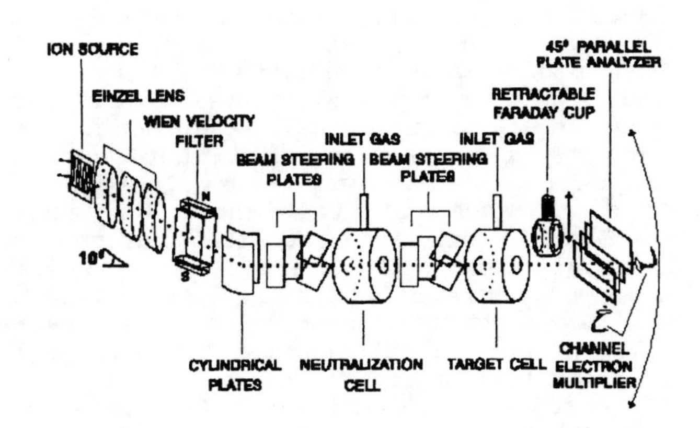

FIGURE 1. Schematic diagram of the apparatus.

II.- EXPERIMENT

The experimental arrangement has been described in detail elsewhere [2, 3], so only a brief survey of the experimental setup will be given here. A schematic diagram of the apparatus is shown in Fig. 1. Kr^+ ions were formed in a colutron-type ion source and accelerated in the energy range of 0.3 to 5.0 keV. The selected ion beam was velocity analyzed with a Wien filter, passed through cylindrical plates to deviate it by 10°, and through a series of collimators before entering the gas target cell. The cell is a cylinder of 2.5 cm in length and diameter, with a 1-mm entrance

CP475, *Applications of Accelerators in Research and Industry*, edited by J. L. Duggan and I. L. Morgan

aperture, and a 2-mm wide, 6-mm long exit aperture. All apertures and slits had knife edges. The target cell was located at the center of a rotatable, computer controlled vacuum chamber that moved the whole detector assembly which was located 47 cm away from the target cell. A precision stepping motor ensured a high repeatability in the positioning of the chamber over a large series of measurements. The detector chamber housed a Harrower-type parallel plate electrostatic analyzer, located at 45° with respect to the incoming beam direction, with two channel electron multipliers (CEMs). The Kr^0 atoms formed by electron capture passed straight through the analyzer through a 1-cm orifice on its rear plate, and impinged on a CEM so that the neutral counting rate could be measured. Separation of charged particles occurred inside the analyzer, which was set to detect the Kr^+ species with a second CEM. This flux was used as a measure of the stability of the beam during the experiment. To measure the angular distributions, a 0.36-mm diameter pinhole was located at the entrance of the analyzer. This geometry permitted the measurement of neutral atoms, the directions of which make an angle of up to ±4° with respect to the incoming beam direction. Path lengths and apertures gave an overall angular resolution of the system of 0.1°. The target thickness was ≈10^{13} atoms/cm^2 in order to ensure a single collision regime. Absolute gas pressures in the cell were measured with a capacitance manometer. The absolute differential cross section was calculated from the relation:

$$\frac{d\sigma}{d\Omega}=\frac{I(\theta)}{nLI_0} \quad (1)$$

where I_0 is the number of Kr^+ ions incident per second on the target, n, the number of Ar atoms per volume unit, L, the effective length of the scattering chamber, and $I(\theta)$, the number of Kr^0 counts per unit solid angle per second detected at laboratory angle (θ) with respect to the incident beam direction. The total cross section was derived by integrating the differential cross section over the solid angle dΩ:

$$\sigma=2\pi\int_0^{\pi}\frac{d\sigma}{d\Omega}\sin(\theta)d\theta \quad (2)$$

Extreme care was taken when the absolute differential cross section was measured. The reported value of the angular distributions was obtained by measuring it with and without gas in the target cell with the same steady beam. Then point-to-point subtraction of both angular distributions was carried out to eliminate the counting rate due to neutralization of the Kr^+ beam on the slits and the Kr^0 from background distributions. The Kr^+ beam intensity was measured before and after each angular scan. Measurements not agreeing to within 5% were discarded. Angular distributions were measured on both sides of the forward direction to assure they were symmetric. The estimated rms error is 15%, while the cross sections were reproducible to within 10% from day to day.

III.- RESULTS AND DISCUSSION

Characteristic angular distributions for single electron capture of Kr^+ in Ar are shown in Fig. 2 as a function of the projectile energy (the estimated error is of the size of the data point). In order to have a direct comparison of the data obtained at different values of the energy, the reduced variable $\tau=E_0\theta$ was calculated; E_0 is the impact energy and θ the scattering angle. The reduced cross section is the quantity $\rho(\tau)=(d\sigma/d\Omega)\theta\sin(\theta)$. An estimated value of R_c, the internuclear distance where the electron capture process takes place, was obtained from τ by using an exponentially shielded Coulomb potential [7]. This calculation gave a value of R_c=

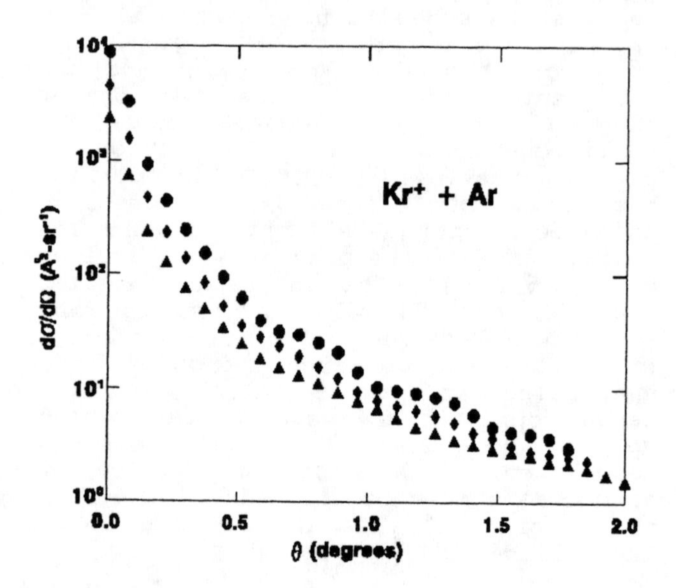

FIGURE 2. Measured absolute differential cross sections for single electron capture. (●) 5.0 keV; (▲) 1.0 keV; (▼) 0.3 keV.

1.882 a_0. Our results of the total cross sections for single electron capture are shown in Fig. 3. For a preliminary understanding of the trend of the total cross sections, the charge transfer cross section was discussed in terms of the model for near-resonant charge transfer proposed by Olson [6]. Since no theoretical potentials for $KrAr^+$ have been published so far, the total single electron cross sections were calculated using the following parameters: the crossing distance R_c= 1.882 a_0 calculated above; the coupling matrix element H_{12}=0.198 a.u. was calculated through the expression [8]: $H_{12}(R_c)=R^*\exp(-0.86R^*)$, where $R^*=(\alpha+\gamma)R_c/2$. We used $\alpha^2/2$=15.759 eV as the effective ionization potential of the target and $\gamma^2/2$=13.996 eV as the ground state electron affinity; and the first derivatives of the potential difference between the two electronic states involved in the charge exchange $|\Delta V'|$= 4.26 a.u. a_0^{-1}, which was fitted to get the same value of the maximum cross section predicted by Massey's criterion, together with the universal reduced cross section of Olson [6]:

$$\sigma_{10}=4\pi R_c^2 G(\lambda) \qquad (3)$$

where $G(\lambda)$ is the tabulated integral

$$G(\lambda)=\int_1^\infty e^{-\lambda x}(1-e^{-\lambda x})x^{-3}dx \qquad (4)$$

and

$$\lambda=\frac{2\pi H_{12}^2(R_c)}{v\hbar|\Delta V'(R_c)|} \qquad (5)$$

(v is the projectile velocity).
The results of this calculation are shown in Fig. 3 (solid line). Although the Olson model calculations [6] are not expected to be highly reliable, the calculation of σ_{10} is seen to agree in magnitude with the present measurements. In this case, it is clear that the calculations with the Olson model reproduce well the energy dependence of the cross section. Also plotted in Fig. 3 are the reported values of Maier [1] at low energies. The shape of the total cross section (the oscillatory behavior in the total cross section) suggests that distinct mechanisms are important in determining the single electron capture process in Ar. It is interesting to observe structures in the cross section between 50 and 165 eV. It thus appears desirable that a detailed theoretical analysis be carried out to further check this behavior.

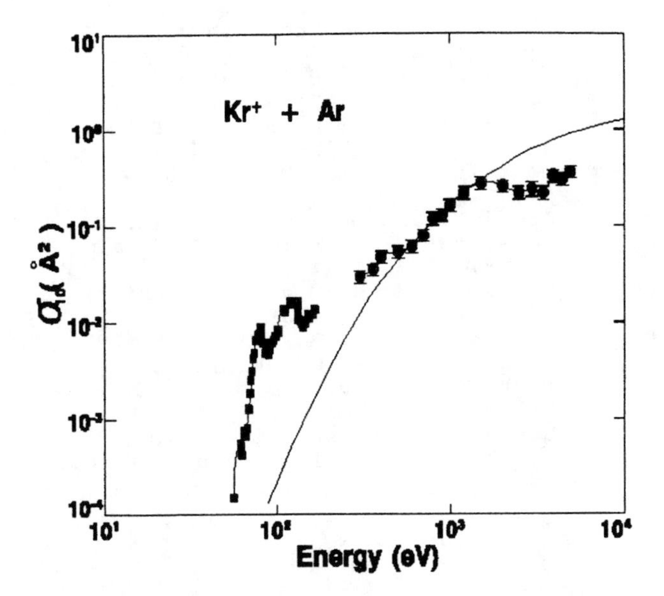

FIGURE 3. Total cross sections. (●) present measurements; (■) from Ref. 1; (___) Olson model.

The results of the present work can be summarized as follows:
a) Differential and total cross sections for single electron capture of Kr^+ on Ar are reported.
b) The reduced differential cross sections suggests the existence of a curve crossing at R_c= 1.882 a_0. It thus appeared desirable that a detailed theoretical analysis be carried out to confirm this critical transition region around R_c= 1.882 a_0.
c) The total cross sections for single electron capture show an oscillatory behavior.
d) The magnitudes of the present total cross sections for single electron capture are in good agreement with the semiempirical model of Olson [6].
e) We believe that the low-energy maximums in the cross section occurs when the reactant and the product of the reaction are excited, while that the high energy data occurs with the reactant and product in the ground states.

ACKNOWLEDGMENTS

We are grateful to B. E. Fuentes for helpful suggestions and comments. The authors wish to thank Sr. Anselmo Gonzalez for his technical assistance. This work was supported by the DGAPA-IN100392 and CONACyT 3659P-E9607.

REFERENCES

[1] W. B. Maier II, J. Chem. Phys. 69, 3077 (1978).

[2] H. Martínez and J. M. Hernández Chem. Phys. 215, 285 (1997).

[3] H. Martínez, J. M. Hernández, P. G. Reyes, E. R. Marquina and C. Cisneros, Nuclear Instrum. and Methods B124, 464 (1997).

[4] M. Tsuji, N. Kaneko, M. Furusawa, T. Muraoka, and Y. Nishimura, J. Chem. Phys. 98, 8565 (1993).

[5] M. Tsuji, N. Kaneko, and Y. Nishimura, J. Chem. Phys. 99, 4539 (1993).

[6] R. E. Olson, Phys. Rev. A2, 121 (1970).

[7] F. T. Smith, R. P. Marchi, K. G. Dedrick, Phys. Rev. 150, 79 (1966); F. T. Smith, R. P. Marchi, W. Aberth and D. C. Lorents, Phys. Rev. 161, 131 (1967).

[8] R. E. Olson, F. T. Smith and E. Bauer, Appl. Opt. 10, 1848 (1971).

ENHANCED PRODUCTION OF NONEQUIVALENT ELECTRON CONFIGURATIONS $3ln'l'$ ($n' \geq 6$) IN SLOW keV Ne^{10+} + He COLLISIONS: EXPERIMENT AND THEORY

J.-Y. Chesnel,[1,2] C. Harel,[3] B. Sulik,[4] F. Frémont,[1] B. Pons,[3] H. Jouin,[3] H. Merabet,[1] C. Bedouet,[1] X. Husson,[1] and N. Stolterfoht[2]

[1] *Centre Interdisciplinaire de Recherche Ions-Lasers, Unité Mixte (n° 6637) CEA-CNRS-ISMRA-Université de Caen, 6 Bd Maréchal Juin, F-14050 Caen Cedex, France*

[2] *Hahn-Meitner Institut GmbH, Bereich Festkörperphysik, Glienickerstrasse 100, D-14109 Berlin, Germany*

[3] *Centre de Physique Théorique et de Modélisation de Bordeaux, Laboratoire des Collisions Atomiques, Université de Bordeaux I, 351 Cours de la Libération, F-33405 Talence Cedex, France*

[4] *Institute of Nuclear Research of the Hungarian Academy of Sciences, P.O.Box 51, H-4001 Debrecen, Hungary*

Mechanisms for double-electron capture producing projectile doubly-excited states in Ne^{10+} + He collisions are studied. Emphasis is given to slow collisions with projectile energies of a few keV. At these impact energies the production of configurations $3lnl'$ of *nonequivalent* electrons ($n \geq 6$) is dominant. The creation of these nonequivalent electron states $Ne^{8+}(3lnl')$ with $n \geq 6$ originates from dielectronic processes involving electron-electron interaction. The present close-coupling calculations are compared with our recent measurements of *absolute* double-capture cross sections. A fairly good agreement between theory and experiment is found, especially for the total double capture as well as for the significantly populated configurations $3l6l'$.

INTRODUCTION

Charge exchange is at the heart of many important processes in atomic collisions (1-8). Several experimental and theoretical data are available for multicharged ion-atom collisions in the range usually referred to as the low-energy range (1-8). For example, for Ne^{10+} ions, low projectile energies are typically about 100 keV. Little is known about collisions involving bare ions at energies lower than 10 keV. For basic research, the investigation of charge transfer mechanisms for *double*-electron capture at very low energies is essential, however

In slow ion-atom collisions, double-electron capture can result in the creation of configurations $nln'l'$ involving doubly-excited states of the projectile. For the creation of the states $nln'l'$, we consider two kinds of mechanisms referred to as *monoelectronic* and *dielectronic* processes (6,9). These mechanisms are recalled in terms of the potential-curve diagram shown in Fig. 1 which is relevant for the Ne^{10+} + He collision. This figure illustrates resonance conditions at curve crossings (6). As noted in Refs. (10,11), it is useful to distinguish two categories of configurations $nln'l'$, namely, those involving (near-) equivalent n values ($n' \approx n$) and those involving nonequivalent n values ($n' >> n$). Hence, referring to Fig. 1, we first consider the production of the near-equivalent electron configurations $3l4l'$. The entrance channel crosses the channel labeled asymptotically $Ne^{9+}(4l)$ + He^+, where a single-electron transition into the orbital $4l$ may occur. When the internuclear distance decreases to 2.5 a.u., a second transition may produce the states $3l4l'$. Hence, two-step monoelectronic processes produced by nucleus-electron interactions may create (near-) equivalent electron configurations. In addition, *two*-electron transfers involving the electron-electron interaction may also populate the (near-) equivalent electron configurations (Fig. 1).

In the creation of *nonequivalent* electron configurations, such as $3lnl'$ ($n \geq 6$) in Ne^{10+} + He collisions, monoelectronic processes are unlikely to occur, as shown by model calculations in Ref. (6). These states (e.g., $3l7l'$) can be produced via a direct process of correlated double capture (CDC). In Fig. 1, this process corresponds to the crossings between the entrance channel and the double capture $Ne^{8+}(3lnl')$ + He^{2+} channels. Likewise, after a monoelectronic transfer into either the $4l$ or $5l$ orbital, a correlated transfer and excitation (CTE) process may also populate the configurations $3lnl'$ ($n \geq 6$). The CDC and CTE mechanisms have in common that they involve simultaneous *two-electron* transitions (Fig. 1), where the interaction between the electrons plays an essential role (9,10).

Here, we focus the attention on double-electron capture in ion-atom collisions involving keV impact energies. At keV energies, the projectile velocity is at least 10 times smaller than the mean velocity of the target electrons. Hence, the electron orbitals evolve in an extremely adiabatic manner and this adiabaticity may give rise to characteristic features in the capture mechanisms. Between adiabatic potential

CP475, *Applications of Accelerators in Research and Industry*, edited by J. L. Duggan and I. L. Morgan

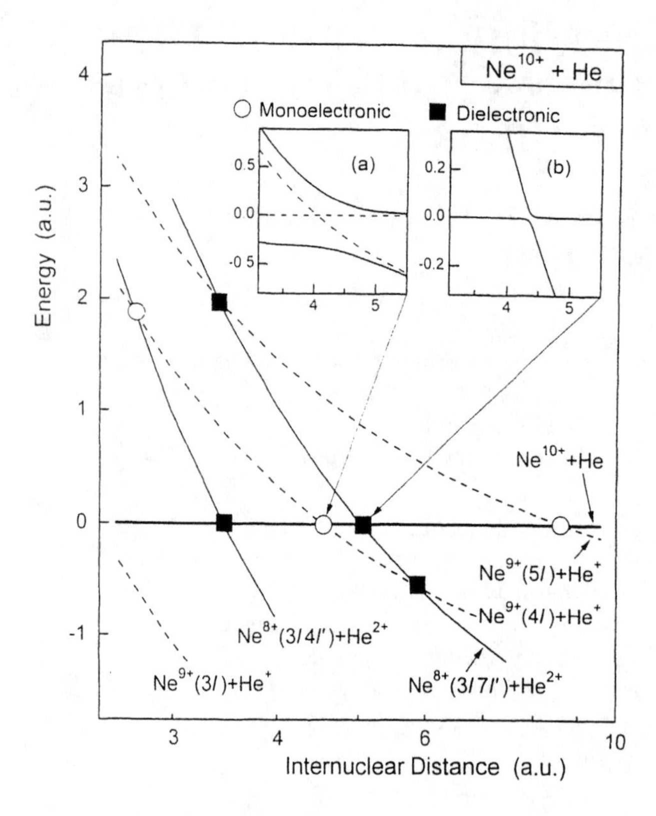

FIGURE 1. Diagram of approximate potential curves for the system Ne^{10+} + He. A limited number of curves relevant for the production of configurations $3lnl'$ is shown. Dashed and solid lines correspond to single and double capture, respectively. Crossings where monoelectronic and dielectronic transitions are expected are indicated by circle and square symbols, respectively. In insets (a) and (b) examples of adiabatic potential curves in the vicinity of monoelectronic and dielectronic crossings, respectively, are shown. These insets depict avoided crossings. A more detailed diagram is presented in Ref. [10].

curves, avoided crossings occur (see insets of Fig. 1), where transitions can take place (12). Because of the high adiabaticity, transitions at strongly avoided crossings as in inset (a) of Fig. 1 are expected to be significantly reduced at keV energies, increasing the selectivity in the production of final-state configurations. Thus, considering monoelectronic and dielectronic processes, the question arises as to which process is favored in collisions under strong adiabaticity conditions.

In this paper, our recent study (10) of 1- to 150-keV Ne^{10+} + He collisions is first reviewed. In this study (10) the projectile velocity dependence of double-electron capture was investigated experimentally by means of Auger electron spectroscopy. Specific effort was devoted to cross-section measurements at impact energies below 10 keV, which are significantly lower than those studied earlier (6,13). The main goal was to analyze the role of dielectronic processes in very slow collisions (10). To compare our measurements (10) with theory, newly calculated cross sections are presented in this work. The present calculations were performed using the close-coupling method within the framework of a straight-line impact-parameter approximation (14). From the comparison between theory and experiment, specific test calculations are suggested to determine the relative importance of the different dielectronic processes, such as CDC and CTE.

AUGER ELECTRON SPECTRA

Figure 2 shows typical Auger spectra measured for Ne^{10+} + He collisions at impact energies of 1, 10 and 150 keV (10). Details about the experimental method are given in Ref. (10). The observed peaks originate from the Auger decay of doubly-excited states due to $3lnl'$ and $4lnl'$ ($n \geq 4$). Figure 2 indicates that in 150-keV Ne^{10+} + He collisions Auger electron emission originates primarily from the near-equivalent electron configurations $4l5l'$, $3l4l'$ and $3l5l'$. Moreover, the decay of the overlapping components $4l4l'$ and $3lnl'$ ($n \geq 8$) contributes significantly to Auger electron emission. In contrast, for projectile energies of 1 and 10 keV, Auger emission is predominantly due to the decay of the nonequivalent electron configurations $3lnl'$ ($n \geq 6$). Moreover, higher-resolution L-Auger spectra presented in Ref. (10) clearly show that Auger emission from the nonequivalent electron states $3lnl'$ ($n = 6$ - 9) populated at 1 and 10 keV is significantly larger than that due to the overlapping components $3lnl'$ ($n \geq 10$) and $4l4l'$.

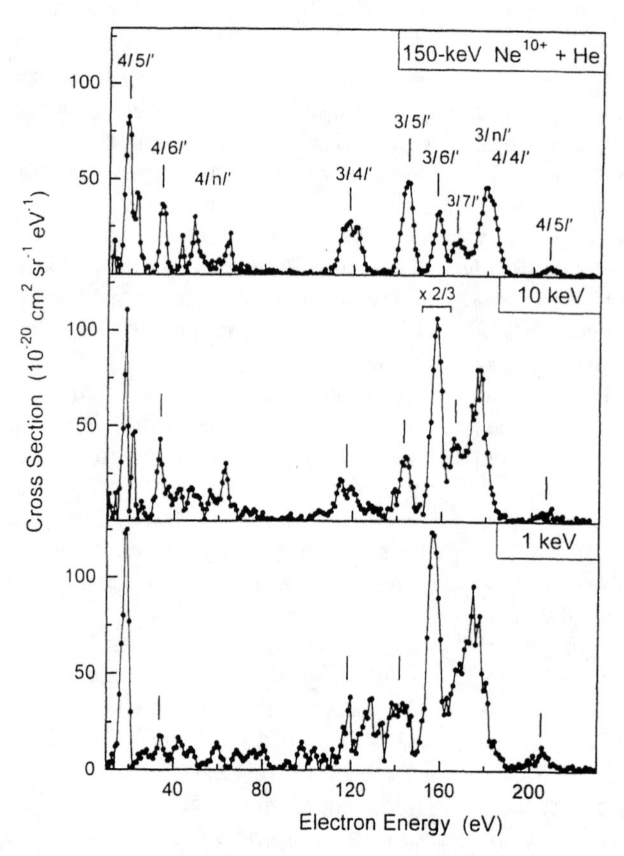

FIGURE 2. Spectra of Auger electrons produced in Ne^{10+} + He collisions at projectile energies of 1, 10 and 150 keV. The emission angle is 40° with respect to the beam direction. The peaks result from the Auger decay of the states $3lnl'$ and $4lnl'$ ($n \geq 4$) of Ne^{8+} ions. (From Ref. [10]).

RESULTS AND DISCUSSION

As described in Ref. (10), double-electron-capture cross sections were determined from the Auger spectra (Fig. 2) by using available Auger yields (11,15-17). The results for the Ne^{10+} + He system are shown in Fig. 3. When the impact energy decreases from 150 to 1 keV (i.e., at projectile velocities ranging from 0.53 to 0.04 a.u.) significant changes occur in the population of the configurations $3lnl'$ and $4lnl'$ ($n \geq 4$). At the highest energy, double-electron capture creates predominantly the (near-) equivalent electron configurations $3lnl'$ and $4lnl'$ (n = 4 and 5). The cross sections associated with the (near-) equivalent electron configurations are rather constant as the collision energy decreases, while the cross sections for producing the nonequivalent electron configurations $3lnl'$ with $n \geq 6$ increase noticeably (Fig. 3). At few-keV impact energies the production of the nonequivalent electron configurations $3lnl'$ ($n \geq 6$) becomes dominant (Fig. 3). The corresponding cross sections increase by a factor larger than three when the impact energy varies from 150 to 10 keV.

To discuss the observed variations of the cross sections for producing the nonequivalent electron states, it is instructive to consider the mechanisms responsible for their production. In our previous work (6) calculations using the

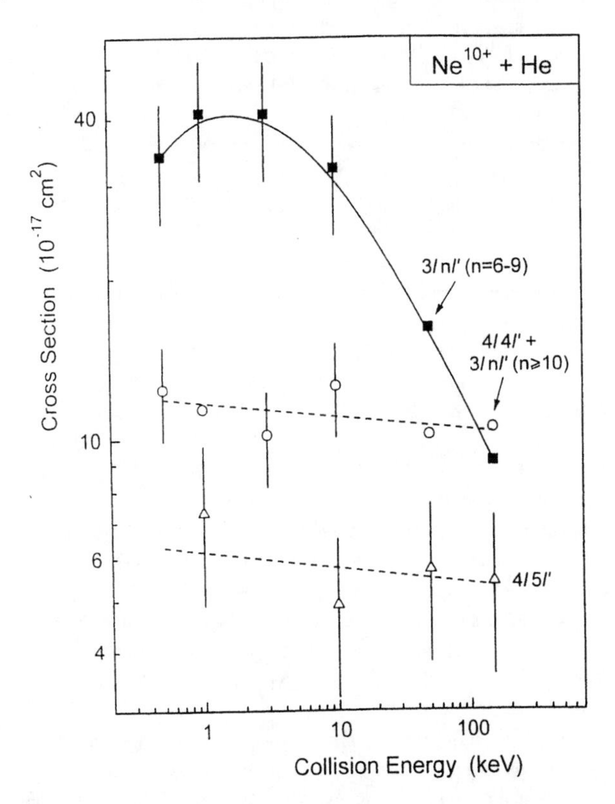

FIGURE 3. Cross sections for producing the nonequivalent electron configurations $3lnl'$ (n = 6 - 9) in Ne^{10+} + He collisions in comparison with those obtained for the near-equivalent electron configurations $4l5l'$ and the overlapping states $4l4l'$ and $3lnl'$ ($n \geq 10$). The curves are to guide the eye. (From Ref. [10]).

Landau-Zener model (18) indicated that the states $3lnl'$ ($n \geq 6$) are created by dielectronic processes, such as CDC and CTE (10). Hence, in view of the double capture processes, the experimental results (Figs. 2 and 3) show that the dielectronic processes gain importance when the collision energy decreases down to the few-keV range. Note that the dielectronic processes may also contribute to the production of the (near-) equivalent electron states (Fig. 1). Since cross sections for populating the (near-) equivalent electron states are rather unchanged at very low energies (Fig. 3), the contribution of the monoelectronic processes is likely to be reduced at few-keV impact energies. Consequently, the relative importance of the monoelectronic and dielectronic processes in Ne^{10+} + He collisions is reversed when the collision energy decreases from 150 to 1 keV.

A qualitative understanding of the energy dependence of dielectronic processes is provided when discussing the capture mechanisms with respect to approximate adiabatic potential curves, as those shown in the insets in Fig. 1. Details about these curves are given in Ref. (10). At the monoelectronic avoided crossing presented in inset (a) the minimum distance between the curves is at least ten times larger than the corresponding energy difference at the dielectronic crossing depicted in inset (b). Since dynamic couplings decrease with decreasing collision energy, the probability for transitions at strongly avoided crossings, such as *monoelectronic* crossings [see inset (a)], becomes weak at very low impact energies. Consequently, dielectronic processes are likely to gain importance with respect to monoelectronic transitions.

A quantitative analysis of the individual capture mechanisms requires a theoretical treatment of the double-electron capture. Taking into account the dielectronic interaction, we calculated the total cross sections for producing the configurations $3lnl'$ and $4lnl'$ ($n \geq 4$) in Ne^{10+} + He collisions. The framework of the theoretical method is the same as that described in Ref. (14). In a straight-line impact-parameter treatment (14), the total electronic wave function is expanded onto a set of configurations built with product of one-electron diatomic-molecular orbitals (OEDM), that are solutions of a one-electron two-center Hamiltonian. The electronic momentum transfer is taken into account through the same two-electron common translation factor as in (14). In the present calculations we have used a basis set of 25 dielectronic configurations, allowing the treatment of the n = 4, 5 and 6 single-capture channels (the n = 5 channel is predominantly populated by single-electron capture) as well as the $3lnl'$ and $4lnl'$ (n = 4 - 7) double-capture channels. The introduction of the above expansion in the Schrödinger equation leads to a set of coupled equations that is numerically integrated using a vectorized version of the PAMPA code (19). The results for the total double-electron capture and for the predominantly populated configurations $3l6l'$ are compared with the experimental data in Fig. 4.

There is a fairly good agreement between the *ab initio* calculations and the experimental results. Note that the present comparison is done for *absolute* cross sections. The experimental energy dependence of the cross sections is well reproduced by the theory. Moreover, the calculations for the total double-electron capture agree with experiment in the entire impact-energy range investigated here. For the configurations $3l6l'$, the differences between theory and experiment do not exceed a factor of 2 (Fig. 4). It should be pointed out that differences of a factor of 2 occur also for the states $4l4l'$. However, the global good agreement between the present calculations and the experimental data suggests that additional test calculations based on the same theoretical method may lead to a quantitative evaluation of the relative importance of each capture mechanism (e.g., the relative contribution of CTE versus that of CDC). Such additional test calculations are under progress. The corresponding results will be compared in details with our experimental data (10) as well as with recent results obtained by means of recoil-ion-momentum spectroscopy (15).

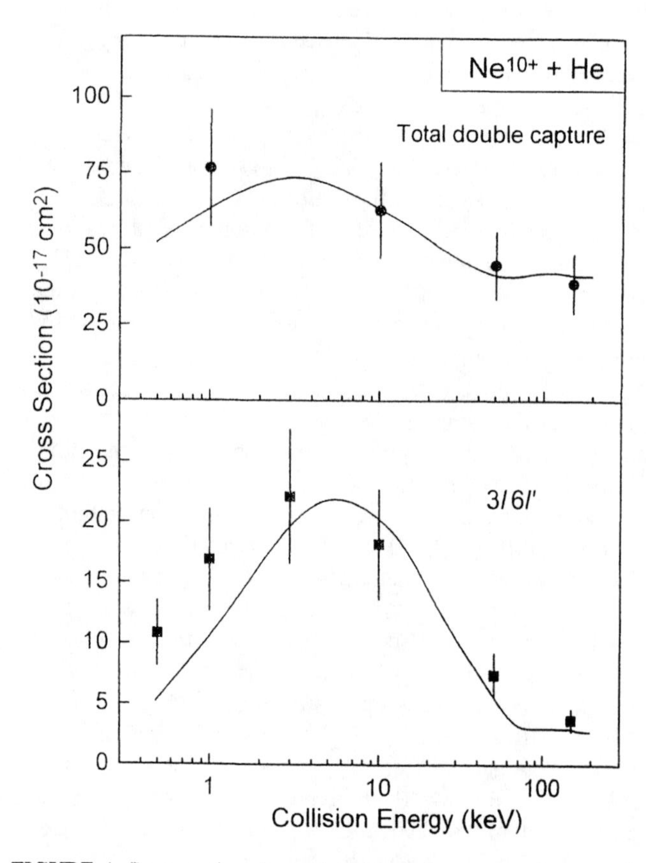

FIGURE 4. Cross sections for total double-electron capture (upper part) and for producing the nonequivalent electron configurations $3l6l'$ (lower part) in 0.5- to 150-keV Ne^{10+} + He collisions. The theoretical results (curves) are shown in comparison with the experimental data (solid symbols). The experimental uncertainties are about 20%.

Summarizing, the collision-energy dependence of double-electron capture in slow Ne^{10+} + He collisions is studied in this paper. Collisions under strong adiabaticity conditions are investigated, where the projectile velocity is one order of magnitude smaller than the Bohr velocity. The total contribution of the dielectronic processes is shown to increase strongly when the projectile energy decreases. Close-coupling calculations of double-capture cross sections are compared with experimental data, showing a fairly good agreement. Further calculations are planned to determine the individual contributions of the different capture mechanisms.

This work was performed within the framework of the Collaboration Research Program PROCOPE. One of us (J.Y.C.) acknowledges the support of the Alexander-von-Humboldt Foundation, Germany. Concerning the present calculations, C.H., B.P., and H.J. are indebted to support by the Centre National Universitaire Sud de Calcul (Montpellier) and by the Conseil Régional d'Aquitaine.

REFERENCES

1. Bordenave-Montesquieu A., Benoit-Cattin P., Gleizes A., Marrakchi A. I., Dousson S., and Hitz D., *J. Phys. B* **17**, L223-L227 (1984).
2. Stolterfoht N., Havener C. C., Phaneuf R.A., Swenson J. K., Shafroth S. M., and Meyer F. W., *Phys. Rev. Lett.* **57**, 74-77 (1986).
3. Winter H., Mack M., Hoekstra R., Niehaus A. and de Heer F. J., *Phys. Rev. Lett.* **58**, 957 (1987).
4. Van der Hart H. W., Vaeck N., and Hansen J. E., *J. Phys. B* **27**, 3489-3514 (1994).
5. Bachau H., Roncin P., and Harel C., *J. Phys. B* **25**, L109-L115 (1992).
6. Fremont F., Merabet H., Chesnel J.-Y., Husson X., Lepoutre A., Lecler D., and Stolterfoht N., *Phys. Rev. A* **50**, 3117-3123 (1994).
7. Schmeisser C., Cocke C. L., and Mann R., *Phys. Rev. A* **30**, 1661-1671 (1984).
8. Vernhet D. *et al.*, *J. Phys. B* **21**, 3949-3968 (1988).
9. Stolterfoht N., *Physica Scripta* **T51**, 39-46 (1994).
10. Chesnel J.-Y., Sulik B., Merabet H., Bedouet C., Frémont F., Husson X., Grether M., Spieler A. and Stolterfoht N., *Phys. Rev. A* **57**, 3546-3553 (1998).
11. Chesnel J.-Y. *et al*, *Phys. Rev. A* **58**, 2944-2961 (1998).
12. Stolterfoht N., *Progress in Atomic Spectroscopy, Part D*, ed. by Beyer H. J. and Kleinpoppen H., (Plenum Publishing Corp., 1987), p. 415.
13. Chesnel J.-Y. *et al*, *Phys. Rev. A* **53**, 4198-4204 (1996).
14. Harel C., Jouin H., and Pons B., *J. Phys. B* **24**, L425-L430 (1991).
15. Flechard X., Duponchel S., Adoui L., Cassimi A., Roncin P., and Hennecart D., *J. Phys. B* **30**, 3697-3708 (1997); and private communication.
16. Abdallah M. *et al.*, *Proc. of the XIV International Conference on the Application of Accelerators in Research and Industry*, Denton, Texas, 1996, edited by J. L. Duggan and I. L. Morgan, AIP Conf. Proc. No. 392 (AIP, New York, 1997), p. 209; and private communication.
17. Merabet H., Cremer G., Fremont F., Chesnel J.-Y., and Stolterfoht N., *Phys. Rev. A* **54**, 372-378 (1996).
18. Landau L. D., *Phys. Sovietum* **2**, 46 (1932) and Zener C., *Proc. Roy. Soc. A* **137**, 696 (1932).
19. Salin A., *Comput. Phys. Commun.* **62**, 58 (1991).

Calculation and Interpretation of Electron Momentum Densities

S. B. Trickey
Quantum Theory Project, Departments of Physics and of Chemistry, Univ. of Florida, Gainesville, FL 32611-8440

Intrepretations of scattering experiments in terms of electron momentum densities (spectral, total) commonly are formulated in a ground-state independent electron picture. Most calculations on extended systems and increasingly many on moleucles are based on Density Functional Theory, which raises subtle, significant technical and interpretation questions. We give a terse review of the relationship between DFT orbitals and Dyson orbitals, the significance of the Lam-Platzman correction, sensitivity to choice of approximate exchange-correlation model, and the role of the difference between the DFT and actual Fermi surface. Specific unsolved problems and related probes are formulated.

I. DISTINCTIONS AMONG ORBITALS

Measurement of the spectral (hence in principle also the total) electron momentum density by (e,2e) and (γ,eγ) experiments is of growing interest and importance. Certain oft-ignored or dismissed technical distinctions among the electron orbitals involved actually are physically significant for both interpreting the experiments and for using and understanding Density Functional Theory (DFT), kinetic theory of stopping, and so-called Hartree-Fock-Slater calculations of generalized oscillator strengths.

Within the plane–wave impulse approximation, the measured cross section in (e,2e) and (γ,eγ) experiments is related directly to the square of the Dyson orbital momentum density[1] for the excitation "j" (hence, spectral density)

$$\rho_{D,j}(\mathbf{p}) = \hat{\Phi}_j^*(\mathbf{p})\hat{\Phi}_j(\mathbf{p})$$

where the caret indicates the Fourier transform (from real space) and "D" denotes Dyson. (Throughout ρ will be momentum densities, and real-space number densities will be n.) The Dyson orbital (electron detachment amplitude)[2] is

$$\Phi_j(\mathbf{r},\sigma) \propto \langle \Psi_j^{N-1} | \Psi_0^N \rangle_{N-1}$$

where the subscripted expectation value indicates integration over $N-1$ electron coordinates, Ψ_0^N is the N-electron ground state, and Ψ_j^{N-1} is the $N-1$ electron state generated by removing an electron from natural spin orbital φ_j.

The natural spin orbitals give the first-order reduced density matrix in orbital-diagonal form[3]:

$$\begin{aligned}
&\gamma(\mathbf{x}';\mathbf{x}) \\
&= N \int d\mathbf{R}ds\, \Psi_N^*(\mathbf{x}';\mathbf{R},\mathbf{s})\Psi_N(\mathbf{x};\mathbf{R},\mathbf{s}) \\
&= \sum_i n_i\, \varphi_i^*(\mathbf{x}')\varphi_i(\mathbf{x})
\end{aligned}$$

The combined space-spin coordinate is $\mathbf{x} = (\mathbf{r},\sigma)$. The NSO's are orthonormal, with occupation numbers obeying $0 \le n_i \le 1$ and the electron (number) density for spin σ is $n_\sigma(\mathbf{r}) = \gamma(\mathbf{r},\sigma;\mathbf{r},\sigma)$

In terms of the Dyson orbitals, which are not orthogonal,

$$\gamma(\mathbf{x}';\mathbf{x}) = \sum_j \Phi_j^*(\mathbf{x}')\Phi_j(\mathbf{x})$$

Comparison with the NSO expansion of γ above shows that each Φ_i is a specific linear combination of the NSO's.

The total electron momentum density is

$$\rho(\mathbf{p}) = \hat{\gamma}(\mathbf{p};\mathbf{p}) = \sum_i \rho_i(\mathbf{p}) = \sum_i n_i\, \hat{\varphi}_i^*(\mathbf{p})\hat{\varphi}_i(\mathbf{p})$$

In general

$$\rho_i(\mathbf{p}) \neq \rho_{D,i}(\mathbf{p})$$

though in the familiar "target Hartree-Fock approximation" equality holds[1].

The situation is more subtle in Hohenberg-Kohn-Sham Density Functional Theory[4]. The HKS equation (compressed spin labeling) is

$$\left\{ -\frac{\hbar^2}{2m}\nabla^2 + V_\sigma[\rho(\mathbf{r})] \right\} \phi_{i,\sigma}(\mathbf{r}) = \mathcal{E}_{i,\sigma}\phi_{i,\sigma}(\mathbf{r}),$$

where the spin number density is

$$n_\sigma(\mathbf{r}) = \sum_i f_{i,\sigma} |\phi_{i,\sigma}(\mathbf{r})|^2 .$$

Casida[5] has shown that the HKS exchange-correlation potential v_{XC} is the best local approximation possible for the Dyson self-energy operator (which is non-local). The quality criterion is extremalization of Klein's energy expression for the Dyson equation.

CP475, *Applications of Accelerators in Research and Industry*,
edited by J. L. Duggan and I. L. Morgan

II. OPTIMUM APPROXIMATE ORBITALS

Casida as well as Duffy et al.[6] also discussed the extent to which his result might justify the assumption that the HKS orbitals might be the best (or, failing that, high quality) approximations to the Dyson orbitals. If they were, then the "target Kohn-Sham approximation"

$$\rho_{D,i,\sigma}(\mathbf{p}) \approx \rho_{i,\sigma}^{HKS}(\mathbf{p})$$
$$\rho_{i,\sigma}^{HKS}(\mathbf{p}) = f_i|\hat{\phi}_{i,\sigma}(\mathbf{p})|^2$$

also would be validated. With only intuitive support, this approximation has seen long use in, for example, calculation of the total momentum density for Compton profiles[7] and stopping powers[8]. On the basis of arguments rooted in Casida's result, Duffy et al.[6] gave important numerical results about the differences arising from a selection of practical DFT approximations.

However, the argument itself, though suggestive, is not entirely rigorous (in fairness, neither Casida nor Duffy et al. claimed otherwise) for reasons that seem to have gone unconsidered. One is that, as usual with variational stability, a second-order error in the local v_{xc} that approximates the Dyson self-energy operator best corresponds to a first-order error in the orbitals. Second is the Lam-Platzman correction[9,10], which commonly is omitted. The LP correction is the shift of the momentum density of the HKS auxiliary system relative to the real system

$$\rho_\sigma(\mathbf{p}) = \rho_\sigma^{HKS}(\mathbf{p}) + \Delta\rho|_{LP}(\mathbf{p})$$

In turn there are two problems in combining the LP correction with the Casida result. Even if the real-space HKS orbital ϕ_j were the best local operator eigenfunction approximation to the Dyson orbital Φ_j, the LP correction shows that in general the same relationship does not necessarily follow for the Fourier-transformed orbitals. Furthermore, the LP correction applies to the total momentum density, not orbital by orbital. Since the LP correction is derived by the adiabatic connection formulation of DFT (adiabatically adding the contribution of the total momentum operator to the basic Hamiltonian, then extracting the momentum distribution via the Hellmann-Feynman theorem), it is not at all evident how to obtain an orbital-by-orbital analogue to LP or even that such an analogue exists.

III. NUMERICAL TESTS

A straightforward but somewhat difficult to execute numerical test of the approximate equivalence of the HKS and Dyson orbitals would be to study the overlap $\langle\hat{\phi}_j(\mathbf{p})|\hat{\Phi}_j(\mathbf{p})\rangle$. Lacking that, the numerical comparison by Duffy et al. of valence momentum densities from configuration interaction and approximate DFT calculations supports the assumed equivalence. Note that this support addresses primarily the issue of individual orbitals and that there seems to be less validity for the inner orbitals. That fact is consistent, at least, with the occurence of problems in computed Compton profiles (next paragraph), which use the total EMD of course.

How much difference in $\rho(\mathbf{p})$ does the LP correction make? For Compton profiles, the result is remarkably system-independent. For bulk Cu, Bauer and Schneider[10] found a rather small but noticeably beneficial effect. Recently Králik et al.[11] computed the LP correction for the Compton profile of bulk Si and found a similarly small effect, which however put the DFT results in very close agreement with their high-accuracy variational Monte Carlo results. Both show a significant discrepancy with respect to experiment at high momentum transfers. Similarly, Filippi and Ceperley's[12] comparison of Compton profiles for bcc Li computed via approximate DFT and Quantum Monte Carlo methods shows that LP brings the two in close agreement but that there remains significant disagreement with experiment.

How much influence does the choice of approximate DFT v_{XC} model have on the calculated HKS momentum density and how do such calculations compare with correlated wave-function calculations? Numerical comparisons by Duffy et al.[6] show that the differences are small (typically less than or equal to the experimental error bars) for the valence SEMD's of the molecules treated. For inner valence orbitals particularly, there are notable discrepancies between all calculations and experiment. The largest differences are at low momenta.

IV. SPECIFIC PROBLEMS AND OPPORTUNITIES

The discrepancies between experiment and calculation for both total and spectral EMD's are an obvious problem. On the purely theoretical side, a known problem worth mentioning is that the LP correction for any local v_{XC} model is isotropic. It is also worth reiterating that there is so far no obvious way to achieve an orbital-by-orbital counterpart to the LP correction. Having this would be equivalent in principle to knowing the difference $\Phi_j - \phi_j$ in real space, which suggests the difficulty involved. Two distinct routes for possible progress on this problem are excited state DFT and the procedure given by Jansen[13].

Assuming availability of high-quality orbital mean excitation energies (and the validity of the associated assumptions)[14], the gas-solid differences in proton stopping predicted purely from kinetic theory of stopping[15] would provide a useful check of computed EMD's since in the sense that the kinetic theory cross-section expression is an integral over the EMD which differs substantially from the Compton profile. Predictions for Li[16] await experimental test.

Recently we[17] we have begun calculational investigation of a suggestion by Mearns and Kohn[18] regarding the difference between the DFT and actual Fermi Surfaces. Briefly, the discontinuities in the EMD are mapped to the Fermi Surface (FS). Therefore, in the EMD's for all momenta lying in each of several planes, the presence and relative position of secondary peaks relate directly to the FS. Comparison of high precision measurement with calculations therefore provides another test of the target Kohn-Sham approximation.

Finally, the unresolved matter of the precise physical content of the HKS orbitals [beyond the obvious point that they construct $n_\sigma(\mathbf{r})$] leaves open the opportunity to devise clever potentials for the generation of better approximations to the Dyson orbitals. Duffy et al.[6] tested two such schemes and found both lacking. Similarly, there is no real justification for the old scheme of calculating dipole oscillator strengths from "Hartree-Slater" orbitals[19], another variant on the idea of a target Kohn-Sham approximation, but that lack is also opportunity for finding effective operators that generate superior orbitals in the sense of yielding better oscillator strengths.

ACKNOWLEDGMENT

Discussions with R.J. Mathar, N.Y. Öhrn, J.R. Sabin, and J. Wang, correspondence with M.E. Casida and E.R. Davidson are acknowledged with thanks.

[1] E.R. Davidson, Can. J. Phys. **74**, 757 (1995)
[2] R. Longo, B. Champagne, and Y. Öhrn, Theor. Chim. Acta **90**, 397 (1995)
[3] P.-O. Löwdin, Phys. Rev. **97**, 1474 (1955)
[4] References to original literature are in E.S. Kryachko and E.V. Ludeña, *Energy Density Functional theory of Many-Electron Systems* (Kluwer, Dordrecht, 1990); *Density Functional Theory of Many-Fermion Systems*, S.B. Trickey special ed., Adv. Quantum Chem. **21** (1990); R.M. Dreizler and E.K.U. Gross, *Density Functional Theory* (Springer, Berlin, 1990); R.G. Parr and W. Yang *Density Functional Theory of Atoms and Molecules* (Oxford, NY, 1989)
[5] M.E. Casida, Phys. Rev. A **51**, 2005 (1995).
[6] P. Duffy, D.P. Chong, M.E. Casida, and D.R. Salahub, Phys. Rev. A **50**, 4707 (1994); P. Duffy, Can. J. Phys. **74**, 763 (1996).
[7] J.R. Sabin and S.B. Trickey, J. Phys. B **8**, 2593 (1975).
[8] J.Z. Wu, S.B. Trickey, J.R. Sabin, and D.E. Meltzer, Nucl. Instr. Meth. B **56/57**, 340 (1991).
[9] L. Lam and P.M. Platzman, Phys. Rev. B **9**, 5122 (1974)
[10] G.E.W. Bauer, Phys. Rev. B **27**, 5912 (1983); G.E.W. Bauer and J.R. Schneider, Phys. Rev. B **31**, 681 (1985)
[11] B. Králik,P. Delaney, and S.G. Louie, Phys. Rev. Lett. **80**, 4253 (1998)
[12] C. Filippi and D.M. Ceperley, preprint
[13] H.J.F. Jansen, Phys. Rev. B **43**, 12025 (1991)
[14] J.R. Sabin and J. Oddershede, Phys. Rev. A **26**, 3209 (1982); *ibid.* **29**, 1757 (1984)
[15] P. Sigmund, Phys. Rev. A **26**, 2497 (1982)
[16] J. Wang, R.J. Mathar, S.B. Trickey, and J.R. Sabin, J. Phys.: Cond. Matt. (submitted)
[17] S.B. Trickey, R.J. Mathar, J. Wang, and J.R. Sabin, 18th Werner Brandt Workshop, June 1998 (unpublished)
[18] D. Mearns and W. Kohn, Phys. Rev. B **39**, 10669 (1989)
[19] J.L. Dehmer, M. Inokuti, and R.P. Saxon, Phys. Rev. A **12**, 102 (1975)

Zero Degree Electron Emission in H^+ Collisions with Gas Targets

M. B. Shah, K. O. Lozhkin, J. Geddes, H. B. Gilbody and C. McGrath

Department of Pure and Applied Physics, The Queen's University of Belfast, Belfast BT7 1NN, UK

A recently constructed apparatus permits measurements of the electron velocity distributions at zero degrees for fast ions in collision with target atoms and molecules. A 260mm diameter hemispherical analyser with a half angle acceptance of 1.5° and an energy resolution $\Delta E/E$ of 0.012 is used to measure the velocity distribution of the emitted electrons. The analyser and the surrounding interaction region are double shielded against the earth's magnetic field by use of μ-metal enclosures. Measurements show that electrons with > 2eV pass energy travel through the analyser with no appreciable deflection due to the residual magnetic field. For 40keV H^+ projectiles in collision with H_2 and He targets the electron energy spectra at zero degree emission angle show that it is dominated by a well-defined electron capture to the continuum peak. Existence of saddle-point electrons at zero degrees is not confirmed.

INTRODUCTION

Angular and velocity distributions of electrons emitted in ion-atom collisions are able to provide invaluable insights into the mechanisms of ionization and have been a subject of intense experimental and theoretical activity in recent years (cf 1). The observed energy spectra of the emitted electrons reveal structures that can be identified with different emission mechanisms (2, 3). The three main features are the broad soft electron peak, the electron capture to the continuum (ECC) peak present at emission angles close to zero degrees and the binary peak. In attempts to describe these features theoretical models have been developed where 'one centre' and 'two centre' effects are considered. The term 'one centre' refers to the ejection of an electron into the continuum of either the target or projectile alone while 'two centre' refers to electron emission in the combined fields of both target and projectile. Soft electrons (electron velocity $v_e \approx 0$) arise from glancing collisions and small momentum transfers. For projectiles with high velocity v_p and low charge state, the electrons are emitted essentially in the field of the target ion and consequently are best described within the 'one-centre' model. The ECC electron peak ($v_e \sim v_p$) arises from electrons that travel alongside the receding projectiles at asymptotically large distances. While a 'one centre' calculation involving the projectile predicts the existence of this feature, the influence of 'two centre' effects must be taken into account especially during the initial part of the ion-atom interaction. The binary peak ($v_e \sim 2v_p$) results from a two-body (binary) hard collision of the fast projectile with an effectively stationary target electron. For high impact velocities ($v_p >>$ the electron orbital velocity) 'one centre' theories such as the Born or impulse approximations give excellent predictions.

A further ionization process that gives rise to 'saddle-point' electrons has been predicted (4). This mechanism is based on the idea that some of the electrons find themselves on the moving saddle point of the Coulomb potential between the projectile ion and the target ion. Electrons that have the same velocity as the saddle point feel no force and are lifted to the continuum as the projectile and target separate. For singly charged ions on atomic targets the 'saddle point' electrons have velocity $v_e \sim v_p/2$. This process provides an ideal test-bed of direct 'two-centre' effects during a collision. However many theoretical and experimental results suggest that this process remains speculative (5, 6).

In the present work we have recorded the velocity distribution of the electron spectra at zero degrees with respect to the incident ion beam for 40 keV H^+ incident on helium and molecular hydrogen. Our velocity range v_e/v_p from 0.27 to 2.3 covers the regions associated with the formation of 'saddle point', ECC and binary electrons. We also discuss the elaborate precautions that we took to ensure that present spectra are free from many of the difficulties encountered in the past (2).

EXPERIMENTAL APPROACH

A schematic diagram of the apparatus is shown in Fig. 1. The fast momentum analysed H^+ ion beam passed through the differentially pumped target chamber where the gas pressure was low enough to maintain single collision conditions. The ion beam and electrons emitted from the target at zero degrees with respect to the ion beam travelled 50 mm from the target before entering a large hemispherical analyser spectrometer. The fast ions, which were not deflected significantly in the spectrometer, passed through an aperture in the wall of the spectrometer and were collected in a Faraday cup. The electron energy spectra were recorded using the hemispherical spectrometer.

Screening from the earth's magnetic field was provided by a double layer of μ metal. The inside of the main

CP475, *Applications of Accelerators in Research and Industry*,
edited by J. L. Duggan and I. L. Morgan

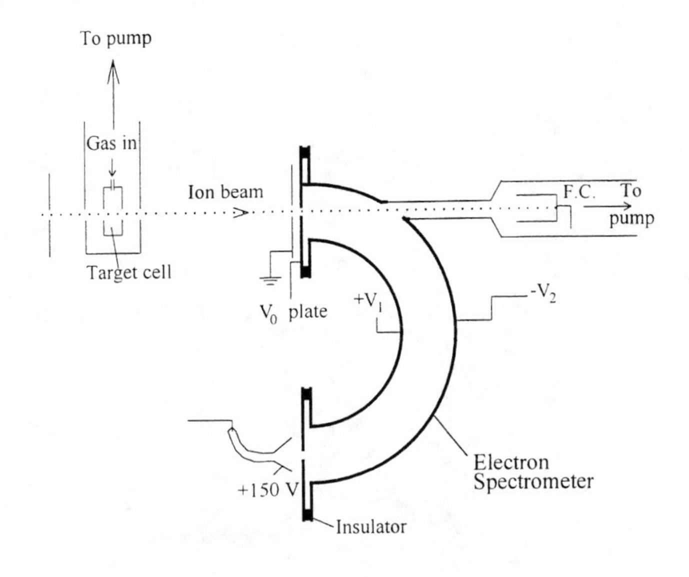

FIGURE. 1. Schematic diagram of the apparatus

experimental chamber was screened by a 2 mm thick layer. The target, the drift region from the target to the hemispherical analyser and the analyser itself were separately screened by an inner layer of 2 mm thick μ metal. Tests showed that 2eV electrons formed in the target within the acceptance angle of the analyser were transmitted with negligibly small attenuation.

Electrons exiting the spectrometer were detected by a channeltron in a pulse counting mode. The intensity of the projectile beam was monitored by a Faraday cup (F.C.).

The electron energy spectra at zero degrees were obtained by measuring the electron distribution when gas was introduced in the target cell and subtracting it from the background signal of electrons which arose when the main chamber was flooded to a pressure that corresponded to the value attained when gas was introduced into the target cell.

In the development of the experimental arrangement much attention was given to the collimination of the ion beam, the overall design of the target cell and the operation of the energy analyser.

The ion beam

The impact of the ion beam on any surface in the vicinity of the hemispherical analyser gave rise to a large background electron signal that could exceed the real signal. This problem was minimised by careful collimation of the ion beam. The H^+ ion beam was passed through a collimator 0.5 mm in diameter placed 500 mm upstream from the target cell. A further collimator with a 2mm aperture was placed 50 mm upstream from the target. The apertures in the target region and the entrance to the hemispherical analyser were all 3 mm in diameter. A series of deflection plates, placed before the target, enabled the ion beam to be directed cleanly through the target without making collisions with the entrance or exit apertures.

The target

In a recent study (7) where a gas jet from a needle was used to provide the target, the measured spectra of low energy electrons emitted at 10 degrees were dependent on the spacing between the projectile beam and the exit tip of the needle. The low energy part of the spectrum was enhanced as the distance between the needle tip and the ion beam was increased. This observation was attributed to the increase in the interaction volume due to the divergence of the gas beam. However it is not clear why only the low energy electron signal should be affected. Effects of this nature were avoided by the use of a small target gas cell (1 cm in length). The region outside the cell was differentially pumped (Fig. 1). This arrangement allowed the target gas to be well defined while the pressure in the experimental chamber was ~ 10^{-8} torr. Consequently the scattering of the low energy electrons did not degrade the electron spectra.

The hemispherical analyser

The hemispherical analyser used to record the electron velocity distributions had a mean diameter of 260mm. The spectrometer was operated at a fixed pass energy (V_1 and V_2 fixed) by raising the base plate of the spectrometer to the required voltage V_0. An earthed plate was placed close to the base plate to prevent penetration of this voltage into the drift region and distortion of the electron trajectories was minimised. A lens system, which could give rise to a collection efficiency that depended on the emission angle and energy of the electrons, was not used to collect and focus electrons into the spectrometer. The acceptance angle of the analyser was 1.5 ± 0.2 degree and the energy resolution ($\Delta E/E$) was 0.012 (FWHM). Tests showed that for 40keV H^+ incident ions, identical spectra were obtained when the spectrometer was used at fixed pass energies (V_1 and V_2 fixed) of 8 or 16eV with V_0 ramped. Further with V_0 earthed and the analyser used in the variable energy pass mode, an identical spectra was obtained after theoretical allowance was made for the variation in transmission with electron energy..

Passage of the fast primary ion beam through the background gas within the hemispherical analyser gave rise to electrons that were detected. Separate pumping of the hemispherical analyser greatly reduced this background signal.

RESULTS AND DISCUSSION

Figures 2 and 3 show our measured spectra at zero degrees for 40keV H^+ incident ions in H_2 and helium respectively. The spectra are completely dominated by the ECC peak. The shape of the ECC cusp shows the normal asymmetry observed in the measurements of other workers (8-14) where the differential cross sections at energies above the peak fall off more steeply than those at lower energies. The energy resolution of the spectrometer (0.16eV FWHM for the data in Figures 2 and 3) is much smaller than the steeply falling high energy tail (0.5eV HWHM). Hence the energy dependence of the ECC peak is faithfully reproduced by our measurements.

The binary peak centred at $v_e = 2v_p$ is negligibly small at the present projectile energy because the projectile velocity is of the same magnitude as the velocity spread associated with the target electron. The existence of structure associated with 'saddle-point' electrons (which should appear at the $v_e = v_p/2$) is not confirmed. Often experimental evidence for the existence of 'saddle point' electrons has been retracted later (7). Past efforts have concentrated on emission at angles greater than zero degree. In these cases it has been suggested (4) that the structure observed at $v_p/2$ electron velocity could arise from the interplay between the prominent peaks of ECC at zero degrees and the soft electrons at 90 degrees.

Another reason for the appearance of $v_p/2$ structure could be the role played by the target gas geometry as highlighted earlier (7). Here an artificial enhancement of

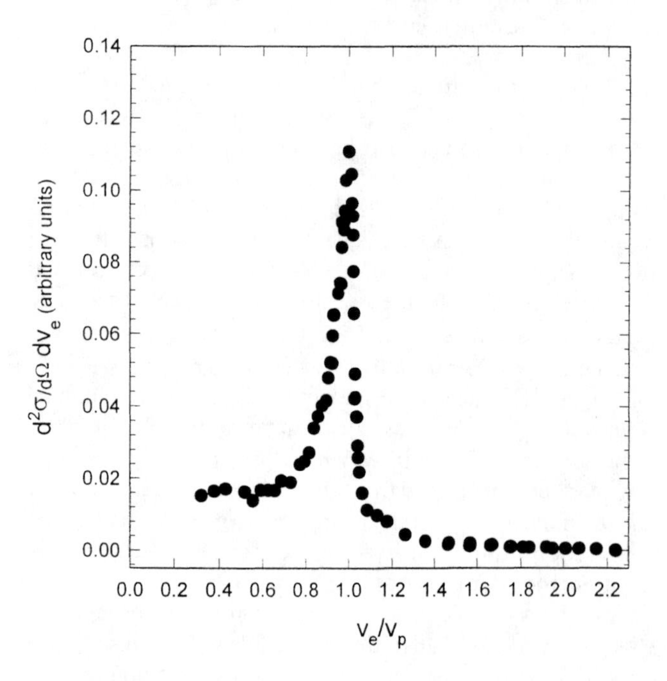

FIGURE 2. Electron emissions at zero degree for 40 keV H^+ ion impact in H_2.

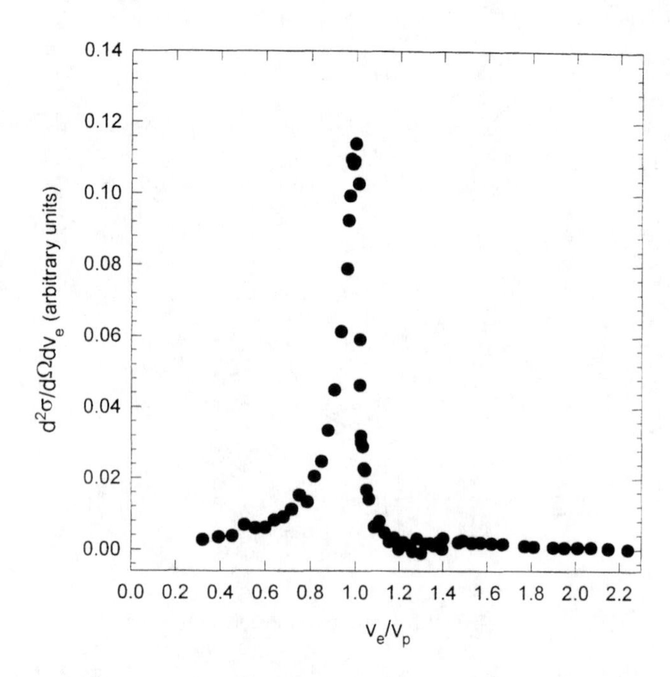

FIGURE 3. Electron emissions at zero degree for 40 keV H^+ io impact in He.

the low energy region of the spectrum was observed as the target volume increased. In our initial experimental arrangement we had used a crossed beam set up in which target beam prepared in second differentially pumped chamber was crossed by the projectile beam. Then the only gas load introduced in the main chamber was that due to the dumping of the target beam. The ensuing pressure rise in the main chamber and the spectrometer was such that we also found similar enhancements of low energy electrons and in some runs small structures appeared at electron velocities approximately $v_p/2$. Our final closed target cell arrangement together with the differentially pumped regions ensures that the target gas is well confined and the measured spectra do not show any evidence of enhanced contributions at low energy. The spectra at zero degrees for 40keV H^+ impact energy are reproducible and do not show the presence of the saddle-point electrons.

The soft electron contribution ($v_e \sim 0$) were not recorded in the present experiment because our spectrometer could not measure electrons with energies below 2eV with a good degree of confidence. However, it is abundantly clear from the present spectra that most of the contribution to electron emission at zero degree comes from the ECC peak.

ACKNOWLEDGEMENTS

This work is part of a large programme supported by the U.K. Engineering and Physical Sciences Council. One of us (C. McGrath) is also indebted to the Department of

Education, Northern Ireland for the award of a Research Studentship.

REFERENCES

1 Stolterfoht N, DuBois RD and Riwarola R, Springer Series on Atoms and Plasmons (Springer, Heidelberg), (1997).
2 Suarez S, Two-centre effects in ion-atom collisions, AIP Conference Proceedings, ed. by Gay TJ and Starace AF, Vol362, p29, (1996).
3 Richard P, Two-centre effects in ion-atom collisions, AIP Conference Proceedings, ed. by Gay TJ and Starace AF, Vol362, p69, (1996).
4 Olson RE, Phys Rev, A27, p1871, (1983).
5 McCartney M, Phys Rev, A52, p1213, (1995).
6 Gay TJ, Two-centre effects in ion-atom collisions, AIP Conference Proceedings, ed. by Gay TJ and Starace AF, Vol362, p19, (1996).
7 Suarez S, Garibotti C, Bernardi G, Focke P and Meckback W, Phys Rev A48, p4339, (1993).
8 Bernardi G, Suarez S, Fainstein PD, Garibotti CR, Meckback W and Focke P, Phys Rev A40, p6863, (1989).
9 Robdro M and Andersen FD, J. Phys. B: At. Mol. Phys., Vol 12, p2883, (1979).
10 Pregliasco RG, Garibotti CR and Barrachina RO, J. Phys. B: At. Mol. Opt. Phys., Vol 27, p1151, (1994).
11 Gulyas L, Sarkadi L, Palinkas J, Kover A, Vajnai T, Szabo Gy, Vegh J, Beneyi D and Elston SB, Phys Rev A45, p4535, (1992).
12 Dahl P, J. Phys. B: At Mol. Phys, Vol 18, p1181, (1985).
13 Gibson DK and Reid ID, J. Phys. B: At Mol. Phys, Vol 19, p3265, (1986).
14 Andersen LH, Jensen KE and Knudsen H, J. Phys. B: At Mol. Phys, Vol 19, pL161, (1986).

Solid State Effects in Binary Encounter Electron Emission by Swift Heavy Ions (13.6 - 77 MeV/u)

H. Rothard[1], G. Lanzanò[2], E. De Filippo[2], D. Mahboub[3], A. Pagano[2]

[1]*CIRIL (Centre Interdisciplinaire de Recherches Ions Laser, CEA-CNRS-ISMRa), BP 5133, Rue Claude Bloch, F-14070 Caen Cedex 05, France*
[2]*INFN, Sez. di Catania and Dipartimento di Fisica, Corso Italia 57, I-95129 Catania, Italy*
[3]*INFN-LNS and Dipartimento di Fisica, Corso Italia 57, I-95129 Catania, Catania, Italy*

We report on binary encounter electron ejection from thin solid foils (C, Al) by highly charged heavy ions ($Ar^{17+,18+}$) at high energies (13.6 - 77 MeV/u). For the collision systems 13.6 MeV/u Ar^{17+} on C and 77 MeV/u Ar^{18+} on Al, comparison has been made with a recent relativistic theory. This theory, based on the Electron Impact Approximation, describes the shape of the binary encounter electron peak and the angular dependence of the cross section well for thin targets only. For thicker targets, solid state effects (in particular, electron transport) influence the shape of the binary encounter electron peak initially given by the target atom Compton profile. New experiments were recently performed with Ni^{28+} (45 MeV/u).

INTRODUCTION

Electron Ejection is a fundamental consequence of energetic atomic collisions and an important probe for atomic collisions with swift projectiles such as ions, atoms and clusters. Experimental data (in particular differential electron spectra and angular distributions) are an important test for the theoretical description of ionization [1]. Furthermore, radiation effects due to electronic energy deposition by swift particles in condensed matter are closely related to ionization and the subsequent electron transport and secondary electron cascade multiplication. The deposited energy is distributed along and around the ion track, and after relaxation and thermalization, the energy initially deposited in electronic excitation may finally result in creation of defects, nuclear tracks, damage and even modification of material properties [1,2].

Specific effects can be observed with heavy ions due to the high charge involved and resulting high ionization cross sections, strong induced perturbation and large electronic energy loss. The "simplest" ionization mechanism is a two body collision between projectile and target electrons [3]. The corresponding so-called "binary encounter electrons" (BEE) should be ejected with an emission angle θ dependent velocity of

$$v_{BE} = 2\, v_p \cos(\theta) \qquad (1)$$

if the interaction with the target nucleus, binding energy and relativistic effects are neglected (v_p denotes the projectile velocity, at relativistic velocities, the momentum $p = \gamma v$ with $\gamma = (1-(v/c)^2)^{-1/2}$ is the relevant quantity). Since electrons are bound to the target nucleus in different shells, the observed distribution of BEE at fixed angle is not a δ-function as described by equation (1), but a distribution centred near v_{BE} which reflects the initial one dimensional momentum distribution of the bound electrons of the target ("Compton profile").

BEE spectra in ion-atom collisions were extensively studied at rather low impact velocities and interesting effects have been observed concerning the influence of the projectile on the shape, position and intensity of the BEE Peak. For example, enhancements of the emission intensity for partially stripped ions compared to bare ions and a splitting of the peak at certain emission angles were observed [3,4]. These results were explained in terms of elastic scattering of target electrons in a screened projectile potential which may lead to quantum mechanical diffraction structures in the angular distribution (Ramsauer-Townsend effect, recently reviewed by Lucas et al [4]). We refer the reader to recent reviews of electron ejection in swift ion-atom collisions [3,4] for a complete literature survey. Electron emission from solid targets by swift ions is subject of ref. [5].

CP475, *Applications of Accelerators in Research and Industry*, edited by J. L. Duggan and I. L. Morgan

Recently, BEE emission at higher velocities induced by Ar ions was studied experimentally with thin solid foil targets at projectile energies of 13.6 MeV/u by Rothard, Jakubassa-Amundsen and Billebaud [6], at 35 and 95 MeV/u by De Paola et al. [7], at 77 MeV/u by Lanzano et al. [8] and at 400 MeV/u by Azuma et al. [9]. A relativistic theory based on the electron impact approximation (EIA) has been developed by Jakubassa-Amundsen [10].

BEE FROM THIN FOILS: COMPARISON OF THEORY AND EXPERIMENT

Let us briefly summarize the relativistic EIA as described in detail by Jakubassa-Amundsen [10]. In this quasielastic scattering approximation, ionization takes place via electron transfer to the projectile continuum. The active target electron scatters elastically from the projectile field, and the corresponding cross section $d\sigma_e^{rel}(k,\theta)/d\Omega$ is then folded with the electron's momentum distribution $\varphi_i^T(\mathbf{q})$ in its initial state. In the numerical calculations, Hartree-Fock bound state wave functions and experimental atomic binding energies were used. The explicit formulae are given in ref. [10].

We compare the result of the calculation (dotted lines) to experiment in Fig. 1, which shows experimental spectra from thin Al foils taken in forward direction (in the direction of the beam, $\theta = 0°$, at 13.6 MeV/u, and close to the beam direction, $\theta = 3°$ at 77 MeV/u). The spectra at 13.6 MeV/u (also the ones of fig. 3) were measured by means of a magnetic spectrometer [6], the spectra at 77 MeV/u by a time of flight method [8] with the multidetector ARGOS. Both experiments were performed at GANIL (Caen).

The forward spectra for 13.6 MeV/u projectiles shows the BEE peak at about 30 keV. The BEE peak width is well described by theory at 13.6 MeV/u, but the experimental spectrum is broader than theory at 77 MeV/u. Note the quite important shift of the experimental BEE peak to lower energies (from approximately 170 down to 130 keV) at 77 MeV/u. The origin of this discrepancy is still unclear [8].

The energy integrated single differential cross sections are plotted as a function of the emission angle in Fig. 2 for Ar^{q+} projectiles and C or Al targets (as indicated). We also included experimental data at 95 MeV/u from DePaola et al. [7] and the result of the EIA calculation (dotted line). At 13.6 MeV/u (C target) and 77 MeV/u (Al target), experimental and EIA SDCS agree well for all emission angles. At 95 MeV/u, the general experimental increase of the SDCS with θ is fairly well reproduced by theory, but strong deviations occur at $\theta = 0°$ at 95 MeV/u. See refs. [6-8] for a detailed discussion of these findings.

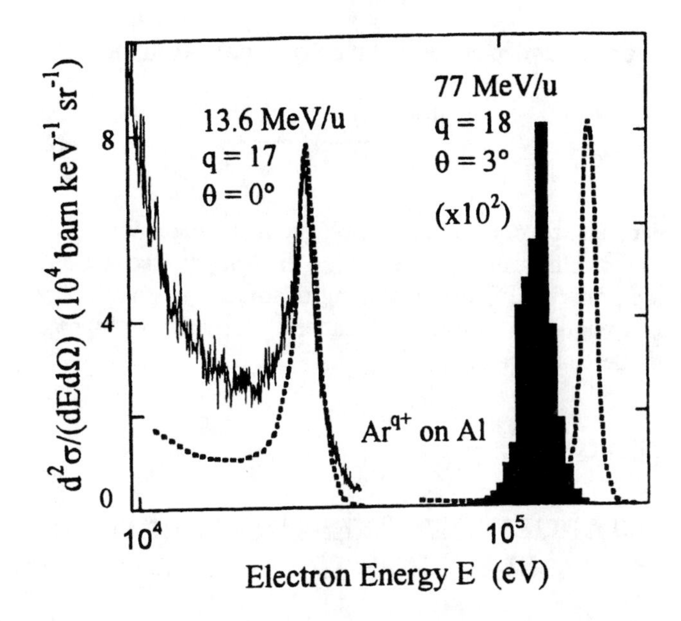

FIGURE 1: Electron spectra from the collision system Ar on Al foil. Comparison of the EIA calculation [10] (dotted lines) to experimental spectra at 13.6 MeV/u (thickness 33 μg/cm², from ref. [6]) and at 77 MeV/u (thickness 90 μg/cm², from Lanzanò et al. [8]). The double differential cross sections $d^2\sigma/(dEd\Omega)$ (lin. scale) are plotted as a function of the electron energy (log. scale).

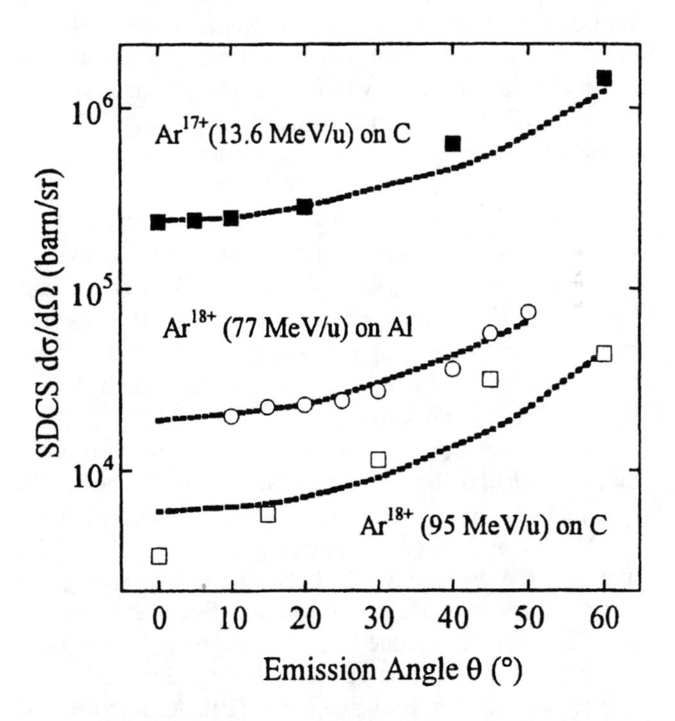

FIGURE 2: Angular dependence of the single differential cross section for the indicated collision systems. Included are data taken at the French heavy ion accelerator GANIL at 13.6 MeV/u (C, 4.4 μg/cm²) [6] and 77 MeV/u u (Al, 90 μg/cm²) [8]. The data at 95 MeV/u (C, 80 μg/cm²) are from DePaola et al [7]. Dotted lines: EIA calculation [10].

The angular dependence of the SDCS roughly follows the

$$\frac{d\sigma_{BE}(\theta)}{d\Omega} \sim \frac{1}{\cos^3\theta} \qquad (2)$$

dependence one would expect from a simple two-body Rutherford scattering between a free target electron and the projectile nucleus including relativistic kinematics but non relativistic Rutherford scattering formula as proposed by DePaola et al. [7].

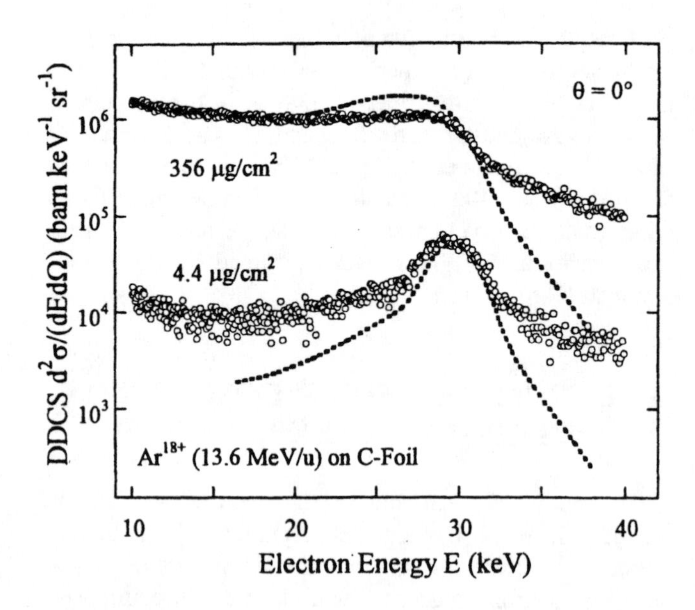

FIGURE 3: Double differential electron spectra $d^2\sigma/(dEd\Omega)$ as a function of electron energy E measured with a very thin C foil (4.4 μg/cm²) and a "thick" foil (356 μg/cm²) in forward (beam) direction (θ = 0°) under perpendicular impact of Ar^{18+} at 13.6 MeV/u [6]. Dotted line: EIA and transport calculation [14].

TARGET THICKNESS DEPENDENCE: ELECTRON TRANSPORT

As can already be seen from fig. 1, the shape of BEE spectra from solid targets may differ from calculations made for collisions with single atoms. In particular, a peak broadening is observed which can be attributed to electron transport (secondary scattering of the electrons on their way to the surface towards the detector) within the foil target [6,11,12]. On the other hand, if the initial peak shape for very thin foils is well described by theory, studies as a function of target thickness allow to observe the evolution of electron emission from single collisions (as in atomic collisions with low density gas targets) up to multiple collisions (as in the bulk of solids, where electron transport phenomena become important).

Fig. 3 shows 0° spectra for a very thin C foil (4.4 μg/cm²) and a "thick" foil (356 μg/cm²). In terms of the inelastic mean free path λ for fast electrons in solids, the thickness of the thin foil is about 0.7 λ, and that of the thick foil 250 λ. We mention that the inelastic electron mean free path can be approximated by: $\lambda(E) = A\,E^n$ with approximately A = 0.1 and n = 0.81 for carbon, if E is measured in eV and λ in Å [13].

We observe how the overall intensity increases. Note that in particular the low energy wing is oncreased for the 356 μg/cm² target. The electrons from deep within the solid suffer both angular scattering and energy loss. This broadens the distribution (angular and energy straggling) and shifts the maximum to lower energies. The change of the BEE peak shape due to such effects can be described by transport theory. The initial peak shape for thin targets is given by the EIA as shown above [10]. At high electron energies, the SELAS approximation, i.e. the separation of energy loss and angular scattering, can be applied [11]. The result of such a treatment of electron transport in solids [14] is shown in Fig. 3 (dotted lines). The observed peak shape is well reproduced by theory [14].

CONCLUSION

BEE from thin foils probe the very first consequence of ion-solid interaction, that is the event of primary ionization. Already at this stage, "high charge effects" (deviations from first order theory, multiple ionization) and the complex dynamics of electrons within the combined (screened non-Coulomb) projectile-target potential render this phenomenon complicated [3]. Nevertheless, the relativistic EIA theory describes the shape and the angular dependence well for thin targets both at 13.6 MeV/u and 77 MeV/u [6,8].

For thicker targets, solid state effects (electron transport) influence the shape of the binary encounter electron peak initially given by the target atom Compton profile. Such effects have to be taken into account in atomic collision experiments even if single collision conditions for the primary interaction (e.g. ionization) are fulfilled. Not only the electronic structure may be different for free atoms and atoms bound in solids, but also the observable spectra of e.g. electrons may be different from the primary ones due to transport effects in condensed matter [6,11-13].

From an experimental physicists point of view, for BEE emission studies at high energies, the application of multidetector arrays such as ARGOS as used up to now only in nuclear physics experiments [8] allows the simultaneous measurement of fast electron velocity spectra in the entire angular range. This allows to study BEE emission in a very efficient way. Even multiple coincidences (electrons, recoiled atoms, scattered projectiles and fragments) become possible [8]. Also, absolute quantitative measurements of cross sections (double or single differential, DDCS or SDCS) are now possible. A preliminary analysis of a recent experiment performed at INFN-LNS (Catania) yields a SDCS of 65 kb/sr for Ni^{28+} (45 MeV/u) on C at an emission angle of 6°. This value can be compared to the calculated SDCS values of fig. 2.

ACKNOWLEDGEMENTS

We would like to thank Doris H. Jakubassa-Amundsen for the excellent collaboration and fruitful discussions.

REFERENCES

1. *Ionization of Solids by Heavy particles,* Ed. R.A. Baragiola, Ed., Plenum, New York , 1993, NATO ASI Series **B306**.
2. *Proc. of the 3rd Int'l Conf. on Swift Heavy Ions in Matter SHIM-95*, Eds. N. Angert, A. Bourret and J.P. Grandin, Nucl. Instrum. Meth. **B107** (1996).
3. N. Stolterfoht, R.D. Dubois and R.D. Rivarola, *Electron Emission in Heavy Ion-Atom Collisions*, Springer Series on Atoms and Plasmas **20** (1997).
4. M.W. Lucas, D.H. Jakubassa-Amundsen, M. Kuzel and K.O. Groeneveld, Int. J. Mod. Phys. **A12**, 305 (1997).
5. H. Rothard, Scanning Microsc. **9**, 1 (1995).
6. H. Rothard, D.H. Jakubassa-Amundsen and A. Billebaud, J. Phys. B: At. Mol. Opt. Phys. **31**, 1563 (1998).
7. B.D. DePaola, Y. Kanai, P. Richard, Y. Nakai, T. Kambara, T.M. Kojima and Y. Awara, J. Phys. B: At. Mol. Opt. Phys. **28**, 4283 (1995).
8. G. Lanzanò, E. DeFilippo, S. Aiello, M. Geraci, A. Pagano, Sl. Cavallaro, F. LoPiano, E.C. Pollacco, C. Volant, S. Vuillier, C. Beck, D. Mahboub, R. Nouicer, G. Politi, H. Rothard and D.H. Jakubassa-Amundsen, Phys. Rev. **A58** (1998).
9. T. Azuma, T. Ito, K. Komaki, T. Tonuma, M. Sano, A. Kitagawa, E. Takada and H. Tawara, Nucl. Instrum. Meth. **B132**, 245 (1997).
10. D.H. Jakubassa-Amundsen, J. Phys. B.: At. Mol. Opt. Phys. **30**, 365 (1997).
11. G. Schiwietz, J.P. Biersack, D. Schneider, N. Stolterfoht, D. Fink, V.J. Montemajor and B. Skogvall, Phys. Rev. **B41**, 6262 (1990).
12. R.A. Sparrow, R.E.Olson and D. Schneider, J. Phys. B28, 3427 (1995).
13. M. Jung, H. Rothard, B. Gervais, J.P. Grandin, A. Clouvas, and R. Wünsch, Phys. Rev. **A54**, 4153 (1996).
14. D.H. Jakubassa-Amundsen and H. Rothard, submitted to Phys. Rev. **B** (1998).

Double Electron Knockout to Study Electron Correlation

J. H. Moore, M. A. Coplan, J. W. Cooper

University of Maryland, College Park, Maryland 20742

J. P. Doering, B. El Marji

Johns Hopkins University, Baltimore, Maryland 21218

Electron-impact double ionization probes the role of multi-electron interactions in the dynamics of electron scattering, as well as the nature of electron correlation in a target atom. Electron impact double ionization of magnesium is being studied with the ultimate goal of measuring the pairwise joint momentum distribution of atomic electrons. A multi-detector apparatus permits 120 triple-coincidence measurements to be performed simultaneously. (e,3e) and (e,(3-1)e) experiments have been carried out at incident-electron energies of 400 to 3500 eV with ejected-electron energies between 35 and 100 eV. Resonant (Auger) and nonresonant (direct) double ionization have been studied. Analysis of the data in terms of the net momentum of the ejected electrons and the momentum of the residual doubly-charged magnesium ion has revealed the existence of underlying symmetries during the course of the collision and unexpectedly large residual ion momentum. Current models of double ionization fail to account for many of the experimental observations.

For most of this century, the independent-electron model has been the basis of the descriptions of atomic electronic structure and the interaction of electrons with atoms. As many branches of physics focus upon the so-called many-body problem, atomic physicists are considering the range of validity of the independent-particle model and are searching for a physical picture of the correlated motion of atomic electrons. The corresponding experiments involve the investigation of phenomena, such as multiple ionization, that are inherently the consequence of the collective or cooperative motion of atomic electrons. Conceptually, the most straight-forward experiment is electron-impact double ionization.

The use of knockout experiments to investigate structure has its origins in nuclear physics and, in deference, electron-impact double ionization is referred to as an (e,3e) experiment. If two electrons could be ejected from an atom by electron impact in a billiard-ball-like collision, the relative momentum distribution of the targeted pair could easily be determined from a measurement of the momenta of the incident and the three outgoing particles. The relative momentum distribution of a pair of atomic electrons describes completely their correlated motion. At present it is not known how to carry out an experiment in which the effects of the scattering dynamics can be separated from those of the structure of the target. Most theory and all the experiments have been directed towards a better understanding of the double ionization process.

In a complete (e,3e) experiment, the energies and directions of the incident, scattered, and ejected electrons are measured with the three outgoing electrons detected in (triple) coincidence. For a given incident energy, there are eight independent variables (typically two angles for each outgoing electron and the energy of two of the outgoing electrons with the third determined by energy conservation); the measured cross section is eightfold differential. The cross section for double ionization is very small so data are accumulated slowly. Only a small portion of the eight-parameter space can be investigated with any one apparatus. At present, the situation is akin to the proverbial blind men's investigation of the elephant.

(e,3e) processes can be roughly divided into two regimes according to the momentum transferred by the incident electron to the target. Photoionization is mimicked by high-energy, low-momentum-transfer collisions in which the incident electron is scattered through a small angle. This is the dipole regime to be contrasted with the regime of high-momentum-transfer collisions that are suggestive of knockout processes. The first (e,3e) measurements were carried out in the dipole regime on rare gas atoms with incident-electron energies of about 5 keV, scattering angles less than 0.5°, and ejected-electron energies of a few eV. (1) We have chosen to investigate relatively higher momentum-transfer collisions in which the incident- and ejected-electron energies are sufficiently large to minimize final-state

CP475, *Applications of Accelerators in Research and Industry*,
edited by J. L. Duggan and I. L. Morgan

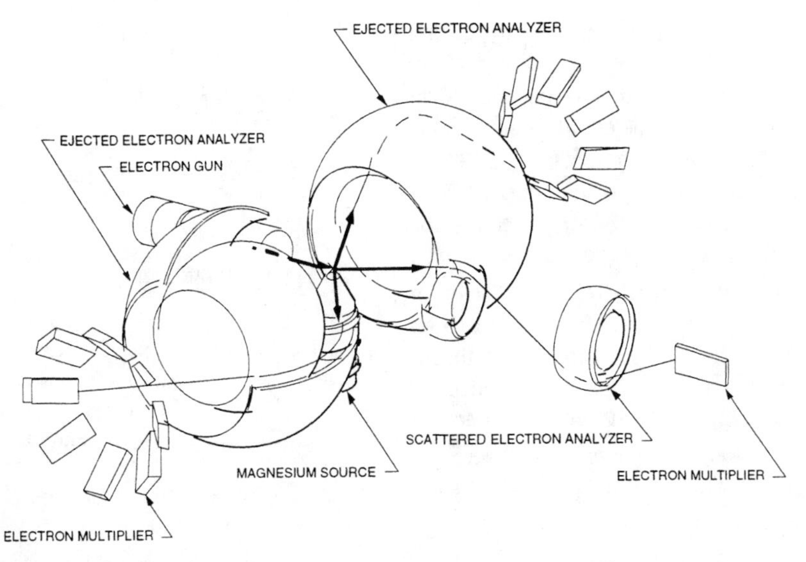

Figure 1. Electron-impact double-ionization spectrograph. Shown are the electron gun, magnesium source, the three electrostatic energy analyzers, and the electron detectors. The scattering angle is 18°.

coulomb interactions. The target has been magnesium, a quasi-two-electron atom in which the motion of the two (outer) *3s*-electrons is believed to be highly correlated

The (e,3e) cross section is small in virtually all scattering regimes. The statistical exigencies of triple-coincidence measurements place strict limits on the incident-beam current and target density required for optimum signal-to-noise. As a consequence (e,3e) apparatus are multiplexed devices employing arrays of detectors so that many triple coincidence measurements can be carried out at one time.(2) Our apparatus is illustrated in Fig. 1. Truncated spherical electrostatic analyzers establish the energies of the ejected and scattered electrons that are detected. A single electron multiplier detects incident electrons scattered at 18°. There are 16 ejected-electron detectors. As many as 120 (e,3e) measurements can be performed simultaneously.

The mechanism of electron-impact double ionization has been described by a number of models. Important parameters include the momentum transfer vector, $\mathbf{K} = \mathbf{p_o} - \mathbf{p_s}$, where $\mathbf{p_o}$ and $\mathbf{p_s}$ are respectively the incident- and scattered-electron momentum vectors; the net momentum of the ejected pair of electrons, $\mathbf{P} = \mathbf{p_1} + \mathbf{p_2}$, where $\mathbf{p_1}$ and $\mathbf{p_2}$ are the momenta of the ejected electrons; $\mathbf{p_{ion}} = \mathbf{K} - \mathbf{P}$, the momentum of the doubly-charged ion core; and the plane of the ejected pair of electrons defined by $\mathbf{p_1} \times \mathbf{p_2}$. The mechanism of double ionization has been described in terms of single and multiple interactions of the incident electron with the target. Perhaps the most simple is the *composite-particle* model that treats the two target electrons as a single entity that is knocked out with momentum **P**. In this event the distribution in **P** would be expected to be centered about **K**. The *shake-off* model describes another single-hit mechanism in which the incident electron ejects one electron and leaves behind an excited atom that subsequently decays to emit a second electron. In this case one expects $\mathbf{p_1}$ to be aligned with **K** and $\mathbf{p_2}$ to be isotropically distributed.(3) Alternatively the incident particle is considered to interact once to excite the target atom and the correlation between a pair of electrons is responsible for their subsequent simultaneous ejection. (4, 5) This two-electron shake-off model would be expected to be applicable to photoionization or electron impact in the dipole regime. The *two-step* models involve sequential binary interactions in which the incident ejects first one and then another electron or, alternatively the incident electron strikes one target electron that subsequently strikes a second electron. The kinematics of the two-step models is such that at least one and usually both ejected electrons move into the forward hemisphere relative to the incident-electron direction.

In our (e,3e) experiments on magnesium the incident energy has been varied between 422 eV and 1052 eV with ejected electrons detected at energies of 55 eV to 100 eV.(6, 7) We have also carried out (e,(3-1)e) experiments in which the scattered electron goes undetected thus averaging over the momentum-transfer direction.(8) We have also observed resonant double ionization *via* the Mg $L_{2,3}M_1M_1$ Auger process in low-momentum transfer (e,3e) experiments.(9) In general our results provide no obvious support for any of the models described above. Although certain symmetries have been observed in the angular distributions, these are not what was anticipated. In the Auger experiments we observe an angular correlation between the ejected pair but no preference for one electron to move in the momentum-transfer direction. In the high-momentum-transfer experiments the net-ejected-electron-momentum vector **P** shows no strong tendency to align with **K**. Ejected electrons, especially ejected electrons of the same energy, tend to be emitted in diametrically opposite directions (for equal energies this corresponds to **P** = 0). For two ejected electrons at 90° to one another, the double-ionization cross section is large when the plane of the two ejected electrons contains the incident direction (i.e., $\mathbf{p_1} \times \mathbf{p_2}$ perpendicular to $\mathbf{p_o}$) without regard for the direction of **P**. The most striking observation is the relatively large cross section for both electrons to move in the backwards direction. In both our (e,3e) and (e,(3-1)e) geometries it is possible to calculate at least a lower limit for $\mathbf{p_{ion}}$ (= **K** - **P**), the momentum of the residual ion core. Relatively large cross sections are found for collisions for which the magnitude of $\mathbf{p_{ion}}$ exceeds 4 a.u. (10)

The large residual-ion momentum observed in electron-impact double ionization implies either a significant transfer of momentum from the incident electron to the core--an inefficient process given the disparity between the mass of the electron and core--or ejection of the two electrons with large *initial* net momentum, that is, two *3s*-electrons that are near each other and close to the atomic core. If the latter interpretation is correct, double ionization in our kinematic regime may prove to be a valuable tool to investigate electron motion close to the nucleus. In any event it is clear that double ionization is fundamentally different from single ionization and there is a great deal more to be learned.

ACKNOWLEDGMENTS

Research supported by NSF Grant PHY-95-15516.

REFERENCES

1. see for example, C. Schröter, B. El Marji, A. Lahmam-Bennani, A. Duguet, M. Lecas, and L. Spielberger, *J. Phys. B*, **31**, 131-143, 1998; A. Lahmam-Bennani, A. Duguet, A. M. Grisogono, and M. Lecas, *J. Phys. B*, **25**, 2873-2884, 1992.
2. M. J. Ford, J. P. Doering, J. H. Moore and M. A. Coplan, *Rev. Sci. Instr.*, **66**, 3137, 1995.
3. B. El Marji, C. Schröter, A. Duguet, A. Lahmam-Bennani, M. Lecas, and L. Spielberger, *J. Phys. B*, **30**, 3677, 1997.
4. F. W. Byron and C. J. Joachain, *Phys. Rev. A*, **164**, 1, 1967.
5. R. J. Tweed, *J. Phys. B*, **6**, 398, 1973.
6. B. El Marji, J. H. Moore, J. P. Doering, M. A. Coplan, J. W. Cooper, *Bull. Am. Phys. Soc.*, **43**, 1337, 1998.
7. M. J. Ford, J. H. Moore, M. A. Coplan, B. El Marji, J. P. Doering, *Bull. Am. Phys. Soc.*, **42**, 1747, 1997.
8. M. J. Ford, B. El Marji, J. P. Doering, J. H. Moore, M. A. Coplan, J. W. Cooper, *Phys. Rev. A*, **57**, 325-330, 1998.
9. M. J. Ford, J. P. Doering, M. A. Coplan, J. W. Cooper, J. H. Moore, *Phys. Rev. A*, **51**, 418-423, 1995.
10. M. J. Ford, J. H. Moore, M. A. Coplan, J. W. Cooper, J. P. Doering, *Phys. Rev. Lett.*, **77**, 2650-2653, 1996.

Investigation of Multi-Electron Processes in 60 keV O^{6+} + Ar Collisions Using a Triple Coincidence Technique

H. Merabet, H. M. Cakmak[1], A. A. Hasan, E. D. Emmons, T. Osipov[2], R. A. Phaneuf, and R. Ali

Department of Physics, University of Nevada, Reno, NV 89557 USA

Multiple electron capture processes in 60 keV O^{6+} + Ar collision system have been investigated by means of time-of-flight triple-coincidence measurements of Auger electrons, scattered projectile and target ions. This technique is particularly suited for the investigation of electron capture processes involving capture of three or more electrons. From the measured sub-partial Auger spectra, it can be seen that double electron capture dominantly populates the configurations (3,n) with n=3-5. Triple electron capture is found to dominantly populate the configurations (3,3,3) and (3,3,4), while quadruple electron capture populates the (3,3,3,3) configuration.

INTRODUCTION

Studies of multi-electron capture processes in highly charged ion-atom collisions have attracted considerable interest during the last two decades. Significant progress has been made in investigating single and double electron capture processes using the methods of translational energy gain spectroscopy (1-3), recoil ion momentum spectroscopy (4,5), photon spectroscopy (6,7) and Auger electron spectroscopy (8,9). On the contrary, little is known about multi-electron processes. Indeed, from a theoretical point of view, understanding multi-electron processes is a two-fold problem. Firstly, the different mechanisms involved in the collision process that lead to the production of multiply excited states must be recognized and described. Secondly, the radiative and non-radiative properties of the resulting multiply excited states must be known. Moreover, quantum mechanical or semiclassical treatments of collisions involving more than two electrons are prohibitively difficult due to the large number of channels involved.

Although the classical over-barrier model has been extended (10) to account for multiple electron capture, the extended model (ECB) does not take into account the electronic correlation which has been found to play an important role in many cases for double electron capture (11,12). In fact, the ECB model cannot accommodate many of the possible reaction channels such as the simultaneous capture of two or more electrons, and the population of highly Rydberg states (12). This model stops at giving the final capture state distribution on the projectile and possible target excitation. In order to further account for the final collision products, relaxation schemes for the multiply excited states must be invoked.

During the last six years, the Groningen group has made significant contributions (13,14) toward understanding Auger electron spectra obtained in multiple electron capture processes by means of the coincident detection of Auger electrons and target ions. They demonstrated that many of the Auger lines attributed to doubly excited states must have been derived from multiply excited projectile states through successive autoionization processes.

It is important to note that Auger electron spectroscopy in singles mode does not give adequate information on multiple electron capture processes. This is due to the fact that the spectra would contain contributions from doubly, triply, quadruply and possibly quintuply excited states that make the analysis of the spectra rather difficult. However, the partial Auger spectra corresponding to the different target ion charge states are much easier to analyze than singles spectra. One should also realize that these spectra can be further simplified if the final projectile charge state is also determined.

In the present paper we report triple-coincidence measurements of Auger electrons, scattered projectile, and target recoil ions in 60 keV O^{6+} + Ar collisions. Such measurements provide sub-partial Auger spectra corresponding to specific final projectile and recoil ion charge states.

EXPERIMENTAL SETUP

The measurements were performed at the multicharged ion research facility at the University of Nevada, Reno (UNR). The 60 keV O^{6+} ion beam was provided by the UNR 14 GHz electron cyclotron resonance (ECR) ion source, and guided to the collision chamber where it crossed a supersonic Ar gas jet at 90°. The jet furnished a well-localized target with an effective target density of about 0.1 m Torr. After the collision, the target recoil ions were

---

[1]Present address: Department of Physics, Celal Bayar University, Manisa, Turkey.

[2]Present address: Department of Physics, Kansas State University, Manhattan, KS 66506.

CP475, *Applications of Accelerators in Research and Industry*, edited by J. L. Duggan and I. L. Morgan

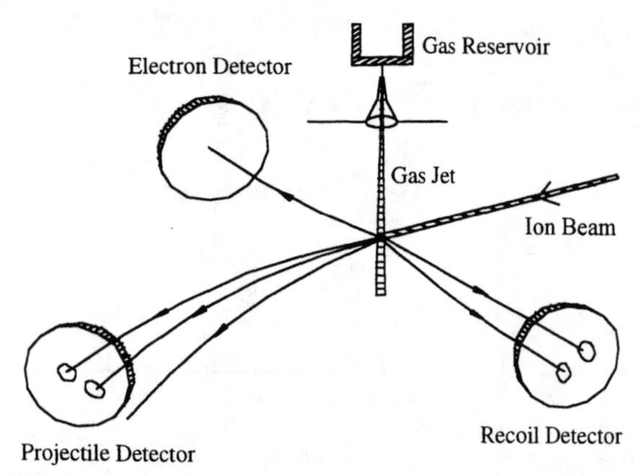

FIGURE 1: A schematic of the experimental setup.

extracted transversely to the ion beam direction by a uniform electric field (≈ 10 V/cm), traveled through a time-of- flight (TOF) spectrometer and were then detected by a microchannel plate detector. The charge exchanged projectiles were charge analyzed downstream from the collision chamber by a parallel-plate electrostatic deflector and detected by a two-dimensional position sensitive microchannel plate detector (2D-PSD) (Fig. 1).

Coincident TOF measurements provided the recoil ion charge states while the impact positions on the projectile 2D-PSD provided the final projectile charge states. Auger electrons ejected at 90° relative to the incident beam traveled through a TOF electron spectrometer, located opposite to the recoil ion TOF spectrometer with its entrance aperture close to the interaction region, and were detected by another microchannel plate detector. TOF coincidence measurements of the electrons and the charge exchanged projectiles provided the TOF of the electrons (Fig. 1).

The collision chamber was differentially pumped such that high ion beam purity was maintained prior to entering the chamber and after exiting it and entering the electrostatic analyzer. Typical pressures of about 5×10^{-9} and 1×10^{-8} Torr were maintained up- and downstream of the collision chamber, respectively, while the collision chamber residual pressure was about 1.2×10^{-7} Torr during the experiment. Double collisions were estimated to be less than 0.5 %. A fast timing signal derived from the electron detector was used to start a time-to-digital converter (TDC) which was stopped by a signal from the projectile detector. Another TDC was started by a fast signal from the recoil detector and stopped by the projectile signal. Analog-to-digital (ADC) converters read the position signals from the projectile detector. The time and position information were then read by a computer and stored in list mode for further processing. The true triple-coincidence rate was about 0.65 Hz for a primary ion beam current of 10 pA.

RESULTS AND DISCUSSION

Fig. 2(a) is a scatter plot representing coincidences between recoil ions and Auger electrons or photons. Since the electron detector views the interaction region, any photons with energy higher than about 12 eV that are emitted toward the detector will be detected. The projection of Fig. 2(a) onto the vertical axis gives the recoil ion TOF spectrum shown in Fig. 2(c). The equivalent of a singles Auger electron spectrum results from the projection of Fig. 2(a) onto the horizontal axis as shown in Fig. 2(b).

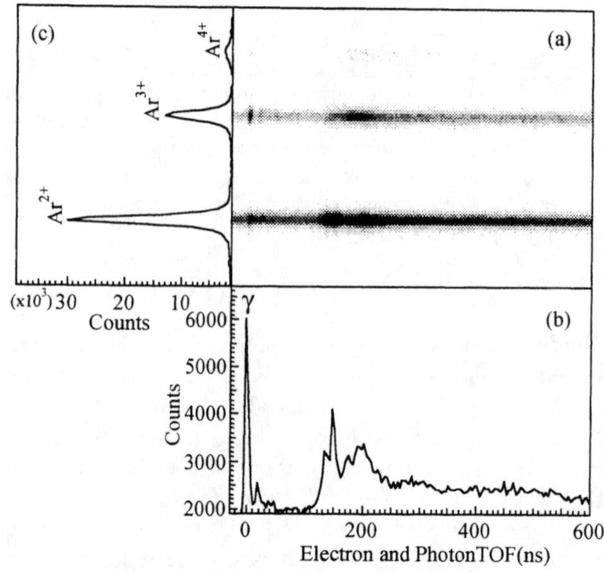

FIGURE 2: (a) A scatter plot representing coincidences between recoil ions and Auger electrons or photons, (b) singles Auger electron/ photon spectrum (the label γ indicates photons), and (c) recoil ion TOF spectrum.

It is evident from Fig. 2(a) that the singles Auger electron spectrum resulted from processes involving loss of two to four electrons by Ar, thus rendering the interpretation of the singles electron spectrum a difficult task. However, partial Auger electron spectra specific to each recoil ion charge state can make the analysis much easier. Such spectra are obtained by placing 2D-windows corresponding to the different recoil ion TOF in Fig. 2(a), and projecting the events within each window onto the horizontal axis. Partial spectra so obtained are shown in Fig. 3. Clearly, the partial spectra present noticeable differences. Fig. 3(a) shows the partial Auger electron spectrum corresponding to double electron capture. The Auger line identification is carried out using the Hartree-Fock code by Cowan (15). According to the ECB model, the reaction window associated with capturing the two outermost electrons overlaps the configurations with principal quantum numbers (3,4) and (3,5) and therefore should be dominantly populated. Their subsequent autoionization gives rise to *L*-Auger lines. Indeed, we observe appreciable populations of these two configurations. There is, however, a substantial population of the (3,3) configuration, which can be accounted for within the framework of the ECB model only if the assumption of double electron capture accompanied by target excitation is invoked. Such a process is a three-electron process whereby the three outermost electrons are assumed to have been molecularized. The projectile then captures the third and either the first or the second electrons, while the residual target ion recaptures the other

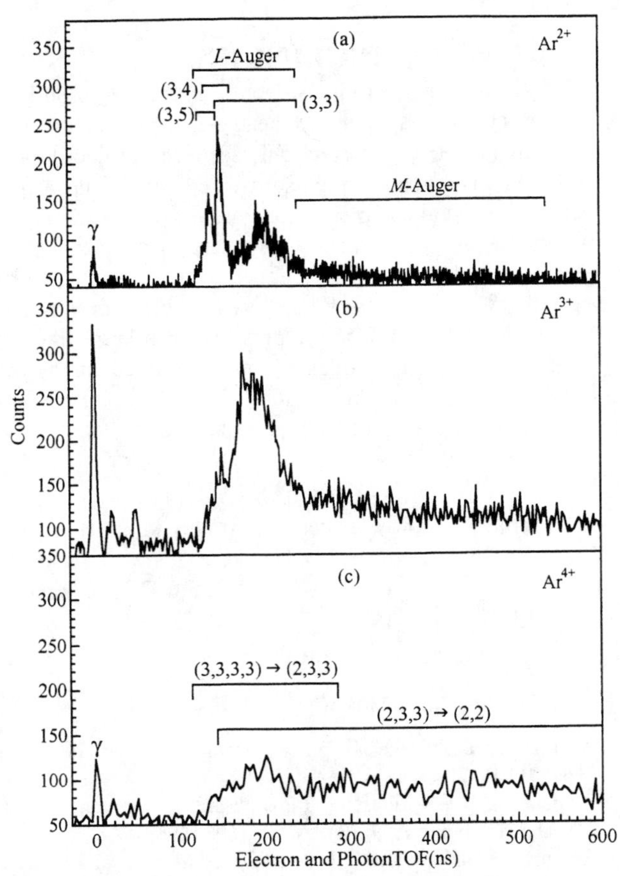

FIGURE 3: Partial Auger electron/photon spectra corresponding to the different recoil ion charge states.

electron into an excited level. de Nijs *et al.* (14) argued in favor of such a process in the case of the 60 keV C^{6+} + Ar collision system. While this assumption explains the presence of Auger lines derived from the (3,3) configuration, a comparison of the ECB predicted relative intensities of the (3,3) configuration to the (3,4) and (3,5) configurations with the experimentally determined ones shows that the ECB model underestimates the importance of the (3,3) configuration by about a factor of four.

In addition to the *L*-Auger lines, we also observe weak *M*-Auger lines in Fig. 3(a), an indication that the configurations (4,n) with n ≥ 5 are weakly populated. In general, our findings concerning double electron capture processes for this collision system are similar to those of de Nijs *et al.* (14) for the C^{6+} + Ar collision system. It seems that the electron capture process is dominated by the incoming projectile charge state in both systems, while the projectile core effect is negligible; at least in the case of O^{6+}.

The partial Auger spectrum corresponding to triple electron capture is shown in Fig. 3(b). Inspection of Fig. 4, however, shows that the triply ionized recoil ions are found in coincidence with projectiles that changed their charge state by one or two units. Hence, the partial Auger spectrum is still a composite one, and therefore can be further reduced according to the final projectile charge states. This

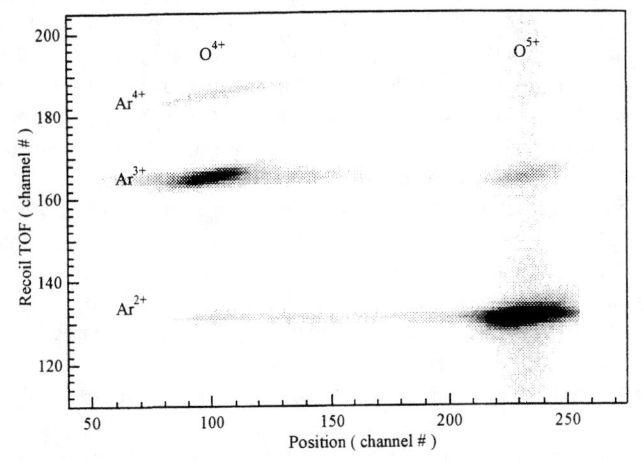

FIGURE 4: A scatter plot representing coincidences between scattered projectiles and recoil ions.

reduction results in the sub-partial Auger spectra shown in Fig. 5. These sub-partial Auger spectra provide further insights into the population and relaxation pathways of triply excited states. It is important to note here that the small concentration of events in Fig. 4 corresponding to the combination (Ar^{2+}, O^{4+}) represents true double electron capture events, where both electrons stabilize radiatively, and are then found in coincidence with photons only.

The ECB model predicts a reaction window for the capture of the three outermost electrons that overlaps the configurations (3,3,3), (3,3,4), and (3,4,4). The autoionization of the first two configurations results in singly excited daughter configurations, thus leading to the retention of two electrons by the projectile. Auger lines derived from these configurations should then be found in the sub-partial spectrum corresponding to the (Ar^{3+},O^{4+}) combination. Substantial population of these configurations can indeed be inferred from Fig. 5(a). The quasi-continuum electron distribution beyond about 260 ns can be accounted for using the ECB model only if the triple electron capture is assumed to be accompanied by target excitation. The capture may then proceed into the configurations (2,3,3) and (2,3,4) resulting in the observed distribution.

On the other hand, the configuration (3,4,4) is expected to predominantly autoionize to the (3,3) doubly excited configuration, which in turn autoionizes, resulting in the retention of one electron only by the projectile. Indeed, the sub-partial spectrum, shown in Fig. 5(b), corresponding to the (Ar^{3+},O^{5+}) combination exhibits Auger lines resulting from the two autoionization steps. Worthy of particular notice is the appearance of a *K*-Auger line in the (Ar^{3+},O^{5+}) sub-partial spectrum. This is attributed to the presence of a small metastable (1s2s) component in the O^{6+} ion beam. One may argue that the relative ratio of the (Ar^{3+},O^{5+}) to the (Ar^{3+},O^{4+}) events is enhanced due to the metastable component, and that the Auger lines in the (Ar^{3+},O^{5+}) channel are not necessarily derived from the initial population of the (3,4,4) configuration. However, examination of Fig. 3(b) shows that *K*-Auger electrons constitute only about 1.5% of the total electrons detected in coincidence with Ar^{3+}, whereas a separate measurement of the recoil ion charge state fractions in coincidence with the final projectile charge states yielded an (Ar^{3+},O^{5+}) fraction of 25% of the total events leading to the production of triply ionized recoils. A metastable component of a few percent can not account for this fraction. In fact, even assuming a 5% metastable component, we still infer about a 20% fractional population for the (3,4,4) configuration. This calls into question the conclusion by Nakamura *et al.* (16), who also studied the 60 keV O^{6+} + Ar collision system using high resolution Auger electron spectroscopy in singles mode, that the (3,3) doubly excited states are not affected by Auger cascades from triply excited states.

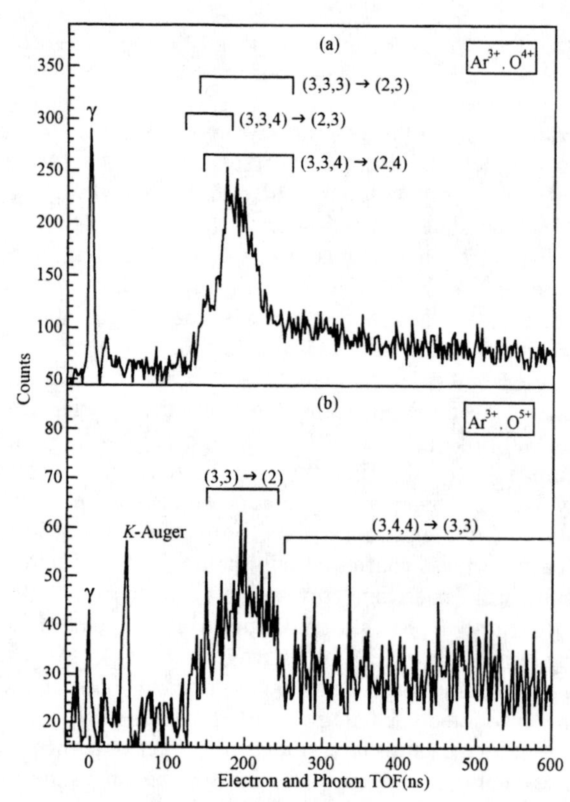

FIGURE 5: Sub-partial Auger electron spectra corresponding to triply ionized recoil ions and projectiles that retained (a) two electrons and (b) one electron.

Quadruple electron capture is predicted by the ECB model to proceed into the (2,3,3,3) and the (3,3,3,3) configurations. Disregarding the metastable ion beam component, the former configuration autoionizes once leading to the retention of three electrons by the projectile. We did not attempt to observe such events. The latter configuration autoionizes to the (2,3,3) continuum limits. The resulting configurations then autoionize to the (2,2) limits, resulting in the retention of two electrons by the projectile. Fig. 4 shows that Ar^{4+} recoil ions are found in coincidence with O^{4+} only. Furthermore, the partial Auger spectrum corresponding to quadruply ionized recoils, shown in Fig. 3(c), does indeed contain Auger lines consistent with the initial population of the (3,3,3,3) configuration .

CONCLUSIONS

We have investigated multi-electron processes in 60 keV O^{6+} + Ar collision system by means of triple-coincidence measurements of Auger electrons, scattered projectile and recoil ions. The measurements provided sub-partial Auger spectra corresponding to specific final projectile and recoil ion charge states. Such spectra provide better understanding of multi-electron processes. The experimental findings were compared with the predictions of the ECB model. While the presence of a large fraction of the observed Auger lines is well accounted for by the model, the predictions concerning the relative intensities are not in good agreement with the experiment.

AKNOWLEDGMENT

H. Merabet acknowledges the travel support of the ACSPECT Corporation, Reno, Nevada. This work has been supported in part by the Nevada NSF and DOE Chemical Physics EPSCoR programs.

REFERENCES

1. Schmeisser, C., Cocke, C. L., and Mann, R., *Phys. Rev. A* **30** 1661 (1984).
2. Andersson, R.L., Pederson, J.P.O., Bárany, A., Bangsgaard, J.P., and Hvelpund, P., *J. Phys. B* **22**, 1603 (1989).
3. Roncin, P., Gaboriaud, M. N., and Barat, M. , *Europhys. Lett.* **16**, 551 (1991).
4. Cassimi, A., Duponchel, S., Flechard, X., Jardin, P., Sortais, P., Hennecart, D., and Olson, R. E., *Phys. Rev. Lett.* **76**, 3679 (1996).
5. Abdallah, M. A., Wolff, W., Wolf, H. E., Sidky, E., Kamber, E. Y., Stöckli, M., Lin C. D., and Cocke, C. L., *Phys. Rev. A.* **57**, 4373 (1998).
6. Bliman, S., Bonnet, J.J., Hitz, D., Ludcec, T., Druetta, M., and Mayo, M., *Nucl. Instrum. Methods Phys. Res. B* **27**, 579 (1987).
7. Martin, S., Bernard, J., Chen, Li , Denis, A. , and Désesquelles, J., *Phys. Rev. A* **52**, 1218 (1995).
8. Holt, R.A., Prior, M.H., Randall, K.L., Hutton, R., McDonald, J., and Schneider, D., *Phys. Rev. A* **43**, 607 (1991).
9. Stolterfoht, N., Sommer, K., Swenson, J.K., Havener, C.C., and Meyer, F.W., *Phys. Rev. A* **42**, 5396 (1990).
10. Niehaus, A., *J. Phys. B* **19**, 2925 (1986).
11. Stolterfoht, N., *Phys. Scr.* **T46**, 22 (1993); **T51**, 39 (1994).
12. Chesnel, J.-Y., Merabet, H., Frémont, F., Cremer, G., Husson, X., Lecler, D., Reiger. G., Speiler, A., Grether, M., and Stolterfoht, N., *Phys. Rev. A* **53**, 4198 (1996).
13. Posthumus, J. H., and Morgenstern, R., *Phys. Rev. Lett.* **68**, 1315 (1992).
14. de Nijs, G., Hoekstra, R., and Morgenstern, R., *J. Phys. B* **29**, 6143 (1996).
15. Cowan, R. D., *The Theory of Atomic Structure and Spectra* (University of California Press, 1981).
16. Nakamura, N., Ida, H., Matsui, Y., Wakiya, K., Takayanagi, T., Koide, M., Currell, F.J., Kitazawa, S., Suzuki, H., Ohtani, S., Safronova, U.I., and Sekiguchi, M., *J. Phys. B* **28**, 4743 (1995).

MEASUREMENTS OF PHOTOELECTRON ANGULAR DISTRIBUTIONS USING A NEW MAGNETIC ANGLE-CHANGING TECHNIQUE

G. C. King, D. Cubric, and F. H. Read

Dept of Physics and Astronomy, University of Manchester, Manchester, M13 9PL, UK

A new magnetic angle-changing technique is reported for the measurement of angular distributions in photoelectron spectroscopy. This technique employs a localised and shaped magnetic field that allows the angular measurements to be made without any movement of the electron spectrometer. The technique has been used to measure angular distribution parameters for the rare gases over energy regions containing autoionization resonances. The measurements have been made as continuous functions of both photon energy and photoelectron ejection angle, which highlights the dramatic effect of these autoionising resonances on the angular distribution parameter.

INTRODUCTION

There continues to be a great deal of theoretical and experimental interest in the study of autoionising resonances in the rare gases because they are crucial to the understanding of the coupling of bound states to the continuum. A variety of experimental techniques has been used to study these autoionising resonances including photoabsorption, photoelectron spectroscopy and ion mass spectroscopy. The use of photoelectron spectroscopy has a number of important advantages and can give more complete information about the photoionization process. The autoionising resonances invariably decay to more than one final ion state and the use of photoelectron spectroscopy allows these various channels to be isolated and individual partial ionisation cross sections to be measured. Measurements of angular distributions of photoelectrons also provide particularly sensitive and detailed information about autoionising resonances. As predicted by Dill (1) there are rapid changes in the angular distribution of the ejected photoelectrons across the resonance features. Indeed, Dill showed that these variations are a particularly sensitive probe of the photoejection dynamics.

In conventional measurements of the angular asymmetry parameter, β, a rotatable angle-resolving spectrometer is used to measure the intensities of photoelectrons of a particular energy at typically two angles of emission with respect to the electric field vector of the photon beam, e.g. 0° and 90°. In the present work a new magnetic angle-changing device has been used to obtain the angular information. Here a localised and shaped magnetic field is used to change the direction of the outgoing photoelectrons without changing their point of origin, i.e. from the photon beam-gas beam interaction region. We have used the technique to measure angular asymmetry parameters for the rare gases over energy regions containing autoionising resonances. These include measurements of angular distributions for the (ns^2np^5) $^2P_{3/2}$ ionic states over the energy ranges containing the $nsnp^6n'p$ autoionising resonances and also measurements of autoionising states lying between the $^2P_{1/2}$ and $^2P_{3/2}$ ionic states.

APPARATUS AND EXPERIMENTAL METHOD

The present apparatus (2) consists of a tuneable source of VUV radiation, an electron spectrometer to measure the energies of the photoelectrons and a localised and shaped magnetic field placed at the photon beam-gas beam interaction region. A schematic diagram of the apparatus is shown in figure 1. The localised magnetic field (3,4) is produced by an arrangement of three pairs of co-axial solenoids, the common axis of which lies along the direction of the photon beam. There is a gap between the pairs of solenoids to allow passage of the gas beam and also the emitted photoelectrons, which are detected in the plane perpendicular to the axis of the solenoids. The arrangement of the solenoids and also the resultant action on the photoelectrons is illustrated in figure 2. This shows a cross section through the solenoid coils and the variation of the magnetic field with radial distance from the interaction region. The magnetic field becomes essentially zero at a radial distance of about 25 mm so there is no disturbance to the operation of the electron spectrometer. This rapid cancellation of the magnetic field is achieved by

CP475, *Applications of Accelerators in Research and Industry*,
edited by J. L. Duggan and I. L. Morgan

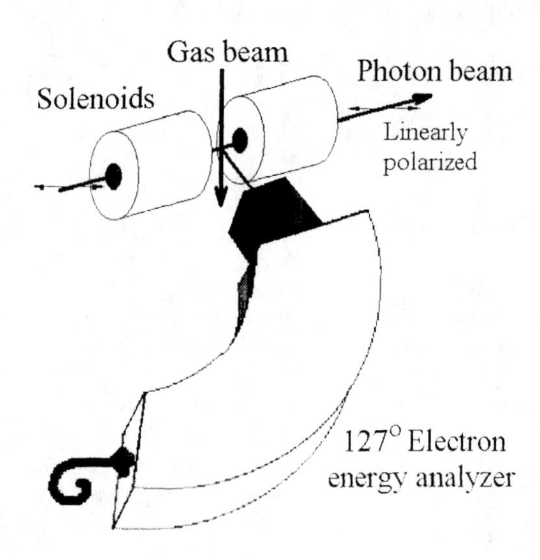

FIGURE 1 Schematic diagram of the photoelectron spectrometer.

making the overall dipole magnetic moment of the solenoid system equal to zero by setting solenoid currents to satisfy the condition

$$\sum_i n_i I_i R_i^2 = 0$$

where i labels the solenoid, n is the number of turns, I is the current, and R is the radius of the solenoids layer.

Figure 2 also shows examples of the resultant changes in directions of emitted photoelectrons of various kinetic energy. Figure 2 illustrates the important point that the photoelectrons change their out-going direction when they move radially away from the interaction region in the field-free region. This property arises because of the axial symmetry of the magnetic field for which a component of the generalised momentum is a conserved quantity. By suitably changing the solenoid currents, the change in angle for electrons of a specific energy can be varied in a controlled way, and so their angular distribution can be measured. The present technique has the valuable advantage that angle-resolved measurements can be made with a stationary electron spectrometer and mechanical movement in the vacuum system is avoided.

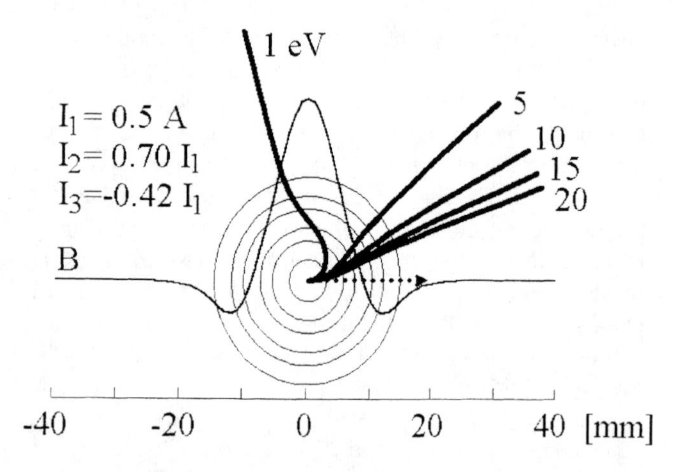

FIGURE 2 A cross section through the solenoid coils. The radial magnetic field profile is plotted for the coils for which the dipole magnetic moment is equal to zero and several electron trajectories for different energies are presented.

The electron spectrometer consists of a 127° cylindrical deflector analyser with a twin lens optical system and has been described elsewhere (5). This spectrometer provides the necessary range of photoelectron kinetic energy at an energy resolution of typically 60 meV. The angular resolution is ±5 degrees. The experiments were carried out at the Daresbury Laboratory Synchrotron Radiation Source using a 5-metre MacPherson monochromator. This provided photons over the energy range of interest with a resolution of typically 15 meV. The photon flux was monitored with a photodiode having an aluminium oxide surface. The signal from the diode was used to normalise the data for the variations of photon flux with energy and time.

RESULTS AND DISCUSSION

We have used the magnetic angle-changing technique to measure variations in asymmetry parameters, β, across autoionization resonances in the rare gas atoms argon, krypton and xenon. These include measurements of angular distributions for the $(ns^2np^5)\ ^2P_{3/2}$ ionic states over the energy ranges containing the $nsnp^6n'p$ autoionising resonances and also measurements of autoionising states lying between the $^2P_{1/2}$ and $^2P_{3/2}$ ionic states. The technique is ideally suited for this since it provides complete pictures of angular distributions as continuous functions of both ejection angle and photon energy.

As an example, data obtained in krypton over the energy region between the $^2P_{1/2}$ and. $^2P_{3/2}$ ionic states is shown in figure 3. The contour plot, shown at the top of figure 3, shows the detected photoelectron yield as continuous functions of both photon energy and ejection angle. (The ordinate axis gives the solenoid currents, which in turn are related to the ejection angle.) To obtain such a spectrum the photon energy hν and the collection energy, E_c, of the photoelectron spectrometer are varied synchronously according to the expression $E_c = h\nu - BE$ where BE is the binding energy of the electronic orbital. This yields a constant ion state spectrum corresponding to the partial ionisation cross section. Then the angular distribution is measured at a succession of fixed photon energies, which are incremented by a small amount, typically 10 meV, to obtain the 2-dimensional spectrum. Figure 3 (middle) shows the measured partial ionisation cross-section for the $^2P_{3/2}$ state of krypton. It was obtained

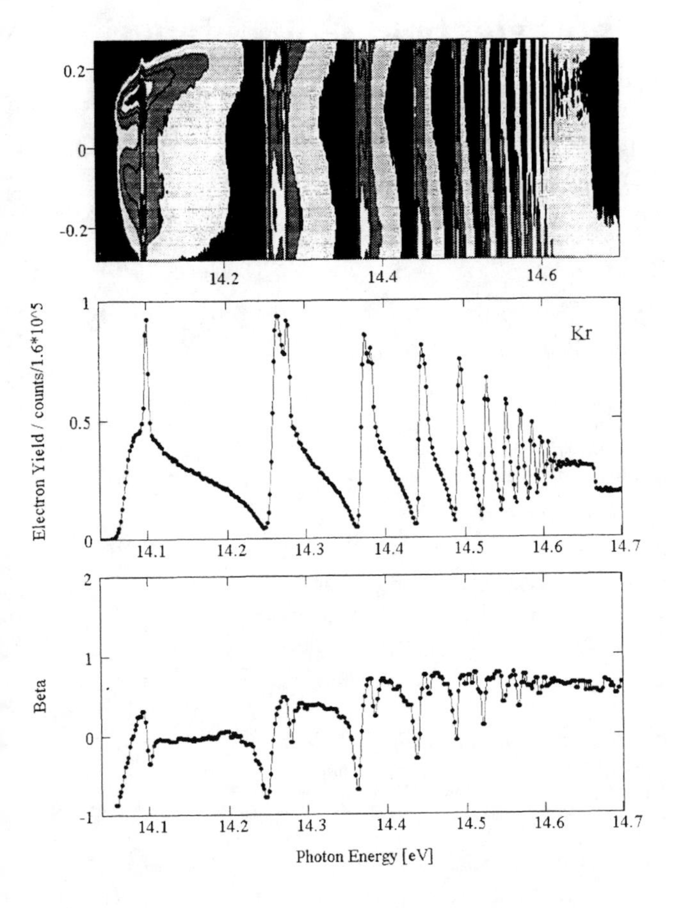

FIGURE 3 Photoelectron yield contour plot (top), partial cross section (middle) and angular distribution parameter β (bottom) of krypton photoelectrons in the region of the ((n=2)s', nd' n=6 to ∞) autoionizing resonances.

by taking a cut of the 2D spectrum at θ=60° corresponding to the magic angle for the measured value of the polarisation, P=0.65. The spectrum shows sharp structure due to the ((n=2)s',nd', n=6 to ∞) autoionising resonances Figure 3 (bottom) shows the variation of β with photon energy that has also been deduced from the 2D data. It may be noted that the β parameter, at a particular photon energy, was deduced using not only two values of ejection angle, but rather using the full angular range of the measurements. This increases the statistical accuracy of the results. It is interesting to see that the β parameter spectrum shows up features that are not observed in the ionisation cross-section; the β parameter is a more sensitive probe since it involves a ratio of cross-sections.

A full interpretation of the β parameter variation based on experimental data alone is not straightforward. The resonances in this region are not isolated. Furthermore, there are significant spin-orbit and configuration-interaction effects. Indeed, the present data emphasises the need for theoretical calculations to enable the interpretation of the data.

SUMMARY

A new magnetic angle-changing technique has been described that uses a localised and shaped magnetic field to change the outgoing directions of photoelectrons in photoelectron spectroscopy. This enables angular measurements to be made with a stationary electron analyser. Our experiments have confirmed the high performance and potential of the technique. Indeed it largely eliminated several possible sources of systematic error. For example photon energy drifts are minimised because there is minimum delay between measurements at different angles. The studies have provided valuable new data about the role and characteristics of the autoionising resonances in the noble gas atoms. They have also highlighted the need for more theoretical studies to understand the role of these resonances.

REFERENCES

1. Dill,d. Phys Rev A 7 1976 (1976).
2 Cubric, D., Thompson, D.B., Cooper, D.R., King, G.C. and Read, F.H., J. Phys. B: At. Mol. Opt. Phys., **30**, L857-64 (1997).
3. Zubek, M., Gulley, N., King, G.C. and Read, F.H., J. Phys. B: At. Mol. Opt. Phys., **29**, L239 (1996).
4. Read, F.H. and Channing, J.M., Rev. Sci. Instrum., **67**, 2372 (1996).
5. Hall, R.I., McConkey A., Ellis, K., Dawber, G., Avaldi, A.,MacDonald, M. A., and King, G.C. Meas. Sci. Tecnnol., **3** 316 (1992).

Angular Correlation in Auger Photoelectron Coincidence Spectroscopy from the Cu(111) Surface

R. Gotter[2], A. Attili[1], L. Marassi[3], D. Arena[5], R. A. Bartynski[5], D. Cvetko[2], L. Floreano[2], S. Iacobucci[4], A. Morgante[2], P. Luches[3], A. Ruocco[1], F. Tommasini[2], and G. Stefani[1]

1 *INFM Unita' di Roma 3 and Dipartimento di Fisica "E. Amaldi". Universita' di Roma Tre, via della Vasca Navale 84, I-00146 Roma, Italy*
2 *Laboratorio Nazionale TASC-INFM, Trieste, Italy*
3 *INFM Unita' di Modena and Dipartimento di Fisica Universita' di Modena, Modena, Italy*
4 *CNR Istituto di Metodologie Avanzate Inorganiche, Montelibretti, Italy*
5 *Department of Physics and Astronomy, Rutgers University, Piscataway NJ*

The feasibility of angle resolved APECS (*Auger Photo-Electron Coincidence Spectroscopy*) on solids is demonstrated with an experiment performed at the ALOISA beamline (ELETTRA, Trieste) on the $L_3M_{45}M_{45}$ Auger transition of the Cu(111) surface. This beamline, with its multicoincidence detection system based on several individual electron analysers, was designed in order to perform such experiment with good efficiency.
The correlation effects displayed by the measured angular distribution suggest inadequacy of the *two step* model that is usually adequate for non coincidence experiments performed at the same energy and on the same orbital.

INTRODUCTION

Electron-electron coincidence spectroscopies have been extensively applied to atomic and molecular physics since early seventies and ever since then they have grown in number and relevance. To apply this methodology to solids and surfaces has been a major target since early days, but the average time needed to complete a coincidence experiment has hampered its attainment. This time is usually too long when compared with the speed at which a clean surface gets contaminated, even in a state of art ultra high vacuum system. This handicap has been surmounted during the past decade by the help of new multicoincidence, high luminosity apparatuses and high brilliance light sources. Once again, the implementation of innovative instrumentation opens up new applications to established methodologies. The sizeable number of papers dealing with application of coincidence spectroscopies to solids does show that the time is ripe for seeing more and more work in this field.

Similarly to what happens for atoms and molecules, these experiments can be used to study the dynamics of electron interaction with and within a solid. This will be the focus of this talk and, among the spectroscopies, we shall concentrate on Angle Resolved Auger-Photoelectron Coincidence experiments (AR-APECS). In such ionisation experiments the kinematics is fully determined and two main issues are and have been always at stake :

a) what is the dominant mechanism that leads to ejection of electron pairs from a solid surface

b) what information is yielded by these "exotic" spectroscopies that is not already available from currently used spectroscopies.

In the past few years APECS has gained relevance in the field of electron spectroscopies, in particular whenever it is necessary to discriminate electron correlation (many body) effects from single particle behaviour. In the past, experimental studies on the correlation between photoelectron and Auger electron have been performed with APECS measurements exclusively resolved in energy. These studies have shown that the correlation in energy of the two final electrons is relevant to the point that the energy is conserved by the pair as a whole, not by the two electrons independently. The effect was seen on the 3p orbital of copper, in an experiment performed with an overall energy resolution narrower than the natural lifetime broadening of the core hole[1]. This is a confirmation of the correlated behaviour of the two final electrons, whenever the core hole lifetime is shorter than the two hole final state lifetime. These are indeed the conditions under which a single step model, rather than the more usual two step, is to be used to describe the formation and relaxation dynamics of the core hole.

CP475, *Applications of Accelerators in Research and Industry*, edited by J. L. Duggan and I. L. Morgan

Besides the possibility to investigate the ionisation dynamics, the unique capability of APECS to disentangle signals originated from different sites and/or from overlapping spectral features and its high surface sensitivity make it very attractive for surface analysis [2].

Only recently the angular correlation of the final electron pair of APECS has been investigated by theory [3] and experiments [4]. Both studies were limited to gas phase and to low kinetic energies where continuum final state effects dominate. The extension to higher energies, where initial state effects are relevant, and to solids, where diffraction from the crystal lattice is expected to modulate the differential cross section, is a major task. On the other hand, APECS experiments offer the challenge of disentangling, if possible at all, momentum correlation modulations from the diffraction ones, the latter being due to diffraction from the crystal lattice of the two final unbound electrons.

In the present paper we report the results of the first angle resolved APECS experiment performed on a solid sample.

EXPERIMENTAL

In order to demonstrate the feasibility of angle resolved APECS on solids, we have used the experimental apparatus of the beamline ALOISA (*Advanced Line Overlayer Interface Surface Analysis*), installed at the ELETTRA Synchrotron Radiation Facility (Trieste). This is a multipurpose beamline for surface science studies and operates in the range 200-8000eV. Due to the multicoincidence detection system, this is, at present, the only apparatus capable of efficiently performing such experiments.

For the soft x-ray region (photon energies lower than 1.7 KeV) the radiation produced by an undulator is monochromatized by a plane mirror plane grating monochromator. The configuration is essentially a slitless Hunter one [5], in which the dispersive system is a modification of the SX-700 [6]. In the photon energy range of the experiment (1400 eV) the resolving power can be as good as 5000. However, in these first measurements a reduced resolving power was used, because the main aim was not an optimal energy resolution but an optimisation of the luminosity of the apparatus.

The incoming photon beam hits the surface of the sample mounted on a six-degree of freedom manipulator (three for the translation and three for the rotation). The incident angle of the SR beam can be varied from 0 to 10 degrees with respect to the sample surface, while by rotating the sample around the x-axis, which is the light propagation axis, is possible to change continuously the polarisation of the incident electromagnetic field with respect to the sample surface.

Seven individual electron analysers, arranged in two arrays, collect the electrons emitted from the surface, discriminating both the angle of emission and the kinetic energy of the particles. Each analyser consist of a retarding electron optics that forms an image of the illuminated area of the sample onto the entrance slit of an energy dispersive element, at the exit of which an electron multiplier is positioned. The electron optics can be operated in two different modes, high resolution (HR) and low resolution (LR), depending on the accepted solid angle at the target (2.4×10^{-4} and 7.2×10^{-4} srad respectively). The dispersive element is a hemispherical sector (reduced to 176.4° to decrease fringing field effects) with a mean radius of 33 mm. Time walk dispersions of 10 ns in the HR mode, and 2 ns in the LR mode were measured.

The presence of several independent analysers is necessary in order to obtain a reasonably large accepted solid angle of the detection system while maintaining a good discrimination in energy and angle. Acquiring simultaneously coincidences under different kinematic conditions reduces the acquisition time necessary to obtain the desired statistical uncertainty.

The angular scan is made possible by two independently rotatable frames on which the electron analysers are mounted. A so called bimodal frame hosts two of them and is able to rotate around the z-axis, while a so called axial frame, hosting the remnant five, can rotate around the x-axis. The whole system (bimodal plus axial frame) can rotate rigidly around the x-axis through the rotation of the experimental chamber. Combining these rotations with the degrees of freedom of the sample allowed by the manipulator, it is possible to reach almost every configuration of the electron momentum components in the frame of reference fixed to the surface of the sample studied.

ANGLE RESOLVED APECS SPECTROSCOPY

The feasibility of an AR-APECS investigation has been demonstrated with an experiment performed on the $L_3M_{45}M_{45}$ transition of the Cu(111) surface with a photon beam of 1400 eV and overall energy resolution of 2 eV. The analysers of the bimodal frame were used to detect $2p_{3/2}$ photoelectrons, while those hosted on the axial frame were tuned on the Auger transition. The two analysers tuned on the photoelectron were positioned on a maximum and on a minimum of the diffraction pattern respectively.

Under these conditions ten different pairs of coincident electrons are measured simultaneously. A true coincidence counting rate between 10^{-2} and 10^{-1} Hz for each pair has been obtained. The true to accidental ratio was close to

1 :1. This has permitted to achieve, with an acquisition time of few hours, a statistics adequate for angular distribution measurements.

In the present experiment, the energy is well above threshold of the $2p_{3/2}$ core hole and lifetime is longer than the lifetime of the final state. The two step model should be appropriate to describe the ionisation and the following Auger de-excitation process. Hence, neither energy nor angular correlation between the two final electrons should be expected.

The overall energy resolution of the experiment is too poor to observe energy correlation effects. Conversely, the angular resolution is good enough to reveal correlation effects, if present.

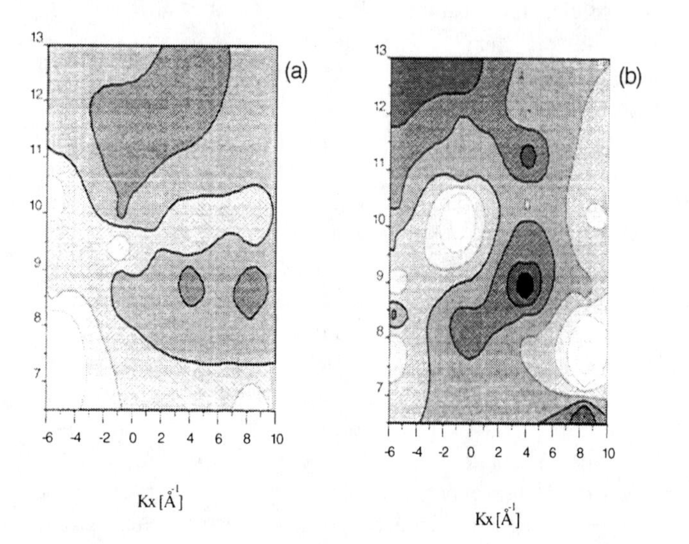

Figure 1 . Contour plot (b) of the true APECS coincidence count rate, measured at a minimum of the Cu L_3 photoelectron diffraction pattern, compared with the incoherent product (a) of photoelectron and Auger $L_3M_{45}M_{45}$ angular distributions. X and Y scales are orthogonal components of the projection onto the surface of the Auger electron momentum. Light grey means high count rate ; the peak to valley maximum ratio is 3.

The most relevant finding of this investigation is that the APECS angular distribution exhibits a shape not always reducible to the incoherent combination of the individual angular distributions of the two final electrons. A discrepancy between the two step uncorrelated model and the experiment has been found when the photoelectron is at a minimum of the diffraction pattern. This is clearly shown in figure 1, where the contour plot of the measured coincidence rate is plotted alongside with the product of the single Auger and photoelectron angular distributions, measured contemporary to the coincidence signal. When the photoelectron is on the photodiffraction maximum, the coincidence angular pattern is indistinguishable, within error bars (roughly 20%), from the product of the correspondent Auger and photoelectron distributions

The two step model, that is commonly accepted for non coincident experiments performed at the same energy and on the same process of core hole creation and relaxation, doesn't explain a difference between coincident and non coincident angular distributions.

To properly describe the experiment, theory should be done within a single step model framework, taking into account electron correlation and diffraction from the crystal lattice of the emitted electron pair. In such a model, the intermediate ionic state is seen as a continuum distribution over all possible core hole states plus the photoelectron wavefunction, and the whole process is treated as a resonance in the double photoionisation.

It has been recently predicted [7] that the emission of electron pairs following the absorption of a single photon by a solid should obey propensity rules expressed by the scalar product of the photon polarisation vector and the vector momentum of the centre-of-mass of the pair of final electrons. Consequently, diffraction of the pair from the lattice occurs when the centre-of-mass momentum changes by a lattice reciprocal vector during the photoemission. The effect, that has been actually observed in an angle integrated double photoionisation experiment on the valence band of Cu(001) and Ni(001) [8], is expected to be present irrespective of the degree of localisation of the orbital. Further investigations with AR-APECS are needed in order to show whether or not this quasi-particle behaviour is present also in the case of core-hole ionisation.

CONCLUSIONS

AR-APECS coincidence experiments on surfaces have been demonstrated to be possible

They possess some unique characteristic not shared with other electron spectroscopies. Besides the enhanced discrimination and sensitivity already known for APECS, there is the capability to investigate new aspects of electron correlation in bound and continuum states. Indeed, the present experiment shows that the APECS angular distribution measured for the 2p ionisation of Cu, doesn't immediately speak in favour of the two step model that is usually adopted to describe core ionisation-relaxation processes.

ACKNOWLEDGEMENTS

The authors are grateful to Progetto Coordinato CNR, Comitato Nazionale scienze Fisiche, to MURST : "Progetti di ricerca di rilevante interesse nazionale" and to NATO Grant CRG970175 for partial support .

REFERENCES

1. Jensen et al., *Phys. Rev. Lett.* **62** (1989) 71
2. E. Jensen et al., *Phys. Rev.*, B**41** :1246, (1990)
3. L. Végh, J.H.Macek, *Phys. Rev.*, A**50**, 5, 4031 (1994)
4. J. Viefhaus et al., *Phys. Rev. Lett.* **80**, 8, 1618 (1998)
5. D. Cvetko et al., *Proceedings SPIE '97*, (1997)
6. M. Domke et al., *Rev. Sci. Instr.* **63**, 80 (1992)
7. J. Berakdar, *Phys. Rev.* B **58**, xx (1998)
8. J. Berakdar, S.N. Samarine, R. Herrmann, J. Kirshner, *Phys. Rev. Lett.* (1998) in press

Saturation Effects in Electron Loss[1]

G. M. Sigaud[2], W. S. Melo[3], A. C. F. Santos, M. M. Sant'Anna[3], E. C. Montenegro

Departamento de Física, Pontifícia Universidade Católica do Rio de Janeiro
Caixa Postal 38071, Rio de Janeiro, RJ 22452-970, Brazil.

The electron loss of dressed ions by heavy neutral atoms can be highly non-perturbative, in which concerns one of the two competing mechanisms which govern electron loss, namely the screening contribution. The behavior of the total electron loss cross sections with the target atomic number, Z_2, shows a strong saturation as Z_2 increases. PWBA calculations present such a behavior for the electron-electron contribution (antiscreening) but not for the screening, since this saturation is related to a non-perturbative regime. In this work, we compare measured total electron loss cross sections of He^+, C^{3+} and O^{5+} ions, with energies ranging from 1.0 to 3.5 MeV, by H, He, Ne, Ar, Kr and Xe atoms, with calculations for the screening contribution based on the free-collision model, as well as with other models, showing that the inclusion of other competitive channels is needed for a better description of this process.

INTRODUCTION

It has been well established that the loss of an electron of a dressed energetic ion by a neutral target atom is governed by two competing mechanisms, the screening and the antiscreening modes [1,2]. In the case of light targets, both modes are conveniently described by first-order models, such as the plane wave Born approximation (PWBA) [2,3]. However, for heavier targets the situation is not so simple and clear, since the dominant Z_2^2 dependence predicted by the PWBA, where Z_2 is the target atomic number, is not observed in experiments, as, for example, in the data presented by Sant'Anna *et al* [4]. The total electron-loss cross sections for intermediate-velocity He^+ projectiles measured by these authors as a function of the target atomic number present a strong saturation for $Z_2 \geq 10$. First-order calculations for the antiscreening contribution show this behavior, which is not present, however, in those for the screening.

This fact can be understood because there can be a significant contribution to the screening mode from collisions occurring at small impact parameters. This means that the interaction between the projectile active electron and the screened target nucleus can be highly non-perturbative [5]. On the other hand, since the antiscreening contribution is due to a sum of several electron-electron interactions, a perturbative treatment is convenient for its appropriate description.

More recently, Sigaud *et al* [6] reported of a similar saturation with increasing atomic number of the target in measurements of the cusp yields of the electron loss to the continuum (ELC) process for 1.0 MeV He^+ projectiles by atomic and molecular gaseous targets.

Such a saturation for the screening was already observed experimentally in the excitation of highly-charged projectiles in collisions with neutral targets [7-11]. Theoretical attempts to describe those data have been made using either second-order or non-perturbative models, quite successfully in some cases [7,8,11-15].

A non-perturbative approach, using coupled-channel calculations, to the screening contribution to the total electron loss cross sections was presented by Grande *et al* [16] for the He^+ data of Ref. [4]. This method held much better results in comparison with first-order treatments, but was still unable to explain completely the experiment for the heaviest targets.

In this paper, the experimental total electron-loss cross sections of He^+, C^{3+} and O^{5+} projectiles in collisions with H, He, Ne, Ar, Kr and Xe targets from Refs. [4,17,18] are compared with calculations for the screening contribution using the free-collision model [19], together with first-order calculations for the antiscreening [20]. For the He^+ projectile, comparison with the coupled-channel calculations of Grande *et al* [16] is also presented. The purpose is to study the dependence of the total electron loss cross section with the target atomic number in order to analyze the roles played in this process by the different participants in the collision, namely, the projectile and target electrons and nuclei.

---

[1]Work supported in part by CNPq, FINEP, FAPERJ and MCT (PRONEX).
[2]Corresponding Author.
[3]Present address: Instituto de Física, Universidade Federal Fluminense, Brazil.

TOTAL ELECTRON LOSS AND THE TARGET ATOMIC NUMBER

Two models for the calculations of the screening contribution to the total electron loss are used in the comparison with the experimental data. An extension of the free-collision classical-impulse approximation [21] presented by Riesselmann *et al* [19] was employed for the three projectiles studied, while, in the case of He^+, the calculations by Grande *et al* [16] using the coupled-channel method are also presented.

In the free-collision model, the ionization of the active electron of the projectile is treated as the scattering by the target atom of a free electron, whose velocity is the vector sum of the velocity of the projectile center of mass in the laboratory frame and that of the electron in the projectile frame. The momentum transferred to the electron during the collision must be such that it acquires enough energy to be ionized, which implies a lower limit for the free-electron scattering angle, that is, an upper limit to the impact parameter of the collision. The total loss cross section is, then, obtained by integrating the electron-scattering differential cross section over all the scattering angles greater than the minimum value, and over all the possible orientations of the electron velocity in the projectile with respect to the projectile velocity in the laboratory.

The application of this model to evaluate the screening contribution to the loss is restricted to considering only the elastic scattering of the electron, since, in this mode, the target electrons remain in the ground state during the collision.

Riesselmann *et al* [19] applied this model to calculate the electron loss cross sections of H(1s) and the single- and double-electron loss of H^- by gaseous targets. They restricted the calculations to projectile velocities greater than the root-mean-square velocity of the active electron in the projectile, a procedure that was also used here. A very important feature of the application of this model to the screening mode then arises from this approximation, since it leads to a threshold at the same projectile velocity of the onset of the antiscreening. However, it limits the applicability of the model to energies greater than 0.4, 1.4 and 4.0 MeV for He^+, C^{3+} and O^{5+} projectiles, respectively.

As mentioned above, in the case of the electron loss of He^+ projectiles another method to evaluate the screening contribution was also considered, namely the coupled-channel calculations presented by Grande *et al* [16]. In this non-perturbative treatment, the ion-atom collision is described by the many-body time-dependent Schrödinger equation, where the full electronic wavefunction is expanded in terms of unperturbed eigenfunctions of the projectile. Those authors obtained the total electron loss cross sections by adding, in a weighted manner to take into account of unitarity, the screening contribution obtained using this method to the antiscreening contribution calculated with the extended sum-rule method of Montenegro and Meyerhof [20].

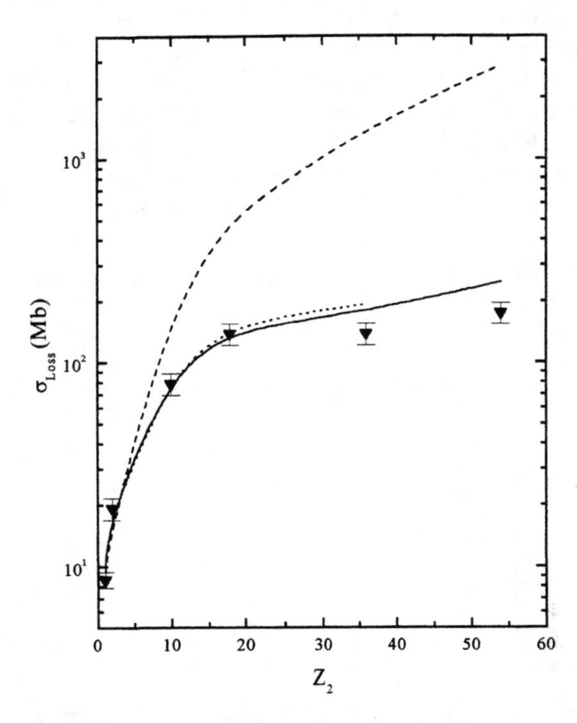

FIGURE 1. Total electron loss cross sections of 1.5-MeV He^+ as a function of the target atomic number. Experiment: ▼, from Ref. [4] (He-Xe) and from Ref. [18] (H). Theory: dashed curve: sum of screening (PWBA) and antiscreening (Ref. [20]); dotted curve: from Ref. [16]; solid curve: sum of screening (free-collision model, this work) and antiscreening (Ref. [20]).

In Fig. 1 the experimental total electron loss cross sections of 1.5 MeV He^+ projectiles from Refs. [4,18] as a function of the target atomic number are compared to calculations of the screening contribution within three different models, namely, the PWBA, the free-collision model and coupled-channel method. In all three cases, the antiscreening contribution is obtained using the procedure of Montenegro and Meyerhof [20]. The coupled-channel calculations of Grande *et al* were performed only up to the Kr target because, as pointed out by those authors, this method presents an intrinsic difficulty, namely, an unrealistic capture of the He^+ electron into an occupied state of the target [16].

Several features can be readily noted. First, the saturation with increasing target atomic number, which appears in the experimental data, also occurs in the theoretical total cross sections when the screening contribution is calculated using either the free-collision model or the coupled-channel method. As mentioned before, this does not occur when first-order models are employed. Second, the results obtained in the free-collision model for He^+ projectiles are very close to the coupled-channel calculations, being in very good agreement with experiment within the error bars up to the Ar target. For heavier targets, there are some discrepancies, which can be as high as 30% for the Xe target, due to the constraint, assumed within the free-collision model, of the non-ionization of the target atom during the collision [22].

In Figs. 2 and 3 the theoretical total electron loss cross sections, with the screening contribution calculated within the PWBA and the free-collision model, as functions of the target atomic number, are compared to data for 3.0 MeV C^{3+} and 3.5 MeV O^{5+} projectiles from Refs. [17,18], respectively. Once more, in both cases the antiscreening contribution is also obtained using the procedure of Montenegro and Meyerhof [20].

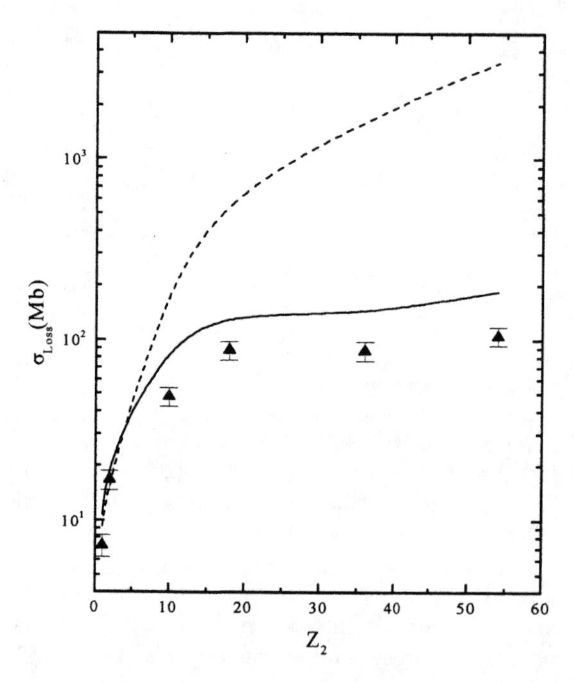

FIGURE 2. Total electron loss cross sections of 3.0-MeV C^{3+} as a function of the target atomic number. Experiment: ▼, from Ref. [17] (He-Xe) and from Ref. [18] (H). Theory: dashed curve: sum of screening (PWBA) and antiscreening (Ref. [20]); solid curve: sum of screening (free-collision model, this work) and antiscreening (Ref. [20]).

Here again, there is a strong saturation in the experimental results in both cases; for the O^{5+} projectile, a kind of structure beyond the error bars can even be noted in the region of $Z_2 \approx 36$. All these trends are followed by the calculations using the free-collision model. However, these calculated cross sections lie systematically above those from the experiment, the discrepancies ranging from 35% to 80% for C^{3+} and factors 3 to 10 for O^{5+}, in contrast with the good agreement presented by the He^+ and also by the H^0 and H^- projectiles [19].

These discrepancies can be attributed to the competition of the electron loss with at least two other processes which

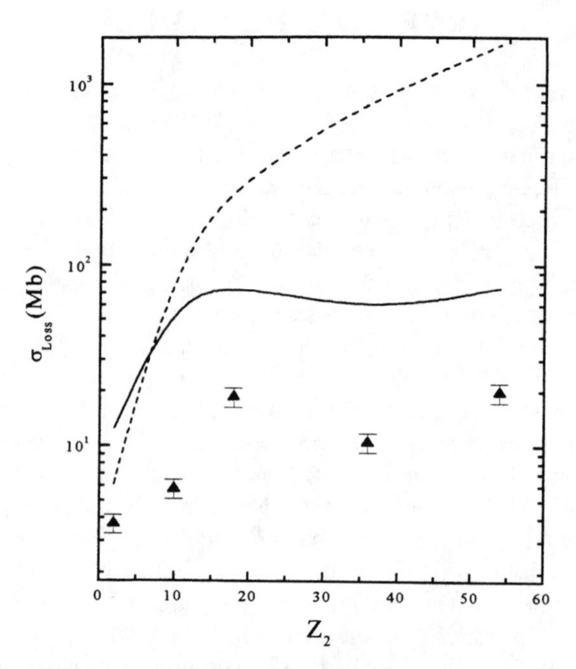

FIGURE 3. Total electron loss cross sections of 3.5-MeV O^{5+} as a function of the target atomic number. Experiment: ▼, from Ref. [17]. Theory: dashed curve: sum of screening (PWBA) and antiscreening (Ref. [20]); solid curve: sum of screening (free-collision model, this work) and antiscreening (Ref. [20]).

may occur as exit channels of the collision, namely, the single or multiple *ionization of the target*, which can be followed by the *capture by the projectile* of one – or more – of these emitted electrons. For the lowest velocities studied here, these processes are very important and can be up to two orders of magnitude higher than the electron loss [17]. Thus, the coupling of these channels with the loss must be considered in the analysis of the data [23].

There are no complete theoretical models to describe the direct target ionization and the electron capture in intermediate-velocity collisions between highly-charged ions and heavy neutral atoms. However, Montenegro *et al* used with success the independent electron model (IEM) [24] to describe the electron loss of C^{3+} and O^{5+} by H_2 and He targets in the same energy range as considered here [23]. Following the procedure presented by these authors, the total electron loss cross section by a neutral target, with N active electrons in the outer shells – which are the electrons most likely to participate in the collision – for a projectile which has one active electron, can be written as

$$\sigma_{loss} = [1 - P_C(0) - P_I(0)]^N \sigma_{screen} + \\ + [1 - P_C(0) - P_I(0)]^{(N-1)} \sigma_{anti}, \tag{1}$$

where $P_C(0)$ and $P_I(0)$ are, respectively, the electron capture and direct ionization probabilities at zero impact parameter. σ_{screen} and σ_{anti} are the calculated screening and antiscreening contributions to the one-electron loss, respectively.

Even the use of this rather simple approach is not an easy task for heavy atoms, since there are still no reliable methods to evaluate the probabilities which appear in Eq. (1). Nevertheless, in order to verify whether the IEM is a reasonable approach to describe the problem, Eq. (1) was employed to fit the values of σ_{screen}, calculated in the free-collision model, and first-order σ_{anti} to the experimental σ_{loss}, using the combined probability $P_C(0) + P_I(0)$ as a fitting parameter, for the C^{3+} and O^{5+} projectiles. The values of N were taken as 2 for He and 8 for the other targets. The estimated values for the combined probability lay between 0.06 and 0.8 and their qualitative behaviors were as expected, with a steep decrease with increasing collision energy – which is characteristic of the capture channel – and a slow decrease with the target atomic number, which is due to the use of the same number of target active electrons for $Z_2 \geq 10$.

CONCLUSIONS

A comparison between experimental and theoretical total electron loss cross sections for He^+, C^{3+} and O^{5+} projectiles by gaseous targets from H to Xe was presented. The calculations for the screening contribution were performed using the PWBA and an extension of the free-collision model for all systems, and the coupled-channel method for the He^+ projectile. Both these non-perturbative models show a much better agreement with experiment than the PWBA, presenting the same saturation with increasing target atomic number observed in the data.

However, despite the qualitative agreement with experiment, there remain, mainly in the case of C^{3+} and O^{5+} projectiles, quite large quantitative discrepancies, which increase with the projectile charge. These differences were assigned to the competition of other exit channels to the collision, i.e., the direct target ionization and the electron capture by the projectile. A rough estimate of the contributions from these alternative channels within the IEM showed that the inclusion, in a unitarized way, of their coupling is necessary to a proper description of the problem.

REFERENCES

1. McGuire, J. H., Stolterfoht, N., and Simony, P. R., *Phys. Rev.* **A24**, 97 (1981).
2. Montenegro, E. C., Meyerhof, W. E., and McGuire, J. H., *Advances in Atomic and Molecular Physics* **34**, 249-300 (1994).
3. Montenegro, E. C., Meyerhof, W. E., McGuire, J. H., and Cocke, C. L., *The Physics of Electronic and Atomic Collisions – XIX International Conference* (ed. by Dubé, L. J., Mitchell, J. B. A., McConkey, J. W., and Brian, C. E.), New York: AIP, 1995, AIP Conference Proceedings **360**, 515-524.
4. Sant'Anna, M. M., Melo, W. S., Santos, A. C. F., Sigaud, G. M., and Montenegro, E. C., *Nucl. Instrum. Meth.* **B99**, 46 (1995).
5. Walters, H. R. J., *J. Phys. B* **8**, L54 (1975).
6. Sigaud, G. M., Jorás, F. S., Santos, A. C. F., Montenegro, E. C., Sant'Anna, M. M., and Melo, W. S., *Nucl. Instrum. Meth.* **B132**, 312 (1997).
7. Brendlé, B., Gayet, R., Rozet, J. P., and Wohrer, K., *Phys. Rev. Lett.* **54** , 2007 (1985).
8. Wohrer, K., Chetioui, A., Rozet, J. P., Jolly, A., Fernandez, F., Stephan, C., Brendlé, B., and Gayet, R., *J. Phys. B* **19**, 1997 (1986).
9. Xu, X. –Y., Montenegro, E. C., Anholt, R., Danzmann, K., Meyerhof, W. E., Schlachter, A. S., Rude, B. S., and McDonald, R. J., *Phys. Rev.* **A38**, 1848 (1988).
10. Sulik, B., Stolterfoht, N., Ricz, S., Kádár, I., Xiao, L., Schiwietz, G., Grande, P., Köhrbrück, R., Sommer, K., and Grether, M., *The Physics of Electronic and Atomic Collisions – XVIII International Conference* (ed. by Anderson, T., Fastrup, B., Folkman, F., and Knudsen, H.), Book of Abstracts p. 528 (1993).
11. Chabot, M., Wohrer, K., Chetioui, A., Rozet, J. P., Touati, A., Vernhet, D., Politis, M. F., Stephan, C., Grandin, J. P., Macias, A., Martin, F., Riera, A., Sanz, J. L., and Gayet, R., *J. Phys. B* **27**, 111 (1994).
12. Thumm, U., Briggs, J. S., and Schöller, O., *J. Phys. B* **21**, 833 (1988).
13. Mukoyama, T., and Lin, C. D., *Phys. Lett.* **141A**, 138 (1989).
14. Rodriguez, V. D., and Miraglia, J. E., *Phys. Rev.* **A39**, 6594 (1989).
15. Mukoyama, T., and Lin, C. D., *IX International Conference on the Physics of Highly Charged Ions* (ed. by Mokler, P. H., and Lüttges, S.), Book of Abstracts p. 113 (1998).
16. Grande, P. L., Schiwietz, G., Sigaud, G. M., and Montenegro, E. C., *Phys. Rev.* **A54**, 2983 (1996).
17. Melo, W. S., Sant'Anna, M. M., Santos, A. C. F., Sigaud, G. M., and Montenegro, E. C. (to be published).
18. Sant'Anna, M. M., Melo, W. S., Santos, A. C. F., Sigaud, G. M., Montenegro, E. C., Meyerhof, W. E. and Shah, M. B., *Phys. Rev.* **A58**, 1204 (1998).
19. Riesselmann, K., Anderson, L. W., Durand, L., and Anderson, C. J., *Phys. Rev.* **A43**, 5934 (1991).
20. Montenegro, E. C., and Meyerhof, W. E., *Phys. Rev.* **A43**, 2289 (1991).
21. Dmitriev, I. S., and Nikolaev, V. S., *Sov. Phys. – JETP*, **17**, 447 (1963).
22. Voitkiv, A. B., Sigaud, G. M., and Montenegro, E. C. (to be published).
23. Montenegro, E. C., Sigaud, G. M., and Meyerhof, W. E., *Phys. Rev.* **A45**, 1575 (1992).
24. McGuire, J. H., *Electron Correlation Dynamics in Atomic Collisions,* New York: Cambridge University Press, 1997, ch. 4.

FLOW OF EJECTED ELECTRON CURRENT IN ION-ATOM COLLISIONS

E. Y. Sidky

James R. Macdonald Laboratory, Department of Physics, Kansas State University, Manhattan, KS 66506-2604

Ab initio calculations for the complete momentum space wave function are performed for the first time for proton-Hydrogen collisions in the 10-25 keV/amu range. Interference between the bound and ejected components of the wave function are seen. The interference fringes are accounted for by modeling the ejected electron component of the wave function as a free expansion.

INTRODUCTION

With advances in the computational ability of modern computers many research efforts in atomic theory have focussed on numerically exact solutions for the quantum-mechanical three body problem. In particular much attention has been given recently to the ionization channel in keV energy ion-atom collisions [1–3]. Interest in this system has been sparked by detailed experimental data which map out the momenta – in three dimensions – of ejected electrons [4,5]. On the theoretical side ionization in ion-atom collisions addresses the fundamental issue of how does one describe the motion of a light charge, the electron, in the field of two moving charges, the target and projectile nuclear cores. Since the Coulomb interaction has such a long range, numerical solution of the Schrödinger equation for ion-atom collisions must cover an enormous range in internuclear distance to get results comparable with experiment. Typically, for ground state targets, internuclear distances of hundreds of atomic units are reached in calculations of the collision system.

To attain integration of the Schrödinger equation to such large internuclear separation one must include the gross features of the electron current during and after the collision. For the excitation process one knows that the average motion of the electron matches the motion of the target nucleus, while for capture the average flow goes with the projectile ion. Accordingly, plane wave factors or other electron translation factors (ETFs) have been employed to account for the background motion of the electron, see reviews [6–8]. For ionization parameterization of the basic motion of the electron is not as well-studied. One approach has been to go to scaled coordinates to accommodate the ionization channel [3,9]. However, here, we study a recent proposal, based on classical trajectory Monte Carlo calculations, suggesting that one should treat the electron flow as free expansion after ejection from the collision region [10]. We demonstrate, with an *ab initio* quantum mechanical calculation, that parameterization of free expansion of the ejected electron component of the total wave function does indeed account for oscillatory features arising from the ionization channel. Atomic units are used throughout.

THEORETICAL METHOD

For the quantum mechanical calculation of the proton-Hydrogen collision system, we employ the standard semi-classical approach [8]. The internuclear motion is treated classically taking the projectile path as linear with constant velocity $\vec{v}$. The electron sees a time-dependent potential and its wave function solves the following Schrödinger equation:

$$i\frac{\partial}{\partial t}\psi(\vec{r},t) = \left(-\frac{1}{2}\nabla^2 - \frac{1}{|\vec{r}|} - \frac{1}{|\vec{r}-\vec{R}|}\right)\psi(\vec{r},t), \qquad (1)$$
$$\vec{R} = \vec{v}t + \vec{b}.$$

$\vec{R}$ is the internuclear separation. Time $t = 0$ is when the projectile ion is at closest approach to the target. The coordinate system is the natural frame: the x-axis is parallel to the projectile velocity, the y-axis is parallel to the impact parameter and the z-axis is perpendicular to the collision plane. The method for solving eq. (1) is presented in detail in ref. [11]. Briefly stated here are the essential points necessary for this work.

The wave function is written as a two-center expansion in momentum space:

$$\Phi(\vec{p},t) = \sum_{l,m} \tilde{T}_{l,m}(p,t)Y_{l,m}(\hat{p}) + e^{-i(\vec{p}\cdot\vec{R}-\frac{1}{2}v^2t)}\sum_{l,m}\tilde{P}_{l,m}(q,t)Y_{l,m}(\hat{q}), \qquad (2)$$
$$\vec{q} = \vec{p} - \vec{v}.$$

The two-center expansion reduces the number of angular harmonics representing the electron current. For results shown here six harmonics (up to and including all

CP475, *Applications of Accelerators in Research and Industry*, edited by J. L. Duggan and I. L. Morgan

d-states) on each center was sufficient [12]. The plane wave factor in eq. (2) reflects the fact that probability current flowing with the projectile is localized about the projectile proton in configuration space. Momentum space is used, since the wave function must tend to zero for large momenta; even for large t. Thus all radial functions in eq. (2) obey:

$$\tilde{T}_{l,m}(p,t) \to 0 \text{ as } p \to \infty, \qquad (3)$$
$$\tilde{P}_{l,m}(q,t) \to 0 \text{ as } q \to \infty.$$

A study of the proton-Hydrogen system, for the energy range from the experimental works [4,5], is presented in ref. [12]. Here, we concentrate on results from one energy and one impact parameter: $E = 25$ keV ($v = 1.0$ a.u.) and $b = 1.2$ a.u.

RESULTS AND DISCUSSION

In Fig. 1 is shown a cross section in the collision plane of the amplitude of the momentum space wave function at a distance 29 a.u. past closest approach ($vt = 29$ a.u.). The amplitude is shown to highlight features that are more difficult to see in a density plot.

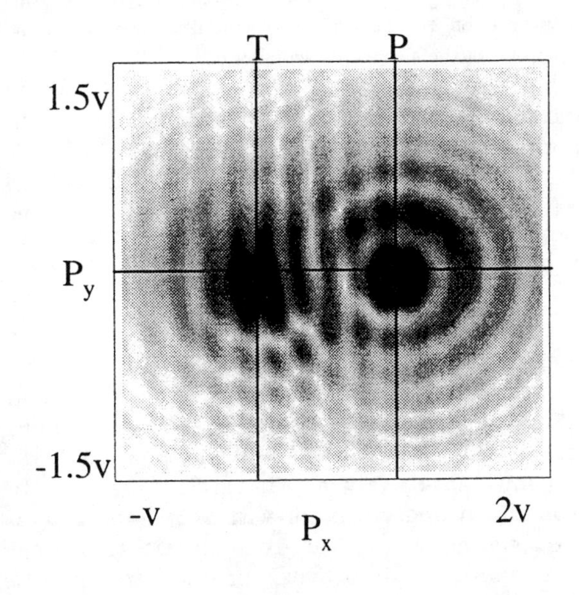

FIGURE 1. Momentum space wave function cross section, $|\Phi(p_x, p_y, p_z = 0, t = 29)|$, for a 25 keV proton on Hydrogen collision at impact parameter $b = 1.2$ a.u. The target and projectile longitudinal momenta are symbolized by T and P respectively. The electron momentum is written in terms of the projectile velocity.

Three sets of interference fringes are apparent in Fig. 1: nearly vertical stripes and two sets of concentric circles centered on both target and projectile velocities. The origin of the vertical stripes is due to the internuclear separation in configuration space, and it is already taken into account with the exponential multiplying the projectile expansion in eq. (2). The concentric rings appear in the numerical solution for the radial functions $\tilde{T}_{l,m}$ and $\tilde{P}_{l,m}$. They result from interference between the expanding ejected electron component and the bound component of the momentum space wave function. As will be seen the rings can be parameterized by taking into account free expansion of the ejected electron cloud.

Illescas and Riera propose an exponential factor $\exp(iU(\vec{r},t))$ to parameterize the overall motion of the ejected electron cloud in configuration space [10]:

$$U(\vec{r},t) = \frac{vr^2}{2R}. \qquad (4)$$

Again v is the projectile velocity and R is the internuclear separation. This form satisfies Galilean invariance, and gives an electron current speed proportional to the distance away from the coordinate origin. (Not surprisingly, the exponential factor from ref. [10] is the same as the one introduced by Soloviev and Vinitsky [9] to counteract artificial flow induced by transforming to scaled coordinates, where the internuclear separation is always unity.) Furthermore, ref. [10] states that the exponential can be factored out of the total wave function:

$$\psi(\vec{r},t) \to exp(i\frac{vr^2}{2R})\xi(\vec{r},t). \qquad (5)$$

Only the component of the configuration space wave function that is far from both nuclear centers will be affected by the transformation, leaving the bound states largely unaffected for large R. The transformation corresponding to eq. (5) in momentum space is complicated by the quadratic dependence on r in $U(\vec{r},t)$.

Fourier transformation of the right-hand side of eq. (5) gives the equivalent transformation in momentum space:

$$\Phi(\vec{p},t) \to exp(-i\frac{Rp^2}{2v}) \otimes \Xi(\vec{p},t). \qquad (6)$$

p-independent prefactors have been omitted. The essential differences from eq. (5) is the convolution in place of the product and the reversed roles of R and v. Having R in the numerator of the phase in eq. (6) implies that the spatial frequency of the rings seen in Fig. 1 increases with internuclear separation. Eq. (6) has only interpretive value since the implied numerical deconvolution is not practical. To perform the unfolding of the exponential factor from the momentum space wave function, the wave function is transformed to configuration space where the exponential from eq. (5) is divided out, and the result is brought back to momentum space. Since the wave function eq. (2) is a two-center expansion the p in the transformation eq. (6) is taken as the distance from the target and projectile velocities for their respective expansions. As the interference effect in Fig. 1 is the same for all harmonics, the transformation is shown only for the s-harmonic on each center, $\tilde{T}_{00}$ and $\tilde{P}_{00}$.

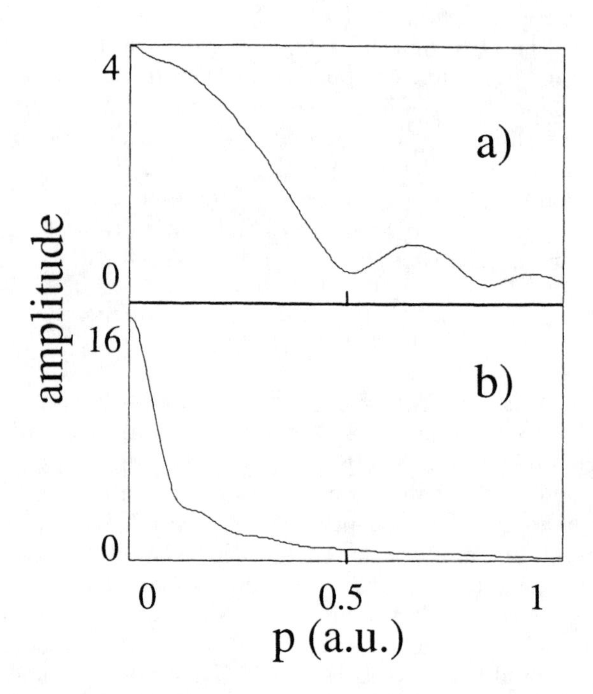

FIGURE 2. a) S-wave amplitude of the target momentum space wave function cross section at $vt = 29$ a.u.. b) The same S-wave after deconvolution of the exponential factor representing free expansion of the electron wave function.

Fig. 2a shows $\tilde{T}_{00}$ for the same energy and impact parameter as the distribution shown in Fig. 1, and Fig. 2b $\tilde{T}_{00}$ after deconvoluting the exponential in eq. (6). The same is illustrated in Fig. 3 for $\tilde{P}_{00}$ centered around the projectile. In both Figs. 2a and 3a oscillations are present in the radial amplitude which correspond to the innermost fringes in Fig.1 centered on the target and projectile respectively. After taking out the overall expansion of the ejected electron component one sees from Figs. 2b and 3b that the oscillations in amplitude of the total wave function disappear. Noting the scale of the y-axis on Figs. 2b and 3b, it is clear that much of the radial distribution is shifted towards zero momentum. (Transformations eq. (5) and eq. (6) preserve the wave function normalization.) Physically, this means that there is only slight deviation of the ejected electrons from a free expansion. The tail, extending out from the low momentum peak, is now composed of only bound states, since they are not affected by eq. (6), thus interference between ionization and excitation or capture is eliminated. Fig. 3b shows structure at low momentum, but for $q > 0.3$ a.u. the amplitude is smooth, which is the desired effect of parameterizing the free expansion.

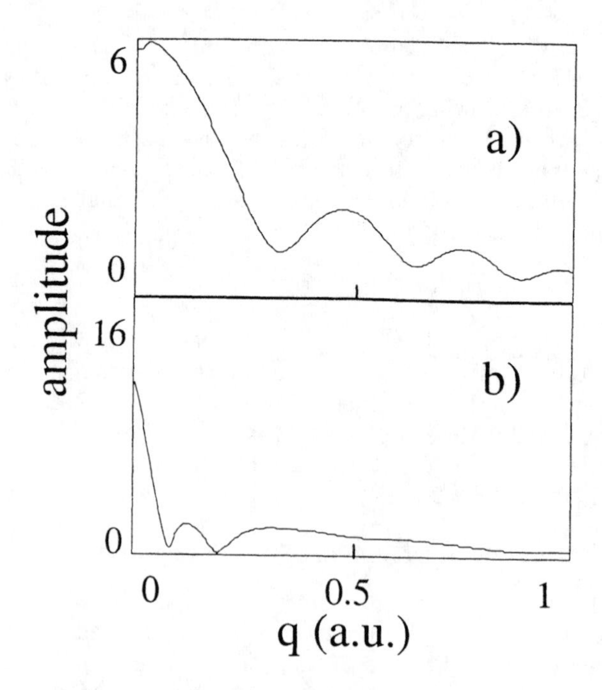

FIGURE 3. a) S-wave amplitude of the projectile momentum space wave function cross section at $vt = 29$ a.u.. b) The same S-wave after deconvolution of the exponential factor representing free expansion of the electron wave function.

Illescas and Riera find that, classically, the principal motion of ejected electrons in ion-atom collisions is free expansion even in the presence of the long-range Coulomb field from the target and projectile protons [10]. We demonstrate this is also the case quantum mechanically.

Both configuration space and momentum space present different numerical challenges in integrating the time-dependent Schrödinger equation. The electron wave function scales with R in configuration space, and accordingly, as seen in Fig. 1 and eq. (6), oscillations are present which scale with R in momentum space. Either one must have a very large integration box, or have correspondingly high resolution with a small integration box. Accounting for the background motion of ejected electrons allows one to achieve numerical solutions to the Schrödinger equation in momentum space for large internuclear separation without resorting to ultra-high radial resolution.

ACKNOWLEDGMENT

I would like to thank CD Lin for helpful discussions. This work was supported by the Division of Chemical Sciences, Office of Basic Energy Sciences, Office of Energy Research, U.S. Department of Energy.

---

[2] Wells J C, Schultz D R, Gavras P and Pindzola M S, *Phys. Rev. A* **54**, 593 (1996).
[3] Ovchinnikov S Y, Macek J H and Khrebtukov D B, *Phys. Rev. A* **56**, 2872 (1997).
[4] Dörner R, Khemliche H, Prior M H, Cocke C L, Gary J A, Olson R E, Mergel V, Ullrich J and Schmidt-Böcking H, *Phys. Rev. Lett.* **77** 4520-4523 (1996)
[5] Abdallah M A, Cocke C L, Wolff W, Wolf H, Kravis S D, Stöckli M and Kamber E, Phys. Rev. Lett. **81**, 3627 (1998).
[6] Delos J B, *Rev. Mod. Phys.* **53**, 287 (1981).
[7] Errea L F, Harel C, Jouin H, Méndez L, Pons B and Riera A, *J. Phys. B* **27**, 3603 (1994).
[8] Fritsch W and Lin CD, *Phys. Rep.* **202**, 1 (1991).
[9] Soloviev E A and Vinitsky S I, *J. Phys. B* **27**, L557 (1985).
[10] Illescas C and Riera A, *Phys. Rev. Lett.* **80** 3029 (1998).
[11] Sidky E Y and Lin CD, *J. Phys. B* **31**, 2949 (1998).
[12] Sidky E Y and Lin CD (in preparation)

THREE BODY EFFECTS IN LOW ENERGY (e,2e) PROCESSES

J. Rasch†, Colm T. Whelan‡

Institut de Physique, Laboratoire de Physique Moléculaire et des Collisions,

Technopôle Metz 2000, Rue Arago, France

‡Department of Applied Mathematics and Theoretical Physics,

University of Cambridge, Silver Street, Cambridge, CB3 9EW, England

Within the last two years a number of highly refined measurements have been performed on H targets which have yielded accurate absolute data for a range of energies and geometries[1] and it would appear that the experimental situation for this, the simplest of atomic targets is now resolved. The theoretical situation is however far from satisfactory and in this paper we will analysis some of the main approaches and characterise their strengths and their weaknesses. We have developed a numerical method which allows us to evaluate triple differential cross sections(TDCS) using the most complex position dependent analytic ansatz wave function and we will present results, using this for low energy (e,2e) processes. We will see that this approach fails when incident channel effects, such as target polarization are likely to be strong.

1 THE BBK APPROACH

In this section we consider the method of analytic ansatz wave functions as applied to electron impact ionization. The application of this type of wave function to (e,2e) processes goes back to a paper by Brauner, Briggs and Klar (BBK) in 1989. Quite generally the TDCS can be written

$$\frac{d^3\sigma}{d\Omega_f d\Omega_s dE} = (2\pi)^4 \frac{k_s k_f}{k_0} \left(|f+g|^2 + 3|f-g|^2\right), \quad (1)$$

where the prior form of the direct amplitude can be written as $f = \langle \Psi_b^- | V_a | \Phi_a \rangle$ with incident electron wave vector $\mathbf{k}_0$, and where $\mathbf{k}_s$ and $\mathbf{k}_f$ are the wave vectors of the exiting electrons. Φ_a is the initial state of the system, i. e. $e^{i\mathbf{k}_0 \cdot \mathbf{r}_f} \psi_0(r_s)$, where ψ_0 is the ground state wave function of the hydrogen atom. V_a is the initial channel interaction potential

$$V_a = \frac{1}{|\mathbf{r}_s - \mathbf{r}_f|} + \frac{-1}{r_f}$$

and Ψ_b^- is the exact solution of the Schrödinger equation developed from the final state.

If we now choose an ansatz wave function that is an exact solution of a particular model Hamiltonian $H_0 \Psi^-_{\text{ansatz}} = E \Psi^-_{\text{ansatz}}$ it can be shown that

$$f(\mathbf{k}_s, \mathbf{k}_f) = \langle \Psi^-_{\text{ansatz}} | V_a | \Phi_a \rangle - \langle \Psi^-_{\text{ansatz}} | (H - H_0)(E - H + i\eta)^{-1} V_a | \Phi_a \rangle .$$

where by neglecting the second term we get

$$f(\mathbf{k}_s, \mathbf{k}_f) \approx f(\mathbf{k}_s, \mathbf{k}_f)_{\text{ansatz}} = \langle \Psi^-_{\text{ansatz}} | V_a | \Phi_a \rangle .$$

A more thorough discussion of the derivation can be found in[2]. It should be clear that any ansatz wave function is only a good model if the neglected second term of the amplitude is small compared to the first one.

A popular choice for the ansatz wave function is

$$\begin{aligned}\Psi_f^-(\boldsymbol{r}_1, \boldsymbol{r}_2) \approx & \frac{1}{(2\pi)^3} e^{i\mathbf{k}_s \cdot \mathbf{r}_s} e^{i\mathbf{k}_f \cdot \mathbf{r}_f} \\ & \times N_s \, {}_1F_1 \left[i\alpha_s, 1, -i(k_s r_s + \mathbf{k}_s \mathbf{r}_s) \right] \\ & \times N_f \, {}_1F_1 \left[i\alpha_f, 1, -i(k_f r_f + \mathbf{k}_f \mathbf{r}_f) \right] \\ & \times N_{sf} \, {}_1F_1 \left[i\alpha_{sf}, 1, -i(k_{sf} r_{sf} + \mathbf{k}_{sf} \mathbf{r}_{sf}) \right] \quad (2)\end{aligned}$$

where $N_j := \Gamma(1 - i\alpha_j) \exp(-\alpha_j \pi / 2)$ and $\alpha_j = Z_j / |\mathbf{k}_j|$ Brauner *et. al.* [3] originally suggested taking $Z_f = Z_s = -1$, $Z_{sf} = 1$.

1.1 DYNAMICALLY SCREENED 3 BODY WAVE FUNCTION

The original Brauner *et. al.* [3] choice lead to values of the TDCS for hydrogen which where markedly too small at low impact energies. In order to rectify this and include more of the dynamics into the ansatz functions several different effective charges have been proposed. The simplest form was proposed by Berakdar and Briggs [4] who took

$$\begin{aligned} Z_s &= Z_f = -\frac{4Z - \sin(\vartheta_{sf}/2)}{4}, \\ Z_{sf} &= 1 - \sin^2(\vartheta_{sf}/2). \end{aligned} \quad (3)$$

The approach of Berakdar [5, 6] is based upon the idea that the best way to construct Ψ_f is to approximate the 3 body wavefunction by a wavefunction Ψ_f^{ansatz} that is an eigensolution of an approximate Hamiltonian H_0 such that the terms from the total Hamiltonian not included in H_0 can be neglected asymptotically. This lead him to

CP475, *Applications of Accelerators in Research and Industry*,
edited by J. L. Duggan and I. L. Morgan

position dependent effective charges of the form

$$
\begin{aligned}
Z_f &= -Z + \left[\left(\frac{3+\cos^2 4\gamma}{4}\right)^2 \frac{r_{sf} r_f^2}{(r_s + r_f)^3}\right] \\
Z_s &= -Z + \left[\left(\frac{3+\cos^2 4\gamma}{4}\right)^2 \frac{r_{sf} r_s^2}{(r_s + r_f)^3}\right] \\
Z_{sf} &= 1 - \left[\left(\frac{3+\cos^2 4\gamma}{4}\right)^2 \frac{r_{sf}^2}{(r_s + r_f)^2}\right] \qquad (4)
\end{aligned}
$$

At this point it is important to point out that one can only calculate TDCS involving Ψ_f^{DS3C} using a fully numerical method. However, calculations have been performed for (e,2e) on H using a "momentum" version or asymptotic version of Ψ_f^{DS3C}. In this method one assumes that classically in the asymptotic region the three particles can be converted in to velocity coordinates where in atomic units we can write $r_{s,f,sf} = k_{s,f,sf}\ t$, where t is time. It should be noted that for equal energy sharing the momentum dependent effective charges reduce to (3).

2 THE DWBA APPROACH

In the following discussion, we will compare our results with a variant of the distorted wave Born approximation, (DWBA). The DWBA, in contrast to the BBK, starts from the post form where it can be shown that the **exact** scattering amplitude may be written

$$f = \langle \zeta_1^-(\mathbf{k}_f, \mathbf{r}_f)\zeta_2^-(\mathbf{k}_s, \mathbf{r}_s)\,|V_b - V_1 - V_2|\,\Psi_a^+\rangle \qquad (5)$$

where $V_1(\mathbf{r}_f)$, $V_2(\mathbf{r}_s)$ are effective potentials associated with the outgoing electrons. As a first approximation to Ψ_a we could choose,

$$\Psi_a^+ = \zeta_0^+(\mathbf{k}_0, \mathbf{r}_f)\psi(\mathbf{r}_s) \pm \zeta_0^+(\mathbf{k}_0, \mathbf{r}_s)\psi(\mathbf{r}_f) \qquad (6)$$

where $\pm$ denotes spin singlet$^+$ / triplet$^-$ static exchange wavefunction for electron scattering by the H atom in the state ψ.

This approximation is discussed in great detail elsewhere, Whelan *et. al.*[7] Walters *et. al.*[8] and Rasch[9], here we will only make a few remarks. Firstly the wavefunctions, ζ_1^-, ζ_2^- are really quite general and once we have the exact wavefunction Ψ_a^+ on the left hand side of equation (5) we will get the exact scattering amplitude. However, when we make the approximation (6) we have, in effect, produced a first order approximation in the $1/r_{sf}$ potential. Now, as pointed out by Whelan *et. al.*[10], in this case we have recovered our post-prior equivalence which was characteristic of the full problem and one can interpret $\zeta_0^+(\mathbf{k}_0, \mathbf{r}_f)\psi(\mathbf{r}_s)$ as being associated with the incident channel and $\zeta_1^-\zeta_2^-$ as being associated with the final channel. The picture one has is of the incoming electron scattering in the incident channel distorting potential, in this case the static exchange

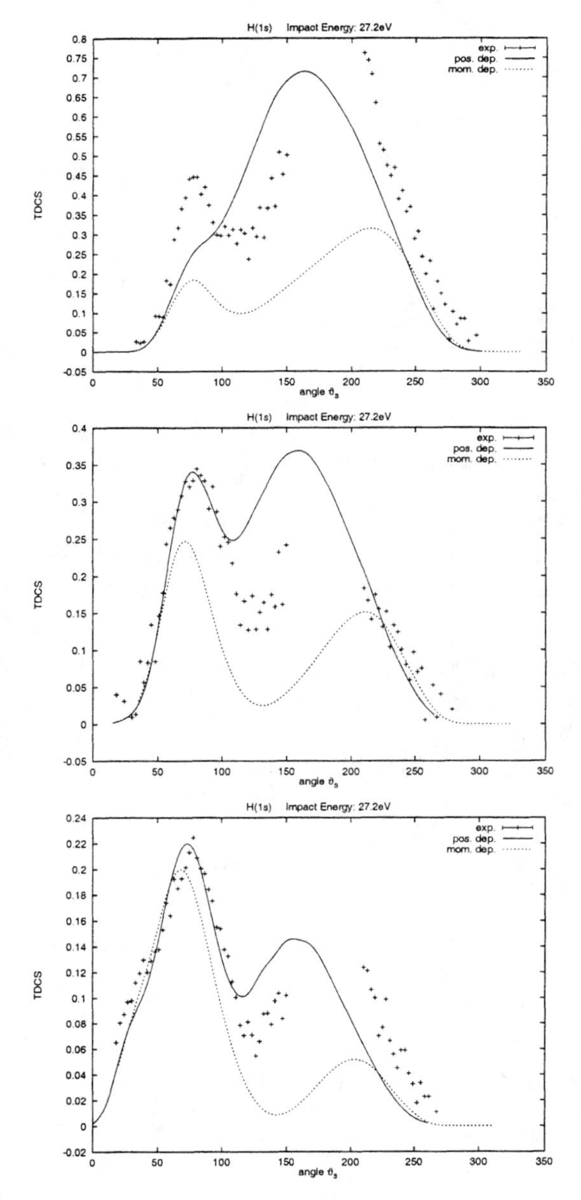

Figure 1: TDCS in coplanar energy sharing geometry, experimental data is internormalised but relative therefore all exp. data is scaled by the same overall factor to give the best fit to theory. Shown are three figures in asymmetric geometry at an impact energy of E_0 = 27.2eV, E_s = E_f = 6.8eV, angle of one of the two outgoing electrons is fixed at 345° (top), 330° (middle), 315° (bottom). Curves in all figures: (a) Berakdar position dependent effective charge calculation (solid); (b) momentum dependent effective charge calculation (dashed).

potential of the hydrogen atom, the ionizing electron-electron interaction occuring once and the two exiting electrons elastically scattering in the potentials V_1 and V_2 on their way to the detectors. So after making the approximation, (equation (6)), the character of $\zeta_1^-\zeta_2^-$ changes. They are no longer "projectors" which act on Ψ_a^+ to pick out the exact scattering amplitude, they are

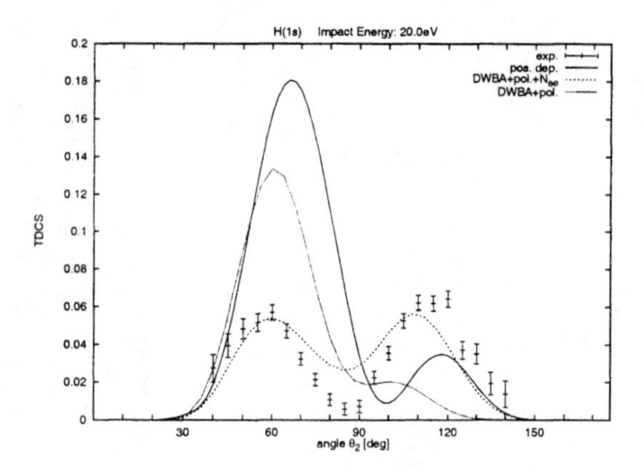

Figure 2: Coplanar symmetric geometry is shown with an impact energy of $E_0 = 20$eV, $E_s = E_f = 3.2$eV. Curves: Berakdar position dependent effective charge calculation (solid), DWBA with polarisation and Gamow factor (dashed), DWBA with polarisation (dotted).

now an intrinsic part of the approximation and it is important to include as much physics as possible in the choice of V_1, V_2.

For example, if $E_s \approx E_f$ then a reasonable choice would be,

$$V_1(\mathbf{r}_f) = -\frac{1}{r_f}, \qquad V_2(\mathbf{r}_s) = -\frac{1}{r_s}. \tag{7}$$

with this choice of DWBA direct and exchange scattering amplitudes become

$$f = \left\langle \zeta_1^-(\mathbf{k}_f)\zeta_2^-(\mathbf{k}_s) \left| \frac{1}{r_{sf}} \right| \zeta_0^+(\mathbf{k}_0)\psi \right\rangle, \tag{8}$$

$$g = \left\langle \zeta_1^-(\mathbf{k}_s)\zeta_2^-(\mathbf{k}_f) \left| \frac{1}{r_{sf}} \right| \zeta_0^+(\mathbf{k}_0)\psi \right\rangle. \tag{9}$$

We remark that a feature of this approximation is that ζ_0^+ will be different for singlet and triplet scattering, this will result in different direct and exchange amplitudes which we will denote by f^t, g^t, f^s, g^s with t and s denoting triplet and singlet. This is discussed in detail elsewhere, see Curran and Walters[11], Rasch[9]. The TDCS is then given by

$$\frac{d^3\sigma}{d\Omega_s d\Omega_f dE_s} = \frac{(2\pi)^4 k_s k_f}{4k_0}\left(|f^s + g^s|^2 + 3|f^t - g^t|^2\right). \tag{10}$$

Despite its simplicity this approximation has been used to good effect to describe those (e,2e) measurements where interactions with the nucleus play a significant role; for example in the study of multiple scattering effects in coplanar symmetric geometry, Whelan *et. al.*[7], the identification of strong interference effects in out of plane geometry, Rasch *et. al.*[12], and in the treatment of inner shell ionization processes[13]. Indeed the relativistic generalisation has proved spectacularly successful in the study of deep inner shell ionization of heavy metal targets, see e.g. Whelan *et. al.*[14], Nakel & Whelan[15].

The DWBA as formulated above does not include any effect from the post-collisional electron-electron interaction, nor indeed from higher order effects such as polarisation in the incident channel. One logical way to correct for these defects would be to expand Ψ_a^+ beyond the first order. However, even to take this expansion to second order presents serious technical difficulties, (see Rouet *et. al.*[16]; Madison *et. al.*[17]; Walters[18]). Of course the post-prior equivalence would be lost and with it the intuitively appealing idea of "incident" and "final" channels. One heuristic approach that has proved extremely useful has been to try to include higher order effects in the ζ_0 by means of a polarisation potential and final sate interactions by means of a multiplicative factor, see Whelan *et. al.*[20, 7] and Rasch[9]. This approach contains something of the philosophy of the DWBA and allows one to retain the concept of initial and final channel effects. In this approximation the TDCS is given by

$$\frac{d^3\sigma}{d\Omega_s d\Omega_f dE_s} = N_{ee}\frac{d^3\sigma^{DWBA}}{d\Omega_s d\Omega_f dE_s} \tag{11}$$

where N_{ee} is either the Gamow factor, $N_{ee} = \gamma/(e^\gamma - 1)$, with $\gamma = 2\pi/|\mathbf{k}_s - \mathbf{k}_f|$ or the Ward & Macek factor[21], in the calculations presented here we use the Gamow factor. These factors act to take some account of of the final channel electron-electron repulsion, in that they impose the correct kinematical character on the TDCS. For example, $N_{ee} = 0$ when $\mathbf{k}_s = \mathbf{k}_f$. For further discussion see Whelan *et. al.*[20] and Röder *et. al.*[1].

In Figures 1 and 2 we compare the various forms of the analytical ansatz approach with experimental data. It is seen that it is an improvement when using the positional dependent form of Berakdar[22]. However, all variants largely fail in those cases where the DWBA are most sensitive to incident channel effects (see Figure 2).

3 CONCLUSION

All the theoretical approximations fail to adequately reproduce experiments for low energy (e,2e) processes. The analytic ansatz wave function performs only poorly in regimes where incident channel 3 body effects are thought to be important, while the convergent close coupling approach has difficulty in dealing with final state interactions. The ansatz approach is difficult to improve since each attempt to include more Physics forces one to generate a new trial wave function. The convergent close coupling approach has the advantage that one may continuously add more pseudo states, but its asymmetric character and the nature of 3 body interactions in the final channel means that it will not be possible to simply extend this procedure to convergence. It is time for a new approach!

4 ACKNOWLEDGEMENTS

We gratefully acknowledge support from the Royal Society, the EU (ERB4001GT965165). This work was (partially) supported by the National Science Foundation through a grant for the Institute for Theoretical Atomic and Molecular Physics at Harvard University and Smithsonian Astrophysical Observatory. We are grateful for the use of the Hitachi parallel high performance computer, Cambridge, U.K.

References

[1] J. Röder, J. Rasch, K. Jung, Colm T. Whelan, H. Ehrhardt, R. J. Allan, and H. R. J. Walters. "On the role of Coulomb 3 body effects in low energy impact ionization of H(1s)". *Phys. Rev. A*, 53(1):225–33, 1996.

[2] S. P. Lucey, J. Rasch, and Colm T. Whelan. On the use of analytic ansatz wavefunctions in the study of (e,2e) processes. *Proc. Roy. Soc. A, in press*, 1998.

[3] M. Brauner, J.S. Briggs, and H. Klar. Triply-differential cross sections for ionization of hydrogen atoms by electrons and positrons. *J. Phys. B*, 22:2265, 1989.

[4] J. Berakdar and Briggs. J. S. Three-body Coulomb continuum problem. *Phys. Rev. Lett.*, 72:3799, 1994.

[5] J. Berakdar. Approximate analytical solution of the quantum-mechanical 3-body Coulomb problem. *Phys. Rev. A*, 53(4):2314–26, 1996.

[6] J. Berakdar. Parabolic-hypersherical approach to the fragmentation of three-particle Coulomb systems. *Phys. Rev. A*, 54(2):1480–6, 1996.

[7] Colm T. Whelan, H. R. J. Walters, and X. Zhang. (e,2e) and All That ! In Colm T. it et. al. Whelan, editor, *(e,2e) and Related Processes*, pages 1–32. Kluwer academic publishers, Netherlands, 1993.

[8] H. R. J. Walters, Colm T. Whelan, and X. Zhang. *(e,2e) and related processes.* Kluwer academic publishers, Netherlands, 1993.

[9] J. Rasch. *(e,2e) processes with neutral atom targets.* PhD thesis, Cambridge University, 1996.

[10] Colm T. Whelan, H. R. J. Walters, J. Hannsen, and R. M. Dreizler. High-energy electron-impact ionization of H(1s in coplanar asymmetric geometry). *Aust. J. Phys.*, 44(1):39–58, 1991.

[11] E. P. Curran and H. R. J. Walters. Triple differential cross sections for electron impact ionization of atomic hydrogen – a coupled pseudostate calculation. *J. Phys. B*, 20:333, 1989.

[12] J. Rasch, Colm T. Whelan, R. J. Allan, S. P. Lucey, and H. R. J. Walters. Strong interference effects in the triple differential cross section of neutral atom targets. *Phys. Rev. A*, 56(2):1379–83, 197.

[13] Zhang X., Colm T. Whelan, H. R. J. Walters, R. J. Allan, P. Bickert, W. Hink, and S. Schönberger. (e,2e) cross-sections for inner-shell ionization of argon and neon. *J. Phys. B*, 25(20):4325–35, 1992.

[14] Colm T. Whelan, H. R. J. Walters, S. Keller, H. Ast, J. Rasch, and R. M. Dreizler. Inner shell (e,2e) processes. *Can. J. Phys.*, 74:804–810, 1996.

[15] W. Nakel and Colm T. Whelan. Relativistic (e,2e) processes. *Physics Reports*, 1997. in press.

[16] F. Rouet, R. J. Tweed, and J. Langlois. The effect of target atom polarisation and wavefunction distortion in (e,2e) ionization of Hydrogen. *J. Phys. B*, 29:1767, 1996.

[17] D. H. Madison, I. Bray, and I. E. McCarthy. Exact Second Order Distorted Wave Calculation for Hydrogen Including Second Order Exchange. *J. Phys. B*, 1991.

[18] H. R. J. Walters. Perturbative methods in electron – and positron – atom scattering. *Physics Reports*, 116:1, 1984.

[19] S.J. Ward and J.H. Macek. Wave functions for contiuum states of charged fragments. *Phys. Rev. A*, 49:1049, 1994.

[20] Colm T. Whelan, R. J. Allan, J. Rasch, H. R. J. Walters, X. Zhang, J. Röder, K. Jung, and H. Ehrhardt. Coulomb three-body effects in (e,2e) collisions: The ionization of H in coplanar symmetric geometry. *Phys. Rev. A*, 50:4394, 1994.

[21] S. J. Ward and J. H. Macek. Wave-functions for continuum states of charged particles. *Phys. Rev. A*, 49(2):1049–56, 1994.

[22] J. Berakdar. Approximate analytical solution of the quantum-mechanical three-body Coulomb continuum problem. *Phys. Rev. A*, 53:2314, 1996.

[23] Colm T. Whelan, R. J. Allan, and H. R. J. Walters. PCI, polarisation and exchange effects in (e2e) collisions. *in "Journal de Physique", Volume 3, Colloque 6*, pages C6–39, 1993.

[24] J. Rasch. *(e,2e) processes with neutral atom targets.* PhD thesis, University of Cambridge, 1996.

[25] M. Streun, G. Baum, W. Blask, J. Rasch, I. Bray, S. Jones, D. H. Madison, H. R. J. Walters, and Colm T. Whelan. Low energy spin polarized (e,2e) processes with Li targets. *submitted to J. Phys. B*, 1998.

Experimental Studies of Electron Emission Produced by Fast, Grazing Ion-Surface Interactions

S. B. Elston and R. Minniti

Department of Physics and Astronomy, University of Tennessee, Knoxville, Tennessee 37996-1200

Recent measurements of electron emission from grazing collisions of carbon ions, in various charge states and having velocities of a few atomic units, with atomically clean and flat surfaces are presented and discussed. The absolute yield corresponding to convoy electron emission, apparently accelerated by the projectile-induced image potential, agrees well with a simulation by Reinhold and Burgdörfer, but the spectral distribution agrees less well. Experimental data is also available for the angular distribution of emission. Studies of the influence of the projectile charge and the extent to which charge state equilibration occurs are also examined.

INTRODUCTION

Electron emission produced by fast, multicharged ions in grazing collisions with surfaces provides an interesting arena in which atomic and condensed matter physics come into contact. Here, experimental and theoretical techniques developed in the study of ion-atom collision processes can be applied to a broad area of current technological as well as scientific interest. In the present paper, we review a comparison between observations of kinetic electron emission (1) and the predictions of a recent, highly detailed microscopic simulation of emission (2). We also consider the effect of unexpected projectile charge dependencies and independencies.

So-called kinetic processes (3) dominate emission in the collision velocity regime of interest here, in contrast to the potential emission processes more common in low velocity collisions (4), and lead to three main spectral features. A low-energy secondary-emission (cascade) tail results from degradation of higher energy electrons through multiple inelastic and elastic scattering in the bulk of the target. A broad, prominent peak in the forward direction corresponds roughly to the well-known convoy-electron peak (CEP) found in ion-solid foil transmission experiments (5), and is closely related to the more fundamental binary ion-atom collision processes of target-electron capture and projectile-electron loss to projectile-centered continuum states (6). In the ion-surface case, this peak is dramatically shifted and broadened because of interaction between the emitted electron and the dynamic surface image or "wake" potential that forms as the surface electron density rearranges in response to the passing Coulomb projectile. The present experiments were in part undertaken with a view toward using the CEP as a probe of the surface wake formation process. Finally, a feature corresponding to the ion-atom collision binary-encounter (BE) or direct ionization process appears as a broad, high-energy shoulder that overlaps to some extent with the CEP.

The simulation employs classical trajectory Monte Carlo (CTMC) methods to follow emitted electrons, uses a detailed microscopic description of both core- and valence-surface-electron densities, and provides a careful treatment of the transport and multiple scattering of electrons near the surface and of surface-image interactions. Several discrepancies noted between the simulation results and previous measurements (7) led to the present investigations and to a determination of the absolute yield of emission contained within (a fraction of) the convoy electron peak.

EXPERIMENTAL METHOD

Most of the experimental details associated with the measurements presented here have been described previously (1). Beams of carbon ions with total kinetic energies between 2.4 and 6.0 MeV, and charge states ranging from q/e =+1 to +4, were produced by the Oak Ridge EN Tandem Van de Graaff Accelerator Facility and collimated to a diameter of 0.2 mm and an angular divergence less than 0.015° (half-angle). The beams were directed onto silicon (100) target surfaces, which had been prepared according to a technique reported by Swartzentruber *et al.* (8) and shown by them to reliably produce atomically clean and flat surfaces. Surface angles of incidence (measured to the surface in the plane containing the incident velocity and the surface normal) were sufficiently small to ensure grazing, non-penetrating collisions, typically $\theta_{in} \leq 0.25°$. Substantially larger incidence angles result in surface penetration as the fraction of projectile kinetic energy associated with motion perpendicular to the surface exceeds the average repulsive wall provided by the surface layer of target ion cores. This penetration can be distinguished by the onset of large-angle projectile scattering, and a critical angle for the onset of penetration is observable (9).

The apparatus, shown schematically in Fig. 1, was connected to the beam source by two stages of differential UHV pumping, permitting base vacuums in the main scattering chamber of 1×10^{-10} Torr. A key feature of this apparatus is the ability to image emission angle information by means of a microchannel plate-amplified, position-sensitive detector (PSD) viewing the output of a 160° spherical sector electrostatic electron spectrometer through the spectrometer exit aperture. This feature permits high resolution,

CP475, *Applications of Accelerators in Research and Industry*,
edited by J. L. Duggan and I. L. Morgan

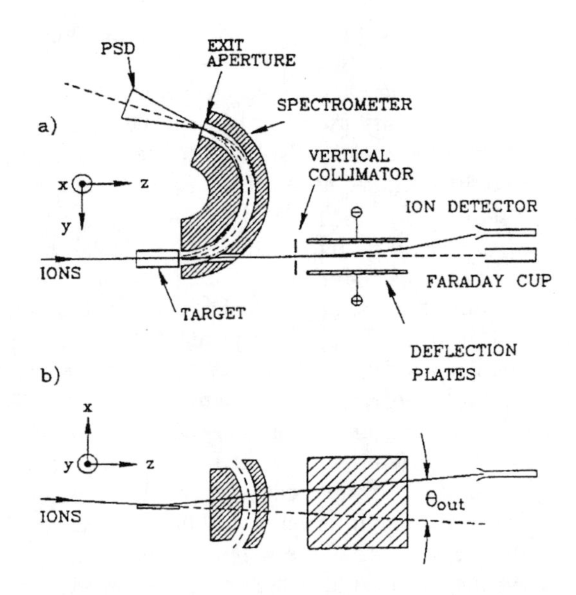

FIGURE 1. Schematic diagram of the apparatus. The surface target is centered at the entrance focus of the spherical sector spectrometer. The nominal energy resolution of the spectrometer, set by the exit aperture diameter, is $\delta E/E \sim 1\%$. The PSD views a ~ 10° wide (half-angle) sample of the angular distribution of emission from the target, which is closely reproduced by the focusing properties of the spectrometer at the exit aperture, with a resolution (also set by the exit aperture diameter) $\delta\theta/\theta \sim 0.3°$. Horizontal (x-) translation of the ion detector, shown in view (b), combined with vertical collimation, shown in view (a), permits measurement of projectile ion scattering angle distributions, while electrostatic deflection of transmitted projectiles (view (a)) permits charge state distribution measurements.

simultaneous emission angle- and energy-resolved measurements over a broad range of emission angles and, through scanning the pass-energy of the spherical sector electron spectrometer field, of emission energy. The spectrometer is rotatable about the entrance focal region to permit measurement of angular distributions of electron emission beyond the range provided for by the PSD at a single spectrometer orientation. A hole provided in the outer sector of the spectrometer permits passage, for subsequent analysis, of projectiles that have been specularly reflected from the surface target, making the present design particularly useful for measuring emission near the specular reflection direction. Projectiles exiting the scattering region are vertically collimated by a horizontal slit (see the upper (a) view of Fig. 1), electrostatically deflected to provide projectile charge state selectivity, and detected by a discrete-dynode electron multiplier. The lower view (b) of Fig. 1 illustrates that adjustable positioning of the ion detector permits measurement of the projectile scattering angular distribution in addition to providing charge state analysis. A second key feature of the apparatus is that of specular reflection selectivity, which permits rejection of events associated with projectiles scattered to large angles. Large scattering angles are likely to result from hard collisions due to surface penetration (at terrace edges, for example) or due to surface contaminants which cause local departures from surface regularity. This selectivity is achieved by positioning the ion detector to intercept ions scattered into the specular reflection direction and requiring a specified differential time-of-flight coincidence between electrons detected by the spectrometer/PSD and ions detected by the ion detector. This coincidence method can also help to guard against contamination of spectral measurements due to scattering from residual defects and contaminants and against time-dependent target degradation due to vacuum-environment exposure and radiation damage.

Finally, this coincidence detection technique permits determination of absolute electron emission yields by evaluating the ratio of detected electrons to detected projectiles and applying corrections for detector efficiencies. This has the advantage of avoiding the difficult problem of determining the fraction of incident beam striking the portion of the target viewed by the spectrometer, since for a perfect surface target, every relevant incident (grazing) projectile should be specularly reflected.

RESULTS AND DISCUSSION

Electron emission spectra observed for two instances of incident projectile velocity are displayed in Fig. 2. These are obtained by integrating the emission angular distribution over the full-width-at-half-maximum (FWHM) of the convoy peak, as detailed in Ref. 1, where the data of Fig. 2 are thoroughly discussed. The emission yield is made absolute by means of the coincidence detection method outlined above. When the total yield is obtained by integration over the FWHM of the CEP in the spectra of Fig. 2, one obtains, from the experiment, ~ 0.5 e^-/ion and ~ 0.25 e^-/ion at projectile (ion) velocities of 3.5 a.u. and 4.5 a.u. (corresponding to incident kinetic energies of 3.6 and 6.0 MeV), respectively. In comparison, the simulation predicts ~ 0.7 and ~ 0.65 e^-/ion, respectively, for the same integration intervals. This level of agreement for the total yield is considered to be reasonable given the complexity of the surface emission production and transport process, and is taken to indicate that the basic processes modeled in the simulations are indeed the important contributions to emission. However, in comparing the normalized spectra of Fig. 2 and considering the comparisons made in Ref. 7, we observe that there are several discrepancies worthy of further consideration.

The simulation results for E_{proj}=3.6 MeV indicate a high-energy shoulder due to binary-encounter (BE) ionization of valence electrons. A similar shoulder is not observed in the experiment. In this spectral region, CTMC dynamics should be accurate and dynamical screening should be unimportant. This discrepancy may indicate that the momentum density of the valence electrons (the Compton profile) is not accurately described in the simulation, which models the valence electrons as pure atomic orbitals (2). Thus measurements of the emission spectra in the BE region may provide a sensitive probe of the Compton profile of surface valence electrons.

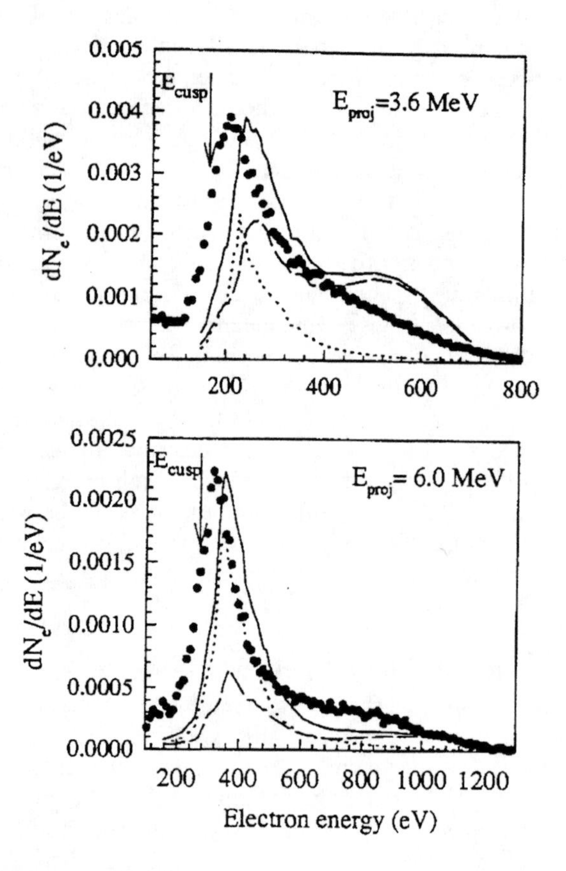

FIGURE 2. Absolute emission yield spectra from 3.6 and 6.0 MeV C ions on Si (001) surfaces at surface-incidence angles of 0.15° (solid points) versus theory (from Ref. 1, solid line), peak normalized to experiment. The emission has been integrated over emission angles corresponding approximately to the FWHM of the angular distribution. The yield scale applies to experiment; actual theory yield within the convoy peak FWHM is 2.8x that shown here for the 6 MeV result, and 1.5x that shown for the 3.6 MeV result. The calculated contributions of valence and core electrons are shown by the dashed and dotted lines, respectively.

The shift of the CEP toward higher emission energy and away from the binary ion-atom "cusp" location indicated in Fig. 2 is conjectured (1) to be the result of an underestimate of the yield in the low-energy side of the cusp. This underestimate could arise as a consequence of the classical treatment of electron scattering in the combined surface-barrier and dynamical screening potential, which has been shown to result in rainbow scattering (7). The low-energy side of the CEP corresponds to the dark side of the rainbow, and while classically forbidden, might be significantly populated in a full quantum treatment. On the other hand, it is possible that the dynamical screening is simply overestimated in the simulation, perhaps as result of an overestimate of the effective charge of the (ensemble-average) projectile near the surface. Accounting for the former possibility would produce an even greater discrepancy between experiment and theory with respect to the total yield while repairing the peak shift. Alternatively, reducing a hypothetical overestimate of effective projectile charge might be expected to reduce the yield and bring the experimental yields closer to the simulation results.

Because the effective projectile charge state near the surface may be a contributing factor in some of the observed differences between the measurements and the simulation predictions, we now outline the dependence of the spectra and yields on the incident and exit projectile charges. In essence, we find that the observed total yields and spectra are independent of the exit projectile charge state for all collision velocities studied to date. In addition, the exit projectile charge state is independent of the incident projectile charge state at all collision velocities studied. In other words, the projectiles, as a group or ensemble, achieve the condition known in ion-solid collisions as charge state equilibrium as a result of dynamic balance between probabilities for capture and loss of electrons.

Final charge state equilibrium not withstanding, we do observe (10) a dependence of the convoy electron peak shift upon the *incident* projectile charge (but *not* on the exit charge), at the highest velocity studied (v_{proj}=4.5 a.u., corresponding to E_{proj}=6.0 MeV). We interpret this to mean that, at least in this case, although charge state equilibrium occurs in the exit, specularly reflected, beam, this state is not achieved close to the surface where convoy production occurs or during the critical phase when the convoy electron is propagating in the potential responsible for shifting and/or shaping the CEP.

A final result regarding projectile charge state distributions is in order, and is presented in the case of 3.0 MeV carbon projectiles in Fig. 3. We find that the distribution of projectile exit charge states observed with grazing incidence on a silicon (100) surface target is skewed toward higher charges than the same beam transmitted by an amorphous carbon foil target of equilibrium thickness (20 $\mu g/cm^2$). Accordingly, the mean charge in the grazing incidence case is higher than that for bulk foil transmission, as seen in Fig. 4. While ideally one would compare with a silicon foil target in transmission, we are limited here to accounting for the difference in target Z. We find that available data and models (11) for the distribution of projectile charge states, following transmission in bulk media, suggest that in this collision velocity regime, a silicon foil should produce a lower mean charge than carbon, apparently because the higher electron count associated with higher Z favors capture over loss. The only data we have been able to locate that is nearly relevant is for 8.8 MeV oxygen ions transmitted through carbon ($\langle q_{out}\rangle$=5.61) and silicon ($\langle q_{out}\rangle$=5.35), and suggests that inclusion of the target Z dependence will only make the observed difference between grazing incidence and bulk equilibrium distributions larger. Our observation, that carbon ions glancing from silicon surfaces produces a higher mean charge than carbon ions transmitted through equilibrium thickness carbon foils, is thus remarkable and is likely attributable to an ion-surface interaction effect.

It may be that any charge state equilibration that occurs at distances close to the surface, where capture is most

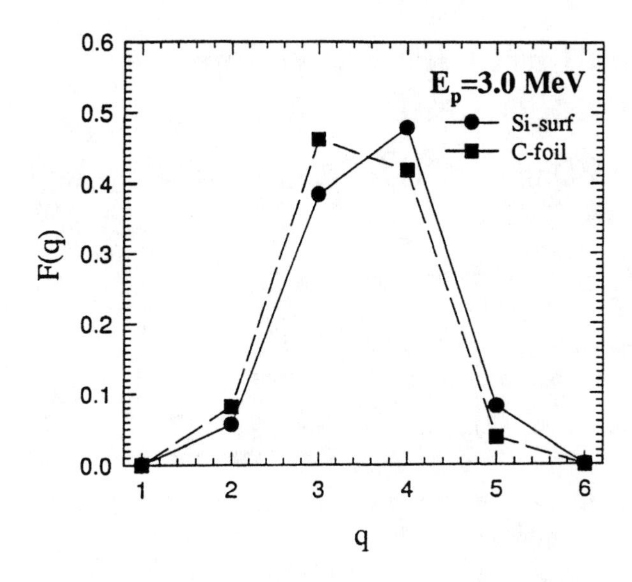

FIGURE 3. The distribution of charge state fractions observed for specularly reflected carbon ions exiting a silicon (100) surface with 3.0 MeV carbon ions incident, compared with the distribution for similarly incident ions after transmission through a 20 $\mu g/cm^2$ carbon foil.

likely to occur and the scattering environment most closely resembles that of bulk media, involves lower effective projectile charge than does the scattering environment further away, where the final shape of the CEP is determined. While only conjecture at this point, this suggests that further investigation of the emission process should carefully assess the effect of projectile charge state evolution as well as vary the target species and span a broad collision velocity range.

CONCLUSION

We have reported absolute measurements of the high-energy spectrum of electron emission produced by fast, grazing collisions of carbon ions with atomically clean and flat silicon (100) surfaces, and have found the yield of convoy electrons to agree reasonably well with a CTMC simulation by Reinhold and Burgdörfer (2). Details of residual discrepancies in the binary encounter emission regime of the spectrum suggest that further measurements may serve to probe surface valence electron Compton profiles. In addition, a detailed explanation of the convoy electron peak shape and location may provide new insight into rainbow scattering and charge equilibration processes occurring during fast ion-surface interactions.

ACKNOWLEDGMENTS

The authors acknowledge many fruitful discussions with C. O. Reinhold and J. Burgdörfer. Thanks are due to the Silicon Products Department of Texas Instruments, Inc. for the gift of target wafers, and to the staff and management of the EN Tandem Van de Graaff Accelerator Facility at Oak Ridge National Laboratory. This work was supported in

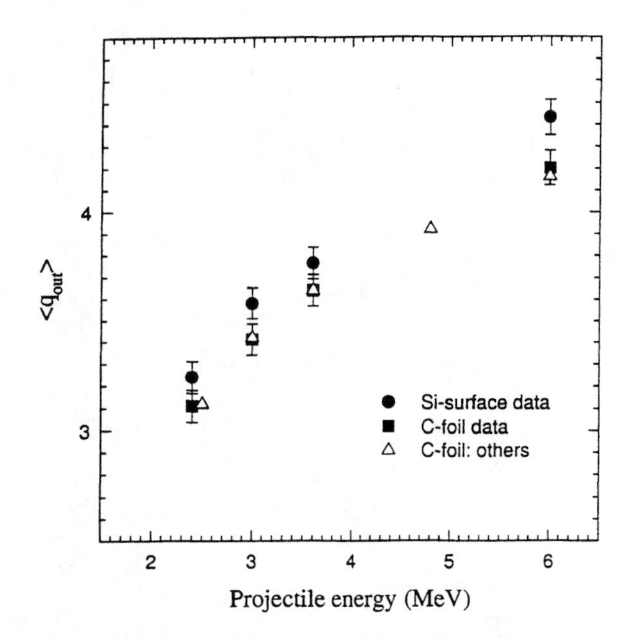

FIGURE 4. The mean charge of exit carbon projectiles measured after grazing collisions with a silicon (100) surface, compared with that for carbon projectiles transmitted through a 20 $\mu g/cm^2$ carbon foil. The foil results from Ref. 11 are also shown for comparison.

part by the National Science Foundation, and by the U. S. Department of Energy, Office of Basic Energy Sciences, under contract No. DE-AC05-96OR22464 with Lockheed Martin Energy Research Corporation.

REFERENCES

1. Minniti, R., Elston, S. B., Reinhold, C. O., Lim, J. Y., and Burgdörfer, J., Phys. Rev. A **57**, 2731 (1998).
2. Reinhold, C. O., and Burgdörfer, J., Phys. Rev. A **55**, 450 (1997).
3. Baragiola, R. A., Nucl. Instrum. Methods Phys. Res. B **78**, 223 (1993).
4. Varga, P., and Winter, H., *Electron Emission II,* Berlin: Springer-Verlag, 1992.
5. Harrison, K. G., and Lucas, M. W., Phys. Lett. **33A**, 142 (1970).
6. Breinig, M., *et al.*, Phys. Rev. A **25**, 3015 (1982).
7. Reinhold, C. O., Burgdörfer, J., Kimura, K., and Mannami, M., Phys. Rev. Lett. **73**, 2508 (1994).
8. Swartzentruber, B. S., *et al.*, J. Vac. Sci. Technol. A **7**, 2901 (1989).
9. Minniti, R., Ph. D. Dissertation, University of Tennessee, 1997 (unpublished).
10. Lebius, H., Minniti, R., Lim, J. Y., and Elston, S. B., Phys. Rev. A **54**, 4171 (1996).
11. Shima, K., Mikumo, T., and Tawara, H., Atomic Data and Nuclear Data Tables **34**, 364 (1986), and references therein.

Electron-Impact Excitation of Multicharged Ions: Merged Beams Experiments

M. E. Bannister*, Y.-S. Chung†, N. Djurić‡, G. H. Dunn‡, A. C. H. Smith§, B. Wallbank‖, and O. Woitke‡

*Physics Division, Oak Ridge National Laboratory, Oak Ridge, Tennessee 37831-6372

†Department of Physics, Chungnam National University, Gung-Dong 220, 3005-764 Daejon, South Korea

‡JILA, University of Colorado and National Institute of Standards and Technology, Boulder, Colorado 80309-0440

§Department of Physics and Astronomy, University College London, London WC1E 6BT, United Kingdom

‖Department of Physics, St. Francis Xavier University, Antigonish, Nova Scotia, Canada, B2G 2W5

Abstract. Electron-impact excitation cross sections for several multicharged ions have been measured near threshold using the merged electron-ion beams energy loss (MEIBEL) technique. This technique allows the investigation of optically-allowed and forbidden transitions with sufficient energy resolution, typically about 0.2 eV, to resolve resonance structures in the cross sections. Results from the JILA/ORNL MEIBEL experiment on allowed transitions in several multicharged ions demonstrate the ability of various theoretical methods to predict cross sections in the absence of resonances. Comparisons of R-matrix calculations and measured cross sections for transitions in Mg-like Si^{2+} and Ar^{6+}, however, indicate that theory must continue to evolve in order to more accurately predict cross sections involving significant contributions from dielectronic resonances and interactions between neighboring resonances.

INTRODUCTION

Inelastic electron-ion collisions play a significant role in the transport properties of many laboratory and astrophysical plasmas. Cross sections and rate coefficients for these processes are crucial for modeling and diagnosing plasmas in research areas such as controlled fusion, plasma processing, lighting discharges, and astrophysics. Theoretical efforts have produced much of the existing data for electron-impact excitation of ions. The excitation process can be direct or a result of resonant dielectronic capture followed by autoionization that leaves the target ion in an excited state. In optically-forbidden transitions, resonant enhancements to the cross section are significant and can dominate direct contributions by an order of magnitude or more near threshold (1).

Nearby dielectronic resonances often interfere (2) through direct configuration interaction (CI) and indirect interactions with a common continuum, resulting in significant changes in the excitation cross sections as calculated in the close-coupling R-matrix (CCR) formulation. It has also been found (2) that the overall resonance structure is extremely sensitive to the exact positions of the individual resonances. Only recently have experimentalists been able to provide theorists with critical benchmark cross sections for transitions dominated by resonances.

EXPERIMENTAL TECHNIQUE

The present experimental method of measuring absolute total cross sections for electron-impact excitation of multicharged ions relies on detecting electrons that have lost most of their energy during inelastic collisions with ions. This merged electron-ion beams energy-loss (MEIBEL) technique has been used by researchers in the JILA/ORNL collaboration (3) and at the Jet Propulsion Laboratory (4), but the details given here will be specific to the former group. The JILA/ORNL MEIBEL technique (5), with the apparatus shown in Fig. 1, employs trochoidal analyzers with longitudinal magnetic fields and transverse electric fields to merge and demerge an electron beam with an ion beam extracted from an electron-cyclotron resonance ion source. The demerger serves as an energy analyzer, separating inelastically scattered electrons from unscattered or elastically scattered electrons and deflecting them onto a calibrated position sensitive detector. The unscattered primary electrons and those elastically scattered at small angles are collected in a Fara-

CP475, *Applications of Accelerators in Research and Industry*, edited by J. L. Duggan and I. L. Morgan

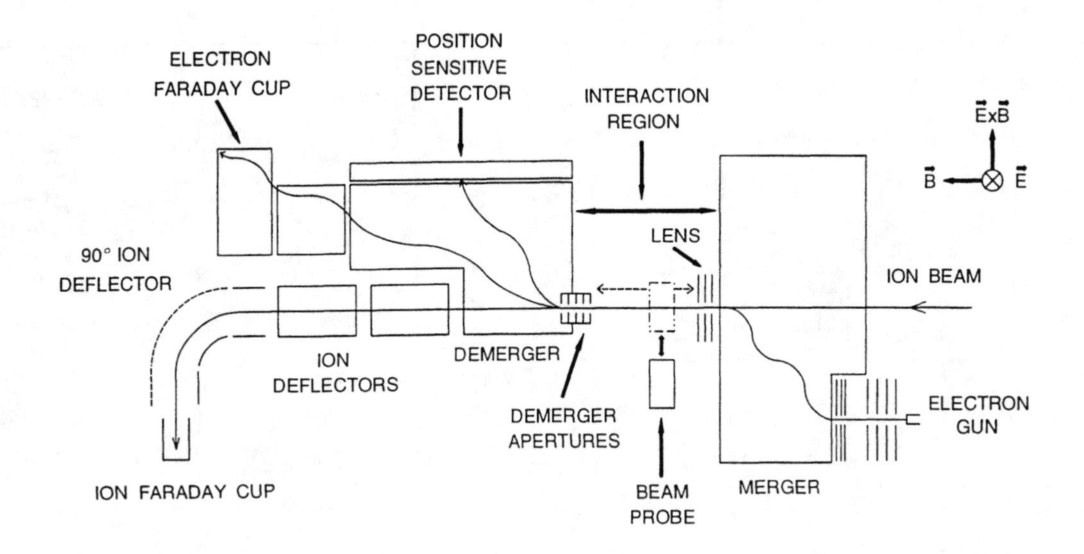

FIGURE 1. Schematic drawing of the JILA/ORNL merged electron-ion beams energy loss (MEIBEL) apparatus. See the text for a description.

day cup, since they are deflected less by the trochoidal fields. Electrons elastically scattered through large enough angles to reach the detector are blocked by a series of apertures at the entrance of the demerger. In order to separate the true signal from the backgrounds caused by the passage of the beams through the apparatus, both beams are chopped in a phased four-way sequence with the detector counts directed into four histogramming memories. The signal is then obtained by appropriate addition and subtraction of the four outputs. By measuring the beam overlaps at several points along the merge path using a two-dimensional video beam probe (6), the cross sections are put on an absolute scale. A three-dimensional trajectory modeling program (7) may be used to correct the measured cross sections at higher interaction energies.

The MEIBEL technique has distinct advantages over the traditional crossed-beams fluorescence method. Since the MEIBEL method involves the detection of low energy electrons, it may be applied both to excitation to radiating and also to non-radiating states, whereas the fluorescence technique requires excitation to a radiating state. Secondly, the MEIBEL technique features the complete collection of the signal electrons with detection efficiencies of 50-70%, bettering the capabilities of the fluorescence method by more than an order of magnitude. Finally, the merged-beams geometry yields an energy resolution of typically 0.2 eV, which is about two times better than that of most fluorescence experiments. These advantages are demonstrated in the measurements of spin-forbidden transitions to metastable states shown in the next section. The disadvantage of the MEIBEL technique is the limited energy range over which cross sections can be measured.

RESULTS

The dipole-allowed $2s \rightarrow 2p$ transition in Li-like B^{2+} has been investigated (9) with the MEIBEL technique, with the results shown in Fig. 2. The theoretical predictions of both Bartschat and co-workers (10, 11), an R-matrix with pseudo-states (RMPS) calculation, and Clark and Abdallah (12), a distorted-wave (DW) calculation, are shown convoluted with a 0.24 eV FWHM Gaussian representing the experimental energy distribution. The agreement between the experimental data and the RMPS calculation is striking, whereas the DW calculation appears to overestimate the cross section by about 25%.

The MEIBEL technique was also used to measure cross sections (13) for excitation of the dipole-allowed $3s^2\,{}^1S \rightarrow 3s3p\,{}^1P$ transition in Ar^{6+} as shown in Fig. 3. The 8-state CCR calculations of Griffin *et al.* (1) and the DW calculations of Clark *et al.* (14), each convoluted with a 0.24 eV FWHM Gaussian representing the experimental energy distribution, are also shown in Fig. 3. There is excellent agreement between all three sets of data for this transition. The measurement of this allowed transition nails down the absolute energy scale of the experiment, which is crucial for the measurement of the spin-forbidden transition in Ar^{6+} discussed below.

Since spin-forbidden transitions usually have small non-resonant contributions, the cross sections are often dominated by dielectronic resonances as discussed in the introduction. This is clearly demonstrated by the experimental excitation cross sections (15) for the $3s^2\,{}^1S \rightarrow 3s3p\,{}^3P$ transition in Si^{2+} that are shown in Fig. 4 along with separate 12-state CCR calculations of Baluja *et al.* (16) and Griffin *et al.*

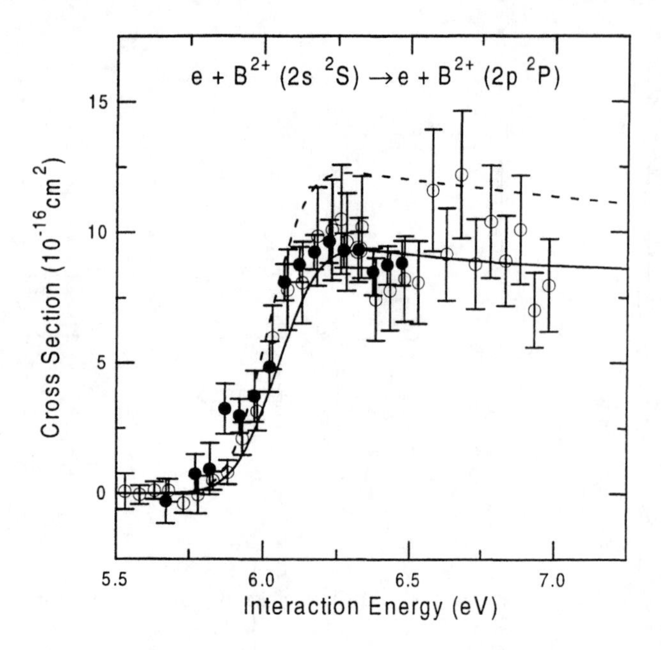

FIGURE 2. Cross sections for electron-impact excitation of the 2s → 2p transition in B^{2+}. The circles represent the MEIBEL results of Ref. (9) with relative error bars at the 90% confidence level. The outer error bars on the point at 6.3 eV represent the total expanded uncertainty. The solid curve is the RMPS calculation of Refs. (10) and (11) and the dashed curve is the DW calculation of Ref. (12), both convoluted with a 0.24 eV FWHM Gaussian representing the experimental energy distribution.

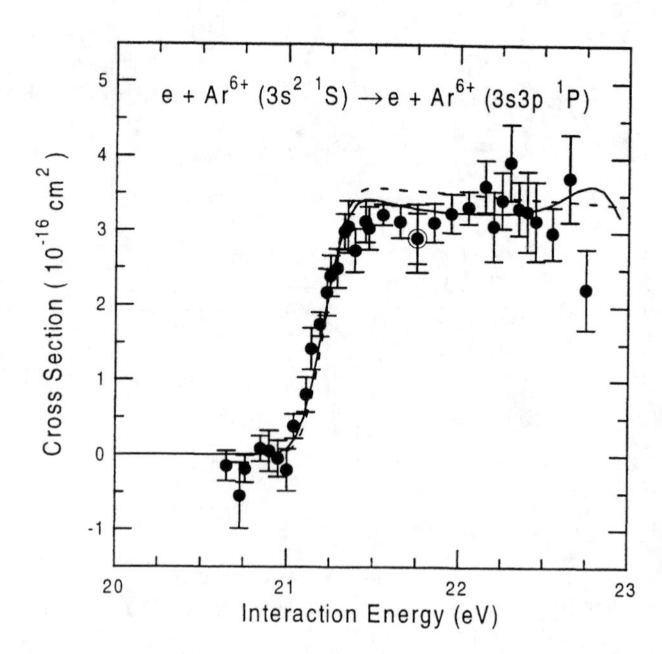

FIGURE 3. Cross sections for electron-impact excitation of the $3s^2\,^1S \rightarrow 3s3p\,^1P$ transition in Ar^{6+}. The circles are the MEIBEL results of Ref. (13) with relative error bars at the 90% confidence level. The outer bars on the point at 21.75 eV represent the total expanded uncertainty. The solid curve is the 8-state CCR calculation of Ref. (1) convoluted with a 0.24 eV FWHM Gaussian representing the experimental energy distribution. The dashed curve is the DW calculation of Ref. (14) convoluted with the same Gaussian.

(1), which are each convoluted with a Gaussian of 0.24 eV FWHM representing the experimental energy distribution. All three sets of data agree very well on the large resonance peak near 6.7 eV that dominates the non-resonant contributions to the cross section by almost an order of magnitude. The second resonance peak predicted by theory could not be investigated due to ion energy limitations in the experiment that prevented measurements beyond 7.6 eV.

As predicted by the 8-state CCR calculations of Griffin *et al.* (1), the experimental excitation cross sections (13) for the spin-forbidden $3s^2\,^1S \rightarrow 3s3p\,^3P$ transition in Ar^{6+} are also dominated by dielectronic resonances. The measurements are shown in Fig. 5 along with the CCR calculations (1) convoluted with a Gaussian of 0.24 eV FWHM. There is very good agreement on the resonance feature near 15.5 eV. However, a discrepancy exists for the peak near 14.4 eV, indicating that the theory has difficulty calculating the precise positions of the contributing resonances and their interference. Comparison of the CCR predictions with those of the independent-processes isolated-resonance distorted-wave (IPIRDW) approximation (17), which neglects interactions between resonances, suggests that the lower-energy peak may be strongly influenced by interference effects in the CCR calculation.

CONCLUSIONS

The MEIBEL technique is a powerful tool (18) for investigating near-threshold electron-impact excitation of ions. It is particularly useful for spin-forbidden transitions which are dominated by dielectronic resonances. The close-coupling R-matrix (CCR), R-matrix with pseudo-states (RMPS), and distorted-wave (DW) methods are all fairly successful in predicting cross sections for dipole-allowed transitions in the absence of significant resonance contributions. The agreement between experiment and theory is not as good when resonances dominate, as for spin-forbidden transitions, and varies greatly even for different resonances in the same transition. The experimental cross sections discussed here serve as crucial benchmarks for the close-coupling R-matrix theory and suggest that some refinements are required for the calculations to accurately reproduce the resonance positions and cross section contributions.

ACKNOWLEDGMENTS

The authors thank N. R. Badnell, T. W. Gorczyca, D. C. Griffin, M. S. Pindzola, J. S. Shaw, and K. Bartschat for providing unpublished cross section calculations. This research

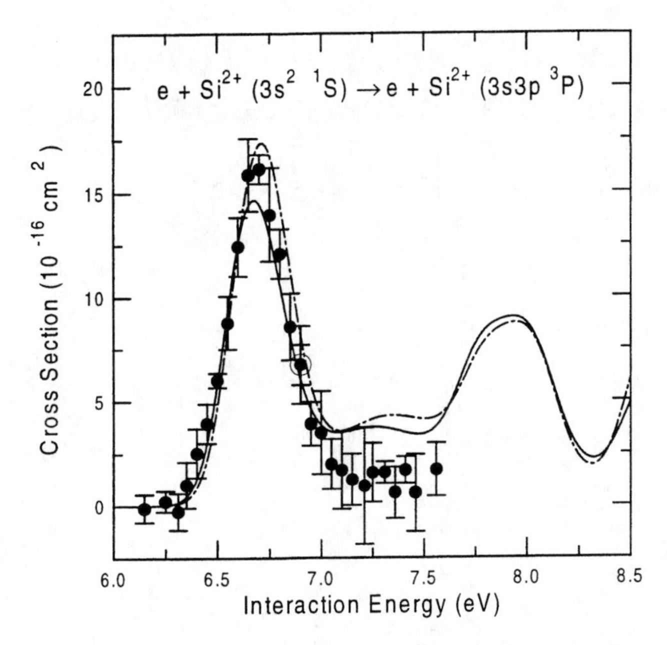

FIGURE 4. Cross sections for electron-impact excitation of the $3s^2\,^1S \rightarrow 3s3p\,^3P$ transition in Si^{2+}. The circles are the MEIBEL results of Ref. (15) with relative error bars at a 90% confidence level. The outer bars on the point at 6.9 eV represent the total expanded uncertainty. The curves are convolutions of a Gaussian (0.24 eV FWHM) with CCR calculations from Ref. (1) [solid] and Ref. (16) [dash-dot].

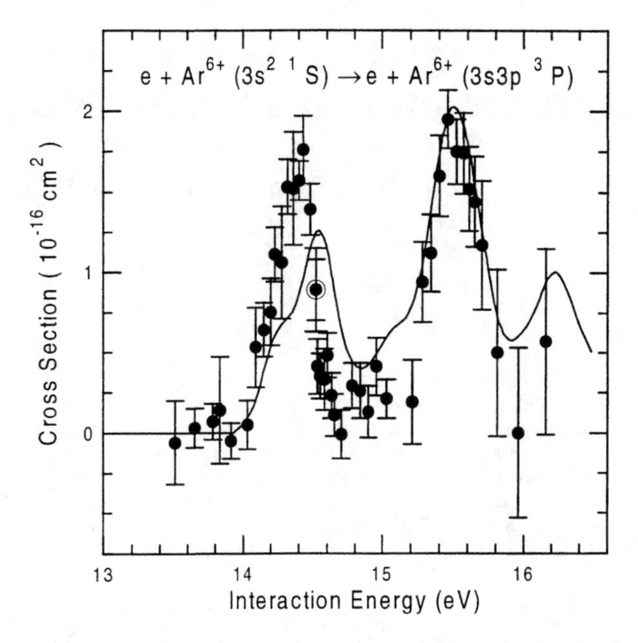

FIGURE 5. Cross sections for electron-impact excitation of the $3s^2\,^1S \rightarrow 3s3p\,^3P$ transition in Ar^{6+}. The circles are the MEIBEL results of Ref. (13) with relative error bars at a 90% confidence level. The outer bars on the point at 14.5 eV represent the total expanded uncertainty. The solid curve is a convolution of a Gaussian (0.24 eV FWHM) with CCR theory from Ref. (1).

was supported by the Office of Fusion Energy Sciences of the U. S. Department of Energy, under Contract No. DE-AC05-96OR22464 with Lockheed Martin Energy Research Corp. and Contract No. DE-A102-95ER54293 with the National Institute of Standards and Technology.

REFERENCES

1. D. C. Griffin, M. S. Pindzola, and N. R. Badnell, *Phys. Rev.* A **47**, 2871-2880 (1993).
2. D. C. Griffin, M. S. Pindzola, F. Robicheaux, T. W. Gorczyca, and N. R. Badnell, *Phys. Rev. Lett.* **72**, 3491-3494 (1994).
3. E. K. Wåhlin, J. S. Thompson, G. H. Dunn, R. A. Phaneuf, D. C. Gregory, and A. C. H. Smith, *Phys. Rev. Lett.* **66**, 157-160 (1991).
4. S. J. Smith, K.-F. Man, R. J. Mawhorter, I. D. Williams, and A. Chutjian, *Phys. Rev. Lett.* **67**, 30-33 (1991).
5. E. W. Bell, X. Q. Guo, J. L. Forand, K. Rinn, D. R. Swenson, J. S. Thompson, G. H. Dunn, M. E. Bannister, D. C. Gregory, R. A. Phaneuf, A. C. H. Smith, A. Müller, C. A. Timmer, E. K. Wåhlin, B. D. DePaola, and D. S. Belić, *Phys. Rev.* A **49**, 4585-4596 (1994).
6. J. L. Forand, C. A. Timmer, E. K. Wåhlin, B. D. DePaola, G. H. Dunn, D. Swenson, and K. Rinn, *Rev. Sci. Instrum.* **61**, 3372-3377 (1990).
7. SIMION 3D Version 6.0; David A. Dahl, Idaho National Engineering Laboratory.
8. D. W. Savin, L. D. Gardner, D. B. Reisenfeld, A. R. Young, and J. L. Kohl, *Phys. Rev.* A **51**, 2162-2168 (1995).
9. O. Woitke, N. Djurić, G. H. Dunn, M. E. Bannister, A. C. H. Smith, B. Wallbank, N. R. Badnell, and M. S. Pindzola, *Phys. Rev.* A **58** (to be published, 1998).
10. P. J. Marchalant, K. Bartschat, and I. Bray, *J. Phys.* B **30**, L435-L440 (1997).
11. K. Bartschat, private communication.
12. R. E. H. Clark and J. Abdallah, Jr., *Phys. Scr.* **T62**, 7-18 (1996).
13. Y.-S. Chung, N. Djurić, B. Wallbank, G. H. Dunn, M. E. Bannister, and A. C. H. Smith, *Phys. Rev.* A **55**, 2044-2049 (1997).
14. R. E. H. Clark, N. H. Magee, J. B. Mann, and A. L. Merts, *Astrophys. J.* **254**, 412-418 (1982).
15. B. Wallbank, N. Djurić, O. Woitke, S. Zhou, G. H. Dunn, A. C. H. Smith, and M. E. Bannister, *Phys. Rev.* A **56**, 3714-3718 (1997).
16. K. L. Baluja, P. G. Burke, and A. E. Kingston, *J. Phys* B **13**, L543-L545 (1980).
17. N. R. Badnell, D. C. Griffin, T. W. Gorczyca, and M. S. Pindzola, *Phys. Rev.* A **50**, 1231-1239 (1994).
18. Further experimental details and tabulations of the data are available at the World-Wide Web site http://www-cfadc.phy.ornl.gov/meibel/.

Theoretical Studies of the Effects of Electron-Electron Correlation in the Photoionization of Small Molecular Systems

Robert R. Lucchese, Mona Wells, Shaleen K. Botting

Department of Chemistry
Texas A&M University
College Station, Texas 77843-3255

Photoionization cross sections were computed using correlated initial and final target states obtained with standard *ab initio* techniques and using a numerical representation of the continuum scattering orbitals. We will focus on ionization processes where electron-electron correlation effects are important. The first example is the resonant ionization of CO at energies just below the threshold for C $1s^{-1}$ ionization. We will examine the lifetime of the resonant state and the pathways for the Auger decay of this state. A second example is the inner valence photoionization of C_2H_2 where we will consider the energy dependence of photoionization cross sections leading to satellite states in comparison to the energy dependence of the photoionization cross sections leading to the corresponding main hole ionic states.

INTRODUCTION

One application of synchrotron radiation has been to the study of molecular photoionization. Theoretically, there has been much progress made in the study of small molecular systems. In particular, linear systems have been studied in great detail with nonlinear polyatomic systems receiving less study due to the extensive computations required for these systems. The level of calculation has increased from single-channel frozen-core Hartree-Fock, to multichannel frozen-core Hartree-Fock (1), to multi-channel configuration interaction (MCCI) (2) and multi-channel complete active space (MCCAS) type calculations (3).

THEORY AND METHODS

Here we will describe two calculations on small linear molecular systems which have been performed using a single-center expansion technique combined with the Schwinger variational principle. The MCCI wave function for a system with a continuum electron is written as (2)

$$\Psi_{\text{MCCI}} = \sum_{i=1}^{N_c} \Phi_i(\xi_i) = \sum_{i=1}^{N_c}\sum_{j=1}^{N_b} C_{ij}\psi_j(\xi_i), \tag{1}$$

where ξ_i is the ith channel scattering state, and Φ_i, represents the configuration interaction (CI) wave function of the residual ion in channel i. N_c is the number of channels and N_b is the number of configuration state functions (CSF) in the expansion. The notation $\psi_j(\xi_i)$ implies a spin-adapted N-electron CSF, not a simple product of ψ_j and ξ_i.

By inserting this wave function into the Schrödinger equation, a matrix form of the equation can be obtained which then can be used to determine the channel scattering functions ξ_i. The form of the scattering potential found in the channel Schrödinger equation (2) and its evaluation (4) have been discussed in detail previously. When this approach is applied to photoionization, the channel transition moment may be obtained, where final variational expression for the channel transition moment, $I_i^{\text{L(V)}}$, is given by

$$I_i^{\text{L(V)}} = \left\langle R_i^{\text{L(V)}} \middle| \xi_i^0 \right\rangle + \sum_{\alpha,\beta} \left\langle R_i^{\text{L(V)}} \middle| \underline{G}_c \underline{V}_Q \middle| \phi''_\alpha \right\rangle \left\langle \underline{V}_Q - \underline{V}_Q \underline{G}_c \underline{V}_Q \right\rangle_{\alpha\beta}^{-1} \times \left\langle \phi''_\beta \middle| \underline{V}_Q \middle| \xi_i^0 \right\rangle, \tag{2}$$

where the quantity $R_i^{\text{L(V)}}$ is a function of the electronic dipole operator, $\mu_i^{\text{L(V)}}$. The superscript L(V) denotes that this equation may be solved either for the length form or for the velocity form. The potential $\underline{V}_Q$ in Eq. (2) is a Phillips-Kleinman pseudo-potential and $\underline{G}_c$ is the multi-channel Coulomb Green's function matrix given by

$$\left(\underline{G}_c\right)_{ij} = G_c(E_i)\delta_{ij}, \tag{3}$$

where the Coulomb Green's function is defined as

$$\left(-\frac{1}{2}\nabla^2 - \frac{1}{r} - E_i\right)G_c(r,r') = -\delta^3(r-r'). \tag{4}$$

CP475, *Applications of Accelerators in Research and Industry*,
edited by J. L. Duggan and I. L. Morgan

In the mixed form the doubly differential photoionization cross section of a given channel is proportional to the product of the length and velocity forms of transition moments from Eq. (2) and is given by

$$\frac{d^2\sigma^{\mathrm{M}}}{d\Omega_{\hat{k}_i} d\Omega_{\hat{n}}} = \frac{4\pi^2\hbar\omega}{c}\mathrm{Re}\left\{\left(I_i^{\mathrm{L}}\right)^* I_i^{\mathrm{V}}\right\}, \quad (5)$$

The final equation for differential cross section is obtained by integrating Eq. (5) over all spatial orientations of the molecule in the laboratory frame.

AUGER DECAY OF CO

The first study we will consider here is the photoexcitation and subsequent decay of the C $1s \rightarrow 2\pi^*$ state of CO (5). This state occurs at an energy just below the threshold for the creation of the C $1s^{-1}$ core hole state of CO. For this system we represented the initial and final states in complete active space configuration interaction (CASCI) wave functions using and active space of 10 electrons in 10 orbitals. We then performed a series of five channel calculations with four closed channels coming from ion states used to represent the resonant C $1s \rightarrow 2\pi^*$ state and a fifth open channel leading to one of the valance ion states to which the resonant state decays through an Auger decay process. In preliminary calculations we found that the decay rates to different final valence ion state were the same if we included several open channels or if we summed the rate from a series of calculations where only one open channel was included. Including 31 valence channels we found a total width of 60.0 meV. Adding an estimated radiative decay width of 3.9 meV to this non-radiative decay width leads to a total width of 63.9 meV which is in reasonable agreement with the experimental estimate of 85±3 meV of Shaw *et al.* (6).

Using the relative rates computed here it is also possible to predict the dexcitation electron spectrum. We find that there are two types of lines in agreement with the experimental data (7). We find participator Auger decay going to low energy ion states, where the electron in the excited π^* orbital participates in the Auger process. Decay to higher energy ion states often leaves an electron in the π^* orbital. The decay into such states is referred to as a spectator decay process since the electron in the π^* orbital is not involved in the decay process.

INNER VALENCE IONIZATION OF C_2H_2

We have also studied the inner valence ionization of C_2H_2. In C_2H_2 the $2\sigma_g^{-1}$ ion state is strongly correlated with configurations involving excitations into the π^* orbital (8). This configuration mixing leads to the presence of four strongly excited satellite peaks in the photoelectron spectrum at approximately (9) 26.8, 28.0, 30.0, and 31.0 eV just above the that of the main $2\sigma_g^{-1}$ peak which is at 23.5 eV. The intensities of the ionization leading to these states may be related to the intensity of the ionization leading to the main $2\sigma_g^{-1}$ line. If all of the intensity for the excitation of these lines comes from the configuration mixing with the $2\sigma_g^{-1}$ state, then at high energy this intensity ratio should approach a ratio of the corresponding spectroscopic intensity factors (SIF) which are defined as

$$I_{\mathrm{SIF}} = \sum^{c} \left|\left\langle \Psi_{s,c}^{\mathrm{MCCI},N-1} \middle| a_q \middle| \Psi_{s,(0)}^{\mathrm{MCCI},N} \right\rangle\right|^2, \quad (6)$$

where $\Psi_{s,c}^{\mathrm{MCCI},N-1}$ is the target ion state, $\Psi_{s,(0)}^{\mathrm{MCCI},N}$ is initial state, and a_q is an annihilation operator for an electron in orbital q.

The ion states were computed using 9 electrons in set of 13 active orbitals. These states were represented by multi-configuration self-consistent field (MCSCF) wave functions which were computed using a state averaged MCSCF procedure with 17 ion states. The orbital occupations were restricted leading to a reduction in the size of the calculation without a significant increase in the error in the energies of the states when compared to more detailed calculations (10).

Figure 1 shows the ratio of the four satellite component cross sections to the cross section of the $2\sigma_g^{-1}$ main hole for photon energies ranging from 40 to 1000 eV. Each panel in Fig. 1 also has a straight line across the panel which is the ratio of SIFs matching the same states used in the cross section ratios. Ratios are given both for single channel calculations where only the asymptotic ion state is included and multi-channel calculations where 11 ion channels are included. The cross section ratios for both satellite 1 and satellite 2 converge to the SIF ratios in a manner falling within the Becker and Shirley (11) nomenclature for shake-up satellites. Moreover, at lower energies where correlation effects are important there is structure imposed by resonant features. There is a high energy σ_u shape resonance, which has been previously characterized (12), and which is more strongly manifest in the $2\sigma_g^{-1}$ main line than in the satellite 1 line, leading to a dip in the ratio at ~50 eV. In the satellite 2 ratio, however, the same σ_u shape resonance leads to a peak at ~60 eV. We can also see that the correlation introduced in the channel coupling causes the ratios to converge to their high energy limits more slowly than when there is no channel coupling.

The behavior of the cross section ratios for satellites 3 and 4 seen in Fig. 1 is quite different from that seen for the first two satellites. Here the ratios do not converge to the SIF ratios and we cannot assign a purely $2\sigma_g^{-1}$ origin to these satellite features. The slight humped feature at ~200 eV is however characteristic of ionization leading to the $3\sigma_g^{-1}$ state. This feature suggests that satellites 3 and 4 borrow their intensity from both the $2\sigma_g^{-1}$ and $3\sigma_g^{-1}$ channels.

CONCLUSIONS

Current computational capabilities allow for a detailed investigation of the effects of electron-electron correlation in photoionization processes in small molecular systems. In our study of the ionization of CO, we saw that we could obtain an accurate estimate of the lifetime of the resonant autoionizing state. In the study of the photoionization of C_2H_2 we found that the SIFs can be used to determine relative cross sections only for selected satellite states. Furthermore, we found that for the satellites where the SIF ratio is approached asymptotically, correlation effects cause this ratio to be obtained only at photon energies above 500 eV.

ACKNOWLEDGMENTS

Funding from The Welch Foundation under grant A-1020 is gratefully acknowledged.

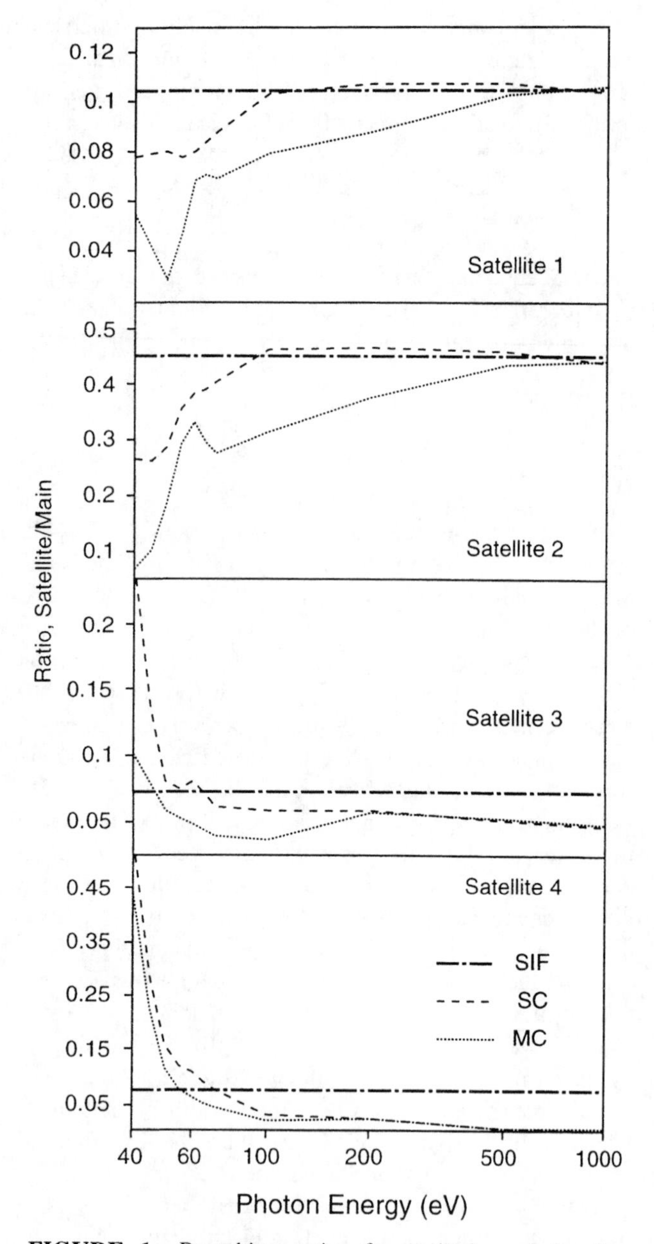

FIGURE 1. Branching ratios for excitation of the four satellite states of the $2\sigma_g^{-1}$ ion state of C_2H_2. SIF is the ratio computed from the spectroscopic factors, SC is the result from a single-channel calculation, and MC is the result from a multi-channel calculation.

REFERENCES

1. R. R. Lucchese, K. Takatsuka, and V. McKoy, Phys. Rep. **131**, 147-221 (1986).
2. G. Bandarage and R. R. Lucchese, Phys. Rev. A **47**, 1989-2003 (1993).
3. R. E. Stratmann, G. Bandarage, and R. R. Lucchese, Phys. Rev. A **51**, 3756-65 (1995).
4. R. E. Stratmann and R. R. Lucchese, J. Chem. Phys. **102**, 8493-505 (1995).
5. S. K. Botting and R. R. Lucchese, Phys. Rev. A **56**, 3666-74 (1997).
6. D. A. Shaw, G. C. King, D. Cvejanovic, and F. H. Read, J. Phys. B **17**, 2091 (1984).
7. W. Eberhardt, E. W. Plummer, C. T. Chen, and W. K. Ford, Aust. J. Phys. **39**, 853 (1986).
8. M. Wells and R. R. Lucchese, (to be published).
9. S. Svensson, E. Zdansky, U. Gelius, and H. Ågren, Phys. Rev. A **37**, 4730-3 (1988).
10. M. S. Moghaddan, S. J. Dejardins, A. D. O. Bawagan, K. H. Tan, Y. Wang, and E. R. Davidson, J. Chem. Phys. **103**, 10537-47 (1995).
11. U. Becker and D. A. Shirley, Physica Scripta **T31**, 51-66 (1990).
12. D. Lynch, M.-T. Lee, and R. R. Lucchese, J. Chem. Phys. **80**, 1907-16 (1984).

Electron Angular Distributions Measured with Simultaneous Analysis of Emerging Projectile Charge-State

P.Focke, G.Bernardi and W.Meckbach

Centro Atomico Bariloche and Instituto Balseiro, RA-8400 Bariloche, Argentina

The experimental setup at the Centro Atomico Bariloche for measurement of energy and angular distributions of electron emitted by atomic collisions was further developed including charge-state analysis of the emerging projectiles. The simultaneous detection of the ejected electrons and emerging projectiles provided a more detailed information about the collision dynamics. This arrangement was used for the study of three different collision processes: a) "electron transfer into the continuum" for the neutral projectile H, b) "electron loss" process of H incident on Ar, and c) "transfer ionization" for an incident p beam. We present a description of the experimental setup and the main results obtained.

INTRODUCTION

The study of distributions of electrons emitted in ion-atom collisions provides valuable information for the understanding of the fundamental mechanisms underlying the dynamics of the electron-ion interaction. This information is obtained from measuring the double differential cross section, $d\sigma/dEd\Omega$, that is differential in energy E of the electron and its angle of emission θ relative to the beam direction. The extensive theoretical and experimental research performed on the energy and angular distributions permitted identification of several different production mechanisms[1,2]. In many collision systems the emission of electrons is a superposition of several channels, for instance if the incident projectile carries electrons (dressed projectile), then emitted electrons can result from ionization of the target, the projectile or even both collision partners. Other cases may result from multielectron emission with nude projectiles such as transfer ionization. A separation of at least part of these contributions in the measured distributions can be done if the charge state of the emerging projectile is simultaneously identified. This represents a further step in providing additional information for the understanding of the atomic collision dynamics. It was with this aim that we expanded our available facility for measuring electron distributions incorporating a coincidence technique. We applied this approach to the study of some simple cases. For one collision system we measured, at the emission angle of $\theta = 0°$, the "electron transfer into the continuum" processes. For the other case, we used a neutral H projectile and identified the emission where at least one electron comes from the projectile. A third system studied focussed on the mechanism of "transfer ionization" using a nude H+ projectile and measured electron distributions associated with an emerging neutral H.

In the following we describe briefly the experimental setup and the main results obtained.

EXPERIMENT

A schematic of the apparatus is shown in figure 1. A detailed description has been given elsewhere[3,4] and for the present purpose we highlight only some relevant features. A beam of light ions, for the present cases H+, is provided by the Cockcroft-Walton accelerator of the Centro Atomico Bariloche. The accelerator has an ion source of the radio frequency type and the range of accelerating voltages that provides useful beam is from 20 to 300 kV. The projectile beam before entering the experimental line is momentum analyzed by a 90° bending magnet.

The experimental line consists basically of three sections. In the first part, the beam preparation section, there are two sets of adjustable slits used to define the collimation and to control the intensity of the beam. In the middle of the section there is a differentially pumped charge-exchange chamber used to produce a neutral beam. With gas in that chamber, normally air, the projectiles emerge in a mixture of charged and neutral components. The charged beam being eliminated downstream by an electrostatic deflector. The pressure, under normal operating conditions of the gas cell, is about 1×10^{-6} Torr in the beam transport region.

The second section is the interaction region consisting of a chamber with the electron analyzer. Before entering the chamber, the beam passes through a second electrostatic deflector that strips it from an accompanying electron component. The electron analyzer is of the cylindrical mirror type with second order focusing [5]. The analyzer, mounted on the chamber lid, can be rotated around a vertical axis perpendicular to the beam direction. The analyzer axis, tilted 42.3° respect to the horizontal plane, passes through the intersecting point of the beam and rotation axis. This point is the location of the collision region and object focus of the analyzer. Small slots are cut in the plates to allow the projectiles go through without

CP475, *Applications of Accelerators in Research and Industry*,
edited by J. L. Duggan and I. L. Morgan

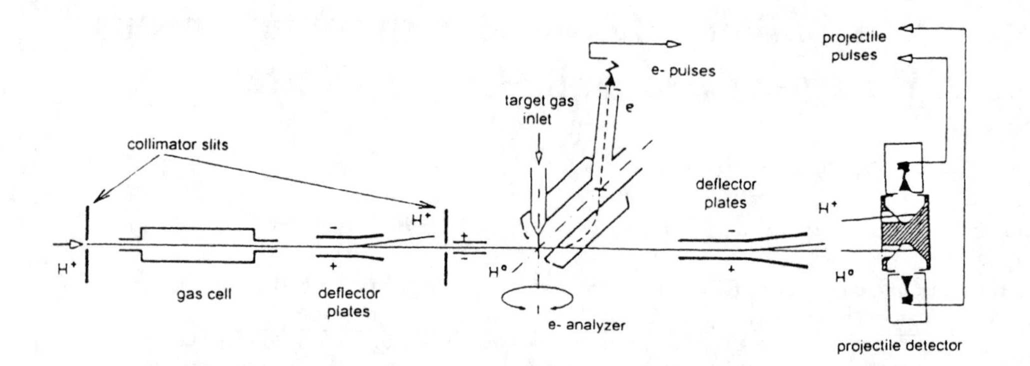

Figure 1. Schematic view of the experimental setup for measuring electron emission spectra.

hitting the walls. With this arrangement it is possible to measure complete angular distributions in a continuous range from $\theta = 0$ to $180°$, a salient feature is that it includes the interesting region of angles around the beam direction. At the image point there is an aperture that, together with the interaction region size, determines the analyzer resolution $R = \Delta E/E$, which can be set from typically 1% to 6%. The analyzed electrons are detected by a channeltron multiplier located a certain distance from the image point. In front of the funnel entrance there are a set of interchangeable apertures that define the angular acceptance θ_0 of the instrument, ranging from 0.25 to $2.5°$. The gas target is provided by an effusive flow from a 0.25 mm inner diameter hypodermic needle located along the instrument rotation axis. Under normal operation, the needle tip is positioned 0.2 mm away from the projectile beam border. Under this working condition we get an efficient use of the needle beaming property obtaining an increase in signal by a factor of 50 with respect to a homogeneous distributed target. The chamber is pumped with a 1500 l/s oil diffusion pump that maintains a base pressure of 1×10^{-7} Torr, and increases to 1×10^{-5} Torr when using the beam target. The interaction region and analyzer are shielded from the earth's magnetic field utilizing a set of three mutually perpendicular Helmholtz coils reducing it to less than 5 mG.

In the third section of the setup, the projectiles that emerge from the collision region go through an electrostatic deflector that separates the different charge states into well defined beams, for the present case H and H+. These beams are recorded by two channeltron detectors operating in conjuction with a secondary electron converter consisting of an Al surface. This detection system allows the handling of high count rates and prevent the channeltrons from being exposed directly to the projectile beams. In addition, it allows the simultaneous registration of two charge-state signals. For the relative detection efficiency for neutral and charged projectiles we could not see any detectable difference when using an H and H+ beam. The vacuum in this section was typically 2×10^{-6}Torr.

We measured angle resolved electron energy spectra with and without analyzing the emerging projectile charge state. For the former, "singles" or non coincident spectra the data acquisition is performed using a standard multiscaling technique. At a fixed angle an energy spectrum is obtained scanning the voltage V applied to the analyzer, changing the voltage after a certain amount of charge is collected (or a number of particles) in one of the ion detectors (under certain conditions it was possible to insert a Faraday cup instead). To include the charge state information a coincidence technique is used. Electrons emitted in the collision are energy analyzed and the pulses are fed as a "start" signal to a time-to-amplitude converter. The "stop" signals are provided by the pulses of one of the projectile detectors. In this case, for each energy, a time spectrum is recorded in a multichannel analyzer. Afterwards a complete set of such spectra is processed to extract the real coincidence events and assembled into an energy spectrum. Since the pulses of the three detectors are simultaneously counted, the electron energy spectrum can be normalized to the total number of projectiles or projectiles of a particular charge state. Typical count rates in the different detectors were, for instance, for incident H+: H detector $\approx1\times10^5$ cts/s, H+ detector $\approx$ 2 to 10 % of the H counts, and electrons $\approx$ 2 cts/s. The "dark" count rate was $\approx$1 cts/min and typical measuring times were about 45 min for a given electron energy. For some collision systems, the time spectra for e-H+ and e-H coincidences showed a small background of random events outside a prominent coincidence peak. It was possible to combine the signals from both projectile detectors into one "stop" signal yielding a time spectrum with two well separated coincidence peaks. This procedure made possible simultaneous acquisition of data for both charge state channels resulting in an saving of overall measuring time.

MEASUREMENTS

1. Electron Transfer Into the Continuum

As one of the applications of our setup we measured electron energy spectra at the emission angle $\theta = 0°$ for a 15 to 200 keV incident H beam on an Ar target. Our interest was the study of the "cusp electrons", that is electrons with emission velocities close to the projectile velocity v_P [6]. Electrons emitted in this region are described by the "transfer into the continuum" mechanisms. Coincidence spectra were measured with emerging H and H+ to observe the distributions for the "electron capture into the continuum" (ECC) and "electron loss to the continuum" (ELC) processes. In figue 2 we show spectra for different targets. For both channels we observed cusp-shaped distributions. A prominent peak is seen too for the e-H coincidences that corresponds to a capture process into the continuum for a neutral projectile[7]. The shape of the ECC peak is seen to be symmetric and target independent. On the contrary the ELC cusp is much broader displaying a slight asymmetry towards the low energy wing.

The relative yields for both processes, defined as the value of the normalized coincidence counts at the cusp top to total (singles) cusp height, showed that the contribution of the ELC process is becoming dominant at projectile energies above 200 keV. At lower energies, < 50 keV, it is nearly energy independent reaching a value of 40 % for the ECC contribution[4], showing the importance of the ECC channel contribution at lower collision energies.

The presence of a cusp-shaped peak is known to be related to the long range nature of the Coulomb interaction between the electron and the ionic projectile in the final state. As for a neutral outgoing projectile there is no such field, it is expected that some type of dipole like interaction should be in effect instead in this case. Under certain circumstances this dipole field can lead to the formation of a cusp-shaped peak. In general, this is the case if the cusp observed can be associated with an excited H projectile. In particular for an n = 2 state, due to the degeneracy between the s and p states, a strong enough dipole field of the excited H projectile can develop. This is not possible for the ground state 1s. That this can be the case is provided by some experimental facts consistent with this assumption. We have observed a change in the yields when using an incident H beam completely quenched of metastables 2s states relative to a partially quenched one[4]. For a beam that carries such excited states we observed higher relative ECC yields. The description given here is valid within the frame the "final-state interaction" theory; that decouples the effect due to the final state electron-projectile interaction from the actual ionization process or from any electron production mechanism [8]. A consequence of the final-state theory is that analogies are expected to show up for the present ECC process and the cusp observed in other collision processes, for example in H^-detachment and in transfer ionization. We note that alternative models have been proposed to account for the ECC cusp associated with a neutral projectile [9].

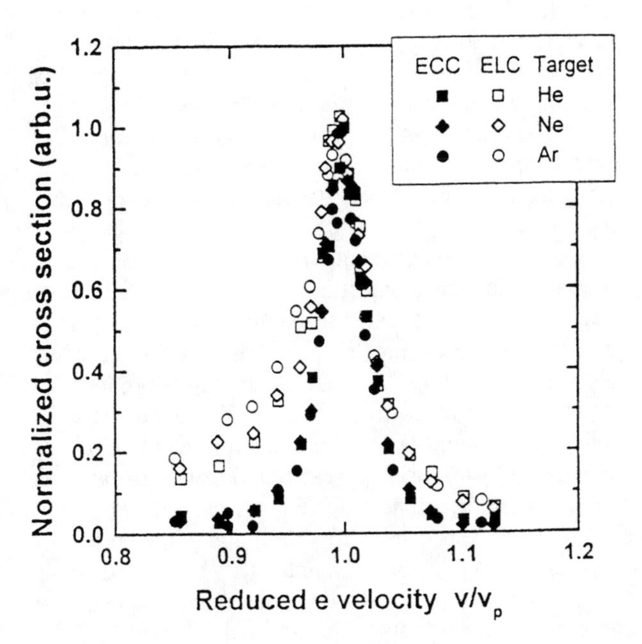

Figure 2. Set of electron energy spectra measured at $\theta = 0°$ for 75keV H incident on different target gases. The spectra have been normalized to 1 at the peak top.

The situation for the ELC cusp is better known. It results from a postcollisional Coulomb interaction between the stripped electron and the residual ionized projectile. It is generally assumed that ELC peaks are symmetric[6]. The present observation of a skewness toward lower electron energies can be interpreted as due to a contribution of a simultaneous target ionization channel. It is expected that at intermediate collision energies for multielectronic targets such contributions can be important[10].

2. H Projectile Ionization

Using an H incident beam of 78 keV on an Ar target we measured electron distributions as a function of energy and angle in the complete range from $\theta = 0$ to180° [11]. In figure 3 we show some of the spectra obtained. We measured distributions for "singles" and in coincidence with emerging H+. Under this latter condition we selected the channels that correspond to projectile ionization (or electron loss) and simultaneous projectile and target ionization.

For the singles spectra at $\theta = 0°$ we observed a sharp cusp. This cusp results from two electron transfer into the continuum mechanisms: the electron loss from the H projectile[12] and capture of a target electron by the neutral H [7]. For both processes the cusp shape is a characteristic feature of the long range final-state interaction between the electron and emerging projectile as described previously. With increasing θ the cusp extends into a broad peak positioned near the energy T and visible until $\theta \approx 40°$. Here T is the energy of an electron moving with the projectile speed v_P. This peak has been clearly observed for higher projectile energies and up to $\theta = 180°$ [13]. At $\theta > 90°$ a different more complex structure is observed consisting of a peak splitting with a minimum in place where a peak maximum was visible at lower angles. At higher emission angles this

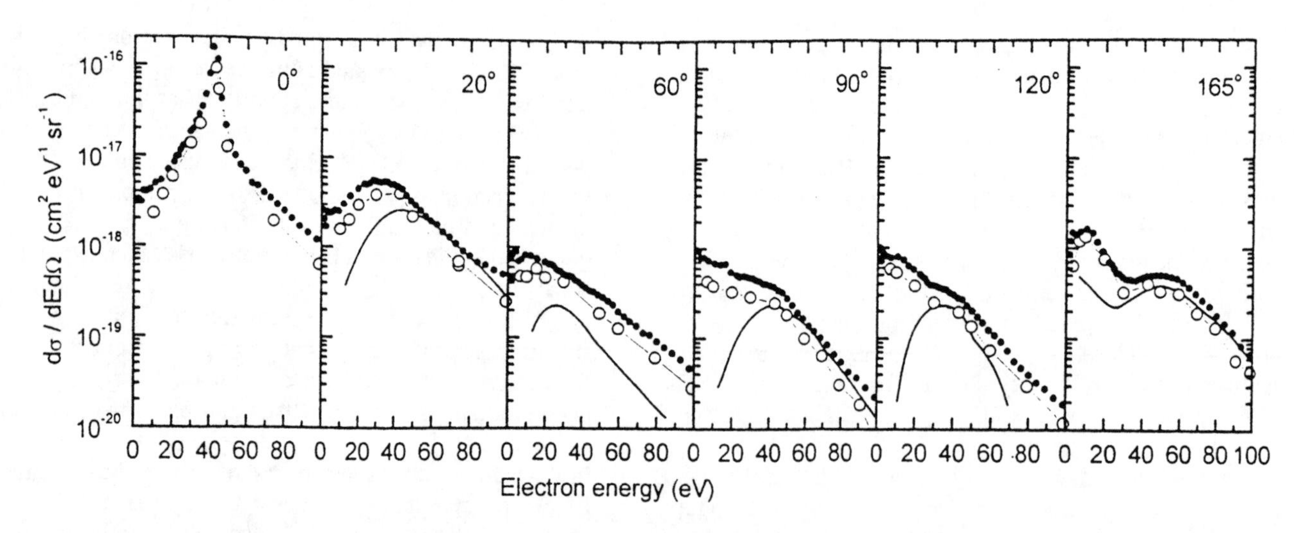

Figure 3. Double differential cross sections at different emission angles for 78 keV H incident on Ar. Measured total electron emission (full circles) and in coincidence with emerging H+ (open circles). Full line is the calculated cross section according to the electron scattering model.

structure is becoming more clearly visible. The coincidence data show in general nearly the same trend as the singles spectra. From this we conclude that the main features seen in the singles spectra are not obscured by the single target-ionization channel. The angular distribution measured for electrons emitted with an energy T shows two shallow minima at $\theta = 70°$ and $\theta = 135°$.

The structure for $\theta > 10°$ can be understood in the frame of the "elastic scattering model" (ESM)[14] that considers the projectile electron as a nearly free particle scattered elastically off the screened potential of the target atom. That the electron is initially in a bound state is taken into account considering its momentum distribution given by the Compton profile and binding energy B. That model describes the main feature of a broad peak with its maximum located at an energy $T - B$. Calculations performed by Stolterfoht [11] of double differential cross sections following the idea of the ESM showed a reasonable agreement with the measured spectra and accounted for most of the present features observed. The calculation was performed using a well modeled potential for the neutral Ar atom. From this the appearance of the maximum, the minima in the angular distributions and the splitting of the peak at larger angles could be described as an interference effect between different partial waves of the scattered electron.

The electron loss mechanism with its characteristic interference pattern effect was extensively studied at higher projectile energies, and this kind of phenomena are known as the "generalized Ramsauer-Townsend effect"[15]. This effect is not only limited to electron detachment collisions but has also been observed in a quite different electron emission mechanism that provided useful complementary information. In energetic collisions with ionic projectiles the doubly differential cross section shows a broad peak known as the "binary-collision peak", produced in this case by electrons emitted from the target. Here too a complex structure was reported consisting of localized intensity minima at small emission angles and double peak structures (splitting), again interpreted as interference effects of the emitted electrons[16].

3.Transfer Ionization

Other system studied was the two-electron process of transfer ionization (TI). Measurements were performed on a simple atomic system with the purpose to provide data useful for theoretical work concernig the role of electron correlation in ionizing collisions. We measured the TI produced by 100 keV H+ incident on Ar in which one of the emitted electrons is bound to the projectile as H and the second electron is free[17]. This measurement provided double differential cross sections at an intermediate collision energy where reported data of this type are scarce.

In this kind of experiment where the TI ionization cross section is relatively small it is important to account for the occurrence of double collision events in the background gas[18]. The main source is the incident H+ beam, which contains a small amount of accompanying H produced by electron capture in the beam line before reaching the target. The incident H+ and H projectiles can have charge exchange collision prior to reaching the charge-state analyzer yielding wrong detection events. Such double collision processes can be important due to the large cross sections involved by charge exchange. These undesirable effects can be circumvented by maintaining good vacuum conditions in the transport line and reducing the target thickness to a value where a linear dependence of count rates dominates. The spectra thus obtained are shown in figure 4 together with singles spectra. We observe that the

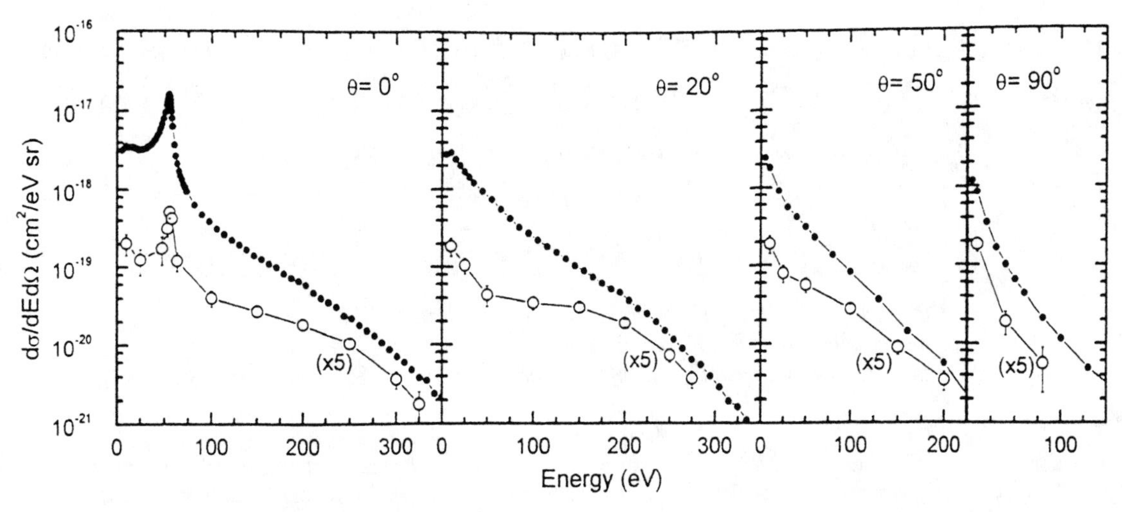

Figure 4. Measured double differential cross sections for 100 keV H+ incident on Ar in the forward direction: (full circles) total electron emission, (open circles) transfer ionization. Full circle data for θ = 50° are from ref.[19].

TI shows a general trend like that of the total emission. In the forward angles the contribution from binary collision electrons is more clearly seen in the TI spectra. At θ = 0° a cusp is clearly visible[18], that again is associated to an emerging neutral projectile and according to the picture of the final-state interaction theory [8] may have many features common to the previously mentioned ECC with incident H. For emission at the backward direction, θ > 90° the TI spectra are roughly isotropic. At low electron energies the TI cross section is almost isotropic, while the total electron emission (singles spectra) show a decrease with increasing angles. The data show that in such of collisions one of the electrons is captured into a bound state while the other electron is emitted predominantly by close-collision contributions. In TI the low energy electrons, resulting from soft collisions, apparently are emitted without being influenced by the neutral outgoing projectile. Only those small quantity moving nearly with the projectile velocity v_P are affected presenting a cusp-shaped peak.

ACKNOWLEDGMENTS

We thank R.O. Barrachina, K.O. Groeneveld and N. Stolterfoht for their comments and discussions that helped enlighten many of the present topics. We also are grateful to I. A. Sellin for a critical reading and to D. Fregenal and S. Suarez for their cooperation during the measurements. We acknowledge the CONICET for financial support.

REFERENCES

[1] Rudd M.E. and Macek J., *Case Stud. At. Phys.* **3**, 48 (1972)
[2] Stolterfoht N., DuBois R.D. and Rivarola R.D., *"Mechanisms for electron emission in heavy ion-atom collisions"* prepared as a book at Springer Verlag, Heidelberg (1997)
[3] Bernardi G., Suarez S., Fregenal D., Focke P. and Meckbach W., *Rev. Sci. Instrum.* **67**, 1761 (1996)
[4] Focke P., Bernardi G., Fregenal D., Barrachina R.O. and Meckbach W., *J. Phys. B: At. Mol. Opt. Phys.* **31**,289(1998)
[5] Risley J.S., *Rev. Sci. Instrum.* **43**, 95 (1972)
[6] Breinig M. *et al.* , *Phys. Rev. A* **25**, 3015 (1982). Groeneveld K.O., Meckbach W., Sellin I.A. and Burgdorfer J., *Comments At. Mol. Phys.* **4**,187 (1984)
[7] Sarkadi L., Palinkas J., Kover A., Berenyi D. and Vajnai T., *Phys. Rev. Lett.* **62**, 527 (1989)
[8] Barrachina R.O.: *J. Phys. B: At. Mol. Opt. Phys.* **23**, 2321 (1990); *Nucl. Instrum. Methods B* **124**, 198 (1997)
[9] Jakubassa-Amundsen D.H., *J. Phys. B: At. Mol. Opt. Phys.* **22**, 3989 (1989). Szoter L., *Phys. Rev. Lett.* **64**, 2835 (1990). Kunikeev Sh.D. and Senashenko V.S., *Sov. Phys. JETP* **75**, 452 (1992)
[10] Horsdal-Pedersen H. and Larsen L., *J. Phys. B: At. Mol. Opt. Phys.* **24**, 4099 (1979)
[11] Focke P., Bernardi G., Meckbach W. and Stolterfoht N., *J. Phys. B: At. Mol. Opt. Phys.* **31**, 3893 (1998)
[12] Drepper F. and Briggs J., *J. Phys. B: At, Mol. Phys.* **9**, 2063 (1976). Meckbach W., Vidal R., Focke P., Nemirovsky I.B. and Gonzalez-Lepera E., *Phys. Rev. Lett.* **52**, 621 (1984)
[13] Wilson W.E. and Toburen L.H., *Phys. Rev. A* **7**,1535 (1973)
[14] Burch D., Wieman H. and Ingalls W.B., *Phys. Rev. Lett.* **30**, 823 (1973)
[15] Kuzel K. *et al.*, *Phys. Rev. Lett.* **71**, 2879 (1993). Lucas M.W., Jakubassa-Amundsen D.H., Kuzel M. and Groeneveld K.O.,*International Journal of Modern Physics A***12**,305(1997)
[16] Kelbch C. *et al.*, *Phys. Lett. A* **139**, 304 (1989). Wolf W. *et al.*, *J. Phys. B: At. Mol. Opt. Phys.* **28**, 1265 (1995)
[17] Bernardi G., Focke P. and Meckbach W., *Phys. Rev. A* **55**, R3983 (1997)
[18] Vikor L. *et al.*, *J.Phys.B: At. Mol. Opt.Phys.* **28**,3915(1995)
[19] Rudd M.E., Toburen L.H. and Stolterfoht N., *At. Data Nucl. Data Tables* **23**, 405 (1979)

Measurement of the Degree of Polarization for the Radiative Decay of He^+ (np) (n=2 and 3) and He (1snp) $^1P^0$ states Following Electron Impact on He

R. Bruch, H. Merabet, M. Bailey[$], and A. Shevelko[&]

Department of Physics, University of Nevada, Reno, NV 89557 USA
[$]Atmospheric Science Center, Desert Research Institute (DRI), Reno NV 89506 USA
[&]Optical Division, Lebedev Physical Institute, Russian Academy of Sciences, Moscow, Russia

New experimental results are presented on the degree of linear polarization of the extreme ultraviolet (EUV) emission of the neutral and ionized helium following electron impact excitation and ionization-excitation of He atoms. The polarization of the photon emission from decay of the He (1snp) $^1P^0$ states with wavelengths of 51.7 to 58.4 nm has been measured for electron impact energies ranging from 30 to 1500 eV. In addition, the polarization for the He^+ (2p) $^2P^0$ and He^+ (3p) $^2P^0$ states with wavelengths of 30.4 and 25.6 nm, respectively, has been obtained from threshold (66 eV and 73 eV) to 1500 eV. Furthermore, a comparison of the experimental results with theoretical calculations is provided.

INTRODUCTION

The helium atom presents the simplest, strongly bound, two electron system, and is therefore ideally suited for the study of few body collision dynamics and electron correlation effects. Investigations of electron impact excitation, ionization, and simultaneous ionization-excitation of helium with the subsequent detection of photons or scattered electrons have provided important tests for theoretical models of electron correlation and scattering dynamics during the collision process (1-4). The characteristics of helium have also been investigated via proton and ion impact experiments (5-6) and more recently with photoionization studies utilizing synchrotron sources (7-8).

The measurement of the polarization or angular distribution of the photons emitted by excited atoms and ions yield important information pertaining to the magnetic substate populations which can be used to test theoretical approaches at a deeper level than that accessible through total or differential cross section measurements. Both polarization and angular distribution techniques have experimental difficulties, but of the two, polarization analysis is generally the simplest to perform if a suitable polarization analyzer is available in specific wavelength regions.

The degree of polarization of radiation emitted by atoms and ions following electron impact contains important information concerning the excitation of individual magnetic sublevels (9). When electrons have an anisotropic or beamlike velocity distribution, particular sublevels are preferentially populated, and the target it left in an anisotropic state. The radiation can be visualized to arise from an ensemble of excited atoms or ions with electric dipoles oriented along the beam axis, and two mutually vertical dipoles of equal strength that are both perpendicular to the beam axis. The degree of linear polarization or polarization fraction is then given by

$$P = \frac{I_{\parallel} - I_{\perp}}{I_{\parallel} + I_{\perp}}. \qquad (1)$$

where $I_{\parallel}$ is the intensity of radiation with electric field vectors oriented along the beam axis and $I_{\perp}$ is the intensity of radiation with a transverse electric field vector (perpendicular to the plane formed by the incident electron beam and the direction of observation). The measurement and analysis of the degree of polarization or the angular distribution of the photon emission can reveal the anisotropy or structure of the emitting source.

The present paper provides the first comprehensive experimental results on the degree of linear polarization of radiation in the EUV from ionized-excited helium (HeII) following electron impact. These measurements have been performed using a molybdenum-silicon (Mo/Si) MLM polarimeter (10) whose reflection characteristics have been optimized for radiation with a wavelength of 304 Å at a grazing incidence angle of 50°. In addition to the 304 Å emission, the polarization of the 256 Å emission from the HeII (3p) $^2P^0 \rightarrow$ (1s) ^2S transition has been measured as well. The same polarimeter has also been used to measure the polarization of HeI emission for the transitions,

$$e^- + He(1s^2)^1S \rightarrow He\ (1snp)^1P^0 + e^-$$
$$\hookrightarrow He^+ (1s^2)^2S + h\nu \qquad (2)$$

These transitions involve the emission of radiation with wavelengths ranging from λ=517 to 584 Å for n=2-5, respectively. This radiation is dominated by the 584 Å

CP475, *Applications of Accelerators in Research and Industry*,
edited by J. L. Duggan and I. L. Morgan

emission arising from the n=2 level. Furthermore theoretical calculations are performed and compared with the experimental data.

RESULTS AND DISCUSSION

The apparatus used in this study consists of three main components: the polarimeter; the electron gun, target cell and Faraday cup; and a 1.5 meter grazing incidence monochromator. A PC controlled data acquisition system was used to operate the system and to record the data. For more details, a complete description of this experimental set up is given by Bailey *et al.* (11).

The pressure dependence of the emission from neutral and ionized-excited helium have been measured with the polarimeter. From our results it is obvious that the polarization fraction determined for an electron impact energy of 60 eV is approximately constant for pressures ranging from 0.05 to 1 mTorr. A target pressure of 0.25 mTorr was selected for the polarization measurements of the HeI $(1snp)^1P^o$ series. The corresponding pressure dependence of the polarization fraction for the 304 Å radiation has been studied as well. Here the polarization fraction exhibits a much more gradual change with pressure than the HeI results and reaches a constant value for pressures at 1 mTorr and below. Hence a gas pressure of 1 mTorr was used for both the $(2p)^2P^o$ and $(3p)^2P^o$ decays of HeII studied in this work.

Moreover, the statistical error of the intensities measured are found generally to vary from 0.5% to 2% over most of the range of impact energies. However, for the lowest energy threshold measurements, it was as high as 8%. Since the sum and difference of intensities are involved, the relative error of the polarization fraction increases rapidly with decreasing intensity. The relative error also becomes larger when the polarization fraction is small, reaching a maximum when it is close to zero. While the statistical error was the largest contributor to the total error of the polarization fraction, other factors were included as well.

A. HeI $(1snp)\,^1P^0$ Polarization

The HeI $(1snp)\ ^1P^0$ polarization data measured in this work for electron impact energies ranging from 30 to 350 eV are shown in Fig. 1. Included in this figure are the previous experimental results of Hammond *et al.* (12) which extend down to 0.8 eV above threshold at 22 eV. The measured polarization is observed to drop rapidly from unity close to threshold to a value of approximately 0.48 at a 25 eV. The polarization rises again to a secondary maximum value of approximately 0.57 at 40 eV and then decreases gradually with increasing impact energy.

The results from this work are in excellent agreement with the results of Hammond *et al.* which were measured with errors of less than 1%. Included in the figure are the Born approximation calculations of Vriens *et al.* (13) which show very good agreement with both experimental sets of data even down to an impact energy of 80 eV. This is a somewhat surprising result since the Born approximation

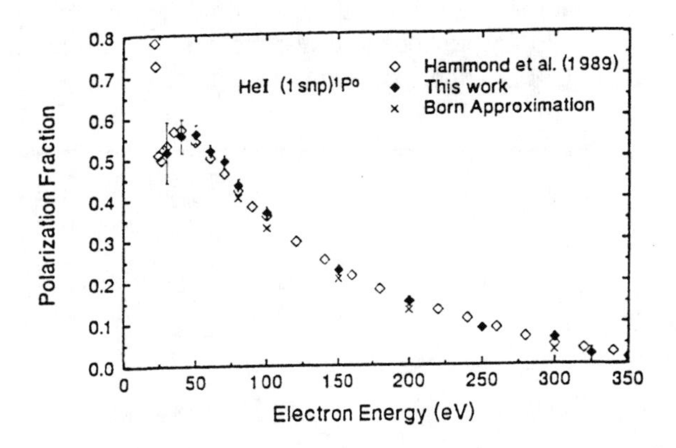

Figure 1: . Polarization of HeI $(1snp)^1P^o \rightarrow (1s^2)^1S$ radiation as a function of electron impact energy for energies ranging from 22 to 350 eV. ♦, this work; ◊, data from Hammond *et al.*; x, Born approximation.

fails to correctly predict the total cross sections for the excitation of $(1snp)\ ^1P^0$ states at low energies[11].

The results from this work and from Hammond *et al.* (12) have not been corrected for cascade effects. Hammond *et al.* have calculated the effects of cascade on the observed polarization data for the HeI (1snp) $^1P^0$ decay using the cascade corrections of Donaldsen *et al.* (14). This correction increases the polarization by about 20% at an impact energy of 30 eV but by only 10% at 100 eV. These corrections become negligible above approximately 300 eV. In contrast Steph *et al.* (15) have performed a cascade free measurement of the HeI (1snp) $^1P^0$ polarization using an electron-photon angular correlation technique for impact energies ranging from 30 to 500 eV. Their results indicate that the polarization correction is only half as large as that estimated by Hammond *et al.*, however the corrections of Hammond *et al.* assume that the cascade contribution to the total emission cross section is unpolarized. This assumption may not be correct. In either case, the cascade correction is small enough that the corrected polarization data retains good agreement with the Born approximation predictions.

The polarization data near threshold exhibits two interesting features. First, as previously mentioned, the threshold model for single excitation of helium predicts a threshold polarization value equal to one, i.e. $\sigma(1) = 0$ at threshold corresponding to a pure s wave for the outgoing electron (12,15). This predicted result has apparently been verified experimentally by Norén *et al.* (16). Second, the pronounced dip in the polarization at an impact energy of 25 eV appears to coincide with the peak of the cross section for the production of metastable helium $(1s2s)^1S$ and $(1s2s)^3S$ states measured by Mason and Newell (17). A similar feature is observed in the HeII polarization data from this work at the threshold energy for the double ionization of helium (see section B). From these and other observations of polarized emission (16), it is clear that the polarization can be noticeably affected by the presence of competing excitation channels. The intermediate to high energy impact data from this work are displayed in Fig. 2. along with the previously mentioned data of Hammond *et al.* and the Born approximation calculations (13).

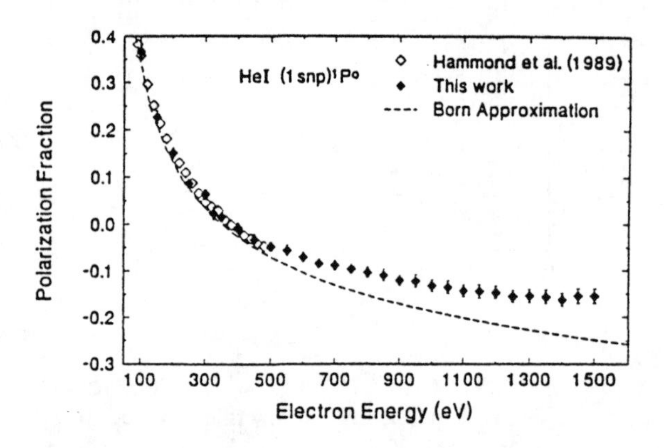

Figure 2. Polarization of HeI (1snp)^{1}P^o → (1s^2)^{1}S radiation as a function of electron impact energy for energies ranging from 100 to 1500 eV. ♦, this work; ◊, data from Hammond et al.; --, Born approximation.

The high energy impact data from this work appear to be approaching an asymptotic value of approximately **P**(^{1}P^0) = -0.17 at an impact energy of 1500 eV, while the Born approximation results are still decreasing at this energy. The Born approximation results reach an asymptotic limit at impact energies well above 3000 eV (13).

B. HeII (np) ^{2}P^0 Polarization

The He$^+$(2p) ^{2}P^0 and He$^+$(3p) ^{2}P^0 polarization results from this study for the full range of impact energies are shown in Fig. 3. and Fig. 4. The (3p) ^{2}P^0 polarization results appear to be slightly higher than those for the (2p) ^{2}P^0 level for impact energies less than approximately 400 eV (see Fig. 5), however the two data sets are nearly identical above this energy. This appears to indicate that the polarization for the (np) ^{2}P^0 series of He$^+$ is also approximately independent of the principle quantum number in a manner similar to the (1snp) ^{1}P^0 series of neutral helium.

The He$^+$(2p) ^{2}P^0 and He$^+$(3p) ^{2}P^0 polarization results in these figures have not been corrected for cascade effects since the cross section measurements for many of the excited states of He$^+$ necessary to accurately calculate these corrections have not yet been performed. However, due to the much shorter lifetimes of the excited HeII states in comparison with those with the case of HeI, cascade effects are expected to be smaller than for HeI (3).

Both the He$^+$ (2p) ^{2}P^0 and He$^+$(3p) ^{2}P^0 results from this work exhibit threshold polarizations of approximately **P**(^{2}P^0) = 0.2 with the polarization decreasing gradually for higher impact energies. These threshold values present clear evidence for the contribution of partial waves with L > 0. Also evident in both of these polarization measurements is a small dip at an impact energy of approximately 79 eV which happens to coincide with the threshold energy for the

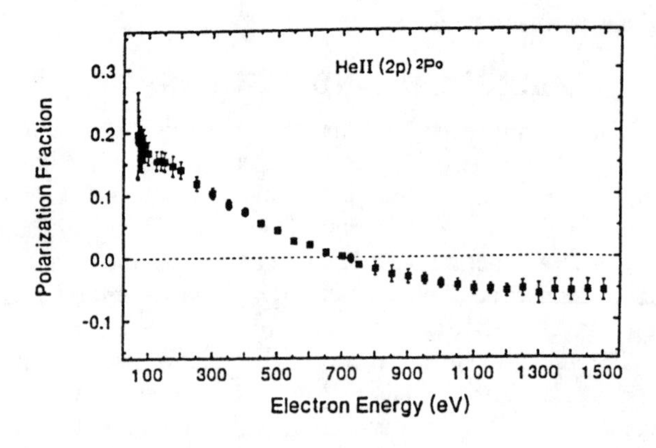

Figure 3: Polarization of HeII (2p)^{2}P^o → (1s)^{2}S radiation as a function of electron impact energy for energies ranging from 66 to 1500 eV.

double ionization of helium. While the He$^+$(2p)^{2}P^0 and He$^+$(3p) ^{2}P^0 data are similar within the stated error limits, the (3p) ^{2}P^0 results appear to be slightly higher for impact energies less than approximately 400 eV.

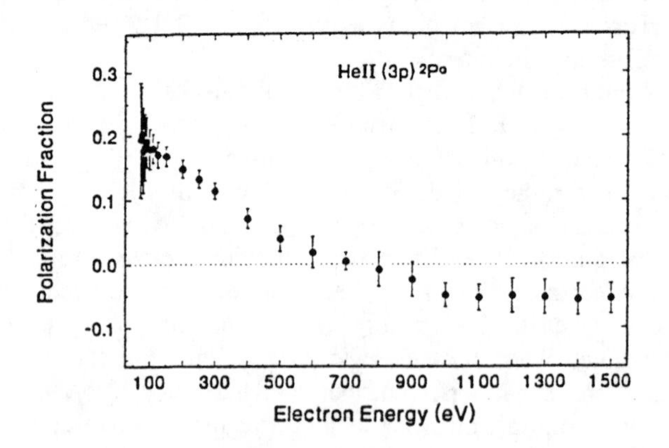

Figure 4: Polarization of HeII (3p)^{2}P^o → (1s)^{2}S radiation as a function of electron impact energy for energies ranging from 73 to 1500 eV.

The threshold calculation of the (2p) ^{2}P^0 polarization by Götz *et al.* (4) using a DWBA + R-Matrix method was found to have an approximately constant value of 0.22 for impact energies ranging from 65.4 to 72.9 eV. This value is in good agreement with the threshold polarization measurements in this work. In the DWBA + R-Matrix method of Götz *et al.*, the projectile electron is described by a distorted wave, calculated in the static ground state potential of the He atom while the interaction between the ejected electron and the residual ion is treated by an R-Matrix close-coupling expansion method. In this calculation, the physical 1s, 2s, and 2p orbitals of He$^+$ have been included to construct the corresponding target states in a close-coupling treatment of the ejected-electron-residual-ion collision system. In addition, two pseudo-orbitals, 3s and 3p, have also been included to improve the description of the He ($1s^2$) ground state. While this calculation does not include the possibility of exchange between the projectile and target electrons, the parameterization yields surprisingly good agreement with the experimental data from this study.

The $He^+(2p)$ $^2P^0$ threshold measurements performed were initially obtained with a 25% VYNS transmission filter (11), this filter provides approximately a 1000:1 transmission ratio for 304 Å radiation versus the dominant 584 Å radiation from neutral helium. However a review of the electron impact cross section data for the excitation of the (1s2p) $^1P^0$ and (2p) $^2P^0$ levels of neutral and ionized helium reveals that the ratio of these cross sections at energies just slightly above the $(2p)^2P^0$ threshold exceeds 1000:1 as illustrated in Fig. 5. Hence a 25% transmission filter will transmit roughly equal amounts of the (1snp) $^1P^0$ and (2p) $^2P^0$ emission intensities and in our view is not adequate for an accurate determination of the threshold

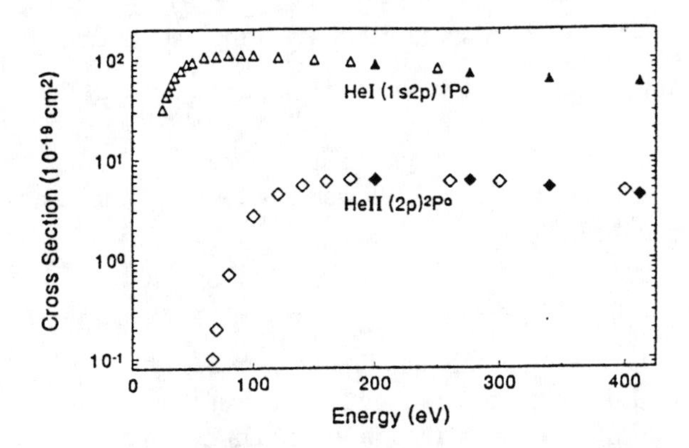

Figure 5: Cross section data for the electron impact excitation of the $(1s2p)^1P^o$ and (2p) $^2P^o$ levels of neutral and ionized helium. Δ, data from $(1s2p)^1P^o$ (2) and ▲, Bailey *et al.* (18); ◊, (2p) $^2P^o$ data from Forand *et al.* (3) and ♦ Bailey *et al.* (18).

polarization and associated alignment parameters of the L-line in He^+. The 10% VYNS transmission filter used in this experiment provided a transmission ratio of approximately 100,000:1 so that the polarization measured with this filter contained approximately only a 1% contribution from the (1snp) $^1P^0$ emission.

The DWXA calculations of Itikawa *et al.* (19) for electron impact excitation of ground state ionized helium presented in Fig. 6. shows reasonable agreement with both the experimental results obtained from this work and the DWBA + R-Matrix calculation of Götz *et al.* near threshold, though this calculation represents a fundamentally different process. These e^- + He^+ calculations are not expected to yield correct results for the (e^-, e^-, $h\nu$) reaction of simultaneous ionization-excitation since electron correlation effects between the projectile and ejected target electron are not included in such calculations. Also at energies near threshold for ionization-excitation, a perturbed potential of the ion due to the interaction of the two slow outgoing electrons is expected to play an important role in determining the final angular momentum states of both the excited ion and the outgoing electrons (20). While the calculations of Itikawa *et al.* do not take these factors into account because of the fundamental difference between the two collision mechanisms, some interesting conclusions may be inferred from a comparison of these calculations with those of Götz *et al.* and the present experimental data.

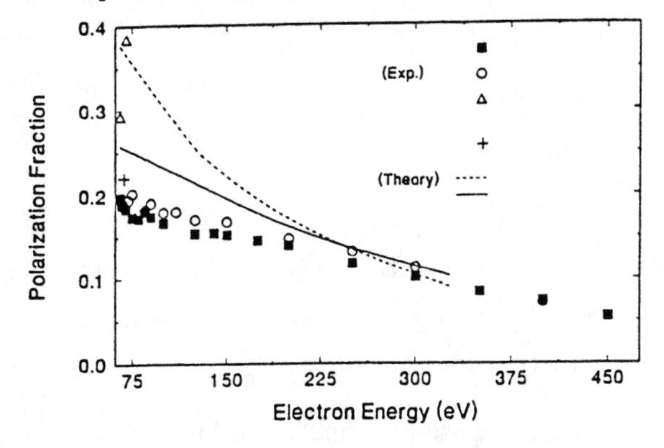

Figure 6: Polarization of HeII $(np)^2P^o \rightarrow (1s)^2S$ radiation as a function of electron impact energy for energies ranging from threshold to 450 eV. Experiment: ■, $(2p)^2P^o$ and o, $(3p)^2P^o$, this work ; Δ, $(2p)^2P^o$, Götz *et al.* (4). Theory : +, DWBA + R-matrix, Götz *et al.* ; dashed line, DWBA (e^- + He^+), Haug *et al.* (21) ; solid line, DWXA, Itikawa *et al.* (19).

Finally, It would be of great interest to compare these experimental data with proton impact on He. Indeed, experiments are currently being made to investigate proton and molecular ion impact excitation and ionization-excitation of helium where the effects of charge exchange and multi-centered projectile excitation will be explored.

ACKNOWLEDGMENTS

This Project has been supported by NATO Grant No. CRA 930032 and Acspect Corporation, Reno, Nevada.

REFERENCES

1. McGuire, J.H., *Electron Correlation Dynamics In Atomic Collisions*, (Cambridge University Press, 1997).
2. Westerveld, W.B. *et al.*, *J. Phys. B.* **12** 115 (1979).
3. Forand, J.L. *et al.*, *J. Phys. B:* **18** 1409 (1985).
4. Götz, A. *et al.*, *J. Phys. B.* **29** 4699 (1996).
5. Hippler, R., and Schartner, K.H., *J. Phys. B.* **7** 618 (1974).
6. Träbert, E. *et al.*, *Nucl. Instr. Meth. B* **23** 151 (1987).
7. Samson, J.A.R. *et al.*, *Phys. Rev. A.* **25** 848 (1982).
8. Jiménez-Mier, J. *et al.*, *Phys. Rev. Lett.* **57** 2260 (1986).
9. Eminyan, M. et al., *J. Phys. B* **7** 1519 (1974).
10. Alexandrov, Yu. M., *Rev. Sci. Instrum.* **60** 2124 (1989).
11. Bailey, M. *et al.*, submitted, *Phys. Rev.* **A.** (1998).
12. Hammond, P. *et al.*, *Phys. Rev. A.* **40** 1804 (1989).
13. L. Vriens and J.D. Carrière, *Phys. Scp.* **49** 517 (1970).
14. F.G. Donaldson, F.G. *et al.*, *J. Phys. B* **5** 192 (1972).
15. Steph, N.C., and. Golden, D.E., *Phys. Rev. A.* **26** 148 (1982).
16. Norén, C. et al., *Phys. Rev. A.* **53** 1559 (1996).
17. N.J. Mason, N.J. *et al.*, *J. Phys. B.* **20** 1357 (1987).
18. Bailey, M., *Ph.D. thesis, University of Nevada, Reno* (1997).
19. Y. Itikawa, Y. *et al.*, *Phys. Rev. A.* **44** 7195 (1991).
20. Heideman, H. G. M. et al., *Phys . B.* **13** 2801 (1980).
21. Haug, E., Sol. Phys. **71,** 77 (1981).

Multiply excited ion formation and Auger stabilization

S. Bliman*# and M. Cornille **

* *Université de Marne la Vallée. Champs sur Marne*
77454 Marne la Vallée Cedex 2 - France
 # LSAI - UPR, Université Paris-Sud, 91405 Orsay - France

** *DARC, Observatoire de Paris, UPR 176 CNRS*
92195 Meudon Cedex - France

Until recently charge exchange collisions, single and double capture (SC and DC) considered projectile ions in ground states. It is well known that long lived metastables do exist and participate in SC and DC processes (He-like 1s 2s $^3S_1$ and Ne-like 1s $2s^2$ $2p^5$ 3s $^3P^0_{0,2}$ metastable ions are long lived). The DC by these ions ends in the formation of triply excited ions: Be-like and Mg-like triply excited ions respectively. The analysis of the Auger spectra shows that the stabilization of these ions takes places via two Auger electron cascade and/or one Auger electron and implies the complementary part that is fluorescence.

INTRODUCTION

The existence of multiply excited ions has long been known. Recently using charge exchange collisional approaches and/or XUV inner shell allowed photo excitation has to understand their formation. Actually, the analysis of their stabilization via Auger decays to different available contenua is possible [1],[2].

Experiments were performed to observe X and VUV radiation following the excitation mechanisms : (inner shell excitation and ionization) Auger stabilization was studied separately [3]. These two different approaches were not correlated.

Theoretical calculations gave level energies, decay rates and fluorescence yields for target inner shell excitation and ionization [4], [5].

In view of a clarified presentation of the results of the analysis of our observations, we will first introduce a very general definition of what is meant by multiply excited ions or atoms, seen from energy considerations. If we conisder an ion and/or an atom, in its ground state, we may define its first E_{i1} ionization energy. If a level has an excitation energy larger than E_{i1}, then the system is doubly excited and is likely to stabilize via the emission of one Auger electron. This definition may be extended and generalized: if the excitation energy E_{exc} is larger than the sum of (n-1) ionization energies, the ion is n fold excited and may decay via (n-1) Auger electron emissions. This concept is illustrated for example by the K shell ionization of an argon atom. The K hole excitation energy is of order 3.2 keV and leaves Ar^+. In fact this excitation energy is larger than:

$$E_{K\,hole} > \sum_{q=0}^{12} E_{ionis.}\, q \rightarrow q+1 \;(\;\approx\; 2.650\ \text{keV})$$

In this case, not only has fluorescence been observed [6] but also Auger electron emissions were predicted and observed [7], [8] even though, it was not possible to assign the continua to which these decays took place.

More recently, in the study of the interaction of highly ionized ions at low kinetic energies with surfaces, « hollow ions » were observed using X-ray spectroscopy [9]. The complementary part, namely Auger stabilization has not really been explained : in fact a huge number of low energy electrons were detected but no assignement of their decay to a specific continuum has been attempted [10]. This is easy to understand since the initial state of the interaction is not energetically assigned and no continua are identified.

The availability of multiply charged ion sources [11] renewed interest led to studies multiple charge transfer in highly charged ions colliding with many electron targets [10]. Double electron capture by ground state ions focused great attention and the Auger spectroscopy of the doubly excited systems was largely developped [12],[13]. Multiple capture were mostly studied by coincidence techniques where the projectile final charge state was related to the target charge state [14]. This could not give information about the levels that had been populated and about their stabilization.

CP475, *Applications of Accelerators in Research and Industry*,
edited by J. L. Duggan and I. L. Morgan

Recently, more graduated approaches to create multiply excited systems were initiated. The formation of « hollow lithium atoms » by multi-photon excitation pumping showed that the states populated were above the second ionization limit from the gound state $1s^2\ 2s\ ^2S_{1/2}$. They decay via two Auger steps ending in Li^{2+} (1s) $^2S_{1/2}$. [These conference proceedings F. Wuilleumier et al.]. Another approach to creating multiply excited ions consists in inducing double electron capture in charge exchange collision of metastable ions (He-like or, Na-like long lived metastable ions for example) with two electron targets (He, H_2). In the next section, taking advantage of the similar collisional features in S.C. by ground state and metastable Ar^{8+} from He, assuming that in D.C., the metastable behaves like the ground state projectile, we will give the different characteristics of the collision products and their Auger stabilization.

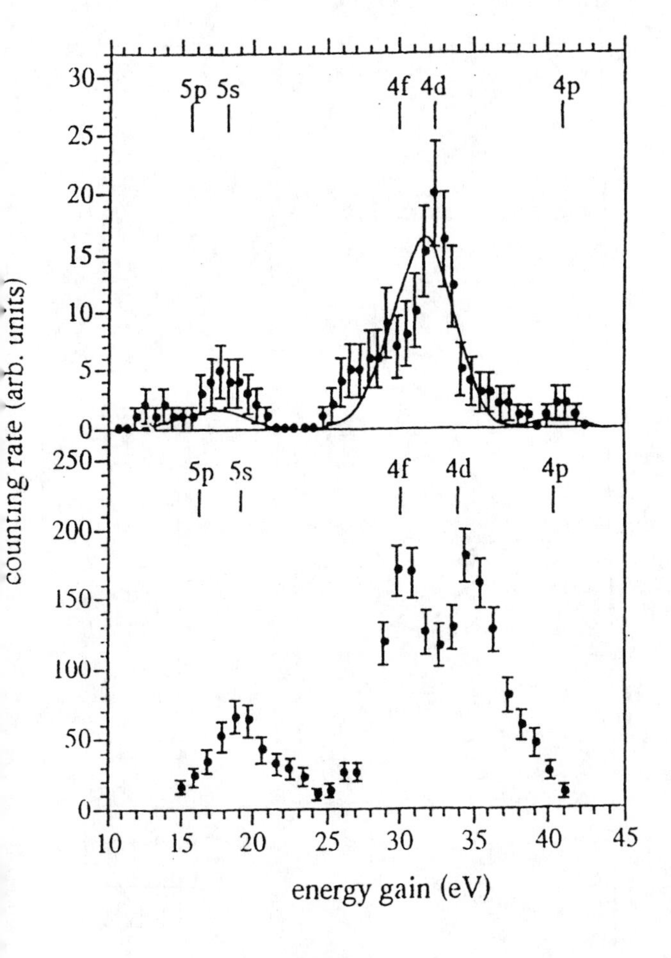

Fig 1. Energy-gain spectra for Ar^{7+} ions produced in Ar^{8+}+ /He collisions at θ=0°. Upper part : ground-state projectiles $Ar^{8+}(2p^6\ ^1S_0)$; lower part : metastable excited projectiles $Ar^{8+*}(2p^5\ 3S)^3\ P_{0,2}$.

FORMATION AND DECAY OF MAGNESIUM LIKE TRIPLY EXCITED Ar IONS

A beam of Ar^{8+} ions is extracted from a ECR ion source. The energy of the ions is 10 keV/charge. After mass and charge selection, the ions are passed in a collision cell, differentially pumped, containing He gas at a pressure of order 2-4x 10^{-5} Torr. In these conditions the single collision condition is fullfiled. The beam contains a metastable fraction $(1s^2\ 2s^2\ 2p^5\ 3s)\ ^3P^0_{0,2}$ of order 5% of the total beam current. The metastability life time is of order 0.3 millisecond [15].

The experimental techniques used to observe the collisions

Ar^{8+} $(2p^6)\ ^1S_0$ + He → Ar^{7+} $(2p^6 4l)^2\ L_J$ + He^+ (1)

→ Ar^{6+} $(2p^6 3l\ 4l')$ + He^{2+} (2)

Ar^{8+} $(2p^5 3s)\ ^3P^0_{0,2}$ +He → $Ar^{7+}(2p^5\ 3s\ 4l)^{2,4}\ L_J$+$He^+$ (3)

→ Ar^{6+} $(2p^5\ 3s\ 3l\ 4l')$ + He^{2+} (4)

are X-VUV spectroscopy, Auger spectroscopy and translational energy gains spectroscopy (TES).

The TES approach reveals that for both projectiles the energy gains for single capture are the same, pointing to the same capture levels 4l, specifically. $2p^6\ 4l\ ^2L_J$ for the ground state projectile and $2p^5\ 3s\ 4l\ ^{2,4}L_J$ for the metastable projectile. This is shown in Fig.1. This evidences the usefullness of the experimental scaling rule [16],[17] predicting the most probably populated level n (principal quantum member) in single capture state projectile.

The double capture (D.C) by the ground is clearly characterized in the X-VUV spectrum where the final transition Ar^{6+} $(2p^6 3s3p)\ ^1P_1$ → Ar^{6+} $(2p^6 3s^2)\ ^1S_0$ at 585.3Å (end of the cascade) is seen. More over, the scaling rules point the DC to Ar^{6+} $(2p^6 3l\ 4l')$. This is confirmed by the optical transitions observed at 281.4Å $(2p^6\ 3s\ 4s\ ^1S_0 \rightarrow 2p^6\ 3s\ 3p\ ^1P^0_1)$ and at 350.4 Å

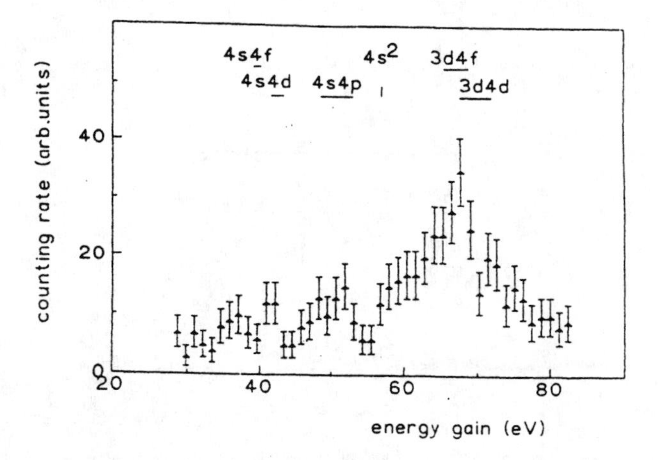

Fig 2. Energy-gain spectrum for Ar^{6+} ions produced in $Ar^{8+}(2p^6)\ ^1S_0$+He collisions. Scattering angle 0°

$(2p^6\,3s\,4f\,^1F^0_3 \rightarrow 2p^6\,3s\,3d\,^1D_2)$. An important feature is that all the $(2p^6 3s\,4l')\,^{1,3}L_J$ decay radiatively : they are all located below the first ionization limit of Ar^{6+} from the ground state [18]. The TES spectrum of the D.C. by the ground state projectile (Fig.2) shows the full energy gain spectrum of Ar^{6+} peaking an Ar^{6+} $(2p^6\,3d\,4f)$ (ioni-zation limit) [18] with a Rydberg tail associated with the fraction of the D.C. that shares its decay among radiation and Auger electron emission. This TES spectrum shows the fraction were both electrons remain on the ion. The complementary part, the Auger spectrum is shown in Fig.3 where a nice Rydberg tail is seen with a part superimposed which origin is remains to be explained.

The total Auger spectrum showing the low energy part as in Fig.3 and the high energy part above 100 eV is presented in Fig.4. This high energy part is unambiguously attributable to the Auger decay of sodiumlike core excited Argon ions in states Ar^{7+} $(1s^2\,2s^2\,2p^5\,3l\,n'l')$.

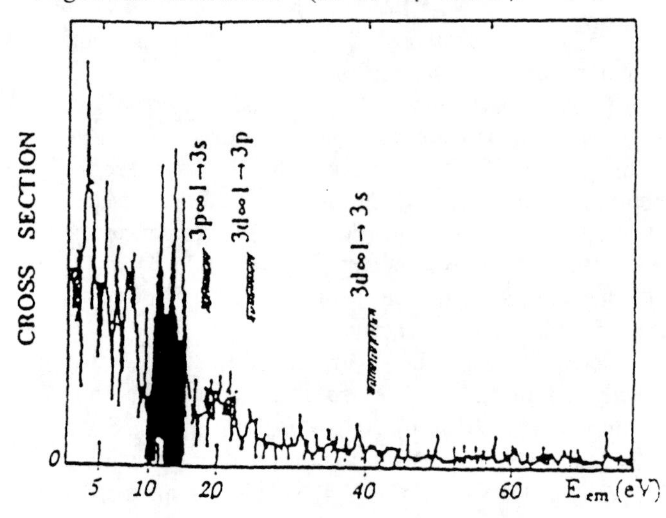

Fig 3. Auger spectrum for the energy range 0-45 eV in the emitter frame. The region of interest for the first Auger electron emission of triply excited Ar^{6+} ions is shaded.

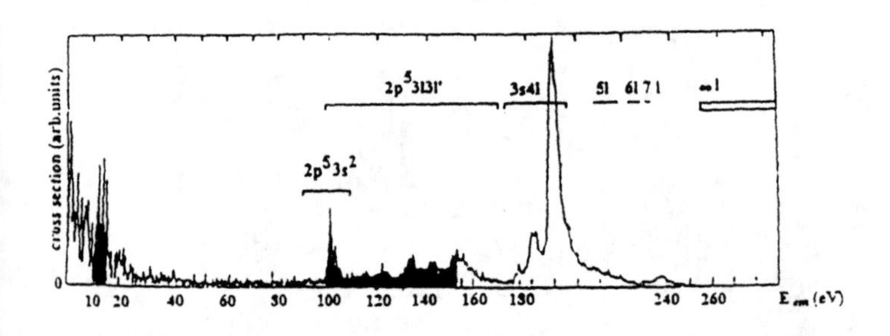

Fig 4. Auger spectrum for Ar^{6+}+ He. The regions of interest for the Auger stabilization of triply excited Ar^{6+} ions are shaded.

A careful analysis and identification of the different lines in the high energy part of the Auger spectrum evidences the transitions $(2p^5\,3s\,4l)\,^2L_J$ due to S.C. by the metastable projectile (energy range 180-200 eV). A week Rydberg tail, extends to the series limit $(2p^5\,3s\,nl)\,^2L_J$ at approximately 252.eV. These are based on the sets of calculated atomic data [15] where it is shown quartet states decay by fluorescence, if cascades from the quartet states populate lower ^4L energy levels, these would decay by photon emission.

In these conditions, we are left with two regions of the Auger spectrum open for analysis and identification (see shaded regions in Fig.4).

The first step into the analysis is to assume that the D.C. collision features are the same for both the ground state and metastable projectile. For the ground state, the TES spectrum [Fig.2] shows that the D.C. peaks at 70 eV energy gain onto the configuration 3l4l'. In these conditions assuming that D.C. on the metastable projectile (process 4) takes place with the same energy gain, makes it possible to locate approximate' these levels on the energy scale of Ar^{6+} from the gro ıd state. The result of that is shown diagramatically in ig.5. This gives to see that some of these triply excited l.vels populated by D.C.

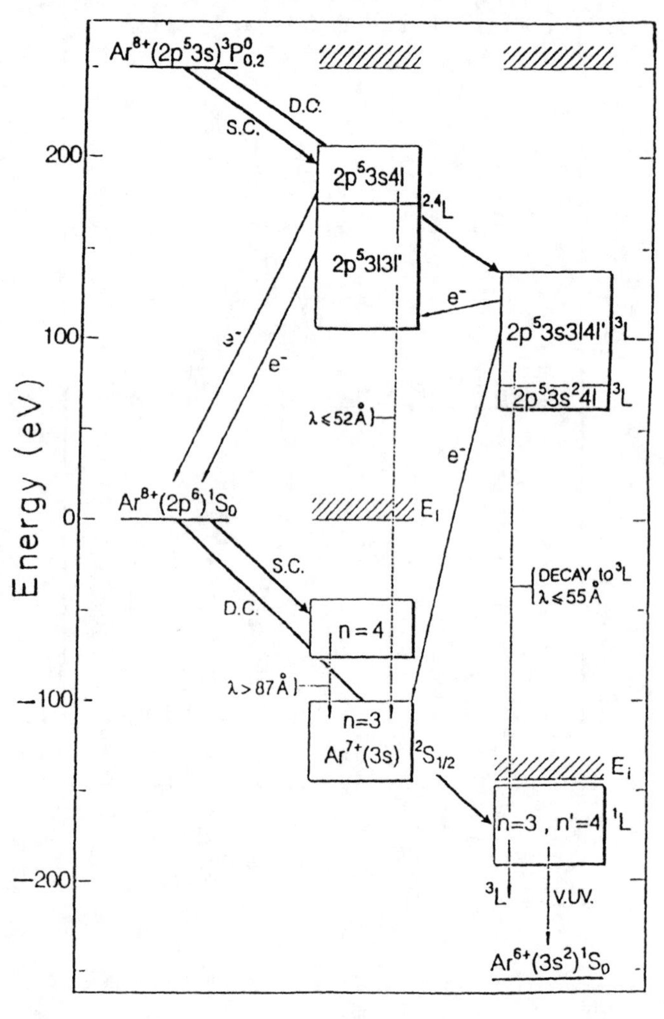

Fig 5. Level diagram of the Ar^{8+} + He system. Left column : entrance channels ; central column : single-capture exit channels to Na-like Ar ; right column : double-capture exit channels to Mg-like Ar.

are likely to find a decay channel into the sodium like core excited levels ($2p^5$ 3l 3l'). Before going further and to support this assumption, an atomic set of levels characteristics (energy mostly) were calculated using Cowan's code [19]. The results confirmed the estimates from the above assumption. Then inspecting the levels position, we deduce that the triply excited states ($2p^5$ 3s 3l 4l'), as far as energetically located in the energy scale above the $2p^5$ 3s 3l' levels stabilize via a first Auger step of low energy to the intermediate continua $2p^5$ 3l 3l' giving a series of closely packed peaks (shaded region in the low energy part of Fig.4) according to

$$Ar^{6+} (2p^5\ 3s\ \ 3l\,4l') \rightarrow Ar^{7+} (2p^5\, 3l\, 3l')\ ^2L + e^-$$

This second step is followed by the second step

$$Ar^{7+} (2p^5\ 3l\,3l')\ ^2L \rightarrow Ar^{8+} (2p^6)\ ^1S_0 + e^-$$

This step is associated with the shaded region seen in the energy region above 100 eV (Fig.4).

In terms of Auger lifetimes, it seems that at least for the first step (low energy), the transitions are probably such faster than the higher energy ones. Typical lifetimes for the high energy transitions calculated are of the order 10^{-13} -10^{-14}s, while the low energy ones should range around 10^{-15} -10^{-16}s. The transitions are skeched by arrows in Fig.5.

CONCLUSION

We have shown a new way to create triply excited ions by a double capture collisionnal approach. They could be on metastable ion. The Auger stabilization has been shown to be two step cascade.

The experimental approach can be extended to other ions, in long lived metastable states (He like ions in 1s 2s ^3S, we have shown a new way to create triply excited ions by a double capture approach. The Auger stabilization has been shown to be a two step cascade.

The problem is to be able to show that energetically there exists a connexion to doubly excited levels in the adjacent charge state to be the intermediate continua for the decay. Theoretically, this is a very difficult problem that has to be solved for each specific system and accurate treatments are scarce.

An interesting aspect is that the understanding of these phenomena give views to new experiments such as inner shell photoexcitation of ions : they could be in there G.S. or in metastable states.

REFERENCES

1. Diehl S. et al., *J. Phys. B: At. Mol. Opt. Phys.* **30,** L595 (1997).
2. Bliman S. et al., *Phys. Rev. A* **56**, 4683 (1997).
3. Carlson T.A. et al. *Phys. Rev. A* **140**, A 1057 (1965).
4. Bhalla C.P., *Phys. Rev. A* **8**, 2877 (1973).
5. Larkins F.P., *J. Phys. B.* **9**, L29 (1971).
6. Cauchois Y., and Senemaud C., *Wavelengths of X ray emission lines and absorption edges* (Pergamon, New York 1978).
7. Kayaishi T. et al.,. *J Phys. B* **27**, L115 (1994).
8. Koike F., in Proc. of Oji Int. Seminar on Atomic and Molecular Photoionization, Tsukuba, Japon, Sept. 1995, p. 283, Universal Academy Tokyo, 1996.
9. Briand J.P. et al. *Phys. Rev. Lett.* **65**, 159 (1990).
10. J. Limburg et al. *NIM Phys. Rev. B,* **98**, 436 (1995).
11. Proc. 4th Intern. Conf. On Ion Sources 1991, RSI 63, April 1992.
12. Mack E.M., PhD Thesis, University of Utrecht, The Netherlands (1987).
13. Boudjema M., PhD Thesis, Université d'Alger, Algerie, 1990 (unpublished).
14. Martin S. et al. *Physica Scripta* **T73, 149** (1997)
15. Bliman S. et al. *Phys. Rev. A* **46**, 1321 (1992).
16. Mann R., Folkmann F. and Meyer H.F., *J. Phys. B,* **14**, 1161 (1981).
17. Hutton R. et al. *Phys. Rev. A*, **39**, 4902 (1989).
18. Marseille P. et al. *J. Phys. B*, **20**, 5127 (1987).
19. Cowan R.D. *The theory of atomic structure and spectra* (University of California Press, Berkeley, 1981) and Computer Codes.

Branching Ratios in Dissociative Recombination Measured in a Storage Ring

L. Vikor[a], A. Al-Khalili[a], H. Danered[b], N. Djuric[c], G. H. Dunn[c], M. Larsson[a], A. Le Padellec[a], S. Rosén[a], M. af Ugglas[b]

[a] *Department of Physics, Stockholm University, Box 6730, S-113 85 Stockholm, Sweden*
[b] *Manne Siegbahn Laboratory, Stockholm University, S-104 05 Stocholm, Sweden*
[c] *JILA, University of Colorado, Campus Box 440, Boulder, CO 80309-0390 USA*

Short review of branching ratios measurements in dissociative recombination process of polyatomic molecules by means of storage ring is given, with the main advantages of this experimental technique.

INTRODUCTION

Dissociative recombination (DR) is a process in which molecular ions recombine with electrons and dissociate into neutral fragments (1, 2). It takes place in plasma of (dark) interstellar molecular clouds, inner coma of comets, planetary ionospheres, and in the ionized layers of Earth's upper atmosphere.

In the DR process molecular ions are neutralized and destroyed, and for a polyatomic ion it is important to know probabilities for producing different sets of neutral fragments, known as branching ratios. For modeling the chemistry of interstellar molecular clouds, the branching ratios data are necessary, but in the long list of gas phase reactions, given at the University of Manchester Astrophysics Group homepage (http://saturn.ma.umist.ac.uk:8000/) for branching ratios are usually given only assumed values in the lack of measured or calculated data.

The DR branching ratios are basically difficult to measure, because the DR products are neutral particles, and electric and magnetic field cannot be used to separate the different decay channels. The theoretical determination of the branching ratios is also difficult. It is a complex problem, since a large number of potential energy surfaces have to be considered, and only model calculations can be performed. In his theoretical model Bates (3) assumes that the dissociation which requires the least number of valence bonds to be rearranged is favored. Herbst (4) used a statistical phase space theory to make predictions of DR branching ratios.

Before storage ring measurements, the DR branching ratios were measured mostly by spectroscopic techniques, for thermalized polyatomic ions in flowing afterglow plasmas (5). These methods are quite complicated and the complete set of branching ratios cannot be obtained in one measurement. The uncertainties from normalization of data from different measurements increase the uncertainty of the obtained branching ratios.

By use of the single-pass merged-beam technique (6), measurements were improved in a sense that complete set of DR branching ratios is obtained in one measurement, and branching ratios can be obtained for a range of energies. However, densities of merged electron and ion beams are limited, for the electron beam in order to avoid the possible space-charge effect, and for the ion beam by ion source production of a particular molecular ion. This results in long time for a data collection. The problem is also that in this type of experiments the ion-electron interaction occurs a short time after ion extraction from a source, and ions are typically vibrationally excited, which makes it difficult to compare obtained results with a theory and with results of other experiments.

A storage ring, which in spirit is a combination of merged-beam and ion trap technique, with use of a grid technique (which will be explained later in the text) becomes an excellent tool for measuring branching ratios in dissociative recombination. It provides energy dependent DR measurements, down to an interaction energy that corresponds to low temperatures in interstellar molecular clouds or inner coma of comets. It also provides interaction with vibrationally cold ions, and from one measurement complete branching ratios in DR process can be obtained.

EXPERIMENTAL TECHNIQUE

Branching ratios in DR were measured at two storage rings up to now: ASTRID at the University of Aarhus, Denmark, and CRYRING at Stockholm University, Sweden. The main idea of the measurement will be given on the example of the CRYRING storage ring, ilustrated in Fig. 1 (7). Molecular ions produced in ion source MINIS are injected into the ring, accelerated, and kept circulating by a periodic structure of bending and focusing magnets. The pressure in the ring is very low, about 10^{-11} Torr, and ions can be stored for a time

CP475, *Applications of Accelerators in Research and Industry*,
edited by J. L. Duggan and I. L. Morgan

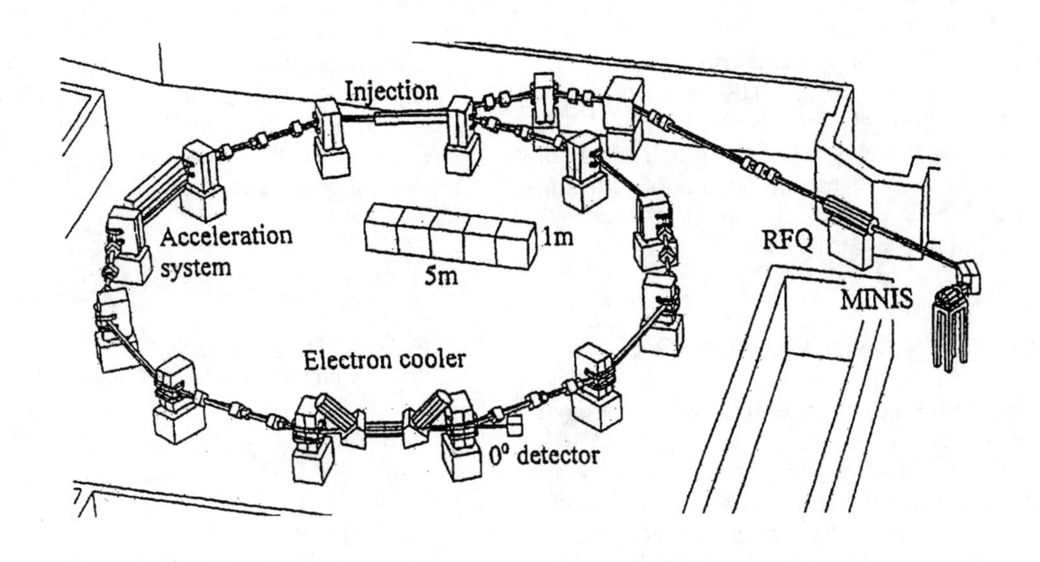

FIGURE 1 The CRYRING facility at the Manne Siegbahn Laboratory, Stockholm

longer than vibrational relaxation time for most of molecular ions. In the section of the storage ring labeled as "electron cooler", the ion beam is merged along the distance of about 1 m with an electron beam, which serves as a target. For every injected ion beam, interaction repeats many times during the beam circulation in the ring, and the data gathering is as many times faster than in the single-pass merged beam experiment. By tuning the electron energy, the interaction energy can be changed (center-of-mass energy). Following the electron cooler, the ions are bent by a magnet and continue to circulate in the ring, while the neutral products from the DR process as well as those produced in collisions with the rest-gas follow a straight line and hit an energy-resolving surface barrier detector. In the energy spectrum of neutral fragments the peak formed by DR usually dominates.

Peak position scales with the mass of the corresponding particles impinging the detector and the background from the collisions of molecular ion with the rest-gas appears as peaks at lower energies. High ion beam energy and ultra high vacuum reduces the background.

In order to determine the branching ratios, a grid with a known transmission is inserted in front of the detector. The probability for neutral fragments to pass the hole is equal to grid transmission T, and the probability to be stopped is equal to 1-T. Particles stopped by the grid do not contribute to the signal from the detector, and the DR signal splits over a series of peaks. The distribution of events into the peaks depends on the branching ratios and on the grid transmission. The spectrum shown in Fig. 2 illustrates the grid effect. This is the spectrum of neutral fragments from DR of the NH_2^+ ion at 0 eV

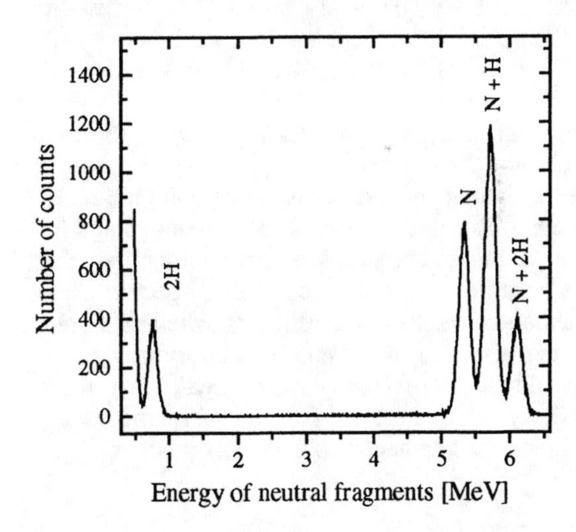

FIGURE 2. The spectrum of neutral fragments in dissociative recombination of the NH_2^+ ion at 0 eV interaction energy, obtained with a grid in front of the detector.

collisional energy, obtained with a grid in front of the detector. Without the grid, the DR products would appear at the full beam energy, that was 6.1 MeV. The peaks are well separated, because of the high energy of the ion beam. From the integrated number of counts in the peaks and with the known grid transmission T, branching ratios can be calculated.

RESULTS AND DISCUSSIONS

Up to now DR branching ratios have been measured for these molecular ions: H_3^+, H_2D^+, H_2O^+, H_3O^+, CH_2^+, CH_3^+, CH_5^+, NH_2^+, NH_4^+ (8, 9, 10, 11, 12,13).

For the investigated molecular ions two- and three-body DR fragmentation channels were energetically opened, and for most of the ions, DR was dominated by three-body break-up. For the H_3^+ molecular ion at low energy (< 0.1 eV) 75% of the DR events finish in the three-body H + H + H channel. For the H_2O^+, CH_2^+ and NH_2^+ ions also three-body channel was found to be dominant: 68% in O + H + H (10), 63% in the C + H + H (11), and 66% in N + H + H (13). However, for the molecules with larger number of atoms the obtained branching ratios are more different. The difference can be seen comparing the respective two-body branching ratios. For DR of H_3O^+ 33% goes into the two-body H_2O + H channel (10), for CH_5^+ only 5% goes into the two-body CH_4 + H channel (12), and for NH_4^+ 69% into the two-body NH_3 + H channel (13). Herbst and Lee (4) suggested that the dominance of the three-body channels (for CH_5^+ and H_3O^+) could be explained by secondary fragmentation of vibrationally or electronically excited molecular products of the two-body channels.

The fragmentation patterns are at this point not really understood. Regularity seems to be missing, since for some of the more complex ions greater fragmentation occurs, while in other complex species, less fragmentation results. A question is whether all the generated products are result of the primary or of further dissociation. One may anticipate that these results may guide efforts to obtain a general theoretical explanation for fragmentation in the DR process.

ACKNOWLEDGMENTS

We would like to acknowledge the Göran Gustafsson Foundation, the Swedish Natural Science Research Council and the Swedish Institute. G. D. and N. D. acknowledge support in part by the Office of Fusion Energy, U.S. Dept. of Energy under contract DC-A102-95ER54293 with the National Institute of Standards and Technology.

REFERENCES

1. Bates, D. R., *Phys. Rev.* **78**, 492- 93 (1950).
2. Larsson, M., *Annu. Rev. Phys. Chem.* **48**, 151-79 (1997).
3. Bates, D. R., *J. Phys. B: At. Mol. Opt. Phys.* **24**, 3267-84 (1991).
4. Herbst, E., Lee, H.-H., *Astrophys. J.* **485**, 689-696 (1997).
5. Adams, N. G., Herd, C. R., Geoghegan, M., Smith, D., Canosa, A., Gomet, J. C., Rowe, B. R., Queffelec, J. L., and Moralis, M., *J. Chem. Phys.* **94**, 4852-57 (1991).
6. Mitchell, J. B. A., Forand, J. L., Ng, C.T., Levac, D. P., Mitchell, R. E., Mul, P. M., Clayes, W., Sen, A., and McGowan, J. Wm., *Phys. Rev. Lett.* **51**, 885 –88 (1983).
7. Strömholm, C., Semaniak, J., Rosén, S., Danared, H., Datz, S., van der Zande, W., and Larsson, M., , *Phys. Rev. A* **54**, 3086-94 (1996).
8. Datz, S., Sundström, G., Biedermann, Ch., Broström, L., Danared, H., Mannervik, S., Mowat, J. R., and Larsson, M., *Phys. Rev. Lett.* **74**, 896-899 (1995).
9. Larsson, M. *Astron. Astrophys.* **309**, L1-L3 (1996).
10. Vejby-Christensen, L., Andersen, L. H., Heber, O., Kella, D., Pedersen, H. B., Schmidt, H. T., Zajfman, D., *Astrophys. J.* **483**, 531 (1997).
11. Larson, Å., Le Padellec, A., Semaniak, J., Strömholm, C., Larsson, M., Rosén, S., Peverall, R., Danared, H., Djuric, N., Dunn, G., and Datz, S., 1998, Astrophys. J. 505, to be published.
12. Semaniak, J., Larson, Å., Le Padellec, A., Strömholm, C., Larsson, M., Rosén, S., Peverall, R., Danared, H., Djuric, N., Dunn, G., and Datz, S., *Astrophys. J.* **498**, 886-95 (1998).
13. Vikor, L., Al-Khalili, A., Danared, H., Djuric, N., Dunn, G. H., Larsson, M., Le Padellec, A., Rosén, S., af Ugglas, M., to be published.

The Ionization and Fragmentation of H_2 and CH_4 and their Deuterated Variants by Photon Impact

R A Mackie, A M Sands, R Browning, K F Dunn and C J Latimer

Department of Pure and Applied Physics, The Queen's University of Belfast, Belfast BT7 1NN, N Ireland, United Kingdom

The kinetic energy spectra of fragment protons and deuterons produced in the dissociative photoionization of H_2, D_2 CH_4 and CD_4 by VUV photons within the energy range 12-50 eV have been measured. Cross sections for negative ion production have also been determined. It is shown that, in addition to direct photoionization, intermediate superexcited states play an important role in these photoionization processes.

INTRODUCTION

In addition to being of fundamental interest, the absorption of VUV photons by molecular hydrogen and methane is important in interstellar clouds, planetary atmospheres and plasma physics[1,2]. Superexcited molecular states, which have several different decay channels including dissociative autoionization and ion-pair formation can play an important and often dominant role in the absoption process.

In recent work[3,4,5] we have shown that such states make a significant contribution to the formation of energetic protons in the photodissociative ionization of H_2.

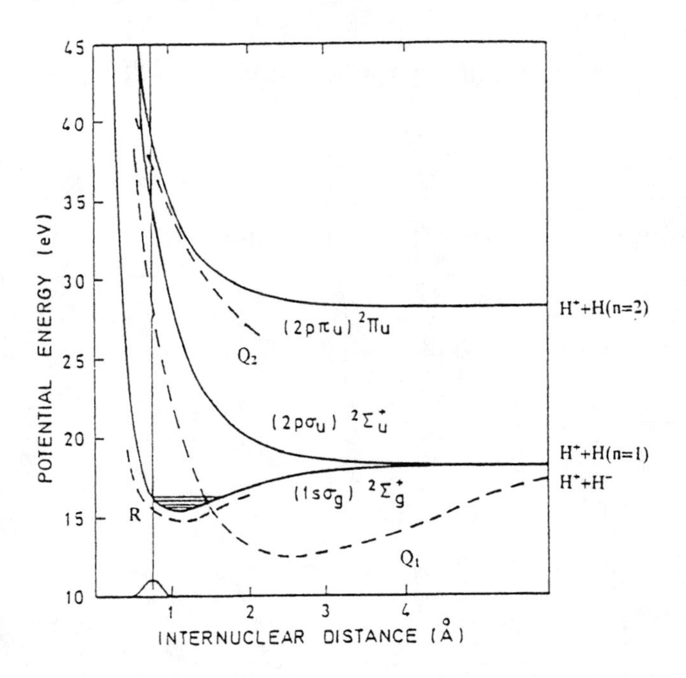

FIGURE 1. Potential energy curves of H_2 and H_2^+ involved in the current work. R, Rydgerg states; Q_1 and Q_2 doubly excited states.

Furthermore differences in the fragment energy spectra of the light and heavy isotopes was shown to give detailed information on the autoionizing processes involved, since heavier nuclei separate more slowly than lighter.

In the present work we have extended these studies to concentrate on states of Π symmetry, by making measurements perpendicular to the $\underline{E}$ vector of the synchroton radiation. In addition the negative/positive ion pair channel, which is of special interest since it can yield precise values for electron affinities and the potential energy curves of the precursor excited molecular states, has been studied. These processes have been observed previously[6,7] to have a rotational state dependence (ortho and para-hydrogen give different results) and are believed to result from predissociation of neutral Rydberg states although the detailed identity and nature of these states is not yet clear[8]. Similar fragment ion energy spectra measurements have been made in CH_4 and CD_4 where the dynamics of fragmentation are of great interest since methane is the smallest molecule which may be expected to dissociate in a statistical manner[9].

EXPERIMENTAL APPROACH

The experiments were carried out at the Daresbury Laboratory (UK) synchrotron using experimental arrangements and procedures almost identical to that described in our earlier work[3,10]. The VUV photon beam (~ 10^{11} hv/s) was horizontally polarized (~ 80%) and contained < 5% second order radiation. The photon beam emerged from a capillary light guide and was crossed at 90° by a low pressure gas jet of target molecules. Three different types of experiment were performed.

To determine the energy spectrum of the fragment ions formed in the gas jet target a rotatable wide aperture parallel plate electrostatic analyser ($\Delta E \simeq 0.8$ eV) viewed the field free interaction region perpendicular to the E vector of the radiation. Spectra were corrected for the

CP475, *Applications of Accelerators in Research and Industry*, edited by J. L. Duggan and I. L. Morgan

transmission factor of the apparatus. For these measurements, no mass analysis of the fragment ions was performed since molecular ions are known to be formed with near thermal energies and in the present work we are interested in energetic fragment protons ($\geq$ 1.5 eV).

To determine cross sections for the production of mass analysed product ions (positive or negative), the interaction region was viewed from the opposite direction by a quadrupole mass spectrometer. Mass descrimination effects were eliminated by normalizing the data, in a subsiduary ion beam experiment, to well established proton impact data[11].

To determine partial photoionization cross sections for the formation of energetic fragment ions, ions were extracted by a carefully chosen weak electric field of 2.25 V/cm and then further accelerated to form a beam of 15 eV energy which then passed through a retarding grid system before further acceleration and detection by a channel electron multiplier. Under these conditions fragment ions with energies > 1.5 eV were only collected over an angle of approximately $\pi/10$ and so the measurements could again be made perpendicular to the radiation E vector.

RESULTS AND DISCUSSION

Fragmentation of H_2 (D_2) via the $^2\Pi_u$ ion state

At photon energies $\geq$ 36 eV, energetic protons arising from direct Franck-Condon excitation to the repulsive $^2\pi_u$ ionic state might be expected to dominate the spectrum (see Fig. 1). Fig. 2 shows the fragment proton and deuteron spectra obtained at 90° in H_2 and D_2 at a photon energy of 36.5 eV. The large group centred around an energy of 3.0 eV can readily be attributed to this process.

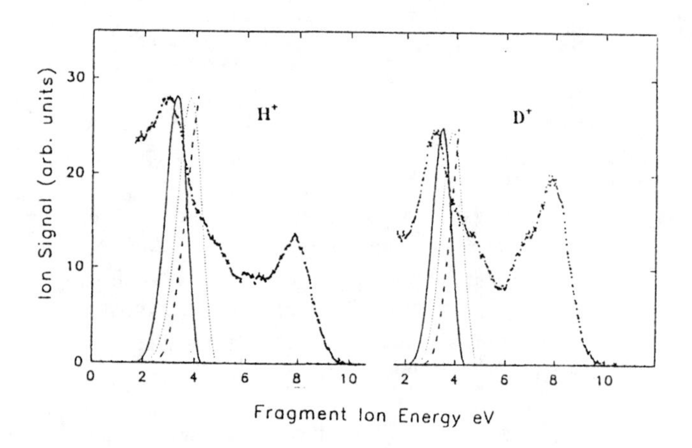

FIGURE 2. Fragment proton and deuteron energy spectra at 90° in H_2 and D_2 at 36.5 eV. The broken curves given the reflection approximation for excitation to the $^2\Pi_u$ state. The dotted curve shows the effect of incorporating the energy bandpass of the ion analyser while the full curve also incorporates a simple Airy function shift.

Now the most widely used method for describing such bond continuum transitions is the reflection approximation in which the continuum wavefunction is replaced by a normalized δ-function at the classical turning point[12,13]. Fig. 2 shows the results of such a calculation[5] and agreement with experiment is disappointing, even after modifying the calculation to incorporate the known transmission function of the ion energy analyser. Recently however it has been shown[14] that the final state wavefunction is well approximated by an Airy function which has a first maximum which is displaced from the classical turning point by 1.019 $[\hbar^2/2\mu v^2]^{1/3}$ where μ is the moleuclar reduced mass and v' is the steepness of the final state potential. As can be seen in Fig. 2, which this simple shift is incorporated into the calculations, agreement with the present data is greatly improved.

In addition to the $^2\Pi_u$ feature, a significant group of higher energy ions at ~ 8.1 eV is apparent. In accord with our earlier work[5] these energetic fragments clearly arise via excitation to the $^1\Pi_u$, Q_2 superexcited state followed by autoionization. The maximum allowed energy of fragment protons is ½(36.5 – 18.1) = 9.2 eV. This indicates that there is a preferred energy loss of ~ 2.2 eV in the autoionizing process from this state which is taken up by the outgoing electron, which in turn indicates that autoionization to the $^2\Sigma_u$ state of H_2^+ occurs with maximum probability at an internuclear separation of 1.8 a_o where the separation of the two potential energy curves is ~ 2.2 eV.

Ion-pair formation in H_2 and D_2

Figs. 3 and 4 show the cross sections obtained for the ion pair production process in normal H_2 and D_2 at 300 K at a wavelength resolution of 0.025 nm ($\Delta E \cong$ 6 meV). Figure 1 shows the results for a hydrogen target. Several

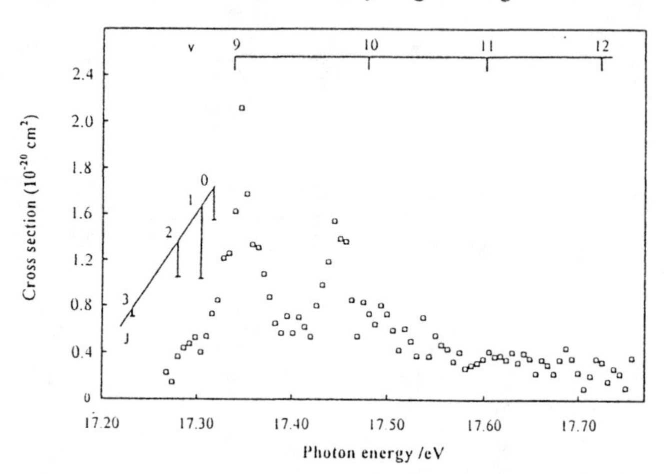

FIGURE 3. Ion-pair formation in the photoabsorption of normal hydrogen. Theoretical thresholds for excitation from different rotational levels (J) of the ground state are shown with the length of the lines indicating the room temperature populations. Approximate energies of predissocating Rydberg states converging to H_2^+, v ≥ 9 are also shown.

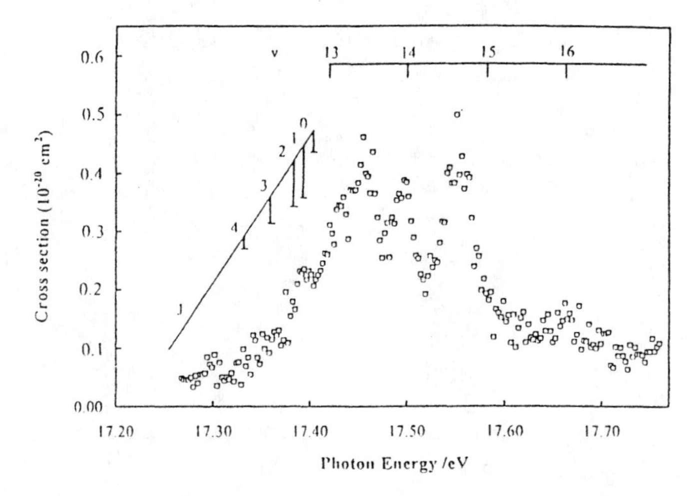

FIGURE 4. Ion-pair formation in the photoabsorption of normal deuterium. Thresholds and energy levels as in Fig. 3.

thresholds are observed corresponding to excitation from different rotational levels. These correlate well with the threshold values shown in the figure, based on the thermodynamic threshold at 18.0766 eV and an electron affinity for H atoms of 0.7542 eV. The structure in the observed cross section is in accord with a mechanism involving predissociating unresolvable Rydberg states converging to H_2^+ ($X^2\Sigma_g$) (see Fig. 1) which have vibrational quantum numbers v = 9, 10 and 11, and total angular momenta N = 0,2. The value of the photoionization cross section for the dissociation continuum (~ 0.4 x 10^{-20} cm^2) is in harmony with the only previous measurement[7], an upper bound only, of 0.5 x 10^{-20} cm^2. The corresponding results for a deuterium target (Fig. 4) also show that the position and size of the threshold features are in good agreement with the theoretical predictions and the structure at higher energies clearly relates to Rydberg states converging to D_2^+ ($X^2\Sigma_g$), v ≥ 13, N = 0, although perturbations due to configuration interaction seem likely. The fact that the D^- cross sections are approximately 33% of the H^- clearly indicates that a predissociation process is involved[15] in a complicated and not fully understood, dissociation mechanism[8].

Fragmentation of CH_4 and CD_4

Fig. 5 shows the fragment proton and deuteron spectra obtained at 90° in CH_4 and CD_4 at a photon energy of 46.6 eV. Two main groups of ions are apparent centred around 2.1 eV and 3.5 eV. There is a significant isotope effect. Partial photoionization cross sections for the formation of energetic > 1.8 eV and > 3.1 eV in hydrogen are shown in fig. 6. For the lower energy group thresholds are observed at 22.0 and 29.0 eV. Now CH_4 has the $(1a_1)^2(2a_1)^2(1t_2)^6$ electron configuration, where $1a_1$ and $2a_1$ correspond to the 1s and 2s arbitals of the carbon atom respectively. Photoelectron spectroscopy and (e, 2e) experiments[16] show

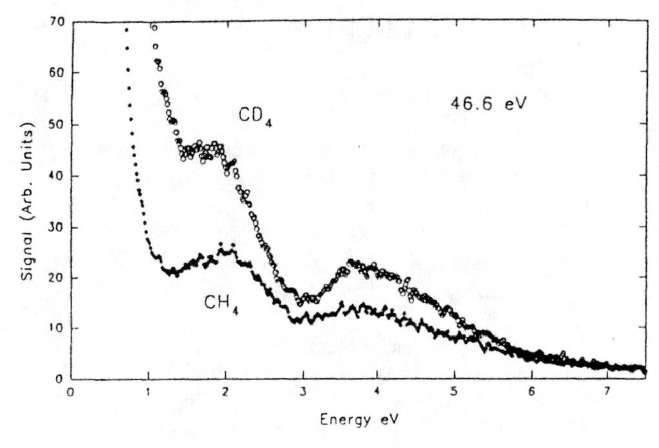

FIGURE 5. Fragment proton and deuteron energy spectra at 90° to the radiation E vector in CH_4 and CD_4 at a photon energy of 46.6 eV.

that the $(2a_1)^{-1}$ state has a threshold at 22.4 eV, indicating that the H^+ ions are arising from this state. At higher energies two, as yet unclassified, doubly excited states X_1 and X_2 are observed. The X_2 states, possibly $(1t_2)^{-2}(3a_1)^1$, corresponds to the second threshold for the formation of H^+ ions (E > 1.8 eV) observed in our work. Energetic considerations show that the final decay channel is CH_3 + H^+ + e (rather than CH_2 + H^+ + H + e, which is also possible)[17]. The higher energy group of photons with E > 3.1 eV has a single threshold at 35.0 eV. This onset correlates with the appearance energy of CH_3^+ in double ionization[18] and the steady increase above this is in accord with the additional availability of double ionization reaction channels.

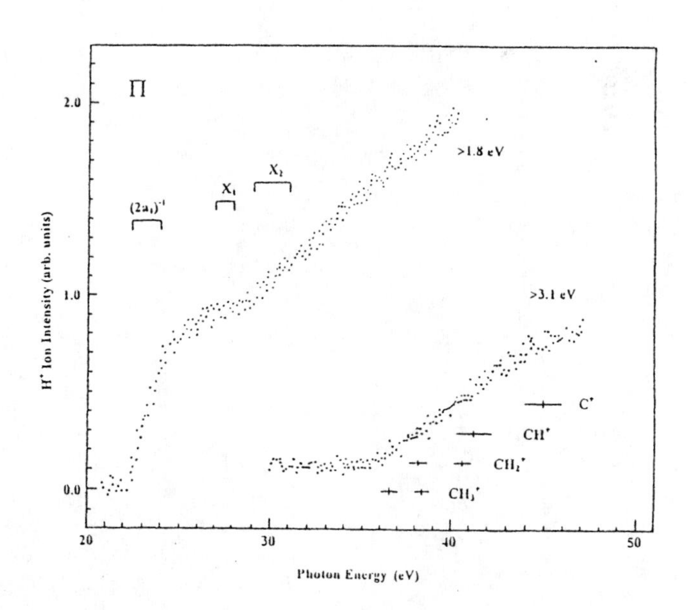

FIGURE 6. Partial photoionization cross sections for the formation of energetic fragment protons with E > 1.8 eV and E > 3.1 eV at 90° in CH_4. Vertical lines indicate thresholds for the states indicated.

REFERENCES

1 Shimamura I, Noble C J and Burke P G, Phys Rev A **41**, 3545, 1992.
2 Fox J L in Dissociative Recombination, Eds B R Rowe, J B A Mitchell and A Canosa (Plenum, New York), 1993.
3 Latimer C J, Dunn K F, Kouchi N, MacDonald M A and Geddes J, J Phys B: At Mol Opt Phys **26**, L595, 1993.
4 Geddes J, Dunn K F, Kouchi N, MacDonald M A, Srigengan V and Latimer C J, J Phys B: At Mol Opt Phys **27**, 2961, 1994.
5 Latimer C J, Geddes J, MacDonald M A, Kouchi N and Dunn K F, J Phys B: At Mol Opt Phys **29**, 6113, 1996.
6 McCullough K E and Walker J A, Chem Phys Lett **25**, 439, 1974.
7 Chupka W A, Dehmer P M and Jivery W T, J Chem Phys **63**, 3929, 1975.
8 Berkowitz J in VUV and Soft X-ray Photoionization, Ed U Becker (Plenum, New York), 1996.
9 Field T A and Eland J H D, J Electron Spec Rel Phenom **73**, 105 (1995).
10 Latimer C J, Dunn K F, O'Neill F P, MacDonald M A and Kouchi N, J Chem Phys **102**, 722, 1995.
11 Browning R and Gilbody H B, J Phys B: At Molec Phys **1**, 1149, 1968.
12 Tellinghuisen J in Advanced in Chemical Physics, vol LX, Ed K P Lawley (Wiley-Interscience, New York), 299, 1985.
13 Gislason E A, J Chem Phys **58**, 3702, 1973.
14 Schinke R, Photodissociation Dynamics (CUP, Cambridge) 1993.
15 Berry R S and Nielsen S E, Phys Rev A **1**, 395, 1970.
16 Van der Wiel M J, Stoll W, Hamnett A and Brion C E, Chem Phys Lett **37**, 240, 1976.
17 Samson J A R, Haddad G N, Masuoka T, Pareek P N and Kilcoyne D A L, J Chem Phys **63**, 3929, 1975.
18 Hatherly P A, Stankiewicz M, Frasinski L J and Codling K, Chem Phys Lett **159**, 355, 1989.

Differential Cross Section Measurements in Ion-molecule Collisions

Song Cheng

Department of Physics and Astronomy, University of Toledo, Toledo, Ohio 43606

A 14 m long beam line dedicated to study very small scattering angles in ion-molecule collisions has been set up in the University of Toledo Heavy Ion Accelerator (THIA) Laboratory. Together with position sensitive detectors for both the projectile and the recoil particles' detection, the beam line can be used to measure the projectile forward scattering angles of up to 2.5 milliradians (mrad) with a 0.025 mrad resolution, in coincidence with information on the recoil particles such as recoil charge states, energy, momentum and the molecular orientations.

INTRODUCTION

In studying of collisions of ions with molecular targets, many important quantities need to be measured in order to compare experimental data with theoretical expectations. Two of these quantities for the projectile are the forward scattering angle and the projectile momentum change. As for the recoil, the similar quantities are the recoil energy and the directions, which, in the case of molecular fragmentation, bear signatures of the molecular orientation at the moment of the collision. The more quantities that are measured, the more stringent tests the experimental data provide for the theories. Otherwise, the un-measured quantities need to be integrated over all possible values in theories and thus possibly mask important insights into understanding the collisions. In this sense, differential cross sections are more valuable than total cross sections.

For a heavy projectile with energies in the order of 1.0 atomic unit per mass (e.g. 25 keV proton), the forward scattering angle is typically of order of 1 mrad[1]. To obtain a meaningful differential cross section measurement for the forward scattering angle, a resolution of 0.01 mrad (1% of 1 mrad) or better is needed. Such a small angular resolution requires a long beam line that will allow the scattered projectile to travel a relatively long distance before it hits the detector. An additional requirement is that the projectile needs to be carefully collimated in order to achieve the required high angular resolution. On the recoil side, the main concern is how to effectively collect recoil particles and derive important parameters such as the recoil energy, momentum, and the molecular orientation in the case of the fragmentation. In this paper, we describe a setup which aims to meet these requirements and show an example using the setup to measure the forward scattering angle in coincidence with the target molecular orientations in the collision of 100 keV proton with hydrogen molecules. For a convenience of discussion, this example is used throughout this paper.

In this example, a theoretical calculation[2] predicts that for a fixed hydrogen molecular orientation, the forward scattering angle of the proton projectile shows an oscillatory structure in the differential cross section for the electron capture process (Fig. 1). Thus an experimental setup that can verify this prediction should be able to: (1) determine the hydrogen molecular orientation at the moment of the collision; (2) measure the forward scattering angle with a sufficient resolution to show the structure; and (3) separate the electron capture process

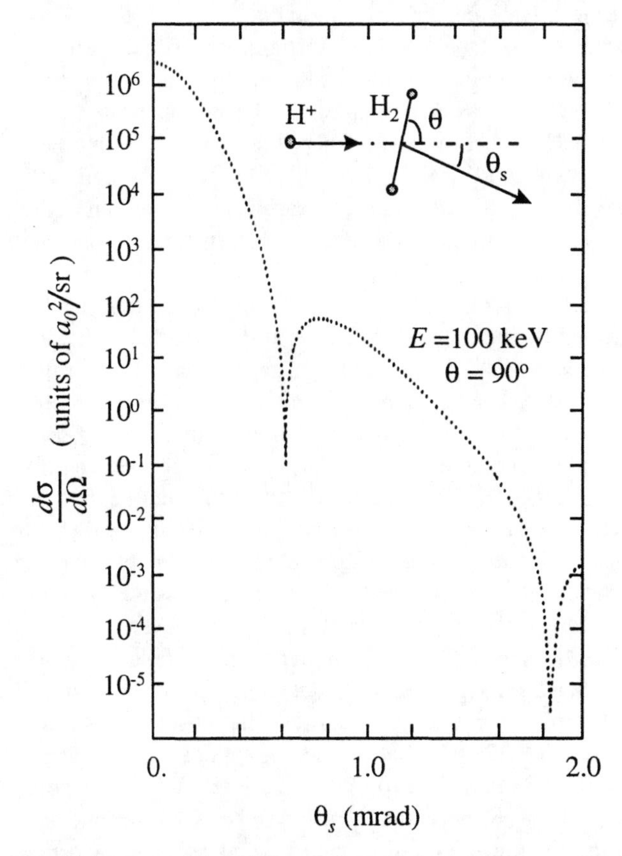

FIGURE 1. Differential cross section for electron capture process for 100 keV proton colliding with hydrogen molecules whose molecular orientation is fixed at θ = 90° (see reference 2).

CP475, *Applications of Accelerators in Research and Industry*,
edited by J. L. Duggan and I. L. Morgan

from the other scattering processes. All three requirements are generally needed in order to provide a stringent test of theories in ion-molecule studies.

BEAM LINE DESCRIPTION

The projectile ion beam is generated as follows: positive ions generated in a conventional ion source are extracted by a potential in the range of 6 to 20 kV applied to the ion source, mass analyzed by a bending magnet, and are accelerated to energies up to 250 keV. Then the ion beam is directed down a long beam line by an electric field deflector system. A schematic diagram of the long beam line is shown in Fig. 2. The beam is first collimated by the slits S1 to define a reference point for the rest of the beam line. The beam current is reduced from a couple of μA down to the nA scale after passing the slits S1 with a slit width in the order of 100 μm. Before traveling down the long beam line, the beam is slightly steered horizontally to the right by the deflector DP1 and if necessary, vertically by the deflector DP2. Steering the beam slightly in the horizontal direction directs the beam a little off the center to hit the slits S2 that is 6.5 meters downstream. Due to this long flight path, some projectile ions will interact with the background gas and capture or lose electrons. For example, for the proton beam, there will be H^0 and H^- impurities near the slits S2. With a help from two additional horizontal deflectors DP3 and DP5 near the slits S2, the charged beam is bent twice and only the H^+ is directed through a collimator of 0.1-mm diameter (the AP1 inside the collision chamber) to enter the collision chamber. The H^0 impurity beam is not affected by the deflections and is stopped on the right side of the slits S2. This pre-collision separation of the various charge state beams in the incident projectile beam is very important for this collision system since the charge exchange processes have a large cross sections in this energy range[3]. The pure charge state incident beam then enters the collision chamber to interact with the target. The combination of deflectors DP2 and DP4 provides a vertical steering to accommodate a possible minute misalignment that the collimator AP1 may have.

Inside the collision chamber, target molecules are introduced through a jet at 90 degrees with respect to the beam from the side opposite to the position sensitive detector (PSD2). The jet is made of a glass capillary array from the Galileo Electro-optics Corporation[4]. The pore diameter of an individual capillary tube is 25 μm and the length is 2 mm long. The size of the glass capillary array is about 1.25 mm. After interactions with the projectile ions, the molecular ions are extracted and accelerated to the PSD2 by an electric field set up by two plates G_1 and G_2 that are separated apart by 2.4 mm (see the exploded view of the extraction region in Fig. 3). The voltages on the plates, V_1 and V_2, are chosen in such a way that the electric field between the plates is strong enough to accelerate all molecular ions and molecular fragments in all directions after the collision towards the PSD2. Too strong electric field in this region is avoided since it would also deflect the projectile beam considerably. For hydrogen molecules, the maximum initial energy of the molecular fragments are about 10 eV. So at least 20 V potential difference should be maintained across the two plates, assuming the projectile interacts with the target right in the middle of the plates. In reality, the voltage difference should be larger to accommodate the beam size of the projectile and the case in which the projectile is a little off the center of the two plates.

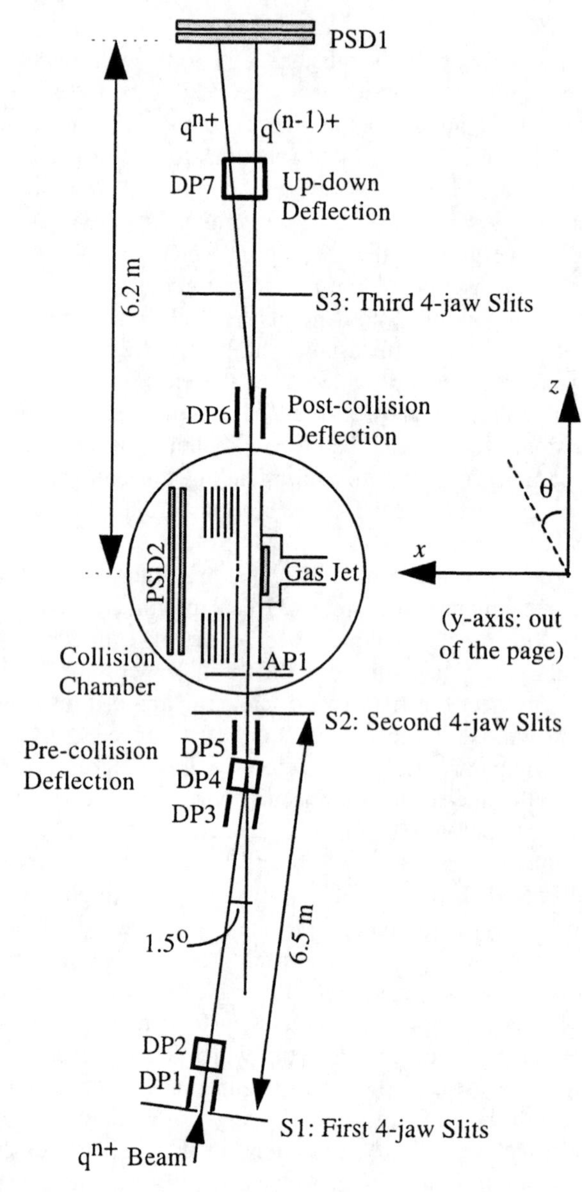

FIGURE 2. A schematic of the long beam line setup. DP1, DP3, DP5, and DP6 are horizontal deflectors. DP2, DP4, and DP7 are vertical deflectors. AP1: a 0.1-mm diameter aperture. PSD1 and PSD2: two-dimensional position sensitive detectors.

There is a 6.3 mm diameter aperture with a high transmission grid on the plate G_2 to limit the PSD2 solid angle so that it can detect only recoil fragments from collisions that happen in the vicinity of the gas jet. A trade-off exists when deciding the diameter of the aperture and it will be discussed further later in this paper. The extracted molecular ions or fragments after passing the aperture are accelerated by a uniform electric field (l_2 in Fig. 3) and then are allowed to drift in a field-free region (l_3 in Fig. 3) before hitting the two-dimensional position sensitive detector. A measurement of the position (y,z) that ions hit the PSD2 and the time-of-flight that ions take from the jet to the PSD2 is sufficient to allow a calculation of the molecular orientation of the target molecule at the moment of the collision.

A second position sensitive detector PSD1 is placed about 6.2 meters downstream in the forward direction at the end of the beam line to detect the scattered projectile beam. The beams of different charge states are separated by the post-collision deflector DP6 so that they hit the PSD1 at different locations horizontally. The relative counts at these horizontally separated groups represents the percentage of beam that changes charge states after beam interacts with the target gas jet. The position spread within each group with respect to the center of the group itself is proportionally related to the forward scattering angle θ_s as defined in the Fig. 1. At about the midpoint between the post-collision deflector DP6 and the PSD1, there is a vertical deflector DP7 which alters the beam trajectories slightly in the vertical direction. The purpose

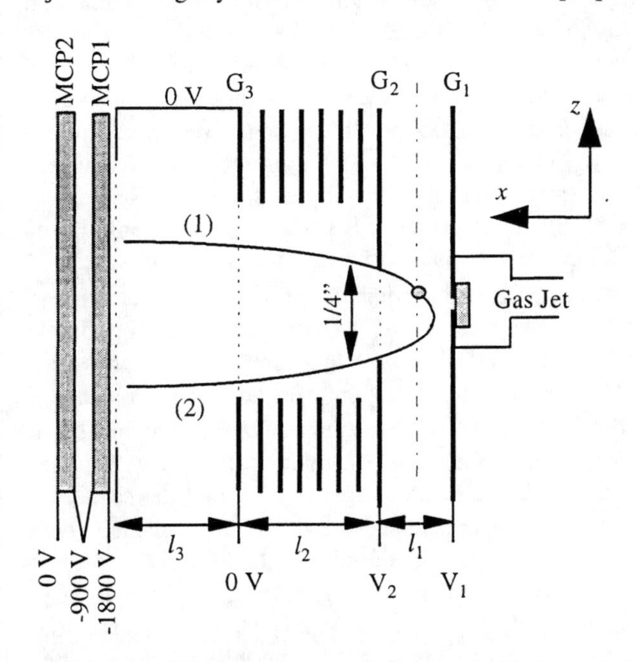

FIGURE 3. An exploded view of the extraction region and the detector PSD2. MCP: micro-channel plates. l_1: the scope of the extraction field set up by the V_1 and the V_2. l_2: the scope of the acceleration field set up by the V_2 and the ground. l_3: the scope of field-free drift region. A 1/4" aperture is on the plate G_2.

of the DP7 is to estimate the contributions to the beam impurities developed *after* the projectile beam passes the slits S2 and interacts with the background gas before it interacts with the gas jet. Since for the beam that changes charge states during the path from the deflector DP6 to the deflector DP7, it will be separated vertically on the PSD1. The relative counts at these vertically separated groups represents the percentage of beam that changes charge states when the beam travels through the background gas of "target thickness", which is equal to the background gas density multiplied by the separation from the deflector DP6 to the deflector DP7. The background gas densities in locations *before, in,* and *after* the collision chamber are monitored or estimated so that the background signal can be accounted for in the data. The coincidence between timings from the PSD1 and the PSD2 relates the forward scattering angle with the molecular orientation of the target molecules at the moment of the collision.

Placing the PSD1 at about 6.2 meters downstream from the target gas jet is to gain a sufficient angular resolution needed to observe the narrow interference structure as shown in the Fig. 1. For the PSD1, that has a linear resolution of about 0.15 mm, the angular resolution that can be achieved by this arrangement is about 0.024 milliradian, or about 0.0014°, which should be sufficient to resolve the oscillatory structure shown in the figure. The size of the position sensitive detector PSD1 is about 20 mm in radius. That means it can cover the forward scattering angle up to 3.2 milliradians if the beam is let to hit in the middle of the detector. In reality, a reliable coverage is about 80 % of 3.2 milliradians, to be 2.5 milliradians.

In order not to deteriorate the 0.024 milliradian angular resolution achieved on the PSD1, the incoming projectile beam has to be collimated at least in the same order of magnitude, or about 0.012 milliradian. To satisfy this condition, the projectile is collimated by two slits S1 and S2 that are separated by 6.5 meters upstream. In order to close these slits down to a zero width, the opposite slits are designed in such a way that they will not touch each other when closed down to a zero width and will overlap about one millimeter if they are continued to close down. No absolute calibration or alignment is needed for these two precision-micrometer-driven slits and the zero slit width is obtained by watching the projectile counting rates to go to zero as the slits are closed down. A 0.012 milliradian divergence of projectile calls for slit openings of 80 microns for both the slits S1 and the slits S2. The real openings of the slits are actually dictated by the beam current that the forward PSD1 can handle. For example, the THIA accelerator routinely delivers about 5 microamperes of proton beam onto the slit S1 with a beam diameter of about 5 mm. However, a position sensitive detector can normally only handle about 10,000 particles per second. So the beam current has to be reduced down to less than a few femto-amperes when the beam reaches the PSD1. Therefore, very narrow slit widths are needed to cut the beam intensity down to an acceptable range, resulting in slit openings of approximately 50-60 microns.

APPLICATION TO p + H_2 COLLISION

We would like to use this setup to measure the forward scattering angle θ_s in coincidence with the molecular orientation for electron capture process in p + H_2 collisions. The proton energy is chosen to be 25 keV and 100 keV in order to directly test the theoretical predictions by Deb *et. al.*[2]. Their predictions can be summarized as: (1) for a fixed molecular orientation, the forward scattering angle θ_s shows an oscillatory structure; (2) as the projectile energy changes, or the fixed orientation θ changed to other than 90°, the oscillatory structure changes accordingly.

The main difficulty in this application is how to determine the molecular orientation at the moment of the collision. For a sufficiently high energy proton impact, there is a certain probability[5,6] that, after the electron capture event, hydrogen molecular ions are left in one of the dissociative states. For example, for 25 keV proton impact on H_2, the total capture cross section σ_C is about 5 $\times 10^{-16}$ cm^2 (ref. 7), and the cross section for transfer ionization σ_{TI} ($p + H_2 \Rightarrow H^0 + H_2^{++} + e$) is about 1.0×10^{-17} cm^2 (ref. 5), and the cross section for transfer excitation σ_{TE} ($p + H_2 \Rightarrow H^0 + (H_2^+)^*$) is about 1.0×10^{-16} cm^2 (ref. 5). Therefore, when electron capture events happen, 22% of the recoiling hydrogen molecular ions will be left on the dissociative states which will dissociate into H^+ and H^0 in the case of $(H_2^+)^*$, or dissociate into two H^+ ions in the case of H_2^{++}.

When the molecule ions undergo dissociation, the momenta of the two partners are equal in magnitude and opposite in direction (the thermal energy of the hydrogen molecules can be neglected when compared with the dissociation energy carried away by these two partners, which in the order of a few eV to 10 eV each), and the directions of the momenta determine the molecular orientation. Since the collision time is much shorter than the rotation time of hydrogen molecules (which is in the order of 10^{-12} second), the molecules can be considered 'frozen' at the moment of the collision. The post-collision interaction between the projectile and the dissociated ions has little effect on the trajectory of the dissociated ions[8]. Therefore, the probability of finding the dissociated particles in a neighborhood of a fixed orientation is proportional to the solid angle $d\Omega$ associated with the orientation. For example, in the case of $\theta = 90°$ and $\Delta\theta = \pm 2.5°$, the probability is about 4.3%, assuming that molecular fragments in all directions are collected and the azimuthal symmetry is valid.

In order for the PSD2 to collect the dissociated particles of kinetic energy of a few eV (up to about 10 eV) and moving in all directions, V_1 is set to 491 V, and V_2 to 453 V. The lengths l_1, l_2, and l_3 are set to 2.4 mm, 19.6 mm, and 20.8 mm respectively to maximize the time-of-flight resolution for the fragments[1]. Measurements of the time-of-flight of the fragments and the position (y,z) where the fragments hit the PSD2, theoretically, are enough to determine the molecular orientation. However, the determination of the orientation gets complicated due to the fact that the gas jet is not an ideal point source but rather has an extended spread along the beam axis. In this case, the molecular fragmentation can happen anywhere along the beam seen by the PSD2. Therefore, the origin of the collision can not be defined and the molecular orientation can not be calculated. There is no solution to this for a given jet. To cure the problem to some extent, an aperture is placed on the plate G_2 to restrict the view of the PSD2 to the region near the jet. The influence of the aperture on the fragments to be detected depends on which channel the H^+ comes from: the $(H_2^+)^* \Rightarrow H^+ + H^0$ channel or the $H_2^{++} \Rightarrow H^+ + H^+$ channel. For the $H^+ + H^0$ channel, only the H^+ is detected. The origin of H^+ needs to be known in order to calculate the molecular orientations. In this sense, an aperture with a smaller diameter is preferred. However, a smaller diameter aperture needs a stronger electric field in the extraction region l_1 to ensure that all fragments pass the aperture without being blocked. A stronger electric field would deteriorate the time-of-flight resolution and in turn deteriorate the orientation measurement. For the $H^+ + H^+$ channel, it does not matter where the origin is, since the momentum conservation dictates that if one H^+ comes with an angle θ, then another must come with $180° - \theta$. In this sense, one would prefer to have no aperture at all. Based on the initial velocity of the H^+ fragments, the geometry of the extraction region, and the diameter of the PSD2 (40 mm), a compromise of 6.36 mm diameter is used for the aperture in this setup.

An estimate of the coincidence rate can be made based on the known cross sections. As stated earlier, for 25 keV proton impact on H_2, the total capture cross section $\sigma_C = 5 \times 10^{-16}$ cm^2, $\sigma_{TI} = 1.0 \times 10^{-17}$ cm^2 for the transfer ionization, and $\sigma_{TE} = 1.0 \times 10^{-16}$ cm^2 for the transfer excitation. From the relation $Y = \sigma(nL)N$, where N is the projectile rate; (nL), the product of the target density and the scattering length; is the target thickness; σ, the cross section for a specific process; and Y, the rate of the specific process, it is reasonable to assume the (nL) to be 2×10^{14}/cm^2 to keep the single collision condition valid. Under this condition, the total capture rate is $Y_c = 0.1N$ and the rate for transfer ionization is $Y_{TI} = 0.002N$, and for transfer excitation, $Y_{TE} = 0.02N$. Since the projectile detector PSD1 can handle about 10,000 particles per second, and we are only interested in the charge exchange processes, the projectiles that are not involved in charge exchange process (H^+ in this case) are blocked from hitting the projectile detector by the slits S3, and only the H^0 ions are allowed to hit the detector. Assuming that, after interaction with the target, the projectile H^0 rate is 10,000/s, then about 2,000/s are from the transfer excitations (of cross section 1.0×10^{-16} cm^2) and 200/s are from the transfer ionization (of cross section 1.0×10^{-17} cm^2). All these transfer excitation and transfer ionization events lead to the dissociation of hydrogen molecules. The events whose molecular orientations are at $\theta = 90° \pm 2.5°$ regardless the azimuthal angle are about 86/s (4.3% of the 2000/s) and 8.6/s (4.3% of the 200/s) respectively for the transfer excitation and transfer ionization.

The coincident events can be dramatically increased by the following consideration. When we examine the differential cross section shown in the Fig. 1, we realize that the oscillatory structures lie totally beyond the scattering angle $\theta_s > 0.4$ milliradian and only account for about 1% of the total cross section. That is to say that 99% of the electron capture events have forward scattering angles $\theta_s < 0.4$ milliradian and do not contribute to the oscillatory structures. Therefore if we block the events that hit the projectile detector PSD1 in the center within the radius of $0.4 \times 10^{-3} \times 6.2$ m = 2.4 mm, we can reduce the projectile counting rate by a factor of 100 without throwing away any events which contribute to the oscillatory structures. In this way, the beam intensity can be increased 100 fold, which in turn increases the molecular fragmentation rates by 100 fold without saturating the detector. An easy solution to block the beam from hitting the center of the detector is to use a channel plate that has a hole in the center, which is commercially available from the channel plate manufacturer, Galileo Electro-optics Corporation[4].

The molecular orientation of $\theta = 0°$ which is parallel to the beam direction can not be measured by the above discussed procedures for the $H^+ + H^+$ channel. In this case, the time-of-flights for the both H^+ are the same and will be counted as one. However, the use of the isotopic HD molecule will overcome this difficulty. Though each fragment still has the same initial momenta, their initial velocities are quite different, which means their time-of-flights are quite different also.

ACKNOWLEDGMENTS

The author would like to thank Dr. Cocke for discussions on the feasibility of this experiment and Dr. Kamber for encouragement which stimulated me to present this paper in this meeting.

REFERENCES

1. Cocke, C. L. and Olson, R. E., *Physics Reports* **205**, 153-219 (1991).
2. Deb, N. C., Jain, A., and McGuire, J. H., *Phys. Rev. A* **38**, 3769-3772 (1988).
3. Stier, P. M. and Barnett, C. F., *Phys. Rev.***103**, 896-907 (1956).
4. Galileo Electro-optics Corporation, Glass Capillary Array Data Sheet No. 7100, and Microchannel Plate Data Sheet No. 9000.
5. Lindsay, B. G., Yousif, F. B., Simpson, F. R., and Latimer, C. J., *J. Phys. B: At. Mol. Phys.* **20**, 2759-2771 (1987).
6. Cheng, S., Cocke, C. L., Kamber, E. Y., Hsu, C. C., and Varghese, S. L., *Phys. Rev. A* **42**, 214-222 (1990).
7. Tawara, H., Kato, T., Nakai, Y., *Atomic Data and Nuclear Data Tables* **32**,235-303 (1985).
8. Dunn, G. H. and Kieffer, L. J., *Phys. Rev.* **132**, 2109-2117 (1963).

Molecular Dynamics in Attosecond-duration Intense-field Pulses Using a Tandem Accelerator

D. Mathur

Tata Institute of Fundamental Research, Homi Bhabha Road, Mumbai 400 005, India

External perturbation of intramolecular electric fields can result in structural changes that influence energy, entropy and molecular dynamics. We have probed the ionization and fragmentation dynamics of triatomic and polyatomic molecules induced by intense, transient fields generated in fast (50-100 MeV) collisions with Si^{q+} (q=3-8) ions, and by using a laser. In the former case, at an impact parameter of ~3Å, the target molecule experiences an attosecond-duration field whose magnitude (~0.05 V/Å) matches that induced by intense (10^{13} W/cm^2) laser light. The temporal and directional properties of the two fields are very different: the linearly-polarized light yields a fixed field direction whereas the ion-induced field is time-dependent. Moreover, the ion-induced fields are of attosecond duration whereas the laser-induced ones last for picoseconds. We explore how the dissociative ionization patterns measured in our experiments reflect these temporal and directional differences.

INTRODUCTION

We have explored the possibility of studying the behavior of molecules in intense electric fields of attosecond duration. We denote external fields as being intense when their magnitude matches intramolecular coulombic fields, so that the resulting ionization and dissociation dynamics lie in the regime of non-perturbative physics. In the context of molecular systems, it is useful to compare interaction times to typical rotational and vibrational time periods (tens of picoseconds and tens of femtoseconds, respectively). Intense electric fields of such duration are presently generated by tightly focussing light from intense lasers that produce light pulses of picosecond and femtosecond duration. Can molecular dynamics be probed in intense fields that are of even shorter duration? Can fast beams of highly charged ions be used for such purpose? The correspondence between ion-induced fields and light-induced ones is not straightforward, but the prospect of broadening horizons for intense-field-induced dynamics to sub-femtosecond timescales is alluring enough for us to probe some of the similarities and differences between ion- and light-induced fields, and the effect these might have on the morphology of field-induced dissociative ionization (DI) of simple molecules like water and benzene.

Exposure of molecules like H_2O to high voltages, to surfaces and interfaces, or to certain solutes, can give rise to local and far-ranging structural effects that influence the energy, entropy and molecular dynamics. H_2O is also an apt species for our studies from the viewpoint of untangling the effects of, on the one hand, the magnitude and, on the other, the directional properties of the applied electric field. On the other hand, benzene represents a prototypical molecule with ring structure and presents interesting possibilities for studies of the morphology of field-induced DI of polyatomic systems.

METHODOLOGY

We collided H_2O and C_6H_6 with 50-100 MeV $Si^{3+,8+}$ ions in order to generate intense electric fields of *attosecond* duration while longer-duration fields in our experiments were generated using 35-ps wide light pulses (wavelength 532 nm) from an Nd:YAG laser whose output was focussed to 30μm. H_2O or C_6H_6 vapor was introduced into the field-irradiation zone as an effusive beam, and positive ions were analyzed by time-of-flight (TOF) spectrometry. In exploring the correspondence between the two types of fields, we consider different ion-velocity regimes:

- In the case of relativistic ion beams (energy ≈GeV/nucleon, $\beta=v/c\approx1$), the Weizsacker-Williams equivalent photon picture [1] equates the effect of the fields to a pair of orthogonally directed photon pulses. Such equivalent-light pulses have recently been applied to studies of atomic ionization [2].
- Our experiments were conducted at lower impact energies ($\beta\approx0.04$) where the electric and magnetic field components associated with the ion beam do not form an electromagnetic wave as the latter component is deficient by β. So, the Poynting vector gives the effective intensity (I) experienced by the target molecule. At light intensities of 10^{13} Wcm^{-2}, the corresponding field is ~0.05 V/Å.
- In the limit of zero-velocity ions (β=0), the interaction is Coulombic: a field of 4.8 V/Å results if Si^{3+} is placed at an impact parameter of 3Å from the target molecule.

CP475, *Applications of Accelerators in Research and Industry*,
edited by J. L. Duggan and I. L. Morgan

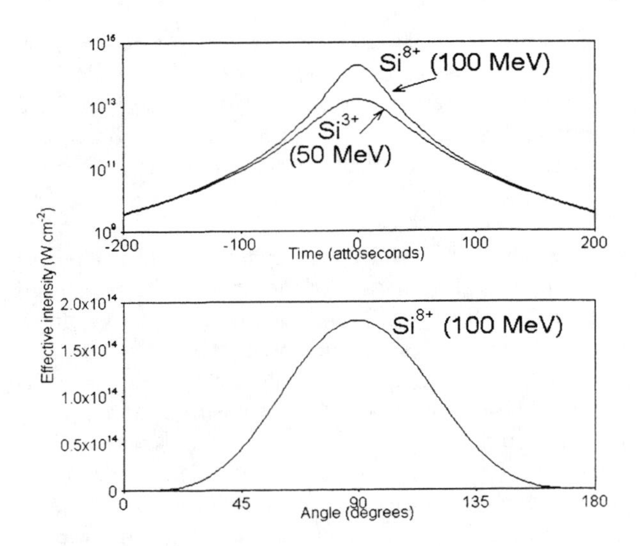

FIGURE 1. Calculation of field intensities (top panel) generated by fast Si^{3+} and Si^{8+} at an impact parameter of 3Å. The time dependence of the effective intensity has a half-width of 30 as. The corresponding light-induced field intensity has a gaussian temporal profile with a half width of 35 ps. The bottom panel shows the ion field's directional properties expressed in terms of the angle between the field vector and the target molecule. 0^0 is in the direction of the projectile ion beam.

Using MeV beams, the variation of I with time (fig.1) shows a sharply-peaked function with a half-width of about 30 attoseconds; the peak intensity depends on impact parameter b. For b~3Å, the ion-induced field is equivalent to 10^{13}-10^{14} W cm^{-2} (similar in magnitude to our laser-induced fields). As a function of the angle that the direction of the field vector makes with the center of the H_2O target, however, the variation is slow, with a half-width of 90°. This constitutes one essential difference between laser- and ion-induced fields; *although in terms of magnitude the two fields may be considered equivalent, the directional properties in the two cases are very different.* We recall that the magnitude of the laser-induced field also exhibits a time dependence that is governed by the gaussian temporal profile of the laser pulse.

RESULTS AND DISCUSSION

Before considering the field-induced DI patterns obtained with H_2O and C_6H_6 targets, we consider the quantification of I for ion-induced fields. A range of impact parameters come into play in the recoil ion spectrum that we measure (fig. 2) and this range manifests itself in the mean recoil energy imparted to each product. For the parent H_2O^+ ion in our TOF spectrum, the width translates to a recoil energy of 40 meV. A lower limit of b can be deduced from this using the established method [3] based on Olson's classical trajectory Monte-Carlo technique [4] to deduce the impact parameter dependence of multiple ionization transition probabilities in fast-ion collisions. A value of 3Å was obtained as the lower limit for b in this case. Collisions that occur at lower values of b give rise to recoil energies far in excess of 40 meV and can be discriminated

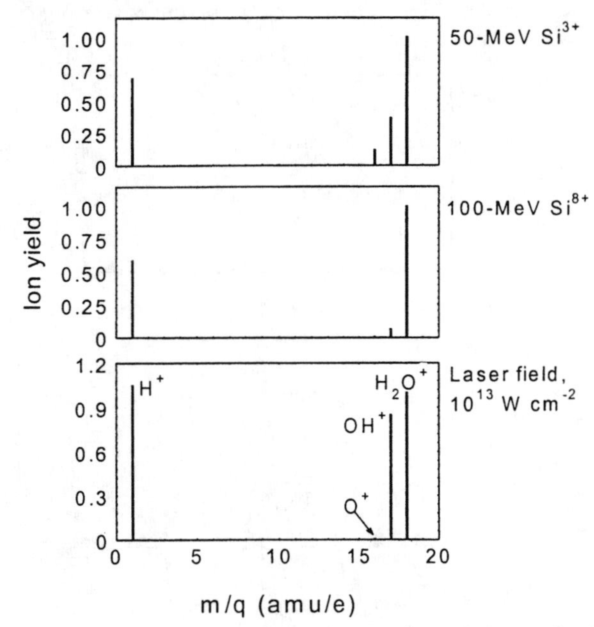

FIGURE 2. Single ionization and fragmentation of water by fast ion impact and in an equivalent light field.

by enhancing the angular resolution with which recoil ions are monitored in an orthogonal direction to the fast projectile beam. The veracity of this method for deducing the lower-limit of b was confirmed by measurements of total cross sections for formation of low-energy Ar^{q+} (q=1-10) recoil ions made in the same apparatus.

A difference between the two types of interactions lies in the possibility of electron capture and loss processes in the ion-impact case; these may complicate the deduction of b-values for recoil production; measurements conducted in our apparatus confirmed that for Si^{q+} (q=3-12), direct ionization dominates. The contribution of single ionization processes to the DI patterns obtained with $Si^{3+,8+}$ projectiles and with a laser field are shown in fig. 2. The relative intensities of H^+, OH^+ and O^+ in the ion-impact spectra indicate a decrease in the degree of fragmentation for Si^{8+} collisions: the O^+ yield reduces to <0.5% of the parent ion peak from ~20% obtained with Si^{3+}. With picosecond light, the significant feature in the DI pattern is that the H^+ fragment is somewhat more prolific than the parent ion, with almost total suppression of O^+. The fragment ion yields, except for O^+, are a little enhanced in the case of laser-field data but the almost-zero yield of O^+ ions is somewhat akin to the DI pattern obtained with Si^{8+}. The ion-induced DI pattern shown in fig.3 manifests multiple ionization of water. Inner-shell excitation is known to be almost independent of projectile charge state; for valence-shell electrons, the cross section falls with collision energy. In our experiments we did not probe the relative importance of valence- and inner-shell processes, but we expect that Si^{8+} ions might give rise to somewhat less valence-shell ionization than Si^{3+} on the basis of the

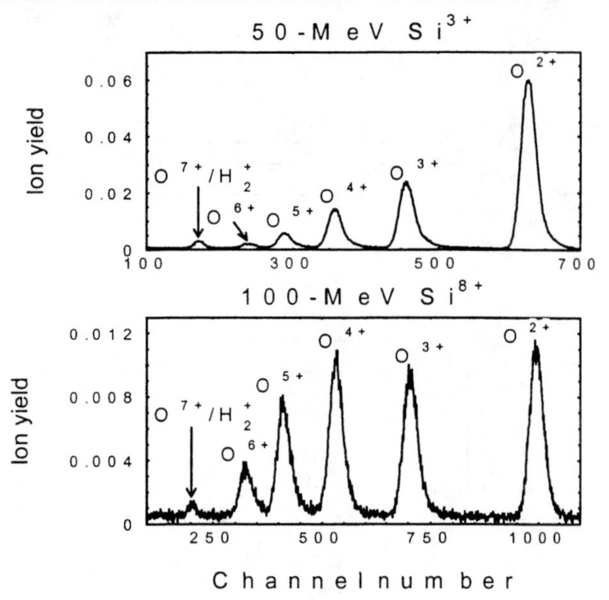

FIGURE 3. Mass spectra of slow recoil ions produced interactions of water with fast Si^{3+} (top) and Si^{8+} (bottom).

doubling of the impact energy. On the other hand, the extent of multiple ionization would be expected to increase with collision energy, resulting in enhancement of O-ions in high charge states. Our data show some contrasting facets: (i) The overall yield of O^{q+} (q>1) relative to H_2O^+ is, somewhat counterintuitively, an order of magnitude less for higher-energy Si^{8+}- H_2O collisions. (ii) On the other hand, the distribution of q-values show that the yield of O^{q+} (q>1) obtained with Si^{3+} decreases with q whereas for Si^{8+} there is a substantial enhancement of the higher charge states relative to O^{2+}. The morphological features of the data in figs. 2 and 3 bring to light the following features:

- The dynamical interactions are obviously complex even when attosecond timescales imply purely electronic interactions on timescales that are shorter than rotational and vibrational periods. Both the intensity of the applied fields as well as their time duration have important implications for the dynamics.
- Analysis of the DI patterns in terms of zero-field structural properties of water are inappropriate because of the gross distortions of potential energy surfaces that the intense fields would induce.
- Application of fields that last for, say 30 attoseconds, implies an energy uncertainty of ~22 eV, making irrelevant interpretations that involve conventional molecular states!

Further work is clearly necessary in order to gain insight. Nevertheless it is legitimate to ask whether the lack of similarity of DI patterns obtained at two equivalent field magnitudes is due to differences in the directional properties of the field, or to differences in the interaction times?

It is interesting to consider the implications of differences in the directional properties of the light- and ion-induced fields in the context of the water molecule. As noted above, H_2O molecules were exposed to *linearly-polarized* laser light, with the corresponding field having a well-defined, time-independent direction whereas the direction of the ion-induced field changed in the course of the interaction (fig. 1). It is known that the interaction of initially randomly-oriented molecules with linearly-polarized light can result in their spatial alignment prior to DI [5] due to the interaction of the laser's electric field vector with the anisotropy of molecular polarizability, α; the resulting induced dipole moment exerts a torque on the molecules such that diatomics and linear triatomics tend to align their internuclear axes along the field direction. At laser intensities $I=10^{13}$ Wcm^{-2}, the field-molecule interaction energy overwhelms the field-free rotational energy and the extent of alignment depends on α and I. For H_2O, however, angle-resolved experiments at these intensities [5] show that both O^+ and OH^+ possess isotropic angular distributions. The relative magnitudes of the components of the induced dipole moment which lie parallel and perpendicular to the molecular symmetry axis influence the angular distributions; calculations show that although the parallel component of the dipole moment increases with field magnitude, and the perpendicular component decreases marginally, the perpendicular component dominates up to field magnitudes of ~0.1 a.u., much larger than the fields in our measurements (~0.04 a.u.). The interaction potential under such conditions has minima at $\pm 90^o$ and, as a result, would induce H_2O to spatially orient in a direction which is perpendicular to the applied field [5]. When the field is applied along the symmetry axis of H_2O, the perpendicular component is zero and the molecule would tend to align in the field direction. Consequently, when a field is applied to a randomly-oriented ensemble of H_2O molecules, as the interaction potentials have minima in orthogonal directions, the net torque experienced by the molecules is negligible. These molecules are, therefore, not aligned in any preferred direction when they interact with linearly-polarized electric fields. *This leads us to postulate that the directional properties of the applied fields do not significantly determine the intense-field DI dynamics of H_2O.* Consideration of interaction times might, therefore, be pivotal in attempts to gain insight into the differences between ion- and light-induced DI patterns in the water molecule.

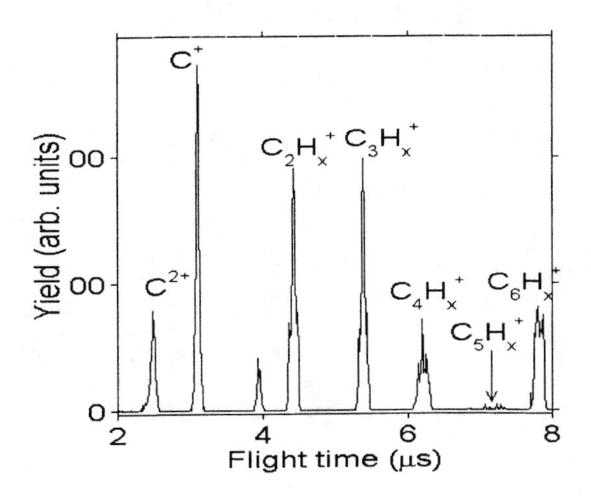

FIGURE 4. Mass spectrum of C_6H_6 upon irradiation with 532-nm laser light of intensity 1 x 10^{13} W cm^{-2}.

We have also probed the morphology of field-induced DI of C_6H_6. The following morphological features emerge from the TOF spectra we measured in light fields: the light-induced fragmentation pattern is very different to that obtained in electron impact mass spectrometry. Whereas the $C_6H_6^+$ ion dominates the electron-impact spectrum, and fragments smaller than C_2H_x. contribute only marginally to the overall fragmentation pattern, our light-field-induced spectrum (fig. 4) shows C^+ as the dominant fragment, with the benzene parent ion being only a relatively minor constituent of the mass spectrum. The intense laser field clearly gives rise to much more fragmentation. Our picosecond fragmentation pattern is very similar to that obtained using nanosecond laser light of different wavelengths. Qualitatively, we postulate that the DI pattern is determined by charge transfer type of couplings between adjacent molecular orbitals accessed in the course of unexpectedly fast ladder switching that appears to occur while the laser pulse is on. Our data indicate that such internal conversion between different excited states within the ladder appears to occur on a timescale that may be significantly shorter than 35 ps. Our measurements also indicated that the gross features of the mass spectra were only weakly dependent on laser intensity, at least over the range 1-8 x 10^{13} Wcm^{-2}.

The ion-induced DI pattern in benzene (fig.5) shows global similarities to the light-induced one. The $C_6H_6^+$ parent ion peak is again a relatively minor constituent of the mass spectrum. All the $C_nH_x^+$ fragments observed in fig.5 are also seen in fig.4. The notable difference in the ion-induced spectrum is the prominent presence of H^+ fragments. There is also evidence of multiple ionization of molecular fragments. It is probable that these are manifestations of lower-impact parameter interactions where the effective field intensity is somewhat larger than the corresponding light field obtained in our picosecond laser experiments. In order to explore further, it would clearly be of interest to extend the laser experiments to higher intensities, and using shorter (femtosecond) pulses. Both these aspects are currently being probed in new experiments in our laboratory.

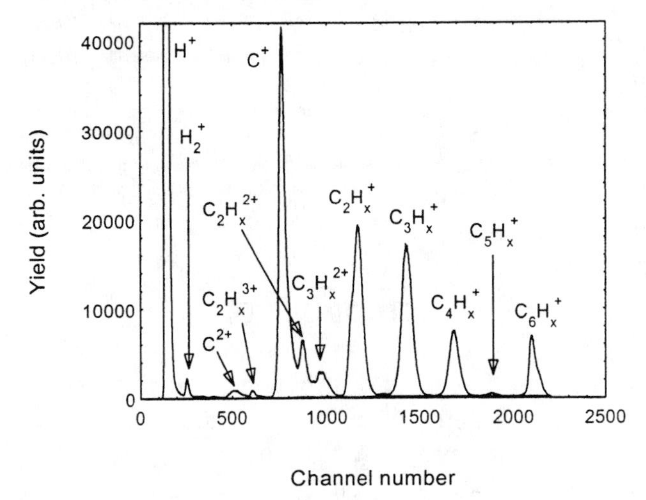

FIGURE 5. Mass spectrum of C_6H_6 obtained with 50-MeV Si^{3+}.

SUMMARY

We have explored the possibility of using fast beams of highly-charged ions to irradiate molecules with intense, attosecond-long electric fields. Such fields are similar to laser-induced ones in that both have time-dependent magnitudes (gaussian for laser-induced fields, and dependent on impact parameter and velocity for ion-induced ones). Comparison of DI patterns obtained using such fields reveal intriguing similarities and differences that ought to encourage further explorations. Recent work on more complex polyatomics is presented elsewhere [6].

ACKNOWLEDGMENTS

The efforts of my colleagues V.R. Bhardwaj, F.A. Rajgara, K. Vijayalakshmi, V. Kumarappan, and of A.K. Sinha (Nuclear Science Centre), are gratefully acknowledged.

REFERENCES

1. Jackson, J.D., *Classical Electrodynamics*, New York: Wiley, ch.15.
2. Moshammer, R., et al., Phys. Rev. Lett. **79**, 3621 (1997).
3. Tonuma, T., et al., J. Phys. B **17**, L317 (1984).
4. Schlachter,A.S., et al., Phys. Scr. **T3**, 143 (1983).
5. Bhardwaj, V.R., et al., J. Phys. B **30**, 3821 (1997).
6. Bhardwaj, V.R., et al., Phys. Rev. A **58**, 3849 (1998).

Dissociative ionisation of simple molecules by photons and fast ions

K F Dunn[1], P A Hatherly[2], C J Latimer[1], M B Shah[1], C Browne[1], A M Sands[1], B O Fisher[2], M K Thomas[2] and K Codling[2]

[1]Department of Pure and Applied Physics, The Queen's University of Belfast, Belfast, BT7 1NN
[2]J J Thomson Physical Laboratory, University of Reading, Reading RG6 6AF, U.K.

Ion pair production and their kinetic energy distribution in the charge transfer of 3-20 keV H^+ and He^+ ions with simple molecules has been studied. Multiple ionisation fragmentation of SF_6 by 1 MeV protons using the triple coincidence technique has allowed the different fragmentation channels to be identified unambiguously along with the kinetic energies released. The data is compared with that obtained with 54 and 695 eV photons using the PEPIPICO technique.

INTRODUCTION

The kinetic energy distribution of H^+ fragments from collision induced fragmentation of H_2 has been extensively studied experimentally. In particular it has been shown[1,2] that a broad spectrum of energetic H^+, from 2-10 ev, is produced in heavy particle collisions. In recent work in our laboratory we have studied the production of energetic fragment ions and ion pairs produced in the single and double ionisation of simple molecules by H^+ and He^+ impact in the energy range 3-30keV. Within this energy range, while several channels are possible, the energetic ions arise predominantly from charge transfer[3].

$$A^+ + BC \rightarrow A^+ + B^+ + C + e \quad (1)$$
$$\rightarrow A + B^+ + C \quad (2)$$
$$\rightarrow A^+ + B^+ + C^+ + 2e \quad (3)$$
$$\rightarrow A + B^+ + C^+ + e \quad (4)$$
$$\rightarrow A^- + B^+ + C^+ \quad (5)$$

EXPERIMENTAL APPROACH

A schematic diagram of the apparatus is shown in figure 1. A 3-30 keV-ion beam was crossed at 90° by a low pressure gas jet of target molecules at the centre of a ramp voltage labelled region. Fragment ions, at a prescribed energy and perpendicular to the gas jet and ion beam were selected using two parallel plate analysers which view the interaction region

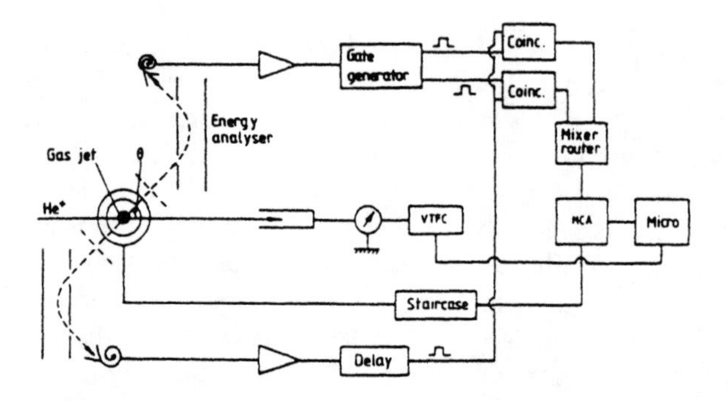

FIGURE 1.A schematic diagram of the apparatus.

from opposite directions to allow detection of ion pairs with equal and opposite momentum. The technique ensures that the transmission factor of the analysers is not required and that the thermal motion of the gas target has a negligible effect.

RESULTS AND DISCUSSION

Uniquely in the case of a hydrogen target there exists only one state of the doubly charged ion H^+H^+, a pure Coulomb potential leading to a "Coulomb explosion". While an energy spectrum for such proton pairs produced by 15 keV H^+ shows the expected single peak at ~9eV for 15 keV He^+ impact two peaks are now present at ~5 and 9eV. The low energy group was found to be insensitive to projectile velocity and so cannot be explained as a simple

CP475, *Applications of Accelerators in Research and Industry*, edited by J. L. Duggan and I. L. Morgan

breakdown of the Franck-Condon principle. A two step process therefore seems likely. Both the $H_2^+(^2\Sigma, \nu)$ states and the H_2^* Rydberg states have the appropriate internuclear separation for an intermediate step. However, Franck-Condon transitions to such states produce a range of vibrational states which would require a vibrational period of $\sim 10^{-14}$s to reach the required internuclear separation. Since the collision time is less than $\sim 10^{-15}$s this possibility can be excluded.

To further explore this phenomenon we have investigated the double ionisation of deuterium by both H^+ and He^+. The consequences of increasing the nuclear mass are (a) the reduced mass of the nuclei will narrow the potential well and hence the energy distribution of the fragment ion pairs and (b) the nuclei will separate more slowly. The experimental results now show a low energy peak for both H^+ and He^+ projectiles, see for example figure 2.

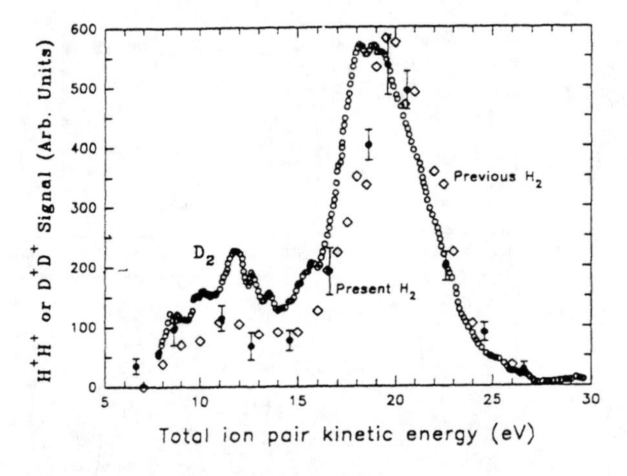

FIGURE 2. An energy spectrum of ion pairs produced in 15 keV He^+ - H_2 ,D_2 collisions .The present He^+ - H_2 results o are compared with our previous measurements .

Savage et al suggested a two step process to explain the observed peaks. The H_2^+ $(^2\Sigma, \nu)$ states have the correct internuclear separation to give the observed low energy peak. However, given the time scale constraints it must be populated via autoionisation of one of the doubly excited autoionisation states of H_2 eg $^1\Sigma_g$ $(2p\sigma_u^2)$ which are significantly excited in such collisions.

The double ionisation of H_2 and D_2 has also been studied by Beckord et al[5] in the energy regime 50-350 keV using a coincidence technique. Since they used an electron as the start pulse of the coincidence they observed reactions of the type (3) and (4). However their low energy peak at ~3.5eV for 200 keV He^+ on H_2 , figure 3, was attributed not to the target, but to background gas. They concluded from an analysis of the TOF spectra of the low energy pairs and residual gas analysis that the proton pairs originated from water vapour in the background gas.

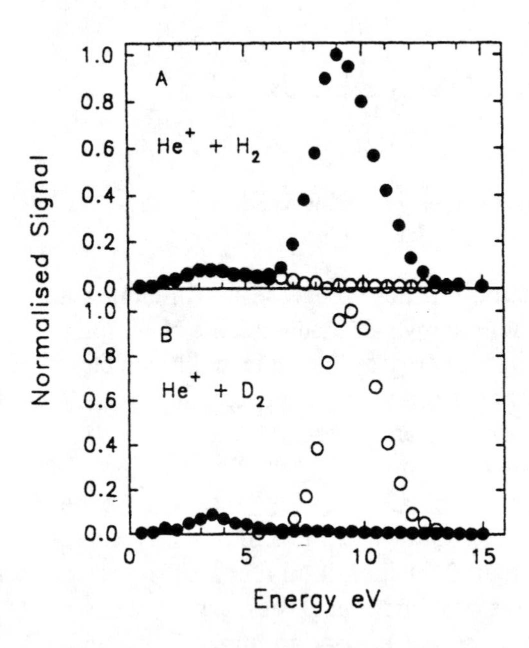

FIGURE 3. Energy spectra of ion pairs for a) 200 keV He^+ - H_2 o and background gas o and b) 200 keV He^+ on D_2 o and coincident H^+ pairs from background gas o (Beckord et al 1994)

Subsequently we have repeated our previous measurements using a diffusion pump and liquid nitrogen trap to minimise background water vapour and passing the target gas through a liquid nitrogen bath to remove possible impurities. Our previous work (Fig 2) was readily reproducible.

Residual gas analysis showed our two principal background impurities to be nitrogen and water. We therefore also studied ion-pair production from nitrogen and water vapour. Figure 4 shows ion pair data for 15 keV He^+ collisions in

H_2, N_2 and H_20. From time of flight information the ion pair observed for water is H^+H^+. The size of the ordinate gives the relative cross

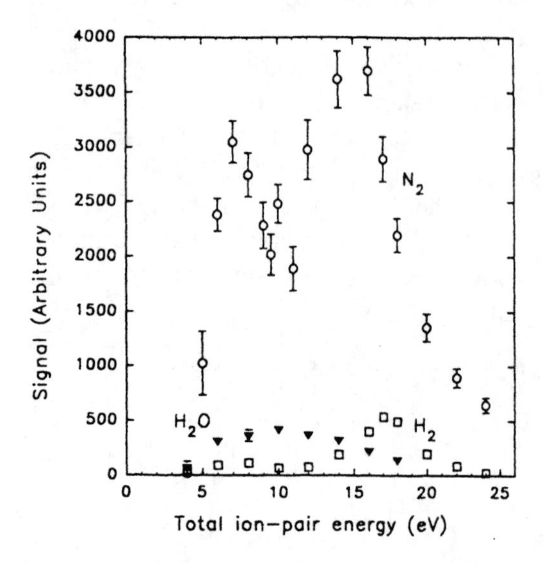

Figure 4. Energy spectra of ion pairs produced in 15 keV He^+ - N_2, H_2O and H_2 collisions.

section for these processes. While the water vapour shows a broad peak the N^+N^+ ion pair kinetic energy spectrum is highly structured, the cross section is large and dependent on incident ion species and energy.

The dissociation of N_2^{++} has been extensively studied both experimentally[6,7] and theoretically[8]. Yousif et al identified much of the structure in terms of quasi bound dissociating states of N_2^{2+}. The discrete energy groups at 7, 10 and 12 eV we have identified with the $^1\Pi_u$,$^3\Pi_g$ and $^1\Sigma_u$ states. The width of the highest energy peak, extending to 25 eV, suggests that this arises either via charge transfer ionisation of N_2 into autoionising dissociative states or from the $^3\Delta_u$ state .

As the degree of ionisation increases and we move from diatomic to multiatomic molecules, the traditional methods used to study fragmentation ionisation become increasingly inadequate. Groups at Reading[9] and Oxford[10] Universities faced with these problems developed the modern electron ion-ion triple coincidence method PEPIPICO which allows unambiguous correlations to be made between ion pairs following multiple ionisation. They used a time of flight apparatus whose ion drift tube had been configured to meet the Wiley-MacLaren condition that the TOF of an ion depends not only on its mass/change ratio but is also proportional to its momentum parallel to the drift tube axis. An electron provides the start pulse and the coincident ions arrive at times t_1 and t_2 .

The Reading group has studied photodissociative ionisation of simple molecules using both VUV radiation exciting outer valence electrons and soft X-ray excitation of inner

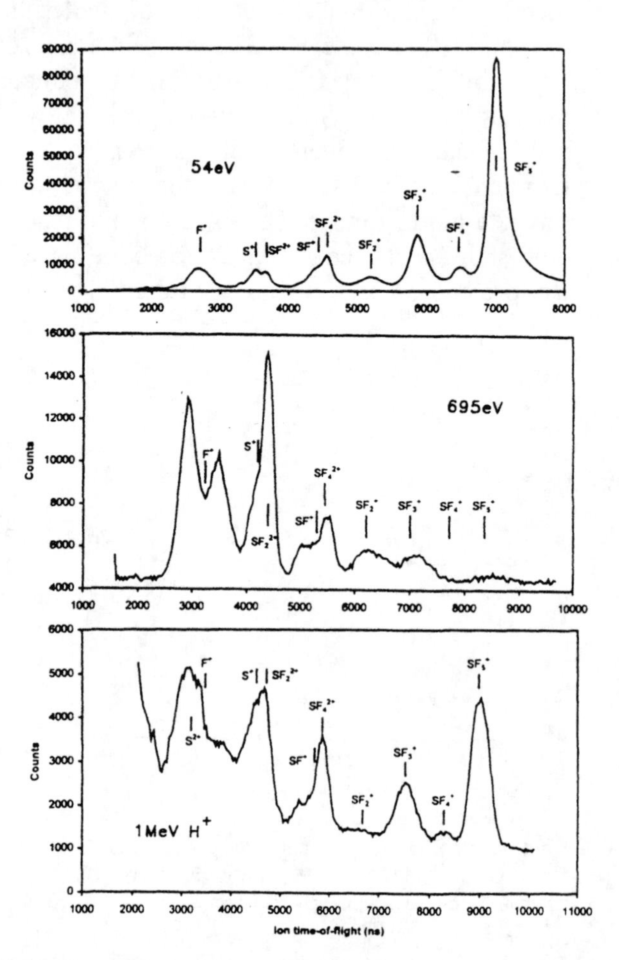

FIGURE 5. TOF spectrum for SF_6 following photoionisation at 54eV , 695 eV [F(1s)] and 1MeV proton impact.

shells. We have combined our resources to study dissociative ionisation by 1 MeV protons. Although apparently different in principle to photoionisation, proton impact shows many similarities since the passage of a 1 MeV proton near a molecule appears as a fast (~fs) electromagnetic pulse which may be resolved into a large range of frequency components. While several molecules were studied we will discuss only SF_6.

Simple TOF spectra, figure 5, show the mass of the fragments present but contain no definitive information on ion correlation. Noticeable, however, is the alternating intensity of SF_n^+, ions with odd numbers of fluorines having high intensities. Although the parent ion is not stable the spectra show the presence of the stable double ions SF_4^{2+} with S^{2+} also present after proton excitation. To obtain information on the ion pairs the data is processed to display counts vs (t_1+t_2) to identify the ion pairs or counts vs (t_2-t_1) to reveal the kinetic energy releases.

The heaviest ion pair corresponds to $F^+ + SF_3^+$ while following both F(1s) or proton impact the atomic pairs F^+ , S^+ and F^+, F^+ are produced. For SF_3^+ and F^+ both 54 eV and proton ionisation show peaks at 5 eV with a maximum energy of ~10 eV suggesting the same final state of the doubly charged ion. In the case of F(1s) excitation the energy extends to 25 eV. In the other channels studied the greatest similarity is between the F(1s) ionisation and proton impact. This is reasonable since proton impact may contribute both valence double and core excitation-like effects depending on the impact parameter.

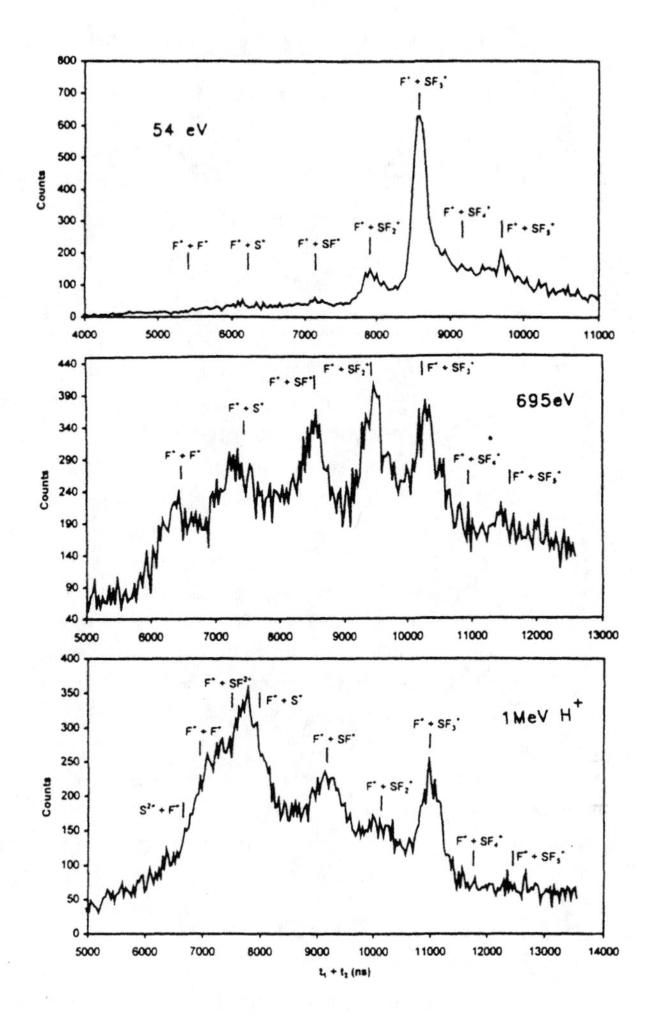

FIGURE 6. Plots of $t_1 + t_2$ for SF_6 following photoionisation at 54ev, 695ev and 1MeV proton impact.

REFERENCES

1. Edwards AK, Woods RM, Davis JL and Ezell RL, *Phys.Rev.A*,**42**,1367,1990
2. Savage OG , Lindsay BG and Latimer CJ, *J Phys B :At Mol Phys* , **23**, 4313-4320, 1990
3. Afrosimov VV,Leiko GA, Mamaev Yu A, and Panev MN,*Sov.Phys.-JETP*,**29**,648.1969
4. Geddes J, Dunn KF, Kouchi N, MacDonald MA, Srigengan V and Latimer CJ, *J Phys B :At Mol Phys*, **27**, 2961-70, 1994
5. Beckord K, Becker J, Werner U and Lotz HO,*J Phys B: At Mol Phys*, **27**, L585-L589, 1994 .
6. Yousif FB , Lindsay BG and Latimer CJ, *J Phys B :At Mol Phys* ,**23**, 495-504, 1990
7. Lundquist M, Edwardsson D, Baltzer P and Wannberg B , *J Phys B :At Mol Phys*, **29**, 1489-99, 1996
8. Bennett FR,*Chem.Phys*.,**190**,53,1994
9. Frasinski LJ,Stankiewicz M, Randell KJ , Hatherly PA and Codling K, *J Phys B*, **19**, L819-L821,1986
10. Eland JHD, Wort FS and Royds RN, *J. Electron Spectrosc. Relat. Phenom.* , **41**,297,1986

Ion Storage Ring Measurements of Dielectronic Recombination for Astrophysically Relevant Fe^{q+} Ions

D. W. Savin*, N. R. Badnell† T. Bartsch‡, C. Brandau‡, M. H. Chen§, M. Grieser**, G. Gwinner**, A. Hoffknecht‡, S. M. Kahn*, J. Linkemann**, A. Müller‡, R. Repnow**, A. A. Saghiri**, S. Schippers‡, M. Schmitt**, D. Schwalm**, and A. Wolf**

*Columbia Astrophysics Laboratory, Columbia University, New York, NY 10027, USA
†University of Strathclyde, Glasgow, G4 0NG, United Kingdom
‡Institut für Kernphysik, Strahlenzentrum der Justus-Liebig-Universität, D-35392 Giessen, Germany
§Lawrence Livermore National Laboratory, Livermore, CA 94550, USA
**Max-Planck-Institut für Kernphysik, D-69117 Heidelberg, Germany

Iron ions provide many valuable plasma diagnostics for cosmic plasmas. The accuracy of these diagnostics, however, often depends on an accurate understanding of the ionization structure of the emitting gas. Dielectronic recombination (DR) is the dominant electron-ion recombination mechanism for most iron ions in cosmic plasmas. Using the heavy-ion storage ring at the Max-Planck-Institute for Nuclear Physics in Heidelberg, Germany, we have measured the low temperature DR rates for Fe^{q+} where $q = 15$, 17, 18, and 19. These rates are important for photoionized gases which form in the media surrounding active galactic nuclei, X-ray binaries, and cataclysmic variables. Our results demonstrate that commonly used theoretical approximations for calculating low temperature DR rates can easily under- or over-estimate the DR rate by a factor of ~ 2 or more. As essentially all DR rates used for modeling photoionized gases are calculated using these approximations, our results indicate that new DR rates are needed for almost all charge states of cosmically abundant elements. Measurements are underway for other charge states of iron.

INTRODUCTION

Heavy-ion storage rings, coupled with electron cooling techniques, are an important laboratory tool for studying electron collisions with highly charged ions (Müller & Wolf 1997). Of particular interest for astrophysics is the ability of storage rings to study low energy dielectronic recombination (DR), which at the low electron temperatures predicted for photoionized cosmic plasmas (Kallman *et al.* 1996) is the dominant electron-ion recombination process for most ions (Arnaud & Rothenflug 1985; Arnaud & Raymond 1992). Using storage rings, those DR resonances important in photoionized plasmas can be measured by merging the electron and ion beams. Using a co-linear geometry one can achieve near zero eV relative collision energies.

Storage rings are unique for their ability to study low energy DR of highly charged ions with a narrow energy resolution. Common laboratory techniques for measuring DR such as electron beam ion traps (Beiersdorfer *et al.* 1992) and tokamak plasmas (Bitter *et al.* 1993) can produce highly charged ions, but cannot simultaneously reach near zero relative velocities. Crossed electron-ion beams techniques also can achieve neither the required low relative velocities nor a narrow enough energy resolution for resolving DR resonance structure (Müller *et al.* 1987; Savin *et al.* 1996). A merged electron-ion beams technique is the only way to achieve the desired near zero eV collision energies. DR measurement can be carried out using a single-pass technique which merges an electron and ion beam for some distance, demerges them, and then separates the recombined ions from the primary beam and detects both beams separately (e.g., Andersen *et al.* 1992). One disadvantage of this method is the low signal rate. Also, for ions with partially-filled shells, data analysis is usually complicated by the presence of metastable ions in the beam.

Ion storage rings are the optimal method for measuring low energy DR. Storing the ions, one can accumulate ions and thus increase the signal rate. For our Fe^{17+} measurement (Savin *et al.* 1997), we stored currents of 30-50 μA as compared with typical ion currents of $\lesssim 1$ μA in single-pass experiments (Andersen, Bolko, & Kvistgaard 1990). Ions can typically be stored on the order of tens of seconds which is long enough for essentially all metastable ions to relax to their ground state before begining measurements. Electron cooling techniques can be used on the stored ions (Poth 1990). Cooling reduces the energy spread of the ions, and more ions can be stored. Using an adiabatically expanded electron beam (Pastuszka *et al.* 1996) in combination with the merged beams geometry results in typical electron energy spreads of $k_B T_\perp \sim 18$ meV transverse to

CP475, *Applications of Accelerators in Research and Industry*,
edited by J. L. Duggan and I. L. Morgan

the beam velocity and $k_B T_\| \sim 0.18$ meV longitudinally (Savin *et al.* 1997). With this narrow energy spread one can resolve a large number of individual DR resonances and determine their energies and strengths accurately.

ASTROPHYSICAL MOTIVATION

Photoionized plasmas form in planetary nebulae, H II regions, cold nova shells, stellar winds, and the intergalactic medium (IGM) and in the media surrounding active galactic nuclei (AGN), X-ray binaries, and cataclysmic variables. Spectroscopic observations of these objects can address fundamental questions in astrophysics. For example, observations of the IGM yields information on the chemical evolution of the universe (Giroux & Shull 1997) and AGN, which are thought to contain supermassive black holes, can be used to study General Relativity (Nandra *et al.* 1997).

Meeting the need of accurate atomic data is extremely timely. The upcoming launches of the *Advanced X-Ray Astrophysics Facility* (*AXAF*; Markert 1993) and the *X-Ray Multimirror Mission* (*XMM*; Brinkman 1993) will put in orbit satellites which will collect X-ray spectra from extrasolar objects of a higher quality and resolution than has been achieved in the past. These spectra are expected to revolutionize the field of X-ray astrophysics. Unambiguous interpretation of the collected spectra, though, will be impeded by errors and uncertainties in the atomic data base.

Models of photoionized plasmas require accurate DR rates for hundreds of ions. Laboratory measurements can provide only a fraction of the needed rates and modelers must rely on theoretical calculations. In the past, theorists calculated DR rates for a select number of ions along an isoelectronic sequence and then interpolated along the sequence for a needed DR rate. However, atomic structure does not scale smoothly along an isoelectronic sequence. Thus it is necessary explicitly to calculate all the needed DR rates. Laboratory measurements must be selected to test these calculations in the most efficient manner possible.

We have chosen to study DR of iron because iron is the highest-Z cosmically abundant element and an important constituent of almost all astronomical plasmas. Iron plays a pivotal role in determining the line emission and thermal and ionization structure of photoionized plasmas (Kallman *et al.* 1996; Hess, Kahn, & Paerels 1997). Also, due to its high Z, iron is the last cosmically abundant element to lose all its electrons. As such, iron can be used spectroscopically to probe extreme conditions where all other important elements have already been stripped to bare nuclei.

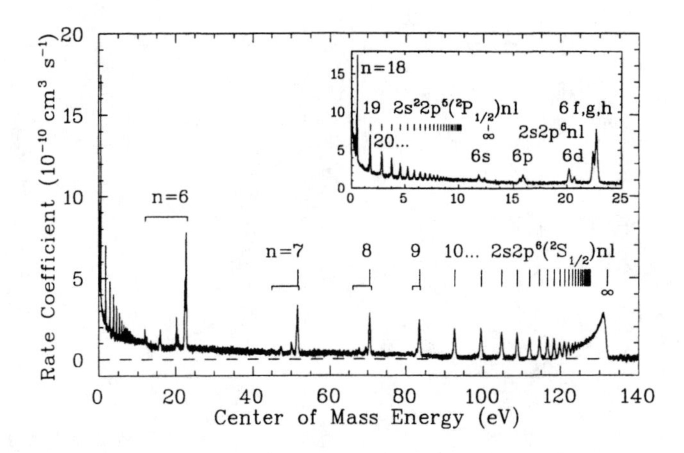

Figure 1: Measured Fe^{17+} to Fe^{16+} recombination rate coefficient versus collision energy. The nonresonant "background" is due to radiative recombination and charge transfer with residual gas in the storage ring.

EXPERIMENTAL PROGRAM

Using the heavy-ion test storage ring (TSR; Habs *et al.* 1989) at the Max Planck Institute for Nuclear Physics in Heidelberg, Germany, we have undertaken to measure the low temperature DR rates for a wide range of iron ions. These measurements will provide a comprehensive set of benchmark measurements for a number of important isoelectronic sequences. To date we have measured DR for Fe^{15+} (Linkemann *et al.* 1995), Fe^{17+} (Savin *et al.* 1997), Fe^{18+}, and Fe^{19+} (Savin *et al.*, in preparation).

DR is a two-step electron-ion recombination process that begins when a free electron collisionally excites an ion, via an $nl_j \to n'l'_{j'}$ excitation of a bound core electron, and is simultaneously captured. DR is complete when this state emits a photon which reduces the energy of the recombined system to below its ionization limit. Low energy DR usually involves a $\Delta n = 0$ excitation of a core electron.

DR measurements using a storage ring are carried out by merging, in one of the straight sections of the ring, an ion beam with an electron beam. After demerging, any recombined ions formed are magnetically separated from the stored ions and directed onto a detector. The relative electron-ion collision energy can be precisely controlled and the recombination signal is measured as a function of this energy. Detailed descriptions of DR measurement techniques using storage rings have been given elsewhere (e.g., Kilgus *et al.* 1992; Lampert *et al.* 1996).

Our recent measurement of Fe^{17+} DR (Savin *et al.* 1997) is shown in Figure 1. Fe^{17+} can undergo $\Delta n = 0$ DR via two different channels,

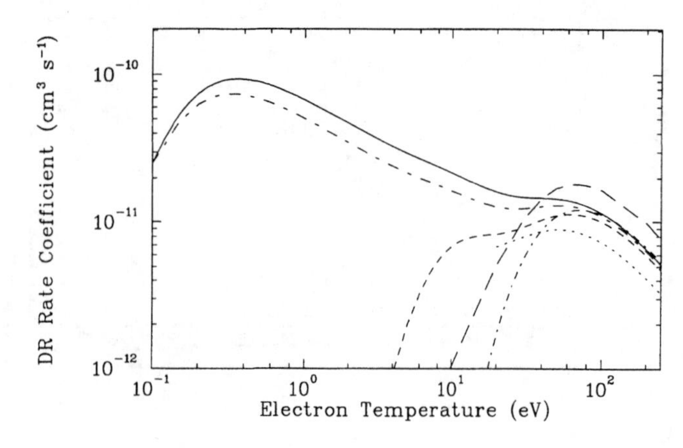

Figure 2: Fe^{17+} to Fe^{16+} Maxwellian-averaged $\Delta n = 0$ DR rate coefficients. See text for explanation.

$$Fe^{17+}(2s^2 2p^5[^2P_{3/2}]) + e^- \rightarrow \left\{ \begin{array}{ll} Fe^{16+}(2s^2 2p^5[^2P_{1/2}]nl) & (n \geq 18) \\ Fe^{16+}(2s2p^6[^2S_{1/2}]nl) & (n \geq 6). \end{array} \right. \quad (1)$$

The radiative stabilization of either of the above Fe^{16+} autoionizing states to bound configurations of Fe^{16+} leads to DR resonances for electron-ion collision energies between 0 and 132 eV.

We have integrated the measured DR resonance strengths and energies with a Maxwellian electron velocity distribution to yield a total Fe^{17+} $\Delta n = 0$ DR rate coefficient as a function of $k_B T_e$ (Figure 2, upper solid line). The estimated total experimental uncertainty in our inferred DR rate is better than 20%. Various theoretical DR rates are also shown in Figure 2. At $k_B T_e \sim$ 15 eV, near where Fe^{17+} is predicted to peak in fractional abundance in photoionized gas with cosmic abundances (Kallman *et al.* 1996), our measured DR rate is a factor of $\sim$ 2 larger than the calculations of Roszman (1987; long dashed curve), Chen (1988; dotted curve), and Dasgupta & Whitney (1990; short dashed curve). These theoretical rates all tend rapidly to zero at $k_B T_e < 20$ eV because they have not included DR via $2p_{1/2} \rightarrow 2p_{3/2}$ fine-structure core excitations.

At higher plasma temperatures, where the $2p_{1/2} \rightarrow 2p_{3/2}$ channel is unimportant, significant discrepancies exist between our inferred rate and the calculations of Chen and Roszman. Chen underestimates the DR rate by a factor of $\sim$ 1.5. This may be partly due to approximations which ignore DR via core excitations to levels with $l > 8$. Including these additional DR channels would increase the calculated DR rate. Roszman overestimates the DR rate by a factor of $\sim$ 1.6. This may be partly due to the extrapolation technique to high n levels used to calculate the DR rate. If the extrapolation was initiated for $n < 18$, as is likely, this leaves out an important autoionization channel which reduces the DR rate. However, it is likely that Roszman also made many of the same approximations as Chen and so the true source of the discrepancy is unclear. The low temperature rate of Roszman goes to zero faster than that of Dasgupta & Whitney because Roszman calculated that DR via $2s2p^5nl$ configurations becomes energetically possible at $n = 7$ whereas Dasgupta & Whitney find DR starts at $n = 6$. Our measurements show that DR via this channel is allowed for $n \geq 6$.

The agreement between our inferred rate and that of Dasgupta & Whitney is probably serendipitous. They made many of the same approximations as Chen and also began their extrapolations at $n = 16$. It is unclear, but it may be that the agreement is a result of approximations roughly canceling one another out. What is clear is that comparisons only of rate coefficients cannot be used to distinguish definitively between the various theoretical techniques. Measurements of DR resonance strengths and energies are needed to provide benchmarks for the detailed atomic physics which goes into calculating DR rates.

Using the data shown in Figure 1, we have extracted DR resonance strengths and energies for comparison with theory (Savin *et al.*, in preparation). A direct comparison of these results with the work of Chen, Roszman, and Dasgupta & Whitney is not possible because they do not present resonance strengths resolved as a function of n. We have carried out a new calculation which includes the $2p_{1/2} \rightarrow 2p_{3/2}$ channel and accounts for DR for $l \leq 12$. The calculated resonance strengths for DR via the $2p_{1/2} \rightarrow 2p_{3/2}$ agree with experiment to within $\lesssim 30\%$ which is larger than the estimated 20% total experimental uncertainty limits. Our measurements agree with the new calculations to within $\lesssim 20\%$ for DR via the $2s \rightarrow 2p$ channel. Using the new calculations we have generated a DR rate coefficient which agrees well for $k_B T_e \gtrsim 60$ eV, but differ by $\sim$ 30% below 60 eV (Figure 2, lower solid curve; Savin *et al.* 1997).

ASTROPHYSICAL IMPLICATIONS

Photoionized plasmas are most commonly modeled using the DR rates of Aldrovandi & Péquignot (1973), Shull & van Steenberg (1982), Nussbaumer & Storey (1983), Arnaud & Rothenflug (1985), and Arnaud & Raymond (1992). But for a few exceptions, all of these DR rates have been calculated using either pure *LS*-coupling or the Burgess formula approximation (Burgess 1965). Neither approximation accounts for DR via fine-structure core excitations ($nl_j \rightarrow nl_{j'}$). If an ion forms at a temperature $k_B T_e \lesssim \Delta E_{fs}$, where ΔE_{fs} is the energy of the fine-structure core excitation, existing DR rate coefficients can under-estimate the DR rate by rather large factors. For ions which form at $k_B T_e \gtrsim \Delta E_{fs}$, the approximations used for most of the existing DR rates could easily result in under- or over-estimating the DR rate by a significant factor.

In conclusion, there are major uncertainties in the low temperature DR rates for most ions with partially filled shells. Determining the magnitude of the effect these uncertainties will have on modeling and interpreting spectra of photoionized plasmas requires reliable DR rates for all the relevant charge states. Our laboratory studies of iron ions will help to provide many of the needed DR rates; and for those charge states not measured, our work will provide valuable benchmarks to test the various techniques for calculating DR along isoelectronic sequences.

ACKNOWLEGDGMENTS

We thank the staff and technicians of the TSR group for support during the beam time. This work was supported in part by NASA High Energy Astrophysics X-Ray Astronomy Research and Analysis grant NAG5-5123. Travel and living expenses for DWS were supported by NATO Collaborative Research Grant CRG-950911. The experimental work has been supported in part by the German Federal Minister for Education, Science, Research, and Technology (BMBF) under Contract Nos. 06 GI 475, 06 GI 848, and 06 HD 854I. Work performed at Lawrence Livermore National Laboratory was under the auspices of the US Department of Energy (contract number W-7405-ENG-48).

REFERENCES

Aldrovandi, S. M. V. & Péquignot, D. 1973, Astron. Astrophys., **25**, 137; 1976, Astron. Astrophys., **47**, 321.

Andersen, L. H., Bolko, J., & Kvistgaard P. 1990, Phys. Rev. A **41**, 1293.

Andersen, L. H., Pan, G.-Y., Schmidt, H. T., Pindzola, M. S., & Badnell, N. R. 1992, Phys. Rev. A **45**, 6332.

Arnaud, M. & Rothenflug, R. 1985, Astron. Astrophys. Suppl. Ser., **60**, 425.

Arnaud, M. & Raymond, J. 1992, Astrophys. J., **398**, 394.

Beiersdorfer, P., Phillips, T. W., Wong, K. L., Marrs, R. E., Vogel, D. A. 1992, Phys. Rev. A **46**, 3812.

Bitter, M., Hsuan, H., Hill, K. W., & Zarnstorff, M. 1993, Physica Scripta **T47**, 87.

Brinkman, A. C. 1993, in *UV and X-ray Spectroscopy of Laboratory and Astrophysical Plasmas*, ed. E. Silver & S. Kahn, (Cambridge University Press: Cambridge), 469.

Burgess, A. 1965, Astrophys. J., **141**, 1588.

Chen, M. H. 1988, Phys. Rev. A, **38**, 2332.

Dasgupta, A. & Whitney, K. G. 1990, Phys. Rev. A, **42**, 2640.

Giroux, M. L. & Shull, J. M. 1997, Astron. J. **113**, 1505.

Habs, D. et al. 1989, Nucl. Instrum. Methods, **B43**, 390.

Kallman, T. R., Liedahl, D., Osterheld, A., Goldstein, W., & Kahn, S. 1996, Astrophys. J., **465**, 994.

Kilgus, G., Habs, D., Schwalm, D., Wolf, A., Badnell, N. R., & Müller, A. 1992, Phys. Rev. A, **46**, 5730.

Lampert, A., Wolf, A., Habs, D., Kilgus, G., Schwalm, D., Pindzola, M. S., & Badnell, N. R. 1996, Phys. Rev. A, **53**, 1413.

Liedahl, D. A., Kahn, S. M., Osterheld, A. L., & Goldstein, W. H. 1990, Astrophys. J. Lett., **350**, L37.

Linkemann, J. *et al.*, 1995, Nucl. Instrum. Methods **B98**, 154.

Markert, T. H. 1993, in *UV and X-ray Spectroscopy of Laboratory and Astrophysical Plasmas*, ed. E. Silver & S. Kahn, (Cambridge University Press: Cambridge), 459.

Müller, A., Belić, D. S., DePaola, B. D., Djurič, N., Dunn, G. H., Mueller, D. W., & Timmer, C. 1987, Phys. Rev. A **36**, 599.

Müller, A. & Wolf, A. 1997, in Accelerator-Based Atomic Physics Techniques and Applications, ed. J. C. Austin & S. M. Shafroth (AIP Press: New York), Ch. 5.

Nandra, K., George, I. M., Mushotzky, R. F., Turner, T. J., & Yaqoob, T. 1997, Astrophys. J. **477**, 602.

Nussbaumer, H. & Storey, P. J. 1983, Astron. Astrophys., **126**, 75.

Pastuszka, S., Schramm, U., Griesser, M., Broude, C., Grimm, R., Habs, D., Kenntner, J., Miesner, H.-J., Schüßler, Schwalm, D., & Wolf, A. 1996, Nucl. Instrum. Methods **A369**, 11.

Poth, H. 1990, Phys. Rep. **196**, 135.

Roszman, L. J. 1987, Phys. Rev. A, **35**, 2138.

Savin, D. W., Gardner, L. D., Reisenfeld, D. B., Young, A. R., & Kohl, J. L. 1996, Phys. Rev. A. **53**, 280.

Savin, D. W., Bartsch, T., Chen, M. H., Kahn, S. M., Liedahl, D. A., Linkemann, J., Müller, A., Schippers, S., Schmitt, M., Schwalm, D., & Wolf, A. 1997, Astrophys. J. Lett. **489**, L115.

Shull, J. M. & Van Steenberg, M. 1982, Astrophys. J., Suppl. Ser., **48**, 95; **49**, 351.

Excitation and ionization of exotic and non-exotic atoms in heavy-ion collisions

D. Trautmann, Z. Halabuka, T. Heim, K. Hencken, H. Meier
Department of Physics and Astronomy, University of Basel, CH-4056 Basel, Switzerland

and

G. Baur
Institut für Kernphysik (Theorie), Forschungszentrum Jülich, D-52425 Jülich, Germany

We have applied the semi-classical formalism to the excitation and ionization of neutral and highly ionized atoms. By extending this method to the case of relativistic projectile energies, we open up an exciting field of applications, with entirely new types of processes becoming possible. The recent production and detection of a new type of an "exotic atom", namely antihydrogen, provides a striking example where the formation proceeds through bound-free pair production with antiprotons. Bound-free pair production is also of practical importance for relativistic heavy-ion colliders, constituting one of the dominant beam loss processes. A further application of this formalism pertains to Coulomb excitation and break-up of pionium, i.e., "exotic atoms" consisting of a $\pi^+\pi^-$-pair in a bound state. Cross sections for various transitions and break-up processes also in the screened Coulomb field of a nucleus are evaluated analytically and numerically.

I. GENERAL FORMALISM

The semi-classical formalism has been established [1–3] as a powerful tool to investigate excitation, as well as ionization processes even at relativistic energies. The technical issues arising in this method's application are well understood and manageable, as we will show in our contribution. The formalism is not limited to "conventional" systems (atoms and electrons), but it applies equally well to exotic aggregates, such as the recently detected antihydrogen, or bound states of $\pi^+\pi^-$-pairs (pionium), to mention but two.

The semi-classical approximation (SCA) describes an arbitrary collision process in terms of a *trajectory* associated with the projectile and characterized by an *impact parameter* b. The cross section for the transition—excitation or ionization—of the target system from an initial state $|i\rangle$ to a final state $|f\rangle$ is obtained by integrating over the impact parameter:

$$\sigma_{fi} = 2\pi \int_0^\infty b \, \mathrm{d}b \, P_{fi}(b), \tag{1}$$

where $P_{fi}(b)$ denotes the impact parameter dependent transition probability given by

$$P_{fi}(b) = \frac{1}{2l_i + 1} \sum_{m_f, m_i} |a_{fi}(b)|^2 \tag{2}$$

(summed over un-observed magnetic quantum numbers). In first order of the interaction, the amplitude takes the form of the matrix element

$$a_{fi}^{(1)}(b) = \langle f | H_{\text{int}}(\vec{R}_b(t)) | i \rangle \tag{3}$$

where $\vec{R}_b(t)$ describes the projectile's trajectory specified by the impact parameter in the target's rest frame (Fig. 1). The target system may consist of an ordinary (non-exotic) atom, or it may represent an exotic "atomic" complex of oppositely charged particles other than electrons and nuclei. Furthermore, the projectile energies can range from non-relativistic values to the extreme relativistic regime.

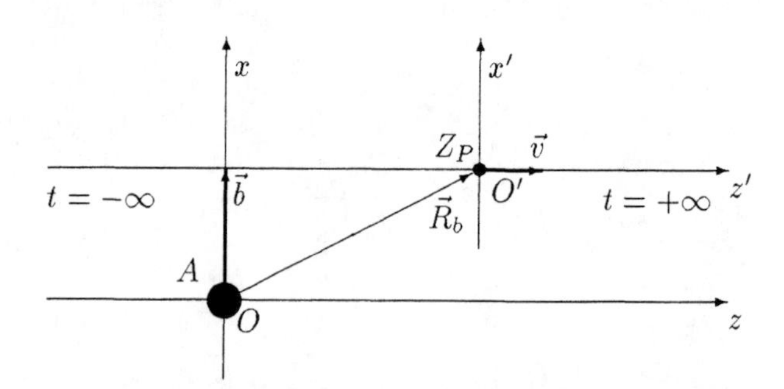

FIG. 1. Semi-classical picture of a projectile with charge Z_P moving on a trajectory with impact parameter b past a target "atom" A consisting of particles of opposite charge.

II. EXCITATION AND IONIZATION OF ATOMS IN ION-ATOM COLLISIONS

In the case of ions colliding with exotic or non-exotic "atomic" systems, the interaction Hamiltonian H_{int} is given by the Coulomb interaction between the target system's components and the screened or unscreened field of the moving projectile's charge, described by the Liénard-Wiechert potential or its non-relativistic limit, as appropriate. Similarly, $|i\rangle$ and $|f\rangle$ are the initial and final states of the "atomic" system. In most cases, e.g. for the ionization of K- or L-shell electrons, it is sufficient to use relativistic or non-relativistic, bound or continuum

CP475, *Applications of Accelerators in Research and Industry*,
edited by J. L. Duggan and I. L. Morgan

hydrogenic wave functions. However, in the case of ionization from the M-, N- or even higher shells, one has to use numerical Dirac-Hartree-Fock-Slater wave functions.

The literature [4,5] provides many examples showing excellent agreement between experiment and this theory applied to ionization and excitation of inner-shell electrons. Besides these "conventional" systems, we have also employed this approach to describe exotic systems, as demonstrated in the following sections.

III. PRODUCTION OF ANTIHYDROGEN

A first "exotic" application of the formalism comprised in (1)–(3) pertains to the recently detected atom of antihydrogen [6,7]. The synthesis of an antiproton and a positron is achieved in [7] by passing relativistic antiprotons through a hydrogen gas target to produce electron-positron pairs. If a positron's velocity matches that of the antiproton, they may combine to form an atomic complex. In lowest order the formalism for calculating the production of antihydrogen is identical to the one used for studying bound-free pair production in ion-ion collisions. By crossing symmetry the matrix element for bound-free pair production is equivalent to the one for ionization. This analogy is illustrated in Fig. 2.

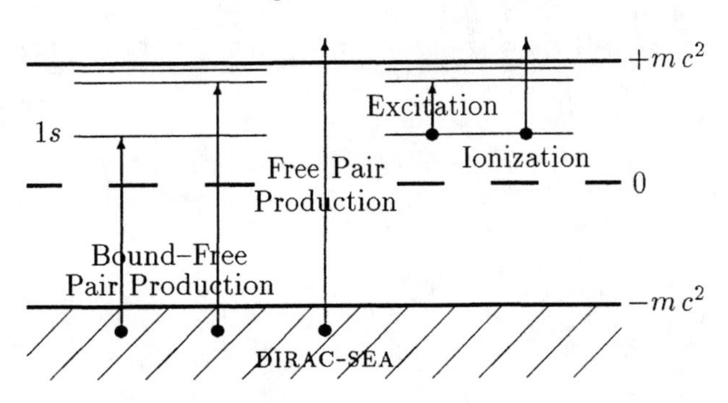

FIG. 2. Various transitions in the Dirac-sea picture.

In contrast to the case of ionization, the relevant momentum transfer for the production of antihydrogen, i.e. the energy required to lift an electron from the Dirac-sea of negative energy (a positron) into a bound state, is on the order of $\omega \approx 2m_e$ (we set $\hbar = c = 1$ in this section). Therefore, we use exact Dirac-Coulomb wave functions to describe the initial and final states $|i\rangle$ and $|f\rangle$ of the antihydrogen atom (actually, its positron) in (3).

We calculate the cross section for antihydrogen production in SCA, or equivalently in plain-wave Born approximation (PWBA) [8]. The Hamiltonian $H_{\rm int}$ entering (3) is given by the electromagnetic interaction between antihydrogen and the (hydrogen) gas target. Due to the large values for ω, screening effects are negligible for this case, as long as $\gamma = (1-\beta^2)^{-1/2} \lesssim 200$ [9].

With an unscreened potential, the total SCA cross section for pair production with electron capture in a heavy ion collision is given by [10]

$$\sigma_{\rm tot} = 8\pi \left(\frac{Z_P\alpha}{\beta}\right)^2 \int_{m_e}^{\infty} \mathrm{d}E_i \int_{q_0}^{\infty} \frac{s\,\mathrm{d}s}{[s^2-(\beta q_0)^2]^2} \times \sum_{\kappa_i}\sum_{m_f,m_i} \left|\langle f\,|(1-\beta\alpha_3)\mathrm{e}^{\mathrm{i}\vec{q}\vec{r}}|\,i\rangle\right|^2 . \qquad (4)$$

(Note that in the antihydrogen's rest-frame the *target gas* becomes the *projectile*, indicated by Z_P.) Here $\vec{q}$ denotes the momentum transfer from projectile to target whose magnitude is $|\vec{q}| = s$, and $q_0 = \omega/\beta$. The third component of the Dirac matrices is α_3. In our calculation, $|i\rangle$ is the wave function of an electron with negative energy E_i in the continuum. Similarly, $|f\rangle$ is a Dirac-Coulomb wave function of a K-shell electron.

Using relativistic wave functions with their two-component structure yields different radial form factor components

$$J^\ell(E_i,s) = \int_0^\infty r^2\,\mathrm{d}r\, j_\ell(sr) \times [g_{\kappa_f}(r)g_{E_i,\kappa_i}(r) + f_{\kappa_f}(r)f_{E_i,\kappa_i}(r)] , \qquad (5)$$

$$I_\ell^\pm(E_i,s) = \int_0^\infty r^2\,\mathrm{d}r\, j_\ell(sr) \times [f_{\kappa_f}(r)g_{E_i,\kappa_i}(r) \pm g_{\kappa_f}(r)f_{E_i,\kappa_i}(r)]. \qquad (6)$$

In (6), g_κ and f_κ denote the large and small components of the Dirac-Coulomb wave function, respectively.

For purposes of illustration we present in Fig. 3 the

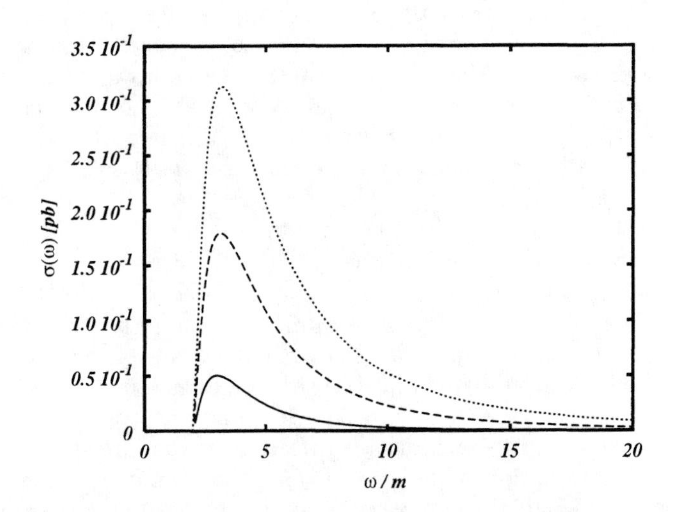

FIG. 3. The cross section for antihydrogen production as a function of energy transfer $\omega = (E_f - E_i)/\beta$. The solid line shows the cross section for $\gamma = 3$ (Lorentz factor of the projectile in the antihydrogen's rest frame). The dashed line shows the cross section for $\gamma = 6$, and the dotted line for $\gamma = 10$.

cross section for antihydrogen production versus energy transfer ω. The minimum energy transfer required to create a bound-free pair with a positron in the $1s$ state is $\omega_{\min} = (1+\sqrt{1-Z_T^2\alpha^2})m_e \approx 2m_e$ (with $Z_T = 1$). The diagram shows that the cross sections peaks near threshold, and that the peak moves slightly to larger ω for higher values of the Lorentz factor γ due to contributions from higher multipoles.

IV. EXCITATION AND IONIZATION OF PIONIUM

In another—seemingly unrelated—example, we apply the SCA formalism to the excitation and break-up of pionium. In the experiment "DIRAC", recently proposed and accepted at CERN [11], collisions of 24 GeV/c protons with target nuclei will produce $\pi^+\pi^-$-pairs that may combine to form a bound complex called "pionium", i.e., a $\pi^+\pi^-$-atom, most likely in an s-state. While moving through the target material, the pionium may annihilate into $\pi^0\pi^0$-pairs, if they are still in an s-state, or they may get excited and eventually break up into "atomic" pairs due to the electromagnetic interaction with the target atoms or nuclei. The pionium's lifetime in its ground state is dominated by the annihilation process, governed by strong interaction. Measuring this lifetime to better than 10% accuracy thus provides a crucial test for chiral perturbation theory [12] predicting a lifetime of $\tau = (3.7\pm0.3)\cdot 10^{-15}$ sec for the ground state [13]. This annihilation time is much shorter than the average time for radiative transitions. Spectroscopy of pionium can therefore only be studied through $\pi^+\pi^-$-pairs emerging from the breakup of such atoms in the target [14]. Thus, in order to extract the pionium lifetime with a precision of 10% we need to determine the electromagnetic excitation, de-excitation, and breakup cross sections of $\pi^+\pi^-$-atoms in the Coulomb field of a target atom or nucleus to a precision of a few percent [11]. We show that the theoretical calculation of these cross sections to such high precision is indeed possible.

In the rest-frame of the pionium, the target material's ion can be treated as a classical particle moving on a straight-line trajectory $\vec{R}_b(t) = (b, 0, \beta ct)$ at nearly the speed of light, while the pionium at the origin is treated quantum mechanically. As we are only interested in pions forming atom-like complexes, their relative velocity must be small (of order $v_\pi/c \approx \alpha$). Hence non-relativistic hydrogenic wave functions are perfectly appropriate for the initial and final states $|i\rangle$ and $|f\rangle$ of the pionium. On the other hand, the complex charge distribution of the target atoms is taken into account by including a *screening function* in the scalar potential experienced by the pionium. In our calculations, we use screening functions that reproduce exactly expectation values of powers of the radial variable obtained with full Dirac-Hartree-Fock-Slater wave functions for the heavy ion or atom [15]. Neglecting magnetic terms (estimated to contribute no more than 0.4%), H_{int} reduces to the interaction of the pions with the scalar potential of the ion with charge Z_P, given in the pionium's rest frame by

$$\Phi(\vec{r},t) = \frac{Z_P e}{2\pi^2}\sum_{k=1}^{N} A_k \int \frac{\exp\left[\mathrm{i}\vec{s}\cdot\left(\vec{r}-\vec{R}_b(t)\right)\right]}{s^2+\alpha_k^2-(\beta s_z)^2}\,\mathrm{d}^3 s, \quad (7)$$

where the screening parameters A_k and α_k are taken from [15]. Performing the integration over coordinate space implied in the matrix element (3) with the two pions positioned at $\pm\vec{r}/2$ from the pionium's center-of-mass, we obtain the impact parameter dependent transition amplitude in first order of the scalar interaction

$$\begin{aligned} a_{fi}^{(1)}(b) &= \frac{2Z_P\alpha}{\mathrm{i}\beta}\sqrt{4\pi(2\ell_f+1)(2\ell_i+1)}\,(-1)^{m_f} \\ &\times \sum_{\ell,m} \mathrm{i}^{\ell-m}\sqrt{2\ell+1}\begin{pmatrix} \ell_f & \ell & \ell_i \\ 0 & 0 & 0\end{pmatrix} \\ &\times \left[1-(-1)^{\ell}\right]\begin{pmatrix} \ell_f & \ell & \ell_i \\ -m_f & m & m_i\end{pmatrix} \\ &\times \sum_{k=1}^{N} A_k \int_0^\infty s\,\mathrm{d}s \frac{B_{\ell m}(b,q_0,s)}{s^2+\alpha_k^2-(q_0\beta)^2} F_{fi}^{\ell}\left(\frac{s}{2}\right), \end{aligned} \quad (8)$$

again with $q_0 = (E_f - E_i)/(\beta\hbar c)$, and with the straight-line trajectory factor [16]

$$\begin{aligned} B_{\ell m}(b,q_0,s) &= \Theta(s-q_0)Y_{\ell m}\left(\cos^{-1}\left(\frac{q_0}{s}\right),0\right) \\ &\times J_m\left(b\sqrt{s^2-q_0^2}\right) \end{aligned} \quad (9)$$

containing the step-function and a Bessel function. In the radial form factors in (8),

$$F_{fi}^{\ell}(k) = \int_0^\infty r^2\,\mathrm{d}r\, R_f(r)\, j_\ell(kr)\, R_i(r), \quad (10)$$

R_i and R_f stand for bound state or continuum wave functions of the pionium, as appropriate. These form factors can easily be evaluated using standard methods described in [17].

Integrating the squared amplitude over the impact parameter b yields the inelastic cross section $\sigma_{fi}^{(1)}$ for the transition—excitation or breakup—between states $|i\rangle$ and $|f\rangle$. The total cross section for excitation and breakup from the state $|i\rangle$ is obtained by summing over bound and integrating over continuum final states:

$$\sigma_{\text{tot},i}^{(1)} = \sum\!\!\!\!\!\!\int_f \sigma_{fi}^{(1)}, \quad (11)$$

or by using the completeness relation for the set of final states:

$$\sigma^{(1)}_{\text{tot},i} = 16\pi \left(\frac{Z_P\alpha}{\beta}\right)^2 \int_{q_0}^{\infty} s\, ds \left[1 - F^0_{ii}(s)\right] \times \left[\sum_{k=1}^{N} \frac{A_k}{s^2 + \alpha_k^2 - (\beta q_0)^2}\right]^2 . \quad (12)$$

(Note that the elastic cross section vanishes.) The determination of cross sections and transition probabilities thus reduces to the accurate and fast calculation of the radial form factors.

Fig. 4 illustrates the suitability of our approach under quite general conditions. It shows the dependence of the total inelastic cross section on the kinetic energy of the heavy ion (in the rest-frame of the pionium). In order to remove the dominating (but trivial) factor $1/\beta^2$ in (12), the cross section is divided by this factor. As can be seen from the figure, $\beta^2\sigma_{\text{tot},i}$ is essentially constant in the energy range of interest to experiment DIRAC, i.e., between 2 GeV and 10 GeV.

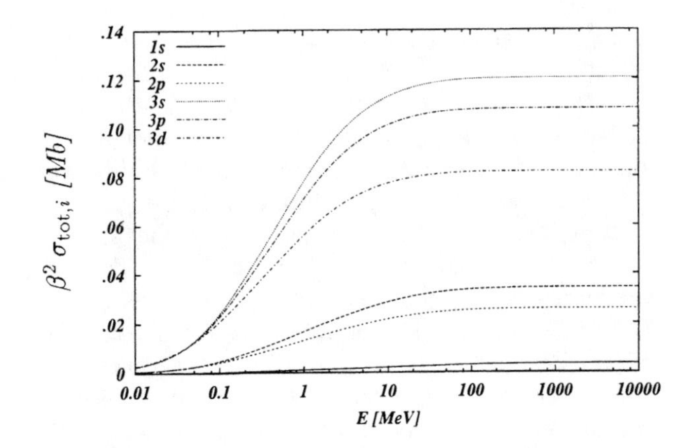

FIG. 4. Total inelastic cross section from different initial states i (as indicated in the figure), multiplied with the (dominating) factor β^2, over a wide range of projectile energies (from 10 keV to 10 GeV). Calculations for Ti target.

V. CONCLUSION

We demonstrated with two explicit examples the suitability of the semi-classical approximation to ion-ion or ion-atom collisions for non-exotic as well as exotic systems. Specifically, we applied this formalism first in the calculation of the cross section for relativistic antihydrogen production. We presented cross sections in excellent agreement [8] with the experimental results obtained at Fermi-Lab [7]. Similar calculations for electron capture in extremely relativistic ion-ion collisions with arbitrary charge have been performed and shall be presented in a forth-coming paper. A current extension of this application is the formation of muonic atoms with *muons* stemming from pair production (in analogy to the positron in the case of antihydrogen). In this case, the larger amplitude of the muon wave function near the nucleus must be taken into account by using an extended charge distribution for the nucleus, rather than a pure Coulomb potential.

In a second "exotic" application, we calculated excitation and break-up cross sections for pionium ($\pi^+\pi^-$-atoms). The experiment "DIRAC", to be performed at CERN, aims at measuring the pionium's lifetime in its ground state to high precision, providing a crucial test for chiral perturbation theory. Attaining the required precision hinges on an accurate calculation of all electromagnetic processes competing with the strong interaction. The SCA method satisfies this requirement. We calculated the necessary form factors accurately and efficiently. On-going developments in our group concentrate on including magnetic interaction terms, as well as higher-order perturbation contributions.

---

[1] J. Bang and J.M. Hansteen, *K. Dan. Vidensk. Selsk. Mat. Fys. Medd.* **31** (1959) 13
[2] L. Kocbach, *Z. Phys. A* **279** (1976) 233
[3] D. Trautmann and F. Rösel, *Nucl. Instr. Meth.* **169** (1980) 259
[4] Z. Halabuka, W. Perger and D. Trautmann, *Z. Phys. D* **29** (1994) 151 and references contained therein
[5] M. Kavcic, Z. Smit, M. Budnar and Z. Halabuka, *Phys. Rev. A* **56** (1997) 4675
[6] PS210 Collaboration, W. Oelert, spokesperson; G. Baur *et al.*, *Phys. Lett. B* **368** (1996) 251
[7] G. Blanford *et al.*, *Phys. Rev. Lett.* **80** (1998) 3037
[8] H. Meier, Z. Halabuka, K. Hencken, D. Trautmann and G. Baur, *Eur. Phys. J. C* **5** (1998) 287
[9] A.H. Sørensen, *Phys. Rev, A* (1998) submitted
[10] J. Eichler and W.E. Meyerhof, *Relativistic Atomic Collisions,* (Academic: San Diego, 1995)
[11] L.L. Nemenov et al., *Proposal to the SPSLC: Lifetime measurement of $\pi^+\pi^-$-atoms to test low-energy QCD predictions,* CERN/SPSLC 95-1, SPSLC/P 284
[12] H. Leutwyler, in *Proc. XXVI Int. Conf. on High Energy Physics,* Dallas, 1992, edited by J.R. Sanford, AIP Conf. Proc. No. 272 (AIP, New York, 1993) p.185
[13] J. Gasser and H. Leutwyler, *Ann. Phys.* (N.Y.) **158** (1984) 142
[14] L.G. Afanasyev and A.V. Tarasov, *Phys. At. Nucl.* **59** (1996) 2130
[15] F. Salvat, J.D. Martinez, R. Mayol and J. Parellada, *Phys. Rev. A* **36** (1987) 467
[16] P.A. Amundsen and K. Ashamar, *J. Phys. B* **14** (1981) 4047
[17] D. Trautmann, G. Baur and F. Rösel, *J. Phys. B* **16** (1983) 3005

Looking for polarization bremsstrahlung in the midst of inner-shell ionization: Analysis of the x-ray spectrum in electron interactions with thin-film targets

C. A. Quarles and S. Portillo

Department of Physics and Astronomy, Texas Christian University, Fort Worth, Texas 76129

Calculations of the total bremsstrahlung spectrum including polarization bremsstrahlung (PB) for high energy electrons on atoms have been made recently by A. V. Korol, A. G. Lyalin and A. V. Solovy'ov. This has motivated us to look for the PB effect for 25 and 50 keV electrons on a variety of thin-film targets including C, Al, Cu, Ag, Tb and Au. PB is predicted to be a significant increase in radiated photon intensity at energies below the target K and L absorption edges. A good model of the thick-target bremsstrahlung background due to electrons elastically scattered into the detector window and a good understanding of the Ge and Si(Li) detector response are crucial for interpretation of the data. We have used a geometry in which the detector-window background is significantly reduced from that in prior experiments. We have analyzed the photon spectra from above 4 keV to the kinematic endpoint. The data are very well fit by normal bremsstrahlung alone. No PB contribution is seen in the data. Finally, we conclude with an argument why, in fact, we should not expect any PB effect when charged particles interact with solid film targets.

INTRODUCTION

Normal Bremsstrahlung is the radiation by a charged particle when scattered in the Coulomb field of a target atom. Polarization Bremsstrahlung (PB) is the radiation by the target atom when polarized by the scattering of a charged particle. Reality should include both effects. The question is to what extent does one or the other amplitude dominate the cross section. Interference can be expected when the radiated photon energy is near one of the characteristic energies of the target. Prior bremsstrahlung spectrum experiments with electrons in the 5 to 500 keV energy range incident on thin-film targets have been in very good agreement with the prediction of normal bremsstrahlung without any contribution from PB.

Very recently, A. Korol, A. Lyalin and A. Solovy'ov [1] have done the first calculation of the PB effect for the photon spectrum produced by electrons in an energy range amenable to experiment with solid state x-ray detectors. The results of ref. 1 are based on treating the normal bremsstrahlung component in a DWBA approximation and the polarization component in the Born approximation. An important result of the calculation is the prediction of an increase in the photon intensity radiated at photon energies less the target's K absorption edge (and an even larger increase below the L edge, etc.) These calculations provide the first definite theoretical predictions on the PB effect from electrons incident on atoms that can be compared with experimental data.

In previous experiments the level of background from bremsstrahlung produced by electrons which elastically scatter in the target into the detector window has been of the same order or larger than the predicted PB effect. So, we have tried design a detector geometry to reduce the background and increase the sensitivity to the predicted PB effect.

The background can be reduced by placing a well-collimated detector at some distance from the target and then inserting a low Z absorber about half-way between the target and the detector to absorb the elastically scattered electrons.[2] The insertion of the absorber, which is just thick enough to stop the 50 kV electrons, produces thick-target bremsstrahlung seen by the detector. But since the detector is some distance from the absorber, the solid angle is kept small. The thick-target bremsstrahlung produced in the absorber can be calculated rather well and can be included in the analysis of the bremsstrahlung data when compared with the bremsstrahlung theory. The data presented below are consistent with a level of background about an order-of-magnitude less than previous experiments and well below the predicted level of PB.

Data has been taken at two incident electron energies, 25 and 50 keV on thin-film targets of Al, Cu, Ag, TbF_3, Au and UF_4. With the exception of the Al target that was self-supporting, the targets were on backing of 15 $\mu g/cm^2$ carbon. The data for Ag and Cu at 50 and 25 keV are shown in Figure 1. The data for the other targets at 25 keV are shown in Figure 2. The data at 25 keV were taken with a thinner scattered electron absorber foil (4.7 mil kapton) and so can be extended to lower radiated photon energy than the data for 50 keV.

In Figure 1 and 2 the data are compared with the normal bremsstrahlung prediction. The fitting procedure includes both the thick-target absorber background and the detector response. Very good fits to the data are obtained for both the Ge and the Si(Li)

CP475, *Applications of Accelerators in Research and Industry*,
edited by J. L. Duggan and I. L. Morgan

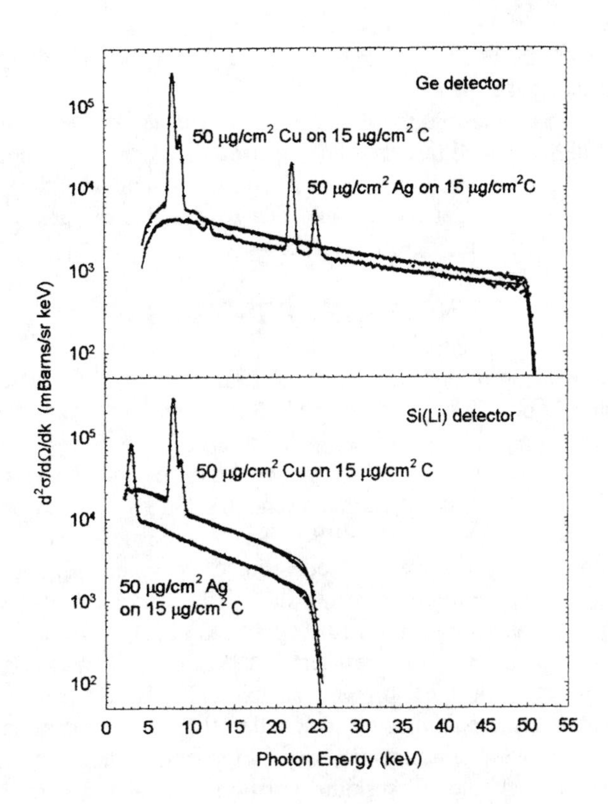

Figure 1. Cross section for Bremsstrahlung from 50 and 25 keV electrons on Cu and Ag targets taken with Ge and Si(Li) detectors. Data is fitted to model with normal bremsstrahlung theory including detector response and thick target background.

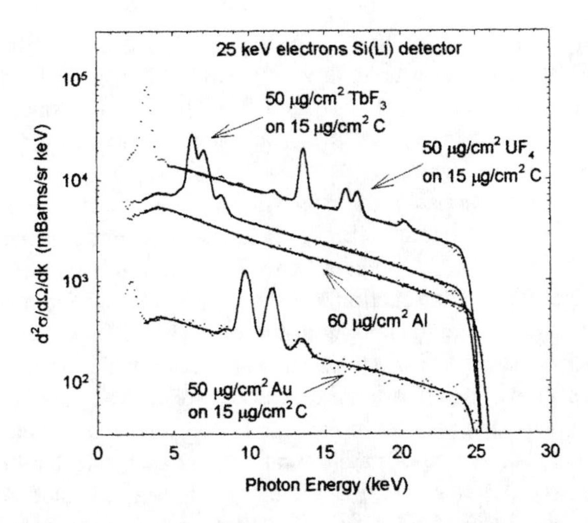

Figure 2. Cross section for bremsstrahlung for 25 keV electrons on various targets taken with the Si(Li) detector. Data is fitted to model including normal bremsstrahlung, detector response and thick-target background above 4 keV photon energy.

detectors. The model includes the efficiency of each detector, the absorption in the electron absorber foil and the detector window, and the Ge detector K x-ray escape effect very well. The fitting procedure has been described in detail elsewhere.(2) It is a Monte Carlo model which constrains the parameters of the fit that describe the detector response (two parameters) and the remaining small thick-target background (one intensity parameter). Intensity parameters are varied to fit the characteristic x-ray peaks to Gaussian peak shapes localized at the known x-ray energies. The energy calibration of the system is fitted by two parameters, the energy per channel and the zero-energy channel.

COMPARISON WITH PB PREDICTIONS

The PB effect can be tested by calculating the difference between the data and the fit to normal bremsstrahlung. We will concentrate on the data for Cu and Ag since there are predictions for these atoms.(1) In Figures 3 and 4, this difference is compared to the difference between the total bremsstrahlung prediction of ref. 1 and the normal bremsstrahlung. Thus the theoretical prediction includes the contribution from the polarization amplitude as well as the interference between the two amplitudes. In both figures the effect of absorption in the electron absorber and detector window on PB is also shown as the solid line.

Looking at Figure 3 for 50 keV electrons, it is clear that the data are inconsistent with a significant polarization bremsstrahlung effect. This is especially notable for Ag target, but can be seen at the lower photon energies, from 4 to 8 keV for Cu as well. A significant increase in photon intensity due to PB is predicted for both Cu and Ag, but not observed. The major polarization bremsstrahlung effect is expected below the K edge, or below the characteristic x-ray peaks in the data. The increase in photon intensity is predicted to continue into the region dominated by the window absorption. With the absorbers necessary to reduce background, we are able to measure photons down to about 4 keV. However, the increase expected as photon energy decreases to 4 keV is not seen in the data.

In Figure 4, the result for the difference between the data and normal bremsstrahlung at 25 keV is shown and compared with the PB theory of 50 keV. One would expect that the PB effect would be even larger at 25 keV and that the PB would scale approximately inversely with the incident particle velocity. So we consider the comparison with the 50 keV Ag to be conservative. The data are consistent with normal bremsstrahlung theory. Clearly there is no PB effect for the Ag data. The data for Cu is less clear since the fit to normal bremsstrahlung is not as good. There is a peak in the difference data at about 6 keV, below the K x-ray region. A similar but smaller effect is noticeable in the 50 keV data. This effect is seen only for the Cu target and is more likely due to contamination in the target than PB. (Some contamination was seen in the Cu target in He^+ data mentioned below.) The effect does not have the expected PB photon energy dependence at either energy.

The results for the difference between the data and the model with normal bremsstrahlung for 25 keV electrons on targets of Au, Al, TbF_3 and UF_4 are shown in Figure 5. We have displaced each curve by a scale constant of 1000 to separate the data for each target. The peaks in the UF_4 data are due to a small Ta contamination in the target which was not included in the fit.

It would be interesting to have atomic PB theory such as that for Cu and Ag to compare with these data. However, it is clear from the figure that no significant deviation from the normal bremsstrahlung model is observed in any of the targets studied.

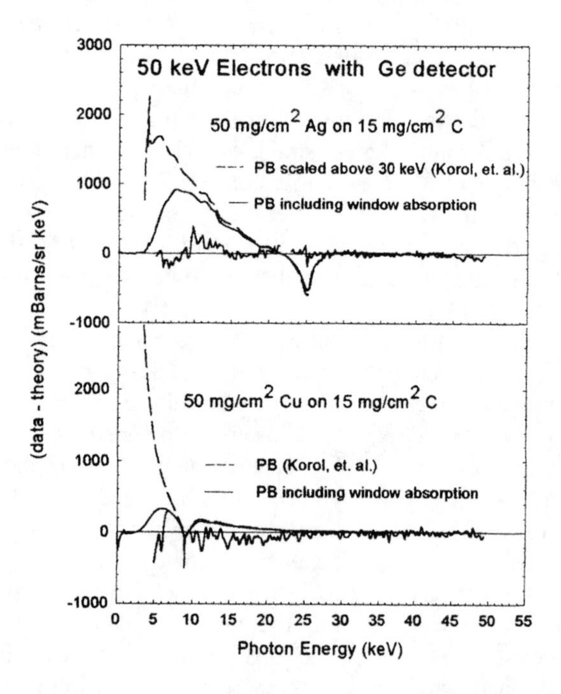

Figure 3. Difference between data and normal bremsstrahlung for 50 keV electrons on Ag and Cu thin-film targets compared to prediction for PB(1). The effect of absorption in the scattered electron absorber and detector window on the predicted PB is shown as the solid line.

Recently data has been taken on the 2 MeV van de Graff at UNT for 1 MeV and 1.5 MeV He^+ on the same Cu and Ag targets and with the same Si(Li) detector used in the electron experiment. With heavy particle projectiles, there is essentially no normal bremsstrahlung, so the PB effect should be more easily observed. PB is expected to be of the same order as that predicted for the electrons or perhaps even larger since the velocity of the projectile is lower than that of the 25 keV electrons. The energy of the projectile was chosen to reduce the normal bremsstrahlung from secondary electrons. The full results will be published elsewhere, but the initial conclusion is that no PB effect is seen in the heavy particle bombardment of either thin-film or thick solid targets.

In conclusion, the data are consistent with no PB effect at all from solid targets. The electron data can be described by a model with only normal bremsstrahlung including a small thick-target background due to the electron absorber and the expected detector response.

SO, WHERE IS THE PB?

At this point we would like to suggest that perhaps the comparison of our data from solid targets with the atomic model of Korol, Lyalin and Solovy'ov may not be appropriate. We have been focusing on a "single interaction" model. This model has proven to be very effective in describing normal bremsstrahlung, but it may be inadequate to describe the PB effect. There is a completely different perspective on radiation of charged particles interacting with bulk materials in which polarization of the medium plays the central role. V. L. Ginzburg provides a good summary of this perspective.(3) Cerenkov radiation is perhaps the best known example of radiation from a polarizable medium, even when the charged projectile is moving with constant velocity. Transition radiation is another example. Transition radiation is usually discussed at much higher projectile energy. But it should be present at low energy and essentially has the same origin as the PB effect from atoms that we have been searching for.

Consider the PB effect at a large impact parameter. There is essentially no acceleration of the projectile in this case, but the PB effect is there nevertheless for an isolated atom. The PB effect occurs in a solid, however, only if there is a variation in the polarization density of the medium such as at a boundary between a uniform medium and the vacuum. Perhaps the PB effect from a single isolated atom can be thought of as the extreme case of a variation in the "medium."

When the impact parameter with a particular atom in a solid is small enough for the normal bremsstrahlung amplitude to be significant, the PB amplitude is still affected by the interaction of the atom with the other polarized atoms as well as with the projectile. It is easy to visualize the screening effect on a single atom due to the polarization of neighboring atoms in the solid. The obvious effect is to suppress the polarization due to the projectile alone. Thus we can imagine that in an exact calculation there would be no net PB effect at all in a continuous polarizable medium. Certainly this expectation would be in agreement with experiment at the present level of sensitivity.

SUMMARY AND CONCLUSIONS

We have found no experimental evidence for a PB contribution in electron interactions with solid targets. We have also found no evidence for a PB contribution at the level expected from the single atom theory in the interaction of a heavy projectile such as He^+ with either thin-film or thick solid targets.

We have suggested that calculations of the PB bremsstrahlung spectrum for a charged projectile with an atom may not be appropriate for comparison with data from a solid, either a thin- or thick-film. The problem is not so much with the thickness of the target. It is still appropriate to treat normal bremsstrahlung as a single interaction of the electron with a single atom in the target. Rather, the problem is in treating the polarization contribution as a "single" particle interaction. By its very nature the polarization contribution is a collective or many body effect that depends not only on the properties of the incident charged projectile but significantly on the interaction with neighboring atoms in the solid which are, of course, also polarized by the projectile. We suggest that the PB effect in solids may be better discussed theoretically from the perspective that has been used to discuss radiative effects such as transition radiation.(3)

Finally, it would be interesting to look for an atomic PB effect in the electron interaction with a dilute gas or gas beam target such as 5 - 25 keV electrons on Kr and other noble gases. In this case the polarization interaction with neighboring atoms may be small enough for the single atom model of Korol, Lyalin and Solovy'ov to be applicable. The modest amount of earlier data with gas jet targets appears to be in good agreement with normal bremsstrahlung.(4) However, it may be time to reconsider this case with the reduced background experimental setup and the data analysis tools described here.

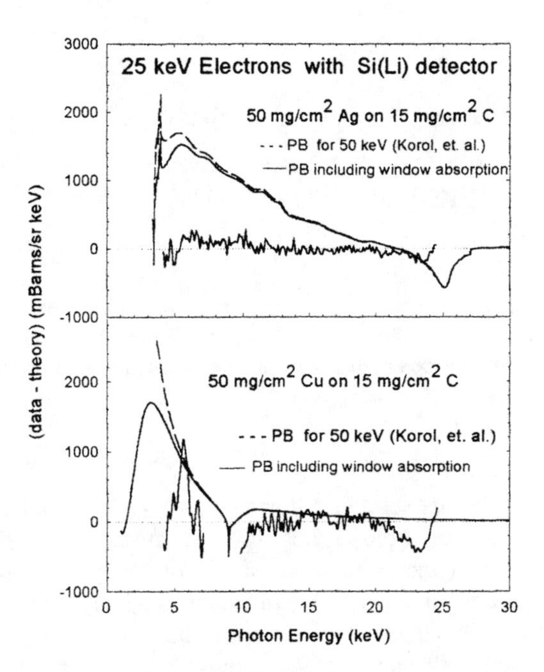

Figure 4. Difference between data and normal bremsstrahlung for 25 keV electrons on Cu and Ag. The PB theory for 50 keV is shown for comparison. The effect of absorption in the scattered electron absorber and detector window on PB is shown as a solid line.

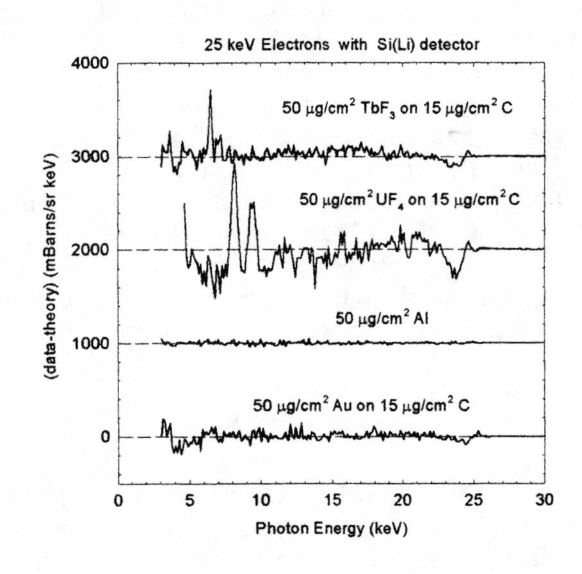

Figure 5. Difference between data and normal bremsstrahlung for several other targets.

ACKNOWLEDGEMENTS

We thank the TCU Research Fund for support. We appreciate the help of Mike Murdock and David Yale in the TCU machine shop. We are grateful to Mike Ellis for conversion of the Monte Carlo fitting program to Visual Basic 5.

REFERENCES

[1] A. V. Korol, A. G. Lyalin and A. V. Solovy'ov., J. Phys. B: At. Mol. Opt. Phys. **30** (1997) L115; Abstracts TU074,TU077 and TU078, 20th ICPEAC Abstracts, Vienna Austria, 23-29 July 1997, eds. F. Aumayr, G. Betz, And H.P. Winter; and private communication (1997).

[2] C. A. Quarles, *Accelerator-Based Atomic Physics Techniques and Applications,* eds. S.M. Shafroth and J. C. Austin, AIP Press 1997, p. 237-278.

[3] V. L. Ginzburg, Physics-Uspekhi **39** (10) (1996) 973-982.

[4] Lee Estep and C. A. Quarles, Physica **145C** (1987) 369; and references therein.

The Emission of Charged Particles with eV Energies from Hot Graphite

R. Sears, Q.C. Kessel, E. Pollack and W.W. Smith

Department of Physics and the Institute of Materials Science, The University of Connecticut, Storrs, CT 06269

Thermal desorption spectroscopy of graphite (grafoil) has been investigated by measuring the the energies of the emitted ions with a hemispherical electrostatic analyzer under ultra-high vacuum conditions. At 850 °C the masses of most of the emitted ions are found to be in the range of 48*u* to 60*u* by time-of-flight techniques. The present data show that under certain conditions, the ions may be emitted with energies above those expected for thermal emission. It is not clear whether these energies are the result of local charging of the surface or surface chemistry.

The unexpected emission of charged particles with eV energies from a hot graphite surface is reported. Although the emission of "positive and negative electricity" is a well known phenomena (1), the present data show, that under certain conditions, ions may be emitted with energies above those expected for thermal emission. Energy storage and energy transfer to and from graphite has long been of interest (2,3), and has been related to the concentration of defects, such as vacancies, interstitials and impurities. The excitation of phonons and plasmons in various forms of carbon are also familiar phenomena (4). Energy related effects are important to take into account when graphite is used at high temperatures, such as in fission reactors (as a moderator) and fusion reactors (as divertors). Because of this, the emission of particles from graphite surfaces has also been investigated by a number of techniques, including thermal desorption spectroscopy (TDS) (4), direct recoil spectroscopy, and elastic recoil detection (5). Penetration, trapping and the self-sputtering behaviors of ions impinging on graphite surfaces have also been investigated and show that impurity ions may be held strongly between the basal planes of graphite (6).

INTRODUCTION

The technique used to obtain some of the present data is similar to TDS, except that the desorbed particles are energy analyzed by a 150 degree, 100 mm radius, hemispherical electrostatic analyzer and detected with a channeltron electron multiplier. The hot graphite surface emitting the charged particles consists of a piece of grafoil (7), a form of graphitic carbon, mounted with its back side facing a hot filament. The electrotatic analyzer detects ions emitted from the front side and the temperature of this side is measured with an optical pyrometer. The vacuum is maintained on the 10^{-10} torr range with a cryopump. Figure 1 shows an outline of the apparatus. The sample is placed on a carrier which fits, in turn, on the manipulator. The heated sample may be rotated to face either the electrostatic analyzer or the residual gas analyzer.

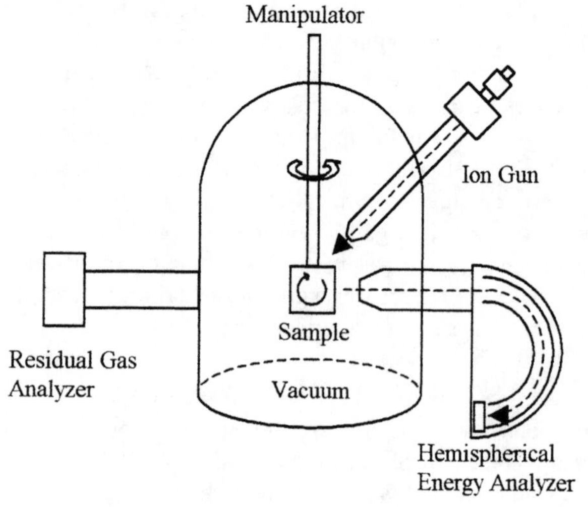

FIGURE 1. The experimental arrangement.

THE EXPERIMENT

Figure 2 shows two energy spectra of the emitted positive ions obtained at temperature of 1010°C. Below 600 °C, peaks are not observed in the spectra, but as the temperature increases peaks develop in the energy spectrum and, initially, shift toward higher energies as the temperature is increased. The emission intensity increases rapidly with temperature. Also the relative intensities of the peaks depend upon the time and temperature history of

CP475, *Applications of Accelerators in Research and Industry*,
edited by J. L. Duggan and I. L. Morgan

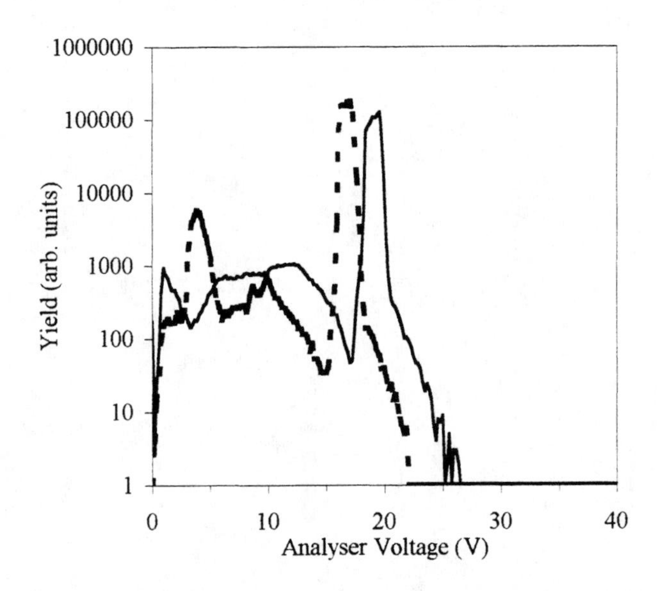

FIGURE 2. The intensity of emitted ions plotted as a funciton of analyzer voltage. The intense peaks correspond to approximate energies of 15eV. The data corresponding to the dashed line were taken approximately 30 minutes after the data of the solid line. Both were taken at a sample temperature of 1010°C. Note the emergence of a lower energy peak at a voltage corresponding to roughly 3.7eV.

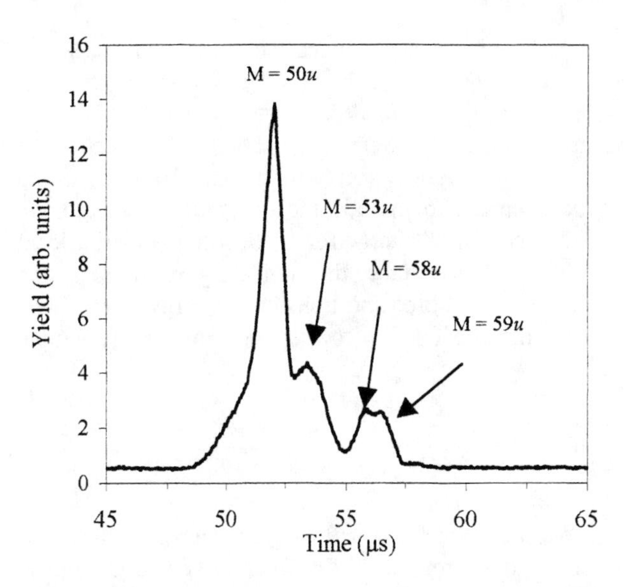

FIGURE 3. Time-of-flight spectrum of the dominant peak in Fig. 2. The peaks in this spectrum correspond to masses in the range of 48 to 60*u*.

the sample. For example, Fig. 2 shows a spectrum taken immediately upon reaching 1010°C (solid line) and another taken 30 minutes later. The energy of the prominant peak is seen to shift to lower values, while a second peak develops at a low energy. Scans made by moving the sample in the plane of the surface indicate that the emission is not uniform accross the surface. The electrostatic analyzer measures only the ionized particles' energies and gives no information about the corresponding masses, so time-of-flight (TOF) techniques were used to determine the approximate masses of the ions in the main peak. These results are shown in Fig. 3. This TOF spectrum was obtained by applying a square voltage pulse to the sample and shifting the apparent energy of the peak away from the window of acceptance of the spectrometer. In this manner a timing pulse was generated and a TOF spectrum obtained. The mass assignments are estimated to be correct to about 1 or 2*u*, the uncertainty being due to an uncertainty in $t = 0$ for the time spectrum. A separate TOF procedure, in which square voltage pulses with different amplitudes are applied to the sample, results in similar values for the masses and a value of $q = 1$ for the charge state of the particles. The range of masses identified in Fig. 3 are appropriate for a number of combinations of C_nH_m in association with elements with such as O and S (as noted below, the sample may contain up to 450 ppm of S).

DISCUSSION

Neither the origin of the ions nor the source of energy for this charged particle emission is known. Grafoil is not a well characterized material in terms of its crystal structure and may be considered to be a collection of oriented graphite crystallites. Graphite is a unique layered crystal with very different electrical and thermal conductivities parallel and perpendicular to its basal planes. The foil does have a thermal conductivity of 140 W/m·K along its width and length and only 5 W/m·K through its thickness (7) (to be compared with values of 190 -390 W/m·K and 1-3 W/m·K, respectively for highly oriented pyrolytic graphite (8)). The premium grade of grafoil has a carbon content of 98% and may contain up to 450 ppm of sulfur. It probably contains forms of carbon, other than graphite, as well. Atoms, including C, are easily intercalated between the basal planes (9). The fast particles observed might be ejected from between the basal planes, from a surface or a crystal boundary and might be from the sample itself, or from a hydrocarbon, other molecule, or atom adsorbed onto the surface. The source of the energy is the more important question. How can the thermal energy in the sample be converted to the electron volts of kinetic energy the particles are observed to have? Color centers produced by radiation and thermal luminescence are well known, but result in the emission of photons, the photons being emitted when defects are annealed at higher temperatures. Plasmons have energies of tens of eV and the typical bulk plasmon energies for graphite are about 6 and 25 eV. Electron energy loss spectra for forms of carbon ranging from diamond to C_{60}, show not only these energies but additional peaks at energies of about 15 and 35 eV (10). On the other hand, while fast particles can deposit energy into a solid by a variety of means, thermal, creation of defects and plasmons, etc., thermal energy in the bulk, by itself, would not be expected to provide the energies observed here. Possible sources of energy for these ions might be local charging of the surface, the annealing of stresses and defects and the subsequent transfer of energy to the ions, or exothermic chemical processes occurring on the sample surface.

ACKNOWLEDGMENTS

We have greatly benefited from conversations with our colleagues, Professors Best, Budnick, Fernando, Otter, Sinkovic and Dr. Laine, in particular. Michael Newman assisted with the assembly of the apparatus and participated in the preliminary measurements. This research was supported by awards from the Research Corporation (Q.C.K.), the Connecticut Space Grant Consortium under NASA grant NGT-40037 (R.S.), The University of Connecticut Research Foundation, and Connecticut Innovations, Inc. This last grant was made possible through a collaboration with Advanced Technology Materials, Inc., of Danbury CT.

REFERENCES

1. Richardson, O.W., *The Emission of Electricity From Hot Bodies*, London, Longmans, Green and Co. 1916.
2. Prosen, E. J. and F. R. Rossini, J. Res. Natl. Bur. Standards **33**, 439 (1944).
3. Dienes, G. J. and G. H. Vineyard, Radiation Effects in Solids, Interscience publishers, Inc., New York 1957.
4. Siegle, R., J.A. Davies, J.S. Forester and H.R. Andrews, Nucl. Instr. and Meth. D **90**, 606 (1994).
5. Ahmad, S., M.N. Akhtar, A. Qayyum, B. Ahmad, K.Babar, and W. Arshed, Nucl. Inst. and Meth. in Phys. Res. B **122**, 19 (1997).
6. Choi, W., C. Kim and H. Kang, Surf. Sci **281**, 323 (1993).
7. Grafoil is the tradename of graphite based paper manufactured by UCAR Carbon Company, Inc.
8. Pierson, H. O., *Handbook of Carbon, Graphite, Diamond and Fullerenes*, Park Ridge, NJ, Noyes Publications, 1993, p. 157.
9. Dresselhaus, M. S., and G. Dresselhaus, Adv. Phys. **30**, 139 (1981).
10. Hirai, H., K. Kondo, N. Yoshizawa, and M. Shiraishi, Chem. Phys. Lett. **226**, 595 (1994).

A Fully Characterized Multilayer Mirror (MLM) Polarimeter in the EUV Range for Application in Atomic and Surface Collision Experiments

M. Bailey

Atmospheric Science Center, Desert Research Institute (DRI), Reno NV 89506 USA

H. Merabet, R. Bruch, and A. Shevelko[&]

Department of Physics, University of Nevada, Reno, NV 89557 USA

[&]Optical Division, Lebedev Physical Institute, Russian Academy of Sciences, Moscow, Russia

A new multilayer mirror (MLM) polarimeter has been designed, constructed and optimized to analyze polarization in the extreme ultraviolet (EUV) wavelength range. In particular, a MLM polarimeter with a resolving power of about 6 at 304 Å (Lyman α of He^+) has been developed and fully characterized for detailed polarization studies at wavelengths of 256 Å (HeII (3p →1s) and 304 Å (HeII (2p →1s). The MLM has also been used as a single flat surface mirror polarimeter for the analysis of longer wavelength, from 517 to 584 Å for HeI (1snp) $^1P^o \rightarrow (1s^2)\ ^1S$ transitions.

INTRODUCTION

In investigations of laboratory and or astrophysical experiments in the extreme ultraviolet (EUV) wavelength range (10 nm< λ <100 nm), we are frequently challenged by spectroscopic and polarization measurements. In these types of investigations the greatest challenge generally is the analysis of very weak radiative sources. Therefore the main objective of the present work is to create a new compact optical device which consist of a multilayer mirror that can be used as an efficient polarimeter to measure the degree of polarization in the EUV wavelength range.

The excitation process in atomic and molecular collision physics generally leads to anisotropic excited magnetic substates giving rise to polarized radiative transitions (1). Therefore a compact efficient EUV polarimeter would be an ideal tool for studing accelerator based collisional processes.

In this work extensive prototype polarization measurements, following electron impact on He, have been performed using a molybdenum/silicon multilayer mirror polarimeter whose reflection and polarization characteristics have been optimized for the 304 Å emission of HeII. The degree of linear polarization was determined by measuring the intensity of photons with electric field vectors parallel ($I_{\parallel}$) and perpendicular ($I_{\perp}$) to the plane formed by the incident electron beam and the direction of observation (see Fig 1.). These intensities were observed at 90° to the incident electron beam and are related to the integrated magnetic substate cross sections (2) *i.e.*,

$$P = \frac{I_{\parallel} - I_{\perp}}{I_{\parallel} + I_{\perp}}. \qquad (1)$$

Such measurement of the polarization or angular distribution of the photons (3) emitted by excited atoms and ions yields important information pertaining to the magnetic substate populations which can be used to test theoretical approaches at a deeper level than that accessible through total or differential cross section measurements. Moreover such MLM devices have important applications for astrophysical applications, detailed study of collision and plasma processes, as well as for the accurate characterization of polarization sensitive detection equipment.

Because of the poor reflectivity of metals in the EUV, reflection analysis of weak optical sources has generally not been feasible for performing accurate polarization measurements of radiation below approximately 500 Å. For the Lyman-α emission of HeII, the only method available until recently has been the measurement of the angular distribution of the radiation intensity which can be related to the degree of polarization. Such a measurement has recently been performed by Götz *et al.* (4).

However, due to the more advanced optical properties of MLM's, it is now possible to construct enhanced reflection analyzers with good polarizability and reflectivity. Analysis with those MLM involves a more direct method for the measurement of the polarization, avoids some of the problems encountered with angular distribution experiments, and according to Fano and Macek (5), can provide a more sensitive means for measuring polarization.

CP475, *Applications of Accelerators in Research and Industry*, edited by J. L. Duggan and I. L. Morgan

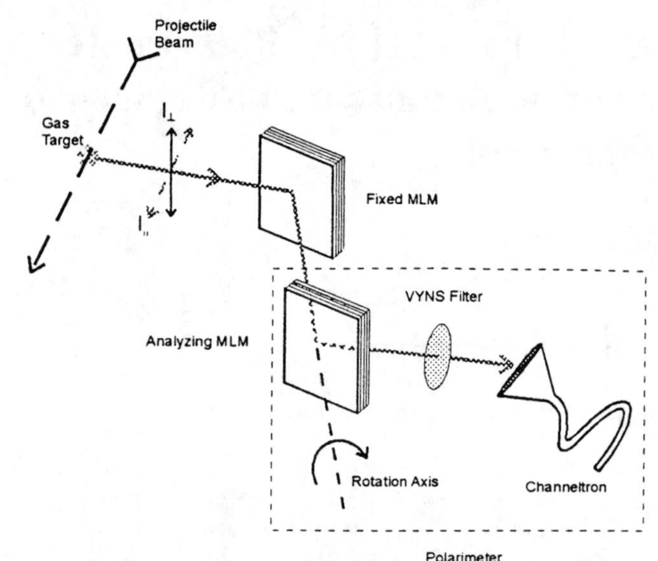

FIGURE 1: A Schematic view of the crossed polarimeter set up used to measure the intrinsic polarizability of the Mo/Si.

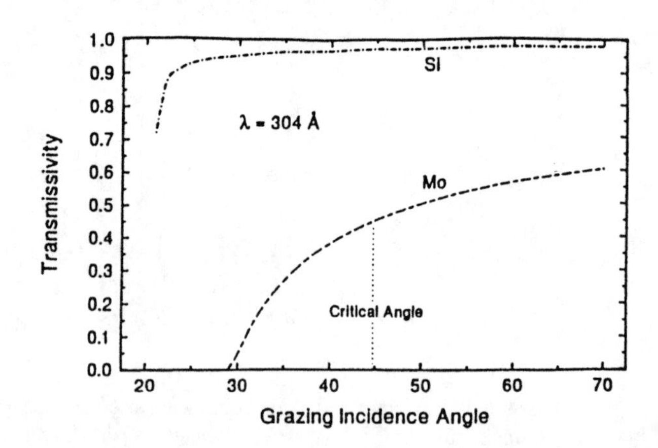

FIGURE 2: Transmissivity of the Mo and Si layers of a single period of the MLM polarimeter as a function of grazing incidence angle for 304 Å radiation.

In the present paper we primarily describe the optical characteristics of our new MLM polarimeter for the EUV wavelength region. The optical filters used as well as the internal polarizability of our MLM device are also presented. Furthermore a short description of the experimental set up is provided and finally some representative results of our polarization measurements are presented.

OPTICAL PROPERTIES OF MULTILAYER MIRROR POLARIMETER

A multilayer mirror consist of a layered structure in which the refractive index varies periodically with depth and therefore selectively reflects electromagnetic radiation at particular wavelengths. Such layers generally consist of quarter wave thicknesses of materials with a difference in refractive indices as large as possible, similar to optical reflection coatings. At normal incidence the layer thicknesses are equal to 1/4 λ. The path length differences together with the phase shifts occurring upon reflection from the interfaces leads to constructive interference for specific wavelengths, and the intensity of the reflected waves depends on the magnitude of the contrast between the refractive indices of the layer materials. However in the EUV wavelength region, nearly all optical materials are absorbing and reflectivities for most metals decrease sharply for wavelengths below approximately λ = 500 Å. Reasonable reflectivities can be obtained for grazing incidence conditions but at these angles the polarizability is small or even negligible. This disadvantage of low reflectivity of metal surfaces for EUV wavelengths can be overcome with multilayer interference structures such as MLMs, although some absorption at EUV wavelengths cannot be avoided.

The MLMs developed in this study consists of alternating thin layers of molybdenum and thicker layers of silicon. While not periodic in the traditional sense, they are periodic in the total thickness of a layer pair. This type of MLM polarimeter has been optimized for λ= 304 Å radiation incident at a grazing angle of 50°. The Bragg condition is satisfied for

$$m\lambda = 2\, d \sin\theta \qquad (2)$$

where d is the lattice spacing, λ is the incident wavelength m is the order of reflection, and θ is the grazing angle of incidence.

We have found that the Mo/Si MLM device doesn't act as a multilayer interference structure for the 584 Å radiation from neutral helium (5), corresponding to the HeI (1s2p) $^1P^o \rightarrow (1s^2)\ ^1S$ transition. In this case the reflection will be exclusively specular from the top Mo layer.

For shorter wavelength EUV radiation, the transmissivity of the multilayer structure is critical, therefore the reflectivity and transmissivity of each period has to be maximized while minimizing absorption. The transmissivity of the Mo and Si layers of a single period of the MLM as a function of grazing incidence angle for radiation with wavelengths of 304 Å and 256 Å have been calculated and are shown in Fig. 2. and Fig. 3. As it can be seen the transmissivity is low below the critical angle because the reflectivity is high, and above the critical angle the transmissivity increases as the incidence angle increases towards the normal. Hence silicon makes a good spacing material for these wavelengths. Constructive interference due to Bragg reflection for a particular period thickness, d, can only be satisfied for specific wavelengths incident at specific angles (see Fig. 2.). Therefore MLMs are somewhat dispersive and have a wavelength resolution

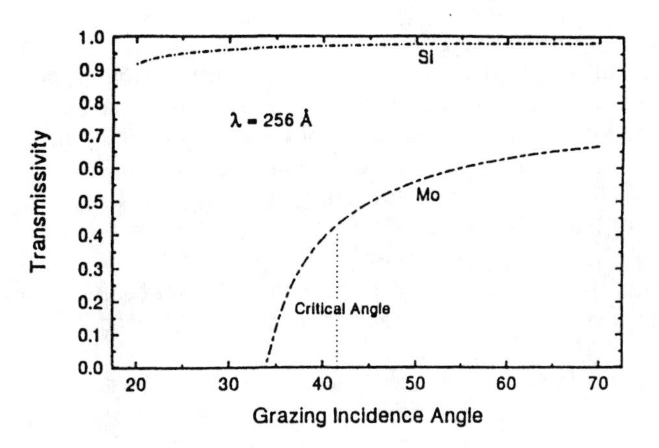

FIGURE 3: Transmissivity of the Mo and Si layers of a single period of the MLM optical device as a function of grazing incidence angle for 256 Å radiation.

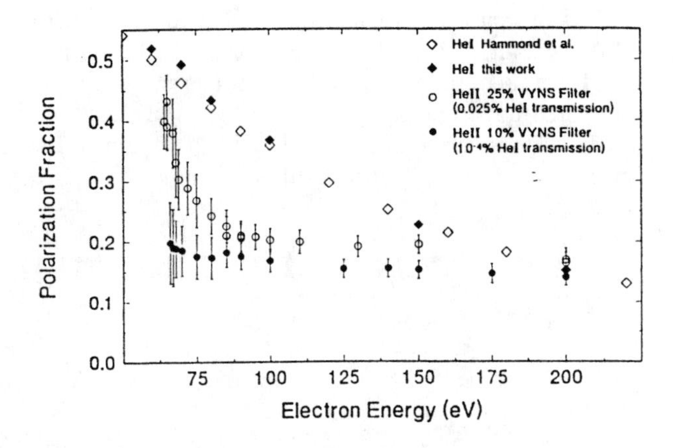

FIGURE 4: Threshold measurements of Lyman-α of HeII (2p→1s) polarization results obtained with 10% and 25% VYNS transmission filters. The HeI $(1snp)^1P^o \rightarrow (1s^2)\ ^1S$ polarization results are presented for comparison.

that ranges from 0.1 Å at λ= 10 Å to 7 Å at λ = 150 Å to 50 Å at λ = 300 Å (6). The resolution of a MLM for EUV wavelengths is approximately related to the number of periods which contribute to the reflectivity,

$$\lambda/\Delta\lambda = mN \qquad (3)$$

where N is the effective number of periods and m is the order of the reflection (7).

In the EUV region around λ = 300 Å, absorption is significant and it is not advantageous to construct MLMs with a large number of layers since only the top 5 to 7 periods will substantially contribute to the total reflectivity. Calculations of the design parameters for a MLM require a computer program of modest complexity and a few examples of these have been published (9).

Other factors can affect the performance of a MLM device besides interface roughness, such as the chemical stability of the layer materials. Some materials used for MLMs diffuse over time leading to the formation of interlayers at the interface boundaries, especially if the MLM is exposed to high temperatures (5,9).

Due to possible oxcidation, the Mo/Si MLMs used in this study were periodically cleaned with pure ethanol and immediately placed in the vacuum chamber. The intensities measured with the MLM in the earlier phases of this work, when compared with measurements obtained after cleaning, indicate changes in reflectivity consistent with the results of Underwood and coworkers (10).

OPTICAL FILTERS

In order to accurately measure the EUV polarization of HeII radiation following electron impact in He, it is necessary to suppress the dominant HeI emission. To obtain the necessary selectivity between these two spectral regions a filter material known as VYNS-3 (11) was obtained. VYNS-3 is a copolymer material containing a 90% vinyl chloride, 10% vinyl acetate mixture and is produced by Union Carbide (VYNS-3 is a specific formula of this copolymer) (12).

VYNS filters have superior transmission characteristics for the HeII EUV wavelengths when compared to the aluminum-formvar combination or other polymer materials. For example, VYNS filters transmitting 25% of the incident 304 Å radiation from helium have transmission ratios of approximately 1000:1 when compared with the transmission of 584 Å radiation (13) while measurements with a 10% VYNS filter have shown a 100,000:1 ratio transmission. This improved suppression of the 584 Å radiation is most likely due to the chlorine content of the polymer. Both carbon and chlorine exhibit much stronger absorption for 584 Å than for 304 Å radiation, but chlorine is superior to carbon in this respect (14). VYNS filters are relatively easy to fabricate. In order to make filters with 304 Å transmissions of approximately 25%, a 0.5% solution by weight of VYNS-3 dissolved in 1,2-dichloroethane was prepared (VYNS-3 is a white crystalline powder with a density of 1.37 to 1.39 g/cc (15). Furthermore the use of such a filter has the advantage to suppress the backgound produced by electrons, ions and metastables. Hence we can have a much better signal for the dominant excitation lines in the measured spectra (16).

The transmission of each filter was determined with the 1.5 m grazing incidence monochromator (16,17) by placing the filter in front of the entrance slit and comparing the measured line intensities of the relevant wavelengths with those measured without the filter. These transmissions measurements were also used to determine the filter thicknesses.

POLARIZABILITY OF THE Mo/Si MLM

The mirror polarizability, is given in terms of the reflection coefficients for the different orientations of the incident electric field vector by the Eq. 1. A direct measurement of the polarizability of the Mo/Si MLMs used in this study was possible because two identical MLMs were provided by the Lebedev Physical Institute. With these two identical MLMs, a crossed polarimeter configuration (18) can be utilized to determine the polarizability by two straightforward relative intensity measurements when the polarization of the incident radiation is known. In a crossed polarimeter setup, one of the MLMs is set in a fixed orientation and illuminated with a source of known polarization. The second MLM and polarimeter are then used to analyze the polarization of the radiation reflected from the first MLM. The difference between the known polarization and the polarization of the radiation reflected from the first MLM can be used to extract the instrumental contribution. A schematic of this setup is shown in Fig. 1. More details about this matter are given in Ref. (5,16).

POLARIZATION RESULTS AND DISCUSSION

The apparatus used in this study consists of three main components, 1: the polarimeter, 2: the electron gun, target cell and Faraday cup, and 3: a 1.5 meter grazing incidence monochromator. A PC controlled data acquisition system was used to operate the system and to record the data. A complete and detailed describtion of this experimental set up is given by Bailey *et al.* (17).

The polarization results performed in this work present the first comprehensive, electron impact, EUV measurements for the HeII $(2p)^2P^o$ and $(3p)^2P^o$ states of ionized-excited helium, and greatly extend the impact energy range of polarization data for the previously studied $(1snp)^1P^o \rightarrow (1s^2)\ ^1S$ series of neutral helium. In this section we just show an illustration of the experimental results using our new developed MLM polarimeter. More details of the polarization study will be given elsewhere (16). The polarization data measured in this work are in excellent agreement with the earlier results of Hammond *et al.* (19) using a reflecting gold mirror which were measured with errors of less than 1%. Furthermore the $(2p)^2P^o$ and $(3p)^2P^o$ threshold measurements from this work shown in Fig. 4. were performed with a 10% VYNS transmission filter. The $(2p)^2P^o$ threshold measurements in this study were initially obtained with a 25% VYNS transmission filter, and as was previously stated, this filter provides approximately a 1000:1 transmission ratio for 304 Å radiation versus the dominant 584 Å radiation from neutral helium. However a review of the electron impact cross section data for the excitation of the $(1s2p)^1P^o$ and $(2p)^2P^o$ levels of neutral and ionized helium reveals that the ratio of these cross sections at energies just slightly above the $(2p)^2P^o$ threshold exceeds 1000:1 (16). At Threshold, we had to carry out measurements using the 10 % VYNS transmission filter in order to suppress the HeI contribution to less than 1% when compared to HeII polarization (see Fig. 4).

This fully characterized polarimeter can be used for the measurement of the polarization and spectral characteristic of a variety of accelerator beam experiments producing EUV radiation. The measurement of the degree of polarization and intensity of polarized beams of radiation is significant in the analysis of hot dense plasmas, ion-atom collision experiments, surface diagnostics, EUV astronomy of the sun and stars, investigations of surface analysis with electrons, photons, atoms, molecules and ions, nanotechnology and fundamental understanding of EUV optics and instrumentation.

The MLM polarimeter will be utilized for A^{q+} + He experiments to study the magnetic substate population of the He (1snp) ^1P and He^+ target states as a function of ion beam energy, projectile charge and composition.

ACKNOWLEDGMENTS

We would like to express our gratitude to N. N. Salashchenko and Yu. Ya Platonov for fabricating the MLMs at the Institure for Applied Physics in Russia. We are indebted to James Manson for his detailed instructions concerning the construction of the VYNS filters. This Project has been supported by NATO Grant No. CRA 930032 and Acspect Corporation, Reno, Nevada.

REFERENCES

1. Mehlhorn, W., *Phys. Lett. A* **26** 166 (1968).
2. Percival, I.C., and Seaton, M.J., *Philos. Trans. R. Soc. London, Ser. A* **251** 113 (1958).
3. Götz, A. *et al.*, *J. Phys. B: At. Mol. Opt. Phys.* **29** 4699 (1996).
4. U. Fano and J.H. Macek, *Rev. Mod. Phys.* **45** 553 (1973).
5. Bailey, M. *et al.* , submitted, *Appl. Opt.* (1998).
6. Regan, S.P. *et al.*, *Rev. Sci. Instr.* **68** 1002 (1997).
7. Underwood, J.H. *et al.* , *Appl. Opt.* **20** 3027 (1981).
8. Davis, J.C. *et al.*, (*Lawrence Berkeley Laboratory, CA*, 1993).
9. Barbee, T.W. *et al.* , *Appl. Opt.* **32** 4852 (1993).
10. Underwood, J.H. *et al.*, *Appl. Opt.* **32** 6985 (1993).
11. *Union Carbide Material Safety Data Sheet, Union Carbide Chemicals and Plastics Company, Inc., Solvents & Coatings Materials Division, Danbury CT* (1992).
12. Bender, H.A. *et al.*, *Appl. Opt.* **32** 6999 (1993).
13. Manson, J.E. *Appl. Opt.* **12** 1394 (1973).
14. Henke, H., Gullikson, E.M., and Davis, J.C., *Atomic Data and Nuclear Data Tables* **54** 181 (1993).
15. Ichikawa, Y. *et al.*, *Appl. Opt.* **26** 3671 (1987).
16. Bailey, M., *Ph.D. Thesis, University of Nevada, Reno* (1997).
17. Bailey, M. *et al.*, submitted, *Phys. Rev.* **A.** (1998).
18. Dhez, P., *Nucl. Instr. Meth.* **A261** 66 (1987).
19. Hammond, P. *et al.*, *Phys. Rev.* **A. 40** 1804 (1989).

Precision Atomic Spectroscopy with an Integrated Electro-Optic Modulator

C. Koehler, D. Livingston, J. Castillega, A. Sanders, and D. Shiner

Department of Physics, University of North Texas, Denton, TX 76203

We have explored the use of recently developed high speed integrated electro-optic modulators as a tool for precision laser studies of atoms. In particular, we have developed a technique using a high speed modulator as a key element and applied it to the study of the fine structure of the 2^3P state of atomic helium. This state has been of long standing interest in atomic physics and its study has been the aim of several recent experiments using various precision techniques. We present our method and results, which indicate an order of magnitude improvement in precision over current results is possible.

INTRODUCTION

Electro-optic modulators have been and continue to be used in a variety of ways in atomic spectroscopy (1) and laser metrology (2). Recently, new types of electro-optic modulators, integrated electro-optic modulators, have become commercially available from a number of sources (3). These modulators are fabricated with integrated circuit technology and combine small microwave guides (~ 10 μm transverse dimension) with single mode nonlinear optical waveguides (4,5). They were developed for and are now primarily used in high speed fiber optic signal transmission systems. Our interest in these integrated electro-optic modulators arises because their exceptional speed (20 and 40 GHz models are available), efficiency (microwave power of 20-30 dBm for a modulation index of 1.8) and bandwidth ($0 \rightarrow f_{max}$) allow for flexible laser frequency modification and control. They should help make feasible new approaches in precision laser studies of atoms. In fact we are developing a technique using a high speed modulator as a key element. We report on that technique and demonstrate its strengths and weaknesses within the context of an experimental study of the fine structure of the 2^3P state of atomic helium (6,7). This state is under fairly intense study by experimenters using a variety of precision techniques and approaches involving lasers (8,9), interferometry (10), and microwave technology (11). Thus a comparison of these experiments should serve both as a proving ground for precision atomic methods and for facilitating a comparison of the relative strengths and drawbacks of various techniques (see Fig. 1).

The current interest in the 2^3P state of atomic helium arises for several reasons. Helium is the simplest multi-electron atom and as such has long been a proving ground for theoretical methods in atomic physics. Unlike hydrogen, analytic solutions to a realistic zeroeth order problem are not known for helium, and so approximation methods must be employed from the outset. Nevertheless, methods have been developed with remarkable precision (12,13) and helium now serves as one of the most sensitive systems for the study of the electron-electron interaction (14,15). Particular interest in the 2^3P state comes from its comparatively large fine structure splitting (~32 GHz) and its small natural linewidth (~1.6 MHz), facilitating a precision measurement. Furthermore, a combination of theoretical and experimental progress suggests that the 2^3P state of the helium atom might be used to provide a newly competitive "atomic physics" source of the fine structure constant, alpha. An uncertainty of at most a few kHz is required for this possibility, while an uncertainty of 300 Hz is very desirable. The 300 Hz precision would lead to a precision in alpha comparable to its best current determination, that from the electron g-factor, and thus would allow a full comparison of theory and experiment in that system (16,17).

Our experiment on the fine structure of the 2^3P state is designed with this 300 Hz precision as a goal. If reached this would constitute an order of magnitude improvement over the best current measurements on this state. We will in this paper describe our technique and our experiment, and then present data and results which support our approach.

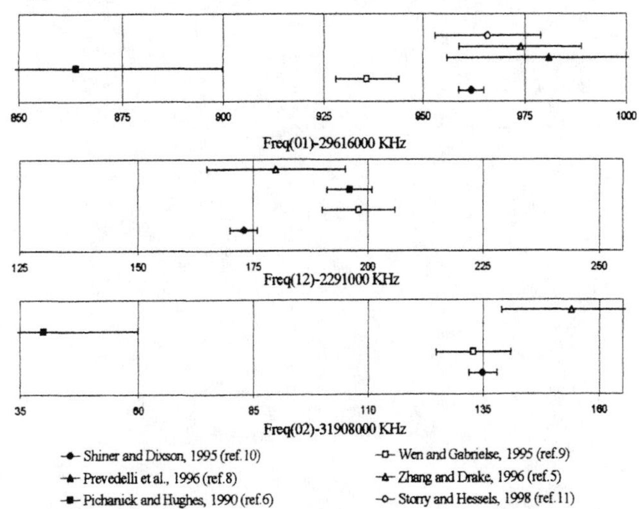

FIGURE 1. Comparisons of Theoretical and Experimental Fine Structure Intervals for the 2^3 P_J State of Helium.

CP475, *Applications of Accelerators in Research and Industry*, edited by J. L. Duggan and I. L. Morgan

TECHNIQUE

Our approach is to use a narrow linewidth tunable laser to induce the 1.083 μm allowed electric dipole transitions between the metastable 2^3S state and the 2^3P fine structure states. Our earlier measurements (7,10) took differences in transitions wavelengths to determine fine structure intervals. Since we are now interested only in the fine structure splittings themselves, which are a factor of ~ 10^4 smaller than the overall transition frequencies, a more direct frequency determination is desirable. This would avoid the wavelength metrology which limited the precision of our previous results. Our approach for doing this is to approximately center the infrared laser between the J=2 and J=0 transitions, to electro-optically modulate the laser at appropriate microwave frequencies, and to use the laser frequency side bands to alternately drive the fine structure transitions. The 2^3P state along with other low lying levels in atomic helium are shown in Fig. 2. The optical layout is shown in Fig. 3. This approach has several important advantages. (1) The difference in frequency between the sidebands is determined by a very well defined microwave source independent of the carrier laser frequency or jitter. (2) The laser frequency spectrum used to excite the two transitions is identical, thus eliminating an important possible source of asymmetric treatment of the two transitions. (3) Quick and convenient computer controlled synthesizer tuning between transitions can be performed, minimizing possible sources of error from, for example, various drifts in the experimental system. (4) Large modulation indices (β=1.84) can be used to give good sideband power (34% of carrier) and stability against rf power changes.

The important limitations in the use of integrated electro-optic modulators involve maximum laser power (10-50 mW for λ = 630-1500 nm) and specified wavelength range (± 20 nm). With regard to wavelength range, our experience suggests a much larger range is possible than manufacturers guarantee. This is important since stock modulators are usually available at 1300 nm and 1550 nm (and occasionally at others like 820 nm and 1060 nm). We currently use a 1300 nm modulator at λ=1083 nm. The fiber is slightly multimode at this wavelength but wrapping the fiber adequately attenuates the 01 mode (trying the same approach when λ=633 nm was not satisfactory). Laser powers of ~ 10 mW are sufficient for our experiment. If substantially higher powers are required, the modulator can still be used to calibrate and control frequency intervals, though not to directly drive the transitions.

EXPERIMENT

We have built an external cavity diode laser as the laser source for the experiment. We use an SDL 1083 nm DBR diode laser (18) and obtain optical feedback from a mirror mounted on a 1 meter long cervit spacer (similar efforts, but a different configuration are reported in (8). Its short term frequency stability is probably 20 kHz, based on extrapolation of our data on shorter cavities. In any event it is much less than the 200 kHz resolution of our best interferometer. This laser is frequency locked to a 3 meter reference cavity, and this cavity is in turn stabilized to an iodine stabilized He-Ne laser (19) having 300 Hz stability with 100 sec averaging times.

We summarize the atomic preparation and detection for the measurement with reference to Fig. 4. The principle is based on the Rabi, molecular beam, magnetic resonance technique, except that an allowed electric dipole transition is used in the "C" region (instead of the magnetic resonance). A beam of metastable helium atoms is prepared in the 2^3S_1 state by electron bombardment and passes into a magnetic deflection apparatus with the usual "A" and "B" deflection magnets and a homogeneous magnetic field of variable strength (5-30 Gauss) in the "C"

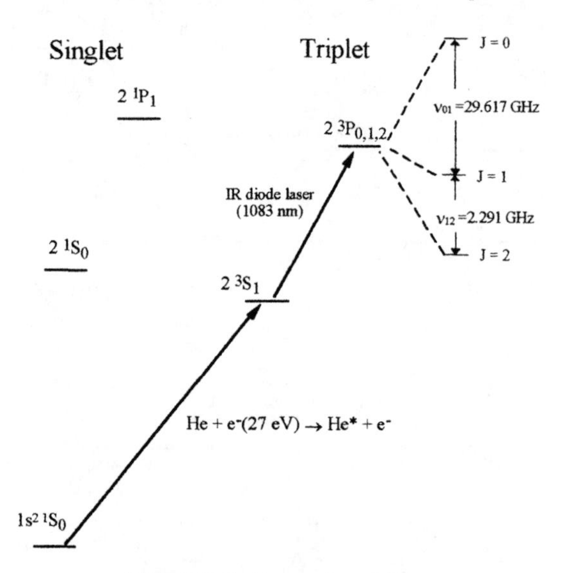

FIGURE 2. Energy Level Diagram

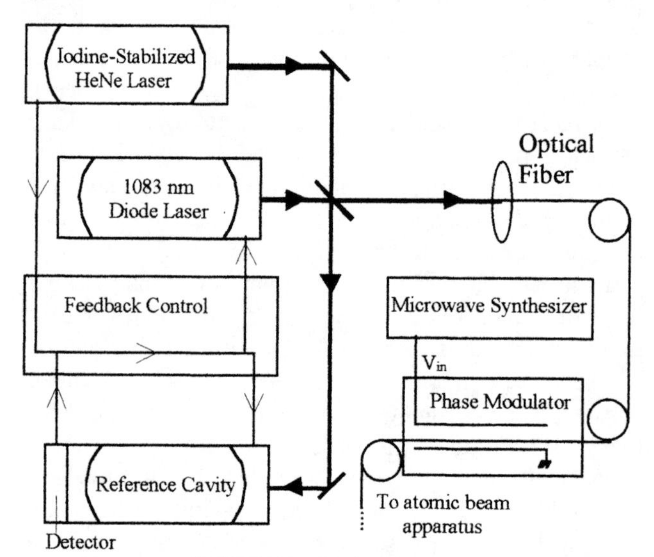

FIGURE 3. Optical Layout

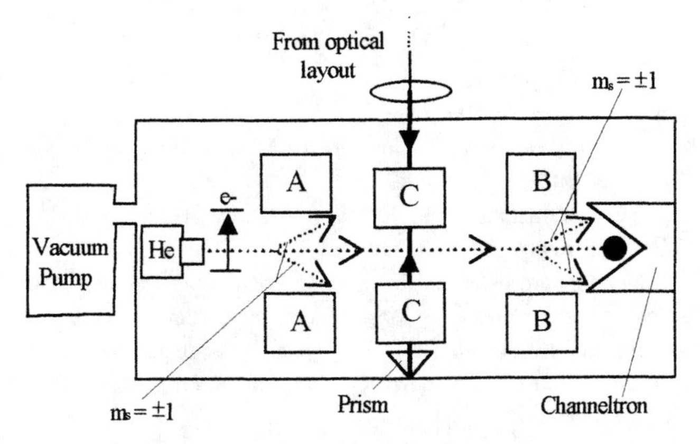

FIGURE 4. Atomic Beam Apparatus

region. Transitions among magnetic substates in 2^3S_1 are caused if the laser is tuned to the 1.08 μm $2^3S_1 \rightarrow 2^3P_{0,1,2}$ transitions. Because the selected 2^3S_1 m=0 substate of the beam is well collimated and undeflected in the beams machine, there is only a small broadening caused by residual Doppler effects and shifts are canceled by a retroreflecting prism. Our detection method involves using a channeltron and pulse counting electronics which gives the necessary sensitivity for counting individual metastables.

DATA AND RESULTS

We have observed the relevant fine structure transitions: 2^3S_1 (m=0) -> 2^3P_2 (m=-1,0,+1), 2^3S_1 (m=0) -> 2^3P_1 (m= -1,+1) and 2^3S_1 (m=0) -> 2^3P_0. We lock the IR laser carrier frequency to a point approximately midway between the J=0 and J=2 transitions and step the microwave frequency under computer control. We provisionally define the line center as the average of two frequencies on either side of the transition which give equal count rates. We take the differences in line center frequencies to give the fine structure interval, with the only correction being for the well known second order Zeeman shifts. These are calculated with negligible error using the magnetic field dependent transitions to determine the magnetic field strength. We have performed a number of long data runs. We can run the experiment on average for about 10 hours unattended. The counting error (purely statistical square root of N fluctuations) after one of these data runs is about 300 Hz in the fine structure intervals. The data analysis shows that various small drifts (for example in the laser carrier frequency, the laser intensity, the atomic beam intensity, in the magnetic field, etc.) are sufficiently small in the sense that they are being averaged out in our method without degrading the statistical precision of our results. Figure 5a shows the stability of the various parts of our experiment is not currently a limiting factor, since repeated measurements lead to consistent results assuming only counting errors. This conclusion allows us to focus primarily on studying various possible sources of systematic error.

Our initial systematic studies have focussed on the J=2 to J=1 interval, both the m=-1 to m=-1 and the m=+1 to m=+1 transitions. We do this since a recent laser experiment (9) agrees well with the older microwave measurements (6) and

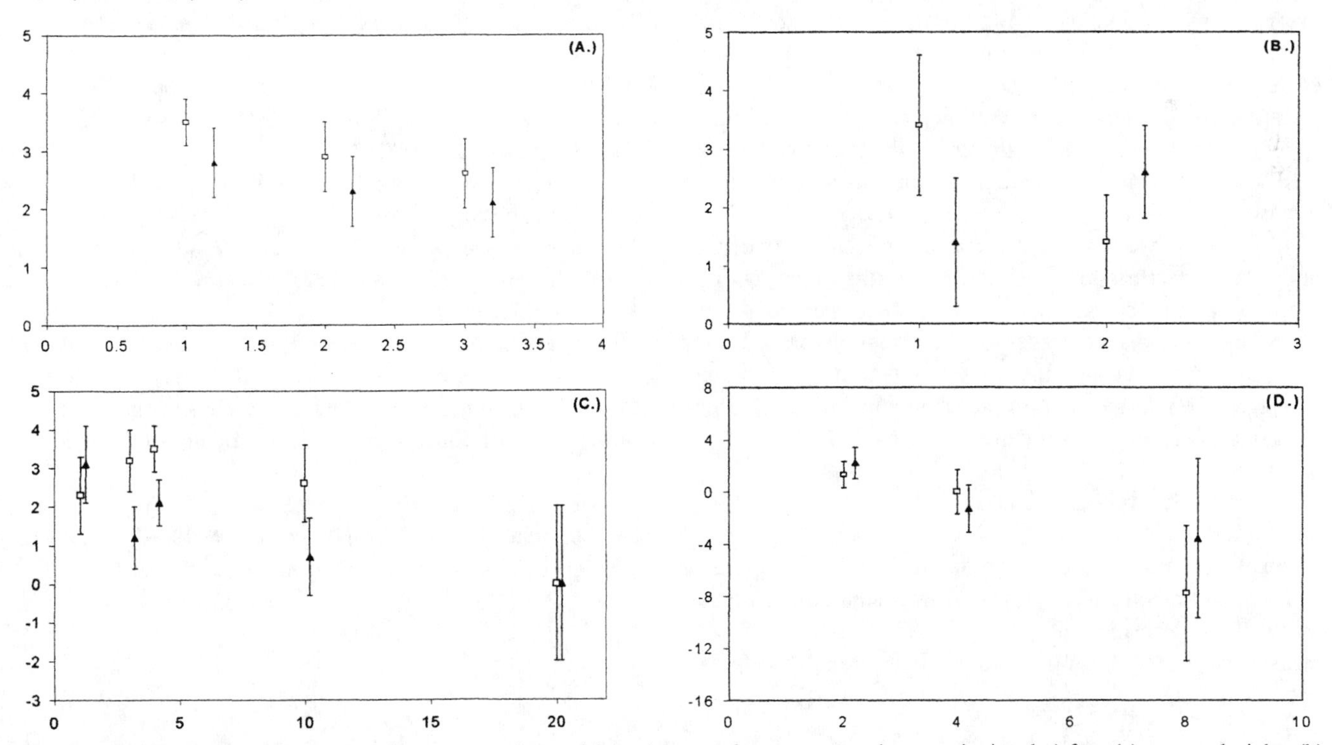

FIGURE 5. The measured J=2 to J=1 fine structure interval (m=+1 -> m=+1 squares, m=-1 -> m=-1 triangles) for: (a) repeated trials, (b) ideal and nonideal microwave modulation index, (c) various laser powers (arb. units), and (d) various laser step sizes (MHz). The vertical axis is in kHz with a fixed ~ 2 GHz frequency subtracted.

not with our previously reported results on this interval using wavelength metrology (7). To reach our goal the observed linewidth (~ 2 MHz) must be split by one part in 10^4. Thus drifts in the modulator efficiency must be kept less than this over each transition. To this end, we are able to operate the modulator, because of its excellent efficiency, with the optimum modulation index of ~ 1.84. This makes the first order side bands very insensitive to microwave power changes (a 1% change in rf power gives a .01% change in optical side band power). As a test, Fig. 5b shows data collected with the ideal modulation index and with a small modulation index (which is sensitive to microwave power changes and would amplify the effect). The consistency shows that this effect is negligible.

We wish to check that the applied laser power does not shift the observed transition frequencies. This is shown in the Fig. 5c where results measured with various laser powers (up to 20 times the usual power) are in agreement.

We also with to check that the transition intervals are independent of the step size chosen in the data collection. Figure 5d shows the interval determined from data collected ± 1MHz, ± 2 MHz and ± 4 MHz from the line center. The line half width is ~ 1 MHz so this data goes well into the wings of the transition (with a subsequent reduction in statistical precision).

We perform Doppler tests by collecting data with and without a retroreflected laser beam present. These tests currently limit any Doppler errors to less than 1 kHz.

An important remaining test is to collect data at different magnetic field settings to check consistency and verify the magnetic field corrections. Additionally, greater statistical precision is needed on several systematic checks since current tests yield results at the ~ 1 kHz level. To this end we are currently finishing a rebuilding of the atomic beam apparatus to allow for easier tuning of the magnetic field as well as better alignment, state selection and atomic beam intensity.

In summary, we have demonstrated the usefulness of an integrated electro-optic modulator in the context of precision atomic spectroscopy. We have discussed some advantages and disadvantages of these modulators. With the techniques and supporting data presented here, an order of magnitude improvement in precision appears feasible in the study of helium fine structure.

ACKNOWLEDGMENTS

We thank William Burns for discussions on integrated electro-optic modulators. This work was supported in part by the Texas Advanced Research Program and a precision measurement grant from the National Institute of Standards and Technology.

REFERENCES

1. See, for example: Wijngaarden, W. A., *Adv. At. Mol. Opt. Phys.* **36** 141-182 (1996).

2. See, for instance: Ye, J., Ma, L.-S., Day, T. and Hall, J. L., *Optics Letters* **22** 301-303 (1997).

3. Vendors are listed in, for instance, the Physics Today or Laser Focus Buyers Guides. Most of our measurements were done using a 1300 nm 20 GHz phase modulator from the Optoelectronics & Electronics Div. of the Sumitomo Osaka Cement Co., Tokyo, Japan.

4. See, for instance, Tamir, T. et al., *Guided-Wave Opto-electronics*, New York, Plenum Press (1995).

5. Burns, W. K. in *Guided-Wave Opto-electronics*, Tamir, T. et al., eds., New York, Plenum Press (1995) p231-236.

6. Background on the interest in this state and a summary of their landmark measurements on it is given by F. M. J. Pichanick and V. W. Hughes in *Quantum Electrodynamics*, edited by T. Kinoshita (World Scientific, Singapore, 1990).

7. Shiner, D., Dixson, R. and Zhao, P., *Phys. Rev. Lett.* **72** 1802-1805 (1994).

8. Prevedelli, M., Cancio, P., Giusfredi, G., Pavone, F. S. and Inguscio, M., *Optics Comm.* **125,** 231-234 (1996).

9. Wen, J., Ph.D. thesis (G. Gabrielse advisor), Harvard University, 1995 (unpublished).

10. Shiner, D. and Dixson, R., *IEEE Trans. Instrum. Meas.* **44**, 518 (1995).

11. Storry, C. H., and Hessels, E. A., *Phys. Rev. A.* **58**, R8-R11 (1998).

12. Drake, G. W. F., in *Long-Range Casimir Forces: Theory and Recent Experiments on Atomic Systems*, Frank S. Levin and David A. Micha, Eds., Plenum Press, New York, 1993, pp. 107 - 217.

13. Baker, J. D., Freund, D. E., Hill R. N., and Morgan III, J. D., *Phys. Rev. A* **41**, 1247-1273 (1990).

14. Zhang, T., Yan, Z., and Drake, G. W. F., *Phys. Rev. Lett.* **77**, 1715-1718 (1996).

15. Zhang, T. and Drake, G. W. F., *Phys. Rev. A* **54**, 4882-4922 (1996); Zhang, T., *Phys. Rev. A* **53**, 3896 (1996); Zhang, T. *Phys. Rev. A* **54**, 1252 (1996).

16. Taylor, B. N., "The Fine Structure Constant," *Units and Fundamental Constants in Physics and Chemistry,* Landolt-Bornstein Numerical Data and Functional Relationship Series, O. Madelung Ed. in Chief, Springer-Verlag, Berlin, 1993, pp. 3.125-3.131.

17. Kinoshita, T., *Rep. Prog. Phys.* **59**, 1459-1485 (1996).

18. C.L. Bohler and B.I. Marton, *Opt. Lett.* **19**, 1346-1348 (1994).

19. Model 100, Winters Electro-Optic, Inc., Longmont CO.

Photodetachment Cross Sections: Use of a Saturation Technique in Laser-Ion Beam Interactions

D. H. Lee[1)], D. J. Pegg[1)], and D. Hanstorp[2)]

[1)] *Department of Physics, University of Tennessee, Knoxville, Tennessee 37996, USA*

[2)] *Department of Physics, Göteborg University and Chalmers University of Technology, S-412 96 Göteborg, Sweden*

Cross sections for the single photodetachment of negative ions have been investigated using a saturation technique combined with fast ion-laser-beam electron spectroscopy. A crossed beam apparatus has been employed in the measurements. Both spatial and temporal profiles of the pulsed laser beam are accounted for in the calculation of the expected photoelectron yields. The saturation dependence of the electron yield on photon flux has been used to determine state-selective partial cross-sections for photodetachment of C^- and Ca^- ions at a photon energy of 2.08 eV.

I. INTRODUCTION

Photodetachment and photoionization processes are of fundamental interest in atomic physics. They provide a good testing ground for understanding electron correlations, which are particularly important in the cases of weakly-bound systems such as negative ions. In negative ion photodetachment studies, crossed and merged laser-ion beam geometries have been widely used. These beam methods can isolate a particular channel by selectively detecting, via energy analysis, either product of the photodetachment process: the ejected photoelectrons or the excited residual atoms. In the present experiment, photoelectron spectroscopy was combined with a crossed ion-laser beam geometry to measure *partial* photodetachment cross sections for selected channel.

In this article we report on photodetachment cross sections measured by studying the saturation dependence of the photoelectron yield on the photon flux. The advantage of this method is that a cross section can be obtained from merely the shape of the corresponding saturation curve without further knowledge of quantities that are difficult to measure, such as the efficiency for detection of photoelectrons and the geometrical overlap of the laser and ion beams. In addition, the method can be used to study ions with more than one bound state, such as C^-.

The saturation method was first employed in photodetachment by Hall *et al.* [1], and later by Blondel and co-workers [2] and Stapelfeldt *et al.* [3], in measurements of multiphoton detachment cross sections. Hotop and Lineberger [4] used the technique to measure single photon detachment cross sections. We have also previously used the saturation method to determine the single photon detachment cross sections for Li^- [5], C^- [6], and Ca^- [7]. Balling *et al.* [8] reported cross section measurements on the Cu^- ion using the same technique. All the above experiments involved fast beams of negative ions. The same method has been applied in the measurement of photodetachment cross sections using quasi-stationary negative ions in Penning [9] and Paul [10] traps and photoionization cross sections using gaseous and vapor atomic targets [11].

II. EXPERIMENT

The experiment was performed at the Oak Ridge National Laboratory. Figure 1 shows a schematic of the crossed laser-ion beam apparatus. A detailed description is given elsewhere [6]. A beam of positive ions was extracted from an ion source and accelerated to a selected energy in the range 5 to 100 keV. A small fraction of this beam was converted to the negative ions via two sequential electron captures in a Li vapor cell. The negative ion component was deflected 10° with respect to the incident beam direction. The ion beam was collimated by use of two 1.6 mm diameter apertures. The ion beam current was monitored using a Faraday cup placed downstream of the interaction region. A linearly polarized photon beam, derived from a flashlamp-pumped pulsed dye laser, perpendicularly intersected the ion beam. A lens was used to focus the laser beam into the interaction region. The laser pulse energy was measured with a power meter after the laser beam had traversed the interaction region.

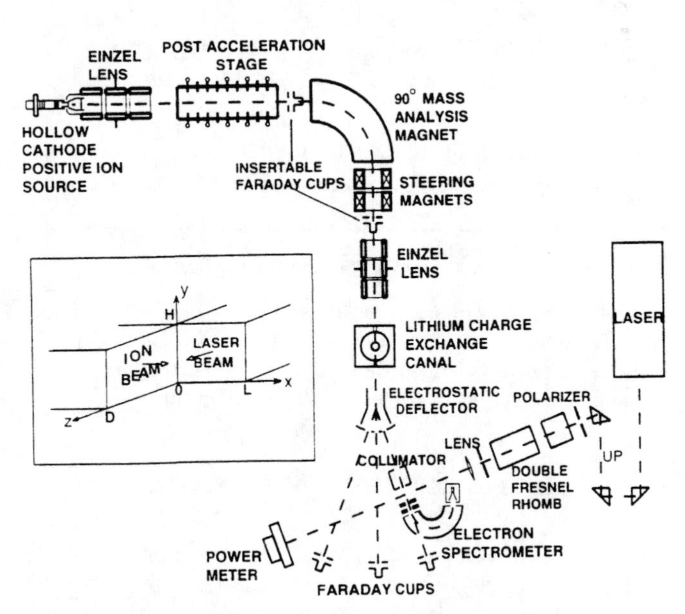

FIG. 1. A schematic of the crossed-laser-ion beam apparatus. Inset: A diagram of the interaction region showing the crossed laser and ion beams.

CP475, *Applications of Accelerators in Research and Industry*, edited by J. L. Duggan and I. L. Morgan

Photoelectrons detached in this region were collected in the forward direction (the direction of motion of the ion beam) and energy analyzed by use of a spherical sector spectrometer. Figure 2 shows typical spectra of photoelectrons detached from 80-keV beams of C^- and Ca^- ions at a photon energy of 2.08 eV. The electron yields were obtained by fitting the spectral peaks to Gaussian lineshapes. Corrections were made, when necessary, to account for counting losses associated with the detector dead time. The resolved peaks are kinematically shifted (they correspond to electrons ejected into the forward direction) and are associated with the photodetachment channels (1,2,3) shown in the energy level diagrams in the figures.

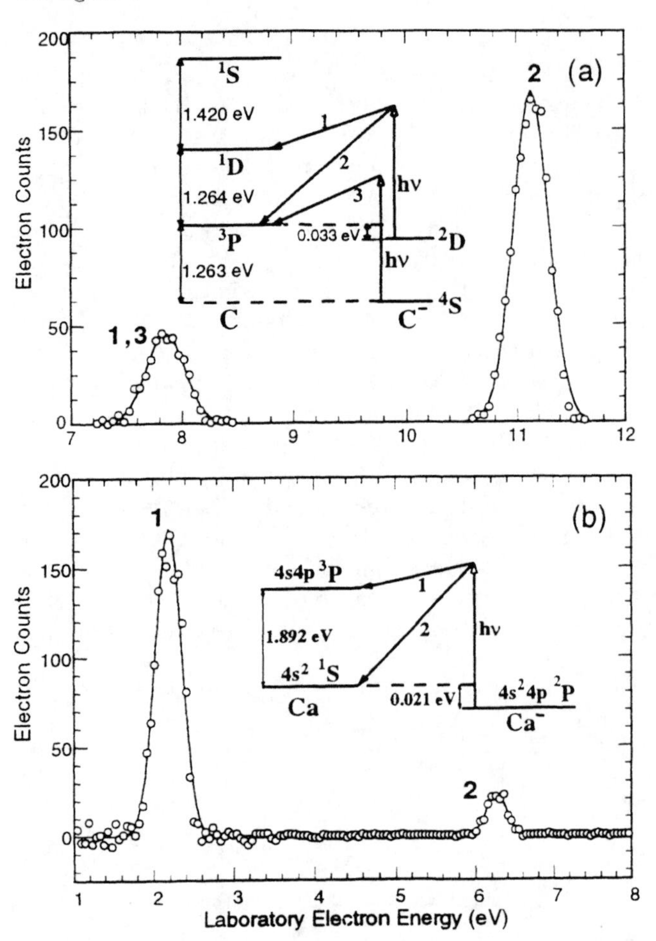

FIG. 2. Spectra of electrons ejected in the forward direction from a 80-keV beam of (a) C^- and (b) Ca^- ions using a photon energy of 2.08 eV. Peak 1, 2, and 3 correspond to the channels designated in the energy level diagram.

III. PHOTOELECTRON YIELD ANALYSIS

First we derive an expression for the expected yield of photoelectrons in the photodetachment process. In the analysis, the measured data is fit to this theoretical form. In a spatially and temporally uniform laser field of photon flux Φ, the probability p that an ion will undergo photodetachment when passing through the laser field is given by

$$p = 1 - e^{-\sigma\Phi\tau}, \tag{1}$$

where σ is the photodetachment cross section and τ is the interaction time which, in the present case, is determined by the transit time of the ion through the laser field. The transit time τ, which is typically a few nanoseconds, is much smaller than the laser pulse duration T_0 of several microseconds. Thus, the photon flux varies little temporally within the interaction time of each ion with the laser field. The ion does, however, experience a spatial variation of the photon flux during its passage through the interaction region. The photon flux can be written in terms of a spatial distribution function $g(x,y)$ and a temporal distribution function $f(t)$ in the following manner

$$\Phi(x,y,t) = \frac{W}{\hbar\omega} g(x,y) f(t), \tag{2}$$

where W represents the laser pulse energy and $\hbar\omega$ is the energy of each photon. The functions $g(x,y)$ and $f(t)$ are normalized according to: $\int_0^H \int_0^L g(x,y)dxdy = 1$ and $\int_0^{T_0} f(t)dt = 1$. Shown in Fig. 3 are the measured (a) temporal and (b) spatial distributions of the laser beam used.

One can use the photon flux expression above, to write an equation for the expected number of single photodetachment events, Z, per laser pulse

$$Z = \frac{1}{T_0} \int_{T=0}^{T_0} \int_{y=0}^{H} \int_{x=0}^{L} \rho D\{1 - exp[-\sigma \frac{W}{\hbar\omega} \int_T^{T+\tau} f(t) \times g(x+vt,y)dt]\} dxdydT, \tag{3}$$

where v is the velocity of the ions traveling in the x direction, D is the size of the interaction region in the z direction (see Fig. 1 inset) and ρ is the ion beam density, which is assumed constant. The spatial distribution of ions in the beam is, in general, Gaussian. We used a set of small apertures to select only those ions near the beam axis. In this manner we produced, to a good approximation, a beam whose ions are uniformly distributed in space in the interaction region.

The photoelectron yield, Y, defined in the present case as the total number of electrons detected for a given number of laser pulses, n, can be written as

$$Y = nZ \frac{F(\Delta\Omega,\alpha,\beta)}{4\pi} \eta, \tag{4}$$

where η represents the collection and detection efficiency of the electron spectrometer. The quantity $F(\Delta\Omega,\alpha,\beta)$, described in Ref. [7], is an angular function determined by the solid angle of collection $\Delta\Omega$, the asymmetry parameter β characterizing the photoelectron emission, and the polarization angle α defined as the angle between the laser polarization vector and the electron collection direction. Assuming a uniform photon flux Φ_0, Eq. (4) can be reduced in the unsaturated limit $\sigma\Phi\tau \ll 1$ to the form

$$Y = n\rho V\sigma\Phi_0\tau\frac{F(\Delta\Omega,\alpha,\beta)}{4\pi}\eta. \quad (5)$$

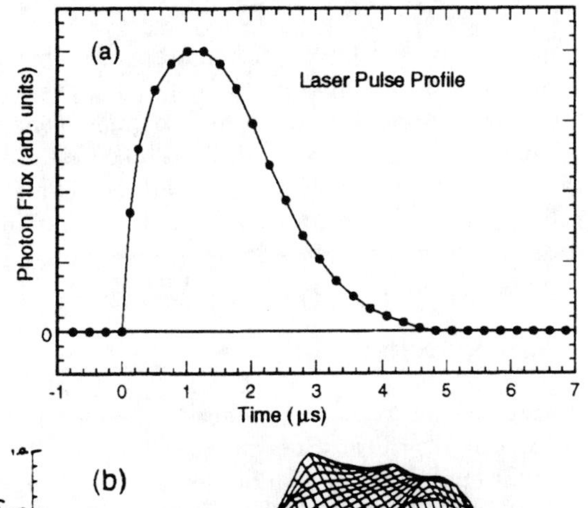

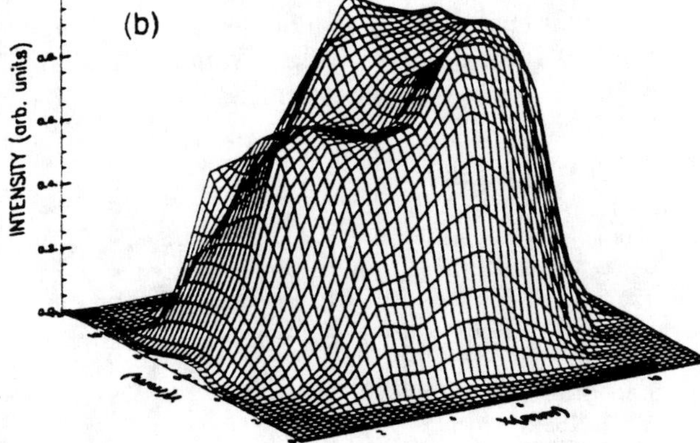

FIG. 3. Temporal (a) and spatial (b) profiles of the laser pulse used in the present work. The spatial profile was measured prior to focusing of the laser beam.

In general, photons in a laser pulse are non-uniformly distributed in space and time as shown in Fig. 3. The temporal distribution $f(t)$ was found to be a skewed Gaussian with a FWHM of about 2.2 μs. The spatial distribution $g(x,y)$, which was further selected by a 4-jaw slit, was found to be more complex but variations in intensity within the cross-sectional area of the laser beam in the interaction region were found to be less than $\pm 20\%$. For the present case of $\tau \ll T_0$, two different spatial functions $g(x,y)$ were tried in the analysis: the measured function and a uniform function. It was found that in this single photon experiment the photoelectron yield was relatively insensitive to the form of the spatial function. For example, when the measured and uniform functions were used, a difference in yield of less than 1% was obtained. For ease of analysis, we chose the simpler uniform spatial distribution function in the present work.

Assuming a uniform spatial function and the condition $\tau \ll T_0$, the electron yield of Eq. (4) can be simplified to

$$Y = n\rho V\frac{F}{4\pi}\eta\frac{1}{T_0}\int_0^{T_0}\{1 - exp[-\sigma\frac{W}{\hbar\omega A}f(T)\tau]\}dT, \quad (6)$$

where V is the geometrical interaction volume defined by the crossed laser and ion beams and A $(= LH)$ represents the cross-sectional area of the laser beam at the site of the interaction.

In order to extract the cross section from the yield of electrons recorded as a function of the measured laser pulse energy $\tilde{W}$ (in relative units, $W = \kappa\tilde{W}$), we can cast Eq. (6) in a more operational form

$$Y = b\frac{1}{T_0}\int_0^{T_0}\{1 - exp[-a\frac{\tilde{W}}{\hbar\omega}f(T)\tau]\}dT, \quad (7)$$

where a $(= \sigma\kappa/A)$ and b $(= n\rho V\eta F/4\pi)$ are two dimensionless parameters that can be obtained from the least-squares fit of the data to the above theoretical form. The quantity $\kappa\tilde{W}/\hbar\omega$ represents the true number of photons in a single laser pulse. The quantities κ, the proportionality constant of the laser power meter, and A, the cross sectional area of the laser beam, were not directly measured in the experiment. Instead we chose to eliminate their dependence by use of a reference ion, whose photodetachment cross section is known. Thus, curves representing the saturation of the production of photoelectrons in the photodetachment process were taken in pairs: one for the ion of interest X^- and the other one for the reference ion, which in the present experiment was D^-. The cross section of interest $\sigma[X^-]$ can be determined from a knowledge of the reference cross section $\sigma[D^-]$ and the corresponding fitting parameters $a[X^-]$ and $a[D^-]$ as follows: $\sigma[X^-] = (a[X^-]/a[D^-])\sigma[D^-]$.

IV. RESULTS AND DISCUSSION

Shown in Fig. 4 are typical saturation data measured for fast negative ions of D, C, Li, and Ca. As a test of the our apparatus and the saturation method we first determine the cross section for photodetachment of Li^- via the ^2Skp channel using D^- (photodetached via the ^2Skp channel) as the reference ion. At a photon energy of 2.08 eV, the D^- photodetachment cross section has been calculated to be 33.5(5) Mb [12]. Normalization of the measured Li^- data to the corresponding D^- data results in a Li^- cross section of 72(12) Mb. This value is in agreement with previous experimental [5] and theoretical [13–15] results.

The saturation method was then applied to the problem of determining the cross sections for photodetachment of $C^-(^2D)$ via $^3Pks,d$ channels and $Ca^-(^2P)$ via the ^3Pkp and $^1Sks,d$ channels. We again used the photodetachment of D^- via the ^2Skp channel as the reference. The measured cross sections, which were reported in Refs. [6,7], are listed in Table I along with the theoretical values.

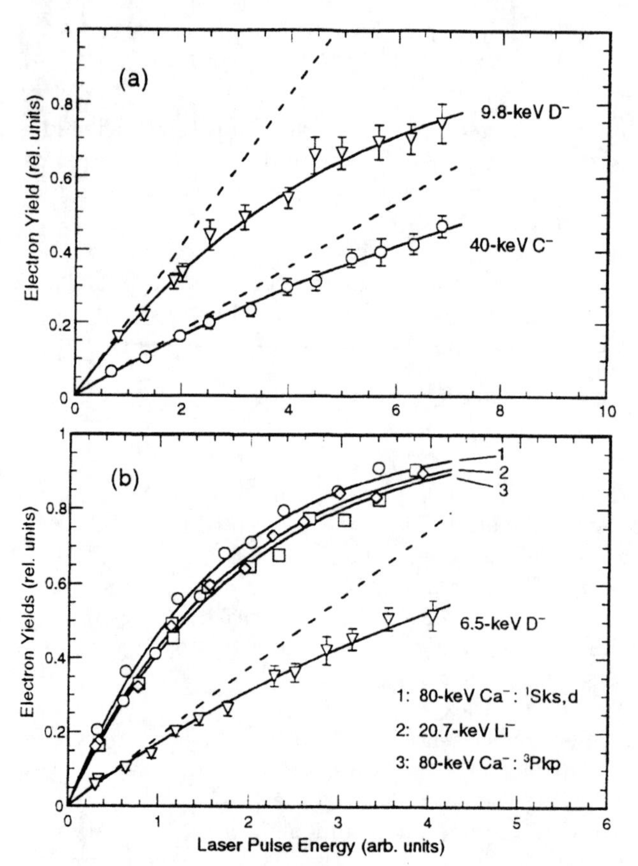

FIG. 4. Saturation data for D^-, C^- [for the channel 2 in Fig. 2-(a)], Li^-, and Ca^- [for both channel 1,2 in Fig. 2-(b)] measured at a photon energy of 2.08 eV and various ion beam energies [data symbols are, respectively, triangle in (a) and (b), circle in (a), diamond in (b), circle and square in (b)]. Solid curves are the best fits for the saturation data. For comparison, the dashed lines represent the dependence of electron yield on laser pulse energy if the photodetachment process were not saturated [as described by Eq. (5).]

TABLE I. Partial cross sections σ(Mb) for the photodetachment of the C^- and Ca^- ions at a photon energy 2.08 eV.

Ion	Transition	σ[Expt.]	σ[Theor.]
C^-	$^2D \rightarrow {}^3Pks,d$	13(2)[a]	13.8[b]
Ca^-	$^2P \rightarrow {}^3Pkp$	61(5)[a]	157[c], 109[d]
	$^2P \rightarrow {}^1Sks,d$	66(7)[a]	153[c], 107[d]

[a]This work.
[b]Reference [15].
[c]Reference [16] (length form).
[d]Reference [16] (velocity form).

The C^- result agrees with a recent R-matrix calculation of Ramsbottom *al.* [15]. In the case of the Ca^- results, the measured cross sections are about a factor of two smaller than theory at 2.08 eV. The measured ratio $\sigma(^1Sks,d)/\sigma(^3Pkp)$ of 1.08(14) is, however, in agreement with the corresponding theoretical ratios. The measurement of Kristensen *et al.* [17] at an energy of 1.39 eV agrees with the length form of the calculation of Yuan and Fritsche [16] but not with the velocity form. This measurement seems to have been made sufficiently far below the Ca(^3P) threshold that the opening of the ^3Pkp channel has a negligible effect on the $^1Sks,d$ channels. We can only conclude that the discrepancy between experiment and theory at 2.08 eV is due to an overestimation in the calculation of the degree of mixing of the $(^1Skd)^2D$ and $(^3Pkp)^2D$ final state channels in the vicinity of the Ca(^3P) threshold.

V. CONCLUSION

In summary, a crossed laser-ion beam apparatus coupled with electron spectrometer has been employed to measure partial photodetachment cross sections using a saturation method. This technique has been demonstrated in the cases of C^- and Ca^- photodetachment cross section studies. The primary advantage is that the partial cross section can be determined from the fitting parameter describing the degree of saturation alone.

ACKNOWLEDGMENTS

This research is supported by the US Department of Energy, Office of Basic Energy Sciences, Division of Chemical Sciences. We would also like to thank the Physics Division, Oak Ridge National Laboratory(ORNL) for the use of their facilities.

REFERENCES

[1] J.L. Hall, E.J. Robinson and L. M. Branscomb, Phys. Rev. Lett. **14**, 1013 (1965).
[2] For example, C. Blondel, R.J. Champeau, M. Crance, A. Crubellier, C. Delsart and D. Marinescu, J. Phys. B **22**, 1335 (1989).
[3] H. Stapelfeldt, C. Brink and H.K. Haugen, J. Phys. B **24**, L437 (1991).
[4] H. Hotop and W.C. Lineberger, J. Chem. Phys. **58**, 2379 (1973).
[5] J. Dellwo, Y. Liu, C.Y. Tang, D.J. Pegg and G.D. Alton, Phys. Rev. A **46**, 3924 (1992).
[6] D.H. Lee, W.D. Brandon, D.J. Pegg and D. Hanstrop, Phys. Rev. A **56**, 1346(1997).
[7] D.H. Lee, D.J. Pegg and D. Hanstrop, Phys. Rev. A **58**, 2121 (1998).
[8] P. Balling, C. Brink, T. Andersen and H.K. Haugen, J. Phys. B **25**, L565 (1992).
[9] N. Kwon, P.S. Armstrong, T. Olsson, R. Trainham and D.J. Larson, Phys. Rev. A **40**, 676 (1989).
[10] R.J. Champeau, A. Crubellier, D. Marescaux, D. Pavolini and J. Pinard, J. Phys. B **31**, 249 (1998); *ibid* **31**, 741 (1998).
[11] For example, M.V. Ammosov, N.B. Delone, M. Yu. Ivanov, I.I. Bondar, A.V. Masalov, Adv. At. Mol. and Opt. Phys. edited by D. Bates and B. Bederson, Academic Press, 1992, Vol.29, p33 (and references therein).
[12] P. Decleva, A. Lisini and M. Venuti, J. Phys. B **27**, 4867 (1994).
[13] D.L. Moores and D.W. Norcross, Phys. Rev A **10**, 1646 (1974).
[14] R. Moccia and P. Spizzo, J. Phys. B **23**, 3557 (1990).
[15] C.A. Ramsbottom, K.L. Bell and K. A. Berrington, J. Phys. B **27**, 2905 (1994).
[16] J. Yuan and L. Fritsche, Phys. Rev. A **55**, 1020 (1997).
[17] P. Kristensen, C.A. Brodie, U.V. Pedersen, V.V. Petrunin and T. Andersen, Phys. Rev. Lett. **78**, 2329 (1997).

Laser Induced Radiative Recombination with Polarized Photons *

E. Justiniano[†], G. Andler[‡], P. Glans, M. Saito[#], W. Spies, W. Zong, and R. Schuch

[†]Department of Physics, East Carolina University, Greenville, NC 27858-4353, USA.

[‡]Manne Siegbahn Laboratory, Stockholm University, S-104 05 Stockholm, Sweden.

Department of Atomic Physics, Stockholm University, S-104 05 Stockholm, Sweden.

[#]Laboratory of Applied Physics, Kyoto Prefectural University, Kyoto 606, Japan.

Laser induced radiative recombination (LIR) to the $n = 3$ states of deuterium was studied at the storage ring CRYRING. The aim of this work is to better understand the origin of the below threshold gain observed in such experiments. A beam of 23 MeV/amu D^+ ions was stored and phase-space cooled in the storage ring. The rate of radiative recombination into the $n = 3$ states of deuterium was enhanced over the spontaneous rate by overlaping linearly polarized laser radiation from a one-stage optical parametric oscillator (OPO) system to the ion-electron interaction region in the cooler. To probe for directional effects in the below threshold gain we rotated the polarization direction of the laser with respect to the direction of an induced transverse electric field obtained by displacing the center of the electron beam from that of the ion beam.

Introduction

A substantial number of advances in atomic physics in this decade have been due to the development of inovative techniques for trapping, storing, and cooling ions and atoms. Heavy-ion cooler rings have provided a uniquely well suited environment to study several aspects of electron-ion recombination [1, 2]. The detailed knowledge of electron-ion recombination reactions and their rates is needed for the modeling of astrophysical and laboratory plasmas [3, 4]. From the basic science point of view, the study of these phenomena yields knowledge on the structure and decay modes of many-electron systems. The information, thus obtained, provides stringent experimental benchmarks with which to gauge our understanding of relativistic and quantum-electrodynamical corrections as well as of the role played by electron-electron correlations.

In the electron cooler of a storage ring, positive ions and electrons may spontaneously recombine in a binary collision where a free electron is captured to a bound state in the ion, while a photon is simultaneously emitted balancing energy and momentum in the reaction. This process, radiative recombination (RR), may be regarded as the time inverse of photoionization [5, 6]. If these reactions take place in an intense laser field, recombination rates into specific final states of the ion may be enhanced by stimulated photon emission leading to a gain for the rate of stimulated population of a given state over the spontaneous reaction rate. Experimental evidence for this process, known as laser induced recombination (LIR),is still relatively recent [7, 8, 9]. LIR is a resonant process, as the laser wavelength must match the corresponding wavelength of the free-bound RR transition. One outcome of LIR experiments has been the observation of below-threshold gain [7, 10, 11] which was originally attributed to the Stark states in the continuum induced by external fields. However, this explanation was found not to be entirely satisfactory.

The region of the LIR gain spectrum exhibiting below-threshold gain is expected to be influenced by three body recombination and by collective screening effects. Below-threshold gain may also be related to the enhancement recently found in RR studies that have yielded measured recombination rates at very low relative velocities that are much larger than what theory predicts [12, 13]. This puzzling phenomenon has drawn much attention both theoretically [14, 15, 16, 17] and experimentally [18, 19, 20, 21]. For ions with a complex electronic structure, such as Ar^{13+}[18], Au^{28+} [12], and Pb^{53+}[22], most of the enhanced rates may be attributed to DR resonances occuring at very low relative energies. However, for bare ions DR is not possible and the enhanced rates found for these ions still poses an open and challenging problem. Schramm *et al.* [23] have recently reported evidence that for C^{6+} ions, LIR shows a similar enhancement to that of spontaneous RR at very low detuning energies of the merged ion and electron beams.

*Supported by the Swedish Natural Science Research Council (NFR).

CP475, *Applications of Accelerators in Research and Industry*,
edited by J. L. Duggan and I. L. Morgan

All of this indicates the importance of fully understanding the LIR below threshold intensity.

In order to find further evidence for the origin of the below threshold gain we have investigated LIR into the $n = 3$ states of deuterium using polarized laser radiation from a one-stage optical parametric oscillator (OPO) system and intentionally enhancing the space-charge field to detect the influence of possible directional effects in the below-threshold gain. The results show that the energy resolution is limited by the broad linewidth of the OPO. However, if we account for the linewidth our results are in good agreement with a theoretical model based on an average-field induced threshold shift due to the space-charge of the electron beam.

Experiment

For the most part, the experimental setup used here has been described before [11, 25] and we will only highlight some of its features. The experiments were performed with the CRYRING storage ring at the Manne Siegbahn Laboratory of Stockholm University. The ions, produced from a Penning ion source, were injected into the ring, after pre-acceleration to 300 keV/amu in an RFQ, and accelerated to 23 MeV/amu prior to storage. During electron cooling, the ions were merged over an effective length of $\ell = 0.8$ m with a velocity matched electron beam, confined by a solenoidal magnetic field of 0.03 T to a diameter of 40 mm. Figure 1 shows a schematic view of the the section of the ring where this experiment takes place. The electron densities were about 2×10^7 particles/cm^3 and the electrom beam temperatures, were $kT_\perp = 10 - 20$ meV (transverse) and $kT_\parallel = 0.10 - 0.15$ meV (longitudinal).

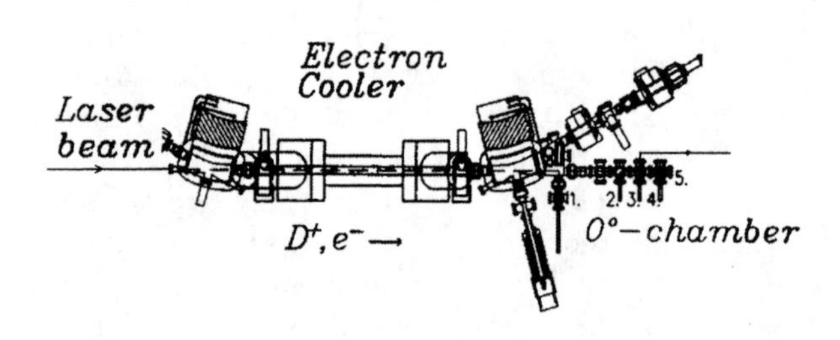

Figure 1: The section of CRYRING where the experiment takes place. The scrapers at 1 and 2 are used to overlap the laser beam with the ion and electron beams. The quartz crystal used to deflect the laser beam out of the vacuum is at 3. The surface barrier detector is at 4, and at 5 a position sensitive microchannel plate detector is available for beam diagnostics.

The laser beam was tuned to laboratory wavelengths in the range 644.7–662.6 nm and aligned to overlap with the co-moving ion and electron beams at the cooler with the help of two horizontal and two vertical scrapers. Because of the Doppler shift, these wavelengths translate into a scan (827.6–805.3 nm) about the $n = 3$ threshold wavelength of 820.36 nm [26]. The laser beam was produced by a Spectra-Physics Quanta-Ray MOPO-710 where a BBO crystal is pumped by the 355 nm third harmonic of a Spectra-Physics Quanta-Ray GCR 270-30 Nd:YAG. The laser beam was focused such that, at the cooler, its cross-sectional area was ~ 0.06 cm^2 and a weak reflection of the beam prior to entering the ring was used to trigger a fast photodiode. The laser intensity is estimated at around 80 MW/cm^2 in the interaction region. Since our OPO consists only of a power oscillator, its linewidth is relatively broad (~ 2 nm), and wavelength dependent [25].

The rates of recombined ions were measured at the exit of the first bending magnet after the electron cooler by a surface barrier detector (SBD)with unity detection efficiency [21]. To reach the SBD the ions had to traverse a thin (400 μm) quartz crystal used to deflect the laser beam out of the vacuum system. A time-to-amplitude converter (TAC), started by the pulses from the photodiode and stopped by the signals from the SBD, was used to determine coincidences between the detected recombined particles and the laser pulses. Thus, laser induced recombination transitions could be separated from their spontaneous counterparts.

In this experiment the partly polarized radiation of the high-intensity laser radiation from the OPO system was completely linearly polarized using a Glan polarizer. The direction of the electric vector ($\vec{\epsilon}$) could be chosen with a polarization rotator. A well defined direction of the external electric field from the space charge of the electron beam was selected by displacing the center of the electron beam by about 3 mm horizontally. This induces a transverse electric field of about 17 V/cm. The polarization direction of the laser photons was then rotated in steps of 45° relative to this electric field.

Results

The LIR gain is defined as the ratio between the induced rate (R_n^{ind}), with the laser on, and the spontaneous RR rate (R^{spo}), with the laser off. We have calculated theoretical gain curves using the following expression:

$$G_n(\varepsilon) = \frac{R_n^{ind}}{R^{spo}} = \sum_{\ell=0}^{n-1} G_{n\ell}^0 \int_{\phi=0}^{\pi} \int_{\theta=0}^{2\pi} \sigma_{n\ell}(\chi) \sin\phi \cdot \int_{E_\gamma = E_0 - \varepsilon_{sp} \sin\frac{\psi}{2}}^{\infty} f(\varepsilon_t, \phi, \theta) g(E_\gamma)\, dE_\gamma\, d\phi\, d\theta \quad (1)$$

where ε is the energy with respect to the threshold energy (E_0), E_γ corresponds to the photon energy, $\sigma_{n\ell}(\chi)$ is an anisotropy parameter obtained from the differential photoionization cross-section, $f(\varepsilon_t, \phi, \theta)$ describes the flattened velocity distribution of the electrons ($kT_\perp = 20$ meV and $kT_\parallel = 0.15$ meV were used), and $g(E_\gamma)$ is a normalized Gaussian which accounts for the linewidth of the laser. χ denotes the angle between the polarization vector and the direction of the electric space-charge field. The angles χ and ν can be expressed in terms of θ and ϕ, and $\varepsilon_t = E_\gamma - E_0 + \varepsilon_{sp} \sin \frac{\nu}{2}$ where ε_{sp} accounts for the fact that an external electric field lowers the ionization threshold [7]. The factors $G^0_{n\ell}$ contain the R^{spo} part and some other constants. These factors also include the laser intensity. However, since the laser intensity is well-above the saturation level, the saturation intensities for the 3s, 3p, and 3d levels are used.

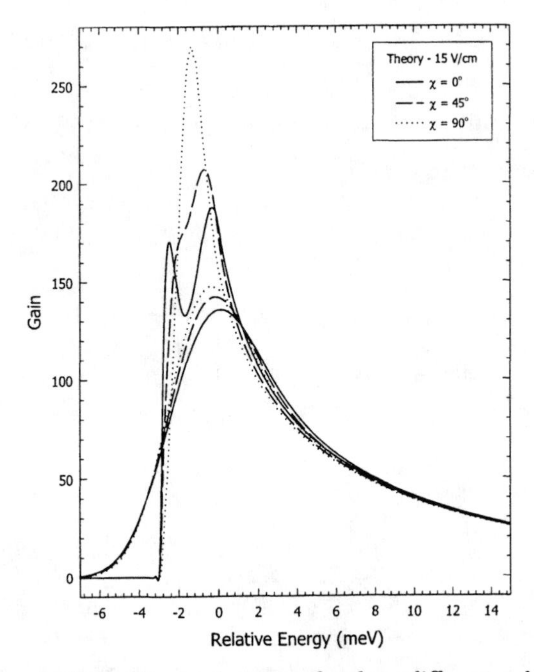

Figure 2: Theoretical gain curves for three different polarization angles (χ) before convolution with the laser bandwidth of the OPO (the lines with a sharp onset at -3 meV) and after convolution (lines with a smooth onset begining at -7 meV).

In Figure 2 theoretical gain curves for an assumed electric field of 15 V/cm at three different polarization angles, χ are shown. The three curves having a sharp onset at about -3 meV were obtained assuming a narrow laser bandwidth whereas the other curves show the gain after convolution with the actual bandwidth of the OPO laser. In the case of narrow laser bandwidth our theory predicts a significant dependence on the angle χ, with a double-peak structure for $\chi = 0°$. However, after convoluting with the actual bandwidth of the OPO the curves are almost identical.

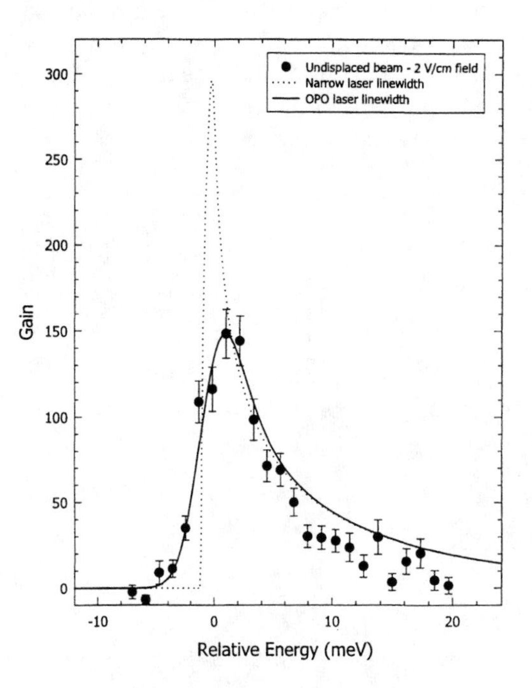

Figure 3: Gain spectrum obtained for an undisplaced electron beam leading to an applied field of 2 V/cm. Theory curves calculated following (1) for this value of the external field.

Two sets of gain curves were measured: one with the ion beam in the center of the electron beam and one with the ion beam displaced 3 mm from the center of the electron beam. In either case, the induced recombination appear in a narrow time window of the time (TAC) spectra, whereas the spontaneous recombinations are distributed homogenously over the full TAC spectra. The experimental gain curves are obtained from the following equation:

$$G_n(E) = \frac{R_n^{ind}}{R^{spo}} = \frac{(N - \bar{N_b}\Delta C)/T}{\bar{N_b}/\delta t} \qquad (2)$$

where N is the number of counts in an interval ΔC around the induced time peak, $\bar{N_b}$ is the averaged number of counts in the spontaneous background, δt is the width of one time channel, and T is the time during which the laser intensity stays above saturation. The latter quantity is not well known, and a value was chosen for each set of data to give maximum gain amplitudes in approximate agreement with theory.

In Figure 3 – no displacement – an electric field ($\vec{F}$) of 2 V/cm was used in the theory. This value was obtained in a similar, earlier study at CRYRING [11]. In Figure 4 the experimental results for four different polarization angles, χ are shown. The lines are the calculated gain curves from fig. 2 for polarization angles of 0° and 90° and an electric field of 15 V/cm which corresponds to a displacement of 3 mm of the centers of the beams.

The agreement between our experimental data and the theoretical model is good. However, from our results it is clear that the both the statistics and the bandwidth has to be improved in order to reveal if the gain is polarization dependent as our theoretical model predicts.

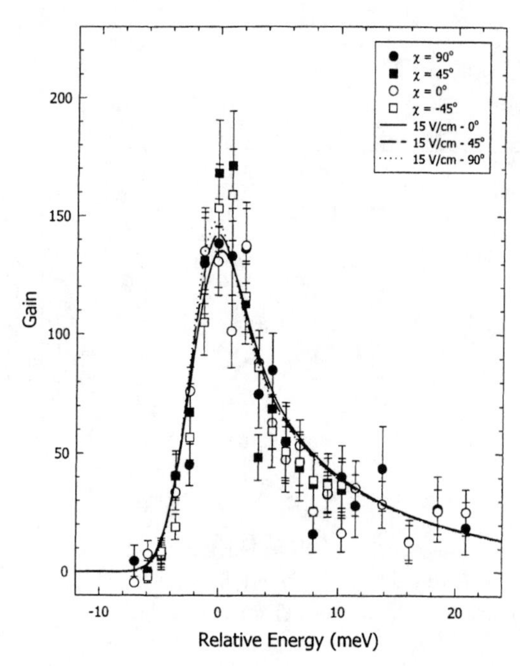

Figure 4: Energy dependence of the measured gain for four different polarization angles (χ) compared to calculated gain curves for polarization angles of $0°$ and $90°$.

Summary

We have presented a measurement of laser induced recombination into the $n = 3$ states of deuterium using linearly polarized laser radiation. By displacing the center of the electron beam in the cooler with respect to that of the coasting ion beam we were able to arbitrarily control the external electric field present in the interaction region and varied the angle between the polarization vector of the laser light and the external electric field. We find that although theory would predict a clear dependence of LIR on this angle, the combination of the relatively broad linewidth and the experimental statistics does not allow its direct observation yet. One of the authors (EJ) acnowledges partial financial support from the Royal Swedish Academy of Sciences.

References

[1] R. Schuch, in *Review of Fundamental Processes and Applications of Atoms and Ions*, C.D. Lin ed., World Scientific, Singapore, 1993, and references therein.

[2] M. Larsson, Rep. Prog. Phys. **58**, 1267 (1995).

[3] K.P. Kirby, Phys. Scr. **T59**, 59 (1995).

[4] H.P. Summers and W.J. Dickson *Applications of Recombination*, NATO ASI Series B, Physics, (New York: Plenum Press) (1992).

[5] M. Stobbe, Ann. Phys., Lpz. **7**, 661 (1930).

[6] H.A. Bethe and E. Salpeter, *Quantum Mechanics of One- and Two- Electron Atoms*, Plenum, New York, 1977.

[7] U. Schramm, *et al.*, Phys. Rev. Lett. **67**, 22 (1991).

[8] F. B. Yousif, *et al.*, Phys. Rev. Lett. **67**, 26 (1991).

[9] S. Borneis, *et al.*, Phys. Rev. Lett. **72**, 207 (1994).

[10] U. Schramm, *et al.*, Hyperfine Interactions **99**, 309 (1996).

[11] S. Asp, *et al.*, Nucl. Instrum. Methods B **117**, 31 (1996).

[12] A. Müller, *et al.*, Phys. Scr. **T37**, 62 (1991).

[13] A. Wolf, *et al.*, Z. Phys. D **21**, S69 (1991).

[14] Y. Hahn and P. Krstic, J. Phys. B: At. Mol. Opt. Phys. **27**, L509 (1994).

[15] Y. Hahn and J. Li, Z. Phys. D **36**, 85 (1996).

[16] M. Pajek and R. Schuch, Hyperfine Interactions **108**, 185 (1997).

[17] G. Zwicknagel, *et al.*, Hyperfine Interactions **99**, 285 (1996).

[18] H. Gao, *et al.*, Phys. Rev. Lett. **75**, 4381 (1995).

[19] H. Gao, *et al.*, Hyperfine Interactions **99**, 301 (1996).

[20] U. Schramm, *et al.*, Hyperfine Interactions **99**, 309 (1996).

[21] H. Gao, *et al.*, J. Phys. B: At. Mol. Opt. Phys. **30**, L499 (1997).

[22] S. Baird, *et al.*, Phys. Lett. B **361**, 184 (1995).

[23] U. Schramm, *et al.*, Abstracts of Contributed Papers XX ICPEAC, vol.2, TU144 (1997).

[24] D.J. McLaughlin and Y. Hahn, Phys. Rev. A **43**, 1313 (1991).

[25] E. Justiniano, *et al.*, Hyperfine Interactions **108**, 283 (1997).

[26] C.E. Moore, *Atomic Energy Levels*, Volume 1, NBS Circular 467, USGPO, Washington DC, 1949.

Reverse Energy-Pooling a in K-Na Mixture

J.O.P. Pedersen*

Ørsted Laboratory, Niels Bohr Institute for Astronomy, Physics and Geophysics, Universitetsparken 5, DK-2100 Copenhagen Ø, Denmark

Experimental rate coefficients for reverse heteronuclear energy-pooling collisions between excited potassium atoms and ground state sodium atoms are presented at thermal energies. The reactions are exothermic and very high rates were observed showing that reverse exothermic energy-pooling is an order of magnitude more efficient than the corresponding forward endothermic energy-pooling reactions.

INTRODUCTION

Collision processes where electronic excitation energy is transferred have been studied both theoretically and experimentally for decades. A particularly interesting process is energy pooling (EP) where two excited atoms collide and produce one highly excited atom and one ground state atom (1). In the experimental studies the colliding atoms are typically prepared in the excited state using optical excitation, and the highly excited atoms, populated by the energy-pooling collisions, are detected through their fluorescence.

The first reported study of heteronuclear energy-pooling was performed by Allegrini et al. (2) in a mixed vapor of potassium and sodium upon irradiation by two cw dye lasers tuned to the D resonance lines of sodium and potassium. The fluorescence from the high-lying levels of potassium and sodium was analyzed using an intermodulation technique to isolate the contribution due only to the heteronuclear processes.

This intermodulation technique has been used in the present work to study the inverse process of energy pooling: reverse energy-pooling (REP), in which an atom prepared in a highly excited state collides with an atom in the ground state, and the result of the collision are two excited atoms.

For experiments dealing only with atomic levels, the identification of REP by fluorescence detection is complicated in a homonuclear system, since the lower states can be populated both by the REP process and by radiative decay of the optically excited atom in the entrance channel. A heteronuclear system, on the other hand, offers the advantage that fluorescence from the atomic species not directly prepared by the laser excitation can be detected.

THE EXPERIMENT

Figure 1 shows a schematic overview of the setup. A capillary glass cell containing a K-Na alloy is heated, creating a mixed alkali-metal vapor in the cell. By propagating two laser beams through the capillary cell, highly-excited potassium atoms are produced, some of which will collide with ground-state sodium atoms and transfer a part of their excitation energy to the sodium atoms. The presence of excited sodium atoms gives rise to characteristic fluorescence light, which is collected with a lens, analyzed by a monochromator, and detected with a photomultiplier. The photomultiplier signal is amplified and filtered to extract the part stemming from the collisions between Na and the laser excited K atoms and finally stored on a PC.

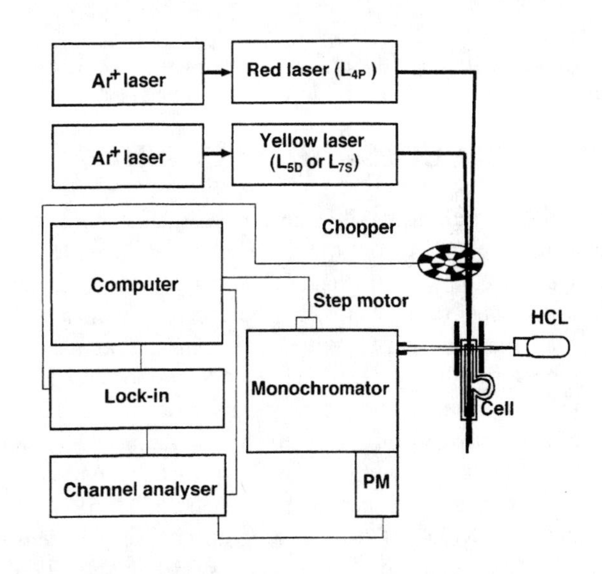

FIGURE 1. Experimental setup. HCL is a hollow-cathode lamp used for measuring ground-state densities and PM the photomultiplier.

In Figure 2 the level schemes of the involved atoms are shown. The excitation of the desired potassium levels is accomplished by two broadband dye lasers pumped by two

* Present address: Danish Center for Earth System Science (DCESS), Niels Bohr Institute for Astronomy, Physics and Geophysics, Juliane Maries Vej 30, DK-2100 Copenhagen Ø, Denmark

CP475, *Applications of Accelerators in Research and Industry*, edited by J. L. Duggan and I. L. Morgan

argon-ion lasers. The first dye laser, L_{4P}, is operating around the $K(4S_{1/2})$-$K(4P_{1/2})$ transition at 769.9 nm and the second, L_{5D} or L_{7s}, on the $K(4P_{3/2})$-$K(5D_{5/2})$ transition at 583.2 nm or the $K(4P_{1/2})$-$K(7S_{1/2})$ transition at 578.2 nm. These combinations are used because they give the best signal-to-noise ratio in the experiment, however all other combinations exciting the K(5D) and K(7S) states through the K(4P) state were attempted

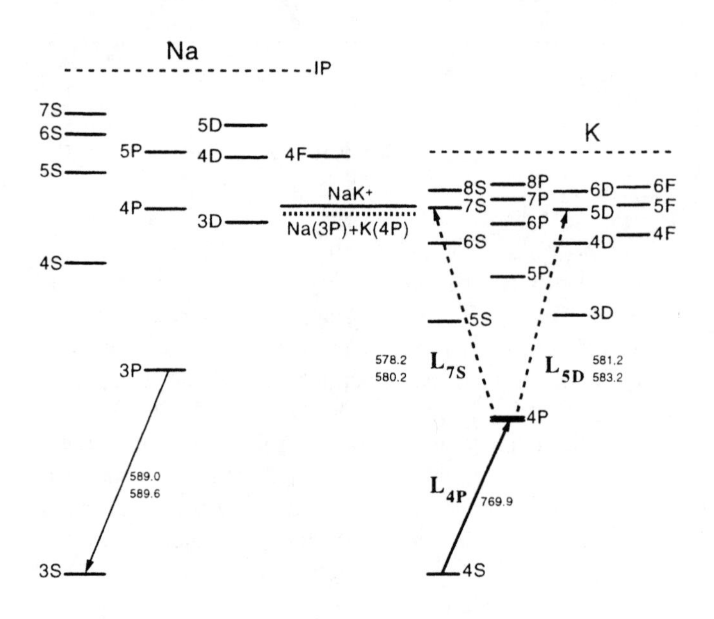

FIGURE 2. Sketch of the excitation and observation scheme. The energy of the exit channel is shown by the dotted line.

RESULTS AND DISCUSSION

A set of rate equations can be established for the two excitation schemes from which the rate coeficients for REP can be calculated. The relations include the ground-state densities and ratios between flourescence intensities for relevant transitions. The ground-state densities are measured by absorption spectroscopy using a K-Na hollow-cathode lamp. The intensity ratios necessary to evaluate the rate constants have been obtained by scanning the monochromator in the range (588.8-590.5 nm) around the $Na(3P_J)$-$Na(3S_{1/2})$ transitions and in the range (580.0-582.0 nm) around the K(7S,5D)-K(4P). Appropriate neutral density filters have been used since intensities differ by about three orders of magnitude.

Figure 3 shows the measured values of rate coefficients calculated from the rate equations. The rate constants contain contributions from both exit channels $K(4P_J)$ which cannot be separated in the experiment, because they are involved in the laser-excitation path. The rate constants for the two $Na(3P_J)$ exit channels are almost equal for the highest temperatures in both cases of excitation, while for decreasing temperatures the ratio between the rate constants for the $Na(3P_{3/2})$ and the $Na(3P_{1/2})$ exit channels increases slightly. In both cases of laser excitation a temperature dependence of the rate coefficients is observed, which, however, is more pronounced for 5D-excitation than for 7S-excitation.

In the case of 5D-excitation we obtain averaged rate coefficients at the mean temperature T= 465 K as 1.1×10^{-9} cm^3s^{-1} and 0.8×10^{-9} cm^3s^{-1} for the $K(4P_J) + Na(3P_{3/2})$ and $K(4P_J) + Na(3P_{1/2})$ exit channels, respectively. The cross sections, related to the rate coefficients by the interatomic mean velocity, are then 125×10^{-16} cm^2 and 95×10^{-16} cm^2, respectively. For 7S-excitation we obtain 2.2×10^{-9} cm^3s^{-1} and 1.4×10^{-9} cm^3s^{-1} for the $K(4P_J) + Na(3P_{3/2})$ and $K(4P_J) + Na(3P_{1/2})$ channels, respectively. The related cross sections are 270×10^{-16} cm^2 and 175×10^{-16} cm^2, respectively.

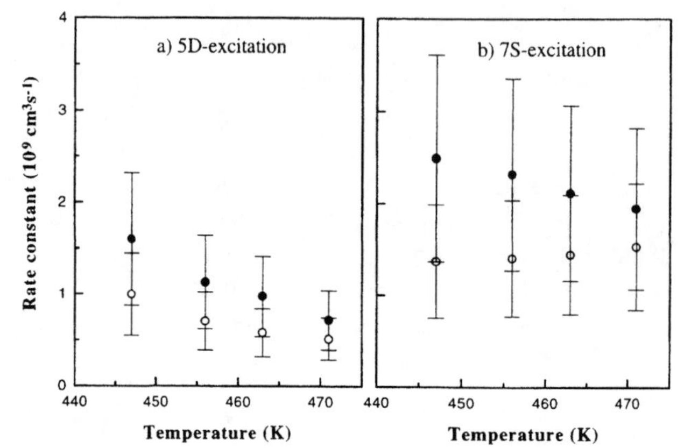

FIGURE 3. Measured rate costants for the REP processes for the 5D (left) and 7S (right) excitation. The open circles are rates for $Na(3P_{1/2})$ and the closed circles are for the $Na(3P_{3/2})$ population.

In (4) a Landau-Zener model was used to calculate the cross section for the K(5D)+Na(3S) reactions. The theoretical treatments did not include the fine-structure interaction in the calculation of the potential energy curves, and for this reason we can only compare the theoretical results with the total experimental cross section for the reverse energy pooling obtained by summing the cross sections for the two exit channels.

The result shows that the model reproduces the experimental result within a factor of two, which is a resonable result for a semiclassical calculation. The model also predicts a much larger cross section for the K(4P)+Na(3P) exit channel than for any other. Also this result supports our observation. An advantage of the present experiment is that dealing with such large cross section, the experiment can be performed at sufficiently low atom density to neglect the fine-structure mixing collisions.

The temperature dependence is relatively pronounced for the $K(4P_J) + Na(3P_{3/2})$ channels in the case of 5D-excitation and less strong for the channel $K(4P_J) + Na(3P_{1/2})$, while the dependence is much weaker for the two channels with 7S-excitation. Several effects can, as discussed in (4), cause a decrease of the rate coefficients with increasing temperature.

In Figure 4 previous results for EP in alkali metals (6) have been compared with the present reults. The rate constants are displayed versus the energy defect. The data fit into the general picture noted by Gabbanini et al (6) that the order of magnitude of the rate coefficients depends on the fundamental parameter $\Delta E/k_BT$, and in the form of an exponential decay function.

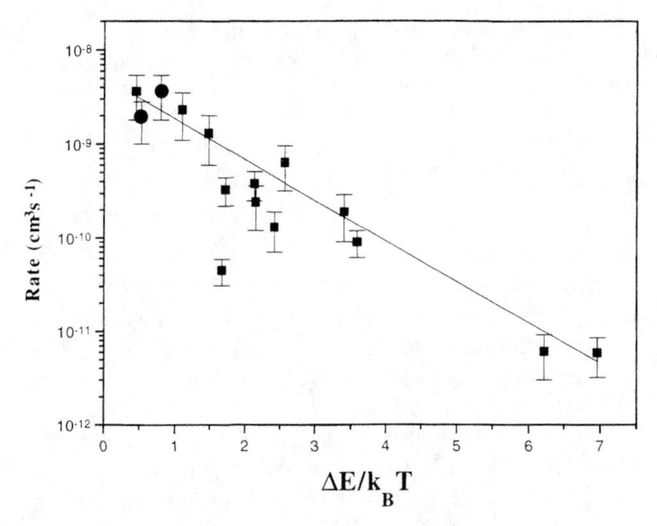

FIGURE 4. Plot of the rate constant experimental values vs. $\Delta E/k_BT$ for the exothermic cases. The points measured in this work are indicated with a circle and the line represents an exponential decay fit.

ACKNOWLEDGMENTS

This work is supported by the Danish Natural Science Research Council, Ib Henriksens Foundation (Copenhagen), and by the EEC-HCM program (contract number ERBCHRXCT 930344). The author thanks his coworkers in ref. (4) and also thanks the Carlsberg Foundation for financial support.

REFERENCES

1. For a recent review see Bicchi, P., Riv. Nouvo Cimento **20**, No. 7, 1 (1997).
2. Allegrini, M., Bicchi, P., Gozzini, S., and Savino, P., Opt. Commun. **36**, 445 (1981).
3. Gozzini,S., Abdullah, S.A., Allegrini, M., Cremoncini, A., and Moi, L., Opt. Commun. **63**, 97 (1987).
4. Guldberg-Kjær, S., De Filippo,G., Milosevic, S., Magnier, S., Pedersen, J.O.P., and Allegrini, M., Phys. Rev. A **55**, R2515 (1997).
5. De Filippo, G., Guldberg-Kjær, S., Milosevic, S., Pedersen, J.O.P., and Allegrini, M., Phys. Rev. A **57**, 255 (1998).
6. Gabbanini, C., Gozzini, S.,Squadrito, G., Allegrini, M., and Moi, L., Phys. Rev. A **39**, 6148 (1989).

Molecular Dissociation in Dilute Gas*

S. N. Renfrow[†], J. L. Duggan, and F. D. McDaniel

Ion Beam Materials Research Laboratory, Sandia National Laboratories, MS 1056, PO Box 5800, Albuquerque, NM 87185
Ion Beam Modification and Analysis Laboratory, Department of Physics, University of North Texas, Denton, Texas 76203

The charge state distributions (CSD) produced during molecular dissociation are important to both Trace Element Accelerator Mass Spectrometry (TEAMS) and the ion implantation industry. The CSD of 1.3 - 1.7 MeV SiN^+, $SiMg^+$, $SiMn^+$, and $SiZn^+$ molecules have been measured for elements that do not form atomic negative ions (N, Mg, Mn, and Zn) using a NEC Tandem Pelletron accelerator. The molecules were produced in a Cs sputter negative ion source, accelerated, magnetically analyzed, and then passed through an N_2 gas cell. The neutral and charged breakups where analyzed using an electrostatic deflector and measured with particle detectors. Equilibrium CSD were determined and comparisons made between molecular and atomic ion data.

INTRODUCTION

The study of ionic charge states dates back to the first of this century, and it continues to be an important topic. Scientists working on fusion research must have accurate data on charge changing cross sections. These data are necessary in the design of fusion chambers to prevent performance degradation of the plasma (1). Accelerator designers and users need charge state data to optimize ion beam production and the acceleration and transport of the beams. Other research areas where these data are important include: beam-foil spectroscopy (2), hyperfine interactions, and atomic physics experiments involving few-electron systems (3). Atmospheric effects following above ground nuclear blasts also are strongly dependent on charge state and charge changing cross sections (4). One of the most sensitive analytic techniques that is available today is Accelerator Mass Spectrometry (AMS) (5). At the University of North Texas, a Trace-Element AMS (TEAMS) system (6) has been developed as a complement to secondary ion mass spectrometry (SIMS). Accurate charge state data for both atomic and molecular ions is essential for the high sensitivities that are possible with these techniques. The goals of this measurement are to improve understanding of the theoretical models and to provide the ability to determine impurity levels without the use of standards.

Studies of molecular beams and the subsequent molecular dissociation have been ongoing since the mid 1970's. When molecular projectiles are incident upon a foil or gas target at MeV energies, most of the binding electrons of the projectile molecule are stripped off within the first few $\mu g/cm^2$ or monolayers of the target. The resulting charged nuclei rapidly separate due to their mutual Coulomb repulsion, converting their initial electrostatic potential energy into kinetic energy of the relative motion. This dissociation process has been termed a Coulomb explosion (7). Because of this fact, fast molecular-ion beams provide a unique source of energetic projectile nuclei that are correlated both spatially and temporally. The foil-induced dissociation of fast molecular ions has been used by several groups to provide new information about molecular-ion structures (8), the charge states of fast ions inside and outside solids (9,10), the interactions of such ions with the solid (11,12), as well as other atomic collision phenomena. More recent experiments with molecular beams include measuring lifetimes of metastable highly charged molecular ions (13,14) and stopping powers of molecules in solids (15).

The purpose of the present investigation is to measure the charge state distributions after molecular dissociation for a variety of atoms that do not form negative ions. This research is of direct interest in quantifying charge-state fractionization for use in Trace Element Accelerator Mass Spectrometry (TEAMS) and for purely fundamental investigation (6). The research will be used to accurately measure the impurity levels of atoms that do not form negative ions well, but atoms that form more abundant negative molecules (16). Negative atomic or molecular ions are needed for injection into a tandem accelerator. Furthermore, the data on molecular dissociation can be used to improve the understanding of SIMS Relative Sensitivity Factors (RSF's), which are conversion factors for impurity levels of different elements when compared to matrix elements.

EXPERIMENTAL DETAILS

In the Ion Beam Modification and Analysis Laboratory (IBMAL) at the University of North Texas, there are a variety of techniques and equipment available to perform fundamental atomic physics studies and material analysis. The IBMAL has a National Electrostatics Corporation (NEC) 9SDH-2 3 MV Tandem Pelletron accelerator which has the ability to accelerate a wide assortment of atomic or molecular ions. The source

CP475, *Applications of Accelerators in Research and Industry*,
edited by J. L. Duggan and I. L. Morgan

of negative ions by Cs sputtering (SNICS) was used in the present experiments. See Figure 1.

The molecule of interest was produced in the SNICS source from in-house produced cathodes. The ion beam was mv/q analyzed with the 30° and 90° magnets and the desired mass was injected into the accelerator. Here m,v and q are the mass velocity and charge of the ion, respectively. The beam passed though a set of electrostatic steerers and an einzel lens, which aligns and focuses the beam on to the terminal electron stripper canal.

The ion beam was accelerated to the terminal and stripped to the $q = 1+$ state in the 60 cm long nitrogen gas cell, which had a minimum of nitrogen gas to allow as many molecules to survive as possible. Negative atomic ions injected into the accelerator normally lose one or more electrons due to collisions with the molecules of the N_2 gas in the stripper channel. The N_2 pressure within the channel is of significant importance for the intensity of the accelerated ions. In the case of molecular ions, it was found that the N_2 pressure has to be roughly one order of magnitude smaller than for atomic ions to get any significant beam intensity. This is quite obvious because a large N_2 density increases the number of break up processes and internal excitations of the molecule which in turn can lead to disintegration. The low N_2 density necessary for a molecular beam represents a N_2 thickness that is much smaller than the thickness necessary to establish a charge state equilibrium for the molecular constituents. A consequence is that events where more than two electrons are stripped in a molecule-N_2 collision are rather scarce. The maximum cluster energy is thus limited to $2eV_T$ + the injection energy (eV_I), where V_T is the terminal voltage and V_I the injection voltage. Optimum molecular beam intensities were obtained when the N_2 stripper gas inlet was almost closed, so that a pressure of 6.2 x 10^{-7} torr was measured at the high energy end of the accelerator. It should be noted that the stripping processes for the molecular ions occur predominantly within the terminal. This conclusion can be drawn from the fact that a magnetic scan of the beam produced no satellite peaks in front of the molecular peak.

After exiting the accelerator, the molecular beam was focused with the electrostatic quadrupole. The $q = 1+$ molecular beam was magnetically mv/q analyzed into the 30° atomic beam line by the HVEC magnet to separate it from any atomic beams. The molecular beam was then passed through a second gas cell, with effective length of 14.4 cm containing nitrogen, which is pressure controlled by feedback from a capacitance manometer. The beam then passes through a normalization detector which consists of a 9 mm aperture followed by a 80 lines/inch gold mesh which has 85% transmission. A small portion of the beam is forward scattered off the gold into an annular particle detector. The rest of the beam passes through the detector's 13 mm hole. Each set of data was taken for the same number of normalization counts. The resultant molecular fragments were analyzed with the 3° ESA for their energy/charge. The resulting elemental fragment ions were measured in either a Faraday cup or a particle detector approximately 2 m from the 3° ESA. The ESA was scanned under computer control to obtain the E/q spectra.

A particle detector was also located 0.5 m from the exit of the 3° ESA; it was used to measure neutrals by setting the ESA to 30 kV which deflects all charged particles into the ESA plate. Another particle detector was 3 m from the exit of the 3° ESA to measure the charged fragments as the ESA is scanned. Negatively charged breakups products were scanned by reversing the polarity of the ESA but none where found for any of the breakups.

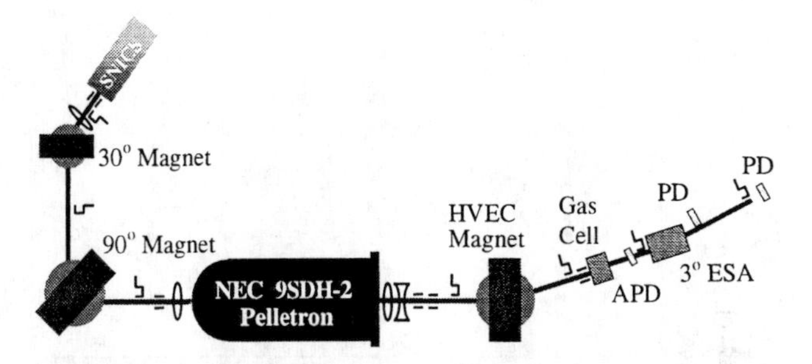

FIGURE 1. The schematic of the experimental setup used for charge state measurements. APD is the annular particle detector for normalization and PD is the particle detector for final charge state counting.

MEASUREMENTS

First, the equilibrium thickness of the N_2 gas was determined for AlO and SiZn. The thickness x of the gas target in $\mu g/cm^2$ is determined by the following equation (14).

$$x = \frac{mlP}{RT} \tag{1}$$

Where m is the target mass (amu), l is the effective gas cell length (cm), P is the gas pressure (torr x 133), R = 8.31 J/mol K, and T is temperature (K). From Kim's thesis (14), the gas cell effective length is 14.4 cm, which gives an N_2 thickness of 1.12 $\mu g/cm^2$ for a $P = 50x10^{-3}$ torr and $T = 288$ K.

The measured gas cell pressure needed to achieve the equilibrium charge state distribution for AlO is reached between 10 and 20 mtorr. At 20 mtorr, the nitrogen gas target thickness is 0.43 $\mu g/cm^2$. Figure 2 shows that the SiZn fragments reach equilibrium CSD at approximately 20 mtorr of N_2. Therefore, an N_2 gas pressure of 30 mtorr was used for all other CSD measurements, which corresponds to a gas target thickness of 0.65 $\mu g/cm^2$.

Figure 3 shows the charge state fractions of AlO, SiN, SiMg, SiMn and SiZn. The solid lines are data of the fragments and the dashed lines are of the atomic ion at the same velocity as the fragment. The dotted lines are a semi-empirical fit published by R.O. Sayer (17) for CSD of high energy atomic ions in dilute gases. The fitting parameters where originally for higher energy ions (0.04 to 10 MeV/amu), therefore the slight disagreement between the present data and Sayer model is expected.

As stated earlier, SiCa was to be measured, but data for SiCa could not be used because of interference from other undetermined molecules which completely concealed any SiCa signal. Also for the SiMg data a large beam of SiC_2 was present which adds uncertainty to the Si fragment data.

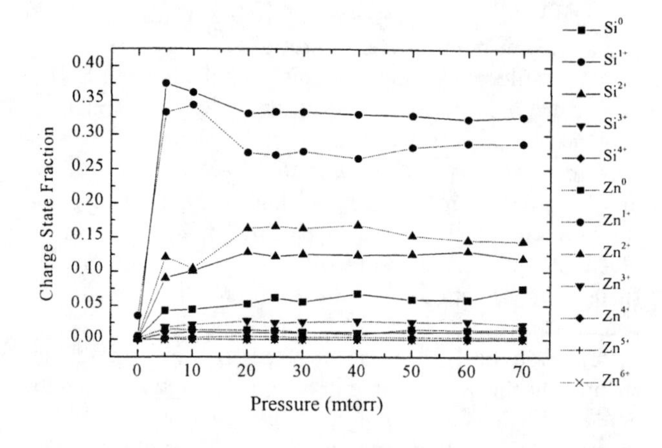

FIGURE 2. Charge state distributions as a function of N_2 gas cell pressure for the fragments of 1.32 MeV.

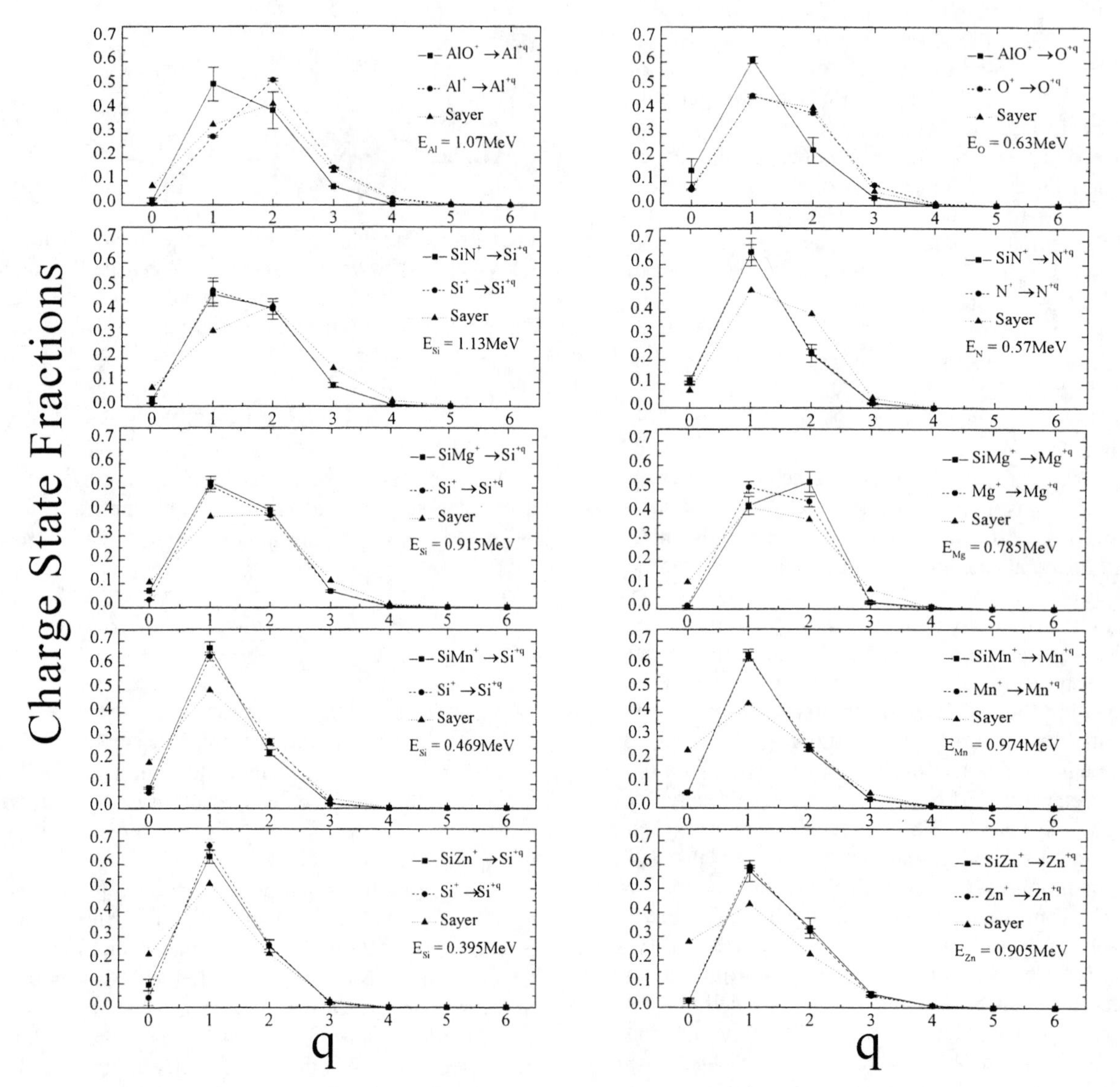

FIGURE 3. Equilibrium charge state fractions of AlO, SiN, SiMg, SiMn, and SiZn after breakup compared to atomic ions of the same elements.

RESULTS AND CONCLUSIONS

Measurements were made of the CSD resulting from molecular breakup and compared to that obtained for atomic ions at the same velocity. Elements (N, Mg, Mn, and Zn) chosen for study are those of interest for Trace Element Accelerator Mass Spectrometry (TEAMS) because they do not readily form negative ions necessary for injection into the tandem accelerator. AlO was chosen as an additional molecule to be studied because it was known to be more tightly bound than the SiX molecules and also it may exist in a 3+ charge state for many microseconds (14).

After production of the SiX^- or AlO^- molecules in the ion source and acceleration of the molecular ions to the terminal, they were stripped of electrons. Only the SiX^+ or AlO^+ molecules were selected after acceleration and passed into a 30° beam line for breakup and analysis of the charge states of the fragments. The SiX^+ or AlO^+ molecular ions were passed through an N_2 gas cell to simulate the conditions in the terminal of the accelerator during molecular breakup.

In order to make realistic comparisons to atomic charge state distributions, atomic ions of the fragments were produced in the terminal by electron stripping, e.g. Si^+, N^+, Mg^+, Mn^+, Zn^+, Al^+, and O^+. The terminal voltage was chosen so that the singly charged ions were accelerated out of the accelerator at the respective fragment ion velocity produced in the molecular breakup. Since the two fragments have the same velocity after breakup, the kinetic energy of the molecule is shared between the two fragments according to their masses.

The measured charge state fractions for SiX molecular fragments were found to be essentially the same as that measured for the atomic ions at the same velocity. Both charge state distributions peak at the same charge state and have approximately the same magnitude. The semi-empirical calculations of Sayer (17) predict lower equilibrium charge state distribution peaks than both the atomic ions and the equilibrium molecular fragment ions. The Sayer calculations predict a higher charge state distribution for some ions that may be due to the fact that the calculations were for higher energy ions.

The average charge state for each ion was determined by multiplying each charge state by the charge state fraction for that charge state. The average charge state difference for the atomic ions and the fragment ions was found to be negligible for the four SiX molecular breakups. For the AlO molecular breakup, the difference was 0.237 of a charge state for Al and 0.60 of a charge state for O. This difference may be due to the more tightly bound AlO molecule compared to the SiX molecules.

At present, TEAMS requires standards to analyze unknown impurity concentrations. The users of SIMS have developed Relative Sensitivity Factors (RSFs) for almost all-possible contaminants in Si, GaAs, and other semiconductor materials, for both ions and molecules. A TEAMS-RSF would use the SIMS-RSF (16) multiplied by a charge state factor and a gas cell attenuation factor as given in Equation 2.

$$\rho_i = \frac{I_i e^{n\sigma_i l} A_m CS_m}{I_m e^{n\sigma_m l} A_i CS_i} RSF \tag{2}$$

where ρ_i is the impurity atom density (atoms/cm^3); subscript m is for the matrix element; subscript i is for the impurity element; I is secondary ion intensity (counts/s); A is the isotope abundance fraction; CS is the charge state fraction; n is the gas number density (cm^{-3}); σ is the cross section for collision (cm^2); l is the path length in the gas (cm); RSF is the relative sensitivity factor (atoms/cm^3).

*Work supported in part by the National Science Foundation, the State of Texas Coordinating Board – Texas Advanced Technology Program, the Office of Naval Research, and the Robert A. Welch Foundation.

†Present Address: Ion Beam Materials Research Laboratory, Sandia National Laboratories, MS 1056, PO Box 5800, Albuquerque, NM 87185

REFERENCES

1. E. Nardi and Z. Zinamon, Phys. Rev. Lett. **49**, 1251 (1982).
2. D.J. Pegg, *Atomic Physics: Accelerators* (P. Richard ed.) Vol 17, 529-606 (Academic Press, 1980).
3. R. Marrus and P.J. Kohr, Adv. At. Phys. **14**, 181 (1978)
4. H.D. Betz, *Applied Atomic Collisions in Physics Vol 4 Condensed Matter* (S. Datz ed.), 2-42, (Academic Press, New York, 1983).
5. D. Elmore and F.M. Phillips, Science **236**, 543 (1987).
6. F.D. McDaniel, S.A. Datar, B.N. Guo, S.N. Renfrow, Z.Y. Zhao, and J.M. Anthony, Appl. Phys. Lett., 3008-3010 vol. 22 num. 23 (1998).
7. J. Remillieux, Nucl. Instr. Meth. **170**, 31 (1980).
8. D.S. Gemmell, Chem Rev. **80**, 301 (1980).
9. I. Plesser, E.P. kanter, and Z. Vager, Phys. Rev. A **29**, 1103 (1984).
10. A. Breskin, A Faibias, G Goldring, M. Hass, R. Kain, Z Vager, And N. Zwang, Phys. Rev Lett. **42**, 369 (1979).
11. Z. Vager and D.S. Gemmell, Phys. Rev. Lett. **37**, 369 (1976).
12. M.F. Steuer, D.S. Gemmell, E.P. Kanter, E.A. Johnson, and B.J. Zabransky, Nucl. Instr. Meth. **194**, 277 (1982).
13. I. Ben-Itzhak, I. Gertner, and B. Rosner, Phys. Rev. A **47**, 289 (1993).
14. Y.D. Kim, Accelerator Mass Spectrometery Studies of Highly Charged Molecular Ions, Dissertation 1994.
15. D. Ben-Hamu, A. Baer, H. Feldman, J. Levin, O. Heber, Z. Amitay, Z. Vager and D. Zajfman, Private communication July 31,1997.
16. R.G. Wilson, F.A. Stevie and C.W. Magee, *Secondary Ion Mass Spectrometry* (John Wiley & Sons, 1989).
17. R.O. Sayer, Rev. de Phys. App. **12**, 1543 (1977).

Secondary Electron Emission from SnTe(001) and KCl(001) Surfaces Induced by Specularly Reflected Protons

G. Andou, S. Ooki, K. Nakajima and K. Kimura

Department of Engineering Physics and Mechanics, Kyoto University, Kyoto 606-8501, Japan

The production process of secondary electrons at surface is investigated utilizing specular reflection of MeV protons at SnTe(001) and KCl(001) surfaces. The secondary-electron yield induced by 0.5-MeV protons specularly reflected from SnTe(001) is about 30 electrons/proton, while the yield is 100 - 160 electron/proton for KCl(001), indicating enhancement of the secondary-electron production process in front of insulator surfaces. The enhancement can be ascribed to the almost complete conversion efficiency of the excited surface plasmons into electron-hole pairs and the large band gap, which results in efficient production of free electrons over the vacuum level by single electron excitation process.

INTRODUCTION

Secondary-electron emission (SEE) induced by ion impact on solids has been extensively studied (1). There are two different mechanisms of the SEE: When the potential energy of the incident ion is larger than twice the work function of the target, potential electron emission (PEE) can occur. The other mechanism, called kinetic electron emission (KEE), is a process of direct transfer of kinetic energy from the incident ion to the target electrons. The mechanism of KEE is explained by a so-called three step model (2). The three steps are production of excited electrons over the vacuum level inside solids, transportation to the surface, and transmission through the surface barrier. Although these three processes are analyzed separately in theoretical study, the separation is usually difficult in experimental study. If the excitation of electrons takes place outside solids, the electron is ejected directly to the vacuum without other processes and the production process can be studied separately from other processes. This can be done by utilizing specular reflection of fast ions from single crystal surfaces. When a fast ion is incident on a single crystal surface with a grazing angle smaller than a critical angle [≈ 11 mrad for 0.5-MeV H^+ on SnTe(001)], the ion is reflected from the surface without penetration inside the crystal (3, 4). Secondary electrons induced by the ion are produced outside the crystal and not subject to the transportation and transmission processes. In the present paper, we report on the measurement of the secondary-electron yield induced by specularly reflected 0.5-MeV protons from SnTe(001) and KCl(001) surfaces. The production processes are discussed with particular emphasis placed on the surface-plasmon-assisted process.

EXPERIMENTAL

Details of the experimental procedure are described elsewhere (5). Briefly, A beam of 0.5-MeV protons from the 1.7-MV Tandetron accelerator was collimated by a series of apertures to less than 0.1 × 0.1 mm^2 and to a divergence angle less than 0.3 mrad. The beam was incident on target crystals at glancing angles θ_i of 1 - 7 mrad. The specularly reflected protons were detected by either a silicon surface barrier detector or a magnetic spectrometer. The secondary electrons induced by the protons were measured by a microchannel plate (MCP, effective diameter ϕ = 20 mm and placed at ~ 10 mm in front of the target surface) in coincidence with the reflected protons. The MCP was biased at + 500 to 700 V to collect all secondary electrons emitted from the target. The pulse height distribution of the MCP signals was registered by a pulse height analyzer.

A single crystal of KCl cleaved along (001) in air was mounted on a goniometer in a UHV scattering chamber. The surface was kept at 250°C during the measurement to prepare and maintain a clean surface (6) and also to avoid surface charging. A single crystal of SnTe(001) was prepared *in situ* by epitaxial growth on the KCl(001) at 250°C. Although the crystal structures of SnTe and KCl are the same (NaCl-type) and the lattice constants are almost the same (0.633 nm and 0.629 nm for SnTe and KCl respectively), electronic structures are different. SnTe is a narrow gap semiconductor with a band gap 0.19 eV, while KCl is a typical insulator.

RESULTS AND DISCUSSION

An example of the observed pulse height distribution of MCP signal measured in coincidence with the reflected proton is displayed together with the distribution measured without the proton beam in Fig. 1. The pulse height I of the MCP signal is proportional to the number of secondary-electrons detected (7). The signals measured without the

CP475, *Applications of Accelerators in Research and Industry*, edited by J. L. Duggan and I. L. Morgan

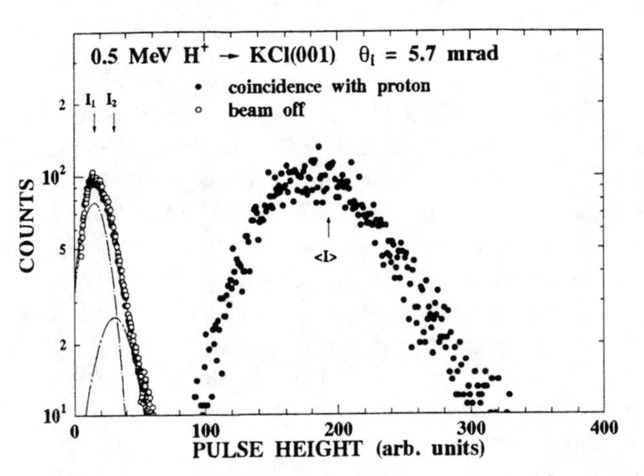

FIGURE 1. Pulse height distribution of MCP signals detected in coincidence with the reflected protons when 0.5-MeV protons are incident on KCl(001) at θi = 5.7 mrad. The distribuion measured without the proton beam is also shown.

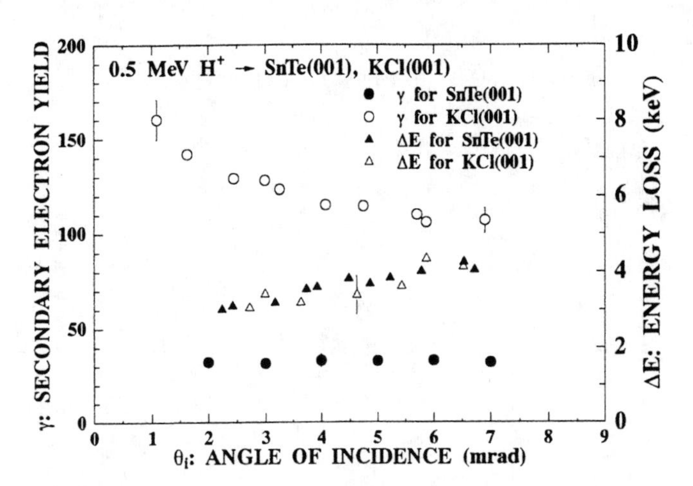

FIGURE 2. Secondary-electron yields produced during specular reflection of 0.5-MeV protons from SnTe(001) and KCl(001). Observed energy losses of the reflected protons are also shown for comparison.

proton beam correspond to detection of single electrons or pairs of electrons which are created by an ion pump evacuating the chamber. The secondary-electron yield γ (the mean number of secondary electrons produced by a single proton) can be derived from the observed distributions using the equation, $\gamma = \langle I\rangle/(\varepsilon I_1)$, where $\langle I\rangle$ is the mean value of the pulse height distribution, I_1 the pulse height for single electron detection and ε [≈ 0.6 (8)] the efficiency of the MCP.

Figure 2 shows the observed secondary-electron yield for both SnTe(001) and KCl(001) as a function of θ_i. The energy loss of the reflected proton is also shown for comparison. Although the observed energy losses are almost the same, the secondary-electron yield for KCl is 3 - 4 times larger than for SnTe. It is known that secondary-electron yield for insulator is enhanced, sometimes by one order of magnitude, compared to metals or semiconductors (9 - 13). This is attributed to the reduced electron scattering due to a large band gap and also a large escape probability due to a lower surface barrier. The present enhancement for KCl, however, cannot be ascribed to these effects, because the electrons observed here were produced outside the crystal. The observed enhancement indicates that the electron excitation process itself is enhanced in front of the insulator surface, although the electron excitation process in insulators has usually been assumed to be suppressed due to a large band gap (13). In order to see the origin of the enhancement of the secondary-electron-production process, more detailed information is extracted from the observed results.

Introducing a position-dependent secondary-electron production rate $P(x)$, i.e., the number of secondary electrons produced by a proton per unit path length travelling parallel to the surface at a distance x from the surface, the secondary electron yield is given by integrating $P(x)$ along the proton trajectory,

$$\gamma(\theta_i) = \int_{traj} P(x)\, dz, \tag{1}$$

where the trajectory lies in the x-z plane. Using the surface continuum potential $V(x)$ the equation of the trajectory is written as

$$\left(\frac{dx}{dz}\right)^2 = \frac{V(x_m(\theta_i)) - V(x)}{E}, \tag{2}$$

where $x_m(\theta_i)$ is the closest approach distance to the surface and E the proton energy. Substituting Eq. (2), Eq. (1) can be written as

$$\gamma(\theta_i) = 2\sqrt{E}\int_{x_m(\theta_i)}^{\infty} \frac{P(x)}{\sqrt{V(x_m(\theta_i)) - V(x)}}dx. \tag{3}$$

This is an integral equation of the Abel type and the solution is given by

$$P(x) = -\frac{1}{2\pi E}\frac{dV(x)}{dx}\{\gamma(0)\sqrt{\frac{E}{V(x)}} + \int_0^{\pi/2}\frac{d\gamma(\theta_i)}{d\theta_i}\Big|_{\theta_i=\sqrt{\frac{V(x)}{E}}\sin(u)} du\} \tag{4}$$

Using Eq. (4), the production rate can be derived from the observed secondary electron yield. The position-dependent

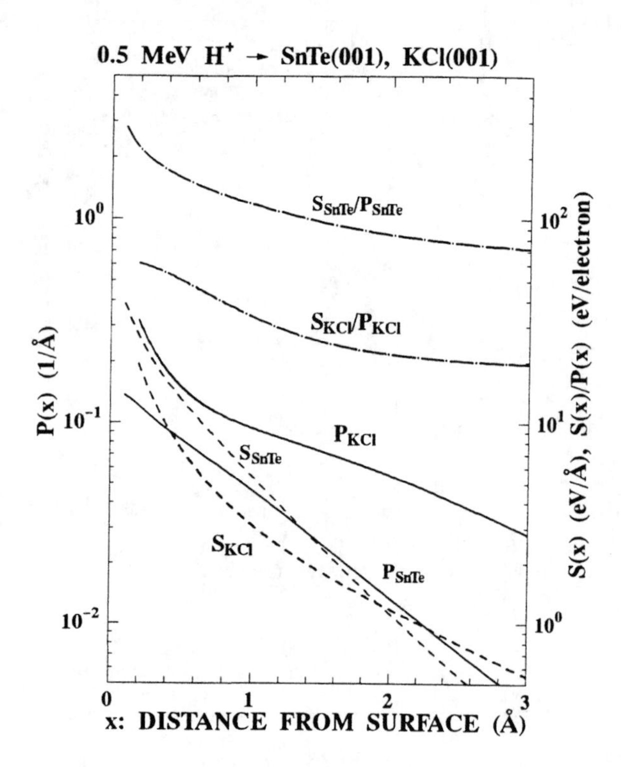

FIGURE 3. Position-dependent secondary-electron production rates P(x) for 0.5-MeV protons at SnTe(001) and KCl(001) surfaces. Position-dependent stopping powers S(x) and the ratios S(x)/P(x) are also shown for SnTe(001) and KCl(001).

stopping power $S(x)$ can be derived from the observed energy loss $E(\theta_i)$ using an equation similar to Eq. (4).

Figure 3 shows the obtained position-dependent secondary-electron production rates for 0.5-MeV protons on SnTe(001) and KCl(001) together with the position-dependent stopping powers. Both the production rate and the stopping power decreases with increasing distance from the surface. Although the secondary-electron production rate is usually assumed to be proportional to the stopping power in phenomenological theories (2), the observed production rate is not proportional to the stopping power for both SnTe and KCl. This is because the dominant process of the secondary-electron production depends on the distance x from the surface. While the surface-plasmon-assisted process is dominant at larger x, bulk plasmon process and single electron excitation process are dominant at small x. The corresponding energy losses for single electron emission are different for these processes.

The ratio of the stopping power to the secondary-electron production rate $S(x)/P(x)$, i.e., the corresponding energy loss for single electron emission, is also shown in Fig. 3. The ratio decreases with increasing x and approaches to constant values of 70 and 20 eV/electron at $x > 2$ Å for SnTe and KCl, respectively. In this large x region, surface-plasmon-assisted process is exclusively dominant for the secondary-electron production. The surface plasmon can decay either via electron-hole pair production or via photon emission. At an ideal smooth surface, decay of the surface plasmon into a photon is prohibited because the laws of energy and momentum conservation are not satisfied at the same time. Real surfaces, however, are not perfect and the conservation law of the transverse momentum is violated, which allows the decay of surface plasmons to photons (14). Thus the conversion probability of surface plasmons to secondary electrons depends on the surface conditions.

Using surface plasmon energies, 11 and 10 eV for SnTe and KCl (15), respectively, and assuming that half of the electrons produced by surface plasmon decay are ejected into the vacuum with the other half impinging into the solid, the conversion probabilities of the surface plasmons into electron-hole pairs are estimated to be ~ 30 and ~ 100 % for SnTe and KCl, respectively, indicating that the KCl surface is much smoother than SnTe. This is consistent with observations performed using atomic force microscopy. While a number of pyramidal hillocks were observed on the SnTe(001), wide terraces (several hundred nm wide) separated by atomic-height steps were observed on the KCl(001). The large enhancement of the secondary-electron production process for KCl(001) at large x is thus ascribed to the large conversion probability of the surface plasmon into an electron-hole pair. It should be noted that even at smaller x the secondary-electron production process is enhanced for KCl compared to SnTe. This might be explained by the small electron affinity of KCl (0.4 eV) which makes single electron excitation very efficient for the production of excited electrons over the vacuum level.

CONCLUSIONS

We have measured secondary electron yields induced by 0.5-MeV protons specularly reflected from SnTe(001) and KCl(001). The position-dependent secondary-electron production rate is derived from the observed yield as a function of the distance x from the surface. The probability of surface plasmon decay to an electron-hole pair is estimated to be ~ 30 % for SnTe(001) and ~ 100 % for KCl(001). The large difference in the obtained probabilities is ascribed to the surface conditions.

ACKNOWLEDGEMENTS

We are grateful to the members of the Department of Nuclear Engineering at Kyoto University for the use of the Tandetron accelerator.

REFERENCES

1. See, for example, M. Rösler and W. Brauer, in *Particle Induced Electron Emission I*, edited by G. Hohler and E.A. Niekisch, Vol. 122 of Springer Tracts in Modern Physics (Springer, Heidelberg, 1991), p. 1.
2. D. Hasselkamp, in *Particle Induced Electron Emission II*, edited by G. Hohler and E.A. Niekisch, Vol. 123 of Springer Tracts in Modern Physics (Springer, Heidelberg, 1991), p. 1.
3. G.S. Harbinson, B.W. Farmery, H.J. Pabst, and M.W.

Thompson, Radiat. Eff. **27**, 97 (1975).

4. K. Kimura, M. Hasegawa, Y. Fujii, M. Suzuki, Y. Susuki, and M. Mannami, Nucl. Instrum. Methods Phys. Res. B **33**, 358 (1988).
5. K. Kimura, S. Ooki, G. Andou, K. Nakajima, and M. Mannami, Phys. Rev. A **58**, 1282 (1998).
6. M. Prutton, *Surface Physics* (Clarendon, Oxford, 1983) p. 9.
7. T. Schenkel, A.V. Barnes, M.A. Briere, A. Hamza, A. Schach von Wittenau, and D.H. Schneider, Nucl. Instrum. Methods Phys. Res. B **125**, 153 (1997).
8. M. Galanti, R. Gott, and J.F. Renaud, Rev. Sci. Instrum. **42**, 818 (1971).
9. C. Baboux, M. Perdrix, R. Goutte, and C. Guilland, J. Phys. D **4**, 1617 (1971).
10. W. Krönig, K.H. Krebs, and S. Rogashewski, Int. J. Mass. Spectrum. Ion Phys. **16**, 243 (1975).
11. L.A. Alonso, R.A. Baragiola, J. Ferron, and A. Oliva-Florio, Radiat. Eff. **45**, 119(1979).
12. H.Jacobsson and G. Holmén, Phys. Rev. B **49**, 1789 (1994).
13. J. Schou, in *Ionization of Solids by Heavy Particles*, edited by R.A. Baragiola (Plenum Press, New York, 1993) p. 351.
14. H. Raether, *Surface Plasmons on Smooth and Rough Surfaces and on Gratings*, Vol. 111 of Springer Tracts in Modern Physics (Springer, Heidelberg, 1988).
15. A. Akkerman, A. Breskin, R. Chechik, and A. Gibrekhterman, in *Ionization of Solids by Heavy Particles*, edited by R.A. Baragiola (Plenum Press, New York, 1993) p. 359.L. Marton and L.B. Leder, Phys. Rev. **94**, 203 (1954).

Energy Loss of Heavy Ions in Solids at Energies Below 5 MeV/u

Annu Sharma, Shyam Kumar and A. P. Pathak*

Department of Physics, Kurukshetra University, Kurukshetra 136 119 INDIA
** School of Physics, University of Hyderabad, Hyderabad 500 046 INDIA*

A comparative study of various energy loss formulations has been made by comparing the calculated stopping power values with the corresponding experimental values for projectiles $6 \le Z \le 92$ in targets $4 \le Z \le 79$ in the energy region ~0.1-5.0 MeV/u. Some systematic and interesting trends have been observed in case of LSS theory. The limitations of TRIM and SRIM calculations have been demonstrated. Stopping power calculations using Hubert et al formulation have been extended successfully beyond its recommended range of validity i.e. 2.5 - 500 MeV/u down to energies as low as 0.5 MeV/u for projectiles $Z \le 29$ in targets upto copper.

INTRODUCTION

Passage of heavy ions through matter and understanding of various interaction processes continue to be of central importance in many branches of physics (1). In recent days, surface modification and characterization of materials and devices are being done using heavy ion beams (2,3). Interpretation of such data requires reliable and precise values of stopping power for heavy ions. Determination of stopping power at low energies poses more problems because of the complexity of the phenomenon of charge capture by projectile and in the absence of an analytical expression of effective charge as a function of projectile speed, the choice becomes limited to an empirical or semi-empirical effective charge parametrization. Experimentalists mostly rely on the existing semi-empirical formulations. This method is not fool proof and there is a continuous need to check the reliability of such formulations through comparison with the available experimental data. There are reports which indicate that the stopping power values for heavy ions deviate significantly from often used semi-empirical models (4,5).

PRESENT WORK

Over the past few years, our group is actively engaged in stopping power measurement experiments using 15 MV Pelletron accelerator facility at the Nuclear Science Centre, New Delhi, INDIA. We have covered several ion species from Z_1= 8-29 in carbon absorbers. The comparison with the calculations indicated some new and interesting trends (6). Our observations coupled with the availability of new experimental data (7-14) have motivated us to undertake a relative comparison of some widely used stopping power formulations over a broad spectrum of projectile-target combinations.

In the present work, a comparative study of different stopping power formulations e.g. LSS theory (15), Ziegler et al. formulation (16) and Hubert et al. formulation (17) has been made by comparing the calculated stopping power values with the corresponding experimental values as available in the literature (5-14, 18, 19) for a variety of projectiles such as C, N, O, Al, Ti, Cu, Ag, I, Xe, Au, Pb and U in various targets like C, Al, Cu, Ag and Au in the energy range ~0.1 - 5.0 MeV/u. The present study has been undertaken in order to check the validity and systematics of these formulations which in addition to their applicability may provide a scope for future research from both theoretical and experimental point of view.

RESULTS AND DISCUSSION

Experimental Validity of LSS Theory

At low ion velocities, $v \le v_0 Z_1^{2/3}$ (v_0 - Bohr's velocity, Z_1 atomic no. of the projectile ion), the LSS theory (15) predicted a linear trend with velocity for the electronic stopping power. It was considered appropriate to compare the experimental data with the LSS predictions using reduced ion velocities and reduced stopping power (20). The LSS values for one particular absorber then fall on a straight line intersecting the origin irrespective of the nature of the projectile ion when reduced stopping power is plotted against the reduced ion velocity. Figure-1 presents such a plot for various heavy ions in an aluminium absorber. The similar plots for carbon, copper, silver and gold absorbers are not shown due to paucity of space. The following systematic trends are observed. For C and Al absorbers, the LSS theory gives good agreement with the experimental stopping power data only within the narrow region of $0.6v_c$-$0.8v_c$ ($v_c = v_0 Z_1^{2/3}$) irrespective of the nature of the projectile. For $v < 0.6v_c$, it underestimates the stopping power and for $v > 0.8v_c$, it starts overestimating the stopping power. For Cu absorbers, the LSS predictions are generally in agreement with the experimental data. Exceptions are Al

CP475, *Applications of Accelerators in Research and Industry*,
edited by J. L. Duggan and I. L. Morgan

and Cu ions at velocities above $v \sim 0.8v_c$ and Au ions at velocities below $v \sim 0.4v_c$. For Ag and Au absorbers, the LSS theory considerably underestimates the corresponding experimental values. Exceptions are Cu ions at velocities $v > 0.6v_c$.

From the above except a few exceptions, it is obvious that for lower Z absorbers like C, Al etc.,the LSS predictions agree with the experimental data only in the limited ion velocity range (~ $0.6v_c$ to $0.8v_c$), for medium Z absorbers like Cu, the LSS values are normally in agreement with data and for high Z absorbers like Ag, Au etc., the LSS formulation significantly underestimates the experimental values; irrespective of the nature of the projectile ion.

Such an observation about the LSS theory is reported for the first time to the best of our knowledge. This trend is one of the main theoretical limitations of the LSS theory. It seems that there is a need to include the dependence of target atomic number in the LSS formula

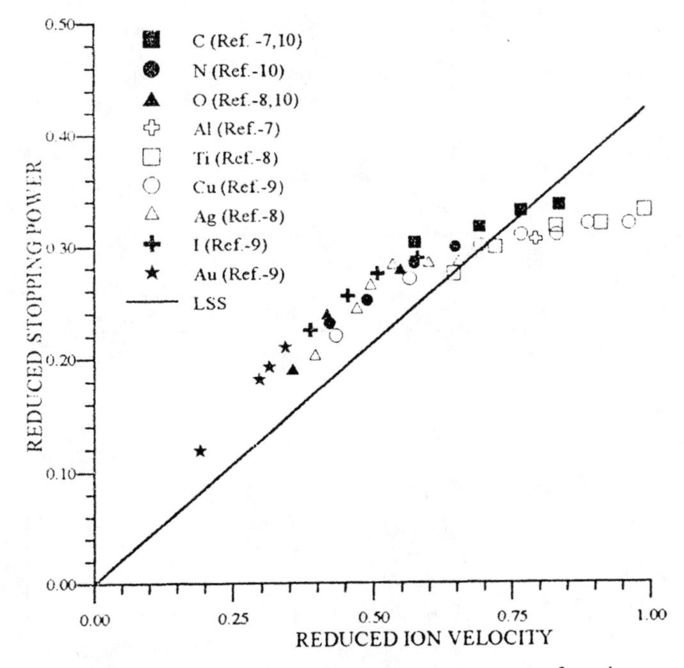

FIGURE 1. The reduced stopping power (in MeV cm^2 mg^{-1}) is plotted as a function of reduced ion velocity ($v_{red}= v / (v_0 Z_1^{2/3})$) in aluminium absorber. Solid line represents LSS values.

more accurately, since for the same projectile at the same velocity, the theory behaves differently in different absorbers.

Comparison with Calculations based on Ziegler et al. Formulation

Calculated stopping power values adopting Ziegler et.al. (16) formulation using TRIM-92 (21) provide a good overall agreement (normally within 10 %) with the experimental data for most of the projectile target combinations considered. Major exceptions are Cl, K, Ca, Sc, Ti, Cr, Mn and Fe ions at energies below 0.2 MeV/u in carbon absorber, where TRIM-92 significantly overestimates (as high as upto 25%) relative to the data (5).

In addition to the above, the following two major discrepancies as explicitly depicted in table-1 are observed in the case of TRIM-92.

1. For the same ion velocities, the calculated stopping power values around Xe show an oscillatory behaviour with maxima at Xe (table-1, Sr. No. 1-6). Such an unexpected behaviour leads to enormously high stopping power values for Xe projectiles as high as from 13% to 53% than the corresponding experimental values (18) in the energy range 0.23-0.41 MeV/u in carbon absorber (table 1, Sr. No. 7-9). Such a behaviour is dominant at lower energies.
2. For high Z projectiles like U, Pb, Xe, Kr etc. in the energy range ~3-5 MeV/u, calculations based on TRIM show an interesting trend. The calculated values considerably underestimate the experimental results (19) for low Z absorbers like Be, C, Al etc.; in contrast, calculations based on this formulation overestimate the experimental values for high Z absorbers like Au, U etc. This behaviour is clearly indicated in table-1 (Sr. No. 11-19). These deviations reduce with increasing projectile energies (4).

Calculated stopping power values using SRIM-96 code recently developed by Ziegler (22) are normally within a few percent than those obtained using TRIM-92 thus providing good overall agreement with the experimental data. In case of SRIM-96, the discrepancy mentioned at (1) above is removed whereas (2) remains as it is, as is clearly shown in table-1. This may be perhaps on account of incorporation of additional experimental data in the fittings contained in SRIM-96 code.

Application of Hubert et al. Formulation

The Hubert et al. (17) formulation is based on the scaling of heavy ion stopping powers relative to those of alpha particles. It has been established in our previous studies (4) that the Hubert et al. (1989) formulation which is recommended for application in the energy range 2.5-500 MeV/u leads to stopping power values in good agreement with the experiment for the energies ~3-400 MeV/u. Further, the calculations from this formulation are consistent with the experimental stopping power data even for those projectile-target combinations viz. heavy projectiles like U Pb, Xe, Kr etc. in light targets e.g. Be, C, Al etc and heavy targets like Au, U etc in the energy range ~ 3-5 MeV/u for which the TRIM-92 and SRIM-96 calculations predict quite significant deviations as is clearly depicted in table-1 (Sr. No. 11-19). Lately

TABLE 1. Experimental and calculated stopping power values for $53 \le Z \le 92$ projectiles in different targets.

Serial No.	Energy MeV/u	Projectile	Target	Experimental Stopping Power* S_{exp} MeV/mg/cm^2	Computed Stopping Power TRIM-92	SRIM-96	Hubert et al
1	0.10	I	C	-	21.96	20.21	-
2	0.10	Xe	C	-	40.67	20.16	-
3	0.10	Cs	C	-	18.11	20.19	-
4	0.40	I	C	-	49.94	49.51	-
5	0.40	Xe	C	-	61.40	51.10	-
6	0.40	Cs	C	-	48.13	51.58	-
7	0.225	Xe	C	35.7	54.9(53)	36.6(3)	-
8	0.283	Xe	C	43.1	57.6(33)	42.2(-2)	-
9	0.414	Xe	C	54.9	61.5(13)	52.2(-5)	-
10	0.443	Xe	C	56.5	62.2(10)	53.9(-5)	-
11	3.20	Kr	C	57.73	50.99(-12)	50.2(-13)	57.74(0)
12	3.13	U	C	156.0	127.2(-18)	133.0(-15)	154.2(-1)
13	3.96	Pb	Be	175.0	113.0(-35)	111.6(-36)	176.8(+1)
14	4.60	U	Al	125	105.0(-16)	104.0(-17)	121.6(-3)
15	4.66	Xe	Al	67.5	61.2(-18)	59.4(-11)	68.5(+1)
16	4.91	U	Au	46.5	51.0(10)	49.7(7)	45.7(-2)
17	4.92	U	Bi	43.0	50.6(18)	50.1(17)	45.28(5)
18	4.85	U	U	41.5	46.8(13)	46.7(13)	42.62(3)
19	4.95	Xe	Au	25.7	29.5(15)	28.2(10)	27.4(+7)

*Pape et al. (18), Bimbot et al. (19); quoted experimental error is ~7%

TABLE 2. .Results of Stopping Power calculation based on Hubert et al formulation compared with experimental stopping power values for different projectile-target combinations.

Energy MeVu	Projectile	Target	Stopping Power MeV mg^{-1} cm^2 Expt**.	Calculated	Energy MeV/u	Projectile	Target	Stopping Power MeV mg^{-1} cm^2 Expt**.	Calculated
0.66	C	C	6.32	7.05(11)	0.91	Mn	C	35.54	39.03(10)
1.50	C	C	4.93	5.22(6)	0.58	Cu	C	37.85	34.16(-10)
0.65	C	Al	4.66	4.99(7)	2.77	Cu	C	42.58	45.28(10)
1.75	C	Al	3.67	3.78(3)	2.75	Cu	Al	33.58	33.17(-1)
0.42	C	Cu	3.15	2.84(-10)	1.10	Cu	Cu	20.68	18.88(-9)
1.62	C	Cu	2.76	2.77(0)	3.26	Cu	Cu	21.78	22.7(4)
0.73	O	C	9.32	10.20(9)	0.54	Li	CR-39	3.06	2.92(-5)
2.98	O	C	5.92	5.90(0)	0.57	C	CR-39	7.61	8.42(11)
0.73	O	Al	7.19	7.13(-1)	0.56	O	CR-39	11.51	12.21(6)
1.33	O	Al	6.51	6.64(2)	0.56	Na	CR-39	17.65	17.49(-1)
0.49	Al	C	18.0	17.78(-1)	0.67	Si	CR-39	23.35	22.81(-2)
2.85	Al	C	13.2	13.80(5)	0.51	Li	LR-115	2.94	2.79(-5)
0.78	Al	Al	12.8	12.08(-6)	0.50	B	Mylar	5.96	6.39(7)
0.98	Al	Cu	8.65	8.08(-7)	0.58	N	Mylar	8.65	9.69(12)
3.68	Al	Cu	6.59	7.01(6)	1.28	C	Ag	2.07	2.30(11)
0.81	Cl	C	24.48	26.12(7)	1.57	C	Ta	1.60	1.68(5)
0.69	Ti	C	31.7	31.47(-1)	1.55	C	Au	1.47	1.53(5)
0.92	Ti	C	33.85	34.72(3)	1.83	Al	Ag	6.75	6.35(-6)
2.17	Ti	C	30.3	32.84(8)	1.25	Al	Ta	4.82	4.54(-6)
0.80	Ti	Al	22.9	20.42(-10)	1.44	Al	Au	4.60	4.37(-5)
2.07	Ti	Al	24.8	23.87(-4)	2.48	Cu	Ta	14.37	13.17(-10)
0.79	Cr	C	34.6	35.75(4)	2.34	Cu	Au	19.10	17.5(-8)

** Kumar et al. (5), Abdesselam et al. (7-10), Raisanen and Rauhala, (11). Raisanen et al. (12), Rauhala and Raisanen (13), Rauhala et al. (14). The experimental values are reported to have an error of ~6%, 5% and 3% respectively for references 5, 7-10 and 11-14. Percentage deviations of the calculated stopping power values from the corresponding experimental values, i.e. $[(S_{cal}-S_{exp})*100] / S_{exp}$ are quoted in the parenthesis with each entry in the tables.

availability of experimental data at energies below 3 MeV/u has paved way to determine the validity of this formulation in the lower energy domain. Table-2 presents the calculated stopping power values based on the application of the Hubert et al. (1989) formulation at energies lower than its quoted validity alongwith the corresponding experimental values (5, 7-14) for projectiles $3 \leq Z \leq 29$ in materials viz. C, Al, Cu, CR-39 ($C_{12}H_{18}O_7$), LR-115 ($C_6H_9O_9N_2$), Mylar ($C_{10}H_8O_4$), Ag, Ta and Au in the energy range ~ 0.5-3.0 MeV/u. The percentage deviation of the calculated stopping power from the corresponding experimental value, i.e. $[(S_{cal.}-S_{exp.})*100] / S_{exp.}$ has been quoted in the parenthesis with each entry in the table. From table-2, it is clear that the calculated stopping power values using the Hubert *et al.* formulation at energies lower than its reported validity provide good agreement normally within acceptable limits with the experimental values for projectiles $Z \leq 29$ in elemental targets upto Cu and polymeric SSNTD materials like CR-39, LR-115 and Mylar for energies as low as 0.5 MeV/u. However for heavier targets like Ag, Ta and Au etc, this parametrization remain valid down to energies ~1.5 MeV/u at least for projectile ions upto Cu. Ouichaoui et al. (23) arrived at a similar conclusion for S and Br ions in targets Al and Cu predicting the validity of this parametrization up to energies as low as 1.5 MeV/u.

ACKNOWLEDGEMENTS

The authors gratefully acknowledge the help provided by Dr J. Raisanen and Dr E. Rauhala, University of Helsinki, Finland. The authors are also grateful to Dr J. F. Ziegler (IBM Research, Yorktown Heights, NY, USA) for kind cooperation. One of the authors (A. S.) is thankful to Kurukshetra University, Kurukshetra for financial assistance.

REFERENCES

1. Ahlen S. P., *Rev. Mod. Phys.*, **52**, 121-173. (1980).
2. Feldman L. C. and Mayer J. W., *Fundamentals of surface and thin film analysis,* New York, North Holland, (1986).
3. Rauhala E. *In Chemical analysis by Nuclear Methods* New York, 1994, pp. 253
4. Sharma S. K., Kumar S., Yadav J. S. and Sharma A. P., *Appl. Radiat. Isot.* **46**, 39-52 (1995).
5. Kumar S., Sharma S. K., Nath N., Harikumar V., Pathak A. P., Kabiraj D. and Awasthi D. K., *Radiation Effects and Defects in Solids.* **139**, 197-206 (1996).
6. Kumar S., Sharma S. K., Nath N., Harikumar V., Pathak A. P., Hui S. K., Kabiraj D. and Awasthi D. K., *Vacuum* **48**, 1027-1029 (1997).
7. Abdesselam M., Stoquert J. P., Guillamue G., Hage-Ali M., Grob J. J. and Siffert P., *Nucl. Instrum. Meth. Phys. Res.* **B61**, 385-393 (1991).
8. Abdesselam M., Stoquert J. P., Guillamue G., Hage-Ali M., Grob J. J. and Siffert P. *Nucl. Instrum. Meth. Phys. Res.* **B72**, 293-301 (1992a).
9. Abdesselam M., Stoquert J. P., Guillamue G., Hage-Ali M., Grob J. J. and Siffert P., *Nucl. Instrum. Meth. Phys. Res.* **B72**, 7-15 (1992b).
10. Abdesselam M., Stoquert J. P., Guillamue G., Hage-Ali M., Grob J. J. and Siffert P., *Nucl. Instrum. Meth. Phys. Res.* **B73**, 115 (1993).
11. Raisanen J.and Rauhala E., *Phys. Rev.* **B41**, 3951-3958 (1990).
12. Raisanen J., Rauhala E., Fulop Zs., Kiss A. Z., Somorjai E. and Hunyadi I., *Radiat. Meas.* **23**, 749-752 (1994).
13. Rauhala E. and Raisanen J., *Phys. Rev.* **B37**, 9249-9253 (1988).
14. Rauhala E., Raisanen J., Fulop Zs., Kiss A. Z. and Hunyadi I., *Nucl. Tracks Radiat. Meas.* **20**, 611-614 (1992).
15. Lindhard J., Scharff M. and Schiott H.E., *Mat. Fys. Medd. Dan. Vid Sels,.* **33**, 14 (1963).
16. Ziegler J.F., Biersack J.P., and Littmark U., *The Stopping and Ranges of Ions in Matter*, New York, Pergamon Press, (1985), Vol. 1.
17. Hubert F., Bimbot R. and Gauvin H., *Nucl. Instrum. Meth. Phys. Res.* **B36**, 357-363 (1989).
18. Pape H., Clerc H. G. and Schmidt K. H.*Phys.A* **286**, 159 (1978).
19. Bimbot R., Gardes D., Geissel H., Kitahara T., Armbruster P., Fleury A. and Hubert F., *Nucl. Instrum. Meth. Phys. Res.* **B108**, 231 (1980).
20. Harikumar V., Pathak A. P., Sharma S. K., Kumar Shyam, Nath N., Kabiraj D. and Avasthi D. K., *Nucl. Instr. and Meth.* **B108** 223 (1996).
21. Ziegler J. F., TRIM 92: *TheTransport of Ions In Matters*, IBM Research, 28-O, Yorktown Heights, NY 10598, USA, (1992).
22. Ziegler J. F., SRIM 96: *The Stopping Range of Ions In Matter*, IBM Research, 28-O, Yorktown Heights, NY 10598, USA, (1996).
23. Ouichaoui S., Rosier L., Hourany E., Bimbot R., Redjdal N. and Beaumevielle H., *Nucl. Instr. and Meth.* **B95**, 463-469(1995).

Atomic Physics with the Scanning Tunneling Microscope

M. Kleber, C. Bracher, and M. Riza

Physik-Department T30, Technische Universität München, James-Franck-Straße, D–85747 Garching, Germany

Backscattering of atomic beams above a given surface yields information similar to the one obtained from scanning the same surface with a scanning tunneling microscope (STM): In both cases the experimentally accessible quantity is the local density of states (LDOS) $n(\mathbf{r}, E)$ of the surface. For the case of backscattering, the LDOS at the turning point of the atom is an important ingredient of the potential between atom and surface. In experiments performed with an STM, the LDOS at the apex of an atomically sharp tip can be determined directly. Probing surfaces locally by an STM allows for the study of basic phenomena in atomic physics, with tunneling of electrons in three dimensions being a central issue.

Introduction

One of the main differences beween solid-state physics and atomic physics is that solid-state physics relies heavily on the use of (single-particle) energy level densities[1] whereas atomic physics is primarily interested in fundamental processes involving a rather small number of electrons. However, there are cases where both fields come into contact. For example, if an atom or ion is scattered from a surface at low energies it will "probe" the local electron density distribution, i. e., the local density of energy levels. In this context, Harris and Liebsch [2] have shown that the repulsive part of the interaction between helium and a metal surface is determined approximately by the unperturbed local density of states of the metal electrons at the position of the He nucleus.

To some extent, the adiabatic limit of a slowly moving atom in the vicinity of a solid surface can be realized by an atomically sharp tip in front of the same surface. Such an arrangement is the basic ingredient of a scanning tunneling microscope (STM, see Figure 1).

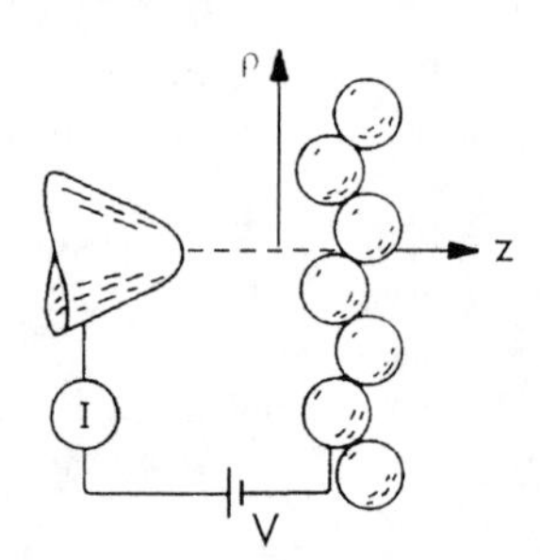

Figure 1. Schematic plot of the scanning tunneling microscope. A weak external electric field allows the electrons to tunnel through the vacuum barrier between tip (on the left) and surface of a sample (on the right).

The discussion presented here stresses the atomic physicist's point of view: We will first divide the macroscopic apparatus into an atomic system (tip and surface layers) whose quantum properties are relevant, and the macroscopic remainder (circuitry and positioning components) appropriately described in classical terms. After identifying the atomic part of the problem we discuss how the electrons tunnel in real space out of the tip (or into the tip). In this way we will be able to understand the spatial resolution of an STM and how to use it for problems in atomic physics.

Dividing the STM in a microscopic and a macroscopic part

In stationary operation electrons travel along a macroscopic wire which narrows down in diameter until it ends into an atomically sharp tip[2]. There, at sufficiently low bias between tip and surface, an electron must tunnel if it wants to cross from tip to surface. From a quantum mechanical point of view electron tunneling is a scattering phenomenon where the electron is scattered across the tunnel junction.

In order to translate this picture into mathematics we divide the whole system into a macroscopic part (circuitry and positioning components) and an atomic system (tip and surface layers) whose quantum properties are relevant (see Figure 2). The Hamiltonian for the total system is split accordingly,

$$H = T + U(\mathbf{r}) + U_{mp}(\mathbf{r}). \tag{1}$$

Apart from the kinetic energy, H contains a sum of two potentials, with $U(\mathbf{r})$ being nonzero only in the atomic system whereas $U_{mp}(\mathbf{r})$ is different from zero in the macroscopic remainder[3]. Since the process of scanning the surface is

[1] The basic concept for the calculation of thermodynamic properties of complex systems, including solids, requires only the knowledge of energy-level densities. See, for example [1]

[2] For the realization of a point-like tip we refer to [3].

[3] For reason of simplicity we assume here the potentials to be local.

CP475, *Applications of Accelerators in Research and Industry*, edited by J. L. Duggan and I. L. Morgan

slow compared to the electronic motion, the STM can be regarded to operate under stationary conditions. The corresponding Schrödinger equation then reads

$$[E - T - U(\mathbf{r})]\psi(\mathbf{r}) = U_{mp}(\mathbf{r})\psi(\mathbf{r}), \tag{2}$$

with $\psi(\mathbf{r})$ being a current-carrying wave function. By introducing the quantum mechanical propagator (or Green function) $\mathsf{G}(\mathbf{r}, \mathbf{r}'; E)$ which is a solution of

$$[E - T - U(\mathbf{r})]\mathsf{G}(\mathbf{r}, \mathbf{r}'; E) = \delta^{(3)}(\mathbf{r} - \mathbf{r}') \tag{3}$$

we can solve (2) for $\psi(\mathbf{r})$,

$$\psi(\mathbf{r}) = \int \mathrm{d}^3 r' \; \mathsf{G}(\mathbf{r}, \mathbf{r}'; E) \, U_{mp}(\mathbf{r}') \, \psi(\mathbf{r}'). \tag{4}$$

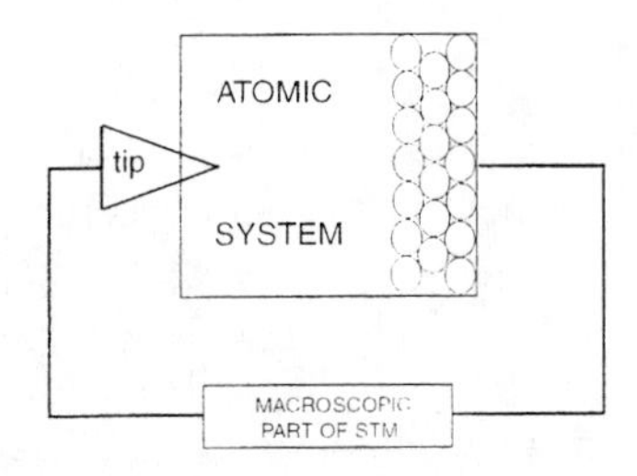

Figure 2. Cut of the STM in a microscopic atomic system and a macroscopic remainder.

There is no compelling principle of how to divide the total STM device into a microscopic part and a macroscopic remainder. We know however that all tunneling electrons must pass through the apex of the tip. Therefore we draw the dividing line between microscopic and macroscopic world in close vicinity of the sharp end of the tip. Mathematically this amounts to expanding the integrand in (4) about the position $\mathbf{r}'$ of the apex of the tip which we choose to be at $\mathbf{r}' = \mathbf{0}$. As a result of such an expansion we can cast (4) in the form

$$\psi(\mathbf{r}) = \sum_{lm} C_{lm} \mathsf{G}_{lm}(\mathbf{r}, \mathbf{r}' = \mathbf{0}; E). \tag{5}$$

In (5), the $\mathsf{G}_{lm}(\mathbf{r}, \mathbf{0}; E)$ are the multipole Green functions which are closely related to harmonic polynomials (or regular solid harmonics) [4]. Here we content ourselves with presenting the explicit form of the monopole term, $l = m = 0$:

$$C_{00} = 2\sqrt{\pi} \int \mathrm{d}^3 r \; U_{mp}(\mathbf{r})\psi(\mathbf{r}),$$
$$\mathsf{G}_{00}(\mathbf{r}, \mathbf{0}; E) = \frac{1}{2\sqrt{\pi}} \mathsf{G}(\mathbf{r}, \mathbf{0}; E). \tag{6}$$

Therefore in the monopole approximation we have

$$\psi(\mathbf{r}) = \psi_{00}(\mathbf{r}) = C \cdot \mathsf{G}(\mathbf{r}, \mathbf{0}; E) \tag{7}$$

with $C = C_{00}/(2\sqrt{\pi})$. Equation (7) tells us that, in monopole approximation, the desired scattering wave function $\psi(\mathbf{r})$ is proportional to the propagator $\mathsf{G}(\mathbf{r}, \mathbf{r}' = \mathbf{0}; E)$ which is known to be a relative probability amplitude for a particle to arrive at point $\mathbf{r}$ if it is created at point $\mathbf{r}'$ [5]. In our case the "creation" of the electron takes place at $\mathbf{r}' = \mathbf{0}$ where the apex of the tip is located.

The tip as an atomic tunneling source for electrons

Equation (7) tells us that the desired wave function $\psi(\mathbf{r})$ is proportional to the Green function which carries an electron from the apex of the tip at $\mathbf{r}' = \mathbf{0}$ to an arbitrary point of the sample (Figure 1). With the Green function satisfying (3) we can use (7) to verify that $\psi(\mathbf{r})$ will obey

$$[E - T - U(\mathbf{r})]\,\psi(\mathbf{r}) = \sigma(\mathbf{r}), \tag{8}$$

where $\sigma(\mathbf{r}) = C\delta^{(3)}(\mathbf{r})$ is a source field. Equation (8) represents the stationary inhomogeneous Schrödinger equation with a source term which arises from coupling the atomic system to the macroscopic environment of the STM (see Figure 2). Because of (5) the strength of the source contains the wave function and therefore should be determined self-consistently, at least in principle. Instead of calculating the strength C of the source it is more practical to relate C to the experimentally observed tunneling current J. From

$$\operatorname{div}\mathbf{j}(\mathbf{r}) = \frac{\hbar}{m} \mathfrak{Im}\left[\psi(\mathbf{r})^\star \nabla^2 \psi(\mathbf{r})\right] \tag{9}$$

the total current is found to be [6]

$$J = -\frac{2}{\hbar}|C|^2 \mathfrak{Im}\big[\mathsf{G}(\mathbf{r} = \mathbf{0}, \mathbf{r}' = \mathbf{0}; E)\big]. \tag{10}$$

Equation (10) relates the tunneling current to quantum propagation in such a way that all path amplitudes must be summed for the closed loops that start and end at the position $\mathbf{r}' = \mathbf{0}$ of the source (i. e. , the apex of the tip). Of course we don't need to perform the sum if we can calculate $\mathfrak{Im}[\mathsf{G}(\mathbf{0}, \mathbf{0}; E)]$ directly. Note that

$$\mathfrak{Im}\big[\mathsf{G}(\mathbf{r} = \mathbf{0}, \mathbf{r}' = \mathbf{0}; E)\big] = \pi n(\mathbf{r} = \mathbf{0}, E)$$

is proportional to the density of states, $n(\mathbf{r} = \mathbf{0}, E)$, at the position $\mathbf{r} = \mathbf{0}$ of the tip. Taking into account the different occupation probabilities $f(E)$ in tip and sample, equation (10) is modified to $J \longrightarrow J \cdot [f(E) - f(e + eV)]$ with V being the applied voltage [7].

Example: Field emission from a point-like tip

The problem of stationary emission of electrons from a point-like tip at $\mathbf{r}'$ can be solved analytically for a homgeneous electric field. Aligning the direction of the corresponding driving force $\mathbf{F}$ along the positive z axis we have

$U(\mathbf{r}) = -Fz$. By solving (3) we obtain the desired Green function

$$G_{\text{ret}}(\mathbf{r},\mathbf{r}';E) = \frac{m}{2\hbar^2|\mathbf{r}-\mathbf{r}'|} \times$$
$$\times \left[\text{Ai}'(a_-)\,\text{Ci}(a_+) - \text{Ai}(a_-)\,\text{Ci}'(a_+)\right] \quad (11)$$

with

$$a_\pm = -\sqrt[3]{\frac{m}{4\hbar^2F^2}}\left[F(z+z'\pm|\mathbf{r}-\mathbf{r}'|)+2E\right]. \quad (12)$$

Here we have introduced the complex Hankel-type Airy function

$$\text{Ci}(s) = \text{Bi}(s) + i\text{Ai}(s) \quad (13)$$

in terms of two real Airy functions $\text{Ai}(s)$ and $\text{Bi}(s)$ as defined in Abramowitz and Stegun [8]. In (11) we have imposed outgoing-wave boundary conditions, hence the appearance of the retarded Green function $G_{\text{ret}}(\mathbf{r},\mathbf{r}';E)$[4]. Though correct, the solution (11) does not particularly illuminate on the physics of tunneling. Therefore we use the semiclassical approximation to (11),

$$\psi(\mathbf{r},E) \sim G_{\text{ret}}(\mathbf{r},\mathbf{r}'=\mathbf{0};E)$$
$$\propto \exp\left\{-\frac{\sqrt{m}}{3\hbar F}\left[F(r-z)-2E\right]^{3/2}\right\}\times$$
$$\times \exp\left\{\frac{i\sqrt{m}}{3\hbar F}\left[F(r+z)+2E\right]^{3/2}\right\} \quad (14)$$

which is correct to exponential order. It is not difficult to derive from (14) interesting properties of tunneling in three dimensions (3D) [9]. In this case we have $E < 0$ which corresponds to a tunneling source with quasibound electrons sitting in a potential well and waiting for the opportunity to tunnel[5] The main escape path will be in direction of the force ($z-$direction in Figure 3). Small deviations from the escape path are possible both in configuration space and momentum space. However such deviations are not favorable for tunneling: Any travel in lateral direction, i. e., in a direction orthogonal to the escape path makes the traversal path across the barrier longer. Diverting some of the initial momentum in lateral direction will also reduce the momentum in escape direction and, therefore, the probability for penetrating the barrier. From what has been said one expects for example both Δx and Δp_x to be as small as possible. The uncertainty principle establishes a lower bound on $\Delta x \Delta p_x$. In fact, by use of (14) it is not difficult to show that the tunneling wave assumes minimum uncertainty

[4]If the electrons were to tunnel into the tip (reverse bias) we would need the advanced Green function.

[5]The case $E > 0$ corresponds to an electron source which emits ballistic electrons in the absence of a barrier.

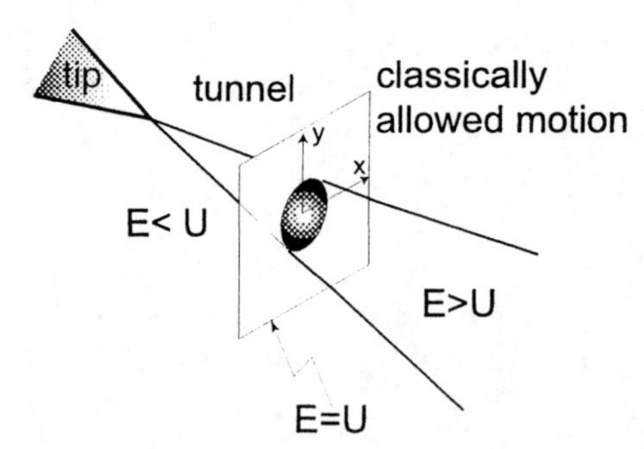

Figure 3. Tunneling spot (shaded) at the turning surface.

$$\Delta x \Delta p_x = \frac{\hbar}{2} \qquad \Delta y \Delta p_y = \frac{\hbar}{2} \quad (15)$$

at the exit of the tunnel [9] . In other words, in 3D we will have a tunneling spot of minimum uncertainty at the turning surface between classically allowed and forbidden motion. (see Figure 3). Minimum uncertainty means a Gaussian density (and current) profile in lateral direction, both in configuration and momentum space. The size of the tunneling spot depends on the barrier crossing time which for field emission turns out to be

$$\tau = m\frac{\Delta x}{\Delta p_x} = \frac{\sqrt{2m|E|}}{F}. \quad (16)$$

It is pleasing to note that τ equals the bounce time (or Büttiker-Landauer time) which is expected to represent the natural semiclassical time scale of motion. From (15) and (16) we can determine the Gaussian distributions at the tunnel exit. However, by expanding (14) we find the Gaussian density profile to be valid along the entire escape path $z < 0$. Therefore, also the current density assumes a Gaussian profile, given by [10]

$$j_z(\mathbf{r}) \propto \exp\left\{-\frac{\kappa\,(x^2+y^2)}{2z}\right\}, \quad (17)$$

where $\hbar\kappa = \sqrt{2m|E|}$ represents the binding momentum at the tip site. Although the results shown here are strictly valid only for pointlike s-wave tunneling sources in the presence of a homogeneous field, we expect them to hold also for more complicated electric field configurations, including guided motion. In Figure 4 we compare expression (17) with the results of a very detailed calculation by Lang et al. [11]. The agreement between the simple model (17) and the model of Lang et al. is better than qualitative.

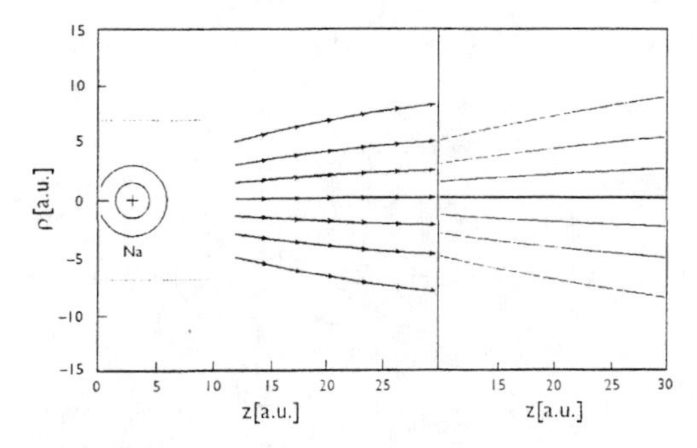

Figure 4. Current-density streamlines from a single-atom point source for an external electric field $F = 1.3\,\text{eV}/\text{Å}$. The left side of the figure depicts the results from the model of Lang et al. [11]. In the right hand part of the figure the corresponding streamlines are plotted as obtained from Eq. (17) with $|E| = 2.35$ eV. The streamlines enclose $\frac{1}{2}$, $\frac{2}{9}$, and $\frac{1}{18}$, respectively, of the total current.

Applications

Due to limited space we present just a few aspects of experiments where 3D-tunneling is essential.

• Squeezing of the tunneling spot

The spatial resolution of the STM will be improved by decreasing Δx and Δy at the expense of increasing Δp_x and Δp_y. Electrons tunneling out of tip states with higher multipoles will yield a sharper image of the surface as compared to electrons which are emitted from lower multipole states [12].

• Backbending of the tunneling "spotlight"

If electrons are ejected by a nonstationary field, tunneling is still a useful concept even though it is to some extent a matter of taste whether or not to speak of tunneling. With ω being the frequency of the applied (laser) field and F its driving force, the Keldysh parameter $\gamma = \omega\sqrt{-2mE}/F$ [13] is the border line ($\gamma \approx 1$) between the tunneling ionization regime ($\gamma \ll 1$) and the multiphoton regime ($\gamma \gg 1$). In an alternating electric field, the emitted electrons may return to the emitting ion (see Figure 5), giving rise to a variety of processes. For example, by scattering off the ion an electron can gain considerable energy, resulting in high-order above threshold ionization. Alternatively, the electron may recombine with the ion and emit its entire kinetic energy plus the binding energy as a single photon. This gives rise to generation of high harmonics. Put into mathematics, such simple pictures for the behavior of atoms in strong laser fields have been shown to work amazingly well [14, 15].

Figure 5. Schematic plot of a closed-loop travel of an electron in a time-dependent field.

• Interference of tunneling trajectories

Tunneling trajectories will interfere only under very special circumstances. For example Madhavan et al. [16] have identified a Kondo resonance in an STM experiment. Their finding can be interpreted as a Fano resonance for an interacting discrete single copper atom coupled to the electronic continuum of a gold surface [16].

Tunneling interference occurs also in laser induced ionization. In this case there can be interference of electrons that tunnel out of atoms at different times within one cycle of the field [17].

References

[1] R. P. Feynman, *Statistical Mechanics* (Benjamin/Cummins, Reading, 1982).

[2] J. Harris and A. Liebsch, J. Phys. C. **15**, 2275 (1982).

[3] H. W. Fink, Phys. Scr. **38**, 260 (1988).

[4] A. R. Edmonds, *Angular Momentum in Quantum Mechanics* (Princeton University Press, Princeton, 1974).

[5] R. P. Feynman and A. R. Hibbs, *Quantum Mechanics and Path Integrals* (McGraw-Hill, New York, 1965).

[6] C. Bracher, M. Riza, and M. Kleber, Phys. Rev. B **56**, 7704 (1997).

[7] J. Tersoff and D. R. Hamann, Phys. Rev. B **31**, 805 (1985).

[8] M. Abramowitz and I. A. Stegun, *Handbook of Mathematical Functions* (Dover Publications, New York, 1972).

[9] C. Bracher *et al.*, Am. J. Phys. **66**, 38 (1998).

[10] B. Gottlieb, A. Lohr, W. Becker, and M. Kleber, Phys. Rev. A **54**, R1022 (1996).

[11] N. D. Lang, A. Yacoby, and Y. Imry, Phys. Rev. Lett. **63**, 1499 (1989).

[12] J. C. Chen, Phys. Rev. B **42**, 8841 (1990).

[13] L. V. Keldysh, Zh. Eksp. Fiz. **47**, 1945 (1964).

[14] A. Lohr, M. Kleber, R. Kopold, and W. Becker, Phys. Rev. A **55**, R4003 (1997).

[15] W. Becker, A. Lohr, M. Kleber, and M. Lewenstein, Phys. Rev. A **56**, 645 (1997).

[16] V. Madhavan *et al.*, Science **280**, 567 (1998).

[17] G. G. Paulus *et al.*, Phys. Rev. Lett. **80**, 484 (1998).

SECTION II

NUCLEAR PHYSICS AND RADIOACTIVE ION BEAM FACILITIES AND EXPERIMENTS

PARITY VIOLATION IN THE COMPOUND NUCLEUS

G. E. Mitchell,[1] J. D. Bowman,[2] B. E. Crawford,[1] P. P. J. Delheij,[3] C. A. Grossmann,[1] T. Haseyama,[4] J. Knudson,[2] L. Y. Lowie,[1] A. Masaike,[4] Y. Matsuda,[4] S. Penttilä,[2] H. Postma,[5] N. R. Roberson,[6] S. J. Seestrom,[2] E. I. Sharapov,[7] D. A. Smith,[2] S. L. Stephenson,[8] Yi-Fen Yen,[2] and V. W. Yuan[2]

[1]*North Carolina State University, Raleigh, North Carolina 27695 and Triangle Universities Nuclear Laboratory, Durham, North Carolina 27708*
[2]*Los Alamos National Laboratory, Los Alamos, New Mexico 87545*
[3]*TRIUMF, Vancouver, British Columbia, V6T 2A3, Canada*
[4]*Department of Physics, Kyoto University, Kyoto 606-8502, Japan*
[5]*University of Technology, Delft, 2600 GA, the Netherlands*
[6]*Duke University, Durham, North Carolina 27708 and Triangle Universities Nuclear Laboratory, Durham, North Carolina 27708*
[7]*Joint Institute for Nuclear Research, 141980 Dubna, Russia*
[8]*Gettysburg College, Gettysburg, Pennsylvania 17325*

Measurements have been performed on the helicity dependence of the neutron resonance cross section for many nuclei by our TRIPLE Collaboration. A large number of parity violations are observed. Generic enhancements amplify the signal for symmetry breaking and the stochastic properties of the compound nucleus permit the strength of the symmetry-breaking interaction to be determined without knowledge of the wave functions of individual states. A total of 15 nuclei have been analyzed with this statistical approach. The results are summarized.

INTRODUCTION AND MOTIVATION

In recent years two developments have encouraged a new approach to the study of the weak interaction in nuclei: the observation of very large enhancement factors for parity violation in low-energy neutron resonances, and the realization that the stochastic properties of the compound nucleus simplify the analysis. Study of the weak interaction has contributed significantly to our understanding, in large part because of the clear signals for weak interaction effects. The weak interaction is both an object of fundamental research and a tool for the study of strongly interacting systems.

Work on the weak interaction in nuclei illustrates this dual role. One goal is the understanding of the effective weak interaction in nuclei, which is a basic problem of many-body theory. The weak interaction is also used to investigate strongly interacting hadronic systems.

This duality is reflected in the study of parity violation in the compound nucleus. Study of the helicity dependence of the scattering of slow neutrons offers a new way to investigate the effective weak interaction in nuclei. Earlier work required detailed calculations for the wave functions. In the new approach we utilize the stochastic properties of the compound system – the parity violating observable is a random variable – and a statistical analysis is adopted.

First the enhancement factors and the new approach are briefly reviewed. Then the experimental system is described and the methods of analysis discussed. The results are then presented and summarized.

ENHANCEMENT FACTORS AND STATISTICAL APPROACH

The classic approach to the determination of the effective parity-violating interaction in light nuclei requires that the wave functions be known with high precision (1,2). Sushkov and Flambaum (3) predicted that parity violating signals would be strongly enhanced near neutron threshold. This was confirmed experimentally at Dubna (4,5) by measurement of the helicity dependence of the total neutron cross section. The observable is the longitudinal asymmetry ϵ: the difference of the transmitted intensities for the two helicity states divided by the sum. A non-zero value of ϵ indicates parity violation. The signal is observed at p-wave resonances that are mixed by the parity-violating interaction with

CP475, *Applications of Accelerators in Research and Industry*, edited by J. L. Duggan and I. L. Morgan
 1-56396-825-8/99/$15.00

neighboring states of the same spin and opposite parity (s-wave resonances). The key to the enhancement is that in heavy nuclei the compound states are very close together (mean spacing D of order 10-20 eV) and that very large s-wave states are mixed into very weak p-wave states (near threshold the s-wave penetrability is much larger than the p-wave penetrability). The first effect is usually called "dynamic" enhancement and the second "kinematic" enhancement. The combination of these enhancement factors can be as large as 10^6. This changes the expected magnitude of parity-violating effects from the value of 10^{-7} in nucleon-nucleon scattering to the percent level.

The observation of very large parity-violating effects is interesting but of little value by itself. The primary disadvantage of the neutron scattering data – the high degree of complexity of the resonance wave function – can in fact be converted into a major advantage. The fluctuation properties of neutron resonances agree with random matrix theory. The matrix elements v that mix the compound nuclear states are considered as random variables. The distribution of these matrix elements yields the mean-square matrix element $\overline{v^2}$, and from this the weak spreading width $\Gamma^{\downarrow}$. The spreading width is adopted to remove the dependence on level density, and gives a direct measure of the strength of the effective parity-violating interaction in nuclei. Thus the major advantage: the statistical approach yields physical information without making any statements about the wave functions of individual resonances. This general approach is discussed in several recent reviews (6-9). The key remaining theoretical issue is the relation between $\Gamma^{\downarrow}$ and the effective parity-violating nucleon-nucleon interaction (10).

EXPERIMENTAL SYSTEM AND PROCEDURE

The TRIPLE Collaboration has performed parity-violation measurements on a number of nuclei. The 800-MeV proton beam from the Los Alamos Neutron Scattering Center (LANSCE) is chopped and then stacked and accumulated in the Proton Storage Ring (PSR). The stored proton beam is directed (at 20 Hz) toward a tungsten spallation target. The neutrons are then moderated to epithermal energies and collimated.

The TRIPLE apparatus and procedure was described by Roberson *et al.* (11). An overview of the TRIPLE experimental system is shown in Fig. 1. The neutron beam is polarized by transmission though a polarized proton target. The protons are polarized in frozen ammonia by the dynamic nuclear polarization method (12). The neutron polarization was about 70%. The spin direction of the neutrons is reversed rapidly by an adiabatic spin flipper (13). The neutron detector system consists of 55 liquid scintillator cells optically coupled to photomultiplier tubes (14).

Figure 1. Overview of the polarized neutron flight path at LANSCE.

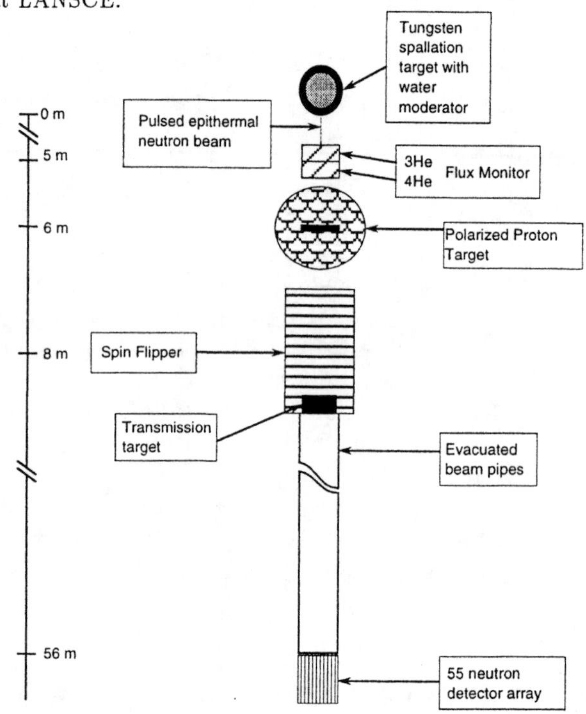

Data are collected for 200 beam bursts. The neutron helicity state is changed according to an eight-step sequence designed to reduce the effects of gain drifts and of residual magnetic fields (11). Each spin flipper state lasts 10 s. After 20 eight-step sequences, the data are stored for later analysis. The large number of small data sets, or runs, are analyzed separately.

ANALYSIS METHOD

In favorable cases, the parity violation can be observed in the raw data. The transmission spectra for the two different helicity states for the 63-eV resonance in ^{238}U are shown in the top portion of Fig. 2. The parity violation is apparent by inspection.

The analysis proceeds as follows. First the neutron resonance parameters are determined with the computer program FITXS (15). This multilevel program includes broadening from the neutron beam, the target-beam interaction, and the detector. The resonance parameters are held fixed while the longitudinal asymmetry P (the difference of the p-wave resonance cross sections for the two helicities divided by the sum) is determined for each resonance for each run. Results for different runs are combined as illustrated in the bottom portion of Fig. 2. The mean value of P and the uncertainty ΔP are determined from the distribution.

Figure 2. Top: ^{238}U transmission spectra for the two helicity states near the 63.4-eV resonance. The parity violation is apparent by inspection. Bottom: histogram of the asymmetries obtained for each of 157 runs for the resonance shown at the top of the figure.

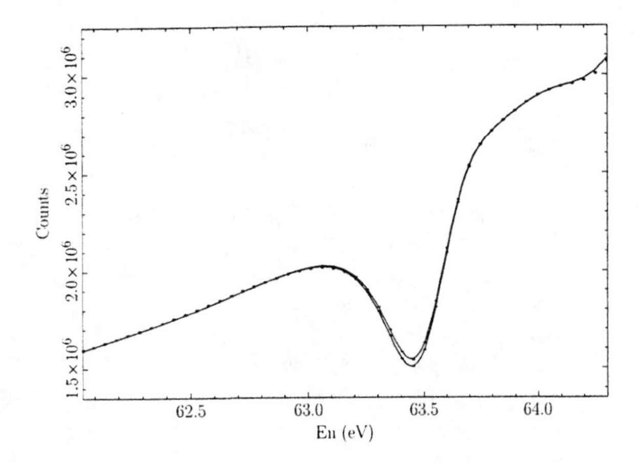

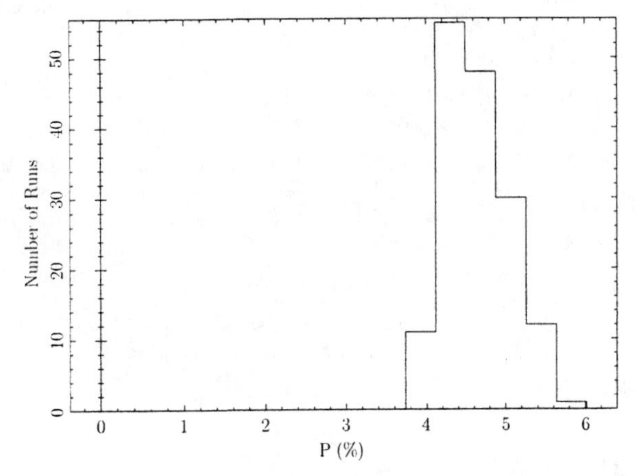

The parity-violating matrix elements v are assumed to be Gaussian random variables. Since there may be several s-wave resonances that contribute to the parity violation at a given p-wave resonance, and therefore several matrix elements (between the p-wave state and the various s-wave states), there are too many unknowns to determine the values of the individual matrix elements. However, since all of these matrix elements are from the same distribution, it is possible to determine the variance $\overline{v^2}$. The maximum likelihood method is used to determine the most likely value of the variance. For spin zero targets, the analysis is straightforward. For targets with spin the analysis becomes much more involved. The details of the analysis are given by Bowman *et al.* (16).

RESULTS

Most of the TRIPLE data consists of transmission measurements of the helicity dependence of the neutron total cross section. Following early measurements on ^{232}Th and ^{238}U, the system was redesigned and rebuilt with improved polarization and detection efficiency. The early measurements yielded a major surprise: for ^{232}Th all of the parity-violating asymmetries had the same sign. This violated the statistical prediction and led to a large amount of theoretical effort. All attempts at explanation that involved distant states required a single particle weak matrix element that was 100-1000 times the expected value. Therefore a crucial issue was whether this non-statistical phenomenon was real, and if real, generic or confined to ^{232}Th. The answer is given in Table I. A remeasurement of ^{232}Th (17) confirmed the non-statistical distribution of the parity-violating asymmetries: all ten statistically significant parity-violating effects have the same sign. However, the effect is probably not generic, since for the other nuclides the signs are much closer to random. Most of the recent attempts to explain the sign effect utilize a local doorway state, but the origin of this doorway state remains unexplained (27-32).

Table 1. Relative signs of parity violations observed by the TRIPLE Collaboration. Effects with statistical significance greater than three standard deviations are included.

Target	Reference	Number	P +	P −
^{81}Br	(18)	1	1	0
^{93}Nb	(19)	0	0	0
^{103}Rh	(20)	4	3	1
^{107}Ag	(21)	8	5	3
^{109}Ag	(21)	4	2	2
^{104}Pd	(20)	1	1	0
^{105}Pd	(20)	7	4	3
^{106}Pd	(22)	1	0	1
^{108}Pd	(20)	1	1	0
^{113}Cd	(23)	3	2	1
^{115}In	(21)	6	3	3
^{117}Sn	(20)	6	3	3
^{121}Sb	(15)	5	3	2
^{123}Sb	(15)	1	0	1
^{127}I	(15)	7	5	2
^{131}Xe	(24)	1	0	1
^{133}Cs	(15)	1	1	0
^{139}La	(25)	1	1	0
^{232}Th	(17)	10	10	0
^{238}U	(26)	5	3	2
total		73	48	25
total excluding Th		63	38	25

One central issue that can addressed directly is the mass dependence of the weak spreading width. Our values are listed in Table II. The data are qualitatively consistent with a constant spreading width. However many of these results are preliminary and have large uncertainties.

It would be quite valuable to measure the weak spreading width for light nuclei. Parity-violation experiments with charged particle resonances in light nuclei are now underway (33).

Table 2. Weak spreading widths obtained by the TRIPLE Collaboration.

Target	Reference	$\Gamma^{\downarrow}$ (10^{-7} eV
^{93}Nb	(19)	≤ 1.0
^{107}Ag	(21)	$5.40^{+3.57}_{-1.99}$
^{109}Ag	(21)	$1.30^{+2.50}_{-0.74}$
^{104}Pd	(20)	$2.53^{+10.9}_{-1.70}$
^{105}Pd	(20)	$1.29^{+2.54}_{-0.83}$
^{106}Pd	(22)	$0.49^{+1.16}_{-0.29}$
^{108}Pd	(20)	$2.33^{+7.71}_{-1.49}$
^{113}Cd	(23)	$16.4^{+18.0}_{-8.4}$
^{115}In	(21)	$0.94^{+0.94}_{-0.39}$
^{117}Sn	(20)	$0.86^{+1.94}_{-0.54}$
^{121}Sb	(15)	$6.45^{+9.72}_{-3.66}$
^{123}Sb	(15)	$1.23^{+15.0}_{-0.96}$
^{127}I	(15)	$2.05^{+1.94}_{-0.93}$
^{232}Th	(17)	$4.7^{+2.7}_{-1.8}$
^{238}U	(26)	$1.35^{+0.97}_{-0.64}$

SUMMARY

The study of parity violation in the compound nucleus has made impressive advances. Large enhancement factors were both predicted and observed. The TRIPLE Collaboration has measured approximately 70 parity violation effects for many nuclei in the mass 100 and 230 regions. Except for ^{232}Th the data seem to agree with the statistical model. With the ansatz that the matrix elements are random variables one can determine the rms matrix element v and the weak spreading width $\Gamma^{\downarrow}$ without knowing the wave functions of the individual states. The non-statistical sign correlation observed in ^{232}Th is ascribed to nuclear structure effects. The key remaining experimental issue is the mass dependence of $\Gamma^{\downarrow}$. The key theoretical issue is the establishment of the connection between the weak spreading width $\Gamma^{\downarrow}$ and the effective parity-violating nucleon-nucleon interaction in nuclei.

ACKNOWLEDGMENTS

This work was supported in part by the U. S. Department of Energy, Office of High Energy and Nuclear Physics, under grants No. DE-FG02-97-ER41042 and DE-FG02-97-ER41033, and by the U. S. Department of Energy, Office of Energy Research, under Contract No. W-7405-ENG-36.

REFERENCES

1. Adelberger, E. G. and W. C. Haxton, *Ann. Rev. Nucl. Part. Sci.* **35**, 501 (1985).
2. Desplanques, B., *Phys. Reports* **297**, 1 (1998).
3. Sushkov, O. P. and V. V. Flambaum, *JETP Lett.* **32**, 352 (1980).
4. Alfimenkov, V. P. *et al.*, *JETP Lett.* **35**, 51 (1982).
5. Alfimenkov, V. P. *et al. Nucl. Phys.* **A 398**, 93 (1983).
6. Bowman, J. D., G. T. Garvey, Mikkel Johnson, and G. E. Mitchell, *Ann. Rev. Nucl. Part. Sci.* **43**, 829 (1993).
7. Frankle, C. M., S. J. Seestrom, N. R. Roberson, Yu. P. Popov, and E. I. Sharapov, *Phys. Part. Nucl.* **24**, 401 (1993).
8. Flambaum, V. V. and G. F. Gribakin, *Prog. Part. Nucl. Phys.* **35**, 423 (1996).
9. Mitchell, G. E., J. D. Bowman, and H. A. Weidenmüller, *Rev. Mod. Phys.* (to be published).
10. French, J. B., V. K. B. Kota, A. Pandey, and S. Tomsovic, *Ann. Phys. (N.Y.)* **181**, 198 (1988).
11. Roberson, N. R. *et al.*, *Nucl. Istrum. Methods* **A 326**, 549 (1993).
12. Penttila, S. I. *et al.*, *High Energy Spin Physics*, edited by K. J. Heller and S. L. Smith (AIP Conf. Proc. No. 343, New York, 1995), p. 532.
13. Bowman, J. D., S. I. Penttila, and W. B. Tippens, *Nucl. Instrum. Methods Phys. Res.* **A 369**, 195 (1996).
14. Yen, Yi-Fen *et al.*, *Time Reversal Invariance and Parity Violation in Neutron Resonances*, edited by C. R. Gould, J. D. Bowman, and Yu. P. Popov (World Scientific, Singapore, 1994), p. 210.
15. Matsuda, Y., Ph. D. thesis, Kyoto University, 1998.
16. Bowman, J. D., L. Y. Lowie, G. E. Mitchell, E. I. Sharapov, and Yi-Fen Yen, *Phys. Rev. C* **53**, 285 (1996).
17. Stephenson, S. L. *et al.*, *Phys. Rev C* **58**, 1236 (1998).
18. Frankle, C. M. *et al.*, *Phys. Rev. C* **46**, 1542 (1992).
19. Sharapov, E. I. *et al.*, to be published.
20. Smith, D. A. *et al.*, to be published.
21. Lowie, L. Y., Ph. D. thesis, North Carolina State University, 1996.
22. Crawford, B. E., Ph. D. thesis, Duke University, 1997.
23. Seestrom, S. J. *et al.*, to be published.
24. Szymanksi, J. J. *et al.*, *Phys. Rev. C* **53**, R2576 (1996).
25. Yuan, V. Y. *et al.*, *Phys. Rev. C* **44**, 2187 (1991).
26. Crawford, B. E. *et al.*, *Phys. Rev C* **58**, 1225 (1998).
27. Auerbach, N., J. D. Bowman, and V. Spevak, *Phys. Rev. Lett.* **74**, 2638 (1995).
28. Flambaum, V. V. and V. G. Zelevinsky, *Phys. Lett.* **B 94**, 277 (1995).
29. Hussein, M. S., A. K. Kerman, and C.-Y. Lin, *Z. Phys.* **A 351**, 30 (1995).
30. Auerbach, N., V. V. Flambaum, and V. Spevak, *Phys. Rev. Lett.* **76**, 4316 (1996).
31. Desplanques, B. and S. Noguera, *Nucl. Phys.* **A 598**, 139 (1996).
32. Sushkov, O. P., *Phys. Rev. Lett.* **77**, 5024 (1996).
33. Mitchell, G. E. and J. F. Shriner, Jr., *Phys. Rev. C* **54**, 371 (1996).

New Measurements of Spin-Dependent n-p Cross Sections

B. W. Raichle[1,4], C. R. Gould[2,4], D. G. Haase[2,4], M. L. Seely[2,4*], J. R. Walston[2,4†], W.Tornow[3,4], W. S. Wilburn[3,4‡], S. I. Penttilä[5], and G. W. Hoffmann[6]

[1]*Department of Physical Sciences, Morehead State University, Morehead, KY 40351*
[2]*Physics Department, North Carolina State University, Raleigh, NC 27695*
[3]*Physics Department, Duke University, Durham, NC 27708*
[4]*Triangle Universities Nuclear Laboratory, Durham, NC 27708*
[5]*Los Alamos National Laboratory, Los Alamos, NM 87545*
[6]*University of Texas, Austin, TX 78712*

We report on new measurements of the spin-dependent neutron-proton total cross-section differences in longitudinal and transverse geometries ($\Delta\sigma_L$ and $\Delta\sigma_T$ respectively) and between 5 and 20 MeV. These transmission experiments involve a polarized neutron beam and polarized proton target. The polarized neutron beam was produced as a secondary beam via charged-particle induced neutron-production reactions. The proton target was cryogenically cooled and dynamically polarized. These data will be used to extract ε_1, the phase-shift parameter which characterizes the strength of the tensor interaction at low energy.

I. INTRODUCTION

We report on new polarized beam-polarized target transmission measurements of the spin-dependent n-p total cross-section differences between 5 and 20 MeV and in both the longitudinal ($\Delta\sigma_L$) and transverse ($\Delta\sigma_T$) geometries [1,2]. These spin-dependent observables are sensitive to the tensor component of the n-p interaction and insensitive to other phase shifts which contribute to low energy n-p scattering [3]. The tensor force gives rise to mixing of the $^3S_1$ and $^3D_1$ angular momentum states and is characterized by the phase-shift parameter ε_1.

II. FUNDAMENTALS OF THE MEASUREMENT

The transmitted flux Φ of a beam of pure spin-up or spin-down neutrons ($P_n = \pm 1$) incident on a target of pure spin-up or spin-down protons ($P_T = \pm 1$) is

$$\Phi = \Phi_0 e^{-x\sigma_{p(a)}} \qquad (1)$$

where Φ_0 is the incident neutron flux, x is the target thickness in units of b^{-1}, and σ_p (σ_a) are the cross sections for a neutron and proton with spin axes parallel (anti-parallel) and are defined

$$\sigma_p = \sigma_0 - \frac{1}{2}\Delta\sigma_{T(L)} \qquad (2)$$

$$\sigma_a = \sigma_0 + \frac{1}{2}\Delta\sigma_{T(L)} \qquad (3)$$

where σ_0 is the unpolarized cross section, and with $\Delta\sigma_T$ and $\Delta\sigma_L$ defined as the differences in cross sections for a beam and target polarized anti-parallel and parallel to each other.

Transmission of a neutron beam with polarization P_n incident on a proton target with polarization P_T is the weighted sum of the transmission of a beam of pure spin-up neutrons ($P_n = 1$) passing through the target (containing both spin-up and spin-down protons) plus the transmission of a beam of pure spin-down neutrons ($P_n = -1$) passing through the target, and is given by

$$\Phi = \Phi_0^+[e^{-N_p^+ x\sigma_p} e^{-N_p^- x\sigma_a}] + \Phi_0^-[e^{-N_p^+ x\sigma_a} e^{-N_p^- x\sigma_p}] \qquad (4)$$

where the fraction of spin-up (N_p^+) or spin-down (N_p^-) protons is given by $N_p^\pm = \frac{1}{2}(1 \pm P_T)$ and the number of incident spin-up (Φ_0^+) or spin-down (Φ_0^-) neutrons is similarly given in terms of P_n.

The measured transmission asymmetry ϵ_n due to reversing beam polarization from $-P_n''$ to $+P_n'$ is defined

$$\epsilon_n = \frac{N(+P_n') - N(-P_n'')}{N(+P_n') + N(-P_n'')} \qquad (5)$$

where N is the number of counts recorded by the detector, which after some algebra can be rewritten

$$\epsilon_n = \frac{(P_n' + P_n'')\sinh(y)}{2\cosh(y) + 2(P_n' - P_n'')\sinh(y)} \qquad (6)$$

---

*Present address: Thomas Jefferson National Accelerator Facility, Newport News, VA 23606
†Present address: Avanti Corporation, Research Triangle Park, NC 27709
‡Present address: Los Alamos National Laboratory, Los Alamos, NM 87545

CP475, *Applications of Accelerators in Research and Industry*, edited by J. L. Duggan and I. L. Morgan

using the shorthand notation $y = \frac{1}{2}P_T x \Delta\sigma_{T(L)}$.

Targets used in these measurements meet the criteria $y < 0.02 \ll 1$ and the TUNL polarized ion source is such that $P'_n \approx P''_n$ so that Equation 6 simplifies to

$$\epsilon_n = \frac{1}{2}P_n P_T x \Delta\sigma_{T(L)}. \quad (7)$$

Notably, the transmission asymmetry is independent of incident neutron flux and the unpolarized cross section.

III. DESCRIPTION OF THE EXPERIMENTAL APPARATUS

Target polarization was achieved via dynamic nuclear polarization (DNP). Dynamically polarized targets operate at a higher temperature than statically polarized targets making them less susceptible to effects of beam heating. In addition, dynamic polarization allows rapid (in $\approx$ 30 min) reversal of target spin, which is crucial for canceling the effects of instrumental asymmetries.

The target was cooled to 0.5 K in a 2.5 T magnetic field by a ^{3}He evaporation refrigerator of a PSI design. The refrigerator has a cooling power of 15 mW at 0.5 K with a ^{3}He flow rate of 0.6 mm/s. The ^{4}He cryostat, target insert, and ^{3}He refrigerator are shown in Fig. 1. Dynamic pumping was with $\approx$ 7 mW of 69 GHz microwaves.

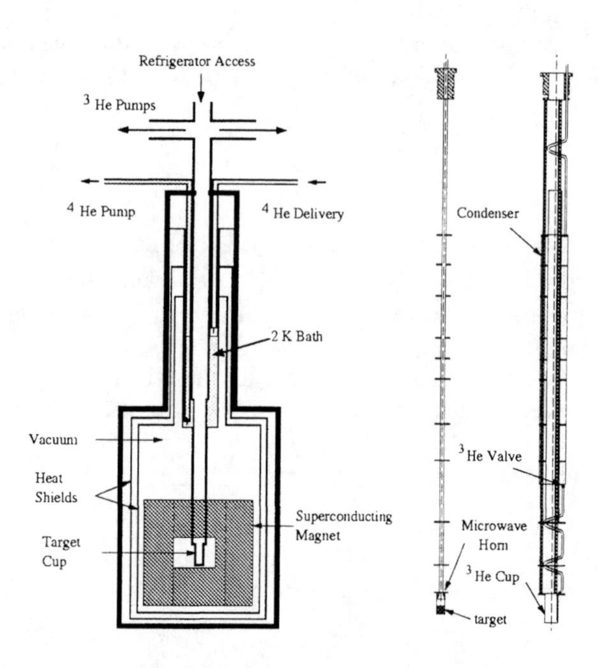

Figure 1. Schematic of the ^{4}He cryostat (left), and target insert and ^{3}He fridge (right). Beam enters the target cup from above the page. The fridge and insert are top loaded into the cryostat.

The target material was propanediol ($C_3H_6(OH)_2$, density 1.27 g/cm^3) frozen into 1 mm diameter beads, giving a hydrogen concentration of 5×10^{22} H/cm^3. The propanediol was chemically doped with EHBA-Cr^V complex to provide free electrons for DNP. Nominal target thickness was 0.06 b^{-1} with a typical $P_T = 65\%$. Target beads filled a 1.4 cm $\times$ 1.4 cm $\times$ 1.4 cm kel-F cup.

The polarized neutron beam was produced as a secondary beam via polarization-transfer reactions. The $^3\mathrm{H}(\vec{p},\vec{n})^3\mathrm{He}$ reaction was used for neutron production near 2 MeV. The $^2\mathrm{H}(\vec{d},\vec{n})^3\mathrm{He}$ reaction was used for neutron production above 5 MeV. The $^3\mathrm{H}(\vec{p},\vec{n})^3\mathrm{He}$ production target was a 1.1 mg/cm^2 thick layer of tritiated titanium evaporated on a copper beamstop. The $^2\mathrm{H}(\vec{d},\vec{n})^3\mathrm{He}$ production target was a 6.0 cm long $\times$ 1.9 cm diameter gas cell filled to 3 atm deuterium gas.

A beam of polarized protons or deuterons was produced by the TUNL atomic beam polarized ion source. Typical proton or deuteron vector polarization was of order 60%. Acceleration was provided by a 10 MV FN tandem Van de Graaff accelerator. Typical beam current on target was of order 1 μA.

Neutrons were detected at 0° by a 12.7 cm diameter $\times$ 12.7 cm long liquid organic scintillator coupled to a photomultiplier tube powered by a 14-stage transistorized base. The 0° detector was collimated to a 2.5 msrad solid angle which illuminated a 1.2 cm $\times$ 1.2 cm cross section of the target.

Due to small changes in the tensor polarization for spin-up and spin-down deuterons, a second detector to monitor incident neutron flux from the $^2\mathrm{H}(\vec{d},n)^3\mathrm{He}$ reaction was placed between the neutron production cell and the polarized target. This detector consisted of a small (11.1 mm $\times$ 22.2 mm $\times$ 25.4 mm thick) liquid organic scintillator coupled to a photomultiplier tube via a 1 m lightpipe.

Figure 2 shows the experimental arrangement of the production cell, polarized target, collimator, and neutron detectors.

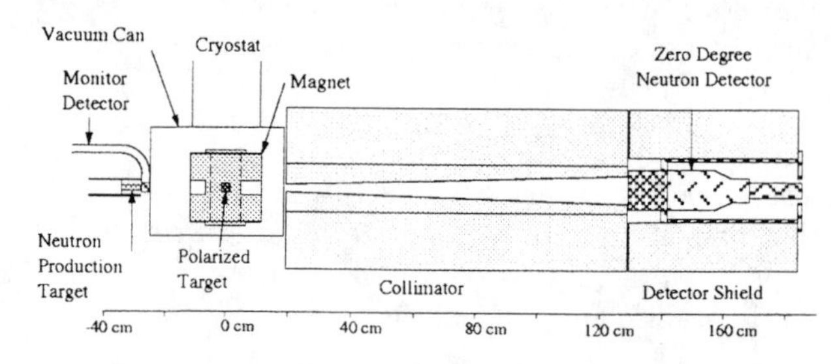

Figure 2. Schematic of the experimental setup from the neutron production cell to the 0° detector.

IV. EXPERIMENTAL PROCEDURE

$\Delta\sigma_T$ or $\Delta\sigma_L$ can be determined from measurements of a neutron transmission asymmetry, neutron beam polarization, and the product of target polarization times thickness according to Equation 7.

A. Neutron Asymmetry Measurements

A polarized beam of neutrons was incident on the polarized proton target, and the transmitted flux was measured at 0°. Transmission was measured in the transverse (vertical) and longitudinal geometries. The number of transmitted neutrons were counted with neutron beam polarization parallel (N_+) or antiparallel (N_-) to the target proton polarization. In principle, the transmission asymmetry could then be calculated:

$$\epsilon_n = \frac{N_+ - N_-}{N_+ + N_-}. \tag{8}$$

To minimize sensitivity to drifts in time to second order, the neutron beam polarization was reversed at 10 Hz in the 8-step sequence $+--+-++-$, and target polarization was reversed every 4 hrs. Pulse-shape discrimination and an energy threshold were utilized to discriminate gammas and low-energy unpolarized neutrons in the main neutron detector. No energy information was recorded with neutron events, however a separate pulsed-beam measurement verified good time-of-flight separation. An energy threshold was set in the monitor detector.

B. Beam Polarization Measurements

Neutron beam polarization was determined from measurements of charged-particle beam polarization and knowledge of polarization transfer coefficients. Charged-particle beam polarization was determined by measuring a left/right scattering asymmetry from p-^{4}He elastic scattering or the ^{3}He($\vec{d}$,p) reaction. The scattering chamber was located several meters upstream from the neutron production cell. The polarimeter cell was 2.54 cm in diameter and contained 1 atm of helium or deuterium gas.

C. Target Polarization Times Thickness Measurements

Target polarization times thickness ($P_T x$) was determined absolutely by neutron transmission, and nuclear magnetic resonance (NMR) was used to continuously monitor target polarization during $\Delta\sigma_{T(L)}$ and $P_T x$ measurements.

At low energy $\Delta\sigma_{T(L)}$ are well determined by potential models and kinematics, and so Equation 7 can be solved for $P_T x$. In this way, measuring the neutron transmission asymmetry ϵ_n allows one to calculate $P_T x$. Measurements of ϵ_n were made at $E_n = 2$ MeV and 0.87 MeV for the transverse and longitudinal geometries respectively. used. Neutron beam energy was calibrated using a ^{12}C resonance.

Relative target polarization was monitored using a PC based continuous NMR.

V. NEUTRON ASYMMETRY ANALYSIS

In practice, the measurement of a transmission asymmetry is susceptible to many systematic effects. We consider effects arising from the following sources: beam polarization dependence of the neutron production reactions, beam current asymmetries, and count-rate dependent neutron detector efficiencies.

Following Walston [2] we write Equation 1 as

$$\Phi = \Phi_0 e^{-\sigma_0 x}[1 + \frac{1}{2} P_n P_T \Delta\sigma_{T(L)} x]. \tag{9}$$

Incident neutron flux Φ_0 is given by

$$\Phi_0 = kI[1 + A_z P_z + \tfrac{1}{2} A_{zz}(0°) P_{zz}] \quad \text{(L)} \tag{10}$$

$$\Phi_0 = kI[1 + A_y P_y - \tfrac{1}{4} A_{yy}(0°) P_{yy}] \quad \text{(T)} \tag{11}$$

which depends on the charged-particle beam current I and vector P_i and tensor P_{ii} polarizations, and the production reaction analyzing powers A_i and A_{ii}. The parameters A_i are small, and non-zero only if there are spin or detector misalignments. We model the detector efficiencies by including linear and quadratic terms

$$N = (\alpha + \beta\Phi)\Phi \tag{12}$$

where N represents the number of neutrons reported as counted by the detector.

Calculating the logarithmic derivative of N and examining small changes in neutron count rate due to reversing the vector polarization of the beam (assuming $|P_i^+| \approx |P_i^-|$) gives the asymmetry

$$\epsilon_N = (\epsilon_I + \epsilon_n + \kappa\epsilon_{P_i} + \kappa'\epsilon_{P_{ii}})m \tag{13}$$

where the measured neutron transmission asymmetry ϵ_N is expressed in terms of asymmetries due to charged-particle beam current (ϵ_I) and vector (ϵ_{Pi}) and tensor (ϵ_{Pii}) polarizations and the spin-dependent cross section (ϵ_n). κ and κ' are respectively functions of A_i and A_{ii}, and m is given by

$$m = \frac{1 + 2\frac{\beta}{\alpha}\Phi}{1 + \frac{\beta}{\alpha}\Phi} \tag{14}$$

which parameterizes the non-linear detector gain. We note that the *measured* neutron asymmetry ϵ_N scales linearly with the *true* asymmetry ϵ_n, but includes false asymmetries related to beam effects. The parameter m is associated with a count-rate dependent efficiency.

A graphical method was used to extract the true asymmetry ϵ_n from the measured asymmetry ϵ_N. Data were collected in 800 ms units during which time the beam polarization was reversed in the 8-step sequence. Measured neutron and beam current asymmetries were calculated for each spin-flip sequence. A plot of ϵ_N binned vs. ϵ_I was then created, an example of which is shown in Fig. 3. The intercept of a least-squares fit to the data is $\epsilon_N(\epsilon_I = 0)$ and the slope is m. The true asymmetry can then be expressed

$$\epsilon_n = \frac{\epsilon_N(\epsilon_I = 0)}{m} - \kappa\epsilon_{P_i} - \kappa'\epsilon_{P_{ii}}. \qquad (15)$$

We have found m values that increase with beam energy and deviate from unity by 30% at 20 MeV.

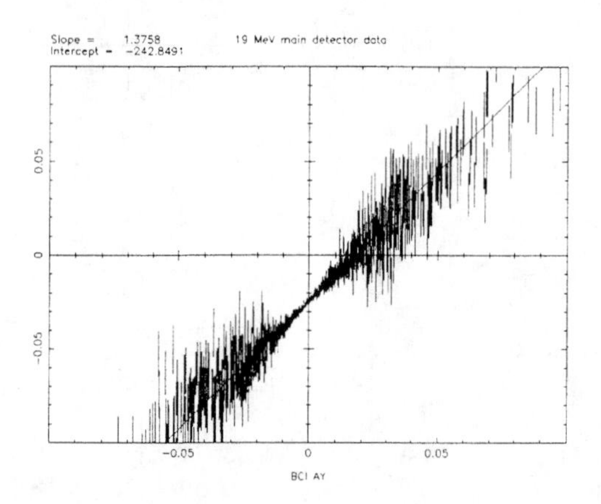

Figure 3. Measured neutron asymmetry ϵ_N binned in beam current asymmetry ϵ_I. Error bars include both statistical and systematic uncertainties. The intercept is $\epsilon_N(\epsilon_I = 0)$ and the slope is m.

The target polarization was then reversed from $+P_i$ to $-P_i$ and as a result the sign of ϵ_n is reversed. The vector and tensor polarization terms are made small by use of a monitor detector and are further canceled by noting that these terms are unchanged when the target polarization is reversed. The difference in ϵ_n for target polarization states gives

$$\epsilon_n^+ - \epsilon_n^- = \left[\frac{\epsilon_N(\epsilon_I = 0)}{m}\right]^+ - \left[\frac{\epsilon_N(\epsilon_I = 0)}{m}\right]^- \qquad (16)$$

noting the kappas cancel and that m is not target spin dependent but depends on count rate and will in general be different for any given measurement. We define the average spin-dependent asymmetry $\bar{\epsilon}_n$ such that

$$\bar{\epsilon}_n = \frac{1}{2}[\epsilon_n^+ - \epsilon_n^-] \qquad (17)$$

which is found from the slopes and intercepts of two graphs.

VI. RESULTS

Measurements of $\Delta\sigma_T$ were made between $E_n = 11$ and 18 MeV from 12/95 through 1/96 [1]. $\Delta\sigma_L$ measurements covered an energy range from $E_n = 5$ through 18 MeV and were made between 8/96 and 6/97 [2]. Results are plotted in Fig. 4. The 5 and 7 MeV values of $\Delta\sigma_T$ are from Wilburn [4]. Also included in these figures are the recent Prague measurements [5] and the Nijmegen partial-wave analysis PWA93 [6].

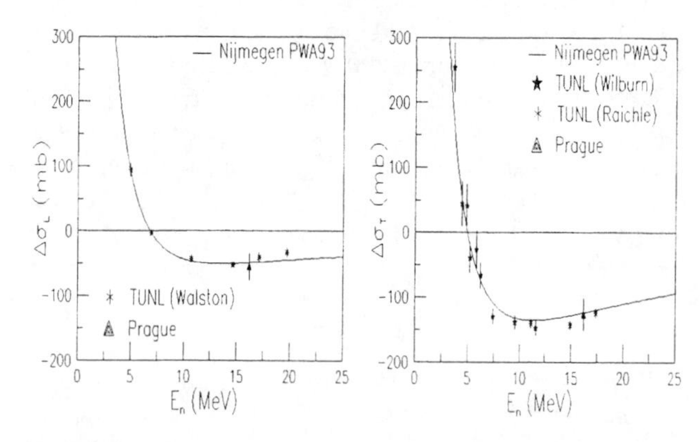

Figure 4. Plot of experimental values of $\Delta\sigma_L$ and $\Delta\sigma_T$. Error bars include statistical and systematic uncertainties.

VII. SUMMARY

We have performed polarized neutron beam-polarized proton target transmission measurements to determine the spin-dependent n-p total cross-section differences $\Delta\sigma_L$ and $\Delta\sigma_T$. There is general agreement between the trend of the TUNL data and theory. The use of a dynamically polarized reduced systematic errors considerably, and the present results have significantly smaller uncertainties than previous measurements at TUNL and at Prague. These new data will be used in a phase-shift analysis to extract the parameter ε_1 which characterizes the strength of the n- p tensor interaction at low energy.

VIII. ACKNOWLEDGEMENTS

We gratefully acknowledge the support of the U.S. Department of Energy, Office of High Energy and Nuclear Physics, under Grant Nos. DEFG05-91-ER40619 and DEFG05-88-ER40441. We also acknowledge the support of the Robert A. Welch Foundation.

[1] B. W. Raichle, Ph.D. thesis, North Carolina State University, 1997.

[2] J. R. Walston, Ph.D. thesis, North Carolina State University, 1998.

[3] W. Tornow *et al.*, in *Physics with Polarized Beams and Polarized Targets*, edited by J. Sowinski and S. Vigdor (World Scientific, Singapore, Indiana, 1989), pp. 75–93.

[4] W. S. Wilburn *et al.*, Physical Review C **52**, 2351 (1995).

[5] J. Brož *et al.*, Zeitschrift für Physik A **354**, 401 (1996).

[6] V. G. J. Stoks, R. A. M. Klomp, M. C. M. Rentmeester, and J. J. de Swart, Physical Review C **48**, 792 (1993).

An Algorithm for Computing Thick Target Differential p-Li Neutron Yields Near Threshold

C. L. Lee and X.-L. Zhou

Department of Nuclear Engineering,, Massachusetts Institute of Technology, Cambridge, MA 02139.

The ^{7}Li(p,n)^{7}Be reaction is a good source of neutrons for accelerator boron neutron capture therapy (BNCT). Both reactor and accelerator neutron sources produce fast neutrons, which must be moderated since BNCT uses epithermal neutrons. Near-threshold BNCT uses proton energies only tens of keV above the reaction threshold, which reduces the thick target neutron yield but also produces neutrons closer to epithermal energies, so that less moderation is required. Accurate methods for calculating near-threshold differential neutron yields from thick targets of lithium, as well as certain low weight lithium compounds, were developed for BNCT source design. Neutron yields for proton beams up to 2.8 MeV will be presented. Good agreement with yields from several targets will be demonstrated.

INTRODUCTION

Boron neutron capture therapy (BNCT) is a binary therapy in which a compound containing ^{10}B is preferentially introduced to a treatment site, such as a glioblastoma multiforme brain tumor. The patient is then irradiated with an epithermal (roughly 1 eV to 10 keV) neutron beam. These neutrons slow down in the head and thermalize at the ^{10}B-loaded tumor, producing a large number of ^{10}B(n,α)^{7}Li reactions whose high linear energy transfer (LET) products deposit dose that is highly localized to the treatment site. Near-threshold BNCT, a novel accelerator based BNCT concept studied under a collaboration between the Massachusetts Institute of Technology (MIT), Idaho State University (ISU), and the Idaho National Engineering and Environmental Lab (INEEL), uses protons with energies only tens of keV above the ^{7}Li(p,n)^{7}Be reaction threshold of 1.88 MeV (1,2). This reduces the thick target neutron yield relative to a higher (and more commonly used) proton beam energy, such as 2.5 MeV, but the neutrons that are produced near threshold have substantially lower kinetic energies than those produced with higher proton energies. This is advantageous for BNCT, since these lower energy neutrons require less moderation to reach the desired epithermal energy range.

In order to model near-threshold BNCT target assemblies, accurate thick target neutron yields, as a function of energy and angle in the lab, are needed. It is desirable that the method for calculating these yields also be capable of calculating neutron yields for higher proton energies in a self-consistent manner, in order to facilitate comparisons between proton energies of interest. In addition, since the low melting point (181°C) and high reactivity of lithium may limit its practical usefulness as an accelerator target for medical applications, accurate modeling of certain lithium compounds is desirable. A calculational method was developed that satisfies these criteria, and this method was implemented into a computer code designed for a high degree of user-friendliness. This paper will outline the major aspects of the method, present differential neutron yields for several proton energies of interest in accelerator BNCT, and provide several examples of experimental verification of these yields.

THEORY

It may been shown that differential thick target yields for (p,n) reactions, $d^2Y/d\Omega dE_n$, may be calculated using the following expression (3):

$$\frac{d^2Y}{d\Omega dE_n}(\theta,E_n)=\frac{fN_o}{eA}\frac{\frac{d\sigma}{d\Omega'}\frac{d\Omega'}{d\Omega}\frac{dE_p}{dE_n}}{\frac{1}{\varrho}\frac{dE_p}{dx}} \tag{1}$$

where $d\sigma/d\Omega'$ is the center-of-mass (CM) differential cross section for the (p,n) reaction, $d\Omega'/d\Omega$ is the transformation Jacobian from the CM frame to the lab frame, dE_p/dE_n is the transformation Jacobian between proton and neutron energies, N_o is Avogadro's number, f is the fraction of lithium atoms that are ^{7}Li, e is the electronic charge, A is the atomic mass of lithium, and $1/\rho\ dE_p/dx$ is the mass stopping power of the lithium target. Tabulated experimental mass stopping powers are used for protons

CP475, *Applications of Accelerators in Research and Industry*,
edited by J. L. Duggan and I. L. Morgan

striking lithium metal (4). For proton energies above 1.95 MeV, Liskien and Paulsen's extensive set of tabulated $^7Li(p,n)^7Be$ cross section values (5) are used in Equation (1). For near-threshold proton energies, however, it is necessary to use a simple, analytical form for the cross section, which has been shown to very accurately agree with experiment (6).

Note that Equation (1) implies the differential neutron yield is a pointwise function of neutron angle and energy. It is also important to note that the Jacobian terms in Equation (1) become infinite as the proton energy approaches threshold, 1.88 MeV, while the CM cross section approaches zero. This indeterminate expression cannot be evaluated with tabulated cross sections, but the analytical form that is unique to the $^7Li(p,n)^7Be$ reaction near threshold, when combined with the analytical expressions for the Jacobians in the differential yield expression, produces a finite, non-zero value for the differential neutron yield at threshold (7). By assuring a smooth transition from the analytical form for the cross section in the near-threshold region to the tabulated form used at high proton energies, a self-consistent set of neutron yields is produced.

It is straightforward to modify this method to generate neutron yields for lithium compounds. In Equation (1), the only terms that are dependent on target material are the atomic mass A, which is changed to the molecular mass M of the lithium compound, and the mass stopping power, for which tabulated values for compounds may be determined using the additivity rule. An additional term n, the number of lithium atoms in a unit cell, is necessary in the denominator.

These techniques, described in greater detail elsewhere, were implemented in a computer code, **LIYIELD** (7). This code was designed to produce differential thick target neutron yields for natural lithium, LiH, Li_2O, LiOH, Li_3N, and LiF targets, for proton energies from threshold to 5 MeV. Since the program was written primarily for near-threshold applications, only the reaction leading to the ground state of ^7Be is included. The output file of differential yields may be written in pointwise format or else in the energy and angular bin format appropriate of a source (sdef) file for the Monte Carlo code MCNP (8).

Figure 1 shows results of calculations using **LIYIELD** for three proton beam energies of interest in current accelerator BNCT research. The representative near-threshold energy 1.91 MeV given in Fig. 1 (a) demonstrates the kinematic forward collimation that is observed for all proton energies below 1.92 MeV. In addition, proton energies of 2.3 to 2.5 MeV are used in most accelerator-based BNCT neutron source designs, so the yields given in Fig. 1 (b) and (c) are of interest to many BNCT researchers.

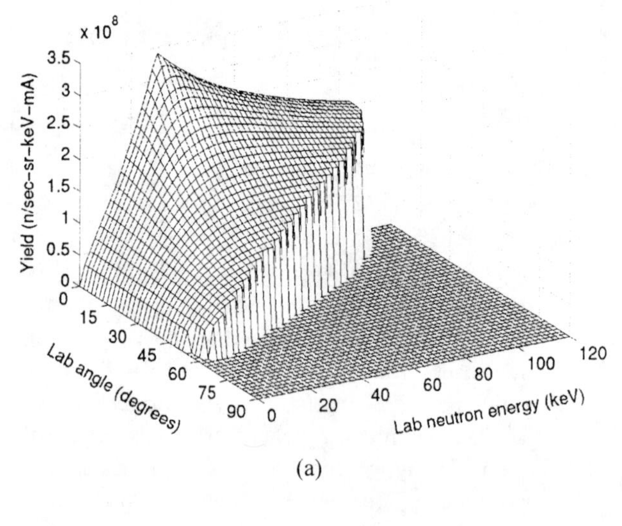

(a)

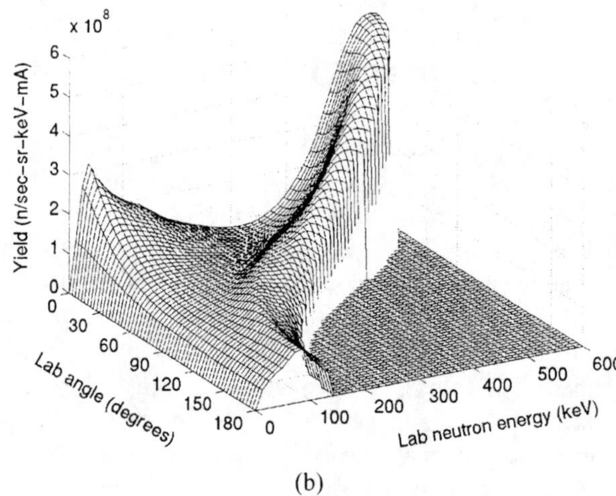

(b)

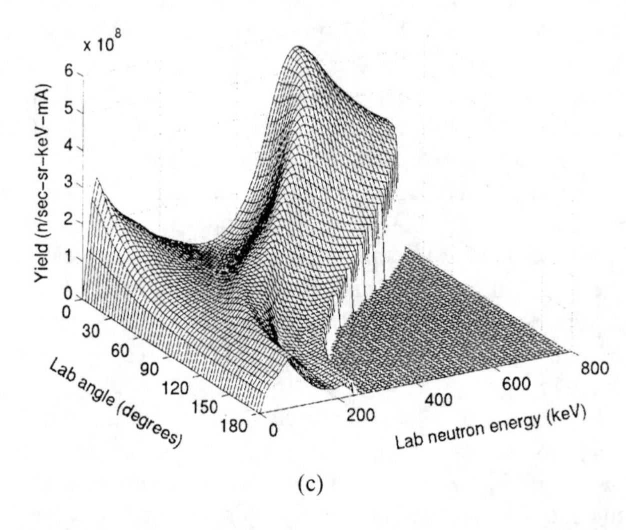

(c)

FIGURE 1. Calculated differential thick target neutron yields from protons striking natural lithium metal. Incident proton energies are (a) 1.91 MeV, (b) 2.3 MeV, and (c) 2.5 MeV.

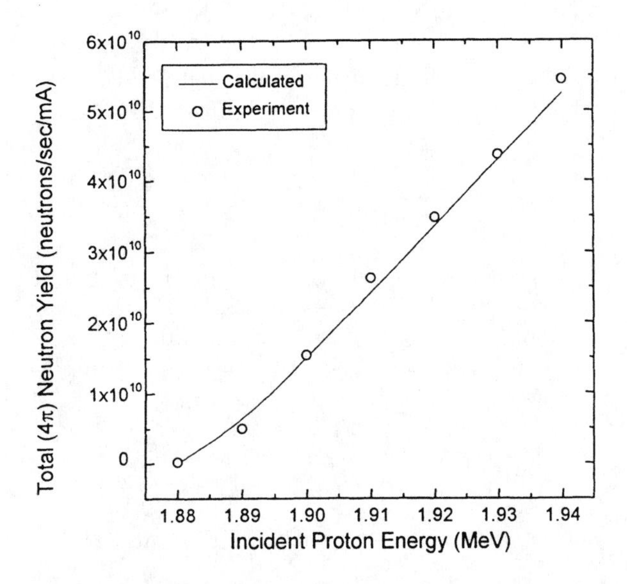

FIGURE 2. Total neutron yields for near-threshold proton beams on natural lithium metal. Good agreement is seen between calculated and experimental yields.

EXPERIMENTS AND RESULTS

Experimental verification of the total (4π) neutron yields for natural lithium metal are shown in Fig. 2. All neutron yield measurements were performed at Idaho State University. A major concern in these measurements was the formation of a corrosion product layer on the lithium surface, which could seriously impact the yield measurement. In order to remove this effect, the lithium target was formed inside the accelerator beam tube under vacuum. A piece of lithium was placed in a small wire cage at the based of a stainless steel (type 304) backing inside a Van de Graaff beam tube (9). The cage was placed below the proton beam area to prevent interference with the beam once irradiation began. The wire leads were attached outside the tube to a Variac voltage controller. Once a vacuum was established in the beam line, the Variac voltage was increased, vaporizing the lithium in the cage and depositing it on the stainless steel backing. Deposition times were increased until yield measurements reached a plateau, indicating that the proton beam was slowing past the reaction threshold in the target. This criterion was satisfied when the Variac voltage remained on for 10-15 minutes.

The total neutron yield was measured using a 4π detector employing 12 18-inch (45.72 cm) long ^{3}He thermal neutron detectors. The counter was designed to have a constant neutron detection efficiency for neutron energies up to 100 keV (10). The end flange of the beam line was placed at the midpoint of the central hole of the counter, and a paraffin plug was placed in the other end. The counter was calibrated before each measurement using a standard AmBe source. The relative error of each data point in Fig. 2 is about 5%, primarily due to fluctuations in the energy of the proton beam.

Additional experiments were performed to measure total neutron yields for LiF (Fig. 3) and Li_2O (Fig. 4). As with the experimental data for lithium metal given in Fig. 2, the relative error for all experimental points is about 5%. The agreement between calculation and experiment verifies not

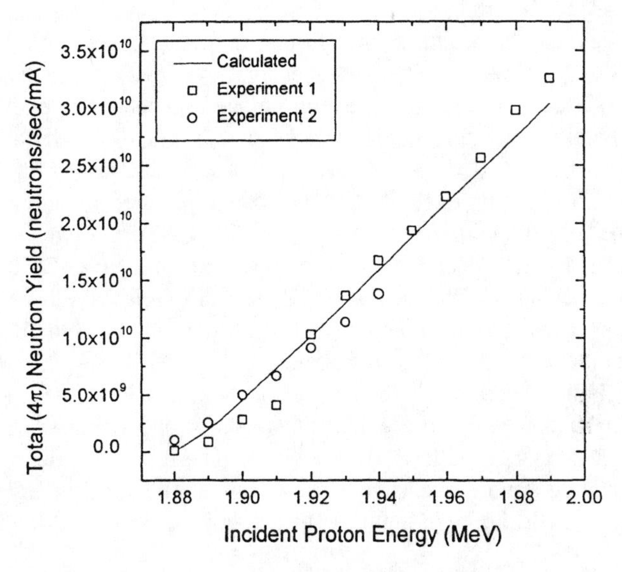

FIGURE 3. Total neutron yields for near-threshold proton beams on LiF. Two experiments were performed for this target, and reasonable agreement with calculated yields was found.

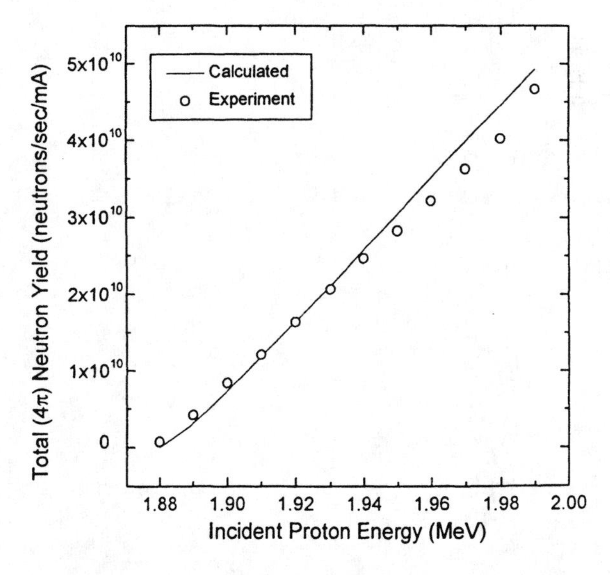

FIGURE 4. Total neutron yields for near-threshold proton beams on Li_2O. There is good agreement between calculated and experimental yields.

only the capabilities of the calculational technique, but also the validity of the additivity rule for stopping powers for these compounds.

CONCLUSIONS

In summary, a calculational method has been developed to generate differential thick target neutron yields for the $^7Li(p,n)^7Be$ reaction for proton energies from threshold to 5 MeV. Yields may be produced from lithium metal targets, as well as certain low mass compounds of lithium. Output files may be produced in a format immediately usable by MCNP. The algorithm used by the **LIYIELD** computer code takes advantage of several unique aspects of the p-Li reaction near threshold, such as the excellent analytical representation of the CM differential cross section in this region. The time required by the code to produce an output file of differential neutron yields in 1°, 10 keV intervals is less than 30 seconds on HPUX 9000 and Sun SPARC 20 workstations. The neutron yields near threshold have been experimentally benchmarked for lithium metal, lithium oxide, and lithium fluoride targets.

Future work includes improvements to **LIYIELD** to make the code more applicable to the broad interests of accelerator-based BNCT researchers. These improvements include a more extensive array of output file formats and greater flexibility in input file formatting.

ACKNOWLEDGMENTS

This work was supported by the INEEL University Research Consortium Grant No. DE-AC07-94ID13223.

REFERENCES

1. Kudchadker, R. J., Kunze, J. F., and Harmon, J. F., *Proceedings of the 7th International Symposium on Neutron Capture Therapy for Cancer*, Zurich, Switzerland, 1996.
2. Zhou, X.-L., and Lee, C. L., *Appl. Rad. Isotopes* **48**, 1571 (1997).
3. Ritchie, A. I. M., *J. Phys. D* **9**, 15 (1976).
4. Janni, J. F., *At. Data Nucl. Data Tables* **27**, 147 (1982).
5. Liskien, H., and Paulsen, A., *At. Data Nucl. Data Tables* **15**, 57 (1975).
6. Gibbons, J. H., and Macklin, R. L., *Phys. Rev.* **114**, 571 (1959).
7. Lee, C. L., *The Design of an Intense Accelerator-Based Epithermal Neutron Beam Prototype for BNCT Using Near-Threshold Reactions*, Massachusetts Institute of Technology Ph.D. thesis, 1998.
8. Briesmeister, J. F., ed., *MCNP – A General Monte Carlo N-Particle Transport Code, version 4A*, Los Alamos National Laboratory: Publication number LA-12625-M, 1993.
9. Bartholomay, R. W., *BNCT Benchmark Experiments for Near Lithium Threshold Target Performance*, Idaho State University Master's thesis, 1998.
10. Kudchadker, R. J., *A Precision Neutron Detector for (p,n) Reaction Measurements*, Idaho State University Master's thesis, 1992.

Excitation of the $^{180}Ta^m$ Isomer in (γ,n) Reactions

A. P. Tonchev, J. F. Harmon, B. D. King

Physics Department, Idaho State University, Pocatello, ID 83209

The yield has been determined for the excitation of $^{180}Ta^g$ ($J^\pi = 1^+$) from the $^{181}Ta(\gamma,n)$ reaction by measurements of γ rays emitted following the electron capture and β decays of 8.15 h ground state. The probability $\sigma_m / (\sigma_m + \sigma_g)$ for the production of the $^{180}Ta^m$ isomer ($J^\pi = 9^-$) after γ absorption was deduced. The role of initial and final spin on reaction yield, along with the relevance for nucleosynthesis of $^{180}Ta^m$, is discussed.

INTRODUCTION

During the last decade ^{180}Ta has to be one of the most intensely studied isotopes. The natural form of this isotope is an isomer, because the ground state $^{180}Ta^g$ has a half-life of 8.1 hr, while that the isomer is ≥ 1.2 10^{15} yr. The isomer has spin 9^- and excitation energy 75.3 keV, and the ground state has spin 1^+. According to the Nilson diagram, the odd proton belongs to the $9/2^-$ [514] orbital, and the neutron to the $9/2^+$ [624] orbital, and the isomer spin is the sum of these angular momenta. Due to the large difference between the spins of the isomeric and ground states (ΔI = 8) and to the absence of levels located below to the isomeric state, with close spin values, the ^{180}Ta nucleus is practically stable in its isomeric state. Because of this, ^{180}Ta has become the subject of study in nuclear physics as a natural isomeric target [1-5]. High-spin targets may have significant absorption cross sections in comparison with usual nuclei, due to nuclear structure, and possible alteration of deformation, and the level density of the compound nucleus. It is very important to establish the existence of some selective population of the levels in the residual nucleus with structure similar to that of the initial nucleus.

In addition to its unusual nuclear structure, ^{180}Ta has important astrophysical consequences [6-11]. This element was found to be the rarest stable isotope in the universe, with abundance of 1.2 10^{-4}. The ratio of $^{180}Ta/^{181}Ta$ is about an order of magnitude smaller compared to other rare odd-odd nuclei like $^{50}V/^{51}V$ = 0.24 or $^{138}La/^{139}La$ = 0.09. It is evident from Fig.1 that ^{180}Ta is bypassed by the slow (*s*) neutron-capture process which proceeds through the stable hafnium isotopes. Furthermore, it is shielded from the β^- and β^+ decay paths following the rapid (*r*) neutron-capture process.

From the preceding discussion, it is apparent that in order to decide among the various proposed ^{180}Ta production mechanisms an important consideration is the relative production cross sections for the short-lived and long lived ^{180}Ta states. In the present experiment we measured the production cross section of the $^{180}Ta^g$ from $^{181}Ta(\gamma,n)$ reaction, in energy range from threshold to 13 MeV, using the activation technique. The cross section for the production of the long-lived $^{180}Ta^m$ has been determined by subtracting the $^{180}Ta^g$ cross section from the total ^{180}Ta cross section. These measurements are compared with the results of a statistical model evaporation calculation. The cross section measurements will allow new information necessary for calculating the abundance of these isotopes in nature to be obtained and will also help in drawing conclusions about the conditions in which nucleosynthesis has taken place (matter density, temperature).

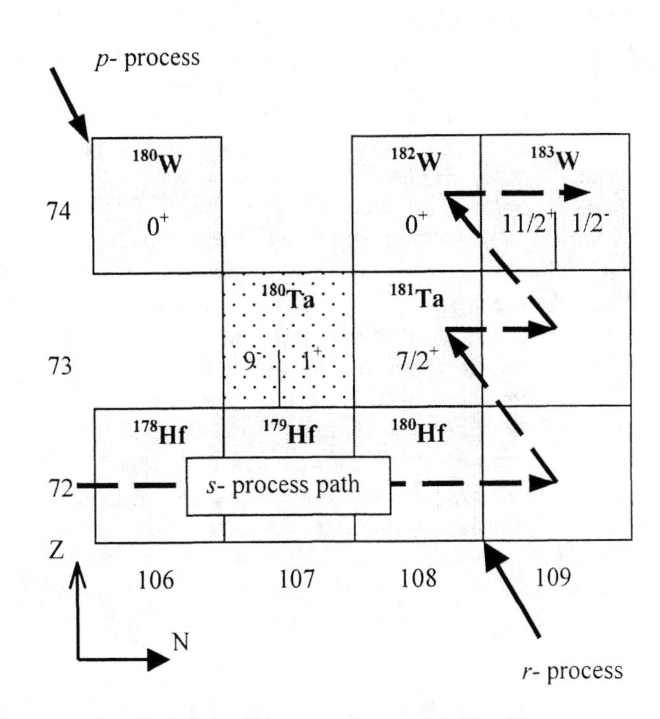

FIGURE 1. Region of the periodic table relevant to our investigation. The *s*-, *r*-, and *p*-process are shown.

CP475, *Applications of Accelerators in Research and Industry*, edited by J. L. Duggan and I. L. Morgan

EXPERIMENTAL PROCEDURE

Measurements of the $^{180}Ta^g$ production cross section were performed with the linear electron accelerator at the Idaho Accelerator Center at Idaho State University. To produce bremsstrahlung radiation, a convertor is placed in the beam. For these experiments the convertor was a 0.551 cm diameter, 0.178 cm thick alloy of 80 % tungsten and 20% copper. Calculations show that for a 9 MeV electron beam, the convertor stopped about 99.6% of the incident electrons. The linear accelerator used could produce electron beam energies up to 13 MeV. The average current of the accelerated electrons usually amounted to about 10 μA.

All measurements of the tantalum samples are carried out with gold foils as a standard which is activated simultaneously. Ta and Au monitor foils were arranged coaxially with the bremmstrahlung target. In this way most systematic uncertainties can be avoided.

Table 1 presents the main decay characteristic of the nuclei in the ground state under study [13]. Isomeric state of ^{180}Ta is too long-lived ($T_{1/2} \geq 1.2\ 10^{15}$ yr) that no γ or β decay was observed. In Table 1 the main γ lines from $^{196}Au^g$ monitor after (γ,n) reaction are also listed.

Sample exposure time was varied up to one hour and then removed to a remote counting room. Delayed *X*-rays and γ rays were observed with LEGe detector equipped with 0.3 μm Be window and having an energy resolution of 600 eV (at 122 keV). The detector efficiency was determined by using the a of samples of standard γ radiation. In this way residual activity from the ground states of tantalum and gold was counted 2 and 6 hours for $^{180}Ta^g$ and $^{196}Au^g$, respectively.

ANALYSIS AND RESULTS

Experimentally, the isomeric ratios are determined in those cases where the nuclei in ground and isomeric states are radioactive, and the values of σ_g and σ_m are measured directly in the same experiment. If the final nucleus is stable in the ground state or very long-lived in isomeric state (as in $^{180}Ta^m$), it is more convenient to obtain the values of the ground state excitation probability.

For a particular γ-ray line the number of events *S* registered in the LEGe detector during the activity determination are given by the expression

$$S = AK_{\gamma}\varepsilon_{\gamma}I_{\gamma}e^{-\lambda t_c}(1 - e^{-\lambda t_m}), \qquad (1)$$

where ε_γ is the LEGe efficiency, I_γ the absolute decay intensity for the investigated γ line (Table 1), and λ is the decay constant. T_c, and t_m are the times of decay and measurement. K_γ gives the correction factor for γ-ray absorption, dead time and geometric factors. For disk shaped samples of thickness *d* one obtains $K_{\gamma ab}=1/d\mu(1-e^{d\mu})$. The γ absorption coefficients μ were taken from Ref. 12.

TABLE 1. Photons from the (EC, β^-) decay of $^{180}Ta^{g*}$ and $^{196}Au^g$

Emitted photons	E_γ [keV]	I_γ [%]
Hf $K_{\alpha 2}$	54.611	20.4 ± 0.8
Hf $K_{\alpha 1}$	55.790	35.7 ± 1.3
W $K_{\alpha 2}$	57.981	0.17 ± 0.03
W $K_{\alpha 1}$	59.318	0.29 ± 0.05
Hf K_β	63.20	15.0 ± 0.6
Hf $K_{\beta 1}$	67.155	0.097 ± 0.016
Hf $K_{\beta 2}$	69.342	0.025 ± 0.004
γ Ta EC	93.331	4.51 ±0.16
γ Ta β^-	103.60	0.81 ± 0.24
γ Au EC	355.68	87 ± 0.01
γ Au EC	332.98	22.9 ± 0.006

*$T_{1/2}$ = 8.152 h

The number of activated nuclei *A* can be written as

$$A = N\,\phi\,\sigma f_i \qquad (2)$$

Here, $\phi = \int\phi(t)dt$ is the time integrated bremsstrahlung spectrum [14], *N* the number of nuclei in the sample, and σ the (γ, n) cross section [16]. The factor

$$f_i = \int_0^{t_i} \phi(t)\,(1 - e^{-\lambda(t-t_i)})\,dt / \int \phi(t)\,dt \qquad (3)$$

accounts for the decay of activated nuclei during irradiation time t_i. For a constant γ flux, f_i reduced to

$$f_i = \frac{1}{\lambda t_i}(1 - e^{-\lambda t_i}) \qquad (4)$$

As the measurements are carried out relative to the ^{197}Au as a standard, the γ flux ϕ cancels out the first order, if the number of activated nuclei is normalized to gold:

$$\frac{A_{Ta}}{A_{Au}} = \frac{\sigma_{Ta}}{\sigma_{Au}}\frac{N_{Ta}}{N_{Au}}\frac{f_{Ta}}{f_{Au}} \qquad (5)$$

The $^{197}Au(\gamma,n)^{196}Au^g$ reaction is convenient to use as a standard, owing to the absence of resonance structure near the threshold energy for those nucleus, the intense γ lines from its decay, the possibility of including the neutron contribution to the measured yields, and the simplicity of preparation of the target. Ground state activity of $^{196}Au^g$ represents also the commutative yields from both high-lying isomeric states $^{196}Au^{m1}$ ($T_{1/2}$=8.1 s) and $^{196}Au^{m2}$ ($T_{1/2}$=9.7 h) if this activity was measured after enough cooling time. In our case $^{196}Au^g$ measurement was done one week after the irradiations.

The resulting dependence of the photonuclear reaction yield on the maximum energy of the bremsstrahlung radiation serves as the basis for determining the reaction

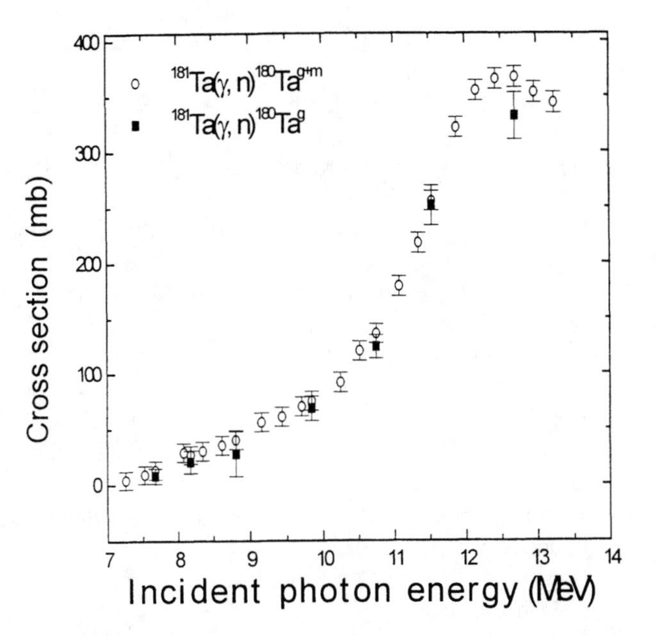

Figure 2. Circles represent total ^{180}Ta production cross section and squares represent ^{180}Ta^{g} production cross section from ^{181}Ta(γ,n) reaction.

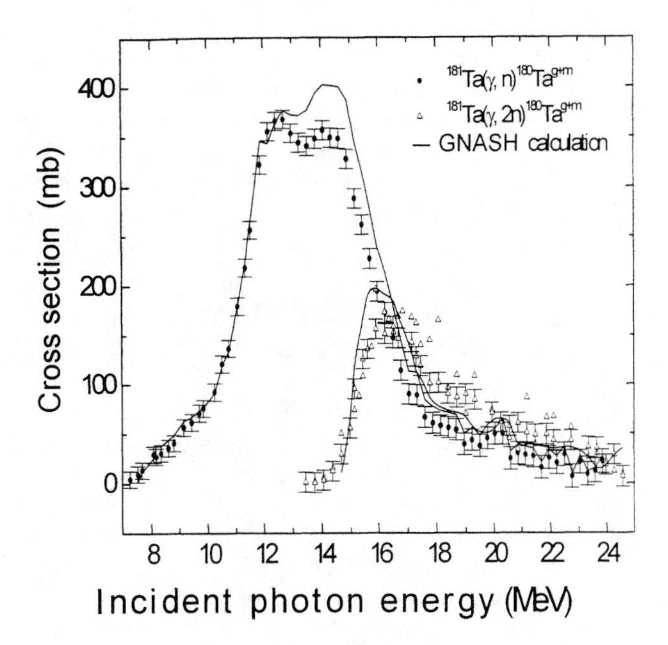

Figure 3. Comparisons of GNASH calculations of ^{181}Ta(γ,n) and ^{181}Ta(γ,2n) cross sections with the experimental data.

cross sections and the excitation function. As in previous papers [17-19], to reconstruct reaction cross section, an iterative procedure for minimizing directed divergences was employed [20]. The step used in reconstructing the cross section was 1 MeV.

In Fig.2 the experimental results of ^{180}Ta^{g} cross section after (γ,n) reaction are presented. The experimental data [16] from ^{181}Ta(γ,n)^{180}Ta^{g+m} are also indicated in the same figure. The ^{180}Ta^{g} cross section may be subtracted from the ^{180}Ta^{g+m} cross section to obtain the cross section for the production of ^{180}Ta^{m}. Because of the small differences between σ(^{180}Ta^{g+m}) and σ(^{180}Ta^{g}) and relatively large experimental uncertainties, we interpret these differences as upper limits on the ^{180}Ta^{m} cross section. These results are summarized in Tab. 2 along with values for σ(^{180}Ta^{m}) which were calculated using a Hauser-Feshbach theory described below.

TABLE 2. Results for the ^{180}Ta^{g+m} and ^{180}Ta^{m} production cross sections for the ^{181}Ta(γ,n) reaction.

Photon Energy (MeV)	σ(^{180}Ta^{g+m}) (mb)	σ(^{180}Ta^{m})^{a} (mb)	σ(^{180}Ta^{m})^{b} (mb)
9.85	75.3±9	<1	0.4
10.75	136±9	2	3
11.53	256±9	7	4
12.43	365±9	12	7

^{a} Obtained by subtracting σ(^{180}Ta^{g}) from σ(^{180}Ta^{g+m}).
^{b} Obtained from GNASH calculations described in text.

THEORETICAL CALCULATIONS

The excitation functions of reactions induced by gamma rays are calculated according to the statistical model of nuclear reactions in the Hauser-Feshbach theory, which takes into account the laws of conservation of the total angular momentum and parity. The new version of GNASH code [21] was used to calculate the absolute photonuclear reaction cross section and deexcitation cross section of discrete states in compound or residual nuclei. It is assumed that in a given range of Z and A (intermediate and heavy nuclei) and bombarding γ energy (below 20 MeV), reactions with compound-nucleus formation and establishment of thermal equilibrium in the nucleus dominate. Direct and semidirect processes give a comparatively small contribution. These assumptions are also used to calculate the isomeric ratio.

The first test of these calculations was photonuclear modeling of (γ,n) reactions on ^{181}Ta. In Fig. 3 we show measurements [16] of the ^{181}Ta(γ,1n) excitation function and the photoneutron yield cross section, compared with GNASH calculations. These calculations are seen to describe the measurements well. To calculate the isomer production sections, experimental information of discrete levels in residual (^{180}Ta^{m}) are needed. In Ref. 22 γ-decays into 9^{-} isomer rotational band was identified. This experimental information together with the existing discrete levels was completed in a γ-cascade filling isomeric or ground state. The last column on Fig. 3 listed the theoretical calculations of isomeric cross section after (γ,n) reactions. The differences between GNASH calculations and experimental information are more significant at the region close to neutron threshold reactions. This is not a surprise if we take into account the large experimental uncertainties of restoring the cross section in this energy range. Theoretical calculations were extended to the maximum of Giant Dipole Resonance where (γ,n) reactions reach their maximum. The maximum values of $\sigma(^{180}Ta^{m})$ = 9 mb were obtained at 14.6 MeV. At this energy, isomeric ratios (σ^{m}/σ^{g}) were calculated to be 0.03, which means 3 % of (γ,n) cross section leads to filling the high-spin isomeric state.

CONCLUSIONS

The possibilities of isomer production of the long-lived $^{180}Ta^{m}$ isomer via (γ,n) reactions was investigated at energies E_{γ}=8-13 MeV. The $^{180}Ta^{m}$ cross section was obtained by subtracting the $^{180}Ta^{g}$ cross section from the total $^{180}Ta^{g+m}$ cross section. At this energy interval, ground state production dominated (by 97 %) after the photoneutron reaction. But the $^{181}Ta(\gamma,n)^{180}Ta^{m}$ cross section may have yielded the observed $^{180}Ta^{m}$ abundance. This might explain the solar $^{180}Ta^{m}$ abundance.

REFERENCES

1. Belov, A.G., Gangrsky, Yu.P., Tonchev, A.P., and Zuzaan, P.,*Hyperfine Interactions* **107,** 167-173 (1997).
2. Karamian, S.A., Collins, C.B., Carroll, J.J., and Adam, J., *Phys. Rev.* C **57,** 1812-1816 (1998).
3. Karamian, S.A., De Boer, J., Oganessian Yu.Ts. et al., *Z. Phys.* A **356,** 23-29 (1996).
4. Norman, E.B., Renner, T.R., Grant, P.J., *Phys. Rev.* **26,** 435-440 (1982).
5. Collins, C.B., Eberhard, C.D., Glesener, J.W., Anderson, J.A., *Phys. Rev.***37,** 2267-2269 (1988).
6. Kellogg, S.E., Norman, E.B., *Phys. Rev.* **46,** 1115-**1131** (1992).
7. Schlegel, C., von Neumann-Cosel, P., Neumeyer F., et al., *Phys. Rev.* C **50,** 2198-2204 (1994).
8. Nemeth, Zs., Kappeler, F., Reffo, G., *Astrophys J.* **392,** 277-283 (1992).
9. Howard, W.M., Meyer, B.S., Woosley, S.E., *Astophys. J.* **373,** L5-L8 (1991).
10. Hainebach, K.L., Schramm, D.N., Blake, L.B., *Astophys. J.* **205,** 920-930 (1976).
11. Rayet, M., *Advances in nuclear astrophysics,* Paris:Editions Frontieres, 1986, pp. 585-593.
12. Storm, E., and Israel, H.I., *Nucl. Data Tables* **A7,** 566-681 (1970).
13. Firestone, R.B., *Tables of Isotopes*, New York: Wiley Press, 1996.
14. Kondev, Ph.G., Tonchev, A.P., Khristov, Kh.G., and Zhuchko, V.E., *Nucl.Instr. and Methods in Phys.Res.* B **71,** 126-131 (1992).
15. The data were taken from the Brookhaven National Laboratory SCISRS file.
16. Bergere, R., Beil, H., Veyssiere, A., *Nucl. Phys* A **121,** 463-481 (1968).
17. Tonchev, A.P., Gangrsky, Yu.P., Belov, A.G., and Zhuchko, V.E., *Phys. Rev.* C **58,** 1-7 (1998).
18. Tonchev, A.P., Harmon, J.F., King, B., *Nuc. Instr. and Methods in Phys. Res.* A, in press, (1998).
19. Gangrsky, Yu.P., Tonchev, A.P., Balabanov, N.P., *Phys.Part. Nucl.* **27,** 428-452 (1996).
20. Zhuchko, V.E., *Sov. J. Nucl. Phys.* **25,** 124-125 (1977).
21. Arthur, E.D., "Comprehensive Nuclear Model Calculations: The GNASH Preequilibrium Statistical Model Code," presented at the ICTP Workshop on Applied Nuclear Theory and Nuclear Model Calculations for Nuclear Technology Applications, Trieste, Italy, February 15-March 18, 1998.
22. Dracoulis, G.D., Mullins, S.M., Byrne, A.P., et al., *Phys. Rev.* C **58,** 1444-1466 (1998).

PARITY VIOLATION IN 2s–1d NUCLEI USING THE (p,α) REACTION

B. E. Crawford,[1] G. E. Mitchell,[1] N. R. Roberson,[2] J. F. Shriner, Jr.,[3], L. K Warman,[3] W. S. Wilburn[4]

[1] *North Carolina State University, Raleigh, North Carolina 27695 and Triangle Universities Nuclear Laboratory, Durham, North Carolina 27708*
[2] *Duke University, Durham, North Carolina 27708 and Triangle Universities Nuclear Laboratory, Durham, North Carolina 27708*
[3] *Tennessee Technological University, Cookeville, Tennessee 38505 and Triangle Universities Nuclear Laboratory, Durham, North Carolina 27708*
[4] *Los Alamos National Laboratory, Los Alamos, New Mexico 87545*

In order to extend our knowledge of the weak spreading width to the nuclear mass 30 region, parity-violation measurements have begun at the Triangle Universities Nuclear Laboratory using the $^{31}P(p,\alpha_o)^{28}Si$ reaction. Longitudinally polarized protons are scattered from ^{31}P nuclei at several MeV, and a large solid angle silicon-strip detector detects the alpha particles at backward angles. By rapidly reversing the spin direction of the protons, we measure the helicity dependence of the alpha yield. A discussion of the working system and preliminary results are given.

INTRODUCTION AND MOTIVATION

Progress in understanding the weak interaction in the nuclear medium is being made by the study of parity violation in the compound nucleus. Following theoretical predictions that parity violation would be enhanced in compound nuclear resonances (1), a group at Dubna observed very large parity violations in several nuclei (2-3). Subsequently the TRIPLE Collaboration has developed techniques which allow measurements of a number of parity violations per nuclide. Using statistical analysis methods, the root-mean-squared (RMS) symmetry-breaking matrix element is determined for each nucleus studied. It is convenient to express the RMS matrix element in terms of a weak spreading width $\Gamma^{\downarrow}$. The TRIPLE collaboration has measured $\Gamma^{\downarrow}$ in two mass regions, $A\approx110$ and $A\approx230$. An interesting remaining challenge is to determine the mass dependence of $\Gamma^{\downarrow}$ (4-6).

We describe an experiment now being performed at the Triangle Universities Nuclear Laboratory (TUNL) to determine $\Gamma^{\downarrow}$ in the mass $A\approx30$ region using the (p,α_0) reaction. This mass region is well separated from the 110 and 230 regions already studied, and extensive high resolution spectroscopic information exists for many nuclei in this region. With this information, predictions have been made of the relative enhancement of parity violation for many resonances in a number of 2*s*-1*d* shell nuclei (7-9). The calculations indicate that the parity-violating longitudinal asymmetries should show strong dependence on energy and angle, with magnitudes ranging from 10^{-3} to 10^{-7}.

The current experiment uses longitudinally polarized protons incident on a ^{31}P target. Four segmented detectors detect elastically scattered protons and alpha particles at 16 polar angles covering a solid angle of nearly 2π. A bias voltage is applied to the target so that one can scan resonances without changing the beam optics. By reversing the helicity of the protons, the longitudinal parity-violating asymmetry in the resonant alpha cross section is determined. The initial goal is a sensitivity of 10^{-4}. Measurements at this level require the reduction of systematic errors to 10^{-5}, and this has been the focus of our early experimentation. We describe the experimental apparatus and preliminary measurements of systematic effects.

EXPERIMENTAL SYSTEM

Longitudinally polarized H^- ions are produced at the Atomic Beam Polarized Ion Source at TUNL and are accelerated to the terminal of the FN Tandem Van de Graaff accelerator. At the terminal the beam passes through a 2 $\mu g/cm^2$ carbon foil which removes the two electrons. The positively charged H^+ beam then accelerates through the same potential in the second half

CP475, *Applications of Accelerators in Research and Industry*, edited by J. L. Duggan and I. L. Morgan

of the Tandem. The energy of the accelerated beam is determined by a pair of 90° bending magnets. Signals from current slits at the output of the 90°-90° magnet system are fed back to the terminal of the Tandem to regulate the beam energy. The feedback is used in two ways. The low frequency components (< 1 Hz) control the corona discharge of the terminal. Higher frequency components control the bias on the stripper foil, where the fast feedback amplifier is set to roll off frequencies higher than 3 Hz.

The beam is directed to the target by a 60° bending magnet. On the final leg of the beam, signals from current slits are fed back to steerers that control the position of the beam on target. Measurements of the beam in this leg show horizontal variations of $\sim 6\ \mu$m and vertical variations of $\sim 30\ \mu$m. Before entering the target chamber, the beam is collimated to 4.3 mm in diameter 33 cm upstream from the target. Two anti-scattering collimators are positioned between this collimator and the detector. The target is positioned on a 4-position target rod that can be biased up to 10 kV, allowing the proton energy to be changed. One of the target positions contains a tuning ring with a 3.2 mm aperture. The beam is focused such that less than 2% of the full beam current ($\sim$ 600 nA) strikes the tuning ring.

FIGURE 1. Schematic for a 16-strip silicon detector.

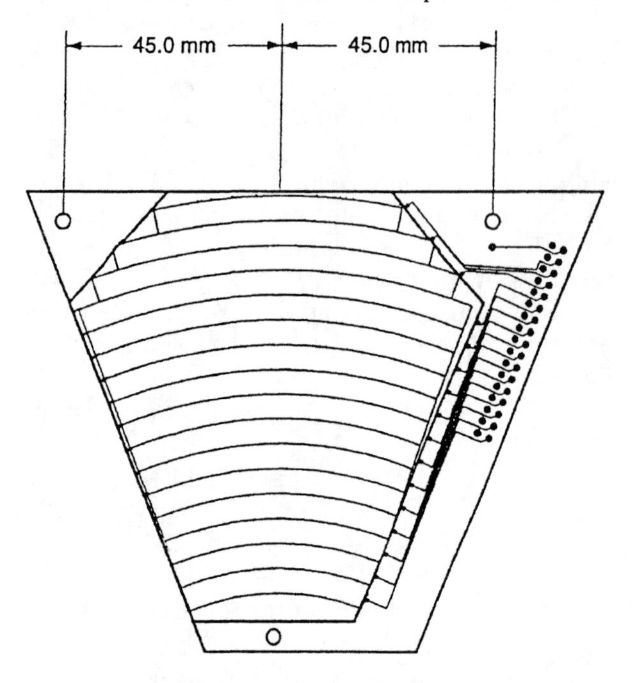

A Micron (10) manufactured silicon strip detector measures elastically scattered protons and alpha particles. The detector consists of four wedge-shaped sections approximately 9 cm long and 9 cm wide at the widest part. Each wedge shaped detector is 300 μm thick and has 16 strips. Figure 1 shows a schematic of one of the 16-strip detectors. The four detectors fit together to form a square pyramid, whose axis is the beam axis. The detector covers 16 polar angles with each strip 2–5° wide in the angular range 110°–166° with respect to the beam direction. Each detector wedge covers one fourth of 2π in azimuthal angle.

FIGURE 2. Sample spectrum from the 10th strip of the UP detector. The two peaks (divided by 30) on the left are from elastic protons from C (left) and ^{31}P (middle). The peak on the right is from alpha particles from the $E_p = 3.3963$ MeV resonance in p+^{31}P.

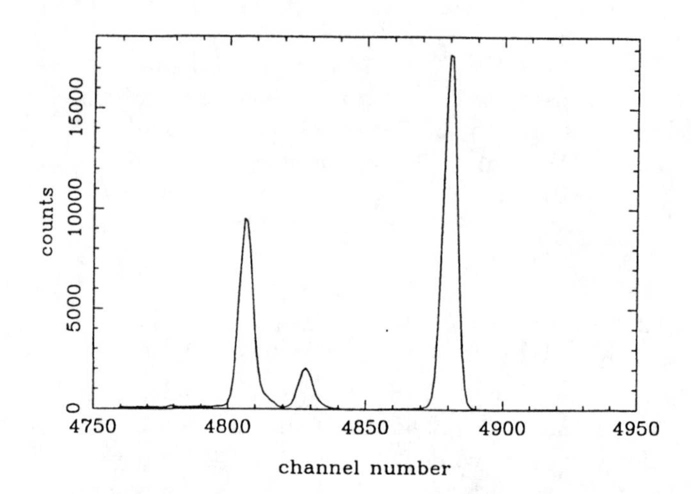

The detector signals are processed by 16-channel preamplifiers which produce fast timing signals (50 ns rise time, 100 ns width) and charge-integrated energy signals (600 ns width). This type of detector and preamplifier combination was developed by Wuosmaa *et al.* (11). The timing signals from each detector are discriminated by two 16-channel constant fraction discriminators (CFDs). One CFD threshold is set below the proton signal, and the other is set above the proton signal but below the alpha signal. Since the elastic proton rate is significantly higher than the alpha rate, which is of primary interest, the proton CFD signal is divided by 30 to limit the data acquisition count rate. This proton signal is combined with the alpha signal to trigger a 16-channel analog-to-digital converter (ADC) (Phillips 7166H). If two strips from a single detector produce signals very close in time, the resulting gate from the CFD could trigger the ADC to sample the wrong part of one of the signals. To avoid this, the multiplicity output of the CFDs is used to veto any gates to the ADC where the multiplicity is greater than one. The ADC data are read via CAMAC by a Microprogrammed Branch Driver (MBD) and bussed to a VAX 3200, where the data are written to disk. Typical count rates are 2 kHz for each of the four detectors, with dead times of 15 μs per event. The dead time for each detector is monitored by scaling a 50 MHz pulser in coincidence with the busy signal from each ADC. Figure 2 shows the well resolved proton and alpha peaks in a sample ADC spectrum from one strip of one of the silicon strip detectors.

FIGURE 3. Asymmetries for a transversely polarized proton beam at 3.396 MeV. The top plot shows the transverse asymmetry for the UP (circles) and DOWN (dots) detectors. The bottom plot shows the sum of the transverse asymmetries from the UP, DOWN, LEFT, and RIGHT detectors.

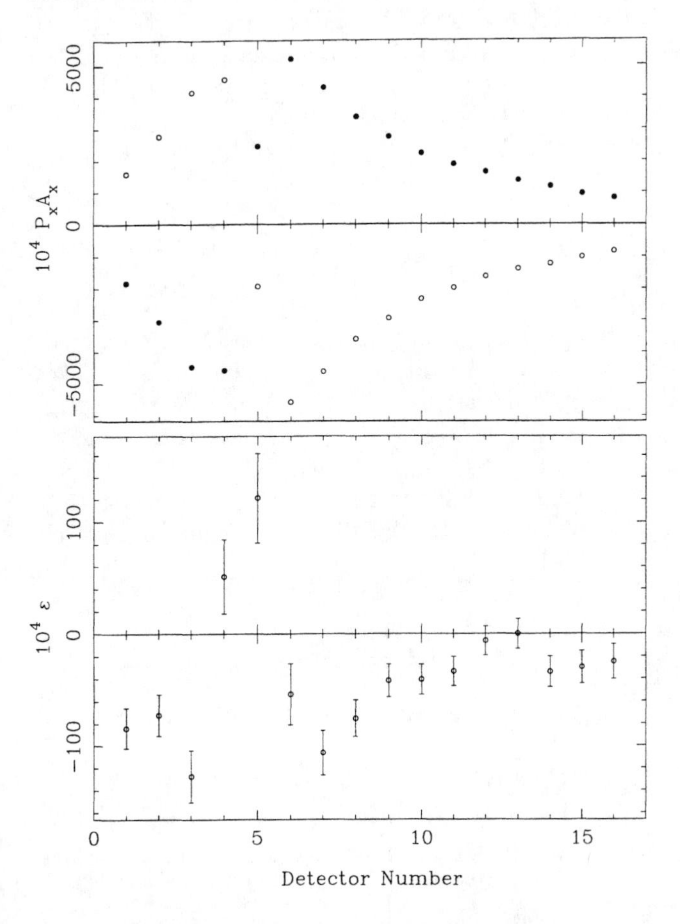

The proton polarization is reversed in the polarized ion source. RF-fields in the ABPIS are used to cause atomic transitions in the H^- ions which in turn cause proton spin flips. The RF fields are switched at 10 Hz to continually switch between the two polarization states. A Wien filter at the output of the ABPIS is used to rotate the proton polarization to the desired angle with respect to the beam direction. Signals from the RF-field transition units are read by the data acquisition system so that the proton polarization direction is known for each ADC event. The beam current is measured by a 2 inch diameter Faraday cup with suppressor ring that is located 36 cm downstream from the target.

TESTS OF SYSTEMATIC ERRORS

Our primary initial focus has been directed at possible systematic errors. Wilburn *et al.* (12) give an expression for the measured asymmetry in the alpha yield,

$$\epsilon = \frac{N^+ - N^-}{N^+ + N^-}, \tag{1}$$

where $N^\pm$ are the alpha yields for + and − helicity protons, respectively. The authors consider ϵ as a function of polar angle, θ, after averaging over all four detectors (UP, DOWN, LEFT, RIGHT) and ignoring helicity-dependent variations higher than first order, and obtain

$$\epsilon(\theta) = P_z A_z + (P_x \delta A_x + P_y \delta A_y)/2 + \frac{\delta \mathcal{L}}{\mathcal{L}}. \tag{2}$$

The first term is the longitudinal polarization times the parity violating asymmetry we wish to measure. The last term results from any helicity-dependent beam luminosity (beam current times target thickness) and is removed by normalizing the alpha yield by the backscattered protons detected in the silicon-strip detectors. The second term contains any residual transverse components in the beam polarization (P_x and P_y) and helicity-dependent difference in the transverse analyzing powers, A_x and A_y. In the ideal case this term vanishes, but misaligned detectors could cause nonzero values for δA_x or δA_y.

FIGURE 4. Proton yield in arbitrary units for the E_p = 3.2762 MeV (p, p_o) resonance in $p+^{31}$P. The yield has been normalized for beam current and solid angle and has been summed over the four back most angles for all four detectors. The positive (negative) helicity data are shown as dots (squares). The data were taken in 250-eV steps.

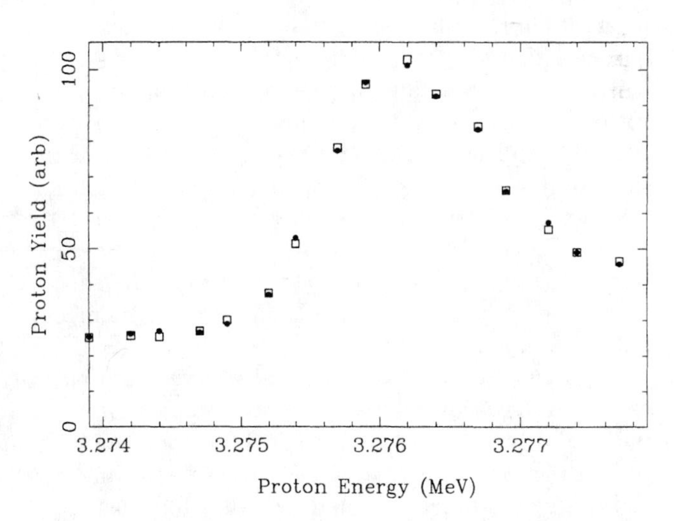

The effect of the second term can be seen in Figure 3. These data were taken with the proton beam polarized transversely to the beam direction in the plane of the LEFT and RIGHT detectors. In the upper plot the UP and DOWN detectors show a large transverse analyzing power. (Data from the LEFT and RIGHT detectors, not shown here, show a small analyzing power resulting from residual transverse polarization in the UP-DOWN plane.) Averaging over the four azimuthal directions, one obtains an ϵ signal two and a half orders of magnitude smaller than A_y. Since during the parity violation experiments the beam is polarized longitudinally, the transverse components are typically 100 times smaller than in this present test, thus reducing the second term in Eq. 2 to a few times 10^{-4}. In addition, with accurate measurements of the transverse analyzing powers, the longitudinal data can be corrected for any transverse components, thus reducing the false asymmetry by at least another order of magnitude.

The effect of a correlation between helicity state and proton energy is not included in Eq. 2. Since many of the resonances of interest are quite narrow, a energy shift could cause a yield shift. If the energy shift is helicity dependent, the yield shift would appear as a false asymmetry. The magnitude of this effect was measured by scanning narrow resonances with both helicity states. The result of a scan of the 3.2762 MeV (p, p_o) resonance in ^{31}P is shown in Figure 4. The normalized proton yields shown are summed for the four back most strips of all four detectors. Preliminary analysis of these resonances indicates that the energy shift is of the order of 10 eV.

The beam energy resolution is also apparent from Figure 4. Since the resonance has an intrinsic full width of 0.4 keV (13), the observed width is essentially the beam resolution - about 1 keV. We should be able to improve the resolution by a factor of two or three.

Time drifts pose another source of systematic errors. The primary tool for addressing this problem is a fast polarization reversal system that follows an "8-step sequence". The sequence is given by $\rightarrow\leftarrow\leftarrow\rightarrow\leftarrow\rightarrow\rightarrow\leftarrow$, where the arrows point in the polarization direction with respect to the beam direction. This procedure cancels to second order any long term time-ordered drifts such as in the gains of the preamplifiers or in target thickness. We believe these effects to be canceled to the 10^{-6} level. Other systematics are also under consideration, but they should be smaller than the effects considered here. One example is higher order moments in the beam polarization. Finally, the major advantage of the segmented detector and the energy variation is having the flexibility to make detailed comparisons with theoretical predictions. Although the theory can only predict the average magnitude of the effect, the angular and energy dependence should be well determined.

SUMMARY

Experiments have begun at TUNL to measure parity violation in the mass A = 30 region. The current experiment measures the asymmetry in the alpha yield from the $^{31}P(p,\alpha_o)^{28}S$ reaction for longitudinally polarized protons. The experimental apparatus employs a nearly 2π silicon strip detector array to measure elastically scattered protons and alpha particles. The segmented detectors covering nearly 2π in solid angle, allow measurement at 16 polar angles in four regions of azimuthal angle. Systematic effects produced by residual transverse components are reduced to the 10^{-5} level by correcting each detector yield with measured analyzing powers and then averaging over all four azimuthal regions. Helicity-dependent energy shifts have been measured to be ~ 10 eV. Time drifts are reduced by a fast spin flip procedure. Measurements of parity violation in the compound nucleus utilizing charged particle resonance reactions appears to be a promising method to determine the mass dependence of the weak spreading width.

ACKNOWLEDGMENTS

The authors would like to thank C. A. Grossmann for his assistance with data collection. This work was supported in part by the U. S. Department of Energy, Office of High Energy and Nuclear Physics, under grants No. DE-FG02-97-ER41042, DE-FG02-97-ER41033 and DE-FG02-96-ER40990, and by the U. S. Department of Energy, Office of Energy Research, under Contract No. W-7405-ENG-36.

REFERENCES

1. Sushkov, O. P. and V. V. Flambaum, *JETP Lett.* **32**, 352 (1980).
2. Alfimenkov, V. P. *et al.*, *JETP Lett.* **35**, 51 (1982).
3. Alfimenkov, V. P. *et al. Nucl. Phys.* **A 398**, 93 (1983).
4. Bowman, J. D., G. T. Garvey, Mikkel Johnson, and G. E. Mitchell, *Ann. Rev. Nucl. Part. Sci.* **43**, 829 (1993).
5. Frankle, C. M., S. J. Seestrom, N. R. Roberson, Yu. P. Popov, and E. I. Sharapov, *Phys. Part. Nucl.* **24**, 401 (1993).
6. Mitchell, G. E., J. D. Bowman, and H. A. Weidenmüller, *Rev. Mod. Phys.* (to be published).
7. Shriner, J. F., Jr. and G. E. Mitchell *Phys. Rev. C* **49**, R616 (1994).
8. Mitchell, G. E. and J. F. Shriner, Jr., *Nucl. Instrum. Methods B* **99**, 305 (1995).
9. Mitchell, G. E. and J. F. Shriner, Jr., *Phys. Rev. C* **54**, 371 (1996).
10. Micron Semiconductor, INC., 252 East Semoran Blvd., Suite 907, Casselberry, Florida 32707.
11. Wuosmaa, A. J. *et al. Nucl. Instrum. Meth.* **A 345**, 482 (1994).
12. Wilburn, W. S., G. E. Mitchell, N. R. Roberson and J. F. Shriner, Jr., in *Applications of Accelerators in Research and Industry*, edited by J. L. Duggan and I. L. Morgan, AIP Conf. Proc. No. **392** (AIP, New York, 1997), p. 297.
13. Fang, Dufei, Ph.D. Dissertation, Fudan University (1987).

Monitoring of Alpha-Beam Properties by the $^{nat}Ti(\alpha,x)^{51}Cr$ Reaction : New Measurements and Critical Compilation

A. Hermanne[1], M. Sonck[1], S. Takács[2], F. Szelecsényi[2] and F. Tárkányi[2]

[1]Vrije Universiteit Brussel, Cyclotron Department, Brussels, B1090, Belgium.
[2] Institute of Nuclear Research of the Hungarian Academy of Sciences, Debrecen, H4001, Hungary.

Cross sections for reactions induced by alpha particles on ^{nat}Ti foils and leading to the formation of ^{47}Sc; ^{51}Cr and ^{48}V were determined from the reaction thresholds up to 42 MeV. Our experimental values are compared to literature values and a good accordance is found. The $^{nat}Ti(\alpha,x)^{51}Cr$ reaction is particularly useful for monitoring α-beams in the 10-20 MeV region but above 20 MeV the $^{nat}Ti(\alpha,x)^{47}Sc$ reaction or the $^{nat}Ti(\alpha,x)^{48}V$ reaction are more suited.

INTRODUCTION

A systematic investigation of monitoring reactions (nuclear reactions induced by charge particles on metals for determining energy and intensity of accelerator beams) on the same target material for the different light ion beams (p, d, ^{3}He and α) allows expansion of reliable databases needed for correct use of these monitor reactions. In earlier publications we compiled, critically evaluated and remeasured a number of absolute cross sections (1, 2, 3). Knowledge of these excitation curves also allows optimization of beam parameters and calibration curves for nuclear analytical studies by thin layer activation (TLA).
The light, corrosion resistant and high strength metal titanium has a wide range of applications in mechanical constructions, air- and spacecraft technology and research tools. Its physical and chemical properties, its unrestricted availability make titanium an ideal target material for beam parameter monitoring and wear and corrosion studies.
A survey of the available data for monitoring low and middle energy charged particles (4) shows that the status of these data is not satisfactory. We reported earlier on the excitation functions of p, d, α and ^{3}He induced reactions on ^{nat}Ti in the low energy region (1, 2, 5, 6). In this work we investigate the cross sections of reactions leading to ^{47}Sc, ^{51}Cr and ^{48}V induced by α beams up to 42 MeV
The errors ascribed by authors to their measured cross sections vary mostly between 10 and 15% but data points of different investigations are often systematically shifted in energy over a wider significant range. We hence thought it to be essential to critically compile and evaluate the existing experimental data, to remeasure consistently over the whole energy range with special attention to all possible errors, and to set up a recommended data set for the most promising monitoring reactions using ^{nat}Ti.

EXPERIMENTAL TECHNIQUES

The cross sections were measured using the stacked foil activation method. The experimental procedures, evaluation of data and error estimation are similar to those reported earlier (1, 7, 8, 9, 10). Only the specific features of the present measurements are discussed here.
Seven stacks containing high purity natural Ti foils (> 99.9% pure) were irradiated in the external beam of the cyclotrons at Brussels (CGR-560) and Debrecen (MGC-20E). The composition of the stacks and the irradiation parameters are given in Table 1.
The target holders were Faraday-cups equipped with a secondary electron suppresser. Total charge on target was derived from the Faraday cup measurements using a digital integrator without additional beam current monitoring. The activity of the irradiated foils was determined by high-resolution gamma spectrometry without chemical treatment. Source to detector distance was at least 10 cm to avoid dead-time and pile-up effects and to minimize efficiency calibration errors (geometry effects). Cross sections were calculated using the activation formula. The stack composition and measuring method assures automatic correction of possible recoil effect errors.
The decay data summarized in Table 2 were taken from Browne and Firestone (11) while the Q-values of the most probable contributing reactions were calculated using the nuclear masses given in this reference. Our previous measurements on contributions of (n,pxn) reactions to the activation (mediated by neutrons generated by interaction with collimator and target material) show that these processes are negligible in the energy range studied (12).

TABLE 1. Target composition and irradiation parameters.

Series/ Number of foils	Foil thickness (μm)	Incident energy (MeV)	Mean beam current (nA)	Irradiation time (s)
a / 17	21.6	42.1	470	3600
b / 29	12.0	40.0	2000	720
c / 19	12.0	29.5	300	7200
d / 15	12.5	26.5	50	600
e / 15	12.5	26.5	100	2400
f / 14	12.5	26.8	100	3600
g / 4	12.5	18.0	100	3600

CP475, *Applications of Accelerators in Research and Industry*,
edited by J. L. Duggan and I. L. Morgan

TABLE 2. Decay data of the investigated radionuclides.

Nuclide	Half life	Eγ(keV)	Iγ(%)
^{51}Cr	27.7d	320.1	9.8
^{48}V	15.98d	983.5	100
		1312.0	97.5
^{47}Sc	3.34d	159.4	68

The total error on cross sections (10%-16%) is estimated by combining the contributing sources in quadrature : uncertainty on number of target atoms (5%), nuclear data (5%), charge (7%), efficiency calibration of spectrometer (3-7%), statistical and peak fitting errors (1-10%).
The effective particle energy in each foil of the stack was calculated using the polynomial approximation of (13). For estimation of the error on the energy the cumulative effects (alpha incident energy, foil thickness and stopping straggling) have been taken into account.

RESULTS AND DISCUSSION

Table 3 contains the numerical values of our measured cross sections for the production of ^{51}Cr, ^{48}V and ^{47}Sc.
Levkovskii (17) reported cross sections measured for some of the individual contributing reactions on highly enriched targets of one or more of the isotopes ^{46}Ti, ^{47}Ti, ^{48}Ti and ^{49}Ti and we calculated from his data the approximate total production cross sections using the natural isotopic composition of Ti (^{46}Ti:8.2%, ^{47}Ti:7.4%, ^{48}Ti:73.8%, ^{49}Ti:5.4%, ^{50}Ti:5.2%). As cross section data for all mother isotopes were not available for each daughter nuclide studied in (17), the derived total cross sections are always underestimated, be it mostly only by small amounts.
In most cases good overall agreement exists between all the new data series and the earlier works.

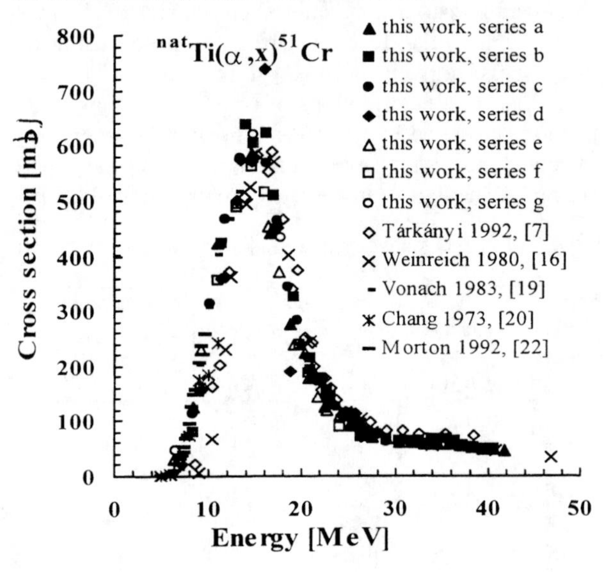

FIGURE 1 : Cross section for the $^{nat}Ti(\alpha,x)^{51}Cr$ process.

$^{nat}Ti(\alpha,x)^{51}Cr$

As this reaction leads to an isotope often used in medicine and tracer experiments 9 different authors have published earlier data sets (6, 16, 17, 18, 19, 20, 21, 22, 23). It was already earlier recognized as a possible monitor reaction (14). Except for the measurements of Levkovskii (17), who used enriched ^{48}Ti and ^{49}Ti targets (and hence underestimation for the computed total production cross section at energies above 22 MeV), the results of all other authors were obtained on natural Ti.
The $^{48}Ti(\alpha,n)^{51}Cr$ reaction with Q= - 2.8MeV on the most abundant isotope is dominant in the energy region studied. Contributions of reactions on another isotope (^{49}Ti) is only evidenced by the measurements presented in (17). The 9 previous sets of data and our 7 new series are generally in good agreement concerning the energy scale, shape and position of the maximum. Discrepancies exist on the maximal value of the excitation function and some slight energy shifts are observed. A more detailed discussion and selection of acceptable data for evaluation (represented in Fig. 1) is included in the section on applications.

$^{nat}Ti(\alpha,x)^{48}V$

All available results are presented in Fig.2. We assessed the activity after total decay of the ^{48}Cr parent. Good agreement between our two data series and the data from (18) is observed while the same remarks as for the previous nuclide are valid for the data of (17): lower values especially at higher energies. Man King Go and Markowitz (15) measured the cross sections of the $^{46}Ti(\alpha,pn)^{48}V$ reaction using enriched ^{46}Ti targets. We renormalized their data to obtain approximate total production cross sections at low energies supposing an abundance of 8.2% for ^{46}Ti in natural Ti. This leads to erroneous results if reaction channels on other isotopes of Ti open which is certainly

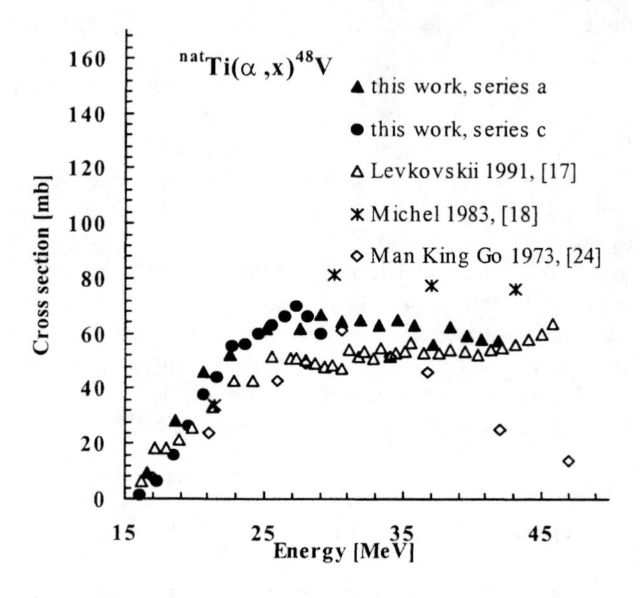

FIGURE 2 : Cross section for the $^{nat}Ti(\alpha,x)^{48}V$ process.

Table 3 : Cross sections (mbarn) of the reactions leading to ^{51}Cr, ^{48}V and ^{47}Sc in function of energy (MeV)

Energy MeV	σ (^{51}Cr) mb	Energy MeV	σ (^{51}Cr) mb	σ (^{48V}Cr) mb	σ (^{47}Sc) mb	Energy MeV	σ (^{51}Cr) mb	σ (^{48}V) mb	σ (^{47}Sc) mb
6.36 ± 1.1	30.4 ± 2.7	17.60 ± 0.5	370 ± 33			25.53 ± 0.4	94.4 ± 11	62.9 ± 7.6	1.1 ± 0.1
6.57 ± 1.0	45.0 ± 5.4	17.79 ± 0.3	432 ± 52			26.20 ± 0.3	72.0 ± 9		
6.90 ± 1.1	18.0 ± 2.0	18.45 ± 1.2	343 ± 39	15.6 ± 1.9		26.28 ± 0.3	94.4 ± 13		
7.22 ± 1.1	20.7 ± 3.0	18.66 ± 0.5	189 ± 23			26.30 ± 0.3	70.3 ± 10.0		
8.33 ± 0.6	80.4 ± 9.6	18.70 ± 1.9	277 ± 39	28.4 ± 4.2	0.04 ± 0.01	26.30 ± 0.3	80.3 ± 7.2		
8.36 ± 2.5	115 ± 13	19.03 ± 0.4	326 ± 39			26.43 ± 0.4	83.0 ± 9.8	66.3 ± 8.0	1.6 ± 0.2
8.36 ± 0.9	125 ± 19	19.13 ± 0.4	240 ± 22			27.30 ± 0.3	78.6 ± 9.3	69.7 ± 8.4	2.4 ± 0.3
9.10 ± 0.9	160 ± 19	19.57 ± 1.1	281 ± 32	26.6 ± 3.2		27.50 ± 1.5	69.0 ± 9.9	61.5 ± 9.0	3.2 ± 0.5
9.41 ± 1.0	160 ± 19	19.86 ± 0.5	241 ± 29			27.79 ± 0.3	76.4 ± 9.2		
10.21 ± 2.2	311 ± 35	20.32 ± 0.4	223 ± 20			28.15 ± 0.3	72.8 ± 8.7	66.1 ± 7.9	3.3 ± 0.4
11.00 ± 0.7	423 ± 38	20.65 ± 1.0	215 ± 24	37.5 ± 4.5		28.99 ± 0.3	62.2 ± 7.5	60.0 ± 7.2	3.8 ± 0.5
11.00 ± 0.7	356 ± 38	20.68 ± 1.8	178 ± 25	45.9 ± 7.2	0.08 ± 0.04	28.99 ± 0.4	68.3 ± 8.2		
11.30 ± 0.9	358 ± 43	20.80 ± 0.5	188 ± 23			29.02 ± 1.3	67.8 ± 9.7	66.6 ± 9.7	5.0 ± 0.7
11.85 ± 1.9	467 ± 53	20.95 ± 0.3	193 ± 23			30.48 ± 1.0	62.7 ± 9.0	64.0 ± 9.3	6.7 ± 0.9
13.00 ± 0.8	500 ± 55	21.69 ± 0.8	173 ± 20	44.0 ± 5.3		31.77 ± 0.5	62.9 ± 7.6		
13.03 ± 0.6	495 ± 44	21.71 ± 0.4	144 ± 13			31.89 ± 0.8	62.0 ± 8.8	64.8 ± 9.3	9.0 ± 1.3
13.35 ± 1.7	575 ± 65	22.00 ± 0.4	178 ± 21			32.67 ± 0.3	65.9 ± 7.9		
14.56 ± 0.6	584 ± 52	22.53 ± 1.7	125 ± 18	52.5 ± 7.7	0.33 ± 0.05	33.25 ± 0.7	58.9 ± 8.4	63.0 ± 9.1	10.9 ± 1.5
14.60 ± 0.7	560 ± 62	22.69 ± 0.7	141 ± 16	55.5 ± 6.7	0.17 ± 0.02	34.58 ± 0.6	58.1 ± 8.3	65.0 ± 9.5	13.6 ± 1.9
14.74 ± 1.5	577 ± 65	23.05 ± 0.4	143 ± 17			35.86 ± 0.5	56.7 ± 8.1	62.7 ± 9.1	15.6 ± 2.2
14.82 ± 0.5	605 ± 73	23.20 ± 0.4	132 ± 16			36.06 ± 0.5	54.3 ± 6.5		
16.01 ± 0.6	740 ± 81	23.67 ± 0.6	125 ± 14	55.9 ± 6.7	0.34 ± 0.04	37.12 ± 0.4	54.2 ± 7.8	56.0 ± 8.2	16.5 ± 2.3
16.22 ± 0.5	623 ± 75	24.10 ± 0.4	90 ± 11			38.34 ± 0.4	52.9 ± 7.6	62.6 ± 9.1	18.3 ± 2.6
16.57 ± 2.1	441 ± 62	24.61 ± 0.5	107 ± 12	60.0 ± 7.2	0.69 ± 0.08	39.53 ± 0.3	50.2 ± 7.2	59.4 ± 8.7	18.4 ± 2.6
17.02 ± 0.3	508 ± 61	24.70 ± 0.3	114 ± 14			39.60 ± 0.3	48.1 ± 5.8		
17.02 ± 0.4	508 ± 61	25.10 ± 1.7	91 ± 13	61.9 ± 9.1	1.40 ± 0.20	40.69 ± 0.3	48.6 ± 7.1	58.1 ± 8.6	17.9 ± 2.5
17.27 ± 1.3	462 ± 52	25.10 ± 0.3	91 ± 10			41.84 ± 0.3	44.5 ± 6.4	57.3 ± 8.3	17.4 ± 2.4

the case at higher energy.
From the threshold at about 15 MeV and the slow slope without clear maximum up to 35 MeV it is clear that different reactions take place on the ^{46}Ti and ^{47}Ti nuclides until the reaction on ^{48}Ti takes over above 30 MeV with a steep increase in cross section.

$^{nat}Ti(\alpha,x)^{47}Sc$

The shape of the excitation curve and its threshold for this nuclide (Fig. 3) indicates that the major contributing reaction at low energy is $^{48}Ti(\alpha,\alpha p)^{47}Sc$ (Q= - 11.4MeV).

Other reactions on low abundant nuclides contribute to the broad maximum situated at an energy higher than expected. This is also supported by the Levkovskii data (17) constructed from results on only ^{48}Ti and ^{49}Ti.
Our data series a shows, in contradiction with results for the other isotopes studied, a slight shift to lower energy.

MONITORING OF α– PARTICLE BEAMS.

The IAEA-CRP proposes to evaluate the $^{nat}Ti(\alpha,x)^{51}Cr$ reaction as a candidate for monitoring α-particle beams (14). The compilation in fig. 1 of all available data, including our 7 new data series, shows that the status of the excitation curve is satisfactory and that an evaluation by comparison to theoretical calculations can be performed after critical selection of experimental values
The effective thresholds and cross section values at low energies are best represented by the data of (19) and (22) obtained by measurements on Van de Graaff accelerators.
The series with cross sections values below threshold or with a too low maximal were rejected (16, 17, 21, 23).
After this selection the remaining data set consisting of 6 series of independent values was sent for evaluation.
The recommended values for this reaction, to be published in the IAEA-TECDOC, can be considered for monitoring up to an energy of 20 MeV. Reactions on ^{nat}Ti leading to ^{48}V and ^{47}Sc nuclides seem more appropriate for energies

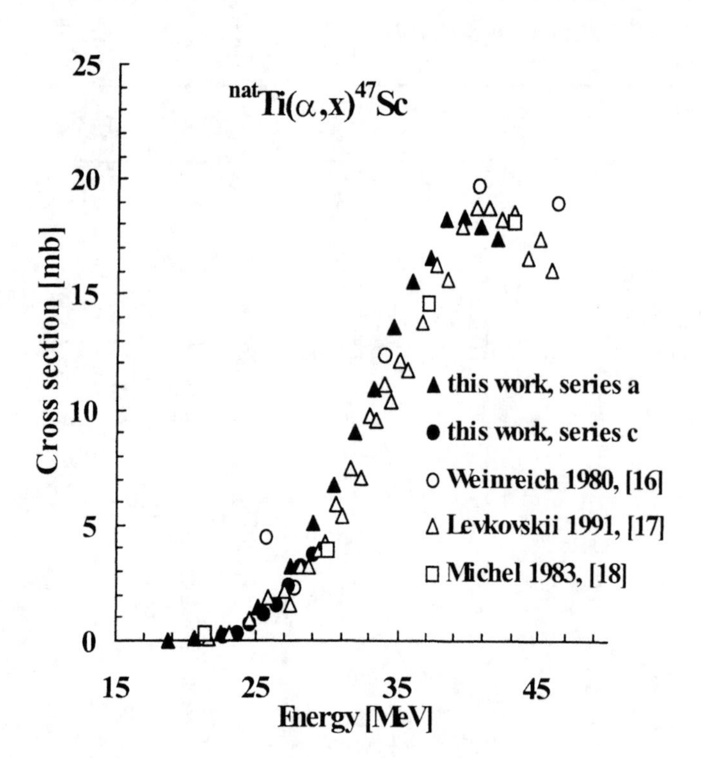

FIGURE 3 : Cross sections of the $^{nat}Ti(\alpha,x)^{47}Sc$ process.

up to 50 MeV. However too few data sets exist for these nuclides to make meaningful evaluations at this time.

CONCLUSIONS

The present work details a comprehensive study on some excitation functions of alpha particle induced reactions on natural titanium for energies up to 42 MeV. The new data are generally in good agreement with the literature data and reasons for discrepancies could be indicated.
The status of the data set for $^{nat}Ti(\alpha,n)$ was critically compiled and an evaluation is underway. The recommended values to be issued by the IAEA will be useful for monitoring energy and intensity of alpha-beams up to 20 MeV. For higher energy, reactions leading to ^{48}V or ^{47}Sc seem more appropriate but additional measurements and intercomparisons have to be made before an evaluation is meaningful.
The cross section data obtained for ^{51}Cr and ^{48}V give valuable information for activation practice and can used for the construction of calibration curves needed in LTA wear measurement and allowing estimation of radiation dose to accelerator parts and isotope production systems.

ACKNOWLEGMENT

This work was done in the frame of an IAEA-CRP "Development of a Reference Charged Particle Cross Section Data Base for Medical Radioisotope Production". Financial support was received from the FWO-Vlaanderen, the Hungarian Academy of Science and from OZR-VUB.

REFERENCES

1. F. Tárkányi, F. Szelecsényi and P. Kopecky, *App. Rad and Isot.*, **42**, 513, (1991).
2. S. Takács , M. Sonck, B. Scholten, A. Hermanne and F. Tárkányi , *Appl. Rad. and Isot.*, **48,** 657, (1997).
3. M. Sonck, A. Hermanne, F. Szelecsényi, S. Takács and F. Tárkányi, *Appl. Rad. and Isot*,. **49,** 1533, (1998).
4. O. Schwerer and K. Okamoto, IAEA (NDS)-218/GZ, (1989).
5. P. Kopecky, F. Szelecsényi, T. Molnár , P. Mikecz and F.Tárkányi , *Appl. Rad. and Isot.*, **44**, 687, (1997).
6. F. Tárkányi, F. Szelecsényi and S. Takács, *Acta Radiologica*, **376**,72,(1992).
7. F. Tárkányi, F. Szelecsényi and P. Kopecky, *Proc.of Int.Conf. on Nuclear Data for Science and Technology, Julich*, (ed. S. M. Qaim, Springer, Berlin, 1992) p. 529.
8. S. Takács, L. Vasváry, F. Tárkányi, *Nucl. Instrum. and Methods B*, **89**, 88, 1994.
9. A. Hermanne, N. Walravens and O. Cichelli, *Proc. of Int. Conf. Nuclear Data for Science and Technology, Jülich*, ed. S. M. Qaim, Springer Verlag, Berlin, 1992), p.616.
10. M. Sonck, A. Hermanne and J. Van hoyweghen, *Appl.Radiat. Isot.*, **47,** 445, (1996).
11. E. Browne and R. B. Firestone , *Table of Radioactive Isotopes*,(ed. V.S. Shirley), Wiley, London, 1986.
12. A. Hermanne, M Sonck, S. Takács, F. Szelecsényi and F. Tárkányi, *Proc. Int. Conf. on Nuclear Data for Science and Technology, Trieste (eds. G. Reffo, A. Ventura and C.Grandi), Conf. Proc. Italian Physical Society*, **59**,.1262, 1997.
13. J.F. Ziegler, *Stopping and Ranges in All Elements, Vol. 4:Helium*, Pergamon Press, Oxford, 1977.
14. F. Tárkányi and P. Oblozinsky, *Proc. Int. Conf. on Nuclear Data for Science and Technology, Trieste (eds.G. Reffo, A.Ventura,C. Grandi), Conference Proceedings of Italian Physical Society*, **59**, 1629, 1997.
15. Man King Go and S. Markowitz, *Phys. Rev. C* **7**, 1464,(1973)
16. R. Weinreich, H. J. Probst and S. M. Qaim, *Int. J. Appl. Radiat. Isot.*, **31**, 223, (1980) and private communication.
17. V. N. Levkovskii, *Activation Cross Sections by Medium Energy (E=10-50MeV) Protons and Alpha-Particles (Experiment and Systematics).* Inter-Vesi, Moscow, 1991.
18. R. Michel, G. Brinkmann and R. Stück, *Radiochimica Acta*, **32**, 173, (1983).
19. H. Vonach, R. C. Haight and G. Winkler G, *Phys. Rev. C*, **6,** 2278, (1963).
20. C. N. Chang, J. J. Kent, J. F. Morgan and S. L. Blatt, *Nucl. Instr. Meth.*, **109**, 327, (1973).
21. A. Iguchi, H. Amano and S. Tanaka, *J. Atomic Energy Society of Japan*, **2,** 682, (1960).
22. A. J. Morton, S. G. Tims, A. F. Scott, V.Y. Hansper, C.I.W. Tingwell and D.G. Sargood, *Nucl. Phys. A*, **537**, 167, (1992).
23. Xiufeng Peng, Fuqing He and Xianguan Long, *Nucl. Instr. Meth -B*, **140**, 9, (1998).

Extraction of the $^1S_0$ Neutron-Neutron Scattering Length from a Kinematically-Complete n-d Breakup Experiment at TUNL

D.E. González Trotter, W. Tornow, C.R. Howell, F. Salinas, R.L. Walter[1] and H. Witała[2]

1. Triangle Universities Nuclear Laboratory, Box 90308, Durham NC-27708-0308.
2. Institute of Physics, Jagellonian University, PL-30059, Cracow, Poland.

The $^1S_0$ neutron-neutron (*nn*) scattering length's currently accepted value (a_{nn}=-18.6±0.3 fm) is derived exclusively from two π^--d capture-reaction experiments, in disagreement with the average -16.7±0.5 fm extracted from kinematically-complete *nd* breakup experiments. This discrepancy may be due to deficiencies in the analyses of n-d breakup data and/or three-nucleon force (3NF) effects. A kinematically-complete n+d→n_1+n_2+p breakup experiment at an incident neutron energy of 13.0 MeV was performed recently at TUNL. The value of a_{nn} was extracted from the direct comparison of experimental and rigorously-calculated theoretical *nd* breakup differential cross sections at four production angles of the *nn* pair. Using modern nucleon-nucleon potential models in the three-nucleon cross-section calculations we obtained a_{nn} =-18.7±0.6 fm, in agreement with the π^--d result. We found no significant effect due to 3NFs on our a_{nn} value.

INTRODUCTION

The principle of charge-symmetry that implies the equality of the nuclear force between protons and between neutrons, was proposed in the early 1930s by W. Heisenberg (1). This symmetry was later found to be approximate, and in the formalism of quantum chromodynamics it is broken due to the up-down quark mass splitting and due to electromagnetic differences like electric charges and magnetic moments. Charge-symmetry breaking (CSB) arises in the meson-exchange picture of the nucleon-nucleon (NN) interaction due to neutron-proton (*np*) mass differences and electromagnetic interactions.

The $^1S_0$ scattering length parameter is very sensitive to variations in the NN potential. A ~1% variation in the potential leads to a ~10% change in the magnitude of the scattering length (2) having a "magnifying glass" effect through which the extent of CSB in the $^1S_0$ part of the NN interaction can be clearly established. Free proton-proton (*pp*) scattering experiments have resulted in a scattering length of $a_{pp}^N = -17.3 \pm 0.4$ fm (3). The measurement of a_{nn} has been more difficult because of the lack of a neutron target. The earliest efforts were centered on kinematically-incomplete n+d→n+n+p (*nd* breakup) final-state interaction (*FSI*) experiments, where only the outgoing proton energy spectrum was observed. A combination of problems in accounting for instrumental effects and the simplicity of the theoretical models utilized to analyze the experimental data led to a large variance in the a_{nn} values obtained (4). On the other hand, kinematically-complete *nd* breakup experiments yielded an average $a_{nn} = -16.7 \pm 0.5$ fm and π^-+d→n+n+γ capture experiments gave an average $a_{nn} = -18.6 \pm 0.3$ fm. It has been suggested that the discrepancy between these last a_{nn} values is due to the action of three-nucleon forces (3NFs) in the exit *nd* breakup channel.

In the last few years it has become possible to rigorously solve the three-nucleon (3N) Faddeev equations using modern NN interactions, even including 3NFs (5). Armed with these new powerful theoretical methods we performed an experiment-oriented theoretical analysis of kinematically-complete *nd* breakup data acquired in an experiment performed at TUNL.

In the present paper we show new cross-section data from the *nd* breakup reaction at a bombarding lab energy E_n= 13 MeV. Values for a_{nn} were extracted by direct comparison of experimental *FSI* cross sections and theoretical calculations produced by a sophisticated Monte-Carlo simulation which included instrumental effects of the experimental setup. We follow with a presentation of an outline of the experimental setup, data analysis and results.

EXPERIMENTAL SETUP

The goal of this experiment was to measure the absolute *nd* breakup cross section for *nn-FSI*

CP475, *Applications of Accelerators in Research and Industry*,
edited by J. L. Duggan and I. L. Morgan

configurations, where the two neutrons in the exit channel leave with zero relative momentum.

A neutron beam was produced using the $^2H(d,n)^3H$ reaction employing a deuterium gas cell at 7.8 atm. The gas cell consisted of a 3.0 cm in length × 1.0 cm in diameter copper cylinder with a 0.25 mil-thick Havar foil entrance window and a 0.051 mm-thick gold beam stop. The neutron beam was collimated through a double-truncated rectangular copper/polyethylene collimator, resulting in a homogeneous flux of neutrons on a deuterated center detector (CD) target 170 cm from the gas cell. A concrete/lead/paraffin wall shielded the array of neutron detectors from the deuterium gas cell's radiation not emitted near 0° (see Fig. 1.)

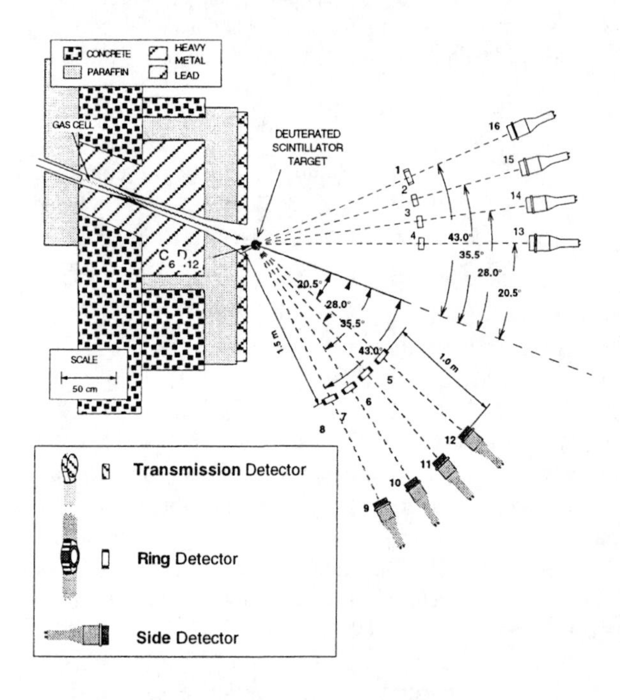

Figure 1. Layout of the target and neutron detectors in the Shielded Source area at TUNL.

The CD consisted of a 6.0 cm in height × 4.0 cm in diameter cylindrical active volume of NE-232 deuterated liquid scintillator enclosed in a glass cylinder of 0.22 cm wall thickness. Neutron detectors were placed at 150 cm (ring) and 250 cm (side) from the CD, with ring-side detector pairs placed at θ_{nn} = 20.5°, 28.0°, 35.5° and 43.0° with respect to the primary neutron beam axis. Ring detectors consisted of a cylindrical NE-213 liquid scintillator volume 7.6 cm in inner diameter, 13.44 cm in outer diameter and 4.0 cm in depth. The active volume was enclosed in a glass and aluminum case. Side detectors consisted of 6.34 cm in diameter × 5.08 cm in length cylindrical volumes filled with BC-501 liquid scintillator enclosed in an aluminum capsule with a glass window as one of the flat surfaces.

In an *nn-FSI* event, the breakup proton was detected in the CD. Neutron 1 was detected by a ring detector, and neutron 2 was detected by the side detector *directly behind it*. The hole in the ring detector subtended the same solid angle as the side detector behind it, so the flux of *nd* breakup neutrons detected by a side detector was minimally attenuated by the ring detector in front of it (see Fig. 1.)

The energies E_1 and E_2 of neutrons 1 and 2 were determined using the time-of-flight technique. The start signals were provided by a concidence between the anode signals of the CD and one of the neutron detectors. The anode signal from the same neutron detector provided the stop signal. The proton's energy E_3 was determined from the pulse-height (PH) signal from the CD. Pulse-shape discrimination was applied to the anode signals from the neutron detectors to reject almost all events due to gamma rays. A software threshold of one-third the PH of the Compton edge of a ^{137}Cs gamma-ray source ($\frac{1}{3}\times$ Cs) was used on the neutron detectors. A PH threshold of $\frac{1}{11}\times$ Cs was used on the CD.

There were ten experimental runs of about ten days each, for a total of two thousand one hundred hours of net data accumulation. The statistical uncertainty at the peak *nn-FSI* cross sections was smaller than ±5% for all configurations studied.

DATA ANALYSIS

Luminosity

For each run, the *nn-FSI* event yield had to be normalized by the luminosity

$$\beta_r Q_r = N_n^r N_d, \qquad (1)$$

where N_n^r is the number of neutrons illuminating the target during a run r, N_d is the density of deuterium atoms in the CD and Q_r is the charge deposited by the deuteron beam on the gas cell beam stop during run r.

The luminosity was determined from the *nd* elastic-scattering yield

$$Y_{el}^r(\theta) = \frac{d\sigma_{lab}}{d\Omega}(\theta)\alpha(E_{el})\varepsilon(E_{el})\beta_r Q_r \Omega, \qquad (2)$$

where α(El) is the transmission probability of *nd* elastic scattering neutrons with energy E_{el}, $\varepsilon(E_{el})$ is the absolute detection efficiency of the neutron detector and Ω is the angle subtended by the neutron detector. The *nd* elastic cross sections $\frac{d\sigma_{lab}}{d\Omega}(\theta)$ were provided by Witała et al.(6).

The yields were extracted from the CD PH for each run, and corrected for multiple-scattering effects using a Monte-Carlo (MC) simulation.

Experimental *nn-FSI* Cross-Section Determination

Breakup event yields as a function of the point-geometry locus (S-curve) length were obtained for each *nn-FSI* configuration. The ideal S-curve is the locus of kinematically-allowed events in energy space given by the incoming neutron energy and angle θ_{nn} of the outgoing neutrons in the *nd* breakup reaction. The origin of the curve is defined at the point where E_2=0 MeV. The length of the ideal point-geometry locus for a given *nn-FSI* configuration was discretized in steps of 0.5 MeV and their associated energies converted to momenta.

The experimental events were projected onto the nearest ideal locus point by minimizing the distance in momentum space

$$K = \sqrt{\sum_{i=1,2,3} (k_i^c - k_i^{\exp})^2}\ , \qquad (3)$$

where k_i^c are the momentum-space coordinates of discrete points along the ideal S-curve and $k_i^{\exp}$ are the momentum coordinates of the experimental events. For each point along a point-geometry S-curve we got a breakup yield $Y_r(S)$ for every run r. After averaging the yields from all runs using the luminosity as a normalizing factor and correcting for cross-talk between front and back neutron detectors, we obtain a yield $Y_{net}(S)$. The absolute cross section as a function of S-curve length is given by

$$\frac{d^3\sigma}{d\Omega_1 d\Omega_2 dS}(S) = \frac{Y_{net}(s)}{\kappa(S)\Omega_1\Omega_2\Delta S}\ , \qquad (4)$$

where $\Delta S = 0.5\,\text{MeV}$, Ω_i are the solid angles subtended by the neutron detectors, and $\kappa(S)$ is the MC-calculated attenuation and efficiency correction factor.

nd Breakup Monte-Carlo Simulations

Correction Factors

The experimental data are distorted by the energy spread of the neutron beam (±200 keV), finite size of target and neutron detectors, CD PH resolution, TOF resolution, detection efficiency of the neutron detectors and neutron flux attenuation due to the construction materials of the target and detectors.

For each history (or event) in the MC simulation we randomly selected points within the CD and two neutron detectors in a *nn-FSI* configuration and calculated an attenuation factor

$$\overline{\alpha}(E_1,E_2,E_3) = \alpha(E_1)\alpha(E_2)\ , \qquad (5)$$

which is the product of the individual attenuation factors for each neutron traveling from the CD to its respective neutron detector. Similar factors were calculated for the neutron detection efficiency

$$\overline{\varepsilon}(E_1,E_2,E_3) = \varepsilon(E_1)\varepsilon(E_2)\ . \qquad (6)$$

Due to the finite CD energy resolution we introduced a weighting factor $\omega(E_3)$ based on a Gaussian distribution.

For each history the efficiency and attenuation factors were multiplied together and projected onto the ideal S-curve weighted by $\omega(E_3)$, obtaining the correction factor $\kappa(S)$.

Finite-Geometry Theoretical Cross Section

For each *nn-FSI* configuration we created a rigorously-calculated library of absolute cross sections spanning the range of the finite geometry given by the CD and neutron detectors, neutron beam energy spread, and value of a_{nn} from -17.0 fm down to -20.0 fm in steps of 0.5 fm. The MC simulation made use of these libraries as bases to calculate cross sections for each history. These cross sections were weighted by efficiency, attenuation and CD energy resolution factors and projected onto the ideal S-curve to obtain a finite-geometry theoretical cross section that could be directly compared with the experimental cross section (see Fig. 2.)

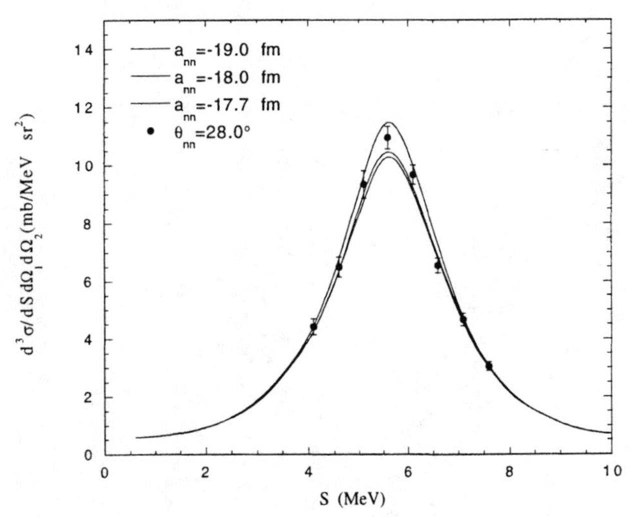

Figure 2. Experimentally-determined *nn-FSI* cross section as a function of S-curve length for θ_{nn} = 28.0° plotted along with finite-geometry theoretical cross section curves for various values of a_{nn} obtained from the MC simulation of the experiment using cross-section libraries based on $^1S_0$-modified versions of the Bonn B potential.

RESULTS AND CONCLUSIONS

Three realistic NN potentials (Bonn B (7)(8) , CD Bonn (9) and NijmI (10)) were used in the creation of different *nn-FSI* libraries to investigate the potential-dependence of the analysis, leading to a 0.1 fm variance in the extracted value of a_{nn}. A χ^2 analysis of the experimental data using a_{nn} as a free parameter in the comparison of experimental and finite-geometry theoretical cross sections gave an average $a_{nn} = -18.7 \pm (0.1)_{stat} \pm (0.6)_{syst}$ fm from the four *nn-FSI* configurations studied.

The systematic error originates from a combination of uncertainties in the attenuation factors, efficiencies, *nd* elastic cross sections, detector dimensions and distance measurements. These errors were combined to a ±5% uncertainty in the absolute cross section for each *nn-FSI* configuration.

As mentioned in the introduction, the difference between a_{nn} values extracted from older kinematically-complete *nd* breakup and π^-+d→n+n+γ capture experiments was suggested to originate in the action of 3NFs in *nd* breakup. One signature of a 3NF influence would be the observation of a systematic angular dependence of a_{nn}. We found no strong systematic dependency of a_{nn} on θ_{nn} (5). Moreover, our result agrees with the π^-+d→n+n+γ capture experiments, indicating that 3NF effects are negligible in the presently investigated region of three-nucleon phase-space.

ACKNOWLEDGMENTS

This work was supported in part by the U.S. Department of Energy, Office of High Energy and Nuclear Physics, under grant No. DE-FG02-97ER41033. The 3N calculations reported in this work were performed on the Cray T916 of the North Carolina Supercomputing Center at the Research Triangle Park, North Carolina.

REFERENCES

1. Heisenberg ,W. *Z. Phys.*, **77**, 1 (1932).
2. de Téramond, G. F., "Charge Symmetry of the Nuclear Interaction and the N-N Scattering Parameters", presented at the Symposium Workshop on Spin and Symmetries, Vancouver, Canada, June 30-July 2, 1989.
3. Miller, G. et al. *Phys. Rep.*, **194**, 1-116 (1990).
4. Tornow, W. et al., *Few-Body Systems*, **21**, 97-130 (1996).
5. Glöckle, W. et al., *Phys. Rep.*, **274**, 107-286 (1996).
6. Witała, H. et al., *Nucl. Phys.*, **A491**, 157-172 (1989).
7. Machleidt, R. et al., *Phys. Rep.*, **149**, 1-89 (1987).
8. Machleidt, R. et al., *Adv. Nucl. Phys.* **19**, 189-376 (1989).
9. Machleidt, R et al., *Phys. Rev. C*, **53**, 1483-1487 (1996).
10. Stoks, V. G. J. et al., *Phys. Rev. C*, **49**, 2950-2962 (1994)

Measurement of the Parity Violating Asymmetry A_γ in $\vec{n} + p \rightarrow d + \gamma$

W.S. Wilburn,[1] A. Bazhenov,[2] C.S. Blessinger,[3] J.D. Bowman,[1] T.E. Chupp,[4] K.P. Coulter,[4] S.J. Freedman,[5] B.K. Fujikawa,[5] T.R. Gentile,[6] G.L. Greene,[1] G. Hansen,[3] G.E. Hogan,[1] S. Ishimoto,[7] G.L. Jones,[6] J.N. Knudson,[1] E. Kolomenski,[2] S.K. Lamoreaux,[1] M.B. Leuschner,[8] A. Masaike,[9] Y. Masuda,[7] Y. Matsuda,[9] G.L. Morgan,[1] K. Morimoto,[7] C.L. Morris,[1] H. Nann,[3] S.I. Penttilä,[1] A. Pirozhkov,[2] V.R. Pomeroy,[8] D.R. Rich,[3] A. Serebrov,[2] E.I. Sharapov,[10] D.A. Smith,[1] T.B. Smith,[1] W.M. Snow,[3] R.C. Welsh,[4] F.E. Wietfeldt,[6] V.W. Yuan,[1] and J. Zerger,[4]

[1] *Los Alamos National Laboratory, Los Alamos, NM 87545*
[2] *Petersburg Nuclear Physics Institute, Petersburg, Russia*
[3] *Department of Physics, Indiana University, Bloomington, IN 47405*
[4] *Department of Physics, University of Michigan, Ann Arbor, MI 48109*
[5] *Department of Physics, University of California, Berkeley, CA 94720*
[6] *National Institute of Standards and Technology, Gaithersburg, MD 20899*
[7] *Physics Department, University of New Hampshire, Durham, NH 03824*
[8] *Physics Department, Kyoto University, Kyoto, Japan*
[9] *KEK National Laboratory, Tsukuba, Japan*
[10] *Joint Institute for Nuclear Research, Dubna, Russia*

The weak pion-nucleon coupling constant H_π^1 remains poorly determined, despite many years of effort. The recent measurement of the ^{133}Cs anapole moment has been interpreted to give a value of H_π^1 almost an order of magnitude larger than the limit established in the ^{18}F parity doublet experiments. A measurement of the gamma ray directional asymmetry A_γ for the capture of polarized neutrons by hydrogen has been proposed at Los Alamos National Laboratory. This experiment will determine H_π^1 independent of nuclear structure effects. However, since the predicted asymmetry is small, $A_\gamma \approx 5 \times 10^{-8}$, systematic effects must be reduced to $< 5 \times 10^{-9}$. The design of the experiment will is presented, with an emphasis on the techniques used for controlling systematic errors.

INTRODUCTION

The hadronic weak interaction is an ideal place to study the interplay between the weak and strong nuclear forces. The weak force is well described by the standard electroweak model, and weak processes involving only leptons can be exactly calculated. Quarks, however, interact via the strong force as well, effectively modifying the manifestation of the weak force between them. Since QCD has not been solved for the non-perturbative regime characteristic of low energies, the parameters of the weak hadronic interaction must be determined from experiment.

The weak nucleon-nucleon interaction can be parameterized by a potential model of the form

$$V_{pnc} = \sum_{\mu=\pi,\rho,\omega} \sum_{\Delta I=0,1,2} H_\mu^{\Delta I} V_\mu^{\Delta I}, \qquad (1)$$

where $H_\mu^{\Delta I}$ is the weak coupling constant corresponding to the exchange of a π. ρ, or ω meson and an exchange of isospin of ΔI. Only one coupling H_π^1 is allowed for π exchange. Most theoretical calculations of this coupling, whether from symmetry considerations [1], QCD sum rules [2], or chiral perturbation theory [3, 4], give a value in the range $1 \leq H_\pi^1 \geq 10 \times 10^{-7}$. In contrast the best experimental determination, from the measurement of circular polarization in the decay of ^{18}F, gives an upper limit of $H_\pi^1 \leq 0.28 \times 10^{-7}$ [5]. In addition, the recent measurement of the anapole moment of ^{133}Cs [6] has been interpreted to give a large value of $H_\pi^1 = 2.26 \pm 0.50(\text{expt}) \pm 0.83(\text{theor}) \times 10^{-6}$ [7]. This interpretation, however, has been disputed [8].

Because of the difficulty in interpreting the results from measurements in nuclei, a measurement in the nucleon-nucleon system is necessary to definitively determine H_π^1, free from nuclear structure assumptions. While most parity-violating experimental observables

CP475, *Applications of Accelerators in Research and Industry*, edited by J. L. Duggan and I. L. Morgan

are sensitive to a linear combination of several weak couplings, the directional asymmetry A_γ in the emission of gammas from np capture, given by

$$A_\gamma = -0.045\left(H_\pi^1 - 0.02H_\rho^1 + 0.02H_\omega^1 + 0.04H'^1_\pi\right), \tag{2}$$

is (to the few-percent level) only sensitive to H_π^1. A measurement of A_γ is therefore a measurement of H_π^1. A previous measurement of A_γ has been performed [9], though not with sufficient precision to obtain a non-zero result. We are proposing an experiment to measure A_γ with a statistical precision of 10% of the predicted value, $A_\gamma \sim 5 \times 10^{-8}$, with negligible systematic error [10].

EXPERIMENTAL DESIGN

In this section we describe the conceptual design for the proposed measurement of A_γ in $\vec{n} + p \to d + \gamma$. The apparatus, shown schematically in figure , consists of a cold neutron source, followed by a neutron polarizer, and a liquid para-hydrogen target, surrounded by an array of gamma detectors. Neutrons from the spallation source are moderated by a liquid hydrogen moderator. The source is pulsed, thus allowing measurement of neutron energy through time-of-flight techniques. The neutron guide transports the neutrons from the moderator through the biological shield with high efficiency. The neutrons are then polarized in the vertical direction by transmission through polarized ^{3}He gas. The neutron spin direction can be subsequently reversed by the radio-frequency resonance spin flipper. The use of this type of a spin flipper, which is possible at a pulsed neutron source, reduces the systematic error associated with the $\vec{\mu}_n \cdot \nabla B$ force, where $\vec{\mu}_n$ is the neutron magnetic moment. The neutrons are captured in the target, which consists of liquid para-hydrogen. This state of hydrogen is required, since neutrons depolarize quickly in ortho-hydrogen, while those with energies below 15 meV retain their polarization in para-hydrogen. Gammas emitted in the capture process are detected in the CsI(Tl) detectors surrounding the target. The parity-violating asymmetry causes an up-down asymmetry in the angular distribution of the gamma-rays for vertical neutron spin. When the neutron spin is reversed, the up-down gamma asymmetry reverses. The parity-violating asymmetry in gamma flux,

$$\frac{d\omega}{d\Omega} = \frac{1}{4\pi}\left(1 + A_\gamma \cos\theta_{s,\gamma}\right), \tag{3}$$

is a measure of H_π^1, as discussed in the introduction.

SYSTEMATIC ERRORS

We distinguish between statistical and systematic errors. The experiment is designed to measure the directional asymmetry of the emission of gamma rays with the neutron spin direction. A source of systematic error produces a signal in the detector that is coherent with the state of the neutron spin; for example, the current in a magnet used to flip the neutron spin might be picked up by the gamma detector, or a guide field might steer the neutron beam up-down as the spin is changed from up to down. A source of statistical error produces a detector signal that is not correlated with the neutron spin direction; for example fluctuations in the number of detected gamma rays due to counting statistics or drifts in amplifier offsets. The size of statistical errors is important when discussing systematic errors, because it is important to be able to diagnose systematic errors in a time that is short compared to the time it takes to measure the directional γ asymmetry. Systematic errors can be further classified according to whether they are instrumental in origin and are present whether or not neutrons are being detected or arise from an interaction of the neutron spin other than the directional γ asymmetry in the $\vec{n} + p \to d + \gamma$ reaction, for example the parity-allowed asymmetry $\vec{s}_n \cdot (\vec{k}_n \times \vec{k}_\gamma)$. Finally, it is important to isolate and study experimentally potential sources of systematic errors. For example we can search for false asymmetries from activation of components of the apparatus due to the capture of polarized neutrons by emptying the liquid hydrogen target. We can monitor in situ effects such as the parity allowed $\vec{s}_n \cdot (\vec{k}_n \times \vec{k}_\gamma)$ correlation in $\vec{n} + p \to d + \gamma$ that produces left-right asymmetries.

It is not possible to give a complete list of sources of instrumental systematic errors. Many come to mind: the influence of magnetic fields on detector gains, shifts in the mains voltage as power supplies are turned on and off, leakage of control signals into preamplifiers, etc. It is essential to be able to tell whether such effects are present in a short time, to learn where they come from, and fix them. These effects are not associated with the neutron beam. There are two types of instrumental asymmetries; additive couplings and gain shifts. Additive couplings will be diagnosed by running the experiment with the beam off and looking for a non-zero up-down asymmetry. The electronic noise is 1/100 of counting statistics. In the presence of electronic noise only, achieving an accuracy of 0.1×10^{-8} (the statistical error in A_γ will be 0.5×10^{-8} in one year of data) will require a running time $5^2/100^2$ of 1 year, $\approx$ 1 day.

In order to search for gain shifts we will illuminate the detectors with light from light emitting diodes. The level of illumination will produce a photo-cathode current 10 times larger than that due to neutron capture where we expect the number of photo-electrons per 2.2 MeV gamma from CsI(Tl) will be $\approx$ 500. The time to measure a gain shift of 0.1×10^{-8} will be $5^2/(10 \times 1000)$ of 1 year $\approx$ 1 day. We will be able to

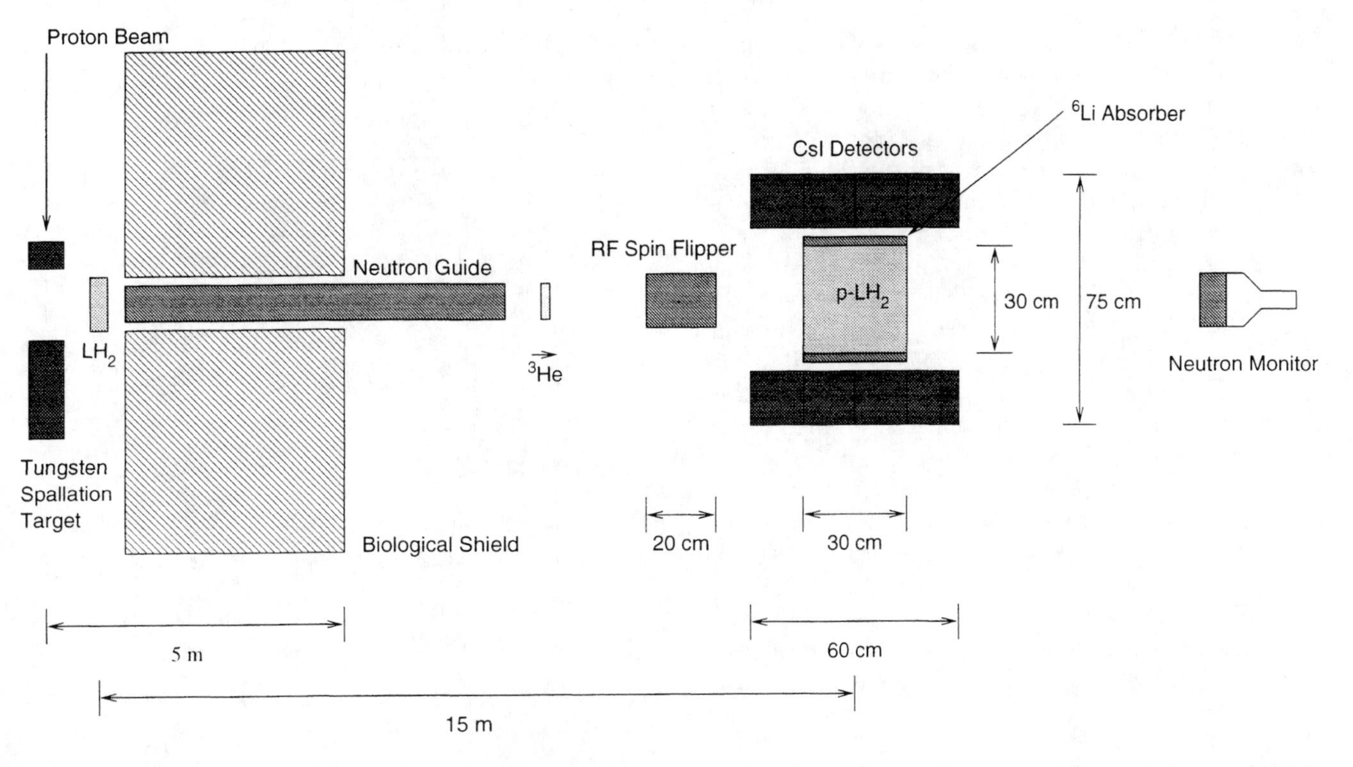

Figure 1: The conceptual design for the proposed experiment, showing the most important elements (not to scale). Approximate sizes and distances are indicated for some features.

diagnose and eliminate instrumental systematic errors before we take beam, and without a complete apparatus. We will be able to check for problems during periods when the beam is off.

The most important experimental tool we have to isolate a parity violating signal in this experiment is the neutron spin flip. It is therefore absolutely essential that the process of flipping the neutron spin have a negligible effect on all other properties of the apparatus. In this section we discuss some of the ways that this idealization may fail, and our estimates for the size of the resulting systematic effect.

In our considerations above we assumed that the spin flip process is "perfect", that is, that the only difference between the flip/no-flip states of the experiment is that the neutron polarization is reversed. In practice this condition cannot be met. We now relax these assumptions and consider the consequences. We will concentrate on two methods of neutron spin reversal: use of a RF magnetic field on the neutron beam and reversal of the polarization direction of the ^{3}He polarizer.

One method of spin reversal consists of reversing the polarization direction of the ^{3}He target. The ^{3}He spin can be reversed by an adiabatic fast passage or adiabatic reversal of the magnetic holding/guide field. The magnetic field (at the polarizer the fully polarized ^{3}He nuclei create a field of about 2 Gauss) due to the reversed magnetic moments of the polarized ^{3}He nuclei in the neutron polarizer causes a change in the static magnetic field at the location of the gamma detectors. This change is about 1×10^{-6} Gauss. Coupled with the measured change in the gamma detector efficiency 2×10^{-5} per Gauss, this gives a negligible efficiency change of 2×10^{-11}.

The other method of neutron spin reversal is effected by turning on and off the ≈ 30 kHz magnetic field in the spin flipper. This field, although closer to the detectors than the ^{3}He cell, can be shielded very effectively because the skin depth of the 30 kHz RF field in aluminum is 0.5 mm. In addition, the intrinsic detector efficiency should be less sensitive to an RF field than a DC field. Care must be taken to insure that there is no spurious electronic pickup induced by the RF switching. We intend to forestall this problem by switching the RF power into a dummy coil when the neutron spin is not being flipped.

We will reverse the neutron spin on a 20 Hz time scale using the RF spin flipper with a $+--+-++-$ pattern. This pattern eliminates the effects of first and second order drifts. The neutron spin will be reversed every few hours by reversing the polarization direction of the ^{3}He polarizer. Finally, we will reverse the direction of the holding/guide field every few hours. Instrumental effects arising from the state of the RF spin flipper, the ^{3}He cell, the holding/guide field, or from other parts of the apparatus will have different dependences on the different reversals. These different dependences can be used to identify the source of potential instrumental systematic errors. Any instrumental or spin-dependent systematic error that depends on the ^{3}He state, the spin flipper state, or the holding field state would be eliminated by averaging over different reversal methods.

In this section we consider systematic errors arising from interactions of the polarized neutron beam itself. This type of false effect is potentially the most difficult to eliminate. Fortunately, these effects are all small, $\ll 10^{-8}$, and do not require heroic efforts to eliminate. In order to produce a false asymmetry, an interaction must occur after the spin is reversed by the RF spin flipper, otherwise the effect of the interaction would be averaged out by the eight-step reversal sequence. The interaction must involve the inner product of the neutron spin vector and some vector made up of the vectors and scalars from the initial and final states. At least one quantity from the final state that deposits energy in the detector must be involved. We have tried to identify all possible Cartesian invariants that satisfy these conditions and evaluate the associated false asymmetries. We evaluated invariants that produced asymmetries $\approx 10^{-10}$ more carefully than asymmetries $\ll 10^{-10}$. Different potential sources of false asymmetry produce effects that depend on time of flight (neutron energy) in a characteristic fashion. The $\vec{n}+p \to d+\gamma$ directional asymmetry, A_γ, produces an up-down pattern (for neutron spin up-down) that is independent of neutron energy up to an energy of 15 meV. Above 15 meV, the neutrons depolarize in the para-hydrogen and the asymmetry vanishes.

SUMMARY

We are proposing an experiment to measure A_γ with a statistical precision of 10% of the predicted value, $A_\gamma \sim 5 \times 10^{-8}$, with negligible systematic error [10]. This measurement will determine the weak pion-nucleon coupling H_π^1 independent of nuclear structure assumptions. The experiment is designed to measure A_γ to 10% of its predicted value with negligible systematic error.

ACKNOWLEDGMENTS

This research was supported in part by the U.S. Department of Energy and the U.S. National Science Foundation.

REFERENCES

References

[1] B. Desplanques, J.F. Donoghue, and B.R. Holstein, Ann. Phys. **124**, 449 (1980).

[2] E.M. Henley, W.-Y.P. Hwang, and L.S. Kisslinger, nucl-th/9809064.

[3] D.B. Kaplan and M.J. Savage, Nucl. Phys. **A556**, 653 (1993).

[4] U.-G. Meißner and H. Weigel, nucl-th/9807038.

[5] E.G. Adelberger and W.C. Haxton, Ann. Rev. Nucl. Part. Sci. **35**, 501 (1985).

[6] C.S. Wood, S.C. Bennett, D. Cho, B.P. Masterson, J.L. Roberts, C.E. Tanner, and C.E. Wieman, Science **275**, 1759 (1997).

[7] V.V. Flambaum and D.W. Murray, Phys. Rev. C **56**, 1641 (1997).

[8] W.S. Wilburn and J.D. Bowman, Phys. Rev. C **57**, 3425 (1998).

[9] J.F. Caviagnac, B. Vignon, and R. Wilson, Phys. Lett. **B67**, 148 (1977).

[10] W.M. Snow, W.S. Wilburn, J.D. Bowman, M.B. Leuschner, S.I. Penttilä, V.R. Pomeroy, D.R. Rich, E.I Sharapov, and V. Yuan, nucl-ex/9704001.

Neutron Capture Measurements on Unstable Nuclei at LANSCE

J.L. Ullmann, R.C. Haight

LANSCE-3, Los Alamos National Laboratory, Los Alamos, NM 87545

M.M. Fowler, G.G. Miller, R.S. Rundberg, J.B. Wilhelmy

CST-11, Los Alamos National Laboratory, Los Alamos, NM 87545

Although neutron capture by stable isotopes has been extensively measured, there are very few measurements on unstable isotopes. The intense neutron flux at the Manual Lujan Jr. Neutron Scattering Center at LANSCE enables us to measure capture on targets with masses of about 1 mg over the energy range from 1 eV to 100 keV. These measurements are important not only for understanding the basic physics, but also for calculations of stellar nucleosynthesis and Science-Based Stockpile Stewardship. Preliminary measurements on ^{169}Tm and ^{171}Tm have been made with deuterated benzene detectors. A new detector array at the Lujan center and a new radioactive isotope separator will combine to give Los Alamos a unique capability for making these measurements.

INTRODUCTION

Neutron capture is one of the first reactions studied in nuclear physics, and measurements have been made on virtually all stable targets. Most cross sections on unstable targets have been inferred from theoretical calculations. This is a difficult reaction to calculate, however, because details of the nuclear structure of the compound nucleus near the neutron binding energy are important. Evidence for this is the order of magnitude differences in thermal neutron capture cross sections for adjacent and near-adjacent nuclei.

In addition to fundamental scientific curiosity, several applied programs have a need for neutron capture cross sections in the keV region. Several facilities at Los Alamos combine to provide a unique capability to make these measurements. A description will be given for a new detector array being designed and built at the Los Alamos Neutron Science Center (LANSCE) to measure capture on small quantities of unstable and stable nuclei, on the order of a milligram, and present some preliminary measurements of capture on ^{171}Tm.

NEUTRON CAPTURE REVISITED

Defense Applications

One of the main drivers for this renewed interest in capture reactions is in understanding nuclear device performance. The extensive data associated with 50 years of nuclear testing is being systematically reviewed using modern analysis and computational techniques, with a goal of understanding nuclear weapons physics. One of the principal methods of diagnosing device performance has been through the use of radiochemical tracer isotopes. The high neutron density during an explosion can result in multiple reactions on a single nucleus, driving it far from stability. In order to calculate the observed isotopic yields, accurate cross sections are required for the nuclei involved, many of which are unstable.

s-Process Nucleosynthesis

A second area of interest is in studying the stellar synthesis of elements heavier than iron. This is believed to occur primarily through neutron capture and β decay. These reactions take place in a "r" (rapid) process that occurs in the explosive environment of supernovae, and in a "s" (slow) process of sequential capture along the valley of β stability that occurs in the "asymptotic giant branch" stage of evolution of low to medium mass stars, or in red giants. The overall mechanism of the s process appears to be understood, and solar and stellar abundances are in general well reproduced (1,2).

The flow of the s-process is illustrated in Fig. 1, for masses near A = 151. Of particular interest is the branching of the flow at unstable nuclei, where the competition between neutron capture and β decay can shed information on the stellar environment where the s process occurred. If the β decay rate in the hot stellar environment is the same as in our laboratory, the neutron density at the s-process site can be inferred from observed abundances and the neutron capture cross section. The most accurate abundances to use are those of nuclei that are formed only by the s-process.

CP475, *Applications of Accelerators in Research and Industry*,
edited by J. L. Duggan and I. L. Morgan

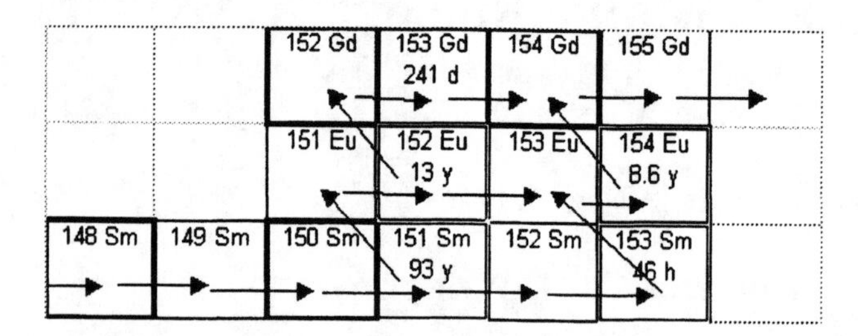

FIGURE 1. S-process flow near mass 151. The branching at ^{151}Sm is of particular interest. Heavy outlines indicate "s-only" isotopes, and double outlines indicate unstable isotopes.

Neutron densities and stellar temperatures extracted from branch-point analysis are given in Ref. 1. They appear quite consistent with a ^{22}Ne(α,n) neutron source at 30 keV and a neutron density of 4 x 10^8 cm^{-3}. However, this consistency is based on calculated values for neutron capture cross sections on unstable nuclei. A comparison of calculated (3) and measured (4) capture cross sections for the stable Sm isotopes shows that the even-even isotopes are calculated to 10% while the even-odd isotopes show poorer agreement. A minimum 25% uncertainty is often assumed for the calculated values of the capture cross section, and clearly more accurate measurements are needed.

Interestingly, although the measured parameters point to a 30 keV ^{22}Ne(α,n) neutron source for the s-process, detailed stellar models favor the ^{13}C(α,n) reaction, with an average energy of 10 keV, as the main neutron producing reaction driving the s process (1). Models predict that the s process occurs during a 20-year pulse of ^{13}C burning followed by a 1-year pulse of ^{22}Ne burning, which alters the observed abundances. Thus while cross sections must be measured over the entire 10 keV Maxwell velocity distribution characteristic of ^{13}C burning, the effect of this episode on the observed stellar abundances must be determined through detailed stellar models.

Transmutation

A third area of growing interest is field of accelerator-driven transmutation technology, in which neutrons produced by an accelerator-driven spallation source are used to transmute long-lived radioactive waste from reactors into shorter-lived, easier-to handle isotopes. A number of capture cross sections need to be accurately known, including capture on fission fragments such as ^{99}Tc and ^{135}Cs, and on lesser actinides such as ^{234}U and ^{242}Pu. Although some measurements do exist, it is likely that more experiments will be required.

LOS ALAMOS CAPABILITIES

The ability to make these measurements requires a confluence of capabilities that exists only at Los Alamos. The first of these is the intense neutron spallation source of the Manual Lujan Jr. Neutron Scattering Center (5). The high neutron fluence available in the critical keV region allows us to make direct measurements on samples with mass 1 milligram or less. Next is the ability to safely handle and produce radioactive targets. Los Alamos National Laboratory supplies many of the radiopharmaceutical isotopes distributed by the U.S. Department of Energy, and most of these are produced at LANSCE by proton spallation on heavy targets. The Nuclear and Radiochemistry group at Los Alamos (CST-11) maintains hot cells for handling the production targets and chemically separating isotopes of interest.

In addition to chemical separation, a magnetic isotope separator, the Radioactive Species Isotope Separator (RSIS) has been specifically designed and built to separate highly radioactive isotopes. The RSIS will be housed entirely in a hot cell. Test operation of the separator has shown separation factors between adjacent isotopes of about 10^4. As an example, it will take about 1 to 3 weeks to produce 1 mg of ^{170}Tm, depending on the efficiency of the ion source, which ranges from about 10 to 30% for Tm.

Finally, a new detector array is being constructed. This will be discussed further below.

PRELIMINARY MEASUREMENTS

A preliminary experiment to measure neutron capture cross sections on a radioactive target used the isotope ^{171}Tm, which has a half-life of 1.9 yr. This isotope was chosen for several reasons. First, milligram quantities of isotopically pure ^{171}Tm were produced by reactor irradiation of ^{170}Er and chemical separation by high-performance liquid chromatography, without the use of an isotope separator. Second, the low-energy of the decays (97 keV β, 66 keV γ) presented the lowest radiological hazard of the possible targets. The target was electroplated on a 12.7 μm Be foil which had a 70 nm layer of titanium vapor-deposited on the surface. This was covered by a second 12.7 μm Be foil.

The experiment was done on Flight Path 4 of the Lujan Center at LANSCE, which views the "high intensity" water moderator. The detectors were two C_6D_6 scintillators, each 12.5 cm diameter by 7.5 cm thick, mounted on an RCA 4522 phototube. The detectors were mounted inside a 30 cm x 30 cm x 10 cm thick Pb shield inside a larger polyethylene house. The flight path was 8.08 m. The results of the measurement are shown in Fig. 2. The results are compared to calculations (6) made with the GNASH code, which underpredicts the data by a factor of 2 to 5. A target of stable ^{169}Tm was also measured, and the results agreed very well with the

GNASH calculations, which used parameters determined from an earlier measurement of capture on ^{169}Tm by Macklin (7).

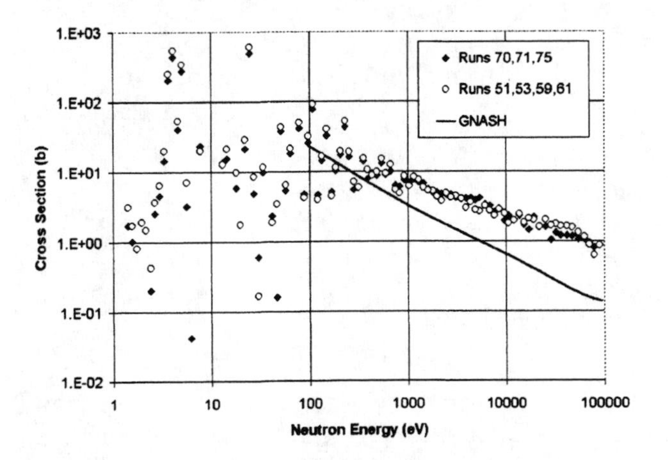

FIGURE 2. Cross section measured for the radioactive target ^{171}Tm, compared to the GNASH theoretical calculation (6).

In this configuration, the signal from the ^{171}Tm was only about 10% above the background from a blank Be foil assembly. Comparison of the background to the count rate with no foil indicates that neutron scattering was responsible for part of the gamma background. The background will be reduced in future measurements by decreasing the mass near the detector array and improving the shielding against neutrons from adjacent flight paths.

DANCE

The results of the preliminary experiment were encouraging and led to a proposal for a new detector array and flight path at the Lujan Center. The new detector will be located at 20 m on Flight Path 14, which views the newly installed "upper tier" water moderator. This moderator will have about half the flux of the high-intensity water moderator on Flight Path 4. However, planned improvements to the accelerator ion source and beam delivery are expected to eventually put up to 200 μA on the spallation target, about a 3X increase over the present proton current. By using different collimation which will view the entire moderator surface, the neutron flux at 20 m will actually be greater than on Flight Path 4 at 8 m.

In addition, a new detector array, the Device for Advanced Neutron Capture Experiments (DANCE) is being designed, and prototype detectors and collimation will be tested at LANSCE in fall, '98. The array will, by necessity, have a fast response and be highly segmented. A 2 Ci target, which was the activity of the ^{171}Tm target reported above, has 7.4 x 10^{10} decays/second, or 740 decays in a 10 ns bin. To get one count per detector in a 10 ns period, assuming 20% efficiency, would require 148 detectors. This count rate will present interesting challenges. A capture event produces about 3 to 10 gamma rays. The segmentation of the detector array will enable multiplicity to be used to help discriminate against decay gamma rays.

The leading candidate for detector material is BaF_2, which has a 0.6 ns component that peaks in the ultraviolet at 220 nm. It also has a slow component that produces five times as much light, so filters or phototubes with special photocathodes will be required. We note that scintillators containing Cs or I have a large neutron capture cross section even at 10 keV, and would therefore produce an unacceptable background. Neutron absorbers, used successfully by others, do not appear to be effective above 1 keV (8).

DISCUSSION

We are designing and building a new detector array for measuring neutron capture on small quantities of unstable isotopes. These measurements will provide valuable cross sections for applied programs and nuclear astrophysics. A partial array is expected to be operational in 1999, and we are very excited about the prospects for physics with this detector.

ACKNOWLEDGEMENTS

We would like to acknowledge P.E. Koehler for pioneering these measurements at LANSCE. This work was supported by the USDOE Science-Based Stockpile Stewardship program, and made beneficial use of the neutron beams at LANSCE.

REFERENCES

1. Wallerstein, G., et al., Rev. Mod. Phys. **69**, 995 (1997).
2. Kappeler, F., Beer, H., Wisshak, K., Rep. Prog. Phys. **52**, 945 (1989). Kappeler, F., Gallino, R. Busso, M., Picchio, G., Raiteri, C.M., Astrophys. J. **354**, 630 (1990).
3. Toucan, K.A., et al., Phys. Rev. C **51**, 1540 (1995).
4. Wisshak, K., et al., Phys. Rev C **48**, 1401 (1993).
5. For a description of the Manual Lujan Jr. Neutron Scattering Center, see http://www.lansce.lanl.gov.
6. Young, P.G., private communication.
7. Macklin, R.L., Drake, D.M., Malanify, J.J., Arthur, E.D., Young, P.G., Nucl. Sci. Eng. **82**, 143 (1982).
8. Wisshak, K., et al., "The Karlsruhe 4π BaF_2 Detector," KFK 4652, 1989.

Studies of the D State of ^{6}Li Using the FSU Polarized ^{6}Li Beam

K.D. Veal,[1,2] C.R. Brune,[1,2] W.H. Geist,[1,2] H.J. Karwowski,[1,2] E.J. Ludwig,[1,2] A.J. Mendez,[1,2] E.E. Bartosz,[3] P.D. Cathers,[3] T.L. Drummer,[3] K.W. Kemper,[3] A.M. Eiró,[4] F.D. Santos,[4] B. Kozlowska,[2,5] H.J. Maier,[6] and I.J. Thompson[7]

[1] *Department of Physics and Astronomy, University of North Carolina at Chapel Hill, Chapel Hill, NC 27599-3255*
[2] *Triangle Universities Nuclear Laboratory, Durham, NC 27088-0308*
[3] *Department of Physics, Florida State University, Tallahassee, FL 32306-3016*
[4] *Centro de Fisica Nuclear da Universidade de Lisboa, Lisboa, Portugal*
[5] *Institute of Physics, University of Silesia, Katowice, Poland*
[6] *University of Munich, Garching, Germany*
[7] *Department of Physics, University of Surrey, Guildford GU2 5XH, U.K.*

One way to quantify the D-state component of the wave function of a nucleus is by the quantity η, the ratio of the D- and S-state asymptotic normalization constants. Analyses of the analyzing powers from transfer reactions induced by polarized ions have been useful for the determination of η in the $A = 2-4$ systems. In an effort to determine η for the $d+\alpha$ relative motion in ^{6}Li we have measured analyzing powers for ($^6\vec{\text{Li}}$,d) reactions on ^{58}Ni and ^{40}Ca at E(^{6}Li) = 34 MeV. The experiments were performed at Florida State University using the Optically Pumped Polarized Lithium Ion Source. We compared the data with the results of well-constrained DWBA calculations assuming a direct α-particle transfer mechanism. With η the only free parameter in the calculations, a best fit to the tensor analyzing power data results in an average value of $\eta = +0.0003 \pm 0.0009$, much smaller than previous determinations.

INTRODUCTION

Few-nucleon systems continue to be a subject of considerable importance in nuclear physics. They provide excellent grounds for testing the description of complex systems on the basis of knowledge of the interaction between nucleons. One aspect of the nucleon-nucleon force that is of continuing interest is the tensor interaction, which is responsible for a large fraction of the binding energy of light nuclei, such as the deuteron, triton, ^{3}He, and ^{4}He (1) and is responsible for the presence of D states in the wave functions of these systems.

Measurements of D-states in the wave functions of light nuclei are generally given in terms of the ratio of the D- and S-state asymptotic normalization constants η. For large r, the S-state ($l = 0$) and D-state ($l = 2$) radial wave functions behave as (2)

$$u_l(r) = \frac{N_l}{\beta r} W_{-\xi, l+\frac{1}{2}}(2\beta r)$$

where N_l is a normalization constant, $W_{-\xi,l+\frac{1}{2}}$ is a Whittaker function, ξ is the Sommerfeld Coulomb parameter, and β is the wavenumber. As such, the quantity η, defined as $\eta = N_2/N_0$, is a measure of the relative strength of the D state in the wave function in the asymptotic region outside of the nuclear interaction. Values of η for the $A = 2$–4 nuclei have been established to within a few percent for the deuteron (3), within 10% for the triton and ^{3}He (4,5), and about 20% for the d + d configuration of ^{4}He (6).

The case for ^{6}Li, however, is less well understood. Cluster configurations of d + α and t + ^{3}He can combine to form $l = 0$ or 2 states and still maintain J$^\pi$ = 1$^+$ for the ^{6}Li ground state. While a D state could exist between the triton and ^{3}He, most experimental and theoretical investigations of the ^{6}Li D state have focused on the d + α configuration.

Contrary to the lighter nuclei, η for the d + α relative motion in ^{6}Li is not well determined, even as to its sign. In fact, the only direct evidence for a D state in the wave function in ^{6}Li is its small negative quadrupole moment $Q = -0.083$ fm^2 (7). Nishioka *et al.* (8) suggested that Q could originate from a delicate cancellation of the deuteron quadrupole moment with a D state of relative motion between the deuteron and α particle in ^{6}Li. The small D-state admixture which reproduced Q resulted in $\eta = -0.0112$. Three-body (αnp) models are generally able to describe very well properties of ^{6}Li such as the binding energy, the magnetic moment, and the charge radius, but systematically predict $Q > 0$ and $\eta > 0$ (3,9). A six-body calculation (10) using realistic two- and three-body forces recently predicted $\eta = -0.07 \pm 0.02$, but underpredicted the binding energy and overpredicted the negative quadrupole moment by an order of magnitude.

Experimentally, the situation is similar where determinations of η have widely varied. The S- and D-state asymptotic normalization constants were found in an analysis of d–α scattering (11) resulting in $\eta = +0.005 \pm 0.014$. Polarized ^{6}Li scattering analyses (12,13) have

CP475, *Applications of Accelerators in Research and Industry*,
edited by J. L. Duggan and I. L. Morgan

suggested that $\eta < 0$ while a polarized ^{6}Li breakup study (14) indicated $\eta > 0$.

In the present work, we have determined η from an analysis of the tensor analyzing powers (TAPs) A_{zz} and A_{xz} from ($^6\vec{\text{Li}}$,d) reactions on ^{58}Ni and ^{40}Ca. The magnitude and sign of the TAPs from these reactions scales with the magnitude and sign of η. Similar methods have been successful in D-state studies of the A = 2–4 nuclei from polarized-deuteron induced transfer reactions (4,5,6). We have recently published the ^{58}Ni data (15) and our main conclusions; the present article provides additional information including some of our ^{40}Ca data.

EXPERIMENTS

The experiments were performed at Florida State University using the optically-pumped polarized lithium ion source (16). The $^6\vec{\text{Li}}$ ions were accelerated to 34 MeV using the Super FN Tandem accelerator into an 85-cm-diameter scattering chamber. The setup in the scattering chamber, shown in Figure 1, consisted of two pairs of Si ΔE-E detector telescopes placed symmetrically on each side of the beam. The telescopes consisted of a total of 4 to 6 mm of Si with 1- to 2-mm Si ΔE detectors. A Ta foil was placed in front of each detector to stop elastically scattered ^{6}Li ions. Each detector telescope had an angular acceptance of either ±2 or 2.2° and subtended a solid angle of 4.7 or 6.2 msr. Additionally, a small detector was placed at 135° with respect to the incident beam direction to monitor the target. The ^{58}Ni targets (>99.76% enriched) were self-supporting rolled foils with thicknesses ranging from 0.8 to 2.0 mg/cm^2. The ^{40}Ca targets (natural abundance) had a thickness of 0.9 mg/cm^2 and were sandwiched between 0.3-mg/cm^2 layers of Au. The polarization was monitored via ^{4}He($^6\vec{\text{Li}}$,α)^{6}Li scattering (17,18). Typical beam polarizations on target were $p_z \approx -0.6$ and $p_{zz} \approx -1.1$.

Transitions to both the 0^+ ground state and the 2^+ first-excited state were well resolved for each of the reactions studied here. However, due to the small cross sections for these reactions, only forward angle data (≤ 45°) were measured. In addition to the TAP data, we also measured the vector analyzing power A_y for these reactions as well to assist in modeling the reaction mechanism.

ANALYSIS

In our analysis we assumed that the reactions take place via a direct α-particle transfer mechanism. Thus we have used the Distorted-Waves Born Approximation (DWBA) to model the reactions. The input parameters for the calculation of the distorted waves came from global optical model parameterizations of ^{6}Li and deuteron scattering. The bound-state wave functions (both S and D state) of the deuteron with the α particle were described by a Woods-Saxon (WS) effective potential with a geometry of $R_{\alpha d}$ = 1.90 fm and $a_{\alpha d}$ = 0.65 fm (19), where the depths were adjusted to reproduce the separation energy. The bound-state system of the α particle with the target nucleus was described by a WS potential with $R_{\alpha T} = 1.25\,A^{1/3}$ fm and $a_{\alpha T}$ = 0.65 fm. With these parameters, the oscillations in the calculations and the data were out of phase. Therefore, to improve the agreement between the calculations and the data, we decreased the deuteron optical model real radius potential by 12%.

A sample of the agreement between the calculations and cross section and vector analyzing power (VAP) data is shown in Figure 2. The overall agreement between the direct transfer calculations and the cross section and VAP data is very good. The remaining analysis is to determine the value of η which results in the best fit to the TAP data.

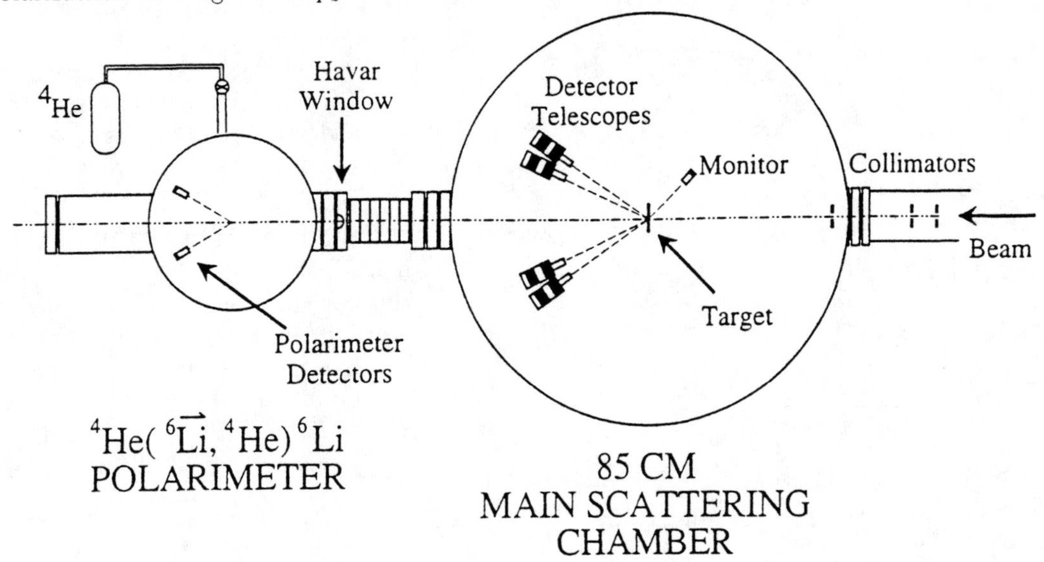

FIGURE 1. A schematic diagram of the scattering chamber for the ($^6\vec{\text{Li}}$,d) reaction experiments.

Treating η as the only remaining free parameter in the calculations, we determined a best-fit value for η for each of the eight TAP angular distributions via χ^2 minimization. We show in Figure 3 the TAP A_{zz} for both the ground and first-excited state transitions from the ^{40}Ca($^6\vec{\text{Li}}$,d)^{44}Ti reaction at 34 MeV. The solid curves correspond to the best fit calculation for each analyzing power while the dashed (dotted) curves correspond to η = +0.015 (−0.015). The uncertainty in the minimization, reflecting the uncertainty in our TAP data, was taken to be the difference in η between χ^2_{min} and χ^2_{min} + 1.

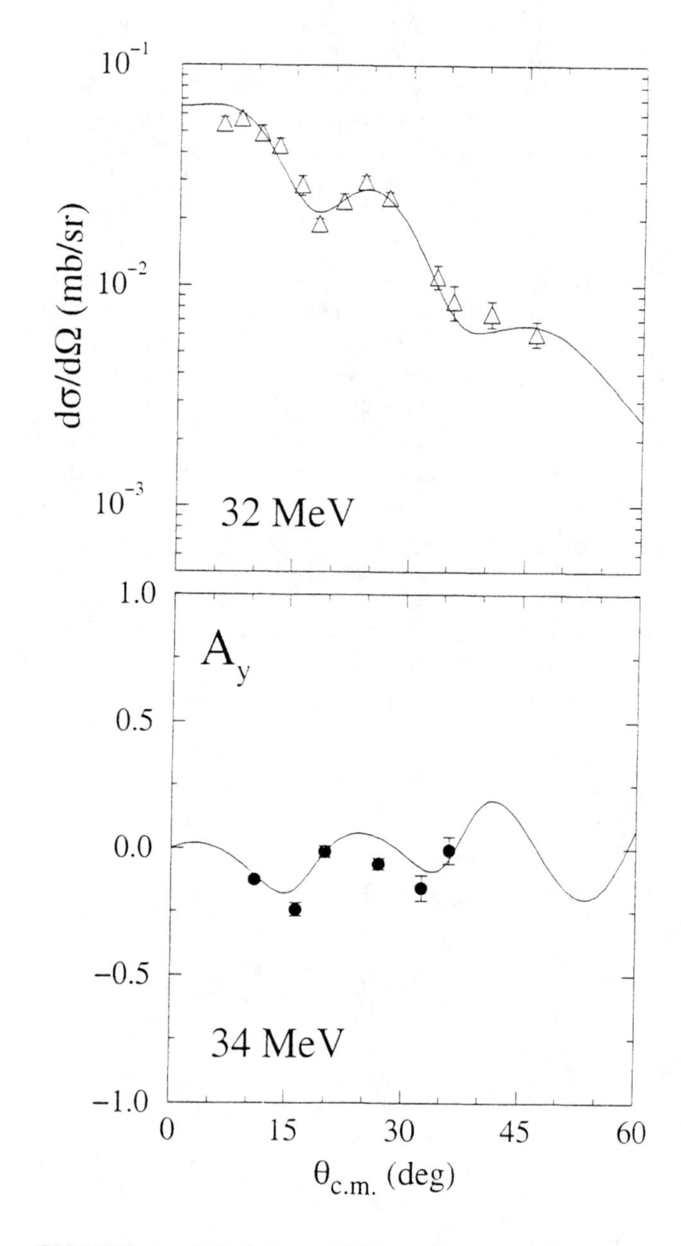

FIGURE 2. Comparison of DWBA calculations of the cross section and VAP for the ^{40}Ca($^6\vec{\text{Li}}$,d)^{44}Ti reaction leading to the 2^+ first-excited state at 1.08 MeV. The cross section data [△, Ref. (20)] and calculations are at 32 MeV while the VAP data [•, present work] and calculations are for 34 MeV.

We also investigated the sensitivity of the calculations to the input optical model parameters and bound-state potential parameters. It was found that the calculated TAPs were very sensitive to the parameters of the α + target bound-state geometry. These parameters ($R_{\alpha T}$ and $a_{\alpha T}$) were allowed to vary by ±15% and each time η was redetermined.

The effects of optical model tensor potentials were also investigated. It has been shown (3,4) that the TAPs from polarized-deuteron induced transfer reactions are sensitive to deuteron-nucleus tensor potentials. It was found that this was not the case for the ($^6\vec{\text{Li}}$,d) reactions studied here. However, a ^{6}Li-nucleus optical model tensor potential did have an effect on the calculated TAPs. We included the ^{6}Li tensor potential derived from 30-MeV ^{6}Li scattering from ^{12}C (18) in the DWBA calculations and found that the best-fit value of η changed slightly from that previously determined.

The final uncertainty in each of the eight determinations of η was calculated by adding in quadrature the contributions from the uncertainties in the fitting, the α + target bound states, and the tensor potentials. The final value for η was then taken to be the average value of the eight individual determinations, weighted by the uncertainty in each determination and resulted in

$$\eta = +0.0003 \pm 0.0009.$$

DISCUSSION

The value for η determined in this work is generally consistent that that of Ref. (11) given the size of the error on that determination. However, it is in disagreement with most previous theoretical estimations for η attributing the quadrupole moment Q to the deuteron quadrupole moment and the D-state in the d + α wave function (8). This result suggests that the origin of Q lies elsewhere in the total ^{6}Li wave function and does not arise from a D state between the deuteron and the α particle. It is hoped that this work will provide constraints on nuclear structure and tensor force effects in future three- and six-body calculations.

ACKNOWLEDGMENTS

We thank P.V. Green, P.L. Kerr, E.G. Myers, and B.G. Schmidt for there assistance with the experiments, and Z. Ayer for contributions in the early stages of this project. This work was supported in part by the US DOE under Grant No. DE-FG02-97ER41041, the NSF under Grant. No. NSF-PHY-95-23974, the Portuguese FCT under Contract Praxis/2/2.1/FIS/223/94, and the Polish KBN under Grant No. 2 P03B 056 12.

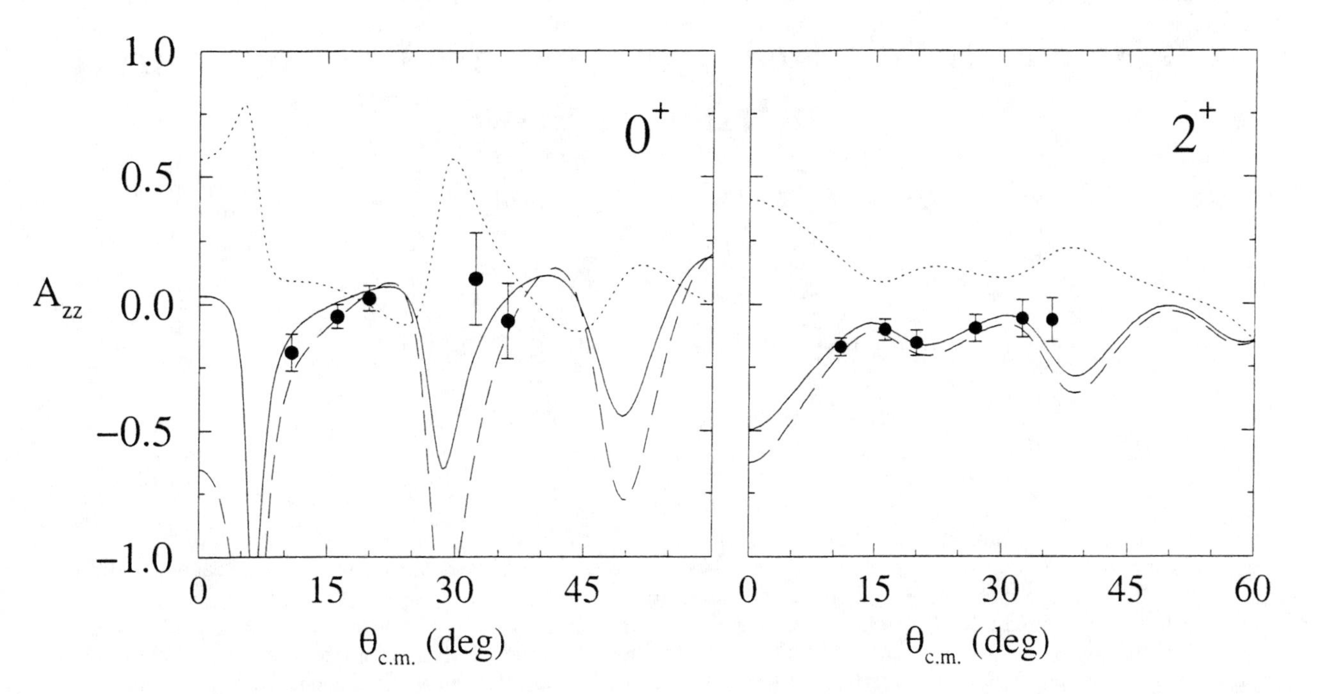

FIGURE 3. A comparison of DWBA calculations of the TAP A_{zz} with data for the ground (0^+) and first-excited (2^+) state transitions of the $^{40}Ca(^{6}\vec{Li},d)^{44}Ti$ reaction. The solid curves correspond to the best-fit value of η for each analyzing power, $\eta = +0.0024$ for the 0^+ transition and $\eta = +0.0114$ for the 2^+ transition. The dashed (dotted) curves correspond to $\eta = +0.015$ (-0.015) and is presented to illustrated the sensitivity of the calculations to η.

REFERENCES

1. Ericson, T. E. O. and Rosa-Clot, M., *Ann. Rev. Nucl. Part. Sci.* **35**, 271 (1985).
2. Eiró, A. M. and Santos, F. D., *Jour. of Phys. G* **16**, 1139 (1990).
3. Lehman, D. R., *Colloque de Phys.* **51**, C6-47 (1990).
4. Kozlowska, B. *et al.*, *Phys. Rev. C* **50**, 2695 (1994).
5. Ayer, Z. *et al.*, *Phys. Rev. C* **52**, 2851 (1995).
6. Karp, B. C. *et al.*, *Nucl. Phys.* **A457**, 15 (1986).
7. Sundholm, D. *et al.*, *Chem. Phys. Lett.* **112**, 1 (1984).
8. Nishioka, H. *et al.*, *Nucl. Phys.* **A415**, 230 (1984).
9. Kukulin, V. I. *et al.*, *Nucl. Phys.* **A586**, 151 (1996).
10. Forest, J. L. *et al.*, *Phys. Rev. C* **54**, 646 (1996).
11. Bornand, M. P. *et al.*, *Nucl. Phys.* **A294**, 492 (1978).
12. Dee, P. R. *et al.*, *Phys. Rev. C* **51**, 1356 (1995).
13. Rusek, K. *et al.*, *Phys. Rev. C* **52**, 2614 (1995).
14. Punjabi, V. *et al.*, *Phys. Rev. C* **46**, 984 (1992).
15. Veal, K. D. *et al.*, *Phys. Rev. Lett.* **81**, 1187 (1998).
16. Myers, E. G. *et al.*, *Nucl. Instr. Meth. Phys. Res. B* **56/57**, 1156 (1991).
17. Mendez, A. J. *et al.*, *Nucl. Inst. Meth. Phys. Res. A* **329**, 37 (1993).
18. Kerr, P. L. *et al.*, *Phys. Rev. C* **52**, 1924 (1995).
19. Kubo, K. I. and Hirata, M., *Nucl. Phys.* **A187**, 186 (1972).
20. Fulbright, H. W. *et al.*, *Nucl. Phys.* **A284**, 329 (1977).

A Dynamically Polarized Hydrogen and Deuterium Target at Jefferson Lab*

J. R. Boyce, C. Keith, J. Mitchell, M. Seely

Thomas Jefferson National Accelerator Facility, Newport, News, VA, 23606

S. Bültmann, D. G. Crabb, & C. Harris

University of Virginia, Charlottesville, VA, 22901

Polarized electron beams have been successfully used at Jefferson Lab for over a year. We now report the successful achievement of polarized targets for nuclear and particle physics experiments using the dynamic nuclear polarization (DNP) technique. The technique involves initial irradiation of frozen ammonia crystals (NH_3 and ND_3), using the electron beam from the new Free Electron Laser (FEL) facility at Jefferson Lab, and transferring the crystals to a special target holder for use in Experimental Halls. By subjecting the still ionized and frozen ammonia crystals to a strong magnetic field and suitably tuned RF, the high electron polarization is transmitted to the nucleus thus achieving target polarization. Details of the irradiation facility, the target holder, irradiation times, ionized crystal shelf life, and achieved polarization are discussed.

INTRODUCTION

New information associated with the nucleus can be obtained from experiments with both a polarized beam and a polarized target. Polarized electron beams have been successfully used at Jefferson Lab for over a year. Target nuclei polarization at Jefferson Lab, however, has only recently been attempted using Dynamic Nuclear Polarization, or DNP (1).

The basic idea behind DNP consists of using properly tuned microwaves in a high magnetic field to transfer the polarization of atomic electrons to the nuclei. Solid ammonia, in the form of NH_3 and ND_3 crystals is a good target material since it is hydrogen- or deuteron-rich and can be readily paramagnetically doped with an unpaired or quasi-free electron. Since the relaxation time of the electron spins is short (on the order of milli-seconds) and that of the nucleons is long (on the order of minutes), a relatively small number of unpaired electronic spins are required to polarize a large number of nuclear spins.

Some of the paramagnetic centers created through irradiation are stably trapped in the ammonia crystals at temperatures as high as 77 K. In effect, the "self life" of these irradiated crystals can be several months if they are stored in liquid nitrogen. SLAC experiments E143 (2-5) and E155 have demonstrated the viability of DNP achieving polarization of up to 85% for protons and 38% for deuterons.

Here we describe our initial efforts to produced such polarized hydrogen and deuterium targets for the nuclear physics research program at Jefferson Lab. We describe the FEL Irradiation Facility, two irradiation procedures: "warm" and "cold," and initial results.

THE FEL IRRADIATION FACILITY

The layout of the FEL accelerator is shown in Figure 1. The space between the beam dump and the wiggler is the irradiation location. For these irradiations, the electron beam from the injector was accelerated to 36 MeV by the cryomodule and directed into the straight-ahead beam dump through the irradiation set-up. The wiggler was not used.

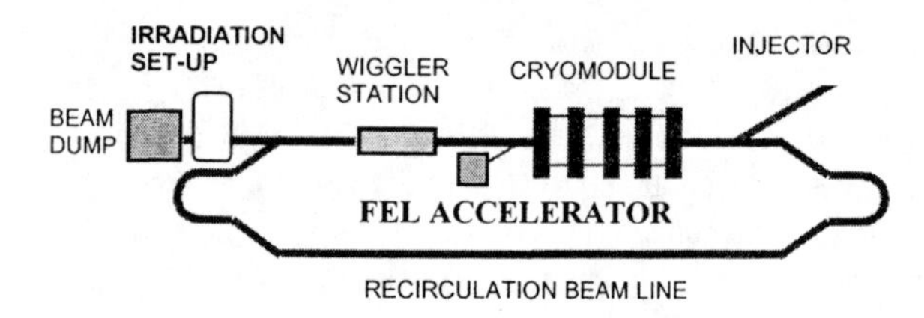

Figure 1. The Jefferson Lab FEL Accelerator layout.

* Supported in part by DOE Contract #DE-AC05-84ER40150

CP475, *Applications of Accelerators in Research and Industry*, edited by J. L. Duggan and I. L. Morgan

The layout of the first irradiation set-up - the "warm" set-up - where the ammonia crystals were in a bath of liquid argon at 87.3 K, is shown in Figure 2. A dewar, modified to have two thin window ports for beam entrance and exit was placed between the beam-line exit and beam dump.

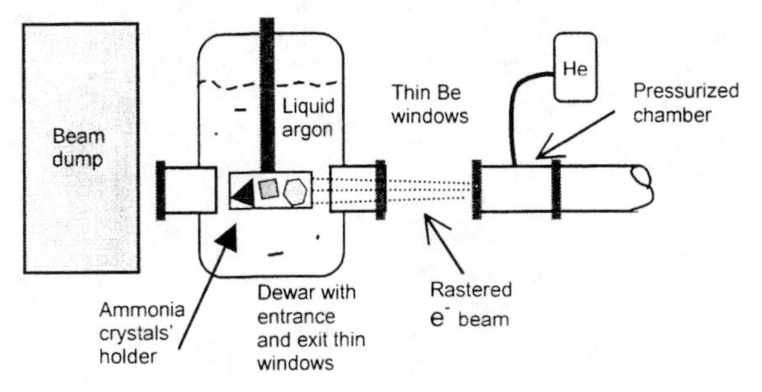

Figure 2. The "Warm" irradiation set-up.

Of major concern when the electron beam is brought out of the normal high vacuum system is the possibility of a breach of the vacuum system due to beam burn through, thus letting room air reach the srf accelerator. To mitigate this situation, a chamber pressurized to 1 atm of helium was placed at the beam exit location. Pressure in the chamber is monitored and interlocked to gate valves that will actuate on a pressure drop. Thus, only a small amount of helium, not room air, would reach the SRF accelerator with vacuum system breach.

A second safeguard against beam burn-through is a raster pattern placed on the beam by air core magnets upstream from the beam's exit chamber. The raster prevents local hot spots from occurring due to the beam and ensures uniform irradiation of the ammonia crystals.

For the "warm" irradiation targets, a total dose of $\sim 10^{17}$ electrons/cm^2 was needed rastered over 4.8 cm^2. This translated into about an 11 hour run with a beam current of 2 μA. Two "warm" irradiations were done: one for crystals of NH_3 and one for crystals of ND_3.

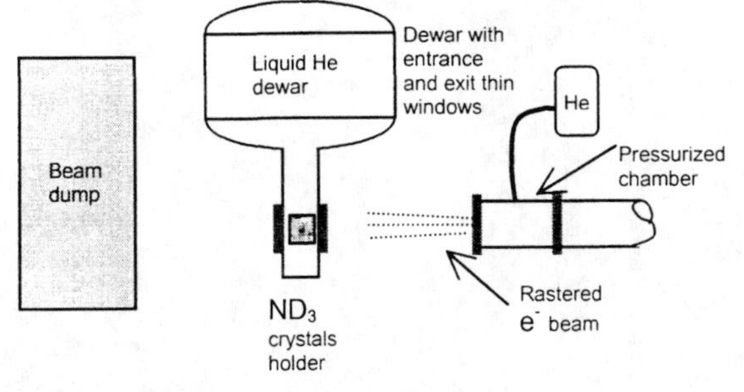

Figure 3. The "Cold" irradiation set-up.

The "cold" irradiation was at liquid helium temperatures (~ 4 K). The "cold" dewar was also modified for minimum beam loss in the dewar and maximum crystal irradiation. Figure 3 is a sketch of the cold set-up. Only ND_3 was used in the "cold" irradiation.

Experience has shown that "warm" irradiation of ND_3 is insufficient for achieving optimal polarization. Previously, "warm" irradiation have been supplemented by additional "in situ" irradiations at 4 K to achieve the desired polarization. We have attempted to get the same results with a liquid helium temperature, or "cold," irradiation at the FEL. This required a dose of one quarter that of the "warm" irradiation. The beam current was reduced to minimize the heat loss, and thus helium boil-off, during the run. The run time for a 1 μA average beam was estimated to be about 5 hours.

The basic run time procedure was to perform three irradiations:

- a "Warm" irradiation of one sample of NH_3,
- a "Warm" irradiation of one sample of ND_3, and
- a "Cold" irradiation of one sample of ND_3.

Seven days of run time were allocated for this irradiation effort: Two for setup and take-down, two for warm runs, one for changing from warm to cold configuration, and two for the cold runs.

RESULTS AND DISCUSSION

In actuality, the entire procedure was successfully executed in four days during a stability test of the FEL facility electron beam. The beam was found to be very stable, delivering the requested beam current without anticipated interruptions.

Other laboratories have been successful in storing irradiated crystals for several months without significant loss of ferromagnetic centers and thus without significant deterioration in polarizability of the target material. Typically polarizations of 90% have been obtained for NH_3 and about 35-40% for ND_3. For the Jefferson Lab irradiated samples, polarization of >90% has been achieved in the NH_3 samples. The Jefferson Lab irradiated ND_3 have been kept in cryogenic storage until the supply of SLAC irradiated samples is fully utilized. It is expected that the ND_3 polarization will be roughly the same as for the SLAC samples, or ~38%.

CONCLUSIONS

Target polarizations can be achieved using the technique of DNP and Jefferson Lab has started a program to utilize this technique to supply polarizable targets for experiments in the nuclear physics research program. This was accomplished during accelerator stability tests of the newly constructed FEL facility.

Proton polarizations of >90% were achieved. Deuteron polarizations have not been verified as of this date but are expected to be on the order of previous attempts by other laboratories: ~38%.

This achievement also illustrates the inter-relationship between basic research and applications research, namely: technology improvements in one field many times benefit the other field.

ACKNOWLEDGMENTS

The authors would like to thank the FEL Team and the Jefferson Lab Operations staff for assistance both prior and during the irradiation runs. Thanks also are due to members of the polarization collaboration group for advice on ammonia crystal handling.

REFERENCES

1. Crabb, D. A. and Meyer, W., *Annu. Rev. Nucl. Part. Sci.,* New York, Annual Reviews, Inc., 1997, 47: pp. 67-109.
2. Abe, K., et al., SLAC-PUB-6580 (1994); Phys. Rv. Lett. **74**, 346 (1995).
3. Abe, K., et al., SLAC-PUB-6734 (1994); Phys. Rv. Lett. **75**, 25 (1995).
4. Abe, K., et al., SLAC-PUB-6997 (1995); Phys. Rv. Lett. **B364**, 61 (1995).
5. Abe, K., et al., SLAC-PUB-6982 (1995); Phys. Rv. Lett. **76**, 587 (1996).

Electron Scattering with Polarized Beams at MIT-Bates Accelerator Center†

M. Farkhondeh, D. Barkhuff, G. Dodson, K. Dow, E. Ihloff, S. Kowalski, E. Tsentalovich, B. Yang and T. Zwart, for the SAMPLE Collaboration‡

MIT-Bates Accelerator Center, Middleton, MA. 01949

A brief introduction to electron scattering with polarized beams as a precise tool for studying electromagnetic structure of nucleons and nuclei will be presented. The status of a parity violating experiment SAMPLE [1] at Bates using a polarized beam to probe the strange quark content of proton will be reported. The polarized source at Bates as used for the SAMPLE experiment is described.

INTRODUCTION

Polarized electron beams are used at MIT-Bates for medium energy electron scattering studies in which the electromagnetic properties of the nucleons and nucleus are probed with the exchange of virtual photons between the electron and target. Polarization provides access to small components of the nuclear response, which arise from the interference between the components of the nuclear currents [2].

In nuclear scattering, the asymmetry between the yield from positive and negative helicity electrons typically ranges between 0.1% to 1%. These asymmetries are quite large compared with the asymmetry that arises from parity violating electron scattering due to the exchange of a Z boson. In recent years, with the development of high quality polarized beams, several parity violation experiments have been carried out or proposed at Bates and other labs in order to test the standard model [3] and to study the strange quark-antiquark ($s\bar{s}$) content of the nucleon [4,5]. The SAMPLE experiment at Bates [4], for example, measures a parity violating asymmetry between the yield of positive and negative helicity electrons scattered at backward angles from an unpolarized liquid hydrogen target. Because the measured asymmetry is extremely small, $\sim 10^{-6}$, it is crucial to suppress any helicity-correlated differences in the properties of the electron beam that may introduce false asymmetries in the measured yield. These include differences in the current, position, angle and energy of the beam incident on the hydrogen target for the two helicities. This parity violating experiment at Bates requires a very high quality beam. For example, the experiment requires helicity-correlated beam position differences to be controlled at the level of a few hundred nanometers at the target for a beam that is few millimeters in diameter. The full specifications for this beam are listed in Table 1. With the maximum 4 Watts peak laser power currently available at the photo-cathode, Quantum Efficiencies (QE) of greater than 0.5% are needed to obtain the SAMPLE peak currents of 12 mA from the gun. So far, this requirement has made it impractical to use high polarization photocathodes that inherently have low QE's. SAMPLE also requires a relatively long lifetime from the source so that a minimum of two Coulombs a day on target may be accumulated for periods of several days between photocathode "activations".

The two major obstacles in running the SAMPLE experiment in the past had been: 1) a source lifetime insufficient to reliably deliver two Coulombs a day, and 2) helicity correlated beam position differences exceeding a few hundred nm [6]. However, we demonstrated substantial progress on both fronts during an extended SAMPLE beam development period last fall in which both the source lifetime and the beam position differences met or exceeded the SAMPLE requirements. Following are descriptions of the polarized injector system and the SAMPLE experiment.

MIT-BATES POLARIZED SOURCE

The polarized injector at Bates consists of a commercial high power CW laser system, a 60 kV diode electron gun, a Wien filter for spin manipulation, a 300 kV acceleration column, a vertical transport line and an achromat with two 45° dipoles. The gun chamber and the associated instrumentation are located inside an elevated Faraday cage which is separated from the ground with corona rings.

Longitudinally polarized electrons are produced by illuminating a GaAs photocathode with circularly polarized laser light at near infrared wavelengths. The maximum theoretical polarization from bulk GaAs is 50%. In practice, this limit falls below 40% due to depolarization of electrons while transporting to the crystal surface. In order to achieve high QE, the surface of the photocathode must be clean and have a Negative

† Work supported in part by the Department of Energy under contract DE-AC02-76ER03069.

‡ For a list of SAMPLE collaboration members visit the SAMPLE web page http://www.npl.uiuc.edu/exp/sample/sampleMain.html.

CP475, *Applications of Accelerators in Research and Industry*,
edited by J. L. Duggan and I. L. Morgan

Electron Affinity (NEA). The surface is heat cleaned at ~600C for ~1 hour. NEA is achieved in the gun chamber by an "activation" process in which Cs and an oxidizer (O_2 or NF_3) are applied. Over time, the surface of the photocathode becomes contaminated with beam induced backstreaming ions and residual gases in the chamber, causing the QE to degrade. Heat cleaning is necessary when the required current can not be maintained at full laser power. This process takes about 10 hours during which no beam is available. While beam is running, cesium is periodically added to the chamber with Cs dispensers to partially restore the QE. Two additional guns identical to the injector gun are available as backups. A typical gun interchange requires only about four days of downtime.

Laser System

The laser system consists of a 30W CW Ar laser pumping a CW Titanium-sapphire laser. The Ar laser is chopped with a synchronous electro-mechanical chopper. The chopper wheel a) reduces the thermal load on the Titanium-sapphire crystal and b) provides an accurate common timing reference for the entire accelerator and the experiment. An electro-optical chopper phase-locked with the mechanical chopper produces 18 μs wide light pulses. With 29 W Ar, peak powers as high as 8 W at ~750 nm are extracted from the Titanium-sapphire. Because of insertion losses in the transport system and the dynamic ranges of various feedback systems only a maximum of ~3.5 W is available to the photocathode. The light is directed by four remotely adjustable transport mirrors to the photocathode through a vacuum port. The circularly polarized light is generated from linearly polarized light incident upon a Helicity Pockels Cell (HPC) acting as a $\lambda/4$ wave plate when appropriate voltages are applied. The helicity of the light and the electron beam is flipped by reversing the polarity of the voltage on the HPC. In the past, for ease of access, the HPC was located on the laser table upstream of the transport mirrors. Later, it was discovered that the mirrors introduced or amplified helicity-correlated position differences that originated from imperfections in the HPC. The HPC was subsequently moved to the accelerator vault and positioned after the last mirror that directs the laser light onto the photocathode. The total flight path between the laser source and the cathode is ~15 m with point to point imaging. The helicity of the electron beam is chosen randomly for each set of 10 consecutive beam pulses and the complement helicities are used for the next ten pulses. At 600 Hz, these "pulse pairs" of opposite helicity are 1/60 second apart. To minimize systematic errors, physics asymmetries are calculated individually for each pulse pair.

SAMPLE EXPERIMENT

The SAMPLE experiment at Bates measures the strange magnetic form factor G_M^s of the proton at low momentum transfer. At Q^2=0, G_M^s is the "strange magnetic moment" (μ_s). The normal magnetic moment of nucleon corresponds to the magnetic coupling to the photon; the weak magnetic moment involves coupling to the Z boson. The neutral weak magnetic moment of the proton can be expressed in terms of known magnetic moments of proton (μ_p) and neutron (μ_n) and an unknown contribution from strange quarks:

$$\mu_p^Z = \frac{1}{4}(\mu_p - \mu_n) - \sin^2\theta_w \mu_p - \frac{1}{4}\mu_s \qquad (1)$$

where θ_w is the weak mixing angle and μ_s is the strange quark magnetic moment. The quantity G_M^s for the proton is measured via elastic parity-violating electron scattering at backward angles. Theoretical predictions for G_M^s in the limit of zero momentum transfer have been summarized in Refs [7,8]. Parity-violating electron scattering is measured in terms of the asymmetry A in the cross section for the scattering of right- and left-helicity electrons from an unpolarized target as shown in Eq. 2:

$$A = \frac{\sigma_R - \sigma_L}{\sigma_R + \sigma_L} \qquad (2)$$

The SAMPLE experiment is a measurement of this asymmetry at backward angles with beam energy of 200 MeV incident on a 40-cm long liquid hydrogen target. The scattered electrons are detected in a large solid angle (~2 sr) Cerenkov detector covering the angular region between 130 and 170 degrees. The average momentum transfer is Q^2 =0.1 $(GeV/c)^2$ and at these kinematics the asymmetry is sensitive to $G_M^s(0) = \mu_s$. The 40 μA beam deposits over 500 Watts of heat into the target, a high flow-rate recirculating liquid hydrogen system. The detector consists of ten mirrors which image the target onto ten 8-inch photomultiplier tubes. There is a remotely controlled light shutter to cover each photomultiplier tube for background measurements. About 1/4 of the data taking runs are with shutters closed. Because the counting rate in each phototube is very high, individually scattered electrons are not counted. Instead, the signal is integrated over the duration of each pulse and normalized to the charge in each burst. In addition, data are taken during "pulse-counting" runs in which beam currents are reduced by several orders of magnitude. In these runs, the individually scattered electrons are detected in coincidence with photons incident on the phototubes. One out of 10 pulses is used as a "tracer bullet" at high enough current to a) monitor the beam in the accelerator and b) to compare the ordinary integrated detector signal to the "pulse-counting" signal. With pulse-counting runs and computer simulation the composition of the detector signals were studied in great detail. The preliminary composition is ~21% non-light, ~55% elastic scattering, and ~22% scintillation light from soft EM radiation from the target. There is also an estimated ~3% Cerenkov light from decaying π^+ and π^0 produced in the target. The measured asymmetry is corrected for the effects of these backgrounds.

Any helicity correlation in beam properties will cause false asymmetries. The raw measured asymmetry A_{raw} must be corrected for these effects. The effects of these corrections are measured periodically by changing the beam parameters (positions, angles and energy) by known amounts and measuring the changes in the yield. The corrected asymmetry A_c is expressed as

$$A_C = A_{raw} - \frac{1}{S}\Sigma_i \frac{\partial S}{\partial \alpha_i} \delta\alpha_i^{LR} \tag{3}$$

where S is the normalized detector yield, α_i is one of the five beam parameters and $\delta\alpha_i$ is the helicity-correlated difference in the beam parameters. The goal has been to make the size of the corrections as small as possible.

Beam Considerations

Due to the linac's low duty factor of less than 1% and a capture efficiency of ~1/3, peak currents exceeding 12 mA are needed from the polarized source. These peak currents, nominal pulse widths of 16μs and a repetition rate of 600 Hz fall in a difficult regime for polarized injector systems. The SAMPLE experiment required about 140 Coulombs of unprecedented quality and stability over 1000 hours at average currents of 40μA. A schematic view of the SAMPLE experiment is shown in Figure 1. A summary of SAMPLE beam parameters is listed in Table 1. The most challenging requirement for SAMPLE was the source lifetime of over three days between activations. This is particularly difficult at high peak currents. One way of reducing the peak current for the same average current is to increase the width of the beam pulse to 25μs from the nominal 16μs (increasing the duty cycle to ~1.5%). By carefully processing the accelerator RF system to wider pulses, it was demonstrated that 40 μA could be delivered to the SAMPLE experiment with injector pulses ~8mA high and ~25μs wide. The helicity-correlated beam position differences of less than a few hundred nanometers are also very demanding. This was particularly problematic for several years when HPC was located on the laser table upstream of the transport mirrors. As discussed earlier, the position differences were significantly reduced when the HPC was moved to its new location.

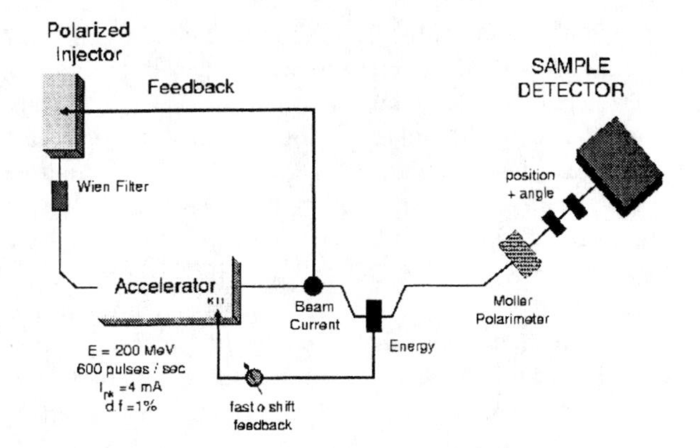

FIGURE 1. A schematic view of the SAMPLE experiment

The stability of the beam must be very high in order to reduce the width of the measured asymmetries. The helicity-correlated intensity differences are maintained below ~10ppm in each half-hour. This is accomplished with an active feedback system by measuring the total charge for each helicity at the end of the linac and adjusting the high voltage (every ~3 minutes) on a Correction Pockels Cell (CPC) in a helicity correlated way. The CPC voltage adjusts the intensity of the light to the photocathode. A λ/10 wave plate installed in front of the CPC determines the sensitivity of the beam intensity to the CPC voltage. The stability of the extracted electron beam depends on the stability in the intensity, mode structure and the pointing of the laser; and on the uniformity of the QE on the surface of the photocathode [6]. A fast feedback system with a gain bandwidth of 100 MHz was developed and implemented to stabilize the electron beam current within each pulse. The differential signal between the toroid signal and a reference level is amplified and fed to a Pockels cell. This feedback system improved the beam current stability by a factor of five.

TABLE 1. SAMPLE Beam Specifications.

Beam Parameter	Specification
Average current on target	~40 uA
Average daily charge on target	~2 C
Beam width on target	~2.5 mm σ
Intensity stability	≤1%
Helicity-corr. Intensity differences*	≤10 ppm
Helicity-corr. Position differences *	≤ 0.3 μm
Time between full activation	≥3 days
Total Coulombs for ^{1}H	~140C

*In 1/2 hour runs.

Performance

In 1995-96, about 1/4 of the required SAMPLE data (30 Coulombs) was collected and a preliminary result was published in 1997 [1]. It was determined that improvements in both the lifetime and the helicity-correlated position differences were needed before the experiment could take additional data. Many improvement of the polarized source and the accelerator system have been implemented in 1997-98 as a result of several beam development tests for SAMPLE. It was demonstrated that long lifetimes could be maintained for several weeks. The 25μs long pulse mode was implemented recently; this significantly improved the photocathode lifetime that already had exceeded the SAMPLE requirements. This past summer, data taking was successfully completed for the hydrogen portion of the experiment. In less than two months, over 130 Coulombs of polarized beam of exceptional quality and

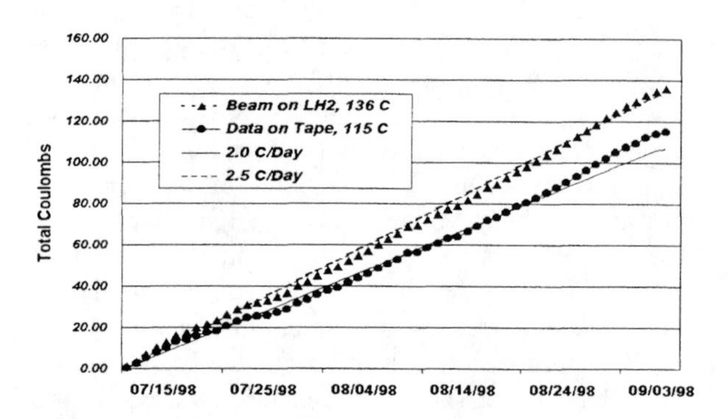

FIGURE 2. The total Coulombs vs. the calendar time shown for the SAMPLE experiment in 1998. The two curves show the delivered beam (▲) and the data on tape (●). The dashed and solid lines represent 2.5C and 2.0 C per day respectively.

reliability was delivered to the SAMPLE experiment. Only three heat cleanings were necessary during this period and lifetimes (1/e) of several hundred hours were achieved. The accumulated charge as a function of time is shown in Figure 2. The beam polarization was measured daily with a Møller polarimeter and averaged about 37% over the duration of the experiment. Helicity-correlated position differences were below 200 nm (see Figure 3). A piezo-based feedback system was also implemented to further reduce the position differences [9]. With the fast beam current stabilizer discussed earlier, a FWHM of ~0.7% in the raw signal and ~0.4% in the asymmetry of the last accelerator toroid were often achieved. Every two days, a helicity λ/2 wave plate was inserted or removed downstream of the HPC to manually reverse the helicity of the beam independent of all electrical signals. Stability in position and intensity of the beam was excellent under this helicity reversal, leading to a negligible correction to the physics asymmetry. The overall stability of the beam was significantly improved by implementing a feedback system to stabilize the accelerator energy. In this system, the beam position is continuously measured in a dispersive region after the linac and the phase of accelerator is adjusted via a fast phase shifter in order to hold the beam energy constant.

Preliminary analysis of the new SAMPLE data indicates that the asymmetry measured this summer is consistent with the results published in 1997 [1] but with significantly smaller statistical uncertainty.

There is a second component to the SAMPLE experiment that is scheduled to take beam next summer [10]. This new measurement of the parity violating asymmetry on deuterium will improve the reliability of the flavor decomposition of the proton magnetism. The deuterium experiment will run at the same kinematics and beam conditions and should reduce the systematic uncertainty of the proton measurement.

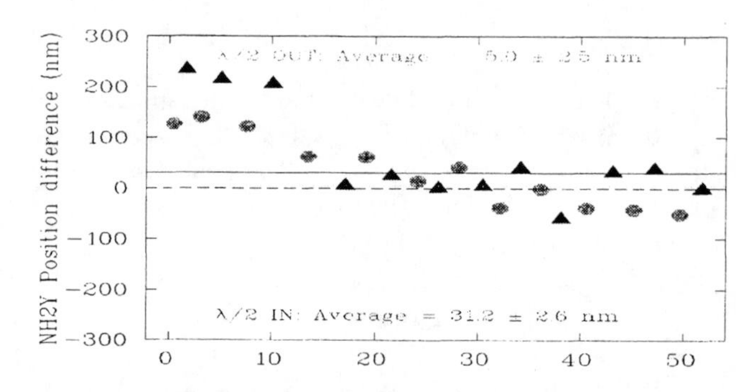

FIGURE 3. A plot of helicity-correlated position differences at the SAMPLE target vs. days into the experiment for the 1998 runs. The preliminary results are shown for the λ/2 plate in (▲) and out (●) averaged over ~ 2 days.

REFRENCES

1. B. Mueller, *et al.*, Phys. Lett. **78**, *3824 (1997).*
2. J. Mandeville, *et al.*, Phys. Rev. Lett., **72**, 3325 (1994).
3. Bates ^{12}C Parity experiment, P. A. Souder *et al.*, Phys. Rev. Lett. **65**, 694 (1990).
4. Bates experiment 89-06 (R. McKeown and D. Beck, contacts).
5. TJNAF experiments E91-017 (D. Beck, spokesperson), E91-010 (P. Souder and M. Finn, spokespersons), and E91-004 (E. J. Beise, spokesperson).
6. M. Farkhondeh *et al.*, Proc. Of 7th Intern. Workshop on Polarized Gas Targets and Polarized Beams, Urbana, IL, August 1997, AIP **421**, p.240.
7. M.J .Musolf *et al.*, Phys. Rep. **239**, I(1994)
8. E.J. Beise *et al.*, in Proceedings of SPIN96 Symposium (Report No. nucl-ex/9610011).
9. B. McKeown and Todd Everett, Kellogg Radiation Laboratory, Caltech, Pasadena, CA 91125.
10. Bates experiment 94-11(M. Pitt and E. Beise, contacts).

Computational Design of High Efficiency Release Targets for Use at ISOL Facilities

Y. Liu and G. D. Alton

Physics Division, Oak Ridge National Laboratory, P. O. Box 2008, Oak Ridge, TN 37831-6368

J. W. Middleton

Engineering Cybernetics, Inc., Houston, TX 77084

This report describes efforts made at the Oak Ridge National Laboratory to design high-efficiency-release targets that simultaneously incorporate the short diffusion lengths, high permeabilities, controllable temperatures, and heat removal properties required for the generation of useful radioactive ion beam (RIB) intensities for nuclear physics and astrophysics research using the isotope separation on-line (ISOL) technique. Short diffusion lengths are achieved either by using thin fibrous target materials or by coating thin layers of selected target material onto low-density carbon fibers such as reticulated vitreous carbon fiber (RVCF) or carbon-bonded-carbon-fiber (CBCF) to form highly permeable composite target matrices. Computational studies which simulate the generation and removal of primary beam deposited heat from target materials have been conducted to optimize the design of target/heat-sink systems for generating RIBs. The results derived from diffusion release-rate simulation studies for selected targets and thermal analyses of temperature distributions within a prototype target/heat-sink system subjected to primary ion beam irradiation will be presented in this report.

INTRODUCTION

Many of the reactions, fundamentally important in nuclear physics and astrophysics, are inaccessible to experimental study using stable/stable beam/target combinations and, therefore, can only be studied with accelerated radioactive ion beams (RIBs). The availability of RIBs offers unique opportunities to further our knowledge about the structure of the nucleus, the stellar processes that power the universe, and the nucleosynthesis burn-cycles responsible for heavy element formation. The well-known on-line isotope separator (ISOL) technique, used for the generation of short-lived radioactive nuclei, involves a multi-step process: the nuclear reaction products must diffuse to the surface of the target material, evaporate from the surface, and effusively flow to the ion source where a fraction of the species of interest are ionized, extracted from the source and mass analyzed prior to post acceleration. The principal means whereby short-lived radioactive species are lost between initial formation and utilization are associated with the diffusion and surface adsorption processes where the hold-up times are long with respect to the life-time of the species of interest. Both diffusion and surface adsorption processes depend exponentially on the operational temperature of the target/ion source, and thus, successful RIB generation requires careful consideration of the physical, chemical, and metallurgical properties of the target material and thoughtful design of highly permeable target matrix systems with short diffusion lengths and controllable target temperatures. In this report, we briefly review our efforts to develop targets that simultaneously incorporate these properties so that useful RIB intensities for a wide range of short-lived species can be generated for use in the astrophysics and nuclear structure physics research programs at the Holifield Radioactive Ion Beam Facility (HRIBF).

HIGHLY PERMEABLE TARGET MATRICES

Efforts are underway at the HRIBF to design fast release targets with controllable temperatures for optimum production and generation of RIBs [1]. The speed of release is dependent on the diffusion constant, the target temperature, and the thickness through which the radioactive particles must pass prior to arriving at the surface of the target material. Therefore, fast release times are synonymous with large diffusion constants, short diffusion lengths (i.e., small target material dimensions), and low enthalpies of adsorption for released products.

Thin Al_2O_3 fibrous targets have been successfully used during on-line production and release of ^{17}F and ^{18}F for use in the HRIBF astrophysics research program [2]. Other refractory fibrous metal oxides, such as ZrO_2, HfO_2 are also being considered for use as target matrices for the production of F isotopes. Still, other fibrous targets such as SiO_2, TiO_2, Ta_2O_5, Y_2O_3 as well as rare-earth can be used. Many other refractory materials such as the metal-carbides are under consideration for RIB applications. We have also chosen low-density carbon fiber, such as carbon-bonded-carbon fiber (CBCF) or reticulated vitreous carbon

CP475, *Applications of Accelerators in Research and Industry*, edited by J. L. Duggan and I. L. Morgan

fiber (RVCF), to serve as the target material plating matrix and thermal transport conduit for heat removal. These fibrous materials are refractory and highly porous (density typically 0.06 to 0.12g/cm^3). Very thin layers of selected target materials, for example, Ni, SiC and UC_2, have been uniformly deposited onto RVCF and CBCF to form composite target matrices using either electroplating (EP) or chemical vapor deposition (CVD) methods [3].

DIFFUSION RELEASE

Diffusion in an isotropic medium is governed by Fick's laws. For the release of radioactive isotopes of decay constant λ, the time-dependent diffusion equation, modified to include the rate of production S and the rate of decay loss E, can be expressed in a Cartesian coordinate representation as follows:

$$\frac{\partial n}{\partial t} = D\left(\frac{\partial^2 n}{\partial x^2} + \frac{\partial^2 n}{\partial y^2} + \frac{\partial^2 n}{z^2}\right) + S(x,y,z,t) - E(x,y,z,t) \quad (1)$$

where n is the concentration of the diffusing substance; D is the diffusion coefficient, assumed to be independent of concentration. For an uniform distribution of radioactive species produced with a primary ion beam of intensity I and charge Z, the production rate S is given by

$$S(x,t) = \frac{\sigma n I L}{ZeV} \quad (2)$$

where e is the charge on the electron, n is the density of interaction nuclei, I is the intensity of the primary ion beam, L is the length of the target material, σ is the cross section for production of the species of interest, and V is the volume over which the species are distributed. The decay loss E is identically equal to

$$E(x,t) = n\lambda \quad \text{where } \lambda = 0.693/\tau_{1/2} \quad (3)$$

for radioactive species with half-life $\tau_{1/2}$,

The rate of diffusion release from a target material can be calculated by solving the appropriate form of Eq. 1. Diffusion in composite RVCF targets coated with selected target materials can be modeled by solving the one-dimensional equation for a planar geometry target. For an initially homogeneous distribution of nuclei in an infinite planar target of thickness l, the fractional release $f(t)$ is given by

$$f(t) = 1 - \frac{2}{l^2}\sum_{n=0}^{\infty}\frac{\exp\{-\mu_n Dt\}}{\mu_n} \quad (4)$$

where $\mu_n = (n+1/2)^2\pi^2/l^2$.

We have assumed that the species do not diffuse into the carbon substrate but are reflected at the boundary and that surface desorption is fast compared to the mean diffusion time. The total release rate per unit surface area can be written in the form

$$R(t) = \frac{2S}{l}\sum_{n=0}^{\infty}\frac{1-\exp\left\{-\left(\mu_n + \frac{\lambda}{D}\right)Dt\right\}}{\mu_n + \frac{\lambda}{D}} \quad (5)$$

The fractional release $f(t)$ of nuclei uniformly distributed within a fibrous cylindrical geometry metal-oxide target material of radius r, can be represented by the summation

$$f(t) = 1 - \frac{4}{r^2}\sum_{n=1}^{\infty}\frac{\exp\{-\alpha_n^2 Dt\}}{\alpha_n^2} \quad (6)$$

where α_ns are the positive roots of Bessel functions of the first kind of zero order, $J_0(\alpha_n r) = 0$. The total release rate per unit length on the fiber material then becomes

$$R(t) = 4\pi S\sum_{n=1}^{\infty}\frac{1-\exp\left\{-(\alpha_n^2 + \frac{\lambda}{D})Dt\right\}}{\alpha_n^2 + \frac{\lambda}{D}} \quad . \quad (7)$$

Figure 1 shows the fractional release, $f(t)$, of Cu from RVCF coated, respectively, with 1 μm and 2 μm of Ni. The fractional release of fluorine from a 6 μm diameter Al_2O_3 fibrous target is shown in Fig. 2. Estimated total diffusion release rates from a number of targets are given in Table 1. Although the diffusion release rates from these highly permeable targets have been calculated in the frame-work of a simple diffusion model, the results show that the diffusion release of short-lived radioactive species such as ^{58}Cu and ^{17}F from these target matrices are sufficiently fast to meet the intensity requirements for research programs at the HRIBF.

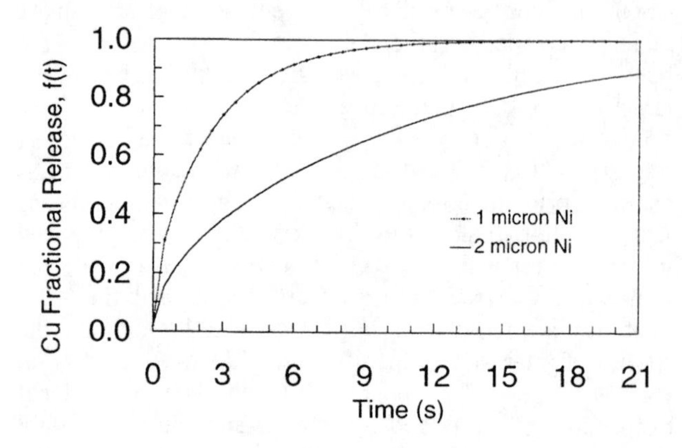

Fig.1. Fractional release of Cu from a composite RVCF matrix coated with 1 μm and 2 μm Ni, respectively. Diffusion coefficient D = 1.53x10^{-9} cm^2/s.

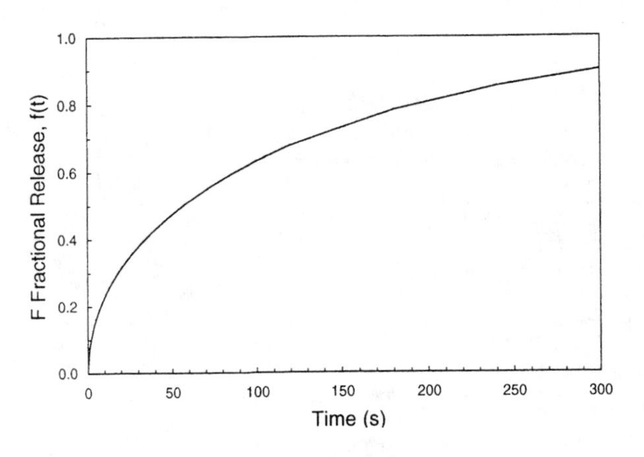

Fig. 2 Fractional release of fluorine from 6 μm diameter fibrous Al_2O_3 target material. Diffusion coefficient: $D = 10^{-10} cm^2/s$.

and equilibrium temperature distributions in target/heat-sink systems.

A prototype target/heat-sink system now under design at the HRIBF is displayed in Fig. 3. The device can be inserted into or removed from the target/ion source through a vacuum inter-lock port. The primary beam, which produces the radioactive species, passes through a thin window and deposits energy in the target material during the slowing down process. The length of the target material is chosen so that the beam exits the target with ~ 4 MeV energy and comes to rest in a C-beam stop that is connected to a Cu heat-sink. The target material reservoir can be resistively heated when necessary to maintain the desired target operational temperatures.

Computational studies have been conducted to simulate the heat transport and temperature distributions in several targets designed for use at the HRIBF. These targets include: fibrous Al_2O_3 for ^{17}F production using low-energy deuteron beams; Ni coated RVCF for ^{58}Cu production using low energy protons; and UC_2 coated RVCF composite targets for production of fission products with 1 GeV protons. The maximum target temperature versus beam power in a 2.4 cm long fibrous Al_2O_3 target irradiated by 20 MeV deuterons, as calculated with ANSYS, is displayed in Fig. 4. The energy of the ion beam was chosen for the production of ^{17}F isotopes via the nuclear reaction $^{16}O(d,n)^{17}F$. The operating temperature of the target is set by the vapor pressure limitation of the ion source system (~ 1650 °C). As shown in Fig. 4, this temperature is already reached with less than 45 W of deuteron beam power, limiting the primary deuteron beam intensities that can be used for the production of ^{17}F. This is due to the fact that these highly permeable fibrous materials have low thermal conductivities. Partially coating the Al_2O_3 matrix with a thin layer of Ir can be an effective means for removal of heat from the target. As also shown in Fig. 4, the maximum beam power can be increased to 150W by adding a 25 μm thick Ir foil spiral rolled between 1.5 mm thickness layers of the Al_2O_3 fibrous matrix. An interesting result, derived from the analysis, is that radiative heat transfer is more effective than conduction even at temperatures less than 1600 °C for low thermal conductivity fibrous target materials such as Al_2O_3.

TABLE 1. Estimated diffusion release rate of selected radioactive species from fibrous and composite targets.

Element	Target	Ion Beam	Prod. Rate/μA	Release Rate/μA	Temp (°C)	D(cm^2/s)
^{17}F	Al_2O_3 Fiber	40 MeV Deuterons	$1.53x10^{10}$	$2.9x10^9$	1430	10^{-11}
^{18}F	Al_2O_3 Fiber	40 MeV Protons	$1.46x10^{10}$	$1.3x10^{10}$	1430	10^{-11}
^{58}Cu	2 μm Ni / RVCF	30 MeV Protons	$5.5x10^8$	$2.1x10^8$	1300	$1.529x10^{-9}$
^{132}Sn	15 μm UC_2 / RVCF	60 MeV Protons	$4.2x10^8$	$2.8x10^8$	2000	$8x10^{-8}$
^{132}Sn	15 μm UC_2 / RVCF	1 GeV Protons	$4.0x10^9$	$2.6x10^9$	2000	$8x10^{-8}$

TARGET/HEAT-SINK SYSTEM

Higher temperatures significantly decrease diffusion release times of particles from targets. Thus, in order to optimize radioactive species release, it is desirable to operate the target to the temperature limit set by the vapor pressure of the target material which does not compromise the ion source efficiency. The limiting vapor pressure of the CERN-type electron-beam-plasma ion source, presently used at the HRIBF, is 2×10^{-4} Torr [4]. It is also important to provide means for removal of beam deposited heat from the target matrix at controlled rates so that, as high as practical, primary beam intensities can be used to produce the species of interest. Several computer codes have been used to simulate heat generation, heat removal, and to estimate equilibrium temperature distributions in target matrices. Heat deposited in the target material by the primary ion beam is simulated by use of the Monte Carlo code, Stopping and Range of Ions in Matter (SRIM) [5]. The heat generation function, derived from SRIM, takes into account the axial energy deposition, dE/dx; effects due to the radial profile of the primary ion beam; and the scattering of the primary ion beam during passage through the target which results in broadening of the beam. The finite element code ANSYS [6] is used to simulate heat transfer, temperature gradients

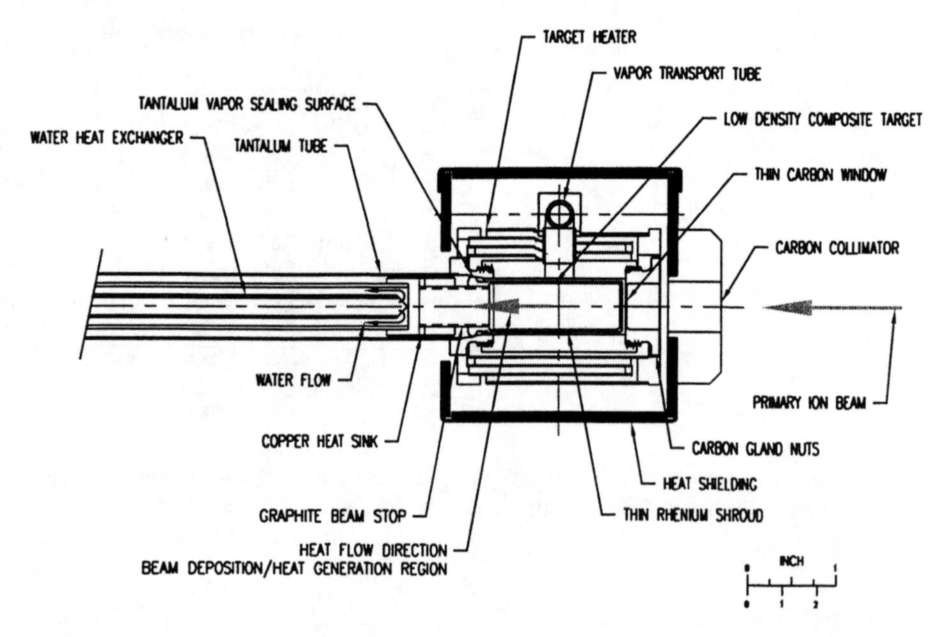

Fig. 3. Schematic drawing of a removable/insertable composite target/heat-sink system showing the components which make up the system now under design at the HRIBF for use in RIB generation. The target material may be particulate, fibrous, or plated onto composite target support disks.

The beam heating results in a non-uniform temperature distribution profile, both radially and axially. It is desirable to smooth out the temperature distribution in the target material. One approach for achieving this effect is to coat the entrance end of the beam-stop with pyrolytic carbon with the high thermal conductivity direction oriented parallel to the face of the beam-stop. Comparisons of the radial and longitudinal temperature profiles with and without pyrolytic carbon show that pyrolytic carbon is effective in smoothing out the temperature distribution in target systems such as Al_2O_3/Ir/RVCF and UC_2/Ir/RCVF.

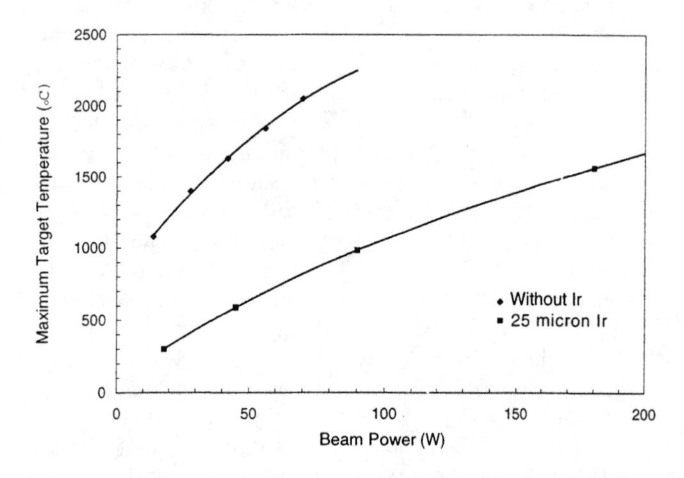

Fig. 4. Maximum target temperature versus beam power in a 2.4 cm long fibrous Al_2O_3 target spiral rolled with 25 μm thick Ir and without Ir, irradiated by 20 MeV deuterons.

ACKNOWLEDGEMENTS

Research at the Oak Ridge National Laboratory is supported by the U.S. Department of Energy under contract DE-AC05-96OR22464 with Lockheed Martin Energy Research Corp.

REFERENCES

1. G. D. Alton, *Nucl. Instr. and Meth. A* **382**, 207 (1996).
2. H. K. Carter, J. Kormicki, D. W. Stracener, J. B. Breitenbach, J. C. Blackmon, M. S. Smith, and D. W. Bardayan, *Nucl. Instr. and Meth. B* **126,** 166 (1997).
3. G. D. Alton, CP392, *Appl. of Accel. in Res. and Ind., Edited by J. L. Duggan and I. L. Morgan*, AIP Press, New York, 1997, p. 429.
4. R. Kirchner and E. Roeckl, *Nucl. Instr. and Meth.* **133**, 187 (1976).
5. SRIM - The Stopping and Range of Ions in Matter, J. F. Ziegler, IBM Research, Yorktown Heights, New York 10598, USA.
6. ANSYS is a finite element computer code designed to solve thermal transport and radiation problems; the code is a product of Swanson Analysis Systems, Inc., Houston, PA 15342-0065.

On-line collection of a radioactive target — a poor man's radioactive beam collider

Ulli Köster

TU München, Physik-Department, 85748 Garching, Bavaria

Abstract. The intense beams of neutron-rich nuclei at reactor-based RIB facilities will allow to produce neutron-rich isotopes of heavy and superheavy elements via fusion reactions. A way to reach the highest N/Z ratios is fusion of very neutron-rich projectiles with neutron-rich (radioactive) target nuclides. Due to the short lifetimes, both projectile and target nuclei have to be produced on-line. Nuclei with a moderate lifetime (minutes or longer) can be implanted by a low-energy ion beam into a collection foil, which is used *simultaneously* as target for the energetic projectile beam. Typical beam energies are 30 to 200 keV for the target beam and 4 A·MeV for the projectiles. Both will be focussed by electrostatic and magnetic multipoles to an overlapping beam spot of 1 mm^2. Depending on the lifetime of the target nuclei, a detection limit of some 3 to 2000 nbarn can be reached, which has to be compared with predicted unhindered fusion cross sections of 10 to 100 μbarn. A carbon foil (about 20 to 100 μg/cm^2 thickness) is appropriate as catcher for tin beams. For more volatile elements (e.g. antimony or tellurium) the foil material has to be adapted to prevent a rapid release of target nuclides from the heated foil.

INTRODUCTION

The second generation Radioactive Ion Beam (RIB) facilities will provide intense beams of exotic nuclei. Using thermal neutron induced fission in reactor-based RIB facilities, beams of very neutron-rich nuclei with intensities of up to 10^{12} s^{-1} could be provided. Such facilities have been proposed for the high-flux reactors ILL in Grenoble: PIAFE (1) and for the FRM-II in Munich: MAFF (2, 3). The latter is now under construction.

Among many other applications the high radioactive beam intensities will allow to produce via fusion reactions more neutron-rich isotopes of the heavy and superheavy elements, thus approaching the region of spherical superheavy nuclei where considerably longer halflives are predicted (4). One possibility is the symmetric or near-symmetric fusion with target and projectile nuclei of comparable size. This way has not yet been explored systematically at sufficient sensitivity, since the macroscopical models requiring an extra-push and an extra-extra-push energy (5, 6) forbid the complete fusion of massive symmetric systems at the interaction barrier. However, the synthesis of the heaviest elements (e.g. (7)) is obviously not described by a macroscopic model which does not include the importance of nuclear structure in the entrance channel (e.g. for N=82) (8). This was reconfirmed by recent results from the investigation of fusion-evaporation residues from the reaction ^{86}Kr on ^{130}Xe and ^{136}Xe (9). Considering the shell structure especially nuclei around the doubly magic ^{132}Sn would be promising projectiles for near symmetric fusion (10).

While new neutron-rich isotopes of heavy elements could be produced in this way, the extremely neutron-rich heavy elements, e.g. the highly interesting isotopes towards ^{264}Fm which, according to a longstanding prediction (11), should mainly decay by symmetric cold fission, will still be out of range by fusion with any combination of stable targets and available radioactive projectiles. A way to reach the required N/Z ratio is fusion of very neutron-rich projectiles with neutron-rich (radioactive) target nuclides. Due to the short lifetimes both, projectile and target nuclei, have to be produced on-line. An appropriate radioactive ion beam collider would not only be very costly, but also require a new separation mechanism as an alternative to the traditional recoil separators which use the favorable reaction kinematics of a fixed target with a strongly forward focussed beam of fusion evaporation products.

RADIOACTIVE ON-LINE TARGET

Alternatively a radioactive target was first proposed for the PIAFE project (12). It can be prepared by implantation of a low-energy radioactive ion beam into a foil. Bombarding this "on-line-target" *simultaneously* with radioactive projectiles minimizes the losses due to radioactive decay of the target nuclei. The reachable luminosity

CP475, *Applications of Accelerators in Research and Industry*,
edited by J. L. Duggan and I. L. Morgan

will be determined by the equilibrium concentration of the target nuclei (equal to the cumulative isotopic yield times their lifetime), the intensity of the projectile beam and the size of the overlapping beam spots. The beam intensities are given by the RIB facility, but the size of the implantation area has to be optimized.

Table 1 shows nuclei suitable for a radioactive target which allow to reach detection limits in the range of nbarn. This has to be compared to unhindered fusion cross sections of 10 to 100 μbarn (10).

In the following we will discuss the technical challenges and physical limitations of such a set-up to be realized at the Munich Accelerator for Fission Fragments (MAFF) (3). This discussion will concentrate on isotopes close to ^{132}Sn, but similar considerations could be applied for fusion reactions in other mass regions, e.g. for neutron-rich isotopes with masses 90 to 100 or even for reactions of neutron-deficient on neutron-deficient isotopes at an accelerator-based RIB facility.

Ion optics

The mass separator has to provide simultaneously two beams of different masses which are extracted from the focal plane by an electrostatic switchyard. This technique is successfully used since long time at ISOLDE (13).

The low-energy ion beam will be provided from an in-pile ion source kept at a positive potential of about 30 kV. An unselective "hot plasma source" will give the best cumulative yields (14). For such a source an emittance of $\varepsilon_{2\sigma} = 15\pi$ mm mrad at 15 keV is quoted (FEBIAD-F with 1.2 mm orifice) (15). Efficient focussing of the beam can be achieved with a multiplet of electrostatic quadrupoles. For better focussing and deeper implantation the target beam can be slightly post-accelerated up to about 200 kV by putting the implantation area on a negative potential. A focal point of about 1 mm^2 could be realized even in case of a non-optimized emittance. Space charge effects can still be neglected for this current density.

The fusion barrier, estimated with the macroscopic Bass formula (16), is around 4 A·MeV for symmetric reactions of tin on tin. For efficient post-acceleration the projectile beam will be charge-breeded to $A/q < 6.3$ in a pulsed ECR ion source (17) and accelerated by a LINAC (18). This one has a design similar to the lead injector LINAC at CERN which gives an emittance of about $\varepsilon_{4rms} = 10\pi$ mm mrad at 4.2 A·MeV output energy (19). The focussing of the projectile beam will be done by a multiplet of strong magnetic quadrupoles. Such a microlens, capable to focus ion beams with up to 200 MeV·q^2/A to sub-μm dimensions, has just been constructed at the Technische Universität in München (20). For our case the requirements concerning the beam size are much less stringent than in real microbeam applications (the current density will be limited by other considerations, see below), but the design of this microlens allows to guide and focus both, target and projectile beam, from the same side onto the target.

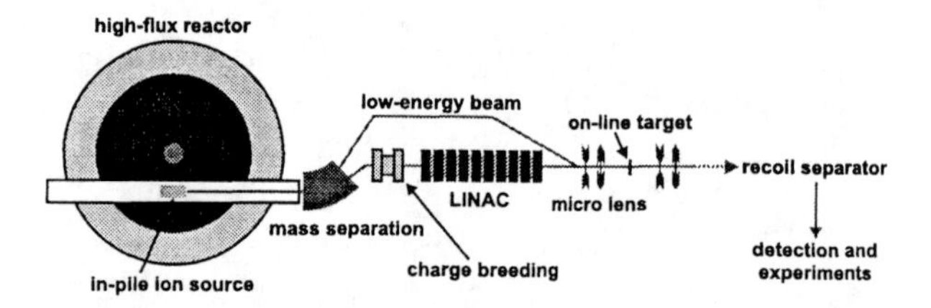

FIGURE 1. Implementation of a radioactive on-line target at the Munich Accelerator for Fission Fragments (MAFF).

While the magnetic rigidity varies between both beams only by a factor of two to six (for 200 keV and 30 keV target beams respectively), the electrical rigidity differs by two to three orders of magnitude. This allows to set the magnetic elements for the high-energy beam and focus the low-energy beam nearly independently by applying additional electrical multipole fields and optionally even a longitudinal acceleration voltage. The electrical multipoles inside the Munich microlens can hold a voltage of up to 4 kV (21) which is largely sufficient to control the low-energy ion beam with up to 200 keV energy. Both beams will be combined with an electrostatic deflector in front of the microlens.

The fusion products will be separated from the primary beam by the MORRIS separator (22). Its first part consists of a Wien-filter type recoil separator similar to SHIP at the GSI (23). With an angular acceptance of about 50 mrad most fusion products will be transmitted even for a highly focussed beam spot.

Collection foil

The key part of an on-line collected target is the collection foil. Its thickness has to be sufficient to be mechanically stable and stop the target beam, but thin enough to avoid too high energy loss of the projectile beam[1] and the fusion products. Foils of low-Z materials (Be, C, ...) with 20 to 100 μg/cm^2 thickness have not only a good ratio of mechanical stability to energy loss, but moreover avoid the losses of target nuclei due to backscattering.

The electrical conductivity of the foil has to be sufficient to avoid a charging-up due to the deposited charge

[1] The fusion excitation function has a narrow width, typically about 10 MeV for asymmetric fusion (24).

Table 1. Possible target nuclides for a radioactive on-line target. The cumulative yields are calculated for ionization with an unselective plasma ion source. The equilibrium concentration and the respective target thickness are given after maximal 5 days of collection on a beam spot of 1 mm^2. A target thicker than 10^{18} atoms/cm^2 is not useful due to excessive energy loss of the projectiles (e.g. limitation for ^{132}Te). The reachable detection limit (for 1 reaction per day) is given for a 1 pnA projectile beam. A ^{126}Sn target could also be prepared off-line, but is given here for comparison. The average and the maximum temperatures for 10% duty cycle are given.

Isotope	Half-life	Cumulative yield (s^{-1})	Equilibrium target size (atoms)	Rad. target thickness (μg/cm^2)	**Detection limit (nbarn)**	20 μg/cm^2 graphite T_{mean} °C	20 μg/cm^2 graphite T_{max} °C	100 μg/cm^2 graphite T_{mean} °C	100 μg/cm^2 graphite T_{max} °C
^{126}Sn	10^5 a	$1.5 \cdot 10^{10}$	$6 \cdot 10^{15}$	135	**3**	200	490	380	850
^{128}Sn	1 h	$7.5 \cdot 10^{10}$	$4 \cdot 10^{14}$	8	**50**	230	500	390	860
^{130}Sn	1.7 / 3.7 min	$3.0 \cdot 10^{11}$	$6 \cdot 10^{13}$	1.3	**320**	330	530	440	870
^{132}Sn	40 s	$1.7 \cdot 10^{11}$	$1 \cdot 10^{13}$	0.2	**2000**	430	580	510	890
^{131}Sb	23 min	$6.5 \cdot 10^{11}$	$1 \cdot 10^{15}$	28	**15**	380	560	470	880
^{132}Te	76 h	$1.1 \cdot 10^{12}$	$1 \cdot 10^{16}$	300	**1.3**	430	580	510	890
^{134}Te	42 min	$1.7 \cdot 10^{12}$	$6 \cdot 10^{15}$	135	**3**	530	640	590	910

of the low-energy beam and emitted secondary electrons. Materials with a specific electrical resistivity $< 100\ \Omega$m will keep the potential difference small enough to prevent any detoriation of the low-energy beam focus. Most metal and graphite foils fulfill this requirement while insulators like diamond-like graphite or pure Al_2O_3 can cause problems.

Foil temperature

Several sources contribute to the heating of the foil:

- The low energy beam will deposit its full kinetic energy in the immediate vicinity of the impact (some nm straggling). With a non-selective ion source the mass yield may reach 600 nA (at mass 134) and thus contribute up to 18 mW heating for 30 keV implantation energy and up to 120 mW for 200 keV implantation energy respectively.
- The stopping power of 4 A·MeV Sn in carbon is about 72 MeV per mg/cm^2 (25), i.e. for a 100 μg/cm^2 foil and 1 pnA of ^{132}Sn the accelerated beam deposits about 7 mW.
- Only a small part of the energy released in the radioactive decay of the target atoms will contribute to a local heating. Most will be deposited further away by the long range beta particles.
- Also the energy loss of the fusion products can be neglected due to their low production rate.

The deposited energy will mainly be emitted by radiation. Neglecting the thermal conduction in the thin foil an upper limit for the foil temperature can be estimated. Depending on the duty cycle of the accelerated beam, the target temperature will fluctuate more or less around the average temperature. Due to the small heat capacity the characteristic time for heating and cooling is in the millisecond range (see (26) for an analytical description).

The heating of the foil has several side effects: by annealing structural damages created by the particle impact the lifetime of the foil can be considerably prolonged. On the other hand an overheating will reduce the lifetime due to evaporation of foil material. Already before, the implanted radioactive isotopes start to migrate from the implantation region to other parts of the foil or even leave it completely. While for tin no large release is expected at the quoted temperatures, the more volatile elements antimony and tellurium start to escape much earlier. Higher target temperatures could be tolerated if the implanted nuclides form an appropriate chemical compound which hinders migration and release (e.g. extremely slow diffusion of Te in Si (27)). In a suitable target matrix the volatilization enthalpy can exceed the sublimation enthalpy considerably, thus keeping the implanted atoms bound even at temperatures above the boiling point (28).

Foil lifetime

To run most of the time "in saturation" after buildup to the maximal concentration of target atoms, the foil lifetime should be at least in the order of days. The irradiation damage mechanisms in carbon foils were studied thoroughly (29) and empirical formulas were proposed to estimate the lifetime of a carbon foil bombarded by energetic heavy ions (30, 31, 32). With a projectile beam of 1 pnA lifetimes of weeks and longer are predicted. However for the low energy beam these formulae cannot be applied since the ions are stopped close to the surface (20 to 110 nm range depending on the implantation energy) while most of the bulk material stays unaffected of the recoil cascades. Also the structural changes due to a high concentration of implanted ions are difficult to evaluate. Therefore the foil lifetime should be tested experimentally with an appropriate stable ion beam.

CONCLUSION

It has been shown that a "poor man's radioactive beam collider" realized by on-line collection of radioactive ions on a target foil could attain a detection limit which is orders of magnitude below the predicted unhindered fusion cross sections. Since the reaction rate depends on the product of the intensities of projectile and target beams, the natural place for such a radioactive on-line target is a reactor-based RIB facility providing by far the highest yields in the region around ^{132}Sn. While carbon foils are suitable for the collection of tin beams, for very volatile elements special foil material has to be chosen to prevent a rapid release of the radioactive nuclides.

The largest uncertainty lies however in the expected fusion cross sections for neutron-rich systems. It is difficult to quantify how the shell stabilization affects the fusion probability and the theories differ significantly in their predictions. Also complementary reaction mechanisms like multinucleon transfer from neutron-rich projectiles to transuranium targets should be considered (33).

The specific requirements of a radioactive on-line target (e.g. the parallel separation of at least two masses) can be foreseen at MAFF without big additional effort as supplement to the huge physics program with stable targets.

ACKNOWLEDGMENTS

I gratefully acknowledge fruitful discussions with members of the MAFF, PIAFE and ISOLDE collaborations. Thanks for support by the Accelerator Laboratory of the TU and LMU München.

REFERENCES

1. PIAFE Project Report, ed. by Ulli Köster and Jean-Alain Pinston, ISN Grenoble, 1998.
2. D. Habs et al., Nucl. Phys. A616 (1997) 39c.
3. MAFF - Munich Accelerator for Fission Fragments, Physics Case and Technical Description, ed. by D. Habs et al., 1999.
4. R. Smolańczuk, Phys. Rev. C56 (1997) 812.
5. W.J. Swiatecki, Phys. Scripta 24 (1981) 113.
6. J.P. Blocki, H. Feldmeier and W.J. Swiatecki, Nucl. Phys. A459 (1985) 145.
7. S. Hofmann et al., Z. Phys. A358 (1997) 125.
8. B. Quint et al., Z. Phys. A346 (1993) 119.
9. C. Stodel et al., GSI Annual Report 1996, p. 17.
10. Yu. A. Lazarev and Yu. Ts. Oganessian, Intern. Workshop on Research with Fission Fragments, Benediktbeuern, ed. by T. von Egidy et al., World Scientific, Singapore, 1997, p. 314.
11. H. Faissner and K. Wildermuth, Nucl. Phys. 58 (1964) 177.
12. Gottfried Münzenberg, GSI Darmstadt, priv. comm.
13. E. Kugler et al., Nucl. Instr. Meth. B70 (1992) 41.
14. U. Köster, O. Kester and D. Habs, Rev. Sci. Instr. 69 (1998) 1316.
15. R. Kirchner et al., Nucl. Instr. Meth. B26 (1987) 235.
16. R. Bass, Nucl. Phys. A231 (1974) 45.
17. Pascal Sortais, CAARI'98, these proceedings.
18. O. Kester et al., LINAC'98, Chicago, Aug. 1998, in press. preprints: http://www.aps.anl.gov/conferences/LINAC98/
19. N. Catalan Lasheras et al., LINAC'96, Geneva, 1996, ed. by C. Hill and M. Vretenar, report CERN 96-07, p. 363.
20. G. Hinderer et al., Nucl. Instr. Meth. B130 (1997) 51.
21. Günther Dollinger, TU München, priv. comm.
22. P. Thirolf et al., Nucl. Instr. Meth. B126 (1997) 242.
23. G. Münzenberg et al., Nucl. Instr. Meth. 161 (1979) 65.
24. S. Hofmann et al., Z. Phys. A350 (1995) 277.
25. J.P. Biersack and L.G. Haggmark, Nucl. Instr. Meth. 174 (1980) 257.
26. D. Marx et al., Nucl. Instr. Meth. 163 (1979) 15.
27. N.S. Zhdanovich and Yu. I. Kozlov, Svoistva Legir. Poluprovodn., ed. by V.S. Zemskov, Nauka, Moscow, 1977, p. 115.
28. Heinz Gaeggeler, CAARI'98, these proceedings.
29. G. Dollinger and P. Maier-Komor, Nucl. Instr. Meth. A282 (1989) 223.
30. G. Frick et al., Rev. Phys. Appl. 12 (1977) 1525.
31. F. Nickel, Nucl. Instr. Meth. 195 (1982) 457.
32. E.A. Koptelov, S.G. Lebedev and V.N. Panchenko, Nucl. Instr. Meth. A256 (1987) 247.
33. Jürgen Friese, in (3).

A Neutron Beam Facility for Radioactive Ion Beams and Other Applications

L.B. Tecchio on behalf of the SPES Study Group

Laboratori Nazionali di Legnaro, Via Romea 4, 35020 Legnaro (PD), Italy

In the framework of the Italian participation in the project of a high intensity proton facility for the energy amplifier and nuclear waste transmutations, LNL is involved in the design and construction of some prototypes of the injection system of the 1 GeV linac that consists of a RFQ (5 MeV, 30 mA) followed by a 100 MeV linac. This program has already been supported financially and the work is in progress. In this context LNL has proposed a project for the construction of a second generation facility for the production of radioactive ion beams (RIBs) by means of the ISOL method. The final goal is the production of neutron rich RIBs with masses ranging from 80 to 150 by using primary beams of protons, deuterons and light ions with energy of 100 MeV and 100 kW power. This project is expected to be developed in about 10 years from now and intermediate milestones and experiments are foreseen and under consideration for the next INFN five year plan (1999-2003). During that period the construction of a proton/deuteron accelerator of 10 MeV energy and 10 mA current, consisting of a RFQ (5 MeV, 30 mA) and a linac (10 MeV, 10 mA), and of a neutron area dedicated to the RIBs production and to the neutron physics, is proposed. Some remarks on the production methods will be presented. The possibility of producing radioisotopes by means of the fission induced by neutrons will be investigated and the methods of production of neutrons will be discussed. Besides the RIBs production, neutron beams for the BNCT applications and neutron physics are also planned.

INTRODUCTION

The international community is showing growing interest in high intensity linacs for scientific, industrial and social applications. Proton linacs with final energies of about 1 GeV and CW operation are proposed for secondary beams production, tritium production, nuclear waste transmutation or energy production in sub-critical accelerator driven reactors. The beam intensities vary for the different proposed applications and are ranging from 10 to 100 mA.

In the framework of the Italian participation in the project of a high intensity proton facility for the energy amplifier and nuclear waste transmutations, LNL is involved in the design and construction of some prototypes of the injection system of the 1 GeV linac that consists of a RFQ (5 MeV, 30 mA) followed by a 100 MeV linac. This program has already been supported financially and the work is in progress. In this context, LNL has proposed a project for the construction of a second generation facility for the production of radioactive ion beams (RIBs) by means of the ISOL method. The final goal is the production of neutron rich RIBs with masses ranging from 80 to 160 by using primary beams of protons, deuterons and light ions with energy of 100 MeV and 100 kW power. This project is expected to be developed in about 10 years from now and intermediate milestones and experiments are foreseen and under consideration for the next INFN five year plan (1999 - 2003). During that period the construction of a proton/deuteron accelerator of 10 MeV energy and 10 mA current, consisting of a RFQ (5 MeV, 30 mA) and a linac (10 MeV, 10 mA), and of a neutron area dedicated to the RIBs production, to the BNCT applications and to the neutron physics, is proposed. The RFQ will be of the same type of the one designed for the high intensity project mentioned above. An intense R&D program on high intensity accelerator techniques and targetry is already in progress.

THE ACCELERATOR

The sequence RFQ-DTL (Drift Tube Linac) is until now the most used scheme for proton linacs in the energy range of 10 - 100 MeV. In our design both DTL and RFQ operate at the main linac frequency of 352 Mhz; in this way we avoid any frequency jump, and the bore hole inside the DTL structure can be kept large enough to have a good margin between beam dimensions and machine acceptance. The RFQ structure is, nowadays, the natural choice for

CP475, *Applications of Accelerators in Research and Industry*,
edited by J. L. Duggan and I. L. Morgan

TABLE 1. Main Parameters of the Accelerator

Parameters	Unit	RFQ	DTL
Input energy	MeV	0.05	5
Output energy	MeV	5	100
Beam current	mA	30	10
RF frequency	Mhz	352.2	352.2
Duty Cycle	%	100	100
Total length	m	5.3	80
Transmission	%	94.6	100
RF power diss.	MW	0.6	8.3
Beam loading	MW	0.15	2.8
Quad. diss.	MW	-	0.6

the low energy part of any linear accelerator. It is very efficient up to the energy of a few MeV and it gives a transmission in excess of 90% of the continuos beam coming from the source at energies of few tens of keV. The acceleration efficiency of the RFQ falls down very rapidly in the range of 1 to 10 MeV and it is mandatory to change the structure. As usual we consider a DTL as a following accelerating segment and the transition has been set at 5 MeV, trading off the RFQ low efficiency at the end of the structure with the higher DTL shunt impedance at its beginnings. The DTL shows a good efficiency up to hundreds of MeV. In Table 1 the main accelerator parameters are summarized.

An intermediate milestone of this project consists of the construction of a proton/deuteron accelerator of 10 MeV energy and 10 mA current and of a neutron area dedicated mainly to the RIBs production, to the BNCT applications and to the neutron physics. In the framework of the high intensity proton linac project a first prototype (in aluminum) of the RFQ accelerator (5 MeV, 30 mA) has been designed and constructed. The prototype has recently been delivered and at present is under RF measurements in the laboratory. After the measurements we will proceed to the construction of the final RFQ, which is planned to be ready in two years. At the same time, in collaboration with the ARGONNE National Laboratory, an intense R&D program dedicated to the study of superconducting cavities for the DTL linac has been initiated. In any case, a low energy proton accelerator (up to 5 MeV) like the above mentioned RFQ is a very suitable accelerator to produce high intensity neutron beams. The reaction $^7Li(p,n)^7Be$ has indeed been proposed as an accelerator-based source of neutrons. This reaction displays a large resonance in the forward direction around 2.3 MeV which extends to about 2.5 MeV. The angular distribution of the produced neutrons shows a pronounced peak at zero degree. The neutron yield (per incident proton) between 0^0 and 30^0 is about 4×10^{-3} (n/p) so that, in our case, an intensity of the order of 2.5×10^{14} (n/s) is expected. The neutron induced fission presents remarkable advantages for the production of neutron rich RIBs with respect to the direct production (proton beam directly on the production target), and exotic beams with intensities of the order of 10^9 - 10^{12} ions/s are possible.

NEUTRON PRODUCTION

High neutron fluxes may generally be obtained by the conversion of primary beams of protons and deuterons into targets of different materials (deuterium, tritium, beryllium,...). The most fruitful reactions are: D(d,xn), Be(d,xn), Th(d,xn), U(d,xn). For the sake of convenience we analyze the reactions that are generally most employed, i.e. Be(d,xn) and U(d,xn).

Independently from the conversion target, the deuterons of 100 MeV/u are the most appropriate projectiles for the production of neutrons. Conversion targets as uranium and beryllium both present characteristics which are suitable for the task, even if they have different angular and energetic distributions. Table 2 shows the mean values of the neutron multiplicity for the different regions of the neutron energy spectrum.

Moreover, the cascade of secondary neutrons produced in the target, which contributes to a great extent to the fission process, must be taken into account. The neutron generation in the Be(d,xn) reaction is the most efficient (secondary neutron multiplicity $\langle M_{n2} \rangle = 3$) in comparison with the generation in the U(d,xn) reaction ($\langle M_{n2} \rangle = 1$) and seems to be the most favorable for the production of exotic beams.

RIBs Production Methods

One of the most used methods for the production of RIBs is the combined process of fission and spallation by means of protons on a target of different chemical-physical nature. In particular, the results of the experimentation at ISOLDE of CERN (1), where the primary beam consists of 600 MeV protons, are well known. The experimental results make us believe that the fission process becomes

Table 2. Mean value $<M_n>$ of the multiplicity of neutrons, in units of emitted nucleons per incident deuteron (100 MeV/u) in the different regions of the energetic spectrum.

Neutron Energy range (MeV)	$<M_n>$ for Be	$<M_n>$ for U238
≥50	0.245	0.080
≥25	0.546	0.224
≥10	0.898	0.385
≥5	1.111	0.554
≥0	1.556	2.210

prevalent at low energies, namely some tens of MeV, whereas the spallation process is dominant at energies higher than 100 MeV. In the 100 MeV region the two processes are in conflict. The simulations carried out by the Monte Carlo codes LAHET (2) and FLUKA (3) provide discordant results on the percentage of the spallation processes for 100 MeV protons on a U238 target, that is 11% and 62% respectively. On the other hand experimentation provides unambiguous data on the production of RIBs, when using protons of different energies as primary beam or neutrons (4). Increasing the proton beam energy, the production region extends more and more towards the direction of the proton-rich isotopes. On the contrary, a great increase of the cross section in the neutron-rich region may be obtained by means of the fission process induced by neutrons on fissile targets, like Uranium and Thorium. The use of the fission reaction induced by neutrons of high energy (100 MeV) appears to be very convenient in order to produce exotic beams of high intensity. Moreover, the conversion of the primary proton/deuteron beam into a neutron flux allows to simplify and partially to solve the problems related to the power dissipation (100 kW) in the production target.

The BNCT Applications

The primary beam, employed for the RIBs production, will not use all the available current. Therefore part of the current can be used to produce, in a dedicated room, thermal neutrons for the treatment of skin melanoma with the BNCT (Boron Neutron Capture Therapy). BNCT may be the best treatment for those skin tumors (melanomas) which are nowadays resistant to ordinary therapy. The 5% of these tumors develop surface metastases which can be treated with BNCT. The 40% of patients with melanoma develop metastases in other parts of the body, such as in the brain. Also brain metastatic tumors may be successfully treated with BNCT.

Up to now only nuclear reactors have been applied as thermal neutron sources for BNCT. However the future hospital-based BNCT centers will need to be associated with small accelerators. The Legnaro BNCT facility aims at finding the optimal therapeutic beam and moreover the dosimetric and microdosimetric procedures for a future accelerator-based BNCT source to be installed in medical centers.

The BNCT treatment could last one hour with thermal neutron fluxes within the range of 10^8 - 10^9 cm^{-2} s^{-1}. Such fluxes can be produced by using thick beryllium targets and 1 mA of protons/deuterons accelerated at 10 MeV. Fast neutrons will be thermalized by using a DO_2 and grafite moderator. ^{6}Li collimators will be used to define the proper size of the thermal neutron beam. The neutron beam will be monitored both in the energy distribution and in its dosimetric and microdosimetric characteristics.

THE R&D PROGRAM

The R&D program on the radioactive beams at LNL is based on the study the production methods of RIBs through the fission induced by thermal and non-thermal neutrons on targets of Th232, U235 and U238. In order to minimize the release time from the target and to maximize the ionization efficiency in the ion source, the technology of the target/ion source system will be studied. This R&D program has been funded by INFN and will correspond to the next two financial years (1999-2000). The experiments will be performed at the Van de Graaff accelerator, by using the 7 MeV, 3 μA deuteron beam for the generation of neutrons, which are subsequently employed in the fission processes. The neutrons will be produced by the following reactions: Be(d,n), D(d,n) and t(d,n).

By using the Be(d,n) reaction at 7 MeV, it is possible to obtain about 1.5×10^{10} n s^{-1} sr^{-1}, at zero degree, of 3.2 MeV average energy. By thermalizing adequately the produced neutrons, a flux of 2 x 10^8 cm^{-2} s^{-1} thermal neutrons may be obtained (5). The high energy (14 MeV) neutrons will be produced by the reaction t(d,n); about 10^9 n/s may be obtained with primary deuterons of 0.5 - 1 MeV energy and 3 μA current (6).

The experimental set-up has already been installed and consists of a bunker housing the beryllium foil for neutron generation and the target/ion source system, followed by a magnetic spectrometer (M/ΔM~800) for the isotopic mass separation and of a detector for isotopes identification. Both the production target and the ion source operate at high temperature (<2500 ^{0}C); the source is a conventional surface ionization source charge 1+; the operation voltage is 20 kV.

A first prototype of the target/ion source system has been successfully tested and the on-line separator has been calibrated with stable ions.

CONCLUSIONS

The availability of a neutron facility for the RIBs production at LNL within the next five/seven years is becoming a rather realistic possibility, but it still depends on the approval of the next five-year plan of INFN. Apart from such considerations, in the framework of the high intensity proton linac program a first prototype (in aluminum) of the RFQ accelerator (5 MeV, 30 mA) has been designed and constructed. The prototype has recently been delivered and at present is under RF measurements in the laboratory. After the measurements we will proceed with the construction of the final RFQ, which is planned to be ready in two years from now. Its own features will fulfill completely the requirements for a neutron facility of average intensity. A parallel R&D program to investigate the feasibility of RIBs production through the fission induced by neutrons on fissile targets has been financially supported and the experimental set-up has already been installed at the CN accelerator of LNL. Preliminary tests of the on-line separator are in progress.

ACKNOWLEDGMENTS

I would like to thank all the colleagues of the Accelerator Division and of the SPES study group for their support during the preparation of this paper.

REFERENCES

1. Ravn, H.L., *Physics Reports* **54**,201 (1979).
2. Prael, R.E., Bozoian, M., *LA-RU-88-3238*, Los Alamos (1998).
3. Ferrari, A., Sala,P., *Proceedings of MC93*, Tallahassee, Florida 1993, Ed. World Scientific, Singapore 1994.
4. Ravn, H.L., et al., *Nucl. Instrum. Meth.* **B88,** 441 (1994).
5. Agosteo, S., et al., *Rad. Prot. Dos.* **70**, 559 (1997).
6. Lee, W.C., et al., *Nucl. Instr. Meth.* **B99**, 739 (1995).

Approaches to Develop Targets For Production of Intense Radioactive Ion Beams

W. L. Talbert[1], D. M. Drake[1], M. T. Wilson[1], J. J. Walker[1] and J. W. Lenz[2]

1) Amparo Corporation, Santa Fe, New Mexico 87504
2) John. W. Lenz and Associates, Waxahachie, Texas 75165

Approaches to develop targets for production of intense radioactive ion beams (RIBs) have been evaluated over the past five years. It is acknowledged that many desired physics objectives using RIBs can be met only by using production beams of energetic protons with currents up to 100 μA. Such beams can be made available at future spallation neutron facilities. The production targets will require active cooling to control operational temperatures due to internal heating caused by the production beam. A target concept has been selected, and calculational analyses of the target concept have been performed to guide the design of a prototype target for an in-beam test of the actual thermal behavior. For this test, a high-power test facility is needed; fortunately, the beam currents required exist at the TRIUMF accelerator facility. An experimental proposal has been approved for such a test.

INTRODUCTION

The future construction of a National Spallation Neutron Source provides an opportunity to consider parasitic use of a small fraction of the neutron production beam for producing radioactive species which can be accelerated and used for studies in new frontiers of physics (1).

For more than 13 years there has been growing interest in the nuclear physics community in developing a facility employing an Isotope Separator On-Line (ISOL) followed by a heavy-ion accelerator to produce intense, accelerated radioactive ion beams (RIBs). Arguments have been made that such a facility requires only existing state-of-the-art technology. However, the anticipated extension of production beam intensities by a factor of up to 100 times those used in existing facilities requires that the methods for development of targets to produce the intense RIBs be carefully validated.

Over the past five years, procedures have been developed to evaluate concepts which could result in high power targets that could be used in beams of up to 100 μA (2). An experiment has been approved at the TRIUMF accelerator facility in Vancouver, B.C. to test a prototype target. An analysis of this test will investigate the validity of the simulation procedures as applied to more general high power target development.

There have been other activities for such high power target development, especially at the Rutherford Appleton Laboratory (3), where a radiatively-cooled tantalum target has been designed to operate at 2200°C. This target, however, has not been tested at high production beam intensities.

In general, RIB production targets will be made of materials for which radiative cooling is not efficient, that is, operated at temperatures lower than required for radiative cooling to dissipate the heat loads resulting from production beams passing through the (thick, of the order of tens of g/cm^2) targets. For this reason, other cooling approaches are required. The test prototype target has been designed for water cooling, and the approach chosen can be adapted to a variety of target materials chosen for specific RIB production.

TARGET ISSUES

The major issues associated with high-power targets are: (a) material compatibilities at high temperatures; (b) release properties of the materials/target design; (c) target material thermal conductivity, and (d) coupling of the target to a cooling system.

Other issues include the heat deposition profiles along the target, and a construction approach that ensures thermal performance as provided by numerical simulations. In addition, an actual target may require auxiliary heating for use at less than maximum (100 μA) production beam intensities.

A prototype target design has been developed for exposure to 500-MeV protons at the TRIUMF accelerator in Vancouver, B.C. to provide an experimental check of the computer simulations for energy deposition rates coupled with thermal analysis. The energy deposition rates were determined using the LAHET Code System (4) and the thermal analysis was performed with the COSMOS/M finite element code (5).

PROTOTYPE TARGET DEVELOPMENT

The development of high-power targets for production of intense RIBs is a step-wise process, starting

CP475, *Applications of Accelerators in Research and Industry*,
edited by J. L. Duggan and I. L. Morgan

with numerical simulations of energy deposition by the production beam and subsequent thermal analysis. Such simulations, if part of a validated approach, can be used instead of the intense empirical effort that is both costly and time consuming.

Several classes of targets have been successfully employed at existing RIB facilities. These include metallic foils, molten metals, and powders. Because of the requirement for good internal thermal conductivity of a high-power target, only the first two classes can at present be suitable for application to RIB production with intense production beams. An intense research effort will be needed to improve the thermal conductivity and other thermal properties of powder or powder-like target materials through introduction of "impurities" such as graphite fibers.

Because of the well-known thermal properties of metals, and the desire to choose a refractory metal for the prototype target material that is of low Z (for minimum activation to assist post-experiment handling) and that can withstand an intermediate temperature (1500°C to 2000°C), vanadium (Z=23) metal foils were chosen as the target material (a maximum temperature of 1700°C for vanadium is assumed, representing a tolerable vapor pressure of the target material for a connected "phantom" ion source). A compatible container material is niobium, which was chosen as a medium-Z metal that is more refractory than vanadium, yet relatively easy to fabricate. The target dimensions are: target material radius, 9.5 mm; target material length, 20 cm; niobium target container, 0.5-mm thick, niobium fins of inner radius 1.0 cm and outer radius of 2.5 cm.

The vanadium target has been modeled as foils of 100 μm thickness, and stacked with a maximum target density of 50% solid. As a first step to provide a uniform temperature distribution along the length of the target, the target density was graded (for example, by adjusting the foil spacing) to be less dense (37.5% solid density) at the entrance to the target, becoming more dense along the length of the target. The graded density can be accomplished by changing the spacing between target foils. This approach was first suggested by the Rutherford Appleton project (3). The result of a power deposition calculation using the LAHET Code System (4) for the graded density target is shown in Fig. 1 for an incident beam current of 100 μA. It is clear that the energy deposition rate per unit length is approximately constant throughout the length of the target. The discrete changes in density are easily seen in the picture, as the density is incremented (there are four density regions).

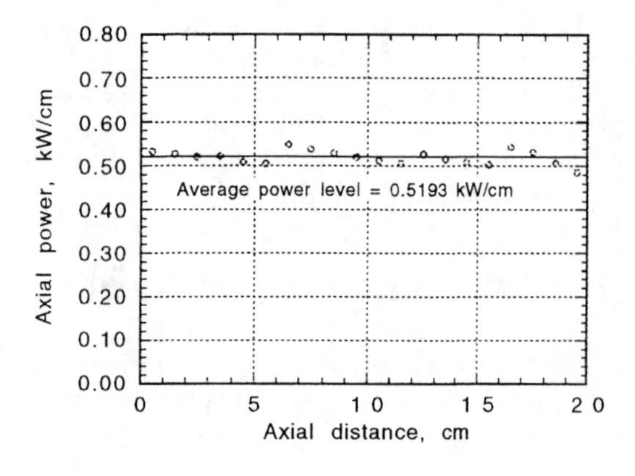

FIGURE 1. Axial power deposition profile in prototype test target

For the prototype target test, a cooling approach has been selected that lends confidence to interpretation of the experimental results. Despite the past attention given to the use of thermal barriers (2) and radiative cooling (3), the approach selected incorporates radial cooling fins distributed along the length of the target. This approach has several advantages. First, it can be modified to accept a range of thermal conditions for a candidate target. Second, the spacing and thickness of the fins can be modified to adjust the cooling rates along the target to match the requirements for cooling to keep the target temperature as uniform as possible.

A schematic representation of a portion of the prototype target (which shows only three of 21 fins) is shown in Fig. 2.

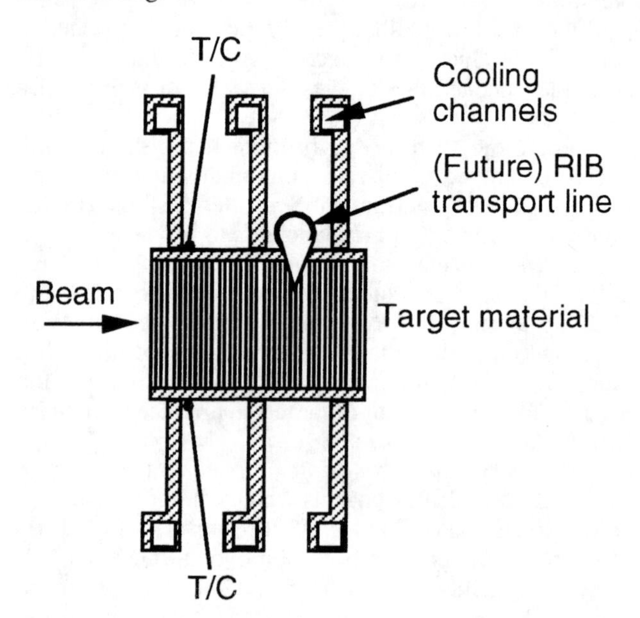

FIGURE 2. Schematic of partial axial segment of the prototype test target (spacing of target material is graduated along length of target).

Construction details for this target are not yet final, but establishing a robust thermal connection between the fins and target material container could include diffusion bonding or one-piece construction. It is assumed that the thermal contact between the target material and target container is achieved through diffusion bonding.

The cooling fins couple the target to cooling lines, shown in the figure as being attached to the outer radius of the fins. Also shown are thermocouple connections for the experiment.

Another cooling approach recently studied is that of using longitudinal fins instead of radial fins. This approach has the possible advantage that thermal stresses resulting from temperature gradients in the fins are not of concern. While the approach looks promising, it is too early in the studies to make judgment of the suitability for the use of longitudinal fins.

Among the issues to be decided are: How many fins -- two, four, eight? A previous study at Berkeley (unpublished) suggested the use of 24 fins, segmented to control axial temperatures. The preliminary studies indicate that longitudinal temperature variations can be accommmodated by adjustment of the fin height, being shorter at the entrance to the target and longer at the end of the target.

A longitudinal fin approach presents the possibility of easier fabrication than for radial fins, where the fin profiles can be made on a lathe, and the body of the target container made from a single rod of material with a milling operation.

ENERGY DEPOSITION

The energy deposition rate for the prototype target has been simulated using the LAHET Code System (4) with the result shown in Fig. 3, assuming an incident beam intensity of 100 μA. The initial radial energy distribution is indicative of the incident beam intensity profile, centered at the target axis with a parabolic radial intensity profile of 2.8-mm extent. Note that the beam diverges along the length of the target, giving rise to an energy deposition profile at the end of the target quite different from that at the front. This effect results in more energy being deposited near the radial edge of the target at the target end and represents a source of thermal non-uniformity, as will be shown later.

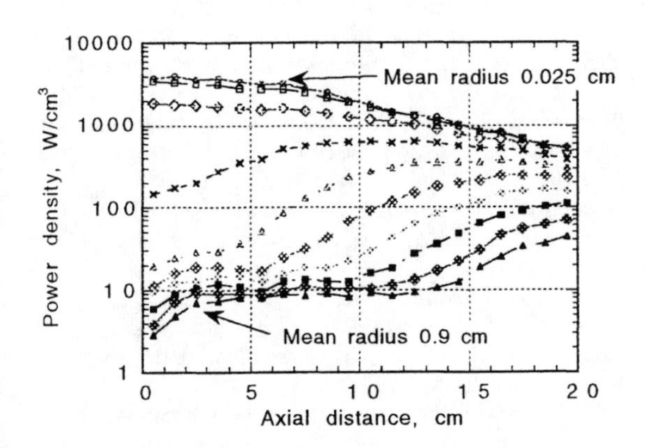

FIGURE 3. Energy deposition profiles in the prototype test target

THERMAL ANALYSIS

The results of the energy-loss simulation are used as thermal energy sources for an equilibrium temperature distribution simulation using the finite element COSMOS/M code system (5). The cooling fin thicknesses used in the thermal analysis were adjusted to achieve a uniform temperature along the axis of the target. The adjusted fin thicknesses ranged from 2.5 mm at the front of the target to 0.6 mm at the end.

The resulting temperature distributions are shown in Fig. 4, where the axial dependencies of the temperatures in the target volume are shown as a function of radius. Note that at the cooling fin locations, there are local temperature minima at the outer radius of the target material due to local heat flux variations.

While the temperature at the target central region is constant, as desired, the temperatures for the outer regions vary significantly (>600°C) along length of target. This provides a good test of the simulations, but results in expected target performance that is not suitable for release

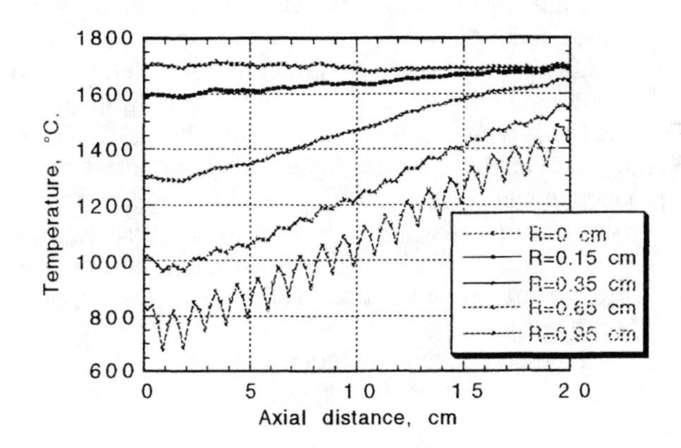

FIGURE 4. Temperature distributions in prototype test target

of produced RIBs, which is considered to be optimized for a favorable geometry if the entire target is uniform in temperature.

PROTOTYPE TARGET TEST

A proposal for a prototype target thermal test has been approved at TRIUMF, subject to availability of a test location able to provide a proton beam intensity of 100 μA. The observable in the test is the outer container sleeve temperature distribution along the length of the target. Figure 5 shows the expected temperatures at the target material radial boundary at five locations along the target length for different incident beam currents.

The prototype target design has been made for 100-μA operation. Note that at reduced incident beam currents, below about 50 μA, the observable temperature gradient is much less pronounced. The test therefore will require an incident beam intensity of ≥50 μA up to 100 μA to validate the simulations used in the target design.

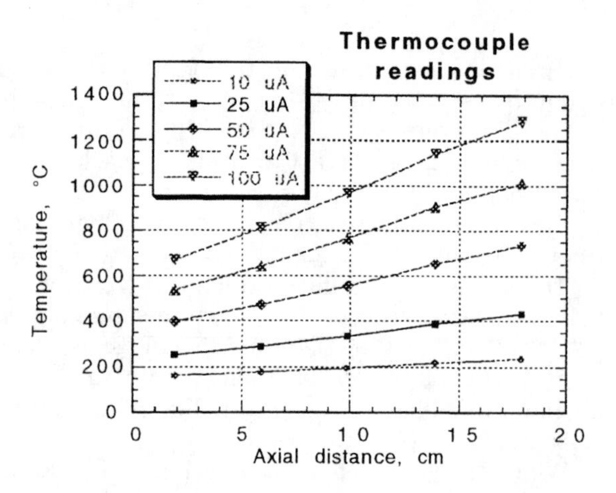

FIGURE 5. Temperature readings of the target boundary as functions of target length and incident beam current

The prototype target test was proposed as being a thermal test only; as such, the target incorporates no transfer line to an ion source, and no released RIBs will be observed. The purpose of the test is strictly to check the simulation approaches to the target design and fabrication method. If the test results in validation of the simulations, then confident application of this simulation approach can be used to design other targets for RIB production which include geometric and thermal properties that favor release of the produced radioactive species.

The prototype target design must be fabricated so that the thermal properties are well known. It is proposed either to diffusion bond the components together to assure robust thermal conductivity conditions or to construct the niobium container and fins out of a single piece of niobium metal, or a combination of the two approaches.

The design of the vanadium target foil assembly is still under discussion. Alternatives to the fabrication of this component include a foil-and-spacer arrangement or a spring-like spacer assembly with foils inserted between turns of the spring. Either approach will be diffusion bonded as a rigid assembly, which is then inserted in a tight fit into the niobium container. The larger thermal expansion dependence of vanadium during heating is expected to result in diffusion bonding of the target assembly to the niobium target container when heated to a uniform temperature of about 1200°C in a vacuum oven prior to the thermal test.

FURTHER DESIGN STUDIES

The prototype test target design results in quite large thermal gradients within the target material, as mentioned above. If it is reasonable to assume that the RIB production rates track the energy deposition profiles within the target material, the fact that target temperatures also loosely track the energy deposition profiles would indicate that the target regions in which RIB activities are produced have proper temperatures for good release. However, in the transport of the radioactive species from the target material to an ion source, the activities will traverse colder regions of the target and may be inhibited from effusion to the ion source.

This is a new feature of RIB production targets. At low production beam intensities, where external heating is required to establish the proper target temperature, the temperature distribution is nearly uniform throughout the target. Thus, all regions of the target have the same effusion rates. Because of the limitations of the less-than-perfect thermal conductivity of even metallic foils, high power targets will naturally sustain (sometimes large) temperature gradients within the target material.

The use of an annular incident beam profile can mitigate such temperature gradients (2), and could also be exploited to incorporate a void in the central target regions that would also assist in radioactive species effusion.

The cooling fin approach adopted in this work was easily adjusted to allow for differences in energy deposition rate profiles to establish uniform axial temperature. Other geometric modifications to the fins and/or use of auxiliary heating need to be explored as possible approaches to improve temperature uniformity of a high power target.

Finally, the useful (although not large) class of molten metal targets offers another route to high power targets. If a molten target is heated intensely and non-uniformly, then convection and evaporation processes within the target can be employed to provide thermal uniformity and cooling, respectively. For high power application, the evaporation rate of the target material could be quite high. For this reason, use of a refluxing chimney that is integral to the transfer line to the ion source could provide control over evaporative loss of the target material and sustain the cooling requirements. Such an approach also has the appeal that it would be self-regulating in temperature

REFERENCES

1. *Scientific Opportunities with an Advanced ISOL Facility*, Proceedings of the Columbus, Ohio Workshop held July 30 - August 1, 1997, November, 1997 (unpublished).
2. Talbert, W. L., Hodges, T. A., Hsu, H.-H. and Fikani, M. M., *Rev. Sci. Instr.* **68**, 3019 (1997) and *Proceedings of 14th International Conference on Applications of Accelerators in Research and Industry*, edited by J. L. Duggan and I. L. Morgan, AIP Press CP392, New York, p.385 (1997)
3. Bennett, J. R. J., *Nucl. Instr. Methods* **B126**, 105 (1997).
4. Prael, R. E. and Lichtenstein, H., *User guide to LCS: The LAHET Code System*, Los Alamos National Laboratory report LALP 91-51 (1991).
5. *The COSMOS/M Code* (Structural Research and Analysis Corporation, Los Angeles, CA. 1988).

The Production and Generation of Radioactive Ion Beams (RIBs)*

A. Schempp

Institut für Angewandte Physik, J.W. Goethe-Universität, D-60054 Frankfurt/M, Germany.

There is a strong interest in physics with radioactive ion beams, which has triggered a number of projects where nuclear physics installations are upgraded with accelerators for low charge state radioactive beams. A survey of the status of these projects is given with emphasis on the low energy linacs, which are limiting the acceptance and transmission of these low current ion beams.

INTRODUCTION

Physics of structure and interaction of nuclear matter has been mainly confined to nuclei which are stable or can be produced with beams of stable nuclei. The valley of stability has been explored extensively and naturally the interest has also been in the hills and the areas hidden behind them. These areas of proton-rich or neutron-rich nuclei exist in the cosmos and therefore are of special interest for astrophysics, studies of the early stages of the universe and for a better understanding of the nuclear structure.

This is the field of Radioactive Ion Beams (RIBs), where a large number of unstable isotopes can be produced and studied. Unstable radioactive beams have been available for about 30 years from the on-line isotope separators (ISOL) e.g. at CERN. The low energy beams available restricted the kind of experiments. The improvement of experimental techniques and the growing interest in this kind of physics has led to a number of proposals and projects to accelerate these beams to energies close to or above the Coulomb barrier to study nuclear reactions.

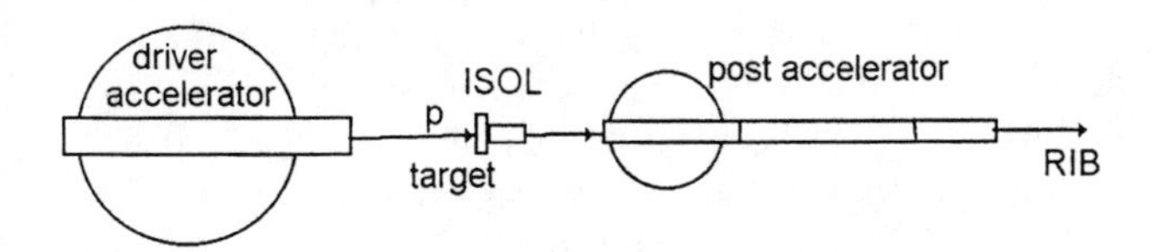

FIGURE 1. Scheme of a ISOL based RIB-facility

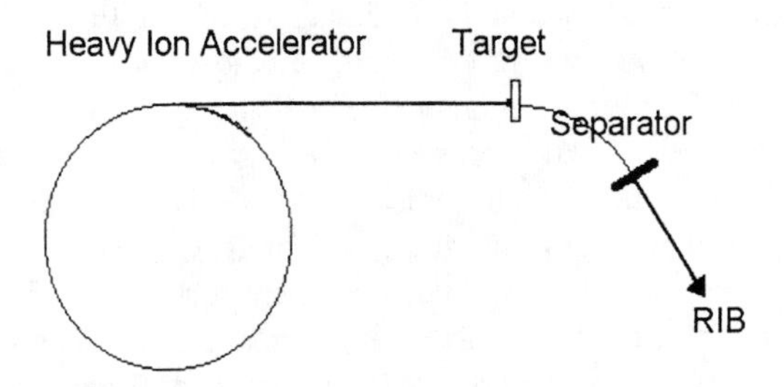

FIGURE 2. Scheme of a Fragment Separator based RIB-facility

As indicated in fig.1 a RIB facility consists of a driver accelerator, which in the case of CERN is the 1GeV PS-Booster, a target ion source separator unit and a low current, high duty factor post-accelerator.

Figure 2 shows the scheme of a parallel path to the generation of these exotic nuclei, the use of fragment separators at HI-facilities with high energy heavy ion beams. This method is complimentary, and the difference in experimental possibilities and the restriction in number of ions and ion species is obvious.

RIB facilities set up beams of separated ions for applying methods successfully used in nuclear physics, for which fixed target experiments with 10^4-10^{10} ions/sec at energies between 1 and 10 MeV/u are typical.

In general these radioactive beam facilities enhance the experimental possibilities and the range of possible beams by orders of magnitude. So the first generation of installations were upgrades of existing facilities to do research in that new field with rather low investments.

After these succesful forerunners special bigger projects were started for the postacceleration of ISOL beams or to feed existing nuclear physics machines with new ion species generated in new ISOL sources with new accompanied driving accelerators.

FRAGMENT SEPARATOR RIBs

Radioactive ions can be produced through fragmentation of very energetic heavy ions. At GSI (Germany) beams up to Uranium are available at energies below 20 MeV/u from the Unilac and up to 2 GeV/u from the SIS synchrotron. The fragmented ions can be studied directly or collected in the ESR storage ring. There, cooling techniques can be applied, which increase the phase space density but also restrict the life time of the ions used. From the ESR they can be reinjected into the SIS for further acceleration or even deceleration down to 5 MeV/u.

At GANIL (France) the primary heavy ion beams are produced by two coupled cyclotrons (K=400) with output energies up to 100 MeV/u. For isotopic selection a system with an achromatic spectrometer and a Wien filter is used.

*Work supported by the BMBF

CP475, *Applications of Accelerators in Research and Industry*, edited by J. L. Duggan and I. L. Morgan

At RIKEN (Japan) the primary beams come from a SS-Cyclotron with an output energy of 20MeV/u for U and 135 MeV/u for light ions. At the NSCL (USA) beams from 50-200MeV/u are delivered by the K=1200 Superconducting cyclotron. Upgrades of these systems by adding better injectors with higher intensities are planed.

ISOL-BASED FACILITIES

The ISOLDE facility at CERN was the first in that field. Their ion source development is the basis for most projects. The ions are produced in a spallation target and are transfered by diffusion to an ion source, from which singly charged ions are extracted and separated. This provides a wide variety of ion beams with easy energy variation. The ions can be changed rather simple and the intensities are high but at low energy.

The first postacceleration of a RIB was achieved at Louvain la Neuve (Belgium) using a K=110 Cyclotron for ^{13}N. The new ARENAS booster accelerator now accelerates to 0.2 MeV/u for A/q > 13 and 0.8 MeV/u for A/q > 6.5.

The HRIBF at ORNL (USA) uses the ORIC cyclotron (K=105) as source of primary protons and an ISOL source on a platform to generate negative ions for injection into the 22 MV Tandem.

Figure 3 shows a layout of the HRIBF, which successfully operates with light RIBs. Upgrade plans with the replacement of the ORIC by a K=200 cyclotron with higher currents and a superconducting Tandem-booster with ΔU=50MV are being studied.

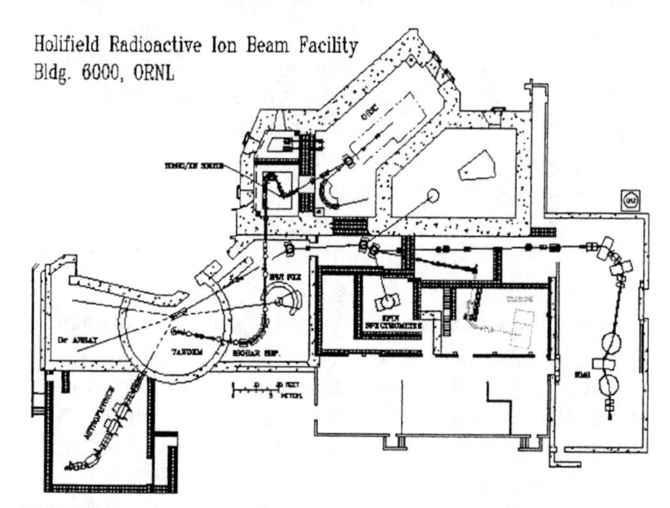

FIGURE 3. Layout of the HRIBF facility

SPIRAL at GANIL is a project where high energy heavy ions are used as the driver for the ISOL target and a K=260 cyclotron is the energy booster (2-29MeV/u).

At INS (Japan) a linac development program had been started to build a high duty factor 25 MHz-RFQ and IH-linacs for RI beams. This first linac project shows the advantage of a linac, the high transmisssion for the delicate particles but also the difficulties to get a high duty factor, which is around 30% at INS.

A new room temperature cw-linac for RIBs is being built for the ISAC project at TRIUMF (Canada). The primary beam from the SS-Cyclotron is feeding the TISOL ion separator.

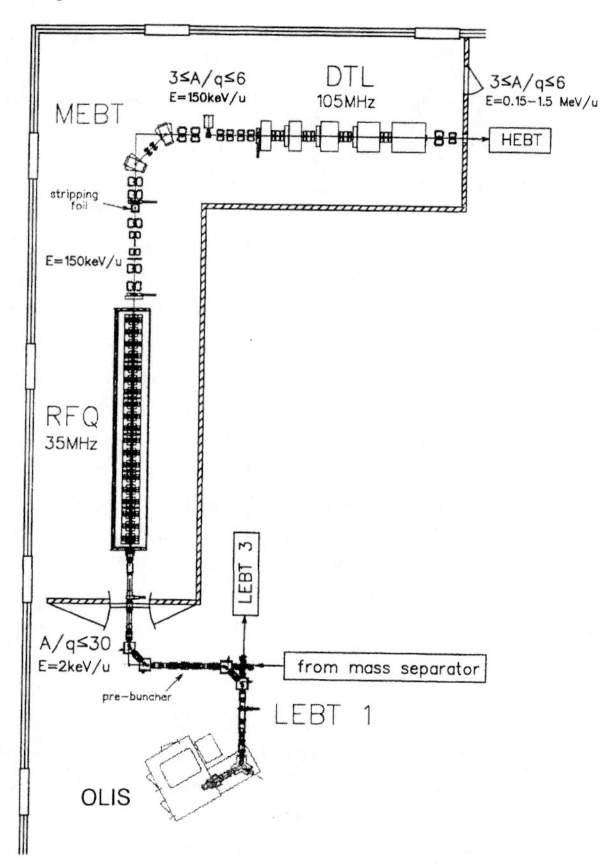

FIGURE 4. Layout of the ISAC facility at TRIUMF

The RI-beams are postaccelerated by a new linac shown schematically in fig. 4. This linac will operate with 100% duty factor, it consists of a 36 MHz RFQ (4-rod -split ring structure) and an IH linac (4 tanks) based on the GSI development but with separate rebunchers. At present the first RFQ-part has been successfully tested with beam.

Another project presently under construction is REX-Isolde at CERN, a collaboration led by LMU-Munich. The concept is quite different. To avoid cost intensive cw-linacs, charge collecting and breeding stages are between the RIB ion source and the linac, which is planned to operated at 10% duty factor. Radioactive ions are collected, cooled and bunched in a Penning ion trap, then transfered to an EBIS ion source, where the charge state is increased to match the minimum A/q=4.5, necessary for the compact linac. The Linac starts with an RFQ at 101.2 MHz based on the successful development on the GSI-HLI injector and the high current injector at MPI Heidelberg. From there also the short IH-cavity and 7-gap spiral resonators are adopted to bring the beam up to 2.2 MeV/u, allowing an energy variation down to 0.85MeV/u.

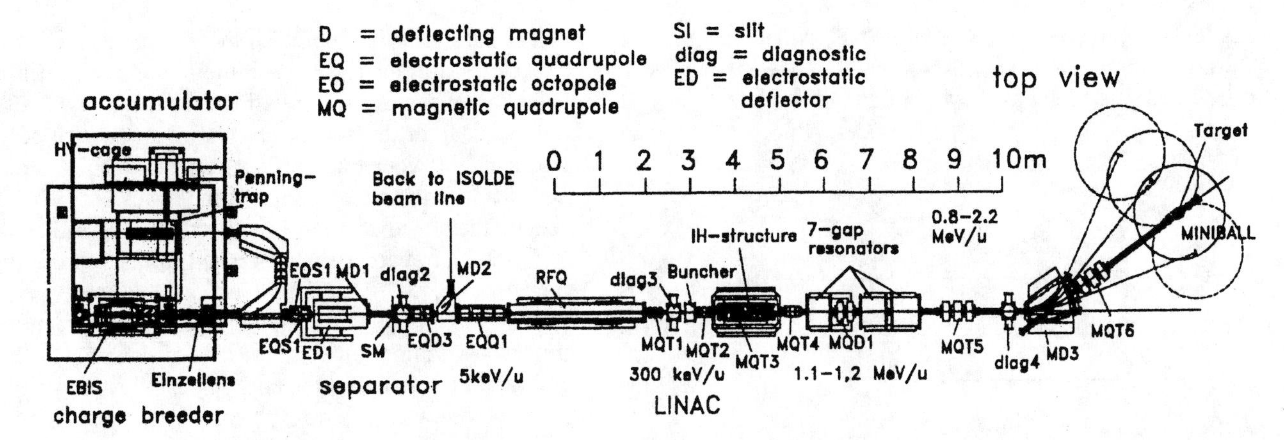

FIGURE 5. Layout of the REX-ISOLDE linac at CERN

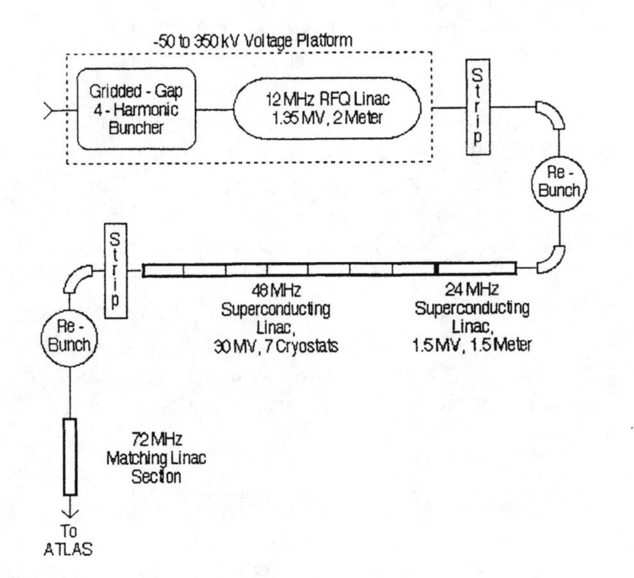

FIGURE 6. The ANL RIB linac upgrade

A new proposal adds a RIB facility to the ATLAS accelerator at ANL. This includes a proposed superconducting production linac of 215 MeV and an ISOL source. On the postaccelerator side ATLAS has to be upgraded as well for lower charge to mass ratios. A new low energy linac with a room temperature RFQ (12 MHz) on a HV-platform for energy variation, and new 24 and 48 MHz superconducting linacs (ΔU=31.5MV) allow the acceleration of RIBs with A/q=132. With two strippers the charge states can be increased to the requirements of ATLAS (A/q=6.6). A layout is shown in fig.6.

PROJECTS

There are a number of new projects which will use existing or planned high current accelerators as production machines for feeding a ISOL source.

At KEK along the JHF project a RIB facility is planned, which uses part of the 3GeV proton beam. This E-arena facility is an extension of the successful INS development of SCRFQ and ICH-linacs.

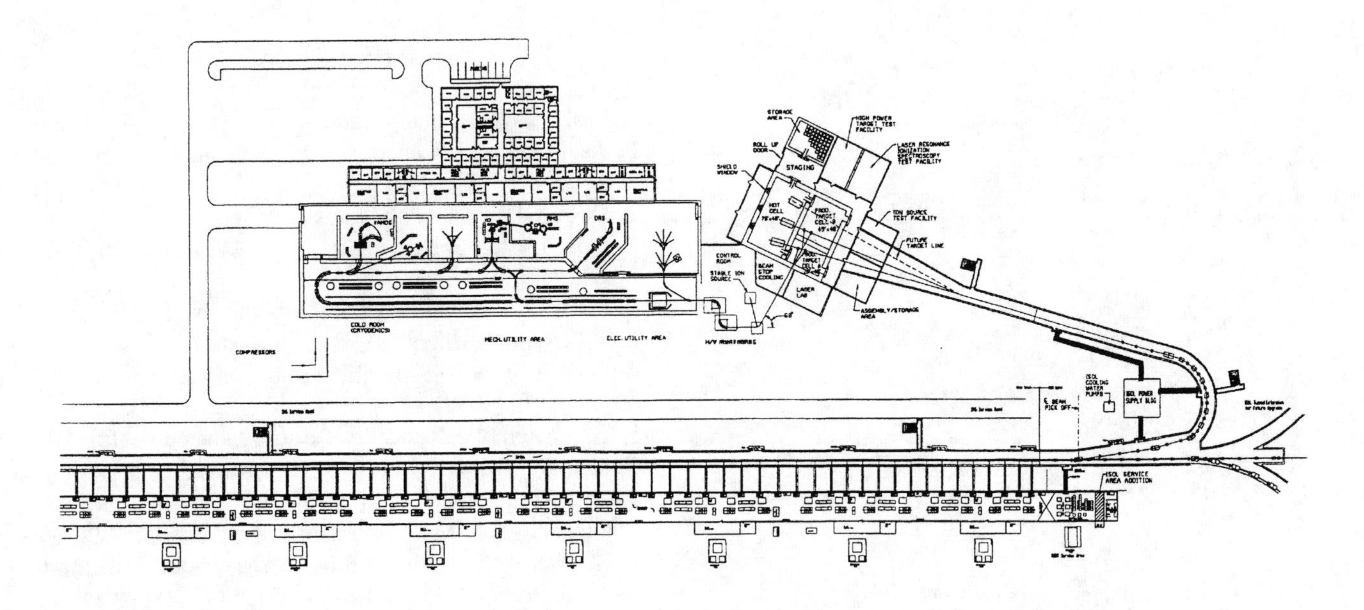

FIGURE 7. NISOL facility at the SNS

At ORNL the construction of the SNS spallation source naturally leads to a parasitic use of that high power beam as driver of a ISOL target as shown in figs. 7,8.

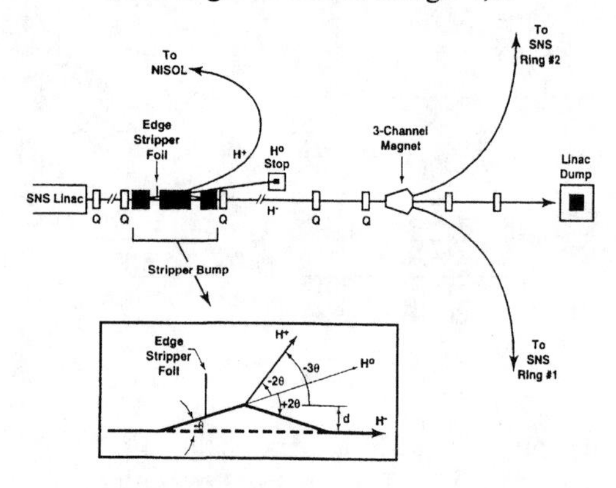

FIGURE 8. Scheme of SNS-NISOL switchyard

The RIB postaccelerator most likely will use a combination of RFQs and low frequency quarter wave structures like ANL accelerating up to 15MeV/u.

Similar concepts are proposed for other high power accelerators like at LANL, RAL, INR and Bejing.

DISCUSSION

The physics issue is clear: to study a large number of nuclei with extreme proton to neutron ratios for new phenomena. The access to RIBs is achieved by complicated accelerator systems, which first produce ISOL type ions on special target/ion sources and secondary beam postaccelerators to deliver beams at different energies for experiments.

The Isospin study demonstrated the possibilities and the complexity of such systems.

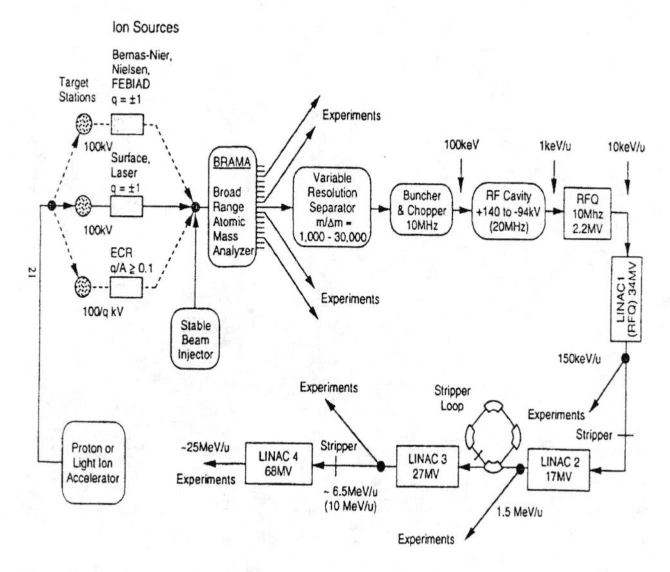

FIGURE 9. The IsoSpin Laboratory Concept [6]

Weak (or expensive) points of these schemes are: the low overall transmission and the requirement of a cw-accelerator. The low transmission is caused by the low charge state from the source and the required strippers to get reasonable accelerator efficiencies. The small number of ions leads to the cw operation of the linacs.

These facts lead to the limitations of light radioactive ions in most projects, being restricted to masses of $A/q <$ 30 or 50.

Work to improve the target RIB source currents therefore is important, because improvements give proportional intensities at the end. The application of ECR-type ion sources to ionize efficiently and possibly to get higher charge states directly from the source has been persued at GANIL, INS/KEK and Louvain la Neuve.

A big step is the REX-concept, a Penning trap and an EBIS charge state breeder, even far from being simple devices, lead to a very compact accelerator, which is pulsed favourably in terms of length and cost of a linac.

A better (high power) production accelerator increases problems with the target but gives larger RIB currents. The use of the new SNS linac leads easily to very high current RIBs. The idea to use d, He beams instead of protons is an attractive feature of the ANL proposal.

A "cheap" production unit is proposed for the proposed fragment accelerators in Grenoble (PIAFE) and at LMU Munich, which will use systems like the REX-linac to accelerate beams from research reactors replacing an ISOL source.

ACKNOWLEDGEMENTS

I would like to thank all the colleagues from other laboratories who provide information for this paper. G.Alton, O. Kester, K. Shepard, R. Laxdal.
Thanks to H.Vormann for checking the manuscript.

REFERENCES

1. Proc. Accelerated Radioactive Beams Workshop, Parksville B.C. (Canada) L.Buchmann, J.Dáuria Eds. TRIUMF Rep. TRI-84-1
2. D.K. Olsen et al., Status Report for the Holifield Radioactive Ion Beam Facility, Proc. Cyclotrons 95, World Scientific (1995) 634
3. P. Bricault, Linear Accelerators for Exotic Ion Beams, Proc.LINAC98, Chicgo 1998
4. D.Habs et al., The REX-Isolde project, Nucl. Instr. Meth. B 139 (1998) 135
5. G. Alton, Advanced Target Concepts for RIB Generation, AIP CP392 (1997) 429
6. Post-Accelerator Issues at the IsoSpin Laboratory, S. Chattopadhyay, J.M. Nitschke Eds., LBL 35533, 1994
7. K. Shepard, J. Kim, A Low Charge State Injector for ATLAS, Proc. PAC 95 , IEEE 95CH35843 (1996) 1128

In-Beam Production and Transport of Radioactive ^{17}F at ATLAS

R.C. Pardo, B. Harss, K.E. Rehm, J. Greene, D. Henderson, C.L. Jiang, J.P. Schiffer, J.R. Specht and B. J. Zabransky

Argonne National Laboratory, Argonne, IL, 60439 USA

Beam currents of radioactive ^{17}F($T_{1/2}$ = 65s) as high as $2x10^6$ s^{-1} have been produced at the ATLAS facility and delivered to target for nuclear physics research. The d(^{16}O,^{17}F)n and p(^{17}O, ^{17}F)n reaction were used to produce the ^{17}F in the energy range of 65-110 MeV with ^{17}F intensities of up to 250 pnA. The target employed is a liquid nitrogen cooled H_2 gas cell, with HAVAR windows, operating at up to $8X10^4$ Pa pressure. A new beam optics geometry consisting of a superconducting solenoid immediately after the production target followed by a single superconducting resonator has significantly improved the total capture efficiency of the transport system. The superconducting solenoid captures the highly divergent secondary beam and refocuses it to improve the beam match into the remainder of the transport system. A single superconducting resonator then 'debunches' the beam, reducing the energy spread by a factor of four. The beam energy can also be varied, using the resonant cavity, without changing the primary beam energy. Detailed discussion of the results, comparison to calculations, and further possible improvements will be presented.

INTRODUCTION

The interest in beams of radioactive isotopes for a wide variety of research in nuclear physics continues to grow. At the ATLAS facility more than 15% of all beam time in the past year was devoted to not naturally occurring radioactive species. The development of a ^{17}F beam ($T_{1/2}$=65 s) has been in progress for a number of years and the first successful experiments with such a beam were reported earlier(1,2,3).

For the production of ^{17}F, we use an in-flight technique exploiting the features of inverse kinematics to produce a well-defined secondary beam. Primary beams of either ^{16}O or ^{17}O bombard a gas cell containing correspondingly either deuterium or hydrogen. The ^{17}F beams are produced via either the d(^{16}O,^{17}F)n or p(^{17}O,^{17}F)n reactions in the energy range of 65-110 MeV. The beam properties are determined by the reaction kinematics and effects of straggling in the gas and the windows of the gas cell.

By normal accelerator standards, these beams are of very poor quality with large energy and angular spreads. The beam emittance can be minimized by using a good transverse and time focus of the primary beam on the production target, but the effect is limited by the significant thickness of the target gas cell. In this paper we report the performance of a new beam optics configuration which improves our ability to capture and transport the secondary beam produced in these inverse reactions and achieves a significant reduction in the energy spread of the secondary beam.

PRODUCTION AND TRANSPORT

Primary Gas Target

In order to achieve high secondary beam intensities, a high pressure gas cell with metal windows is used as the production target which can withstand primary beam intensities up to 250pnA. The gas cell consists of a double-walled 2.5 cm inner diameter cylinder 7.5 cm long. The chamber between the walls can be filled with a circulating cooling liquid such as liquid nitrogen which both stabilizes the target thickness and increases the effective thickness at a given pressure. The windows are 1.9 mg/cm^2 HAVAR™ (a cobalt chromium nickel alloy) foils soldered to a stainless steel ring with an inner diameter of 1.3 cm and mounted on the gas cell using an indium gasket. The pressure used in the gas cell ranged up to $8.5X10^4$ Pa. The lifetime of the HAVAR foils was up to 80 hours depending on the beam current, spot size, and cooling. Cooling the cells to liquid nitrogen temperature is critical to optain the longest window lifetimes. In order to eliminate the need for opening the beam line system after a window failure, two assemblies of three gas cells each were stacked together on linear translator stages.

First Production and Transport Geometry

The gas cell was initially installed in front of a bending magnet leading to the Enge split-pole spectrograph at the ATLAS accelerator at Argonne National Laboratory. The optics of the spectrograph beamline was modified by the addition of a second quadrupole doublet (see Fig. 1) to improve the transport efficiency for the ^{17}F beam which is estimated to have a total normalized emittance ($e_n=\gamma\beta\varepsilon$) of 11.3$\pi$ mm·mr. Even with the addition of a second quadrupole the beamline angular acceptance is approximately 0.55°. The acceptance was further reduced due to the bending magnet dispersion and the energy spread of the reaction products.

In this configuration, ^{17}F beams with energies between 55-100 MeV have been produced. The average reduced ^{17}F beam intensity was 700 (s·pnA)$^{-1}$ and corresponded to a beam transport efficiency of approximately 1.5%. With a primary ^{17}O beam of up to 250 pnA, rates of 2×10^5 ^{17}F/s on the secondary target were achieved into a 1 cm diameter beam spot.

CP475, *Applications of Accelerators in Research and Industry*, edited by J. L. Duggan and I. L. Morgan

1999 The American Institute of Physics 1-56396-825-8

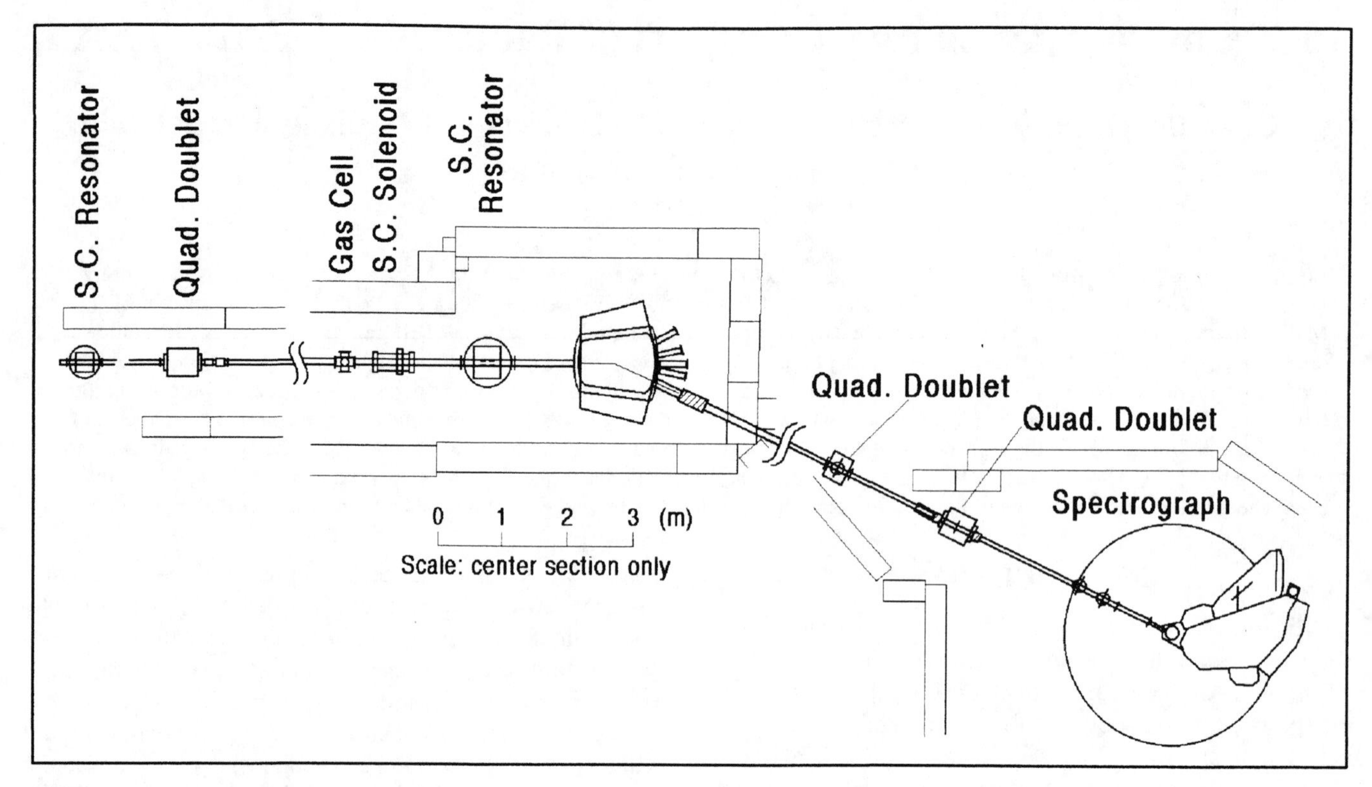

Figure 1. Floor plan of the ATLAS radioactive beam in-flight production facility.

New Production and Transport Geometry

The beam matrix optics code TRANSPORT (4) and a locally developed beam ray-tracing code LINRAY(5), used in the design of ATLAS, were used to developed an improved production and transport system for the ^{17}F beam. LINRAY was modified to include the correct particle distributions and correlations in the reaction process as well as straggling effects in the gas cell windows. LINRAY is a cylindrically symmetric code and therefore does not treat beam transport through quadrupoles correctly, nor the small dispersion induced by the beamline switching magnet. The output from LINRAY through the gas cell, solenoid, and resonator then served as the starting point for TRANSPORT in modeling the remainder of the beam line.

The results of the beam optics studies showed that improved capture and transmission of the ^{17}F ions was possible by the installation of additional focusing elements in the transport system, especially close to the target and the use of RF cavities before and after the target to minimize the longitudinal emittance growth and reduce the energy spread of the outgoing ^{17}F. The addition of a transverse focusing lens near the production target reduces the beam divergence before particles strike an aperture and also controls the size of the beam through the second RF cavity. Due to the large transverse emittance of the secondary beam the optimum transmission achieved is the result of compromises at the various constrictions in the beamline leading to the spectrograph.

The new production and transport configuration developed is shown in Fig. 1. The production target has been moved upstream approximately 5 meters. In this position, the gas cell is between two existing ATLAS superconducting rebuncher resonators and just ahead of a newly installed 2.4T superconducting solenoid. The solenoid is mounted in such a way as to be easily moveable over a 0.63 meter distance for best placement depending on the reaction kinematics. The superconducting solenoid forms a transverse waist after the second resonator. This optics design is a compromise between the need for a sufficiently small ^{17}F beam envelope through the 2.5 cm. diameter aperture of the resonator and the need to minimize the divergence of the beam so that as much beam as possible can be captured and refocused by the two quadrupoles on the spectrograph beamline.

By placing the production target between two RF cavities, it is possible to reduce the longitudinal emittance of the secondary beam and then use the energy-time correlation of the beam to reduce the energy spread with the second resonator. The RF cavity upstream from the production target refocuses the primary beam to a small time width on the production target which minimizes the longitudinal emittance of the secondary beam and maximizes the energy-time correlation necessary to reduce the ^{17}F energy spread. The RF cavity after the production target then 'debunches' the beam, significantly reducing the energy width of the secondary beam.

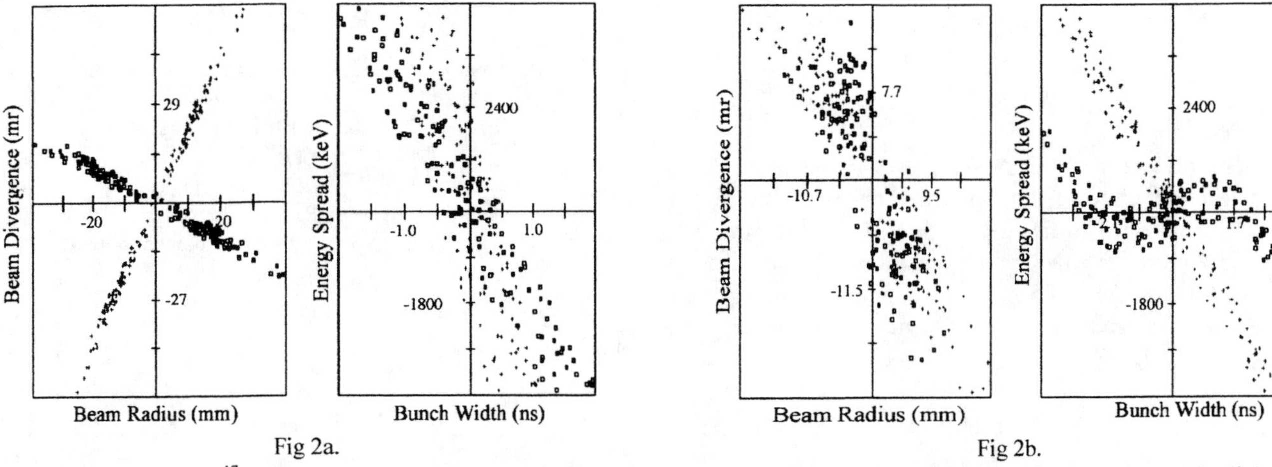

Figure 2. Calculated ^{17}F phase space (a) at the superconducting solenoid and (b) at the second rebunching resonator. The light '+' symbols represent particle coordinates entering the devices while the darker squares represent particles leaving the devices. Note the bimodal transverse distributions and the nonlinear effects of the RF on the longitudinal emittance.

The phase of the second RF cavity can also be adjusted to change the ^{17}F beam energy over a range, in this case, of approximately ±4 MeV while still maintaining a small energy spread. This feature allows the choice of production energy to be chosen partially based on secondary beam yield while allowing excitation functions to be more easily mapped out using the energy variability from the second RF cavity.

Calculations with the program LINRAY allowed the optimum relative position of the gas cell, solenoid, and resonator to be determined. An example of the calculated effects of the new solenoid and resonators on the transverse and longitudinal phase space for the p(^{17}O,^{17}F)n reaction is shown in Fig. 2.

The calculations demonstrate that the use of a 'rebunching' resonator in front of the gas cell to produce a minimum time width at the target for the primary beam is crucial for the best use of the second resonator. The primary beam bunch width without rebunching is approximately 3 ns FWHM. Using only the resonator after the gas cell to reduce the energy spread of the secondary beam reduces that energy spread from ±4 MeV to ±1.7 MeV. When a rebunching resonator is used to refocus the primary beam to a bunch width of ±0.5 ns, the secondary beam's longitudinal emittance is reduced, the energy-time correlation is improved and an energy width of ±0.7 MeV for the ^{17}F beam is predicted for the full beam.

This effect was easily observed in the experiment where we found that using only the second rebunching resonator increased the transmitted beam by only 10-20%, but that using both resonators yielded a 50% increase in beam current compared to optimization with only the superconducting solenoid.

The improvement in the energy width of the ^{17}F beam was also easily observable. The energy spread of the beam arriving at the secondary target without any resonator is approximately 1.6 MeV which is essentially a measure of the acceptance of the beam transport system. With increasing experience in tuning the secondary beam and improved diagnostic information as this project progressed it was possible to reduce the energy spread at the spectrograph target to as little as 0.3 MeV FWHM and routinely to a little less than 1 MeV FWHM. This is consistent with the modeling calculations noting that we have been able to transmit to target only about 25% of the total expected beam, as discussed below.

The gas cell is now significantly closer to the final primary beam quadrupole which produces a waist at the gas cell. If the target where 'thin' one would expect a reduction of the transverse emittance which is effectively a product of the beam spot size and the maximum emission angle for the ^{17}F from the reaction kinematics. Unfortunately because we are forced to use a gas cell in order to accept reasonable primary beams on the production target, the finite extent of the interaction region means that the beneficial effect of a smaller primary beam waist is largely lost. The problem is exacerbated further by the multiple scattering of the primary and secondary beams in the HAVAR windows of the gas cell. Therefore we believe that the transverse emittance of the beam is little changed from that estimated above for the first production geometry used.

Two examples of the secondary beam delivered to the spectrograph are shown in Fig. 3. Components from the primary beam which scatter from the gas cell window frames and other areas constitute the dominant source of tails from the primary beam. Any primary beam degraded in energy sufficiently to match the magnetic rigidity of the desired ^{17}F will be transmitted to target and those components are seen in Fig 3. The best ^{17}F beam purity achieved has been approximately 90%, with 75% a more typical value.

TRANSMISSION RESULTS

The transport of this 'reprocessed' secondary beam to the reaction target at the spectrograph by the remaining beam is now significantly improved. With the new configuration a LINRAY calculations predict a total transport efficiency of 52% or 26% including the stripping fraction. Instead the best efficiency achieved so far is approximately 6%.

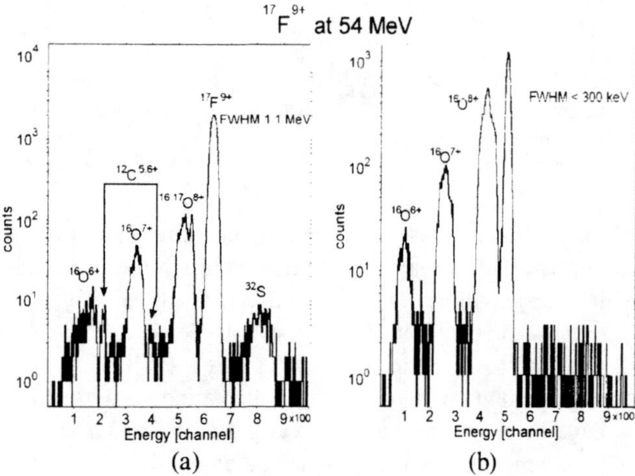

Figure 3. Two examples of beam components arriving at the secondary target. In (a) the energy spread is approximately 1 MeV FWHM and approximately 88% of the beam is ^{17}F. In (b) only 44% of the ions delivered are ^{17}F but the energy spread is less than 300 keV.

A number of factors may contribute to the discrepancy between the observed transmission and the predicted value. First the apertures of the two quadrupoles continue to be significant constraints to the transport efficiency. LINRAY's cylindrical symmetry assumptions cannot properly model a quadrupole. Second, the beam spot at the spectrograph target is significantly larger than for the earlier geometry (12 mm X 8 mm compared to 3 mm X 5 mm.). Such a large size cannot fully be accepted by the spectrograph detector and this reduces our useable efficiency by 50%. These problems could be improved with larger aperture quadrupoles moved to locations reducing the beamline magnification. Unfortunately such quadrupoles are not available to us at this time. Finally realizing the optimum optics calculated in a real experiment is a difficult process and our learning curve may yet yield further improvements in overall transmission.

This work was supported by the U.S. Department of Energy, Nuclear Physics Division, under contract W-31-109-ENG-38.

REFERENCES

1. B. Harss, et al., Proc. of 1996 Lin. Accel. Conf, Geneva, Switzerland, CERN 96-07, 496(1996).
2. B. Harss, et al., Proc. of 1997 Part. Acel. Conf., Vancouver, Canada, 2696(1998).
3. K.E. Rehm, et al, Phys. Rev. Lett., in print.
4. K.L. Brown, Advances Particle Physics., 1, 71(1967).
5. R. Pardo, LINRAY Operating Manual, unpublished.

Recent Experimental and Technical Achievements at ISOLDE/CERN

U. Georg

Div. EP, CERN, CH-1211 Geneva 23

At the on-line mass separator facility ISOLDE ion beams of radioactive isotopes are produced by nuclear reactions induced by bombardment of thick targets with a 1 GeV pulsed proton beam. With the use of a large variety of target materials and ion sources over 600 isotopes of more than 60 elements (Z=2 to 88) have been produced up to now. This great choice offers amplitude of possibilities for experiments in nuclear physics, atomic physics, nuclear solid state physics, nuclear astrophysics, biophysics, nuclear medicine and even particle physics.

A variety of new results have been obtained during the last year like the magnetic moment of ^{11}Be, the beta delayed charged particle emission of ^{9}C, high precision mass measurements with an accuracy up to 10^{-7}, the half life of the waiting point nucleus ^{129}Ag, the direct observation of the diffusion of free hydrogen in III-V semiconductors and monolayer-resolved measurements in ultrathin magnetic multilayers.

INTRODUCTION

Since the move of ISOLDE from CERN's Synchro-Cyclotron (SC) to the Proton Synchrotron Booster (PSB) in 1992 extensive work has been devoted to the development of new beams, i.e., the production of new isotopes, beams of more intensity and the ionization of new elements (1). The later was mainly achieved by the development work done on the laser ion source (LIS) (2). This enabled various experiments to extend their research by benefiting from the chemical selectivity and the higher ionization efficiencies of the LIS. The results mentioned here are partially based on this progress while others show new development by the experiments. The presented results are only a selection of the successful work of the different collaborations that have been performing experiments at ISOLDE within the last year.

NUCLEAR PHYSICS

Laser Spectroscopy

Laser spectroscopy experiments have a long tradition at ISOLDE for the investigation of nuclear parameters like spins, moments and charge radii by measuring the hyperfine splitting and the isotopic shift of whole chains of isotopes using different advanced methods that have been developed over the years (3).

Last year the collinear laser spectroscopy collaboration COLLAPS has taken advantage of the recently found laser ionization scheme for the production of Beryllium ion beams (1,2). This enabled the measurement of the long awaited magnetic moment of ^{11}Be as well as the magnetic moment of ^{7}Be and the isotopic shift of ^{7,9,10,11}Be. The particular interest in the magnetic moment of ^{11}Be is arising from the fact that this nucleus is the only known one-neutron halo nucleus and that it has an anomalous ground-state spin-parity. Several calculations are already existing and the measured value gives information on the relative weights of the $|^{10}Be(0^+)\cdot 1s_{1/2}\rangle$ and $|^{10}Be(2^+)\cdot 0d_{5/2}\rangle$ components of the ^{11}Be ground state wave function (5). The magnetic moment was deduced from β-NMR signals from optically polarized $^{11}Be^+$ implanted into a crystal of metallic Be inside a calibrated 0.3 T magnetic field.

The results on the other isotopes were obtained by standard collinear fast beam laser spectroscopy on Be^+. The value for the magnetic moment of ^{7}Be delivers valuable information for the estimation of electron and proton capture rates in the solar neutrino problem. Furthermore together with the known moment of the stable ^{7}Li the isoscalar and isovector magnetic moments of the lightest T=1/2 mirror pair of the 0p-shell can be derived.

Nuclear Spectroscopy

The IS361-Collaboration has recently investigated the beta decay of ^{9}C. The daughter nucleus $^{9}B^*$ is unbound and therefore disintegrates into one proton and two alpha

CP475, *Applications of Accelerators in Research and Industry*, edited by J. L. Duggan and I. L. Morgan

particles. As ^{9}C is the mirror nucleus of ^{9}Li for who's transition into ^{9}Be very large beta strength had been found the investigation of ^{9}C offers the unique possibility to compare this strength for mirror nuclei, as for all other known candidates in this region like ^{6}He, ^{8}He and ^{11}Li the mirror nucleus is particle unbound.

When performing the experiment, the ^{9}C ions were stopped in a thin carbon foil and the decay products from $^{9}B^{*}$ were registered by two Double Sided Si Strip Detectors (DSSSD) (300 and 500 μm thick) triggered by an annular Si-detector recording the β-decay of the ^{9}C. The DSSSD were complemented by a 1000 μm thick 2500 mm^2 Si-PAD detector and a 700 μm thick 2000 mm^2 Si-detector respectively what allowed the detection of the high energy protons. From the measured energies and the angles of all fragments the proton-α-α correlation could be deduced yielding the excitation energy of ^{9}B (6).

Mass Measurements

Accurate measurements of nuclear masses far from stability are particularly needed for constraining nuclear mass models since their predictions vary quite drastically only a few isotopes beyond measured values. This leads for example to large uncertainties when modelling nucleosynthesis reaction networks. Actually there are two experimental set-ups installed at ISOLDE to extend the number of known atomic masses, the ISOLTRAP collaboration using a Penning trap as spectrometer (7) and the MISTRAL collaboration using a transmission rf-mass spectrometer (8). In both cases the mass measurement is carried out via the determination of the cyclotron frequency $\omega_c=q/m\cdot B$ of ions in a strong magnetic field B.

The ISOLTRAP collaboration has been measuring the masses of long isotopic chains over the past decade at ISOLDE. In the first version of the ISOLTRAP system a stopping-reionization technique was used. In order to overcome the resulting restriction to elements that can be surface ionized, end 1996 a RFQ ion beam buncher has been incorporated, which can catch the ISOLDE beam at 60 keV beam energy and inject it at lower velocity into the Penning trap (9). Due to this new tool the masses of $^{184,\ 185g,m,\ 186,\ 191g,m,\ 193g,m,\ 194,\ 197}Hg$ and of $^{196,\ 198}Pb$, ^{197g}Bi, ^{198}Po, ^{203}At were measured. An accuracy of $\Delta M/M=10^{-7}$ was achieved (10). To resolve ground state and isomeric state of some of the mercury isotopes the trap was operated at a resolving power of up to $5\cdot 10^{6}$.

The MISTRAL spectrometer is a complementary set-up as it can be used for mass measurements of very short-lived (ms) isotopes while ISOLTRAP is limited to longer half-lived (>0.1 s) isotopes. The expected accuracy of measured masses is up to 10^{-7} and the mass resolving power larger than 10^{5}. First and very successful mass measurements have been performed in summer 1998 on $^{23\text{-}30}Na$ with the three heaviest having half-lives of less than 50 ms.

Astro Physics

The search for the neutron-magic ^{129}Ag was the trigger starting the LIS development programme at the new ISOLDE. The decay properties ($T_{1/2}$, P_n and S_n) of this isotope are of particular astrophysical interest as it is one of the few reachable nuclei lying in the r-process path. The LIS is the only possibility to ionize silver atoms in valuable amounts and to suppress isobaric beam contamination from the easy ionizable indium against which the experiment is extremely sensitive. The actual experiment is the detection of β-delayed neutrons. Despite the use of the LIS and a special gating of the ion beam using the time structure of the laser ionization to suppress the ^{129}In contamination the decays of ^{129}Ag represent less than 0.1% of the detected neutrons within the first 300 ms.

The final breakthrough last year has been achieved after taking into account the hyperfine splitting of ^{127}Ag (I=9/2) and shifting the laser frequency with respect to the mean frequency of the stable ^{107}Ag (I=1/2) (11).

The measured half-life of 46^{+5}_{-9} ms is in good agreement with theoretical predictions, but is by about a factor of 3.5 too short with respect to the stellar abundance. Therefore the existence of additional isomeric states beside the investigated ground state is taken into consideration (12).

SOLID STATE PHYSICS

The branch of solid state physics at ISOLDE has continuously been growing over the last decades and is now covering more than 30 percent of the physics programme.

One solid-state collaboration using also the indium suppressed silver beam (^{117}Ag) prepared by the LIS is investigating the diffusion of hydrogen in III-V semiconductors like InP, GaAs and InAs. This diffusion behaviour has been extensively studied over the last years, mainly with respect to the modification of the electronic properties due to the formation of complexes of hydrogen with dopants or other defects. This leads to their passivation and the removal of the corresponding levels in the band gap of the semiconductor (13).

After the decay of the implanted ^{117}Ag into ^{117}Cd the sample is heated to repair damage caused by the implantation into the lattice. The radioactive acceptor ^{117}Cd forms Cd-H pairs that are stable until the ^{117}Cd decays into ^{117}In which sets the hydrogen free and lets it diffuse inside the semiconductor. This diffusion is observed by perturbed $\gamma\gamma$ angular correlation (PAC) spectroscopy.

Here the diffusive jump of a free hydrogen away from the ^{117}In located next to it is investigated within a 100 ns time window as a function of temperature. From the PAC spectra it could be observed that the hydrogen occupies two distinct interstitial lattice sites at low temperature. With increasing temperature these sites are depopulated by the diffusive jumps of hydrogen. From the temperature dependence of the jump rate the diffusion constant can be determined (14).

One experimental set-up at ISOLDE's UHV beam line is the "Apparatus for Surface Physics and Interfaces at CERN" (ASPIC). This collaboration is investigating magnetic multilayers and semiconductor-metal interfaces on an atomic scale by using nuclear probes and performing PAC spectroscopy. The installation allows the preparation of clean surfaces, the growth of monolayers on these surfaces and all analysis techniques needed for the samples (15).

Recently, induced magnetism in Pd, when Pd is in contact with Ni, has been investigated. This is of particular interest as 4d-elements like Pd, which are normally paramagnetic might become magnetic at the surface or close to the interface when they are in contact with ferromagnetic metals. In ASPIC ^{100}Pd/^{100}Rh and ^{111m}Cd/^{111g}Cd were used as nuclear probes. It was found that in a single crystal Pd is polarized up to a distance of at least 7 monolayers from Ni and fluctuations of the induced magnetic moments have been observed (16). Whereas in ultrathin Pd layers on Ni ferromagnetic behaviour of Pd was seen (17).

Also this type of experiments will take advantage of the LIS developments producing stronger Ag beams with less In contamination when investigating the ultrathin magnetic multilayers with the ^{117}Cd/^{117}In nuclear probe produced from the decay of ^{117}Ag.

OUTLOOK

Currently the REX-ISOLDE project is being set-up that will allow post-accelerating the existing radioactive ion beams produced by ISOLDE up to an energy of 2 MeV/u (18). In a first experiment these energetic radioactive ions will be used for a detailed study of very neutron-rich isotopes in the region around the magic neutron numbers N=20 and N=28 via Coulomb-excitation and neutron-transfer reactions.

In addition to the high energy beams also lower-energy pulsed beams of singly or multiply charged ions can be provided for example for weak interaction studies in ion traps or solid state physics experiments using the possibility of deeper implantation.

Another extension of the ISOLDE physics programme will be enabled by the upgrade of the proton beam energy from 1 GeV to 1.4 GeV, which is planned for second half 1999. This will result in higher production rates of the most exotic nuclei far from stability and will open the field for the investigation of isotopes up to now not available at ISOLDE.

REFERENCES

1. Lettry, J. A., "Recent Target and Ion Source Developments at CERN-ISOLDE", contribution to this conference. See also Lettry, J. A., et al., *Rev. Sci. Instr.* **69**, 761 (1998).
2. Fedoseyev, V. N., "The Use of Lasers for the Selective Ionization of Radionuclei for RIB Generation", contribution to this conference.
3. Billowes, J. and Campbell, P., *J. Phys.* G **21**, 707 (1995). See also Otten, E. W., *Treatise on Heavy-Ion Science*, Vol. 8, New York: Plenum Press, 1989, p. 517.
4. Kappertz, S., et al., "Measurement of the Magnetic Moments of ^{7}Be and of the Halo Nucleus ^{11}Be", presented at the 2nd Int. Conf. on Exotic Nuclei and Atomic Masses ENAM98, Bellaire, Michigan, June 23-27, 1998.
5. Suzuki, T., et al., *Phys. Lett. B* **364**, 69 (1995).
6. Tengblad, O., "Beta decay asymmetry in mirror nuclei: A=9", presented at the 9th Nordic Meeting on Nuclear Physics, Jyväskylä, Finland, August 4-8, 1998.
7. Bollen, G., et al., *Nucl.Instr. and Meth* A **368**, 675 (1996).
8. de Saint Simon, M., et al., Phys. Scripta T**59**, 406 (1995).
9. Bollen, G., et al., "A Radio Frequency Quadrupole Ion Beam Buncher for ISOLTRAP", presented at the ENAM98 conference, Bellaire, Michigan, June 23-27, 1998.
10. Bollen, G., et al., "Mass measurements with a Penning trap mass spectrometer at ISOLDE", presented at the ENAM98 conference, Bellaire, Michigan, June 23-27, 1998.
11. Kratz, K.-L., "New nuclear-structure measurements on r-process nuclei", presented at the ENAM98 conference, Bellaire, Michigan, June 23-27, 1998.
12. Kratz, K.-L., et al., "Laser isotope and isomer separation of heavy Ag nuclides: Half-life of the r-process waiting-point isotope ^{129}Ag and structure of neutron-rich Cd nuclides", presented at Int. Conf. on Fission and Properties of Neutron-Rich Nuclei, Sanibel Island, Florida, December 1997.
13. Burchard, A., et al., "First PAC studies on the hydrogen diffusion in III-V semiconductors", presented at the 7th International Conference on Shallow-Level Centers in Semiconductors, Amsterdam, July 17-19, 1996.
14. Burchard, A., "Microscopic study of the hydrogen diffusion in III-V semiconductors", presented at the MRS 1998 Spring Meeting, Symposium H: Hydrogen in Semiconductors and Metals, San Francisco, April 13-17, 1998.
15. Potzger, K., et al., *Nucl.Instr. and Meth* B, in press.
16. Bertschat, H.H., et al., *Phys. Rev. Lett.* **78**, 342 (1997).
17. Bertschat, H.H., et al., *Phys. Rev. Lett.* **80**, 2721 (1998).
18. Kester, O., "Status of the REX-ISOLDE Project", contribution to this conference.

The Production and Acceleration of Radioactive Ion Beams at the HRIBF

R. L. Auble

Physics Division, Oak Ridge National Laboratory, Oak Ridge, TN.

The Holifield Radioactive Ion Beam Facility (HRIBF) includes a K=100 cyclotron (ORIC) which provides high-intensity light-ions for producing radioactive atoms, and a 25 MV tandem electrostatic accelerator which is used to accelerate the radioactive-ions for nuclear structure and nuclear astrophysics research. Ion sources and targets suitable for the production of various radioactive ion beams (RIBs) have been developed. Operational experiences, problem areas, and plans for future beam development are discussed.

INTRODUCTION

The HRIBF is a first-generation Isotope-Separator-On-Line (ISOL) type radioactive ion beam (RIB) facility. In an ISOL-type facility, radioactive atoms of the desired species are produced by bombarding a suitable target with high intensity beams from one accelerator, and the radioactive atoms are then ionized, mass separated, and injected into a second accelerator to produce high quality beams for research.

A major part of our operating schedule since beginning operation in late 1996 has been devoted to restoring the light-ion acceleration capability of the cyclotron, improving the performance of the tandem at low terminal voltages and the development of target/ion-source assemblies suitable for the production of radioactive ions. Progress of the various development programs, and our operational experiences to date, are outlined in this paper.

ACCELERATOR DEVELOPMENT

The accelerator used for the production of radioactive atoms at the HRIBF is the Oak Ridge Isochronous Cyclotron (ORIC). The ORIC is a 1.5 m diameter variable energy cyclotron which began operation as a light ion accelerator in 1962. Although the main magnetic field was designed for a maximum rigidity of about 1.45 Tm, corresponding to a proton energy of 100 MeV, the maximum proton beam energy is presently limited to about 42 MeV by the tuning range of the RF system. Several accelerator improvement projects have been carried out, or are in progress, which are designed to improve the performance and reliability of the ORIC. In collaboration with the cyclotron group at Michigan State University, the central region was modified to improve the extraction efficiency and values of 85% are now routinely achieved for both proton and deuteron beams. This can be compared to typical values of 40%-50% prior to modification. Extracted beam currents are presently limited to 50 μA for protons and deuterons, and up to 200 particle-μA for other beams, by our Accelerator Safety Envelope. With the improved extraction efficiencies, these beam currents can be achieved without excessive power dissipation in, and activation of, the extraction system. At the present time, these administrative limits are not the limiting factor for the production of RIBs. The high power density produced by the relatively low beam energies available from the ORIC results in target temperature limitations which, for the RIB production targets presently in use, limit the beam current on target to about 5 μA for protons and deuterons.

Radioactive atoms produced by beams from the ORIC are ionized, mass separated and charge exchanged to produce negative ions which are then injected into the 25URC tandem electrostatic accelerator. The tandem has proven to be very reliable and economical to operate but several additions and modifications were required to meet the needs of a RIB facility. The most significant change was the replacement of the original corona-point voltage grading system with resistors to allow stable operation at lower terminal voltages. This modification now allows stable operation at terminal voltages as low as 1 MV to meet the beam energy requirements of the nuclear astrophysics research program. In addition, since the intensity of the injected radioactive beam is very low, usually below the detection level of standard Faraday cups, particle detectors have been installed at focal points inside the tandem to aid in tuning the beam through the accelerator. Similar detectors have also been installed on the external beam lines.

RIB PRODUCTION PROCESS

Radioactive atoms are produced by fusion-evaporation reactions between light ion beams from the ORIC and the target atoms chosen to produce the radioactive species of interest. Since the energies available from the ORIC are relatively low, the radioactive species accessible are those which can be produced by simple reactions such as (p,xn) or (d,xn) reactions. The cross-sections for such reactions are usually known and the production rate for a specific product atom can be readily calculated. Typical production rates for the radioactive species being studied at the HRIBF are on the order of 10^{-3} atoms per incident beam particle. Thus, for an incident proton or deuteron beam intensity of 5 μA, radioactive atoms would be produced at a rate of about $3x10^{10}$/s.

The number of radioactive atoms which can be delivered

CP475, *Applications of Accelerators in Research and Industry*, edited by J. L. Duggan and I. L. Morgan

to an experiment is, of course, much smaller than the number produced and can vary over several orders of magnitude depending on the operating conditions in the target/ion source assembly (1). The major loss mechanism for many of the radioactive species studied thus far has been the decay of the radioactive atoms during the time required to diffuse out of the target and drift to the ion source. For chemically active elements, losses will occur due to reactions between the radioactive atoms and the materials used in the target, vapor transport tube, and ionization chamber. Such elements require the formation of nonreactive molecules in order to be extracted from the target and transported to the ion source with any reasonable efficiency.

The target/ion source assembly is the 'heart' of any ISOL-type RIB facility. It must be capable of long term operation in a harsh environment of high temperature and high radiation fields while maintaining high voltage and high vacuum integrity. The design presently in use is the Electron-Beam-Plasma (EBP) positive-ion source (1) which is based on the design of the target/ion source used at ISOLDE (2).

Extensive testing is essential for any ion source prior to its use for the production of radioactive ion beams, and facilities have been built or modified for both off-line and on-line testing. The initial off-line testing of new or modified ion source designs is carried out at the Target Ion Source Test Facility (TISTF) (3). The TISTF allows convenient measurement of ion source performance over a wide range of operating parameters in a controlled, radiation-free, environment.

An ion source test stand has been constructed to assemble ion source components, check dimensional tolerances, ensure the integrity of vacuum seals and cooling water circuits, and measure electrical parameters prior to installation at either the on-line test facility or the RIB Injector. The stand also includes target heating capabilities for outgassing target materials and for efficiency and release time measurements of various radioactive species from irradiated targets (4).

For on-line testing, the UNISOR isotope separator has been modified (5) to allow target/ion source assemblies to be installed and tested with low intensity proton and deuteron beams from the tandem electrostatic accelerator. The UNISOR facility allows measurement of ion source efficiencies for the radioactive species under investigation while minimizing radiological exposure and contamination hazards.

RIB TRANSPORT/INJECTION

The target/ion source assembly is mounted on a high voltage platform, referred to as the RIB Injector, which also supports electrostatic beam optics elements, the first-stage mass-analyzing magnet, and charge-exchange cell. The high voltage platform is designed to operate at up to 300 kV for compatibility with the existing tandem accelerator stable-ion injector. The high voltage platform is presently operated at a voltage which provides 200 keV ions for injection into the tandem. A lower operating voltage was chosen to reduce leakage currents and prevent sparks since studies of the tandem showed that there was no significant difference in beam transport efficiency between 200 and 300 keV. Electronic components are mounted on a separate high-voltage platform which is radiologically shielded from the ORIC beam to prevent radiation damage to solid-state devices. The target/ion source is electrically isolated from the high-voltage platform and can be operated at voltages up to 60 kV although, at the present time, the ion source is normally operated at 40 kV.

Following extraction of the positive ion beam from the EBP ion source, the desired beam species is selected using the first-stage mass analyzing magnet which has a measured mass resolving power, $M/\Delta M$, of approximately 1,000. The beam, now consisting of a single mass but possibly several isobars, is then charge-exchanged to produce the negative ions required for injection into the tandem. At the present time, cesium vapor is used as the charge-exchange medium, but other charge-exchange media are being investigated.

A new beam line was constructed to connect the RIB Injector with the tandem accelerator. All of the focussing elements are electrostatic, and therefore mass independent, allowing initial beam tuning to be done with stable ion beams. Included in the transport line is a high resolution mass-analyzing magnet which was designed to provide a mass resolving power of 20,000 to allow for possible isobar separation. However, to achieve this level of mass resolving power requires very high beam quality. The energy spread introduced by ripple in the RIB Injector high voltage supplies has been measured and found to be less than the design goal of ± 10 eV. However, measurements of the energy-loss straggling introduced by the charge-exchange process, for ^{70}Ge beams and typical charge-exchange vapor densities, gave a FWHM of approximately 42 eV. This effectively limits the mass resolution to about 1 part in 5,000 for beams in this mass region.

Ga and As Beams

A liquid germanium target was developed for the production of arsenic and gallium radioactive ion beams. The target consists of enriched ^{70}Ge contained in a graphite target cell having internal dimensions approximately 9 mm diameter by 4 mm thickness. Radioactive ion beams which have been produced by bombardment of the target with 42 MeV proton beams from ORIC are ^{69,70}As from ^{70}Ge(p,xn) reactions and ^{67}Ga from ^{70}Ge(p,α) reactions.

In the first RIB experiment performed with this target, which was carried out in May-June 1997, the ORIC beam current delivered to the RIB Injector was limited to less than 4 μA due to count rate limitations of the experimental apparatus. At a proton beam current of 4 μA, a maximum of 1.4×10^{6} ions/s of ^{69}As beam was delivered to the experimental station. This corresponds to a total efficiency, defined as the number of accelerated radioactive ions delivered to the experimental station divided by the number of radioactive atoms produced, of 6×10^{-5}. The total efficiency includes the transmission through the exchange cell of 70%-80%, charge-exchange efficiency of approximately 40% for producing As^- or Ga^- ions, transmission through the tandem, including charge state fractionation due to stripping in the terminal, of about 20%, and the target/ion source efficiency. The target/ion source efficiency includes losses in the target and transfer tube plus the ionization efficiency of the ion source for producing As^+ or Ga^+ ions. Measurements at UNISOR (6) have given target/ion source efficiencies up to 0.85% for the production of $^{69}As^+$, but in this experiment, the efficiency appeared to be significantly lower, about 0.11%, for reasons which have not been determined.

Initially, the mass 69 beam was dominated by ^{69}As. However, the release time for As from the molten Ge target is long (6) compared to the half-life of ^{69}As ($T_{1/2}$= 15.2 m) resulting in a buildup of ^{69}Ge and ^{69}Ga in the target and after several days of intermittent operation, the mass 69 beam was dominated by ^{69}Ga. Since the mass difference between ^{69}As and ^{69}Ga is only one part in 10^4, no significant isobar separation could be achieved. The beam line was then tuned for mass 67 and approximately 10^5 ions/s of ^{67}Ga beam was delivered to the experimental station.

A second experiment was carried out in September 1997 with the intent of increasing the ORIC beam intensity and yield of the ^{69}As beam. A beam of ^{69}As was successfully transported to the Recoil Mass Separator but the maximum beam intensity which could be achieved was only 10^5 ions/s for proton beam intensities up to about 6 μA. As the ORIC beam current was increased to 10 μA it was discovered that the yield of ^{69}As decreased rapidly. Post-run examination of the target revealed that most of the Ge had been vaporized and deposited in the vapor transport tube connecting the target and ion source apparently due to excessive temperatures induced by localized beam heating. A new target is now being designed which will incorporate baffles intended to condense and recirculate the Ge and allow operation with higher ORIC beam currents.

Fluorine Beam Development

Aluminum oxide in various physical forms has been investigated at the UNISOR test facility in an effort to develop a target suitable for the production of ^{17}F beams (6). Release measurements on solid alumina targets showed <0.1% release of ^{18}F($T_{1/2}$=1.83 h) but in similar tests of low density fibrous material, composed of 3-μm diameter fibers, >80% release of ^{18}F was observed (7) at a temperature of 1400 °C. Since fluorine is chemically active, it reacts with elements in the target and it was found that 88% of the fluorine extracted from the ion source is in the form of AlF^+ molecules. The best target/ion source efficiency obtained thus far is about 0.2% (8) and efforts are continuing to improve the yield.

In order to be injected into the tandem, the AlF^+ must first be converted to a negative ion. Charge-exchange studies (9) carried out at UNISOR indicated that little or no AlF^- is formed. However, it was determined that the AlF^+ molecule can be dissociated and F^- ions formed with a combined molecular-breakup/charge-exchange efficiency of approximately 10%. Similar charge-exchange efficiencies were observed using both Cs and Mg charge-exchange vapors.

Based on the success of the UNISOR measurements, a first high intensity study of a fibrous alumina target was carried out on the RIB Injector in March-April 1998 using a beam of 28 MeV deuterons from the ORIC. The calculated ^{17}F production rate was 3.7×10^9/s per μA of deuterons. The maximum ion source efficiency was obtained at a deuteron beam current of about 2 μA. At higher deuteron beam currents, the total yield increased but the source efficiency decreased due, at least in part, to a rapid increase in the Al^+ beam emitted by the ion source at higher deuteron beam currents. The maximum yield of $Al^{17}F^+$ observed was $\approx 3 \times 10^6$ ions/s at a deuteron beam current of 4 μA which corresponds to a source efficiency of about 0.02%. Further increases in the deuteron beam current resulted in reduced yields and, because of the decline in source output, no attempt was made to produce a F^- beam. Post-run examination revealed that the target material had suffered significant shrinkage and deterioration.

A second high intensity test of this target material was made in August 1998. In this test, the deuteron beam current was limited to 2 μA to avoid target degradation. The yield of $Al^{17}F^+$ was $\approx 2 \times 10^6$ ions/s. After breakup of the AlF molecule and charge exchange in the charge-exchange cell, the yield of $^{17}F^-$ was measured to be $\approx 5 \times 10^4$ ions/s. This gives an efficiency for molecular dissociation and charge exchange of ≈2.5%, or a factor of 4 smaller than the value obtained in the low intensity measurements at UNISOR. Attempts to transport this small amount of $^{17}F^-$ beam through the tandem were unsuccessful.

At least part of the problem encountered in transporting the fluorine beam can be attributed to the large energy spread and emittance growth introduced by the molecular breakup process. Measurements in the isobar separator of $^{19}F^-$ beams from the breakup of $Al^{19}F^+$, produced by introducing SF_6 into the ion source, gave energy widths of ≈400 eV. The energy width is also more sensitive to the density of the charge exchange vapor, presumably due to the low energy, 16.5 keV, of the ^{19}F atoms following the molecular breakup process. The emittance growth has not yet been measured.

Studies of other metal-oxides having higher working temperatures are in progress in an effort to develop targets capable of withstanding higher deuteron beam currents. A major part of the development program for these target materials is the addition of a vapor feed system to provide a reliable source of aluminum or other metals to form metal-fluoride molecules.

A negative-ion surface ionization source is also under development (10) which would avoid the problems associated with the molecular breakup and charge-exchange processes. Measurements made on a prototype negative-ion source, using SF_6 to produce $^{19}F^-$ beams, gave an ion source efficiency of about 3.5% at the TISTF without target material present, and about 1.5% at the UNISOR facility with target material present. On-line tests at UNISOR, using low intensity 22 MeV deuteron beams from the tandem, have resulted in target/ion source efficiencies as high as 0.02% for the production of $^{17}F^-$ and 0.5% for the production of $^{18}F^-$. The difference in efficiencies for $^{17,18}F^-$ and $^{19}F^-$ is presumably due to losses associated with the decay of the radioactive species due to the hold-up time in the target. A hold-up time of 16.4 minutes (6) has been reported for release of fluorine from alumina fiber targets operated at 1470 °C in an EBP ion source, but later measurements suggest that a fast release component with a hold-up time of about 3 minutes is also present.

FUTURE DEVELOPMENT

The accelerator development plans include replacement or upgrading of the most troublesome power supplies, installation of a recirculating gas stripper in the tandem and developing $^{3,4}He$ beams from the ORIC. The ORIC trim coils, several sections of which are no longer operable due to water leaks, will be replaced if required for accelerating beams needed for RIB production.

In the area of RIB development, the immediate plan is to concentrate on the development of negative-ion sources to eliminate the need for charge exchange, and modification of the liquid germanium target holder to handle higher beam power. A 'batch-mode' target/ion-source is also being fabricated for use with longer lived radioactive species. This source will allow a target to be irradiated with beam from the ORIC, then rotated to a negative-ion cesium-sputter source for production of a RIB beam while a new target is being irradiated by the ORIC beam. The first tests of this source, using Ni targets for the production of a ^{56}Ni beam, are expected to be made late this year.

Low-density uranium carbide targets are being investigated for the production of neutron-rich RIBs through proton induced fission of ^{238}U. Low-intensity studies are being carried out at the UNISOR facility to measure fission fragment release from targets consisting of thin UC_2 coatings on carbon fibers.

ACKNOWLEDGMENTS

The author wishes to acknowledge the skillful development and efficient operation of the ORIC and tandem accelerators by the HRIBF accelerator operations staff, and the dedicated efforts of the ion source development group in the development, testing, and operation of the targets and ion sources used to produce radioactive ion beams at the HRIBF.

Research at the Oak Ridge National Laboratory is supported by the U.S. Department of Energy under contract number DE-AC05-96OR22464 with Lockheed Martin Energy Research Corporation.

REFERENCES

1. G. D. Alton, D. L. Haynes, G. D. Mills and D. K. Olsen, *Nucl. Instr. and Meth.* **A328**, 325-329 (1993).
2. T. Bjørnstad, E. Hagebø, P.Hoff, O. C. Jonsson, E. Kugler, H. L Ravn, S. Sundell and B. Vosicki, *Phys. Scripta* **34**, 578-590 (1986).
3. G. D. Alton, J. Dellwo, S. N. Murray, C. A. Reed, *Physics Div. Prog. Rpt. ORNL-6842, Sept. 1994,* p. 1-25.
4. H. K. Carter, J. Kormicki, J. Breitenbach, S. Ichikawa, P. F. Mantica, G. D. Alton, J. Dellwo, *Physics Div. Prog. Rpt. ORNL-6842, Sept. 1994,* 1-37.
5. H. K. Carter, P. F. Mantica, J. Kormicki, C. A. Reed, A. H. Poland, W. L. Croft, E. F. Zganjar, *Physics Div. Prog. Rpt. ORNL-6842, Sept. 1994,* 1-39.
6. H. K. Carter, J. Kormicki, D. W. Stracener, J. B. Breitenbach, J. C. Blackmon, M. S. Smith and D. W. Bardayan, *Nucl. Instr. and Meth.* **B126**, 166 (1997).
7. J. Kormicki, H. K. Carter, D. W. Stracener, J. B. Breitenbach, J. C. Blackmon, M. S. Smith and D. W. Bardayan, *Physics Div. Prog. Rpt. ORNL-6916, Sept. 1996,* p. 1-46.
8. D. W. Stracener, private communication (1998).
9. H. A. Schuessler and J. Lassen, private communication (1998).
10. G. D. Alton, R. F. Welton, B. Cui, S. N. Murray and G. D. Mills, *Nucl. Instr. and Meth.* **B142**, 578-591 (1998).

The Use of Lasers for the Selective Ionization of Radionuclei for RIB Generation

V.N. Fedoseyev

Institute of Spectroscopy, Russian Academy of Sciences, Troitsk, 142092, Moscow region, Russia

The method of laser resonance step-wise ionization of atoms is used for radioactive ion beams production. Two types of selective laser ion sources (LIS) are applied at on-line mass-separators: a buffer gas cell LIS and a hot cavity LIS. These sources provide both high selectivity and high efficiency of the ionization process. At the ISOLDE/CERN facility short-lived isotopes of Ag, Mn, Zn, Be, Ni, Cu were studied using a hot cavity LIS and a system of pulsed copper vapor and dye lasers. RIBs of a dozen different elements have been generated with the LIS technique, while such possibility exists for about 80% of the total number of chemical elements.

INTRODUCTION

The intensity and physical composition of radioactive ion beams (RIBs), produced at on-line mass separators, are strongly dependent on the type of ion source used to ionize the element of interest. The general requirements of the ion source are the following [1]: the ionization efficiency should be high; the element selectivity large; the intrinsic delay short; the ion beam emittance small; and the energy spread should be minimal. The laser resonance photoionization technique [2,3] fulfills these requirements. In particular, it is outstanding concerning efficiency and selectivity. As the interest in production of exotic nuclei far from stability increases, a demand of high ionization efficiency and high selectivity is becoming more important. Therefore, laser ion sources based on resonant excitation of atomic transitions by wavelength tunable lasers have been implemented at a number of RIB facilities.

Several approaches to the practical realization of the chemically selective laser ion source (LIS) have been suggested and developed. It is possible to divide these sources into three groups according to the physical environment of laser produced ions:

1) laser ionization of free atoms at low pressure in a hot cavity;
2) laser ionization in a cell, filled with a buffer gas;
3) deposition of radioactive species on a cold surface followed by laser ablation and ionization.

Up to now only the first two groups of sources have been implemented in direct connection with target units at on-line mass separators facilities. In particular, RIBs have been generated using hot cavity LISs at the IRIS facility of Leningrad Nuclear Physics Institute since 1990[4]; at CERN-ISOLDE since 1991[5,6]; and at the on-line mass separator at GSI Darmstadt since 1993 [7]. Using the gas cell LIS, the first RIB of ^{55}Ni was produced in 1994 at LISOL (Leuven Isotope Separator On Line) [8].

Laser ion sources for on-line isotope separators have recently been reviewed by Piet Van Duppen [9]. The present paper will, therefore, focus on recent progress made toward the production of RIB using laser ionization techniques. The laser ion source facility at CERN-ISOLDE will be described and shown as an example of a highly efficient and selective tool, used for a broad range of RIB physics experiments. In the following section, a brief description of LIS principles will be given.

BASIC CONCEPTS OF THE LASER ION SOURCE OPERATION

Laser Resonant Ionization

In the process of step-wise resonant photoionization of an atom, the ionization threshold E_{ion} can be overcome by consecutive absorption of several photons of energies $\hbar\omega_k$. In the case of three-step ionization, the excitation of the atom proceeds resonantly if the lasers are tuned to the transitions between atomic states with energies E_i:

$$\hbar\omega_1 = E_1 - E_0, \quad \hbar\omega_2 = E_2 - E_1, \quad \hbar\omega_3 = E_3 - E_2.$$

In the last step of excitation, a transition to the ionization continuum or to an autoionizing state can be used.

Since each element has its unique energy level structure, the step-wise ionization process is highly selective. Using lasers with narrow line-widths, isotopic and nuclear isomeric selectivity can be accomplished.

CP475, *Applications of Accelerators in Research and Industry*,
edited by J. L. Duggan and I. L. Morgan

The efficiencies of the atomic excitation and ionization processes are determined by the transition cross-sections σ_k; the laser beam intensities used in the excitations as well as the spontaneous decay times of excited states τ_k (k = 1, 2). In principle, the ionization efficiency can reach a value close to 100% [12]. Optimal conditions are provided by simultaneous irradiation of the multilevel system with a train of laser pulses of the same duration $\tau_p << \tau_k$ at the resonance frequencies ω_k. For effective excitation and subsequent photoionization of excited atoms, the energy densities of the pulses should satisfy the following saturation conditions:

$$\mathcal{E}_k \geq \mathcal{E}_k^{sat} = \frac{\hbar\omega_k}{2\sigma_k}, \qquad \mathcal{E}_3 \geq \mathcal{E}_{ion}^{sat} = \frac{\hbar\omega_3}{\sigma_3}, \qquad (1)$$

Usually atoms spend a limited transit time t_{tr} in the zone of interaction with the laser radiation. In this case, the probability of ionization is proportional to the value of the duty factor $f \times t_{tr}$, where f is the laser pulse repetition rate.

Gas Cell Laser Ion Source

In the high-pressure buffer gas cell, nuclear reaction products are decelerated during collisions with buffer gas atoms. This idea, combined with the ion-guide technique [11], has been put into the design of the laser ion source at the LISOL facility. In this source (Fig. 1) the nuclear reaction products recoiling out of the target are thermalized and neutralized in a high-pressure noble gas (He, Ar). Then the atoms are selectively ionized by the laser light and are swept out of the cell by the gas flow. Due to the ion storage capacity of noble gas, ionized species are transported to the front end of the isotope separator. The latest version of the gas cell LIS [12] is equipped with the sextupole ion guide (SPIG), which consists of six rods cylindrically mounted on a sextupole structure with an inner diameter of 2.5 mm. An oscillating voltage applied to the rods provides the confinement of the ions within the structure. The buffer gas is pumped out efficiently through the gaps between the rods, while the ions are transported to the extraction electrode with the velocity of the gas jet.

Beams of radioactive Ni, Co and Rh isotopes have been produced at LISOL using selective laser ionization in a gas cell. With the laser system, operating at $f = 400$ Hz, ion source efficiencies of up to 6.6% (Co) have been reached [13]. Ionization of atoms was effected by excitation in two saturated steps with an estimated laser ionization efficiency of 100%. The spatial overlap of the laser beams with the atomic flow (duty factor) was as high as 60%. However, there are some losses of nuclear reaction products inside the cell and during ion transport. In particular, a fraction of the atoms may diffuse to the cell wall and stick to it. It was also found that even minor impurities in the inert buffer gas lead to charge-exchange losses of laser ionized atoms. In order to suppress this loss factor, a gas purification system was used. For fission reaction products, the stopping efficiency is the main loss factor.

FIGURE 1. General layout of the IGLIS and the SPIG. V_{rf} =0-150

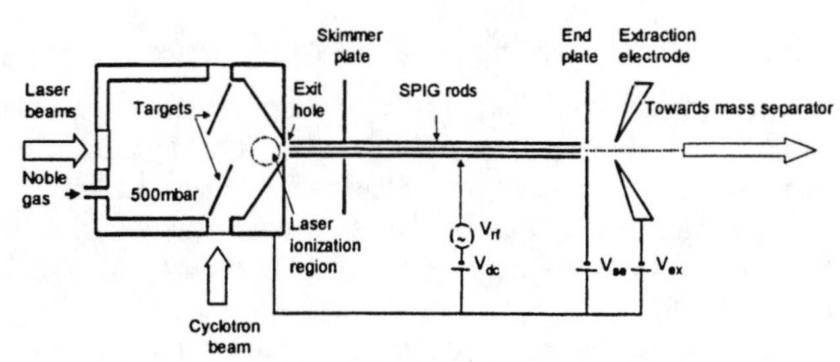

V, f_{rf}=4.7 MHz, V_{dc} and V_{se}=0-300 V, V_{ex}=0-5 kV [12].

The gas purity also strongly affects the selectivity of ionization in the cell. The selectivity S of a laser ion source is defined as the ratio of isotope yield with the lasers tuned to resonance to the yield with the lasers tuned off the resonance. For the gas cell LIS, $S > 300$ was obtained when fusion reaction products (Ni, Co) were ionized, while for Rh fission products S was found to be 50 [13]. The selectivity was mainly determined by the residual ions, that remained in the buffer gas after thermalization of recoil products.

The gas cell LIS is very attractive for studies of short-lived radioactive nuclei of refractory elements because of its short delay time (typically 25 ms). On the other hand, its application is limited to thin targets. RIBS of relatively low intensities have been produced using a laser ion source of this type.

Hot Cavity Laser Ion Source

In a hot cavity where atoms are moving through an ionizing laser beam with thermal velocities, they can pass through the laser beam many times before leaving the cavity. As a consequence, the transit time can be multiplied. This concept has been used in several experiments where high laser ionization efficiencies were needed [14-16]. This technique is well suited for on-line mass-separator applications where the ionization with a laser beam can be effected by directing the laser beam along the axis of a tubular thermal ion source (Fig. 2).

It is well known that ions in a hot metal cavity might be captured by a potential trough created in the plasma, due to strong electron emission from the surface of the cavity wall [17, 18]. In case of thermodynamic equilibrium the cavity plasma potential is defined by the temperature of the wall and by the density of ions, n_{is}, and electrons, n_{es}, near the surface [18]:

$$\Phi_p = \frac{kT}{2e}\ln\left(\frac{n_{is}}{n_{es}}\right), \quad (2)$$

The value of Φ_p may reach −2.2 eV [19], which is enough for preventing ions with thermal energies less than this value from striking the cavity walls. The ions can be extracted out of the cell by a penetrating external electrical field [7, 16], or they can be pushed towards the exit due to the internal field, by heating the tubular cavity with a DC current [4, 6].

For a cylindrical cavity of length L and internal diameter d, the ionization efficiency η is given by the expression:

$$\eta = \frac{f\eta_{photoion}}{f\eta_{photoion} + \frac{dv}{4L^2}}, \quad (3),$$

where $\eta_{photoion}$ is the laser ionization efficiency per pulse and v is the mean thermal velocity of the atoms [6]. This formula is valid if $L >> d$ and the ions are extracted with 100% efficiency.

The temperature of the cavity should be kept high enough to avoid long time adsorption of atoms on its walls. Consequently, thermal ionization of atoms may take place on the hot surface, particularly, if the ionization potential of the atom is low. The value of selectivity S can be as high as 8000 [20] or more. However, atoms with low ionization potentials may be ionized quite efficiently in the cavity, thus substantially reducing the ion source selectivity. Therefore, cavity materials with low electron work functions would help solve this problem.

It is possible to improve the LIS selectivity by making use of the pulsed structure of the ion current, produced by the lasers. The temporal behavior of ion beams has been studied in [6, 21, 22]. A gating technique has been applied for studying neutron-rich ^{129}Ag [23] with the LIS. This technique relies on chopping the ion beam synchronously with the laser pulses, through which the isobaric contamination by ^{129}In could be suppressed by a factor of three.

LASER ION SOURCE IMPACT ON THE ISOLDE FACILITY

The implementation of the LIS at ISOLDE on-line mass separator was triggered by the growing need of isobar selective RIB generation and the demonstrated ability of the laser resonance ionization technique to fulfill conditions needed for nuclear physics experiments. Institute of Spectroscopy has made important contribution to the CERN-ISOLDE research program through their strong background in nuclear physics and LIS developments.

In Fig. 2 the LIS operation is illustrated in simplified form. The laser system consists of copper vapor lasers (CVL), dye lasers, nonlinear crystals and other optical elements for manipulating light beams. Three CVLs with $f = 10$ kHz are triggered simultaneously and provide two beams with a total average power of ≤ 75 W. Two or three dye lasers can be pumped by the CVL. However, one CVL beam can also be used to non-resonantly ionize atoms from a highly excited state. To perform resonance ionization the wavelengths of dye laser radiation are tuned to atomic transitions. Nonlinear frequency doubling and tripling techniques are used to generate ultraviolet radiation necessary for the excitation of specific atomic transitions [24]. The focussed laser beams are merged together

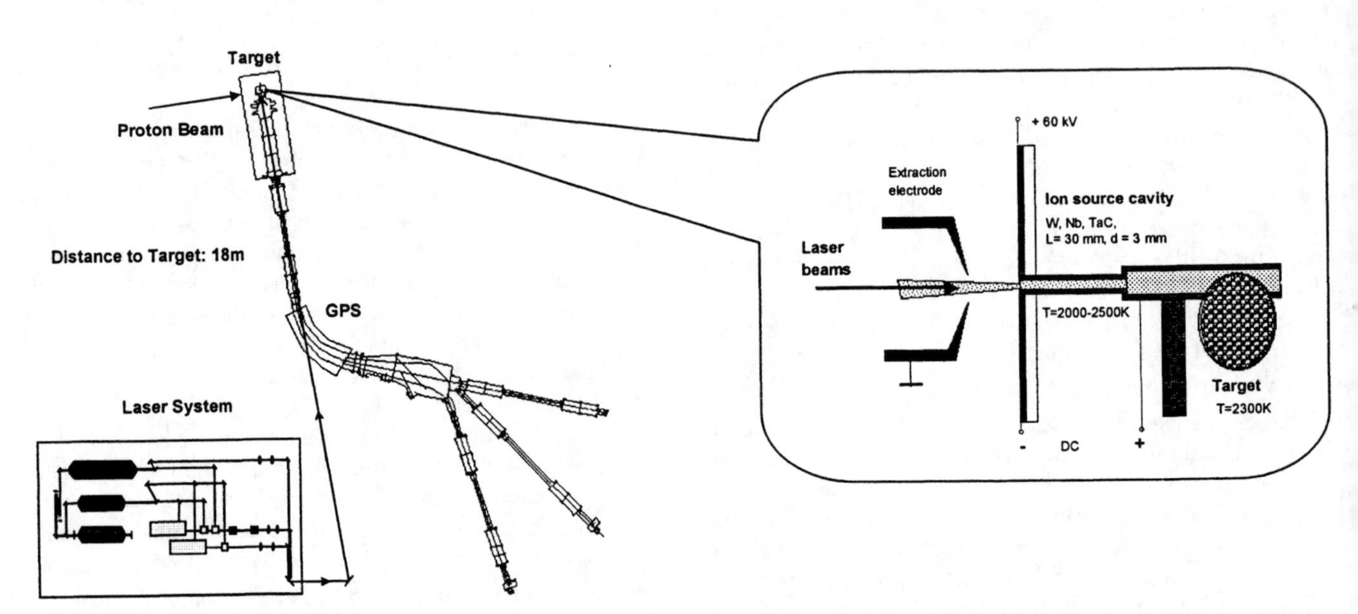

FIGURE 2. RIB generation at ISOLDE/CERN: layout of the laser system, on-line general purpose separator (GPS) and the hot cavity LIS.

and reach the ion source through a quartz window in the vacuum chamber of the separator magnet. There is an additional set of optics for turning laser beams to the front end of the High Resolution Separator (HRS). In this case, the optical path length between the laser setup and the ion source increases from 18 m to 23 m.

Several LIS cavities have been developed at CERN-ISOLDE. In principle, it is possible to use a standard and very reliable surface ionization source, usually made of tungsten. Thus, the same source can be used for generation of RIBs with or without laser ionization. This approach has high selectivity when the isobars have high ionization potentials. For suppression of surface ionization, cavities made of niobium and tantalum carbide have been developed. Increasing the ohmic resistance of the cavity gives a higher internal field gradient, which makes the ion bunches shorter. By applying the ion beam gating technique the selectivity can be even further increased. Therefore, a niobium cavity with reduced wall thickness has been designed [21].

In order to find an optimal ionization scheme for a specific element, preliminary experimental investigations are usually necessary. Although, this can be done at CERN-ISOLDE facility, access is limited due to the heavy experimental program. Therefore, data of other resonance ionization experiments can provide important and useful information. In any case, some test measurements with stable isotopes must be done before applying the LIS for RIB generation. In particular, values of LIS efficiency have been measured off-line, using small samples of stable isotopes. Results of these measurements are summarized in Table 1. Ionization efficiency measurements for nickel failed because of large amount of Ni impurity in the tantalum target container. The LIS has not yet been used for RIB production of all elements. Nevertheless, there are above 10 experiments for which the technique is applicable, which use or are waiting to use this technique for RIB generation.

As an illustration of yields obtained with the LIS, data taken from the generations RIBs of beryllium are presented in Fig. 3. The LIS has been used recently for studying the nucleus ^{11}Be [25] and for measurements of the ^{7}Be$(p,\gamma)^{8}$B cross section [26].

TABLE 1. Ion beams generated with the LIS at CERN-ISOLDE. E_{ion}-atomic ionization energies; values of ionization efficiency η were measured using calibrated samples.

Element	E_{ion}, eV	η, off-line, %	Produced RIB, mass numbers
Be	9.32	7	7, 9-12, 14
Mg	7.65	9.8	
Mn	7.44	19	49 - 69
Ni	7.64		56 - 70
Cu	7.73	6.6	57 - 78
Zn	9.39	4.9	58 - 73
Ag	7.58	14	107 - 129
Cd	8.99	10.4	98 - 128
Yb	6.25	15	

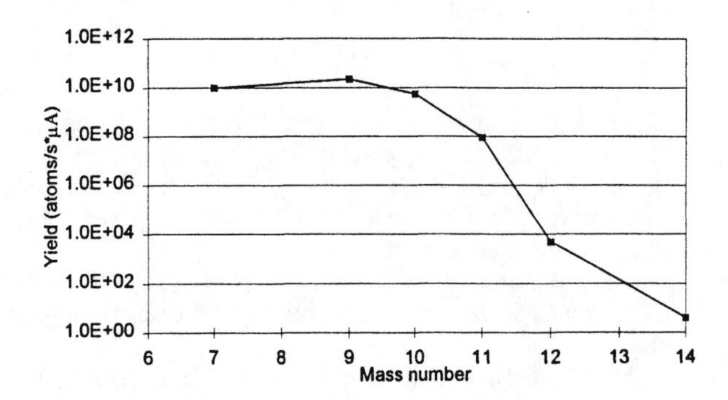

FIGURE 3. Yields of Be isotopes from the UC_2-C target of 53 g/cm^2 with the LIS at CERN-ISOLDE, 28.09.1998.

Laser resonance ionization also provides a unique possibility for nuclear isomer separation. Due to differences in the hyperfine structures of atomic lines, it is possible to tune lasers for isomer selective ionization. This feature of the LIS has considerably facilitated line recognition in the study of ^{122}Ag γ – ray spectra [23].

CONCLUSIONS

At the present time at least three on-line mass separators are generating RIB with the LIS technique. These include: IRIS at Gatchina (Ho, Yb, Tm); LISOL at Leuven (Ni, Co, Rh); and CERN-ISOLDE (see Table 1). Using existing lasers, the method of step-wise resonance ionization can be applied for an even broader range of chemical elements. Efficient ionization schemes are expected for 79 atoms, including all of the metals, with copper vapor laser systems.

LIS efficiency is comparable with efficiencies of other ion sources. With the expected progress in the laser development, it is reasonable to predict further increases in ionization efficiency, although minimizing of the ion losses during extraction also can lead to substantial additional improvements.

Most attention should be given toward the increasing the LIS selectivity, which, in certain cases, is still below desirable levels. Since laser resonance ionization is an extremely selective process, measures to avoid nonselective ion generation in the LIS must also be developed.

The hot cavity LIS is rather attractive due to its simplicity and easy coupling with RIB targets. Since the laser system can be placed far away from the hot zone, it is possible to manipulate the element composition of ion beams remotely. It is especially advantageous to apply this kind of ion source for RIB generation at future large-scale RIB facilities, in particular, high flux nuclear reactors, connected with fission fragment accelerators, where the in-pile ion sources are to be used [27].

REFERENCES

1. Ravn, H.L., and Arrardyce, B.W., *Treatise on Heavy-Ion Science, Vol.8*, ed. by Bromley, D.A., New York: Plenum Press, 1989, pp.363-439.
2. Letokhov, V.S., *Laser Photoionization Spectroscopy*, Orlando: Academic Press, 1987.
3. Hurst, G.S., and Payne, M.G., *Principles and Applications of Resonance Ionization Spectroscopy*, London: Hilger, 1988.
4. Alkhazov, G.D., et al.., *Nucl. Instr. And Meth.*, **A306**, 400-402, 1991.
5. Scheerer, F., et al., *Rev. Sci. Instrum.*, **63**, 2831-2833, 1992.
6. Mishin, V.I., et al., *Nucl. Instr. And Meth.* **B73**, 550-560, 1993.
7. Fedoseyev, et al., "Study of short-lived tin isotopes with a laser ion source", presented at the Seventh International Symposium on Resonance Ionization Spectroscopy, Bernkastel-Kues, Germany, July, 1994. Ed. by Kluge, H.-J., Parks, J.E., Wendt, K., in *AIP Conf. Proc.*, **329**, 465-467, 1995.
8. Vermeeren, L., et al., *Phys. Rev. Lett.*, **73**, 1935-1938, 1994.
9. Van Duppen, P., *Nucl. Instr. And Meth.* **B126**, 66-72, 1997.
10. Balykin, V.I., et al., *Sov. Phys. Usp.*, **23**, 651-678, 1980.
11. Ärje, J., et al., *Phys. Rev. Lett.*, **54**, 99-101, 1985.
12. Kudryavtsev, Y., et al., *Rev. Sci. Instrum.*, **69**, 738-740, 1998.
13. Kudryavtsev, Y., et al., *Nucl. Instr. And Meth.* **B114**, 350-365, 1996.
14. Andreev, S.V., et al., *Opt. Commun.*, **57**, 317-320, 1986.
15. Andreev, S.V., et al., *Phys. Rev. Lett.*, **59**, 1274-1276, 1987.
16. Ames, F., et al., *Appl. Phys.*, **B51**, 200-206, 1990.
17. Latuszynski, A., Raiko, V.I., *Nucl. Instr. And Meth.*, **125**, 61-66, 1975.
18. Huyse, M., *Nucl. Instr. And Meth.*, **215**, 1-5, 1983.
19. Kirchner, R., *Nucl. Instr. And Meth.*, **A292**, 203-208, 1990.
20. Jokinen, A., et al., ISOLDE-collaboration, *Nucl. Instr. And Meth.* **B126**, 95-99, 1997.
21. Jading, Y., et al., *Nucl. Instr. And Meth.* **B126**, 76-80, 1997.
22. Lettry, J., et al., *Rev. Sci. Instrum.*, **69**, 761-763, 1998.
23. Kratz, K.-L., "Laser isotope and isomer separation of heavy Ag nuclides: half-life of the r-process waiting-point nuclide ^{129}Ag and structure of neutron-rich Cd nuclides", presented at the International Conference on Fission and Properties of Neutron-Rich Nuclei, Sanibel Island, Florida, USA, November 10-15, 1997.
24. Erdmann, N., et al., *Appl. Phys.*, **B66**, 431-433, 1998.
25. Kappertz, S., et al., "Measurement of the magnetic moments of ^{7}Be and of the halo nucleus ^{11}Be", presented at the 2nd International Conference on Exotic Nuclei and Atomic Masses, Michigan, USA, June, 1998.
26. Hass, M., et al., "Measurement of the ^{7}Be(p,γ)^{8}B cross-section with an implanted ^{7}Be target", presented at the 2nd International Conference on Exotic Nuclei and Atomic Masses, Michigan, USA, June, 1998.
27. Köster, U., *Rev. Sci. Instrum.*, **69**, 1316-1321, 1998.

SIRIUS: A Proposal for an Accelerated Radioactive Beams Facility at ISIS

P. V. Drumm, CLRC Rutherford Appleton Laboratory, Chilton, Didcot, Oxon, OX11 0QX, U.K.

A Proposal for a future Radioactive Beams Facility has been developed based on ISIS, the world's brightest pulsed neutron source. Radioactive ions are produced by the spallation process driven by a 100 μA, 800 MeV proton beam delivered from the ISIS synchrotron accelerator. This is ten times more powerful than currently available in the world today, and follows on from the successful work performed by the **RIST** project [1] to develop a high power radioactive beam target for such a facility. The proposed design provides for both low (200 keV), and high energy (10 MeV/A) radioactive beams with high to medium mass resolution. A flexible facility layout allows the maximum simultaneous and independent use of the radioactive beams by a number of different users. An overview of the proposed facility is presented.

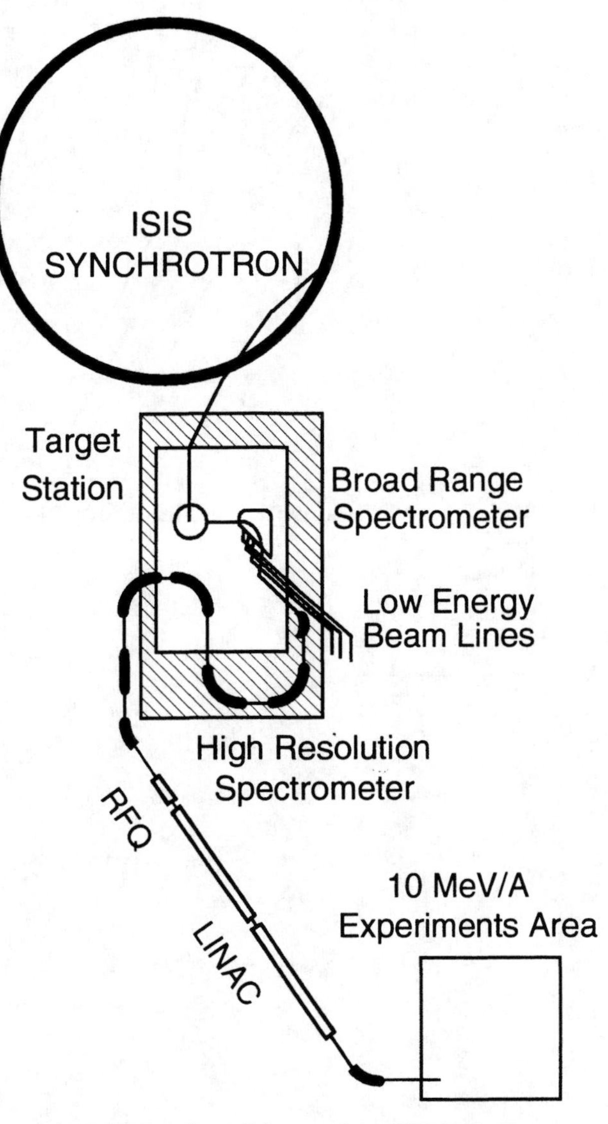

FIGURE 1. Overall layout of the SIRIUS Facility.

INTRODUCTION

The ISIS accelerator currently provides a 200 μA pulsed proton beam at an energy of 800 MeV to a neutron spallation target operating at 50 Hz. Approximately 100 μA of additional beam current could be accelerated in the ISIS synchrotron for use by other applications. One such application would be to generate intense radioactive beams by spallation on a hot target feeding an Isotope Separator On-Line facility (ISOL). An outline design of such a facility, called SIRIUS, has been studied at ISIS to determine the likely costs. The facility would provide mass separated beams at low and high energy, and provide beams simultaneously to a number of experiments. Table 1 lists the major parameters of the facility.

TABLE 1. Specification of the major parameters for SIRIUS

Proton Current	100 μA
Proton Energy	800 MeV
Proton Pulse Intensity	$3.74\text{x}10^{13}$ particles
and Frequency	16 Hz max. variable
Extracted Ion Energy & charge state	200 keV, 1^+.

SIRIUS is designed to provide four mass separated beams from a broad range spectrometer for nearly independent use since, within certain limits, the masses can be independently selected. Three of these beams are available for experiments at low energy (200 keV), while the fourth mass separated beam is available for post acceleration after passing through a high-resolution spectrometer. The proposed accelerator will produce ions at

CP475, *Applications of Accelerators in Research and Industry*,
edited by J. L. Duggan and I. L. Morgan

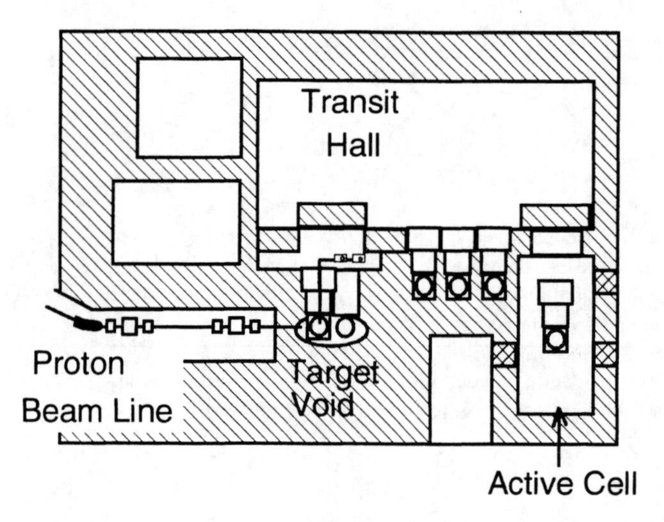

FIGURE 2. Schematic of the SIRIUS target station.

10 MeV/A for a mass range up 240 amu.

TARGET STATION

In the proposed scheme, the ISIS synchrotron would provide beams to both the existing neutron target station and to SIRIUS. Sharing of the beam with the neutron target is made on a pulse-by-pulse basis by the switching of a magnet in the synchrotron ring. The SIRIUS beam line is also furnished with a kicker magnet to switch the beam, again on a pulse basis, to one or other of a pair of target positions within the target station, Figure 2. For maximum flexibility two target positions, separated vertically by about 1.3m, are incorporated into the design but to save on costs the two targets share the same void vessel (i.e. the two targets are not shielded from each other when in position). The loss of flexibility from this is slight since only when targets are being inserted or removed does the proton beam have to be turned off - an eight hour changeover process is anticipated. The targets are held at potential inside a grounded vacuum vessel and supported on a 3m long plug carrying services and the extracted beam line. The plug becomes part of the shielding when the target is installed.

Figure 2 indicates the provision of an active handling cell and active target storage bays. The latter can be used as off-line active test stations. It would be possible to provide off-line beams to the separators from these targets if required.

The design study has investigated the operation of the target at a potential of up to 200 kV. At this voltage, the insulation problems are significant, both in air and of the insulation of the services which must reach the target through the plug. The latter is effected by a set of thick

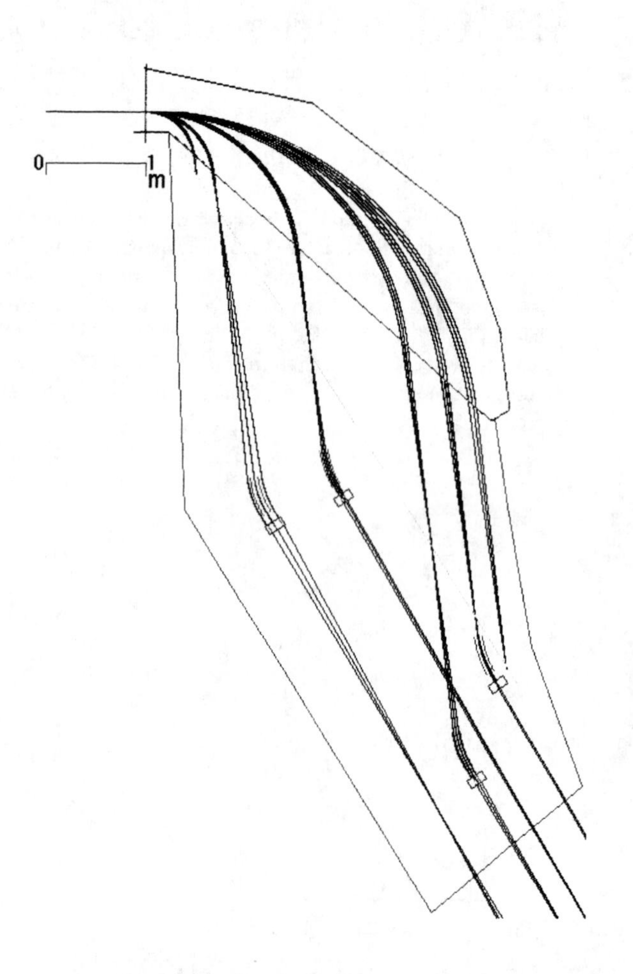

FIGURE 3. Schematic layout of the Broad Range Magnetic Separator. Four mass separated beams are extracted parallel to the focal plane.

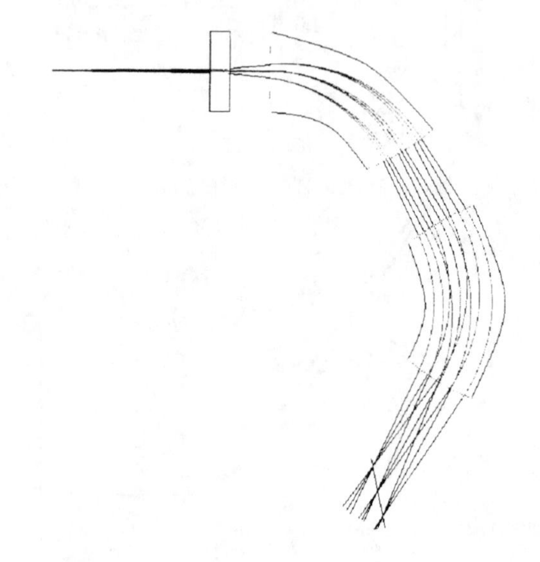

FIGURE 4. Schematic of the high resolution separator capable for 1 in 20,000 resolution.

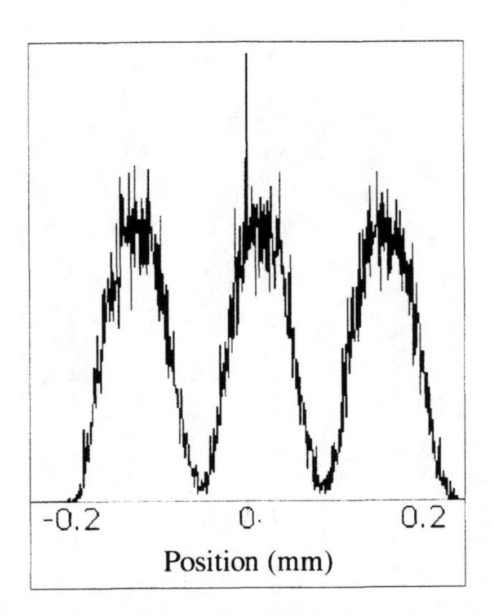

FIGURE 5. Ray traced beams of masses differing by 1 part in 30,000. The image size is ±0.2 mm with an angular divergence of ±10 mr and an energy spread of ±4 eV.

concentric ceramic insulators across which the 200 kV is gradually dropped in a controlled manner. Services at the different potentials (focusing potentials, diagnostics, heating currents etc) can then be placed in the appropriate layer of insulator. The motivation for requiring such a high potential (rather than a more usual 30-60 kV) is the provision of reasonably energetic beams to the low energy experiments where it is particularly useful for implantation studies. The consequence is that the beam lines require much larger voltages for deflection and focusing elements. An alternative approach of using lower energy beams from the target (say 30 keV) reduces the insulation problems and the voltages on the beam line elements significantly. One or more RFQ accelerators could then be used to generate variable and higher energies for implantation experiments.

The radioactive beams emerge from either of the on-line targets and are deflected in to the vertical where they are transported through the roof of the target station and into the magnetic separator vault through a system of electrostatic deflectors and quadrupole doublet lenses. The beams from the two target system merge on top of the target station before entering the first of the magnetic separators.

MASS SEPARATORS

Two magnetic separators are envisaged. The first provides simultaneously a number of beam lines from a fixed field magnet (2), Figure 3. The mass selected into each of these can be independently chosen by optics elements that slide parallel to the focal plane within the vacuum box and provide (as designed) four separated masses. The emerging beam lines are in fixed positions. Three of these beam lines are available for low energy experiments for which a floor above the linac and power supply hall and level with the separator vault is planned.

A second separator is used to obtain isotopically pure beams (M/ΔM ~ 20,000 to 30,000, Figure 4). Figure 5 indicates the ideal performance expected from this separator for beams with M/ΔM of 30,000 and with a modest energy spread (±4 eV). This has been achieved by the use of a defocusing quadrupole in front of the separator and the careful shaping of the entrance and exit faces of the magnets. In reality, controlling these fields would be difficult and they would need to be supplemented by multi-pole focusing elements Beams from this separator are fed into the accelerator or can be used for low energy experiments.

POST ACCELERATION

Initial acceleration of the beam from the high resolution mass separator is achieved with an RFQ placed on a high-voltage platform. The voltage on the platform is adjusted to (mostly) decelerate the incoming beam to an energy of 1 keV/A, and has a range of +200 to -40 keV. The beam is bunched before entering into the RFQ making the device much shorter, saving on power and floor space. Bunching of the beam is necessary since the beam from the target and ion-source is essentially continuous compared to the time scale of the RF frequency of the linac elements. The RFQ is operated at 12 MHz and is expected to cover the entire mass range operating CW at low beam intensities. Following the RFQ is a second buncher, which prepares the beam for injection into the first of many super-conducting linac cavities. Gas stripping at this point is used to raise the charge state of the beam (to A/Q~1/60). The first of the linac cavities are compact devices based on the 4 gap structures in use at *Argonne* (3). Since the beam energy per nucleon from the RFQ is affected by the presence of the HV-platform, the first resonators are adjusted to accommodate this. Independently phasing the resonators gives the maximum flexibility for accelerating the full mass range.

For higher beta values, a two gap (quarter wave) resonator can be used. Stripping foils are positioned at various points in the linac to increase the charge state and efficiency of the acceleration (with a loss of intensity).

A schematic representation of the accelerator is shown in Figure 6.

DESIGN OPTIONS & RISKS

Since the successful completion of the RIST project to design a high power tantalum target (1), there are probably no major uncertainties in the design of the target and target station, except as always the need to develop new targets as a facility develops. The remote handling aspect is considered to be fairly straight forward based on the expertise in target station design and operation available at ISIS.

However, probably the major concern is in achieving the specification of 200 kV required for beams to be used in implantation studies. The target station has been designed (with an additional margin) for operation at up to 300 keV on the extraction. Consequently, the optics required to deliver this beam to the outside world is somewhat unwieldy, requiring some 1.3 m of insulation in air and significant structures to provide the voltage to the target.

The RFQ designed to accelerate all masses in the 10-240 mass range is very ambitious and will need to be proved experimentally.

At a local level, height restrictions, and a desire to keep to proven systems forced the design into a horizontally mounted target scheme. The design of a vertically mounted target system may offer a more flexible solution.

Consideration should be given to the use of higher charge state ion-sources and the consequences for the design and cost of the linac. The present generation of thermal and hot-plasma sources used on on-line isotope separators are mainly optimised for the generation of singly changed ions. The availability at a later date of higher charge states at high efficiency could simplify and reduce the cost of the linac.

PROSPECTS

Within Europe nuclear physics with radioactive beams is developing at a fast pace. Facilities at Louvain-la-Neuve have matured and are being actively constructed at GANIL. Experience with accelerated ions will shortly be obtained at ISOLDE. In the European context the SIRIUS facility with its increased intensities can be regarded as a third generation machine.

ACKNOWLEDGEMENTS

This work has been undertaken in collaboration with colleagues in the CLRC, and UK Universities.

REFERENCES

1 C. J. Densham et al., *Proceedings 14th CAARI*, Denton, Texas, 1996, p377.10

2 J. M. Nitschke, Proc. of the Workshop on Post-Accelerator Issues at the Iso-Spin Laboratory, p 64, 1993.

3 "Concept for an advanced Exotic Beam Facility based on Atlas", 1995.

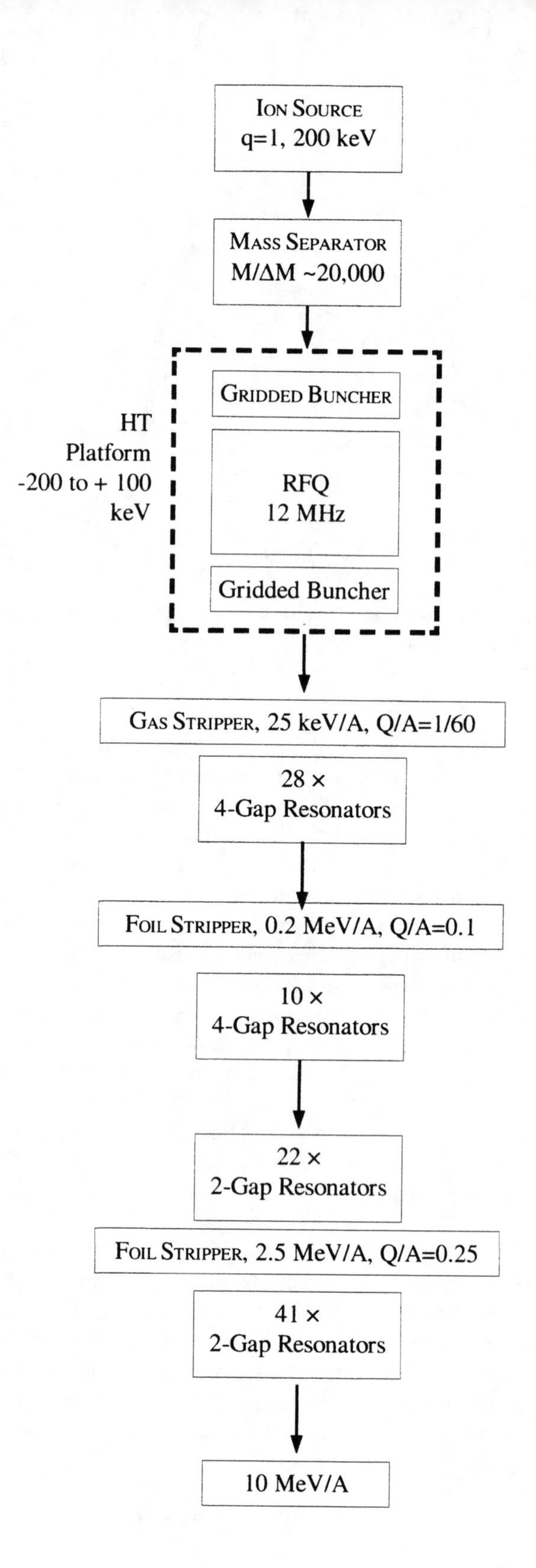

Figure 6. Schematic representation of the SIRIUS RFQ and Linear Accelerator

High Efficiency Release Targets for Radioactive Ion Beams: A Different Approach

M. Gaelens, M. Loiselet, G. Ryckewaert

Centre de Recherches du Cyclotron, Université Catholique de Louvain
Chemin du Cyclotron 2
B-1348 Louvain-la-Neuve, Belgium

At the Radioactive Beam Facility in Louvain-la-Neuve, post-accelerated radioactive ion beams are produced using a high intensity (several hundred μA) primary beam of low energy protons (30 MeV). Together with the use of an Electron Cyclotron Resonance ion source, this requires dedicated target development for each radioactive species to be accelerated, with a range of different problems and methods. Here we present examples of how different production methods are being used.

INTRODUCTION

The Radioactive Ion Beam facility at the University in Louvain-la-Neuve was started in 1989. It uses a two-accelerators configuration coupled by an Electron Cyclotron Resonance ion source. The first accelerator, which is used for the production of the radioactive species, is a low energy (30 MeV) high intensity (300 μA) proton cyclotron, named CYCLONE30. The proton beam is sent on a dedicated target from which the radioactive atoms are extracted through diffusion. The heating necessary for fast diffusion is provided by the beam power (9 kW for 300μA). By molecular flow the atoms are transported to the ECR ion source from which a low energy (10*q keV) beam is extracted. This beam is then post-accelerated by a second cyclotron, CYCLONE, to the required energy. This type of primary beam makes our facility somewhat different from other radioactive beam facilities now being planned and build.

TARGET MATERIAL CHOICE

The schematic target set-up used is shown in Fig. 1. Before the proton beam hits the target, it first passes a collimator and then a 0.5-mm graphite window that separates the target vacuum from the cyclotron vacuum. Activity diffusing from the target is pumped away to the ion source. The whole target assembly is tilted 30° face down. The beam is first tuned on the 45-mm diameter target using a small retractable collimator located before the main collimator. During operation the beam spot is moved over the target by means of a 50Hz wobbling magnet. The target is enclosed in its own cavity, details of which will be discussed later.

With our primary beam (p,n) reactions give the highest yield but some important exceptions exist. Table 1 gives all our present radioactive beams the reactions used and the yield, if known, together with the target material. The yield is defined here as the ratio of radioactive atoms produced inside the target to the number of incident protons.

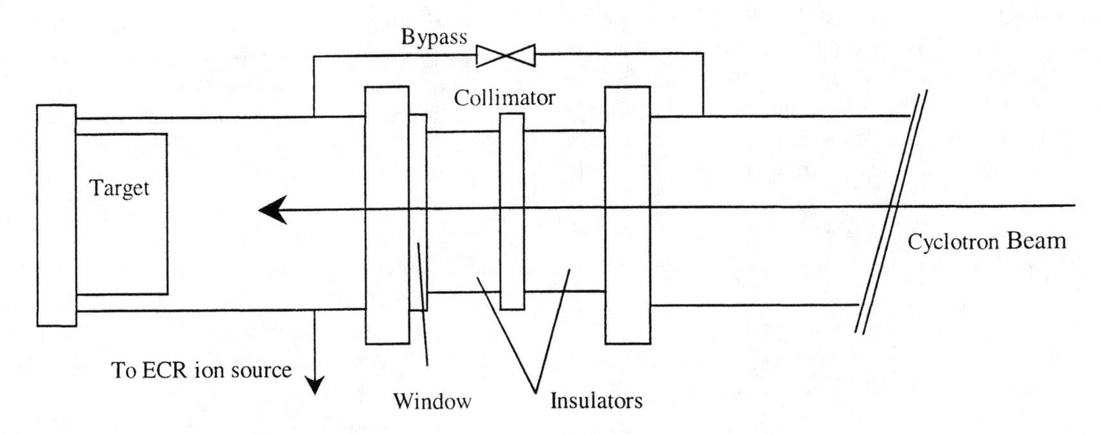

FIGURE 1. Details of the targetbox showing target, window and collimator geometry. An idea of the size is given by the targetdiameter which is 45 mm.

CP475, *Applications of Accelerators in Research and Industry*,
edited by J. L. Duggan and I. L. Morgan

TABLE 1. Available radioactive ion beams together with the targetmaterial, typical primary beam intensity, reactions used and their yield. For the definition of yield, see text.

Isotope	Half-life	Reaction	Yield (30 MeV)	Material	Proton Intensity	Comments
^{6}He	0.8 s	$^{7}Li(p,2p)$	$1\text{-}5\cdot10^{-5}$	LiF	150-200 μA	Liquid
^{11}C	20 min	$^{11}B(p,n)$	-	BN	75-100 μA	Solid, 2 reactions
		$^{14}N(p,\alpha)$	$1.9\cdot10^{-3}$	BN	75-100 μA	
^{13}N	10 min	$^{13}C(p,n)$	$1.6\cdot10^{-3}$	^{13}C	150-200 μA	Enriched, solid
^{15}O	2 min	$^{15}N(p,n)$	$1.5\cdot10^{-3}$	$B^{15}N$	--	Enriched, not used
		$^{19}F(p,\alpha n)$	$1.5\cdot10^{-3}$	LiF	200-250 μA	Liquid
^{18}F	110 min	$^{18}O(p,n)$	$2.8\cdot10^{-3}$	$H_2{}^{18}O$	12 μA	Enriched, 350 μl
^{18}Ne	1.7 s	$^{19}F(p,2n)$	-	LiF	150-200 μA	Liquid
^{19}Ne	17 s	$^{19}F(p,n)$	$1.1\cdot10^{-3}$	LiF	150-200 μA	Liquid
^{35}Ar	1.8 s	$^{35}Cl(p,n)$	-	NaCl	50 μA	Liquid

As can be seen, except for ^{6}He production, a reaction can be found that gives a high yield for the desired radioactive isotope. The target materials used to take advantage of these high-yield reactions are very diverse. They range from solids to enriched liquids and each of them requires its own specific development. The LiF and NaCl targets are in fact powder targets that melt at primary beam intensities above 100 μA. For these liquids, the target container has to be completely closed to prevent evaporation of the material and because of the inclined target orientation. The closing lid has to be strong and accept thermal stresses during inhomogeneous solid-liquid phase transitions and has also to be thin to minimize beam energy loss. Most importantly it has to be as transparent as possible for the radioactive atoms. Currently we use a 0.5-mm graphite disk that fulfills all these requirements. To prevent pressure buildup in the target cavity, the current on the target must be raised slowly to its operating value which requires even for a thoroughly outgassed target about 2 hours.

LiF is our most versatile material. In addition to the production of ^{15}O, ^{18}Ne and ^{19}Ne, it has provided the capability to produce ^{6}He through the $^{6}Li(p,2p)$ reaction which has a rather low yield. Due to the use of a diffusion-based target and a relatively long transfer line from target to ion source (~ 2 m), the short half-life of this isotope (808 ms) is close to the lower limit we can efficiently produce at our facility. Extraction efficiencies out of one version of such a LiF target is still some tens of percent and final beam intensities reach $9\cdot10^{6}$ particles per second ($^{6}He^{1+}$).

CHEMISTRY IN THE TARGET

Extraction of noble gases from the target is rather straightforward from a chemistry point of view. No molecular reactions take place that can hamper or promote diffusion and extraction. However, physical phenomena like surface sticking or trapping can be present and influence the efficiency of the target.

Much more important in our case is the chemistry of extraction of elements other than noble gases. In the case of ^{13}N, the major mode of extraction is in the form of $^{13}N^{14}N$ molecules. For that to occur, we have to add continuously a small flow of $^{14}N_2$ to the target to maintain good extraction efficiency. With this method we achieved extraction efficiencies of about 25%.

Also for the extraction of ^{11}C from the BN target a support gas of O_2 is required because the extracted occurs in the form of CO and CO_2. However, at high temperatures the oxygen starts to react with BN to form layers of B_2O_3 that largely cancels the effect of adding the gas. Such large amounts have to be added that it becomes problematic to maintain good ionization efficiencies in the ECR source. These targets are being studied using externally heated, scaled-down versions. (1)

Another example of chemistry in radioactive atom extraction is the case of ^{15}O. This isotope is produced in the LiF target through the $^{19}F(p,\alpha n)$ reaction which has a yield of $1.6\cdot10^{-3}$ for 30 MeV protons. Early tests to extract it from the target however showed almost zero extraction efficiency. To study the chemistry of this particular case also the externally heated scaled-down target was used in combination with a low intensity primary beam. The gases released from the target were trapped at liquid nitrogen temperature and the activity in the trap was monitored. With this configuration we clearly saw that the extraction of ^{15}O was occurring but at a very slow rate. This is shown in Fig. 2A where the activity coming from the target is given as a function of time. A 2-minute irradiation at 50 nA was performed. The high count rate during this period is due to increased background. When the irradiation was stopped, the decay of ^{19}Ne, also produced in the LiF target through the $^{19}F(p,n)$ reaction (see Table 1), was observed and some activity was extracted with longer half-life.

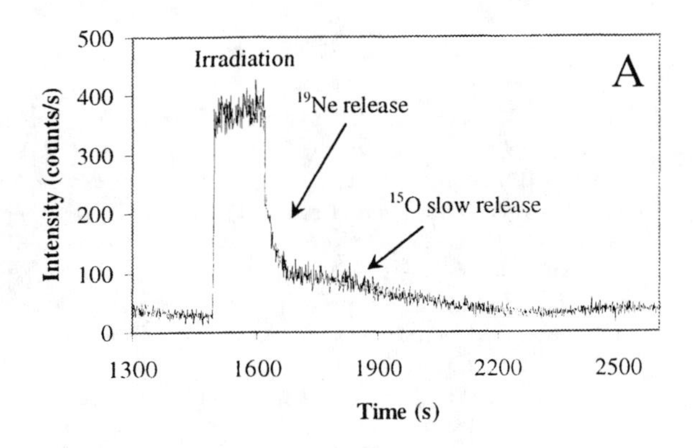

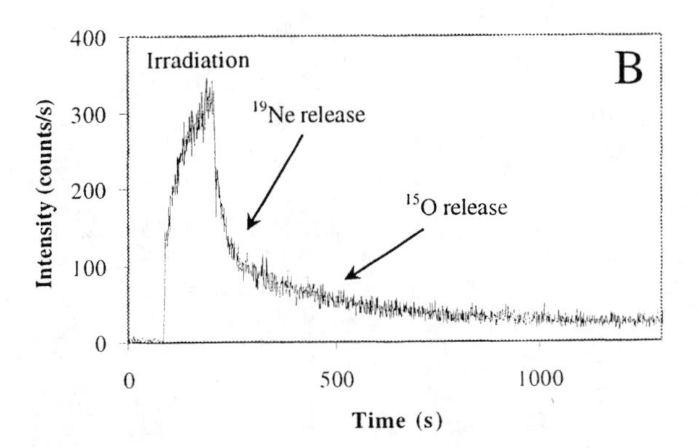

FIGURE 2. Comparison of release of ^{15}O from a LiF target without carbon added (A) and with a 50%/50% mixture of LiF and carbon (B). The target temperature in both cases is 300° C. Timescale offset is arbitrary.

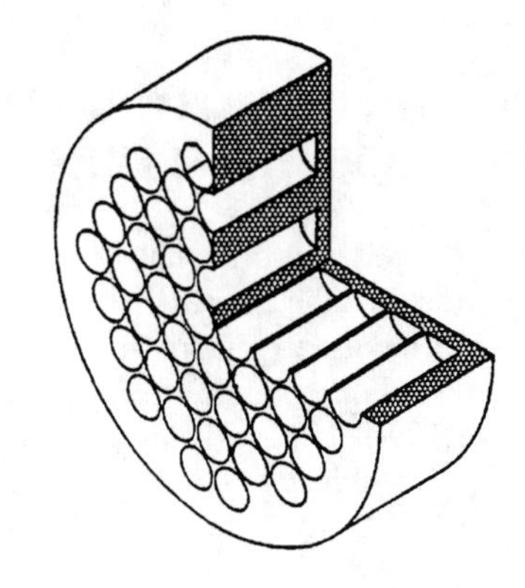

FIGURE 3. The full-size target used for ^{15}O. The diameter is 50 mm, height 17 mm and 48 holes of 5 mm diameter are present.

Isotopes with a short half-life (^{18}Ne, ^{6}He) could not be seen because of the losses in the 6 m transferline from the target to the trap. This activity was ^{15}O but its measured half-life was distorted because during the measurement, several tens of seconds after irradiation stop, there was still oxygen being release from the target. Because of this slow release, most of the ^{15}O was lost in the target. As oxygen has to be extracted in molecular form, at least two different options are readily apparent, namely $^{15}O^{16}O$ and $C^{15}O/C^{15}O_2$ with the carbon provided by the target holder. The $^{15}O^{16}O$ case was rejected early because even with only partially outgassed targets, extraction was very poor, this in contrast with the ^{13}N case where partially outgassed targets provided at first enough nitrogen for efficient ^{13}N extraction. To test for the CO/CO_2, case we mixed carbon powder with the LiF and again looked at ^{15}O extraction. This time the extraction was much faster confirming that CO or CO_2 formation is the primary means of extraction (see Fig. 2B). The speed of diffusion was apparent already during irradiation, as oxygen was arriving at the trap after a few seconds during irradiation. However, at this stage the release efficiency was still very low (< 10^{-3}). For the full-scale model, shown in Fig. 3, we opted for a graphite matrix with numerous holes filled with LiF. In this way, we provided a much larger graphite area than the standard target and reduced the mean distance from within the LiF to the nearest graphite surface. Also this design is structurally more stable and assures a homogenous LiF-Carbon distribution even with LiF in the liquid phase. Once tested, this target proved to be efficient in ^{15}O extraction (up to 12% efficiency) and robust (tested up to 8.1 kW of beam power). The extraction mode at high temperature was determined to be CO rather than CO_2, in accordance with thermodynamical data.

However, every target design has to be studied separately and chemistry often intervenes at unexpected places. For example the design of the carbon matrix for ^{15}O extraction was thought to be a good starting point for a target for the production of ^{14}O. By replacing the LiF with BN for ^{14}O production via the $^{14}N(p,n)$ reaction (yield $1.1 \cdot 10^{-4}$ for 30 MeV protons), the same extraction chemistry of CO formation could be used. However, it appeared that at high temperatures BN layers on graphite prevent the oxidation of the graphite by forming a robust barrier of B_2O_3. In this way any oxygen produced in the BN can not reach the graphite surface and no $C^{14}O$ formation can take place. Indeed, when tested, there was no evidence of ^{14}O escaping from the target. Also other nitrides like AlN and SiN have this same oxidation preventing properties. This chemistry problem leads us to choose CYCLONE as the primary beam producer : it can also deliver ^{3}He beams which allow the use of the $^{12}C(^{3}He,n)^{14}O$ reaction. In this way the ^{14}O is readily extracted as it is imbedded in graphite at the moment of its formation. A target specifically designed for this and used in low-energy (50 kV) ^{14}O experiments will be used (1,2). Post-acceleration will be performed by the new K=44 cyclotron (3).

PRODUCTION OF ISOTOPES WITH LONGER HALF-LIVES

In specific cases where the half-life of the desired isotope is "long" (larger than approximately 1 hour), still other techniques can be used to increase the efficiency of the beam production. For example, we post-accelerated a ^{18}F beam (half-life 109.7 min) by using off-line chemistry (4). As is well known, fluorine is a very reactive element and difficult to extract from targets. Because of its relatively long half-life we developed on off-line chemistry process to convert the ^{18}F produced in an enriched $H_2{}^{18}O$ target into $CH_3{}^{18}F$ (fluoromethane). This chemistry which is totally automated so that high intensities can be handled (up to 500mCi) takes about 45 minutes and is 50 % efficient, excluding decay-losses. The $CH_3{}^{18}F$ is trapped in a liquid nitrogen cooled volume and is transferred from the chemistry processing lab to the ECR ion source under He pressure. There it is released in the source through a thermally controlled needle valve. The whole process of production, chemistry and transfer takes about 3 hours and is timed as to provide a nearly continuous beam for the user with intensities of up to 10^6 pps.

CONCLUSIONS

The fact of having a different approach to produce radioactive ion beams, namely the use of low-energy, high intensity proton beams, results in specific problems during the development of new beams. Targets are very dedicated because the target material choices are limited. Moreover, for elements other than noble gases, the chemistry in the target is important and limits even further the freedom in target design, often requiring specific designs. Isobaric contamination is inherent in our design but is successfully eliminated with the separation capability of the post-accelerator. In all, our approach is somewhat limited in the number of different beams that can be produced, especially far from stability, but very high intensities of post-accelerated radioactive ion beams can be obtained and total efficiencies can be large.

ACKNOWLEDGEMENTS

This report presents results of research funded by the Belgian Program on Interuniversity Poles of Attraction (PAI) initiated by the Belgian State, Federal Services of Scientific, Technological and Cultural Affairs and by the Institut Interuniversitaire des Sciences Nucléaires (IISN).

REFERENCES

1. Decrock P., Huyse M., Van Duppen P., Baeten F., Jongen Y., *Nuclear Instruments and Methods* **B58**, 252-259 (1991)
2. Gaelens M., *Production and Use of Intense Radioactive Ion Beams: ^{14}O as a case study*, Leuven: Ph.D. Thesis, 1996
3. Loiselet M., Barue Ch, Berger G., Breyne D., Colson J.M., Daras Th, Gaelens M., Goffaux H., Lannoye E., Postiau N., Ryckewaert G., Jacobs L., presented at the 15th International Conference on Cyclotrons and their Applications, GANIL, Caen, France, 14-19/06/1998
4. Cogneau M., Decrock P., Gaelens M., Labar D., Leleux P., Loiselet M., Ryckewaert G., *Nuclear Instruments and Methods* **A420**, 489-493 (1999)

Status of the REX-ISOLDE Project

O. Kester, D. Habs, T.Sieber, H. Bongers, K. Rudolph, A. Kolbe, P. Thirolf
Sektion Physik, LMU München, D-85748 Garching, Germany

G. Bollen, I. Deloose, A.H. Evensen, H. Ravn
CERN, CH-1211 Geneva 23, Switzerland

F. Ames, P. Schmidt, G. Huber
Johannes Gutenberg Universität, D-55099 Mainz, Germany

R. von Hahn, H. Podlech, R. Repnow, D. Schwalm
MPI für Kernphysik, D-69029 Heidelberg, Germany

L. Liljeby, K.G. Rensfelt
Manne Siegbahn Laboratory, S-10405 Stockholm, Sweden

F. Wenander, B. Jonsson, G. Nyman
Chalmers University of Technology, Gothenburg, Sweden

A. Schempp, K.-U. Kühnel, C. Welsch
J. W. Goethe-Universität Frankfurt, D-60325 Frankfurt, Germany

U. Ratzinger
GSI, D-64220 Darmstadt, Germany

P. van Duppen, M. Huyse, L. Weismann
K.U. Leuven, B-3001 Leuven, Belgium

A. Shotter, A. Ostrowski, T. Davison, P.J. Woods
University of Edinburgh, GB-Edinburgh EH9 3JZ, Scotland

and the REX-ISOLDE collaboration

The Radioactive beam Experiment (REX-ISOLDE)(1,2,3) at ISOLDE/CERN is under progress and first tests are carried out with some of the structures. The radioactive ions from the online mass separator ISOLDE will be cooled and bunched in a Penning trap, charge bred in an electron beam ion source (EBIS) and finally accelerated in a short LINAC to a target energy between 0.8 and 2.2 MeV/u. The LINAC consists of a radio frequency quadrupole (RFQ) accelerator, which accelerates the ions up to 0.3 MeV/u, an interdigital H-type (IH) structure with a final energy between 1.1 and 1.2 MeV/u and three seven gap resonators, which allow the variation of the final energy. All components of the experiment are now in production or undergo first test measurements. Such measurements are ion capture tests of the trap, electron beam tests of the EBIS, low level measurements and first power tests of the RFQ and the first 7-gap resonator. In this paper the status of the experiment, and the proposed schedule are presented.

INTRODUCTION

REX-ISOLDE is a first generation radioactive nuclear beam (RNB) project aiming at two main goals:

The demonstration of a new concept to bunch, charge-breed and post-accelerate single charged, low energetic ions in an efficient way.

The study of the structure of very neutron-rich Na, Mg, K and Ca isotopes in the vicinity of the closed neutron shells N=20 and N=28 by Coulomb excitation and neutron transfer reactions (4) with a highly efficient γ- and particle-detector array MINIBALL (5).

The experiment dwells on established techniques and state of the art machines, but represents a new way of combining these structures. In contrast to other RNB-facilities, REX-ISOLDE is an experiment which is built up at the existing ISOL-facility ISOLDE at CERN (6). At ISOLDE the technique to produce low energetic radioactive beams and the handling of high radioactivity of the target have been developed over many years. 680 isotopes from 70 elements have been produced with different sources like the surface ionisation source, the plasma source and the new laser ion source (7,8).

In the first experiment the accelerated ions will be used to study the dynamic properties and to examine the shapes of very neutron rich nuclei close to semimagic shells. Due to a rather low final energy of about 2.2 MeV/u the first experiment is limited to light nuclei ($A \leq 50$) to reach the Coulomb barrier. In contrast to the structure of stable nuclei, which are thoroughly investigated experimentally and theoretically, the structure of nuclei close to the neutron dripline

CP475, *Applications of Accelerators in Research and Industry*,
edited by J. L. Duggan and I. L. Morgan

is not well established yet. REX-ISOLDE provides a test of the shell model over a wide range in isospin. Different predictions from the nuclear shell model are obtained, when going from the valley of stability to the neutron rich region. One prediction is the change of the nuclear potential from a box shaped Woods-Saxon potential to a parabolic potential (9). This modification leads to changes of the magic numbers and hence to new predictions of regions of nuclear deformations. Options for extensions to higher and lower beam energies than the region of 0.8-2.2 MeV/u are on hand, which opens up a broad field of future experiments in nuclear spectroscopy, nuclear astrophysics and solid-state physics.

THE CONCEPT OF POST ACCELERATION

The basic concept of REX-ISOLDE is sketched in Fig.1. Singly charged ions coming from ISOLDE with 60 keV energy are retarded by the platform potential of 60 kV of the Penning trap and injected continuously into the trap, where they are accumulated and cooled. After 20 ms which are defined by the required breeding time, bunches of 10 μs length will be re-accelerated to 60 keV and transferred to an electron beam ion source (EBIS). After charge breeding (10-15ms) to a charge-to-mass ratio > 0.22 the ions are injected into a radio frequency quadrupole (RFQ) accelerator via an achromatic mass separator (10,11). Due to the required low injection energy (5 keV/u) of the RFQ the EBIS platform potential has to be lowered from the 60 kV retarding potential down to 20 kV while charge breeding takes place. After 15 ms charge breeding time the ions are expelled out of the EBIS in 100 μs.

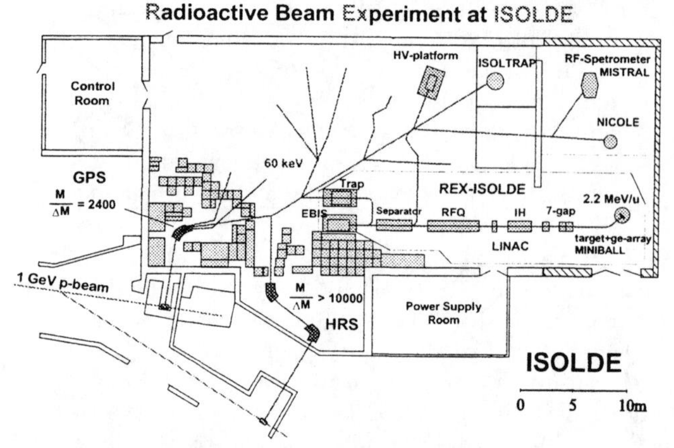

FIGURE 1. Schematic of REX-ISOLDE in the ISOLDE hall

The time structure of the experiment is shown in Fig.2. The 1 GeV proton pulses from the PS booster have a repetition time of 1.2 s. The release time due to the diffusion rate of the ions in the ISOLDE target determines the time dependence of number of ions which reach the Penning trap. The bunching of the pulses from the EBIS is done by the RFQ where the pulses at the exit will have a length of 0.8 ns and a distance of 10 ns. The accelerator consists of a RFQ, an interdigital H-type (IH) drift tube accelerator followed by three seven-gap resonators. The 7-gap resonators and the IH-structure will allow to vary the energy at the target between 0.8 and 2.2 MeV/u to meet the experimental requirements. The LINAC operates at a resonance frequency of 101.28 MHz and max. 10% duty cycle.

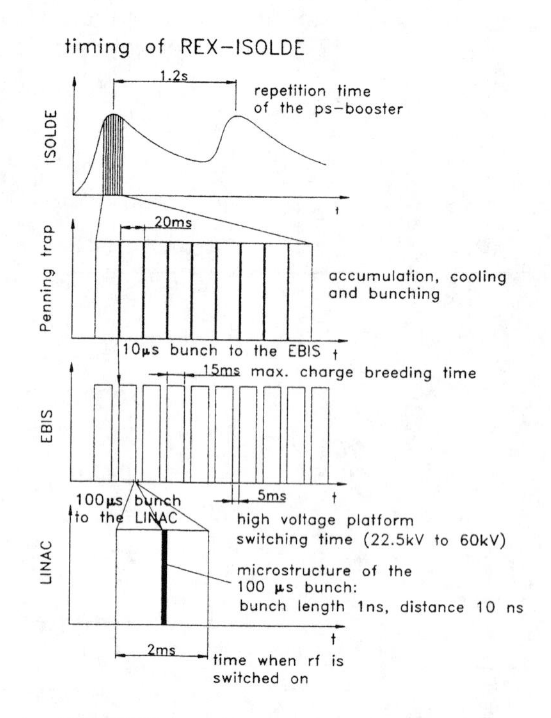

FIGURE 2. Time structure of REX-ISOLDE

The choice of a Penning trap as accumulator for the continous beam of 1+ ions from the ISOLDE mass separator, avoids a cost expensive cw-LINAC and cooling problems in the structures. The charge multiplication in the EBIS allows the acceleration of the light ions to the Coulomb barrier with a LINAC which has a length of only 10m.

PRESENT STATUS OF THE PROJECT

As most of the hardware is now completed or in production, an overview of the project status and a more detailed schedule can be given. It is planned to install the mass separator and the LINAC in the next ISOLDE shut down period 1999. First beams are expected in the middle of 1999, when all power transmitters are delivered.

The Penning trap

Beside the bunching and accumulation of the ions, the cooling of the ions in order to decrease the ISOLDE emittance at 60 keV of 30 π mm mrad down to 5 π mm mrad which is the EBIS acceptance is a crucial task of the trap (12). The ions are injected into the 1 m REX-trap, where they are slowed down by collisions with the atoms of a buffer gas. The energy loss of the ions during a single oscillation through the trap has to be comparable to the initial energy spread of the ISOLDE ions to reach capture efficiencies close to 100%. A special sideband cooling technique of the magnetron motion in the trap (13,14) will be used to separate the radioactive ions from unwanted species. The Penning trap is fully assembled on the high voltage platform including the differential pumping stages which ensure good vacuum conditions in the beam line. High tension tests with the equipment and first tests of ion injection into the trap have been carried out in order to test the injection optic. The electrode structure consisting of gold plated copper rings isolated by ceramic spacers is assembled and installed in the solenoid bore. The central field strength is 3 T and the gas pressure inside the trap 10^{-3} mbar. A plasma ion source has been installed to test the ion capture with a stable 60 keV $^{40}Ar^{+}$ beam in October 1998. In addition all the structures of the experiment may be tested with ions from this source.

The REX-EBIS

The simplicity, efficiency and costs of the accelerator are directly related to the charge state of the ions which have to be accelerated. Fast and efficient charge breeding is required to fulfill these requirements. For low intensities ($<10^{9}$ions/s) the EBIS is the best source for highly charged ions. In contrast to a plasma ion source, an EBIS uses monoenergetic electrons from an electron gun focused by a strong magnetic field to produce highly charged ions (15). The parameters of the REX-EBIS are: A current density of 200 A/cm^2, a beam current of 0.5 A, a 1.5 m Solenoid providing a trap length of 0.8 m with a tip field of 2 T and a beam energy between 5 and 10 kV. According to these parameters the charge to mass ratio of 1/4.5 is reached for Na in 6 ms and for K in 8 ms. In contrast to the Penning trap with 10^{-3} mbar buffer gas pressure, the EBIS requires a pressure lower than 10^{-10} mbar. Therefore several differential pumping stages along the transfer line between trap and EBIS are planed. Concerning a partial residual gas pressure of $5*10^{-13}$ mbar Ar the production of highly charged Ar ions is comparable to 10^6 injected radioactive K ions. However, the intensity of the radioactive ions is much smaller than of the ions from residual gas. The optical design of the transfer line is completed and the electrostatic triplet lenses will be built together with the lenses of the mass separator in Munich. The construction of the kicker-bender of the transfer line is completed as well and will be given to the workshop. After several repairs of the REX-EBIS solenoid by the manufacturer due to damages occurring during quench tests, the EBIS magnet is now shipped back to CERN and will be tested soon. The vacuum system is completed and assembled at Stockholm. The electron gun, the collector and the electrode system are completed and will be shipped to CERN.

The q/A-separator

The yield from ISOLDE for isotopes from Na, Mg, K and Ca can be up to 100 times lower than the amount of residual gas ions from C, N, O and Ar coming out of the EBIS. Due to the potential depression of the electron beam, the extracted ions will have an energy spread, which limits the q/A-resolution of a magnetic separator. In order to handle these problems an achromatic separator system in an "S"-shape form has been designed and is currently in production. In addition the separator beam line has to match the beam behind the mass slit into the acceptance of the RFQ, it has to prepare a 60 keV*q beam which can be transfered back to the ISOLDE main beam line to deliver ions for experiments in solid state physics. This transfer is done in front of the RFQ, which can be seen in Fig.1. The achromatic separator consist of an electrostatic and of a magnetic bender. The electrostatic bender compensates the energy dispersion of the magnetic bender. To reduce the energy dispersion at the mass slit to zero the second order aberations have to be corrected by appropriate curvatures of the pole faces of the magnet. The third order aberations are diminished by an electrostatic octopole, which is placed at the beginning of the system due to space restrictions (10). The resolving power of this system depends on the emittance of the injected beam. In order to provide a resolving power of 150, which is sufficient to separate all isotopes from residual gas ions, the EBIS emittance has to be lower 10 π mm mrad. The electrostatic elements of the separator are under construction or already in the work shop like the quadruplet lens in front of the RFQ. The magnet with a maximum field of 0.2 T is ordered and the platform which carries the lenses and the deflector is under construction. The construction of the deflector and the tank are completed. The separator and the transfer beam line between trap and EBIS will be installed in the beginning of 1999.

The LINAC

The linear accelerator of REX-ISOLDE has to provide a flexible beam energy for the different ion species and energies below and at the Coulomb barrier. Additional to the energy range given above, there is the possibility to trace a beam from the RFQ through the whole LINAC towards the target so that experiments with a 300 keV/u beam can be performed (16). A future upgrade of REX-ISOLDE would require a higher resonator voltage of the IH-structure than the 5 MV which presently requires 60 kW. Due to the high shunt impedance of the structure and the maximum power of the amplifier of about 100 kW such an upgrade is possible.

The 4-rod-RFQ

The 4-rod-RFQ will accelerate the radioactive ions with a mass-to-charge ratio between 3-4.5 from 5 keV/u to 300 keV/u. The required rod voltage is 28-42 kV (17) resulting in a rf power consumption of 30-40 kW. Since the lay-out of the RFQ is very conservative the maximum voltage provides acceleration of ions with charge-to-mass ratios up to 1/6.5.

FIGURE 3. Picture of the REX-RFQ

The mechanical construction of the REX-RFQ is similar to the RFQ1 of the Heidelberg high-current injector (18).

The resonator has a length of 3m and carries 18 support stems. Figure 3 shows the fully assembled REX-RFQ. The water cooled ground plates, the stems and the mini-vane like quadrupole electrodes are mounted in the tank. The capacitive plungers are installed, cabled and tested. The vacuum system has been installed and first vacuum tests have been performed. Low level frequency tuning and flatness measurements have been done. The measured flatness without the end cells is below 1%. The Rp value is 170 kΩm with a quality factor of 3900. To provide an ion beam for tests in Munich a duoplasmatron ion source from the Munich tandem laboratory is used, which delivers 1mA He^{1+} ions. The power amplifier for the RFQ will be delivered in November, when the first beam tests will be performed.

The matching section

To match the beam into the acceptances of the IH-structure a section consisting of two magnetic quadrupole triplet lenses and a rebuncher is required. The first triplet lens focuses the beam through the rebuncher and produces a waist for diagnosis. The second triplet lens matches the beam to the transverse acceptance of the IH-structure. The aperture of the lenses is 3 cm. The design of the matching section is finished and the lenses with a maximum gradient of 60 T/m are ordered. The rebuncher is a three gap split ring resonator with a peak voltage of 50 kV which has to match the phase spread of $\pm 15^o$ to the phase spread acceptance of $\pm 10^o$ of the IH-structure. An 1:1 model of the resonator has been built and tuned to the proper frequency. The model measurements resulted in a quality factor of 3500 and a shunt impedance of 4.5 MΩ/m. The final drift tube assembly and the tank have been manufactured and will be assembled for first power tests.

The IH-structure

The IH-structure accelerates the ions from 0.3 MeV/u to an extraction energy between 1.1 and 1.2 MeV/u. This structure is a short version of the IH-tank1 of the CERN LINAC III (19) and consists of a center frame which carries the drift tubes and two half shells carrying cooling jackets. For the transverse focusing in the first acceleration section an inner tank quadrupole triplet lens is foreseen, which is completed and will be installed in the water cooled vacuum housing. The drift tubes and the stems are completed and will be brazed and copper plated together with the triplet lens. The piston tuners are in production and will be completed until the end of 1998. The IH-tank will be copper plated in the end of the year. Figure 4 shows the components of the IH-vacuum tank. Delivery is scheduled in November. The whole structure has 20 gaps and a total length of 1.5 m. The change of the final energy of the IH-structure can be achieved by adjusting the gap voltage distribution via capacitive plungers. To lower the final energy (1.1 MeV/u) of the IH-structure is nescessary for deceleration of the ions down to 0.8 MeV/u. This change of the final energy via the piston tuners has been examined once more by a detailed MAFIA model of the power resonator. The calculations are in very good agreement with measurements taken from the 1:2 down scaled model of the resonator (20) and show that the design gap voltage distributions can be achieved in the power resonator for both final energies with the proper resonance frequency. With the existing MAFIA model the eigenfrequency tuning of the power resonator via the half shell height can be done accurately and gives advice for the final tuning at the manufacturer.

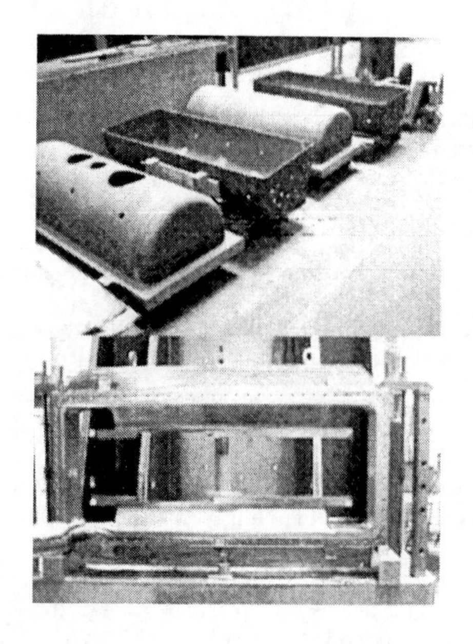

FIGURE 4. Components of the IH-vacuum tank

The 7-gap resonators

The high energy section of the REX-ISOLDE LINAC consists of three 7-gap resonators similar to those built for the high-current injector at the MPI für Kernphysik in Heidelberg (21).

FIGURE 5. Half shell and tank of the first 7-gap resonator

Each resonator has a single resonance structure. It consists of a copper half shell to which three copper arms are attached on each side which are made of hollow profiles which surround the drift tubes and carry the cooling water. The rf-power will be fed into the resonator near one of the three legs of the half shell, where the magnetic flux is maximum. Model measurements show that 1.75 MV resonator voltage can be achieved with 90 kW rf-power. The model measurements are finished and the components of the power resonators are in production or already completed. The tank and the resonant structure of the first 7-gap cavity are shown in Fig.5. The first half shell is brazed, cleaned and polished. The first 7-gap resonator and the vacuum system will be assembled now and with the delivery of the first power amplifier in October 1998 power and beam tests can start.

The MINIBALL array and the target area

The MINIBALL γ-detector array (22) consists of a new generation of Ge-detectors with large full-energy peak efficiency to make an optimum use of the expensive radioactive beams. With its compact arrangement it is mainly suited for detecting events with small γ-ray multiplicities (5). In the final system the array will consist of 6 clusters with 7 individually encapsulated 6-fold segmented Ge-detectors which are a new development of Eurisys. A prototype has been tested and had shown a resolution of 2.1 keV at 1.33 MeV γ-energy. First measurements with a flash-ADC system to determine the position where the γ enters the detector have been performed (23). A new design of the detector arrangement has been done with smaller clusters of detectors (3 and 4 detectors). It is possible to combine a cluster of 3 detectors with a cluster of 4 detectors to a 7 fold cluster, but in a first stage an array with 3-fold clusters can be assembled with a good photo peak efficiency. Currently 3 detectors in Munich, 2 detectors at the MPI-K in Heidelberg and 2 detector in Cologne are operational. The favoured electronics for the MINIBALL is developed by the company XIA and will be comparable to that of the GRETA project in Berkeley, USA. In contrast to former Ge-detector electronics analog circuits are replaced by digital electronics using flash ADC's and DSP's. The new data acquisition system for the MINIBALL has been tested successfully in several experiments at Munich with different detector configurations and will include the new electronics hardware so that for the first MINIBALL measurements at the end of 1999 a powerful detector and acquisition system will be present. The target chamber is completed and has been vacuum tested. Prototypes of the particle detectors (double sided silicon strip detector DSSSD) have been produced and tested with an alpha-source. The pre-amplifiers have been manufactured and tested as well as cablings and connectors. Tests of the multiwire gas chamber behind the target chamber which is the beam position and intensity monitor has been tested in Heidelberg with heavy ions.

ACKNOWLEDGMENTS

This experiment is partly funded by the BMBF under contract No. 06HD802I and No. 06LM868I(4).

REFERENCES

1. Habs, D., et al., *Nucl. Phys.* **A616**, 29c-38c (1997).
2. Habs, D., et al., *Nucl. Instrum. and Meth.* **B126** 218 (1997).
3. Habs, D., et al., *Nucl. Instrum. and Meth.* **B139**, 128 (1998).
4. "Radioactive beam Experiment at ISOLDE: Coulomb Exitation and Neutron Transfer Reactions of Exotic Nuclei", *Proposal to the ISOLDE committee*, CERN-ISC94-25.
5. Habs, D., et al., *Prog. Part. Nucl. Phys.* **38**, 1-13 (1996).
6. Albrow, M., et al., "The ISOLDE Facility at the PS-Booster", CERN-PSCC-89-29
7. Ravn, H., et al., *International Workshop on Research with Fission Fragments*, World Scientific Singapore, 1997, pp. 62-xxx.
8. Mishin, V.I., et al., *Nucl. Instrum. and Meth.* **B73**, 550 (1993).
9. Dobaczewski, J., et al., *Phys. Rev. Lett.* **72**, 981 (1994).
10. Rao, R., et al., "The q/m-Seperator for REX-ISOLDE", proceedings of the CPO5, 1998, to be published in *Nucl. Instrum. and Meth.* **A**
11. Rao, R., Habs, D., Kester, O., Rudolph, K. and Sieber, T., "The q/m-Separator for REX-ISOLDE", presented at the EPAC98, Stockholm, June 22-26, 1998.
12. Ames, F., Bollen, G., Huber, G. and Schmidt, P., "REX-TRAP, an Ion Buncher for REX-ISOLDE", presented at the ENAM98, Michigan, USA, June 1998.
13. Bollen, G., et al., *Nucl. Inst. and Meth.* **A368**, 675 (1996).
14. Raimbault-Hartmann, H., et al., *Nucl. Inst. and Meth.* **B126**, 378-382 (1997).
15. Wolf, B., *Handbook of ion sources*, CRC Press, Boca Raton, 1995.
16. Kester, O., et al., "The REX-ISOLDE LINAC", presented at the EPAC98, Stockholm, June 22-26, 1998.
17. Sieber, T., et al., "Design and Status of the RFQ for REX-ISOLDE", presented at the LINAC98, Chicago, August 23-28, 1998.
18. Madert, M., et al., *Nucl. Instrum and Meth.* **B139**, 427-440 (1998).
19. Warner, D., et al., "CERN Heavy-Ion Facility Design Report", CERN 93-01 (1993).
20. Keil, B., diploma thesis, University of Munich, August 1997.
21. von Hahn, R., et al., *Nucl. Instrum. and Meth.* **A328**, 270-274 (1993).
22. Workshop on Physics with a Germanium-Mini-Ball, MPI für Kernphysik, Heidelberg, May 24-26, 1995.
23. Fischbeck, Ch., diploma thesis, University of Munich, October 1997.

Silicon Strip Detectors for Radioactive Beam Experiments

F. Maréchal[†]

Institut de Physique Nucléaire, IN_2P_3-CNRS, 91406 Orsay Cedex, FRANCE

Abstract. The availability of radioactive beams with reasonable intensities and good optical qualities makes possible the study of direct reactions induced by unstable nuclei. Such experiments can be performed by measuring the energy and angle of light recoiling particles. High resolution position sensitive detectors using silicon strip technology are well suited tools to determine the two-body kinematics of the reaction. The newly built and innovative array MUST, dedicated to such studies, will be presented and its performances discussed in the light of experiments performed at the GANIL accelerator.

INTRODUCTION

Direct reactions can provide a wealth of information on nuclear structure and interaction potentials which are fundamental problems in nuclear physics. Such reactions have been performed extensively with stable nuclei and form the foundation of our current understanding of nuclear systems. The recent avaibility of radioactive beams with sizeable intensities at facilities such as GANIL, GSI, NSCL-MSU and RIKEN opens the possibility of investigating nuclear structure far from stability through nuclear reactions induced by unstable nuclei.

Elastic scattering yields information on the nuclear matter distribution and the effective nucleon-nucleon potentials and therefore should be sensitive to new manifestation of nuclear structure such as halos or neutron skins. Inelastic scattering to low lying collective states gives access to transition probabilities and nuclear deformations, thus is a well suited tool to scan new regions of deformations. Finally, transfer reactions are a tool of choice for the study of nucleon orbitals in nuclei. Such reactions are of primary importance in the light of the profound modifications of shell strucutre already observed in some region of the nuclear chart, for example the weakening of the N=20 and N=28 shell closures for ^{32}Mg and ^{44}S, respectively. These topics represent a few examples of novel physics that can be addressed through the study of reactions between unstable nuclei and light charged particles.

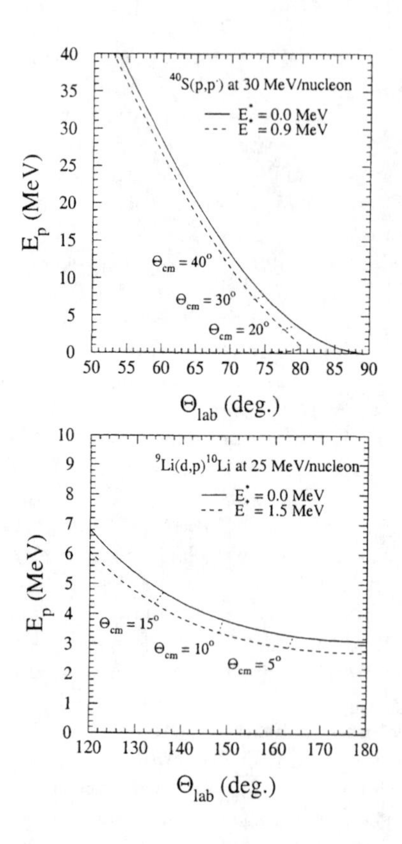

FIGURE 1. Kinematics for the ^{40}S(p,p') reaction at 30 MeV/A (top) and for the ^{9}Li(d,p) reaction at 25 MeV/A (bottom). The figures show the energy-angle correlation for the recoiling light particle in the case of reactions leading to the ground state (solid line) or an excited state (dahsed line). The locations of various center of mass scattering angles are indicated.

DETECTION REQUIREMENTS

One of the best way to obtain information is to perform direct reactions induced by light particles which have no excited states, and for which the interaction potentials are better determined than for heavy ions. Inverse kinematics must then be used, the unstable projectile impinging on the light target which entails specific experimental re-

CP475, *Applications of Accelerators in Research and Industry*, edited by J. L. Duggan and I. L. Morgan

quirements. It is possible to determine the two-body kinematics of the reaction, either by detecting the outgoing heavy nucleus, for example with a spectrometer, or by measuring the energy and angle of the recoiling light particle. In the first case, however, as soon as the projectile is somewhat heavy ($A \geq 20$), due to the kinematic focusing the angular resolution is no longer sufficient to obtain meaningful angular distributions. Moreover, a simple measurement of the outgoing quasi-projectile does not allow one to access unbound states which will decay in-flight by particle emission.

If we take proton scattering as an example, the recoiling light particle method seems to be of more general use, and is in addition more cost-effective. In order to measure down to very forward center of mass angles a low energy threshold is necessary, approximatively 500 keV protons. The energy range of the detector must thus cover from 0.5 up to about 50 MeV for protons. In the case of elastic or inelastic scattering (fig. 1, top), to obtain a reasonable excitation energy resolution, less than 1 MeV, an angular resolution in the laboratory frame of about 0.5° is crucial. For transfer reactions (fig. 1, bottom), the excitation energy resolution is more strongly dependent on the energy resolution of the light particle measurement, which should not be worse than 100 keV. In both cases, identification of the light particles over the entire energy range is also crucial in order to suppress background arising in particular from break-up or fusion reactions. Moreover, a large solid angle must be covered to overcome the low intensities of radioactive beams.

Some experiments with unstable beams have already been performed using this method (1, 2, 3). All of them used a system of telescopes based on silicon strip technology which provides the necessary resolutions, along with particle identification through energy loss, energy and time of flight measurements. In the next section, I will describe the MUST detector which is dedicated to the study of nuclear reactions induced by radioactive beams on light particles. Its aim is to employ the recoiling light particle method under optimal conditions in a wide range of experimental circumstances.

THE *MUST* DETECTOR ARRAY

Description

This new detection system built by an IPN-Orsay, DAPNIA/SPhN-CEA Saclay and DPTA/SPN-CEA Bruyères le Châtel collaboration consists of 8 silicon strip-Si(Li) telescopes. Each telescope consists of a 60x60 mm^2 300μm thick double sided silicon strip detector with 60 horizontal and 60 vertical 1mm wide strips, backed by a 58x58 mm^2 implented Si(Li) diode of approximatively 3mm depleted thickness. Particular care was taken to minimize the entrance window of the Si-strip detectors in order to limit the energy loss of the particles in this dead layer and thus allow for a very low energy threshold. For applications where higher energy particles must be detected, the telescopes are equipped with a third stage made with a 15 mm thick CsI crystal read out by photodiode. The system is modular to adapt to diverse experimental requirements. A picture of one telescope is presented on fig. 2.

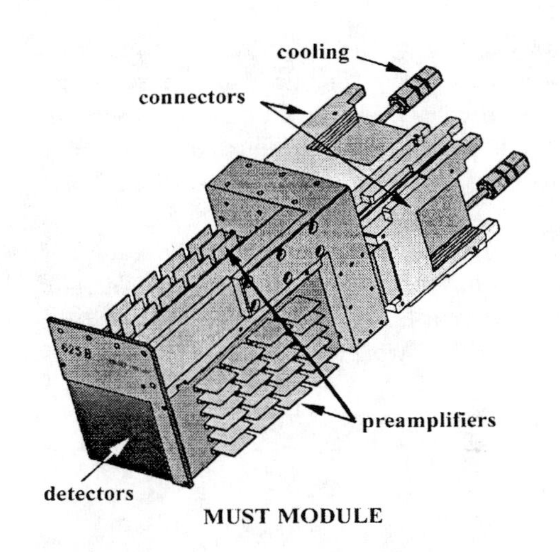

FIGURE 2. Artistic view of one of the eight telescopes.

In order to present the energy ranges and the particle identification methods, let us consider the case of protons. Protons of less than 6 MeV stop in the strip detector and are identified by energy vs. time of flight measurement. From 6 to 25 MeV, protons traverse the strip detector and stop in the Si(Li) detector. They are identified by ΔE-E technique and their total energy is measured as the sum of the energies deposited in the strip and Si(Li) detectors. Above 25 MeV, protons are stopped in the CsI crystal, and once again identified through ΔE-E technique.

Each of the 120 strips of one silicon detector is connected to a preamplifier located under vacuum directly behind the detector. Despite the low power consumption of the preamplifiers, all the modules are cooled by circulation of cold water at 10 °C. This ensures a stable energy resolution for the Si(Li) detector whose resolution otherwise exhibits a strong temperature dependence. Compact custom-made electronics in VXIbus standard have been designed by the electronics department of the Institut de Physique Nucléaire - Orsay. All the necessary functions (discriminators, amplifiers, energy and time digitalization) for one telescope are housed in one VXI module of size D. The trigger functions, the generators for test pulses and the signal processing of the Si(Li) detectors are grouped in a separate module. That means that the 1000 logical and analogical channels of the array are housed in only one crate placed adjacent to the reaction chamber and fully remote controlled by a PC computer, including pulse visualisation on an oscilloscope. These electronics are read out by a dedicated data acquisition system based on a SUN workstation. Details of the electronics can be found in references (4, 5, 6).

Test Experiment

In order to assess the performances of the MUST detector under realistic conditions. a test experiment was performed at the GANIL facility, Caen, France. One telescope was placed in a scattering chamber 15 cm away from a 1 mg/cm^2 CH_2 target. This target was bombarded by a 77 MeV/nucleon ^{40}Ar beam. The aim of the experiment was to measure elastic and inelastic scattering of the protons from ^{40}Ar in inverse kinematics. Therefore, the detector was located at a mean angle of 81°. The angular resolution given by the width of the strips was 0.4°.

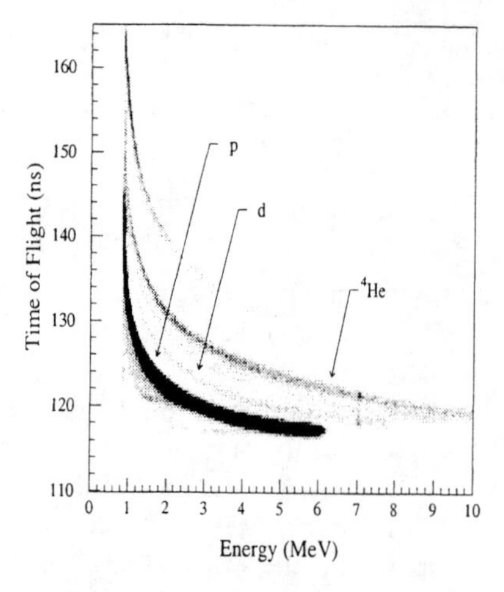

FIGURE 3. E-time of flight identification for particles which stop in the silicon strip detector.

The energy resolution measured for the strip detector was 50 keV, whereas the time resolution measured between the strip detector and the cyclotron RF was 1.2 ns at E_p=1.7 MeV and 0.83 ns at E_p=3.1 MeV. These values are higher than the intrinsic time resolution of the strip detector, because the contribution of the RF signal has not been subtracted. ΔE, E and time of flight measurements allowed identification of light charged particles between 500 keV and 25 MeV proton energy (see fig. 3 and fig. 4). Excellent isotopic identification of the light particles is achieved.

In order to obtain excitation energy spectra and angular distributions, the excitation energy and scattering

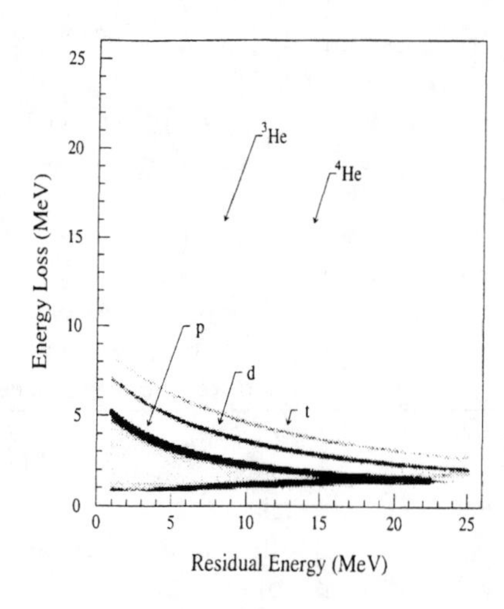

FIGURE 4. ΔE-E identification for more energetic particles which punch through the first detector.

angle in the center of mass frame are deduced from the measured proton energy and angle using relativistic kinematics. Figure 5 shows the excitation energy spectrum for protons with laboratory energies between 6 and 15 MeV. We clearly separate from the ground state the first 2^+ and 3^- excited states known to be located at 1.46 and 3.68 MeV. The excitation energy resolution is 600 keV, which is comparable to the resolution which would be obtained by detecting the scattered projectile in a high performance spectrometer such as SPEG at GANIL, which has a nominal resolution of $\Delta p/p=10^{-4}$ (7), demonstrating the effectiveness of the detection of the recoiling light particle.

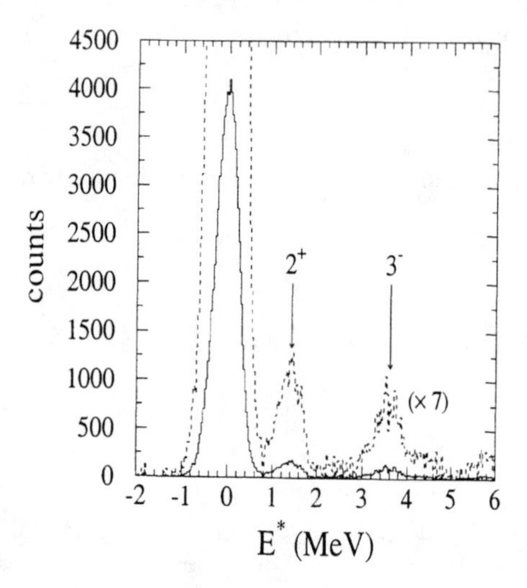

FIGURE 5. Excitation energy spectrum from the ^{40}Ar(p,p') reaction at 77 MeV/nucleon. The dashed curve is the same spectrum multiplied by a factor 7.

The angular distributions of the three states were extracted by subtracting a background independent of excitation energy and are presented in fig. 6. The dashed curves represent coupled channel calculations using the code ECIS (8) and optical potential parameters extrapolated from the potential determined by Sakaguchi et al. at 65 MeV (9). The deduced deformation parameters $\beta_2 = 0.22 \pm 0.01$ and $\beta_3 = 0.24 \pm 0.02$ obtained for the 2^+ and 3^- states respectively are in excellent agreement with the values obtained in previous experiments performed using direct kinematics (10).

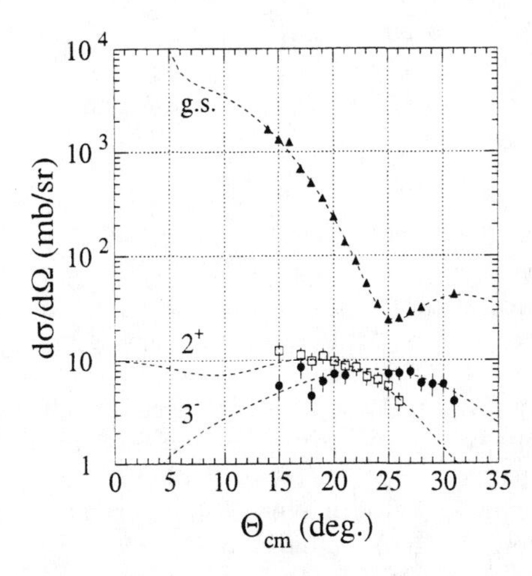

FIGURE 6. Angular distributions for the ground state (triangles), the first 2^+ excited state (squares) and the first 3^- excited state (circles) in the ^{40}Ar(p,p') reaction at 77 MeV/A. The dashed lines are coupled-channel calculations (see text).

These results demonstrate both the reliability of the inverse kinematics method and the capability of silicon strip detectors, such as the MUST array, to yield high quality angular distributions of excited states populated in inverse kinematics reactions.

CONCLUSIONS

The measurement of recoiling light particles is shown to be a powerful method for the study of nuclear reactions in inverse kinematics. Silicon strip arrays, such as the MUST detector system, can achieve isotopic identification of light charged particle and have demonstrated their capability to measure angular distributions of excited states populated in inverse kinematics reactions. The MUST array has recently been successfuly used to measure several reactions induced by radioactive beams at GANIL.

Acknowledgments

I would like to thank my colleagues, physicists, engineers and technicians, from the IPN-Orsay, DPTA/SPN-CEA Bruyères le Châtel, and DAPNIA/SPhN-CEA Saclay whose collaboration has been so valuable during the development of the MUST project.

REFERENCES

†. Present address: Department of Physics, Florida State University, Tallahassee, Florida 32306, USA

1. G. Kraus et al., *Phys. Rev. Lett.* **73** (1994) 1773.
2. A.A. Korsheninnikov et al., *Phys. Rev.* **C53** (1996) R537.
3. J.H. Kelley et al., *Phys. Rev.* **C56** (1997) R1206.
4. F. Maréchal, PhD Thesis, Université d'Orsay (1998), Internal Report IPNO-T-98-02.
5. S. Ottini, PhD Thesis, Université d'Orsay (1998), Internal Report DAPNIA/SPhN-98-01T
6. Y. Blumenfeld et al., to be published in *Nucl. Instr. and Meth.*
7. L. Bianchi et al., *Nucl. Instr. and Meth.* **A276** (1989) 509.
8. J. Raynal, *Phys. Rev.* **C23** (1981) 2571.
9. H. Sakaguchi et al., *Phys. Rev.* **C26** (1982) 944.
10. R. de Leo et al., *Phys. Rev.* **C31** (1985) 362.

BEARS: a Radioactive Ion Beam Initiative at LBNL

J. Powell, F. Q. Guo, P. E. Haustein[†], R. Joosten, R.-M. Larimer, C. M. Lyneis, P. McMahan, D. M. Moltz, E. B. Norman, J. P. O'Neil, M. W. Rowe, H. F. VanBrocklin, D. Wutte, Z. Q. Xie, X. J. Xu and Joseph Cerny

Lawrence Berkeley National Laboratory, University of California, Berkeley, CA 94720, and [†]Chemistry Department, Brookhaven National Laboratory, Upton, NY 11973

BEARS is an initiative to develop a radioactive ion-beam capability at Lawrence Berkeley National Laboratory. The aim is to produce isotopes at an existing medical cyclotron and to accelerate them at the 88" Cyclotron. To overcome the 300-meter physical separation of these two accelerators, a carrier-gas transport system will be used. At the terminus of the capillary, the carrier gas will be separated and the isotopes will be injected into the 88" Cyclotron's Advanced Electron Cyclotron Resonance ion source. The first radioactive beams to be developed will include 20-min ^{11}C and 70-sec ^{14}O, produced by (p,n) and (p,α) reactions on low-Z targets. Tests at the 88" Cyclotron lead to projections of initial ^{11}C beams of 2×10^8 ions/sec ^{14}O beams of 1×10^6 ions/sec. Construction of BEARS is expected to be completed in the spring of 1999.

There is currently extensive world-wide activity in the development and construction of radioactive ion-beam facilities of various types. The availability of beams of unstable nuclei offers exciting new opportunities for research into nuclear structure and nuclear astrophysics. BEARS, or Berkeley Experiments with Accelerated Radioactive Species, is an initiative to develop a radioactive ion-beam capability at Lawrence Berkeley National Laboratory (LBNL).

The basic concept for the initial BEARS system involves the coupling of isotope production at an existing medical cyclotron in building 56 of LBNL with post-acceleration by the 88" Cyclotron. As these accelerators are separated by a distance of about 300 m, isotopes will be transported between the two via a gas-jet capillary (see Fig. 1). Carrier gas would be pumped by a high-throughput Roots blower at the building-88 end. Preliminary tests have shown that a total transport time of less than a minute is easily achieved, and times as short as 20 sec should be possible. Actual construction of this transfer line was started in September of 1998, with completion expected in the spring of 1999.

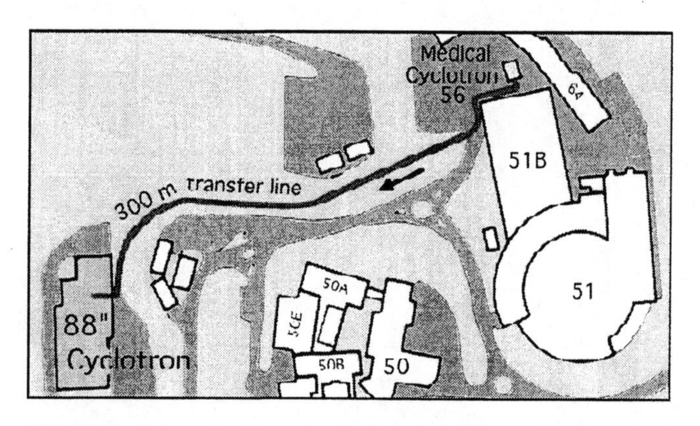

FIGURE 1. Site map transfer line under construction between buildings 56 and 88.

The medical cyclotron of the Medical Isotope Facility (1) produces a 10 MeV proton beam with typical intensities of 40 μA. Several light proton-rich isotopes can be produced, via (p,n) and (p,α) reactions on light-Z targets. Initially, we have focused on the production of ^{11}C ($t_{1/2}$=20 min) and ^{14}O ($t_{1/2}$=71 sec), both produced from a nitrogen gas target. The maximum thick-target yields are approximately 1×10^{11} atoms/sec of ^{11}C and 5×10^9 atoms/sec of ^{14}O (2).

After transport, the radioisotopes are to be injected into one of the two electron-cyclotron-resonance ion sources (ECRIS) of the 88" Cyclotron (3). These sources can reliably achieve good ionization efficiencies at high charge states; however, they require vacuums of less than 10^{-6} torr to operate. Therefore the central technical challenge of BEARS is the coupling of a high-pressure gas transport system to an ECRIS. Two separate techniques have been explored.

Non-gaseous activity can be transported by a gas-jet system via the well-known technique of introducing small aerosol clusters of the order of 0.1 to 1 micron in size (4). The aerodynamics of Laminar flow prevents these aerosols from striking the wall of the capillary, and any activity that attaches to these clusters is transported with high efficiency. The large size and momentum of these aerosols also allows them to be separated from most of the carrier gas via a differentially-pumped skimming system, allowing the possibility of direct coupling to an ion source. Such a system, previously used at LBNL in the isotope separators OASIS (5) and RAMA (6), has the potential for being broadly applicable to many isotopes, but does not work with volatile compounds.

To test this technique, a four-stage differentially-pumped skimming system was constructed and coupled

CP475, *Applications of Accelerators in Research and Industry*,
edited by J. L. Duggan and I. L. Morgan
1999 The American Institute of Physics 1-56396-825-8

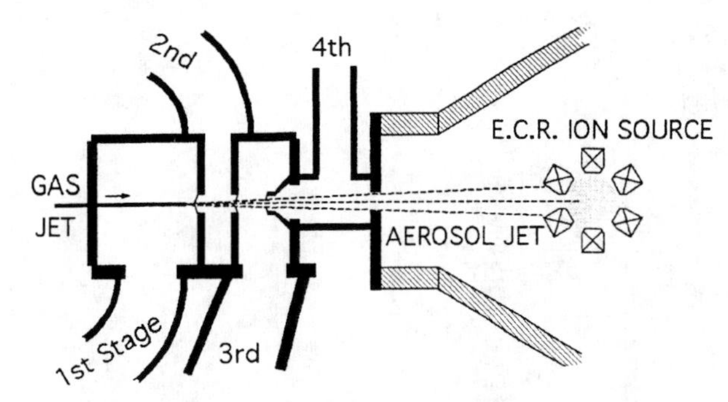

FIGURE 2. Skimming system for injecting aerosols into the ECR ion source.

directly to the ECR, the older of the two ion sources of the 88" Cyclotron. This is shown in Fig. 2. Aerosol clusters and carrier gas enter the first pumping stage in a jet of near-sonic velocity. The heavier clusters exit in a narrower cone than the expanding gas, allowing then to pass through the small holes in the skimmers. Once inside the ECR, the aerosols are caught on heated surfaces, to vaporize the activity. Tests have shown that, with a full gas load, ECR performance is not significantly degraded.

Unfortunately, it was found that this system failed to transport significant amounts of ^{11}C or ^{14}O, the initial BEARS production isotopes. This was traced to the majority of the activity forming gaseous compounds and thus not attaching to the aerosol clusters. The fraction of ^{11}C in a chemical form that could be transported was only on the order of 0.1-0.5%. However, this small amount of activity was successfully injected into the ECR, and an extracted beam of ^{11}C was identified, although at very low intensities.

Most of the produced activity is believed to be in the chemical form of such molecules as $^{11}CO_2$ or $N_2{}^{14}O$ (the CO_2 produced from a small oxygen component in the nitrogen target gas). To separate such molecules, a cryogenic trapping system has been developed. This technique, found to be much more effective for transporting ^{11}C and ^{14}O, is illustrated in Fig. 3. After transport through the capillary, the carrier gas was passed through a coil of 1/8" stainless-steel tubing, about 1.5 m long, submerged in liquid nitrogen. After stopping the gas flow and allowing the remaining nitrogen to be pumped away, the trap was connected to either the ECR or AECR-U, the 88" Cyclotron's second ECRIS (7). The liquid nitrogen was then replaced by an alcohol bath containing dry ice, quickly raising the temperature to 195 K. This temperature increase released most of the trapped activity while keeping contaminants such as water frozen. The released gas was then bled directly into the ECR or AECR-U plasma region through an adjustable needle valve. In initial tests at the 88" Cyclotron, trapping and release efficiencies were 45% for ^{11}C and 65% for ^{14}O and past experience with medical isotope production cyclotrons suggests that efficiencies as high as 90% should be achievable with simple development.

Ionization efficiencies of the two ion sources were determined in several tests at the 88" Cyclotron using small amounts of ^{11}C and ^{14}O. The sources were first optimized for a chosen charge state using ^{12}C or ^{16}O; then the analyzing magnet was adjusted for the mass difference between the stable and radioactive isotopes. The results of these tests are shown in Table 1. Also shown for

TABLE 1. Preliminary ionization efficiencies with the 88" Cyclotron's two ion sources.

Ion	*ECR*	*AECR-U*	*AECR-U with stable ^{12}C and ^{16}O*
$^{11}C^{1+}$	1.1 %		
$^{11}C^{2+}$	0.7 %		
$^{11}C^{3+}$	0.4 %	4 %	
$^{11}C^{4+}$	0.9 %	11 %	24 %
$^{11}C^{5+}$	0.1 %	4 %	14 %
$^{11}C^{6+}$		2 %	
$^{14}O^{3+}$	0.4 %		
$^{14}O^{4+}$	0.4 %		
$^{14}O^{5+}$	0.4 %		
$^{14}O^{6+}$		3.6 %	27 %
$^{14}O^{7+}$		1.2 %	6 %
$^{14}O^{8+}$		0.4 %	

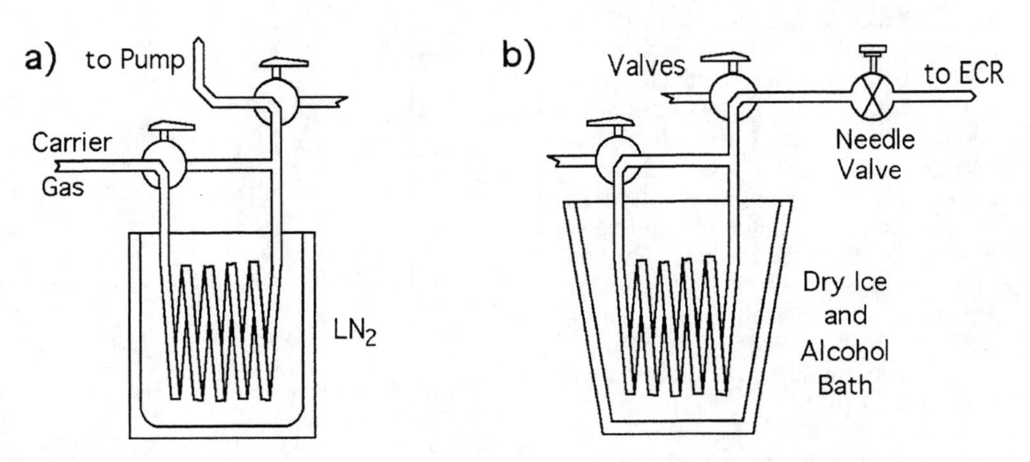

FIGURE 3. Cryogenic trapping system: (a) trapping and (b) release at dry ice temperatures into the ECR ion source.

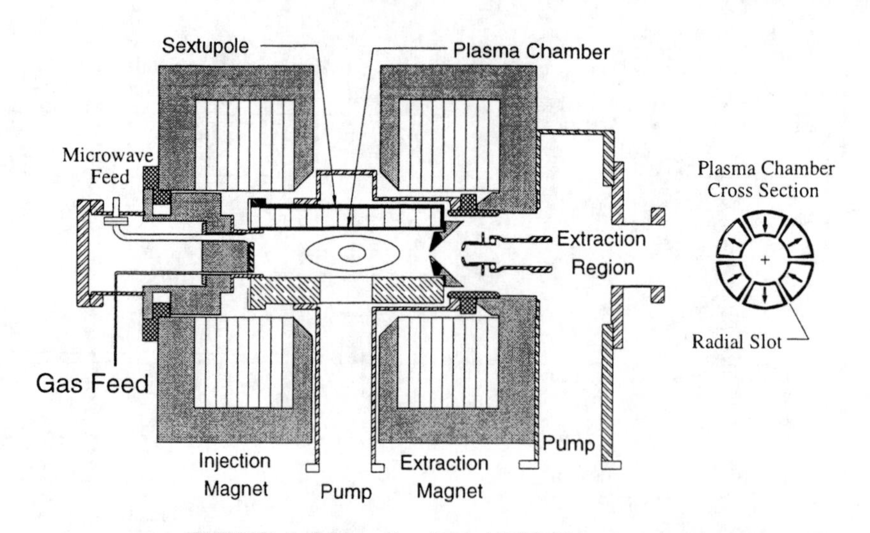

FIGURE 4. Schematic of the AECR-U ion source.

comparison are results for the AECR-U for ^{12}C and ^{16}O measured with a calibrated CO leak.

The AECR-U (a schematic of which is shown in Fig. 4) was revealed to have particularly good ionization efficiencies: up to 11% for $^{11}C^{4+}$ and 2% for fully-stripped $^{11}C^{6+}$. For ^{14}O the efficiency was as high as 3.6%. Although less than half of what was achieved for the stable isotopes, the possible reasons for which will be discussed later, these results indicate that intense radioactive beams at high charge states may be produced using the AECR-U.

All the tests described above were performed using the 88" Cyclotron for isotope production, since the 300 m transfer line from the medical cyclotron has not yet been completed. In order to test the acceleration of radioactive beams before completion of the 300 m gas line, a simple batch-mode transfer system was set up, in which ^{11}C activity was carried down from the medical cyclotron in a transportable cryogenic trap mounted in a well-shielded lead "pig". At building 88, the activity was injected into the AECR-U in the same manner describe above.

For these tests, the 88" Cyclotron was first tuned using $^{22}Ne^{8+}$, after which the cyclotron RF frequency was shifted slightly to the calculated value for $^{11}C^{4+}$. The difference in charge-to-mass ratio between $^{22}Ne^{8+}$ and $^{11}C^{4+}$ is small enough that the other tuning parameters of the accelerator need not be adjusted, yet large enough to be resolved by the RF shift, ensuring minimal neon contamination of the ^{11}C beam. In contrast, ^{11}B can not be separated from ^{11}C in the cyclotron, and a small residual yield of $^{11}B^{4+}$ was observed.

With the AECR-U and 88" Cyclotron optimized for 100 MeV $^{11}C^{4+}$, and a NaI detector monitoring build-up of activity at the beam stop, ^{11}C from the medical cyclotron was bled into the ion source over a period of about ten minutes. Figure 5(a) shows the measured activity during this period and for about one hour afterwards. Also shown, in Fig. 5(b), is the calculated beam intensity extracted from the activity data using the known 20 min half life of ^{11}C. The amount of ^{11}C in the sample was approximately 160 mCi at the point that the valve was opened, and the overall AECR-U plus cyclotron transmission efficiency was 0.6%, about half of what would be estimated using the 11% ion-source efficiency for $^{11}C^{4+}$ and the observed 13% cyclotron transport efficiency for $^{22}Ne^{8+}$.

As seen from Fig 5(b), the ^{11}C beam intensity was around 2 x 10^7 ions/sec during the time that ^{11}C was bled into the ion source. Surprisingly, the beam intensity sharply increased when the ^{11}C bleed valve is closed, then

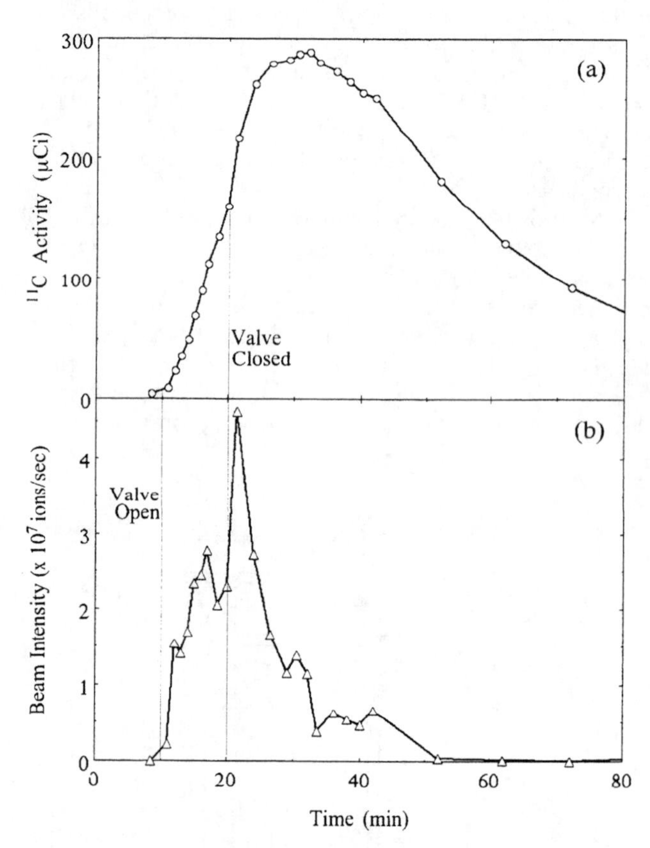

FIGURE 5. First acceleration of a 100 MeV ^{11}C beam: (a) activity build-up and decay at the beam stop and (b) extracted beam intensity in units of 10^7 ^{11}C ions/sec. The vertical lines delineate the time period during which activity was introduced into the ion source through a bleed valve.

trailed off over a long time frame of more than 20 min.

These results are very new and not fully understood; however, they are at least partly caused by one technical difficulty: the residual load of a few non-radioactive gases, such as N_2O, NO, or CO_2, that are cryotrapped out of the target/carrier gas at the same time as the activity. Some simple steps have been taken to reduce this load, but there is still on the order of 0.1 torr-liter of non-radioactive gas that accompanies each sample of activity introduced into the ion source. This gas load is close to the limit that the AECR-U can handle, and its presence is believed to cause the lower ionization efficiencies for ^{11}C and ^{14}O relative to stable species (see Table 1).

The sudden jump in beam intensity, seen in Fig. 5(b) when the bleed valve into the AECR-U was closed, may be due to the drop in gas load resulting in an increase in ion-source efficiency (the pressure in the plasma region of the AECR-U drops within several seconds). In this interpretation, the long trailing-off of the ^{11}C beam, with a half-life of around 7 min, must be attributed to an unusually long hold-up time for ^{11}C in the source. However, this contradicts the measured AECR-U hold-up times for stable gaseous species, including ^{12}C, of less than 20 sec. This apparent long hold-up time for ^{11}C was observed in the earlier ion-source tests, but no such effect was seen for ^{14}O.

Another possibility is that the high gas load affects the plasma in such a way that it no longer recycles ions that stick the walls of the chamber (normally the plasma acts to scrape off these atoms). When the gas load is reduced and the plasma returns to optimum conditions, the plasma may start to recycle the built-up supply of ^{11}C on the walls.

Further tests will be performed to investigate the long ^{11}C hold-up times. In any case, a greater effort will be made to reduce the gas load on the AECR-U through such means as reducing the oxygen content of the target gas and increasing the overall cleanliness of the entire gas-handling system.

Completion of the full BEARS system (including automation of the cryotrap system) is expected some time in the spring of 1999. Assuming only the known production rates of the medical cyclotron and the measured system efficiencies described in this paper, we expect initial $^{11}C^{4+}$ and $^{14}O^{6+}$ beams of 2×10^8 and 1×10^6 ions/sec, respectively. For comparison, the Louvain-la-Neuve facility in Belgium currently reports an available $^{11}C^{1+}$ beam of 10^7 ions/sec and $^{13}N^{1,2,3+}$ beams of $1\text{-}4 \times 10^8$ ions/sec (8). Future development is expected to improve the BEARS system, in particular by increasing the trapping efficiency, optimizing the ion source and cyclotron tuning, and reducing the high AECR-U gas load. With these and other moderate enhancements, $^{11}C^{4+}$ and $^{14}O^{6+}$ beams of 2×10^9 and 3×10^7 ions/sec, respectively, should be achievable.

ACKNOWLEDGMENTS

This work supported by USDOE, Division of Nuclear Physics, under contracts DE-AC03-76SF00098 and DE-AC02-98CH10886.

REFERENCES

1. VanBrocklin, H. F. and O'Neil, J. P., in *Applications of Accelerators in Research and Industry*, ed. by Duggan, J. L. and Morgan, I. L., New York: AIP Press, 1997, pp. 1329-1332.
2. Kitwanga, Sindano wa, et al., *Phys. Rev.* C **42**, 748-752 (1990).
3. Lyneis, C. M., et al., Proc. of the 14th Conf. on Cyclotrons and Their Applications, Cape Town, South Africa, Oct 8-13, 1995, pp. 173-176.
4. Wollnik, H., *Nucl. Inst. and Meth.* **139**, 319-323 (1976).
5. J. M. Nitschke, in Proc. of the Int'l Conf. on the Properties of Nuclei Far From Stability, Leysin, 1970 (CERN report 70-30, 1970) p. 153.
6. Moltz, D. M., et al., *Nucl. Inst.and Meth.* **172**, 507 & 519 (1976).
7. Xie, Z. Q., *Rev. Sci. Instrum.* **69**, 625 (1998).
8. Vervier, J., *Nucl. Phy.* **A616**, 97c-106c (1997).

A High Intensity Electron Beam Ion Trap for Charge State Boosting of Radioactive Ion Beams

R. E. Marrs and D. R. Slaughter

Lawrence Livermore National Laboratory, Livermore, CA 94551.

A high intensity electron beam ion trap under development at LLNL could be adapted for charge state boosting of radioactive ion beams, enabling a substantial reduction in the size and cost of a post-accelerator. We report estimates of the acceptance, ionization time, charge state distribution, emittance, and beam intensity for charge state boosting of radioactive ions in this device. The estimates imply that, for tin isotopes, over 10^{10} ions/s can be ionized to q = 40+ with an absolute emittance of approximately 1 π mm mrad at an energy of 30 × q keV.

INTRODUCTION

The U. S. scientific community is considering options for the construction of a major new facility for the production and acceleration of radioactive ion beams (RIBs). The facility will be based on the ISOL concept in which radioisotopes diffuse out of a high temperature production target, are singly ionized, and pass through a mass separator. The proposed scientific program requires accelerating the radioactive ions to energies up to roughly 10 MeV/u. To reach these energies in a post-accelerator of reasonable size and cost the radioactive ions must be ionized to high charge states either before acceleration or by stripping between stages of acceleration.

One option for reaching high charge states is ionization of (low energy) q = 1+ ions in an electron beam ion source (EBIS) or trap (EBIT). In these devices ions are trapped within the space charge of an electron beam and ionized to high charge states by electron impact. For example, isotopes of tin (Z = 50) could be ionized to q = 40+ or 48+ before acceleration. A high intensity EBIT under development at LLNL may be able to satisfy the expected performance requirements of a future RIB facility with a substantial reduction in the size of the post-accelerator and an improvement in performance compared to ion stripping between acceleration stages.

ELECTRON BEAM ION TRAP

The electron beam ion trap was developed at LLNL to study x-ray emission from very-highly-charged ions. The ions are trapped in the space charge potential of an electron beam and confined axially by voltages applied to drift tubes or trap electrodes (1,2). High charge states are reached by successive ionizing collisions with beam electrons, and the final charge state can be controlled by selecting the electron beam energy and confinement time. The EBIT concept is shown in Fig. 1. The related EBIS was developed as a source of highly-charged-ion beams (3). The EBIT uses a higher electron beam current density and a shorter trap length than the traditional EBIS. However, the EBIT can also be operated as an ion source.

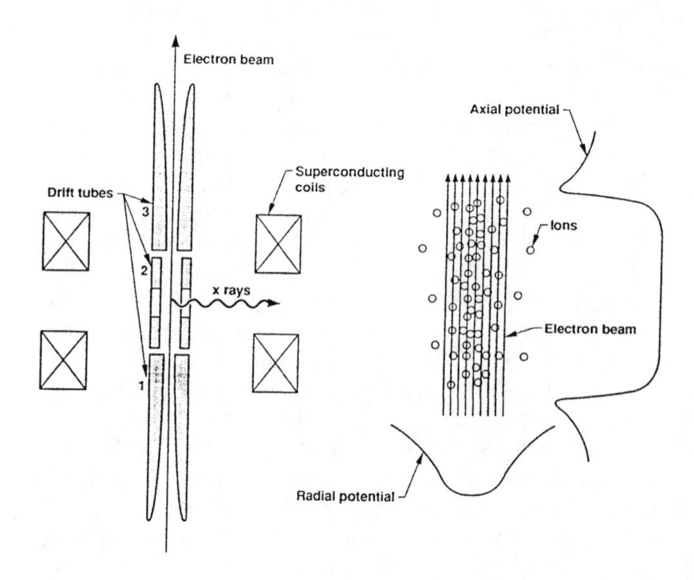

FIGURE 1. Illustration of ion confinement in an EBIT. The magnetic field compresses the electron beam, but has little effect on the ions.

A high intensity EBIT under development at LLNL is expected to increase the x-ray emission rate from trapped ions by a factor of 100 (per cm of beam length), and the

CP475, *Applications of Accelerators in Research and Industry*,
edited by J. L. Duggan and I. L. Morgan
1999 The American Institute of Physics 1-56396-825-8

highly-charged-ion beam intensity by a factor of 1000 over that of existing EBITs. This is accomplished by roughly 10-fold increases in the total electron beam current, the current density, and the length of the trap. Actual parameters for the LLNL high intensity EBIT are given in Table I, and a scale drawing is shown in Fig. 2.

EXPECTED PERFORMANCE

The performance of the LLNL high intensity EBIT operating as a charge state booster can be inferred from calculations and experience with existing EBIT devices. The profile of the magnetically compressed electron beam in such devices is known to be roughly Gaussian with a radius determined by the electron temperature and the magnetic field (if any) at the cathode (4). Normally the magnetic field at the cathode is carefully zeroed with a bucking coil in order to obtain maximum electron beam compression. Electron beam profile measurements in magnetic fields up to 3 T confirm the predictions of theoretical models. For the high intensity EBIT operating at 6 T with a 5-A electron beam from a 6.4-mm radius cathode, theory predicts that 80% of the electron beam is within a radius of 48 μm. The electron beam radius is almost independent of the electron energy; however, we use an electron energy of 30 keV for the estimates presented here.

TABLE 1. Typical parameters for the LLNL high intensity EBIT operating as a RIB charge state booster.

Magnetic field	6 tesla
Electron beam energy	30 keV
Electron beam current	5 A
Beam radius (80% current)	48 μm
Central current density	1.1×10^5 A/cm^2
Trap length	25 cm
Total electron charges	7.9×10^{10}

Ion Acceptance

The acceptance of the EBIT trap is a very important parameter. It must be large enough to accommodate the emittance of the singly charged radioactive ion beam from the production source with minimal losses. Acceptance and emittance have the same units, and both are a measure of the transverse phase space volume of the ion beam. As such, they can be related to the temperature and radius of the ions in an ion source or trap. We calculate the absolute acceptance and emittance for ions entering or leaving the EBIT trap from the formula

$$\varepsilon = \pi R_{80}\sqrt{\frac{T_i}{qeU}} \qquad (1)$$

where R_{80} is the radius in the trap containing 80% of the ions, T_i is the temperature of the trapped ions, q is the ion charge state, e is the elementary charge, and U is the acceleration potential of the ion beam at which the emittance is specified (5). For a Maxwellian distribution of ion velocities at temperature T_i , and a Gaussian shaped radial ion distribution, the phase space defined by Eq. (1) contains 67% of the ions.

Captured ions will have good overlap with the electron beam if the ion temperature is no larger than the space charge potential difference from the axis to the edge of the beam, which is U_e = 450 V for the parameters given in Table I, neglecting ion space charge. (The characteristic ion and electron radii for Gaussian distributions are in fact equal when $T_i = qeU_e$.) The corresponding absolute acceptance for T_i = 450 eV is ε = 5.9 π mm mrad at U = 30 kV. This is comparable to the emittance of the types of ion sources used at ISOL facilities (6).

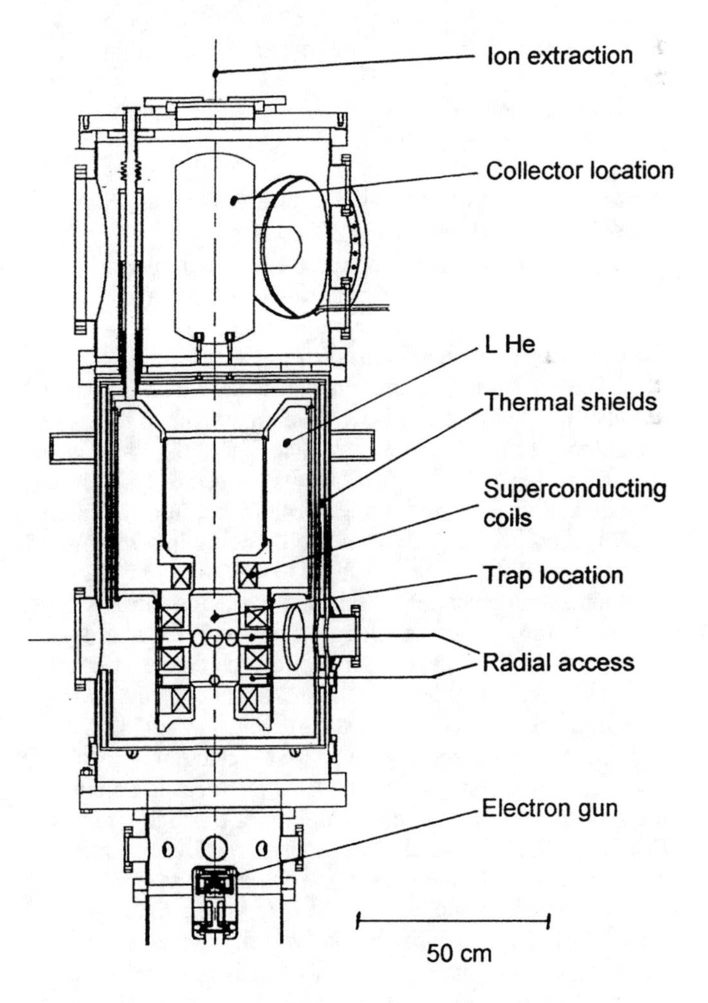

FIGURE 2. Layout of the high intensity EBIT at LLNL.

Note that the acceptance for capture into the trap without requiring good overlap with the electron beam is much larger than the value given above because the 5-mm radius of the trap electrodes is 100 times larger than the electron beam radius, and the 4.5-kV space charge potential at the trap electrode radius is 10 times larger than that at the beam edge. In summary, the capture efficiency for radioactive ions injected into the LLNL high intensity EBIT from ISOL type sources is expected to be close to 100% because of the deep potential well produced by the high-current, highly-compressed electron beam.

Ionization time

The ionization time for stripping singly charged radioactive ions to high charge states is important for two reasons: (1) The ion beam intensity will be reduced for isotopes with lifetimes less than the ionization time, and (2) the ionization rate sets an upper limit on the highly-charged-ion production rate. In an EBIT, ionization to high charge states proceeds primarily by sequential single ionization. The characteristic time for each ionization step is $\tau_i = 1 / \sigma_i \varphi_e$, where σ_i is the ionization cross section and φ_e is the electron flux. The total time to remove N electrons is then $t_N = \sum_{i=1}^{N} \tau_i$. We used Lotz-formula cross sections (7) to calculate the ionization times for tin ($Z = 50$) ions at 30-keV electron energy. The value of the electron flux used was $\varphi_e = 3.4 \times 10^{23}$ cm^{-2} s^{-1}. This corresponds to the average electron current density for electron and ion radial distributions that have a Gaussian profile with the same characteristic radius, and with high-intensity-EBIT parameters as given in Table I.

The total time required to reach the different charge states of tin is plotted in Fig. 3. The time required to ionize tin isotopes from $q = 1+$ to $q = 40+$ (Ne-like) and $q = 48+$ (He-like) is 3.0 ms and 34 ms, respectively. These closed shell configurations are favorable because the jump in ionization potential between shells results in a narrower charge state distribution. In fact, 30 keV is below the K-shell ionization potential for tin; hence the ionization stops at $q = 48+$, and that charge state will have a high abundance after 34 ms. This effect could be used to improve the purity of a ^{100}Sn beam by removing contaminating ^{100}In ($Z = 49$) ions, for which $q = 47+$ is the highest possible charge state in a 30-keV electron beam.

The ionization time can be used to calculate an upper limit for the highly-charged-ion output of the high intensity EBIT used as a charge state booster. If every highly charged ion could be removed as soon at it was produced, the ion production rate would be N_i / t_q, where N_i is the ion capacity of the trap and t_q is the ionization time to reach charge state q. Assuming $q = 40+$ and 50% neutralization of the electron space charge by ions, $N_i = 1.0 \times 10^9$ ions. The upper limit for the production rate of Sn^{40+} ions is then 3.3×10^{11} ions/s. The actual production rate will be lower and is difficult to calculate, but it is probably between 10^{10} and 10^{11} ions/s.

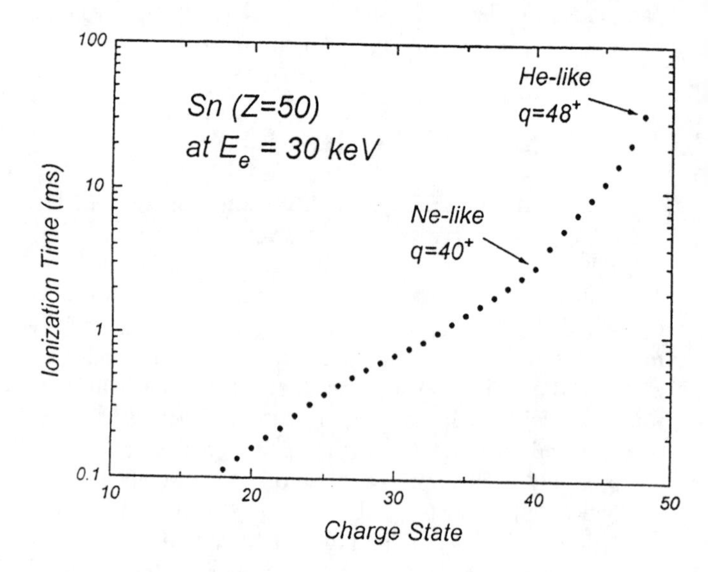

FIGURE 3. Calculated total ionization time to reach the different charge states of tin in a high intensity EBIT.

Charge state distribution

The charge state distribution of the highly charged radioactive ions extracted from an EBIT charge state booster is important because, as with foil stripping, only one charge state is likely to be accelerated and the others will be lost. The charge state distribution in an EBIT is determined by a combination of ionization, recombination, and loss processes. Measurements from existing EBIT and EBIS devices provide a good indication of what can be expected. An abundance of 90% for the $q = 46+$ (Ne-like) charge state of barium has been observed in an EBIT (1). Away from closed shells, an abundance of roughly 25% was observed for the most abundant ($q = 77+$) charge state in a beam of thorium ions extracted from an EBIT (8). EBIS measurements, available for lighter elements such as argon, show similar distributions (9). Thus, experience with existing EBIT and EBIS devices suggests that an ionization balance with 25% or more of the ions in the most abundant charge state can be expected for RIB applications.

Emittance

The stripping of accelerated ions to high charge states with foils or gas can only increase the emittance of the beam. In marked contrast, stripping of ions in an EBIT reduces the beam emittance. Rough measurements of the emittance of beams from existing EBITs and computer modeling suggest that an absolute emittance of roughly 1 π

mm mrad at 30 kV can be expected for the LLNL high intensity EBIT (10, 11). One contribution to the emittance reduction can be seen from the explicit q dependence of Eq. (1). However, this is offset by the Coulomb energy gained by ions as they move toward the beam axis and by electron beam heating of very-highly-charged ions held for long times in the trap. Evaporative ion-ion cooling, in which light (stable) ions are introduced into the trap as a coolant, is used to control the temperature and emittance of highly charged ions in EBIT devices (2).

The decrease in ion emittance in the EBIT has a favorable impact on a RIB post-accelerator. For example, beam losses and radioactive contamination would be reduced, the resolving power of a mass separator would be improved, and RIBs could be focused to micron-size spots.

Duty cycle

Ions can be injected into an EBIT either continuously, at an energy just above the axial trap barrier, or in bunches by lowering the trap barrier during the arrival of the ion bunch. With continuous injection, all injected ions not ionized to higher charge in one transit of the trap will escape the way they entered. Returning to the example of tin, the ionization time for Sn^{1+} ions in the high intensity EBIT is 0.16 μs, a factor of 100 less than the 16-μs bounce time of 500-eV tin ions in a 25-cm trap. Hence, 100% ionization and capture can be achieved with either injection mode. Note that this is not true for EBIS devices that use low electron beam current densities of order 200 A/cm^2.

Charge-boosted ions can be extracted either continuously by allowing them to leak out over the axial trap barrier, or in batches (12, 13). Using batch mode, the EBIT can function as an ion accumulator so that a pulsed RFQ and LINAC can be used as a post-accelerator. Note that the 3.0-ms ionization time to produce Sn^{40+} ions is compatible with a 120 Hz repetition rate.

IMPACT ON RIB FACILITY

The use of an EBIT for charge state boosting will have a large impact on the post-accelerator that follows it. For example, an estimate of the post-acceleration required to accelerate radioactive ^{132}Sn ions to an energy of 7 MeV/u with two stages of conventional stripping indicates that a total accelerating potential of 77 MV is required, and that the expected stripping efficiency is 8% (14). If the ions were stripped to q = 40+ in an EBIT before acceleration, then only 24 MV of acceleration potential would be required.

The REX-ISOLDE project at CERN is using a large Penning trap with buffer gas cooling as an accumulator and emittance reducer for singly charged ions. The ions will be transferred to an EBIS in batch mode for stripping to higher charge states. The capacity of the Penning trap is expected to limit the ion throughput to ~10^7 ions/s (15). The estimates presented here indicate that a high intensity EBIT does not require a Penning trap accumulator.

CONCLUSIONS AND FUTURE PLANS

The estimates presented here indicate that a high intensity EBIT under development at LLNL is an attractive option for charge state boosting of radioactive ion beams in an ISOL-type RIB facility. Other advantages of the EBIT are a significant reduction in the ion beam emittance and the separation of interfering isobars in some cases. These features will facilitate cleaner beams and reduce losses.

An experimental verification of the performance characteristics presented here is crucial for an evaluation of options for a future RIB facility. We propose to use ISOL-type sources for injecting ions into a high intensity EBIT at LLNL in order to obtain measurements of the performance characteristics presented here.

ACKNOWLEDGMENT

This work was performed under the auspices of the U. S. Department of Energy by Lawrence Livermore National Laboratory under Contract No. W-7405-Eng-48.

REFERENCES

1. R. E. Marrs, M. A. Levine, D. A. Knapp, and J. R. Henderson, *Phys. Rev. Lett.* **60**, 1715 (1988).
2. M. A. Levine, R. E. Marrs, J. R. Henderson, D. A. Knapp, and M. B. Schneider, *Phys. Scr.* **T22**, 157 (1988).
3. E. D. Donets in *The Physics and Technology of Ion Sources*, edited by I. G. Brown (John Wiley, New York, 1989), p. 245.
4. D. A. Knapp, R. E. Marrs, S. R. Elliott, E. W. Magee, and R. Zasadzinski, *Nucl. Instrum. Meth. A* **334**, 305 (1993).
5. R. Keller in *The Physics and Technology of Ion Sources*, edited by I. G. Brown (John Wiley, New York, 1989), p. 23.
6. U. Köster, O. Kester, and D. Habs, *Rev. Sci. Instrum.* **69**, 1316 (1998).
7. W. Lotz, *Z. Phys.* **216**, 241 (1968).
8. D. Schneider, M. W. Clark, B. M. Penetrante, J. McDonald, D. DeWitt, and J. N. Bardsley, *Phys. Rev. A* **44**, 3119 (1991).
9. B. Visentin, P. Van Duppen, P. A. Leroy, F. Harrault, and R. Gobin, *Nucl. Instrum. Meth. B* **101**, 275 (1995).
10. R. E. Marrs, *Nucl. Instrum. Meth. B* (in press).
11. R. E. Marrs, D. H. Schneider, and J. W. McDonald, *Rev. Sci. Instrum.* **69**, 204 (1998).
12. M. P. Stockli, *Rev. Sci. Instrum.* **67**, 892 (1996).
13. L. P. Ratliff, E. W. Bell, D. C. Parks, A. I. Pikin, and J. D. Gillaspy, *Rev. Sci. Instrum.* **68**, 1998 (1997).
14. J. A. Nolen, *Rev. Sci. Instrum.* **67**, 935 (1996); *Concept for an Advanced Exotic Beam Facility Based on ATLAS*, edited by E. Rehm (Physics Division, Argonne National Laboratory, 1995).
15. D. Habs et al., *Nucl. Phys.* **A616**, 29c (1997).

Nuclear Astrophysics with RIBs at Oak Ridge National Laboratory

D. W. Bardayan

Physics Division, Oak Ridge National Laboratory, Oak Ridge, Tennessee 37831
and
A. W. Wright Nuclear Structure Laboratory, Yale University, New Haven, Connecticut 06511

The Daresbury Recoil Separator (DRS) and Silicon Detector Array (SIDAR) have been installed at Oak Ridge National Laboratory (ORNL) to perform measurements of reaction cross sections of astrophysical interest using radioactive ion beams (RIBs). For example radioactive ^{17}F beams will be used to determine the ^{14}O(α,p)^{17}F and ^{17}F(p,γ)^{18}Ne stellar reaction rates - both of which are important reactions in the Hot-CNO cycle. The first reactions studied will be ^{1}H(^{17}F,p)^{17}F and ^{1}H(^{17}F,α)^{14}O. These experiments will require ^{17}F beams with intensities of 10^4 - 10^6 ions per second in conjunction with the SIDAR. The ^{1}H(^{17}F,p)^{17}F reaction will be used to probe resonances in ^{18}Ne which contribute to the ^{17}F(p,γ)^{18}Ne stellar reaction rate, while ^{1}H(^{17}F,α)^{14}O will be used to determine the stellar reaction rate of the inverse reaction ^{14}O(α,p)^{17}F. In preparation for these experiments, measurements have been made of the ^{1}H(^{17}O,p)^{17}O and ^{1}H(^{17}O,α)^{14}N reaction cross sections. When higher beam currents of ^{17}F become available, a direct measurement of the ^{1}H(^{17}F,^{18}Ne) resonance strength will be made using the DRS. To test the performance of the DRS for measuring capture reaction cross sections, the well-known ^{1}H(^{12}C,^{13}N) cross section has been measured at an energy similar to those proposed for radioactive beam experiments. Results from these stable beam experiments are discussed.

INTRODUCTION

There are a number of astrophysical events during which hydrogen serves as fuel for (p,γ) fusion reactions under non-hydrostatic equilibrium conditions. These explosive hydrogen burning events, which include novae and X-ray bursts, are among the most energetic explosions (~$10^{38\text{-}45}$ ergs) known in the universe. They are characterized by extremely high temperatures ($10^{8\text{-}9}$ K) and densities ($10^{3\text{-}5}$ g/cm^3), conditions which cause (p,γ) reactions to rapidly (on timescales of ns - min) produce nuclei on the proton-rich side of the valley of stability. Any such nuclei produced with beta-decay half-lives longer than, or comparable to, the mean time between fusion events will become targets for subsequent nuclear reactions. Sequences of (p,γ) reactions on proton-rich radioactive nuclei can therefore occur during these explosions (1), and the observable ashes of such nuclear burning sequences are an important probe of the conditions in these events.

The Hot-CNO (HCNO) cycle (^{12}C(p,γ)^{13}N(p,γ)^{14}O(e$^+\nu_e$) ^{14}N(p,γ)^{15}O(e$^+\nu_e$)^{15}N(p,α)^{12}C) is a primary reaction chain through which explosive hydrogen burning occurs. The energy generation rate of this sequence is limited by the beta-decay lifetimes of ^{14}O and ^{15}O at moderately high temperatures. When the stellar temperatures are high enough (approximately 300 million degrees or higher), the beta-decay of ^{14}O can be bypassed by the ^{14}O(α,p)^{17}F reaction, and the reaction sequence ^{14}O(α,p)^{17}F(p,γ)^{18}Ne(e$^+\nu_e$) ^{18}F(p,α)^{15}O can increase the energy generation rate and alter the abundances of the CNO nuclides. The reaction sequence ^{16}O(p,γ)^{17}F(p,γ)^{18}Ne(e$^+\nu_e$)^{18}F(p,α)^{15}O can alter the CNO nuclide abundances as well, while the sequence ^{14}O(α,p)^{17}F(p,γ)^{18}Ne(e$^+\nu_e$)^{18}F(p,γ)^{19}Ne(p,γ)^{20}Na can provide a path from the HCNO cycle into the rapid proton (rp) capture process. The energy generation rate in the rp-process can be two orders of magnitude larger than the HCNO cycle (1), and at temperatures over one billion degrees, elements more massive than Fe can possibly be formed (2). In order to understand these cataclysmic stellar events, we must know the ^{17}F(p,γ)^{18}Ne and ^{14}O(α,p)^{17}F stellar reaction rates.

Comparisons of astrophysical models with the latest observations require measurements of the important reactions involving radioactive isotopes. By producing high-quality, intense beams of the radioactive ions involved in explosive nucleosynthesis, the Holifield Radioactive Ion Beam Facility (HRIBF) at ORNL (3,4) has the potential to significantly improve the understanding of these spectacular stellar explosions. Radioactive ion beams are produced at HRIBF by an ISOL-type target/ion source (5,6). A high temperature (1100 - 2200° C) refractory target is bombarded by a 0.5 kW light ion (p, d, ^{3}He, or ^{4}He) beam from the K=105 Oak Ridge Isochronous Cyclotron (ORIC). The radioactive reaction products then diffuse out of the hot target material and through a short (10 cm) transfer tube to a modular ion source, where they are ionized and ex-

CP475, *Applications of Accelerators in Research and Industry*,
edited by J. L. Duggan and I. L. Morgan

tracted. Fibrous Al_2O_3 targets (7) are being used to produce ^{17}F via the $^{16}O(d,n)^{17}F$ reactions at 30 MeV. Once produced the radioactive ions are charge-exchanged (if they are positive ions) and then undergo two stages of mass analysis (with $\Delta m / m < 10^{-4}$) before their injection into the 25-MV tandem accelerator, subsequent acceleration, and delivery to the experimental areas.

THE $^{17}F(p,\gamma)^{18}Ne$ REACTION

The $^{17}F(p,\gamma)^{18}Ne$ stellar reaction rate at temperatures characteristic of stellar explosions is thought to be dominated by resonances in ^{18}Ne at excitation energies of 4.520, 4.561, and 4.589 MeV. The 4.561 MeV resonance, however, has only been observed in a measurement of the $^{16}O(^3He,n)^{18}Ne$ reaction (8), but it was not observed in measurements of the $^{20}Ne(p,t)^{18}Ne$ and $^{12}C(^{12}C,^6He)^{18}Ne$ reactions (9,10). From comparison to the isobaric analog nucleus ^{18}O, this resonance may have $J^\pi = 3^+$, which potentially makes it a strong $\ell = 0$ transition in the $^{17}F(p,\gamma)^{18}Ne$ reaction. The $^{17}F(p,\gamma)^{18}Ne$ capture reaction itself - with a cross section $\approx$ 2.5 μb - is too weak to be used to search for the resonance. We, therefore, plan to first measure the $^1H(^{17}F,p)^{17}F$ excitation function at the HRIBF to confirm the existence of this state and measure its properties: resonance energy, width, spin, and parity. Since the cross section for this reaction is large (~ 1 barn), this measurement can be performed with relatively low ^{17}F beam currents (10^4 ^{17}F/s with a 50 μg/cm^2 CH_2 target). Once the resonance is found and higher beam currents of ^{17}F become available, a direct measurement of the resonant $^{17}F(p,\gamma)^{18}Ne$ cross section will be made using the Daresbury Recoil Separator (DRS) now located at ORNL. An intensity of approximately 6×10^7 (0.01 pnA) would be required to make a 10% measurement in 25 days at the dominant resonance.

The elastic scattering excitation function of $^{17}F(p,p)^{17}F$ will be measured with a ^{17}F beam and a polypropylene $(CH_2)_n$ transmission target in inverse kinematics (i.e., $^1H(^{17}F,p)^{17}F$). The scattered protons will be detected in an annular array of single-sided silicon strip detectors in the target chamber downstream of the target location. This silicon detector array (SIDAR) is comprised of 128 segments with 16 radial (from 5 to 13 cm) and 8 azimuthal divisions, similar to the LEDA array used at Louvain-la-Neuve (11). The array subtends 15 to 35 degrees in the lab, allowing detection of the forward-focused protons while passing the ^{17}F beam out of the target chamber. The SIDAR has the advantages of large solid angle coverage and simultaneous measurements of the particle energy and angular distribution. With a beam current of 10^4 ^{17}F/s and a CH_2 foil of thickness 50 μg/cm^2, an average of 300 protons per hour will be detected in the SIDAR. This will allow 3% statistics to be obtained in 4 hours at each beam energy, and several beam energies between 10 and 13 MeV would be used to map out the excitation function. From

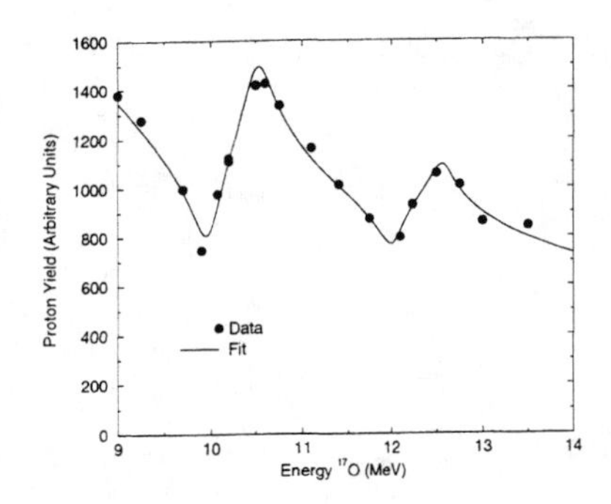

FIGURE 1. The $^1H(^{17}O,p)^{17}O$ excitation function.

the excitation function, the resonance energy and width of the state can be extracted. From the proton angular distribution measured while on resonance, the spin and parity of the state can be deduced.

In preparation for this experiment, a measurement of the $^1H(^{17}O,p)^{17}O$ excitation function has been performed. The energy range of the ^{17}O beam was picked to populate the isobaric analog in ^{18}F of the 3^+ state sought in ^{18}Ne. We also included in our measurement a nearby 2^+ state in ^{18}F. This allowed us to examine the sensitivity of the angular distribution measurement to the spin and parity of the state that we are populating. Proton yields were measured for 19 beam energies from 9 to 13.5 MeV over a period of 4 days with ^{17}O beam currents of about 5×10^6 ^{17}O/s. The normalized proton yields are plotted in Fig. 1 along with a fit to the data. The fit was made using a Breit-Wigner formalism (12), and the fit results are given in Table 1. The proton angular distributions are plotted in Fig. 2 and clearly show our sensitivity to the spins and parities of the states involved. We get a much better fit to the 10.08 MeV angular distribution data if we assume a 3^+ state in the fit; while the 12.24 MeV angular distribution data is fit much better by a 2^+ angular distribution. This agrees with the known spins and parities of these states (13).

A direct measurement of the $^1H(^{17}F,^{18}Ne)$ reaction cross section will be made using the DRS (14). The ^{18}Ne recoils will be detected at the DRS focal plane by a position-sensitive carbon-foil microchannel plate and a ΔE-E gas ionization counter. This method of direct recoil detection in inverse kinematics has several advantages over traditional capture γ-ray detection techniques (15). A high detection efficiency is possible because all recoils are within a cone of half-angle 0.5° with respect to the beam direction, and the signal-to-noise ratio is improved because the detectors are located far from the target. Other detection methods also become possible such as delayed activity detection

TABLE 1. The $^1H(^{17}O,p)^{17}O$ Fit Results.

J^π		Fit Results	Accepted (13)
3^+	E_x (MeV)	6.1605 ± 0.0009	6.1632 ± 0.0009
3^+	Γ (keV)	13.8 ± 0.6	14.0 ± 0.5
2^+	E_x (MeV)	6.2741 ± 0.0014	6.2832 ± 0.0009
2^+	Γ (keV)	11.4 ± 0.9	10.0 ± 0.5

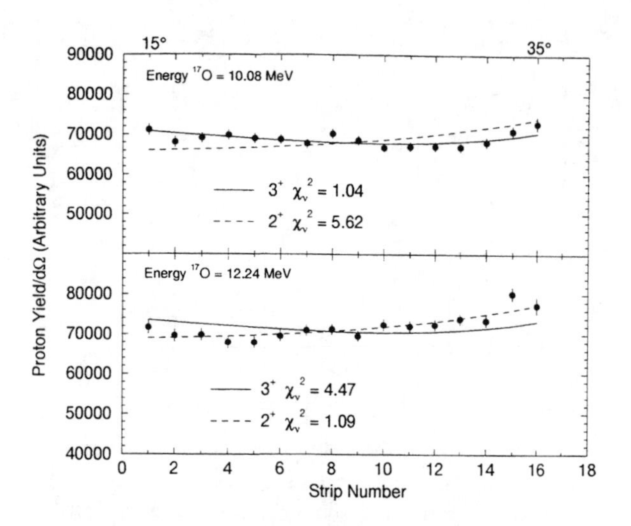

FIGURE 2. The $^1H(^{17}O,p)^{17}O$ angular distributions at the resonance energies. The solid line is a fit assuming that the state is a 3^+ resonance. The dashed line is a fit assuming that the state is a 2^+ resonance. The reduced χ^2 of the fits are shown.

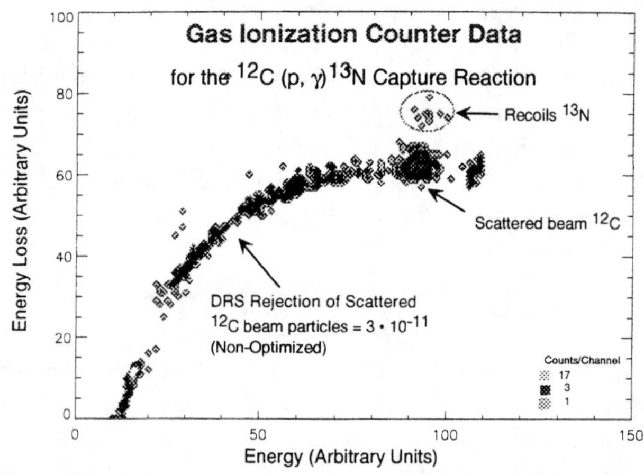

FIGURE 3. Particle identification in the DRS ionization counter.

and recoil-γ coincidences. The recoil detection approach is, however, challenging because: the separator must be located along the beam axis and accepts all of the beam particles; the recoils are only 10^{-10} to 10^{-12} times as intense as the projectiles; and the projectiles and recoils are similar, having nearly identical momentum and differing in velocity and mass by only a few percent. The recoil separator must, therefore, be optimized to collect the ions of interest while simultaneously suppressing the transmission of unwanted scattered beam projectiles. The focal plane detectors must also be capable of distinguishing the recoils from the projectiles, which is quite challenging in view of the low energy (0.4 - 2.0 MeV/amu) and low mass of the particles to be identified.

To test the performance of the DRS for measuring capture reaction cross sections, a measurement of the well-known $^1H(^{12}C,^{13}N)$ cross section has been made. A 0.666 MeV/amu ^{12}C beam was used to bombard a CH_2 target, and ^{13}N recoils were collected at the focal plane of the DRS for 12 hours. The ionization counter spectrum for this capture reaction measurement is shown in Fig. 3. The suppression of scattered beam particles - defined by the ratio of beam particles incident on target to those reaching the focal plane - was 3×10^{-11}, which is within the range needed (10^{-10} to 10^{-12}) for capture reaction measurements. This rejection value is consistent with previous measurements that utilized elastic scattering reactions (e.g. $^{12}C(^{14}N,^{12}C)^{14}N$) (16). Furthermore, the gas ionization counter cleanly separated the ^{13}N recoils from the scattered ^{12}C projectiles. The combined projectile rejection of the DRS and the focal plane detector was, therefore, beyond that needed for capture reaction measurements. The DRS transmission, however, for the recoils was 7%, a factor of 7 lower than expected. Since the CH_2 targets could only withstand 0.5 pnA of beam, the ^{13}N count rate was only a few counts per hour which was too low to be used to optimize the DRS settings. We have found in subsequent tests with scattering and fusion-evaporation reactions that the optical elements of the DRS were not appropriately tuned during this measurement and that the dispersions of the different sections of the DRS were not matched properly. This resulted in the ^{13}N recoils not being focused at the spectrometer focal plane and a reduction in the observed transmission. Optimization of the DRS using fusion-evaporation and scattering reactions is in progress, and measurements of capture reactions with stable beams are planned for the future.

THE $^{14}O(\alpha,p)^{17}F$ REACTION

The rate for the $^{14}O(\alpha,p)^{17}F$ reaction is also quite uncertain. Stable beam spectroscopy measurements indicate that a 1^- state at E_x = 6.150 MeV in ^{18}Ne provides the dominant resonant contribution to the $^{14}O(\alpha,p)^{17}F$ reaction rate at temperatures less than 10^9 K. At these temperatures, the rate depends upon the unknown spectroscopic factors (total and partial-alpha widths) of the 6.150 MeV state, as well as upon the $\ell = 1$ direct reaction component and the interference between them (9). Higher energy states, including those observed at 6.29, 7.05, 7.12, and 7.35 MeV, are expected to make a significant contribution to the reaction rate at temperatures greater than 10^9 K.

A measurement of the $^1H(^{17}F,\alpha)^{14}O$ reaction will be made to determine spectroscopic properties of resonances at E_x = 6.0 - 7.5 MeV in ^{18}Ne. A ^{17}F beam with energies between 39 and 64 MeV will be used with a transmission polypropylene CH_2 target. The recoil alpha particles will be detected in the SIDAR which will be run in "telescope" mode (i.e. a 100-μm-thick detector backed by a 500-μm-thick detector) to discriminate between the alpha recoils and scattered protons and projectiles. The SIDAR will subtend angles 10 to 25° in the lab. Additionally, the ^{14}O recoils will be detected by a smaller annular silicon detector which will be placed behind the SIDAR and subtend angles 3.2 to 6.5° in the lab. This "mini" detector is also

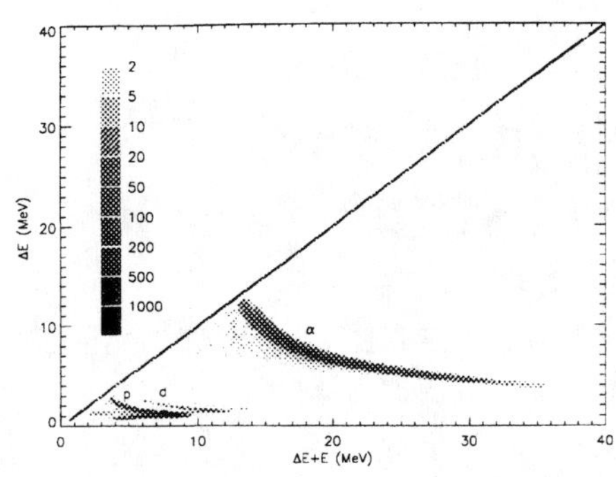

FIGURE 4. Particle identification from the $^{1}H(^{17}O,\alpha)^{14}N$ reaction using the SIDAR.

highly segmented (16 radial and 4 azimuthal divisions) and, therefore, can handle the high counting rates from elastic scattering experienced at such forward angles without significant pileup.

In preparation for this experiment, we have performed a measurement of the $^{1}H(^{17}O,\alpha)^{14}N$ reaction. A 40 MeV ^{17}O beam (5×10^{6} $^{17}O/s$) was used to bombard a 100 $\mu g/cm^2$ CH_2 target for a period of 3 days. The SIDAR particle identification is shown in Fig. 4 where the alpha recoils are cleanly separated from the other reaction products and scattered projectiles. The recoil detector spectrum is shown in Fig. 5, where a peak from ^{14}N recoils was observed. The shaded spectrum was produced by gating on alpha recoils in the SIDAR which reduced the background significantly.

CONCLUSIONS

We have made stable beam measurements of reactions similar to the ones we will measure with a radioactive ^{17}F beam. We have demonstrated our readiness to perform radioactive ion beam experiments and measured reaction cross sections which will contribute background to our radioactive beam experiments. Properties of resonances in ^{18}F have been measured with the $^{1}H(^{17}O,^{1}H)^{17}O$ reaction

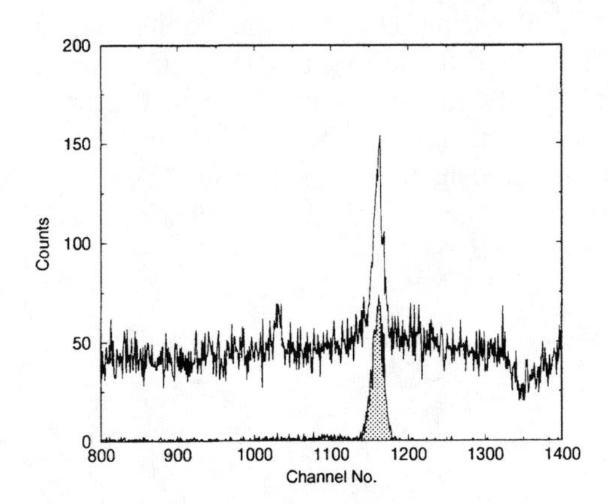

FIGURE 5. The energy spectrum from the recoil silicon detector. The shaded graph was gated on alphas in the SIDAR.

using the SIDAR, and a measurement of the $^{12}C(p,\gamma)^{13}N$ reaction cross section has been made using the DRS. We have, furthermore, made a measurement of the $^{1}H(^{17}O,\alpha)^{14}N$ cross section by detecting ^{14}N and alpha recoils in coincidence in two detector arrays.

ACKNOWLEDGMENTS

Many people have been involved in the DRS and SIDAR commissioning work. These include P. Bertone, J. C. Blackmon, A. E. Champagne, A. A. Chen, T. Davinson, U. Greife, V. Hansper, M. A. Hofstee, A. N. James, B. A. Johnson, P. E. Koehler, R. L. Kozub, Z. Ma, J. W. McConnell, W. T. Milner, P. D. Parker, D. E. Pierce, M. Rabban, D. Shapira, A. C. Shotter, M. S. Smith, F. Strieder, K. B. Swartz, C. Ulrey, D. W. Visser, and P. J. Woods. Research at the Oak Ridge National Laboratory is supported by the U. S. Department of Energy under contract DE-AC05-96OR22464 with Lockheed Martin Energy Research Corp. Additional support is provided by contracts DE-FG02-91ER40609 with Yale University, and DE-FG02-96ER40995 with Tennessee Technological University.

REFERENCES

1. Wallace, R. K., and Woosley, S. E., Ap. J. Suppl. **45**, 389-420 (1981).
2. Schatz, H., *et al.*, Phys. Rep. **294**, 168-263 (1998).
3. Garrett, J. D., *et al.*, Nucl. Phys. **A557**, C701-C714 (1993).
4. Ball, J. B., Nucl. Phys. **A570**, C15-C22 (1994).
5. Alton, G. D., *et al.*, Nucl. Inst. Meth. **A328**, 325-329 (1993).
6. Alton, G. D., *et al.*, Nucl. Inst. Meth. **A382**, 207-224 (1996).
7. Alcen™, Alumina Fiber, RATH Performance Fibers, Wilmington, DE, USA.
8. García, A., *et al.*, Phys. Rev. **C43**, 2012-2019 (1991).
9. Hahn, K. I., *et al.*, Phys. Rev. **C54**, 1999-2013 (1996).
10. Park, S. H., *et al.*, submitted to Phys. Rev. C.
11. Coszach, R., *et al.*, Phys. Lett. **B353**, 184-188 (1995).
12. Blatt, J. M., and Biedenharn, L. C., Rev. Mod. Phys. **24**, 258-272 (1952).
13. Tilley, D. R., *et al.*, Nucl. Phys. **A595**, 1-170 (1995).
14. James, A. N., *et al.*, Nucl. Inst. Meth. **A267**, 144-152 (1988).
15. Smith, M. S., *et al.*, Nucl. Inst. Meth. **A306**, 233-239 (1991).
16. Smith, M. S., "Nuclear Astrophysics and the Daresbury Recoil Separator at the Holifield Radioactive Ion Beam Facility," presented at the International Symposium on Origin of Matter and Evolution of Galaxies 97, Atami, Japan, Nov. 5-7, 1997.

The Physics of Sub-critical Lattices in Accelerator Driven Hybrid Systems : The Muse Experiments in the MASURCA Facility

J.P. CHAUVIN* - J.F. LEBRAT* - R. SOULE* - M. MARTINI*
R. JACQMIN* - G.R. IMEL* - M. SALVATORES*

Commissariat à l'Energie Atomique, Centre d'Etudes de CADARACHE, Direction des Réacteurs Nucléaires, Département d'Etude des Réacteurs, Service de Physique Expérimentale, Laboratoire des Programmes Expérimentaux 13108 SAINT-PAUL-LEZ-DURANCE - France - Tél. (33) 4.42.25.36.55 - Fax (33) 4.42.25.77.08
e-mail : chauvin@macadam.cea.fr

ABSTRACT

Since 1991, the CEA has studied the physics of hybrid systems, involving a sub-critical reactor coupled with an accelerator. These studies have provided information on the potential of hybrid systems to transmute actinides and, long lived fission products. The potential of such a system remains to be proven, specifically in terms of the physical understanding of the different phenomena involved and their modelling, as well as in terms of experimental validation of coupled systems, sub-critical environment/accelerator. This validation must be achieved through mock-up studies of the sub-critical environments coupled to a source of external neutrons. The MUSE-4 mock-up experiment is planed at the MASURCA facility and will use an accelerator coupled to a tritium target. The great step between the generator used in the past and the accelerator will allow to increase the knowledge in hybrid physic and to decrease the experimental biases and the measurement uncertainties.

INTRODUCTION

Since 1991, the CEA has studied the physics of hybrid systems, involving a sub-critical reactor coupled with an accelerator.

These studies have provided information [1] on the potential of hybrid systems to transmute actinides such as Americium, Neptunium, Curium and, possibly long lived fission products.

This potential [2] is to be found in:

- the concentration of wastes in a limited number of dedicated facilities,
- the sub-criticality of such a system, which is a particularly attractive argument in favour of the safety of such concepts and which, more particularly, allows for the introduction of new fuels.

Nevertheless, the potential of such a system remains to be proven. Specifically, the physical understanding of the different phenomena involved and their modelling, including experimental validation of coupled sub-critical systems/accelerators must be shown.

This validation must be achieved through mock-up studies of the sub-critical environments coupled to a source of external neutrons from the accelerator.

In these phenomena, the main neutronic parameters are those characterising the source and its amplification. The spatial distribution of neutron flux and thus the power in a sub-critical environment with a source, essentially depends on the properties of the environment, on its sub-criticality level and on the characteristics of the neutron source (intensity, importance, spatial distribution).

The first studies conducted on hybrid systems thus led the CEA to define an experimental program in the MASURCA reactor. This program, an inherent part of the ISAAC theoretical and experimental study program [3], of which the experimental configuration called MUSE "1 to n" are of increasing importance, started in 1995. These use the maximum potential of the MASURCA reactor, for the validation of the properties and characteristics of the neutronic coupled systems.

CP475, *Applications of Accelerators in Research and Industry*, edited by J. L. Duggan and I. L. Morgan

ISAAC PROGRAM AND MUSE EXPERIMENTAL CONFIGURATIONS IN MASURCA

The CEA, with the MASURCA reactor[4] having begun prospective studies on hybrid systems since 1991, started developing in 1995 an experimental program, the MUSE program in ISAAC, on the validation of the properties and characteristics of these systems [5].

In fact, in an external neutron source supplying a sub-critical multiplying system, the relation between the source characteristics (spectrum, position...) and sub-critical level (k_{eff}) determines in principle the neutron flux level and the overall energy balance of the coupled system. This can be modelled if the source neutron importance is taken into account in a appropriate manner.

Neutronics of sub-critical systems

The neutron flux distribution Φ in a sub-critical core with a source S is the solution of the well-known inhomogeneous equation:

$$A\Phi = M\Phi + S, \qquad (1)$$

(A: net neutron loss operator / M: fission production operator)

The flux of the associated homogeneous systems (i.e. without source) is solution of the equation

$$A\Phi_o = (1/K_{eff})\, M\Phi_o, \qquad (2)$$

According to these equations, we define some neutronic quantities that are proved to be particularly relevant to the description of the sub-critical source driven systems. For instance,

$$K_S = \frac{< M, \Phi >}{< M, \Phi > + S} \qquad (3)$$

represents the ratio of fission neutrons to the total neutron source in the sub-critical system and :

$$\phi^* = \frac{\dfrac{< \Phi_o, S >}{< S >}}{\dfrac{< \Phi_o^*, M\Phi}{< M\Phi >}} = \frac{\overline{\Phi_s^*}}{\overline{\Phi_f^*}} \qquad (4)$$

is the ratio of the importance of external source neutrons to the importance of fission neutrons. This key parameter, which will be measured in the experiment, is particularly sensitive to the energy of the neutrons emitted by the source and indicates the importance of the source on the whole system.

MUSE EXPERIMENTS

The MUSE experiments give insight into the physics of sub-critical cores and provide valuable information for validating the sophisticated methods and data used for the analysis of these systems.

The MUSE [6] experiment has shown the potential of the MASURCA facility to investigate the physics of sub-critical multiplying systems in the presence of an external source. The accuracy of the measurements is very high, and known experimental techniques, used in standard experiments to validate the physics of fast neutron cores, can be applied with confidence.

The short, exploratory MUSE-1 programme has provided some insight into the physical behaviour of the neutron population in the sub-critical system.

Moreover, a relevant integral parameter, i.e. the external neutron source effectiveness ϕ^*, was defined and measured with high accuracy. This point is very relevant, since ϕ^* can be a key parameter in the optimisation of a hybrid system, to improve the energy balance and to provide an optimised neutron importance distribution.

The MUSE-2 [7] experiments were devoted to the experimental study of diffusing materials (sodium and stainless steel) placed around the external source and have been performed in co-operation with CNRS-France.

Good agreement was found between measured and calculated values of radial and axial ^{235}U fission rate distributions and relative importance of external source neutrons (ϕ^*).

MUSE 3

The different experimental configurations have been obtained by loading the MASURCA reactor with an U02-Pu02 fuel containing 25 % of Pu ("PIT" fuel), and Na as coolant. A pulsed source producing 14 MeV neutrons from D-T reactions was introduced in the middle of the core. Several configurations were set up by modifying the level of sub-criticality (fuel unloading) or surrounding the source with a buffer made successively of sodium and lead. The experimental configurations are summarised in the following table:

Configuration Name	Reactivity (pcm)	Number of PIT cells in core
MUSE3-ref	- 110	960
MUSE3-SC1	- 440	944
MUSE3-SC2	- 1100	912
MUSE3-SC3	- 1410	896
MUSE3-Pb	- 3920	1040 + 128 for lead buffer
MUSE3-Na	- 4099	1040 + 128 for sodium buffer

The measurements performed in MUSE-3 experiment are equivalent to those in the MUSE-1 and 2 experiments. In addition, using the pulse mode of the neutron source generation we performed dynamic measurements. A complementary determination of large negative reactivities was obtained by using the source in the pulsed mode. The count rate distribution was recorded on a multichannel analyzer during repeated pulses, and several methods were used to determine the reactivity of the system. Such measurements, cross-checked with those obtained with traditional sub-critical source multiplication methods, have helped us define an alternative method to determine reactivity in a source-driven sub-critical system and the kinetics parameters of the systems.

These experimental results are being analysed using the ERALIB-1 nuclear date library, combined with the ECCO cell code and the ERANOS [8] code system to perform spatial 3D nodal transport calculations. Nevertheless, both the resolution time of the burst (about 3 μs) and the intensity of the pulse (2E8 n/s) should be optimized to decrease experimental biases and measurement uncertainties.

The MUSE-4 experiment is planned in the near future at the MASURCA facility and will use a high intensity deuterium accelerator coupled to a tritium target to solve these problems. The core composition, which is under study at this point, may include lead coolant.

The use of this accelerator called "GENEPI" jointly defined between the CEA and the CNRS enables us to obtain an even higher degree of representativity of the critical MASURCA mock-up versus hybrid systems. It will provide the necessary elements for the study of a pulsed or continuous high-energy source as well as for the study of diffusant materials that can surround it.

CONCLUSION

This experimental programme whose three first configurations have already been achieved, and others are planned, is a parametric study of the physical phenomena occurring in a hybrid system. Each step of this parametric study represented by a specific configuration has allowed the validation, through comparison between experiment and calculation, of the system of codes used in such a medium (ECCO-ERANOS code system).

The next configurations of this programme will allow us to understand more deeply the contribution of the external source through the use of an accelerator. The future will be certainly devoted to the validation of the solutions proposed in the beginning for the demonstration [9], [10].

REFERENCES

[1] M. SALVATORES - I. SLESSAREV - M. UEMATSU - A. TCHISTIAKOV
The Neutronic Potential of Nuclear Power for long-term Radioactivity Risk Reduction.
International Conference on Evaluation of Emerging Nuclear Fuel Cycle Systems.
GLOBAL'95. September 11-14, 1995. Versailles, France.

[2] M. SALVATORES - I. SLESSAREV - A. TCHISTIAKOV
"Nuclear Power Development and hybrid system role"
Second International Conference on Accelerator-Driven Transmutation Technologies and Applications Kalmar, Sweden - June 3-7, 1996.

[3] M. SALVATORES et al.
"French Programs for Advanced Management Options"
Proc. 2nd Int. Conf. on ADS, Kalmar, June 1996.

[4] J.C. CABRILLAT - J.P. CHAUVIN - M. MARTINI
The nuclear fuel cycle backend.
The use of the CEA critical facilities for the assessment of the physics.
Nuclear Engineering and Design 170 (1997).

[5] M . SALVATORES et al.
"The neutronics of a source-driven multiplying medium and its experimental validation at MASURCA"
Proc. of Int. Conf. PHYSOR'96. Mito (Japon), Sept. 1996.

[6] M. SALVATORES et al.
"MUSE 1 : A first experiment at MASURCA to validate the physics of sub-critical multiplying systems relevant to ADS".
Int. Conf. on Accelerator Driven Transmutation Technologies, Kalmar, June 1996.

[7] R. SOULE - M. SALVATORES - R. JACQMIN - M. MARTINI - J.F. LEBRAT - P. BERTRAND - J. BROCCOLI - V. PELUSO
Validation of neutronic methods applied to the analysis of fast sub-critical systems : the MUSE 2 experiments.
Int. Conf. on Evaluation of Emerging Nuclear fuel Systems GLOBAL'97 Yokohama - Japan - Octobre 1997.

[8] J.Y. DORIATH et al.
"ERANOS : The Advanced European System of Codes for Reactor Physics Calculations"
Proc. Int. Conf. Math. Methods and Supercomputing, Karlsruhe, 1993.

[9] C. RUBBIA et al.
"Conceptual Design of a Fast Neutron Operated High Power Energy Amplifier"
CERN/AT/95-44 (1995).

[10] M. SALVATORES - I. SLESSAREV - A. TCHISTIAKOV, G. RITTER
"The potential of Accelerator Driven Systems for transmutation of Power Production using Thorium or Uranium Fuel Cycles"
Nucl. Sci. Eng. (March 1997).

Near Threshold (Gamma, 2e) Studies

G. C. King, N. A. Gulley, G. Dawber, H. Rojas and L. Avaldi

Dept of Physics and Astronomy, University of Manchester, Manchester M13 9PL, UK

Near threshold (γ,2e) measurements have been made in helium to determine the triple differential cross section. The measurements have been made for various energy sharings of the two outgoing electrons. The value of the total excess energy, E, was varied from 0.08 to 0.84eV while the ratio of the energies of the two electrons, E_1:E_2 , was varied over the range from 1 to 20. These measurements have been used to deduce the shape of the correlation function.

INTRODUCTION

In recent years there has been considerable interest in the photo-double ionisation process since it is the archetypal system to study the three-body Coulomb problem. The experimental studies investigate the dynamics of the excess energy sharing between the two outgoing electrons and their associated angular correlations e.g.(1). The triple differential cross section (TDCS) measurement is the most complete picture of the photo-double ionisation process and the best experimental test of it. It is obtained by measuring the energies and ejection angles of both outgoing electrons simultaneously in a coincidence measurement. This can be achieved by using two separate angle-resolving spectrometers as has been done in the present work. The observed shape of the TDCS is determined by a product of factors including kinematical factors which contain the ejection angles, θ_1 and θ_2, of the two photoelectrons, and an angular correlation function that results from the Coulomb repulsion of the two electrons. The correlation function depends on the electron energies E_1 and E_2 and on the mutual angle, θ_{12}, between them. Looking at the photo-double ionisation process close to threshold has the advantage that the expression for the TDCS can be somewhat simplified so that the correlation function can be isolated and more easily determined. The shape of the correlation function has been the subject of debate for many years although most theories agree that, close to threshold, it is a Gaussian centred at θ_{12} = 180°. Furthermore the width of the Gaussian is predicted to vary as a function of the total excess energy, E, shared by the two electrons. The function is given by the relationship $\theta_{1/2} = \theta_o E^{\rho}$, with ρ predicted by Wannier (2) to be 0.25. The purpose of the present study was to determine the shape of the correlation function very close to threshold ($E \rightarrow 0$), and compare this with theory. This was done by measuring TDCS for various values of E and the ratio E_1/E_2.

APPARATUS AND EXPERIMENTAL METHOD

The experimental apparatus used in the present work consists of a photon source of variable energy, an effusive gas jet and a pair of identical electron analysers. These analysers are arranged in a plane perpendicular to, and centred about, the photon beam direction. The photon beam, which was linearly polarised in the horizontal plane, was provided by the TGM beamline at the Daresbury SRS. The photon flux was monitored by an aluminium oxide photodiode placed beyond the interaction region. One analyser was placed at a fixed angle with respect to the direction of the linear polarisation of the beam. The second analyser could rotate about the photon beam direction over a large angular range. An earlier form of the apparatus has been described elsewhere (3). TDCS were obtained by performing a coincidence measurement between the signals from each analyser, using standard coincidence counting electronics. The total and random signals as well as the signals from each analyser and the photodiode current, were all recorded in the data acquisition software, which also controlled the photon energy. A spectrum of coincidences as a function of photon energy, $E_{h\nu}$, for a given excess energy, E, was obtained by scanning $E_{h\nu}$ over the region centred around $E_{h\nu} = E_{th} + E$, where E_{th} corresponds to the photo-double ionisation threshold. The collection energies of the two analysers, E_1 and E_2, were such that $E = E_1 + E_2$. The overall energy width of a coincidence peak obtained in this way, due to the finite energy resolution of both analysers and the incident photons, was 150meV. The total coincidence counting rate was typically between 2 and 0.1 coincidences per second with the true to random ratio in excess of 20; the latter being limited by the available photon flux. Due to these low count rates a large number of scans were taken for each angle.

CP475, *Applications of Accelerators in Research and Industry*,
edited by J. L. Duggan and I. L. Morgan

The angular behaviour of the spectrometer was characterised regularly by observation of the $He^+(N=2)$, and $He^+(N=1)$ photoionisation cross sections as a function of the angle θ_2 of the moving analyser. In the case of $He^+(N=2)$, the distribution is very nearly isotropic ($\beta \sim 0$) in the near threshold region, whereas the $He^+(N=1)$ has an angular distribution of $\beta=2$. Additional measurements were made using the neon $Ne^+2s^{-1}(^2S)$ state which also has a value of $\beta=2$. These measurements allowed the degree of linear polarisation of the photon beam in the energy range of interest to be determined as $S_1 = 0.76 \pm 0.04$.

RESULTS AND DISCUSSION

For measurements of the TDCS it is important to maximise both the angular range and the coincidence signal to obtain as complete a picture as possible. It has been pointed out (4,5) that the strongest signal should arise when one analyser is fixed perpendicular to the polarisation vector. For this reason the first analyser was fixed at 90° with respect to the polarisation axis and the second analyser was rotated over a total angular range of 135° (from 195° to 330°).

TDCS data were taken for various values of total energy and energy sharings (6). The values of E varied from 0.08 to 0.84 eV, while the ratio $E_1:E_2$ varied over the range from 1 to 20. An example of a measured TDCS obtained in He for $E_1 = 0.2$, $E_2 = 0.1$ eV is shown in figure 1.

The observed shape of the TDCS is determined by a product of factors including kinematical factors which contain the ejection angles, θ_1 and θ_2, of the two photoelectrons and an angular correlation function that

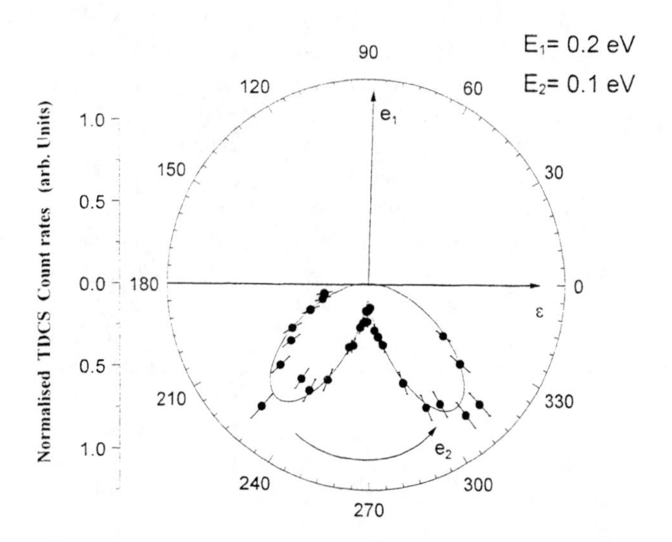

FIGURE 1. Measured TDCS for photodouble ionisation of helium, for electron energies, E_1=0.2 eV and E_2=0.1 eV.

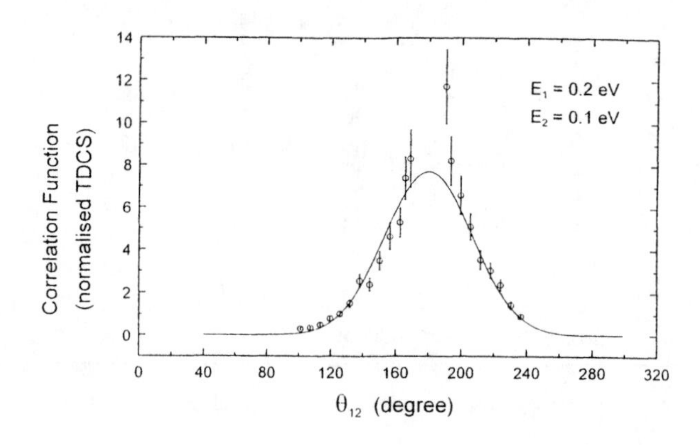

FIGURE 2. Correlation function deduced from the TDCS data for E_1=0.2 eV and E_2=0.1 eV.

results from the Coulomb repulsion of the two electrons (7). As noted above, the correlation function depends on the electron energies, $E_1$and E_2 and on the mutual angle θ_{12} between them. For ionisation close to threshold, the expression for the TDCS can be simplified to

$$\text{TDCS} = |\, a_g\,(E_1, E_2, \theta_{12})\,|^2\,(\cos\theta_1 + \cos\theta_2)^2$$

where $|\, a_g\,(E_1, E_2, \theta_{12})\,|^2$ is the correlation function and $(\cos\theta_1 + \cos\theta_2)^2$ contains the kinematical factors. Since θ_1 and θ_2 are determined, the correlation factor could be obtained from the expression (8)

$$\text{correlation factor} = \frac{\text{measured TDCS}}{\text{kinematical factor}}$$

Using this procedure the correlation function has been deduced for the various measured TDCS. For example the deduced correlation function for the data of figure 1, (i.e. for $E_1 = 0.2$ eV, $E_2 = 0.1$ eV) is shown in figure 2. Note that a good fit is not expected for the correlation function close to 180° since the kinematic factor tends to infinity around this point.

The present work extends the TDCS measurements to within 0.08eV of threshold. From the measurements of the shape of the correlation function as a function of E and the ratio E_1/E_2 two conclusions can be made. As E approaches zero the Gaussian shape provides a better fit to the correlation function while the shape of the correlation function is essentially independent of the ratio E_1/E_2. This was expected and had been predicted by theory e.g. (4, 9). However, the relationship, $\theta_{1/2} = \theta_0 . E^{\rho}$ does not appear to hold. Combining the present data with previous close-to-threshold data (10) θ_0 was determined to be $74 \pm 6°$ but ρ varied dramatically with a 'best fit value' found to be 0.038 ± 0.025. This value of ρ is much lower than the Wannier prediction of 0.25, and therefore further analysis of these values is currently being undertaken. Recent above-threshold studies have also indicated that the exponent is less than 0.25 e.g.(8, 11, 12) but no fixed value for the parameters have yet been determined. The present work demonstrates the trend in the value of θ_{12} , suggested by other workers as threshold is approached.

SUMMARY

Triple differential cross sections have been measured in helium in the near threshold region for total energies as small as 0.08eV. These measurements have allowed the shape of the correlation function to be deduced for various values of total energy and energy sharings. The shape is seen to approach that of a Gaussian as threshold is approached. The measurements have also allowed the constant θ_o and the exponent ρ to be determined. The obtained value of ρ is seen to be consistent with recent, non-threshold measurements.

REFERENCES

1. Schmidt, V., Rep.Prog.Phys. **55**,1483-1659 (1992).
2. Wannier, G.H., Phys. Rev. **90** 817 (1953).
3. Hall, R.I., McConkey A., Ellis, K., Dawber, G., Avaldi, A.,MacDonald, M. A., and King, G.C. Meas. Sci. Tecnnol., **3** 316-324 (1992).
4. Huetz, A., Lablanquie, P., Andric, L., Selles, P, and Mazeau, J., J. Phys.B: Atom. .Mol. Opt. Phys **27** L13-18 (1994).
5. Schmidt, V., Schaphorst, S.J., Krassig, B., Schwarzkopf, and Scherer, N., J. Elec. Spec. and Rel. Phenom., **79** 279-282 (1996).
6. Gulley, N. J., Dawber, G., Rojas, H., Avaldi, L., MacDonald, M. A. and King, G. C., J. Phys.B: Atom. Mol.Opt.Phys. to be published (1998)
7. Huetz, A. L., Selles, P, Waymel, D, and Mazeau, J., J. Phys.B: Atom. Mol. Opt. Phys **24** 1917 (1991).
8. Malaget, L., Selles, P., Lablanquie, P., Mazeau, J.,and Huetz, A., J. Phys.B: Atom. .Mol. Opt. Phys. **30** 263-276 (1997).
9. Maulbetsch F and Briggs J S, J.Phys.B:Atom. Mol. Opt. Phys. **26** 4095 (1994)
10. Dawber G, Avaldi L, Zubek M, McConkey A G, Rojas H, Gulley N, MacDonald M A, King G C and Hall R I, Can. J. Phys. **74** 782-788 (1996)
11. Kazansky, A.K., and Ostrovsky, V. N., Phys. Rev. A **52** 1175 (1995).
12. Reddish, T., Cvejanovic, S., and Shiell, R., SPIG 96 Kotor, Yugoslavia (1996).

Novel, Spherically-Convergent Ion Systems for Neutron Source and Fusion Energy Production*

D. C. Barnes, R. A. Nebel, F. L. Ribe, M. M. Schauer, L. S. Schranck, and K. R. Umstadter

Los Alamos National Laboratory, Los Alamos, NM 87545

Combining spherical convergence with electrostatic or electro-magnetostatic confinement of a nonneutral plasma offers the possibility of high fusion gain in a centimeter-sized system. The physics principles, scaling laws, and experimental embodiments of this approach are presented. Steps to development of this approach from its present proof-of-principle experiments to a useful fusion power reactor are outlined. This development path is much less expensive and simpler, compared to that for conventional magnetic confinement and leads to different and useful products at each stage. Reactor projections show both high mass power density and low to moderate wall loading. This approach is being tested expeimentally in PFX-I (Penning Fusion eXperiment – Ions), which is based on the following recent advances: 1) Demonstration, in PFX (our former experiment), that it is possible to combine nonneutral electron plasma confinement with nonthermal, spherical focussing; 2) Theoretical development of the POPS (Periodically Oscillating Plasma Sphere) concept, which allows spherical compression of thermal-equilibrium ions; 3) The concept of a massively-modular approach to fusion power, and associated elimination of the critical problem of extremely high first wall loading. PFX-I is described. PFX-I is being designed as a small (<1.5 cm) spherical system into which moderate-energy electrons (up to 100 kV) are injected. These electrons are magnetically insulated from passing to the sphere and their space charge field is then used to spherically focus ions. Results of initial operation with electrons only are presented. Deuterium operation can produce significant neutron output with unprecedented efficiency (fusion gain Q).

INTRODUCTION

A new approach is being developed for thermonuclear plasma confinement. In this approach, referred to as Penning Fusion (PF), electrons are contained as a nonneutral plasma beam or cloud in a trap of the Penning type, *i.e.* one with both applied electrostatic and magnetostatic fields. This trap is configured with an access through which electrons are injected and subsequently ejected and collected. Electron space charge forms an electrostatic well for confinement of an ion minority. Thermonuclear ions require a well depth of order 100 kV. Achievable electric fields limit average densities in such a system to less than 10^{18} m^{-3}. An interesting power density results with strong spherical focussing of the ions.

The physics of nonneutral electron confinement imply that thermonuclear conditions may be produced in a PF system with a spherical radius of a few mm. This small size and associated low cost imply a development path orthogonal to that for conventional fusion confinement systems. A range of applications can be addressed sequentially (beginning now!) as system performance improvements are developed and demonstrated. In contrast to traditional approaches, success or failure does not hinge on achieving performance suitable for the ultimate application of stationary fusion electric power generation.

In the short term, PF systems may be developed as sources of energetic neutrons or protons. A typical commercial application of PF as a neutron source is for petroleum well logging. For this application, the small size and high Q (fusion power/input power) offer significant advantages over existing beam-target approaches.

The physics of PF is being studied theoretically and experimentally, and the state of this development is reported here.. In particular, a small experiment called PFX-I (PF eXperiment - Ions) has just begun operation, and its design and initial operation are described.

In the remainder of this paper, the physics principles and scaling laws which fix system configurations are first presented. Next, relevant theory and the results of a previous, electron-only experiment, PFX, are reviewed. Then, the design and initial operation of PFX-I are discussed. Finally, future development directions and conclusions are presented.

* This work is supported by the Office of Fusion Energy Sciences, US Department of Energy

CP475, *Applications of Accelerators in Research and Industry*, edited by J. L. Duggan and I. L. Morgan

PHYSICS PRINCIPLES AND SCALING

Size/Voltage/Power Variation

To analyze the PF concept, an idealized system is first considered. In this idealization, an electrically conducting, grounded spherical shell (radius a) is filled with an electron plasma. In order that spherical symmetry hold, the electron density should depend only on spherical radius r. The simplest such variation is the case when the density n_o is constant in space. In this case, the space charge potential is:

$$\Phi_{sc} = \frac{e\, n_o}{6\, \varepsilon_o}\left(r^2 - a^2\right) \quad . \tag{1}$$

The resulting electric field at r = a is E = - 2 W/a, where $W = e n_o a^2 / 6\, \varepsilon_o$ is the well depth in V. For thermonuclear ions, W ~ 100 kV. Since the wall is an anode (E points away from wall), E approaching 10^8 V/m is possible, so that a > 2 - 5 mm may be sufficient to prevent electrical breakdown at the wall.

An interesting feature of PF systems is the *increase* in fusion output with *decreasing* radius a. This occurs if a varies while W is maintained constant at a thermonuclear value. Then, n_o varies as a^{-2}, and reactivity varies as $a^3\, n_o^2$, provided that the mean ion density squared varies as n_o^2.

To analyze ion density variation, note first that the average ion density is some fraction f_i (typically 0.01 - 0.1) of n_o. Next, because of spherical focussing, there is a gain in output from peaking of the density, which depends on a/r_c, with r_c the radius of the spherical focus. Two modes of spherical ion convergence have been considered. In the first case (1-2) of Spherically Convergent Ion Flow (SCIF), steady, radial ion flow produces a central focus whose density is enhanced by $(a/r_c)^2$, compared to the average. In the second case (3-4) of Periodically Oscillating Plasma Sphere (POPS), a thermal ion cloud on a low adiabat undergoes large-amplitude radial oscillations, periodically compressing the cloud to a r_c, with density enhanced by $(a/r_c)^3$.

From these considerations, the space (and time) averaged power output may easily be shown to be

$$P = F(W)\, f_i^2 \left(\frac{a}{r_c}\right)^{\nu} a^{-1} \quad , \tag{2}$$

where F(W) is proportional to the fusion reactivity ($<\sigma\, v>$) times W^2, and where $\nu = 1$ for SCIF and $\nu = 2$ for POPS (4).

From Eqn. (2) follows the curious goal of making the system size a as small as possible, provided all other system parameters may be maintained at the desired values. Practical considerations of electrical breakdown and precision of construction suggest that an optimum system might have a ~ 5 - 10 mm, W ~ 50 - 300 kV, and a/r_c ~ 100 - 1000. For these parameters, P = 0.2 - 700 W, with fusion gain Q = 0.01 - 30.

The unique scaling of PF systems suggests that their development will be accomplished by beginning with larger, lower-voltage systems for confirming the physics principles and evolve by improved technology of miniaturization and electrical breakdown resistance, with Q and P increasing with each improvement.

Electron Physics Principles

The ideal system considered to this point may be approximated by flowing electrons through the trap. In a real system, the surface of the ideal spherical shell will be interrupted to allow this electron flow. The simplest arrangement is that of a properly shaped, rotationally symmetric conductor (topologically a cylinder) which allows electron flow through its "polar" regions. For convenience, this conductor is maintained at electrical ground, and electrons are supplied from a cathode axially removed from the trap volume and raised to a high negative voltage.

A key feature of the electron physics is that electrons should be maintained in a nonthermal distribution. To see this, note that for *thermal* electrons to have a nearly constant density in the trap, their temperature must be much larger than the well depth, $k_B\, T_e >> e\, W$. At the same time, these electrons are ultimately confined by an applied voltage V_o, and the required excellent electron confinement requires, $V_o >> k_B\, T_e$. Thus, for thermal electrons, $V_o >>>> W$, and an impracticably enormous applied voltage would be required to produce W = 100 kV.

The lowest V_o results when electrons are monoenergetic. As reviewed in the next Sec., this was approximated in

PFX by collecting electrons continuously at a potential close to that of the emitting cathode. If the resulting mean electron residence time τ_e is much less than the electron collision time, a nearly monoenergetic electron distribution will result.

Within the trap volume, electrons should be prevented from reaching the surrounding conductor. Else, a large electron current would flow at a large applied voltage and an impracticably large power input would be required. In PF, this is accomplished by magnetic insulation of the anode (grounded surrounding conductor). Brillouin flow (5) of electrons is assured by arranging for the cathode to be at zero magnetic flux, while an applied B provides radial confinement. A proper combination of B variation and electron inertia will produce the desired magnetic insulation. Some details are discussed subsequently.

Figure 1 shows the arrangement of these required elements of PF, in which magnetic iron is used to provide B shaping near the trap volume.

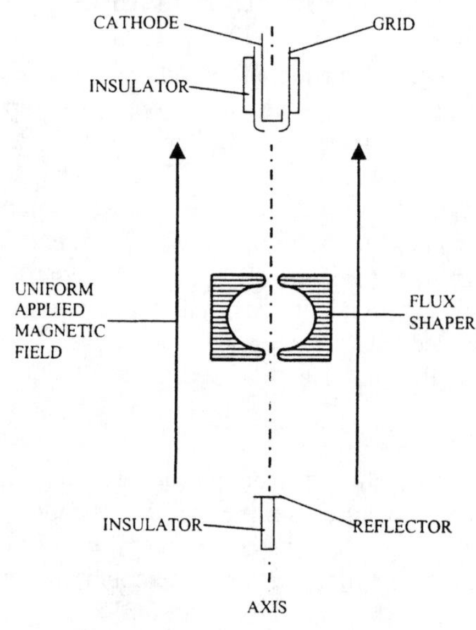

Figure 1. Arrangement of PF system.

PHYSICS BACKGROUND

PFX-I is based on two recent physics advances. First, in a previous spherical Penning-trap, electron-only experiment (PFX) (6-8), spherical focussing of a well-confined, monoenergetic, quiescent electron cloud was demonstrated. PFX was constructed as a high-precision, hyperbolic electrode Penning trap, and operated at liquid helium temperature inside the cold bore of a superconducting magnet. A LaB_6 emitter injected electrons through a 400 μm hole in the lower hyperbolic cathode. These electrons were collected primarily at this lower cathode. The ring anode was operated at voltages up to 10 kV relative to the essentially grounded cathodes. Diagnostics included electrode currents and analysis of trapped and upscattered electrons.

The principal observations were the signature of a steady-state, spherical focus at precisely the theoretical "spherical" point, where the applied voltage and B were such as to produce a spherical well for Brillouin flow electrons. This signature included peaking of the anode current, indicating large angle deflection of the initial electron beam, energy upscattered electrons, and trapped (downscattered) electrons, all of which peaked sharply at the spherical point. Additionally, PFX showed a tendency to self-organize into a spherical focus when the beam current was sufficient, and to exhibit a hysteresis as beam current was varied.

PFX technology is not optimum for ion confinement. First, the cryogenic vacuum does not permit gas handling required to source ions into the trap. Second the hyperbolic trap geometry produces only a shallow central virtual cathode, and leads to very inefficient use of the applied voltage for ion confinement. Finally, electron-electron collisions imply a large power required to maintain the focussed electron distribution, prohibiting large Q operation. For these reasons, PFX-I is designed with unfocussed electrons, and will demonstrate instead ion focussing in the virtual cathode produced by these electrons.

A second physics advance is a theoretically developed ion focussing mode which offers improvements in both power (density) and Q. It has been suggested theoretically (9) that ion-ion collisions in the SCIF mode of ion focussing may limit Q to values near of below unity. Further, Eqn. (2) shows that the reactivity increases only linearly with convergence ratio for SCIF. The ion oscillations studied in the analysis of POPS avoid both of these difficulties. First of all, a striking physics result is that ion oscillations of a Gaussian density profile ion cloud are completely undamped by collisions of by collisionless ion effects. Secondly, reactivity increases quadratically with convergence ratio. A major thrust of PFX-I is to study POPS in a practical system and demonstrate these physics predictions. For this, a uniformity of electron density of order 0.1 - 1 % will be required to access the desired large convergence regime. In the next Sec. the design and initial electrical operation of PFX-I is described.

THE PFX-I EXPERIMENT

PFX-I has been designed to produce a nearly uniform electron density in a trapped electron beam. For efficiency of operation, this beam reflexes axially from a passive second cathode (Fig. 1) with a very large recirculation fraction. The electrostatic potential at the edge of the beam is tailored to be that given by Eqn. (1) by the shape of a surrounding grounded electrical conductor. Shaping of the guiding magnetic field B(z) is chosen to keep the beam density uniform as it slows near the axial potential minimum of the virtual cathode.

The mechanical details of PFX-I are those necessary to operate at applied voltages up to 100 kV. POPS operation requires modulation of the electron cloud at the ion resonant frequency, which will be 50 - 100 MHz for PFX-I parameters. This modulation is accomplished by varying the voltage of the upper cathode by a few hundred V at the required frequency. Thus, a stable, variable-frequency RF source and power amplifier must be floated up to 100 kV to supply the upper cathode. A large (2 m cube) grounded enclosure surrounds a "hot rack" containing the RF equipment and heater supplies for the upper cathode. This hot rack if floated up to 100 kV and all contained equipment operates at relative voltages of a few hundred V. A high-voltage conduit connects the hot rack to an oil-insulated feedthrough for the cathode. Passive electrical components within this "hot deck" provide isolation of the RF from the DC power supplies.

PFX-I currently is undergoing electrical tests, and has produced an electron beam which is collected by a simple cylindrical anode and by the lower reflector electrode for analysis. DC operation at applied voltages over 50 kV has been accomplished. Electron source characterization with applied, uniform B will be completed this Fall.

Subsequent to these electrical tests, the PFX-I anode will be replaced by one with the proper shape, and B shaping will be implemented by magnetic iron structures placed near the ends of the trap region. The required magnetic configuration may be calculated using the equations of paraxial beam theory.

These are, conservation of particles, axial momentum balance, and radial momentum balance

$$\rho^2 v n = I \quad , \tag{3}$$

$$\frac{1}{2} m v^2 - e\Phi = E \quad , \tag{4}$$

$$m v \frac{\partial}{\partial z} v \frac{\partial \rho}{\partial z} = e \frac{\partial}{\partial r}(\Phi + \Phi_B) \quad , \tag{5}$$

where I and E are constants, $\rho(z)$ the beam radius, and Φ_B is the effective potential represented by B(z) for Brillouin flow electrons

$$\Phi_B = -\frac{e B^2 r^2}{8 m} \quad . \tag{6}$$

Combining Eqs. (1) with (3-6), the magnetic field on axis may be found to be

$$B = B_o \sqrt{\frac{1 + z^2/2 z_o^2}{1 + z^2/z_o^2}} \quad , \tag{7}$$

where B_o is the field strength in the midplane and z_o is an axial scale length. In terms of the trap parameters,

$$B_o = \sqrt{\frac{2 m\, n_o}{\varepsilon_o}} \quad , \tag{8}$$

$$z_o = \sqrt{\frac{6 \varepsilon_o W}{e^2 n_o}} \quad . \tag{9}$$

From (7) it may be seen that the required B is a weak anti-mirror, *i.e.* B is strongest in the midplane and varies by only 13% with z. Figure 2 shows the magnetic equipotentials in the right half space z .> 0 associated with the field of Eqn. (7).

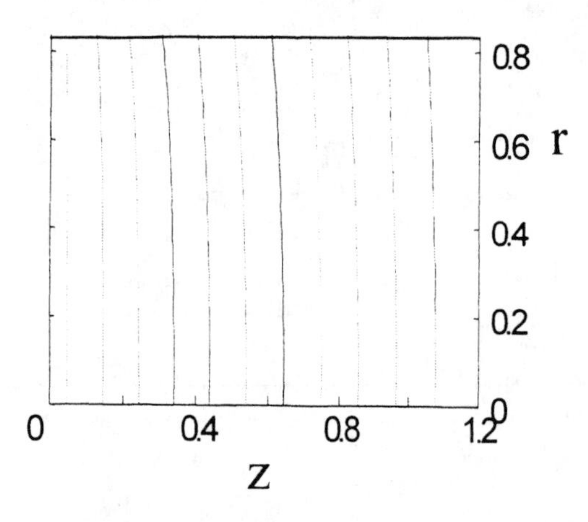

Figure 2. Magnetic equipotentials associated with required B(z).

The PFX-I design will be completed by calculating the paraxial electron motion in the magnetic field of Fig. 2 and specifying Φ on the midplane portion of the edge of the beam according to Eqn. (1). Then the external Φ may be found and an equipotential chosen for the shape of the grounded conductor.

Two other important results are connected with the slowing down of the electron beams near the midplane. First, the amount of slowing is directly related to the well depth W. To maximize W/V_0, the trap radius a should be comparable or larger than the axial scale length z_0. Thus, to optimize well depth for a given applied voltage, the paraxial approximation should be stressed to its limit, perhaps necessitating a more complete calculation. Second, two-stream stability of the two electron beams (including the reflexed beam requires that $W/V_0 < 2/3$. It appears that a careful design will produce more than 1/2 the applied voltage as well depth.

The principal diagnostics for PFX-I are measurement of electrode currents and spectroscopic observation of Stark shift on electron impact excitation of a static neutral gas fill. The latter will be used to measure the space charge electric field, from which the uniformity of electron density may be deduced. Additional diagnostics will look for energetic ions or neutrals produced during POPS operation using Silicon Barrier Diodes or other particle detectors.

PF DEVELOPMENT PATH

As discussed previously, PF development involves improvements in electrical breakdown resistance, particle handling, and increased machine precision, as contrasted to increasing size, cost, and complexity in conventional fusion confinement systems. A huge advantage of this is an affordable development path and a sequence of intermediate products with an individual development cycle which may be financed by applications of previous products.

It appears that PF may be scaled from its present conceptual state to a final fusion power reactor. The reactor concept discussed previously (4) consists of a massively modular arrangement of small ($a < 1$ cm) spherical cells, each producing a modest power output. It should be clear from Fig. 1 that a string of such cells may be aligned axially with the electron beams passing from one cell to the next. Further, since the applied B is uniform at large distances, an array of such linear "fuel rods" may share a common magnet, high-voltage, vacuum system, and heat exchanger.

This design has two remarkable properties. First, the physics may be completely verified at low cost, risk, and with modest shielding requirements in a single cell or a small array of such cells. Second, the large total wall area allows modest levels of wall loading, not too great of an extrapolation from fission reactor parameters. Thus, the materials engineering necessary for a final reactor product may also be accomplished without a multi $B development program. These features are unique for fusion power concepts, and are only possible in a system with nonneutral electron confinement, leading to a plasma size of millimeters instead of meters.

DISCUSSION AND CONCLUSIONS

Penning Fusion appears to offer a unique set of physics and technological challenges and advantages. The critical issues are high electrical breakdown resistance, precision of machine design, construction, and alignment, and high vacuum quality. Because performance increases with decreasing size, individual experiments are small and inexpensive and do not have an unfavorable cost scaling as performance is improved.

PFX-I has been designed and is beginning to investigate these unique physics. Initial operation will focus on electron only (with a low-pressure background static gas fill) operation and will examine how and over what parameter range a nearly uniform electron density may be produced in a quiescent, steady system. With proper electrostatic boundary conditions, such an electron configuration will produce a spherical, harmonic well for ions, and the physics of POPS will be studied in such a system.

Successful ion operation of PFX-I may produce up to 10^{11} deuterium-deuterium neutrons/second, with a fusion gain Q of 10^{-3} - 10^{-2}. Such a result would serve as a proof of principle for the PF concept and pave the wave for a low cost, short-term development program to push PF to its ultimate capabilities, possibly providing a fusion power reactor with affordable and flexible properties.

REFERENCES

1. Farnsworth, P. T., "Electric Discharge Devide for Producing Interactions Between Nucleii," U.S. Patent No. 3,358,402, issued June 28, 1966, initially filed May 5, 1956, rev. Oct. 18, 1960, filed Jan. 11, 1962.
2. Hirsch, R. L., *J. Appl. Physics* **38**, 4522-4534 (1967).
3. Barnes, D. C. and Nebel, R. A., *Phys. Plasmas* **5**, 2498-2503 (1998).
4. Nebel, R. A. and Barnes, D. C, *Fusion Technology* **34**, 28-45 (1998).
5. Lawson, J. D., *The Physics of Charged-Particle Beams*, Oxford, Oxford Univ. Press, 1988, pp. 134-136.
6. Mitchell, T. B., Schauer, M. M., and Barnes, D. C., *Phys. Rev. Lett.* **78**, 58-61 (1996).
7. Barnes, D. C., Mitchell, T. B., and Schauer, M.M., *Phys. Plasmas* **4**, 1745-1751 (1997).
8. Schauer, M. M., Mitchell, T. B., Holzscheiter, M. H., and Barnes, D. C., *Rev. Sci. Instr.* **68**, 3340-3345 (1997).
9. Nevins, W. M., *Phys. Plasmas* **2**, 3804-3819 (1995).

SECTION III

POSITRON SOURCES, EXPERIMENTS AND THEORY

Positron Annihilation in Opals: Evidence of Positronium Formation

J. M. Urban, C. A. Quarles

Department of Physics, Texas Christian University, Fort Worth, Texas 76129

Natural opal samples have been studied with positron annihilation spectroscopy methods: Lifetime spectroscopy (LT) and Doppler Broadening (DB) spectroscopy. Additionally, samples have been examined by the Scanning Electron Microscope (SEM) and the BET method. The mean lifetimes and DB spectra shape of opals have been compared to other rocks by constructing the quotient spectra. Unusual properties of opals have been explained by positronium formation in the three-dimensional tektosilicate framework structure (SiO_2*n H_2O) of the opals.

INTRODUCTION

When we first began studying rocks with positron annihilation (PA) spectroscopy, we focused our attention on sandstones and carbonates because of their significance for the petroleum industry as the potential reservoirs of hydrocarbons. We found that the PA signal depends mainly on the composition of the rock but is also influenced by the physical condition of the rock and its structure. We developed a theoretical description based on consideration of annihilation of the positron with the core and valence electrons in the rock to explain how the differences in composition and structure between sandstones and limestones affect the PA signal (1,2). To further test our understanding of the physical processes and the positron response in rocks we decided to examine opals, which have a composition similar to sandstone but much higher microporosity.

OPALS: ORIGIN, EVOLUTION AND STRUCTURE

Opals are minerals that are composed of hydrated silicon dioxide (SiO_2 * nH_2O) in a framework structure called tektosilicate. The content of water in opals ranges from 3-10%. Opals originate from the silica deposits on the ocean floor. These deposits are accumulations of the siliceous remains of diatoms, radiolarians and other organisms that secrete shells of amorphous silica. Once opals are formed, they undergo a maturation process:

Opal-A $\rightarrow$ Opal-CT $\rightarrow$ micro-quartz $\rightarrow$ quartz

The speed of this process depends on the temperature and pH conditions of the environment. Opals do not crystallize directly to quartz but they form other more open and less dense crystals called tridymites and cristobalites. Both crystals have a structure which can be described in terms of layers formed from SiO_4 tetrahedra linked by four oxygen vertices, with alternate tetrahedra pointing up and down, to form joined hexagonal rings. In addition, water is connected to the crystalline structure producing indefinite 3-dimensional "cages" or frameworks with complex structure. The presence of this internal framework structure is the origin of the high micro-porosity and low density of opals. The maturation of opals leads to increasing crystallization and the loss of water. Finally opals are completely converted into non-hydrated quartz. Quartz has a tetrahedral structure with the silicon in the center (SiO_4). (3)

EXPERIMENTAL DETAILS

The samples have been provided by the Chevron Petroleum Technology Co. Samples have been observed with an optical microscope and by the Scanning Electron Microscope (SEM) (JEOL JSM-6100) to see the surface features and EDX (Energy Dispersions X-Ray analysis) to identify elemental composition.

We did the standard Doppler Broadening (DB) and lifetime (LT) measurements on opals at room temperature and in air. The source and samples were in the sandwich arrangement. For the LT measurements a 30 μCi ^{22}Na source was used. For the DB measurements we used a 10 μCi ^{68}Ge source which significantly reduced the background around the 511 keV line.

For DB measurements we used an Ortec high purity coaxial germanium detector with a resolution about 1.2 keV at 570 keV and a total active volume of 95.1cm^3. The signal from the detector was amplified and recorded in a multi-channel analyzer (MCA) and finally sent to a PC computer. For analysis, the DB spectra were divided into the three regions characterized by S, SW and W-parameters. In this article we will only focus on the S-parameter, which was calculated by dividing the central area of the peak, 511 ± 1 keV, by the total peak area. We applied a step-like function (4) in order to eliminate background.

The apparatus for the LT measurement was a typical fast-timing coincidence setup. We used Hammatsu H1949 photomultipliers coupled to BC-418 plastic scintillators, with an active volume of 6.4 cm^3. The output signals from the photomultipliers were fed to Ortec constant fraction differential discriminators. The time-to-amplitude converter output for each coincident event was recorded in the MCA and finally transferred to the PC computer.

CP475, *Applications of Accelerators in Research and Industry*,
edited by J. L. Duggan and I. L. Morgan

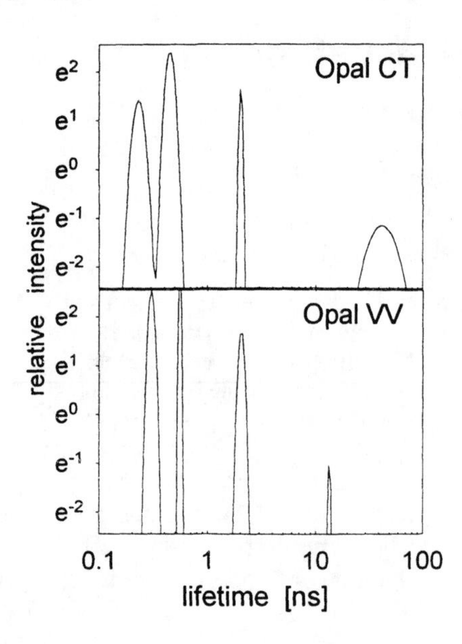

FIGURE 1. Distribution of lifetimes

RELATION BETWEEN LIFETIME AND PORE SIZE. LIFETIME DISTRIBUTION.

The decay rate of ortho-positronium (o-Ps) pick-off, λ_{po}, can be expressed as a function of cavity size (5) with radius R according to the formula:

$$\lambda^{(10)}_{po} = \lambda_a * [1-R/R_0+1/(2\pi)*\sin(2\pi R/R_0)] \quad (1)$$

$$\text{with } \lambda_a=1/4\lambda_S* + 3/4\lambda_T \cong 2/\text{ns} \quad (2)$$

where λ_S = 8 ns^{-1} is the self p-Ps (para positronium) decay rate, $\lambda_T = 7.04*10^{-3}$ ns^{-1} is the self o-Ps decay rate, and $R_0=R+\Delta R$, with ΔR=1.656 Å.

The lifetime, τ_{po}, can be found from:

$$\tau_{po}=1/(\lambda_{po} +\lambda_T) \quad (3)$$

In Fig.1 the measured lifetime distribution of opal CT and opal VV has been calculated with Kansy's lifetime fitting program (6). Only the opal CT sample has a distribution with such a long lifetime. These lifetime results suggest a distribution of the cavity sizes in opal CT and rather uniform cavity sizes in other opal samples. The results are consistent with the optical properties of opals. It has been found that the cavity diameter and refractive index determine the range of color displayed and the luster. Opals which are formed with uniform size and regular packed cavities have a glassy or pearly luster caused by the reflection of the light from the regular crystalline planes (Opal VV and Quartz altered from opal samples). Opals, with cavities of irregular size or irregularly packed do not have a play of color and have a dull luster (Opal-CT) (3).

DISCUSSION OF RESULTS

LT and DB results for opals in different states as powders or solids are presented in Table 1.

Opal-A

Opal-A was a diatomite from J. Mansville Quarry. The sample was a fine white powder with grains smaller than 60 mesh (0.25 mm) and a very low density (0.2 g/cm^3). The Opal-A sample has been examined by use of the BET (Braunauer, Emmet and Teller) nitrogen adsorption method in order to determine its surface area and micro-porosity. The surface area of opal-A, 13.9 m^2/g, was two orders of magnitude higher than other powdered rocks (like sandstones or carbonates). Thus, the total micro-porosity of Opal-A calculated from BET was about 2 %.

Opal-A was the most difficult for straightforward interpretation because it was the only fine powdered sample with the lowest density of the opal samples studied. The origin of the micro-porosity of this sample was two-fold.

First, there were some open micro-pores between the grains. The size of these pores depends on the size and the shape of the grains. This micro-porosity was detected by BET analysis because nitrogen could penetrate these open spaces between grains of powder. Second, there are cavities inside the diatoms formed by the three-dimensional network tektosilicate structure that were not detected by BET but are seen by positron annihilation measurements.

TABLE 1. Lifetime measurement values and S-parameter value for opals and some other rock samples. "s"-solid, "p"-powder, "g"-grain

sample	τ_1[ns]	I_1 %	τ_2 [ns]	I_2 %	τ_3 [ns}	I_3 %	τ_4 [ns]	I_4 %	<τ>[ns]	S
Opal A	0.287	48.5	0.517	39.7	1.96	10.3	16.3	1.55	0.799	0.5659
Opal CT s.	0.206	27.4	0.464	58.3	2.00	9.93	33.1	4.46	1.979	0.5594
Opal CT p.	0.256	37.0	0.488	48.3	2.03	10.3	31.8	4.42	1.947	---
Opal VV p.	0.235	28.4	0.487	54.0	2.06	16.6	13.2	1.08	0.813	0.5593
Quartz a/f Opal s.	0.210	19.2	0.362	69.20	1.03	7.80	2.37	3.80	0.461	0.5339
Quartz a/f Opal p.	0.124	12.0	0.327	74.1	0.93	10.3	2.48	3.63	0.461	0.5339
Quartz Av. p.	0.256	81.7	0.423	17.6	3.33	0.68	---	---	0.306	0.5157
Illite g.	0.271	38.5	0.448	59.8	1.81	1.65	---	---	0.402	0.5289
Dolomite Penfield	0.219	89.2	0.443	10.1	2.52	0.75	---	---	0.259	0.5006

The shortest lifetime value, τ_1=0.287 ns (Tab.1), is attributed to the self p-Ps annihilation, 0.125 ns, and to the delocalized positron trapping in the bulk. This lifetime value was higher than the value in the solid opals because of the lower bulk density of this fine powdered sample (0.2 g/cm^3). The second lifetime value, τ_2=0.517 ns, is attributed to positron trapping effects. This value is also higher than in the other opals due to the intensification of the trapping mechanism on the grain boundaries and in surface states. The third lifetime component had a value of about 2 ns and a relatively high intensity 10-17% in all opal samples. The presence of this lifetime value in all opals and its absence in non-opal samples suggests that this must be connected to the internal structure of the opals. We think that it is caused by the pick-off annihilation of the o-Ps in crystobalites commonly present in the mature opals and observed by other authors (7). The presence of the long lifetime, τ = 17 ns, in Opal-A can be explained under the assumption that some of the pores are closed, not accessible to the oxygen in the air, which acts as a quencher. Therefore the longest lifetime can be caused only by the o-Ps pick-off in the internal cavities within the grains. From the shape of the BET hysteresis isotherm (the difference between adsorption and desorption isotherms) it can be concluded that the pores in Opal-A have a narrow opening with a wide body, which additionally impedes oxygen from penetrating inside the cavities. This 17 ns lifetime distribution is very narrow (see Fig. 1) which means that the cavities have a relatively uniform size with a radius of about 8.7 Å ± 2 Å calculated from formula (2). The lower intensity of this longest component compared with Opal-CT suggests that the framework in Opal-A that is in the first stage of opal maturation is less developed and thus the amount of cavities is limited.

Opal-CT

Opal-CT came from Chico Martinez Creek near Buttonwillow, CA. It contained some asphalt contamination. The Opal-CT sample was solid with brown and black inclusions, dull luster and a brown streak. Opal-CT had a complex internal structure and composition. The surface consisted of a porous, mainly silicate surface. In the spherical pores of different sizes some crystals - mainly pyrite FeS_2 and some cristobalites were seen. Cristobalite crystals on the surface were arranged in the characteristic star shaped clusters. This sample contained also some areas dominated by sulfur and calcium. The structure of Opal-CT, being highly micro-porous, is reflected by the presence of the long lifetimes with the distribution of the longest lifetime linked to the distribution of the micro-pore sizes. This is caused in turn by different numbers of water molecules interconnected in crystobalites and tridymites. The longest lifetime value was 33 ns when no distribution was assumed (Tab.1), but it shifted to 45 ns when a distribution was allowed (Fig.1). The difference is not significant if we consider the fact that, according to the eqn. (2), a change in lifetime values, from 30 to 45 ns, corresponds to the relatively small range of radii sizes from 12-14 Å. Additionally, the existence of different energy states which positronium can populate leads to different lifetime values for the same radii sizes. So, while the lifetime distribution is related to the distribution of the pore sizes, it can also be affected by the physics of the formation of the positronium traps.

We compared the lifetime values in solid and powdered Opal-CT and in quartz altered from opal. The mean lifetime value in opals were slightly higher in the solid than in the powdered material (Tab.1). The effect can be explained in the following way. When this internal structure is partly destroyed in the powdering process it leads to the lowering of a lifetime value connected to the cavity size, τ_4, and its intensity, I_4.

Opal-VV

Opal Virgin Valley (VV) came from Virgin Valley, Humbolt Co., NV. It had white-yellowish subangular-shaped grains, with the size of 1-5 mm, pearly luster and a white streak. The Opal-VV sample examined by SEM was much more uniform in composition, mainly Si and Al, than the Opal-CT sample. Some areas in Opal-VV looked exactly like the amorphous silicate areas with spherical pores filled with crystals in Opal-CT. But these areas were usually smaller in size and richer in crystals than in Opal-CT. The Opal Virgin Valley sample was characterized also by four lifetimes with the longest lifetime value at 13 ns corresponding to a cavity radius of 7.7 ± 0.4 Å. But the intensity of this lifetime was only 1%, the lowest in comparison with the other opal samples. The intensity of the lifetime which is attributed to the pick-off annihilation in crystobalite is higher in Opal-VV, I_3 = 16.6% , than in other opals, suggesting that in the Opal-VV the crystallization process was more advanced with more crystobalites than in the other more amorphous opals.

Quartz a/f Opal

The sample of diagenetic quartz altered from opal (Quartz a/f Opal) had a black solid, glassy luster and a light-brown streak. According to the SEM, this sample had the most crystalline structure of all three samples. But some residual areas which look like the smooth amorphous silicate area in Opal-CT with some small micro-pores were detected also.

The lifetime and S-parameter values for this highly crystallized sample were lower than the values for opals but higher than the values for the quartz and sandstone samples. This sample contains still 2 % of water in the quartz structure. So, this sample was still in the process of diagenesis with a structure intermediate between Opal-A and quartz. The presence of the cavities in the opal-like part contribute to the longest lifetime, τ=2.37 ns, with the intensity, I=3.8%, from the o-Ps pick-off effect. But, the presence of crystoballites characterized by the 2 ns lifetime component in the other opal samples was overshadowed by the dominant quartz crystals with much smaller open volume.

OPALS COMPARED TO OTHER ROCKS

In sandstones and carbonates a linear relationship between the S-parameter and mean lifetime value was observed. In contrast, in opals we do not observe a relationship between the mean lifetime and the S-parameter. The sandstone and carbonate rocks were characterized by low micro-porosity, therefore low positronium formation. Positronium formation in opals is the major cause of the breakdown of the relation between the mean lifetime and S-parameter values. Lifetime is much more sensitive to the internal structure of the materials than the DB measurements, therefore lifetime values in micro-porous opals are very high. DB senses the valence/core annihilation ratio which depends on the annihilation process (whether with free or trapped positrons or positronium) but it does not sense how long it takes for the positron to annihilate, therefore it does not depend on the size of the void or cavity. So, the increase in the lifetime values in micro-porous opals is higher than the increase in the S-parameter value leading to the breakdown in the relation between lifetime and S parameter when significant positronium is formed in the sample.

The quotient spectra have been obtained by dividing the standard DB spectrum of a Silicon crystal (in the 1-1-1 orientation) by the sample spectrum. All spectra have been smoothed and normalized. As we see from Fig. 2, the centroids of the selected spectra are arranged in order according to their S-parameter. The S-parameter and mean lifetime values were higher in sandstone and quartz samples than in carbonates (1,2),(Tab.1). Therefore, the lowest S-parameter and thus the highest ratio (Si-crystal to sample) is obtained for the carbonate (Dolomite Penfield) sample. The quartz altered from opal (q/o) sample lies between the quartz (Quartz Aventurine) and opals, as expected and is the first spectrum where we start to see the curvature at the sides of the spectra that is characteristic of opals. This curvature is caused by a different shape of the opals DB spectra in comparison to the Si-crystal. The DB shape for opals, in the region close to the center (in the range: 511± 1.5 keV, which corresponds to the momentum, $p \approx 6*10^{-3}$ mc) is much narrower than that of the standard sample. This is why we see "wings" on the sides of the centroid in opals. The annihilation of delocalized positrons in the Si-crystal occurs mainly with valence electrons. These electrons being bound to the lattice structure carry some non-zero momentum. In opals there is a significant contribution from

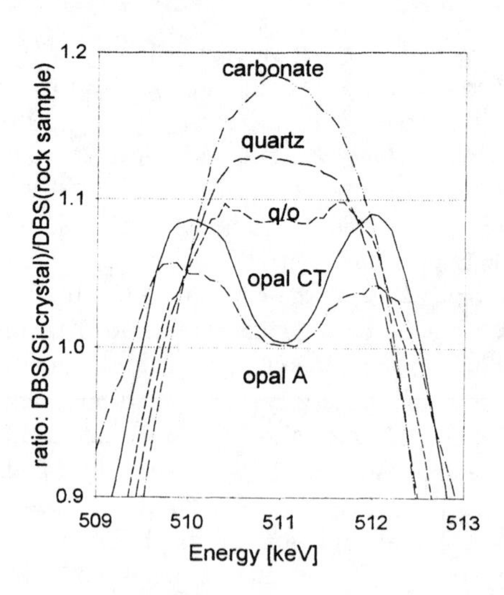

FIGURE 2. Quotient DBS rock spectra

the self para-positronium annihilation which is almost Doppler-free therefore the width of the centroid is narrower. In all the opal samples the narrowest distribution of the momentum around 511± 1.5 keV is observed in the Opal-CT sample in which the positronium formation and self para-positronium annihilation is the most probable because of the most developed internal cavity structure.

SUMMARY AND CONCLUSIONS

Natural opals have unusual properties in comparison to other rocks. By use of the positron annihilation measurements we can calculate the size of the internal cavities and we can determine the stage of maturation in opals from the early amorphous stage through the formation of 3-dimensional cavities and from the first crystallization in crystobalites and tridymites to the completely crystallized quartz structure.

ACKNOWLEDGMENTS

This research has been supported by the Chevron Petroleum Technology Co. and the TCU Research Foundation. We are especially grateful to Dr. G. Salaita for providing the samples and donation of equipment.

REFERENCES

1. Urban, J.M., Quarles, C.A., Acta Phys.Pol., **A 88**, 241 (1995).
2. Urban, J.M., Ph.D. Dissertation, TCU, to be published, 1998.
3. Berry, L.G., Mason, B., *Mineralogy*, San Francisco, W.H.Freeman and Co, 1983, ch.15, pp.393-402.
4. Debertin, K., Helmer, R., *Gamma and X-Ray Spectroscopy*, Amsterdam, Elsevier Sc.Publ., 1988, ch.3, p.162, B8.
5. Eldrup, M., Lichtbody, D., Sherwood, J.N., Chem. Phys., **63**, 51 (1981).
6. Kansy, J., Nucl. Instr. Meth., **A 374**, 235 (1996).
7. Hugenschmidt, C., Holzwarth, U., Jansen, M., Kohn, S., Maier, K.; J. Non-cryst. Sol., **217**, 72, (1997).

THE JAPANESE POSITRON FACTORY

S.Okada, H.Sunaga, H.Kaneko, H.Takizawa, A.Kawasuso, K.Yotsumoto and R.Tanaka

Takasaki Establishment, Japan Atomic Energy Research Institute, 1233 Watanuki, Takasaki, Gunma 370-1292, Japan

The Positron Factory has been planned at Japan Atomic Energy Research Institute (JAERI). The factory is expected to produce linac-based monoenergetic positron beams having world-highest intensities of more than 10^{10} e^+/sec, which will be applied for R&D of materials science, biotechnology and basic physics & chemistry. In this article, results of the design studies are demonstrated for the following essential components of the facilities: 1) Conceptual design of a high-power electron linac with 100 MeV in beam energy and 100 kW in averaged beam power, 2) Performance tests of the RF window in the high-power klystron and of the electron beam window, 3) Development of a self-driven rotating electron-to-positron converter and the performance tests, 4) Proposal of multi-channel beam generation system for monoenergetic positrons, with a series of moderator assemblies based on a newly developed Monte Carlo simulation and the demonstrative experiment, 5) Proposal of highly efficient moderator structures, 6) Conceptual design of a local shield to suppress the surrounding radiation and activation levels.

INTRODUCTION

New beam technologies with accelerators have been emphasized recently. At the Takasaki Establishment, Japan Atomic Energy Research Institute (JAERI), researches on materials science and biotechnology are in progress using a variety of ion beams from an AVF cyclotron, a 3 MV tandem accelerator, a 3 MV single-ended accelerator, a 400 kV ion implanter in TIARA (Takasaki Ion Accelerators for Advanced Radiation Application) facilities. The research fields have stepped into a novel stage of 'quality control on matter itself' at the atomic level, which means both observation and manipulation of the microscopic structures. Positron beam, which is generated as 'slow' (i.e. monoenergetic) positron beam and sometimes accelerated, is a powerful tool for this purpose as well as ion beam. Unlike neutrons and photons, positrons are charged, and therefore the implantation depth in matters can be controlled by varying the injection energy of the beam. Namely positron and ion beams can probe the same positions such as surfaces and interfaces of materials and devices. On the other hand, microscopic features observed and manipulated are different between positron and ion beams, and consequently the perfection could be pursued with the combined use of the two means. An implanted and thermalized positron annihilates with an electron, which is dominantly followed by emission of two photons that maintain the energy and the momentum of the initial state due to the conservation rule. The positron is attracted to atomic vacancies, voids and free volumes, where the nuclear charge densities are lower, due to its positive charge. This results in the prolonged lifetime, which can be measured also by detecting the annihilation radiations. Thus positron can see the electronic structure and the absence of atoms, whereas ion probes the existence of atoms, that is, the atomic structure. Electrons are also charged. When an electron is accelerated into matter, however, it loses its identity quickly through thermalization, and thereafter it becomes useless as a spectroscopic particle. Unlike electrons, slow positrons are repulsed by surfaces of solids due to their positive charge. This endows the positron with high advantage as a topmost surface probe to observe atomic structures by the diffraction (1) and to detect impurity atoms by PAES (Positron Annihilation-Induced Auger Electron Spectro-scopy) (2).

By the combined use of ion beams and positron spectroscopy, we found a unique state of vacancy-hydrogen interaction in silicon. Here proton-implanted silicon was studied by positron lifetime in comparison with He-ion-implanted and electron-irradiated silicons (3). We succeeded in the first observation of RHEPD (Reflection High Energy Positron Diffraction) from a silicon surface by the use of an isotope-based electrostatic slow positron beam, and demonstrated usefulness of positron as a topmost surface sensitive probe (4). Such an isotope-based slow positron beam, however, cannot be applied for advanced analyses like time-dependent observation of transient phenomena and fine structure investigation of sub-micron size local region of materials, because of the insufficient intensity.

We have been promoting design studies for the 'Positron Factory' (5), in which linac-based intense monoenergetic positron beams are planned to be applied for materials science, biotechnology and new fields of basic research. A tentative goal of the slow positron beam intensity is 10^{10}/sec, which is larger by two orders of magnitude than those of existing strongest beams in the world. In this paper, some results of the design studies on a dedicated high-power electron linac and the target system to generate the intense slow positron beam are described.

PLANNED FACILITY

An overview of the planned facility is shown in Fig.1. We have done the conceptual design of the facility, which

CP475, *Applications of Accelerators in Research and Industry*,
edited by J. L. Duggan and I. L. Morgan

includes a high-power electron linac, electron beam lines, a target system, slow positron beam lines and building. The results of the design studies, which are indicated by @ in Fig. 1, are described in the following sections.

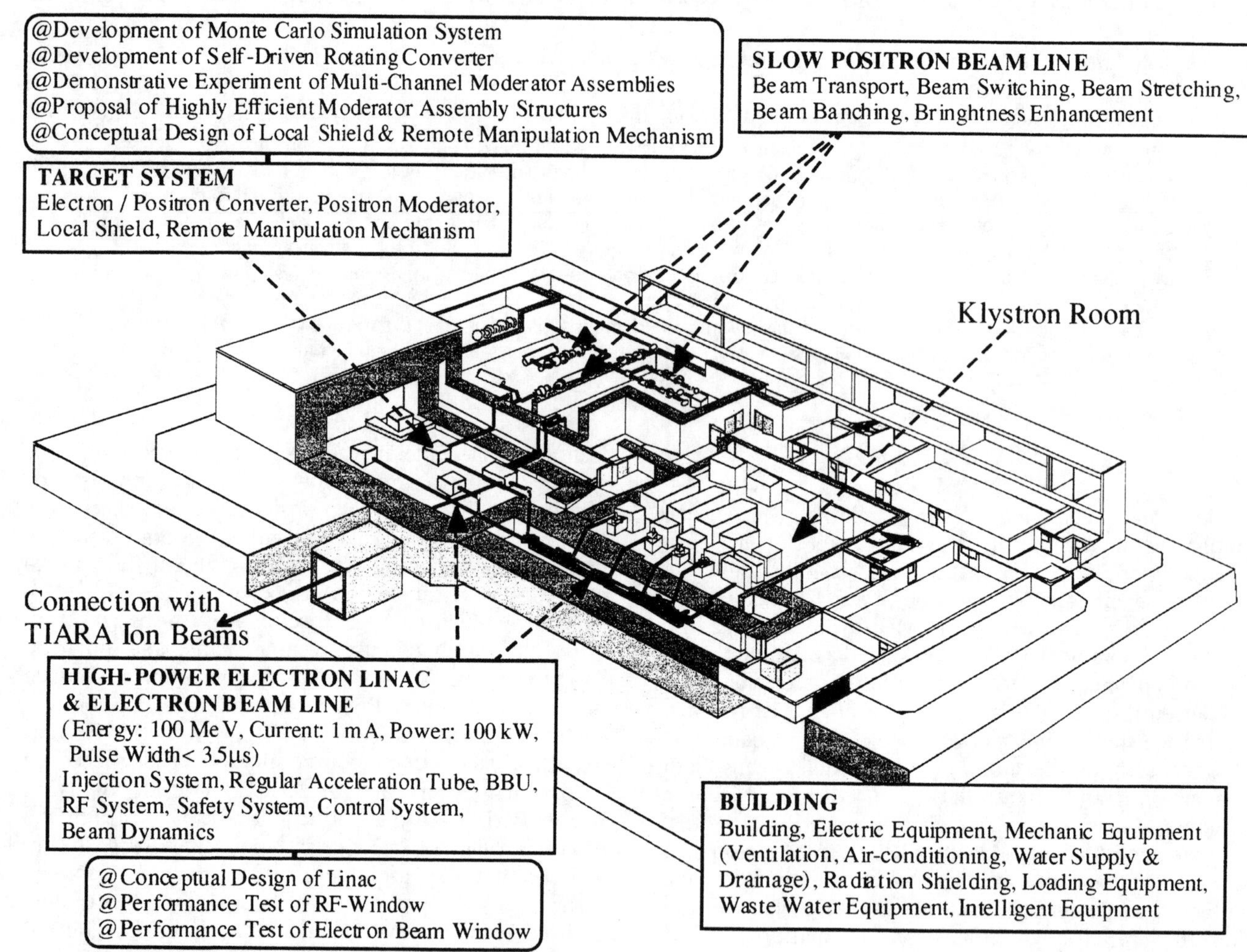

FIGURE 1. An overview of the planned facility and the design study items.

POSITRON BEHAVIOR SIMULATION SYSTEM

We developed a Monte Carlo simulation system named EGS4-SPG (6) for the design studies. In the process to generate a slow positron beam using an electron linac, as shown in Fig. 2, energetic electron beam bombardment from a linac on an electron- positron converter (e.g. tantalum; Ta) causes a cascade shower of Bremsstrahlung and pair production reactions, which results in energetic positron emission. A moderator (e.g. tungsten; W) combined with the converter produces a slow positron beam through a unique emission process due to the negative surface work function for positron.

The EGS4 (7) is a very powerful tool to simulate three-dimensional movements of electrons, positrons and photons in high energy processes like Bremsstrahlung and pair-production. However there is a lower limit of the particle's kinetic energy, under which the EGS4 calculations are less reliable. The limit is called cut-off energy, which is usually around 10 keV. In EGS4, when a particle slows down under the cut-off, the calculation terminates after recording of the three-dimensional coordinate and momentum of the particle and goes to the next incident particle.

In our new simulation system EGS4-SPG, the tracking of positrons does not terminate even under the cut-off. When the kinetic energy of a positron becomes less than the cut-off, the calculation is automatically switched from EGS4 to SPG (5) which had previously been developed by us. The positron is transported from the recorded position in the direction of the recorded momentum by a certain distance z_t, i.e. thermalization length. To determine the value of z_t, the Makhov expression (8) for the probability density function of the thermalization length is used:

$$p(z') = - (d/dz') \exp[-(z'/z_0)^m]. \qquad (1)$$

Here m and z_0 are constants depending on the particle energy and the material. The coordinate z' is along the direction of the positron momentum at the final stage in the EGS4 calculation. This expression is widely used as a stopping profile of positrons having kinetic energies ranging from ~keV to several tens of keV, and the constants can be experimentally determined. Since this formula has a derivative form, it is very suitable to Monte Carlo treatment. The probability distribution function,

$$P(z') = 1 - \exp[-(z'/z_0)^m], \quad (2)$$

can be easily deduced from Eq. (1). The thermalization length z_t, corresponding to a uniform random number R, is obtained from Eq. (2) as follows:

$$z_t = z_0(-\ln R)^{1/m}. \quad (3)$$

After the above transportation, the positron starts to diffuse, and finally is either annihilated or ejected from the surface of the material, according to the probabilities determined empirically. This part in SPG simulates important processes to produce slow positrons. The treatment of particles in SPG is not physically strict. But it is suitable to practical use in design studies of complicated devices using parameters obtained from experiments of simple materials.

Thus EGS4-SPG can trace the three-dimensional positron behavior through from the birth by pair-production in a converter and a moderator to the rebirth by slow positron emission from a moderator.

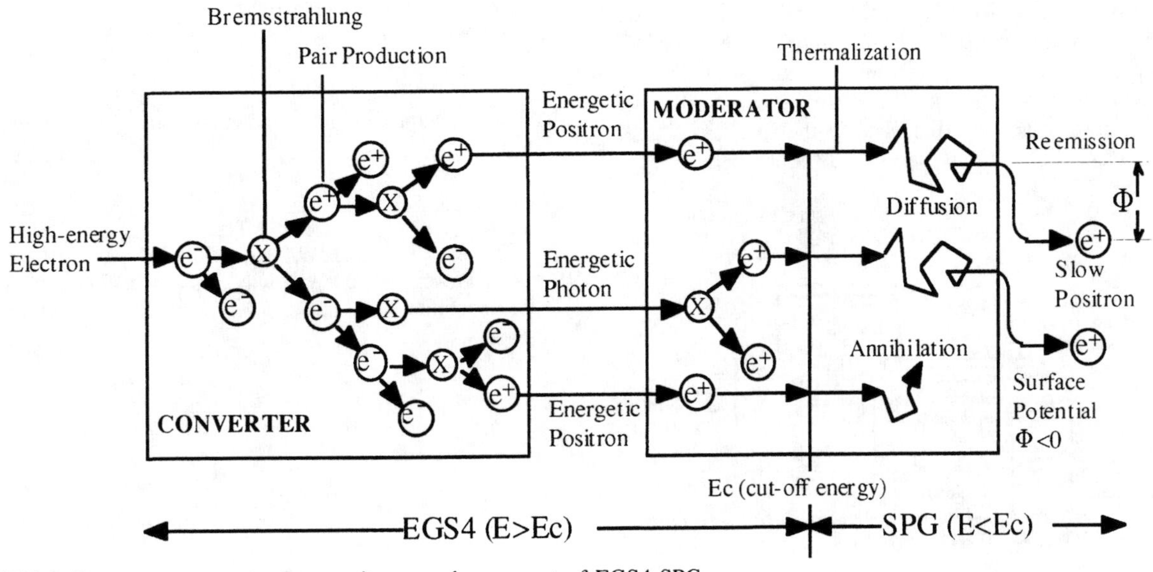

FIGURE 2. Process to generate slow positrons and a concept of EGS4-SPG.

HIGH-POWER ELECTRON LINAC

We performed a design study on a high- power electron linac to generate an intense slow positron beam. An optimum electron beam energy for slow positron generation was estimated to be around 100 MeV by using EGS4-SPG. It was calculated that a tentative goal of the slow positron beam intensity (10^{10}/sec) could be attained with a linac of 100 kW class with the above energy range. A technical survey study confirmed the feasibility of manufacturing such a state-of-the-art linac. Further detailed analyses were carried out concerning thermal deformation of the accelerator structures, beam instability, reliability of the components, down-sizing of the machine and a computer-aided control system. A concept of the linac and the expected specification are shown in Fig.3 and Table 1, respectively.

In addition to the linac itself, the durability of the electron beam window and the RF window of the wave guide circuit from the klystron should be investigated.

The electron beam window will be irradiated by an electron beam of 100 MeV and 1 mA. When the beam size is 10 mm in diameter, the energy fluence rates are estimated to be 100 and 300 W/cm^2 for a titanium and a tantalum window, respectively. Here the thickness of the window is assumed to be 0.1 mm. One side of the window will be in vacuum and the other side will be cooled by water (see the next section).

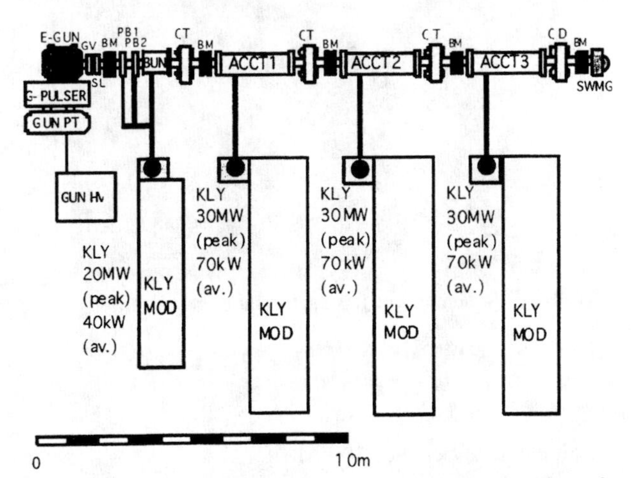

FIGURE 3. A concept of the high-power electron linac for the Positron Factory.

TABLE 1. Specification of the linac.

Beam Energy	100 MeV
Beam Current	1 mA (average)
Beam Power	100 kW (average)
Pulse Repetition	600 pps
Pulse Width	~3.5 μs

We assembled a beam window test stand as shown in Fig.4, and carried out the irradiation test of a titanium and a tantalum window whose thicknesses were both 0.1 mm.. The electron beam energy, the beam diameter and the flow rate of the cooling water are 10 keV, 10 mm and 5.2 lit./min, respectively. The beam currents and the energy fluence rates were 11 mA (103 W/cm^2) for the titanium window and 7.5 mA (78 W/cm^2) for the tantalum window, respectively. These conditions were more severe for the surface parts of the windows in the vacuum sides than an electron beam irradiation of 100 MeV and 1 mA. No significant damages were observed in the irradiation test. Thus the feasibility of the windows was confirmed.

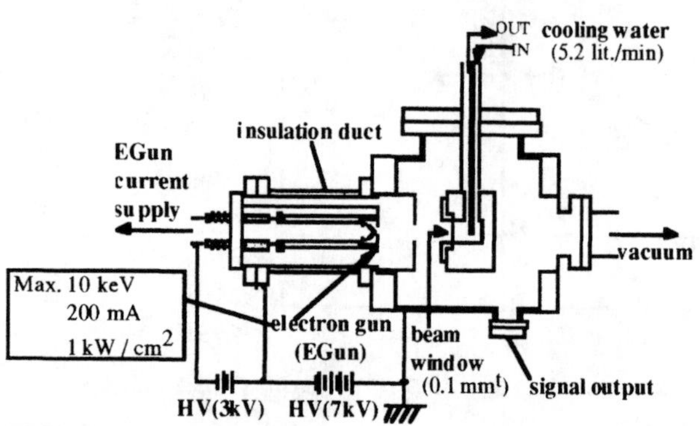

FIGURE 4. Schematic of the electron beam window test stand.

We also made a trial product of the RF window and performed an over-load operation test as shown in Fig. 5. In the wave guide circuit for the linac of the Positron Factory, we plan to divide the RF power into two windows, as shown in the left hand of Table 2. Although such powerful klystrons as in Fig. 3 do not exist in the world, we only need to consider the reliability of the window for half of the power. We used a modified model of Toshiba E3730A klystron, whose specification is shown in the right hand of Table 2. The RF power is sufficient for the test. We carried out a long term operation (continuous 4 hours, total 60 hours) of the RF window. No damages in the window were observed after the stable operation. The temperature rise of the cooling water of the dummy loads was only 0.5 deg. C. Thus the feasibility of the RF window was confirmed.

TABLE 2. Expected and tested conditions of the RF window.

	Expected	Tested
Frequency	2856 MHz	2856 MHz
Peak Power	32 MW/2 windows = 16 MW	36.6 MW
Average Power	72 kW /2 windows = 36 kW	40.1 kW
Pulse Repetition	600 pps	200 pps
Pulse Width	3.5 μs	5.3 μs

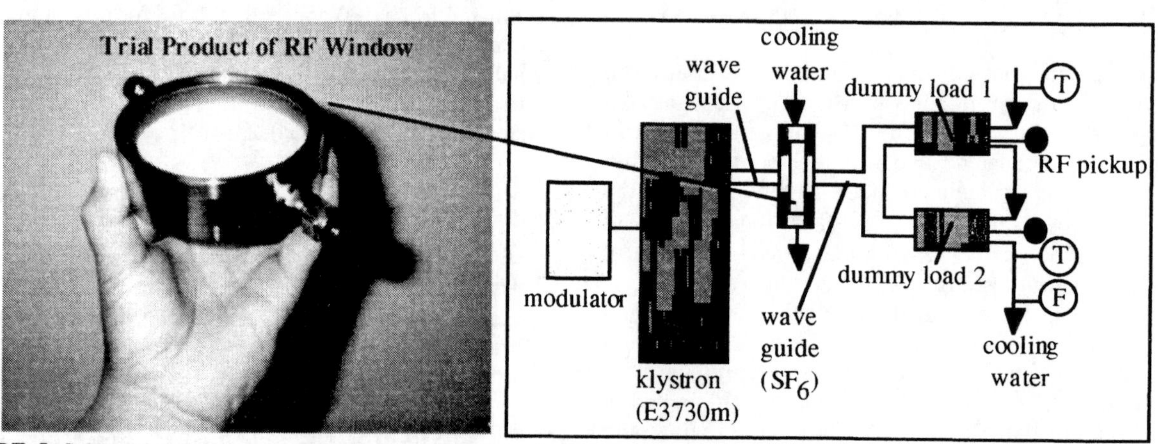

FIGURE 5. Schematic of the performance test of the RF window.

ELECTRON TO POSITRON CONVERTER

If an electron beam of 100 MeV and 100 kW is injected onto a tantalum converter with an optimum thickness of 8.2 mm, a power of 38 kW is deposited. For avoiding the meltdown, it is necessary to divide the converter into several pieces and rotate in a coolant. However a motor and a penetration may not be available due to radiation degradation of insulating and sealing materials. Consequently we have proposed a 'self-driven rotating converter' (9) as shown in Fig.6 which has pivot-type axles and bearings and rotates by a driving force of the coolant itself. The coolant (water) works as a lubricating material as well.

We fabricated a pilot model and confirmed the feasibility by an electron beam irradiation test, three months performance test and a finite element calculation. According to the calculation, the maximum temperature even at the hottest point of the tantalum disks was about 720 °C, which was far lower than the melting point, where the cooling water velocity and the revolution rate of the disk were 470 cm/sec and 300 rpm, respectively.

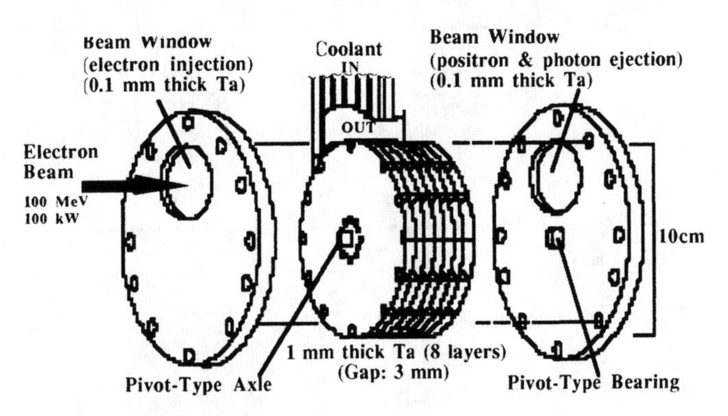

FIGURE 6. A concept of the self-driven rotating converter.

MULTI-CHANNEL MODERATOR ASSEMBLIES

Usually only one slow positron beam is delivered in an operation of one linac. We have proposed 'multi-channel moderator assemblies' (9) to supply multiple slow positron beams simultaneously.

To demonstrate the feasibility of the simultaneous extraction of multi-channel slow positron beams, we fabricated a set of 2 channel tungsten moderator assemblies as shown in Fig.7. The set was composed of 19 tungsten foil layers of 25 μm in thickness. The last layer, which was most distant from the converter, was attached to 1 mm-thick tungsten plates. Slow positrons from the first 9 layers (1st channel) and the second 10 layers (2nd channel) were separately extracted by 2 tungsten mesh grids. Each moderator layer was divided into 3 parts, electrically separated and biased to drift emitted slow positrons by sloping the electric field toward the extraction grids. We observed the slow positron beam profile from the assemblies with a MCP (micro channel plate), using a 100 MeV electron beam from a S-band electron linac at Osaka University.

We also calculated slow positron yields, that is a ratio of the number of slow positrons emitted from each layer of tungsten foils to that of incident electrons onto the tantalum converter, for the same structure as used in the experiment, using EGS4-SPG. The result is shown in Fig.8.

The experimental result is shown in Fig.9. Three peaks were observed in the slow positron beam intensity profile. The largest one was attributed to slow positrons from the first channel which was nearer to the tantalum converter. The second and the third peaks were both attributed to slow positrons from the second channel. It is assumed that back-scattered positrons and pair production reactions by photons give rise to the third peak, because thick tungsten plates were placed at the end of the second moderator assembly. This means that positrons and photons passing through the first and second assemblies still have a potential to generate slow positrons, and also that it will be efficient to place a heavy metal at the end in fabrication of moderator assemblies. The result agrees well with the calculated result. This confirms the feasibility of the proposal and also the validity of the simulation system.

The intensity of slow positrons from the second channel was smaller only by an order of magnitude than that from the first channel. It was concluded that such an extra positron beam will be useful for preliminary or potential researches which are promoted simultaneously with main experiments using the strongest beam.

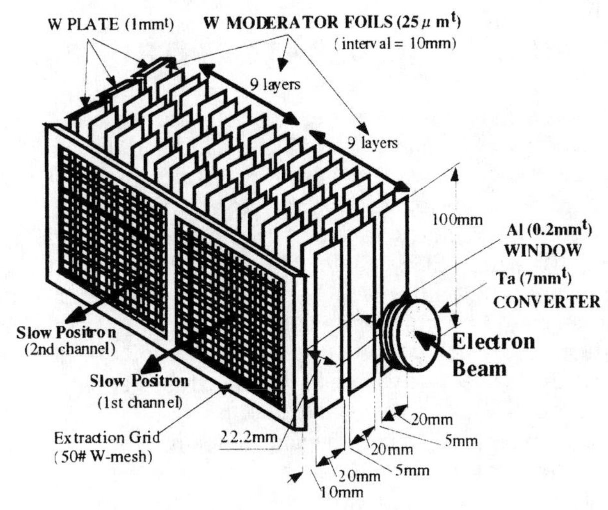

FIGURE 7. Experimental setup of 2-channel moderator assemblies for the demonstrative experiment of the simultaneous extraction of multi-channel monoenergetic positron beams.

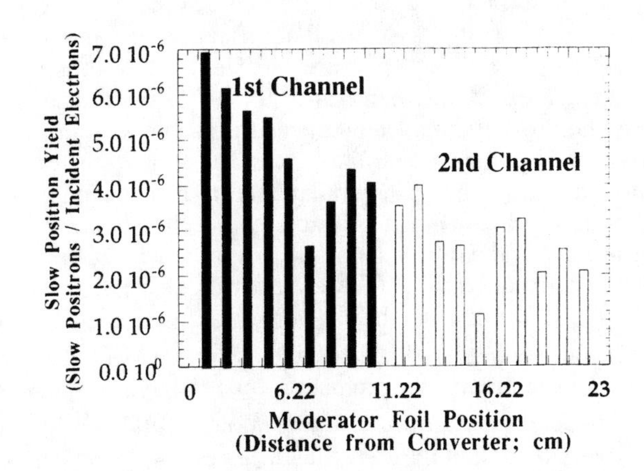

FIGURE 8. Calculated slow positron yields from tungsten foil layers of the two channel moderator assemblies used in the demonstrative experiment as indicated in Fig.7.

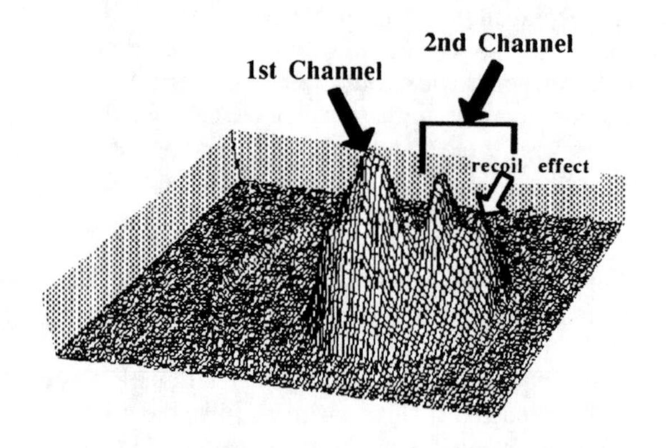

FIGURE 9. The intensity of extracted slow positrons observed with a MCP in the demonstrative experiment of the simultaneous extraction of multi-channel monoenergetic positron beams.

EFFICIENT MODERATOR STRUCTURE

The demonstrative experiment result suggests usefulness of a heavy metal plate for a reflector and importance of the assembly structure. To evaluate the structure effect, we calculated slow positron yields for various moderator assembly structures as indicated in the upper side of Fig.10, using EGS4-SPG. The distance between the converter (8.2 mm-thick Ta) and the moderator was assumed to be 1 cm. Tested structures are as follows: a) a usual one, which consists of ten tungsten foils of 25 μm in thickness vertically placed to the electron beam injection direction (V), b) a set of these foils whose surrounding planes except for the positron and photon injection side and the slow positron extraction one are enclosed by thick tungsten plates i.e. reflectors (V+R), c) a honeycomb-like assembly having an additional set of eleven tungsten foils crossing the above vertical foils (C), and d) a honeycomb-like assembly with reflectors.

The lower side of Fig.10 shows the calculation result. The reflectors cause several tens per cent increase in the yield. Addition of the cross foils enlarges the yield nearly by three times. It is assumed that reflection of particles in the honeycomb-like structure is effective, because the surface area of the moderator foils for slow positron emission is only doubled by the additional cross foils. The yield in the honeycomb-like assembly with reflectors is more than three times larger than that in the usual one.

It is obvious that the structure effect is remarkable especially in the honeycomb-like assembly with reflectors. In the calculation, if a 100 MeV electron beam of 100 kW is applied to this assembly, a very intense monoenergetic positron beam of 10^{12}/sec in intensity is expected. This structure is practically promising for realizing an intense monoenergetic positron beam of more than 10^{10}/sec in intensity, taking empirical efficiencies of beam extraction, transportation and brightness enhancement to reduce the beam size and the energy spread into account.

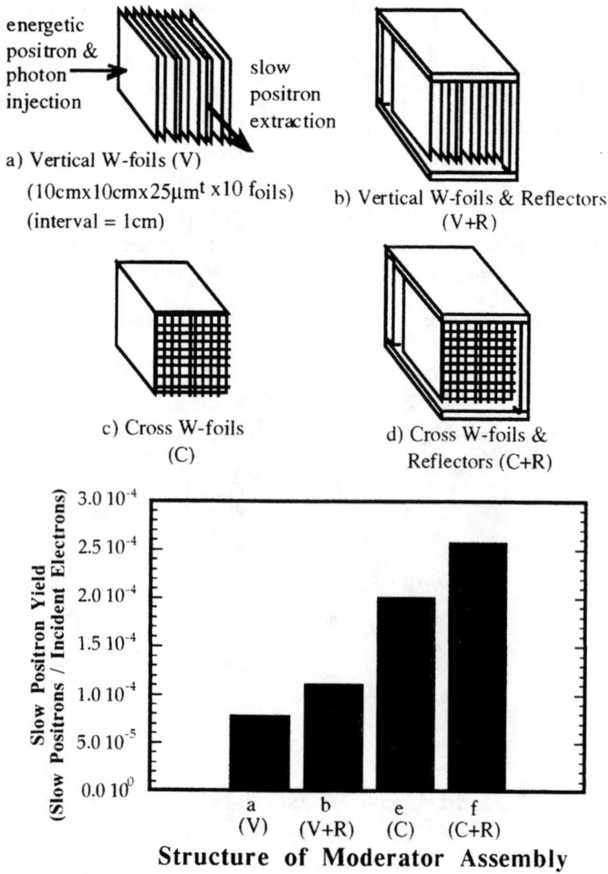

FIGURE 10. Moderator structure effect: calculated slow positron yields with EGS4-SPG (lower side) for moderator assemblies having various structures (upper side).

LOCAL SHIELD & REMOTE MANIPULATION MECHANISM

To suppress the activation of air surrounding the target system and reduce the thickness of concrete walls of the building, we performed a conceptual design of a local shield as shown in Fig. 11, on the basis of the shield calculations. In the design, we intend to reduce the radiation level by two orders with the local shield during the linac operation.

The target system is surrounded by structures consisting of 12 cm-thick steel, 30 cm-thick water and 12 cm-thick steel. The water works as a shield for neutrons and also as a coolant. An additional iron shield of 65 cm in maximum thickness for the forward beam direction will be effective in order to reduce the thickness of the concrete wall. An activated converter and moderator assemblies are manipulated by remote control and transported to containers, using air chucks, a coupler with compressed air, KF-type flanges with clamps and clamp releases with screw drive.

The inside of the local shield is vacuum. Consequently the activation of air in the target room is suppressed to permissible level as shown in Table 3. The calculated result in Table 3 also shows that the saturated activation of the local shield itself is lower than, for example, a typical cyclotron.

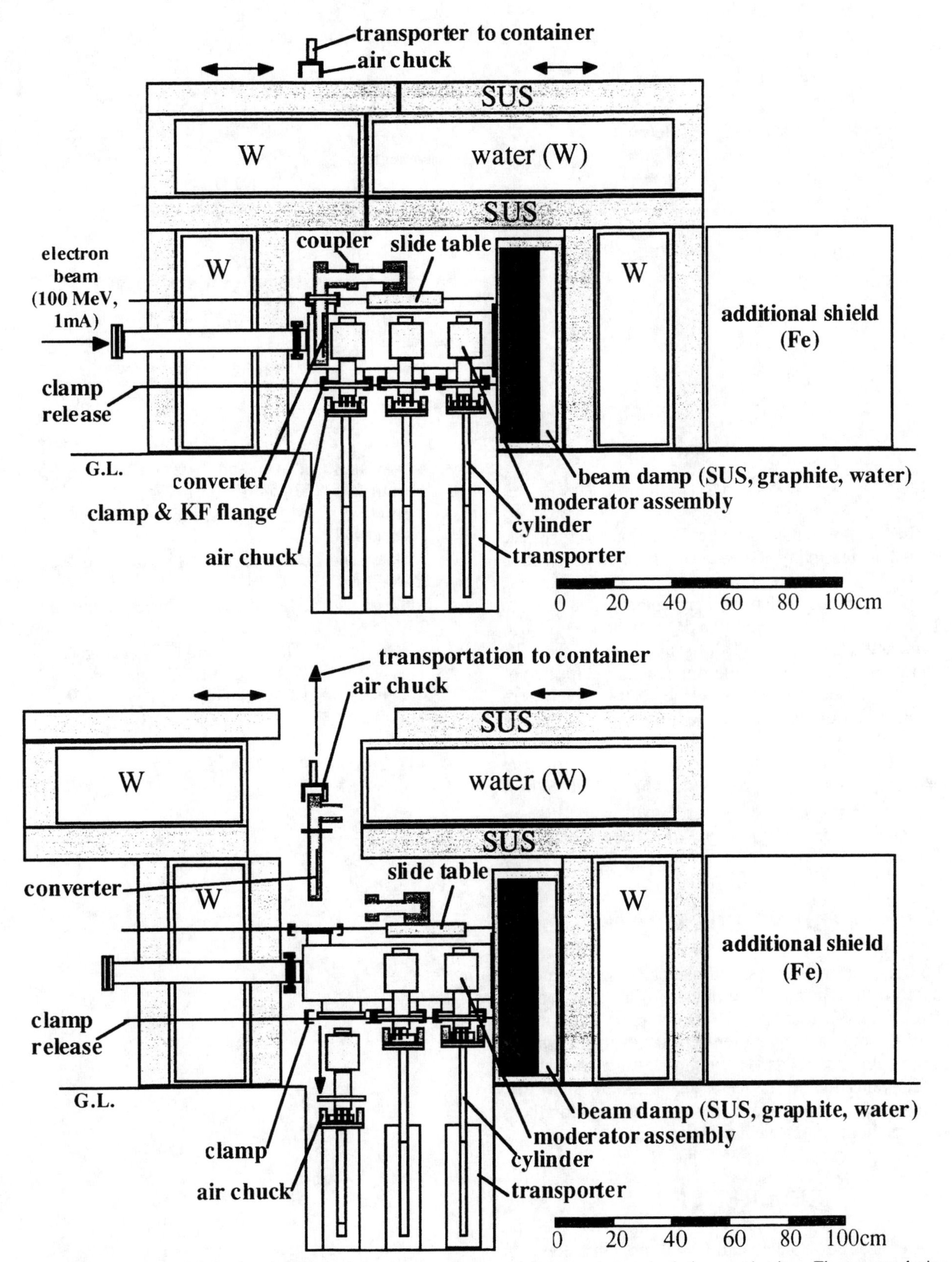

FIGURE 11. A concept of the local shield of the target system and the remote manipulation mechanism. The upper: during the electron bombardment, and the lower: during the manipulation of the converter and the moderator assembly.

TABLE 3. Calculated activation of air, local shield and target.

Activation of Air		Activation of Target System (Dose Rate at 1 m from Local Shield Surface)	
^{3}H (12.3y):	$5.6x10^{-2}$ Bq/cm^{3}	local shield (Fe):	^{55}Fe (2.6y) 0 Sv/h
^{7}Be (53.6d):	$1.1x10^{-4}$ Bq/cm^{3}		^{59}Fe (45.6d) 15 mSv/h
^{11}C (20.3min):	$1.1x10^{-3}$ Bq/cm^{3}		^{46}Sc, ^{54}Mn, ^{51}Cr etc. 26 μSv/h
^{13}N (9.96min):	$5x10^{-2}$ Bq/cm^{3}	converter (Ta):	^{180m}Ta (8.1h) $6x10^{-2}$ mSv/h
^{15}O (123sec):	$6x10^{-3}$ Bq/cm^{3}		^{180}Ta (115d) $4x10^{-1}$ mSv/h
^{16}N (7.14sec):	$2x10^{-7}$ Bq/cm^{3}	moderator (W):	^{182}Ta (115d) $4x10^{-2}$ μSv/h
^{38}C l(37.3min):	$2x10^{-5}$ Bq/cm^{3}		^{183}Ta (5d) $3x10^{-2}$ μSv/h
^{39}Cl (55.5min):	$1.5x10^{-4}$ Bq/cm^{3}		^{181}W (121.2d) 3 μSv/h
^{41}Ar(1.83h):	$1.8x10^{0}$ Bq/cm^{3}		^{185}W (75.1d) 3 μSv/h
			^{187}W (23.9d) $2x10^{-3}$ μSv/h
		water:	^{3}H (12.6y) 0 Sv/h
			^{7}Be (53.3y) $1x10^{-2}$ mSv/h

CONCLUSION

In the design study for the Positron Factory, we confirmed the feasibility of a dedicated high-power electron linac of 100 kW class with a beam energy of 100 MeV, with investigations including performance tests of the electron beam window and the RF window. We also experimentally demonstrated the availability of the self-driven rotating converter and the simultaneous extraction of multi-channel monoenergetic positron beams, which were proposed based on calculations with a newly developed Monte Carlo simulation system. A more efficient moderator structure, which was suggested by the experimental result, is proposed. The world-highest monoenergetic positron beam of more than 10^{10}/sec in intensity will be realized by the use of the above machines and instruments.

ACKNOWLEDGMENTS

The basic idea of the Positron Factory project has been discussed in the Special Committee on Positron Factory Research Plan (Leader: Prof. M.Dohyama). The authors gratefully acknowledge the members for their discussions and encouraging supports. The authors also wish to thank Prof. Hirayama for his advice in using EGS4, and Prof. S.Tagawa, Dr. Y.Honda and their colleagues for their cooperation in the experiment using an electron linac.

REFERENCES

1. W.E.Frieze, D.W.Gidley and K.G.Lynn, Phys. Rev. **B31**, 5628-5633 (1985)
2. R.Mayer, A.Schwab and A.Weiss, Phys. Rev. **B42**, 1881-1884 (1990)
3. A. Kawasuso, H. Arai and S. Okada, Materials Science Forum **255-257**, 548-550 (1997)
4. A.Kawasuso and S. Okada, Phys. Rev. Lett. **81**, 2695-2698 (1998)
5. S.Okada and H.Sunaga, Nucl. Instrum. and Meth. **B56/57**, 604-609 (1991)
6. S.Okada and H.Kaneko, Appl. Surface Science **85**, 149-153 (1995)
7. W.R.Nelson, H.Hirayama and D.W.O.Rogers, The EGS4 Code System, SLAC Report 265 (1985)
8. A.Vehanen and J.Makinen, Appl. Phys. **A36**, 97-101 (1985)
9. S.Okada et. al., Proc. 1994 Int. Linac Conference (Tsukuba, 1994) pp. 570-572

Double Ionization of Noble Gases by Positron Impact

H. Bluhme, H. Knudsen and J. P. Merrison

Institute of Physics and Astronomy, University of Aarhus, DK-8000 Aarhus C, Denmark.

The total double ionization cross sections for helium and neon by positron impact have been measured at energies from threshold to 900 eV. The cross sections exhibit a surprising absence or suppression of positronium (Ps) formation in the near-threshold region. This is a remarkable difference to what is seen in single ionization and in contrast to what has previously been suggested by other experiments. The absolute cross sections for the two targets are compared to a modified Rost-Pattard parametrization. At threshold this model incorporates Wannier theory for double ionization, while it at high energies uses an $E^{-1.5}$ dependence. Good agreement is found, showing that this type of model can be extended from single to double ionization.

INTRODUCTION

In a recent experiment we have studied the total double ionization of helium and neon by positron impact [1]. Included in the cross sections (σ^{2+}_{tot}) are both the direct ionization channel (1) and the ionization channel with positronium (Ps) formation (2), i.e.,

$$e^+ + A \rightarrow A^{2+} + e^+ + 2e^-, \quad (1)$$

$$e^+ + A \rightarrow A^{2+} + Ps + e^-, \quad (2)$$

The two processes have threshold energies at E^{2+}_I and E^{2+}_{Ps}, respectively, where E^{2+}_{Ps} is 6.8 eV lower than E^{2+}_I due to the energy gain for ionization from the binding energy of the positronium. In the measurements we concentrated on low energy impact, and especially the region between E^{2+}_{Ps} and E^{2+}_I was studied. Here the positronium formation channel (2) is the only open channel.

The measurements were done using a pulsed positron beam utilizing random ion extraction so that ions from both (1) and (2) were included in the ion countrates. For more experimental details, the reader is refereed to references [1,2].

RESULTS

Our results (σ^{2+}_{tot}) are shown in Fig. 1(c) and 1(d), together with measurements of the direct double ionization cross section (σ^{2+}_I), which does not include the positronium formation channel (2). In Fig. 1(a) and 1(b) are shown the corresponding cross sections for single ionization (σ^{+}_{tot} and σ^{+}_I), where processes similar to (1) and (2) exist. By comparing Fig. 1(c) to 1(a) and Fig. 1(d) to 1(b) it is obvious that the contribution to σ^{2+}_{tot} from the positronium formation is quite different from the corresponding contribution in single ionization in both gases. The behavior at threshold is very different, with σ^{+}_{tot} showing a sharp onset at threshold (E^{+}_{Ps}), while no such thing is seen in σ^{2+}_{tot} at E^{2+}_{Ps}. Also in single ionization, a large fraction of the total cross section at low energies is attributed to positronium formation, while in double ionization the contribution is hardly discernible.

This kind of behavior was not what had been expected. It would seem natural to believe that capture of an electron by the positron at threshold should be equally as probable in double ionization as in single ionization, since in both cases the positron and an electron leaves the target slowly. Furthermore, previous experiments seemed to suggest this: The difference between the two measurements of σ^{2+}_I in Fig. 1(d) was attributed by Kara et al. [7] to an inadequate discrimination against positronium formation in the experiment by Charlton et al. [9], an explanation, which can be valid only if the positronium channel is strong. Our data shows that Ps formation can not explain the whole difference and the discrepancy between the two σ^{2+}_I remains unresolved. Another experiment on the heavier noble gases, Ar, Kr and Xe [10], has also indicated a significant contribution to σ^{2+}_{tot} from positronium formation, but the emphasis of this experiment was not on the threshold region, and the larger positron energy spread of this experiment may have obscured the results [11].

COMPARISON WITH THEORY

The fact that we observe a suppression of the positronium formation channel (2) and that our cross sections therefore stem nearly completely from the direct ionization channel (1) makes them of interest to theory for direct ionization. In particular the measurements around threshold, which are the first data of its kind, are of great interest to threshold breakup theories like the extended Wannier theory [12]. The basis of this theory is the complete four-particle breakup of the direct ionization

CP475, *Applications of Accelerators in Research and Industry*,
edited by J. L. Duggan and I. L. Morgan

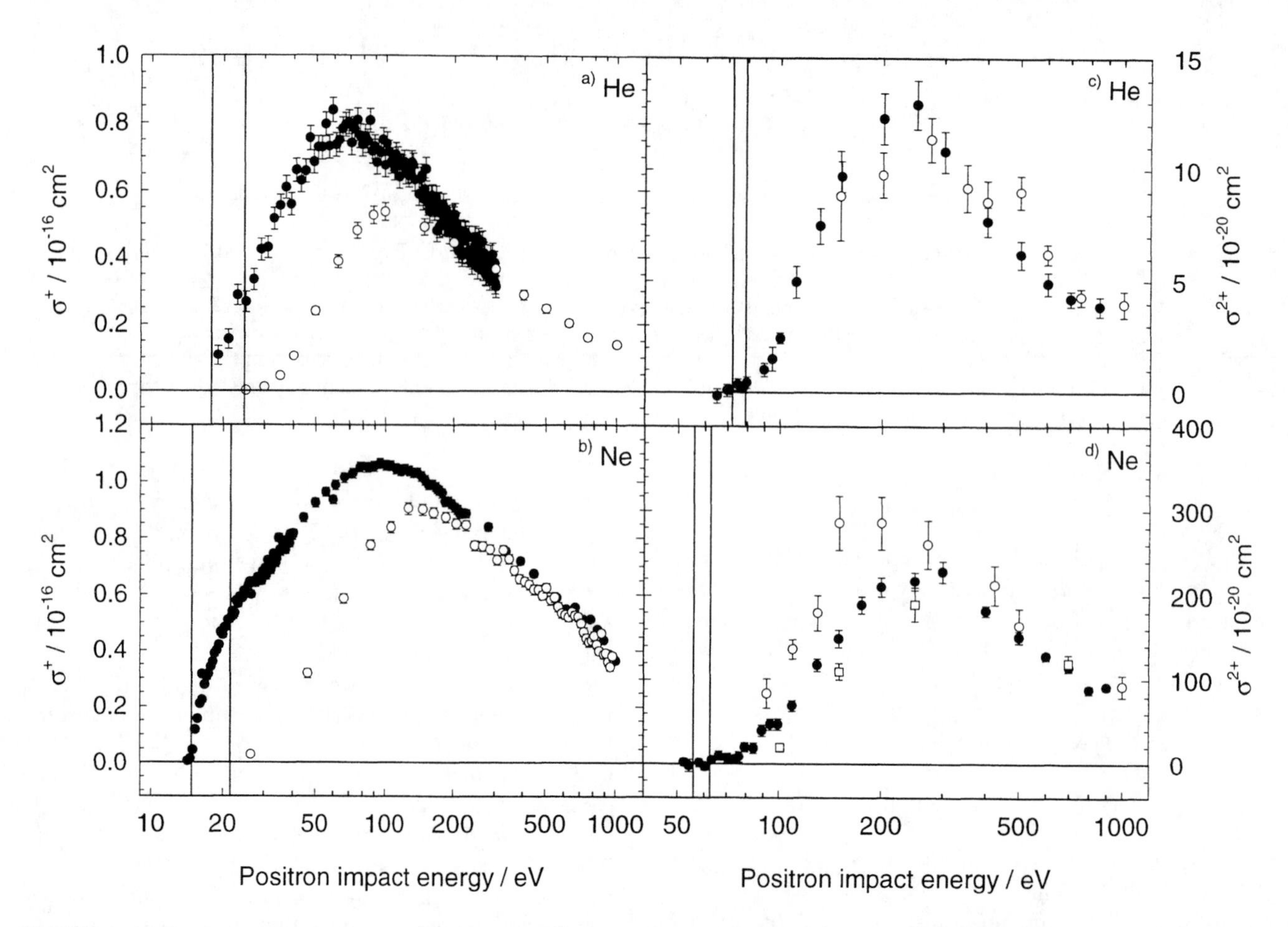

FIGURE 1. (a) Single ionization of He by e^+ impact: (●) σ^+_{tot}, Moxom et al. [3]; (○) σ^+_I, Jacobsen et al. [4]. This cross section was also measured by Moxom et al. [5], but as their results agree with those of Jacobsen et al., they are not shown in this figure. (b) Single ionization of Ne by e^+ impact: (●) σ^+_{tot}, Laricchia [6]; (○) σ^+_I, Kara et al. [7]. This cross section was also measured by Jacobsen al. [4], but as their results agree with those of Kara et al., they are not shown in this figure. (c) Double ionization of He by e^+ impact: (●) σ^{2+}_{tot}, Bluhme et al. [1]; (○) σ^{2+}_I derived from data by Charlton et al. [8] and Jacobsen et al. [4]. (d) Double ionization of Ne by e^+ impact: (●) σ^{2+}_{tot}, Bluhme et al. [1]; (○) σ^{2+}_I derived from data by Charlton et al. [9] and Jacobsen et al. [4]; (□) σ^{2+}_I, Kara et al. [7]. The two vertical lines indicate the relevant values of the threshold energies E_{Ps} and E_I.

channel (1). It predicts that the cross section should behave at threshold as

$$\sigma \propto E^{\alpha}, \tag{3}$$

where E is the positron excess energy above the relevant threshold energy E_I, and α is the so-called Wannier exponent. The first published calculation of α for double ionization by positrons was done by Poelstra et al. on the $e^+ + H^-$ system [13]. Mendez and Feagin have extended the calculations to a neutral target. For a target with infinite atomic mass, they have found a value of $\alpha = 3.613$ [14]. This should be valid for both helium and neon. Recently Kuchiev and Ostrovsky have published an article on threshold breakup of atomic systems into charged fragments [15]. Using a new approach involving so-called scaling configurations, which should be a generalization of the well-known Wannier ridge, they have calculated α for a wide range of systems. Among these is double ionization of a neutral target by positron impact. They found a value of $\alpha = 3.838$. While this value is numerically fairly close to the value of Mendez and Feagin, the difference between them stems from a disagreement in some of the theory's fundamental formulas and possibly in some of the numerical calculations as well [15].

While the validity of Wannier theory is limited to the threshold region, it has been shown by Rost and Pattard [16] that in the case of single ionization it is possible to combine the threshold behavior of Wannier theory with the high energy behavior of the first Born approximation to form a parametrization depending only on the Wannier exponent. This parametrization is universal for the cross sections of all (light) targets for a given projectile over an impact energy range from threshold to high energies. The model involves a scaling of the cross sections into the dimensionless variables $x = E/E_M$ and $y = \sigma/\sigma_M$, in which they all fall on the universal curve. σ_M is the value of the cross section maximum, which is found at energy E_M. Both

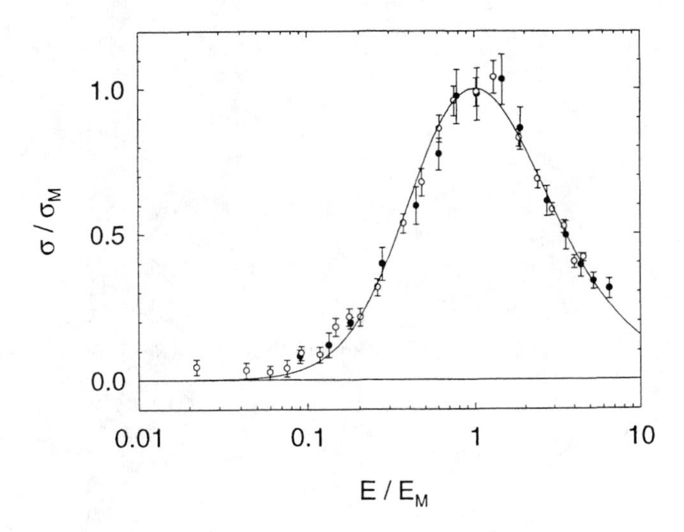

FIGURE 2. The present results scaled relative to E_M and σ_M: (●) helium; (O) neon; (—) modified Rost-Pattard model with α = 3.613 and β = 1.5.

can be found by a fit to the data. The energies E and E_M are again excess energies above E_I. Both the scaling and the curve are equally important aspects of the Rost-Pattard model.

Inspired by the success of the Rost-Pattard parametrization in single ionization, the fact that the Wannier exponent now is available for double ionization and that our data ranges from threshold to high energies, we have tried to modify the Rost-Pattard model to suit double ionization. In Fig. 2 we have transformed our data into the variables x and y as defined above. This causes our data for helium and neon to agree well with each other, showing that the scaling extends to double ionization.

To extend the universal parametrization from single to double ionization one important change has to be made: The high-energy behavior in double ionization is different from the first Born E^{-1} behavior of single ionization. To compensate this we have made a parametrization with a general $E^{-\beta}$ dependence at high energies. It is based on the function

$$f(E) = \frac{1}{(E+E_0)^{\beta}}\left(\frac{E}{E+E_0}\right)^{\alpha}, \quad (4)$$

where $E_0 = E_M\beta/\alpha$. The original Rost-Pattard model for single ionization has $\beta = 1$. The cross section is given by

$$\sigma(E) = \sigma_M \frac{f(E)}{f(E_M)}. \quad (5)$$

When expressed in the dimensionless variables $x = E/E_M$ and $y = \sigma/\sigma_M$, equation (5) reduces to

$$y = \left(\frac{\alpha+\beta}{\alpha\cdot x+\beta}\right)^{\beta}\cdot\left(\frac{\alpha+\beta}{\alpha\cdot x+\beta}\cdot x\right)^{\alpha}. \quad (6)$$

The need for the additional parameter β arises from the different mechanisms, which can lead to double ionization at high energies. At very high energies the governing mechanism is where there is only one interaction between the positron and one of the target electrons, followed by subsequent release of the other electron. This mechanism gives a cross section energy dependence corresponding to approximately $\beta = 1$. At somewhat lower (but still high) energies, a mechanism involving interactions with both electrons come into play. It has an energy dependence corresponding to $\beta = 2$. (See e.g. Knudsen and Reading [17].) For our parametrization we have chosen a value of $\beta = 1.5$ as a compromise between the two. In Fig. 2 we have plotted the universal curve (6). It is based on the Wannier exponent of Mendez and Feagin (α = 3.613). The corresponding curve based on the exponent of Kuchiev and Ostrovsky (α = 3.838) is not plotted since the two curves are virtually indistinguishable due to the small difference in the exponent and our data can therefore not be used to judge which one is best. One observes a good agreement between our data and the curve. We could also have obtained α and β through a fit to the data. This yields α = 2.43 and β = 1.70, but the agreement with the data is not significantly improved.

CONCLUSIONS

Measurements of the total double ionization cross sections for helium and neon have been made. They exhibit a strong suppression of double ionization with positronium formation. The data have been used for a comparison with a modified Rost-Pattard parametrization, showing good agreement. This shows that the Rost-Pattard model can successfully be expanded from single to double ionization.

ACKNOWLEDGMENTS

The authors wish to thank J. M. Feagin for discussions on Wannier theory and for drawing our attention to the Rost-Pattard model, G. Laricchia for providing unpublished data and valuable discussions, M. Charlton and V. N. Ostrovsky for useful discussions, P. Aggerholm for technical assistance, and the Carlsberg Foundation and the Danish Natural Science Research Council for funding.

REFERENCES

1. Bluhme, H., Knudsen, H., Merrison, J. P., and Poulsen, M. R., *Phys. Rev. Lett.* **81**, 73-76 (1998).
2. Bluhme, H., *et al.* (to be published).

3. Moxom, J., Laricchia, G., and Charlton, M., *J. Phys. B* **28**, 1331-1347 (1995).
4. Jacobsen, F. M., Frandsen, N. P., Knudsen, H., Mikkelsen, U., and Schrader, D. M., *J. Phys. B* **28**, 4691-4695 (1995).
5. Moxom, J., Ashley, P., and Laricchia, G., *Can. J. Phys.* **74**, 367-372 (1996).
6. Laricchia, G., (unpublished).
7. Kara, V., Paludan, K., Moxom, J., Ashley, P., and Laricchia, G., *J. Phys. B* **30**, 3933-3949 (1997).
8. Charlton, M., Andersen, L. H., Brun-Nielsen, L., Deutch, B. I., Hvelplund, P., Jacobsen, F. M., Knudsen, H., Laricchia, G., Poulsen, M. R., and Pedersen, J. O., *J. Phys. B* **21**, L545-L549 (1988).
9. Charlton, M., Brun-Nielsen, L., Deutch, B. I., Hvelplund, P., Jacobsen, F. M., Knudsen, H., Laricchia, G., and Poulsen, M. R., *J. Phys. B* **22**, 2779-2788 (1989).
10. Helms, S., Brinkmann, U., Deiwiks, J., Hippler, R., Schneider, H., Segers, D., and Paridaens, J., *J. Phys. B* **28**, 1095-1103 (1995).
11. Paridaens, J., Segers, D., Dorikens, M., and Dorikens-Vanpraet, L., *Nucl. Instrum. Methods Phys. Res., Sect. A* **287**, 359-362 (1990).
12. See e.g., Feagin, J. M., and Filipczyk, R. D., *Phys. Rev. Lett.* **64**, 384-387 (1990); ref. 13.
13. Poelstra, K. A., Feagin, J. M., and Klar, H., *J. Phys. B* **27**, 781-795 (1994).
14. Mendez, J. C., and Feagin, J. M., (private communication).
15. Kuchiev, M. Y., and Ostrovsky, V. N., *Phys. Rev A* **58**, 321-335 (1998).
16. Rost, J. M., and Pattard, T., *Phys. Rev. A* **55**, R5-R7 (1997).
17. Knudsen, H., and Reading, J. F., *Phys. Rep.* **212**, 107-222 (1992).

High Intensity Positron Program at LLNL

P. Asoka-Kumar, R. Howell, W. Stoeffl, and D. Carter
Lawrence Livermore National laboratory, Livermore, CA 94550

Lawrence Livermore National Laboratory (LLNL) is the home of the world's highest current beam of keV positrons. The potential for establishing a national center for materials analysis using positron annihilation techniques around this capability is being actively pursued. The high LLNL beam current will enable investigations in several new areas. We are developing a positron microprobe that will produce a pulsed, focused positron beam for 3-dimensional scans of defect size and concentration with submicron resolution. Below we summarize the important design features of this microprobe. Several experimental end stations will be available that can utilize the high current beam with a time distribution determined by the electron linac pulse structure, quasi-continuous, or bunched at 20 MHz, and can operate in an electrostatic or (and) magnetostatic environment. Some of the planned early experiments are: two-dimensional angular correlation of annihilation radiation of thin films and buried interfaces, positron diffraction holography, positron induced desorption, and positron induced Auger spectroscopy.

INTRODUCTION

Positron annihilation spectroscopy has emerged as a highly sensitive, nondestructive probe to study the nature, concentration, and spatial distribution of defects in materials.[1] Since the early 1970's, considerable efforts went into developing low-energy, monochromatic positron beams that can perform depth-resolved defect analysis of materials. The underlying physical process for converting the positrons emitted from either a radio isotope or a stopping target in an electron linac into a monochromatic beam, known as "moderation," is not very efficient (with a yield of about one positron for every 10^3 positrons entering the moderator material). Therefore, most laboratory beams lack the brightness necessary to realize the full potential of this probe. The present status in our field is analogous to the one experienced by researchers using laboratory X-ray sources before the advent of synchrotron light sources. In this paper, we describe the efforts now underway at LLNL to remedy this situation by building a multiuser experimental program around an existing high current source of moderated positrons.[2]

HIGH INTENSITY POSITRON BEAM

The LLNL primary source of positrons is built at the end of a 100 MeV linac, capable of producing 1 A of electrons at a repetition rate of 330Hz. The energetic electrons are stopped in a water-cooled tungsten target providing a shower of bremstrahlung photons. The pair conversion of the Bremstrahlung photons yields an intense source of positrons. These positrons shine onto a set of well-annealed tungsten foils (25 microns thick) arranged in the form of "Venetian blinds." A fraction of the high energy positrons is slowed down in the tungsten foils and is reemitted as moderated positrons. These positrons are harvested and guided to an experimental hall with a ~ 30 gauss axial magnetic field. The beam transport system contains a curved section to prevent direct line of sight from the source region. The experimental hall and the positron production target are separated by ~ 4.5 m of shielding to ensure a low radiation background environment in the experimental hall. Figure 1 shows the layout of the experimental hall.

The positron beam entering the experimental hall consists of short bursts of high intensity positrons with an energy width of ~ 4 eV. The pulse duration and repetition rates are determined by the pulse structure of the electron linac, with nominal values of 3 μs and 300 Hz, respectively. The intensity of the moderated positrons is determined by the energy, current, and focus quality of the electron beam. With 100 MeV electron beam, every 3×10^5 electrons produce a moderated positron, resulting in a 10^{10} positrons/s beam at full power.

The time structure of the initial positron beam is not well suited for many experiments. Every pulse contains 3×10^7 positrons and will saturate most detection systems. Therefore, the initial beam is captured, stored, and released slowly from a penning trap to produce a time-stretched beam profile. The pulse stretcher consists of 3 elements: a gate electrode at the entrance, a middle storage section (called "floor"), and a barrier electrode at the exit. The voltage on the gate and barrier electrodes is higher than the primary positron energy of ~15 V, except for a short interval when the gate is switched low to let in the pulse of positrons. The positron bunch is thus captured between the gate electrode and barrier electrode. After filling the trap with a positron pulse, the voltage on the

CP475, *Applications of Accelerators in Research and Industry*,
edited by J. L. Duggan and I. L. Morgan
1999 The American Institute of Physics 1-56396-825-8

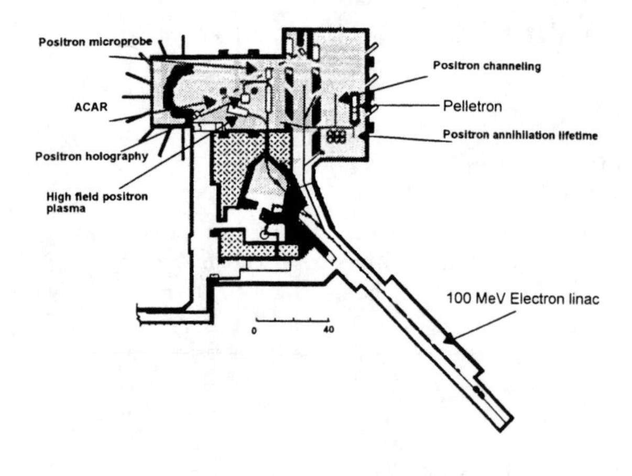

Figure 1. Layout of the intense positron beam at LLNL.

trap floor is raised slowly to spill the positrons over the fixed voltage barrier. The ramp voltage is adjusted to convert the initial 3 μs pulse into a ~ 3 ms pulse at the end of which the next bunch of positrons enters the trap to continue the sequence of events.

The trap efficiency was measured by capturing positrons and holding them for various times. The trap is then emptied by raising the floor and the positron yield is compared to the initial number of positrons. The primary positron intensity is obtained by turning off the barrier voltage. The positron yield is obtained by integrating the annihilation γ-ray signal obtained with a NaI detector. The yield dropped to a 50% level when the positrons were held in the trap for 8 ms, suggesting a positron trap life time of 11.5 ms, at a base pressure of 3.0×10^{-8} torr.

The energy spread of the beam spilling over the barrier is significantly smaller than the starting beam, 20 meV versus ~ 4 eV. The barrier is nominally set at 20 Volt and is followed by an accelerator column consisting of 30 electrodes, equally spaced in distance and voltage to provide a gentle acceleration of 20 eV.

Immediately following the accelerator is a two-gap time-buncher section that is pulsed to produce a nano second wide pulse every 50 ns, and acts as the first pulsing stage of a positron microprobe (see later). Thus, depending on the user requirement, the stretcher-buncher configuration can deliver either a nearly continuous beam or a pulsed beam with well-defined time structure (containing ~ 500 positrons per pulse at 20 MHz) . The stretcher and buncher can also be turned off to keep the time structure of the original linac beam.

EXTRACTION OF POSITRONS FROM MAGNETIC FIELD

Many experimental stations will require an electrostatic environment for optimum performance. Therefore, the positrons have to be extracted out of the guiding magnetic field without significant degradation of the phase space. This is achieved with a magnetic grid consisting of 36 tapered fins pointing toward the center similar to the "spokes" of a wheel. As the spokes approach the center they are terminated at three different distances from the center to have an effective magnetic field termination and > 90% transmission. The magnetic fields are concentrated in the spokes and are guided outward by an iron shield. Tests show that at a distance of 2 mm from the magnetic grid, an internal magnetic field of 30 gauss is reduced to 0.3 gauss.

The motion of positrons crossing the magnetic field through the open space will be nonadiabatic. The positrons are accelerated to 2000 eV with an accelerator section to increase the pitch length compared to the region in which the field lines are terminated. Figure 2 shows the positron beam

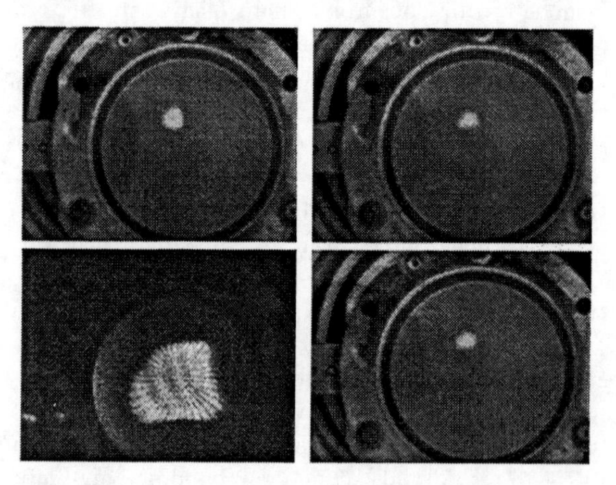

Figure 2. Images of the positron beam captured from a microchannel plate, located 40 cm from the magnetic grid exit. Images are formed with electrostatic lenses.The images correspond to (starting from to left hand corner, clockwise): Primary beam with the trap turned off, Time-stretched positron beam without bunching, Time-stretched and bunched positron beam, Image of the magnetic grid showing the spokes and vane structure. The channel plate active area is 40 mm.

emanating from the grid imaged onto a microchannel plate, located at a distance of 40 cm from the magnetic grid. The image is magnified using electrostatic lenses to show the fine structure of the grid and is collected using the primary positron beam without time stretching. Besides the spokes of the magnetic grid the image also shows the vanes of the starting moderator.

Since we are able to image the primary positron moderator structure through the magnetic grid, it is clear that the scattering of positrons from the magnetic grid is negligible.

Figure 2 also contains images of focused beam spot corresponding to the primary beam, time-stretched beam, and bunched beam. In all three cases the beam spots remained nearly identical showing the excellent transport properties of the beam through the trap, buncher, and magnetic grid systems.

BEAM TRANSPORT

The stretched beam enables the operation of the positron microprobe and a series of positron spectrographic measurements. A switchyard located 40 cm from the magnetic grid can divert the electrostatically transported beam to various experiments. The switchyard accommodates three sets of components: A straight through tube that transports positrons to the positron microprobe, a set of cylindrical plates that deflect the positron beam to a multiuser beam feed line, and a microchannel plate for instantaneous beam characterization and tuning.

The user beam lines are designed with electrostatic elements and the beam is diverted to different end stations using more switchyard stations. Experiments actively pursued include high resolution measurements of electron momentum by angular correlation techniques, positron diffraction and holography, positron induced desorption from the surface, and positron induced Auger electron spectroscopy. These techniques have been developed and tested at other facilities and will be attached to the various end stations.

POSITRON MICROPROBE

LLNL is building a pulsed positron microprobe as a central part of the high intensity positron program. At completion, the microprobe will provide a 100 ps wide bunches of positrons at an intensity of 10^7 positrons/s. The final spot size will be about 1 μm and will have a repetition rate of 20 MHz. This will allow us to perform positron lifetime spectroscopy to determine the defect size and concentration over spatial volumes of 0.025 μm^3. The energy of the positron beam can be varied 1-50 keV to sample the near surface regions (1-10 μm depending on the density of the target material) and buried interfaces.

The positron beam exiting the magnetic grid has the following properties: diameter = 1.2 cm, half-angle of divergence < 1°, energy = 2 keV, a pulse width = 1 ns, and a repetition rate = 20 MHz. Conversion of this beam into a microprobe is challenging and is achieved using several principles.

The most important principle is the brightness enhancement of the beam using successive stages of moderation. The brightness of a focused positron beam is limited by the phase space constraints set by the primary source. Due to the limitations in the primary positron flux (unlike in an electron microscope, where there is an abundant supply of electrons), collimation is not practical for attaining a

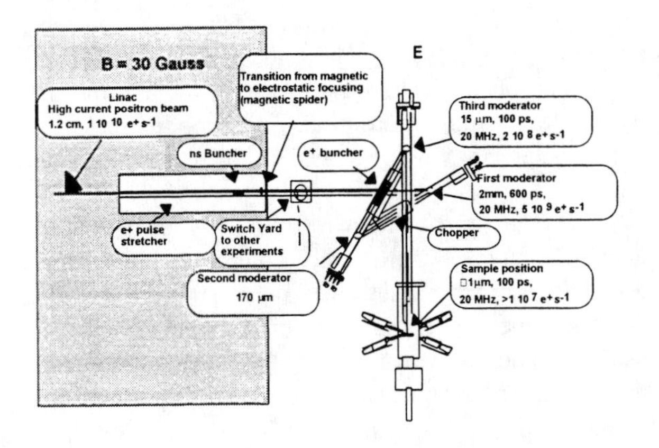

Figure 3. Schematic of the main functional elements of the LLNL 3D-pulsed positron microprobe. The final beam will be ~ 1μm in diameter and pulsed at 20 MHz with a pulse width of < 100 ps. The energy of the beam can be varied

micron beam spot. The brightness of a positron beam can be increased by several orders of magnitude by taking advantage of the nonconservative forces that are present when a beam of fast positrons (~5keV) is slowed down in a target and reemitted as slow positrons.[3] Although, 80% of the beam is lost in the remoderation process, the brightness enhancement of a remoderation stage is orders of magnitude better than that one can achieve using apertures.

In our design, the final spot size of ~ 1μm is achieved by three stages of brightness enhancement, each stage producing a compression of ~ 10 in diameter. Since we have 3 stages of reflection remoderation, the microprobe can be compartmentalized into four straight sections. The angle between each section is chosen to optimize the overall layout of the microprobe (see Figure 3).

To prevent magnetic field causing unwanted beam modifications in an adjacent section, the beam tranport and focusing is performed using electrostatic elements. All electrostatic elements are designed using the ion simulation program SIMION. The first moderator is located 190 cm from the magnetic grid. The length of the column is chosen to produce a time focus at the moderator, and reduce the background in the data acquisition window (see below).

The second column is 50 cm long and produces a spot of 0.15 mm on the second remoderator. This column contains a chopper section, which is a set of parallel plates (4mm×4mm, 6mm separation) designed to let the positron bunch go through while deflecting the unwanted tails (arising from positrons arriving out-of-phase or leaking out of the stretcher) into an absorber. The chopper receives radio frequency voltage timed with respect to the bunches.

The third column (50 cm long) houses the main buncher which will compress the time distribution of the beam from ~ 1ns to 100 ps every 50 ns. The time trigger from the main buncher will be used to obtain the arrival time of the positron bunch at the target. The third column produces a beam spot of 15 μm on the final moderator.

The final column incorporates elements that are necessary to focus the beam to a ~ 1μm spot and to vary the beam energy from 1keV to 50 keV. The final column (113 cm long) contains a buncher section that can be used to provide a tighter time distribution for specialized applications. This column also incorporates a 12-pole deflector lens system to steer the beam across the target. The 12-pole lens also can produce a line focus to scan with greater precision along one dimension.

The components and functions described above have been optimized to yield a set of robust design parameters for the microprobe. The influence of mechanical misallignments of the electrode structures, residual ac magnetic fields, variations in the power supply voltages (ac ripple), and radiation background from successive stages of moderation were all optimized.

The lens elements are distributed to produce an odd number of beam cross-overs in each column, and the cross over positions are adjusted to compensate the beam spot shifts from residual ac magnetic fields. Therefore, a small magnetic field will not alter the position of the electrostatic focus in first order. Simulations show that a one milligauss ac magnetic field broadens the beam spot to 2.6 μm. The microprobe columns are housed in a double walled μ-metal enclosure, and the residual ac magnetic field is expected to be much less than 1 milligauss. It is important to note that a dc background field of 1 milligauss will only shift the final focus spot without degrading the spot size.

The optical column layout and their lengths are optimized to reduce the background radiation in the detector system. Since the bunches are spaced 50 ns apart, there are several bunches in the system at any given instant. The positron annihilations from moderators, the magnetic grid, accelerating grids, and stopper for the chopped beam can all contribute to the background radiation. The length of the optical columns are chosen to produce all these annihilations in a 10 ns window before the target annihilations. Thus, we will have a clean data collection time window of 40 ns that is not disturbed by annihilations from other sources. Since the intensity of 511 keV radiation at the remoderators is several orders of magnitude higher than the target signal itself, this has been an important factor in the layout of various elements.

The location of the radio frequency elements are optimized to reduce the coupling effects. When a bunch receives a radio frequency kick in the chopper, other bunches are located inside well-shielded electrostatic tubes.

Lifetime spectra will be collected using an array of BaF_2 detectors. The detectors will be operated in coincidence to prevent background events. The positron backscattering from the target can cause background signal, and is a serious problem in all existing positron beam lifetime systems. To effectively eliminate this background, we have designed our detector system to record only events emanating from a small volume around the target. Each annihilation event is validated by a pair of detectors arranged at 180 degrees which limits validated events to the volume between the two detectors. If an event occurs in the line of sight between the two coincidence detectors, it could be counted as a good event. The enclosure around the target is large enough to discriminate events that originate from the wall.

When completed, the microprobe will be a unique instrument capable of scanning three dimensional images of defect distributions in materials.

CONCLUSION

A program of new measurements using our intense positron beam is under development at LLNL. The high flux of the LLNL beam and the assortment of spectroscopic tools being developed will enable several unique experiments.

ACKNOWLEDGEMENTS

This work was performed under the auspices of the US Department of Energy by LLNL under contract No. W-7405-ENG-48.

REFERENCES

1. See for example, Proceedings of the 11th International Conference on Positron Annihilation, Eds, Y.C. Jean, M. Eldrup, D.M. Schrader, and R.N. West, Materials Science Forum, **255-257**, Trans Tech Publications, USA (1997).
2. R. Howell, R.A. Alvarez, and M. Stanek, Appl. Phys. Lett., **40**, 751 (1982).
3. A.P. Mills, Jr., Appl. Phys., **23**, 189 (1980).

Bulk Defect Analysis with a High-Energy Positron Beam

J. H. Hartley, R. H. Howell, P.A. Sterne

Lawrence Livermore National Laboratory, Livermore, CA 94551

A program using a positron beam to probe defects in bulk materials has been developed at Lawrence Livermore National Laboratory. Positron annihilation lifetime spectroscopy (PALS) provides non-destructive analysis of average defect size and concentration. A 3 MeV positron beam is supplied by Sodium-22 at the terminal of a Pelletron accelerator. The high-energy beam allows large ($\geq 1\ cm^2$) engineering samples to be measured in air or even sealed in an independent environment. A description of the beam-PALS system will be presented along with a summary of recent measurements.

INTRODUCTION

Positron annihilation lifetime spectroscopy is a powerful tool for the study of defects in materials, but its application has been limited by low data rates and restrictive sample geometry. Samples are sandwiched tightly around a ^{22}Na source, and timing of the positron emission depends on detection of the coincident 1.275 MeV gamma ray. Positron annihilation in the source results in a ~5% systematic contribution to the spectrum that complicates the data analysis.

These difficulties can be greatly reduced by performing PALS measurements with a monoenergetic beam. Sample size and geometry are limited only by the beam size and detector configuration. Data rates are significantly improved because positron implantation can be detected with nearly 100% efficiency. Systematic contributions resulting from positron annihilation in the source are eliminated. In addition, the relatively small phase space of a beam allows better determination of the positron implantation profile.

High-energy positron beams can pass through thin windows before implanting deeply into the sample. This allows *in situ* measurements of engineering samples at temperature, under stress, or in a controlled atmosphere. The bulk PALS measurements are also insensitive to surface conditions, so no special sample preparation is required.

APPARATUS

There are both low and high energy positron beams available at LLNL. New developments in our intense, low-energy beam are found in an accompanying article in this conference.(1) The high-energy beam is derived from a 50 mCi ^{22}Na source moderated by a 1 µm thick tungsten single-crystal foil positioned in the terminal of a 3 MV Pelletron electrostatic accelerator. (See Figure 1.) Approximately $1.5\ 10^5\ e^+s^{-1}$ are emitted from the moderator foil. An additional $4\ 10^5\ e^+s^{-1}$ pass through the foil without thermalizing and are picked up in the accelerating field. The full $5.5\ 10^5\ e^+s^{-1}$ beam can be focused by a magnetic lens to a 3 mm spot at a location 2 m from the exit of the Pelletron for use in PALS. A bending magnet located after the magnetic lens can direct

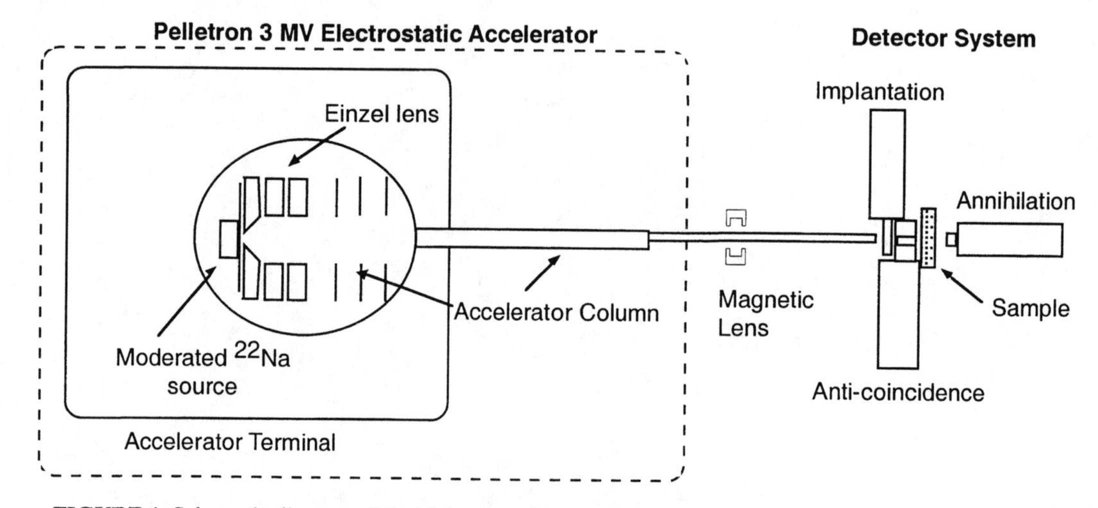

FIGURE 1. Schematic diagram of the high-energy beam-PALS system.

CP475, *Applications of Accelerators in Research and Industry*,
edited by J. L. Duggan and I. L. Morgan
1999 The American Institute of Physics 1-56396-825-8

the moderated beam to other experimental stations.

The basic beam-PALS system utilizes three detectors. Positrons exit the accelerator vacuum system through a 20 μm aluminum window and pass through a 2 mm plastic scintillator that measures the positron implantation time. Positrons lose an average of ~200 keV in the detector, and exit in a diverging beam with a broadened energy distribution. The beam diameter is 1 cm at a sample 4 cm from the detector.

Positrons implant from mm to cm into the sample, thermalize, and annihilate with electrons. Annihilation time is determined from the resulting gamma-rays detected by a BaF_2 detector. Time resolution for this spectrometer is generally less than 250 ps even with broad implantation profiles in large, low-density samples.

Events from positrons that scatter at large angles in the implantation detector or back-scatter out of the sample are rejected by an anti-coincidence detector positioned as an active collimator. The detector can be constructed with a range of apertures to limit the effective beam diameter on a sample. The smallest beam tested to date was 5 mm in diameter and gave no increase in the systematic component of the lifetime spectra over the full-size beam.

This three-detector system is a highly efficient lifetime spectrometer with reduced accidental backgrounds and minimal systematic contributions. In the geometry shown in Fig. 1, 2000 counts per second can be achieved with a 1 cm diameter beam. A systematic contribution appears in the lifetime spectra at an intensity of <0.5%. This is attributable to positrons annihilating in the detector system without being seen by the anti-coincidence detector.

Even this systematic can be eliminated by adding a fourth detector, usually NaI, in coincidence with the BaF_2 annihilation detector. This geometry reduces the counting rate to less than 800 counts per second but restricts counting to annihilation events in the region between the two detectors. This system is modeled on the one used successfully at Stuttgart, Germany.(2)

The four-detector system also allows the use of non-stopping samples. Positrons which pass through the thin sample annihilate more than a meter beyond the coincidence region. The only penalty is a corresponding decrease in the data rate. In low-density samples, the decline in count rate can be balanced against the improvement in timing resolution due to smaller implantation volume.

EXPERIMENTS

The Pelletron beam is being used to perform basic research as well as non-destructive testing of materials. Basic positron annihilation, channeling and transport properties at high energies are studied in single crystals.(3) A collection of annealed, high-purity metal PALS samples has been assembled with single lifetimes from less than 100 ps to over 200 ps. These will serve as standards for comparison to theoretical calculations and testing of data analysis programs.

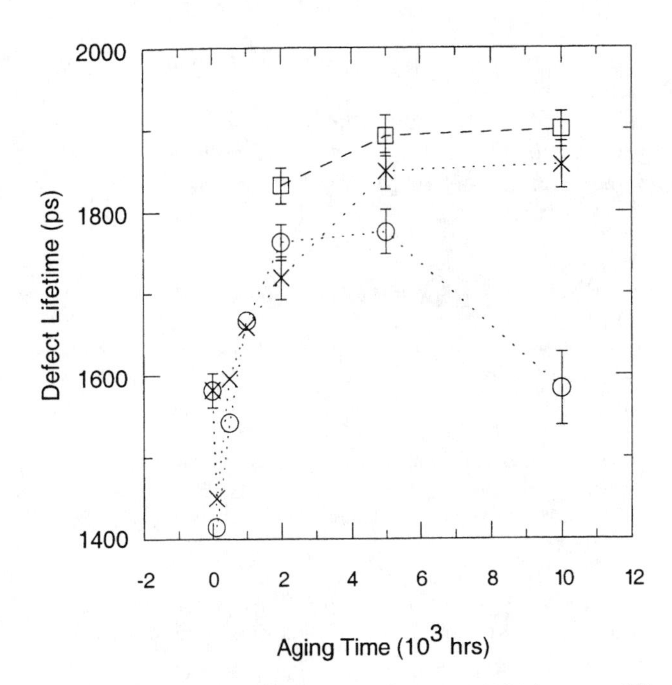

FIGURE 2. Aging of 1908 composite at 200°C in argon (□), air (×), and 80/20 argon/nitrogen (O).

We are using the high-energy beam-PALS system to study aging effects in carbon-fiber resin composites. (See Fig. 2) The composite aging is accelerated by maintaining the samples at elevated temperature in atmospheres of air, argon, or an 80/20 argon-nitrogen mixture.

The lifetime spectra are fit well by two components. The short lifetime (τ_2=370-380 ps) is attributed to positrons annihilating in the bulk, and the long component (I_2=3-5%, τ_2=1.4-1.9 ns) is ortho-positronium in defects. The length of this ortho-Ps lifetime is directly related to average hole volume in molecular solids.(4)

The data show an immediate decrease in the hole volume after the first heating cycle of 120 hrs, followed by a steady increase in hole size that saturates at 2000-5000 hrs. The sample aged in 80/20 atmosphere for 10,000 hrs began to physically disintegrate, so it's not clear that the measured lifetime decrease is physically meaningful. The hole volume changes are being correlated with infrared spectroscopy and mechanical tests to provide a complete description of the changes during aging.

We are also using the high-energy beam to study metal alloys, actinide metals, and polymeric and molecular solids which have been subjected to irradiation, mechanical stress, and/or chemical degradation. The simple geometry and low dose rate allow us to non-destructively evaluate the same samples used in mechanical tests.

ACKNOWLEDGMENTS

We'd like to thank Scott Groves for the aged carbon composite samples, and Alan Hunt and William Patterson for their assistance in operating and maintaining the Pelletron.

Work performed under the auspices of the U.S. Department of Energy by Lawrence Livermore National Laboratory under Contract No. W-7405-ENG-48.

REFERENCES

1. P. Asoka-Kumar, R. H. Howell, W. Stoeffl, *High-Intensity Positron Program at LLNL*, Contribution to this proceedings (1998).

2. W. Bauer, J. Briggmann, H.-D. Carstanjen, S. H. Connell, W. Decker, J. Diehl, K. Maier, J. Major, H. E. Schaefer, A. Seeger, H. Stoll and E. Widmann, *Nucl. Inst. and Meth.* **B50**, 300 (1990).

3. L.V. Hau, A.W. Hunt, J. A. Golovchenko, R. Haakenaasen, and K. G. Lynn, *Materials Science Forum*, **255-257**, 119-123 (1997).

4. Y. C. Jean, in *Positron Spectroscopy of Solids*, A. Dupasquier and A. P. Mills, eds. (IOS Press, Amsterdam, 1995), pp. 563-580.

Concept of an Intense Positron Source at the New Superconducting LINAC "ELBE"

G. Brauer

Institut für Ionenstrahlphysik und Materialforschung, Forschungszentrum Rossendorf, D-01314 Dresden, Germany

R. Ley

Institut für Physik, Johannes-Gutenberg-Universität Mainz, D-55099 Mainz, Germany

H. Schneider and W. Arnold

Strahlenzentrum, Justus-Liebig-Universität Giessen, D-35392 Giessen, Germany

This paper describes intentions to use the superconducting ELBE-LINAC, which is under construction at Forschungszentrum Rossendorf near Dresden/Germany, for the production of an intense slow positron beam of about 2.7 x 10^8 e^+ s^{-1}. Possible experiments in atomic, nuclear, and solid state physics are outlined and especially main applications for materials science - being yet under debate - are mentioned.

INTRODUCTION

Positrons can be created by weak or electromagnetic interaction only. The positrons from weak interaction usually come from radioactive sources like ^{22}Na. From an activity of about 100 mCi (3.7 GBq) about 10^5 moderated slow positrons per second may be extracted. Reactor-based slow positron sources either make use of the weak interaction, e.g. from ^{64}Cu, or electroproduction from the gamma cascade after neutron capture in Cd. About 10^8 slow positrons per second have been reported up to now (Brookhaven, Garching, Delft). In comparison, slow positron intensities two orders of magnitude higher have been obtained by electroproduction using an electron accelerator. From Livermore about 10^{10} slow positrons per second were reported (1, 2), whereas e.g. in Germany the former LINACs at Mainz (3) and Giessen (4-7) delivered up to about 10^8 slow positrons per second.

Modern electron accelerators reach a high degree of reliability, i.e. more than 5,000 h of operation per year. At the Forschungszentrum Rossendorf near Dresden/ Germany a new superconducting LINAC named by the acronym "ELBE" (Electron Source of high Brilliance and low Emittance) is under construction which may be used for the production of an intense slow positron beam (8). Originally planned to deliver a maximum electron energy of about 20 MeV at an average current of 1 mA, it now almost certainly will reach a maximum value of about 40 MeV at the same average electron current. This will allow to produce about 10^8 slow positrons per second. In this paper the concept how to reach this value will be presented and discussed.

POSITRON SOURCE CONCEPT

Bremsstrahlung and pair production

An extremely relativistic electron passing through matter looses its energy mainly by bremsstrahlung. This process can be characterized by the so-called radiation length (rl) which is given by:

$$1\ \text{rl} = (2149\ A) / ((15.80 - \ln Z)\ \rho Z^2) \quad [\text{cm}] \qquad (1)$$

with A being the mass number, Z being the element number, and ρ being the density (g/cm^3) of the target material. One radiation length is the distance over which a fraction of (1 - 1/e) = 63 % of the electron energy is radiated away. The intensity of the bremsstrahlung spectrum is nearly uniform up to 70 % of the endpoint energy - for a comparison of theory with experimental results see ref.(9). The total energy converted into bremsstrahlung within a given target thickness is proportional to the primary energy E_e of the electrons. The gamma quanta from bremsstrahlung loose their energy E_γ by Compton effect and pair production. In a tungsten target the pair production is dominant if $E_\gamma > 6$ MeV and becomes independent of the energy if $E_\gamma > 16$ MeV. The combined probability for primary interactions of bremsstrahlung and the generation of pair production becomes roughly proportional to E_e .

In extremely thin targets, such as entrance windows, only ionization losses are essential. In a thick target, however, the processes of bremsstrahlung and pair production are successively repeated several times, i.e. a shower

CP475, *Applications of Accelerators in Research and Industry*, edited by J. L. Duggan and I. L. Morgan

consisting of gamma rays and electron-positron pairs develops. The number of positrons emerging from a conversion target and their energy spectrum depend crucially on the target thickness and the primary electron energy E_e. As long as the target thickness is small compared with the mean range of the primary electrons, the spectrum of the positrons emerging from the target has a maximum at an energy at about 20 % of E_e (10). The positron spectrum has a steep decrease towards lower energies and a long tail towards higher energies. An increase of the target thickness to a value which is comparable to the mean range of the primary electrons results in a shift of the maximum positron energy to very low energies and simultaneously the total number of positrons reaches a maximum. Theoretical simulations are able to explain the experimental facts at least qualitatively (11-13). With a further increase of the target thickness the maximum of the positron spectrum remains at very low energies but the number of positrons decreases. This is due to the fact that more positrons are annihilated in the rear parts of the target than are created. It is obvious that an optimum thickness d^{opt} for a conversion target exists at which the highest number of positrons with lowest energies will be delivered. Experimental values for d^{opt} have been found by a systematic variation of the target thickness at 9 MeV < E_e < 30 MeV (14), at E_e = 75 MeV (15), and at E_e = 100 MeV (1). All these data can be interpolated by the empirical formula for the optimum thickness of a tungsten target:

$$d^{opt}\,[\mathrm{mm}] = 0.670 + 0.0953\, E_e\,[\mathrm{MeV}] \qquad (2)$$

for 10 MeV < E_e < 100 MeV.

Using d^{opt} values from eq.(2), the observed conversion efficiencies Y for moderated slow positrons as a function of E_e may be interpolated by:

$$Y\,[e^+/10^9\, e^-] = 3.319 \times 10^{-4}\, (E_e\,[\mathrm{MeV}])^{3.327} \qquad (3)$$

for 10 MeV < E_e < 100 MeV.

Thermal Power deposited in the Target

The high energy electrons leave the accelerator beam line through a thin Al window of typical 0.1 mm thickness (about 10^{-3} rl), then are travelling through air for about 20 cm before entering the target chamber through another Al window of the same thickness. Every electron will deposit about 1.4 MeV g^{-1} cm^2 in the aluminium due to ionization. For a current of 1 mA this results in thermal power of about 38 W.

For a thick target the energy deposition is not uniform but reaches a maximum at a certain depth. Detailed calculations about the thermal power distribution as a function of depth can be found in ref.(16). The energy deposition averaged over the volume and the temperature rise have been treated in ref.(17). Due to secondary processes the average energy deposition per electron increases for thick targets to:

$$\Delta E = 2\ [(\mathrm{MeV/e})\,/(\mathrm{g/cm^2})] \qquad (4)$$

The total thermal power P generated by an accelerator current I is given by:

$$P\,[\mathrm{W}] = I\,[\mathrm{A}] \times \Delta E \times d^{opt}\,[\mathrm{cm}] \times \rho\,[\mathrm{g/cm^3}] \qquad (5)$$

Taking the intended ELBE-LINAC characteristics (40 MeV, 1 mA), one obtains at P = 17.4 kW for a W target (density ρ = 19.3 g/cm^3) having the optimum thickness d^{opt} = 0.45 cm.

It is supposed to use a disk-shaped W target of diameter d_2 = 1.5 cm which is in thermal contact with a surrounding water-cooled Cu block. According to ref.(17), the temperature rise ΔT in the target is given by:

$$\Delta T = (0.5 + \ln(d_2/d_1))\,/\,(2\pi\lambda) \times \rho \times \Delta E \times I \qquad (6)$$

where $d_1 \simeq 0.5 \times d_2$ = 0.75 cm is the diameter of the electron beam and λ = 1 W cm^{-1} K^{-1} is the thermal conductivity of tungsten. Inserting all numbers gives:

$$\Delta T = 7.2 \times 10^6 \times I\,[\mathrm{A}] \qquad (7)$$

Inserting the nominal accelerator current of 1 mA would result in a temperature far beyond the melting point of tungsten (3683 K). As a consequence, the accelerator beam has to be expended in diameter or lowered in current in such a way that the fraction hitting the W target disk is low enough to guarantee a ΔT = 2000 K. This can be achieved using I $\leq$ 0.6 mA only or expanding the electron beam to d_1 = 2 cm. According to ref.(17), the thermal power deposited by the expanded beam in the water-cooled Cu block is uncritical.

Using the conversion efficiency of 72 e^+ / 10^9 e^- at 40 MeV, a current of I = 0.6 mA will produce about 2.7 x 10^8 e^+ s^{-1}.

Slow Positron Generation and Transport

The experimental arrangement usually consists of the electron-positron converter target, the W moderator in a vacuum chamber, and the conventional solenoid as a transport system of the positrons to the experimental area (1, 3, 4). Here the 40 MeV electron beam will hit the disk-shaped W target at 30 degrees inclination from perpendicular incidence. The polycrystalline W moderator consisting of several vanes (thickness 25-250 μm, spacing about 5 mm) has to be positioned closely behind the target with their surfaces being oriented parallel to the target normal direction. The moderated positrons will be extracted by a small voltage into the same direction. It is intended to allow biasing of the target-moderator assembly up to 30 keV. The moderated positrons will be magnetically guided inside a 10 m long bent solenoid (about 0.01 T) to an experimental area which is well shielded from the intense γ- and n-radiation arising at the conversion site. The axial position of the beam inside the solenoid is to be controlled by a number of steering coils. At the end of the magnetic transport beam line the positrons will be extracted and transferred into an electrostatic beam line that guides them to a switch. According to the intended further use of the intense slow positron beam after the switch different user facilities are still under debate.

INTENDED USER FACILITIES AND EXPERIMENTS

An intense slow positron beam is of primary interest for investigations in atomic, nuclear and solid state physics, especially in materials science (18-20). Furthermore, applications in chemistry and biology are thinkable but not being considered by us at the moment.

The following experiments are thought to be possible and of high interest in atomic and nuclear physics:

- Elucidation of differences in excitation and ionization processes by particle (electrons, protons) and antiparticle (positrons, antiprotons) impact. Furthermore, multiple ionization of gas atoms and molecules due to positron impact and the estimation of absolute cross-sections for inner and outer shell ionization in connection with stopping power investigations of different materials.
- Efficient production of positronium atoms (Ps) will allow for the performance of stringent tests of predictions from the theory of QED. Ps is unique in this field since it is described by Feynman graphs resulting from annihilation which are not present in hydrogen or muonium. Ground state Ps allows the search for possible exotic decays and the measurement of the branching ratios for rare allowed decays. Precision spectroscopy of Ps in excited states can verify recent calculations of higher order contributions.

A main application in solid state physics would be the use of the intense slow positron beam for 2D-measurements of angular correlation of the annihilation photons in electronic structure investigations, i.e. mainly Fermi surface studies. However, the most important application of an electrostatically guided intense slow positron beam will be in materials science. It is possible to build up a positron micro-beam, i.e. to considerably reduce the positron beam diameter down to the micron range. This may be used for scanning of a sample in x-y direction with depth profiling of defects at a certain position. In addition, the option of beam pulsing should be realized. This will permit the performance of depth-dependent positron lifetime measurements which are essential for the unequivocal identification of vacancy-type defects.

More sophisticated experiments for surface studies of any type at ultra-high vacuum conditions, like positron-induced Auger electron spectroscopy (PAES) or re-emission microscopy, definitely require an extremely low positron energy E with very little energy spread ($\Delta E \leq 0.1$ eV) which may only be achieved via at least one remoderation stage.

More detailed plans regarding the final arrangement of experimental equipment for materials science will be discussed and published elsewhere.

CONCLUSIONS

It has been shown that a realistic number of 2.7×10^8 e^+ s^{-1} may be produced at the ELBE-LINAC using an electron beam of 40 MeV and I = 0.6 mA. Furthermore, possible experiments in atomic physics and applications in materials science were outlined.

Acknowledgements

This work has been supported by Deutsche Forschungsgemeinschaft under grant numbers KON 667/1998 and BR 1250/15-1.

References

1. Howell, R.H., Alvarez, R.A., and Stanek, M., Appl. Phys. Lett. **40**, 751 (1982)
2. Howell, R.H., Cowan, T.E., Hartley, J., Sterne, P., and Brown, B., Appl. Surf. Sci. **116**, 7 (1997)
3. Gräff, G., Ley, R., Osipowicz, A., Werth, G., and Ahrens, J., Appl. Phys. **A33**, 59 (1984)
4. Ebel, F., Faust, W., Hahn, C., Langer, S., and Schneider, H., Appl. Phys. **A44**, 119 (1987)
5. Faust, W., Hahn, C., Rückert, M., Schneider, H., Singe, A., and Tobehn, I., NIM **B56/57**, 575(1991)
6. Schneider, H., Tobehn, I., Ebel, F., and Hippler, R., Phys. Rev. Lett. **71**, 2707 (1993)
7. Hagena, D., Ley, R., Weil, D., Werth, G., Arnold, W., and Schneider, H., Phys. Rev. Lett. **71**, 2887 (1993)

8. Brauer, G., Wendler, W., Büttig, H., Gabriel, F., Gippner, P., Gläser, W., Grosse, E., Guratzsch, H., Dönau, F., Höhnel, G., Janssen, D., Nething, U., Pobell, F., Prade, H., Pröhl, D., Schilling, K.D., Schlenk, R., Seidel, W., Stephan, J., vom Stein, P., Wenzel, M., Wustmann, B., and Zahn, R., Mat. Sci. Forum **255-257**, 732 (1997)
9. Koch, H.W., and Carter, R.E., Phys. Rev. **77**, 165 (1950)
10. Hauptmann, A., Ph.D. Thesis, Inst. für Kernphysik/ Univ. Mainz (1972)
11. Katz, L., and Lokan, K.H., NIM **11**, 7 (1961)
12. Sund, R.E., Walton, R.B., Norris, N.J., and MacGregor, M.H., NIM **27**,109 (1972)
13. Segers, D., Dorikens, M., Paridaens, J., and Dorikens-Vanpraet, L., in: AIP Conf. Proc. **303**, 496 (1994)
14. Bernardini, M., Miller, J., Schuhl, C., Tamas, G., and Tsara, C., Frascati Report LNF **62/66** (1962)
15. Takahane, T., Chiba, T., Shiotani, N., Tanigawa, S., Mikado, T., Suzuki, R., Chiwaki, M., Yamazaki, T., and Tomimasu, T., Appl. Phys. **A51**, 146 (1990)
16. Tabata, T., and Ito, R., Atomic Data and Nuclear Data Tables **56**, 105 (1994)
17. Andreani, A., and Cattoni, A., NIM **129**, 365 (1975)
18. Schultz, P.J., and Lynn, K.G., Rev. Mod. Phys. **60**, 701 (1988)
19. Hulett, Jr., L.D., Brown, B.L., Howell, R.H., Jones, P.L., Lynn, K.G., Okada, S., Smedskjaer, L.C., West, R.N., Gonzales, R., Denison, A.L., Jean, Y.C., Kossler, W.J., Mills, A.P., Jr., Schrader, D.M., Weiss, A.H., and Chen, Y., (eds.), *Application of Positron Spectroscopy to Materials Science*. Adv. Mat. Res. **3**. Zug, Scitec Publ., 1994
20. Dupasquier, A., and A.P. Mills, Jr., (eds.) *Positron Spectroscopy of Solids*. Amsterdam, IOS Press, 1995

SECTION IV

CLUSTERS, FULLERENES, BIOMOLECULES

Energy Deposition in Fullerene-Type Nano-structures

S. Peter Apell[1,2] and John R. Sabin[3,4]

1. Department of Applied Physics , Chalmers University of Technology and Göteborg University, Göteborg, Sweden S-41296
2.Departamento de Física de Materiales, Universidad del Pais Vasco, San Sebastian 20080, Spain
3. Kemisk Institut, Odense Universitet, DK–5230 Odense M, Denmark
4. Quantum Theory Project, Department of Physics, University of Florida, Gainesville, FL 32611, U. S. A.

We consider the energy deposition characteristics of two types of nano-structures with Fullerene-type chemical structure, namely Fullerene spheres and nano-tubes. We make the assumption that these systems can be treated as jellium shells. In the dipole approximation, we find the two systems behave with remarkable similarity. The energy deposition properties depend only on the relative thickness of the shell, and are independent of absolute length scales.

INTRODUCTION

In the past score of years, beginning with the cryogenic calorimetric studies of Andersen (1), it has become possible to study energy deposition processes by swift ions in materials with an accuracy that makes more subtle effects of target structure accessible to investigation. For example, subjects of recent study include deviations from the Bragg rule (2–4), target phase effects (5, 6), surface effects (7), directional and molecular shape effects (8), and projectile isotope effects (9, 10), all of which alter the bulk stopping power of a target by the order of 10% or less

Energy deposition or stopping power is conveniently understood in terms of the Bethe theory (11) which gives the linear energy loss per unit pathlength or stopping power, $-dE/dx$, of a swift ion in matter as

$$-\frac{1}{\tilde{n}}\frac{dE}{dx} = S(v) = \frac{Z_1^2 Z_2 e^4}{4\pi m v^2 \varepsilon_0^2}\, L(v) \tag{1}$$

Here $\tilde{n}$ is the number density of scattering centers, $S(v)$ and $L(v)$ are the stopping cross section and stopping number per scatterer, respectively, and v is the projectile velocity. The charge of the projectile is Z_1, and Z_2 is the number of scattering electrons per scattering center. The interesting physics of the problem is contained in the stopping number, which in Bethe's theory can be written (12):

$$L(v) = \ln\frac{2mv^2}{I_0} \tag{2}$$

The mean excitation energy, I_0, is properly the zeroth moment of the energy weighted dipole oscillator strength distribution (13), and can be looked upon as a materials parameter that indicates how well the target can accommodate energy deposition from the projectile. It is this quantity that we will examine.

Recently, we have been interested in the effect of molecular anisotropies on energy deposition characteristics, and have modeled molecules as jellium spheroids with sharp boundaries and containing a uniform electron density (8). In this case, the mean excitation energy for stopping, I_0, can be related to the plasma frequency of the jellium (14), ω_p, through a depolarization factor (15), n, which depends on the geometry of the ellipsoid alone (8):

$$I_0 = \sqrt{n}\,\hbar\omega_p \tag{3}$$

As this simple model has produced reasonable results for trends in molecular stopping anisotropies(8), we continue with a jellium model, but this time consider systems containing holes, namely Fullerene spheres and nano-tubes.

In this communication, we will consider jellium models of carbon Fullerenes and nano-tubes, and discuss the effect of size and geometry on trends in their stopping characteristics.

FULLERENE SHELLS

In the spirit of our previous work (8), we assume that Fullerenes can be modeled as jellium spheres with a hole: a jellium spherical shell. In this case the system has spherical symmetry and two length scales, the inner (r_{in}) and outer (r_{out}) radii of the jellium shell, which suggests that one should be able to relate the stopping of such a system to that of the filled sphere of radius r_{out} by geometry alone.

The mean excitation energy of the Fullerene in a finite excitation space can be written

$$\ln I_F = \sum_i^{excitations} f_i \ln\hbar\,\Omega_i \tag{4}$$

CP475, *Applications of Accelerators in Research and Industry*, edited by J. L. Duggan and I. L. Morgan

where the $\{\Omega_i = \omega_i/\omega_0\}$ are the allowed excitation frequencies of the system and the $\{f_i\}$ are their dipole oscillator strengths. Here we have invoked the Thomas-Reiche-Kuhn (TRK) sum rule in the form that $\sum_i f_i = 1$ where the sum is over all possible excitations, and $\omega_0 = (4\pi e^2 n_0/m)^{\frac{1}{2}}$ (about 12 eV.) with n_0 the effective electron density of the shell.

In a model such as this which considers only collective modes, a probe passing outside of a very thin shell will excite the surface plasmons, which exhausts the oscillator strength, and thus the surface plasmons alone will be responsible for the energy deposition. There are two dipolar modes for the shell, $\Omega_+(\rho)$ and $\Omega_-(\rho)$, with associated oscillator strengths $f_+(\rho)$ and $f_-(\rho)$, which correspond to the in- and out-of-phase surface plasma modes of the shell and depend upon the ratio of the inner to outer radius of the shell $\rho = r_{in}/r_{out}$. [In the following, we suppress the explicit dependance of the f_i and Ω_i on ρ.] Östling *et al.* (16, 17) have shown that these frequencies and oscillator strengths can be written as

$$\Omega_\pm^2 = \frac{1}{2}\left(1 \pm \frac{Q}{3}\right) \tag{5}$$

and

$$f_\pm = \frac{Q \mp 1}{2Q} \tag{6}$$

where $Q = \sqrt{1+8\rho^3}$. The $\{\Omega_i\}$ and the $\{f_i\}$ are thus determined by the geometry of the spherical shell alone. It should be noted that the plasmon frequencies and oscillator strengths are normalized such that $\Omega_+^2 + \Omega_-^2 = 1$ and $f_+ + f_- = 1$, and that the average Ω^2 is one third for a single sphere (*c.f. e.g.* p.150 of Ref. (18))

$$f_+\Omega_+^2 + f_-\Omega_-^2 = \frac{1}{3} \tag{7}$$

Using these frequencies and oscillator strengths in eq. 4, one obtains a ρ–dependent mean excitation energy

$$I_F(\rho) = \Omega_+^{f_+}\Omega_-^{f_-} \tag{8}$$

for the spherical shell, and therefore stopping (*via* eqs. 1 and 2) becomes dependent on Fullerene geometry alone.

In Fig. 1, we plot the mean excitation energy of Fullerene shells normalized to the value for a filled jellium sphere of the same outer radius *vs.* the ratio of inner to outer sphere radii. We have marked the positions of the Fullerenes that are closest to perfect spheres (*i.e.* those containing 12 pentagons: C_{60n^2}, n = 1, 2, ...). There is only a weak dependence on geometry in the beginning, and therefore we expect only a weak ($\leq$ 10%) dependence on sphere size in the stopping of various carbon Fullerenes. As the differentially thin shell is approached (large Fullerenes if

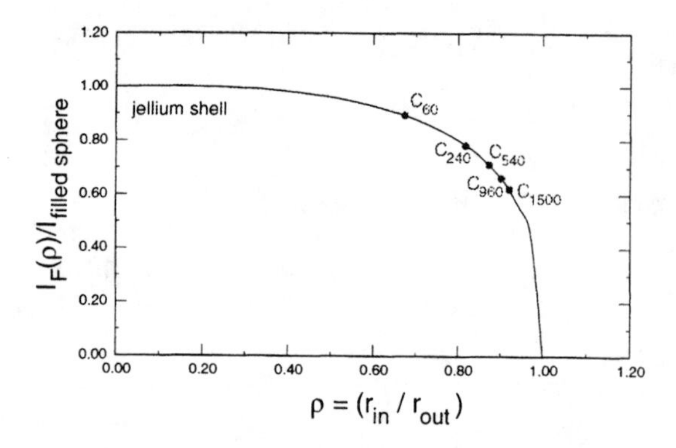

FIGURE 1. Mean excitation ratio for hollow Fullerenes. The hollow Fullerene mean excitation energy is in units of the corresponding filled sphere mean excitation energy as a function of shell thickness, ρ. The positions on the curve of some of the smaller members of the series C_{60n^2}, n = 1, 2, ... are indicated. Note the weak variation for $\rho \leq 0.8$ before it dives for very thin shells.

we posit a constant shell thickness) the variation becomes very large indeed.

We note that here, as with the case of molecules (jellium spheroids) (8), it is geometry alone which determines the mean excitation energy. That is, that the target is spherical and the ratio of inner to outer radii are the only important considerations. Neither the absolute thickness of the shell nor the absolute size of the sphere is relevant, and thus, within the constraints of retardation effects, a homothetic series is expected to have a constant stopping cross section.

CARBON NANO-TUBES

Another structure, similar in chemical composition and bonding and in electron density to the Fullerenes, are the carbon nano-tubes. While Fullerenes have no preferred direction associated with them, as they are, in this model at least, spherical, the length of carbon nano-tubes defines a preferred direction. As momentum can only be transferred from a swift ion in directions perpendicular to its velocity, we will consider only ions traveling parallel to the tube axis. Thus we are concerned with the mean excitation energies in the radial direction.

The surface plasmon frequencies for a cylindrical tube with inner radius r_{in} and outer radius r_{out} can be written

$$\Omega_\pm^2 = \frac{1}{2}(1 \pm \rho^m) \tag{9}$$

where $\rho = r_{in}/r_{out}$ as before. We consider the dipolar excitation corresponding to $m = 1$. For a cylinder, the sum rule in eq. 7 will have one half on the rhs, *viz.*:

$$f_+\Omega_+^2 + f_-\Omega_-^2 = \frac{1}{2} \tag{10}$$

which, when taken together with the TRK sum rule and eq. 9, leads directly to the condition that

$$f_+ = f_- = \frac{1}{2} \tag{11}$$

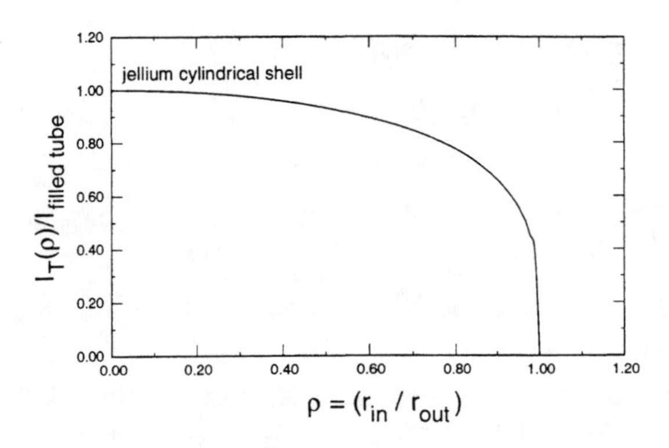

FIGURE 2. Mean excitation ratio for hollow nano-tubes in a direction perpendicular to the tube axis. The hollow tube mean excitation energy is in units of the corresponding filled tube mean excitation energy as a function of shell thickness, ρ. Note the weak variation for $\rho \leq 0.8$ before it dives for very thin shells, and the strong similarity to Fig. 1.

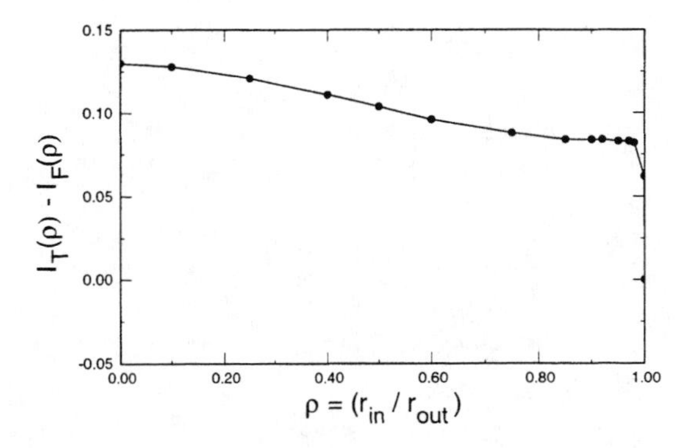

FIGURE 3. Absolute difference in the mean excitation energies of Fullerene shells and nanotubes as a function of the relative wall thickness.

regardless of the value of ρ. The mean excitation energy perpendicular to the tube axis then becomes (from the analog of eq. 4):

$$I_T(\rho) = (\Omega_+ \Omega_-)^{\frac{1}{2}}$$
$$= \frac{1}{\sqrt{2}}(1-\rho^2)^{\frac{1}{4}} \qquad (12)$$

Notice that $I_T(\rho=1) = \sqrt{\frac{3}{2}}\, I_F(\rho=1)$.

In Fig. 2 we present the mean excitation energy for stopping in the radial direction along a nano-tube, normalized to the mean excitation energy of the filled tube $(1/\sqrt{2})$.

SUMMARY

The resemblance between Figs. 1 and 2 is remarkable. Both start at the thick shell limit ($r_{in} = 0$) with the mean excitation energy (normalized to the filled structure) only very weakly dependent on shell thickness, followed by a rapid fall off when the shell becomes thin ($\rho > 0.8$) which, in the systems treated here, correspond to a graphite sheet. It should be noted again, however, that it is not the absolute thickness that matters, rather only the relative thickness with respect to the size of the tube or sphere. The weak dependence of the mean excitation energy of either the Fullerene or the nano-tube on shell thickness for small ρ ($\rho \leq 0.8$) means that in this region it would be very hard to distinguish various sizes of tubes. For larger ρ, however, the mean excitation energy is a strong function of ρ, and different cases should be more easily distinguishable.

In this model, shell thickness is treated as a classical variable, and we allow it to go smoothly to zero. In fact, the thickness of a shell is quantized (19) in units of the atomic size and can not approach zero smoothly. In the case of the classic structure, C_{60}, the relative shell thickness is $\rho = 0.67$, while even for C_{1500}, the thickness has only reached $\rho = 0.92$. Thus, practically, the shells never reach differential thickness, which in any case, would be unphysical.

The general trend, that we find a decreasing mean excitation energy with decreasing thickness, *i.e.* a decrease of I with increasing ρ, is consistent with previous work in Li ultra-thin films (7).

We also note that the normalized mean excitation energies for nano-tubes and spherical Fullerenes are somewhat (1 – 3%) different. In the small ρ region corresponding to thick shells, the normalized mean excitation energies for the Fullerenes are larger than those for the corresponding nano-tubes. For thin shells, the opposite is true, and the two curves cross at $\rho \approx 0.91$.

If we look at the differences in absolute (not normalized to the filled structure value) mean excitation energies for the Fullerenes and nano-tubes, as depicted in Fig. 3, we see that the difference is largest for small ρ, that is, for thicker shells. There follows a weak decrease in the difference as the shell becomes thinner, and a precipitous fall to zero difference for very thin shells. The difference in mean excitation energies in the region of moderate ρ is on the order of 0.1 while the average mean excitation energy in the same region is approximately 0.56, which puts the difference in stopping powers of Fullerene shells and nano-tubes with the same ρ easily within experimental determination.

ACKNOWLEDGMENTS

The authors would like to acknowledge support from the US Army Research Office (Contract #DAA–H04–95–1–0326 to JRS) and to the Swedish Natural Science Research Council (to SPA).

REFERENCES

1. H. H. Andersen, *Studies of Atomic Collisions in*

Solids by means of Calorimetric Techniques, Aarhus University Press, 1974, Doctor's thesis.
2. D. I. Thwaites, Rad. Res. **95**, 495 (1983).
3. D. I. Thwaites, Nucl. Instrum. and Meth. B **69**, 53 (1992).
4. S. A. Cruz and J. Soullard, Nucl. Instrum. and Meth. B **71**, 387 (1992).
5. D. I. Thwaites, Nucl. Instrum. and Meth. B **12**, 84 (1985).
6. J. R. Sabin, J. Oddershede, and P. Sigmund, Nucl. Instrum. and Meth. B **12**, 80 (1985).
7. S. P. Apell, J. R. Sabin, and S. B. Trickey, Phys. Rev. A **56**, 3769 (1997).
8. S. P. Apell, S. Trickey, and J. Sabin, Phys. Rev. A **00**, 0000 (1998), in press.
9. H. Bichsel and M. Inokuti, Nucl. Instrum. and Meth. B **134**, 161 (1998).
10. R. Cabrera-Trujillo, Nucl. Instrum. and Meth. B **00**, 0000 (1998), submitted.
11. H. Bethe, Ann. Phys. (Leipzig) **5**, 325 (1930).
12. U. Fano, Ann. Revs. Nucl. Sci. **13**, 1 (1963).
13. M. Inokuti, Rev. Mod. Phys. **43**, 297 (1971).
14. J. Lindhard and M. Scharff, Kgl. Dan. Vidensk. Selsk.: Mat.-Fys. Medd. **27**, No. 15 (1953).
15. E. Lipparini and S. Stringari, Z. Physik D **18**, 193 (1991).
16. D. Östling, P. Apell, and A. Rosén, Z. Physik D **26**, S282 (1993).
17. D. Östling, P. Apell, and A. Rosén, Europhys. Lett. **21**, 539 (1993).
18. H. A. Bethe, *Intermediate Quantum Mechanics*, Benjamin, New York, 1964.
19. S. B. Trickey, J. Z. Wu, and J. R. Sabin, Nucl. Instrum. and Meth. B **93**, 186 (1994), for an opposite viewpoint, see P. Sigmund, Nucl. Instrum. and Meth. **B95**, 477 (1995).

Applications of Cluster Ion Implantation in Microelectronics Devices

Isao Yamada, Jiro Matsuo, Noriaki Toyoda, and Takaaki Aoki

Ion Beam Engineering Experimental Laboratory, Kyoto University
Sakyo, Kyoto, 606-8501, JAPAN

Ultra-shallow ion implantation by gas cluster ion beams has been demonstrated experimentally and confirmed by molecular dynamics simulations. Implantation of $B_{10}H_{14}$ in Si (100) at 2keV does not cause transient enhanced diffusion (TED) of boron atoms during annealing at 900°C for 10sec. In order to reveal the diffusion mechanism of B atoms, the diffusivity of B atoms in ultra low-energy $B_{10}H_{14}$ ion implantation was measured by Secondary Ion Mass Spectroscopy (SIMS). $B_{10}H_{14}$ ions were implanted at 2, 3, 5 and 10keV. Subsequent annealing was performed at 900°C and 1000°C for 10 sec, respectively. In the case of the 900°C annealing, TED was suppressed as the implant energy decreased and at energy less than 3keV, the TED of B atoms no longer occurred during annealing. High performance 40nm p-MOSFETs with ultra shallow junctions have been fabricated using $B_{10}H_{14}$ cluster ion implantation. The unique characteristics of gas cluster ion beam processes for sputtering has also been applied to very high-rate etching. Yields more than two orders of magnitude higher than those by monomer ions having the same energy and atomic scale smoothing of surfaces to average roughness less than 0.2nm have been demonstrated. This paper will discuss the status of cluster ion beam processes based upon our recent experimental and molecular dynamics simulation results.

INTRODUCTION

As the size of devices on LSI decreases, low-energy ion implantation is essential for the fabrication of very-shallow junctions. However, it is difficult for conventional low-energy ion implanters to obtain high-current beams, because of space charge effects in the beam transport lines. Moreover, it is necessary to suppress diffusion, especially transient enhanced diffusion (TED) during annealing.

Cluster ion implantation has been proposed as a low-energy implantation technique [1]. It is quite easy to obtain high-current beams with low equivalent energy by using cluster ion beams. Cluster impact causes unique damage effects resulting from the localized energy deposition and multiple collisions at near surface regions during impact, and a local amorphous region is formed during penetration of solid surfaces by a cluster [2,3]. Therefore, cluster implantation is expected to form shallow and sharp profiles of defects and to avoid charge-up damage and channeling. In previous work, decaborane($B_{10}H_{14}$) has been used as a type of polyatomic B cluster to reduce the TED of dopant and has been implanted into Si (100) to successfully fabricate very shallow p^+ junctions [1,4]. Since $B_{10}H_{14}$ consists of ten boron atoms, each B atom is implanted with about one-tenth of the total acceleration energy and the ion doping rate of B atoms is ten times larger than the electrical ion beam current indicates [2]. Furthermore, it was found that TED of B atoms was suppressed by low-energy $B_{10}H_{14}$ ion implantation compared with BF_2 ion implantation [4]. These effects offer strong advantages for ultra shallow junction formation [5].

It is necessary to suppress TED of B atoms in order to produce high quality shallow p-type junctions. Recently, the characteristics of diffusion of B atoms implanted into Si at various energies has been reported [6,7]. However, the diffusion during annealing of B atoms implanted at ultra-low energies has not yet been clearly examined and the relation between the TED and characteristics of damage formation induced by cluster ion implantation such as $B_{10}H_{14}$ has also not been examined. In this paper, the characteristics of the diffusion behavior of B atoms in Si implanted with $B_{10}H_{14}$ ion at very low energies (< 1keV/atom), and other important characteristics such as high rate sputtering and surface smoothing, are discussed.

DECABORANE ION IMPLANTATION

Decaborane ($B_{10}H_{14}$; melting point at 99.7 °C; boiling point at 213 °C, 1 atm) is a stable solid at room temperature. A gas sublimed from $B_{10}H_{14}$ solid is fed directly into an ionization chamber and ionized by electron bombardment. Since $B_{10}H_{14}$ is easily fragmented by high-energy electron bombardment, the electron beam energy has to be kept low (60 eV). The ionized $B_{10}H_{14}$ clusters were extracted and accelerated to 2-20 keV. Finally, they are used to bombard Si (100) substrates. After implantation, rapid thermal annealing (RTA) was carried out at 900 °C and 1000 °C for 10 sec. The actual boron dose in the Si was measured by Nuclear Reaction Analysis (NRA)[8]. B depth profiles were measured by SIMS and the diffusion length of B atoms and the diffusivity of B atoms were calculated from the deviation of SIMS profiles before and after RTA.

CP475, *Applications of Accelerators in Research and Industry*,
edited by J. L. Duggan and I. L. Morgan

SHALLOW JUNCTION FORMATION

The suppression of TED is critical in the formation of ultra shallow junctions. Significant effort has been spent trying to understand and model TED. It is reported that TED is caused by the implantation-induced damage, which evolves into a supersaturation of interstitials. In order to reveal the relation between TED and the damage

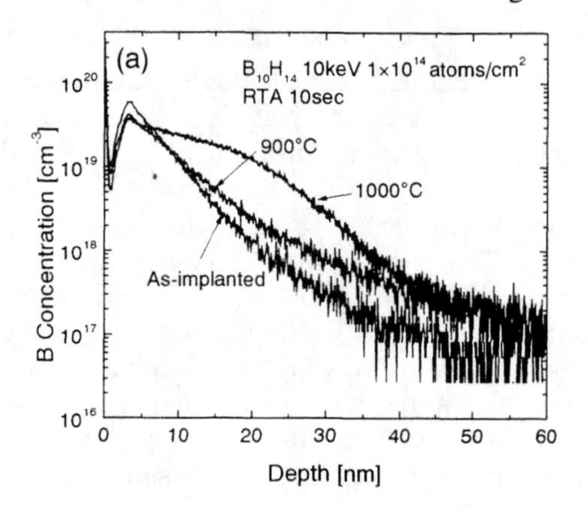

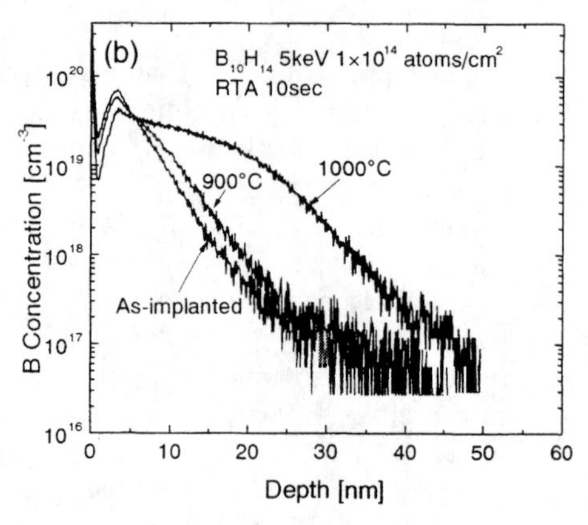

FIGURE 1. Boron diffusion profile of $B_{10}H_{14}$ at the energy of (a)10keV and (b) 5keV before and after RTA

introduced by $B_{10}H_{14}$ ion implantation, the effect of implant energy on the diffusion length was investigated. Figure 1 shows the SIMS profiles of boron atoms implanted with $B_{10}H_{14}^+$ at (a) 10 keV and (b) 5 keV before and after RTA. The TED of B atoms implanted at 5 keV is strongly suppressed compared with that of 10keV because, after 900 °C annealing, the profile of the 5 keV implant is less broad than that of 10 keV after 900 °C annealing. It is estimated from Figs. 1(a) and 2(b) that the diffusion length is 3 nm for 5 keV and 13 nm for 10 keV. The diffusion length is defined as the deviation between as-implanted and annealed profiles at 3×10^{17} atoms/cm^3. Figure 2 shows the diffusion length during annealing at 900 °C and 1000 °C as a function of implant energy. This figure clearly demonstrates that, during annealing at 900 °C, TED is reduced by decreasing the implant energy.

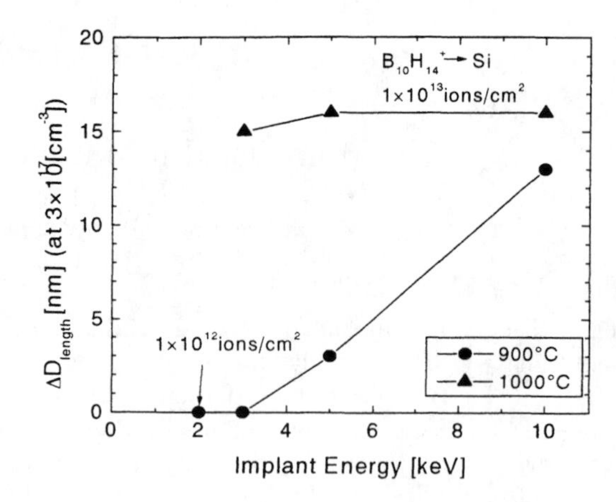

FIGURE 2. Diffusion length of B atoms on implant energies at dopant density of 3×10^{17} cm^{-3}

On the other hand, during annealing at 1000 °C, the diffusion length does not depend on implant energy. These results suggest that the mechanism of B diffusion is different between the annealing temperatures of 900 °C and 1000 °C.

The possible explanation for the suppression of TED with decreasing implantation energy during 900 °C annealing is that the damage is concentrated in a region very close to the surface. In such a case, the surface is believed to act as a sink for interstitials. Consequently, the interstitials introduced by ion implantation are recombined and annihilated at the surface, thereby supressing the TED [6,9]. Figure 3 shows the cross sectional view of the damage in the Si substrate by B_{10} cluster implantation at (a) 10 keV and (b) 3 keV as indicated by molecular dynamics simulations. Solid black points and dark points represent implanted boron atoms and displaced silicon atoms, respectively. This result shows that as the implant energy decreases, displaced atoms reside close to the surface. Since $B_{10}H_{14}$ produces localized damage which is unique to cluster ion bombardment at low energies, the end-of-range (EOR) damage, which is the source of the interstitials for the TED of B [10,11], is reduced.

APPLICATION TO 40 nm P-MOSFET

40nm p-MOSFETs have been fabricated to demonstrate $B_{10}H_{14}$ cluster implantation for shallow source/drain formation. $B_{10}H_{14}$ ion implantation for p-type source/drain (S/D) junctions was performed at an acceleration energy of 30 keV to a dose of 1×10^{13} ions /cm^2, then annealed at 1000 °C for 10 sec. A junction depth of 20 nm has been achieved. For S/D extensions, $B_{10}H_{14}$ ion implantation at 2keV has been carried out to a dose of 1×10^{12}ions/cm^2 followed by annealing at 900 °C for 10 sec. A 7-nm ultra-shallow junction without TED and TD has been achieved [5].

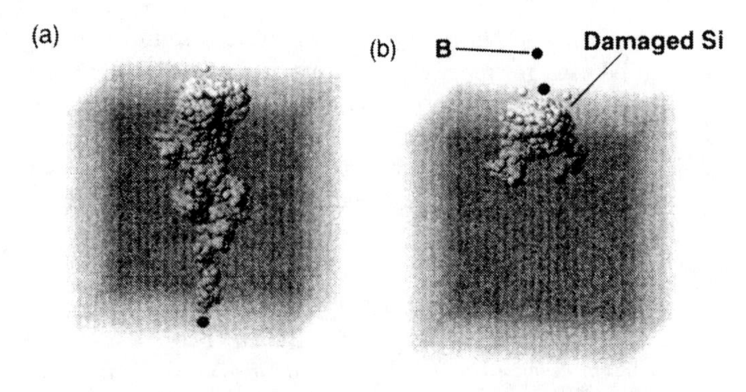

FIGURE 3. Cross sectional view of damage in Si substrates implanted with B_{10} clusters at (a) 10 keV and (b) 3 keV as calculated by molecular dynamics simulations

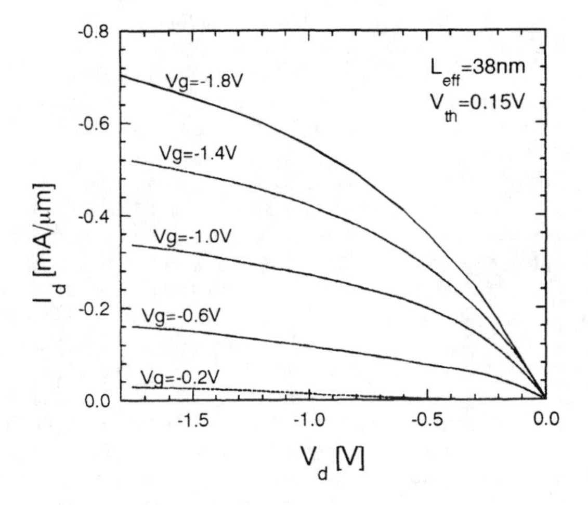

Figure 4. I_d-V_d characteristics of the p-MOSFET

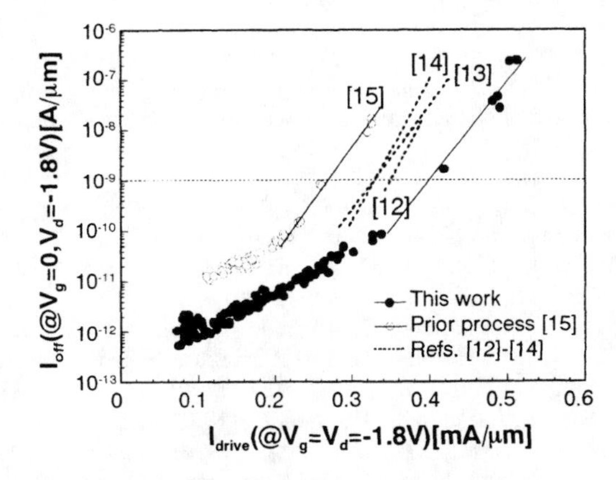

Figure 5. I_{drive} vs. I_{off} of the 40nm p-MOSFET compared with that of recently reported devices

Figure 4 shows I_d-V_d characteristics of the p-MOSFET with a l_{eff} of 38 nm. An excellent saturation drive current of 0.70 ma/μm(@ v_g=v_d= -1.8 v) and a very high gm_{max} of 459 ms/mm was achieved. Figure 5 shows the I_{drive} vs. I_{off} of the 40 nm p-MOSFET compared with that of other recently reported devices. The highest drive current shows 15 % improvement compared with those published data [12-15]. A low s/d series resistance r_{sd} of 760 ohm-m is achieved even if a high sheet resistance (>20 kohm/sq) is used for the extension regions (due to the diminished extension length). The smallest p-MOSFET with a L_{eff} of 38 nm has been demonstrated. Figure 6 shows the SEM image of the device having a poly-Si gate length of 40 nm after sidewall removal. This device has the smallest dimensions and the highest device performance among many developmental devices.

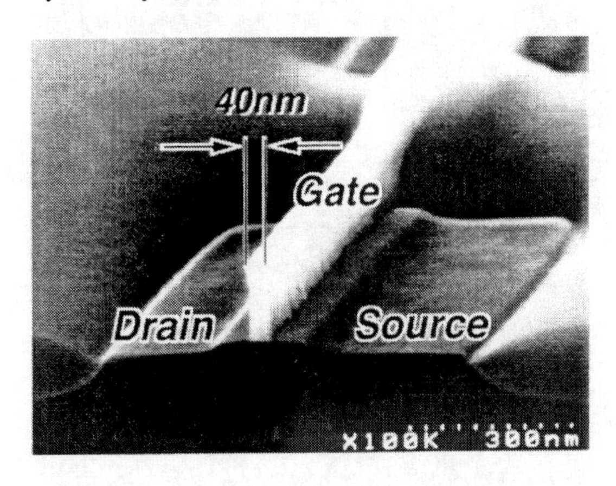

FIGURE 6. SEM image of p-MOSFET having a 40 nm gate after sidewall removal.

APPLICATION TO LOW DAMAGE SPUTTERING

Very high sputtering yields on metal, semiconductor and insulator surfaces due to bombardment with cluster ions have been experimentally observed [1] and this effect has been studied by computer simulation. The sputtering yield Y from Au surfaces due to Ar_n (n=80-200) cluster bombardment at energies of 8-20 keV fits a power dependence $Y \sim n^{2.4}$ on the cluster size n, in good agreement with experiment [16]. The power exponent of this expression, 2.4, is close to the value of 2 which was obtained in a thermal spike model for monomer ions [17].

Cluster ion beam sputtering is characterized not only by high etching rates but also by surface smoothing effects due to a lateral sputtering process. However these characteristics depend on the type of surface interaction process, e.g., physical sputtering or chemical sputtering. Figure 7 shows a comparison of sputtering yields for Si, SiC and W substrates bombarded by Ar monomers, Ar_{3000} clusters and $(SF_6)_{2000}$ cluster ion beams at an acceleration voltage of 20 kV. The sputtering yield is higher when the bombardment is by cluster ions than when it is by an Ar monomer ion beams. Reactive ion sputtering by a SF_6 cluster ion beam was much faster than the physical sputtering by Ar. The effects on resultant surface roughness after etching are also quite different. Physical sputtering by an Ar cluster ion beam produces a smooth surface, and this continues as the bombardment continues. However in the case of SF_6 cluster ions, surface roughness could not be reduced.

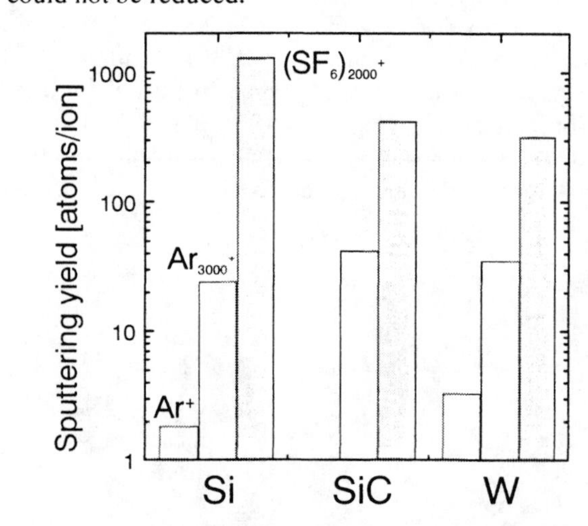

FIGURE 7. Comparison of sputtering yields for Si, SiC, and W substrates bombarded by Ar monomer, Ar_{3000} cluster and $(SF_6)_{2000}$ cluster ion beam at an acceleration voltage of 20kV

Surface smoothing by cluster ion beams has been demonstrated for semiconductors, metals and insulators. For a Cu film deposited on a Si substrate, the original surface contained big hills with an average diameter of 200 nm and the average surface roughness was Ra=5.8 nm. By using Ar cluster ions with dose of $8x10^{15}$ ions/cm^2 at an energy of 20 keV, the surface was smoothed to Ra=1.3 nm. Similar results have been obtained for a YBCO film (from original roughness: Ra=7.9 nm to sputtered surface: Ra=0.5 nm by Ar cluster 20 keV at $2x10^{16}$ions/cm^2), for a SiC film (from Ra=14.5 nm to Ra=0.6 nm by Ar 20 keV at $1x10^{16}$ ions/cm^2), for a CVD diamond film (from Ra=41.3 nm to 8.2 nm at 20 keV $1x10^{17}$ ions/cm^2). These measurements were made with an AFM at 1μm scan size. The potential application of YBCO is for SQUID devices, CVD diamond membrane surface finishing is for SOR and X-ray lithography, metal plated surface smoothing is for non-spherical plastic lens-mold surface smoothing, SiC surface smoothing is for SOR mirrors.

CONCLUSIONS

It has been shown by experiment and simulation that the impact of large clusters on Si is very different from monomer implantation. Local, isotropic deposition of the ion energy can lead to ultra-shallow implantation, high rate sputtering and surface smoothing. Small clusters can also be useful, due to their large mass and mass-to-charge ratio, even though they may not be capable of the fuller range of cluster effects. These properties of cluster beam processing can be used for precisely controlled novel high-throughput shallow implantation and other surface modifications. Processes allowing smaller lateral and vertical dimensions and better reliability are urgently required. Cluster ion beam processing is presented as an advanced approach which will contribute to further progress this field.

REFERENCES

1. I.Yamada and J.Matsuo: *Materials Research Society Symp. Proc.*, Vol. **427**, *"Advanced metallization for Future ULSI"*, Eds. K.N.Tu, J.M.Poate, J.W.Mayer and L.C.Chen, 1996 (Materials Research Society, Pittsburgh, 1997) p. 265.
2. I. Yamada, J.Matsuo, E. C. Jones, D. Takeuchi, T. Aoki, K. Goto, and T. Sugii, *"Range and Damage Distribution in Cluster Ion Implantation"*, *Mat. Res. Soc. Symp. Proc.* Vol. **438** (1997) p. 363.
3. T. Aoki, J. Matso, Z. Insepov, and I. Yamada, *"Molecular Dynamics Simulation of Damage Formation by Cluster Ion Impact"*, *Nucl. Instr. and Meth.*, **B121**, 49 (1997).
4. N. Shimada, T. Aoki, J. Matsuo, I. Yamada, K. Goto, and T. Sugii, *"Reduction of Boron Transient Enhanced Diffusion in Silicon by Low-Energy Cluster Ion Implantation"*, *Mat. Chem. and Phys.* **54**, 80 (1998).
5. K. Goto, J. Matsuo, Y. Tada, T, Tanaka, Y. Momiyama, T. Sugii, and I. Yamada, *"A High Performance 50nm PMOSFET using Decaborane ($B_{10}H_{14}$) Ion Implantation and 2-step Activation Annealing Process"*, IEDM Tech. Dig. (1997) p. 471.
6. A.M. Agarwal, D. J. Eaglesham, H.-J. Gossmann, L. Pelaz, S. B. Herner, D. C. Jacobson, T. E. Haynes, Y. Erokhin, and R. Simonton, *"Boron-enhanced-diffusion of Boron: The Limiting Factor for Ultra-shallow Junction"*, IEDM Tech. Dig. (1997) p. 467.
7. E. J. H. Collart, K. Weemers, and D. J. Gravesteijin, and J. G. M. van Berkum, *"Characterzation of Low-energy (100eV-10keV) Boron Ion Implantation"*, *J. Vac. Sci. Technol.* **B16**, 280 (1998).
8. P. J. Scanlon, G. Farrel, M.C. Ridgway, and R. Valizadeh, *"Nuclear Reaction Analysis of Shallow Boron Implants in Silicon"*, *Nucl. Instr. and Meth.*, **B16**, 479 (1986).
9. A.M. Agarwal, H. -J. Gossmann, D. J. Eaglesham, L. Pelaz, D.C. Jacobson, T. E. Haynes, and Y. E.Erokhin, *"Reduction of Transient Diffusion from 1-5keV Si^+ Ion Implantation"*, *Appl. Phys. Lett.* **71**, 3141 (1997).
10. K. S. Jones, P. G. Elliman, M. M. Petravic and P. Kringhoj, *"Using Doping Superlattices to study Transient-enhanced Diffusion of Boron in Reglown Silicon"*, *Appl. Phys. Lett.* **68**, 3111 (1996).
11. L. H. Zhang, K. S.Jones, P. H. Chi, and D. S. Simons,*"Transient Enhanced Diffusion without {311} Defects in Low Energy B^+-implanted Silicon"*, *Appl. Phys. Lett.* **67**, 2025 (1995).
12. M.Rodder, Q.Z. Hong, M.Nandakumar, S.Aur, J.C.Hu and I-C. Chen, *IEDM Tech. Dig.*, (1996) p. 563.
13. M.Bohr, S.S.Ahmed, S.U.Ahmed, M.Bost, TR. Ghani, J. Greason, R. Hainsey, C.Jan, P. Packan, S. Sivakumar, S. Thompson and S. Yang, *IEDM Tech. Dig.*, (1996) p. 847.
14. L.Su, *et al.*, *Symp on VLSI Tech.*, (1996) p. 12
15 K.Goto, *et al.*, *IEDM Tech .Dig.*, (1996) p. 435.
16. Z.Insepov and I.Yamada, *Nucl. Instr. and Meth.*, **B99**, 248 (1995).
17 .Y. Kitazoe and Y. Yamamura, *Radiat. Eff. Lett.*, **50**, 39 (1980).

Ultra Low Energy Boron Implantation Using Cluster Ions for Decananometer MOSFETs

T. Sugii, K. Goto, T. Tanaka
FUJITSU LABORATORIES LTD., 10-1 Morinosato-Wakamiya, Atsugi 243-0197, Japan

J. Matsuo, I. Yamada
Ion Beam Engineering Experimental Lab., Kyoto University, Sakyo, Kyoto 606-8317, Japan

Two types of applications in ultra low-energy boron implantation with decaborane ($B_{10}H_{14}$) molecules for highly miniaturized devices are presented. One is the formation of a 7-nm-deep junction. This ultra shallow junction was applied to a pMOSFET with a gate length of around 50nm. The other application of this technique is the formation of ultra shallow channels in buried-channel pMOSFETs. We successfully fabricated this type of device, achieving a low threshold voltage and a steep subthreshold.

INTRODUCTION

The pace of MOSFET miniaturization has accelerated over the last few years due to increased demand for processors and system LSIs with higher performance. LSI production with sub-0.2 μm technologies is forthcoming, and current research is concerned with technologies and devices for sub-0.1 μm generations.

In this paper, we review two types of highly miniaturized devices that make use of ultra low energy boron implantation with decaborane ($B_{10}H_{14}$) molecules[1]. One is a decananometer gate-length MOSFET with shallow source/drain extensions [Figure 1(a)]. The ultra low-energy boron implantation technique is used for the formation of ultra shallow junctions at the source and drain extension regions. Achieving a junction depth of less than 30 nm is a prerequisite in sub-0.1 μm devices so as to suppress short-channel effects (SCE). We realized a 7-nm-deep junction with 2 keV $B_{10}H_{14}^+$ implantation and successfully fabricated a decananometer gate-length pMOSFET[2].

The other devices are deep sub-micron buried-channel (BC) MOSFETs[3] [Figure 1(b)] which are commonly used in peripheral circuits in DRAM chips. The critical issues for the use of BC-pMOSFETs below 0.25 um, that is the giga-bit DRAM era, are the low threshold voltage (Vth) and the suppression of SCE, which can be achieved by an extremely shallow counter-doped, p-type layer with a high impurity concentration. Here, the ultra low-energy boron implantation is effective for formation and we fabricated a 20-nm-thick counter-doped layer and demonstrated an SCE-free high-performance 0.18 um BC-pMOSFET.

ULTRA SHALLOW JUNCTION

We have already developed a low-energy $B_{10}H_{14}$ ion implantation technology. Figure 2 shows SIMS profiles

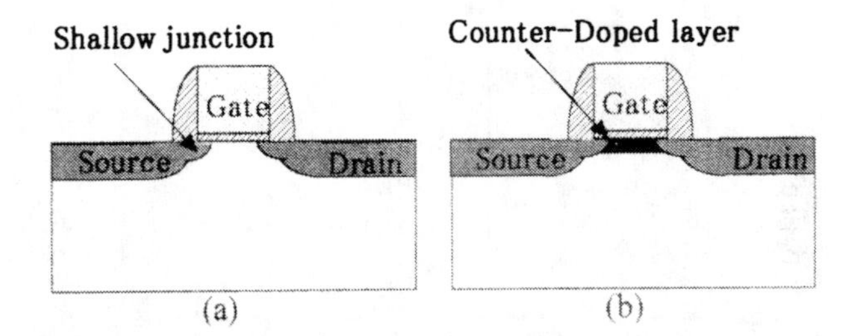

FIGURE 1. Structure (a) is a MOSFET with shallow source/drain extensions. Structure (b) is a buried-channel MOSFET.

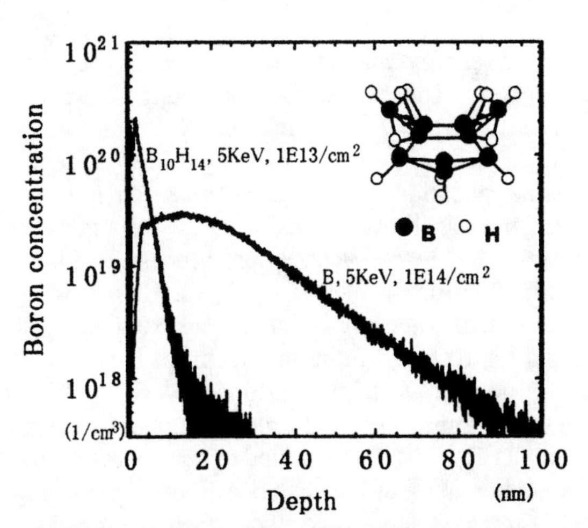

FIGURE 2. Boron profiles of B^+ and $B_{10}H_{14}^+$ implantation.

of $B_{10}H_{14}^+$ implantation and boron one with an acceleration energy of 5 keV and the same effective dose (1×10^{14} 1/cm²). The diagram shows a $B_{10}H_{14}$ structure. $B_{10}H_{14}$ is implanted in clusters, and at the substrate surface, the $B_{10}H_{14}$ clusters resolve into ten boron atoms,

CP475, *Applications of Accelerators in Research and Industry*,
edited by J. L. Duggan and I. L. Morgan

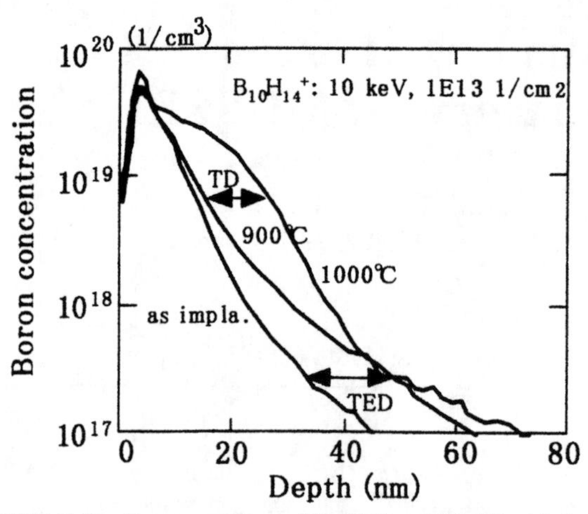

FIGURE 3. Boron profiles of $B_{10}H_{14}^+$ at 10 keV with 1 x 10^{13} 1/cm^2 after implanted and 900C or 1000C annealing for 10 sec.

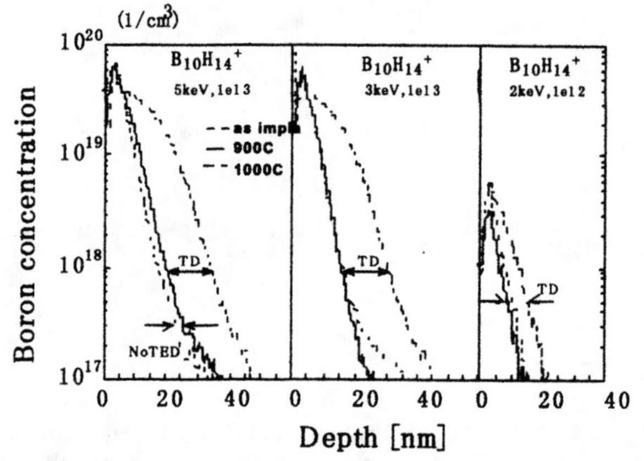

FIGURE.4. Boron profiles implanted by $B_{10}H_{14}^+$ at 5 keV, 3 keV, and 2 keV after 900℃ and 1000℃ for 10 sec.

which leads to implantation with one-tenth lower effective acceleration energy and an effective boron dosage ten times higher. These characteristics suggest possibility that an ultra-shallow junction can be fabricated with enough beam current without invoking charge damage.

For an application of the shallow junction toward any devices, the shallow junction must be maintained throughout the whole fabrication process. Figure 3 shows SIMS boron profiles of $B_{10}H_{14}^+$ at 10 keV with 1 x 10^{13} 1/cm^2 just after implantation and 900C or 1000C annealing for 10 sec. From this figure, it is necessary to suppress not only thermal diffusion (TD) by low temperature annealing but also transient enhanced diffusion (TED). TED was found to be suppressed by the $B_{10}H_{14}^+$ implantation at less than 5 keV as shown in Figure 4. TED remains suppressed even when we increase the dose up to 1 x 10^{14} 1/cm^2 (effective boron dose is 10 times higher). This seems to be due to a surface proximity effect. The interstitial Si atoms created by $B_{10}H_{14}^+$ implantation seem to be effectively terminated at the surface due to the shallow implantation region. A 7-nm-deep ultra shallow junction was obtained by $B_{10}H_{14}^+$ implantation at 2 keV, 1 x 10^{12} 1/cm^2, and 900C, 10 sec annealing, which was the condition used for decananometer gate-length pMOSFET fabrication.

For an application of the shallow junction to channel regions (counter-doped regions) of BC-pMOSFETs in giga-bit DRAM peripheral circuits (Leff: 0.18 um, Tox: 4 nm, Xj: 0.1 um), we investigated appropriate channel impurity profiles using a 2D-device simulator. We varied both the impurity concentration, Na, and the thickness, d, of the counter-doped layer to put the Vth and Drain-Induced-Barrier-Lowering (DIBL) factor into a window, which indicates the values required for pMOSFETs used in the 4-Gbit DRAM generation and beyond. From Figure 5, a 20-nm-thick counter-doped layer with an impurity concentration of 1.5 x 10^{18} 1/cm^3 is necessary to put the device characteristics within the window. Figure 6 shows the SIMS profiles of boron and arsenic at the channel region with optimized process conditions. $B_{10}H_{14}^+$ implantation at 3 keV, 2 x 10^{12} 1/cm^2 and annealing at 950C for 10 sec realized the above mentioned shallow junction.

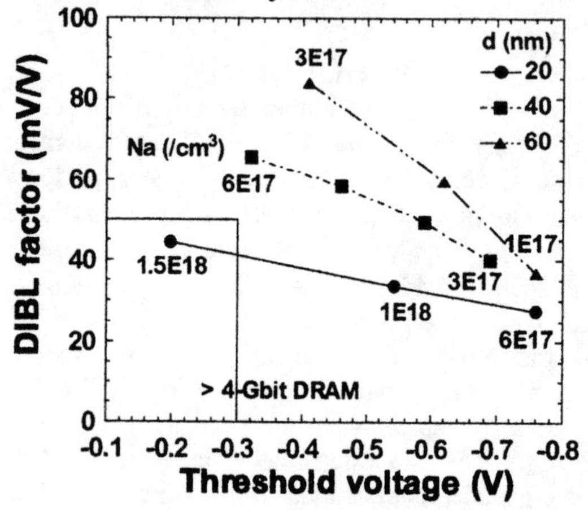

FIGURE 5. Relationship between DIBL and Vth.

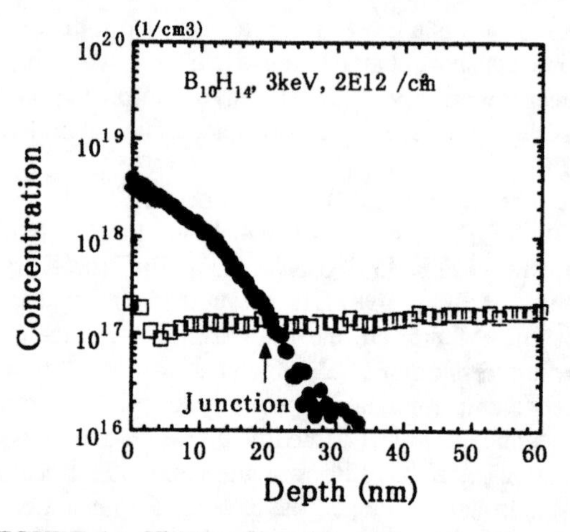

FIGURE 6. SIMS profiles of B and As at the channel.

DEVICE PERFORMANCE

1) Decananometer gate-length pMOSFETs

Figure 7 shows the SEM image of the device under fabrication having a poly-Si gate length of 40 nm. Figure 8 shows the Id-Vd and subthreshold characteristics of the pMOSFET with an effective gate length of 50nm. Figure 8 shows the Id-Vd and subthreshold characteristics of the pMOSFET with an effective gate length (Leff) of 50 nm. An excellent saturation current of 0.5 mA/um and the highest Gmmax of 384 mS/mm was achieved. Figure 9 shows the dependence of Vth and drain current, Id, on Leff. A well suppressed SCE and a continuous increase in Id were observed down to a Leff of 50 nm and below. ΔL=Lpoly-Leff was estimated at about 10 nm by comparing the measured data with SEM and TEM observation. The shallow junction is very effective in reducing the ΔL, which leads not only to reduced SCE but reduced gate-drain overlapped capacitance, Cov. Figure 10 shows the Ion vs Ioff characteristics compared to those of recently reported high-performance pMOSFETs[4-6]. Our work achieves the highest Ion of 0.4 mA/um at a given Ioff of 1 nA/um due to the aggressive scaling of the Leff with reduced SCE. The source/drain resistance, Rsd of 760 ohm-um is low enough for a rather high sheet resistance (> 20Kohm/sq.) for extension regions. This is provided that the length of the extension region is short enough and its lateral boron profile is sharp. Figure 11 shows gate-delay metric of our device together with other reported devices. The result obtained with our decananometer gate-length MOSFET are significant because it guarantees the improvement of the gate delay by scaling a device size at least down to decananometer range with a suppressed SCE by ultra shallow junctions.

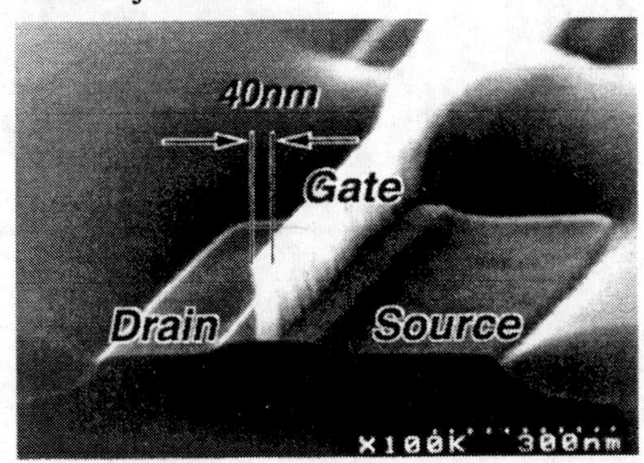

FIGURE 7. SEM image of the device.

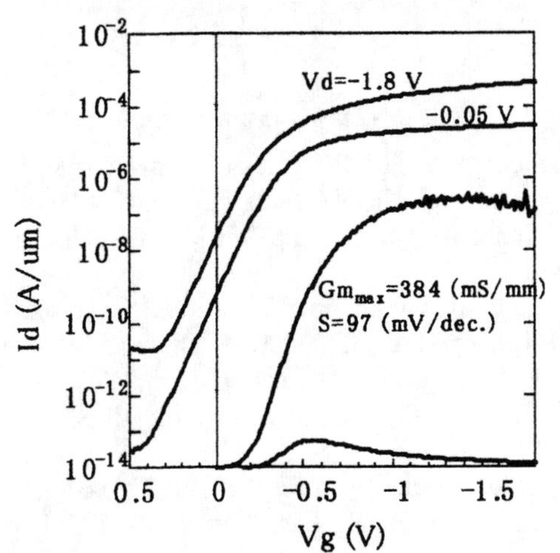

FIGURE 8. Subthreshold characteristics. Leff: 50 nm.

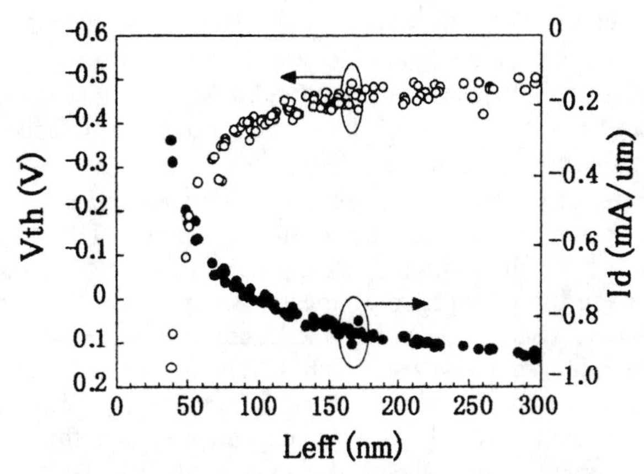

FIGURE 9. Dependence of Vth and drain current on Leff.

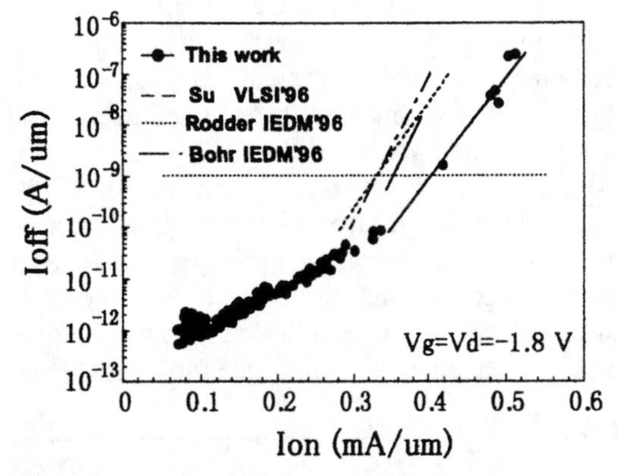

FIGURE 10. Ion vs Ioff characteristics compared to that of recently reported high-performance pMOSFETs[4-6]

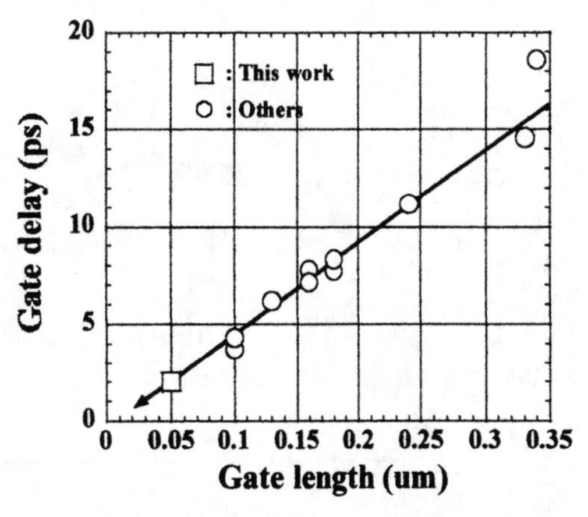

FIGURE 11. Gate delay metric.

2) BC-pMOSFETs

From the results of the above device simulation, we fabricated BC-pMOSFETs using $B_{10}H_{14}$ ion implantation at 3 keV, 3 x 10^{12} 1/cm^2 through the gate oxide. The typical I-V characteristics of a 0.18 um Leff BC-pMOSFET are shown in Figure 12. Figure 13 shows the Vth dependence on the effective gate length for BC-pMOSFETs with different fabrication technologies. Our BC-pMOSFETs with a low temperature process (maximum temp: 950C) suffer less SCE despite low Vth.

To clarify the difference in short channel characteristics, the ΔVth - Vth relationships are shown in Figure 14 with some references[7-9]. Our BC-pMOSFET with a low temperatrure process has the best short-channel characteristics, that is, a low Vth with a small ΔVth. Figure 15 compares the drain current between the low temperature and high temperature process. The drain current for the low temperature process is about twice that for the high temperature process owing to the low Vth. Our BC-pMOSFET with a gate length of less than 0.2 um has achieved a low Vth, a high SCE immunity, and a large drain current, which indicates the applicability of BC-pMOSFETs with $B_{10}H_{14}$ ion implantation and the low temperature process being used in giga-bit DRAMs.

CONCLUSION

Ultra low energy boron ion implantation using $B_{10}H_{14}$ was demonstrated for the fabrication of ultra shallow p-type layers. The technology was applied successfully to decananometer gate length pMOSFETs and low Vth BC-pMOSFETs. Both type of devices show highly suppressed SCE, while demonstrating high current drive ability. The ion implantation technique is very effective and manufacturable and thus is the most promising candidate among various shallow junction techniques for the coming decananometer, multi-gigabit DRAM era.

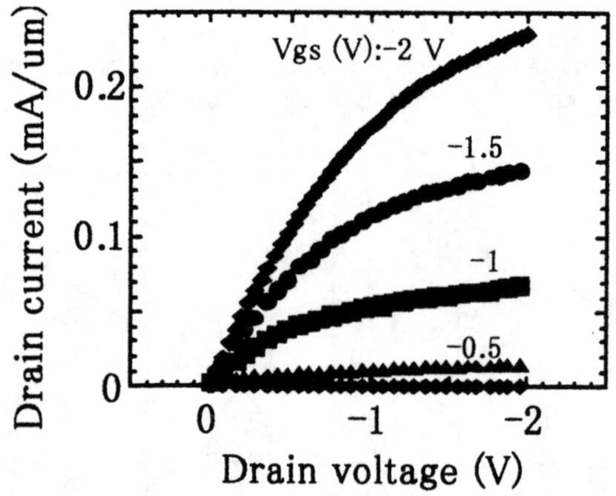

FIGURE 12. I-V characteristics. Leff: 0.18 um

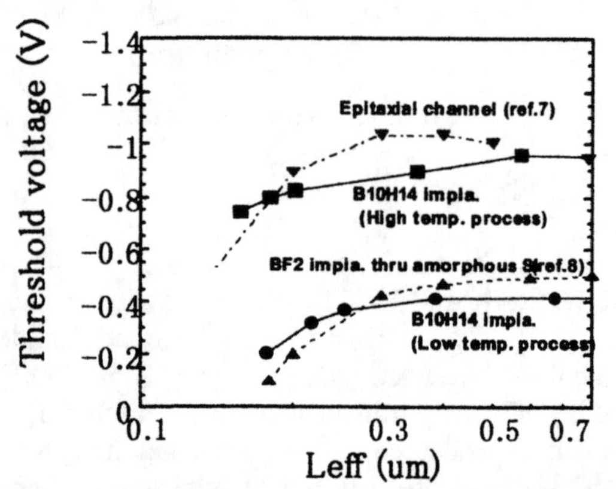

FIGURE 13. Vth dependence on Leff for BC-pMOSFETs with differebt fabrication technologies.

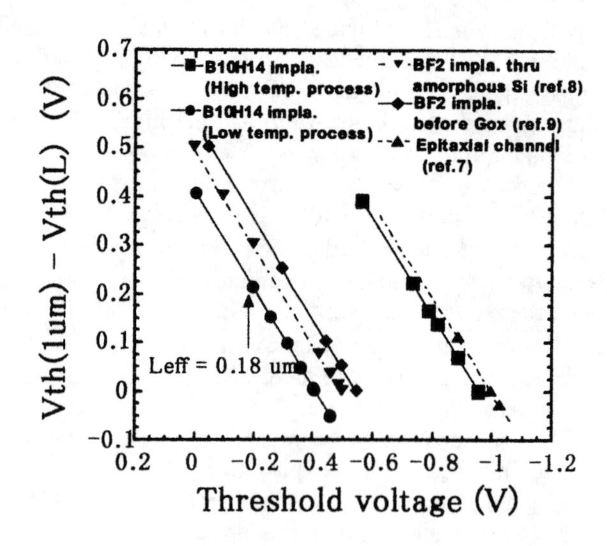

FIGURE 14. (Vth(@1 um) – Vth) vs Vth.

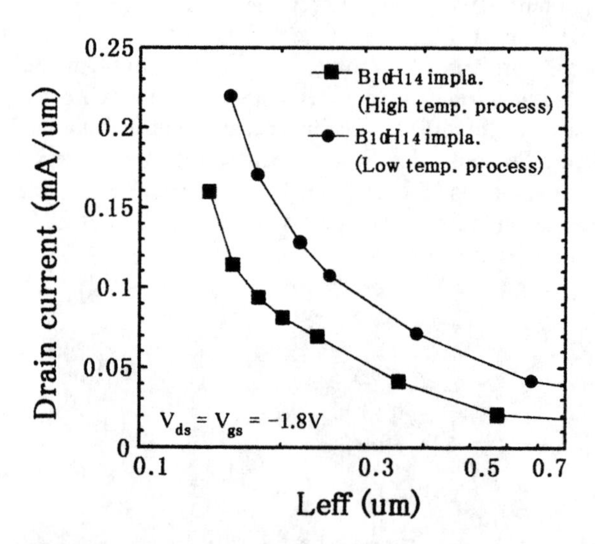

FIGURE 15. Comparison of the drain current of BC-MOSFET fabricated with the low temperature process and high temperature process.

REFERENCES

1. K. Goto, et al., IEDM Tech. Dig., p.435, 1996
2. K. Goto, et al., IEDM Tech. Dig., p.471, 1997
3. T. Tanaka, et al., Symp. on VLSI Tech. Dig., p.88, 1998
4. L. Su, et al., Symp. on VLSI Tech. Dig., p.12, 1996
5. M. Rodder, et al., IEDM Tech. Dig., p.563, 1996
6. M. Bohr, et al., IEDM Tech. Dig., p.847, 1996
7. H. Matsuhashi, et al., Symp. on VLSI Tech. Dig., p.36 ,1996
8. H. Ishida, et al., Tech. Rep. of IEICE, p.61, 1994
9. T. Yoshitomi, et al., Ext. Abs. of 1995 SSDM, p.222, 1995

POLYETHYLENE THIN-FILM GROWTH VIA CLUSTER-SURFACE COLLISIONS

Thomas A. Plaisted[1], John D. Zahrt[2], W. Leigh Young[1], Lifeng Qi[1] and Susan B. Sinnott[1]*

[1]*Department of Chemical and Materials Engineering, University of Kentucky, Lexington, KY 40506*

[2]*Department of Chemistry, Eastern Kentucky University, Richmond, KY 40475*

Molecular dynamics simulations have been performed to investigate the growth of polyethylene thin films via cluster-surface collisions. The substrate is hydrogen-terminated diamond (111) and a beam of four molecular clusters of ethylene impact the substrate at a velocity of 18 km/s. The atoms are characterized by a realistic many-body empirical potential for hydrocarbon systems. The goal of this study is to compare the simulation results to existing experimental data obtained under similar reaction conditions to better understand the process by which solid thin hydrocarbon films grow during molecular cluster-surface impacts.

INTRODUCTION

Controlled thin film growth is crucial for the manufacture of numerous electronic devices and protective coatings. Some of the important quantities are the structure and properties of the thin film and the adhesion between the film and the substrate. Many thin film growth techniques proceed via the acceleration of ionized atoms or molecules at high velocities towards a surface. However, others occur through the acceleration and impact of ionized clusters of atoms or molecules. Cluster-surface collisions are of fundamental interest because they lead to high concentrations of atoms and energy in a very localized part of the surface. Such collisions have therefore been under intense study for some time to better understand the unique chemical and physical processes that are possible following the collision. Examples of systems that have been studied include metal cluster impacts on metal substrates (1-4) and noble gas cluster collisions with ceramic substrates (5-7).

Of special interest is the growth of hydrocarbon thin films, including polymer thin films, through the collision of beams of molecular clusters on relatively rigid surfaces at hyperthermal velocities (8-11). The clusters used experimentally were estimated to contain about 10-20 molecules/cluster and impacted substrates such as mica and glass with incident velocities that ranged from 18 – 26 km/s. When ethylene clusters collided in the lower velocity region, amorphous polyethylene films formed. However, when they impacted with incident kinetic energies corresponding to the higher velocity region crystalline polyethylene films were created.

We have recently studied the nucleation of thin hydrocarbon films on diamond substrates using molecular dynamics simulations (12-14). The studies have shed light on the dependence of nucleation on the reaction conditions, such as molecular reactivity (12), incident velocity (12,13), cluster size (13), and surface reactivity (14). The results detail the chemical reactions that take place on impact. These include the dissociation of molecules into fragments, the addition of molecules and fragments to form short hydrocarbon chains, and the displacement and/or incorporation of surface atoms by the chemical products. Some of the new chemical products scatter away after the collision and some remain on the surface to nucleate a new hydrocarbon thin film, with the amounts varying with changes in the reaction condition variables listed above.

We have also examined film growth through the consecutive impact of two 64-molecule acetylene clusters at velocities of 12 km/s (15). These studies showed that when the second cluster impacted the film that had nucleated from the collision of the first cluster, some of the film was sputtered away and the rest grew through the addition of new chemical products.

We are interested in more closely replicating the experimental conditions with ethylene molecular clusters at higher impact velocities. Therefore in this paper, we investigate the nucleation and growth of polyethylene films on hydrogen-terminated diamond (111), H:C(111), surfaces. Each cluster contained eight ethylene molecules and beams of four clusters impacted H:C(111) at velocities of 18 km/s. This corresponds to an incident kinetic energy of 8664 kcal/mol per cluster. This is more than enough energy to overcome the molecular binding energy of 542 kcal/mol (16) and the enthalpy of polymerization of about 26 kcal/mol (16). We know from our previous studies (12) that the ethylene clusters are less reactive than the acetylene clusters studied in Ref. 15. However, we have also shown (12,13) that higher impact velocities lead to more chemical reactions and surface disruption that contribute to improved thin film nucleation. In addition,

* Corresponding Author

CP475, *Applications of Accelerators in Research and Industry*, edited by J. L. Duggan and I. L. Morgan

smaller clusters have been shown to be more efficient for thin film growth (13). The goal is to understand how the film nucleates and how subsequent cluster impacts affect the growth of the film.

COMPUTATIONAL DETAILS

In these classical molecular dynamics simulations, Newton's equations of motion are integrated with a third-order Nordsieck predictor corrector (17) using a time step of 0.2 fs. The forces on the individual atoms are obtained using a reactive empirical bond order hydrocarbon potential that realistically describes covalent bonding within both the clusters and the surface (18,19). Originally developed to examine the chemical vapor deposition of diamond thin films (19), it has been successfully used to study a variety of other processes at surfaces such as tribochemistry (20), sputtering (21), and surface-catalyzed chemical reactions (22). The long-range van der Waals interactions between the cluster molecules are described with a Lennard-Jones potential that is nonzero only after the covalent interactions have gone to zero as described in Ref. (23). The combined expression for the binding energy of the clusters and surface is therefore:

$$E_b = \sum_i \sum_{j<i} [\, V_r(r_{ij}) - B_{ij} V_a(r_{ij}) + V_{vdw}(r_{ij})]$$

where E_b is the binding energy, r_{ij} is the distance between atoms i and j, V_r is a pair-additive term that takes into account the interatomic core-core repulsive interactions, V_a is a pair-additive term that models the attractive interaction due to the valence electrons, and B_{ij} is a many-body empirical bond-order term that modulates valance electron densities (18,19) and depends on atomic coordination and angles. V_{vdw} is the Lennard-Jones potential. This potential does not account for changes in the electronic structure of the atoms as a result of high-velocity collisions. The assumption taken here and elsewhere (1-7,12-15) is that the impact velocities are low enough that electronic excitations do not play a significant role in the chemical reactions that take place following the collisions.

The three-dimensional molecular clusters are allowed to fully equilibrate at 5 K prior to impact. The H:C(111) surface is equilibrated at 300 K and consists of 4792 atoms in twenty carbon layers, terminated by hydrogen layers on the top and bottom to satisfy the coordination requirements of the carbon atoms. The hydrogen surface layer furthest from the cluster is held rigid. Moving towards the cluster, the next nine carbon layers and all the atoms at the edges of the slab have Langevin frictional forces (17) applied to maintain the surface at a constant temperature and mimic the heat dissipation properties of a real diamond surface. The remaining atoms in the surface, and all of the atoms in the clusters, are allowed to evolve in time according to Newton's equations of motion with no additional constraints. Stringent tests with alternate surface structures were performed to make sure that no energy was scattering from the rigid layers on the bottom of the slab to influence the processes taking place at the surface.

RESULTS AND DISCUSSION

Two series of simulations are considered where beams of molecular ethylene clusters impact the H:C(111) surface at trajectories normal to the surface. The clusters are separated by about 3 Å within the beam. In the fist series of simulations all the clusters in the beam are targeted at the same general area on the surface. In the second series of simulations, the clusters are spread out slightly so that they do not impact the surface in the same general location. In addition, the clusters are randomly oriented relative to each other so that they do not impact the surface with identical orientations. Figure 1 shows selected snapshots from the collision of the second beam of ethylene clusters on the H:C(111) surface. The starting system configuration is illustrated in Fig. 1a.

As each cluster contacts the surface it is flattened out by the impact. Some molecular dissociation occurs that is most noticeable as hydrogen atoms and molecules scattering away from the surface as shown in Fig. 1b. In addition, the force of the collisions imbeds some of the cluster fragments into the surface and creates a small crater in the impact zone. The formation of craters after cluster-surface collisions is a common feature of cluster collisions with metal surfaces that that are generally softer than covalently bound ceramics. However, this is the first report of cratering in diamond due to impact by a molecular cluster.

Some of the molecules that collide, especially in the first two-three clusters in the beam, adhere to the surface while the others scatter back into space as shown in Fig. 1c. However, as they leave they can collide with the other clusters in the beam that are moving towards the surface. Thus some molecule-molecule collisions are observed in space as undoubtedly must take place experimentally. In addition, collisions by subsequent beam clusters sputter away some of the film that has adhered to the surface as was seen previously for another system (15) and as shown in Fig. 1d. Thus the film grows through a repeating pattern of collision, adherence, partial sputtering, chain addition and film growth.

This series of reactions results in a complex sequence of energy transformations as external cluster kinetic

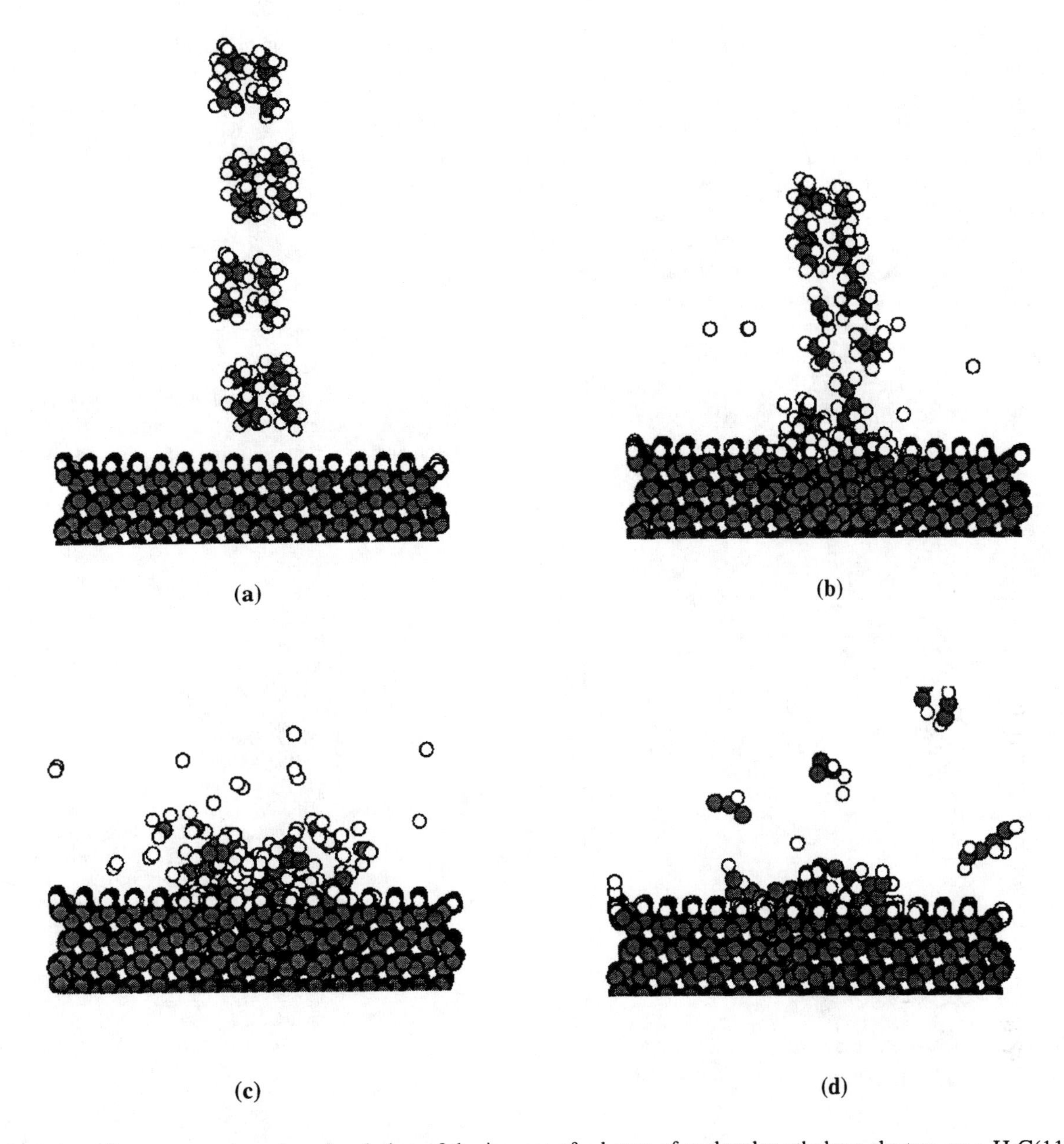

Figure 1: Snapshots from a representative simulation of the impact of a beam of molecular ethylene clusters on a H:C(111) surface. The white spheres are hydrogen atoms and the gray spheres represent carbon atoms.

energy is transferred from one cluster to another and from the impacting clusters to the surface. The extra kinetic energy is also transformed into internal kinetic energy and system potential energy as the chemical reactions proceed. This can be seen in Figure 2 where the total energy of the system is seen to decrease over time as the clusters impact and react with the surface. The total energy of the system is not constant because on impact, much of the cluster's external kinetic energy is transformed into surface kinetic energy that is then dissipated by the Langevin heat bath atoms. Thus these simulations are constant-temperature molecular dynamics. The oscillations shown in Figure 2 between 300 and 520 fs are caused by the cluster-surface and cluster-cluster collisions. The Figure shows how the severity of these oscillations increases, indicating larger energy transformations, as the buildup of molecules on the surface increases, as illustrated in Figure 1. The longest hydrocarbon chains produced through polymerization as a result of the collisions varied between four and twelve carbon atoms in length. About 12-23% of the incident carbon atoms adhered to the surface in a thin film with the amount of sputtering increasing significantly in the case where the molecules impacted the surface in the same general location. This is higher than the 3-10% of carbon adhesion predicted with larger clusters (64 molecules/cluster) impacting H:C(111) at lower impact velocities of 12 km/s (23). Hence these simulations reinforce the trend noted previously that smaller clusters are more efficient for thin film growth (13). Our previous work (23) also indicates that maximum efficiency during the growth process is achieved by increasing the reactivity of the surface. The creation of surface dangling bonds results in nearly 50% of the incident carbon atoms in ethylene molecular clusters adhering to the surface (23). The impact-induced disorder of the surface predicated in these simulations are caused by the initial impacts that also cause the cratering and result in the formation of activated surface atoms that enhance the thin film growth. The film is clearly disordered and polymer-like although the chains are too short to be classified as polymers. Hence the results show good agreement with experimental data albeit the very earliest stages of the experiments.

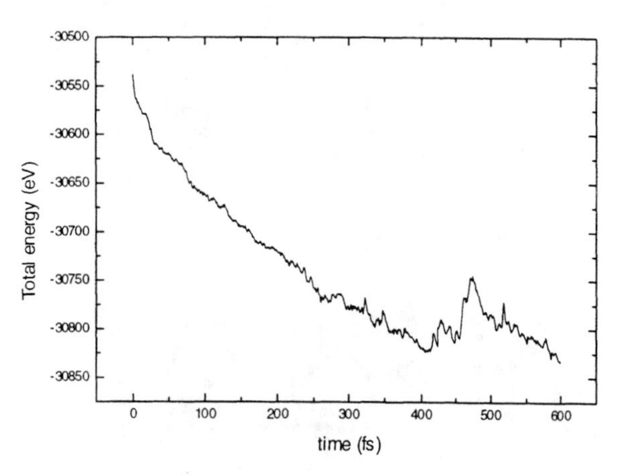

Figure 2: Plot of change in the total energy of the system as a function of time for the simulation shown in Figure 1.

CONCLUSIONS

In this paper we report on preliminary results obtained from classical molecular dynamics simulations used to study the growth of polyethylene through cluster-surface collisions at hyperthermal velocities. The results shed light on the processes that take place during the film growth. These include surface cratering, molecule-surface and molecule-molecule collisions, partial adhesion to the surface, partial sputtering of the adhered molecules when new clusters from the beam collide with the surface, and continued growth of the film from newly adhered molecules and fragments from the beam.

Current efforts are proceeding on two fronts. More realistic beams with tens of clusters impacting comparable surfaces are being modeled to continue to observe the growth of the film. In addition, more trajectories are being performed to yield a better statistical sampling of data. Future efforts will also consider interactions been multiple beams of clusters at normal and other incident angles.

ACKNOWLEDGMENTS

We gratefully acknowledge the National Science Foundation (CHE-9708049) and the Donors of the Petroleum Research Fund, administered by the American Chemical Society, for partial support of this research. L.Q. also acknowledges a graduate fellowship from the Center for Computational Sciences at the University of Kentucky.

REFERENCES

1. Hsieh, H., Averback, R. S. Sellers, H., and Flynn, C. P., *Phys. Rev. B* **45**, 4417, 1992.
2. Vandoni, G., Felix, C., and Massobrio, C., *Phys. Rev. B* **54**, 1553, 1996.
3. Shapiro, M. H., Tosheff, G. A., and Tombrello, T. A., *Nuc. Inst. and Meth.*, **B88**, 81, 1994.
4. Haberland, H., Insepov, Z., and Moseler, *Phys. Rev. B* **51**, 11061, 1995.
5. Cleveland, C. L., and Landman, U., *Science* **257**, 355, 1992.
6. Blais, N.C., and Stine, J.R., *J. Chem. Phys.* **93**, 7914, 1990.
7. Raz, T., Schek, I., Ben-Nun, M., Even, U., Jortner, J., and Levine, R.D., *J. Chem. Phys.* **101**, 8606, 1994.
8. Usui, H., Yamada, I., and Takagi, T., *J. Vac. Sci. Technol. A* **4**, 52, 1986.
9. Fejfar, A., and Biederman, H., *Int. J. Electronics* **73**, 1051, 1992.
10. Gao, H.J., Xue, Z.Q., Wang, K.Z., Wu, Q.D., and Pang, S., *J. Appl. Phys.* **68**, 2192, 1996.
11. Paillard, V., Melinon, P., Dupuis, V., Perez, J.P., Perez, A., and Champagnon, B., *Phys. Rev. Lett.* **71**, 4170, 1993.
12. Qi, L., and Sinnott, S.B., *J. Phys. Chem.* **B101**, 6883, 1997.
13. Qi, L., and Sinnott, S.B., *Nuc. Inst. and Meth.*, **B140**, 39, 1998.
14. Qi, L., and Sinnott, S.B., J. *Vac. Sci. Technol. A* **16**, 1293, 1998.
15. Qi, L., and Sinnott, S.B., *Surf. Sci.* **398**, 195, 1998.
16. Loudon, C.M., *Organic Chemistry*, 1984, ch. 4, pp.91-136.
17. Allen, M. P. and Tildesley, D. J., *Computer Simulation of Liquids*, 1987, ch.3, pp 82-83.
18. Brenner, D. W., *Phys. Rev. B* **42**, 9458, 1990.
19. Brenner, D. W., Harrison, J. A., Shenderova, O.A., and Sinnott, S. B. (unpublished)
20. Harrison, J. A., White, C. T., Colton, R. J., and Brenner, D. W., *Mat. Res. Soc. Bull.* **18**, 50, 1993.
21. Taylor, R. S., and Garrison, B. J., *J. Am. Chem. Soc.* **116**, 4465, 1994.
22. Williams, E.R., Jones, G.C., Jr., Fang, L., Zare, R.N., Garrison, B.J., and Brenner, D.W., *J. Am. Chem. Soc.* **114**, 3207, 1992.
23. Sinott, S.B., Shenderova, O.A., White, C. T., and Brenner, D.W., *Carbon* **36**, 1, 1998.
24. Q., L., Young, W.L., and Sinnott, S.B., *Surf. Sci.* (submitted).

Collisions of Slow Protons with Neutral and Charged Closed-Shell Metal Clusters

M. F. Politis[1], P. A. Hervieux[2], J. Hanssen[2], M. E. Madjet[2] and F. Martín[3]

(1) G.P.S., Université de Paris VI, 2 place Jussieu, 75251 Paris, France
(2) L.P.M.C., Institut de Physique, Technopôle 2000, 57078 Metz, France
(3) Departamento de Química, C-9, Universidad Autónoma de Madrid, 28049 Madrid, Spain

We present a comparative study of $H^+ + Na_{20}$ and $H^+ + Na_{21}^+$ collisions at low velocities. We have found that excitation cross sections are comparable in both cases, while, as expected, capture cross sections are substantially smaller for $H^+ + Na_{21}^+$ than for $H^+ + Na_{20}$. From the calculated energy deposits, we conclude that cluster fragmentation produced after the collision is significantly different in both cases.

INTRODUCTION

Cluster research is a rapidly growing field in which many branches of physics and chemistry are involved (1). In particular, many recent investigations focus on dynamical aspects that take place in collisions with atoms, molecules or surfaces (2). This has been spurred in part by valuable technological developments which have made possible the production of simple mass-selected clusters that were experimentally inaccessible a few years ago. In this respect, very recent experimental works (3) have shown that collisions of metal clusters with highly charged ions is an efficient way to produce multiply charged clusters. There are several mechanisms that may explain the formation of ionic clusters by ion impact : single and multiple electron capture, single and multiple ionisation, capture-ionisation, capture-excitation, etc. While these processes have been extensively studied in ion-atom collisions, very little is known about them in ion-cluster collisions.

In a previous work (4), we have studied the collision of slow protons with Na_{20} clusters using a many-electron quantum dynamical theory. In that work we provided neutralisation and excitation cross sections that could be determined experimentally. However, as production and selection of neutral clusters is rather complicated, most experiments are performed with positively-charged species. Therefore, it seems to us appropriate to perform the same kind of analysis as in Ref. (4) for the isoelectronic closed-shell cluster Na_{21}^+. This ionised cluster can be formed experimentally by photoionisation of the neutral precursor Na_{20} (5). In this paper we will present a comparative study of the two systems, Na_{20} and Na_{21}^+, and we will investigate the role played by the cluster charge in the neutralisation process. We will also provide quantitative capture and excitation cross sections that, we hope, will stimulate future experiments.

THEORETICAL MODEL

As in Ref. (4), we consider the impact velocity range $v_{coll} \sim 0.04 - 0.14$ a.u. (40-500 eV). These values of v_{col} are much smaller than the Fermi velocity of cluster electrons ($v_F \sim 0.6$ a.u. for Na_{20} and Na_{21}^+ (6)) . Hence, as in ion-surface collisions (see (7) and references therein), the electrons can be described as moving in both the effective potential representing the cluster and the projectile potential (this is also the case for the *molecular* method in ion-atom collisions). Closed-shell Na clusters are accurately described by the spherical jellium model (1), which consist in replacing the real ionic core potential by a constant positive background of radius R_C ($R_C = 10.5$ a.u. for Na_{20} and Na_{21}^+). The collision time is roughly $\tau_{col} \sim 2R_C/v_{col}$, which is much shorter than the time required for the interaction between sodium atoms (8). Also, excited clusters resulting from the collision relax with a lifetime $\tau_{rel} \sim 10^{-13}$-10^{-12} s $>> \tau_{col}$ (9), so that dissociation processes resulting from energy relaxation can be ignored during the collision. Consequently, the ionic background of the cluster can be considered as frozen during the collision (10).

In this context, we apply the Kohn-Sham formulation of density functional theory and describe the cluster electron density in terms of single-particle orbitals. The corresponding one-electron potentials, V_C, have been obtained using a local-density approximation with exchange, correlation and a self-interaction correction (LDAXC-SIC, see Ref. (6) for details). The latter correction ensures the correct asymptotic behaviour of the potential, $-(1+q)/r$, where q is the cluster charge. This is crucial in the present study because capture and excitation processes occur mainly at large distances. An important consequence of the quasiseparability of the cluster Hamiltonian is that the total N-electron Hamiltonian $\hat{H}$ can be written as a sum of one-electron effective Hamiltonians,

CP475, *Applications of Accelerators in Research and Industry*,
edited by J. L. Duggan and I. L. Morgan

$$\hat{H} = \sum_{i=1}^{N} \hat{h}(i),$$

with

$$\hat{h} = -\frac{1}{2}\nabla^2 + V_P(|\mathbf{r} - \mathbf{R}|) + V_C(r) \; , \qquad (1)$$

where V_P is the proton Coulomb potential, $V_P = -1/|\mathbf{r} - \mathbf{R}|$, and V_C is the cluster potential. Thus, the *N*-body dynamical treatment reduces to the study of N single-particle problems. This is similar to the *independent electron model* (IEM) of atomic collisions.

We treat the collision in the framework of the impact parameter method, i.e. the projectile follows a straight line trajectory whereas the electrons are described quantum mechanically. In the IEM, each electron is described by a spin-orbital $\psi_i(\mathbf{r},t)$ which satisfies the time-dependent Schrödinger equation

$$\hat{h}\psi_i(\mathbf{r},t) = i\frac{d}{dt}\psi_i(\mathbf{r},t) \qquad i=1,...,N \; , \qquad (2)$$

subject to the initial condition

$$\lim_{t\to-\infty} \psi_i(\mathbf{r},t) = \phi_i(\mathbf{r})\, exp[-i\varepsilon_i t], \qquad (3)$$

where $\phi_i(\mathbf{r})$ is the initially occupied cluster orbital of energy ε_i. The set of N Schrödinger equations (2) has been solved, using the codes of Ref. (11), by expanding the one-electron wave functions in a basis of Born-Oppenheimer (BO) *molecular* states $\{\chi_k(r,R)\}$. This BO states have been obtained by diagonalizing $\hat{h}$ in a *two-centre atomic* basis built from spherical gaussian-type orbitals (GTO) (see Ref. (4)).

The transition probability to a specific final configuration ($f_1, ...f_N$) = $\|\phi_{f_1}\cdots\phi_{f_N}\|$ is given by the (N x N) determinant (12) :

$$P_{f_1,\cdots,f_N} = \det(\gamma_{nn'}) \; ; \; n,n'= 1, ...,N \; , \qquad (4)$$

where $\gamma_{nn'}$ is the one-particle density matrix, $\gamma_{nn'} = \langle f_n|\hat{\rho}|f_{n'}\rangle$, and $\hat{\rho}$ is the density operator which accounts for the time-evolution of the spin-orbitals.

Since the number of active electrons N is large, we have evaluated *inclusive* probabilities which are more easily accessible in experiments. Indeed these probabilities are associated with a series of states in which some levels are occupied irrespective of the occupation of the rest. The inclusive probability $P_{f_1,\cdots,f_q}$ of finding q of the N electrons in the subconfiguration (f_1,...,f_q) while the remaining N-q ones occupy any other states is given by an ordered sum over all exclusive probabilities which include that subconfiguration; it is given by the (q x q) determinant (12) :

$$P_{f_1,\cdots,f_q} = \det(\gamma_{nn'}) \; ; \; n,n'= 1, ...,q \; ; \; q < N \; (5).$$

We have also considered the inclusive probability of finding q occupancies and L-q holes. It is denoted $P_{f_1,...,f_q}^{f_{q+1},...,f_L}$, where occupancies are indicated by subscripts and vacancies by superscripts. As shown in Ref. (12), this probability can be written in terms of probabilities related only to occupancies.

RESULTS AND DISCUSSION

We show in Figs. 1 and 2 the diabatic potential energy curves for the σ states of the $(Na_{20}\text{-}H)^+$ and $(Na_{21}\text{-}H)^{++}$ *quasimolecules*. An inspection of Figs. 1 and 2 suggests that the capture reaction must lead to the formation of H(n = 2) almost exclusively (the H(1s) state lies far below in energy and is not shown in the figure). For simplicity, the molecular orbitals are labeled using the separate "atom" (SA) notation. In the SA limit, hydrogen orbitals include a subscript H, whereas cluster orbitals do not. It must be noticed that excited H orbitals (n = 2, 3, ...) are in fact Stark hybrids due to the cluster electric field. For the n=2 orbitals of H, we use the notation :

$$\xi_{1,H} = (2s_H + 2p_{o,H})/\sqrt{2}\, , \xi_{2,H} = (2s_H - 2p_{o,H})/\sqrt{2}\, .$$

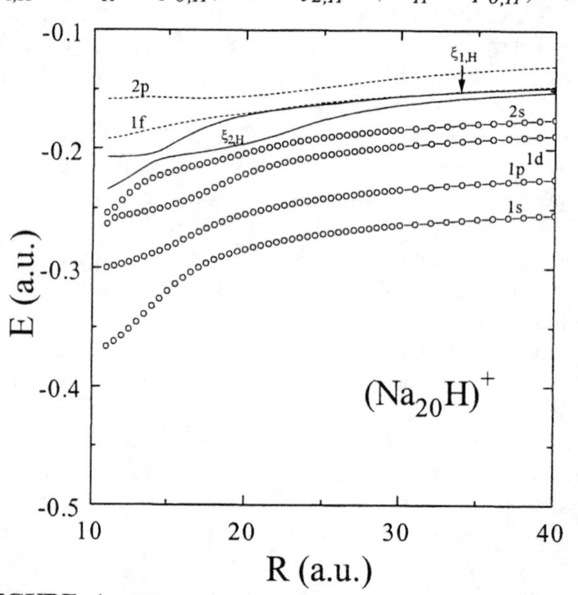

FIGURE 1. $(Na_{20}\text{-}H)^+$ diabatic potential energy curves (σ symmetry). Full lines : H-correlated states (capture channels) ; full lines with circles : occupied cluster states ; dashed lines : empty cluster states.

The *highest occupied molecular orbital* (HOMO) is the 2s orbital. It can be seen that the molecular orbitals denoted 1s, 1p, 1d, and 2s (i.e. those connected to available entrance channels) strongly interact with the $\xi_{2,H}$ orbital. Thus, for both Na_{20} and Na_{21}^+, we have limited the expansion of ψ_i to the following *molecular* states of σ symmetry: the two states dissociating into $\xi_{1,H}$ and $\xi_{2,H}$, the four states dissociating into occupied orbitals of the cluster, namely $1s$, $1p$, $1d$, and $2s$, and the two states dissociating into empty $1f$ and $2p$ orbitals of the cluster. In addition, for Na_{21}^+, we have also included the state dissociating into the $1g$ orbital of the cluster. This amounts to eight states for $Na_{20} + H^+$ and nine states for $Na_{21}^+ + H^+$. These minimal sets of molecular states will allow us to describe the capture reaction as well as cluster excitation.

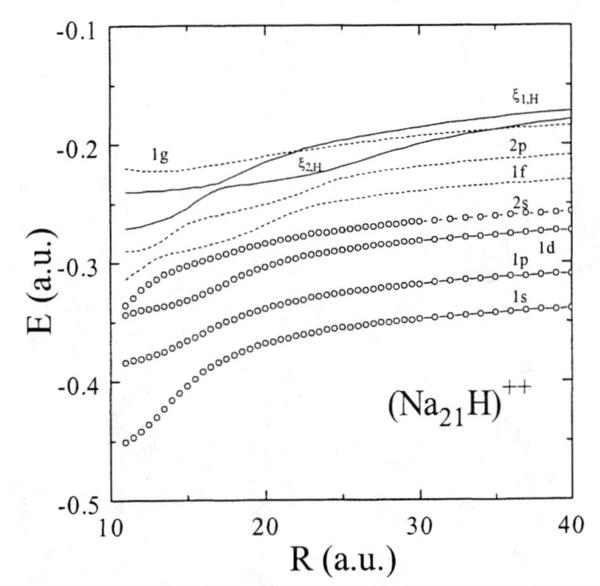

FIGURE 2. $(Na_{21}\text{-}H)^{++}$ diabatic potential energy curves (σ symmetry). Full lines: H-correlated states (capture channels); full lines with circles: occupied cluster states; dashed lines: empty cluster states.

We have evaluated the inclusive probabilities $P_{\xi_{2,H}}$ and $P_{\xi_{1,H}}$. As capture of more than one electron is very unlikely (it leads at most to the formation of H^-, which is a weakly bound anion), $P_{\xi_{1,H}}$ and $P_{\xi_{2,H}}$ can be interpreted as probabilities of finding *one* electron in the projectile. Nevertheless, these probabilities do not correspond to pure single capture reactions, but to a sum of reactions whose common feature is to yield neutral H atoms. They include, in particular, capture-excitation. For Na_{20}, the importance of this many-particle process has been shown and discussed in Ref. (4).

In Fig. 3 we have plotted $bP_{\xi_{2,H}}$ for Na_{20} and Na_{21}^+ as a function of the impact parameter b at several impact energies. It can be clearly seen that the largest contribution to the capture cross section for Na_{20} comes from the region of impact parameters $b \sim$ 20-30 a.u. while, for Na_{21}^+, the capture occurs at smaller distances.

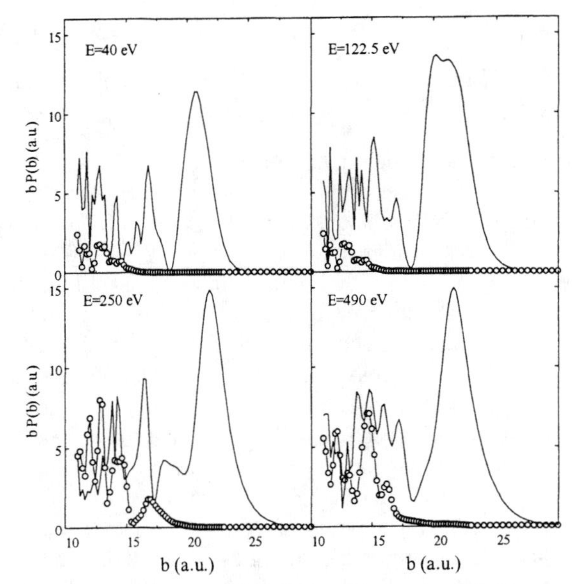

FIGURE 3. Comparison between $bP_{\xi_{2,H}}$ for Na_{20} (full line) and for Na_{21}^+ (line with open circles) as a function of b.

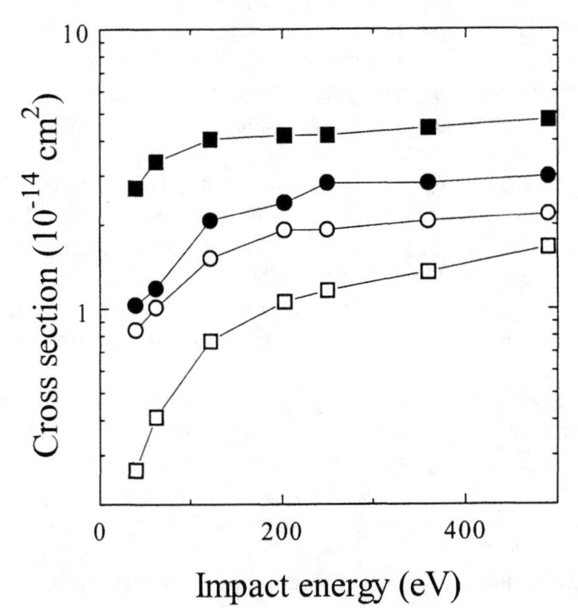

FIGURE 4. Capture and total excitation cross sections for $H^+ + Na_{20}$ and $H^+ + Na_{21}^+$ collisions as functions of impact energy. Circles: excitation; squares: capture; solid symbols: Na_{20}; open symbols: Na_{21}^+.

We define now the *total capture* probability $\hat{P}_{n=2,H} = 2(P_{\xi_{1,H}} + P_{\xi_{2,H}})$. Addition of $P_{\xi_{1,H}}$ and $P_{\xi_{2,H}}$ does not lead to overcounting because the only configurations included in both inclusive probabilities are those with two or more electrons on the projectile which, as discussed above, barely contribute to the total probability. The factor of two appears because the captured electron can have either α or β spin components. Similarly, we define the *excitation* probabilities $\hat{P}_i = 2P_i - (P_i)^2$, where P_i stands for either P_{1f} and P_{2p} for Na_{20} or P_{1f}, P_{2p} and P_{1g} for Na_{21}^+. The total excitation probability is approximately given by $\hat{P}_{exc} = \sum_i \hat{P}_i$ (see Ref. (13)). The corresponding cross sections are shown in Fig. 4.

The capture cross sections are much larger than that observed in H^+-Na collisions (1.5 x 10^{-15} cm^2) (14). For Na_{20} the cross section increases rapidly up to $E \sim 200$ eV and then remains almost flat. This behaviour is qualitatively similar to that observed in H^+-Na collisions, but the flat region is reached much earlier in the present case. Concerning Na_{21}^+, the cross section increases monotonically up to $E \sim 500$ eV. The present results do not allow us to infer the existence of a plateau at higher impact energies. It is worth noticing that for Na_{20}, the excitation cross section is only 2-3 times smaller than the capture cross sections.

The correlation diagram of Na_{21}^+ shows that the main mechanism at low impact energies is the transition from the HOMO to the lowest empty orbital 1*f*. This transition occurs at distances below 20 a.u.. Capture channels are populated through a series of avoided crossings from the orbital 1*f*. Therefore, capture is not directly produced from the HOMO but rather by a series of transitions in which the empty cluster orbitals play the role of intermediate states. Consequently, the capture cross sections are found to be smaller than the excitation cross sections. This is further illustrated by the capture probabilities shown in Fig. 3. In contrast, for H^+-Na_{20} collisions, capture is produced by a direct transition from the initial state to the $\xi_{2,H}$ state (see Figs. 1 and 3). For both systems the total excitation cross section present similar behaviours.

Following the procedure explained in (4), we have evaluated the energy deposit, $E*$, for those clusters which do not undergo capture in the collision. Our results are shown in Fig. 5. Since for the two clusters under consideration, $E*$ is larger than the energy required to evaporate a sodium atom (which is the lowest fragmentation channel) excited clusters will dissociate by ejecting one sodium atom (the dissociation energies are 0.88 eV and 0.93 eV for Na_{20} and Na_{21}^+, respectively). The evaporation times, estimated using the statistical model of Weisskopf (15), are $\tau_{ev} \sim 10^5$ s for Na_{20} and $\tau_{ev} \sim 10^{-1}$ s for Na_{21}^+. As these times are much longer than the typical experimental time of flights measured in the experiments (which are about 10^{-6} s), the decay will not be observed experimentally.

Another point of interest is the fragmentation of the positively charged clusters formed in the collision when capture occurs. It is shown in Ref. (4) that for Na_{20} the captured electrons arise mainly from the 2*s* orbital. Hence, the energy deposited in the remaining Na_{20}^+ ion is much smaller than the dissociation energy (which is 0.85 eV) and, therefore, the Na_{20}^+ cluster will remain stable. In the case of Na_{21}^+ + H^+ collisions, capture electrons also arise form the 2*s* orbital. However the situation is quite different Indeed, the remaining Na_{21}^{2+} (doubly-charged) cluster will break asymmetrically due to the large surface tension of the bulk material and to the mobility of the delocalized electrons. Strong shell effects lead to emission of a charged trimer preferentially. Thus, Na_{21}^{2+} will decay into a large (Na_{18}^+) and a small (Na_3^+) singly-charged fragments. The lifetime of this decay channel is much shorter than the evaporation times discussed above for those clusters that do not contribute to the neutralisation reaction. Consequently, it might be observed experimentally.

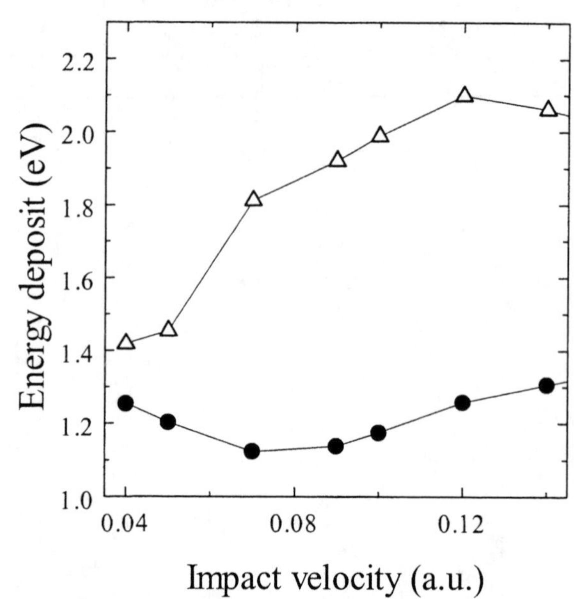

FIGURE 5. Energy deposit in Na_{20}^* (solid circles) and Na_{21}^{+*} (open triangles) as a function of v_{coll}.

CONCLUSIONS

We have evaluated capture and excitation cross sections in H^+ + Na_{20} and H^+ + Na_{21}^+ collisions in the framework of a many-electron quantum dynamical model. As expected, neutralization cross sections are much smaller for the latter than for the former, but, in contrast, excitation cross sections are comparable. From our calculated energy deposits, we have been able to predict fragmentation patterns after the collision. Thus we have concluded that the fragments produced after the neutralization reaction are quite different in both cases. In the near future, we plan to use the present methodology to investigate collisions involving more complicated clusters both neutral and positively charged. Furthermore, the success of the IEM in describing multi-electron capture in ion-atom collisions suggests that the present model is also well adapted to study neutralization of highly-charged ions by metal clusters. Works in these directions are already in progress.

1. Brack M., *Rev. Mod. Phys.* **65**, 677 (1993) ; Heer W. A. de, *ibid.* **65**, 611 (1993).
2. Gotts N. G. and Stace A. J., *Phys. Rev. Lett.* **66**, 21 (1991) ; Goerke A. et al, *J. Chem. Phys.* **98**, 9635 (1993) ; Bréchignac C. et al, *Phys. Rev. Lett.* **72**, 1636 (1994) ; Walch B. et al, *Phys. Rev. Lett.* **72**, 1439

(1994) ; Guissani M. and Sidis V., *Z. Phys. D* **40**, 221 (1997).
3. Chandezon F. et al, *Phys. Rev. Lett.* **74**, 3784 (1995).
4. Politis M.F., Hervieux P. A., Hanssen J., Madjet M. E. and Martín F., *Phys. Rev. A* **58**, 367 (1998).
5. Bréchignac C., Cahuzac Ph., Leygnier J. and Weiner J., *J. Chem. Phys.* **90**, 1492 (1989).
6. Madjet M., Guet C. and Johnson W.R., *Phys. Rev. A* **51**, 1327 (1995).
7. Borisov A. G. et al, *Phys. Rev. B* **54**, 17166 (1996).
8. Ashcroft N. W. and Mermin N. D., *Solid State Physics*, Saunders W. B. Co., Philadelphia 1976.
9. Bréchignac C. et al., *Chem. Phys. Lett.* **189**, 28 (1992).
10. Although cluster fragmentation might be produced in frontal collisions, it barely affects our description because neutralization and excitation occur at long distances.
11. Salin A., *Comp. Phys. Commun.* **62**, 58 (1991).
12. Lüdde H. J. and Dreizler R. M., *J. Phys. B* **18**, 107 (1985).
13. Strictly speaking, in this definition we have not excluded possible overcounting arising from multiple excited configurations. However, we expect that the error introduced in this way would barely affect our conclusions. In particular, it would have a negligible effect in our energy deposit calculations.
14. Allan R. J., *J. Phys. B* **19**, 321, (1986).
15. Hervieux P. A. and Gross D. H. E., *Z Phys. D* **33**, 295 (1995).

Electronic Sputtering from an SiO_2 Target Bombarded by Heavy Ions

N. Imanishi, A. Shimizu, H. Ohta, and A. Itoh

Department of Nuclear Engineering, Kyoto University, Sakyo, Kyoto 606-8501, Japan.

Yields and emission energy distributions of secondary ions were measured for Si and SiO_2 targets bombarded by Si and Cu ions over an energy range between 1 and 5 MeV, where the atomic-collision process changes from a nuclear to an electronic one. Singly charged cluster ions as well as multiply charged monoatomic ions were observed in the case of the SiO_2 target. Dominant species of the cluster ions were $Si(SiO_2)_x^+$, $SiO(SiO_2)_x^+$, and $SiO_2(SiO_2)_x^+$ (x is an integer). The results obtained for the emission energy distribution are summarized as follows: The energy distribution of Si^+ emitted from the SiO_2 target is very narrow compared with that for the Si target; the emission energy is much higher for Si^{q+} (q=1 - 3) than for the cluster ions; the energy distribution of the cluster ions does not depend on the cluster size nor the projectile energy.

INTRODUCTION

Secondary ion emission in an MeV-energy region where the atomic-collision process changes from a nuclear to an electronic one has been studied especially for frozen gases and organic molecules (1 and references cited therein). The observed secondary-cluster-ion yield shows a nonlinear dependence on the electronic stopping power. The nonlinear dependence is partially interpreted by models including mechanisms such as thermal spike, Coulomb explosion, ion track, and pressure pulse (2-6). A pressure-pulse model based on a molecular-dynamics calculation has gained success especially for the desorption of large organic molecules induced by fast heavy ions (5). Emission velocity and kinetic energy distributions have also been measured for frozen gases and organic molecules, and have given insight into some aspects of the formation mechanism of the cluster ions (5, 7, 8).

Studies for tightly bound chemical compounds are scarce, however, except for alkali halides. Therefore, we have focused on the secondary ion yield dependence on the electronic stopping power for an SiO_2 target that has attracted much interest because of its technological importance to silicon devices. In our previous experiments, we observed positively charged cluster ions up to a mass-to-charge ratio of 850 u/e and their yields were characterized by a power function of the electronic stopping power, of which exponent varies with cluster size (9, 10).

The aim of the present report is to extend the measurement to emission velocity and energy distributions of individual secondary ions produced from a silicon-oxide target bombarded by Si and Cu ions at impact energies of 1 to 5 MeV. A silicon target was used as a reference. Observed data are compared with the results for some organic compounds. Thus, any role of the electronic collision will be revealed for the tightly bound chemical compounds.

EXPERIMENTAL

Measurements of the yields and emission energies of secondary cluster ions were carried out using a conventional time of flight (TOF) technique. An Si or a Cu ion beam from the Kyoto University 1.7-MV tandem Cockcroft-Walton accelerator was collimated to a spot of 2 mm in diameter and incident on a target at an angle of 70° with respect to the surface normal (9,10). The targets prepared were a silicon wafer and a 400-nm thick SiO_2 layer epitaxially grown on another silicon wafer. The direct current beam was chopped every 100 μs to a width of 20 ns. The resulting secondary atomic and cluster ions sputtered from the target were extracted to a direction of 90° with respect to the beam axis by two 30-mm-diam grid-type accelerating electrodes set at distances of 10 mm and 15 mm from the beam spot on the target. Then, the ions were focused with an einzel lens and steered with a parallel plate deflector onto a 10-mm-diam entrance slit of a channel electron multiplier (Ceratron) (11). When voltages of 1000 V, 900 V, and 0 V were, respectively, applied to the target and the two accelerating electrodes, sputtered ions with emission energies up to 35 eV could be guided to the detector irrespective of their emission angles. The flight path length was about 482 mm and the time differences between the start signal of the incident beam and stop signals of secondary ion detection were analyzed by a multichannel time-to-amplitude converter. Mass spectra of secondary ions were measured

CP475, *Applications of Accelerators in Research and Industry*,
edited by J. L. Duggan and I. L. Morgan

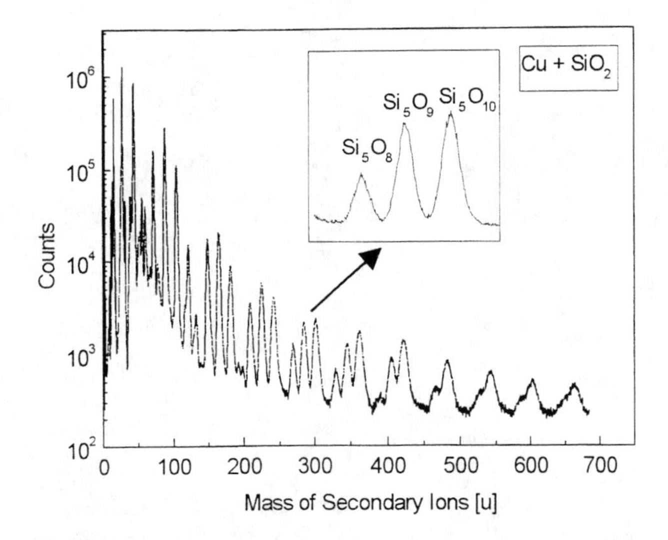

FIGURE 1. An example of the obtained mass spectra of secondary ions for the SiO_2 target bombarded by 3-MeV Cu ions. Dominant species found were $Si_nO_m^+$ species with m being equal to $2n$, $2n$-1, and $2n$-2 (see the insertion).

for Si and Cu projectiles in the energy region of 1 - 5 MeV. The vacuum chamber was baked beforehand at about 400 K for 2 days. The pressures of 10^{-6} Pa and 2×10^{-6} Pa were achieved for the isolated and the experimental conditions, respectively. The front surface of each target was purified by bombarding intensely with Si or Cu ions for 300 s before each 1000 s interval of mass spectra acquisition. The beam intensities were monitored before and after each run. In the case of the SiO_2 target, the ranges of the incident ions were longer than the thickness of the SiO_2 layer. Thus, the incident ions passed through the layer and stopped in the electrically conductive Si wafer. This successfully prevented the target from accumulating electrical charge (12). That is, the intensity of each secondary ion was constant at normal extracting voltages of 0 to 100 V which were applied between the target and the first electrode, but decreased drastically by applying a reverse voltage no more than 1 V. The obtained mass spectra were reproducible and hardly depended on the charges of the incident ions.

RESULTS AND DISCUSSION

Secondary Ion Yields

A typical example of the TOF mass spectra of secondary-ion species is shown in Fig.1. The following features previously found by mass spectrometry (9, 10, 12) were reproduced by the present TOF method; dominant species found were $Si_nO_m^+$ species with m being equal to $2n$, $2n$-1, and $2n$-2, but oxygen rich species of $m\geqq 2n+1$ were not observed. The intensity of the cluster ions inversely depends on the size of the clusters. The yields of secondary ions

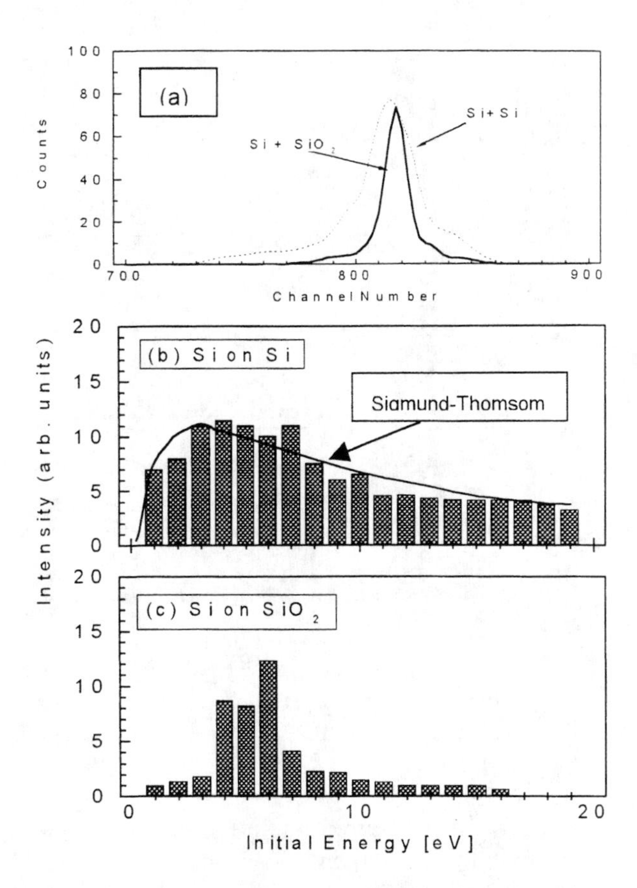

FIGURE 2. Emission energy distributions of secondary Si^+ ions sputtered from the Si and SiO_2 targets bombarded by 3 MeV Si ions. (a) Mass spectra, (b) energy distribution for the Si target, and (c) that for the SiO_2 target. The solid line shown in (b) represents the theoretical distribution deduced from the Sigmund-Thompson linear cascade process.

steeply increase with increasing incident energy in the present energy region, where the electronic stopping power is dominant and increases with increasing incident energy and vice versa for the nuclear stopping power (13). The steep increase of the yields therefore reflects the increasing tendency of the electronic stopping powers of the Si-SiO_2 and Cu-SiO_2 projectile-target systems. The intensity of the large cluster ions increases more steeply with increasing electronic stopping power than that of the small clusters.

Emission Energies of Secondary Ions

Figure 2 compares the emission energies of secondary Si^+ ions sputtered from the Si and SiO_2 targets bombarded by 3-MeV Si ions. In deducing the emission energy distribution from the mass spectra, it was carefully taken into account that the accelerating electrode was not parallel to the target surface and that ions emitted normal to the target surface were deflected by the nonuniform electric field. In the case of the present experimental conditions, an apparent broadening of approximately 2 eV is induced in the emission energy distributions. As shown in Figs. 2-b and 2-c, a great

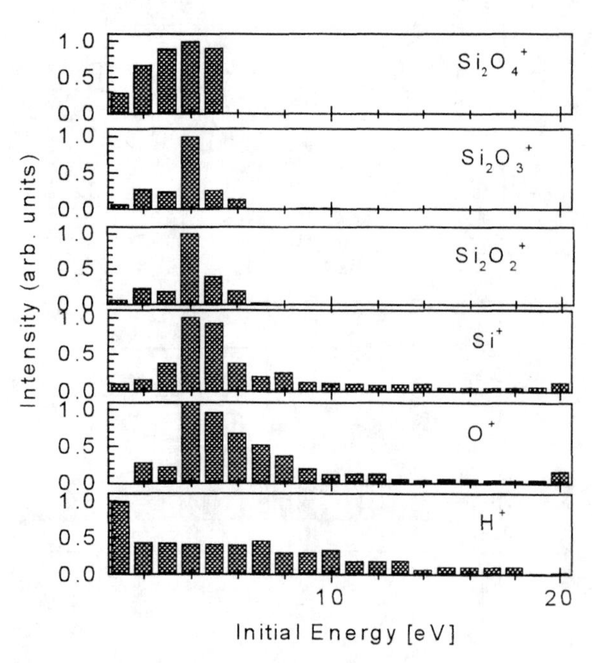

FIGURE 3. Emission energy distributions of secondary atomic and cluster ions sputtered from the SiO_2 targets bombarded by 3-MeV Cu ions.

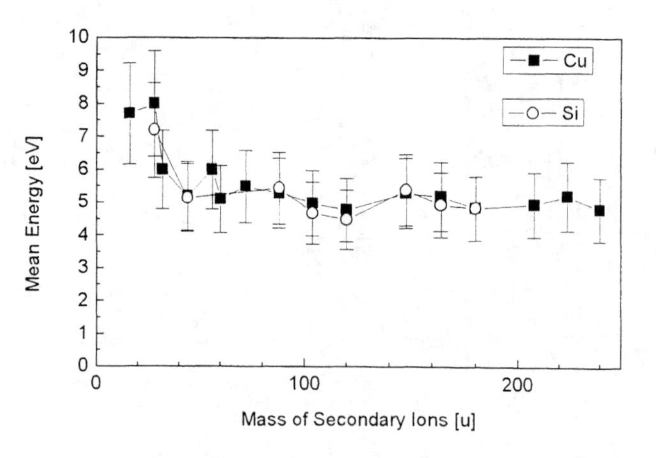

FIGURE 4. Mean energies of secondary ions sputtered from the SiO_2 target bombarded by 5-MeV Si and 5-MeV Cu projectiles.

difference was found between the two energy distributions deduced from the mass spectra shown in Fig. 2-a for the Si and SiO_2 targets. In the case of a conductive target, the electronic sputtering is not expected to occur, and it is generally accepted that the sputtering should proceed through the process of the nuclear collision cascade. The Sigmund-Thompson distribution (14, 15), well accepted for nuclear sputtering, is successfully applied to reproduce the measured energy distribution of the Si-Si system, as shown in Fig. 2-b. Conversely, the energy distribution for the SiO_2 case is very narrow compared with that for the Si target. It is rather difficult to apply the Sigmund-Thompson linear cascade theory to the result of the Si-SiO_2 system. In our previous experiment, the incident energy dependence of the atomic ion yield was found to be different for the Si and SiO_2 targets; the former depends on the nuclear energy depositions S_n in a form of $S_n^{0.78}$, but the latter depends on both the nuclear and electronic energy deposition S_n and S_e as $S_n^{0.78}S_e$ (12). It was concluded then that in the case of the SiO_2 target, a large fraction of the charged secondary atomic ions are produced through the simultaneous process of ionization and recoil caused by single projectile/target-atom collisions in the regime dominated by electronic collisions. The fact that the energy distribution of the Si-SiO_2 system is different from that of the Si-Si system possibly reflects the feature of the single collision. The obtained mean energies of the Si^+, Si^{2+}, and Si^{3+} are 7, 15, and 15 eV, respectively and are much higher than the values of about 5 eV obtained for other cluster ions which will be discussed later. The high average energies reflect the large number of recoiling secondary atoms produced by the incident particles.

The emission energy distributions including cluster ions are shown in Fig. 3 for the SiO_2 target bombarded by the 3-MeV Cu ions. The obtained distributions for the cluster ions are very narrow compared with those of the atomic ions. In the latter cases the distribution is narrower than the shapes expected from the nuclear cascade theory but they still have tails extending to higher energies. The tails probably reflect that the contribution from the linear cascade is still present even in the high-energy region.

The mean energies are compared in Fig. 4 for a wide range of emitted ions for Si and Cu projectiles. Except for the atomic ions of O^+ and Si^+, the mean energies are approximately 5 eV for all of the cluster ions, although the uncertainties in these are about ±20%.

Mean Radial Velocity

Mean radial velocities defined with respect to the target normal were measured by a generally accepted procedure described in Refs. (8,16,17). In the analysis of the data it was carefully taken into account the nonuniform acceleration electric field. Figure 5 shows the obtained mean radial velocities and the mean radial kinetic energies of the secondary ions emitted by 5-MeV Cu ions. The value of mean radial velocity does not depend on the mass of the secondary ions irrespective of whether they are atomic or cluster ions, and the value is almost zero. This last shows that the secondary ions are emitted symmetrically about the direction normal to the target surface. The obtained mean radial kinetic energy decreases slightly with increasing mass, and the average values are about 0.9 eV. By combining these with the values of the mean kinetic energy, we conclude that a large fraction of the secondary ions are emitted within the narrow angle of approximately 25° from the target normal. These results contrast with the data obtained for organic

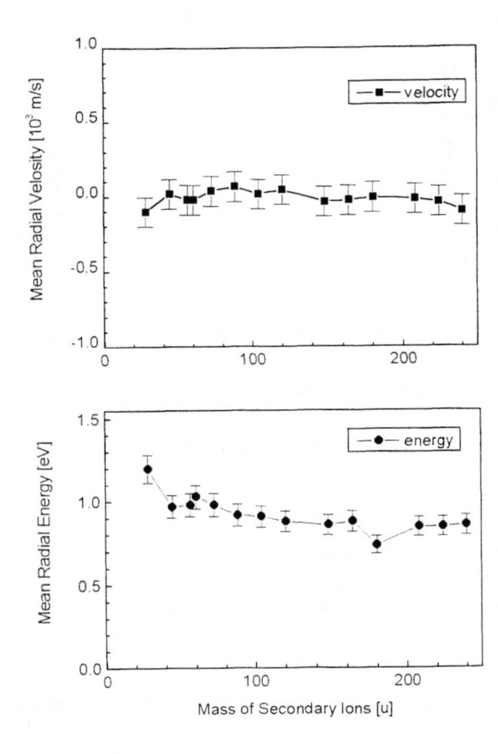

FIGURE 5. The mean radial velocity and kinetic energy are plotted versus the mass of secondary ions.

molecules bombarded by MeV ions. That is, Papaléo et al. observed that the emitted C_nH_m (n and m are integers) have periodically changing mean radial velocities and mean kinetic energies depending on the value of m (8). They claimed that pure carbon clusters C_n^+ are formed in the hot track region and hydrocarbon ions with a higher degree of hydrogenation are formed in an outer track region with lower energy density through the process of recombination reactions of H with C-atoms. The different loci of these chemical events result in the periodic dependence of the radial emission velocity and energy on m. Fenyö et al. found that very large organic ions are emitted in directions far from the target surface normal and that the emission angles are in agreement with the values obtained by the pressure-pulse model (5). Thus, sputtering from targets with high chemical binding energies probably differs from that for weakly bound targets such as organic materials and molecular liquids.

In the case of the cluster ions from the SiO_2 target, the yields depend on the electronic stopping power and it was found that the relative yields of the observed cluster ions depend strongly on the cluster size and are approximately proportional to the power function of Y_0^m (9, 10). Y_0 represents the yield of SiO_2, the main component of the sputtering yield and the exponent m is equal to $P^{0.38}$. The size parameter P is defined as a number of atoms in the cluster divided by 3 and represents that the cluster ions are formed from SiO_2 molecules and SiO_2 constituents. This feature qualitatively supports the sputtering process of the ion track model proposed by Hedin et al. (6). That is, unlike C_nH_m, the cluster ions are not formed by the coagulation of SiO_2 molecules in an outer track region, but are formed in the near-surface region due to the multiple breaking of bonds directly by the incident ions and indirectly by the shower of secondary electrons inside the target. The process should occur irrespective of the direction of the track and result in the emission of the cluster ions in random directions.

CONCLUSIONS

The secondary-ion yields and the emission energies have been measured for Si and SiO_2 targets bombarded by Si and Cu projectiles at energies between 1 and 5 MeV, where the electronic collision becomes more dominant than the nuclear collision in energy deposition. Singly charged cluster ions as well as multiply charged monoatomic ions were observed. Dominant species of the cluster ions were $Si(SiO_2)_x^+$, $SiO(SiO_2)_x^+$, and $SiO_2(SiO_2)_x^+$ (x is an integer). The energy distribution of Si^+ emitted from the SiO_2 target is very narrow compared with that for the Si target and can not be reproduced by the Sigmund-Thompson model. The mean emission energy is much higher for Si^{q+} (q=1 - 3) than for the cluster ions. The mean radial velocity does not depend on the size of clusters and is approximately zero. These observed results show that the atomic ions are produced through the simultaneous process of recoil and ionization in the Si-SiO_2 system, and that the cluster ions are formed near the track region of an incident particle and forced to move in random directions.

ACKNOWLEDGMENTS

This work was done with the Experimental System for Ion Beam Analysis at Kyoto University. We thank K. Yoshida, K. Norizawa, and M. Imai for their useful advice and technical supports during the experiments. It has been supported in part by a Grant-in-Aid for Scientific Research from the Ministry of Education, Science, Sports, and Culture of Japan.

REFERENCES

1. Sundqvist, B. U. R., ed. Behrisch, R., and Wittmaack, K., *Sputtering by Particle Bombardment III*, Berlin, Heidelberg: Springer, 1991, p.257, and refs. cited therein.
2. Bitensky, I. S., and Parilis, E. S., *Nucl. Instr. Meth.* B **21**, 26 (1987).
3. Johnson, R. E., and Brown, W. L., *Nucl. Instr. Meth.* **209/210** 469 (1983).
4. Johnson, R. E., *Inst. J. Mass Spectrum. Ion Process*, **78**, 357 (1978).

5. Fenyö, D., Sundqvist, B. U. R., Karlsson, B. R., and Johnson, R. E., *Phys. Rev.* B **42**, 1895 (1990).
6. Hedin, A., Håkansson, P., Sundqvist, B. U. R., and Johnson, R. E., *Phys. Rev.* B **31**, 1780 (1985).
7. Ens, W., Sundqvist, B. U. R., Håkansson, P. Hedin, A., and Jonnson, G., *Phys. Rev.* B **39**, 763 (1989).
8. Papaléo, R. M., Brinkmalm, G., Fenyö, D., Eriksson, J., Kammer, H.-F., Demirev, P., Håkansson, P., and Sundqvist, B. U. R., *Nucl. Instr. Meth.* B **91**, 667 (1994).
9. Imanishi, N., Kyoh, S., Takakuwa, K., Umezawa, M., Akahane, Y., Imai, M., and Itoh, A., "Proceedings of the International Conference on Accelerator Application in Research and Industry", Denton, 1996,p. 507.
10. Imanishi, N., Kyoh, S., Shimizu, A., Y., Imai, M., and Itoh, A., *Nucl. Instr. Meth.* B **135**, 424 (1998).
11. Murata Mgf. Co., Ltd. in Japan.
12. Kyoh, S., Takakuwa, K., Sakura, M., Umezawa, M., Itoh, A., and Imanishi, N., *Phys. Rev.* A **51**, 554 (1995).
13. Ziegler, J. F., Handbook of Stopping Cross Section for Energetic Ions in All Elements (Pergamon, New York, 1980); Also Biersack, J. P., and Ziegler, J. F., TRIM code.
14. Sigmund, P., *Sputtering by Particle Bombardment I*, Berlin: Springer, 1981, p. 9.
15. Thompson, M. W., *Philos. Mag.* **18**, 377 (1968).
16. Fenyö, D., Hedin, A., Håkansson, P., and Sundqvist, B. U. R., *Int. J. Mass. Spectrom. Ion Proc.* **100**, 63 (1990).
17. Brinkmalm, G., Demirev, P., Fenyö, D., Håkansson, P., Kopniczky, J., and Sundqvist, B. U. R., *Phys. Rev.* B **47**, 7560 (1993).

Molecular Dynamics Simulations of Organic SIMS with Cu_n (n=1-3) Clusters

J. A. Townes[1], A. K. White[1], K. D. Krantzman[1] and B. J. Garrison[2]

[1]Department of Chemistry and Biochemistry, College of Charleston, Charleston, South Carolina 29424
[2]Department of Chemistry, The Pennsylvania State University, University Park, Pennsylvania 16802

Molecular dynamics simulations have been performed to study the effect of cluster size on the emission yield and damage cross section in organic SIMS. A model system composed of a monolayer of biphenyl molecules on a Cu(001) substrate was bombarded with Cu_n (n=1-3) projectiles at kinetic energies of 0.100 keV per atom. The yield increases with cluster size, but a nonlinear enhancement in yield is not observed. The yield-to-damage ratio, on the other hand, increases with the use of clusters, indicating that clusters have the potential to improve the sensitivity of SIMS.

INTRODUCTION

Secondary ion mass spectrometry (SIMS) experiments, which have measured the secondary ion emission resulting from the keV bombardment of solids with monoatomic and polyatomic ions, have raised some interesting issues about the processes for energy deposition leading to the ejection of particles from the solid (1-5). In many cases, polyatomic projectiles produce a big enhancement in the secondary ion yield compared to monoatomic projectiles. The yield is defined to have a nonlinear dependence on the number of atoms in the primary projectile when the yield from a polyatomic projectile containing n atoms with total energy E is more than n times greater than the yield from a monoatomic projectile with energy E/n (2a,b). Experiments show that the degree of enhancement will depend strongly on the kinetic energy, mass, size and composition of the primary cluster as well as the characteristics of the target and matrix or substrate. The greatest enhancements in yields are with molecular ions and molecular fragments (2,3) and with multi-layer targets rather than monolayer films (2d).

There are potential problems with the use of polyatomic projectiles that may overshadow their advantages. In conjunction with producing a larger emission yield, experiments by Van Stipdonk, et. al. show that polyatomic projectiles may also increase the damage cross section on the surface and produce a greater number of molecular fragments (2d). However, experiments by Appelhans, et al. (1a) and Groenewald, et. al. (1b) show both an increase in yield and a smaller damage cross section with polyatomic projectiles. The ultimate test to whether polyatomic projectiles may improve the sensitivity of SIMS is in how they increase the emission yield of intact molecules in comparison to the total damage cross section.

Therefore, the key issue is whether there is a more efficient way to deposit energy with keV projectiles that results in a greater number of ejected intact molecules while minimizing surface damage. In order to address this issue, we have perfomed molecular dynamics simulations of the bombardment of organic films with atomic and cluster projectiles. From the simulations, the emission yield and the yield-to-damage ratio is calculated as a function of cluster size. Although a nonlinear enhancement is not observed with the cluster projectiles, the yield-to-damage ratio increases with cluster size.

METHODS

The classical method of molecular dynamics simulations is used to study the system of interest and the details of this method are described extensively elsewhere (6). The model system, shown in Figure 1, is composed of a monolayer of twenty biphenyl molecules on a Cu(001) microcrystallite consisting of nine layers of 286 atoms. The positions of the biphenyl molecules are determined by allowing the adsorbates to equilibrate on the surface at 0 K. A symmetrically equivalent impact zone is defined on the surface and a set of 150 impact points is evenly distributed over the impact zone. The emission yield with each projectile is calculated by averaging over a set of trajectories, where each trajectory has a different aiming point on the surface.

Potentials developed by DePristo's MD/CEM approach (7) are used for the Cu-Cu interactions. Brenner's hydrocarbon potential is used for the C-C, C-H and H-H interactions (8). The Brenner potential was not developed to describe the repulsive region in which hard collisions occur, and therefore, a Molière pairwise potential is used in the repulsive region of the potential (9). A linear interpolation scheme used by Taylor and Garrison

CP475, *Applications of Accelerators in Research and Industry*,
edited by J. L. Duggan and I. L. Morgan

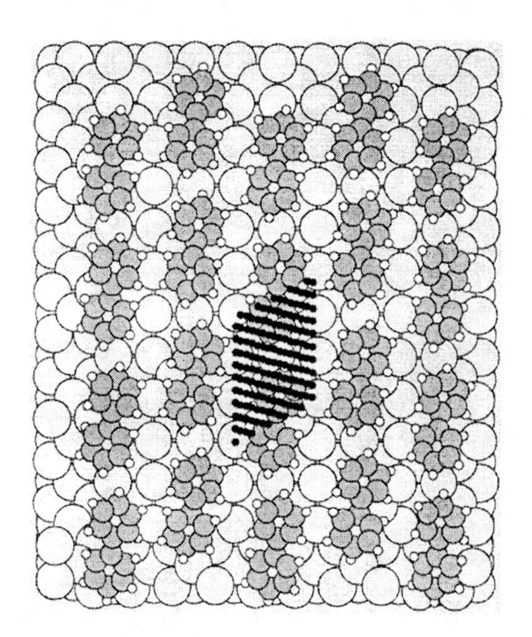

FIGURE 1. Top view of the model system composed of a monolayer of twenty biphenyl molecules on a Cu(001) substrate. The black dots represent the 150 impact points of the bombarding particle.

connects the repulsive and many-body attractive potentials and is described in detail elsewhere (10). A Lennard-Jones potential is used to describe the pairwise Cu-C and Cu-H interactions between the biphenyl molecules and the copper surface, resulting in a binding energy of 2.4 eV (11). A Molière pairwise potential is used in the repulsive region of the potential for these interactions and is connected to the Lennard-Jones potential with a cubic spline polynomial function.

The emission yield is calculated as the number of stable, whole biphenyl molecules ejected from the surface. In order to determine the energy cutoff for stable molecules, simulations were run over a time period of 2 ps for biphenyl molecules with a range of internal energies. From the results of these simulations, it was determined that molecules with internal energies greater than 10 eV are unstable and will fragment before reaching the detector. The total number of damaged molecules is estimated as a sum of the number of ejected, unstable whole molecules and the number of fragmented molecules. From visual inspection of the surface after the bombardment process takes place, the molecules left on the surface intact appear to be undamaged.

RESULTS AND DISCUSSION

The numerical results from the simulations with the Cu_n (n=1-3) projectiles are shown in Table 1. The degree of nonlinear enhancement can be quantified by the enhancement factor. Mathematically, the enhancement factor is defined as $\frac{Y_n(E)}{nY_1(E/n)}$, where $Y_n(E)$ is the yield for the homonuclear cluster at energy E and $Y_1(E/n)$ is the yield for the atomic projectiles at the same velocity (2a,b). An enhancement factor of one indicates that the yield increases linearly with the number of atoms in the cluster. For example, the yield with the Cu_2 cluster at 0.200 keV is simply twice the yield with the Cu atom at 0.100 keV. Table 1 also shows the number of damaged molecules and the ratio of yield to the number of damaged molecules. The yield-to-damage ratio increases with the use of clusters and increases with cluster size.

In Figures 2 and 3, the spatial arrangement of the yield of ejected stable biphenyl molecules and the yield of damaged molecules is shown with the Cu and Cu_3 projectiles, respectively. In these figures, each circle or cone represents one of the twenty biphenyl molecules on the surface. The dark circles represent biphenyl molecules that are not affected by the bombardment. Molecules that are either ejected or damaged are represented by light colored cones. The height of the cone corresponds to the yield of ejected, stable molecules in Fig. 2a and 3a and to the yield of damaged molecules in Fig. 2b and 3b. The impact zone of the incoming projectile encloses the middle two biphenyl molecules.

In Fig. 2a and 2b, the results with the Cu projectile are shown. The Cu projectile leads to the ejection of the two molecules in the impact zone and the immediate neighboring molecules, affecting a total area of 200 $Å^2$. The group of small cones in Fig. 2a represents the number of biphenyl molecules that are ejected intact with internal energies less than 10 eV. In Fig. 2b, the two high cones represent the number of biphenyl molecules in the impact zone that are damaged by the incoming projectile. The incoming projectile damages only a few of the surrounding molecules.

The results with the Cu_3 projectile are shown in Fig. 3a and 3b. Interestingly, the affected area around the impact zone does not increase greatly compared to the area with the Cu projectile. However, a greater proportion of the molecules surrounding the impact zone are ejected intact.

TABLE 1. Yield, Damage, Yield-to-Damage Ratio and Enhancement Factors with Cu_n (n=1-3) clusters

Bombarding Particle	Incident Kinetic Energy (keV)	Yield	Enhancement Factor compared to Cu	Damage	Yield-to-Damage Ratio
Cu	0.100	31	NA	91	0.34
Cu_2	0.200	71	1.1	131	0.54
Cu_3	0.300	97	1.0	164	0.59

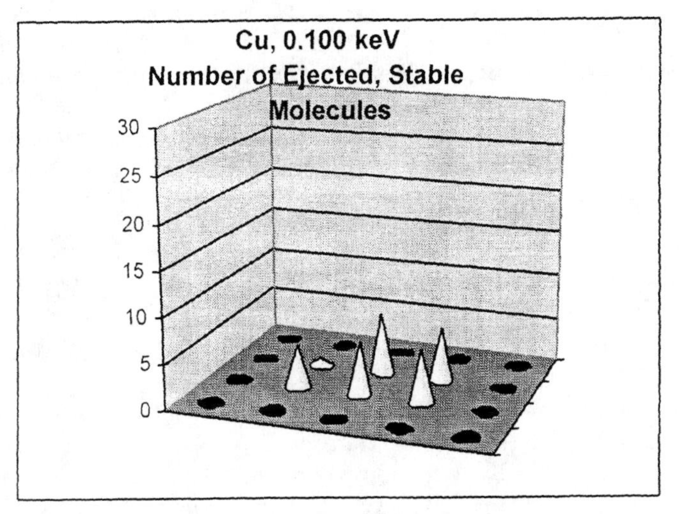

Figure 2a. Spatial arrangement of the yield of ejected stable biphenyl molecules with the Cu projectile.

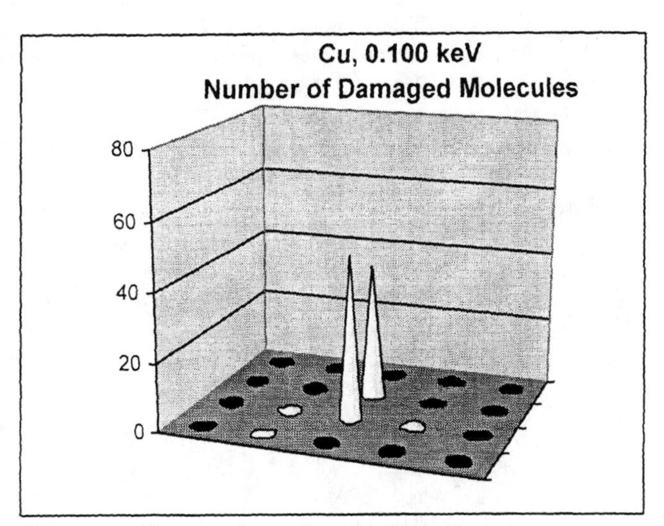

Figure 2b. Spatial arrangement of the yield of damaged biphenyl molecules with the Cu projectile.

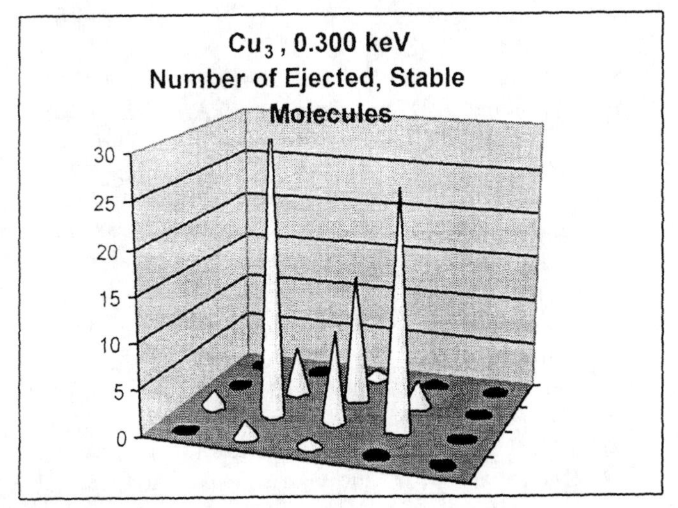

Figure 3a. Spatial arrangement of the yield of ejected stable biphenyl molecules with the Cu_3 projectile.

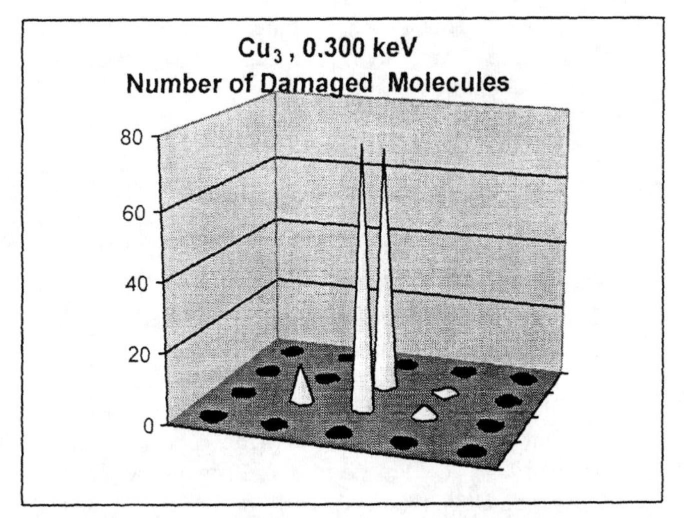

Figure 3b. Spatial arrangement of the yield of damaged biphenyl molecules with the Cu_3 projectile.

In Fig. 3b, the two high cones represent the large number of biphenyl molecules in the impact zone that are damaged. The number of surrounding molecules that are damaged is only a little greater than the number with the Cu projectile. Consequently, the yield-to-damage ratio is larger with the Cu_3 projectile. Previous simulations identified a collaborative mechanism with cluster projectiles (12), in which the cluster atoms initiate adjacent collision cascades that work together to eject the intact molecule from the surface. The collaborative mechanism will lead to a greater proportion of ejected intact molecules, and therefore, to an increase in the yield-to-damage ratio.

CONCLUSIONS

Molecular dynamics simulations of the bombardment of a monolayer of biphenyl molecules on a copper substrate with Cu_n (n=1-3) projectiles have been performed. From the simulations, the yield of ejected, stable molecules and the yield-to-damage ratio has been determined. The greatest contribution to the yield comes from the ejection of molecules immediately surrounding the impact zone. The two molecules in the impact zone are the primary damaged molecules and only a very small amount of the surrounding molecules are damaged.

The yield of ejected, stable molecules increases linearly with cluster size, but a nonlinear enhancement is not observed. However, the yield-to-damage ratio does increase with polyatomic projectiles. With the cluster projectiles, adjacent collision cascades can collaborate to eject the intact molecule from the surface, which leads to an increase in the yield-to-damage ratio.

ACKNOWLEDGMENTS

The financial support of the National Science Foundation, the Petroleum Research Fund and the Research Corporation is gratefully acknowledged. We also thank the use of the computing facilities at the Center for Academic Computing of The Pennsylvania State University and the assistance of Jeff Nucciarone. In addition, we thank Nick Winograd, Michael Van Stipdonk and Anthony Appelhans for insightful discussions about this work.

REFERENCES

1. a) Appelhans, A.D., and Delmore, J.E. *Anal. Chem.* **61**, 1087-1093 (1989). b) Groenewold, G.S., Gianotto, A.K., Olson, J.E., Appelhans, A.D., Ingram, J.C., Delmore, J.E., and Shaw, A.D. *Int. J. Mass. Spectrom. Ion Processes* **174**, 129-142 (1998).
2. a) Blain, M.G., Della-Negra, S., Joret, H., LeBeyec, Y., and Schweikert, E.A. *Phys. Rev. Lett.* **63**, 1625-1628 (1989). b) Benguerba, M., Brunelle, A., Della-Negra, S., Depauw, J., Joret, H., Le Beyec, Y., Blain, M.G., Schweikert, E.A., Ben Assayag, G., and Sudraud, P. *Nucl. Instr. and Meth. B* **62**, 8-22 (1991). c) Van Stipdonk, M.J., Harris, R.D., and Schweikert, E.A., *Rapid Commun. Mass Spectrom.*, **10**, 1987-1991 (1996). d) Harris, R.D., Van Stipdonk, M.J. and Schweikert, E.A., *Int. J. Mass. Spectrom. Ion Processes* **174**, 167-177 (1998).
3. a) Boussofiane-Baudin, K., Bolbach, G., Brunelle, A., Della-Negra, S., Hakansson, P., and Le Beyec, Y. *Nucl. Instr. and Meth.B* **88**, 160-163 (1994). b) Le Beyec, *Int. J. Mass. Spectrom. Ion Processes* **174**, 101-117 (1998).
4. Kotter, F., Benninghoven, A. *Appl. Surf. Sci.* **133**, 47-57 (1998).
5. a) Mahoney, J.F., Parilis, E.S. and Lee, T.D. *Nucl. Instr. and Meth. B* **88**, 154-159 (1994).
6. a) Garrison, B.J. and Winograd, N. *Science*, **216,** 805-812 (1982). b) Garrison, B.J., *Chem. Soc. Rev.,* **21**, 155-162 (1992). c) Bernardo, D.N., Bhatia, R., and Garrison, B.J. *Comp. Phys. Comm.* **80**, 259-273 (1994).
7. a) Stave, M.S., Sanders, D.E., Raeker, T.J., and DePristo, A.E. *J. Chem. Phys.* **93**, 4413-4426 (1990). b) Raeker, T.J. and DePristo, A.E. *Int. Rev. Phys. Chem.* **10**, 1-54 (1991)., c) Kelchner, C.L., Halstead, D.M., Perkins, L.S., Wallace, N.M. and DePristo, *Surf. Sci.* **310** 425-435 (1994).
8. a) Brenner, D.W. *Phys. Rev. B* **42**, 9458-9471 (1990). b) Brenner, D.W., Harrison, J.A., White, C.T., Colton, R.J. *Thin Solid Films* **206**, 220-223 (1991).
9. O'Connor, D.J. and MacDonald, R.J., *Radiat. Eff.*, **34**, 247-250 (1997).
10. Taylor, R.S. and Garrison, B.J. *Langmuir* **11**, 1220-1228, (1995).
11. Allen, M.P. and Tildesley, D.J. *Computer Simulations in Liquids*, Great Britain: Oxford University Press, 1987, pp. 9-11.
12. a) Zaric, R., Pearson, B., Krantzman, K.D. and Garrison, B.J., *Int. J. Mass Spectrom. Ion Processes,* **174**, 155-166 (1998). b) Zaric, R., Pearson, B., Krantzman, K.D. and Garrison, B.J. in Lareau, R. and Gillen, G, (Eds.) *Secondary Ion Mass Spectrometry, SIMS XI*, New York: John Wiley and Sons, 1998, pp. 601-604.

Equipment for Processing by Gas Cluster Ion Beams

Allen Kirkpatrick and James Greer

Epion Corporation, 4R Alfred Circle, Bedford, MA 01730

Gas cluster ions represent emerging new technology for atomic scale processing of surfaces. Applications which have been recognized for gas cluster ion beams include ultra-shallow implantation, surface cleaning, smoothing and planarization, micromachining, reactive formation of films, and ultra-shallow sputtering for high-resolution surface analysis. Gas cluster ion beam systems for manufacturing applications will require substantially higher beam currents than are presently available and other features which have not been associated with laboratory equipment constructed to date. This paper reviews the status of existing gas cluster ion beam equipment and discusses anticipated requirements for future high-throughput processors.

INTRODUCTION

Clusters of gas atoms or molecules held weakly together by van der Waals forces can be formed by homogeneous condensation occurring within the flow of a high-pressure gas expanding through a small nozzle into vacuum. The clusters which can consist of from only a few atoms to many thousands of atoms are produced with broad size distributions determined by the source gas pressure and temperature conditions at the nozzle entrance and by fluid-dynamic conditions within the nozzle. The kinetics of the expansion processes and relationships of cluster size distributions to gas parameters and nozzle configurations are relatively well understood[1,2].

Neutral gas clusters can be ionized and then accelerated through a high potential to produce a beam of energetic gas cluster ions. Accelerated cluster ions can have very substantial energies, but because the total energy of a cluster ion is shared by the large number of atoms of which it is comprised, the energies of individual atoms within cluster ions are typically small. Interactions which can occur between cluster atoms and target atoms during disintegration of cluster ions impacting upon solid surfaces have been found to produce unusual and potentially very useful effects. These effects are expected to lead to advances in methods for atomic scale processing of surfaces and in surface-science instrumentation.

As production applications of gas cluster ion beams emerge from laboratory work now in progress, high throughput processor systems are to be needed. For most of the applications which are being considered, a number of challenging issues will need to be resolved. In particular, practical gas cluster ion beam (GCIB) processors will have to deliver cluster ion currents substantially greater than the highest currents which have been achieved to date.

GCIB EQUIPMENT FOR RESEARCH AND DEVELOPMENT

A number of gas cluster ion beam (GCIB) systems in several configurations have been developed at the Ion Beam Engineering Experimental Laboratory of Kyoto University, where most of the work to date on gas cluster ion interactions with surfaces has originated and where numerous investigations of potential applications have been conducted[3,4,5]. GCIB equipment now also exists in other laboratories in Japan, the U.S. and Europe. All GCIB equipment to date has been designed for R&D use. These systems have provided capabilities to produce and deliver cluster ions from various gases over a wide range of cluster sizes, energies and dose conditions. Other processes, such as concurrent thin film deposition, and system refinements, such as cluster mass filtering, have been incorporated into some of the developmental GCIB equipment. A large amount of knowledge has been generated concerning fundamental cluster interaction effects and possible applications.

Working in collaboration with the investigators at Kyoto University toward commercialization of GCIB technology, Epion Corporation has produced a series of GCIB units which are being used for laboratory research. Figure 1 shows a schematic representation of a general purpose system. Figure 2 illustrates one configuration of the GCIB source components in such a system. It should be noted that the gas loads used for cluster ion generation are relatively large, typically 100 sccm or more, and more than one stage of vacuum pumping is generally required. Differential pumping is used to produce adequately low pressures in the electron bombardment ionization and high potential acceleration regions of the ion source. Vacuum pumps represent a major component of the cost of most GCIB equipment.

CP475, *Applications of Accelerators in Research and Industry*,
edited by J. L. Duggan and I. L. Morgan

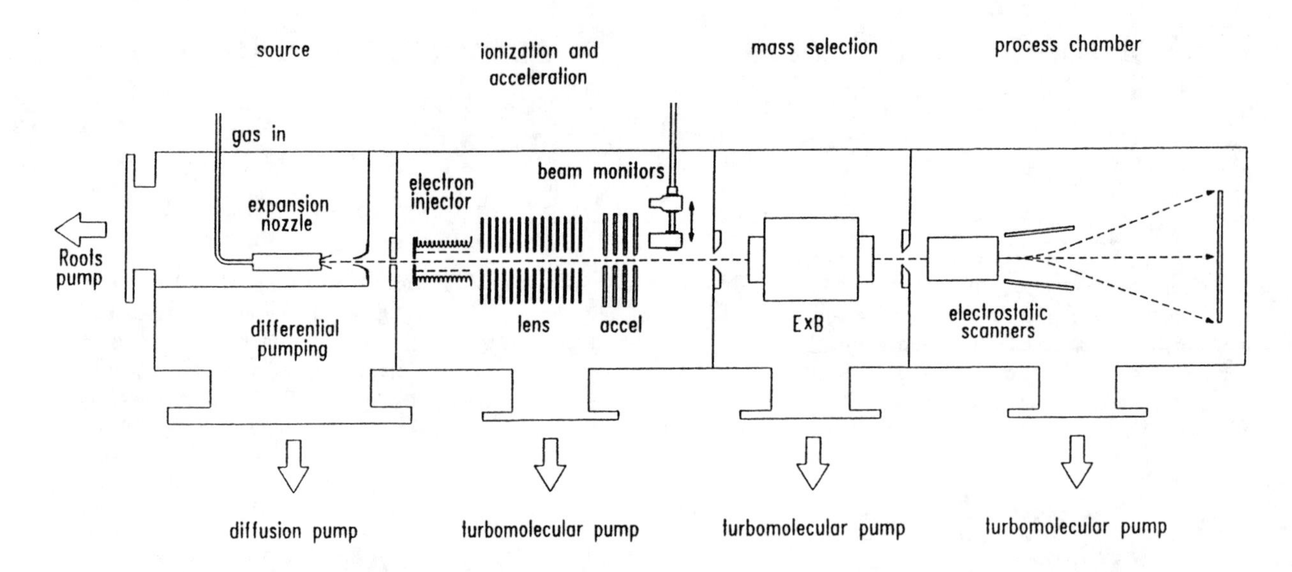

FIGURE 1. Schematic of Laboratory GCIB System

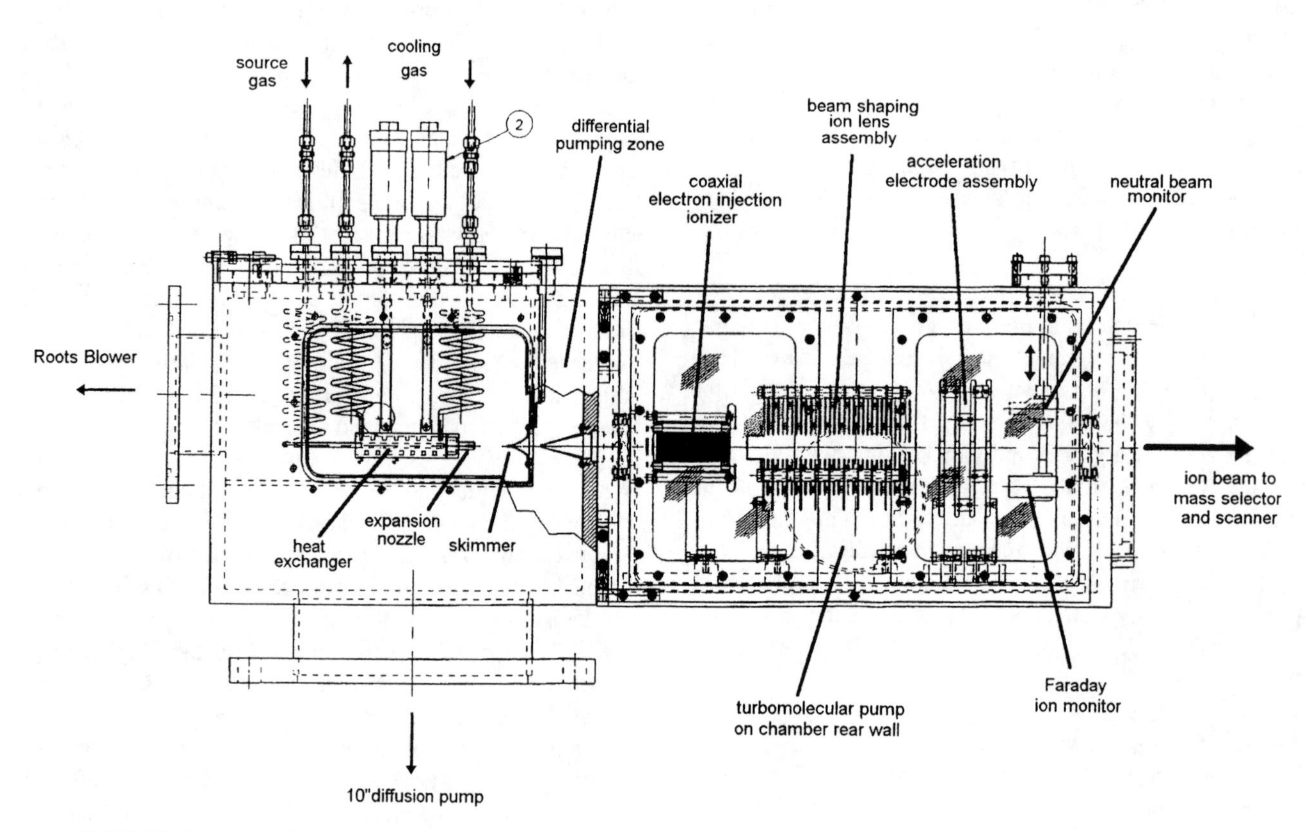

FIGURE 2. Example GCIB Source Configuration

Photographs of two other Epion GCIB units are shown in Figs. 3 and 4. The equipment shown in Figure 3 is a Multi-Beam system which combines two e-beam evaporators and a source able to deliver oxygen cluster ions at up to 30 keV. This system, which has a 20-inch diffusion pump in the main process chamber to facilitate large total gas loads, is being used in Japan for development of low temperature processing for deposition of high-quality transparent conductor oxide films[6]. A 30 keV laboratory system intended for investigations of atomic scale smoothing and for other general research purposes is shown in Fig. 4.

FIGURE 3. Multi-Beam GCIB System

FIGURE 4. Model 3000 Laboratory GCIB System

GCIB EQUIPMENT FOR COMMERCIAL APPLICATIONS

All GCIB equipment to date has been able to provide only a few μamperes of cluster ion current at most. A number of potential applications for GCIB are now under serious consideration. Of these, some, such as ultra-shallow ion implantation or outermost layer sputtering for surface analysis purposes, will require low dose levels ($\leq 10^{12}$ ions/cm^2) or will involve only small process areas. However, most other applications, including surface cleaning, smoothing, polishing, and reactive film deposition, are likely to need dose levels ranging between 10^{14} and 10^{16} ions/cm^2. For these processes to be used in production, it will be necessary to develop methods to deliver cluster ion beams at much higher current levels. Figure 5 shows throughput rate versus beam current for various levels of required process ion dose. For most applications requiring dose levels above 10^{14} ions/cm^2, GCIB currents of hundreds of μamperes will be desirable, if not essential.

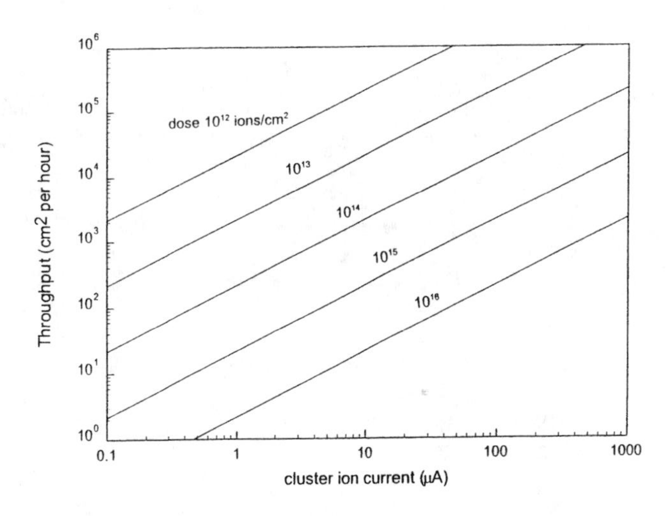

FIGURE 5. Process Area Throughput versus Ion Beam Current as Function of Required Ion Dose

A number of factors are associated with the limitations on maximum ion currents which can be produced in GCIB equipment. Development to obtain appreciably higher usable cluster ion currents involves efforts to increase the efficiency of generation of neutral clusters, to improve the efficiency of cluster ionization, and to reduce ion losses due to space charge effects during transport. Other efforts are being made to engineer improvements to the source configurations, to develop improved approaches to nozzle fabrication, to utilize better combinations of nozzle and skimmer characteristics, and to employ vacuum pump combinations better matched to the total system requirements. More effective methods are also being sought to eliminate from the cluster ion streams the monomer ions and very small cluster ions which are considered to be detrimental in some candidate applications.

SUMMARY

The unique characteristics of the interactions of energetic gas cluster ions with solid target surfaces offer enormous potential for deposition of superior thin films and for shallow surface modification operations. Applications for GCIB are being established from laboratory studies being performed using the commercial GCIB equipment which is now available. Engineering efforts are being conducted to evolve the GCIB equipment which can presently be produced into high throughput processors offering elegant solutions for challenging problems in manufacturing of advanced microdevices.

REFERENCES

1. O.F.Hagena and W.Obert, *J. Chem. Phys.* **56**, 1793 (1972)
2. O.F.Hagena, *Surface Science* **106**, 101 (1981)
3. J.A.Northby, T.Jiang, G.H.Takaoka, I.Yamada, W.L.Brown and M.Sosnowski, *Nuclear Instrum. and Methods in Phys. Res.* **B74**, 336 (1993)
4. I.Yamada, J.Matsuo, Z.Insepov, D.Takeuchi, M.Akizuki and N.Toyoda, *J. Vac. Sci. Technol.* **A14** (3), 1 (1996)
5. I. Yamada and J.Matsuo, "Cluster Ion Beam Processing," *Materials Science in Semiconductor Processing,* **1**, pp27-41 (1998)
6. W.Qin, R.P.Howson, M.Akizuki, J.Matsuo, G.Takaoka and I.Yamada, *Material Chem. & Phys.*, **54**, 258 (1998)

Optical Thin Film Formation by Gas-cluster Ion Beam Assisted Deposition

H. Katsumata[1], J. Matsuo[1], T. Nishihara[1], T. Tachibana[2], K. Yamada[2], M. Adachi[2], E. Minami[1], and I. Yamada[1]

[1] *Ion Beam Engineering Experimental Laboratory, Kyoto University, Sakyo, Kyoto 606-8501, Japan*
[2] *Adachi New Industrial Co. Ltd., ANICS Bldg. 10F, 1-14-20, Itachibori, Nishi-ku, Osaka 550-0012, Japan*

We have developed a gas cluster ion beam assisted deposition system for high-quality optical thin film formation (SiO_2 and TiO_2 *etc.*) with high packing density. Cluster ions can transport thousands of atoms per ion with very low energy per constituent atoms. Consequently, densification of films, which is commonly required for optical coatings, can be achieved without the introduction of increased surface roughness and irradiation-induced defects, which are critical issues for conventional ion assisted deposition processes. In this work maximizing the intensity of gas-cluster ion beam current is discussed based upon a few experiments increasing the neutral cluster beam intensity and designing an ionizer for achieving an efficient transportation of the cluster ion beam. As a result, we successfully obtained a high intensity gas-cluster ion current up to ~30 μA, which is one order of magnitude larger than that obtained so far. TiO_2 films were grown on Si substrates by electron beam evaporation of TiO_2 at ambient temperature under O_2-cluster ion bombardment with acceleration energies (V_{acc}) up to 12 keV. Refractive index, n of the films was increased steeply to n=~2.30 above V_{acc}=4 keV. Water-soaking tests for 12 hrs of the samples revealed that an increase in n values due to moisture absorption becomes smaller with increasing V_{acc}, which suggests that the films become more dense with increasing V_{acc} from optical point of view.

INTRODUCTION

Ion beam assisted deposition (IBAD) has been utilized for the formation of optical thin films aiming mainly to densify them because, as formed by an electron beam evaporation, they exhibit relatively porous and loosely packed structures [1,2]. Such structures have adversely unstable properties with time, which arise from moisture absorption. Optical thin films produced using IBAD are likely to suffer from surface degradation due to ion bombardments. Moreover, in oxide coatings, preferential sputtering of oxygen from growing films, which occurs at higher ion energies over several hundreds of eV, may cause a degradation in stoichiometry [1]. Cluster ions can transport thousands of atoms per ion with very low energy per constituent atoms and can deposit their energy in high density within a very localized region, which not only enhances the chemical reaction between clusters and surface adatoms in the very near surface region [3] but also provides an equivalently low energy irradiation effect [4,5]. Furthermore, clusters bring about a surface smoothing phenomenon due to lateral sputtering effects [6]. It has been demonstrated that even the neutral O_2-clusters without acceleration has higher chemical reactivity than oxygen molecules, and exhibits a surface smoothing effect [7]. These properties, intrinsic to cluster ions, can be useful for the low temperature formation of high quality thin films without introducing irradiation-induced defects. Indeed, high quality In_2O_3 films with low resistivity of $5x10^{-4}$ Ω·cm and high transparency of 80 % to visible light have been formed at ambient temperature by O_2-cluster ion assisted deposition process [8]. In such a process, the gas-cluster ion current density used was 300 nA/cm^2 and the growth area was 3 cm^2, so that the total cluster ion current was 900 nA. However, when this process is applied to the practical coating processes which need both high growth rate and large area fabrication, *e.g.*, 100 cm^2, higher intensity gas-cluster ion current of ~30 μA is required.

In this work, maximizing the intensity of gas-cluster ion beam current was examined by two different approaches. One was performed to increase the neutral cluster beam intensity. For this purpose, a new gas-cluster source was developed. Furthermore, two types of skimmer with different orifice diameters were examined. The other approach was to design an ionizer which can transport cluster ions efficiently to the substrates. As a consequence, the cluster ion beam current of ~30 μA was obtained. With the developed system, TiO_2 films were grown on Si substrates at ambient temperature, and their refractive index, n was measured as a function of the acceleration energy (V_{acc}) of the O_2-cluster ions before and after water-soaking tests.

GAS CLUSTER ION BEAM ASSISTED DEPOSITION SYSTEM

Figure 1 shows a schematic view of the gas-cluster ion beam assisted deposition system developed by us. Optical thin films can be grown by electron beam (EB) evaporation of oxides (SiO_2, TiO_2 *etc.*) under gas-cluster (O_2, Ar *etc.*) ion beam bombardment with V_{acc} up to 20 kV. A new gas-cluster source chamber, which can be easily installed on the existing deposition chamber, was also developed. In the previous type of the system a differential pumping chamber was mounted between the cluster source and deposition chamber [8]. However, it is anticipated that the flow of the

CP475, *Applications of Accelerators in Research and Industry*,
edited by J. L. Duggan and I. L. Morgan

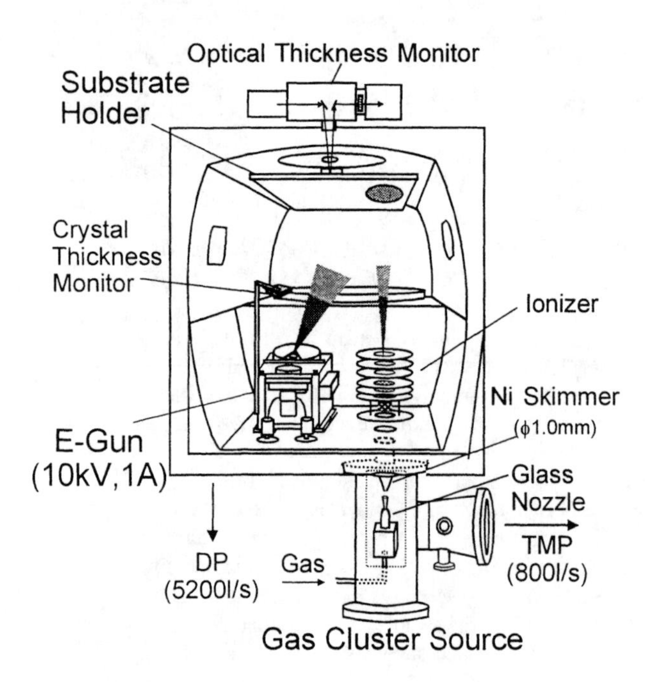

FIGURE 1. schematic view of the gas cluster ion beam assisted deposition system developed in this work

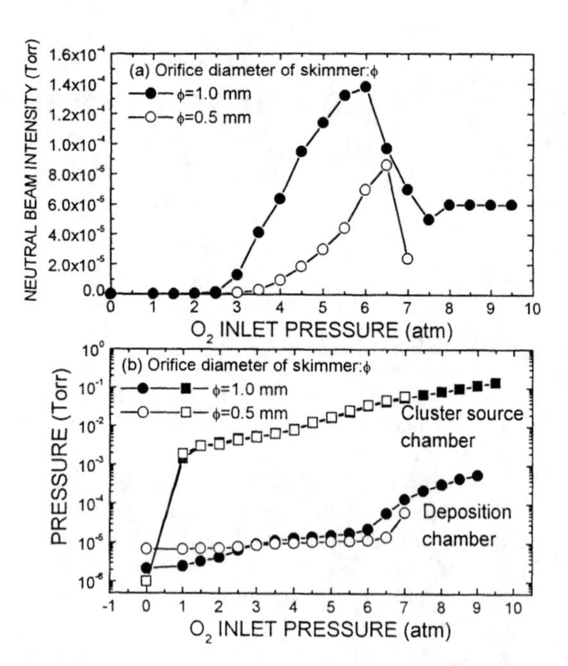

FIGURE 2. O_2-gas inlet pressure applied to the nozzle vs. O_2-cluster beam intensity (a) and the pressures in the deposition and cluster source chambers (b) for two skimmers with different orifice diameters (ϕ) of ϕ=0.5 mm and 1.0 mm

neutral cluster beam is partly disturbed by the differential pumping chamber, which results in a decrease in cluster ion current. Thus, the direct connection of cluster source to the deposition chamber is expected to increase the neutral cluster beam intensity. This arrangement also allows a simplified optical alignment for the neutral cluster beam and reduces the cost of system construction. However, the increase in pressure of the deposition chamber has to be kept low by selecting a suitable skimmer orifice diameter, as discussed in the next section. The deposition chamber is evacuated by a large 5200 l/s diffusion pump to 3×10^{-7} Torr, while the gas-cluster source chamber is evacuated by 800 l/s turbo molecular pump to 1×10^{-6} Torr. The EB evaporator has six crucibles containing the different types of evaporation materials so that multiple layered thin films can be formed.

NEUTRAL GAS CLUSTER BEAM

Gas clusters are generated by expansion of high pressure gases through a Laval glass nozzle into the cluster source chamber. The clusters travel through a Ni skimmer which is used to reduce shock wave effects, and run into the deposition chamber. In order to increase the neutral gas cluster beam intensity, a skimmer with a larger orifice diameter is desirable. However, two competitive processes need to be considered: the requirement to keep a good vacuum of deposition chamber and to increase neutral cluster beam intensity. Figures 2 (a) and 2(b) show the dependence of O_2-gas inlet pressure applied to the nozzle on the neutral O_2-cluster beam intensity and the chamber pressures, respectively, where two types of skimmer with different orifice diameter, ϕ=0.5 and 1.0 mm are compared. The beam intensity, measured with a conventional ionization gauge placed on a beam running pass, increases with increasing inlet pressure up to ~6 atm and then decreases, due to decomposition of clusters which is attributed to the shock wave. Such a tendency of decomposition of clusters can be seen from the pressures in the cluster source chamber above $\sim2\times10^{-2}$ Torr. The beam intensity at 5 atm for ϕ =1.0 mm is 3-4 time higher than that for ϕ =0.5 mm, whereas the pressures in the cluster source and deposition chambers are hardly affected by a difference of ϕ, as seen in Fig. 2(b). Thus, it is concluded that the skimmer with ϕ=1.0mm is preferable to obtain a higher cluster ion beam current.

GAS CLUSTER ION BEAM

An efficient ionization and transportation of the cluster beam and a preferential extraction of cluster ions from ionizer are important in terms of designing the cluster ion source. Figure 3 shows the configuration of the gas cluster ion source. Clusters are ionized inside the cylindrical mesh anode by electron bombardment with an emission voltage up to V_e=300 V and a current up to I_e=200 mA, and then the cluster ions are extracted with an extraction voltage of V_{ext}=0-2 kV. The cluster ions are then focused with an Einzel lens and accelerated up to 20 keV and finally bombard the substrates. Scanning mechanisms of the cluster ion beams are currently under development.

Two fine meshes which are put, respectively, on the anode and extraction electrodes, give a uniform electric

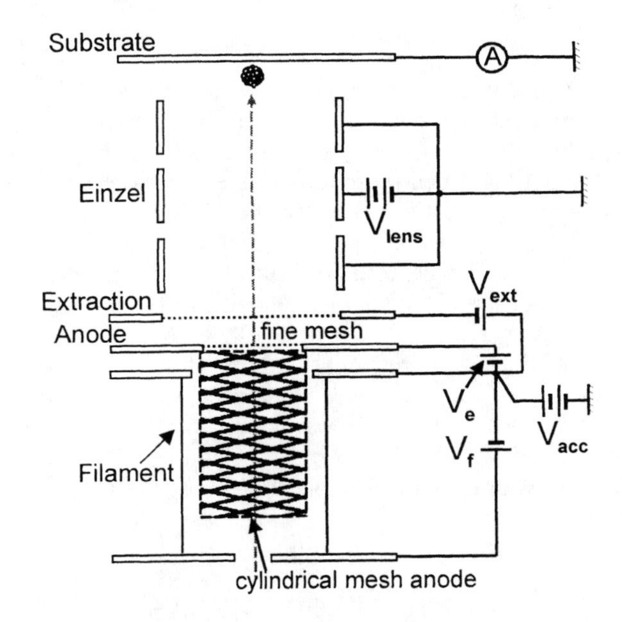

FIGURE 3. Configuration of gas cluster ion source. Neutral gas cluster beam comes from the bottom of ion source

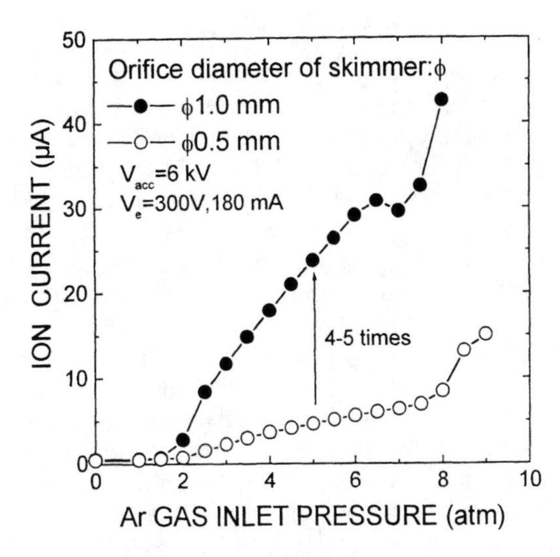

FIGURE 5. The intensity of Ar-cluster ion beam intensity as a function of Ar-gas inlet pressure for two different orifice diameters of skimmer. Sudden increase in ion current above 6 atm is attributed to decomposition of neutral clusters, resulting in an increase in Ar-monomer ion current.

field between the two electrodes. In Fig. 4, Ar-cluster and Ar-monomer ion currents are plotted as a function of total extraction voltage, which is defined as V_e+V_{ext}. The data labeled as "Ar monomer at 2.5 atm" were measured under conditions where few clusters are generated. According to the well-known Child-Langmuir equation which gives a relationship between a space-charge limited current (J_0) and applied voltage (V_0) between two parallel plates, J_0 is proportional to $V_0^{3/2}$. However, our results suggest that the ion current is independent of the extraction voltage in the positive region due to the presence of the two fine meshes.

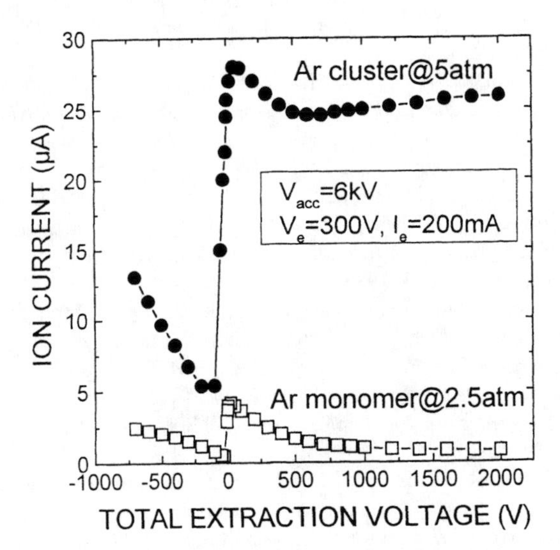

FIGURE 4. Comparison of extraction characteristics between Ar-cluster ion beam and Ar-monomer ion beam. Ar-cluster ions can be preferentially extracted from the ionizer.

This fact contributes not only to a suppression of extraction of monomer ions but also to a preferential extraction of cluster ions because neutral clusters have an initial kinetic energy of less than ~100 eV (at a cluster size=1000) when they are generated by adiabatic expansion. The kinetic energy (K) of a molecule ejected from the nozzle can be expressed as;

$$K=kT_0\cdot\gamma/(\gamma-1), \tag{1}$$

where k is the Boltzmann constant, T_0 the temperature of the supplied gas, and γ the ratio of specific heats. The values of K for single Ar, O_2, and He molecules are calculated to be 64.6, 90.3, and 65.0 meV, respectively. In Fig. 4, the monomer ion current decreases with increasing total extraction voltage whereas the cluster ion current does not significantly decrease, which suggests that the higher extraction voltage of ~2 kV may be useful for a preferential extraction of cluster ions.

Figure 5 shows the Ar-ion beam current as a function of Ar-gas inlet pressure for two skimmers with ϕ=0.5 and 1.0 mm, respectively. The cluster ion current for ϕ=1.0 increases with increasing pressure up to 7 atm and then suddenly increases. Since clusters are known to decompose above 6 atm (Fig. 2 (a)), the sudden increase in ion current can be attributed to a contribution by the monomer ions. By increasing the orifice diameter to 1.0 mm, the intensity of cluster ion current was increased 4-5 times of the original with ϕ=0.5 at 5-6 atm, and, finally, a gas cluster ion current of ~30 μA was successfully obtained. This current value was one order of magnitude higher than that obtained so far in the previous system [8].

TiO_2 FILM FORMATION

TiO_2 films were grown on Si (100) substrates at ambient temperature with the developed system, *i.e.*, electron beam evaporation of TiO_2 under O_2-cluster ion bombardment with V_{acc} up to 12 keV. The O_2-cluster ion current density was less than 1.0 μA/cm^2 which corresponds to 6×10^{15} molecules/cm^2/s (at a cluster size=1000). Prior to the growth, the Si substrates were degreased with organic solvents and were rinsed in high-purity water, and then the native silicon dioxides were etched in buffered HF ($HF:H_2O$=1:50). The TiO_2 growth rate was kept constant at 2.0 Å/s by monitoring it with a crystal thickness monitor. The thickness of grown films was measured by DEKTAK3030™ surface profiler and was ~2000 Å. Refractive index (n) of the grown films was evaluated as a function of V_{acc} by a conventional ellipsometry using a 632.8 nm line of a He-Ne laser. The environmental stability of optical coatings is limited by the porosity of microstructure of the films. In order to investigate the resistance to the moisture absorption, water-soaking test was performed, where the samples were soaked in de-ionized water for 12 hrs. The values of n for these samples were similarly characterized by ellipsometry, and the results are shown in Fig. 6. Although the values of n below V_{acc}=2 kV are n=2.18-2.19, those above V_{acc}=4 kV are increased to n=~2.30. Furthermore, after water-soaking an increase in n values due to moisture absorption becomes smaller with increasing V_{acc}. These results suggest that the films are more dense and the environmental stability of the films increases with increasing V_{acc}. Systematical characterizations associated with the surface roughness, bombardment induced defects, and stoichiometry of the films will be conducted in the near future.

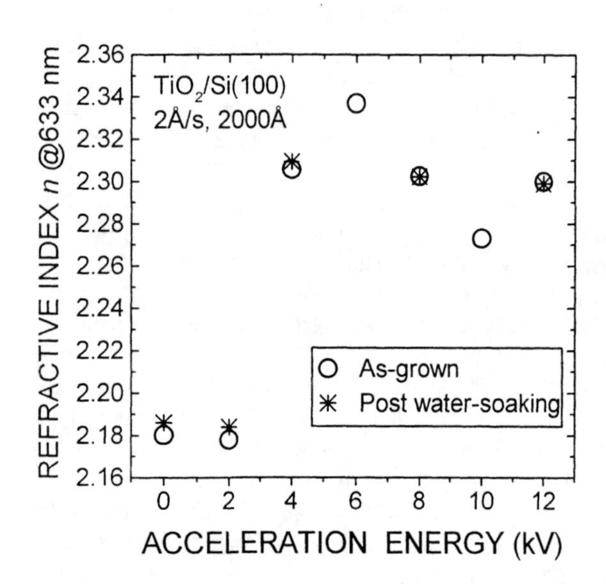

FIGURE 6. Refractive index of two type of films, as-grown and post soakage in de-ionized water for 12 hrs, as a function of acceleration energy of O_2-cluster ions. The average size of O_2-clusters was around 1000.

CONCLUSIONS

A Gas–cluster ion beam assisted deposition system was developed for high quality optical thin film formation. A high intensity gas-cluster ion current of ~30 μA was successfully obtained by following three approaches; 1) to develop a new cluster source chamber which can be easily installed on the existing deposition chamber, and to connect it directly to the deposition chamber, 2) to increase an orifice diameter of skimmer to 1.0 mm, and 3) to design the cluster ion source.

TiO_2 films with refractive indices of 2.18-2.34 were grown on Si (100) substrates at ambient temperature with the developed system. It was found that the films are more dense and the environmental stability of the films increases with increasing acceleration energy of O_2-cluster ions.

ACKNOWLEDGMENTS

The authors would like to thank Dr. A. Perry, visiting professor of Kyoto University, for proofreading this manuscript.

REFERENCES

1. S. Mohan and M.G. Krishna, "*A review of ion beam assisted deposition of optical thin films*," *Vacuum*, vol. **46**, pp. 645-659, 1995.
2. P.J. Martin, H.A. Macleod, R.P. Netterfield, C.G. Pacey, and W.G. Sainty, "*Ion Assisted Deposition of Thin Films*," *Appl. Opt.*, vol. **22**, pp. 178-184, 1983.
3. M. Akizuki, J. Matsuo, I. Yamada, M. Harada, S. Ogasawara, and A. Doi, "*SiO_2 formation at room temperature by gas cluster ion beam oxidization*," *Nucl. Instr. and Meth. B*, vol. **112**, pp. 83-85, 1996.
4. I. Yamada and J. Matsuo, "*Cluster Ion Beam Processing*," *Materials Science in Semiconductor Processing*, Vol. **1**, pp. 27-41, 1998.
5. N. Shimada, T. Aoki, J. Matsuo, I. Yamada, K. Goto, and T. Sugii, "*Reduction of Boron Transient Enhanced Diffusion in Silicon by Low-Energy Cluster Ion Implantation*," *Material Chem. & Phys.*, Vol. **54**, pp. 80-84, 1998.
6. M. Akizuki, J. Matsuo, I. Shin, M. Harada, S. Ogasawara, A. Doi, and I. Yamada, "*Irradiation effects of O_2 cluster ions for lead oxide film formation*," *Nucl. Instr. and Meth. B*, vol. **121**, pp. 166-169, 1997.
7. H. Katsumata, J. Matsuo, T. Nishihara, T. Tachibana, K. Yamada, M. Adachi, E. Minami, and I. Yamada, "*Formation of Oxide Thin Films for Optical Applications by O_2-cluster Ion Assisted Deposition*," *IEEE Proc. 12th International Conference on Ion Implantation Technology*, Kyoto Japan, 1998: in press.
8. W. Qin, R.P. Howson, M. Akizuki, J. Matsuo, G. Takaoka, and I. Yamada, "*Indium oxide film formation by O_2 cluster ion-assisted deposition*," *Material Chem. & Phys.*, Vol. **54**, pp. 258-262, 1998.

Mechanisms of Formation of Metal Cluster Nonlinear Optical Light Guide Structures in $LiNbO_3$ by Ion Beam Implantation

E. K. Williams[a], S. Sarkisov[b], A. Darwish[b], D. Ila[a], D. B. Poker[c], D. K. Hensley[c]

[a]*Center for Irradiation of Materials, PO Box 1447 Alabama A&M University, Normal, AL 35762 USA,*
[b]*Dept. of Natural and Physical Sciences, Alabama A&M University, Normal, AL 35762 USA*
[c]*Solid State Division, Oak Ridge National Laboratory, Oak Ridge, TN 37831, USA*

Metal nanoclusters of Ag, Au and Cu have been produced by ion implantation into x- and z-cut lithium niobate. MeV Ag implantation followed by heat treatment simultaneously produced a waveguide and Ag clusters with a high nonlinear index. Co-implantation of O with Ag at 500° enhanced the ability of the clusters to withstand heat treatment.

INTRODUCTION

Composites of metal nanoclusters and dielectric hosts produced by ion beam implantation are of great interest as efficient optical nonlinear materials. The high third order nonlinear susceptibility of these materials results from the dramatic enhancement of the local optical field in the vicinity of the metal nanoparticles at the wavelength of the surface plasmon resonance. The nuclear damage region produced by bombardment of dielectric hosts with MeV metal ions can also define a light guiding structure with modified refractive index(1). Optical waveguide and composite material can thus be produced simultaneously by the same ion beam resulting in an efficient nonlinear optical device. However, optimal characteristics of such waveguides have not been achieved yet mostly because the atomistic mechanisms of their formation are not yet clear.

Results from three metals (Au, Ag and Cu) that have been implanted into $LiNbO_3$ are reported here. These metals were chosen in part due to their having surface plasmon resonances in the visible regime. This property makes these metals good candidates for nonlinear optical applications and also facilitates study by absorption spectrometry.

Because ion implantation produces damage that can adversely affect the electro-optical properties of a material such as $LiNbO_3$ it is important that it be possible to remove the damage through thermal annealing or demonstrate that the damage is either helpful or benign.

The metal implanted samples have been studied by optical absorption spectrometry(2) to investigate changes in the cluster size due to differing implant and post-implantation heat treatments, by the z-scan technique to measure the nonlinear refractive index(3, 4), by the prism coupling method to investigate the waveguiding properties, by Transmission Electron Microscopy (TEM) to see clusters and implantation damage directly, and by Electron Paramagnetic Resonance (EPR) to study the oxidation state of Ag.

EXPERIMENTAL

Samples of x- and z-cut $LiNbO_3$ were implanted at Oak Ridge National Laboratory in two different energy regimes, MeV and keV. For the z-scan and TEM analysis samples, Ag was implanted at 1.5 MeV to fluences of 2 x $10^{16}/cm^2$ and 1.7 x $10^{17}/cm^2$ at room temperature (RT). Au and Cu were implanted at 2.0 MeV and at room temperature to fluences of 2 x 10^{16} to 8 x $10^{17}/cm^2$.

In the keV regime $^{107}Ag^+$ was implanted at room temperature (RT) and at 500°C into z-cut samples. In order to improve the temperature stability of the Ag nanoclusters $^{16}O^+$ was also implanted. The Ag and O implantations were at energies of 160 and 35 keV, respectively, to fluences of 2, 4 and 8 x 10^{16} ions/cm^2. Energies were chosen based upon SRIM96 (5) simulations to yield equivalent ranges of approximately 50 nm. Three samples were implanted at each temperature: Ag only, Ag followed by O(Ag+O), and O followed by Ag(O+Ag). The Ag and O fluences were equal and the currents rastered over a 5 cm^2 aperture were approximately 20 μA for Ag^+ and 40 μA for O^+. The samples were tilted by ~ 7° to avoid channeling.

The electron paramagnetic resonance measurements were made at room temperature using a Bruker 300 ES spectrometer operating with a microwave frequency of 9.64 GHz and a modulation field at 100 kHz. RBS/channeling spectra were acquired using 2.0 MeV He^+ generated by an NEC 5SDH2 Pelletron. The beam current was 30 to 75 nA and the detector was at 170°. Post-implantation heat treatment at 500°C was done in air.

A Cary 3E UV/VIS spectrometer was used for the optical absorption measurements from which cluster radii were estimated. If the wavelength of the incident light is much larger than the cluster size then the radii of the

CP475, *Applications of Accelerators in Research and Industry*,
edited by J. L. Duggan and I. L. Morgan

clusters can be estimated from the relation $r = AV_f/\Delta\omega_{1/2}$, where A is an empirically determined, metal dependant constant, V_f is the Fermi velocity of the metal and $\Delta\omega_{1/2}$ is the full width at half maximum of the plasmon resonance (2,6). When there is a distribution of sizes of nanoclusters the above relation gives a lower limit on the size of the clusters rather than an average size as the width of the resonance is inversely proportional to the cluster size. TEM analysis is required for an accurate assessment of cluster size distribution.

Light transmission along the implanted layer was studied using the prism coupling technique (7) at 632 nm. The nonlinear refractive index was characterized using the Z-scan technique (3). A Q-switched mode-locked configuration of a 76 MHz pulse rate Nd:YAG laser was used to study the sample at a high peak power density and low pulse repetition rate. A pulse picker selected a single picosecond pulse from the series of pulses in the 350 ns envelope of a Q-switched pulse. At 532 nm it gave a power density of up to 10 GW/cm^2 in each pulse. The repetition rate of the 70 ps pulses was 10 Hz. The average applied power density was a few milliwatts.

RESULTS AND DISCUSSION

For the Cu implanted samples the surface plasmon resonance absorption peaks disappeared immediately upon heat treatment at 500°C, as shown in Figure 1. Also, because the intraband transitions are at a longer wavelength than the plasmon resonance (2) it is difficult to accurately apply the above theory to estimate the Cu cluster size. No z-scan or prism coupling measurements were made on the Cu implanted samples.

In marked contrast to Cu, implanted Au forms stable clusters that grew with heat treatment up to 1000°C. Figure 2 shows the optical absorption spectra from 2.0 MeV Au implanted to 5 x 10^{16}/cm^2 into z-cut $LiNbO_3$. Any surface plasmon resonance from Au clusters present immediately after implantation is obscured by broadband absorption due to implantation damage. As the annealing temperature is raised the resonance increases dramatically. Fits to the data indicate that the lower limit on the cluster size stabilizes at 4.5 nm for heat treatment at 1000°C. Figure 3 shows optical absorption spectra from 2.0 MeV Au implanted to a fluence of 2 x 10^{17}/cm^2. Even at this high fluence no gold clusters are seen prior to heat treatment. Only 30 minutes at 500°C were needed make the clusters visible. Soaking the sample at 1000°C for 1, 4 and 10 hours resulted in a cluster radius of 3.9 nm. Whereas for the 5 x 10^{16}/cm^2 samples the position of the absorption peak shifted from 608 nm after 1 h at 1000°C to 617 nm after 10 h at 1000°C for the high fluence sample the peak position remained stable, but the height decreased. The redshift in peak position is indicative of change in the refractive index of the host, probably due to changes in Au concentration in the regions between the clusters.

TEM analysis of the 5 x 10^{16}/cm^2 Au implanted sample after heat treatment at 1000°C for 10 h shows that the Au clusters are in a ~380 nm thick band 100 nm below the surface. Electron diffraction showed that the Au clusters are indeed crystalline. The facets of the Au crystals are normal to the c-axis. Very few of the crystals are smaller

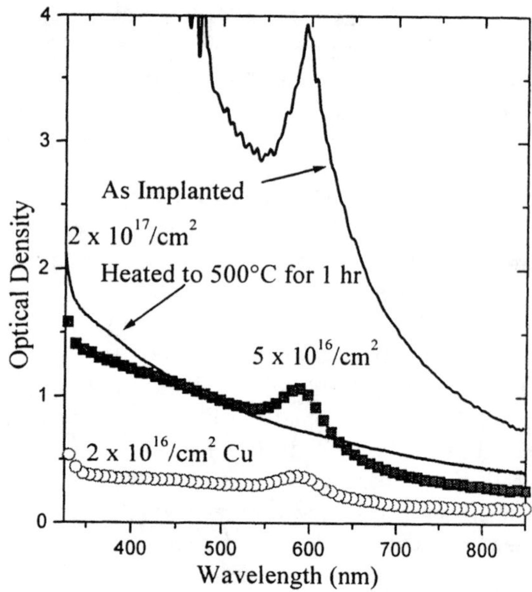

FIGURE 1. Optical absorption spectra of Cu implanted to fluences of 5 x 10^{16}, 2 x 10^{17} and 5 x 10^{17}/cm^2 at 2.0 MeV, before and after heat treatment at 500°C for 1h.

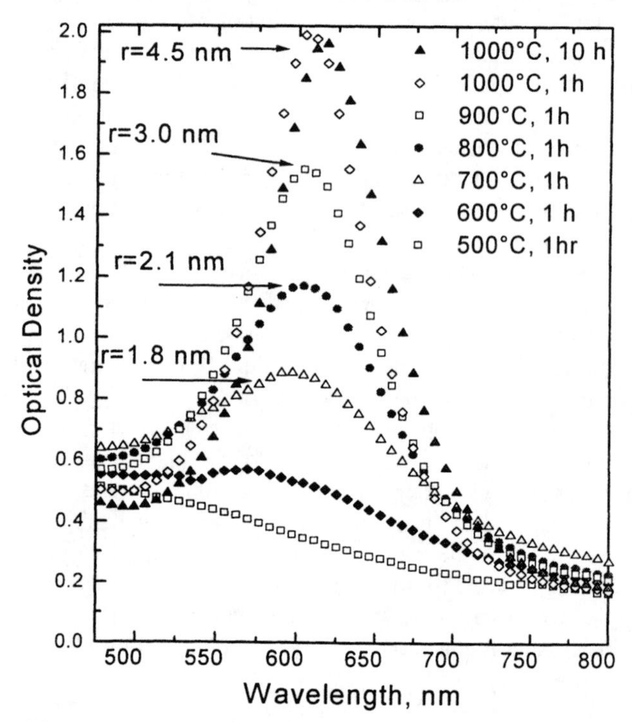

FIGURE 2. Optical density vs. wavelength for z-cut $LiNbO_3$ implanted with 5 x 10^{16} Au/cm^2 at 2.0 MeV and heat treated at 500 to 1000°C for times indicated.

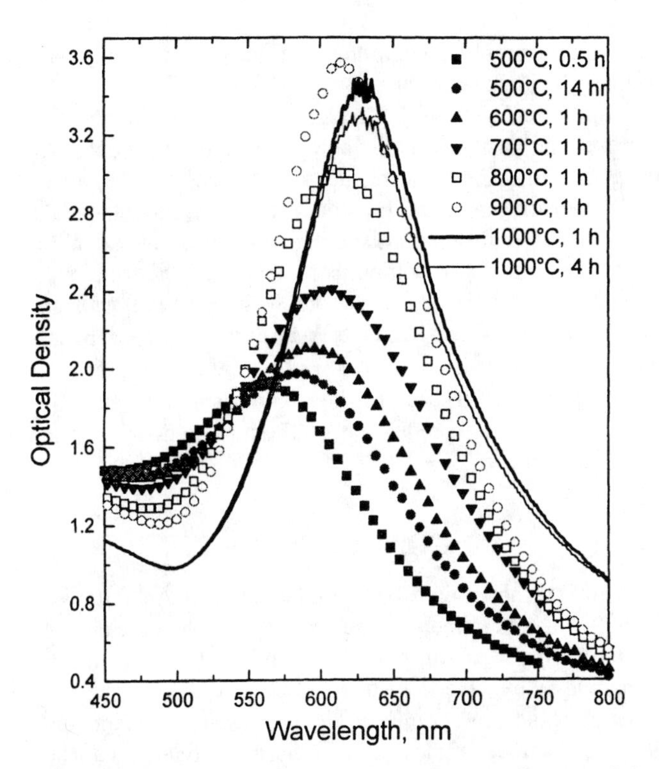

FIGURE 3. Au implanted at 2.0 MeV to a fluence of 2 x $10^{17}/cm^2$ and heat treated in air for the times and temperatures indicated.

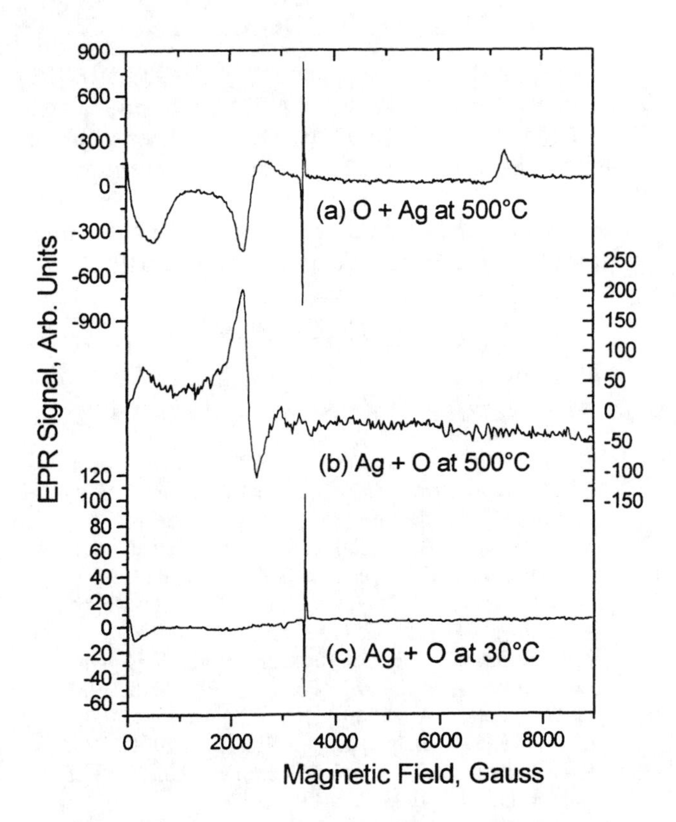

FIGURE 4. EPR spectra from a) O+Ag implanted at 500°C, b) Ag+O implanted at 500°C, and c) Ag+O implanted at room temperature.

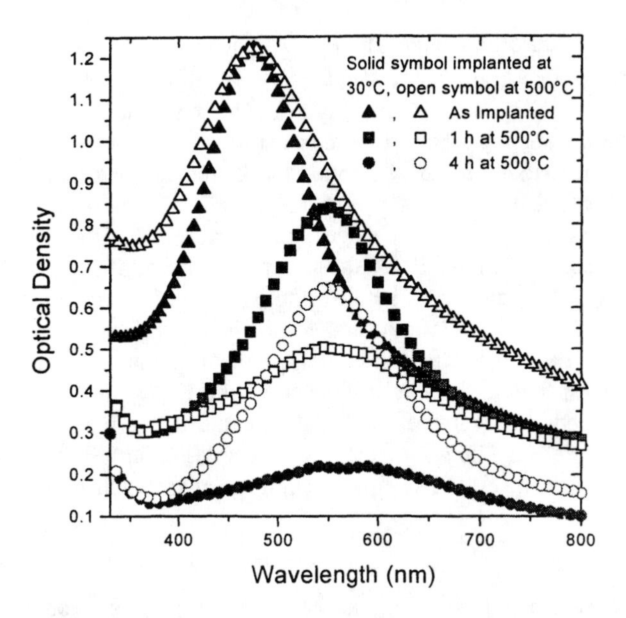

FIGURE 5. Optical absorption spectra from O followed by Ag implanted z-cut $LiNbO_3$ implanted at RT and 500°C and heat treated at 500°C for 1 h and 4 h in air.

than 9 nm in diameter, many are approximately 9 to 11 nm diameter and they range in size up to 40 nm in diameter.

Silver clusters formed in $LiNbO_3$ by ion implantation have been subjected to analysis by EPR, prism coupling and z-scan. Silver clusters are more stable at elevated temperatures than are Cu clusters, requiring heat treatment at 800°C to disappear at approximately the same rate as Cu clusters disappear at 500°C. But the clusters will dissolve at temperatures below 500°C. In order to enhance the temperature stability oxygen was co-implanted with the Ag, to the same fluence and the order of implant was changed. The rationale was that the extra oxygen would bond with the Ag to form AgO or Ag_2O or would assist in regrowth of the $LiNbO_3$ lattice. Ag and O were implanted in the keV regime at room temperature and at 500°C. The purpose of the high temperature implant was to reduce implantation damage. It was expected that a reduction in defects would slow any migration of Ag to those defects and that any clusters that were formed at a high implant temperature would be more stable after post implantation heat treatment.

EPR was used to look for formation of any silver oxides. EPR spectra from three of six samples are shown in Figure 4. The bottom spectrum is from Ag + O at room temperature. The signal is very weak. The central spectrum is from Ag + O implanted at 500°C. Here the effects of the Ag clusters and ions are clearly seen. In the top spectrum, O+Ag implanted at 500°C, the signal at 7500 G is evidence of silver oxide formation, which appear for this sample only.

As expected, the samples implanted at 500°C fared better during 500°C heat treatment than did the room temperature implanted samples. An interesting case is that of the O+Ag implanted samples at 4 x 10^{16}/cm^2 at RT and 500°C. As shown in Figure 5, immediately after implantation the only difference between the two samples is the width of the absorption peak, the narrower peak indicating larger clusters. After 1 h at 500°C the sample implanted at 500°C changes much more than the RT implanted sample. But, after 4 more hours at 500°C the situation is reversed. The RT sample is a mottled, light blue color. The 500°C implanted sample is a uniform bright blue. Lacking microscopic analysis of the crystal one can only speculate that the difference in spectra is due to the difference in degree of implantation induced amorphization and defects between the two samples. Further heat treatment at 800°C results in complete dissolution of the clusters.

Prism Coupling

Prism coupling at 633 nm showed no propagating modes for low energy implants nor for 1.5 MeV Ag implanted at RT until after heat treatment at 500°C for 1 h. The propagation indices of two detected modes are greater than the refractive index of the bulk crystal. The modes can therefore be classified as TE0 and TE1 propagating modes attributed to a light guiding layer on the top of the sample. The increase in refractive index of this layer after heat treatment was due to diffusion of Ag towards the surface which counteracted the decrease in refractive index due to implantation damage remaining after the 500°C heat treatment. The light guiding layer is about 0.75 μm thick, slightly greater than the SRIM96 predicted maximum range of 0.6 μm.

Z-Scan

Figure 6 shows the results of the closed aperture Z-scan for the 1.5 MeV, 1.7 x 10^{16} Ag/cm^2 implanted sample heat treated at 500°C. The estimated nonlinear refractive index (3) of the sample is positive and has a magnitude of approximately 5.0 x 10^{-10} cm^2/W, which compares well to the best known third order nonlinear optical materials (8).

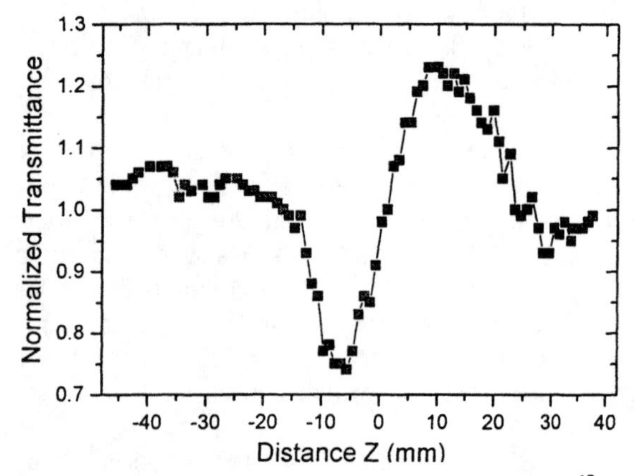

FIGURE 6. Z-scan of $LiNbO_3$ implanted with 1.7 x 10^{17} Ag/cm^2 at 1.5 MeV and heat treated at 500°C for 1 h.

The significant noise component of the Z-scan data is due to intensive light scattering. Because of the noise, we could not apply efficiently the technique of separating slow thermal and fast electronic components of the nonlinear index by fitting the experimental z-scan data with the three-parameter transmission function used by Osborne, et al. (4). For this reason, the estimated nonlinear index more likely combines both components, electronic and thermal. However, when the same sample was measured with an average power of about 200 mW at 575 nm the nonlinear index was reported to be 6.8 x 10^{-8} cm^2/W (9), almost twenty times as large as the newly measured value. The difference is due to reduction of thermal contributions to the measurement.

CONCLUSIONS

We have shown that MeV implantation of Ag followed by 500°C heat treatment produces both a waveguide and an array of small Ag clusters that have a large nonlinear refractive index. Due to its temperature stability Au may be a better choice for nonlinear devices as the implantation damage can be removed by thermal treatment. The nonlinear index needs to be measured for the Au and Ag coimplanted with O samples.

ACKNOWLEDGMENTS

This research was supported by the Center for Irradiation of Materials, Alabama A&M University, and the Div. of Materials Sciences, U.S. Dept. of Energy, under contract DE-AC05-96OR22464 with Lockheed Martin Energy Research Corp., by the Alliance for Nonlinear Optics (NASA Grant NAG5-6532) and by the U.S. Army Research Office under contract DAAH04-96-1-0190.

REFERENCES

1. P.D. Townsend, P.J. Chandler and L. Zhang, *Optical Effects of Ion Implantation*(Cambridge University Press, Cambridge, 1994).
2. U. Kreibig and M. Vollmer, *Optical Properties of Metal Clusters (Springer Series in Materials Science, Vol25)* (Springer Verlag, Berlin, 1995)
3. M. Sheik-bahae, A.A. Said, T.H. Wei, D.J. Hagan and E.W. Van Stryland, IEEE J.Quantum Electronics **26** (1990) 760.
4. D. H. Osborne, R. F. Haglund, F. Gonella, and F. Garrido, Appl. Phys. **B66** (1998) 517.
5. J. F. Ziegler, J. P. Biersack and U. Littmark, *The Stopping and Range of Ions in Solids* (Pergamon Press, NY, 1985)
6. G. W. Arnold, J. Appl. Phys. 46 (1975) 4466.
7. R.Ulrich and R.Torge, *Appl. Opt.* 12 (1973) 2901.
8. R. L. Sutherland, *Handbook of Nonlinear Optics* (Marcel Dekker, Inc, NY, 1996)
9. E. K. Williams, D. Ila, S. Sarkisov, M. Curley, J. C. Cochrane, D. B. Poker, D. K. Hensley, C. Borel, Nucl. Instr. and Meth. B 141 (1998) 268.

Nitride Thin Film Synthesis by Cluster Ion Beam

Hiroshi Saito

Department of Applied Physics, Okayama University of Science, Ridai-cho, Okayama 700-0005, Japan

Small-sized cluster ion beam source using ammonia as source material was constructed for purposes of nitride thin film synthesis. Using this source, we tried nitridation of GaAs and Si. XPS measurements showed that nitridation of GaAs was performed by replacing As atoms with N atoms in temperature range of 550~680°C. In case of Si, nitridation was successful even at room temperature. These results demonstrate that the use of cluster ion beam has great advantage of nitride thin film synthesis.

INTRODUCTION

Metal- and semiconductor-nitrides are expected, on one hand, for materials resistant high temperatures, on the other hand, for photo-electronic devices in ultraviolet-blue region because of their wide band gap nature. Usually nitrides can be synthesized by CVD method only at very high temperatures, sometimes exceeding 1000°C.(1) Synthesis under such high temperatures is difficult to avoid giving unrecoverable damage to underlying structure, and also is impossible to apply for nitridation of metals with low melting point.

In the growth of group III-nitrides by means of MBE, chemically active nitrogen such as the mixture of nitrogen molecules in the triplet excited state, nitrogen atoms and nitrogen ions obtained by RF or ECR discharged nitrogen plasma is usually utilized (1). It is difficult to tell what species in the mixture exerts mainly the synthesis of the material, which leads to the difficulty of finding the solution how to improve film quality. The use of nitrogen or nitrogen-complex cluster beam will be the key technique to overcome the difficulty.

Up to now major application of gas cluster ion beam has been either surface modification such as smoothing and cleaning due to high yield sputtering effect (2,3) or very shallow implantation due to very low energy effect (2,3). We pay attention to another nature of cluster beam, i.e., collective motion during impact between substrate and cluster atoms, and also high density deposition of atoms. These characteristics can be applied to thin film synthesis at low temperatures.

The present report concerns with the construction of small-sized cluster ion beam source intended to install to conventional molecular beam epitaxy chamber. Ammonia is used as source materials. The ammonia cluster beam is applied for nitridation of Si and GaAs. X-ray photo-electron spectroscopy (XPS) measurements showed that nitridation of these materials are successfully achieved.

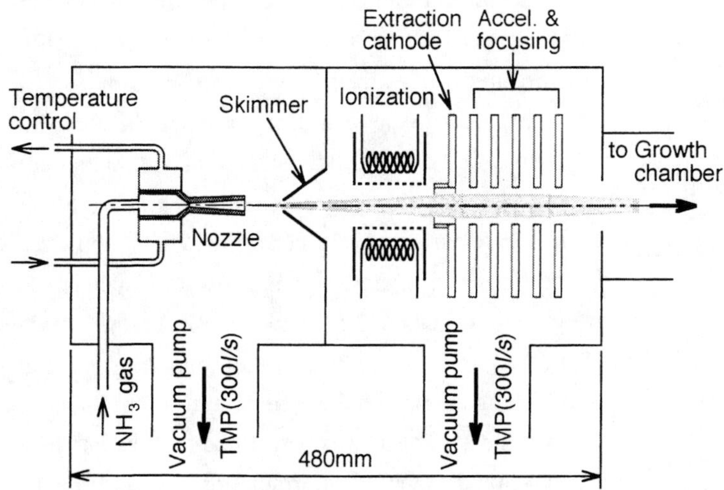

FIGURE 1. Schematic illustration of gas cluster ion source system

CP475, *Applications of Accelerators in Research and Industry*, edited by J. L. Duggan and I. L. Morgan

GENERATION AND CHARACTERISTICS OF AMMONIA CLUSTER ION BEAM

An apparatus presently constructed is schematically shown in Fig.1. The system is consisting of two vacuum chambers; One is for generating clusters and the other is for ionization and acceleration of the clusters. They are separated by a skimmer with an orifice of 0.3mm in diameter. The entire system is intended to be as small as possible to install to conventional molecular beam epitaxy chamber, and has the size of 48cm in length.

Ammonia gas is ejected from the conical nozzle made of fused quartz into vacuum to generate clusters. After passing through the skimmer, the clusters are introduced into the ionization and acceleration chamber. After being ionized by electron bombardment the clusters are extracted by an extraction anode and then accelerated and focused by electrostatic lenses.

The size distribution of the ionized clusters is measured by means of retarding potential method. Example of the results is shown in Fig.2. Figure 2(a) shows ion current as a function of retarding voltage for several amounts of ammonia flow under the extraction voltage of 100V. With increasing ammonia flow rate, the ion current increases. The cluster size is calculated from the tangent slope to the curve shown in Fig2(a) and is shown in Fig.2(b). The abscissa means the number of ammonia molecules in a cluster, i.e. the cluster size and the ordinate the number of that cluster. The peak observed near zero cluster size is due to monomer ions. Under small flow rate the monomer ions dominate the cluster beam. With increasing flow rate, clusters with larger size begin to dominate the beam. Although not shown here, it was found that the ratio of monomer ions occupying the cluster beam decreases with decreasing the extraction voltage. In doping experiments the use of the extraction voltage as low as possible is recommended to minimize the effect of sample damage due to high energy monomer ion irradiation.

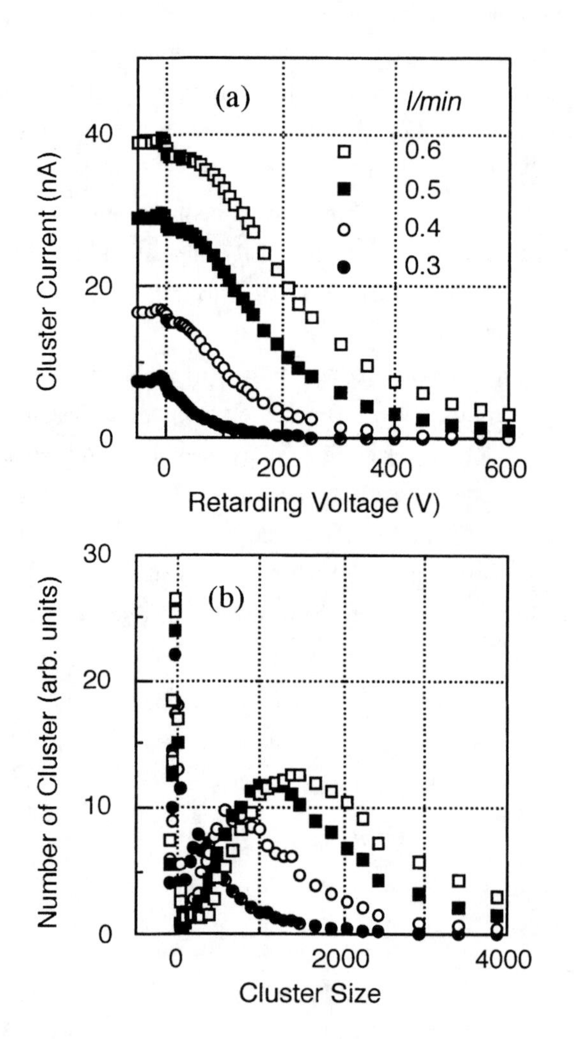

FIGURE 2.
(a) Cluster ion current as a function of retarding voltage for 4 ammonia flow rates.
(b) The size distribution of clusters. The cluster size (abscissa) means the number of ammonia molecules in a cluster.

EXPERIMENTS AND RESULTS

These ammonia clusters are used for nitridation of Si and GaAs substrates. Experiments were performed under the following condition; With an ammonia flow rate of 0.4L/min., a beam with a current density of about $2\mu A/cm^2$ with a peak cluster size of about 10^3 molecules/cluster was obtained. Irradiation with this beam for 2 hours corresponds to the total dose amount up to about $10^{20}/cm^2$. Extraction voltage used was 50V and acceleration voltage was 10kV. No special chemical treatments of both substrate surfaces were performed except only degreased by rinsing in organic solvent prior to installation into the vacuum chamber. In the case of GaAs, the substrate was heated up by exposing to ammonia cluster beam until reaching at prescribed temperature (480°C~740°C), and then the irradiation was performed for 2 hours. In the case of Si, on the other hand, irradiation was performed at room temperature.

After the irradiation, samples were examined by XPS measurements. MgKα ray is used as the X-ray source.

The case of GaAs

We observed a line at about 391eV, which corresponds to the 402eV line (3) due to 1s electron of atomic nitrogen. The energy shift may be owing to the bonding to Ga atom in the solid. Figure 3 shows an example of a detailed XPS spectrum in the energy region from 95 to 220eV for the sample prepared at 740°C. We see several lines from low to high energy direction, i.e., a doublet due to Ga(3p) electron (99 and 102eV) , a doublet due to As(3p) electron (134 and 139eV), a line at 154eV due to Ga(3s) electron, a line at 181eV due to Auger process associated with Ga 3d-2p states, and a broader line at 199eV due to As(3s) electron. All these lines are also observed for unirradiated samples. In addition to these lines, we observe three more lines or sidebands at the high energy side of Ga(3s), Ga(Auger) and As(3s) lines, as indicated by arrows in the figure, named for at the present as Ga(3S)-N, Ga(Auger)-N and As(3s)-N. Intensity of the sidebands and also that of N(1s) line decreases when the sample is etched by Ar ion beam during the XPS measurements. Although not shown in the figure, similar sidebands are also observed for Ga(3d) and As(3d) lines at 13 and 35eV, respectively. It should be noted, however, that no sideband is observed associated with 3p electron both for Ga and As.

Figure 4 summarizes the intensity of the sidebands as a function of N(1s) intensity. Solid lines in the figure indicate least square fits to the data points. One can clearly recognize that the intensity of three sidebands is linearly dependent on N(1s) intensity, suggesting that the sidebands can be ascribed to 3s and 3d electrons of Ga and 3s electrons of As, having a bond to N atoms in the neighborhood. Also plotted in the figure are the intensity of the Ga(3p) and As(3p) lines. The decrease of the As(3p) intensity is proportional to the increase of the N(1s) intensity (dotted line) in contrast to the Ga(3p) intensity, which shows almost flat nature for N(1s) < $3x10^3$, i.e.,except the sample surface.

Results similar to Figs. 3 and 4 are also observed for samples prepared at temperature range of 550~680°C. Samples prepared above and below these temperature ranges, no trace of N signal were obtained. It is found that the sample surface prepared above these temperature range is covered by large amount of Ga droplets due to the preferential desorption of As atoms. As will be stated later, this result suggests that the incorporation of N is performed by replacing As atoms.

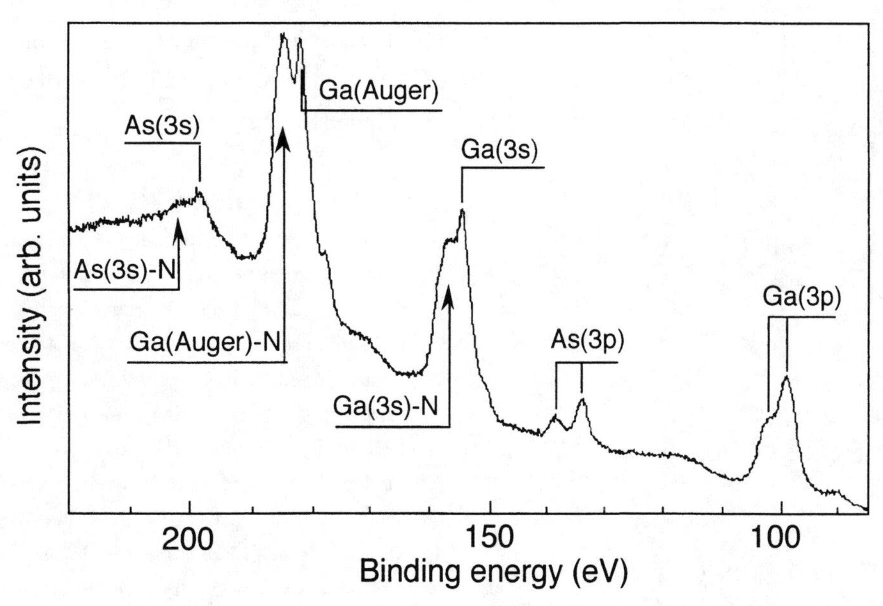

FIGURE 3. XPS spectrum for GaAs prepared at 740°C

The case of Si

We also observed a line at the binding energy of about 396 eV, which can be ascribed to N(1s) electron as in the case of GaAs (3). Furthermore we observed sideband in the high binding energy side of both Si(2s) and Si(2p) lines. The energy separation is 2eV and 3eV, respectively. These two sidebands as well as 396eV line are decreased in intensity when the sample surface is etched by Ar ion beam during the XPS measurements and then disappear completely and also can not be observed for unirradiated sample at all, similar to the case of GaAs. The sidebands are, therefore, attributed to 2s and 2p electrons of Si having a bond to N atom in the neighborhood.

DISCUSSION AND CONCLUSION

We summarize the experimental results for GaAs. No N atoms could be incorporated at lower temperature range where desorption of As atoms does not occur. At higher temperatures where Ga droplets are formed, almost no chemical reaction between N and Ga atoms takes place. These facts suggest that N atoms are incorporated mainly by replacing As atoms. That is, an amount of the reduction of As atoms is in proportion to that of the increase of N atoms, as clearly seen in Fig.4 (dotted line). In contrast, the number of Ga atoms has no relation whether As is replaced by N, in accordance with the behavior of Ga(3p) intensity in Fig.4. Taking into account this consideration and also the tendency of the solid lines in Fig.4, we can conclude that N atoms are not located at the interstitial sites, but just at the As sites.

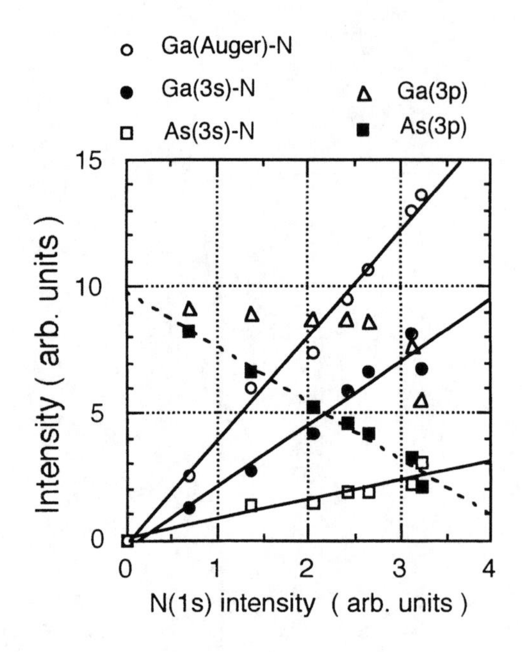

FIGURE 4.
Intensity of three sidebands, Ga(Auger)-N, Ga(3s)-N and As(3s)-N and Ga(3p) and As(3p) lines as a function of N(1s) intensity. Solid and broken lines are least square fits to the data points

There still remain two questions. One is whether As atoms have direct bond with N atoms, because the sideband named As(3s)-N is observed. At present we interpret that the interaction with N is via Ga atom having direct bond with N atom, because As(3s)-N sideband intensity is rather weak. The second is why no sideband associated with 3p electrons is observed. Further investigation is required.

We wish to say that the use of ammonia cluster has a great advantage of nitride synthesis, i.e., the nitridation of Si could be achieved even at room temperature and also the nitridation of GaAs could be possible at lower temperature compared with the conventional CVD method.

ACKNOWLEDGMENTS

The author wish to express hearty thanks to Prof. I. Yamada, Dr. J. Matsuo and Mr. N. Toyoda in Kyoto University for their valuable suggestions for constructing the apparatus and also discussion of experimental results. He is also indebted to Prof. M. Ohishi, Mr. Y. Furuya and Mr. T. Kato in Okayama University of Science for their effort in the construction of the apparatus and experimental works.

REFERENCES

1. A lot of recent works are found in *Proceedings of the 2nd International Symposium on Blue Laser and Light Emitting Diodes*, edited by Onabe K. et al. Tokyo: Ohmusha Ltd., 1998 and also in J. Crystal Growth **184/185** (1998).
2. Yamada I., edited by Duggan J.L. and Morgan I.L., *Application of Acceleration in Research and Industry*, New York: AIP press, 1997, pp. 479-482.
3. Yamada I. and Matsuo J., Materials Science in Semiconductor Processing **1**, 27-41 (1998).
4. Miyazaki E. et al., *Fundamentals and applications of surface science*, Tokyo: Fuji Technosystem Co. Ltd.,1991, ch. 10, pp. 284-294 (*in Japanese*).

Surface Smoothing of CVD-Diamond Membrane for X-Ray Lithography by Gas Cluster Ion Beam

A. Nishiyama[1], M. Adachi[1]
N. Toyoda[2], N. Hagiwara[2], J. Matsuo[2] and I. Yamada[2]

1Central Research Institute, Mitsubishi Materials Corp., Omiya, Japan 330-8508
2Ion Beam Engineering Experimental Lab., Kyoto Univ., Kyoto, Japan 606-8501

Results of the surface smoothing of a CVD-diamond membrane by gas cluster ion beams are presented. An as-deposited diamond membrane with a surface roughness of 400Å Ra was irradiated by Ar cluster ions with a energy of 20keV. A very smooth surface of 30Å Ra was obtained at a dose of 3×10^{17} ions/cm^2. This result can be clarified by computer simulation which shows that the surface smoothing of the diamond membrane was improved by a lateral sputtering of the cluster ions. However, a thin graphite layer was formed on the surface by contamination of monomer ions in the cluster beam, which decreased the transparency of the diamond membrane. A subsequent irradiation with O_2 cluster ions removed these graphite layers.

INTRODUCTION

X-ray Lithography has been developed as a technology which may be applied to the production of 1Gbit-scale DRAMS. X-ray lithograph masks patterns of X-ray absorbers are produced on a membrane material with a high Young's modulus. In order to minimize the distortion of the patterns, SiN and SiC[1] have been used as the mask membrane for X-ray lithography. However, diamond has a higher Young's modulus and high transparency and so is expected to be a superior membrane material.

In general, the following properties are required for a diamond membrane to be applied to the X-ray mask.

(1) Physical and mechanical stabilities under X-ray radiation.
(2) High transparency for X-rays.
(3) High optical transparency for alignment procedures.
(4) Surface roughness to be compatible with pattern resolution.
(5) Mechanical strength to be reliable in mask processing.
(6) Uniformity of residual stress and film thickness.

A CVD-diamond membrane containing an amorphous phase is not stable under X-ray irradiation. A CVD-diamond membrane having good crystallinity shows high radiation stability.[2] A 2 μ m thick polycrystalline diamond membrane with good crystallinity shows an optical transparency of about 50% at a wavelength of 633 nm and the burst strength of 150~350mmHg/cm[3,4].

Surface roughness affects the distortion of patterns and the transparency of the membrane. However, the surface of the as-deposited CVD-diamond is too rough. As a result, several surface smoothing techniques have been studied such as Chemical Mechanical Polishing[5], ion beam modification at grazing angle[6] and plasma etching[7,8].

In this paper, CVD-diamond membranes are irradiated with Ar and O_2 cluster ion beams to smooth their surfaces. The Ar cluster beam acts as a non-reactive beam. The mechanism of surface smoothing by Ar cluster irradiation was compared with the results of Monte-Carlo simulation. In contrast, the O_2 cluster beam acts reactively for diamond ,and different effects for the surface smoothing are expected. The sputtering mechanism with reactive and non-reactive cluster beam is discussed In terms of the experimental results. Sequentially, smooth diamond membranes treated with Ar and O_2 cluster ion beams are shown.

EXPERIMENT

2 μm-thick diamond films were deposited on silicon wafers by microwave plasma CVD using a CH_4 and H_2 gas mixture as the source gas. Silicon wafers were pre-treated to have high nucleation density and high crystallinity. Crystallinity of polycrystalline CVD-diamond films was checked by Raman spectroscopy.

The gas cluster ion beam equipment reported by Matsuo et al.[10] was used to form Ar and O_2 cluster ion beams. Neutral clusters are generated by the cooling effect during adiabatic expansion. Neutral clusters are ionized by electron bombardment with an energy of 150eV.

CVD-diamond membranes were irradiated with Ar and O_2 cluster ions. Sputtering yields were obtained from the sputtered depth measured by a contact surface profiler. Surface morphology and average roughness of the-deposited and irradiated membranes were observed with an Atomic Force Microscope (AFM).

Si substrates were etched from the reverseside using HNO_3+HF mixed-acid solution so that freestanding diamond membranes were obtained. The Transparencies of

CP475, *Applications of Accelerators in Research and Industry*,
edited by J. L. Duggan and I. L. Morgan

the diamond membranes were measured over the wave length range of visible light.

RESULTS AND DISCUSSION

Fig. 1 shows experimental results of surface roughness of a CVD-diamond membrane which has a 330ÅRa initial average roughness. The CVD-diamond surface was irradiated with 20Kev Ar clusters at normal incidence. The surface roughness decreased monotonically with increasing the ion dose down to 30ÅRa at 3×10^{17}ions/cm^2.

In the simulations, as shown in Fig.2, the surface roughness of a Cu target with a 60ÅRa initial average roughness also decreased monotonically with ion dose. The surface roughness of the Cu target was 8ÅRa at 1×10^{16}ions/cm^2.

The values of surface roughness are different between

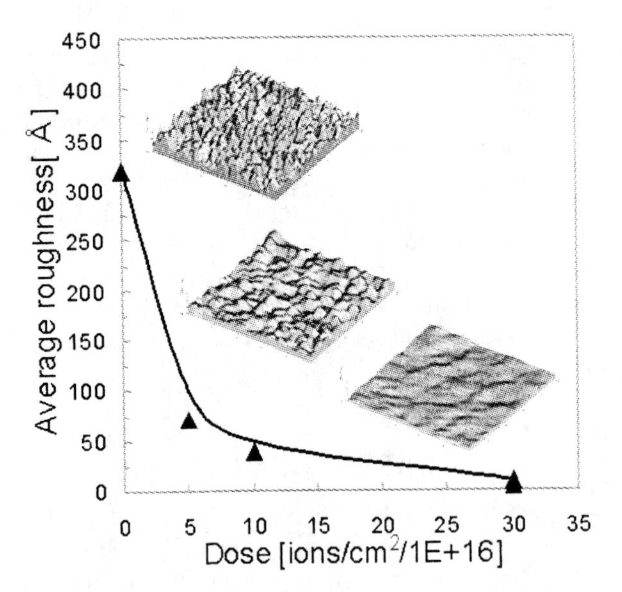

FIGURE 1. Experimental results of diamond surface roughness.

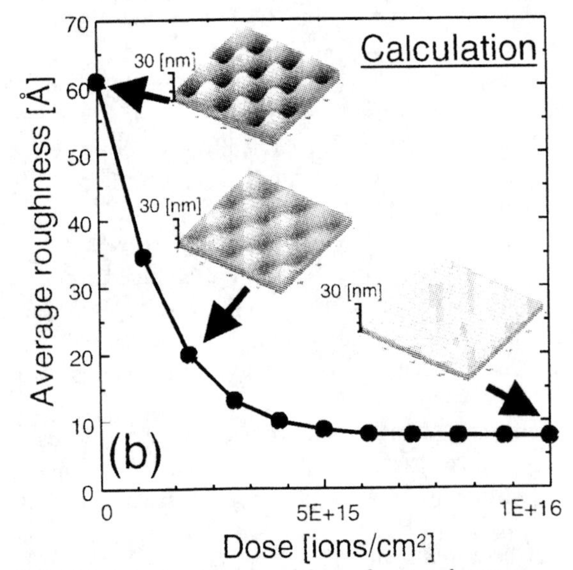

FIGURE 2. Calculated results of Cu surface roughness.

the CVD-diamond and the Cu target. However, this is explained by the difference in their sputtering yields. So, the decreasing curve of average roughness in the simulation can be considered to show good agreement with the experiment. The simulation results confirm that the atomic movement on single cluster ion impact can be described by a single model. The initial surface has hills and valleys expressed by a sine curve. When the surface is sputtered with cluster ion impacts, atoms on the surface are moved to the opposite side from the incident direction, under the condition that the incident angle is larger than the critical angle. As a result, atoms on a hill are moved to a valley, and hills become lower and valleys become shallower. Thus, surface roughness decreases with ion dose.

The surface roughness of diamond membranes is improved through Ar cluster ion sputtering. However, the appearance of the diamond surface became brown after irradiation. It was considered that graphite layers were generated by residual monomer ions in the cluster ion beam and remain on the diamond surface. In contrast, Ar cluster irradiation without monomer ions didn't generate any graphite layer but it's etching rate was too low. Graphite layers influence the optical transmittance of the diamond films which is important for X-ray lithography masks. Fig.3 shows the optical transmittance of the diamond films irradiated with Ar cluster ion beams. The transmittance decreased from 55% to 40% at a wavelength of 600nm.

In order to remove the graphite layers from the diamond surface, the diamond membrane irradiated with Ar cluster ions was chemically treated with a hot H_2SiO_4 + HNO_3 mixed-acid solution. There was a substantial change in color of the irradiated diamond surface but a slight darkish color still remained.

After Ar cluster irradiation, the diamond membrane was then irradiated with reactive O_2 clusters (20keV,1×10^{15}ions/cm^2). The brown color on the diamond surface disappeared with O_2 cluster irradiation. Fig.4 shows the optical transmittance of the same diamond films irradiated subsequently with O_2 cluster ions. The optical transmittance increased to 50%.

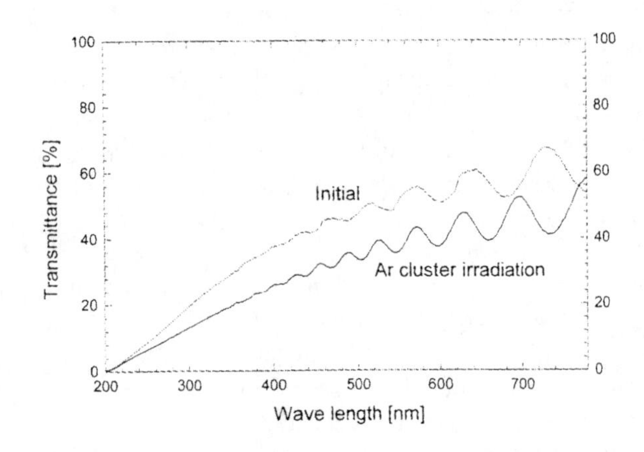

FIGURE 3. The optical transmittance of diamond films irradiated with Ar cluster ion beams.

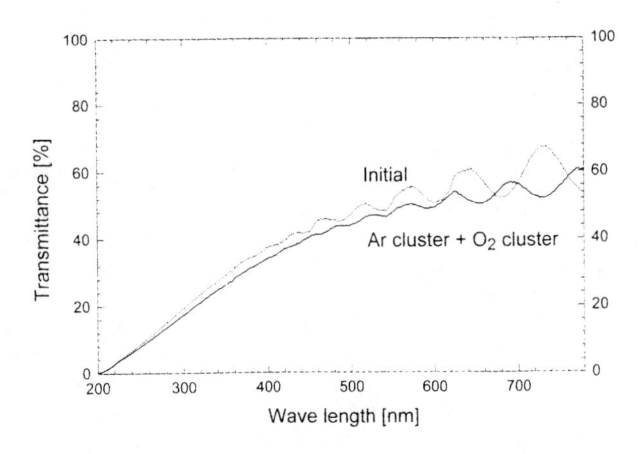

FIGURE 4. The optical transmittance of the diamond films before and after irradiation with an O_2 cluster ion beam.

Single crystalline diamond was irradiated with Ar or O_2 cluster ions in order to confirm the effect of reactive O_2 cluster irradiation in removing carbon from on the diamond. The bonding states of the carbon atoms on the diamond surface were investigated by Raman spectroscopy. Fig.5 shows a Raman spectrum of single crystalline diamond together with spectra after irradiation with Ar or O_2 cluster ions. In the case of initial single crystal diamond, a sharp peak is observed at 1332cm^{-1} with no broad graphite peak. However, there is a broad signal from graphite at around 1600cm^{-1} after Ar cluster ion irradiation indicating that the state of the diamond surface is changed to graphite by residual monomer ions in the cluster ion beam. However, the Raman spectrum after O_2 cluster ion beam irradiation is almost the same as the initial one. As the sputtering yield of carbon with O_2 cluster ions is ten times higher than with Ar, no graphite layer was observed with O_2 cluster ions.

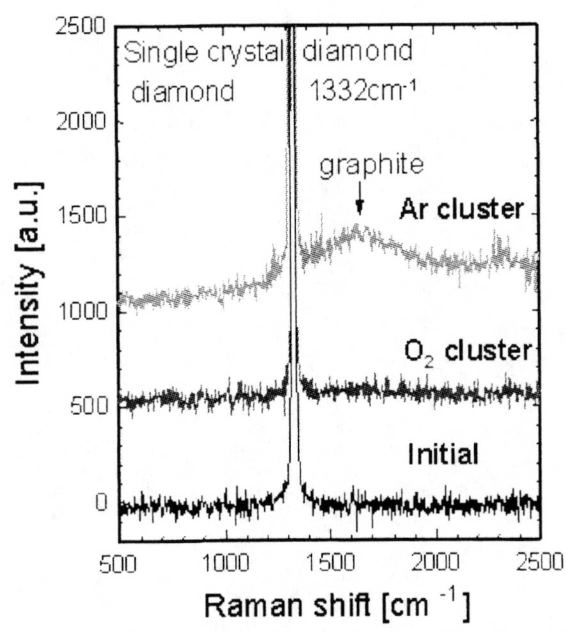

FIGURE 5. Raman spectra from single crystalline diamond and from samples irradiated with Ar or O_2 cluster ions.

These results demonstrate that a qualitatively flat diamond surface without graphite layer can be made by the combination of Ar and O_2 cluster ion beam treatments. Experimentally, the rough diamond surface in the as-deposited condition is irradiated with an Ar cluster ion beam (20keV, 1×10^{17}ions/cm^2) and a smooth surface is obtained from lateral sputtering effects. Subsequently, the graphite layer produced by Ar cluster ions is removed by reactive O_2 cluster ions (20keV,1×10^{15}ions/cm^2).

Fig.6 shows AFM images of an as-deposited diamond membrane (a) and one irradiated subsequently with Ar and O_2 cluster ion beams (b). After the irradiations, the average roughness was reduced from 330ÅRa to 40ÅRa. The optical transmittance of the film was 50%, which is almost same as the initial one.

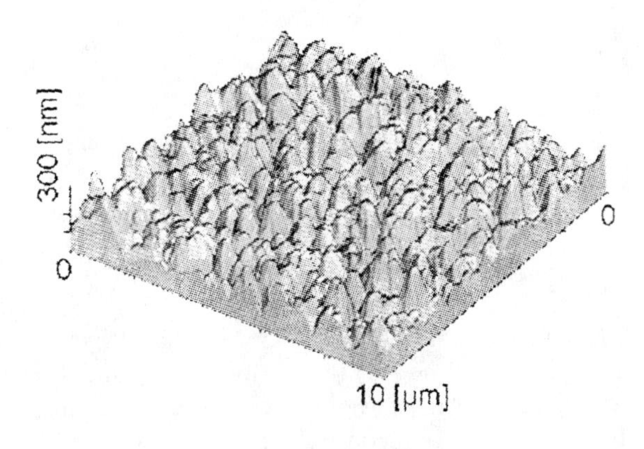

(a) As-deposited diamond. Ra=330Å

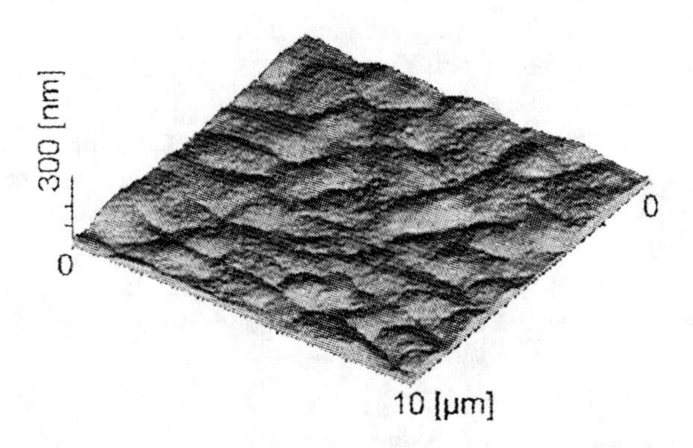

(b) Diamond irradiated with Ar cluster ions (1×10^{17}ions/cm^2) and O_2 cluster ions (1×10^{15} ions/cm^2). Ra=40Å

FIGURE 6. AFM images of a diamond membrane

A mask membrane for X-ray lithography as shown in Fig.7 was successfully manufactured for trial by means of the combination of Ar and O_2 cluster ion irradiations on a 32mm × 32mm area of a 3inch diameter diamond membrane.

FIGURE 7. A mask membrane of diamond for X-ray lithography successfully manufactured for trial.

CONCLUSION

CVD-diamond membranes were irradiated with Ar cluster ion beams. The result of a Monte-Carlo simulation showed good agreement with the experimental results. It confirms the lateral sputtering effect. A disadvantage of Ar cluster ion smoothing is that it generates a graphite layer by residual monomer ions in the cluster beam which reduces the optical transmittance of the diamond membrane.

The sputtering yield of carbon with O_2 cluster ions is enhanced by chemical reactions. Thus, O_2 cluster ions irradiation was very effective in removing graphite layers on the diamond membrane.

As a consequence, it was shown that a smooth diamond surface without a graphite layer could be created in the following way; first, as-deposited diamond films are irradiated with Ar cluster ions and, second, with O_2 cluster ions to remove the graphite layer formed during the first step. The average roughness of diamond treated with this process was 40Å Ra and the optical transmittance at a wavelength of 633 nm was 50%. A very smoothing diamond surface was produced without reducing the optical transmittance, and such material is applicable for an X-ray lithography mask membrane.

REFERENCES

1. M.Kobayashi, M.Sugawara, K.Yamashiro and Y.Yamaguchi, Microcircuit Engineering **11**,237(1990).
2. S.Tuboi, H.Okuyama, K.Ashikaga and Y.Yamashita, J. Vac. Sci. Technol., **B13**, 6(1995)
3. M.F.Ravet and F.Rousseaux, Diamond and Relat. Mater., **5**,812(1996).
4. H.Yoshikawa, N.Kikuchi, H.Yamashita, Y.Matsui and K.Marumoto, Proc. 2nd Int. Conf. on the Appli. of Diamond Films and Relat. Mater., MYU, Tokyo, 445(1993).
5. H.Tokura. C.F.Yang and M.Yoshikawa, Thin Solid Film, **212**,49(1992).
6. S.Ilias, G.Sene, P.Moller, V.Stambouli, J.Pascallon, D.Boucheir, A.Gicquel, A.Tardieu, E.Anger and M.f.Ravet, Diamond and Relat. Mater., **5**,835(1996).
7. H.Buchkremer-Hermanns, C.Long and H.Weiss, Diamond and Relat. Mater, **5**,845(1996).
8. R.E.Rawles, S.F.Komarov, R.Gat, W.G.Morris, J.B.Hudson and M.P.D'Evelyn, Dia. and Relat. Mater., **6**, 791(1997).
9. N.Hagiwara, N.Toyoda, J.Matsuo and I.Yamada,"Monte Carlo Simulation of Surface Smoothing Effect by Cluster Ions", presented at the IIT98 Conference on Kyoto, Japan, June 22-26, 1998.
10. J.Matsuo, H.Abe, G.H.Takaoka and I.Yamada, Nucl. Instr. and Meth. B., **99**,244(1995).

Oxide Film Deposition by Gas-Cluster Ion Assisted Deposition

K. Murai, S.Tamura, M. Kiuchi and N. Umesaki

Osaka National Research Institute, AIST, MITI, Ikeda, Osaka 563-8577, JAPAN

E. Minami, J. Matsuo and I. Yamada

Ion Beam Engineering Experimental Laboratory, Kyoto University, Sakyo, Kyoto, 606-8501, JAPAN

The development of ultra-high quality (UHQ) films is in progress under the NEDO projects*. This project is a collaboration between several companies, national laboratories and universities. A Multi-Beam Gas-Cluster Ion Beam System has been developed for fabricating indium tin oxide (ITO) films. The concept of a multi-beam deposition technique, the description of the Multi-Beam Gas Cluster Ion Beam System as well as our current progress on this project are presented.

INTRODUCTION

Sn doped indium oxide (ITO) films[1] are widely used in flat panel displays (FPD) because the film properties such as transparency and resistivity meet industrial requirements. However, for the higher-speed larger-area opto-electronic device applications, resistivity values of the order of 10^{-5} Ω•cm are required for transparent electrodes. It is quite difficult to achieve this resistivity by conventional deposition techniques, e.g. sputtering and vacuum evaporation, even above 200 °C. Furthermore, the temperature during film formation should be near 100 °C or less in order to reduce cost. It has been demonstrated that the resistivity of ITO films, formed by oxygen gas-cluster ion beam assisted deposition, can achieve 1.8×10^{-4} Ω•cm at room temperature[2] which could not be achieved by other techniques. To apply this technique in industry, a high throughput deposition system is necessary. As a result, a Multi-Beam Gas-Cluster Ion Beam system has been developed within this project.

CONCEPT OF MULTI-BEAM DEPOSITION TECHNIQUE

Normally, a crystalline ITO film has lower resistivity than an amorphous one. The conductivity of ITO results from the substitution of indium (In) by tin (Sn), and from the oxygen vacancies inside the In_2O_3 structure. Therefore, controlling Sn positions and oxygen vacancies is quite important for a low resistivity film.

* The NEDO Regional Consrtium Project, 1997, Project Office: Osaka Science and Technology Center, Utsubo, Osaka, Japan: "R&D of Ultra-High-Quality Transparent Conductive Films Fabrication (FY97-99)", Project leader; I. Yamada.

Figure 1 shows the typical XRD spectra for indium tin oxide (ITO) films fabricated by the sputtering technique. As shown in this figure, the crystallinity of the ITO films changes with the substrate temperature. As seen in Fig.1, the ITO film formed under 200°C is amorphous because the crystallization temperature of ITO is around 200°C. Therefore, it is quite difficult to form crystalline ITO films near 100 °C.

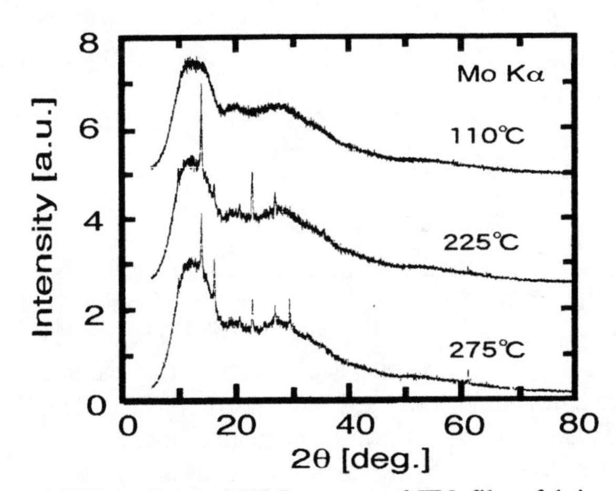

FIGURE 1. Typical XRD spectra of ITO films fabricated by sputtering technique as a function of the substrate temperature. Crystalline ITO are foemed above a substrate temperature of 200 °C.

Considering fabricating crystalline ITO films below 100°C, another excitation beam is necessary in order to excite the deposited film during the deposition process. Figure 2 shows a concept of the multi-beam deposition technique. The additional ion or laser beam irradiates a film deposited by the e-beam, pulsed laser or sputtering

CP475, *Applications of Accelerators in Research and Industry*, edited by J. L. Duggan and I. L. Morgan

deposition methods. The energy of ion beams must be low enough to avoid implantation. Therefore a cluster ion beam source or a low energy ion source is a candidate. The oxygen gas-cluster ion beam[3] may also be useful to control oxygen vacancies inside the films. Since ITO has large absorbance in the infrared and ultra-violet wavelength ranges, the laser beam has to emit in one of these wavelength ranges. Thus, an excimer laser or a CO_2 laser system is a candidate. The grain boundaries and the impurities are reduced by these excitation beams.

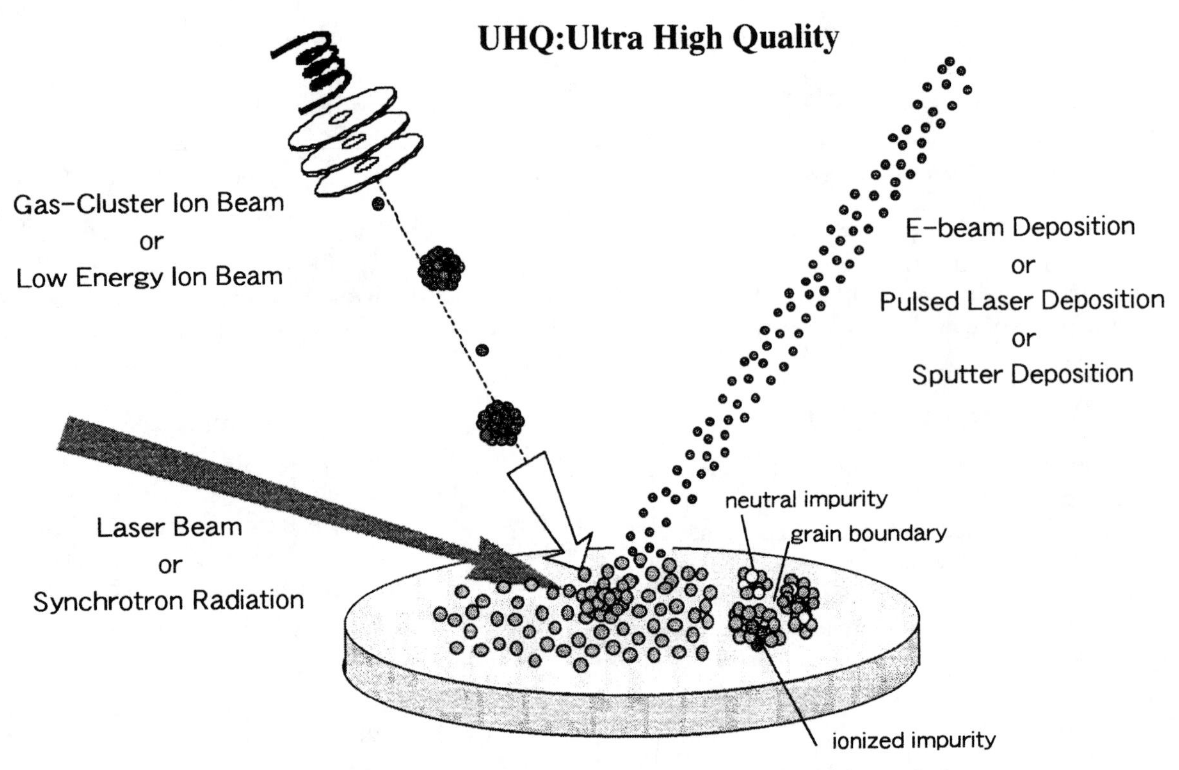

FIGURE.2. Fabrication of UHQ films with multi-beam deposition technique.

A MULTI-BEAM GAS-CLUSTER ION BEAM SYSTEM

Figure 3 shows a schematic of a gas-cluster ion source. Gas-cluster ions are formed from the Laval glass nozzle then ionized by electron bombardment. Then the gas-cluster ions are accelerated and separated from the neutral clusters by a deflector.

Figure 4 shows a schematic diagram of the Multi-Beam Gas-Cluster Ion Beam System developed in our project. The details of this system are also shown in Table 1. As seen in Fig. 4 and Table 1, this system has two e-guns and a gas-cluster ion source inside the chamber. The gas-cluster ion source was fabricated by EPION Corporation. There are two ports for pulsed laser deposition (PLD) and laser annealing. All the components can be computer-controled.

In order to achieve higher deposition rates on a larger area substrate, the gas cluster ion source is scheduled to be upgraded from an ion current of 10 μA to 50 μA. The deposition area will be enlarged to 150 × 150 cm^2. The gas-cluster ion beam can be scanned over this large area.

The laser beam for laser annealing can also be scanned over this area.

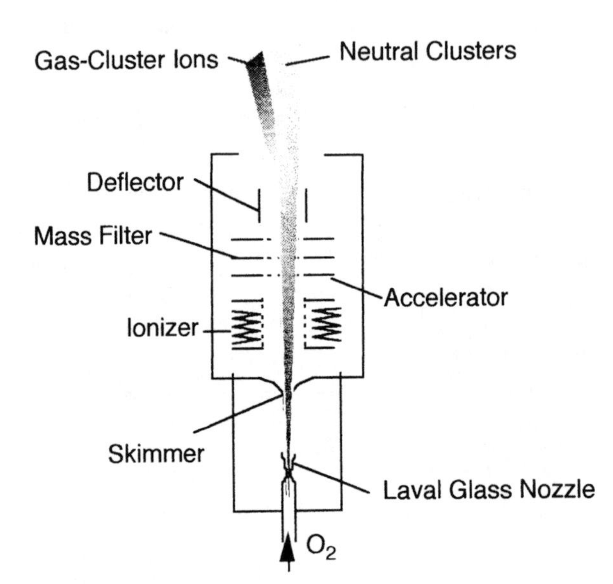

FIGURE 3. Schematic of a gas-cluster ion source.

TABLE 1. Specifications of the Multi-Beam Gas-Cluster Ion Beam System

Component	Performance
Reaction Chamber	**60 × 60 × 90 cm³, SUS304**
Vacuum System	**RP: 2800 liter/min** **DP: 20 inch**
Heater	**maximal substrate temperature: 500°C**
Substrate Size	**100 × 100 cm² (upgrade to 150 × 150 cm²)**
E-gun evaporator	**#1: 100 cc (1 pocket)** **#2: 30 cc (4 pockets)**
Gas Cluster Ion Source	**gas: oxygen** **ion current: 10 μA (upgrade to 50 μA)**
Laser Port	**pulsed laser deposition (PLD)** **laser annealing**

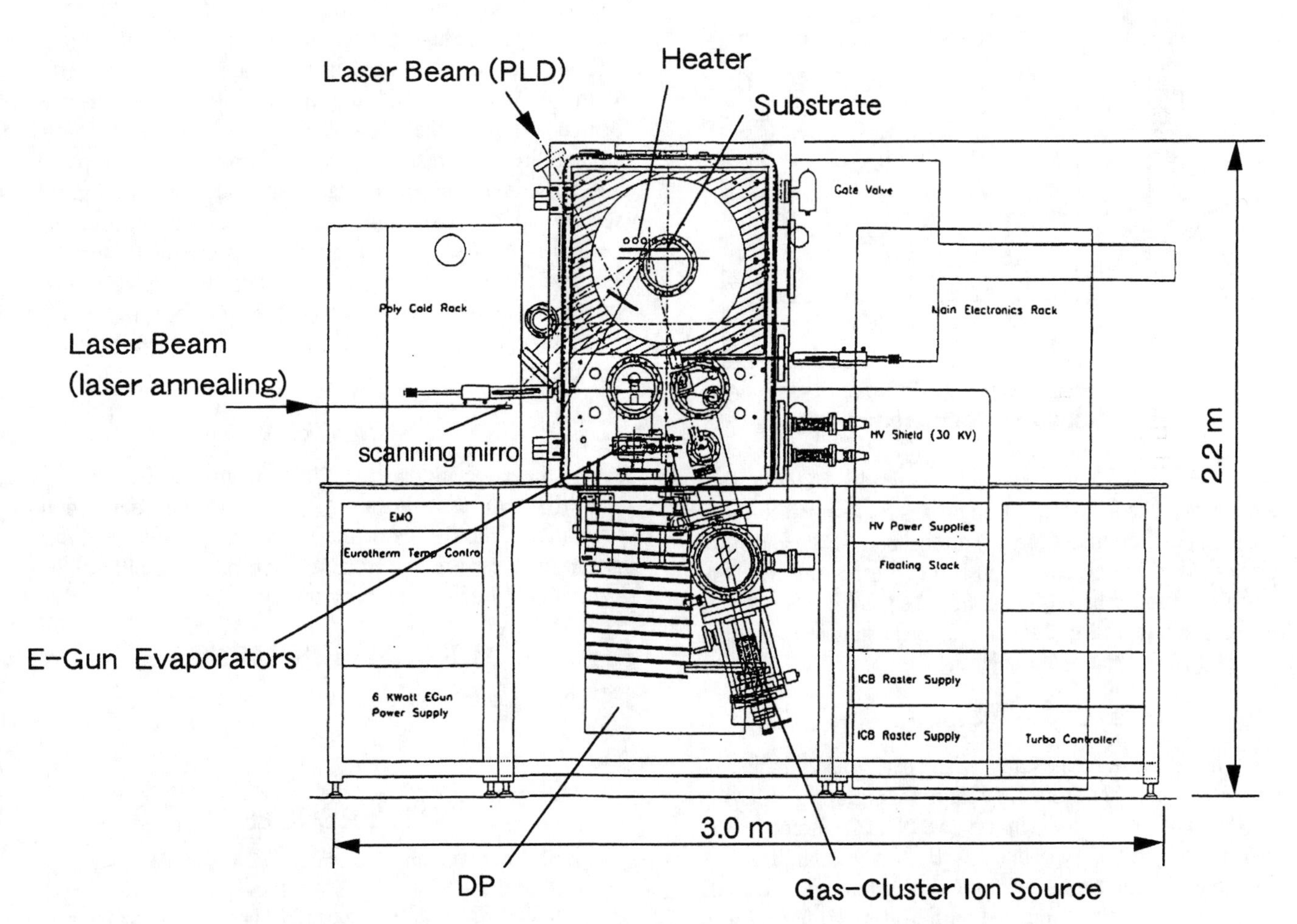

FIGURE 4. Schematic Diagram of a Multi-Beam Gas-Cluster Ion Beam System

TABLE 2. Characteristics of ITO Targets Used in the Experiment.

Target	Density [g/cm³]	Relative Density	Bulk Resistivity [Ω•cm]	Thermal Conductivity [cal/cm•sec•°C]
UHD	7.0	98%	1.3×10^{-4}	0.028
MD	5.0	70%	1.0×10^{-3}	0.008

CURRENT PROGRESS

With the Multi-Beam Gas-Cluster Ion Beam System, we have started to deposit ITO films by e-beam deposition with and without assistance of the gas-cluster ion beam.

Figure 5 shows the current results obtained using the multi-beam system. The resistivities we achieved are plotted as a function of the substrate temperature. The shaded area in this figure shows our goal which has resistivities below 10^{-4} Ω•cm and substrate temperatures below 100 °C.

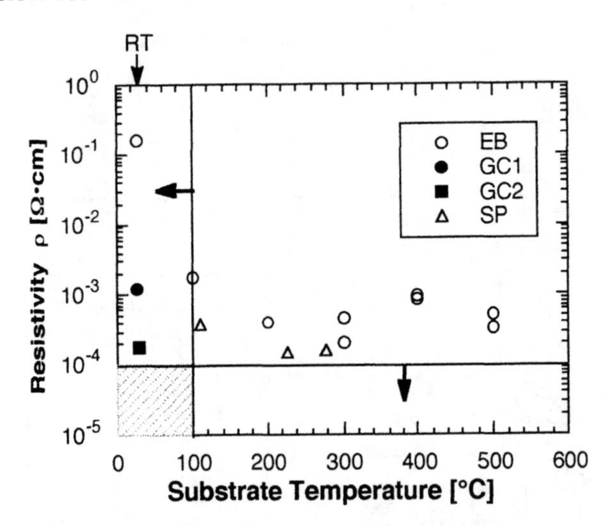

FIGURE 5. Current experimental results. Our goal is shown as a shaded area. Open circles correspond to the e-beam deposition under oxygen atmosphere (EB), and filled circles do to the e-beam deposition with assistance of the gas-cluster ion source (GC1). Filled square corresponds to the data shown in Ref.2 (GC2). Typical sputtering data are also shown by the open triangles (SP).

E-Beam Deposition

The e-beam deposition was performed under an oxygen atmosphere of 2×10^{-4} torr. Ultra-high density (UHD) and medium density (MD) ITO targets were investigated in the experiments. Both of the targets were fabricated by the Japan Energy Corporation. Since UHD has a higher density compared with MD, UHD has a lower heat conductivity than MD. The spitting effect occurred at a deposition rate of more than 5 nm/sec for UHD while it occurred at a deposition rate more than 0.1 nm/sec for MD.

The ITO films deposited without assistance of the gas-cluster ion source (open circles in Fig. 5) have similar resistivities to typical sputtered ITO films (open triangles in Fig. 5).

Gas-Cluster Ion Assisted Deposition

In the gas-cluster ion assisted deposition, oxygen was supplied from the gas-cluster ion source instead of from an oxygen atmosphere as in normal deposition. The oxygen gas-cluster ions were accelerated with 7 kV and separated from the neutral clusters. During the e-beam deposition of ITO at a deposition rate of 0.15 nm/sec, the ion current of 2.7 μA was irradiating over an area of approximately 5 cm², i.e. the ion current density of approximately 5×10^{-7} A/cm². With this condition, a transparent ITO film was formed within the irradiated area even at room temperature. The resistivity of this film was low compared with those made by the e-beam and sputtering depositions (see the filled circle in Fig. 5). These data indicate that the multi-beam deposition technique is useful to achieve our goal shown in Fig.5. Since the experimental conditions studied to date are too few for the many parameters involved, it is necessary to optimize the deposition.

SUMMARY

We have developed the Multi-Beam Gas Cluster Ion Beam System as a result of a multi-beam deposition technique. The details of the multi-beam system and our current progress are reported. We will fabricate ultra-high-quality ITO films using this system.

ACKNOWLEDGMENTS

This project is supported by the New Energy and Industrial Technology Development Organization (NEDO).

REFERENCES

1. R. B. H. Tahar, T. Bam, Y. Ohya, and Y. Takahashi, J.Appl. Phys. **83**, 2631 (1998).
2. W. Qin, R.P. Howson, M. Akizuki, J. Matsuo, G. Takaoka, and I. Yamada, Mat. Chem. and Phys. **54**, 258 (1998).
3. I. Yamada, G. H. Takaoka, M. I. Current, Y. Yamashita, and M. Ishii, Nucl. Inst. and Meth. B **74**, 341 (1993).

Novel Analysis Techniques using Cluster Ion Beams

Jiro Matsuo, Noriaki Toyoda, Masahiro Saito, Takaaki Aoki, Toshio Seki and Isao Yamada

Ion Beam Engineering Experimental Laboratory, Kyoto University,
Sakyo, Kyoto, 606-8501, JAPAN

Ion beams are used intensively for analytical applications. In order to meet the demands of the recent remarkable progress in material science, new ion beam analysis techniques based on large gaseous cluster ion beams are proposed. Cluster ion beams, which provide an equivalent low energy beam, offer many advantages for analytical technique. For instance, sputtering yields of the cluster ions are one or two orders of magnitude higher than monomer ions and smooth, flat surfaces can be maintained during the sputter depth profiling with cluster ions. These advantages are the result of multiple-collision and high local density energy deposition of cluster ions. Gaseous clusters such as argon or oxygen, as utilized in the analysis are generated by supersonic expansion. The beam current of the cluster ions is a few μA which is sufficient for most analyses.

INTRODUCTION

One of the major applications of ion beams is in analysis techniques, such as Rutherford Back Scattering (RBS), Particle Induced X-ray Emission (PIXE), Nuclear Reaction Analysis (NRA), Secondary Ion Mass Spectrometry (SIMS), Auger Electron Spectrometry (AES) and X-ray Photoelectron Spectrometry (XPS). In SIMS, AES and XPS, ion beams are used to sputter surfaces to obtain depth profiles of the elements. There is a strong demand for precise measurement of depth profiles. For instance, SIMS is used most extensively to measure depth profiles of dopants in semiconductor wafers[1-3]. Dopant distribution must be precisely measured as sizes of electron devices shrink.

Shift and broadening of depth profiles are induced by ion beams for sputtering, because of collisional mixing, radiation enhanced diffusion and surfaces roughing[1,3]. These phenomena depend on the ion beam characteristics such as ion mass, energy and incident angle. When the ion energy is high, the depth resolution becomes low, so that low energy ion beams should be used.

Clusters, aggregates of atoms or molecules, are interesting not only as a new state of matter but also as a new beam approach for material analysis and processing. It has been demonstrated that energetic cluster ion beams are very useful for surface processing, providing shallow implantation, high-rate sputtering, surface smoothing, cleaning and film formation as a consequence of the unique irradiation effects [4-15]. For example, when a cluster containing 1000 atoms is accelerated with an energy of 10 keV, each constituent atom carries only 10 eV. Therefore, cluster ion beams are equivalent low energy ion beams with a high current density. The other interesting phenomenon is that many atoms collide with the surface within an area of only several nm^2. Thus, clusters impact a surface with low equivalent energy (low velocity) but with extremely high density.

These characteristics of cluster ion beams are quite suitable for analysis. Some studies of using cluster ion beams in analysis has been carried out. It has been shown that secondary ion emission yields increase with the number of constituents in the cluster ions[16]. This is a positive effect in SIMS. However, metallic cluster ions, which are not suitable for practical analysis application were used in the most of the studies. We have proposed to use large gaseous cluster ions for surface analysis instead of using small metallic clusters.

There are two advantages to using large cluster ions. One is that high depth resolution is expected because of the low equivalent energy. The other is applicability to insulating materials. A cluster ion can transfer many atoms with one charge, so that charging up of insulating materials can be eliminated. This is important not only for inorganic materials but also organic and biomolecular materials.

Here, we address the sputtering characteristics of cluster ions with a view to their applicability in surface analysis and discuss the advantages of this technique.

CP475, *Applications of Accelerators in Research and Industry*,
edited by J. L. Duggan and I. L. Morgan

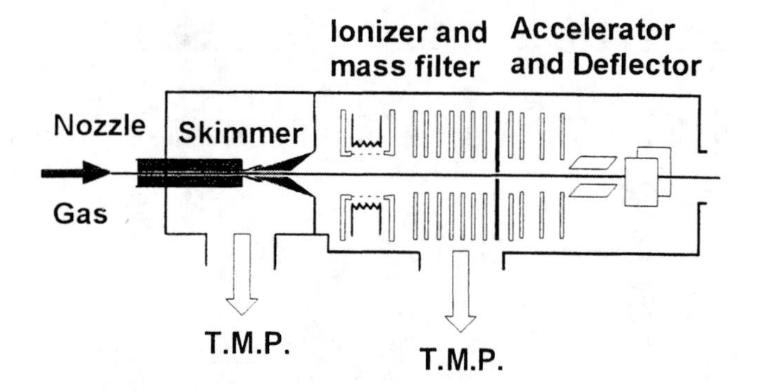

Figure 1 Schematic diagram of the cluster ion source

GAS CLUSTER ION SOURCE

A schematic diagram of the cluster ion source is depicted in Figure 1. This equipment consists of two vacuum chambers which are evacuated by different pumping systems. The technique for the generation of gas cluster ion beams and for the cluster-size separation has been described previously [4,13]. Argon or oxygen cluster beams are formed by adiabatic expansion through a Laval nozzle into the vacuum chamber. The clusters are ionized by electron bombardment, and the size of the cluster ions were selected using the retarding technique.

The clusters were introduced into the ionization chamber through the skimmer. They are ionized by electron bombardment with an ionization voltage of 150 V and an electron current of 45 mA. High current cluster ion beams of a few μA were obtained. The current density was about 1 μA/cm^2 which corresponded to about 1 mA/cm^2, when the cluster size is 1000. This current density is sufficiently high for most analysis applications.

Oxygen ions are widely used in SIMS because of the enhancement of secondary ion emission[2]. In order to increase the oxygen cluster beam intensity, the cluster source was cooled down to 120K with N_2 gas refrigerated by liquid N_2. As a result, the neutral beam intensity of O_2 clusters was increased by approximately a factor of 10 comparing the value at 300K. Oxygen diluted with helium was utilized for the source gas. The cluster beam intensity reaches the maximum value at a 70% mixture ratio. Thus, high intensity oxygen cluster beams were obtained by diluting O_2 gas with He gas and cooling the inlet gases [3,5].

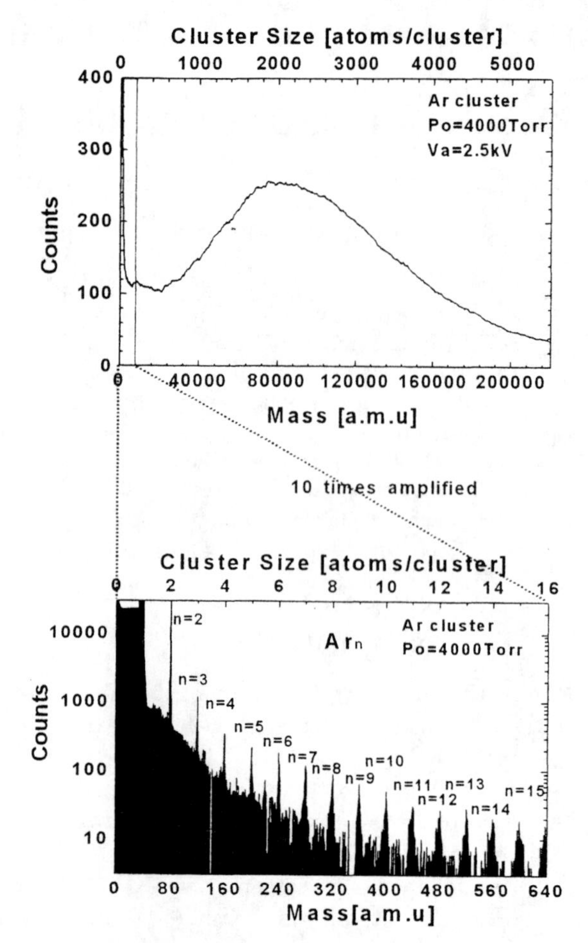

Figure 2 Typical TOF spectra of an argon cluster beam

The cluster size distribution was measured using a Time of Flight (ToF) technique. Cluster beams were ionized by electron bombardment with energy of 70 eV in the Wiley-MacLaren type TOF system. The duration time of the extraction pulse was 10 μs and the repetition rate was 200Hz. Figure 2 shows typical ToF spectra of an argon cluster beam. The argon cluster size distribution goes up to several thousands and the mean cluster size is about 2000. As shown in Figure 2-(a), the mass resolution of the ToF system used in this study is too poor to separate the individual peaks of each cluster size. However as shown in the Figure 2-(b), each different size of argon cluster can be separated, when the cluster size is small (N<30). Many techniques (see. Ref.4) are used to eliminate the monomer and small cluster ions in the cluster source.

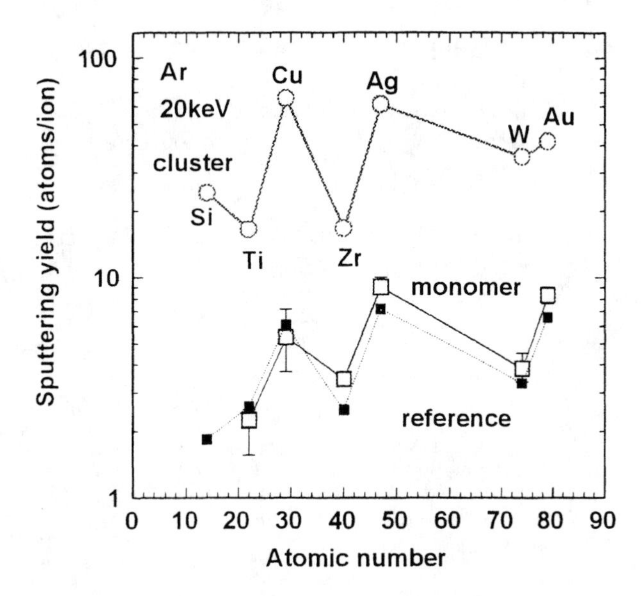

Figure 3 The sputtering yields by monomer and cluster ion, arranged according to target atomic number. The dotted line was the calculated sputtering yields by using the equation proposed by Kanaya et al. Sputtering yields by cluster ion bombardment are proportional to the reciprocal of the sublimation energy of the target atoms.

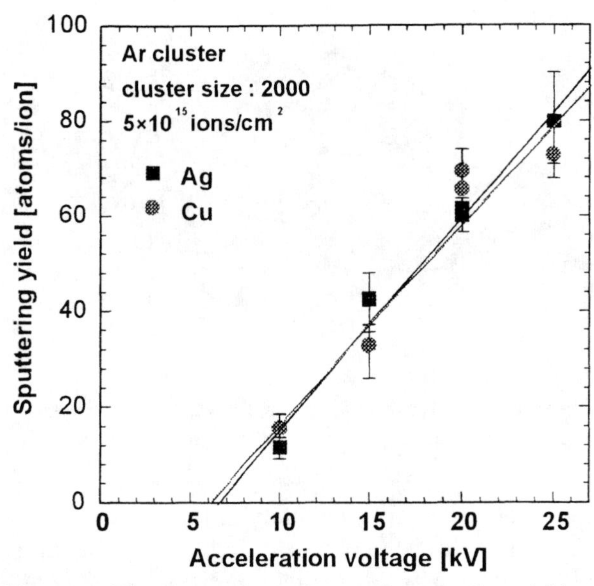

Figure 4 The energy dependence of sputtering yields of Ag and Cu by Ar cluster ions. The threshold energy for the sputtering is 6 keV.

SPUTTERING WITH CLUSTER IONS

Sputtering yields of metals by cluster ion were arranged in Figure 3 according to the atomic number of the target. The sputtering yields were calculated from sputtered depth measured by contact profiler. The sputtering yields of 20 keV monomer ion were also measured with the same procedure as that used for the cluster ion. The dotted line in Figure 3 is the monomer sputtering yield calculated by the equation proposed by Kanaya et al [17]. In the case of monomer ions, quite good agreement was observed between these experimental values and the reported values.

The sputtering yields are about one to two orders of magnitude higher than these of monomer ions. Molecular dynamics calculation shows that multiple collision of cluster atoms near the surface and a high energy density in the central collision zone of the cluster are responsible for very high sputtering yields by cluster ions[18,19].

In the case of monomer ions, the sputtering yields decrease rapidly with ion energy below 1 keV which is the energy used in most advanced SIMS system. Therefore, it is difficult to use very low energy monomer ion beams. However, as shown in Figure 3, several tens of atoms are sputtered by a 20 keV cluster ion which has low equivalent energy.

Figure 4 shows the typical energy dependence of sputtering yields of Ag and Cu by Ar cluster ions. The sputtering yields increase linearly with the ion energy. In this experiment, the threshold energy for the sputtering of both metals is 6 keV, which corresponds to an energy of 3 eV per cluster constituent atom. In the case of a monomer Ar ion, the threshold energies reported for Ag and Cu are 15 and 17 eV, respectively[20]. The threshold energy for a large cluster ion is significantly lower than that of a monomer ion. This result indicates that the mechanism of sputtering by cluster ions is totally different from that of monomer ions. Thus, materials can be sputtered by cluster ions with energies lower than 10 eV which is below the threshold energy of a monomer ion. Atomic mixing which leads degradation of depth resolution can be suppressed by using cluster beams.

Figure 5 shows AFM images of Au surfaces irradiated with Ar cluster ions at an incident angle normal to the surface. The ion energy and ion dose were 20keV and $1x10^{16}$ ions/cm^2, respectively. The Au surface was masked with ultra-fine mesh(2000 mesh/inch). The scanned area of the AFM was 20 μA x 20 μA. The sputtered depth was about 600Å. It should be noted that the surface irradiated with the cluster ions was much smoother than the unirradiated

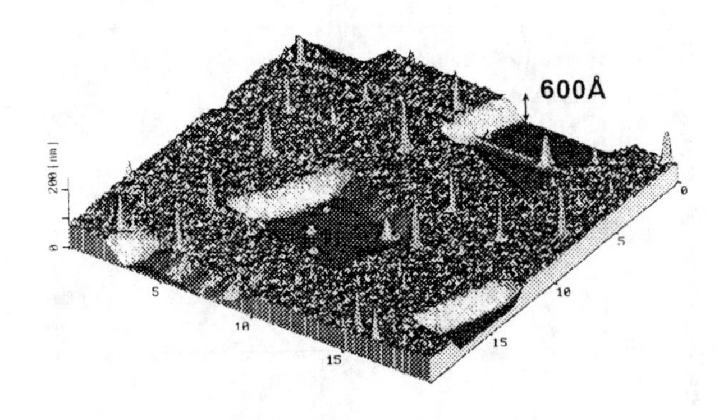

Figure 5 AFM image of Au surface sputtered by Ar cluster ion beams. The surface was covered by a mesh

area, such smoothing effect of cluster ion beams have been reported previously[12].

When a cluster ion containing a few thousand atoms bombards a target, multiple collisions between incident atoms and target atoms occur near the surface, and many atoms are sputtered from the surface in lateral directions. This phenomenon was predicted by molecular dynamic simulations[11,18], and is called "lateral sputtering". This phenomenon is responsible for the smoothing process.

Many materials have been sputtered by Ar cluster ion beams. In any materials, the smoothing effect always keeps the sputtered surface flat during ion bombardment and has been observed even for metal alloys. When metal alloys are sputtered by monomer ions, the surface roughness increases, because of selective sputtering.

No blistering and bubble formation occur with cluster ion bombardment[11], because penetration of the Ar atoms of a cluster is shallow. Thus, the composition and morphology of the sputtered target remain the same during Ar cluster ion bombardment. Depth resolution decreases, when surfaces become rough. Therefore, smoothing effect of cluster ion bombardment is desirable for analysis.

SUMMARY

A novel ion beam analysis technique is proposed by using large gaseous cluster ion beams. The cluster ion beam source and sputtering characteristics are shown. The advantages of using the cluster ion beams are the following;

(1) Equivalent low energy ion beams are utilized.

(2) Charge-up of insulating materials can be eliminated.

(3) High sputtering yields are realized.

(4) Surfaces can be kept smooth during sputtering.

High depth resolution is expected for cluster ion based analysis technique.

REFERENCE

1. C.Montandon, A.Bourenkov,L.Frey,P.Pichler and J.P.Biersack, Rad. Efeects & Defects in Solid, **145**,(1998),p-213
2. A. Bennighoven, C.A.Evans, K.D.McKeegan, H.A.Storms and H.W.Werner(eds.), *Secondary Ion Mass Spectrometry SIMS VII*. John Wiley & Sons, (1990)
3. Z.X.Jiang, P.F.A.Alkemade, J.Vac. Sci. Technol B, **16**, (1998), p.1971
4. J. Matsuo, H. Abe, G.H. Takaoka and I. Yamada, Nucl. Instr. and Meth. B, **99**, (1995),p.244
5. I. Yamada and J. Matsuo, Mat. Res. Soc. Symp. Proc., **396**,(1996), p.149(1996)
6. M. Akizuki, J. Matsuo, S. Ogasawara, M. Harada, A. Doi and I. Yamada, Jpn. J. Appl. Phys. **35**,(1996), p.1450
7. J. Matsuo, W. Qin, M. Akizuki, T. Yodoshi and I. Yamada' Mat. Res. Soc. Symp. Proc., in press (1997).
8. I. Yamada, W.L. Brown, J.A. Northby and M. Sosnowski, Nucl. Instr. and Meth. B, **79**,(1993), p.223
9. J. Matsuo, N.Toyoda and I. Yamada, J.Vac. Sci. Technol. B, **14**, (1996),p.3951(1996)
10. M. Akizuki, J. Matsuo, M. Harada, S. Ogasawara, A. Doi, I. Yamada, Mater. Sci. Eng., **A217-A218**,(1995), p.78
11. T. Aoki, J. Matsuo, Z. Insepov and I. Yamada, Nucl. Instr. and Meth. B, **121**,(1997), p.49
12. I. Yamada, and J. Matsuo, Z. Insepov, D. Takeuchi, M. Akizuki and N. Toyoda, J.Vac. Sci. Technol A, **14**,(1996), p.781
13. M. Akizuki, J. Matsuo, M. Harada, S. Ogasawara, A. Doi, K. Yoneda, T. Yamaguchi, G. H. Takaoka, C. E. Ascheron and I. Yamada: Nucl. Instr. and Meth. B, **99**,(1995), p.229.
14. R. Beuhler and L. Friedman, Chem.Rev., **86** (1986) 521
15. O. F. Hagena, Rev. Sci. Instrum. **63** (1992)2374
16. J.P.Thomas, P.E.Filpus-Luyckx, M.Fallavier and E.A.Schweikert, Phys.Rev. Lett. **55** (1985) p.103
17. K.Kanaya, K.Hojyou, K.Koga and K.Toki, Jpn.J.Appl.Phys., **12** (1973)1297
18. Z.Insepov and I.Yamada, Laser and Ion Beam Modification of Materials: edited by I. Yamada (Elsevier,1994)19
19. H.Hsieh, R.S.Averback, H.Sellers, C.P.Flynn, Phys.Rev. **B45** (1992) 4417.
20. L.Maissel and R.Grang, Handbook of Thin Film Technology (Mcgraw-Hill, 1970)

SECTION V

PIXE, X-RAYS, RBS, ERD, NRA, CPAA, FRAGMENTATION, DESORPTION, SPUTTERING, CODES, AND HANDBOOKS

Secondary Fluorescence in PIXE Channeling Measurements

A. Seppälä[a], R. Salonen[a] and J. Räisänen[b]

[a]*Accelerator laboratory, P.O.Box 43, FIN-00014 University of Helsinki, Finland*
[b]*Department of Physics, University of Jyväskylä, P.O.Box 35, FIN-40351 Jyväskylä, Finland*

The effect of secondary fluorescence on PIXE channeling minimum yields was studied in ZnSe/GaAs heterostructures. The secondary fluorescence enhancement of the Zn K_α line was calculated as a function of proton energy for different ZnSe layer thicknesses. The corresponding yields were measured for an 1800 nm thick ZnSe layer grown on GaAs substrate. An equation for calculating the secondary fluorescence correction to the Zn minimum yield was derived and the corrections to the measured minimum yields were calculated. A notable enhancement in the Zn minimum yield due to secondary fluorescence was observed with proton energies higher than 1.5 MeV.

INTRODUCTION

Ion beam channeling is an effective tool for investigating the properties of crystalline materials. Particle Induced X-ray Emission (PIXE) combined with channeling can be used for studying lattice locations of impurity atoms in crystalline thin films (1).

For the determination of the lattice location of an impurity atom in a host lattice the characteristic X-ray yields are measured as a function of the sample's tilt angle with respect to some crystallographic direction. The lattice location of impurity atoms affects the shape of the angular scan and the minimum yield *i.e.* the yield along the channel direction. The measured yield is compared with the simulated yield for certain lattice location of the studied impurity. Several factors in the measurement system such as beam divergence and sample quality also affect the detected yield. For an accurate analysis all these factors have to be taken into account in the comparison of the simulated and measured data (2).

In this study the generally neglected effect of secondary fluorescence on PIXE channeling results is investigated. The particle-induced X-rays within the matrix may cause secondary fluorescence enhancement of the characteristic X-rays of the studied element. The effect of this phenomenon on the channeling minimum yield of zinc in epitaxially grown ZnSe thin films on GaAs substrates is studied experimentally. An equation for correction of the measured minimum yield is presented. Also calculations to estimate the magnitude of secondary fluorescence in this particular case are presented. To the knowledge of the authors this is the first time this effect has been studied in the present context.

CALCULATIONS

The secondary fluorescence yield for zinc in ZnSe/GaAs heterostructures was calculated theoretically following the procedure presented by Campbell et. al (3). The present PIXE channeling measurement geometry described in the next section was used. A 5° tilt angle between the surface normal and the incoming beam was used in all calculations to avoid channeling effects. The calculations thus correspond to measurements at random direction.

The ionization cross sections were taken from Ref. (3) and the photoelectric cross sections from the XCOM database (4). The attenuation coefficients were taken from the NIST X-Ray Attenuation Database (5). The fluorescence yields for the K shell were obtained from the evaluations of Krause (6). For the relative emission rates of different subshells the theoretically calculated values of Scofield (7) were used. Jump ratios for the subshells were calculated based on the data obtained from the XCOM database (4).

In Fig. 1 the calculated relative secondary fluorescence yields of the Zn K_α line for the ZnSe/GaAs structure are presented as a function of proton energy. The yields for each

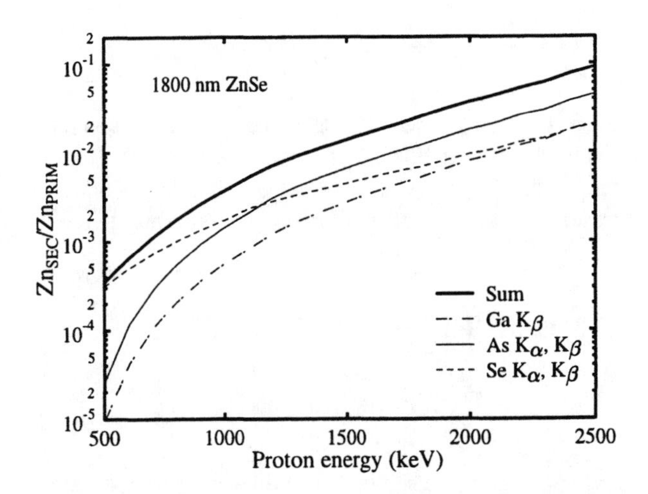

FIGURE 1. The calculated Zn K_α secondary fluorescence yields induced by characteristic X-rays from Ga K_β, As K_α,K_β and Se K_α,K_β as a function of proton energy. The secondary fluorescence yields were normalized with the primary Zn K_α yield calculated in the same geometry. The thickness of the ZnSe layer in these calculation was 1800 nm.

CP475, *Applications of Accelerators in Research and Industry*,
edited by J. L. Duggan and I. L. Morgan

element inducing secondary fluorescence are plotted separately together with the sum of these components. All yields have been normalized with the primary proton induced yield for zinc calculated in the same geometry. The calculations were done for three different ZnSe layer thicknesses: 500, 1800 and 2500 nm. In Fig. 1 the results for the 1800 nm thick layer are shown. No remarkable differences in the sum yields of the three components (Ga, As and Se) were noticed between different layer thicknesses. The relations between the components changed as a function of ZnSe layer thickness, most remarkable differences were observed at low proton energies (500 - 1000 keV).

In the secondary fluorescence calculations the attenuation of Ga and As characteristic X-rays was approximated to be equal in GaAs and ZnSe. The value for GaAs was adopted. In the case of As K_α and K_β the attenuation coefficients are almost the same for GaAs and ZnSe, but a difference exists between these two materials in the case of Ga K_β. The thickness of the ZnSe layer was much smaller than the range of the protons in the GaAs substrate and thus the error in the calculated yields caused by this approximation was small.

To estimate the effect of secondary fluorescence on channeling minimum yields Eq. (1) was derived. This equation is based on the fact that Zn secondary yield is proportional to Ga, As and Se primary yields. Therefore, the ratio of secondary fluorescence yield in the channel direction to the yield in random direction depends on the channeling minimum yields of the elements inducing the secondary fluorescence. This ratio affects the minimum yield of Zn. The modified minimum yield of zinc χ_{Zn}^{sec} and the minimum yield χ_{Zn} are related by the equation

$$\chi_{Zn}^{sec} = \chi_{Zn} + \sum_i \frac{Y_{sec}^i}{Y_{prim}}(\chi_i - \chi_{Zn}), \tag{1}$$

where Y_{sec}^i and Y_{prim} are the secondary fluorescence yield of Zn and the proton induced primary yield of Zn. The index i refers to the element (Ga, As or Se) inducing the secondary fluorescence yield. χ_i is the minimum yield for this element.

MEASUREMENTS AND RESULTS

Experimental Set-Up

The ZnSe samples used in this study were grown at Tampere University of Technology by molecular beam epitaxy on GaAs substrates. The thickness of the undoped ZnSe layer was 1800 nm.

All experiments were performed using the 2.5 MV Van de Graaff accelerator at University of Helsinki. An HPGe detector located at 135 ° angle relative to the incident beam direction was used to detect the X-rays. The sample was mounted into a 3-axis goniometer placed inside a vacuum chamber. The ion doses were obtained using a beam chopper system. The measurement set-up is described in more detail in Ref. (2). The sample alignment was done with 1.5 MeV $^4He^+$ ions using RBS channeling.

PIXE Channeling Measurements

The effect of Zn K_α secondary fluorescence yield on the PIXE channeling minimum yield of Zn was studied experimentally by comparing the minimum yields of zinc and selenium as a function of proton energy. The effect of secondary fluorescence should be detected in the minimum yield of Zn but not in Se. According to the calculations the magnitude of this effect depends strongly on the proton energy. Thus the behaviour of Zn minimum yield should differ from that of Se at higher proton energies. In the case of ZnSe/GaAs heterostructures the characteristic X-rays of arsenic (K_α and K_β), gallium (K_β) and selenium (K_α and K_β) can cause secondary fluorescence of zinc.

The Zn and Se minimum yields in the <100> channeling direction were measured by PIXE with proton energies of 0.75-2.5 MeV. All PIXE yields for the minimum yield determinations were taken from the K_α peaks of zinc and selenium. The statistical uncertainties of the experimental minimum yield values were approximately 2% for Zn and 3% for Se. The results are presented in Fig. 2.

As can be seen the Zn minimum yield increases at the high proton energies though the minimum yield of Se decreases. This might be due to the secondary enhancement of Zn.

The corrected Zn minimum yields, χ_{Zn}, were calculated using Eq. (1). The corrections were based on the presented relative secondary fluorescence calculations for an 1800 nm

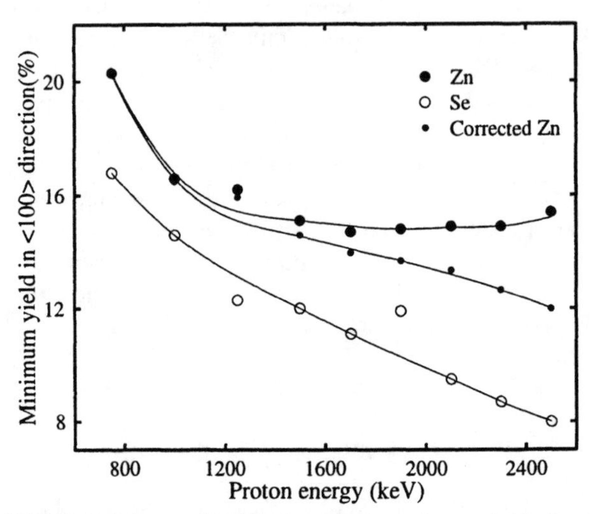

FIGURE 2. Measured PIXE channeling minimum yields for zinc (•) and selenium (o) as a function of proton energy. The sample used in these measurements was 1800 nm thick ZnSe layer on GaAs substrate. The corrected Zn minimum yields calculated using Eq. (1) are also presented. The solid lines were drawn to guide the eye.

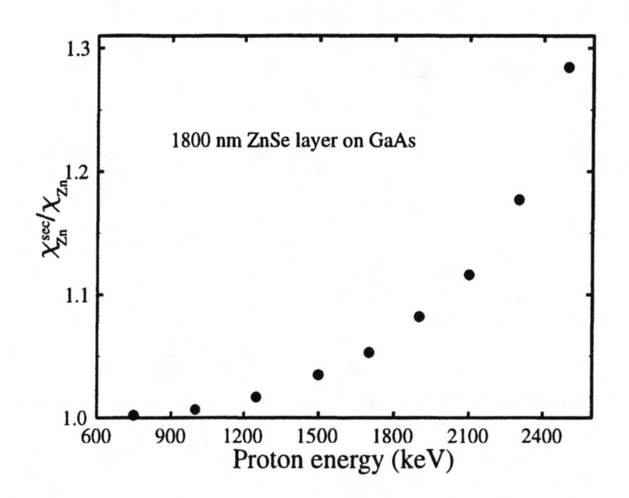

FIGURE 3. The calculated minimum yield ratio, $\chi^{sec}_{Zn}/\chi_{Zn}$, as a function of proton energy. the calculations were performed using Eq. (1) and based on the calculated relative secondary fluorescence yields and the measured minimum yields.

thick ZnSe layer. The minimum yield values of Zn, Ga, As and Se used in the calculations were obtained from the measurements described above. As a function of proton energy the minimum yield values ranged from 73% to 85% for Ga and from 64% to 83% for As. The obtained corrected minimum yields are plotted in Fig. 2.

The ratio of the modified to the correct minimum yield, $\chi^{sec}_{Zn}/\chi_{Zn}$, was determined based on the calculated results above and the measured Zn minimum yields. These results are presented in Fig. 3. It can be noted, that the error caused by secondary fluorescence in the Zn minimum yield for the 1800 nm thick ZnSe layer is more than 10% at high proton energies.

DISCUSSION

According to the calculations the effect of secondary fluorescence was smaller than the experimental error at low proton energies. At high energies the minimum yield enhancement due to secondary fluorescence was over 10-20 % as can be seen from Fig. 3.

In the PIXE channeling measurements at high energies a notable difference between the behavior of zinc and selenium minimum yields was noticed. By calculating the correction for Zn minimum yield using Eq. (1) it has been shown that this difference might be due to secondary fluorescence enhancement. As can be noted from Fig. 2 the behaviour of the Se minimum yield and the corrected minimum yield of Zn as a function of energy are similar within the experimental errors. To confirm this observed effect further studies are needed. In the future, the corrected minimum yields will be compared with simulated PIXE channeling minimum yields.

As can be noted from Eq. (1) the significance of secondary fluorescence in channeling PIXE analysis depends in the case of ZnSe/GaAs heterostructures on the difference between Zn and Ga or As minimum yields. In this heterostructure the other important factor is the ratio of secondary fluorescence yield and primary proton induced yield for zinc. The changes in the Zn minimum yield due to secondary fluorescence induced by Se are small, because the difference between Zn and Se minimum yields is small.

Using Eq. (1) it is possible to estimate the magnitude of the error caused by secondary fluorescence in lattice location determinations. The situation is more complicated if the minimum yield of an impurity is measured, for instance nickel in ZnSe layers. On the other hand it should be noted that the corrections are small if the impurity atoms lay on substitutional sites i.e. the minimum yields from lattice and impurity atoms are almost equal (when the differences in minimum yields due to different masses have been taken into account). In such studies the impurity atom concentration is a factor as important as the layer thickness of ZnSe in the present case.

In conclusion, at high proton energies (>1.5 MeV) in case of ZnSe/GaAs heterostructures the magnitude of secondary fluorescence enhancement has to be estimated for each particular case. The effect on the minimum yield is remarkable only if the difference between minimum yields of the studied element and the element inducing the secondary fluorescence is large. The effect of secondary fluorescence can be minimized by using low energy protons.

ACKNOWLEDGEMENTS

We are thankful to P. Uusimaa at the Tampere University of Technology for growing the ZnSe samples. This work is part of the EPI2 project funded by the Academy of Finland.

REFERENCES

1. Ecker K.H., Quan Z., Schurig T. and Weise H.P., *Nucl.Instr.Meth.B* **118**, 382-387 (1996).
2. Salonen R., Seppälä A., Ahlgren T., Rauhala E. and Räisänen J., *Nucl.Instr.Meth.B* in press.
3. Campbell, J. L., Wang, J.-X., Maxwell, J. A., and Teesdale. W. J.,*Nucl.Instr.Meth.B* **43** , 539-555 (1989).
4. Berger, M. J. and Hubbell, J. H. NBSIR 87-3597, National Bureau of Standards, Maryland, USA (1987).
5. Hubbell, J. H. and Seltzer, S. M. NISTIR 5632, National Institute of Standards and Technology, USA (1995).
6. Krause, M. O, *J.Phys.Chem.Ref.Data* **8**, 307-327 (1979).
7. Scofield, J. H, *At. Data Nucl Data Tables* **14**, 122-130 (1974).

The marginal leakage of some dental cements in humans: a PIXE-microbeam approach

A.Zadro*, G.Cavalleri**, S.Galassini***, G.Moschini***, P.Rossi***, P.Passi*

*Dental School, Department of Dental Materials, University of Padua, Italy
**Dental School, Department of Restorative Dentistry, University of Verona, Italy
***INFN, Laboratori Nazionali di Legnaro, Italy

The marginal leakage and water absorption of dental cements and restorative materials has been investigated by many authors with several techniques, some of which led to valid results. However, no technique could give, by itself, information both on leakage and water absorption, as these measurements usually need different investigations.
PIXE micro beam offers the possibility of investigating these two aspects at the same time, since it is possible to map a proper marker element.
In the present study, cavities were made on 50 extracted human molars, then filled with five different temporary cements (IRM, Cavit W, Kalsogen, Fermit N, SuperEBA). The filled teeth were placed into a 5% silver nitrate solution, and after three days, one, two, three and four weeks were examined. The samples for microPIXE were prepared after embedding the teeth in epoxy resin, and sectioning and grinding them down to a thickness of about 1 mm. The sections were placed on metal holders, and examined with a scanning proton μbeam, in Legnaro (Italy) at the AN2000 LAB of INFN National Laboratories. The beam consisted of 2.4 MeV protons, it had a cross section of 1.5 micron in diameter and typical currents of the order of some μA were used. The maps were obtained by an "ad hoc" software with a McIntosh personal computer. Mapping of silver allowed to evaluate both the marginal leakage and the water absorption for each cement. The samples filled with Cavit W showed a great infiltration, as the tracing element was found in the cement bulk, along the margins and inside the cavity, while those filled with IRM and Kalsogen presented only a deposition of the tracing solution on the cement surface. SuperEBA showed a poor resistance against microleakage, because the marker element was only detected along the cavity margins. Fermit N showed the best marginal integrity, and on its surface no traces of siver were found. In this case the better resistance may be due to the resin present in the composition of the material.

INTRODUCTION

Microleakage of dental restorative materials can damage the fillings and cause caries.
Studies on marginal seal of definitive restorative materials are widely referred in literature; on the contrary, the sealing ability of temporary cements was not studied with the same details, in spite of their importance in the clinical practice.
Many methods were carried out to evaluate the leakage in vitro: colorimeter solutions (basic fuchsin, methilen bleu), radioisotopes, electro-chemical procedures, detection of bacteria and/or particles movements (10). Most bacteria present dimensions approximately of (0.5-1.0) μm, while silver nitrate ions are much smaller (0.059 nm), so that a system able to avoid the leakage of silver ions should also prevent other particles infiltration. The purpose of the present in vitro study was to investigate the sealing property of some of the most employed temporary restorative materials, immersed in a silver nitrate tracing solution, by means of the PIXE-microbeam technique.

MATERIALS AND METHODS

Fifty human molar teeth extracted for periodontal desease were used. The teeth were immersed in formaldehyde 10%, rinsed with saline solution and stored in distilled water. On occlusal surfaces of the teeth cavities of (4x4) mm^2 were performed, standardized by the application of a cardboard template. The pulpal chambers were opened with a diamond cylindrical burr, under water irrigation, and roots canals were reamed and rinsed with a sodium hypochlorite solution and dried with absorbent paper points. The surface of each tooth was painted with 3 layers of transparent nail varnish, and finally coated with boxing wax, to avoid the leaking of the tracing solution through any possible scratch or additional root canal. Acrylic resin was placed on the floor of the cavities, allowing a constant residual space for restorative material of (4x4x4) mm^3. The teeth were subdivided in 5 groups of 10 elements each. The first group was restored with Cavit W, whose composition, as supplied by the manufacturer, is referred in the table 1.

CP475, *Applications of Accelerators in Research and Industry*,
edited by J. L. Duggan and I. L. Morgan

TABLE 1 Cavit W composition

Zinc Oxide
Calcium sulphate
Zinc sulphate
Glycol Acetate
Polyvinylchlorhydrate Acetate
Triethanolamine
Pigments

The second teeth group was restored with Kalsogen (Tab. 2).

TABLE 2 Kalsogen composition

Powder	Zinc Oxide
	Magnesium Oxide
	Zinc Stearate
	Zinc Acetate (Accelerator)
Liquid	Oil of Cloves (or Olive Oil at 15% Volume)
	Acetic Acid (Accelerator)

The third group was filled with IRM (Tab. 3).

TABLE 3 IRM composition

Powder	Zinc Oxide
	Polymethylmetacrylate
Liquid	Eugenol

In the fourth group super EBA cement was used (Tab. 4)

TABLE 4 Super EBA composition

Powder	Zinc Oxide
	Silicon Oxide
	Natural Resins
Liquid	Orthoethossybenzoic Acid
	Eugenol

In the last group Fermit N was used, a light curing material that presents a composition similar to that of the composite restorative resins (tab. 5).

TABLE 5 Fermit N composition

Ethyltriglycolmetacrylate
Silicon dioxide
Polyesterurethanedimetacrylate

Fermit N is however considered a " non definitive" material because it remains slightly elastic to let an easy removal from the teeth , after polymerization.

All the cements were worked carefully respecting the manufacturers' instructions.
The samples were placed in light proof bottles containing a silver nitrate solution at 5% in distilled water. Two samples for each group of teeth were examined at fixed time intervals (3 days, 1 week, 2 weeks, 3 weeks, 4 weeks). Then the specimens were rinsed in distilled water, embedded in light-curing resin after removing the wax, and cut with a diamond rotating disk.
From each tooth two sections were obtained, and then ground with a diamond disk, until a thickness of 1mm was reached. A whole of 100 sections prepared by this

cutting-grinding method were examined with PIXE (Proton Induced X-ray Emission), that allows a quantitative elemental analysis of 1 ppm. The sections, after the removal of the tooth root, were coated with carbon and placed on metallic supports having a hole allowing the passage of the protons beam.
Scansion were carried out on areas ranging from of (0.01x0.01) mm^2 to (2x2) mm^2, providing elemental maps of each sample. The proton beam of 2.4 MeV was produced by a Van de Graaf electrostatic accelerator, focused at a 1.5 micron diameter.

RESULTS

In all the samples filled with Cavit W, a widespread leaking of silver was detected, that after 4 weeks involved the whole layer of the cement. Also the tooth-cement interface was found widely infiltrated by silver (figs. 1-2)
Samples treated with Kalsogen showed a noticeable infiltration of both cement and interface. However, unlike Cavit W, silver didn't infiltrated the whole layer of the cement, neither the whole interface, not even at 4 weeks.
IRM cement showed a limited absorption of silver in the outer layer, with a partial leakage into the interface after 4 weeks (figs. 3-4).

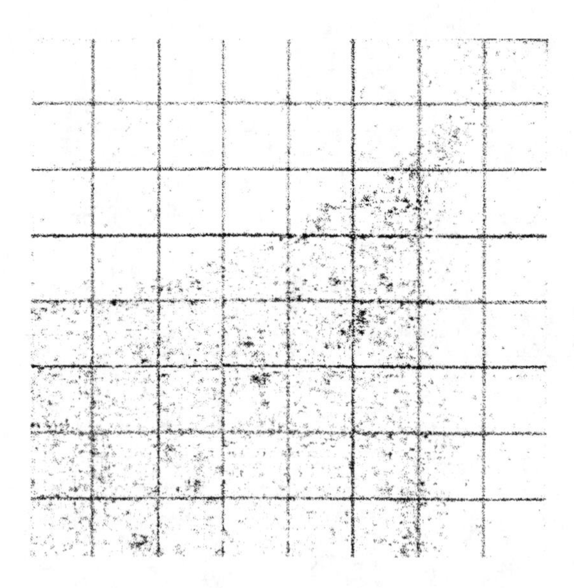

FIGURE 2 The same map of fig. 1, after subtraction of calcium and zinc, displaying silver, that leaked into the cement and the tooth-cement interface.

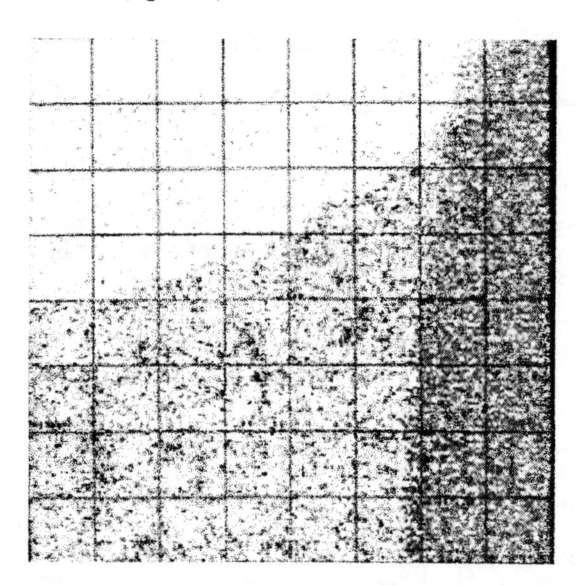

FIGURE 1 PIXE micro-beam map of sample filled with Cavit W cement. Sample immersed in the tracing solution for 4 weeks. Detected elements are zinc (left, corresponding to the cement), calcium (right, corresponding to the tooth), and silver contained in the tracing solution. Silver is masked by the other two elements. Map area = 2 X 2 mm.

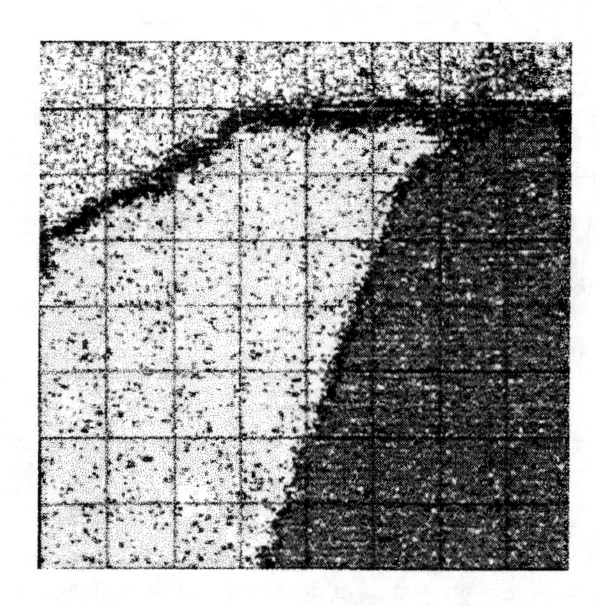

FIGURE 3 IRM cement (zinc oxide-eugenol). Elemental map for zinc (left), calcium (right) and silver.

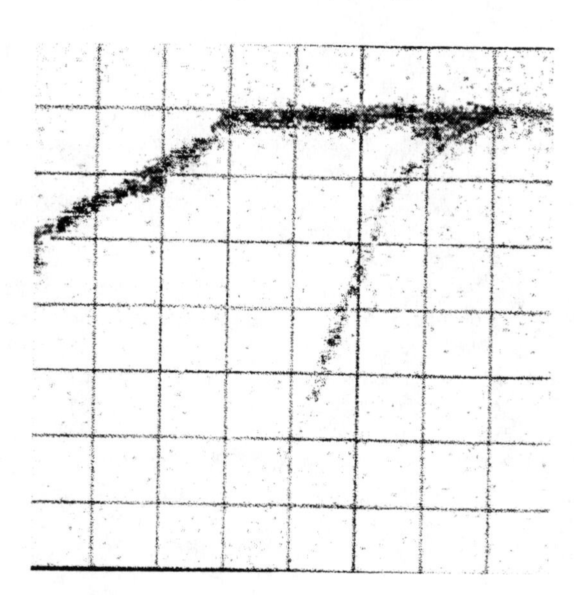

FIGURE 4 The same map of fig.3, showing only silver, that has leaked into the outer layer of the cement, and into the tooth-cement interface.

Super EBA cement had a very limited leakage of silver into the cement, while the infiltration of interface was negligible.
In the samples treated with Fermit N, neither the cement nor the interface were found infiltrated by silver, that concentrated only where fissures or pits of the enamel were present (figs. 5-6).

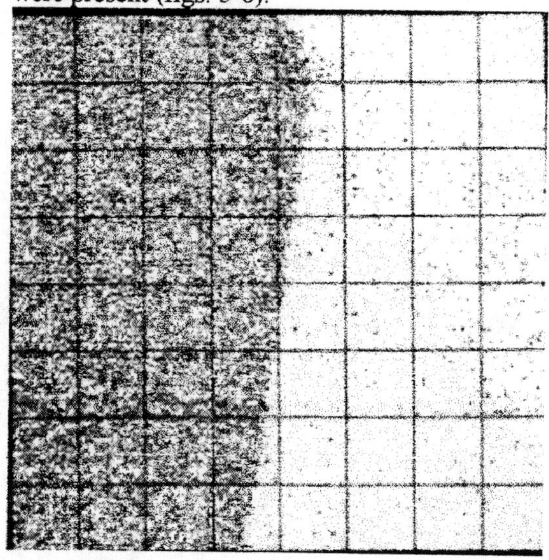

FIGURE 5 Fermit N (acrylic-silicon dioxyde cement). Elemental map for calcium (left), silicon (right) and silver.

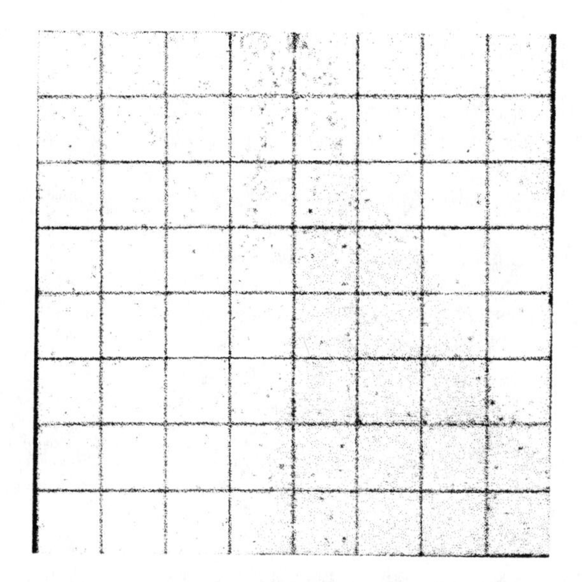

FIGURE 6 The same map of fig.5, showing silver and silicon distribution. Neither cement nor tooth-cement interface are infiltrated by silver, that is concentrated in a small fissure of the enamel. Silicon map was preserved to underline tooth-cement interface.

The tendence to fluid absortion and marginal leakage of these five materials is therefore as follows,
from the worst to the best one: Cavit W-Kalsogen-IRM-Super EBA-Fermit N.

DISCUSSION AND CONCLUSIONS

MicroPIXE seems to be an useful approach in evaluating the fluid absorption and the marginal integrity of dental cements and restorative materials. In the present study, the good results shown by an acrylic composite cement, as Fermit N, could anyway be expected on the basis of previous reports about the characteristics of these materials (1, 2). On the other hand, also the high degree of leakage and infiltration of the zinc oxide-eugenol cements, Cavit W above all, are not surprising, on the basis of their chemical characteristics and the results obtained by other authors (3, 4, 6, 8, 9, 10, 11, 12).
Also the intermediate results obtained with the other three materials could be expected. Indeed, these are eugenol or ortho-ethossybenzoic acid based cements, and fairly soluble in water and saliva (5, 7). However, these evaluations allow to assess the advantages of a microanalytic technique that can be applied to other dental and medical materials, especially if quantitative measurements will be performed.
The possibility to detect very small amounts of the marker (1 ppm) allows to evaluate microleakages that other techniques, as the optical examination of dyes, can not reveal.

REFERENCES

1) Anderson R.W., Powell B.J., Pashley D.H. Microleakage of three temporary endodontic restorations. *J. Endod. 1988, 14, 497-501.*

2) Anderson R.W. Powell B.J., Pashley D.h. Microleakage of temporary restorations in complex endodontic access preparations. *J. Endodon. 1989, 15, 526-9.*

3) Barkhordar R.A., Stark M.M. Sealing ability of intermediate restorations and cavity design used in endodontics. *Oral Surg. 1990, 69, 99-101.*

4) Chohoyed A.a., Bassiouny M.A. Sealing ability of intermediate restoratives used in endodontics. *J. Endodon. 1985, 11, 241-4.*

5) Lamers A.C., Simon M., van Mullem P.J. Microleakage of Cavit temporary filling material in endodontic access cavities in monkey teeth. *Oral Surg. 1980, 49,541-3.*

6) Lee Y.c., Yang S.F., Hwang Y.F., Chueh L.H., Chung K.H. Microleakage of endodontic temporary restorative materials. *J. Endodon. 1993, 19, 516-9.*

7) Lim K.C. Microleakage of intermediate restorative materials. *J. Endodon. 1990, 16, 116-8.*

8) Magura M.E., Kafray A.H., Brown C.E., Newton C.W. Human saliva coronal microleakage in obturated root canals: an in vitro study. *J. Endodn. 1991, 17, 324-331.*

9) Noguera A.P., McDonald N.J. A comparative in vitro coronal microleakage study of new endodontic restorative materials. *J. Endodon. 1990, 16, 523-7.*

10)Tay F.R., Pang K.M., Gwinnet A.J., Wei S.H. A method for microleakage evaluation along the dentin/restorative interface. *Am. J. Dent. 1995, 8,2,4, 105-8*

11) Turner J.E., Anderson R.W., Pashley D.H., Pantera E.A. Microleakage of a temporary endodontic restoration on teeth restored with amalgam. *J. Endodon. 1990, 16, 1-4.*

12) Webber R.T., Del Rio C.E., Brady J.M., Segall R.O. Sealing ability of a temporary filling material. *Oral Surg. 1979, 46, 123-30.*

Elemental Analysis of Soluble and Insoluble Fractions of River-Waters by Particle-Induced X-Ray Emission

H. Yamazaki, K. Ishii, S. Matsuyama, Y. Takahashi, T. Sasaki, H. Orihara,[a] and Y. Izumi[b]

Department of Quantum Science and Energy Engineering, Graduate School of Engineering, Tohoku University, Aramaki-Aza-Aoba 01, Sendai 980-8579, Japan
[a]Cyclotron and Radioisotope Center, Tohoku University, Aramaki-Aza-Aoba, Sendai 980-8578, Japan
[b]Japan Environment Research Co. Ltd., 7-8-13 Shinjuku-ku, Tokyo 160, Japan

A procedure has been developed and tested for PIXE analysis of soluble and insoluble constituents in river-water samples. Three kinds of targets were prepared and analyzed with a PIXE system of 3-MeV proton beam. Insoluble components were filtered on a Nuclepore filter of 0.4-μm pores. For soluble fractions, a target of major components was made from a 0.15-ml filtrate evaporated on a user-made polycarbonate film and trace amounts of heavy metals were preconcentrated in a PIXE-target by means of a combination of dibenzyldithiocarbamate-chelation with subsequent condensation into dibenzylidene-*D*-sorbitol gels. The widespread concentrations (several tenths of ppb to a few tens of ppm) of ~24 elements from Na to Pb were determined simultaneously in a precision sufficient to reveal the elemental distribution between the soluble and insoluble fractions of river waters. Hence, the methodology can be applied to monitor a pollution problem of rivers.

INTRODUCTION

Owing to a greater awareness of the problem of pollution, the composition of rain and surface water has been studied for many years with the analysis of such samples often being carried out by absorptiometry and atomic absorption spectrophotometry. In these analyses, water samples are filtered to remove insoluble components which are then not analyzed. Particle-induced X-ray emission (PIXE) is a rapid and multielemental technique. This technique has been used extensively for analyzing aerosol samples (1,2), but it has not often been applied to water samples because of problems in target preparation (3,4). However, the use of PIXE analysis permits convenient determination of both soluble and insoluble materials in natural waters due to versatility in the form of specimens to be analyzed, providing for some useful information about chemical forms of elements in a water samples.

The PIXE technique is conveniently applied to the analysis of thin samples and offers the possibility of a high absolute sensitivity (nanogram levels). In our previous studies (5), a combination of chelation by dibenzyldithiocarbamate (DBDTC) ions with subsequent condensation into dibenzylidene-*D*-sorbitol (DBS) gels has been developed for preconcentration of trace amounts of heavy metals in water samples, in conjunction with rapid preparation of thin uniform targets containing zirconium as an internal standard. We also have developed a simple preparation method for thin polycarbonate film and used it as a target backing in the PIXE analysis of anions such as sulfate, chromate and arsenate in a wide concentration range (10-2000 ppb).

It is the purpose of the present study to combine these techniques for sample preparation with Nuclepore filtration of insoluble constituents, which results in rapid determination of elemental concentrations in both soluble and insoluble fractions of river waters.

EXPERIMENTAL

Target Preparation

We collected 12 water samples at 2-hour intervals on July 16, 1998 at two locations of the Hirose river, which is a class A river flowing through Sendai city. One sampling point located at the beginning of our town and the other was a 8-km downstream point on the outskirts of the town.

The samples were stored in Nalgene linear polyethylene containers and processed into PIXE-targets within 24 hours. After pH measurement, insoluble components in 30 ml of each sample were filtered under suction with a Nuclepore filter of 0.4-μm pores. The filter was mounted on a Mylar target frame and kept in desiccator for several days. The filtration efficiency was over 98 % for colloids of ferric hydroxide and silver chloride in the range of 20-2000 ppb. The targets for these colloids in a known amount (40 and 400 ppb) were prepared in the same manner and used as an external standard for normalization of PIXE spectra for the insoluble components.

In the preparation of PIXE targets for heavy metals in the soluble fraction of river waters, 2.5 ml of 0.1 % (w/v) DBDTC solution and 25 μl of 1000 ppm Zr in 1M HNO_3, respectively, were added to 25 ml of each filtrate as a chelating agent and as an internal standard, and then the pH of solution was kept around 5 for 4 minutes. The solution gelled immediately after addition of 10 μl of 4 %(w/v) of DBS solution, and the DBS gels containing metal-DBDTC complexes were filtered on a Nuclepore filter of 0.4-μm pores. In our previous study (4), quantitative recoveries of 7

CP475, *Applications of Accelerators in Research and Industry*,
edited by J. L. Duggan and I. L. Morgan

metals (Fe, Co, Ni, Cu, Cd, Hg, Pb) were confirmed up to the concentration of 1 ppm, and the coexistence of Mg, Ca and humic acid in 40 ppm did not interfere with the recovery of the heavy metals.

On the other hand, alkali metals, alkaline earth metals and anionic species in the soluble fraction are not picked up in the preconcentration step described above. The targets for these elements were prepared by depositing 30 μl of filtrate on an user-made polycarbonate film; 10 μl of 1000 ppm Ga in 1M HNO_3 was added to 4 ml of filtrate beforehand. After drying at 60℃, the procedure was repeated four more times to give a total of 150 μl dried on the foil. As we have revealed in our previous study (5), a polycarbonate film of thin and uniform thickness is prepared by dropping a polycarbonate solution in chloroform-benzene mixture slowly on a water surface within a 20-mm aperture of Mylar target frame floating on 50 wt% sucrose aqueous solution. The film offers a good combination of mechanical strength, chemical stability and low X-ray continuum background.

PIXE Analysis

The three kinds of targets were irradiated for 5 minutes in a vacuum chamber by 3 MeV protons (apparent current, 10 nA; beam diameter, 3mm) from a baby cyclotoron at Nishina Memorial Cyclotoron Center, Japan. X-rays from targets were measured with two Si(Li) detectors; one having 300-μm Mylar absorber and high geometric efficiency allows the detection of X-rays > 4 keV, and the other one with a low geometric efficiency is well suited for the detection of elements of the atomic number $Z \leq 20$.

For PIXE spectra analysis, we used a least-squares fitting computer program, which has been developed in our laboratory based on a theoretical approach for the background continuum in PIXE spectra (6). In this program, a background function for elements with atomic number Z=6-30 was obtained as a function of Z and X-ray energy from bremsstrahlung emission cross-sections derived by the theoretical formula based on PWBA and BEA. Then, the spectra of continuous and characteristic X-rays of elements were determined with the background functions and were used to least-squares fit to a measured PIXE spectrum. The detector's intrinsic efficiency and the transmission through absorbers were determined experimentally in the X-ray range of 1-60 keV. The values for the total production cross-sections of X-rays of interest and the correction factors for the relative intensities of multiplets for each characteristic X-ray were obtained from the text book by S.A.E. Johansson and J. L. Campbell (7).

RESULTS AND DISCUSSION

The measurements for 12 river-water samples collected at two locations over 10 hours were done by PIXE, each sample being separated into soluble and insoluble fractions by means of the procedure described before. The detection limits of the PIXE analysis were on the order of several tenths of ppb for elements producing X-rays >5 keV, while the large continuum background of the backings in PIXE spectra incurs an inferior limit, that is, several tens of ppb for elements with atomic number $Z < 20$.

Figure 1 shows typical PIXE spectra obtained from the three kinds of targets; the insoluble components on a Nuclepore filter (in the left-hand side), the deposit of soluble components on a polycarbonate film (in the middle), and the soluble heavy metals preconcentrated into DBS gels on a Nuclepore filter (in the right-hand side). The Ga and Zr peaks are from the internal standards used. The *K* and *L* X-ray lines are observed for 24 elements in the spectra. This indicates that the PIXE technique has the advantages of a truly multielemental character and high speed of analysis with a small amount of samples. The X-ray lines of many heavy metals are clearly detected in the spectrum of the target prepared with the DBDTC-DBS preconcentration. From this result, we can understand why the preconcentration step is indispensable to the PIXE analysis for heavy metals in a low concentration of ng/ml level. In contrast to this, the spectra obtained from the deposit of filtered river-water allows us to determine the concentrations of soluble elements such as alkali and alkaline earth metals, which cannot be picked up in the preconcentration step. An important result is that the PIXE analysis for the three kinds of targets reveals elemental distribution between the soluble and insoluble fractions of river waters.

We now review time-variation in concentrations of major components in the soluble and insoluble fractions (Fig.2). The error bars on the individual data points of Figure 2 are error estimates from the spectral fitting program. It is found from this figure that the elemental concentrations appreciably change with time in a different fashion between the soluble and insoluble fractions. The concentrations of soluble components increased largely on the fourth sample collected at the downstream of the river. On the contrary, the increase of concentrations in the insoluble fraction was observed on the last sample collected at the upstream point. The details for these different observations cannot be discussed because of a short period of collecting the samples. Here it should be noted that PIXE analysis with high sensitivity clearly distinguishes the change of elemental concentrations in the widespread range from 0.1 ppb to 20 ppm (in five orders of magnitude).

The averaged concentrations of the soluble and insoluble components for each six samples collected at two locations are tabulated in Table 1, together with the deviations. Most of alkali metals, alkaline earth metals and anionic species for P, S and Cl are present as soluble components in the range of concentration from 0.5 to 20 ppm. The insoluble fractions include a large amount of Al, Si and Fe, which suggests that the compounds of these elements are insoluble minerals derived from soil. The concentrations of Al and Si, however, are much higher in the soluble fractions than in the insoluble ones, indicating very fine particles of clay and colloidal hydroxides of the elements in the river waters.

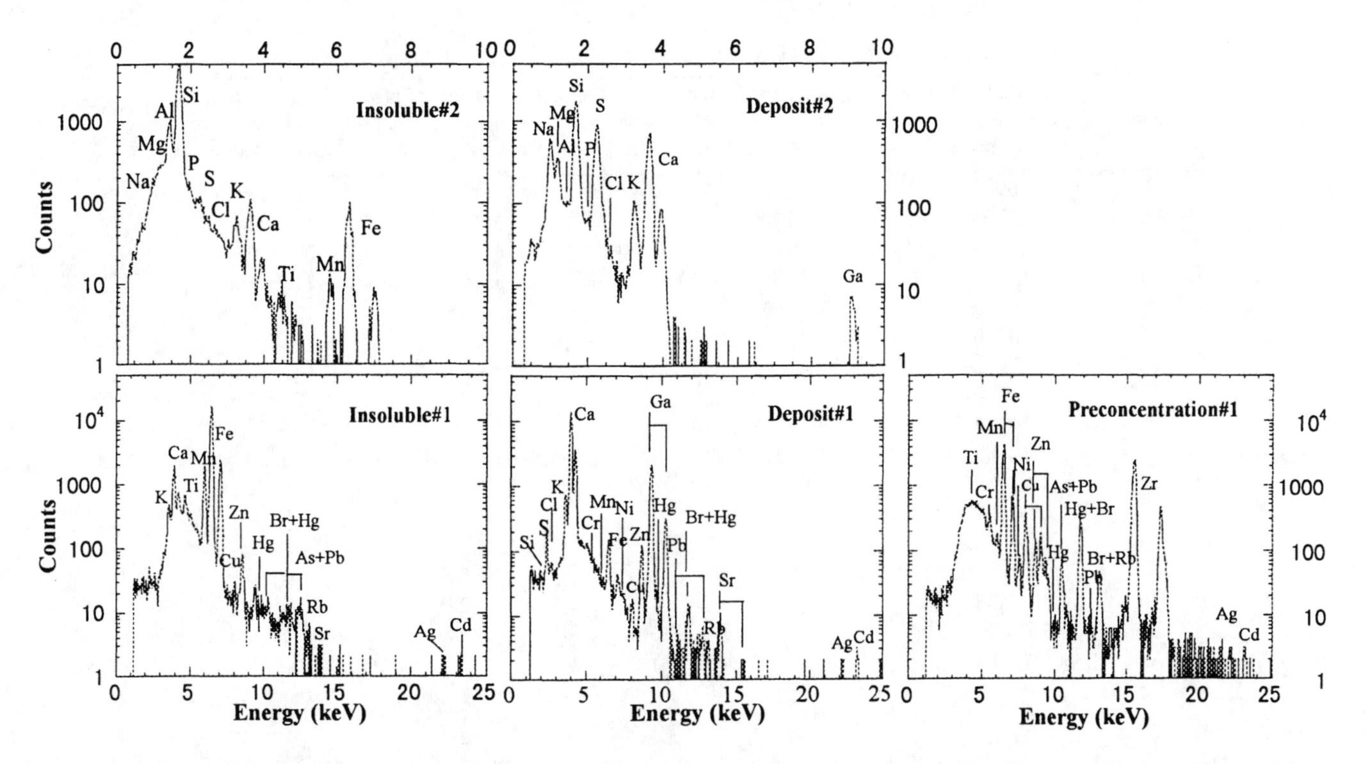

FIGURE 1. Typical spectra of PIXE analyses for the three kinds of targets. The upper spectra were obtained with No.2-detector directed to X-rays of low energy (< 4 keV). The lower spectra were obtained with No.1-detector with 300-μm Mylar absorber.

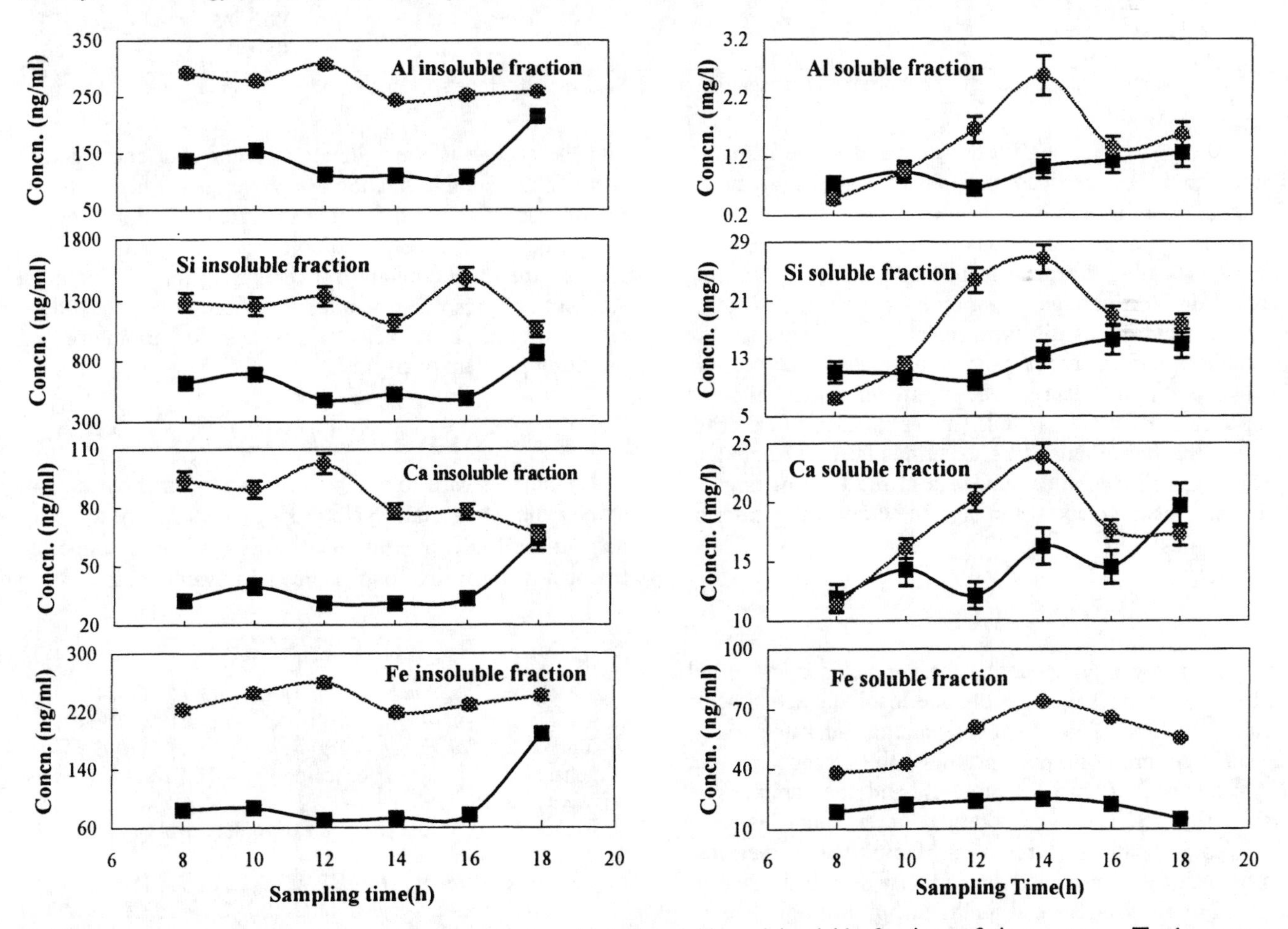

FIGURE 2. Time variation of elemental concentrations in both soluble and insoluble fractions of river waters. ■: the upstream sampling point, ●: the downstream sampling point.

TABLE 1. Concentrations of the Soluble and Insoluble Components of River Waters

Element	Concentration of soluble components (ng / ml)		Concentration of insoluble components (ng / ml)	
	Upstream	Downstream	Upstream	Downstream
Na	6900 ± 1200	8900 ± 3800	42 ± 15	88 ± 11
Mg	2600 ± 410	3800 ± 1200	54 ± 16	108 ± 12
Al	990 ± 220	1410 ± 600	140 ± 45	270 ± 23
Si	12600 ± 2000	17700 ± 6000	610 ± 160	1250 ± 140
P	780 ± 180	1100 ± 390	35 ± 5	69 ± 9
S	8400 ± 1500	11400 ± 2800	34 ± 5	63 ± 9
Cl	490 ± 140	690 ± 330	25 ± 4	41 ± 5
K	1580 ± 250	2420 ± 560	26 ± 6	57 ± 8
Ca	14700 ± 2400	18200 ± 3700	38 ± 14	85 ± 12
Ti	0 ± 0.1	0.2 ± 0.3	4.2 ± 2.6	6.5 ± 0.6
Cr	1.8 ± 0.4	2.0 ± 0.8	0.4 ± 0.1	0.9 ± 0.4
Mn	0.7 ± 0.1	1.1 ± 0.2	12 ± 0.3	32 ± 4
Fe	21 ± 3	56 ± 12	97 ± 41	240 ± 15
Ni	1.3 ± 0.1	1.5 ± 0.3	0.3 ± 0.2	0.6 ± 0.1
Cu	2.8 ± 0.4	3.0 ± 0.2	0.3 ± 0.2	0.5 ± 0.1
Zn	5.4 ± 0.9	6.0 ± 0.2	1.5 ± 0.5	2.9 ± 0.4
As	0.9 ± 0.4	1.1 ± 0.6	0.2 ± 0.2	0.1 ± 0.1
Br	29 ± 2	31 ± 18	0.1 ± 0.1	0.6 ± 0.5
Rb	1.5 ± 0.6	1.8 ± 0.4	0.4 ± 0.3	0.6 ± 0.2
Sr	38 ± 9	55 ± 17	0.2 ± 0.1	0.3 ± 0.1
Cd	1.3 ± 0.6	2.4 ± 0.8	0.3 ± 0.3	0.4 ± 0.3
Ag	2.9 ± 1.1	3.5 ± 1.3	0.2 ± 0.1	0.3 ± 0.3
Hg	0.1 ± 0.1	0.1 ± 0.1	0.4 ± 0.3	1.0 ± 0.5
Pb	0.4 ± 0.4	1.2 ± 0.6	1.6 ± 1.2	3.5 ± 0.7

The mean values and the deviations in the concentrations were calculated for the six samples at each sampling point.

Most of Ti, Mn, Hg, and Pb are included in the insoluble fraction, and the significant portions of other heavy metals are found in both soluble and insoluble fractions.

The elemental concentrations in the soluble fractions are found to be rather independent of the location of collecting samples due to the large deviation with time. On the other hand, concentrations of most elements detected in the insoluble fractions are almost two times higher at the downstream point than at the upstream one, while the composition of the insoluble materials does not differ between the two points. This experimental result naturally implies that clay-colloids condense in the flow of river to form particles large enough to be filtered out with a filter of 0.4-μm pores.

CONCLUSION

In this study, a procedure has been developed and tested for the PIXE analysis of soluble and insoluble constituents in river waters. The target preparation and the PIXE measurement are not time-consuming; the preconcentration step requires less than 10 minutes, and the samples are analyzed by PIXE in which targets are irradiated by a 3 MeV proton beam for 5 minutes. Many kinds of elements can be detected simultaneously in the widespread range of concentrations from several tenths of ppb to a few tens of ppm. The elemental concentrations largely vary with time, and the change in the soluble fraction is not correlated to that of the insoluble fraction. Moreover, some heavy metals of importance as an indication to water-pollution problem are predominant in the insoluble fraction of river waters. Hence, the methodology for preparing targets of both soluble and insoluble components promotes the PIXE analysis to a truly effective means for monitoring a pollution problem of rivers.

ACKNOWLEGEMENT

The authors wish to thank Dr. Sera and Mr. Futatsukawa at Nishina Memorial Cyclotoron Center, Japan for their help in the PIXE measurement. This work was supported financially in part by Tohoku Electric Power Co., Inc.

REFERENCES

1. Akselsson, K.R., *Nucl. Instr. and Meth.*, **B3**, 425 (1984).
2. Kasahara, M., *et al.*, *Nucl. Instr. and Meth.*, **B109**, 471 (1996).
3. Tanaka, S., *et al.*, *Environ. Sci. Technol.*, **15**, 354 (1981).
4. Johansson, E. M. and Johansson, S. A. E., *Nucl. Instr. and Meth.*, **B3**, 154 (1984).
5. Yamazaki, H., *et al.*, *Int. J. PIXE*, **6(3&4)**, 483 (1996); *ibid*, **7(1&2)**, 31, 101 (1997).
6. Murozono, K., *et al.*, , *Int. J. PIXE*, **6(1&2)**, 135 (1996).
7. Johansson, S.A.E. and Campbell, J.L., *"PIXE. A Novel Technique for Elemental Analysis,"* John Wiley & Sons, New York, 1988, pp.313-329.

IBA on functional polymers

M.P. de Jong[1], D.P.L. Simons[1], L.J. van IJzendoorn[1], M.J.A. de Voigt[1], M. A. Reijme[2], A.W. Denier van der Gon[2], H.H. Brongersma[2]

[1]*Research School CPS, Cyclotron Laboratory, Department of Applied Physics, Eindhoven University of Technology, P.O. Box 513, 5600 MB Eindhoven, The Netherlands*

The analysis of element distributions in polymer-based structures using IBA techniques offers the possibility to study a variety of interesting problems, in particular diffusion and reaction phenomena. Indium diffusion in model polymer light emitting diodes (p-LEDs) consisting of a stack Al/poly-(phenylenevinylene)/indium-tin-oxide/glass has been studied with Rutherford backscattering spectrometry (RBS), particle induced X-ray emission (PIXE), X-ray photoelectron spectroscopy (XPS), and low energy ion scattering (LEIS). A second example is provided by the analysis of organic optical gratings, in which the diffusion of labelled monomers during holographic photo-polymerisation of photo-reactive monomer mixtures has been studied with μPIXE using a scanning proton microprobe. Since polymers are sensitive to ion irradiation, a new RBS/ERDA set-up has been constructed that is equipped with a sample holder mounted on a closed cycle helium refrigerator, which enables the cooling of samples to cryogenic temperatures to suppress damage under ion bombardment.

INTRODUCTION

Functional polymers play a key role in a massive field of applications, varying from protective coatings (1) to light emitting diodes (LEDs) (2). In many of these applications, IBA-techniques can give valuable insights into diffusion and reaction phenomena. Two examples of IBA-studies are discussed here, which concern rather different polymer-based structures.

The first example is the study of indium diffusion in model polymer-LEDs (p-LEDs). In a p-LED, the emissive layer consists of a π-conjugated semiconducting electroluminescent polymer (3), for example poly-(phenylenevinylene) (PPV). The electroluminescence takes place through the radiative decay of polaron excitons, which are formed by the recombination of electrons and holes. The electrons and holes are injected into the polymer by two electrodes that sandwich the polymer film. This implies that one of the electrodes has to be transparent to visible light. A transparent electrode that is widely used in p-LEDs is indium-tin-oxide (ITO), which has a work function suitable for hole-injection. However, in several cases indium compounds are found to diffuse from the ITO layer into the polymer film (4-8), which results in undesired electrical doping of the polymer. We have studied the diffusion of indium compounds into PPV with Rutherford backscattering spectrometry (RBS), particle induced X-ray emission (PIXE), X-ray photoelectron spectroscopy (XPS), and low energy ion scattering (LEIS).

The second IBA-study concerns organic optical gratings (9), which are formed by holographic photo-polymerisation of a mixture of mono-acrylate and di-acrylate monomers. The mixture is exposed to an interference pattern of an Ar-laser, resulting in illuminated and dark areas. Because the photo-polymerisation occurs faster for di-acrylates than for mono-acrylates, di-acrylate monomers are depleted from the illuminated areas during the polymerisation reaction, resulting in the diffusion of di-acrylates to the illuminated areas and the diffusion of mono-acrylates to the dark areas. Regions with different indices of diffraction are formed, so that the obtained film behaves as an optical grating. The diffusion of labelled monomers during the lithographic photo-polymerisation reaction has been studied with μPIXE, using a scanning proton microprobe.

Although IBA can be a powerful tool to study polymer structures, it must not be overlooked that polymers are very sensitive to ion beam induced damage. In the case of high-energy (MeV) ion irradiation, significant losses of volatile elements such as hydrogen can occur. To suppress the damage during high energy ion scattering (HEIS) measurements, a set-up has been constructed that is equipped with a sample holder mounted on a closed cycle helium refrigerator, by means of which samples can be cooled to cryogenic temperatures. The first results obtained with this set-up are reported here.

INDIUM DIFFUSION IN MODEL p-LEDs

We studied model p-LEDs that consisted of an ITO electrode on a glass substrate, a PPV layer and, in some cases, a patterned aluminium electrode (see figure 1). Because PPV is insoluble, PPV films can not be deposited directly onto the ITO. A precursor polymer can be used instead, which is converted to PPV after the deposition of a

CP475, *Applications of Accelerators in Research and Industry*,
edited by J. L. Duggan and I. L. Morgan

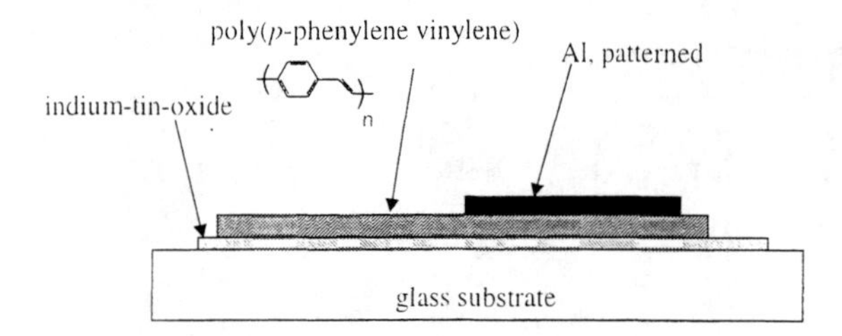

FIGURE 1. Structure of a model polymer-LED.

film. Additionally, soluble PPV derivatives can be obtained by attaching side chains to the conjugated backbone.

Previous studies have pointed out that the ITO layer can erode during the conversion of the chloride sulfonium salt precursor polymer (10) to PPV (see figure 2), resulting in contamination of the PPV layer with indium compounds (4,7). It is suggested that the ITO electrode reacts with the HCl leaving group to form $InCl_3$, which subsequently dopes the PPV. In the alkoxy precursor route (10), HCl is often even added as a catalyst: a so-called acid conversion can be carried out, during which the polymer is heated under an acidic atmosphere created by an Ar/HCl gas flow.

We compared the In contamination levels in model p-LEDs for two different electroluminescent polymers: sulfonium precursor route PPV (Cambridge Display Technology) and OC_1C_{10}-PPV (Philips Research Labs). The latter is soluble and can be spincoated onto an ITO substrate without further chemical and/or heating treatments. The precursor route PPV films were obtained by thermal conversion during 10 hours at 240 °C.

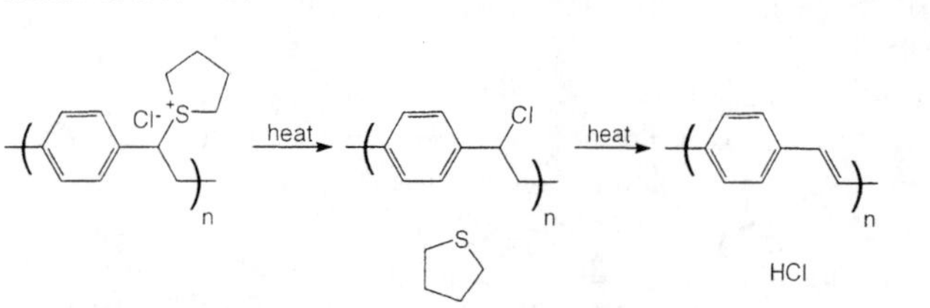

FIGURE 2. Chloride sulfonium salt precursor polymer converts to PPV by heat treatment. Tetrahydrothiophene and hydrochloric acid are eliminated.

RBS measurements on precursor route model p-LEDs showed that the In concentration in the converted PPV was approximately 0.01 at%, homogeneously distributed in depth. The In content increased to about 0.1 at% after 19 hours annealing at 230 °C. Under the patterned Al electrode, annealing resulted in the accumulation of In at the PPV/Al interface, whereas in the uncovered regions the In depth distribution remained homogeneous (see figure 3).

No substantial amount of In was ever found with RBS in the OC_1C_{10}-PPV films, even after 20 hours annealing at 300 °C. This supports the assumption that the erosion of the ITO electrode during conversion of the precursor polymer gives rise to the contamination with In compounds. XPS measurements showed that small traces of In are sometimes

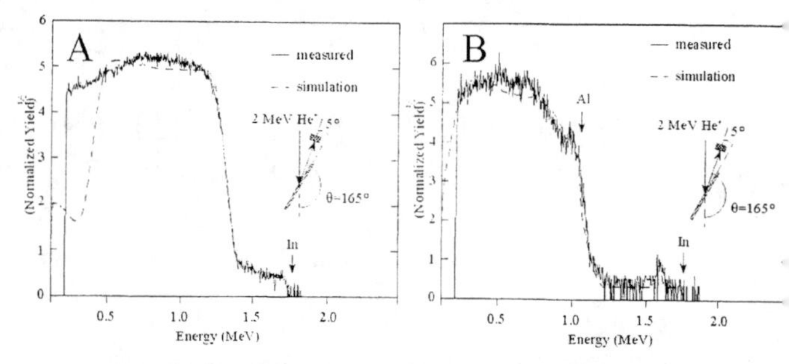

FIGURE 3. RBS spectra of a model p-LED that has been annealed during 19 hours at 230 °C. In the region without Al capping (A), a homogeneous depth distribution of In in the PPV is observed. Under the Al capping (B), In has been accumulated at the Al/PPV interface, as can been seen from the small In-peak at about 1.6 MeV.

present in the annealed OC_1C_{10}-PPV films. The In traces were however accompanied by Cl or F impurities in these cases, which indicates that the presence of halogens in the films can play a role in the In contamination.

To investigate the role played by HCl in the In contamination of PPV films, we exposed OC_1C_{10}-PPV/ITO samples to an Ar/HCl gas flow by leading Ar gas through a solution of HCl in water. The conditions that were used were similar to the conditions for acid conversion of alkoxy precursor route PPV, as described above, but without any heating of the samples. After this treatment, the OC_1C_{10}-PPV films were peeled from the ITO substrate to be examined using the PIXE technique. The same was done with an untreated OC_1C_{10}-PPV film, in which no In or Cl were found. Figure 4 shows lateral scans of the PIXE-yield of In and Cl in the OC_1C_{10}-PPV film that had been exposed to HCl, obtained with a scanning proton microprobe using 3 MeV protons. The measurement shows that islands of indium-chloride are formed, as can be seen from the similar In and Cl distributions. The composition of the islands closely resembles $InCl_3$, as followed from the analysis of a stoichiometric $InCl_3$ salt

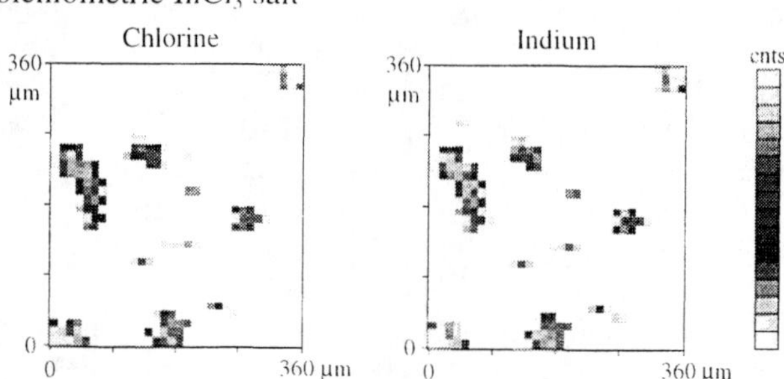

FIGURE 4. Lateral distributions of Cl and In measured with µPIXE in an OC_1C_{10}-PPV film that was peeled of an ITO substrate after 8 minutes exposure to an Ar/HCl gas flow.

LEIS experiments on OC_1C_{10}-PPV/ITO samples exposed to an Ar/HCl flow showed that the indium-chloride resides directly under the outermost atomic layers of the polymer, which demonstrates that the $InCl_3$ has been transported to the polymer surface. After peeling the polymer film from the ITO substrate, the PPV/ITO interface was examined with LEIS. A much larger In and Cl content was found in the ITO than in the PPV. This implies that first the ITO electrode reacts with HCl to form $InCl_3$, followed by the growth of $InCl_3$ islands from the ITO into the polymer film.

The above-described results show that the erosion of the ITO electrode by HCl causes contamination of the overlying PPV film with $InCl_3$. This explains why there is a large difference in the In content in precursor route PPV and OC_1C_{10}-PPV, respectively. The annealing induced accumulation of In at the PPV/Al interface in precursor route model p-LEDs most likely occurs through a reaction between $InCl_3$ and Al that produces atomic In and $AlCl_3$.

HOLOGRAPHIC PHOTO-POLYMERISATION IN ORGANIC OPTICAL GRATINGS

The study of diffusion and reaction phenomena with IBA techniques can also be illustrated by our work on organic optical gratings. The gratings were prepared from a monomer mixture that consisted of 70 mol% di-acrylate monomers and 30 mol% mono-acrylate monomers. The presumed two-way lateral diffusion process during holographic photo-polymerisation (11) of the monomer mixture was studied using a scanning proton microprobe applying the PIXE technique. For this purpose, the mono-acrylate monomers were labelled with a Si containing end-group. The gratings were produced by illuminating a thin film of the monomer mixture, captured between two glass plates, with an interference pattern of an Ar-laser using a 10 μm grating constant. After polymerisation, the films were peeled from the glass plates to be analysed.

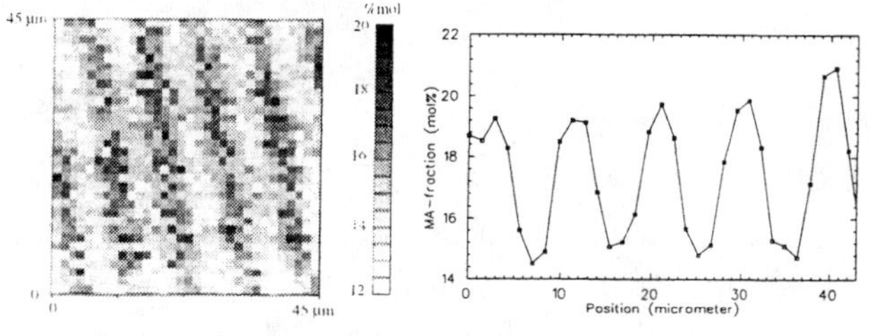

FIGURE 5. mol% distribution of mono-acrylates in an optical grating measured with μPIXE. The right part of the figure corresponds to a line scan, the left part to a surface scan.

Figure 5 shows the mol% distribution of mono-acrylates in an organic optical grating, as calculated from the Si-distribution that was measured with μPIXE. A distinct line pattern can be observed, which matches the 10 μm grating constant of the interference pattern that was used for the holographic photo-polymerisation. Variations in the mono-acrylate fraction of about 5 mol% are measured, which confirms and uniquely quantifies the proposed opposite diffusion of mono-acrylate and di-acrylate monomers.

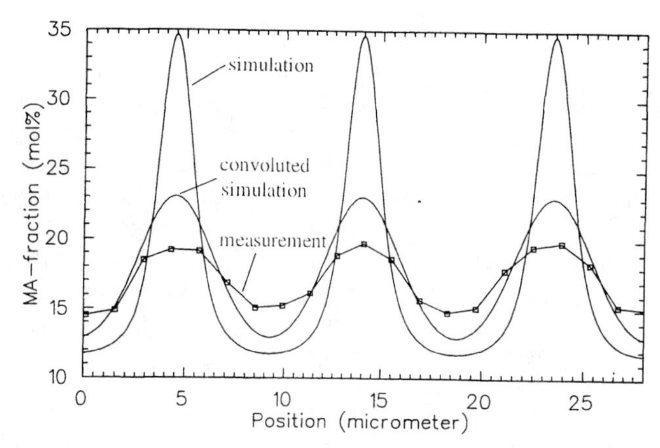

FIGURE 6. The mono-acrylate fraction profile from μPIXE measurement, simulation and convoluted simulation (9) with the beam-spot intensity profile. See text for details.

A comparison of the measured mono-acrylate distribution and a simulation obtained with the photo-polymerisation model proposed by van Nostrum *et al.* (9) is shown in figure 6. Because the proton microprobe produces a finite ion beam with a Gaussian intensity profile, the simulation has to be convoluted with this intensity profile for a good comparison. The difference between the measured concentration profile and the model allows adjusting the free parameters in the model to gain insight in the relevant physical/chemical processes. Important parameters are: the mobility of monomers during polymerisation, the reactivity of the different monomers, the assumed proportionality between reaction rates and light intensity and also the assumed sinusoidal intensity profile of the interference pattern. Although a detailed discussion is beyond the scope of this paper the surplus value of the IBA techniques is provided by the quantified concentration profiles.

SUPPRESSION OF ION BEAM INDUCED HYDROGEN LOSS IN POLYMERS AT LOW TEMPERATURE

It is well known that polymers are sensitive to ion irradiation. Molecular bond cleavage by the incident ions leads to the formation of highly mobile radicals, and volatile molecules are formed in the reactions induced by these radicals (12). This gives rise to the loss of material from polymer films during ion irradiation, which can deteriorate both the quantitativity and the depth resolution in RBS/ERDA experiments.

To suppress the ion beam induced damage during RBS/ERDA experiments, a set-up has been constructed in which the samples can be cooled to cryogenic temperatures. The cooling is accomplished with a two-stage closed-cycle helium refrigerator (APD cryogenics, CSW-204-SL-6.5), which reaches a temperature of about 6 K at the second cooling stage. The sample holder (carrying 6 samples) can be mounted onto the cold end of the cooler by means of a shrink coupling (13) providing good thermal contact at cryogenic temperatures and a loose mechanical contact at room temperature which allows the application of a load lock with magnetic transfer rods. The radiative heat load from the surrounding UHV vacuum chamber is virtually eliminated by the application of extended cold sheets. The samples are brought in thermal contact with the sample holder by means of cryogenic grease (CryCon Grease, APD cryogenics). After applying the grease, the samples are pressed against the surface of the sample holder by small leave springs.

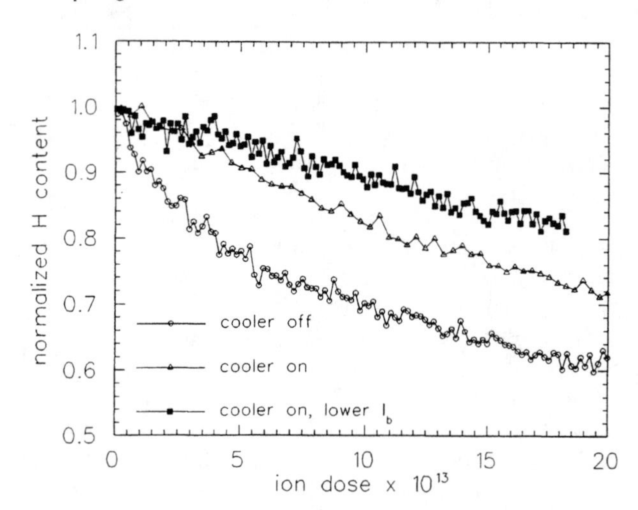

FIGURE 7. The normalized hydrogen content in OC_1C_{10}-PPV films on glass/ITO substrates as a function of ion dose, I_b= 25 nA. The (▪) symbols correspond to I_b= 12.5 nA.

In spite of the fact that the capacity of the cooler is 0.8 W at 8 K and 2.5 W at 10 K, which is higher than the typical heat dissipation of 0.1 W by the ion beam (4 MeV, 25 nA beam), the sample temperature during ion beam analysis critically depends on a number of factors. Most important are: the heat conductivity of the substrate material, the thermal contact to the sample holder and the thermal contact of the shrink coupling.

This paper contains results of the first experiments carried out with the set-up described above. Figure 7 shows the normalized hydrogen content in OC_1C_{10}-PPV during ERDA experiments as a function of the ion dose. The OC_1C_{10}-PPV films were spin coated onto a glass/ITO substrate. The experiments were performed using a 4 MeV He^+ beam, impinging on the sample under a 20° angle. Ejected hydrogen particles were detected at 30° with respect to the incident ion beam. A 16 μm Al stopper foil was placed in front of the detector to stop scattered He particles.

Without cooling, the hydrogen content in the OC_1C_{10}-PPV layer drops to about 80% of the initial content after an irradiation dose of $4 \cdot 10^{13}$ ions. With the cooler turned on, the hydrogen release rate still occurs but is reduced by approximately a factor of three. When the beam current I_b is decreased with a factor of 2 from 25 nA to 12 nA, the hydrogen release rate is further decreased to about a fifth of the release rate at room temperature. Apparently the thermal contact is not sufficient to transport the thermal heat load of the ion beam.

One of the factors that can deteriorate the heat transport away from the OC_1C_{10}-PPV film is the glass substrate on which the polymer film is deposited. Practically all the energy of the impinging ion beam is deposited in the 1 mm thick soda-lime glass substrate, which is not a very good thermal conductor. Therefore, we repeated the measurements for similar OC_1C_{10}-PPV films that were spincoated onto Si substrates, which have approximately a hundred times higher thermal conductivity. The experimental conditions were otherwise exactly similar.

The results of these measurements are shown in figure 8. The difference between the hydrogen release rates in an uncooled versus a cooled sample is now about a factor of ten, with the beam current constant at 25 nA(!) which demonstrates the importance of the heat conductivity of the substrate. Decreasing the beam current to 2.5 nA still leads to a significant improvement demonstrating that heating of the sample still occurs. Further experiments will be carried out to investigate the thermal heat contact between the sample and the holder as well as the efficiency of the shrink coupling.

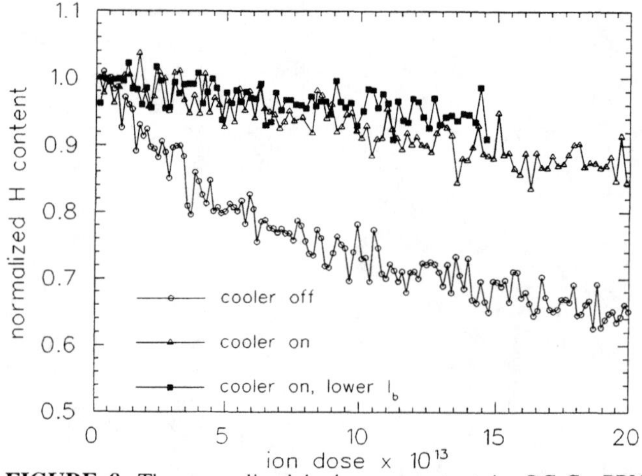

FIGURE 8. The normalized hydrogen content in OC_1C_{10}-PPV films on Si substrates as a function of ion dose, I_b= 25 nA. The (▪) symbols correspond to I_b= 2.5 nA.

So far, we have shown that cooling polymer samples down to cryogenic temperatures during ERDA experiments can result in a factor of ten slower release rate for hydrogen, which is by far the most volatile element in polymers under ion irradiation.

ACKNOWLEDGEMENTS

The authors like to thank Richard Gymer of Cambridge University for manufacturing model p-LEDs, and René van Nostrum for making the organic optical gratings.

REFERENCES

1. T. Donchev, G. Danev, T. Nurgaliev, E. Spassova, V. Tsaneva, R. Chakalova, A. Spasov, *Cz. J. Phys.* **46**, 1525 (1996)
2. J.H. Burroughes, D.D.C. Bradly, A.R. Brown, R.N. Marks, K. Mackay, R.H. Friend, P.L. Burns, A.B. Holmes, *Nature* **347**, 539 (1990)
3. H. Kiess, *Conjugated Conducting Polymers*, Berlin, Springer-Verlag, 1992
4. G. Sauer, M. Kilo, M. Hund , A. Wokaun, S. Karg, M. Meier, W. Riess , M. Schwoerer, H. Suzuki, J. Simmerer, H. Meyer, D. Haarer, *J. Anal. Chem.* **353**, 642 (1995)
5. W. Brütting, M.Meier, M. Herold, S. Karg, M. Schwoerer, *Chem. Phys.* **227**, 243 (1997)
6. M. Meier, S. Karg, W. Riess, *J. Appl. Phys.* **82** (4), 1961 (1997)
7. S. Karg, M. Meier, W. Riess, *J. Appl. Phys.* **82** (4), 1951 (1997)
8. A.R. Schlatmann, D. Wilms Floet, A. Hilberer, F. Garten, P.J.M. Smulders, T.M. Klapwijk, G. Hadziioannou, *Appl. Phys. Lett.* **69**, 1764 (1996)
9. C.F. van Nostrum, R.J.M. Nolte, D.J. Broer, T. Fuhrman, J.H. Wendorff, *Chem. Mater.* **10**, 135 (1998)
10. T.A. Skotheim, R.L. Elsenbaumer, J.R. Reynolds, *Handbook of Conducting Polymers 2th edition*, New York, Marcel Dekker, Inc, 1998, ch. 13
11. W.J. Tomlinson, E.A. Chandros, H.P. Weber, G.D. Aumiller, *Appl. Optics* **15**, 534 (1976)
12. M.P. de Jong, A.J.H.Maas, L.J. van IJzendoorn, S.S. Klein, M.J.A. de Voigt, *J. Appl. Phys.* **82** (3), 1058, (1997)
13. M.P. de Jong *et al.*, to be published

SRXRF Imaging of A Single Brain Cell from A Patient with Parkinson's Disease

K. Takada[1], A.M. Ektessabi[1], S. Yoshida[2]

[1]*Department of Precision Engineering, Graduate school of Engineering, Kyoto University Yoshida Honmachi, Sakyo-ku, Kyoto, 606-8501, Japan*
[2]*Division of Neurological Disease, Wakayama Medical College 27-9 Bancho, Wakayama City, 640-8511, Japan*

Synchrotron radiation X-ray fluorescence spectroscopy (SRXRF) is a potential imaging technique with regard to minimum detection limit, measuring time and being non-destructive on biological samples. These advantages are important for measuring trace elements in biological samples. In this paper, we investigated the distribution of trace elements in the cerebral neurons of the brains of patients with Parkinson's disease (PD), using SRXRF spectroscopy. The cause of PD is unknown but many researchers consider that excessive accumulation of trace metal elements (mainly iron) has strong influence on the generative process of PD. Micro beam imaging (mapping of the elements) with a beam size of 6×8 μm^2, and the energy of 13.5 keV was carried out in a single neuron. The distribution of trace elements in the neurons was successfully obtained in an area of about 100×100μm^2. The same sample was histologically studied with an optical microscope.

INTRODUCTION

In the past decade, synchrotron radiation X-ray fluorescence spectroscopy (SRXRF spectroscopy)[1~3] has been used for imaging trace elements distributions[4~7]. SRXRF spectroscopy is one of the excellent trace element analysis methods having the following advantages[8].

1. The detection limit is low (0.01 ppm level).
2. Measuring time is relatively short (a few seconds).
3. Heat damage is small.
4. The measurement can be done in air.

These points are important advantages for analyzing biological samples. Trace elements have important roll in biological activities. Low minimum detection limit enables us to analyze these trace elements within a tissue.

In this paper, we measured distributions of trace and matrix elements in cerebral neurons obtained from the brain of a patient with Parkinson's disease (PD). Parkinson's disease is one of the major neuro-degenerative diseases. Neuropathological studies have revealed that substantia nigra (SN), which is a part of midbrain, of patients with PD is damaged[9]. The cause of this neuro-degeneration is unknown, but there are indications that iron in SN neurons is related to the neuronal degeneration. Many neuropathologists pointed out that SN neurons of PD patients contain more iron than those of control subjects[10]. The increased iron in affected neurons may be responsible for the degenerative process in PD. Therefore, mapping of iron and other elements in a single neuron has a great significance in trying to understand the role of metal elements in neuro-degenerative process.

For this purpose, substantia nigura neurons of a patient with PD were examined with SRXRF spectroscopy. Using a SR micro beam, we obtained the distribution of the constituent elements in a single neuron. In addition, the same sample was studied histologically using an optical microscope after SRXRF spectroscopy.

MATERIALS AND METHODS

Autopsy specimens of midbrain, including substantia nigra cells, were obtained from a male patient with PD, 72 years of age. The specimen was fixed in 10% formalin, embedded in paraffin and then cut into 8 μm thick sections. Finally, the sections were mounted on a Mylar film for SRXRF analysis.

SRXRF analysis was carried out at the beamline 4A in Photon Factory. SR X-ray was monochromatized with a synthetic multilayer. The X-ray energy was 13.5 keV. Monochromatized X-rays were focused with Kirkpartrick-Baez optics[11]. The cross section of the beam was 6×8 μm^2 on the sample. The sample-stage was moved by x-y step pulsemotors, and distributions (X-ray intensity maps) of Fe, Zn, Ca and S were obtained. The scanning area was 102.5×102.5 μm^2 and was divided into 41×41 pixels. The area of each pixel was 2.5×2.5 μm^2. Each measuring point of the sample was irradiated for 5 seconds. The beam current of the storage ring was about 350 mA. The measurement was done in air. The beamline was equipped with CCD camera in front of the sample holder. The image from this camera gave us visual information of measuring points.

In order to study the presence of other elements, fluorescence X-ray spectra were also obtained at more than 20 points on the same sample. The cross section of the beam was 6×8 μm^2 and each measuring point was irradiated

CP475, *Applications of Accelerators in Research and Industry*, edited by J. L. Duggan and I. L. Morgan

for 10 seconds. The beam current of the storage ring was about 300 mA.

After SRXRF analysis, the sample was deparaffinized and prepared for optical microscopic examination.

EXPERIMENTAL RESULTS

SRXRF imaging was performed on the SN neurons of a PD patient and the distributions of Fe, Zn, Ca and S are shown in Fig. 1a-d. The scale on the right hand side of the map shows the level of the fluorescence X-ray intensity (count). The intensity range of each map was 30~450 counts for Fe, 7~15 counts for Zn, 20~40 counts for Ca and 20~40 counts for S. The concentration of matrix element (Ca and S) in Fig. 1c and Fig. 1d clearly shows that there is a different elemental composition in a certain region in the lower part of the scanning area. Using the CCD camera image, it was confirmed that this region (tissue) was a mass of melanin pigment granules, which is one of the major components of dopaminergic neurons in the substantia nigra (SN). Fig. 1a shows that Fe was selectively accumulated in the melanin pigment granules. Fluorescence X-ray of Fe is one order of magnitude higher than that of the other elements. The map of Zn shows that Zn was concentrated in the left part of the granules.

After performing SRXRF spectroscopy, the area of the sample was studied using an optical microscope. A typical microscopic photograph of the scanning area is shown in Fig. 2. The points (A) and (B) of Fig. 2 show the melanin pigment granules which were released from a dead cell into the tissue. The size of the granules at the point (A) was about 20 μm. Melanin pigment granules in a surviving cell can be seen at the point of (C). From these observations it can be concluded that trace elements (Fe and Zn) didn't accumulate in the melanin pigment granules of the surviving cell. And the melanin pigment granules from a dead cell had a higher accumulation of Fe compared with

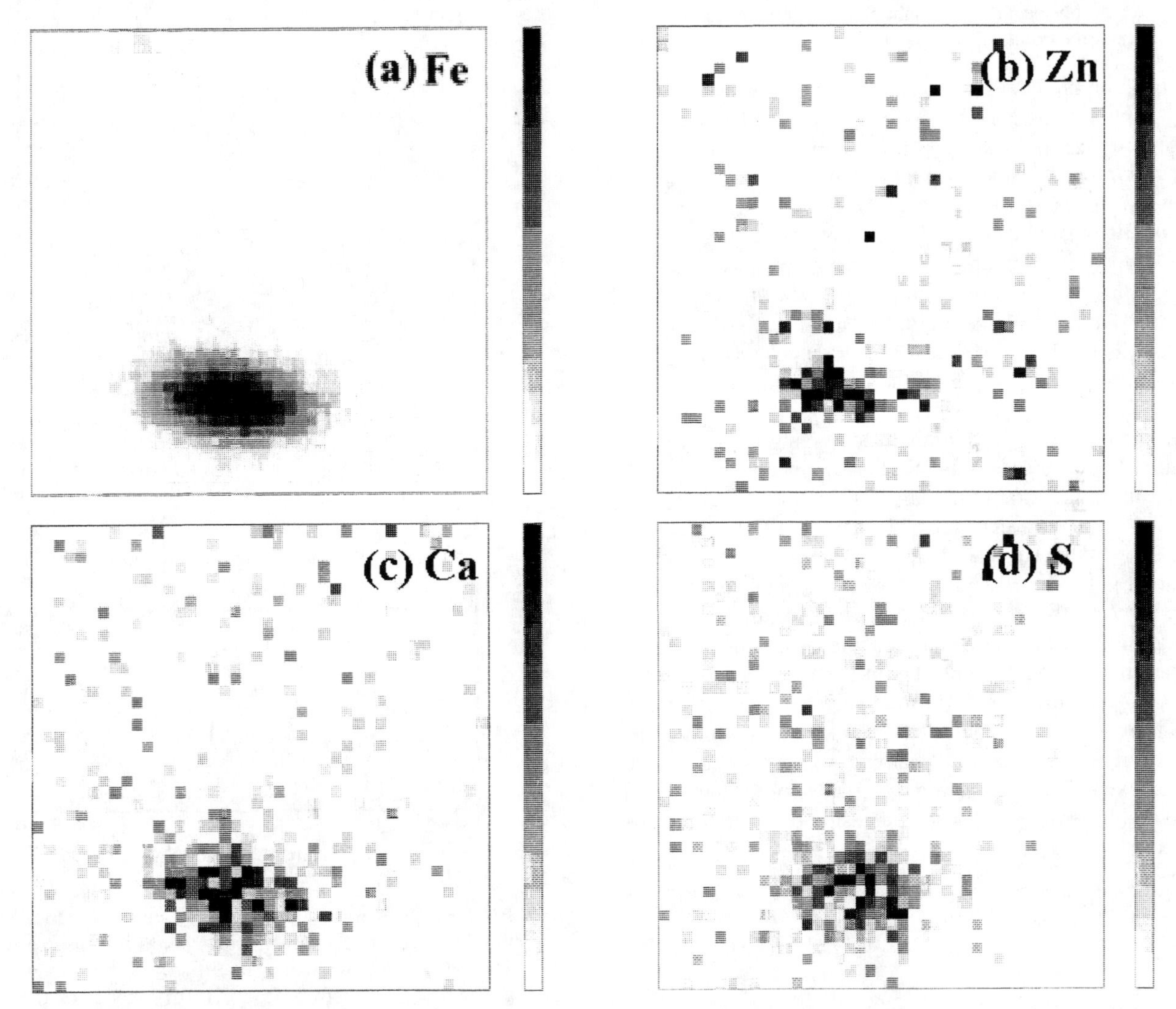

FIGURE 1a-d Map of Fe(a), Zn(b), Ca(c) and S(d) in a single neuron cell in a brain of a patient with Parkinson's disease. The scanning area was 102.5×102.5 μm^2. The scale on the right hand of the map shows level of the X-ray intensity (count). White means high intensity and black means low intensity. The range of intensity is 30~450 for Fe, 7~15 for Zn, 15~40 for Ca and 15~40 for S.

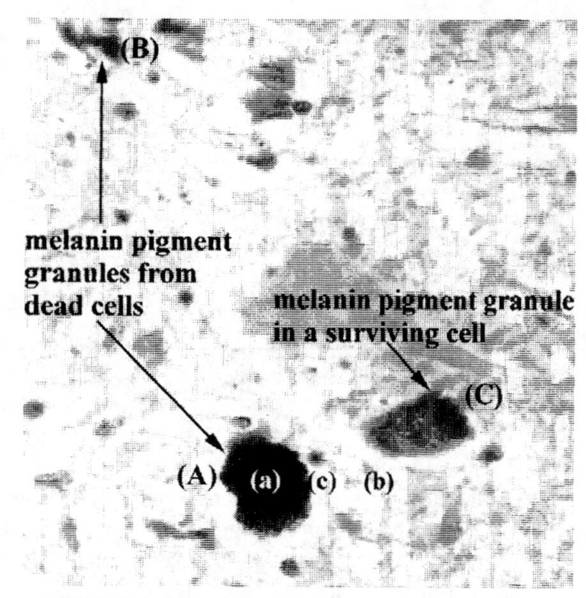

FIGURE 2 Microscopic photograph of the scanning area of the specimen. Melanin pigment granules can be seen at the points of (A), (B) and (C).

the melanin pigment granules in a surviving cell.

Using the optical microscopic images, the measured fluorescence X-ray spectra were grouped into three types; (a) the points inside of the melanin pigment granules released from a dead cell, (b) the control points, which are the points outside of the melanin pigment granules, and (c) the points of the boundary of the melanin pigment granules. One typical spectrum of each group is shown in Fig. 3a-c and measurement points are indicated in Fig. 2 by marks (a), (b) and (c). It is evident that metal elements (Fe, Zn and Cu) were contained inside of the melanin pigment granules from the dead cell with a high concentration. The experiments were done in air hence giving a constant the value of the Ar peaks in all measurements. From each spectrum in the group (a) and (b), peak areas of Fe, Cu, S, Ca and Ar were calculated. Averages, standard deviation and error of these peak areas are shown in table 1. The value of peak area was normalized by the value of the peak area of Ar. The average of the Fe peak area measured inside of the melanin pigment granules was about 11 times higher than that of the control points. The average of the Cu intensity measured inside of the melanin pigment granules was also about twice higher than that of the control points. The errors of the S and Ca peak areas in both groups were large. Even if the errors were accounted for, there was a remarkable difference in the intensity of S in the three different regions.

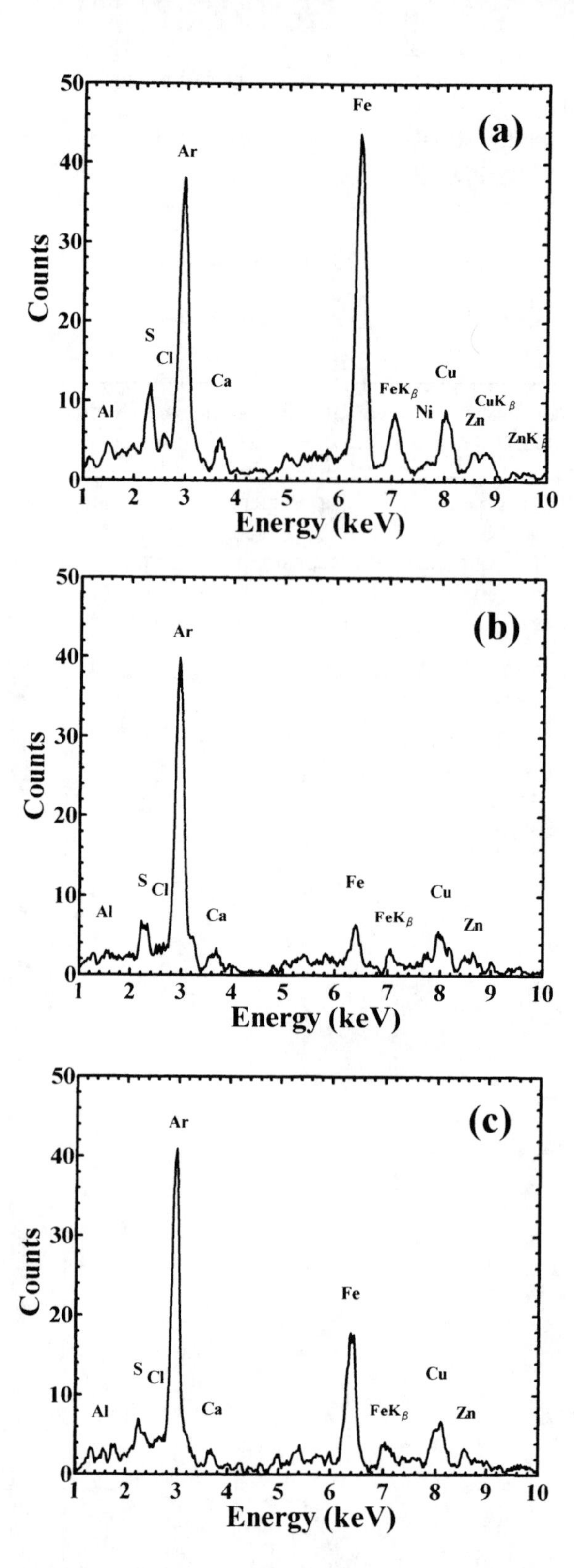

FIGURE 3a, b and c Fluorescence X-ray spectra from the brain tissues of a patient with Parkinson's disease. The measuring point was inside of a melanin pigment granule released from a dead cell (a), outside of the granule (b) and the boundary of the granules (c). The X-ray energy was 13.5 keV.

TABLE 1 Peak area of fluorescence X-ray spectra. The measuring points were inside of the melanin pigment granules released from a dead cell(a) and the other part outside of the melanin granules(b).

Points inside of the melanin pigment granules (a)

	S	Ca	Fe	Cu
Average	68.0	28.0	393.5	96.2
Standard deviation	6.6	10.9	60.3	9.9
Error (%)	9.8	39.0	15.3	10.3

Points outside of the melanin pigment granules(b)

	S	Ca	Fe	Cu
Average	27.0	17.5	35.0	46.5
Standard deviation	8.5	3.7	5.1	13.6
Error (%)	31.4	21.1	14.5	29.2

SUMMARY AND DISCUSSIONS

In this paper, the elemental distributions of the brain cells from a patient with Parkinson's disease is investigated in detail. Fluorescence X-ray spectra and distributions of trace elements in a single neuron were obtained using a micro synchrotron radiation beam. After elemental analysis, histological observation was performed with an optical microscope. Precise correspondence was obtained between the elemental distribution image and the optical microscopic images. These results clearly demonstrated that Fe and Zn were accumulated excessively in melanin pigment granules released from a dead cell.

In order to compare the concentrations of elements in the points inside and those of outside the melanin pigment granules, the peak areas of Fe, Cu, S, Ca and Ar were calculated from fluorescence X-ray in the case of a dead cell. The average of the Fe peak area measured at the melanin pigment granules was about 11 times higher than that at the control points. The average of the Cu peak area measured at the melanin pigment granules was also twice higher than that at the control points. The average values of the Ca and S peak areas were remarkably different in the granules and in the control points.

SRXRF spectroscopy is a highly sensitive and non-destructive trace elemental analysis. Smaller beam (desirably about 1 μm) has to be used in order to obtain better spatial resolution. Photon flux of injected X-ray on samples is proportional to inverse of cross section of the beam. In order to reduce the statistical errors in measurements, we need X-ray intensities one or two order of magnitude higher than the above experiment. Therefore it takes much more time to obtain a better image from a specimen. For example, it took about 160 minutes to image the elements distribution in this experiment. If measurement time had been doubled and the cross section of beam had been 1×1 μm^2, it might be necessary to use as much as about 35 hours. In this way, imaging analysis often involves time difficulties. This problem can be solved by increasing the photon flux from a storage ring. For this reason, it is better to use a brilliant light source for analyzing distributions of trace elements.

In conclusion, excessive and selective accumulation of Fe which was observed in melanin pigment granules released from a dead cell of a patient with Parkinson's disease , may have an important role in the progress of the neurological disorder. Further investigation is needed to establish a clear relation between the level of accumulation and the corresponding pathological effects.

ACKNOWLEDGMENTS

SRXRF experiment was performed at beamline 4A, Photon factory, Tsukuba Japan under project No. 97G178. The authors would like to thank Prof. Atsuo Iida of KEK, Mr. Tomokazu Sano of Kyoto University and Dr. Shinjiro Hayakawa of University of Tokyo for their supports and valuable discussions during this experiment.

REFERENCES

1. Margaritondo G., *Introduction to Synchrotron Radiation*, New York, Oxford University Press, 1988, pp. 3-8.
2. Shaisho H. and Gohshi Y., eds. Shaisho H. and Gohshi Y., *Applications of Synchrotron Radiation to Materials Analysis*, New York, Elsevier, 1996, chapter 2.
3. Valkovic V. and Moschini G., *Rivista Del Nuovo Ciment*, **16**, 1-55 (1993).
4. Kwiatek W.M., *Acta Physica Ponica A*, **86**, 695-703 (1994).
5. Ektessabi A.M., Rokkum M., Johansson C., Albrektsson T., Sennerby L., Saisho and H., Honda S., *Journal of Synchrotron Radiation*, **5**, 1136-1138 (1998).
6. Knor I.B., Naumova E.N., Trounova V.A., Dolbnya I.P. and Zolotarev K.V., *Nuclear Instruments and Methods in Physics Research*, **A369**, 324-326 (1995).
7. Hayakawa S. and Gohshi Y., eds. Shaisho H. and Gohshi Y., *Applications of Synchrotron Radiation to Materials Analysis*, New York, Elsevier, 1996, chapter 3.
8. Sparks Jr. C.J., eds. Winick H. and Doniach S., *Synchrotron Radiation Research*, New York, Plenum, 1980, chapter 14.
9. Kondo T., *Neurological Medicinel*, **42**, 203-212 (1995).
10. Kienzl E., Puchinger L., Jellinger K., Linert W., Stachelberger H. and Jameson R. F., *Journal of the Neurological Sciences*, **134**, 69-78 (1995).
11. Iida A. and Noma T., *Nuclear Instruments and Methods in Physics Research*, **B82**, 129-138(1993).

A Comparison Between PIXE Studies and Electron Microprobe Studies of Rocks From Southern India

G. F. Peaslee, D. A. Carlson, E. C. Hansen, S. S. Hendrickson, R. J. Timmer, A. L. Van Wyngarden, J. D. Wilcox

Chemistry Department and Geological and Environmental Sciences Department, Hope College, Holland MI 49422-9000

The Hope College PIXE facility was used to perform an analysis of zircons, apatites and monazites prevalent in rocks collected in southern India. Thin sections of the samples were prepared for electron microprobe analysis at U. Chicago, and a number of rare earth elements were quantitatively measured. These samples were then fractured and individual crystals were analyzed with 2.3 MeV protons in an internal PIXE irradiation chamber. GUPIX II© software was used to identify and measure trace metals in the samples and the results have been correlated with the electron microprobe data. Whole rock analyses, prepared by making pressed pellets of the powdered rock components were also analyzed by PIXE and XRF. Each of these techniques can be used to address the geological distribution of large ion lithophiles in the mantle, and advantages and disadvantages of each technique will be presented.

INTRODUCTION

The earth's lower crust is a unique laboratory for the study of solid solution geochemistry under extreme pressure and temperature conditions. Trace element analyses of geochemical samples taken from various depths within the earth's crust have been used in a number of studies to identify processes involved in the formation of the continents. One of the most interesting geochemical features of the lower crust is the relative paucity of heat-producing elements, with most of the trace radioactive species being found in the upper crust[1]. This has significant geological implications for continental stability. If there were additional heat sources in lower crustal rocks, it might have kept the lower crust in a molten state instead of producing the relatively stable continents that we enjoy today.

A more general observation is that the lower continental crust is depleted in large ion lithophiles (LIL), which include most of the heat-producing elements. There are several possible theories for this depletion of heat-producing rock from the lower mantle. One possibility is that fluids moving through the lower crust have selectively transported the LIL. To investigate this possibility, among others, the geochemistry of rocks collected in Southern India has been studied for their trace element content[2]. Southern India has been tilted northward and beveled by erosion so that progressively deeper levels of the crust are exposed to the South. At the southern extremes of a 70-km zone on the surface of the Tamil Nadu region of India, rocks that are thought to have originated from as deep as 35 km into the crust are exposed. At the northern extremes of this zone, the rocks are thought to have formed in the upper 20 km of the crust. Several hundred samples have been collected from this region over the past decade and analyzed by Hope College faculty and students[2].

The primary tools used in the analysis of these samples have been the traditional X-ray Fluoresence (XRF) powdered whole rock samples, sent off to commercial labs, combined with Electron Microprobe XRF analysis of individual minerals performed at the University of Chicago Electron Microprobe facility. These techniques are well-established in the field and are generally expensive for an undergraduate research program to afford. With the development of the Hope College Particle-Induced X-ray Emission (PIXE) Facility over the past year, we have explored the possibility of performing further studies of the geochemistry of the southern Indian rock samples with in-house analyses. The preliminary studies have involved the cross-comparison of the techniques for various samples that have been analyzed by several techniques. The results of these preliminary studies are presented here, together with a proposed method of using PIXE to measure uranium content, which is a trace element inaccessible to electron beams.

EXPERIMENTAL METHODS AND RESULTS

The first study involved 16 samples of powdered whole rock analyses from different geographic locations that are thought to correspond to different mantle depths. These samples had previously been sent out for XRF analysis to MSU's Geology Department laboratories, and typical results

CP475, *Applications of Accelerators in Research and Industry*, edited by J. L. Duggan and I. L. Morgan

for an upper-crustal rock versus a lower-crustal rock are shown in Table 1. The exact collection locations of these

Analysis Sample Type	XRF Upper	PIXE Upper	XRF Lower	PIXE Lower
Ti	1610	1120	4790	2640
Fe	10700	10700	43600	43600
Mn	126	95.0	758	802
Ca	16500	12000	33800	31400
K	24000	17000	5810	5020
Cr	21.0	51.1	128	273
Zn	22.0	0.0	47.2	48.0
Rb	20.6	26.8	4.6	0.0
Sr	1320	1400	261	315
Zr	382	283	239	180

Table 1. Concentrations of various trace elements found by PIXE and XRF analyses for two southern Indian rocks. The concentrations are given in units of ppm.

these samples can be found in ref. 2. To perform the PIXE analysis a separate piece of rock from the same sample was powdered in a tungsten ball mill. Commercial cellulose nitrate binder was added to the powdered rock in a mass proportion of 20:1. A pure powdered sample of this binder was shown to be free of heavy metal contaminants in earlier PIXE studies. For each PIXE target made, approximately 0.3 g of this powdered rock mixture was pressed in a hydraulic pellet press at 8000 psi for 30 seconds. The resultant 1-cm diameter pressed pellet was then glued to a thin (500 $\mu g/cm^2$) aluminized polypropylene film, and mounted on a aluminum target frame. These target pellets were thick enough to be self-supporting and did stop the incident beam of protons. The aluminized polypropylene foil was used to reduce charge collection on the electrically non-conductive sample. Additionally, two tungsten filaments were illuminated near the surface of the sample during irradiation to minimize charge build-up on the target.

Approximately 6 to 10 replicate pellets from each sample were irradiated with 3 to 6 μC of 2.3 MeV protons in an evacuated chamber. The resultant x-rays were detected at an angle of 135° from the incident beam, with a 4 mm^2 Si(Li) detector with a thin (0.8 μm) Be window. This detector had a nominal resolution of 200 eV for the 5.9 keV line. Although this detector was ancient, and has since been replaced, significant results were obtained. The signals were processed through a standard 10-bit MCA.

Figure 1 shows sample PIXE spectra from the two samples listed in Table 1. There is a clearly different signature between upper-crustal rock samples and lower-crustal rock samples that was observed for all samples. Each of the PIXE spectra were analyzed offline using GUPIX II[3] software. The absolute normalization of the thick target results in GUPIX was achieved with several NIST standards of Buffalo River Sediment[4], pressed into pellets and irradiated in the same manner as the powdered rock samples. Each of the certified standard element concentrations in the NIST mud were reproduced within 10% using this simple target preparation and analysis technique.

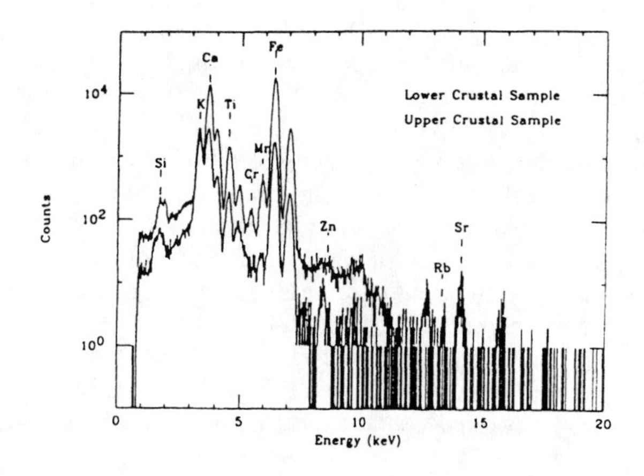

Figure 1: Typical experimental X-ray spectra for two of the samples used to calculate the values in table 1.

The sample concentrations for several elements as identified by GUPIX are also listed in Table 1 for comparison to the XRF results. Figure 2 shows the results for many different elements for all samples which contained measurable quantities of the trace elements. The results obtained from our PIXE analyses are plotted on the abscissa while the XRF data, which are assumed to be accurate, are plotted on the ordinate axis. In each case the axes reflect ratios of elemental concentrations rather than absolute concentrations, to avoid extra uncertainty that arises from integration of the beam current over the time of irradiation.

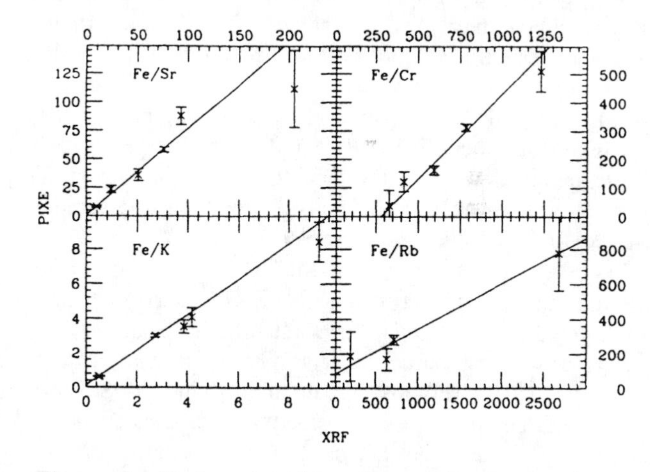

Figure 2: Ratios of various elemental concentrations to iron concentrations determined by PIXE and XRF.

As an extension to this initial project, we have compared PIXE analyses of southern Indian rocks to some electron microprobe analyses performed on the same samples. Although we do not have a microprobe PIXE facility, several rock samples with clearly identified mineral grains were fractured and the relevant minerals mounted in an epoxy base on top of a glass slide. These individual mineral grains were then polished to create a flat surface while mounted in epoxy. A slice of epoxy plus mineral grain was then removed from the glass slide with a rock saw, and mounted to a thin aluminized polypropylene foil with acrylonitrile glue. These foils had already been mounted to standard aluminum target frames and both the glues had previously been measured to be free of any identifiable heavy elements by PIXE analysis. These grains were then exposed to approximately 5 - 15 μC of beam in vacuum at the Hope College PIXE Facility and energy-dispersive x-ray spectra measured with the detector described above. A sample spectrum is shown in the top left panel of Figure 3 for an apatite, with various rare earth peaks identified. The spectra were analyzed off-line with GUPIX II[3] software, and the results are shown in Table 2. Once again ratios of elements (this time normalized to Yttrium concentration) are plotted for the two independent measurement techniques as panels in Figure 3.

Analysis	PIXE	EMP	PIXE	EMP	PIXE	EMP
Sample type	Monazite	Monazite	Apatite 1	Apatite 1	Apatite 2	Apatite 2
Ca			38.9	38.8	31.4	37.2
Y	1.68	1.68	0.027	0.027	0.417	0.417
La	6.99	7.28	0.106	0.096	0.319	0.458
Ce	20.1	23.3	0.213	0.255	0.771	1.075
Nd	8.14	9.86	0.101	0.098	0.342	0.439
Sm	1.50	2.36				
Th	12.4	11.5				

Table 2. Concentrations of various rare earth elements found by PIXE and electron microprobe (EMP) analyses for three large grained rocks. The concentrations are given in units of weight %.

After irradiation at Hope College, the same samples were prepared into grain mount microprobe samples, with more extensive polishing of the surface. These grain mount samples were then optically scanned with a polarizing microscope and particular minerals of monazite [(Ce,La,Nd,Th)PO_4], apatite [$Ca_5(PO_4)_3$] and zircon [$ZrSiO_4$] were identified and mapped for the microprobe. The samples were coated with a thin layer of carbon for conduction. The microprobe analyses for the same samples used in the PIXE study are presented in Table 2.

As a final study to assess the limit of detection for uranium in a typical zircon (SiO_2-ZrO) matrix we have constructed a set of thick target standards. Commercially available silica, zirconium oxide and uranyl acetate powders were mixed with cellulose nitrate binder and pressed into standard target pellets as described above. Using GUPIX analysis of these target standards, we have established a level of detection of approximately 100 ppm for uranium in our new detector and beamline configuration.

DISCUSSION

For the cross-comparison of the PIXE and XRF analyses of powdered whole rock analyses, there is good agreement for every element that is present in sufficient quantity to measure by both techniques. There is dispersion in some data points, which reflects the measurement uncertainties of about 10% in any of our PIXE analyses, plus an additional dispersion that seems to be related to lack of sample homogeneity. At the time these measurements were made,

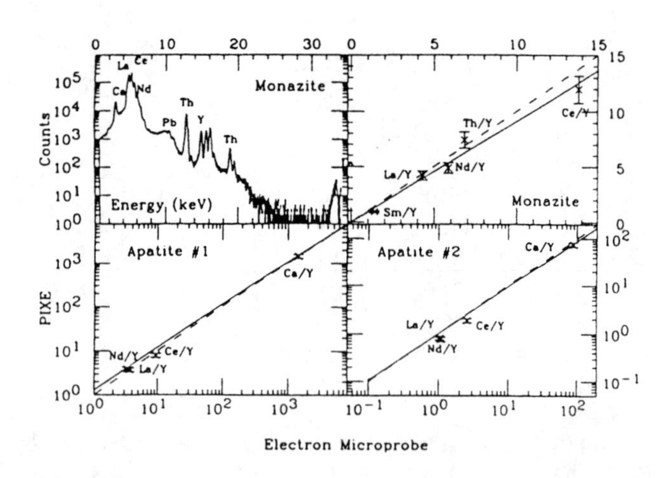

Figure 3: A typical PIXE spectrum of a monazite mineral is shown in the top left panel. The correlation of elemental concentration ratios to yttrium are shown for this sample in the top right panel, plotted against those ratios measured by the electron microprobe. A similar cross-correlation is plotted for two different apatites below. The dashed curves shows the expected correlation for a 1:1 ratio, while the solid curves are the experimental best fits.

our beam spot size was focussed to approximately 1 mm^2, and impossible to make very diffuse. More recent measurements of one of these samples with our improved beamline collimation indicate that sample homogeneity is much less of an issue when the beam spot size is enlarged to 30 mm^2. Sample homogeneity was also seen to improve by a mechanical high-speed shaking of a sample for several minutes prior to pressing it into a pellet. In addition to sample homogeneity, some elements (such as barium) are difficult to peak-fit uniquely, because of the ambiguity of their L-shell x-rays with K-shell x-rays of other elements (such as Ti). Still, even with these preliminary data, there is a definite correlation in the results shown in Figure 2, as indicated by the individual fits to the data. The fact that the slopes are each slightly different for different elements might reflect the differing sensitivity to detection limits experienced in the two x-ray detection methods. These results encourage us to pursue further whole rock analyses, at least for certain elements that are abundant in these rocks.

For the cross-comparison of the PIXE and electron microprobe analyses of grain mounts, the results are again surprisingly good. The microbeam analysis is guaranteed to be exactly of the mineral grain of interest, because electron backscattering fluoresence microscopy is used to identify each bombardment site. The "poor man's" micro-PIXE technique we have employed at Hope College is limited to the optical resolution and mechanical separation of single grains from a destroyed thin section. There was no guarantee that the host rock of each mineral grain was not attached to the edge of each grain of interest, but the rare earth elements of interest seem to be cleanly separated in each of the three samples we have examined to date. Therefore the PIXE analysis reproduces the electron microprobe analyses rather well in these cases, and the slope of each cross comparison curve is close to unity (shown as the dashed lines in Figure 3).

This type of analysis is restricted to mineral grains that are large enough to be mechanically separated from the host rock. However, since the rock samples from southern India have a significant percentage of the mineral zircon scattered throughout the rock, this preliminary work has led to the development of another method to study elements not accessible to even the electron microprobe. Zircons are known to be host minerals for thorium and uranium substitution, and to address the question of lower mantle depletion of these heat-producing elements, it would be advantageous to make a direct measurement of the thorium and/or uranium content of these minerals.

Previous NAA analyses have indicated an expected concentration of 20 - 500 ppm thorium and uranium within a zircon. Since zircons are only present at < 0.01 weight % level in the whole rock, the whole rock concentration of these heavy elements would be in the ppb range. The electron microprobe has very limited sensitivity to heavy elements because of the poor ionization potential of the incident electrons. Protons have much better ionization potentials, even at 2.3 MeV, and direct measurement of these elements in many of the northern rock samples should be feasible if the zircon grains can be separated from the whole rock samples. Since zircons are the most dense non-metallic mineral it should be possible to separate them from most of the surrounding rock in a density column. Further separation of the dense mineral section (>4.0 g/cm^3) of each powdered rock sample can be attempted with magnetic separators, since zircons are non-magnetic and most of the dense metals are magnetic. After this separation, optical microscopy and refractive oils can be used to select zircon grains for PIXE analysis of thorium and uranium content.

CONCLUSIONS

This study has proven the feasibility of using the local PIXE facility to undertake topical geochemical research at Hope College, in that whole rock analyses of powdered samples and thin section analyses can be well reproduced. There are high detection limits for certain elements that have poorly resolved x-ray peaks, and care must be taken to homogenize powdered samples completely, but target preparation is relatively easy. The mechanical separation of minerals from whole rock or from thin sections provides an additional tool to isolate elements of geochemical interest. It is hoped that this technique will allow the direct measurement of thorium and uranium concentrations in zircons extracted from southern Indian rocks.

ACKNOWLEDGEMENTS

This work was supported in part by an REU grant from the National Science Foundation: # CHE-9100801, an interdisciplinary faculty development grant from the Howard Hughes Medical Institute, and a seed grant from the Michigan Space Grant Consortium.

REFERENCES

1. Turcotte, D. L. and Schubert G., *"Geodynamics"*, John Wiley & Sons, (NewYork) 1982. p. 145.
2. Hansen, E. C., Newton, R. C., Janardhar, A. S. and Lindenberg, Sheila, *J. Geology* **103** (1995) 629-651.
3. Maxwell, J. A., Campbell J. L., and Teesdale, W. J., *Nucl. Instr. Meth. Phys. Res.* **B43** (1989) 218-230.
4. NIST Standard Reference Material #2704.

Proton Induced Monochromatic X-Rays : A Technique for Solving Interference Problems in X-Ray Fluorescence Analysis

[1]A. G. Karydas and T. Paradellis

N. C. S. R. «Demokritos», Institute of Nuclear Physics, Laboratory for Material Analysis,
[1] Author to whom all correspondence should be addressed.

With PIXE and EDXRF techniques excellent sensitivities for most of the elements in a wide range of matrix sample compositions can be obtained. Despite this, in some cases, strong interferences originating from the presence in the matrix of an element with high concentration, can limit these sensitivities considerably. A combination of the above techniques, PIXE and XRF, seems to be the most efficient solution to this problem. By choosing the primary target properly, protons can produce an intense, almost monoenergetic exciting X-ray radiation, which in several cases selectively excites the elements of interest in the sample and overcomes the production of X-rays of the element dominating the matrix. The application of this technique to specific interference problems, either in the characterization of thin films deposited onto various substrates (YBaCuO film onto $LaAlO_3$ crystal and a MgF_2 film onto a SiO_2 matrix) or in the determination of trace elements in a high Z thick matrix (copper) is discussed.

INTRODUCTION

Incidence of low energy proton beam (1-3 MeV) on a thick pure target generates an almost monochromatic X-ray radiation with unique characteristics in comparison with conventional X-ray sources. Thick target PIXE can produce yields with intensities of the order of 10^{11}-10^{10} X-rays /μA in the energy range 2-10 keV, which in compact reflection or transmission geometry offer abundant X-ray intensities at the sample position. In addition, an important advantage of proton induced X-rays is their excellent spectral distribution. The bremsstrahlung production mechanism in proton-atom and secondary electron-atom interactions have cross sections which are limited to a few tens of mb/keV or less, while the corresponding ones for the production of characteristic X-rays are at the order of tens or hundreds of barns. Thus, the Proton Induced X-ray Fluorescence (PIXRF) technique has been developed as a complementary technique to the conventional PIXE and XRF techniques, solving not only interference problems, but also achieving low MDL's in the determination of trace elements in biological samples [1, 2]. The advantages and drawbacks of using selective excitation in XRF analysis will be discussed, showing the important features of the PIXRF technique. Also, specific examples showing the way the proton induced X-rays can overcome interference problems between 1-10 keV are given.

IMPROVEMENT IN THE MDL, BY EMPLOYING SELECTIVE EXCITATION IN XRF

The sensitivity in XRF analysis with the use of monochromatic excitation is worsened, mainly when high spectral contributions of particular X-ray lines are present. Below each one of such peaks in the spectrum, due to the incomplete charge collection in the detector, an almost flat background extending to zero energy, is produced with a given intensity relative to the parent peak. For example, in the case where trace elements are analyzed in a biological sample, the background is produced mainly by the scattered radiation in the low Z matrix. On the contrary, in the case where the matrix of the sample is composed from few medium or high Z elements, the background originates from corresponding fluorescent X-ray lines. In both cases, any improvement in the MDL for trace elements determination depends on the possibility to reduce the high intensity X-ray peaks in the spectrum. Thus, for the first case, a polarized exciting radiation or total reflection may be used, while in the second case, the selective excitation is the most efficient approach. It is very useful to examine in a quantitative way the improvement which selective excitation can offer to the MDL, versus the usual monochromatic excitation.

For simplicity, it will be assumed that the sample analyzed (thick target), is composed mainly by an element j and the detection of traces of the element i with $Z_i < Z_j$ is required. In the following the MDL in the XRF analysis of the i element will be compared for the cases of two monoenergetic X-ray beams having the same intensity but different energies: the first one with energy E_1 less than the K-absorption edge energy $U_{K,j}$ of the j element and the second one with energy $E_2 > U_{K,j}$. Without any loss of the generality in our approach, it will be also assumed that the E_1 exciting radiation coincides with the basic characteristic X-ray line of the j element. For the E_1 exci-

CP475, *Applications of Accelerators in Research and Industry*,
edited by J. L. Duggan and I. L. Morgan

tation, the background under the characteristic X-ray peak of the i element is proportional to the intensity $I_s(E_1)$ of the scattered exciting radiation. For the E_2 excitation, it is proportional to the intensity $I_{fj}(E_2)$ of the fluorescent radiation of the j element. Thus, the MDL gain factor in the detection of the i element for the two excitation energies, can be expressed as :

$$R_{MDL} = \frac{(MDL)_2}{(MDL)_1} = \frac{\sqrt{I_{fj}(E_2)}}{I_{fi}(E_2)} \cdot \left(\frac{\sqrt{I_s(E_1)}}{I_{fi}(E_1)} \right)^{-1} \qquad (1)$$

where $I_{fi}(E_1)$ and $I_{fi}(E_2)$ are the intensities of the characteristic X-ray radiation of the i element produced by X-rays of two different energies. The Fundamental Parameter Approach (FPA) states [3] that, for an exciting radiation E_o, the fluorescence and scattered radiation intensity produced from a thick target in reflection geometry can be expressed as

$$I(E_i) \propto c_i \cdot \sigma_i(E_o) \cdot \frac{1}{\mu_s(E_o) + \mu_s(E_i)}$$

where c_i is the concentration of the i element, σ_i is the cross section for the production of the E_i X-ray (characteristic or scattered) and $\mu_s(E)$ is the mass absorption coefficient in the sample for the energy E. Thus, the MDL gain ratio, eq. (1), can be written in the following form :

$$R_{MDL} = R_{fp} \cdot R_{fa} \cdot R_{bp} \cdot R_{ba} \qquad (2)$$

where

$$R_{fp} = \frac{1}{SF_{ji}} \frac{\tau_i(E_1)}{\tau_i(E_2)}, \qquad R_{fa} = \frac{\mu_s(E_2) + \mu_s(E_i)}{\mu_s(E_1) + \mu_s(E_i)}.$$

$$R_{bp} = \sqrt{\frac{c_j \cdot \sigma_j(E_2)/4\pi}{(d\sigma/d\Omega)_s(E_1)}}, \quad R_{ba} = \sqrt{\frac{2 \cdot \mu_s(E_1)}{\mu_s(E_2) + \mu_s(E_j)}}$$

The factor R_{fp} accounts for the for the excitation of the i element at two different X-ray energies and is expressed as a function of the photoelectric cross section τ_i of the i element for the energies E_1, E_2 and by the secondary fluorescent enhancement SF_{ji} of the i element due to the characteristic X-rays of the j element. The factor R_{bp} accounts for the relative production cross section of those X-rays, which cause the background in the spectrum in each case. More specifically, this factor depends on the concentration c_j, the fluorescent cross section $\sigma_j(E_2)$ of element j due to the E_2 exciting radiation and from the scattering (Rayleigh+Compton) cross section $(d\sigma/d\Omega)(E_1)$ of the E_1 exciting radiation in the matrix. The factors R_{fa}, R_{ba} express the relative self-absorption correction for the fluorescent and background radiation. By using the well known exponential approximation of the mass absorption coefficient with the X-ray energy, below and above an absorption edge, these terms can be very well approximated by :

$$R_{fp} = \frac{1}{SF_{ji}} \left(\frac{E_2}{E_1} \right)^{n_i}, \qquad R_{fa} = \frac{J_{K,s}(E_i/E_2)^{n_j} + 1}{(E_i/E_1)^{n_j} + 1}$$

$$R_{ba} = \frac{1.414}{\sqrt{\left[J_{K,s}(E_1/E_2)^{n_j} + 1 \right]}}$$

where $J_{K,s}$ is the absorption jump of the mass absorption coefficient in the sample at the K-edge of the j element.

Two characteristic examples demonstrate the improvement in the MDL. The use of a CuK_α X-ray beam for trace elements quantification in a thick copper matrix, improves the MDL by a factor of 10 compared with other monochromatic X-ray exciting beams of higher energy (figs. 1, 2). In the case of a quartz matrix, the SiK X-ray beam offers still higher gain factor (fig.3).

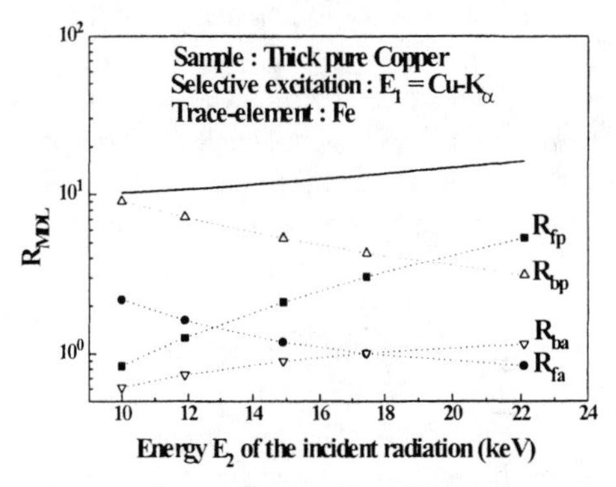

FIGURE 1. Gain factor for the MDL in the XRF analysis of Fe traces in a pure thick copper matrix. Selective excitation with a CuK_α beam is compared with the use of various monochromatic X-ray beams. The partial contribution of each one of the factors involved in equation (2) is also demonstrated.

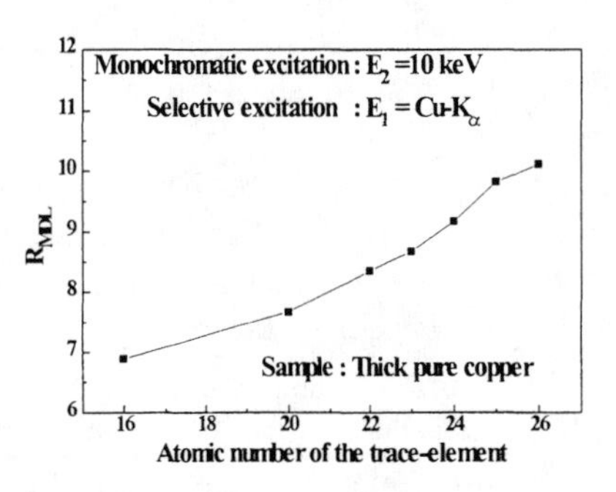

FIGURE 2. Gain factor for the MDL in the analysis of various trace elements in a pure thick copper matrix. Selective excitation with CuK_α beam is compared with a 10 keV excitation.

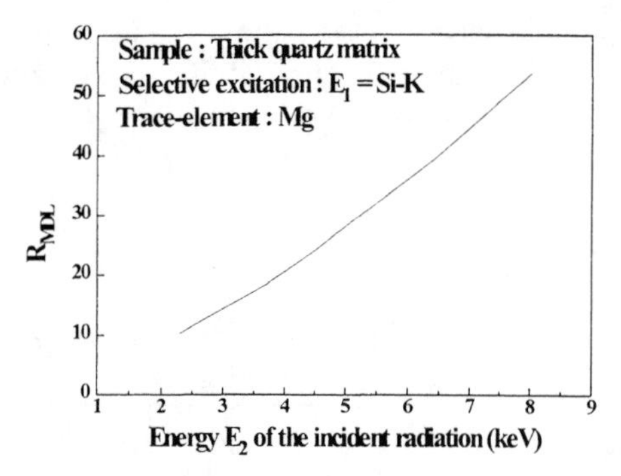

FIGURE 3. The gain factor for the MDL in the detection of Mg traces in a pure SiO_2 matrix. The graph stands for the selective excitation of Mg with a SiK X- ray beam, versus the excitation with various monochromatic beams.

LIMITATIONS IN SELECTIVE EXCITATION - RESONANT RAMAN SCATTERING

The selective excitation in XRF analysis has some inherent limitations. These arise from the fact, that in most of the cases we employ exciting radiation with an energy slightly less than the absorption edge of the major matrix element ($E_o \leq U_{ij}$, where U_{ij} is the absorption energy of the i shell for the j matrix element). This energy condition gives rise to the Resonant Raman Scattering (RRS) [4].

This type of scattering is enhanced significantly according to the law $(U_{i\,j} - E_o)^{-1}$. The RRS is an inelastic process producing a scattered photon and an electron emitted to the continuum. It appears in the spectrum as a continuous distribution, which has onset at the maximum scattered photon energy and extends to zero energy. Under this condition it is very probable, that the RRS distribution may overlap in the spectrum with various characteristic X-ray lines [5,6]. The maximum RRS photon energy is given by $E_{max}=E_o-U_{kj}$, where U_{kj} is the absorption edge of a higher shell that emits the outgoing electron. If i=K and k=L, RRS is characterized as KL type, while if i=L and k=M as LM type. Since the average electron kinetic energy is small (hundreds of eV), the RRS distribution exhibits in the spectrum an almost gaussian profile with the centroid close to the energy E_{max}, having an intense asymmetric tail extending to the lower energies.

In table 1, some interference cases between RRS peaks and the characteristic X-rays of trace elements are presented. The intensity of a fluorescent peak relative to the RRS one, can be calculated from the relation:

$$\frac{I_i(E_o)}{I_{RRS}(E_o)} = \frac{c_i \cdot \sigma_i(E_o)/4\cdot\pi}{(d\sigma/d\Omega)_{RRS}} \tag{3}$$

where c_i is the concentration of the i trace element in the matrix, $\sigma_i(E_o)$ is the cross section for the production of the interfering characteristic X-ray. $(d\sigma/d\Omega)_{RRS}$ is the cross section for the KL-RRS of the incident radiation E_o with the matrix atoms, which satisfy the resonant condition. The RRS cross sections of CuK_α in copper and that of FeK_β in iron are obtained from [4, 7, 8]. Using their known dependence with the energy of the incident radiation and with the average kinetic energy of the emitted electron, the RRS cross sections for other exciting radiations were calculated (table 1). Characteristic interference's between RRS X-ray peaks and proton induced X-rays are shown in fig. 4.

Finally, using eq. (3), it can be easily shown that the intensity of the RRS peaks is so strong, that they correspond to hundreds of ppm of apparent concentration of the interfering element (table 1).

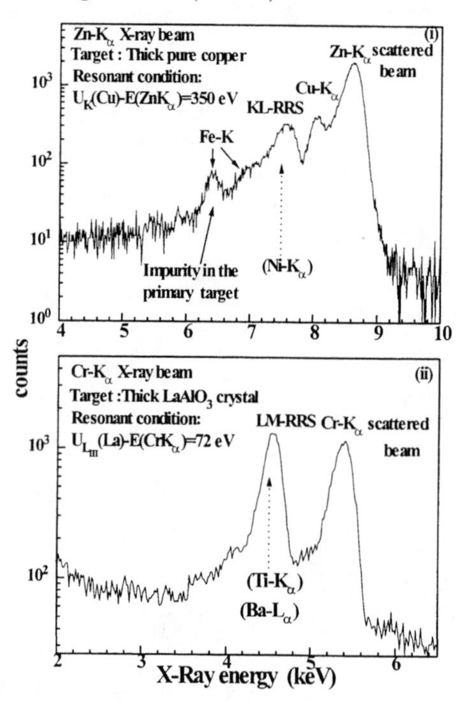

FIGURE 4. Spectra showing the interferences between the RRS peaks and the characteristic X-rays, by employing two proton induced X-ray beams. The respective K_β X-rays are absorbed by the use of filters. Experimental conditions (proton energy, total accumulated charge, filter) : (i) 2 MeV, 2000 μC, Cu 40 mg/cm², (ii) 2 MeV, 3000 μC, V 30.8 mg/cm².

TABLE 1. Examples of elements having characteristic X-rays interfering with RRS peaks. In the last column the apparent concentrations of these elements are shown, in the case where $I_i=I_{RRS}$ in eq. (3).

E_o	Matrix	E_{max} keV	ΔE_R (U_K-E_o) eV	Interfering Element	$\sigma_n/4\pi$ b/sr	$(d\sigma/d\Omega)_{RRS}$ (r_o^2/sr), $r_o^2 \approx 79$ mb	c_i ppm
ZnK_α	copper	7.69	350	Ni	739	6.3	673
CuK_α	copper	7.10	940	Co	716	2.3	254
CoK_α	iron	6.20	187	Mn	680	8.6	999
FeK_α	iron	5.70	712	Cr	725	2.2	240

PROTON INDUCED MONOCHROMATIC X-RAYS FOR SELECTIVE EXCITATION

A typical experimental set-up for the production of proton induced X-rays is based on the reflection geometry. Transmission geometry may also be used offering the flexibility in conducting scattering experiments. Details on these two arrangements may be found in ref. [2,9,10]. For quantification purposes, the calibration procedures are simple and accurate [2]. The FPA equations applied in TTPIXE and XRF analysis permit the extention of the calibration to every element, bombardment energy and primary target with the use of a few standards. In addition, a very important point concerning accurate quantitative analysis is that secondary fluorescence is almost eliminated at selective excitation.

One basic disadvantage of the technique is that protons induce nuclear reactions in the primary target and in turn, the accompanying γ-rays produce a flat background into the Si(Li) detector due to the Compton scattering [2,11]. In order to minimize this source of background, low energy protons are preferred, while the proper choice of the primary target is also important. For example, instead of a copper primary target nickel is more appropriate (fig. 5i) for trace elements determination in a thick pure copper matrix. The total thick target γ-ray yield (E_γ >500 keV) induced by 2 MeV protons is 28 times less in nickel than the one produced by the copper target [12]. Thus, very low MDL for metal trace element can be achieved in this case with a NiK_α X-ray beam instead of a Cu one. From the spectrum of fig. 5i. the following MDL's were found (for a 10 min measuring time) Fe: 3, Mn: 5, Cr: 6, V: 10 and Ti: 14 ppm. In fig. 5ii. a proton induced CrK_α X- ray beam is used to determine the Ba/Y atomic ratio in a YBaCuO film sputtered onto a $LaAlO_3$ substrate. Since intense La characteristic X-rays interfere with the one of Ba, the only possibility for the determination of the above atomic ratio is selective excitation of the film with the CrK_α X-ray beam, employing the energy condition $U_{L3}(Ba) < CrK_\alpha < U_{L3}(La)$. Due to the LM-RRS of the incident beam with La atoms (fig. 4ii.), the contribution to the spectrum from the substrate must be subtracted. The resulting spectrum is shown in fig. 5ii. In fig. 5iii. a proton induced SiK X-ray beam has been used for improving the sensitivity in the analysis of a Mg thin layer deposited onto a quartz matrix. In the spectrum, Na and Al produced from the substrate are seen. The Mg layer thickness was found to be (13.4 ± 1.3) $\mu g/cm^2$, while the MDL for the Mg detection onto a quartz matrix was estimated for a 15 min measurement (2 μA proton current) to be about 42 ng/cm^2.

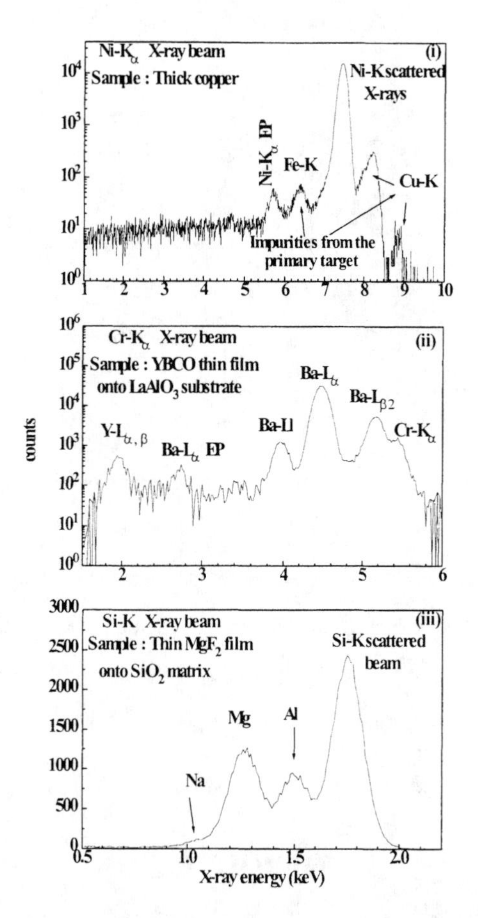

FIGURE 5. Proton induced monochromatic X-rays for solving interference problems: Experimental conditions (proton energy, total accumulated charge, filter) : (i) 2 MeV, 240 μC, kapton 17 mg/cm^2 + copper 8 mg/cm^2, (ii) 2 MeV, 9000 μC, vanadium 30.8 mg/cm^2, (iii) 1.5 MeV, 1800 μC, kapton 3.9 mg/cm^2.

REFERENCES

1. Peisach M., *J. Radioanal. Nucl. Chem.*, **110**, 461(1987)
2. Karydas A. G., Paradellis T., *X-Ray Spectr.*, **22,** 252(1993)
3. Grieken R. and Markowicz A., *Handbook of X-Ray Spectrometry*, edited by Marcel Dekker Inc., 1993, Chapt.1
4. Sparks C. J. Jr., *Phys. Rev. Lett.* **33**, 262(1974)
5. Van Espen P J , Nullens H A. and Adams F. C, *Anal. Chem.*, **51**, 1580(1979)
6. Nishide Y., Sasa Y. and Uda M., *Nucl. Inst.. Methods* **B94**, 271(1996)
7. Bannet Y., Rapaport D. and Freund I., *Phys. Rev.* **A36**, 623 (1977)
8. Suortti P, *Phys. Stat Solid*, B **91**, 657(1979)
9. Karydas A. G., Potiriadis C. and Paradellis T., *X-Ray Spectr.*, **20**, 315(1993)
10. Zeng X., Wu X., Yao H., Yang F. and Cahill T. , *Nucl. Instr. Meth.* **B75**, 99(1993)
11. Uda M., Morito K., Matsui H., Kotani T., Nakamura M. and Ise H., *Proceedings of the 14th Int. Conf. of Appl. of Accelerators in Research and Industry*, p. 539, Denton 1996.
12. Demortier G., *J. Radioanal. Chem.*, **45** 459(1978)

Application of Micro-PIXE Method to Ore Geology

S. Murao[a], S. H. Sie[b], S. Hamasaki[a], V. B. Maglambayan[c] and X. Hu[d]

[a] *Geological Survey of Japan, 1-1-3 Higashi, Tsukuba, Japan 305-8567*

[b] *CSIRO Division of Exploration and Mining, P.O. Box 136, North Ryde, NSW, Australia 2113*

[c] *NIGS, College of Science, University of the Philippines, 1101 Diliman, Quezon City, The Philippines*

[d] *A202 Garden Heights Sakuradai, 989-3 Kashiwa, Japan 277-0005*

Specific examples of ore mineral analysis by micro-PIXE are presented in this paper. For mineralogical usage it is essential to construct a specimen chamber which is designed exclusively for mineral analysis. In most of the analysis of natural minerals, selection of absorbers is essential in order to obtain optimum results. Trace element data reflect the crystallographic characteristics of each mineral and also geologic settings of sampling locality, and can be exploited in research spanning mineral exploration to beneficiation. Micro-PIXE thus serves as a bridge between small-scale mineralogical experiments and understanding of large-scale geological phenomenon on the globe.

INTRODUCTION

Trace element distribution in minerals plays an increasingly important role in both exploration and mineral processing. It contributes to the understanding of ore genesis by preserving the signature of physico-chemical processes in partitioning of elements in the mineral assemblage. In mineral processing precious and rare metals often occur as trace elements and knowledge of the distribution is vital in devising an effective beneficiation strategy. More recently, social expectation for "high-technology metals" and non-toxic "green metals" is driving geologists to gain better understanding of their formation leading to better and detailed exploration model than before. The usefulness of micro-PIXE in mineralogy and geology, with its high sensitivity and spatial resolution for trace element measurements, has been demonstrated in many case histories and reviewed extensively in literature [e.g. 1-3]. Here we present examples of ore mineral analysis mainly done through the collaboration between CSIRO and the Geological Survey of Japan (GSJ).

INSTRUMENTAL SETUP

Even in the study of one field, geologists analyze many kinds of mineral species with micron-size grains and complex textures, and thus stable microbeam is the essential condition for their work. In addition, to pinpoint the analytical area, manual-beam centering is an important function. Also they need a large number of analyses for statistically meaningful results. To achieve this goal the specimen chamber has to be exclusively designed for geoscience applications with following conditions [5, 6].

(1) Transmitted and reflected microscope to observe mineral textures and specimen surface.

(2) Capability to accept various kinds of specimens

CP475, *Applications of Accelerators in Research and Industry*,
edited by J. L. Duggan and I. L. Morgan

(3) Easy system to find analytical points

(4) Quick change of samples and absorbers

(5) Precise measurement of beam charge

(6) Software to calculate the composition of thick target

EXAMPLES OF ANALYSIS

"Ore mineral" is an economic geology terminology, defined as minerals having any kind of useful element at recoverable level. Mineralogically most of them are classified as silicates, oxides and sulfides with various types of structure. But in this report we classify the ore mineral into several categories based on their micro-PIXE analytical properties, and examples (Table 1) are presented following this classification. The analyses were performed at CSIRO using a proton microbeam of 3 MeV that was focused to 10 to 30 micron spatial resolution. The X-rays were detected in an energy dispersive Si(Li) detector at 135-degree angle with respect to the beam direction through various kinds of filters depending on mineral species. The X-ray spectra were deconvoluted for peak areas and converted to concentration using the Geo-PIXE software [7]. Data were collected for a minimum 3 micro-Coulomb beam charge, giving minimum detection limit (MDL) for most elements (Z>26) at 3-50 ppm. The MDL values were defined at 99 % confidence limit.

Light minerals

Most of the light minerals on the earth are silicates such as biotite which usually contain high level (a few percent) of Fe. Some are of special composition such as CaF_2 (fluorite) but Fe is still often detected in such minerals. We tested fluorite and confirmed that it is important to reduce the intensity of the strong X-ray energy of Fe for the analysis. In our experiment, 100 micron aluminum was effective to observe Y, Zr, Rb and Sr that can substitute Ca.

Iron minerals where the Fe K lines dominate

In nature, pyrite, marcasite, pyrrhotite ($Fe_{1-x}S$) and magnetite (Fe_3O_4) are often found as common minerals. Pyrite and marcasite are polymorph of FeS_2 composition. In the case of marcasite, 300 micron thick aluminum filter is used to give a good spectrum. Pyrite and marcasite are iron disulfides and are characterized by anion pairs (S^{2-}-S^{2-}) in the structure. It tends to house trace amount of anions in sulfur's site and cations in iron's site. The analytical result shows two groups of elements, the first with As, Sb and the other with Cu, Zn, Bi and Pb, reflecting such characteristics. As for pyrrhotite, a 300 micron aluminum is also suitable for the measurement. This mineral belongs to NiAs type structure which has the ability to accommodate transition elements in the trigonal bipyramidal holes [8]. The PIXE analysis confirmed this fact to show the concentration of transition elements, Cu, Ni, and Zr. For magnetite, we also use 300 micron aluminum to absorb strong Fe X-rays. Magnetite is of spinel structure that is well known for the flexibility in the range of cations and cation charge combinations [9] and thus it could be an excellent monitor of geological process. In our analysis Cu, As, Zr, Sn, W and Au were detected for two grains from an Indian Sn-W-Cu-As-Au deposit. These examples indicate that the mineral structure strongly control the predominant trace elements in the study area and is thus a useable monitor in geology.

Heavy minerals where K lines of element other than Fe dominate

We selected arsenopyrite (FeAsS) and molybdenite (MoS_2) for the trial. We need 400 micron aluminum for arsenopyrite analysis. The mineral structure includes anion pair, As-S^{2-}, and Fe. The PIXE detected Sb which substitute As-S pair and Ni, Cu and Sn which substitute Fe. For molybdenite we used 300 micron aluminum. The result

TABLE 1. PIXE microprobe analysis of representative ore minerals. Values in ppm unless otherwise noted. Values are given in average with MDLs in parenthesis. "n" indicates the number of analysis.

	Fluorite (n = 2)	Marcasite (n = 3)	Pyrrhotite (n = 6)	Magnetite (n = 2)	Wolframite (n = 9)	Galena (n = 4)
V	25 (15)	-	-	-	-	-
Cr	111 (31)	-	-	-	-	-
Mn	1040 (14)	-	-	-	-	-
Fe	442 (12)	46.6 %	63.5 %	63.7 %	36.4 %	10.2 % (1.7 %)
Cu	-	779 (32)	346 (50)	96 (40)	755 (93)	1.3 % (376)
Zn	41 (9.4)	146 (19)	-	-	134 (59)	638 (189)
Rb	13 (5.0)	-	13 (7.5)	-	-	-
Sr	196 (4.4)	-	42 (7.2)	-	-	-
Y	73 (5.4)	-	-	-	647 (19)	-
Zr	14 (6.4)	-	36 (7.1)	9.4 (6.6)	-	-
Nb	-	-	-	-	36.6 (5.7)	-
In	-	-	-	-	-	-
Sn	837 (6.4)	-	339 (23)	600 (17)	314 (54)	-
As	-	5796 (9.1)	-	-	-	-
Sb	181 (84)	137 (16)	-	-	-	-
Au	-	-	-	79 (31)	-	-
Bi	-	183 (19)	-	-	-	1.3 % (131)
Pb	-	419 (20)	74.7 (23)	-	-	86.6 %
Ag						6910 (52)
W	-	-	-	4620 (77)	45.0%	-
Mo	-	-	-	-	14.2 (6.1)	-

gave trace elements 3110 ppm Cu, 6.6 % Zn, 175 ppm As, 170 ppm Se, 3.8 % Sn, 904 ppm Bi, 1530 ppm Pb and 904 ppm Th, together with 10 % Fe. The MDL of elements of higher Z in molybdenite, compared to other minerals, is poorer due to the tail of Mo K line.

Heavy minerals where L lines of the major elements are dominant

Gold/electrum (alloy of Au and Ag; alloy of more than 20 % Ag is called electrum), wolframite ($(Fe,Mn)WO_4$), and galena (PbS) represent this category. As for gold/electrum many peaks of interest are interfered with by the Au L lines resulting in poor MDL values in the 100 ppm level. In addition, attention must be paid for the artifacts of the detector and filters [9]. In spite of such problems PIXE is still useful in reducing the analytical time, and through the ability of in-situ, grain-by-grain analysis because gold concentrate of enough quantity for bulk analysis cannot be obtained at the beginning of exploration. Usually Bi, Fe, Ni, Cu, Cd, Hg, Pb and Sr are present in grains [10]. It is reasonable to detect such elements in gold/electrum because these but for Sr easily make alloys with Au. Galena grains are usually well analyzed with 200 to 400 micron aluminum filters. In galena both anions and cations are in regular octahedral six-fold coordination [11]. Our analysis indicates that elements of ionic radius which is close to that of six-fold Pb (118 pm) are detected, i. e. Ag (115 pm), and Bi (102 pm). Elements of which energy is higher than Pb L lines (e.g. Ag) give better MDL than other elements. For wolframite, we tried 100 and 300 micron aluminum filters and found that the former gives better analytical results for observing lighter elements such as Fe. In this mineral electronic pile up peaks obscure trace elements with lines between 16 to 20 KeV. In this energy range Y and Nb, both characteristic elements for wolframite, develop some peaks. The MDL for these peaks are in the ppm level, but for many elements in wolframite the MDL is generally poor, typical of minerals in this category.

USING TRACE ELEMENT DATA

One fruitful area to exploit trace element data is the study of ore deposits and formation in geodynamically active regions, in particular subduction areas and island arcs. In Japan such study was first done for an active hydrothermal deposit in the Okinawa Trough [12]. This study showed that rare metal distribution is regulated not only by crystallography but also by geologic environment where each mineral precipitated. Elemental behavior during the fractionation of felsic magma was studied in a polymetallic district of a unique Bi mineralization [13]. The authors selected two kinds of silicates, biotite and hornblende, to monitor the evolution of felsic magma. They found that trace elements in biotite clearly reflect the degree of differentiation of magma in the study area. A similar study has been performed for a world-famous Sb mining area in China. By careful selection of study area and specimens, the authors highlighted the large contrast in trace element distribution between different type of geologic setting, that can be utilized to refine ore deposit model [14].

CONCLUDING REMARKS

As stated above, micro-PIXE has proven to be a powerful tool for delineation of trace element distribution at levels relevant to ore genesis, mineral processing and exploration research. While in general the MDL is good spanning the range 1-100 ppm either not accessible or tedious to obtain by the electron microprobe, other limitations can arise due to the nature of the sample and the response function of the detector. Judicious selection of filters is essential in order to obtain the maximum benefit from the method. The examples discussed in this paper demonstrated the various origins for the control of the distribution of trace elements, in turn providing a handle in applications to new areas of studies.

REFERENCES

[1] Sie, S. H., Ryan, C. G. and Suter, G. F., *Scanning Microscopy* **5**, 977-987 (1991).

[2] Cabri, L. J., Hulbert, L. J., Gilles Laflamme, J. H., Lastra, R., Sie, S. H., Ryan, C. G. and Campbell, I. J., *Exploration and Mining Geology*, **2**, 105-119 (1993).

[3] Campbell, J. L., *Particle Induced X-ray Emission Spectrometry*, John Wiley and Sons, Inc., 1995, ch.6, 313-365.

[4] Sie, S. H., *Nucl. Instr. Meth.* **B130**, 592-607 (1997).

[5] Murao, S and Sie, S. H., *Resource Geology*, **47**, 21-28 (1997).

[6] Ryan, C. G., Cousens, D. R., Sie, S. H., Griffin, W. L., Suter, G. F. and Clayton, E., *Nucl. Instr. Meth.* **B47**, 55-71 (1990).

[7] Vaughan, D. J. and Craig, J. R., *Mineral Chemistry of Metal Sulfides*, Cambridge University Press, 1978, 493 pp.

[8] Putnis, A., *Introduction to Mineral Sciences*, Cambridge University Press, 1992, 457 pp.

[9] Sie, S. H., Murao, S. and Suter, G. F., *Nucl. Instr. Meth.* **B109/110**, 633-638 (1996).

[10] Murao, S., Sie, S. H., Dejidmaa, G. and Naito, K., *Open File Report of GSJ*, No. 220, 24pp (1995).

[11] Ramsdell, L. S., *Am. Mineral.* **10**, 281 (1925).

[12] Murao, S., Sie, S. H. and Suter, G. F., *Nucl. Instr. Meth.* **B109/110**, 627-632 (1996).

[13] Murao, S., Sie, S. H., Nakashima, K., Suter, G. F. and Watanabe, M., *Nucl. Instr. Meth.* **B130**, 671-675 (1997).

[14] Murao, S., Sie, S. H., Hu, X. and Suter, G. F. "Contrasting distribution of trace elements between representative antimony deposits in southern China", *Nucl. Instr. Meth.* **B** (in print).

PIXE Analysis of Low Concentration Aluminum in Brain Tissues of an Alzheimer's Disease Patient

R. Ishihara[1], T. Hanaichi[2], T. Takeuchi[1], A.M. Ektessabi[3]

1 *Dept. of Psychiatry Nagoya University School of Medicine, Nagoya 466- 8550 Japan*
2 *Hanaichi Ultrastructure Research Institute Co. Okazaki, Aichi-prf. 444-2134, Japan*
3 *Graduate School of Engineering, Kyoto University, Kyoto 606-8501, Japan*

An excess accumulation and presence of metal ions may significantly alter a brain cell's normal functions. There have been increasing efforts in recent years to measure and quantify the density and distribution of excessive accumulations of constituent elements (such as Fe, Zn, Cu, and Ca) in the brain, as well as the presence and distribution of contaminating elements (such as Al). This is particularly important in cases of neuropathological disorders such as Alzheimer's disease, Parkinson's disease and ALS. The aim of this paper was to measure the Al present in the temporal cortex of the brain of an Alzheimer's disease patient. The specimens were taken from an unfixed autopsy brain which has been preserved for a period of 4 years in the deep freezer at -80℃. Proton Induced X-ray Emission Spectroscopy was used for the measurement of Al concentration in this brain tissue. A tandem accelerator with 2 MeV of energy was also used. In order to increase the sensitivity of the signals in the low energy region of the spectra, the absorbers were removed. The results show that the peak height depends on the measurement site. However, in certain cases an extremely high concentration of Al was observed in the PIXE spectra, with an intensity higher than those in the other major elements of the brain's matrix element. Samples from tissues affected by the same disease were analyzed using the EDX analyzer. The results are quantitatively in very good agreement with those of the PIXE analysis.

INTRODUCTION

The excessive accumulation of metal ions in brain tissues often plays a toxic role in normal cell functions, even in the cases of essential biological elements such as iron, zinc, copper and calcium. This is even more remarkable in the case of the contaminating element aluminum, which has been thought of for a long time as one of the key neuro-toxic metals in patients with Alzheimer's disease. The concept of a relationship between the neuro-toxicity of aluminum and Alzheimer's disease (AD) began with the development of an animal model. Rabbits exhibiting encephalopathy with neurofibrillaly tangle degeneration after the direct injection of aluminum-containing salts into the cerebral cortex (1) were used. The existence of this experimental model led to a series of analytical comparative studies designed to determine the aluminum content in brain tissues from AD patients. Subsequently, some neurohistopathological findings on the relationship between aluminum and amyotrophic lateral sclerosis (2) and Parkinson's disease (3) have been reported.

In recent years, there have been increasing efforts to measure and quantify the density and distribution of excessive accumulations of essential or other biological elements in the brain tissues. So far, however, this approach has not produced any clear answers on aluminum.

Generally, in an elemental analysis of biological samples, we always encounter major problems in sample preparation techniques. It is necessary to eliminate the loss of metal ions from the biological tissues, or to prevent the samples from being contaminated by other elements in the applied mediums during the sampling procedures.

In this paper, the experimental results are presented based on the measurement of aluminum concentrations in the brain tissues of an AD patient using a proton induced X-ray emission (PIXE) spectroscopy. In these experiments unfixed and freeze-dried specimens, which are expected to have the least loss of metal ions and less contamination by other impurities, were used. Furthermore, the results of the PIXE spectroscopy are compared with and confirmed by those yielded by tests done using the energy-dispersive X-ray micro-analyzer (EDX).

MATERIALS

The autopsy brain materials were obtained from AD patients. For the PIXE analysis, an 84-year-old patient (case 1) and for the EDX analysis, (in both case 1 and case 2) an 86-year-old patient were designated. The temporal cortex was partially divided in preparation for both analyses. These brain tissues were sliced to a 1cm thickness after autopsy, and stored at -80℃ for a period of

CP475, *Applications of Accelerators in Research and Industry*, edited by J. L. Duggan and I. L. Morgan

four years.

SAMPLE PREPARATION

For the PIXE analysis, small blocks of brain tissues were soaked in liquid nitrogen for 1 h. and freeze-dried in a vacuum chamber for 48 h. The blocks were then directly analyzed in bulk form without further processing. For the EDX analysis, the small blocks of brain tissues were immersed in a solution of 2 % glutaraldehyde buffered by 0.1 mol cacodylate, pH 7.4. Thirty minutes later, the blocks were cut into 1 mm^2 pieces and fixed again in the same solution for 2 h. Subsequently, the small tissue blocks were fused in 1.5 mol sucrose, and, after putting each of them on a specimen holder, the specimens were frozen again with slush nitrogen. The refrozen specimens were equipped with ultra-cryokitt, and sliced in 0.3 μm thick sections under the temperature condition of –110℃. These thin-sections were then mounted on copper grids with a carbon film and dried at room temperature.

METHODS OF ANALYSIS

For the PIXE spectroscopy, a tandem type accelerator was used. A proton beam with an energy of 2 Mev, beam current of 1.7 nA, a size of 300×500 μm, and a total dose of 3400 nC was used for measurements. A Si (Li) detector was used to prevent a decrease in the characteristic x-ray signals on the low energy side of the spectra, and no absorber was used. For the same samples, the measurements were done on 5 separate points.

For the EDX microanalysis, a JEOL-2010 electron microscope, equipped with a Noran Voyager III energy dispersive x-ray micro-analyzer with a Si (Li) detector was used. The accelerating voltage was 200 kV, with a beam current of 10 μA. In both cases, the selected areas were analyzed for 200 seconds. Two thin sections from each case underwent microanalysis, and were observed at 3000 to 6000 magnification to confirm the normal condition of the tissue structures, and the regional points of analysis were determined by this electron microscopic observation.

RESULTS

PIXE Analysis

The same sample was measured at five different points. Three distinguished spectra were observed with almost the same highly concentrated levels of aluminum. Three spectra are shown in Fig. 1, marked as (a), (b), and (c). Spectrum (a) shows the concentration of aluminum being much lower than those of the other biological elements present in the brain tissues, (such as phosphorus, calcium and iron). Spectra (b) and (c) show a remarkable decrease in the concentration of the biological elements compared with (a) while the concentration of aluminum is almost the same as that of (a). Spectrum (c) shows a high concentration of aluminum compared with the other biological elements present in the brain tissue at the same point.

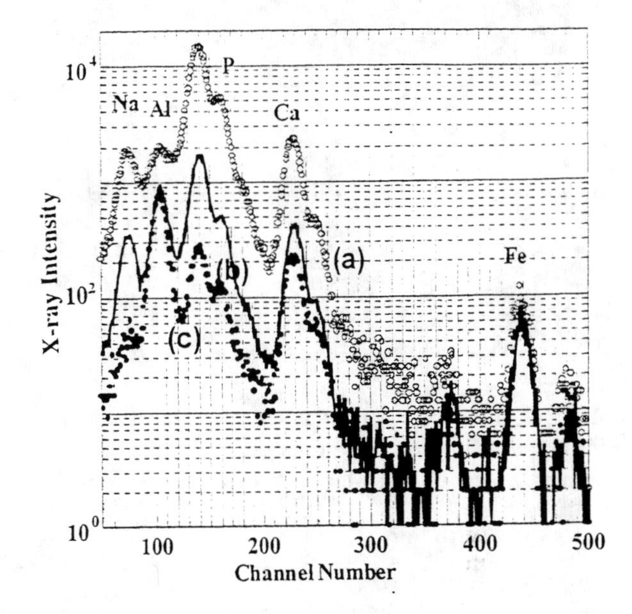

FIGURE 1. PIXE spectrum of the brain tissue in an AD patient

EDX Analysis

The identifications of the tissue structures were carried out based on the following morphological features: in the case of pyramidal neurons, by their outlines, and whether or not they were accompanied by nucleoli in their nuclear region (Fig.2), in the case of the neuropils, by the appearance of network-formations, and in the case of the capillary vessels, by the presence of basal lamina. Depending on the clarity of their appearance, high-density dot-like sediment (deposits) were found. On each section (1×1.2 mm), 5∼6 deposits (0.28∼0.6 μm in diameter) were found, and all of them were analyzed by the EDX analyzer. Extremely high concentrations of aluminum were detected in one out of three deposits found in the brain tissues in both cases. Levels of other biological elements present in the brain tissue such as phosphorous, sulfur and copper, were measured much lower at the same points. No aluminum was detected from outside the deposits. A spectrum of the deposit in the neuronal cytoplasm of case 2 is shown in Fig. 3.

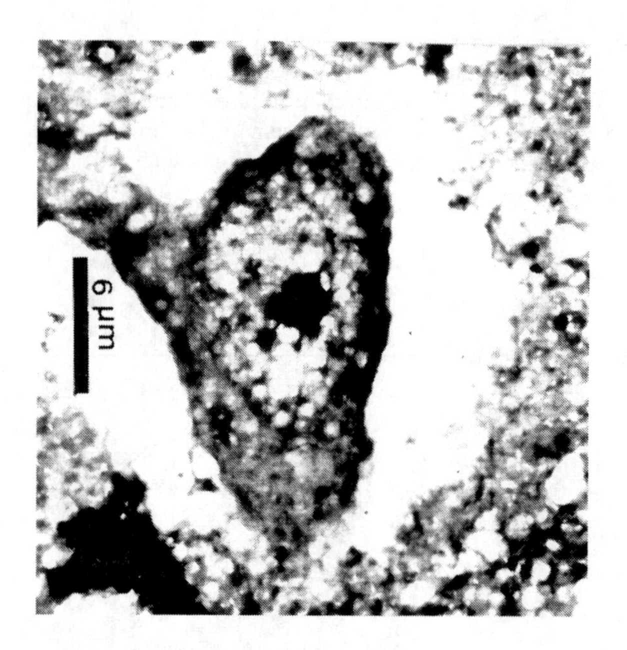

FIGURE 2. Electron micrograph of the freeze-dried thin section of an AD patient's brain tissue. A piramidal neuron including a nucleoli within the nucleus can be seen in the neuropils.

DISCUSSION

Aluminum has been detected in the neuronal nuclear chromatins (4,5), cytoplasms (6), neurofibrillaly tangles (7), senile plaques (8) and pial vessels (9) of the human brain tissues by other investigators, however, the quantity of aluminum concentration is not substantial. In this study the results of the PIXE analysis showed that a much higher peak height of aluminum concentration existed and that a possible marked drop in the levels of the other present biological elements occurred in association with aluminum accumulations in certain regions of our AD patient's brain tissues. We made note as to whether this metal movement occurred in an e extensive area or small points of the brain tissues, then we followed up the results of the PIXE analysis using the EDX analyzer. The results of an EDX analysis corresponded with those of the PIXE, and demonstrated the presence of dot like sediment deposits, a much greater concentration of aluminum.

In view of the tissue sampling conducted using these two methods of analysis, considering that no chemical procedures had been undergone during the PIXE sampling, and that bulk specimens were used, it was expected that the levels of the natural elemental components of the brain wouldn't change. While, for the EDX analysis, since we used thin sections of glutaraldehyde fixed tissue samples, we had to expect some sustained damage during sample preparation. Ultimately, no differences in the detectable levels of aluminum were found between the unfixed and freeze-dried bulk specimens examined using the PIXE spectroscopy, and the glutaraldehyde fixed thin sections analyzed using the EDX.

In general, glutaraldehyde fixes the protein molecules strongly. Therefore, it is suggested that when the aluminum got into the brain tissue, and linked with the protein molecules of the brain matrix and cell organizations, this element stayed on these sites as a result of the strong effects of the glutaraldehyde.

With regards to the other elements, (sodium, potassium and calcium) which seem to exist normally in an ionized state in the brain, and trace elements, (such as chromium, titanium, iron, copper and zinc), the PIXE detected these in all five different sites of the brain sample. Therefore, the unfixed and freeze-dried specimens may have an advantage over the glutaraldehyde fixed specimens when a comparison is being made between the presence of aluminum and the other existing elements in the brain tissue sample, as well as in the correlation between the levels of aluminum and those of the other elements.

CONCLUSION

The results of the PIXE analysis showed a much higher aluminum concentration and a marked drop in the other biological elements in certain regions of the brain tissues of an AD patient. These results were compared to and confirmed by those yielded by tests done using the energy-dispersive X-ray micro-analyzer (EDX).

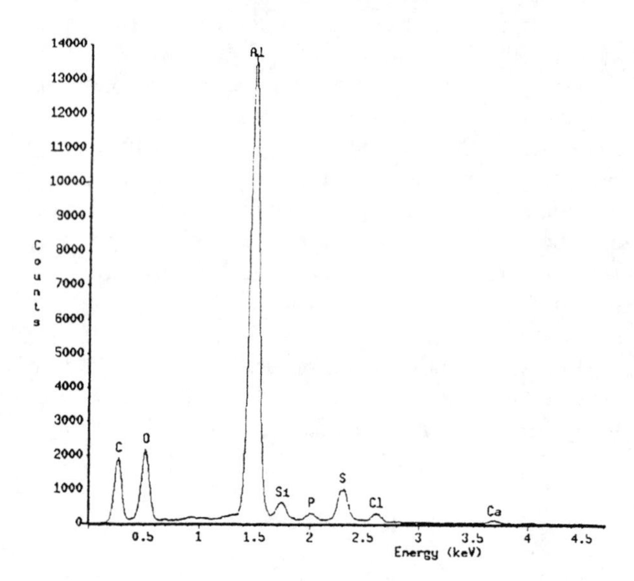

FIGURE 3. EDX spectrum of the deposit in the neurocytoplasm of the brain tissues in AD patient, which shows the highest peak height of aluminum.

ACKNOLEDGMENTS

The authors wish to thank Dr.Y.Kayukawa for his

advice and acknowledge the help of Mr. K. Sato, Mr. Y. Fujita and Mr. S. Gleason

REFERENCES

1. Klatzo, I., Wisniewski, H., Streicher, E., J. *Neropath.exp. Neurol.* **24**, 187-199 (1965)
2. Yoshida, S., *Clin. Neurol.*, **17**, 299-309 (1977)
3. Garruto, P. M., Fukatsu, R., Yanagihara, D., Gajdusek, D. C., *Proc. Natl. Sci. USA* **81**, 1875-1879.March (1984)
4. Crapper, D. R., Quttkat, S., Krishnan, S. S.,Dalton, A., J., DeBoni, U., *Acta Neuropathol.* **50**, 19-24 (1980)
5. Lukiw, W. J., Krishnan, B., Wong, L., Krusk, T. P. A., Begeron, C., McLachlan, D. R. C., *Neuroboiol. Aging*, **13**, 115-121 (1991)
6. Shiraki, H., Yase, Y., *Handbook of Clinical Neurology*,New York, *Elsevier Science Publishers B.V.*,1991,15 (59)
7. Perl, R. C. A., Brody, A. R., *Science*, **208**, 297-299 (1980)
8. Candy, J. M., Oakiey, A. E., Klimowski, J., Carpenter, T. A., Perry, R. H., Atack, J. R.,Perry, E. K., Blessed, G., Fairbrain, A., Edwordson, J. A., *Lancet,* **1**, 354-356 (1986)
9. Crapper, D. R., Krishnan, S. S., Quittkat, S., *Brain*, **99**, 680(1976)

The new dedicated PIXE set-up at the National Environmental Research Institute, Denmark

Kåre Kemp and Peter Waahlin

National Environmental Research Institute
Box 356, DK-4000 Roskilde, Denmark

The Niels Bohr Institute in Copenhagen was in the beginning of the 70'es one of the early places for PIXE. Contributions were made to the theoretical interpretation of the PIXE spectra as well as the practical application. The home-made 4 MV van de Graaff accelerator at the Niels Bohr Institute was an excellent tool for PIXE. The accelerator, which was used for many years, has now found its place on a museum after more than 40 years of active service.
A dedicated PIXE set-up has now been established at the National Environmental Research Institute using a new 1.7 MV Tandem Pelletron (5SDH) from NEC. The main application is elemental analysis of outdoor aerosols. The main work is unsophisticated macro analyses, which do not push the equipment to its limits. This enables automated analysis of about 10,000 samples per year using very limited manpower resources.
The research focuses on the contribution from various source types to the atmosphere over Europe, the North Atlantic and Greenland. Source compositions and their temporal variations are studied.

INTRODUCTION

PIXE (Proton Induced X-ray Emission spectroscopy) is a powerful method for multi-elemental analysis of solid samples (1). PIXE was developed in the beginning of the 1970'es. The need for trace element analyses of environmental samples coincided with available capacity on low energy accelerators, originally designed for nuclear physics studies, and the development of high resolution semiconductor detectors, which enabled separation of the characteristic X-rays from elements with $Z>11$. Used routinely for simple bulk analysis PIXE is a fast, reliable and - compared to other techniques – very cost efficient analysing method. The dissemination of PIXE has however been limited. Possibly due to its origin in Nuclear Physics laboratories, which has endowed it with an air of complicated scientific equipment only to be used at great costs and a lot of trouble.

At the Niels Bohr Institute in Copenhagen preliminary PIXE experiments were started in 1972 at the 4 MV van de Graaff accelerator (2), when the method was still in the early phase. When the capability of the method became evident, the administrative affiliation was moved from the Nuclear Physics department of the University of Copenhagen to the Ministry of Environment. But the old home-made accelerator was used until 1996. During 1996 the set-up was rebuild at the National Environmental Research Institute in Denmark with a new NEC 5-SDH 1.7 MV tandem accelerator. The old accelerator, which was in active service for more than 40 years, has now been revived at a Museum for Electricity.

The new set-up is made with special reference to a non-sophisticated operation, which enables reliable analysis of a large number of samples with a limited use of manpower resources. The set-up and some of the main applications are described in the following.

THE PIXE SET-UP

The 1.7 MV Tandem Pelletron accelerator is equipped with a Duoplasmatron H^- ion source. SF_6 is used as insulation gas in the accelerator tank (Fig. 1). The H^- ions are stripped by N_2 gas in the high voltage terminal. The proton beam is bend in a 90^0 analysing magnet and focused to a spot of approximately 1 mm in diameter on the target. It is possible to have a stable beam of 3.4 protons at a current of around 1 μA at the target position.

A tracer laboratory, where i.a. SF_6 is used as trace gas, is located just one floor above the accelerator room. Levels in the ppt range of SF_6 are measured. The 55 kg of SF_6 in the accelerator caused the people working with the trace analysis a great deal of concern. In order meet this concern the accelerator tank is emptied by cooling the SF6, thus avoiding a mechanical compressor system. This is done through two 25 l steel bottles, which in turn are cooled with LN_2. The two bottles are the standard type used for household butane. A similar system is applied at the University of Central Florida (3). The vapor pressure for SF_6 at LN_2 temperature is around 10^{-4} torr (4). However, in practice it is difficult to reach a pressure below a few torr in the accelerator tank after cooling. The pressure may be further reduced by means of a rotary pump with exhaust to the cooled system. The whole process in emptying and filling the tank takes about one day.

CP475, *Applications of Accelerators in Research and Industry*,
edited by J. L. Duggan and I. L. Morgan

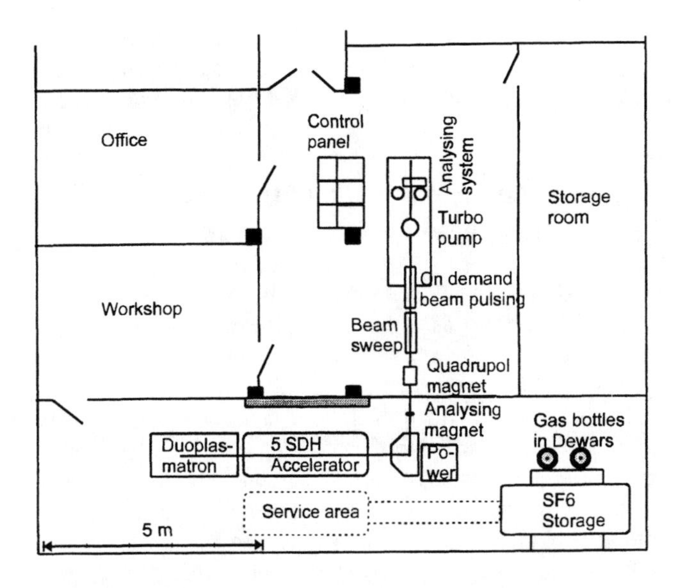

FIGURE 1. Ground plan of the PIXE set-up.

An antiscatter collimator limits the beam size on target. The beam size can be varied from 1 mm in diameter to 1x1 cm^2 by means of eight collimators mounted on a revolving wheel. The beam is swept by to sets of electric deflection plates in order to assure homogeneous beam intensity over the whole beam area. The target is perpendicular to the beam direction and two semiconductor detectors look at the target in an angle of 135^0 to the beam (Fig. 2). The one detector is a 30 mm^2 Si(Li) detector, which is used for the lower and medium energy part of the X-ray spectra, while the other is an 80 mm^2 HP-Ge detector, which is used for the high energy X-rays and at some occasions for γ-rays from the F(p,p') and Na(p,p') reactions. The signals from the Si(Li) detector control an on-demand beam pulsing system with a time resolution of < 0.5 µs. The scattering chamber is further equipped with a particle detector in 135^0 for elastic scattered protons (Fig. 3). An electron gun can be mounted to compensate for charging of electric insulating samples.

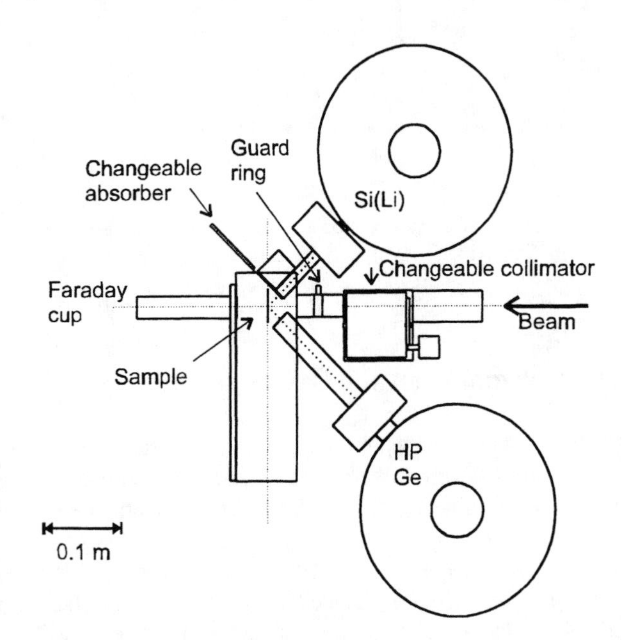

FIGURE 2. Horizontal cut through the scattering chamber.

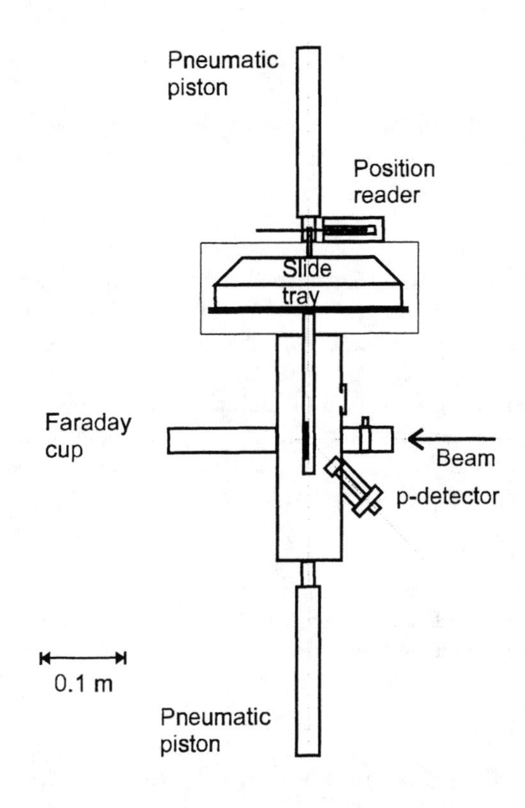

FIGURE 3. Vertical cut through the scattering chamber.

The spectra are recorded by means of two ADC cards in a PC. The spectrum fitting is performed by solving a set of linear equations. Each element is represented by a set of K- and L-lines. The background is defined by a rigid curve that does not follow the steep variations in the spectra.

The samples are in most cases mounted in ordinary slide frames without glass. The sample changing mechanism is build on the frame of a Kodak dias projector for carrousels with 80 slides. The reliability of the sample changer is improved by the use of pneumatic pistons for moving the samples in and out of the beam. A sample carrousel for 30 samples each with a diameter of 2 $cm^{\emptyset}$ can be used as an alternative to the slide tray. The later is mainly used for pellets of e.g. sediments and fly ash. The analysing parameters are varied according to sample type and needed detection limits. The shortest standard analysing time is less than 2 min. It is mainly used for urban aerosol samples, which are bombarded with 150 nA of 2.5 MeV protons. A two-hole "funny filter" absorber in front of the Si(Li) detector limits the count rate. Each sample is exposed to 20 µC. Very "clean" samples are analysed twice using 2 and 3 MeV protons. In the first run the current is around 15 nA and a one-hole "funny filter" is used as absorber in front of the detector, while 300 nA and a thick absorber (e.g. 10 mil Al) are used for the 3 MeV analysis.

The total analysing time may in this case be up to 30 min. The demands for beam energy and current are thus far below the maximal capacity of the accelerator. This ensures a very reliable operation of the accelerator.

APPLICATIONS

The main application is aerosol analysis. The PIXE measurements are an integrated part in the air quality monitoring programs in Denmark. Around 5000 samples per year are analysed in three national networks – a background network, an urban network in several cities and an urban network in the Greater Copenhagen Area. It is essential in these networks that measurements of several pollutants are carried out simultaneous. Beside the elemental composition species like NO_2, SO_2 and O_3 are recorded. PIXE analyses are also performed in the arctic measuring programmes in Greenland. Aerosols are collected both in the far North-east and on the ice cap. The many numbers enables a reliable statistical interpretation of trends and source contributions by means of time series analysis and multivariate statistics.

Several other types of samples - such as sediments, fly ash, plant material and human tissue - are also analysed. Some examples, which illustrate the different types of results, are discussed in the following.

Trend in urban air pollution

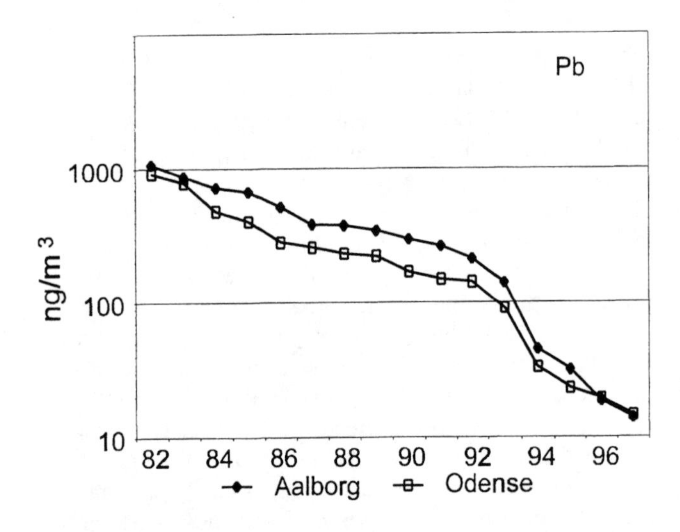

FIGURE 4. Trend for the Pb concentration measured at busy streets in two Danish cities.

Daily measurements of air pollution have been performed continuously in Danish cities since 1982 (5). One of the most striking features has been the result of the removal of lead from petrol during this period. The concentration measured at busy streets were about 1 $\mu g/m^3$ Pb in 1982, i.e. twice the guide line value recommended by WHO. As seen from fig. 4 only 1-2 % remains today where Pb is no longer added to petrol. The present urban level is less than twice the level in rural areas. This means that a major part of the Pb even at busy streets comes from sources outside the city.

Effects on air pollution of the structural changes in Eastern Europe

Two Danish measuring stations have been collecting daily samples since 1979 as a part of an European monitoring network for assessment of the long-range transport of air pollutants. Based on calculations of back trajectory the measurements are divided in eight groups corresponding to transport from North, Northeast, East etc. Based on receptor modelling source apportionment for long-range transport have been calculated using simultaneous results from several measuring stations (6).

Figure 5 shows the trend for long range transported Zn coming from the Western and Eastern part of Europe. There has been a steady decrease from Western Europe as a consequence of better emission control, while the East European contribution was increasing up to and shortly after the structural changes around 1990 and decreasing in the following years.

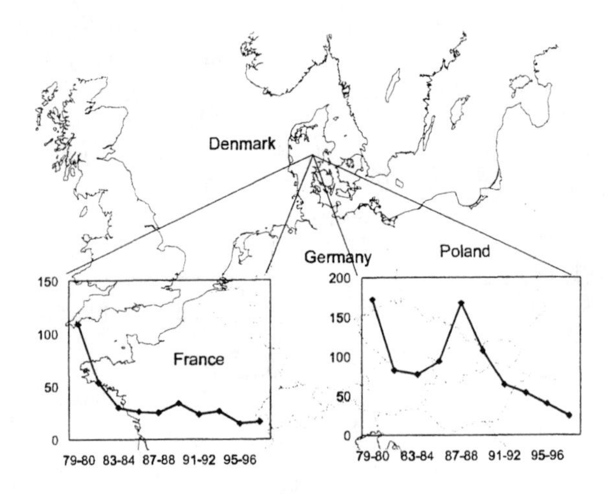

FIGURE 5. Biannual average concentrations of Zn in air coming to Tange in Denmark from South-west and South-east. (unit: ng/m^3).

Measurements on the Greenland Icecap

We have made use of the ability of PIXE to analyse the very small samples from a so-called Streaker aerosol sampler (7). It is possible to use the Streaker for unattended continuos weekly sampling for one year. The particles are collected in two size fractions with a cut point around 2 µm. The streaker has been placed at the Summit of the Greenland ice cap in connection with the major ice core drilling projects. The measured concentrations were in general very low. Receptor model calculations (8) show an anthropogen contribution during the spring 1993 with increased concentrations of e.g. S, K, Zn and Pb (cf. fig 6). Later the same year contributions from forest fires in the Hudson Bay area could be identified by relatively high contributions of K and Zn.

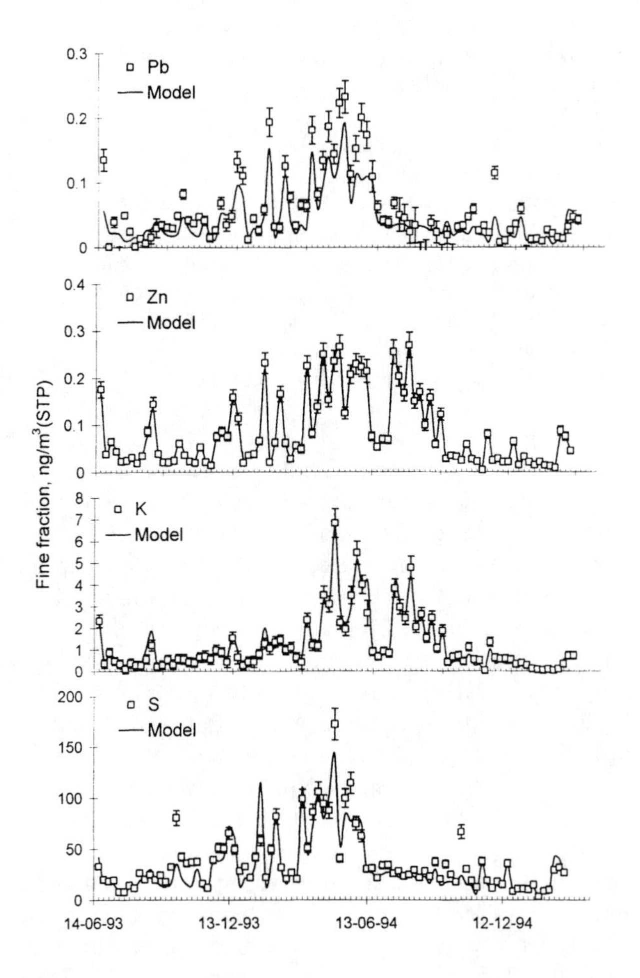

FIGURE 6. Weekly measurements and receptor model calculations of the aerosol composition at the Summit 3300 m A.S.L. during 1993 and 1994 (from (8)).

DISCUSSION

The application of automated PIXE analysis of various types of samples is shown to be very cost-effective tool for elemental analysis of a large number of samples. The total price for analyses in large series is between 5 and 50 US$ per analysis. Each analysis provides results for 15-20 elements. A very smooth and reliable performance of the analysing system is enabled by installation of the new accelerator, which is not pushed to its limits.

The PIXE set-up at the National Environmental Research Institute is in the process of being accredited according to the EN 45001 standard. The accreditation will be necessary in the future for contracts with the state and the Commission of the European Union.

More than 20 years of aerosol analyses have provide data sets, which enables evaluation of long term trends in concentrations and assessments of emissions sources.

REFERENCES

1. Johanson, S.A.E, and Campbell, J.L., PIXE: A novel tecnique for elemental analysis, New York: John Wiley & Sons Ltd, 1988.
2. Folkmann F., et al.,Nucl. Inst. and Meth. **116**, 487-99 (1977).
3. Elias, L., University of Central Florida, Private Communication.
4. Mark, H.F., et al. (eds.), Encyclopedia of Chemical Technology, vol. 9, New York, Interscience, 1966.
5. Kemp, K., Palmgren, F. and Manscher O.H., The Danish Air Quality Programme. Annual Report for 1997, NERI Technical Report No. 245, National Environmental Research Institute, Roskilde, Denmark, 1998.
6. Kemp, K., A Multi-point Receptor Model for Long-range Transport over Southern Scandinavia, Atmospheric Environment **27A**, 823-30 (1993).
7. Kemp, K. and P. Waahlin, Aerosol Sampler for Unattended All-year Sampling at Remote Sites, Proceedings from the Fifth International Symposium on Arctic Air Chemistry, September 8-10, Copenhagen, 1992.
8. Waahlin,P., Kemp, K. and Heidam, N.Z., Year-round Automatic Aerosol Sampling at Summit, Second year technical report for TAGGSI, Program of the EC in Respect of Community in the Field of Environmental Research, Contract EV5V-0412, Bruxelles, 1996.

Stoichiometry of DyFeCo Magneto-Optical Alloys by Combined PIXE/RBS

D. Strivay[1], G. Weber[1], K. Fleury-Frenette[2], J. Delwiche[2]

[1]*Institut de Physique Nucléaire Expérimentale, Sart Tilman B15, Université de Liège, B-4000 Belgium*
[2]*Institut de Chimie, Sart Tilman B6c, Université de Liège, B-4000 Belgium*

DyFeCo alloys offer suitable properties for magneto-optical recording. Their high write and erase sensitivity favors them as materials to design high density multilayer-based recording medium that would be addressed by temperature and field modulation. The samples consist of amorphous thin films of the order of 100 nm deposited by RF magnetron sputtering between an aluminum buffer and a thin aluminum native oxide capping layer onto Si or SiO2 substrates. The stoichiometry of these films has to be known in order to study their magnetic properties. Unfortunately, the analysis of this kind of sample is not obvious neither by RBS or PIXE. Indeed, as Fe and Co are neighbor elements, the RBS spectra do not allow to determine their elementary ratio. On the other hand, PIXE spectra are difficult to analyze due to the overlapping of the different peaks. By combining the two methods it has been possible to obtain the thickness and the stoichiometry of these samples. The experimental setup as well as the computation method used to attain this goal will be also described.

INTRODUCTION

GdTb-Fe and Tb-FeCo-based magneto-optical disks are currently available for data storage. Although not commercially used, some Dy-FeCo alloys also offer suitable properties for magneto-optical recording (1-3). Their performances at room temperature and comparable wavelength are unfortunately slightly less attractive than those of their Gd and Tb analogues (3). Nevertheless, a higher write and erase sensitivity favors them as materials to design a high density multilayer-based recording medium that would be addressed by temperature and field modulation (4).

Dy-(Fe,Co) thin films were to be studied by XMCD (X-ray Magnetic Circular Magnetism) in order to determine the contribution of each element to their integrated magnetic behavior. For magneto-optical alloys, minute changes in composition have been reported to induce large variations in magnetic figures such as the Curie and compensation temperatures (1). Therefore, when monitoring the magnetic properties of such materials, one obviously needs to determine precisely their composition. Both RBS and PIXE were used to this end: to calibrate the deposit process and, thereafter, to control composition.

EXPERIMENT

Sample preparation

DyFeCo films were deposited on Si and SiO_2 substrates covered with an aluminum buffer layer (5-50 nm) to prevent diffusion of bulk contaminants into the magneto-optical layer and epitaxial growth. A 5 nm-thick aluminum capping layer was used to protect the alloys from atmospheric corroding agents. The three layers were deposited by RF magnetron sputtering. This particular process involved a plasma composed of argon ions and electrons moving under a RF voltage. At high frequency, a larger amount of collisions improves the ionization yield and contributes to sustain the plasma at low pressure (5 10^{-3}mbar). A magnetic field also trapped moving electrons to contain the plasma more efficiently in the vicinity of the target surface. The target is coupled to the power supply through an adjustable capacitor in series allowing to generate a negative self-bias potential at its surface. Ions are therefore accelerated towards the target, collide with the latter, and eject material towards the substrate. The substrate is rotated at approximately 1 Hz while the magneto-optical layer is deposited to ensure composition homogeneity over the surface. Such a precaution is required because mosaic targets made of a dysprosium disk partially covered by slices of iron and cobalt foils are used. Sputtering is operated at constant power (0.6 W/cm^2) and impedance varies over the surface of the target. The number of atoms of an element found in the sample, N_S, is therefore linked to that element coverage over the target, θ_t, according to:

$$N_S \propto \theta_t^{n+1}$$

where n is to be determined experimentally.

CP475, *Applications of Accelerators in Research and Industry*,
edited by J. L. Duggan and I. L. Morgan

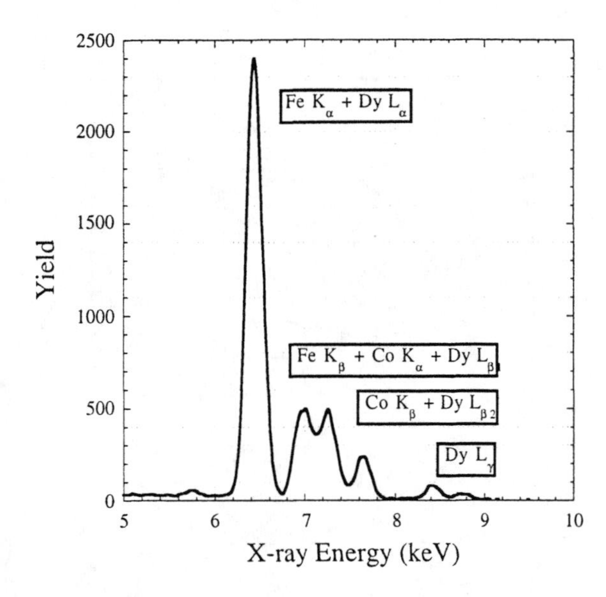

Figure 1. Typical PIXE spectrum of Dy Fe Co films

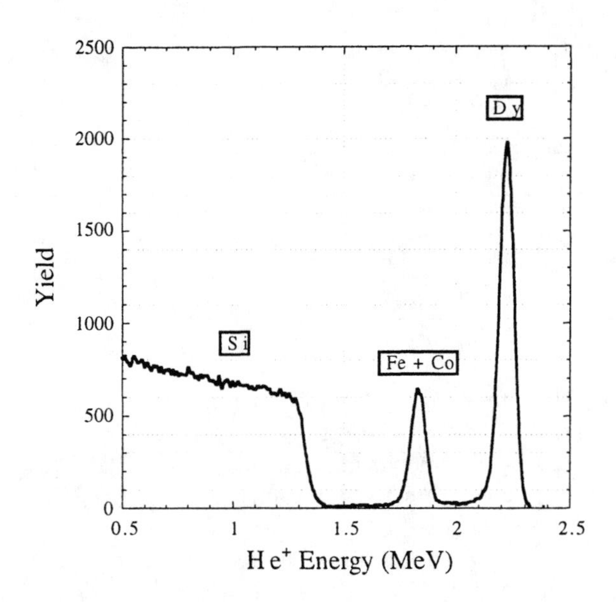

FIGURE 2. Typical RBS spectrum of Dy Fe Co films

PIXE set-up

The PIXE analyses were performed with a 3 MeV proton beam produced by the Van de Graaff accelerator of the "Institut de Physique Nucléaire Expérimentale" of the University of Liège. This beam was extracted through a 6 µm aluminum foil. The incident proton energy on the target was 2.35 MeV with a beam spot of about 1 mm^2. The X-rays were detected by a Si(Li) X-ray detector positioned at 125°, about 30 mm from the target and with a 250 µm Mylar filter. The detector resolution was 0.156 keV at 5.91 keV. A typical PIXE spectrum of a DyFeCo film is shown on Fig. 1. This kind of PIXE spectrum is not easy to analyze due to the overlapping peaks. The spectrum can be separated in four regions: the Fe K_α and Dy L_α peaks, the Fe K_β, Co K_α and Dy $L_{\beta 1}$ peaks, the Co K_β and Dy $L_{\beta 2}$ peaks and finally the Dy L_γ peak.

RBS set-up

The films were subjected to RBS measurements using 2.5 MeV He^+ with a beam spot of about 1 mm^2. The backscattered He^+ were detected by an annular detector with a solid angle of 6.98 msr. Determination of the charge was performed by using a chopper made of a thin gold layer deposited on a light element support. Both sample and gold RBS signals were recorded on the same spectrum. Indeed, the signal is switched from one half to the other when the chopper interrupts the beam. As shown in Fig. 2, the RBS spectrum is also not easy to interpret due to the impossibility to know the Fe/Co ratio without any à priori information.

RESULTS

The elementary composition of the films was first determined by PIXE. The analysis of the PIXE spectra was performed in three steps. First, films containing only one element were analyzed to determine the K_α/K_β of Fe and Co and the ratio of the Dy L peaks. Second, keeping these ratios constant during the fitting process, it has been possible to obtain the area of the Co and Fe K_α and Dy L_α peaks for each sample. Finally, the stoichiometry was obtained by applying correction factors due to the different cross section of X-ray production for each element and to the absorption in the air and in the Mylar filter of the induced X-rays.

The X-ray production cross sections for Fe and Co K rays were computed using the analytical formula of Paul (5), which yields respectively 100.4 and 84.5 barns. The cross section for the Dy L_α production was obtained using the formula

$$\sigma(E, Z) = \frac{e^{\sum_{i=0}^{2}(a_i Z + b_i) \ln^i E}}{c_1 Z + c_2}$$

where the fitting parameters are shown in Table 1. This formula was obtained by fitting experimental values of La X-ray production from various references (6-13). This formula is valid for elements between Ag and U and proton energy below 3.5 MeV. The Figure 3 presents the results of this model compared to the experimental data and to the X-ray production derived from ECPSSR theory

TABLE 1. Coefficients of the fitting formula for the L_α X-ray production cross section of elements between Ag and U and for proton energy up to 3.5 MeV.

a_0	-0.098	a_2	-0.085
b_0	11	b_2	0.068
a_1	0.0357	c_1	0.1003
b_1	0.147	c_2	0.227

(14). The computed Dy L_α production at 2.35 MeV was found to be equal to 80.1 barns.

The X-ray absorptions in the air and in the Mylar filter were computed from the mass attenuation coefficients given by Seltzer (15).

When applying all these correction factors, the stoichiometry of DyFeCo films has been computed (Table 2). The uncertainty on these values is mainly due to the value of the cross sections.

The analysis of RBS spectra were done using RUMP software (16) and by taking the Fe/Co ratio from the PIXE measurements. The layer model was a DyFeCo layer of approximately 50 nm between two Al layers of 5 and 50 nm on a silicium backing. Dy concentration and the film thickness were then easily obtained. As shown by Table 2, the corresponding stoichiometry obtained by RBS is almost identical to the PIXE one.

From these experimental data, N_S was found to be proportional to θ_t, with n equal to 0.39 (Fig. 4). Similar curves can be obtained for Fe and Co.

TABLE 2. Results of the determination of the stoichiometry by PIXE and RBS

PIXE			RBS		
Dy	Fe	Co	Dy	Fe	Co
6.5	93.5	0	6.4	93.6	0
15.3	84.7	0	15.1	84.9	0
17.2	82.8	0	18.2	81.8	0
27.9	72.1	0	27	73	0
30.8	69.1	0.1	32	68	0
6.8	72.2	21.0	7.5	71.1	21.4
17.8	63.4	18.8	19.5	61.6	18.9
19.4	65.5	15.1	20.6	64.1	15.3
24.6	60	15.4	25.3	59.4	15.3
28.6	62.1	9.3	29.5	61	9.5
31.6	61.7	6.7	31.4	61.8	6.8
31.1	55.9	13	32.4	54.8	12.8
32.8	56.7	10.5	33.6	56	10.4
35.4	51.2	13.4	35.1	51.4	13.5

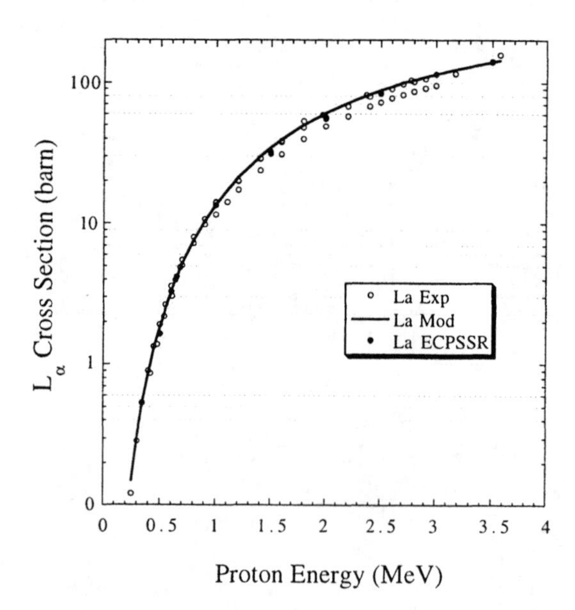

Figure 3. Comparison of experimental Dy L_α (La Exp) cross sections with this general fit (La Mod) and with ECPSSR values (La ECPSSR).

CONCLUSION

Results presented in this paper show that PIXE and RBS are suitable to determine the stoichiometry and the thickness of DyFeCo films. Indeed, although the peak overlapping in PIXE spectrum is high, it is possible to determine the element concentration. From there, RBS analysis becomes straightforward allowing the determination of the film thickness and to confirm the Dy concentration.

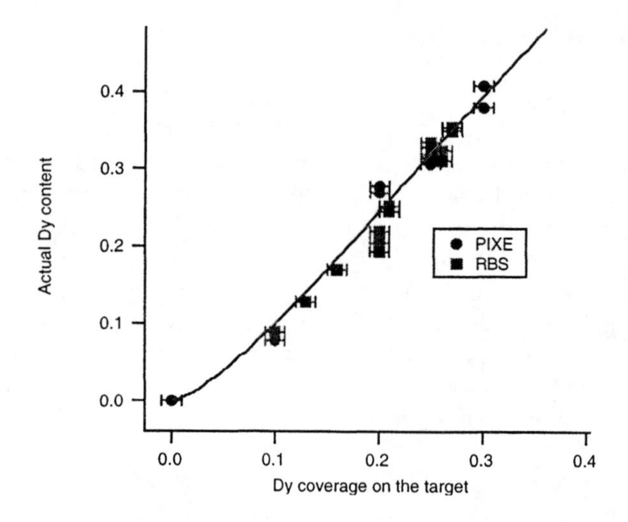

Figure 4. Fitting of the Dy content on the sample as a function of the Dy coverage of the target.

ACKNOWLEDGEMENTS

We are indebted to the Institut Interuniversitaire des Sciences Nucléaires (Belgium) for financial support.

REFERENCES

1. Hansen P., Klahn S., Clausen C., Much G. and Witter K., *J. Appl. Phys.* **69** (5), 3194-3207 (1991).
2. Raasch D., *J. Magn. Soc. Jpn.* **17**, 192-195 (1993).
3. Raasch D., *IEEE Trans. Magn.* **MAG-29**, 34-40 (1993).
4. Hwang W.-K. and Shieh H.-P. D., *J. Appl. Phys.* **81** (6), 2745-2748 (1996).
5. Paul H., *Nucl. Instr. and Meth.* **B3**, 5-10 (1984).
6. Orlic I., Sow C.H. and Tang S.M., *At. Data Nucl. Data Tables* **56**, 159-210 (1994).
7. Sokhi R.S. and Crumpton D., *At. Data Nucl. Data Tables* **30**, 49-124 (1984).
8. Fazinic S., Bogdanovic I., Jaksic M., Orlic I., Valkovic V., *Nucl. Instr. and Meth.* **B94**, 363-368 (1994).
9. Bogdanovic I., Fazinic S., Jaksic M., Orlic I., Valkovic V., *Nucl. Instr. and Meth.* **B109**, 47-51 (1996).
10. Sow C.H., Orlic I., Osipowicz T., Tang S.M., *Nucl. Instr. and Meth.* **B85**, 3133-137 (1994).
11. Padhi H.C., Bhuinya C.R., Dhal B.B., Misra S., *J. Phys. B* **27**, 1105-1114 (1994).
12. John Kennedy V., Augusthy A., Varier K.M., Magudapathy P., Nair K.G.M., Dhal B.B., Padhi H.C., *Nucl. Instr. and Meth.* **B134**, 165-173 (1998).
13. Fazinic S., Bogdanovic I., Jaksic M., Orlic I., Valkovic V., *J. Phys. B* **27**, 4229-4241 (1994).
14. Cohen D.D., Harrigan M., *At. Data Nucl. Data Tables* **34**, 393-414 (1986).
15. Seltzer S.M., *Rad. Res.* **136**, 147-170 (1993).
16. RUMP/GENPLOT. Doolittle L.R., *Nucl. Instr. and Meth.* **B15**, 227-231 (1986). Available from Computer Graphics Service, Lansing, NY.

An Endurance Test of Kapton Foil in In-Air PIXE System

S.Matsuyama, J.Inoue, K.Ishii, H.Yamazaki, S.Iwasaki, K.Goto, K.Murozono, T.Sato

Department of Quantum Science and Energy Engineering, Tohoku University
Sendai 980-8579 Japan

H.Orihara

Cyclotron and Radioisotope Center, Tohoku University
Sendai 980-8578, Japan

We have tested endurance of Kapton foil for beam irradiation, which is used as an exit window of in-air PIXE system. Kapton foil is strong enough to withstand the pressure differential, thin enough to transmit the proton beam with minimal energy loss and does not contain heavier elements. However, little is known about the relation between its lifetime until breakdown and beam parameters such as beam current, beam diameter and energy loss. Endurance tests were made by an in-air PIXE system at Tohoku university. Irradiation was performed by proton beam with 2 or 3 MeV until the foil was broken for various beam parameters. The total charge until breakdown of the Kapton foil was analyzed in consideration of beam current, diameter and energy loss. Total charge until breakdown decrease with increase in the beam current density in case of the beam current density higher than 100 nA/mm^2. On the other hand, total charge until breakdown does not decrease so much in the beam current density up to 100 nA/mm^2. The lifetime of the Kapton foil can be estimated longer than 2 days for the beam current density of 10 nA/mm^2 (7nA in 1mm beam diameter). It is sufficient for the in-air PIXE analysis.

INTRODUCTION

Particle-induced X-ray emission (PIXE) is a useful technique for multielemental analysis with high sensitivity /1/. Generally samples to be analyzed are irradiated in vacuum. Samples which contain moisture cannot be analyzed in the vacuum chamber. Sample size is limited because of the difficulties of placing it in a vacuum chamber. These problems can be solved by extracting the beam into the laboratory atmosphere. We call this arrangement an in-air PIXE analysis. In-air PIXE analysis is attractive especially for biological, botanical and archeological fields than vacuum PIXE.

The easiest way to realize the in-air PIXE analysis is to extract the beam through a thin window into the laboratory milieu. While metallic foils are strong enough to withstand beam damage, their characteristic X-rays from strong background in PIXE analysis. On the other hand, greatly reduced background is achieved using polymer foil. It is known that among polymers, a polyimide such as Kapton (manufactured by Dupont) has the highest radiation resistivity and is suitable for exit window in in-air PIXE system /1,2/. It is a serious problem that the beam exit foil breaks during the experiment, so it is very important to know its lifetime until breakdown. Studies on radiation deterioration of the Kapton foil has been performed from viewpoint of mechanical properties/1,2/. However, there is scarce study on its lifetime as an exit window.

Here, we irradiated the Kapton foil until it was broken and measured total charge through the foil under various beam parameters. We define the total charge through the Kapton foil until breakdown as its lifetime. The average beam current was varied from 10 nA to 1000 nA. Beam size was 0.5 to 3 mm in diameter. The total charge until breakdown of the Kapton foil was analyzed in consideration of current, diameter and energy loss of the beam.

EXPERIMENT

Endurance tests were made using a vertical in-air PIXE (ViaPIXE) system at Tohoku university/3,4/. ViaPIXE is the vertical beam type in-air PIXE system. A single-ended type Dynamitron accelerator in Tohoku University was used in this study. The proton beam from the accelerator is bent by 30 degree using a switching magnet, further bent by 10 degree and led to a beam exit assembly

CP475, *Applications of Accelerators in Research and Industry*,
edited by J. L. Duggan and I. L. Morgan

through a vertical dipole magnet by 90 degree. A fast valve closing system was set just downstream of the switching magnet to protect the accelerator from the vacuum break due to destruction of the exit window. Figure 1 shows the beam exit assembly consisting of a beam viewing port, a Faraday cup, a carbon beam collimator, a beam exit window assembly and an in-air Faraday cup. The exit foil was glued on a copper or aluminum metal ring with a 5 mm inner diameter hole. The exit foil is a Kapton foil of 12.5 or 7.5 μm. The exit window assembly can be changed within a few minutes from the beam exit assembly. Beam current can be measured by the Faraday cup just before the carbon collimator or the in-air Faraday cup (see Fig.1). Beam currents measured by these Faraday cups were consistent with each other.

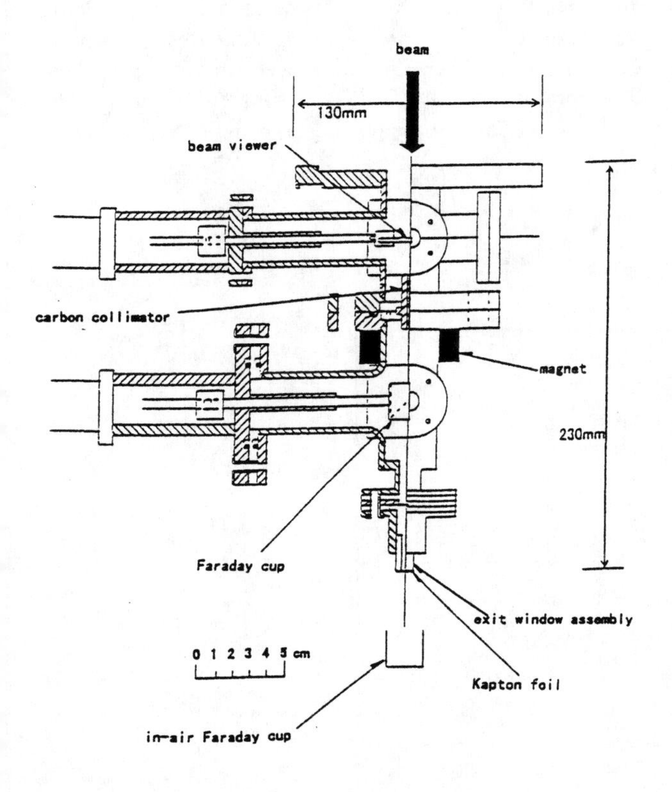

FIGURE 1. A beam exit assembly.

Irradiation with 2 or 3 MeV proton was performed until the foil was broken. The average current varied from 10 to 1000 nA. The beam size was 0.5 to 3 mm in diameter. The beam size larger than 0.5 mm was obtained by the carbon collimator. In order to get uniform beam, the beam was defocused by using a thin (50 mg/mm^2) carbon diffuser foil settled just downstream of a quadrupole doublet lens. An actual beam size was obtained by measuring diameter of a mark burning black on the Kapton foil.

RESULTS AND DISCUSSION

Typical results of the total charge until breakdown as a function of the beam current for the different beam spot size is shown in Figs.2 and 3. Figure 2 shows the results of 7.5 and 12.5 μm Kapton foils for beam spot sizes of 3.2, 4.2 and 3.2 mm^2. While total charge until breakdown depends on the beam spot size, it does not depend on the beam current. Proton irradiation induced chain scission of the polymer and caused the changes in tensile parameters such as tensile strength, elongation and fracture energy/2,5/. Thus, the foil is destructed by radiation damage. Figure 3 shows the result of the 12.5 μm Kapton foil for beam spot size of 0.34 mm^2. Beam current density is about ten times higher than that of the former. In this case, total charge until breakdown decreased with increase in the beam current. Since rise in temperature of the Kapton foil also degrades its malleability and ductility, local heating by ion beam accelerates deterioration of the foil in addition to the radiation damage.

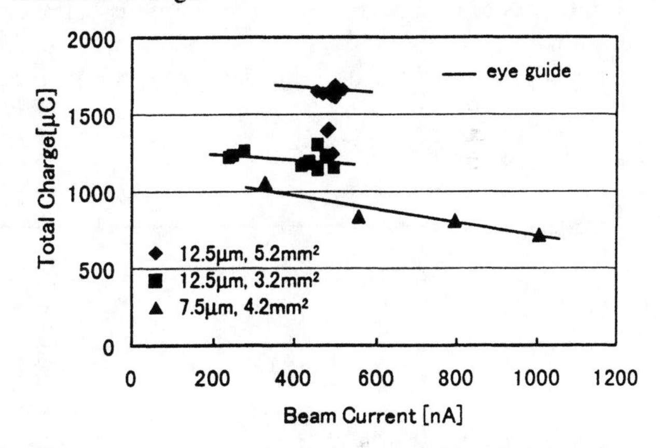

FIGURE 2. The total charge until breakdown as a function of the beam current for 7.5 and 12.5 μm Kapton foil (large spot size).

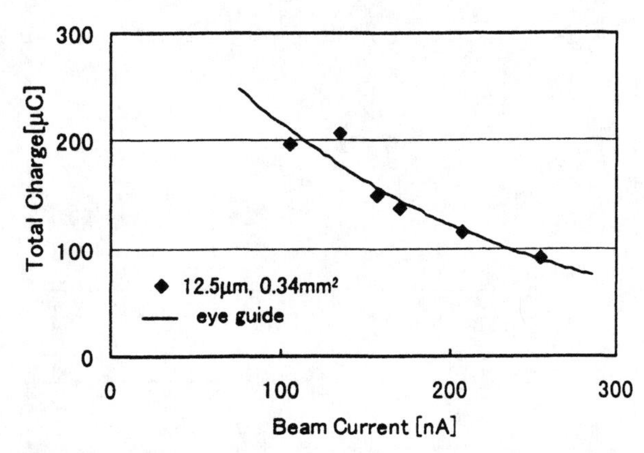

FIGURE 3. The total charge until breakdown of 12.5 μm Kapton foil as a function of the beam current (small spot size).

As shown in Figs. 2 and 3, the total charge until breakdown differs for various beam parameters. In order to research the total charge until breakdown at the different beam parameters, the following normalization may be useful. We introduce the following normalization. The foil gets stress (F) from pressure difference (P) and withstands that stress by tensile strength. In ordinarily, the value of tensile strength at break (F') is larger than that of F. Beam irradiation decrease the value of F' and the foil is broken. The value of F is proportional to the beam spot size (S). The value of F' is proportional to the length of beam circumference (l=2πr, r=beam radius) times foil thickness (t). Furthermore, we introduce the effects of radiation and heat damage into such consideration. The fracture energy of the Kapton foil, which will affect the value of F', is almost in inverse proportion to radiation dose (G) /5/. The fracture energy of the foil decreases slowly with the temperature up to 200°C and decreases sharply with the temperature higher than 200°C. In this study, we assume F' is inversely proportional to temperature (T). While the foil thickness may change continuously with irradiation time, we assume the thickness to be constant. This assumption is sufficient for normalization on our experimental data to evaluate the life time of the foil. Thus, the following equation is derived.

$$F = P \cdot S = \text{constant} \frac{l \cdot t}{G \cdot T} = F' . \quad (1)$$

Radiation dose; total energy deposited to unit volume (G) is defined by energy loss of the proton beam in the foil (dE), total charge (Q), beam spot size (S) and foil thickness (t) :

$$G = \frac{Q \cdot dE}{S \cdot t} . \quad (2)$$

From equation (1) and (2),

$$\frac{Q \cdot dE}{r \cdot t^2} = \text{constant} \frac{1}{T} . \quad (3)$$

Temperature of the foil (T) depends on beam current (I) and energy loss in the foil. We assumed that all of the deposited energy to the foils is converted to heat in disregard of another physical process such as sputtering. Temperature of the foil is estimated by using the following equation based on heat transfer both foil surface and wall.

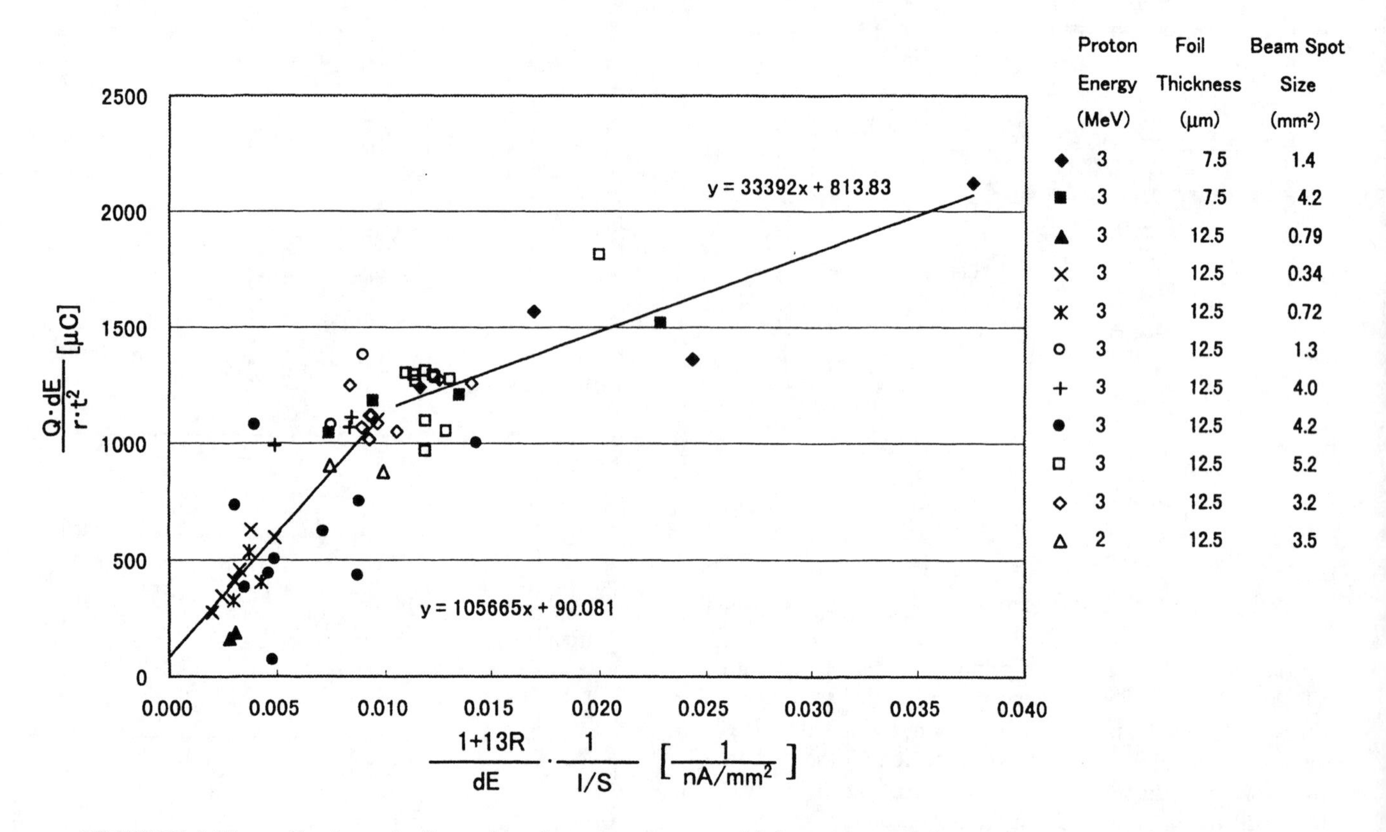

FIGURE 4. Normalized results for various beam parameters with least square fitting curve.

$$I \cdot dE = (h \cdot S + h' \cdot S \frac{2 \cdot t}{r})T, \quad (4)$$

where h : heat transfer coefficient of air,
h' : heat transfer coefficient of the kapton foil.

Since heat conductivity of Kapton foil is 6.5 times higher than that of air, h' is estimated 6.5h. Namely equation (4) become

$$I \cdot dE = (h \cdot S + 6.5 \cdot h \cdot S \cdot 2 \cdot R)T, \quad (5)$$

where R is the ratio of t to r (R=t/r).

The equations (3) and (5) are reduced in the following equation

$$Y = \text{constant} \cdot X, \quad (6)$$

where
$$Y = \frac{Q \cdot dE}{r \cdot t^2},$$

$$X = \frac{(1 + 13R)}{dE} \cdot \frac{1}{I/S}.$$

Figure 4 shows the normalized results for various beam parameters with least square fitting curves. The data are normalized for 3 MeV proton beam with the beam diameter of 2 mm and 12.5μm foil thickness. While Y increases with X, the rate of increase decreases at the region of $X > 0.01$. It shows that temperature of the foil is not so high and does not cause degradation due to heating in the region of low current density. Standard deviations of the data are 31-49 % and 15-24 % for the region lower than 0.01 and higher than 0.01, respectively. Large deviation for the region of $X<0.01$ may be due to temperature fluctuation caused by instability of beam intensity. In the case of PIXE analysis where our usual beam current density is around 10 nA/mm^2 (7 nA in 1 mm beam diameter), the lifetime of the Kapton foil can be estimated longer than 2 days. It is sufficient for in-air PIXE analysis.

CONCLUSION

Endurance of the Kapton foil used as an exit window of in-air PIXE system for beam irradiation has been tested. The total charge until breakdown of the Kapton foil was analyzed in consideration of beam current, diameter and energy loss. The total charge until breakdown decreases with increase in the beam current in the beam current higher than 100 nA/mm^2. On the other hand, in the beam current density lower than 100 nA/mm^2, total charge until breakdown does not decrease so much. The lifetime of the Kapton foil can be estimated longer than 2 days for the beam current density of 10 nA/mm^2. Our results will be useful for the users who want to start an in-air PIXE experiment.

ACKNOWLEDGMENT

The authors are pleased to acknowledge the assistance of Messrs. M.Oikawa and T.Sasaki during the experiments. The authors wish to thank Messrs. R.Sakamoto and M.Fujisawa for their maintenance work for Dynamitron accelerator. The authors also wish to thank Messrs T.Takahashi, K.Komatsu, T.Nagaya and C.Akama for their supports in the fabrication of the beam exit assembly.

REFERENCES

1. Sven A.E.Johansson and John L. Campbell, *PIXE:A Novel Technique for Elemental Analysis*, Singapore, John Wiley & Sons, 1988, chapter 3 and 7
2. Sasuga.T., *Polymer*, **29**, 1562-1568,1988
3. Iwasaki S. et al., *International Journal of PIXE*, **6**, Nos. 1&2, 117-125,1996
4. Iwasaki S. et al., *International Journal of PIXE*, **5**, Nos. 2&3, 163-173,1996
5. David J.T.Hill and Jefferson L.Hopewell, *Radiat.Phys. Chem.*, **48**, No5, 533-537,1996

Irradiation Effects on Ti and SUS-304 Membranes Caused by the Transmission of High-Energy Proton Beam

S. Yamaguchi, S. Nagata, K. Takahiro

Institute for materials Research, Tohoku University, Sendai980-8577, Japan

Ti and SUS-304 membranes are used for the window material of the electron beam processor as well as of the beam line for non-vacuum PIXE. The modification of the window materials during the passage of high-energy charged particles has been examined by experimental simulation using a high-energy proton beam. Polycrystalline Ti membrane of 5 μ m thick and SUS-304 membrane of 6 μ m thick are used as the specimens. Protons of 1 MeV energy are irradiated into the Ti and SUS-304 membranes up to the fluence of 4×10^{17} p/cm^2. Since the projected range of 1 MeV proton on Ti and SUS-304 is larger than 6 μ m, most protons can pass through the membranes. Microstructure and micro-hardness of the proton irradiated specimens are examined by SEM, XRD and Knoop hardness measurements as a function of proton fluence. The results show that the proton irradiation induced recrystallization and softening of the specimen.

INTRODUCTION

The window materials of the electron beam processor as well as of the beam line for non vacuum PIXE are exposed under the irradiation of high-energy electron or ions. In order to assess the durability of the window materials, it is required to examine the modification of metallic membranes induced by the transmission of high-energy charged particles. When a charged particle passes through the membrane, it loses energy due to interaction with electrons (electronic energy loss) is largely dissipated eventually as heat without creating structural defects. On the other hand, the energy loss to the lattice atoms (nuclear energy loss) produce defects by the recoil from the lattice sites if sufficient amount of energies are transferred to the membrane. Thus the irradiation of the charged particles is expected to induce two kind of the competed changes in mechanical property of the membrane; the heat due to the electronic energy loss soften the metals, while the displacement of lattice atoms due to the nuclear energy loss harden the metals. It will be interesting, therefore, to study the modification of microstructure and mechanical property by the passage of high-energy charged particles.

In the present investigation, protons of 1 MeV energy are irradiated into Ti membrane of 5 μ m and SUS-304 membrane of 6 μ m at room temperature. The micro-structure and the micro-hardness of the proton irradiated Ti and SUS-304 membranes are examined by SEM, XRD and Knoop hardness measurements.

EXPERIMENTAL

Specimen used in the present study were polycrystalline Ti membrane of 5 μ m thick and SUS-304 membrane of 6 μ m thick provided from the Nilaco corporation. Protons from 1.7 MV tandem accelerator were irradiated into the specimen target mounted on a Cu sample holder which was cooled by running water. The focused beam swept across the target area by means of a horizontal and vertical scanning system working at high frequencies so that local heating of the target was avoided. The sweep amplitudes were larger than the dimensions of tantalum aperture which limited the beam size immediately in front of the target. Thus, the proton beam was distributed homogeneously over the irradiated area. The flux of protons was about 6 mA/cm^2 typically. Since the projected ranges of 1 MeV proton on Ti and SUS-304 calculated with TRIM program [2] are larger than 6 μ m, most protons (about 99.9%) can pass through the membranes. It was known that the mechanical property was sensitive to the chemical impurities. In order to avoid the contamination due to the recoil mixing between the specimen and the sample holder, the target membrane was folded into double sheet. The membrane which faced to the proton beam directly was used as the specimen. The SEM and XRD measurements were made in order to examine the microstructural evolution of membranes by the proton beam irradiation. Micro-Knoop hardness was determined as the average of 10 indentations on each state of the specimens using a conventional apparatus with a load of 20 g.

CP475, *Applications of Accelerators in Research and Industry*,
edited by J. L. Duggan and I. L. Morgan

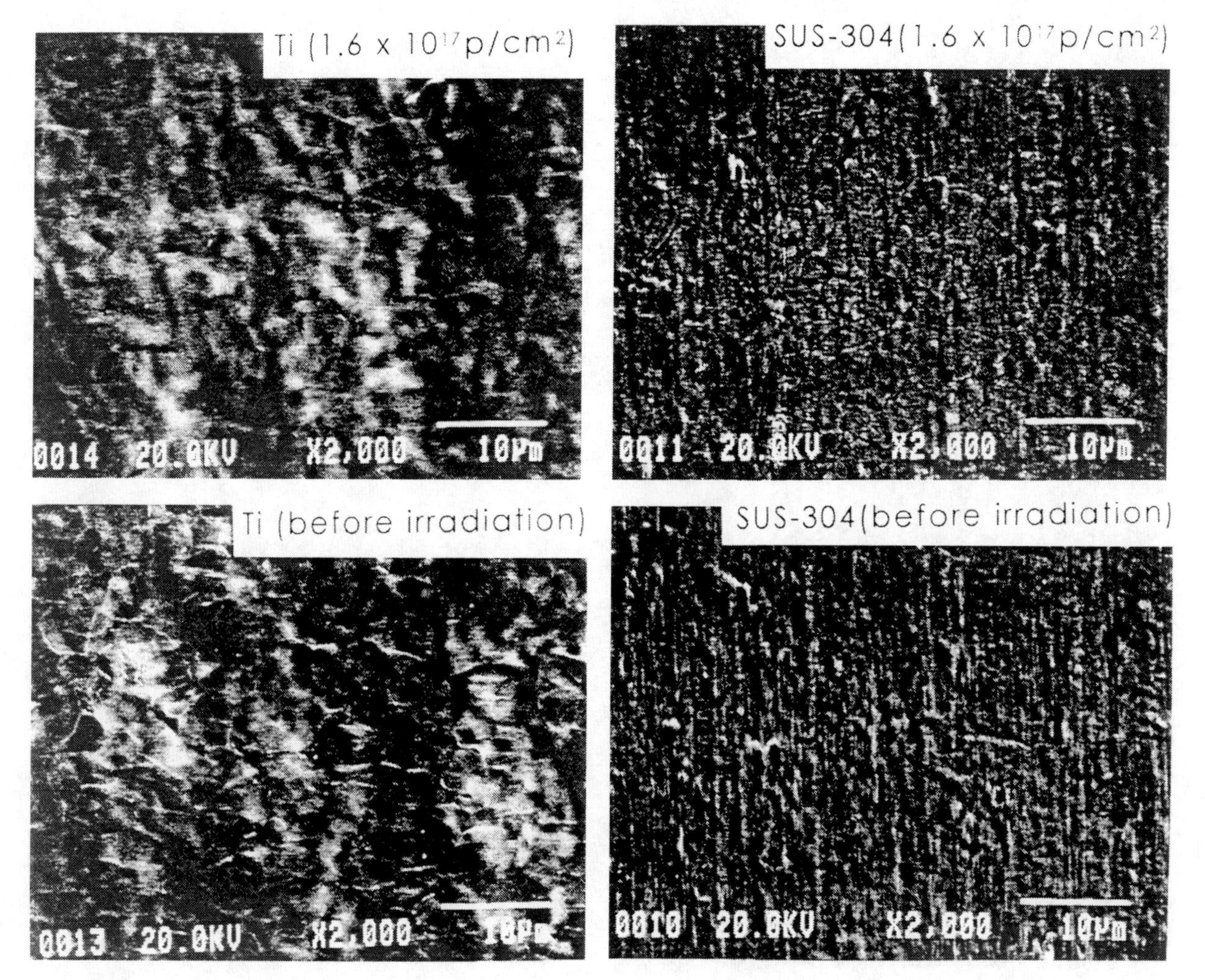

Figure 1 SEM photographs taken from Ti and SUS-304 membranes before and after the proton irradiation.

RESULTS AND DISCUSSION

Figure 1 shows the SEM photographs taken from the Ti and SUS-304 membranes before and after the proton irradiation. After the proton irradiation, grain size of the specimen surface increases appreciably, indicating that the proton irradiation induces grain growth of the polycrystalline membrane.

Figures 2 and 3 show the X-ray diffraction patterns before and after the proton irradiation from the Ti and SUS-304 membranes, respectively. The width of diffraction lines decreases significantly with increasing the irradiation fluence. The narrowing of the diffraction lines may be caused by the grain growth as well as the release of the lattice strain which introduced by the process of making the membrane [3]. These results show that the grain growth by heating due to the ion bombardment takes place preferentially in the experimental condition.

The hardness of the proton irradiated specimens were evaluated by the micro-Knoop hardness. The micro-Knoop hardness is more appropriate than the micro-Vickers hardness for the evaluation of hardness on the thin membrane as less than 6 μ m, since the depth of indentation for the micro-Knoop is shallower than the micro-Vickers.

Figure 4 shows the change of micro-Knoop hardness as a function of the proton irradiation fluence. The micro-hardness is decreased with increasing the irradiation fluence, indicating that the softening by heat originated from the electronic energy loss predominate over the hardening by the displacement of lattice atoms originated from the nuclear energy loss. This is reasonably accepted since the electronic energy loss is about three order magnitude larger than the nuclear energy loss for the transmission of 1 MeV proton through the membrane of 6 μ m thickness [4]; from TRIM calculation [2] the number of displaced atoms produced by the irradiation of 1 MeV proton are only 3.0 and 6.8 atoms/ion for Ti membrane of 5 μ m thick and SUS-304 membrane of 6 μ m thick, respectively.

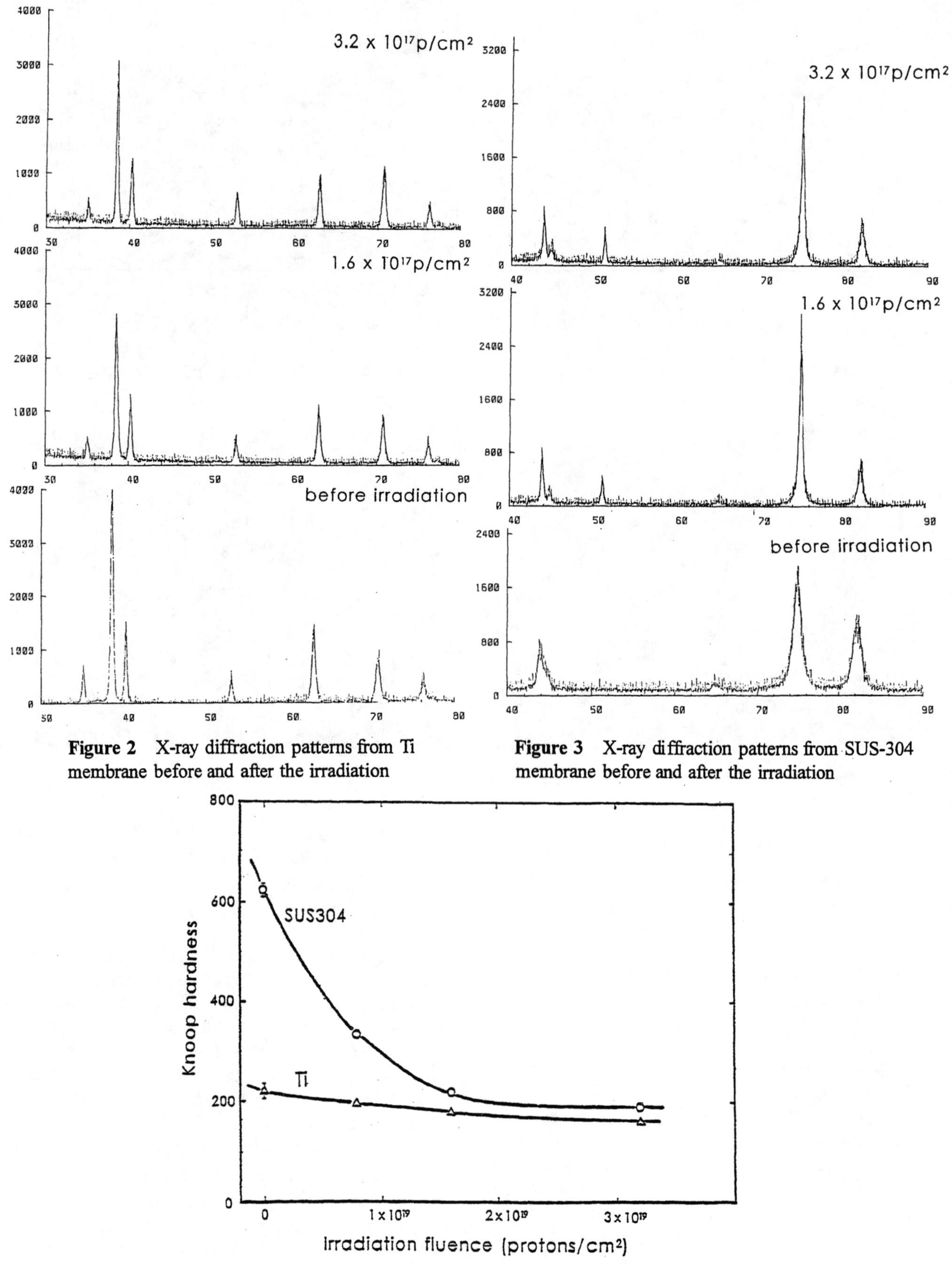

Figure 2 X-ray diffraction patterns from Ti membrane before and after the irradiation

Figure 3 X-ray diffraction patterns from SUS-304 membrane before and after the irradiation

Figure 4 Dependence of micro-Knoop hardness on proton irradiation fluence

The damage production by high-energy light ions in thin metal targets at 4.3 K was studied by Andersen and sorensen [5]. They observed the changes in electrical resistivity as a measure of the radiation damage. The irradiation at room temperature, however, the annealing of damage occurs intensively and reduces the effects of the radiation hardening as seen in the present work.

CONCLUSION

The modification of microstructure and mechanical property produced by the passage of 1 MeV protons were studied by SEM, XRD and micro-Knoop hardness measurements on the Ti and SUS-304 membranes. The transmission of protons induced the grain growth of the polycrystalline membranes. The micro-Knoop hardness of the membranes were decreased by the proton irradiation indicating that the softenung by the heat due to the electronic energy loss predominated over the hardening by the displace ment of lattice atoms due to the nuclear energy loss.

REFERENCE

1] Chr. Lehmann, *Interaction of Radiation with Solid and Elementary Defect Production,* (North-Holland, Amsterdam, 1977) p. 103.

[2] J. F. Ziegler, *Handbook of Ion Implantation Technology,* (North-Holland, Amsterdam, 1992) p.1.

[3] B. E. Warren, *X-ray Diffraction,* (Addison-Wesley, Reading, 1969)), p.235.

[4] H. H. Andersen and J. F. Ziegler, *Hydrogen: Stopping Powers and Ranges in All Elements,* (Pergamon, New York, 1977).

[5] H. H. Andersen and H. Sorensen, Rad. Effects, **14** (1972) 49.

Soft X-ray Appearance Potential Study of Rare Earth-Manganese Compounds

A. R. Chourasia[1] and S. D. Deshpande[2]

[1]*Department of Physics, Texas A & M University - Commerce, Commerce, TX 75429*
[2]*National Research Institute for Metals, 1-2-1 Sengen Tsukuba, Japan 305-0047 and S. S. G. M. College of Engineering, Shegaon (C. Rly.) 444203 India*

Soft X-ray appearance potential spectroscopy (SXAPS) has been employed to study the changes in the electronic structure of RMn_2 compounds (where R = Pr, Sm, Gd and Dy). In this technique the total x-ray emission associated with the thresholds for the excitation of core levels of the atoms in the surface region of the materials is measured. The SXAPS spectra of the Mn $L_{2,3}$ levels in these intermetallics are compared with the corresponding elemental manganese spectrum. The normalized spectra exhibit an increasing trend in the unoccupied density of states at the Fermi level as the atomic number of R increases. This has been interpreted as increasing hybridization between the R 5d and Mn 3d bands. The hybridization is found to influence the magnetic properties of these intermetallics. The core levels are also found to display crystal field splitting that seems to disappear for $DyMn_2$. This correlates very well with the disappearance of the Mn magnetic moment at Dy in these intermetallics.

INTRODUCTION

The rare earth (R) and manganese intermetallics of the type RMn_2 exhibit interesting magnetic properties depending upon the rare earth atom (1-5). These properties are to a large extent due to the competition between the localized R 4f electrons and the itinerant Mn 3d electrons. The 4f moments interact with the 3d moments via a 5d-3d hybridization. The 5d-3d coupling due to local exchange interactions was first considered by Campbell (6) and was corroborated by Brooks et al. (7, 8) using band structure calculations. Duc (9, 10) demonstrated for a large number of R-T compounds (where T is a transition metal) that the R-T exchange interaction can be well understood by applying this 5d-3d hybridization model. The magnetic properties of these intermetallics depend upon the Mn-Mn distance. Huge magnetovolume effects occur (11) which are supposed to be associated with a critical value of the Mn-Mn distance $d_c = 2.66$ Å at ambient temperature, below which the Mn moment is unstable.

The investigation of the electronic structure of these intermetallics is important in order to understand the effect of 5d-3d hybridization on their magnetic properties. In the present investigation we have employed soft x-ray appearance potential spectroscopy (SXAPS) to study the conduction band density of states in the RMn_2 intermetallics. In this technique the derivative of the total x-ray fluorescence yield from the sample surface is measured as a function of incident electron energy. At the threshold of excitation of one of the core levels, both the incident and core electrons are excited to the Fermi energy, E_F resulting in the appearance of an abrupt increase in the total x-ray intensity. The intensity of the soft x-ray emission is determined by the product of the atomic concentration n and the self-convolution of the electronic density of unoccupied states, N(E) above E_F given by (12-14):

$$\int_0^E N(E')N(E-E')dE' \qquad (1)$$

As the matrix element governing the core hole production involves a very short range wave function of the initial core electron state, the technique reveals a localized density of states (DOS). SXAPS is very sensitive to the DOS in the conduction band. Transition metals and rare earths have high unoccupied DOS at E_F making SXAPS a highly suitable technique to study the changes in the electronic configuration of these elements as a result of alloying. The SXAPS theory relates to the self-convolution of the one-electron DOS above E_F. This theory has worked quite well for the 3d transition metals where core level transitions to unoccupied conduction band states occur. In rare earths, the core electron is not excited to the conduction band but to unoccupied 4f orbitals which are quite localized about the excited ion. Therefore, the theory does not adequately explain the SXAPS signal obtained from the rare earths.

In the present investigation we report the SXAPS study on RMn_2 intermetallics where R is Pr, Sm, Gd or Dy. The normalized Mn $L_{2,3}$ level SXAPS spectra in these intermetallics are compared with the corresponding elemental spectrum. The increase in the intensity of the signal with the increase in the atomic number of R is interpreted as an increasing hybridization between the R 5d and Mn 3d bands. The crystal field splitting observed for the core levels is found to disappear for $DyMn_2$. This correlates very well with the magnetic properties at the Mn site in these intermetallics. The results corroborate our earlier studies on these intermetallics (15, 16).

CP475, *Applications of Accelerators in Research and Industry*,
edited by J. L. Duggan and I. L. Morgan

EXPERIMENTAL

In SXAPS electrons from a tungsten filament impact the sample which is biased positively. The potential of the sample is linearly increased with a ramp generator. The soft x-radiation produced is detected photoelectrically. When the target potential approaches the threshold for core level excitation of surface atoms, there is a sharp increase in the photoelectron current. The signal is extracted from the background by taking the data in the first differential mode. This is accomplished by the modulation technique using a phase-lock amplifier. The details of the technique have been described elsewhere (13, 17, 18). The resolution of the spectrometer was 1.0 eV, although changes in the threshold energy could be determined with an accuracy of 0.1 eV.

The samples were prepared by induction melting the high purity (99.95%) starting materials. In order to avoid the formation of R_6Mn_{23} the stoichiometry 1:1.93 was chosen. The as-cast samples were annealed in a high vacuum at 760°C for one week. X-ray diffraction and a. c. susceptibility measurements were used to check the phase purity. The compounds were found to have crystallized in the hexagonal C14 Laves phase type. In the analysis chamber the samples were lightly sputtered by low energy argon ions and annealed. The contamination of the samples was checked by recording the oxygen and carbon K-level spectra. These spectra were barely observable, thus ensuring clean surfaces for examination. The system operating pressure was approximately 10^{-9} Torr.

RESULTS AND DISCUSSION

The normalized SXAPS spectra of the $L_{2,3}$ core levels of Mn in the RMn_2 intermetallics are shown in Fig. 1. These are the plots of the first differential x-ray yield as a function of the sample potential. The spectra represent the average of several runs to improve the signal-to-noise ratio. The spin-orbit splitting for elemental manganese is about 11 eV. Therefore the high energy side of the L_3 structure is slightly superimposed upon the L_2 structure.

In SXAPS the inelastic event occurring leaves both the incident and the core electron in states at E_F. Above this threshold, the electrons can lie in a range of vacant states that conserve energy. The excitation probability reflects the self-convolution of the unoccupied DOS. The strength of the signal therefore represents the DOS at E_F. From Fig. 1, it is seen that as compared to the elemental manganese the signal strength decreases in the RMn_2 intermetallics. This represents a decrease in the DOS at E_F as a result of chemical combination between Mn and R. However, in the series, an increase in the signal strength is observed as we go from $PrMn_2$ to $DyMn_2$. This indicates that the unoccupied DOS at E_F increases as the atomic number of R increases.

In a compound between a rare earth and a transition metal two effects are supposed to take place: (a) a charge transfer among the constituents and (b) a 5d-3d hybridization. The

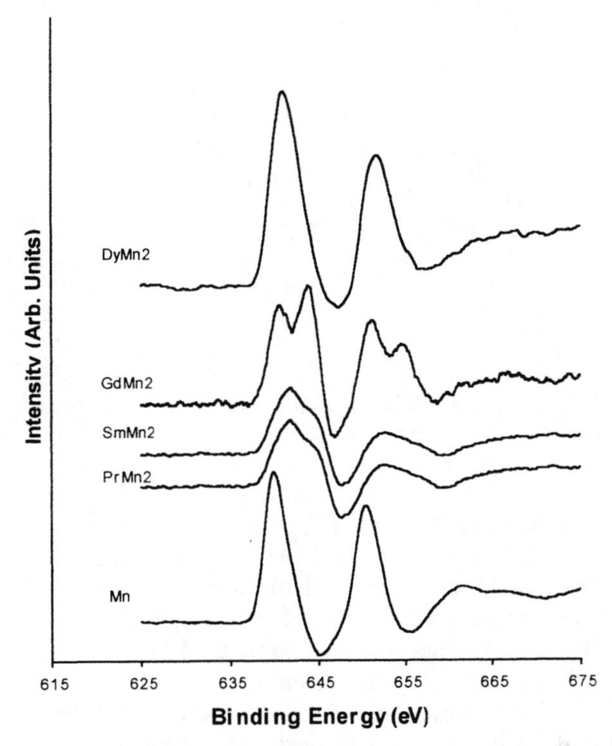

FIGURE 1. Plot of the first differential x-ray yield as a function of electron energy. The $L_{2,3}$ core level spectra of manganese in elemental form and in the RMn_2 intermetallics. The spectra are normalized for equal intensity at 635.0 eV.

Pauling electronegativity for Mn is 1.55 and that for the rare earths is in the range 1.10-1.27. Since Mn is more electronegative, there should be a charge transfer from the rare earths to Mn in these intermetallics. As a result of such a charge transfer, the DOS at E_F for Mn should decrease. This is precisely what we observe for the Mn spectra in these intermetallics. However, the increase in the intensity of the Mn signal among the rare earths emphasizes that R-T charge transfer is not the dominating process upon alloying. This leaves the 5d-3d hybridization as the main effect (19, 20).

In the R-T intermetallics the direct overlap between the 3d and 4f spins is small because the 4f shells are spatially localized. The R-T interactions occur through the intra-atomic 4f-5d and interatomic 5d-3d exchange interactions, as proposed by Campbell (6). Brooks et al. (7, 8) performed self-consistent energy band calculations on various R-3d systems and confirmed Campbell's suggestions. According to their model, the localized 4f spins create a positive localized 5d moment through the 4f-5d exchange and a negative localized moment through the direct 5d-3d exchange. These interactions therefore lead to ferromagnetic ordering if R is a light rare earth, or ferrimagnetic ordering if R is a heavy rare earth.

The 5d-3d hybridization significantly influences the 5d-3d band structure and relative magnitude and sign of the spin polarization at R and T sites. In the R-T intermetallics the pure 5d and 3d bands will hybridize to form hybridized regions at the top of the 3d band and at the bottom of the 5d

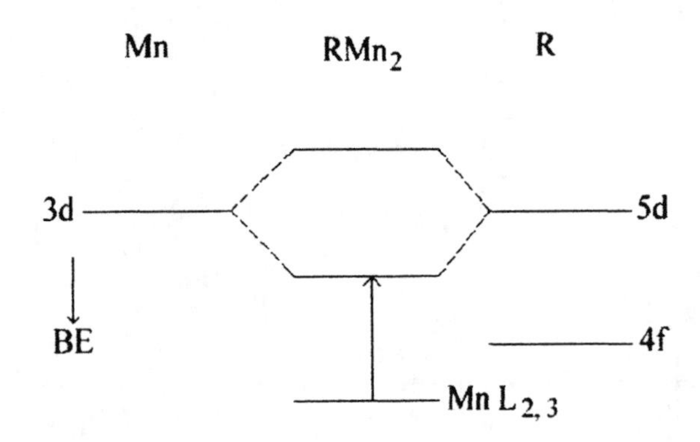

FIGURE 2. Illustration of the simplified hybridization model for the R 5d and Mn 3d levels (not drawn to scale).

band. A schematic of the hybridization picture in R-T intermetallics is shown in Fig. 2. In the case of Mn the 3d band is split into a lower t_{2g} state and an upper e_g state. This splitting lowers the energy of the spin-up 3d states. The extent of hybridization will depend upon the 5d occupation number n_{5d}. The number of electrons in the 5d band is 2.2, 2.1, 1.9 and 1.7 for Pr, Sm, Gd and Dy (21). The reduction of n_{5d} with increasing R atomic number has also been reported for heavy RFe_2 compounds (22). Therefore in the RMn_2 compounds, n_{5d} decreases as the atomic number of R increases. As a result of the 5d-3d hybridization, the unoccupied states in the hybridized band should increase with increase in the atomic number of R. The increase in the signal strength observed in Fig. 1 therefore represents the 5d-3d hybridization. This hybridization increases with the increase in the atomic number of the rare earth in this series. The extent of hybridization of these bands will influence the magnetic properties of these intermetallics. This is in good agreement with model proposed by Brooks et al. (7, 8). The result is consistent with our earlier observations (15, 16) deduced from charge transfer.

The spectra shown in Fig. 1 also exhibit a distinct splitting of about 3 eV for each of the core levels of manganese in the intermetallics. This splitting is due to the crystal field effect. The splitting is seen to decrease with increase in the atomic number of the rare earth and is found to disappear for $DyMn_2$. Wada et al. (11) have studied the magnetic moment at the Mn site as a function of the Mn-Mn distance in these intermetallics. Their study shows that the magnetic moment at the manganese site decreases with the increase in the atomic number of the rare earth. It is found to disappear for $DyMn_2$ and for other heavier rare earths. For heavier rare earths, the Mn-Mn distance is greater than 2.66 Å, the critical distance at which the magnetic moment vanishes. Our data shows that the electrostatic interaction between R and Mn vanishes for distances greater than this value in the RMn_2 intermetallics.

In summary, SXAPS has been employed to investigate the electronic structure of the RMn_2 intermetallics. The strengths of the signal observed for the Mn L2,3 levels indicate an increasing hybridization between the R 5d and Mn 3d bands. The crystal field splitting observed for these levels is found to disappear for heavier rare earths. The results are in complete agreement with the 5d-3d band hybridization and the influence it has on the magnetic properties of these intermetallics.

ACKNOWLEDGEMENTS

The authors wish to thank Dr. G. Wiesinger, Institute for Experimental Physics, Vienna, Austria for providing the samples for the study. The work is supported by Organized Research, TAMU-Commerce.

REFERENCES

1. Yoshimura, K., Shiga M., and Nakamura, Y., *J. Phys. Soc. Jpn.* **55**, 3585-3595 (1986).
2. Shiga M., *Physica B + C* **149**, 293-305 (1988).
3. Cywinski R., Kilcoyne S. H., and Scott C. A., *J. Phys.: Condens. Matter* **3**, 6473-6488 (1991).
4. Ritter C., Cywinski R., Kilcoyne S. H., and Mondal S., *J. Phys.: Condens. Matter* **4**, 1559-1566 (1992).
5. Hauser R., Bauer E., Gratz E., Haufler Th., Hilscher G., and Wiesinger G., *Phys. Rev. B* **50**, 13493-13504 (1994).
6. Campbell I. A., *J. Phys. F* **2**, L47-L50 (1972).
7. Brooks M. S. S., Nordstrom L., and Johansson B., *Physica B* **172**, 95-100 (1991).
8. Brooks M. S. S., Gasche T., Auluck S., Nordstrom L., Severin L., Trygg J., and Johansson B., *J. Magn. Magn. Mater.* **104-107**, 1381-1382 (1992).
9. Duc N. H., *Phys. Status Solidi B* **175**, K63-K67 (1993).
10. Duc N. H., Hien T. D., and Chau N. H., *Acta Phys. Polonica A* **78**, 471-476 (1990).
11. Wada H., Nakamura H., Yoshimura K., Shiga M., and Nakamura Y., *J. Magn. Magn. Mater.* **70**, 134-136 (1987).
12. Park R. L. and Houston J. E., *Surf. Sci.* **26**, 664-666 (1971).
13. Park R. L. and Houston J. E., *J. Vac. Sci. Technol.* **11**, 1-18 (1974).
14. Dev B. and Brinkman H., *Ned. Tijdschr. Vacuumtech.* **8**, 176-184 (1970).
15. Chourasia A. R., Chopra D. R., and Wiesinger G., *J. Elect. Spec. Rel. Phenom.* **70**, 23-28 (1994).
16. Chourasia A. R., Seabolt M. A., Justiss R. L., Chopra D. R., Wiesinger G., *J. Alloys Compds.* **224**, 287-291 (1995).
17. Park R. L., Houston J. E., and Schreiner D. G., *Rev. Sci. Instrum.* **41**, 1810-1812 (1970).
18. Chourasia A. R., *Trends in Vac. Sci. Technol.* **2**, 113-121 (1997).
19. Williams A. R. and Lang N. D., *Phys. Rev. Lett.* **40**, 954-957 (1978).
20. Buschow K. H. J., Bouten P. C. P., and Miedema A. R.,

Rep. Prog. Phys. **45**, 937-1039 (1982).
21. Herbst J. F. and Wilkins J. W., Gschneidner K. A., L. Eyring, and Hufner S. (Eds), *Handbook on the Physics and Chemistry of Rare Earths* **10**, Amsterdam, Elsevier, 1987, 321-360.
22. Brooks M. S. S., Eriksson O., and Johansson B., *J. Phys.: Condens. Matter* **3**, 2357-2371 (1991).

CHARACTERIZATION OF THIN FILMS BY MEANS OF SOFT X-RAY REFLECTIVITY MEASUREMENTS

J. Friedrich

Institut für Angewandte Physik, Heinrich Heine Universität Düsseldorf, 40255 Düsseldorf, Germany.

The suitability of soft x-ray reflectivity measurements for the characterization of thin films and surface layers is discussed. Especially the detection and quantification of oxidized surface layers is of interest for different applications. The evaluations of angular dependent measurements on a Kanigen mirror and on aluminum coated transmission filters for the German X-ray astronomy satellite ABRIXAS are presented. For Nickel films it is shown that a thin carbon coating prevents oxidation. The measurements were performed at the soft x-ray reflectometer at the Hamburg synchrotron radiation laboratory in the energy region 30 eV – 800 eV. In order to resolve the structure of the sample perpendicular to the surface, consistent least square fits of theoretical reflectivity curves have to be required for different photon energies.

INTRODUCTION

Reflectivity measurements are a powerful tool to characterize the optical properties of materials which are described by the optical constants. In the evaluation of those measurements not only the optical constants but also the parameters describing the structure of the sample such as layer thicknesses, surface and interface roughnesses can be determined. Figure 1 shows an example of an angular dependent reflectivity measurement of a single Ni film of 30.4nm thickness in comparison with the calculation. The thickness of the film can be determined due to its characteristic interference pattern. The surface and interface roughnesses lead to a general decrease of the reflectivity and influence the depth of the interference pattern. As the example indicates, precise measurements of the absolute reflectivity over several orders of magnitude are needed in order to obtain accurate values for these parameters (1).

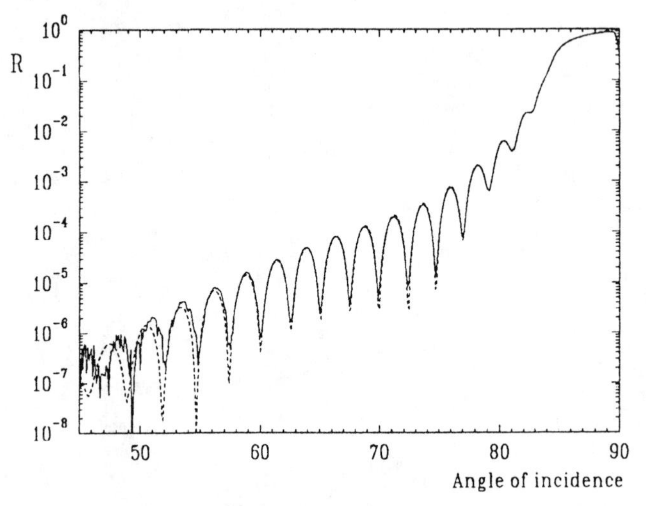

FIGURE 1. Reflectivity vs. angle of incidence at 500 eV of a 30.4nm Ni film on a glass substrate. The measurement (solid curve) is compared with the calculation (dashed curve).

Concerning the monitoring of surface oxidation, photoelectron spectroscopy provides a higher accuracy in the quantification of surface contamination (2). Especially with Auger electron spectroscopy a higher element sensitivity is obtained.

However, an advantage of reflectivity measurements is the flexibility concerning the information depth. By changing photon energy and angle of incidence, the information depth can be changed from ~1nm to several hundred nm, which makes it possible to characterize the surface of a film and to a certain extent the depth profile with one method. In the present paper the suitability of this approach in order to monitor surface oxidation processes is investigated. The possibilities and the limitations of the application of soft x-ray reflectivity measurements are discussed.

EXPERIMENTAL

Beamline

The reflectivity measurements were carried out at the Hamburg synchrotron radiation laboratory HASYLAB with the UHV reflectometer described by Hogrefe et al.(3). It allows computer-controlled independent rotations of the sample and the detector. The measurements were performed with Schottky-type diodes (4,5) (Hamamatsu G1127). Because the sample can be removed from the beam, the necessary normalization was done by moving the detector into the direct beam. Corrections for changes of the incoming photon flux are made by monitoring the total electron yield from the last focusing mirror of the beamline. Depending on the selected photon energy, measurements can be taken over a range of 5 to 6 orders of magnitude. The error of the normalization is less than 1%. The beam divergence is 2 mrad vertically and 5 mrad horizontally.

CP475, *Applications of Accelerators in Research and Industry*, edited by J. L. Duggan and I. L. Morgan

The monochromatized radiation in the soft-x-ray region of 30 to 1500 eV with an energy resolution of $\Delta E/E \cong 1/200$ is supplied by the monochromator Bumble Bee (6,7). By proper choice of the operation modes, the spectral purity of the monochromatized beam can be optimized. Further improvement is achieved by the use of transmission filters. In the energy range covered by the present investigation, the spectral purity was clearly above 99%.

Samples

Measurements on Kanigen, a nickel/phosphorus alloy, were performed at the nonstructured part of a diffraction grating, which was etched into the material. Thus the investigated surface underwent the etching process.
Concerning the analysis on Ni measurements and calculations from a single Ni film of 30.4nm thickness and a Ni/C multilayer consisting of 20 periods of 1.5nm Ni and 4.5nm C, covered by a 1.2nm C layer are presented. The samples were prepared on the smoother side of unpolished float glass ($\sigma < 0.6$nm rms). The deposition was performed with the sputtering facility in the multilayer laboratory of Sinchrotrone Trieste (1). After preparation, the samples remained unprotected in air and an oxidation of the Ni has to be expected (2).
In addition reflectivity measurements on a transmission filter for the CCD of the ABRIXAS satellite (8,9) are presented. These filters consist of a 0.80 µm thick polypropylene foil, which is coated with approximately 60 nm aluminum on both sides. The manufacturing and the characterization of the filters is presented in detail in (10). For the reflectivity measurements presented here the polypropylene foil was put on a Si wafer in order to produce a flat surface.

EVALUATION TECHNIQUE

In case of Kanigen and Ni the evaluation is done by comparing angular dependent reflectivity measurements with the calculation. In general the sample consists of several layers of different material. The reflectivity $\mathbf{R}_0$ of a single layer can be calculated using the Fresnel equation. The reduced reflectivity $\mathbf{R}$ due to scattering from statistically rough surfaces, according to Beckmann and Spizzichino (11) is given by

$$\mathbf{R} = \mathbf{R}_0 \exp-\left(\frac{4\pi}{\lambda}\sigma \cos\theta_i\right)^2. \qquad (1)$$

λ is the wavelength of the radiation, and θ_i the angle of incidence.

The reflectivity of a system with several layers is then calculated with a recursion equation (12,13). Each layer is characterized by 4 parameters, which influence the obtained reflectivity: its thickness d, its roughness σ and its complex dielectric function $\varepsilon = \varepsilon_1 + i\varepsilon_2$. These parameters are determined by least squares fits of the calculated reflectivity curves to the measurements. Fig. 1 shows for example the output of the fit at 500eV for the single Ni film. Since especially in case of systems with multiple layers many parameters have to be determined, parameters have to be determined independently in advance to obtain reliable results. For the energy independent parameters of thicknesses and roughnesses, consistent values in the evaluation of measurements at different photon energies are required. The optical constants of the substrate were determined in advance using an uncoated glass substrate. In case of the more complicated multilayer structure the optical constants of the materials had been kept fixed. The values have been determined by analyzing measurements taken on single films of the materials. The applied procedure is described in detail in.(1,14).

RESULTS AND DISCUSSION

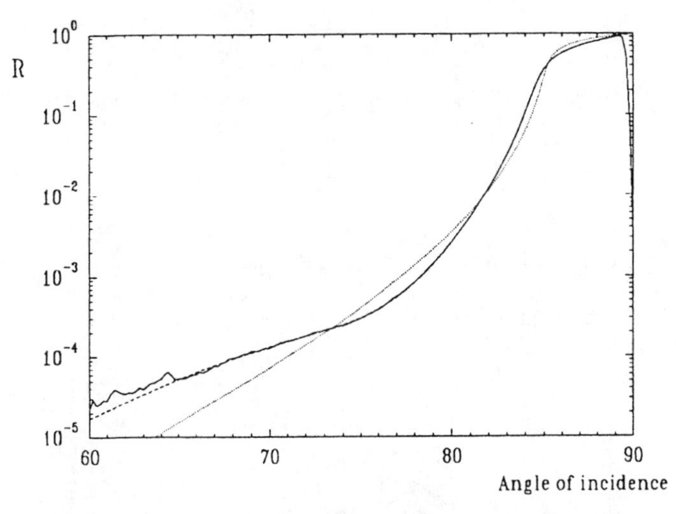

FIGURE 2. Reflectivity vs. angle of incidence at 500 eV of a Kanigen mirror. The measurement (solid curve) is compared with calculations assuming only bulk material (dotted curve) and assuming an additional surface layer (dashed curve).

Figure 2 shows the measurement in comparison with two different calculations of the reflectivity of the Kanigen mirror at 500 eV. The dotted curve denotes the best approximation obtained under the assumption of bulk material, which leads only to a very poor agreement with the measurement. The calculation with an additional surface layer of 2.6nm thickness leads to a very good agreement. This result is confirmed by evaluations at other photonenergies. The surface contamination might result from oxidation or from the application of the etching process.

The very good agreement between measurement and calculation shown in Fig. 1 could only be reached with the assumption of an additional surface layer on top of the Ni film of about 2.4nm thickness. Figure 3 shows the best agreement achieved assuming only a single film. The calculation leads to significantly higher reflectivities around 73 degrees angle of incidence and to significantly lower reflectivity values around 55 degrees. This influence on the calculation can be understood as an additional interference structure with a longer period, which directly indicates the thickness of the structure (Half of the period - from 73 to 55 degree – corresponds to about 8 interference periods of the film.) This simple estimation leads to a thickness of about 2nm which agrees quite well to the value of 2.4nm actually obtained from the fit.

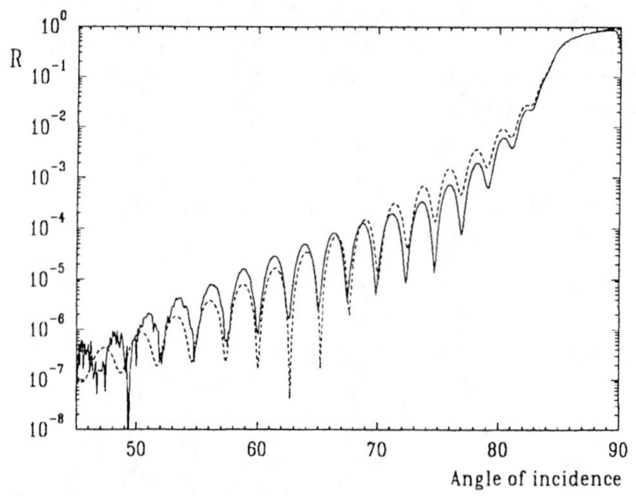

FIGURE 3. Reflectivity vs. angle of incidence at 500 eV of a 30.4nm Ni film on a glass substrate. The measurement (solid curve) is compared with the calculation assuming only a single Ni film (dashed curve).

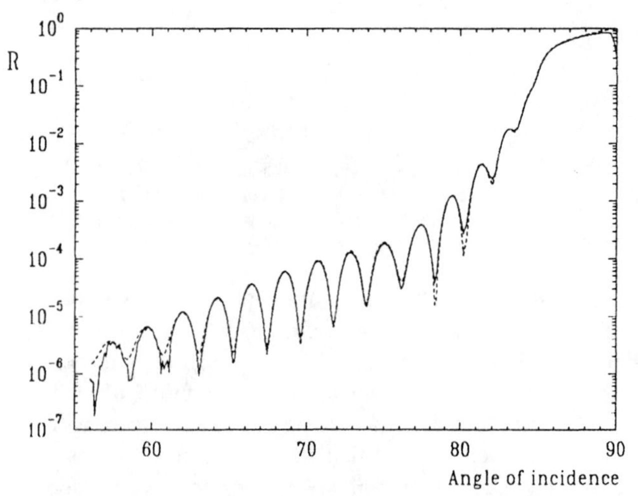

FIGURE 4. Reflectivity vs. angle of incidence at 550 eV of a 30.4nm Ni film on a glass substrate. The measurement (solid curve) is compared with the calculation (dashed curve).

The evaluation of the reflectivity measurement at 550eV (Fig. 4) above the O 1s absorption edge provides more information about the surface layer. Table 1 shows the obtained optical constants of the layer. The larger value of ε_2 at 550 eV means a larger absorption coefficient which indicates the presence of oxygen. Furthermore the agreement between measurement and calculation is less good than at other photon energies. This effect is probably due to the higher impact of the surface layer and due to its imperfect description in the model. In reality one has to assume a gradient rather than a well defined layer.

TABLE 1. Optical constants of the detected surface layer on Ni

Energy (eV)	$1-\varepsilon_1$	ε_2
500	$2.3 \bullet 10^{-3}$	$0.6 \bullet 10^{-3}$
550	$2.9 \bullet 10^{-3}$	$1.2 \bullet 10^{-3}$

Figure 5 shows measurement and calculation of the reflectivity of the Ni/C multilayer. The prominent Bragg peaks are due to the constructive interference of the contributions reflected by all Ni/C period (15). The high frequency interference pattern in between – the Kiessig fringes – correspond to the thickness of the whole multilayer structure.

The excellent agreement of the calculation was obtained using the previously determined optical constants for Ni and C for all layers including the surface layer. This result proves, that the 1.2nm C layer protected the underlying Ni layer from oxidation.

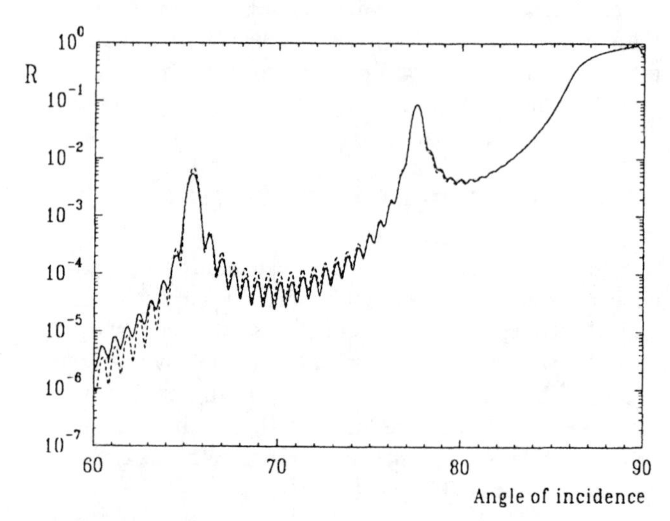

FIGURE 5. Reflectivity vs. angle of incidence at 500 eV of a Ni/C multilayer of 20 periods on a glass substrate. The measurement (solid curve) is compared with the calculation (dashed curve).

The transmission filter for the ABRIXAS satellite turned out to have a roughness exceeding several nm. Thus a proper evaluation of angular dependent measurements concerning surface layers was not possible. In case of Al one expects a 2-5nm thick surface layer of Al_2O_3 to evolve which protects the deeper layers from further oxidation (16,17). However, energy dependent measurements around the O 1s absorption edge can be used to monitor the oxidation process with time. Figure 6 shows measurements at 88, 87 and 86 degree angle of incidence. These measurements will be compared with the same measurements taken after a few months of storage.

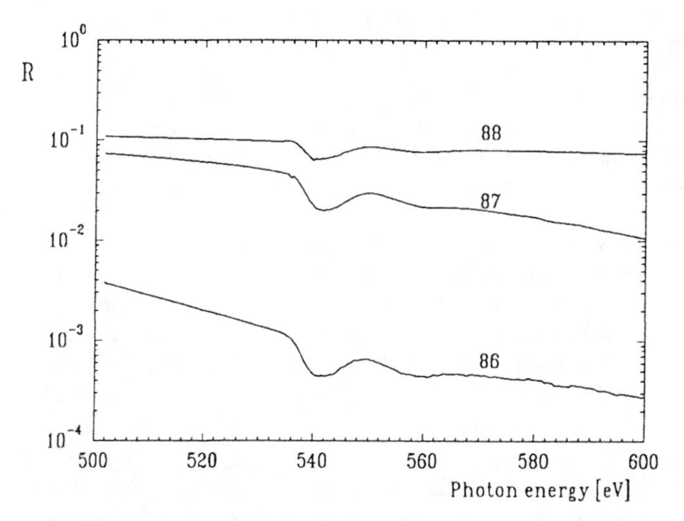

FIGURE 6. Reflectivity vs. photon energy at 88, 87 and 86 degree angle of incidence of a transmission foil similar to the ones planned for the transmission filters of the ABRIXAS satellite.

CONCLUSION

The detection of oxidized surface layers by means of soft x-ray reflectivity measurements is possible for layer thicknesses greater than approximately 1nm. The quality of the results depends strongly on the quality of the sample, especially on its microroughness. However a detailed characterization of thin surface layers, such as the degree of oxidation or the accurate depth profile cannot be expected in general from this method.

In case of rough surfaces, like the transmission foil for the ABRIXAS satellite parameters of an oxidized surface layer cannot be determined accurately. However, energy dependent reflectivity measurements around the O1s absorption edge at grazing angles of incidence can be used to monitor the oxidation process.

ACKNOWLEDGEMENT

The author thanks W. Jark, S. Di Fonzo and B.R. Müller for supplying the Ni-samples, K.-H. Stephan for supplying the special filter foil and I. Diel for his work on the fitting program and the determination of the optical constants of Ni and C.

REFERENCES

1. I. Diel, J. Friedrich, C. Kunz, S. Di Fonzo, B.R. Müller and W. Jark, "Optical constants of float glass, nickel, and carbon from soft x-ray reflectivity measurements"; Appl. Opt., **36**, 6376pp. (1997).

2. J.P. Chauvineau et al., *Application of Thin-Film Multilayered Structures to Figured X-Ray Optics*, Proc. SPIE, San Diego F. Marchall ed. (1990) 245pp.

3. H. Hofgreve, D. Giesenberg, R.-P. Haelbich and C. Kunz, " A new VUV-refectometer for UHV-applications,"Nucl. Instr. and Meth. 208, 415 (1983).

4. J. Barth, E. Tegeler, M. Krisch and R. Wolf, " Characteristics and applications of semiconductor photodiodes from the visible to the x-ray region, " Proc. SPIE 733, 481 (1987).

5. M. Krumrey, E. Tegeler, J. Barth, M. Krisch, F. Schäfers and R. Wolf, " Schottky type photodiodes as detectors in the VUV and soft x-ray range, " App. Opt. 27, 4336 (1988).

6. W. Jark, R.-P. Haelbich, H. Hofgreve and C. Kunz, " A new monochromator for the energy range 5 eV - 1000 eV , " Nucl. Instr. and Meth. 208, 315 (1983).

7. W. Jark,and C. Kunz, "Output diagnostics of the grazing incidence plane grating monochromator BUMBLE BEE (15 - 1500 eV)", Nucl. Instr. and meth. A246, 320(1986)

8. G. Richter, G. Hasinger, P. Friedrich, K. Fritze, J. Trümper, H. Bräuninger, P. Predehl, R. Staubert, and E. Kendziorra, "ABRIXAS, A BRoad-band Imaging X-ray All-sky Survey"; Experimental Astronomy, 159-162(1995).

9. L. Strüder et al.,"The MPI/AIT X-ray-imager-high-speed-pn CCDs for X-ray detection,"Nucl. Instrum. and Meth. A288(1), 227(1990).

10. K.-H. Stephan, H. Bräuninger, F. Haberl, P. Predehl, H.J. Maier, J. Friedrich, D. Schmitz, " Characterization of Thin Film CCD Filters on Board the German Astronomy Satellite ABRIXAS by Soft X-Ray Transmission Measurements, "Proc. 15th Intern. Conf. On the Appl. of Accel. In Research & Industry, to be published

11. P. Beckmann and A. Spizzichino, " The scattering of Electromagnetic Waves from Rough Surfaces," Pergamon Press, Oxford (1963)

12. L.G. Parratt, " Surface Studies of Solids by Total Reflection of X-Rays, " Phys. Rev. **95**(2) 359pp. (1954).

13 O.S. Heavens. " Optical Properties of Thin Solid Films, " Butterworth Scientific Publication, London (1955)

14. J. Friedrich, I. Diel, C. Kunz, S. Di Fonzo, B.R. Müller and W. Jark, "Characterization of sputtered nickel/carbon multilayers with soft x-ray reflectivity measurements "; Appl. Opt., **36**, 6329pp. (1997).

15 E. Spiller, "Low-Loss Reflection Coatings Using Absorbing Materials,"; Appl. Phys. Lett., **20**(9), 365pp. (1972).

16 P.H. Bering, G. Hass, and R.P. Madden, "Reflectance-Increasing Coatings for the Vacuum Ultraviolet and Their Applications,"; J. Opt. Soc. Am., **50**, 586pp. (1960).

17 R.W. Fane and W.F.J. Neal, "Optical Constants of Aluminum Films Related to Vacuum Environment J. Opt. Soc. Am., **60**, 790pp. (1970).

Upgrading RBS Analysis of Single Crystals by Application of Triple Correlation

H. Ellmer, R. Aichinger, and D. Semrad

Institut für Experimentalphysik, Abteilung für Atom- und Oberflächenphysik
Johannes Kepler Universität Linz, 4040 Linz, Austria

Ion accelerators in combination with appropriate goniometers and detector systems have been widely used to investigate single crystals. In most of these measurements the information is obtained from the change in backscattering yield when a low index axis or plane becomes aligned with the beam. From a statistical point of view this decrease is due to a negative correlation between the projectile flux distribution and the target atom positions at the same depth. Using a 'true' 180° detector, an additional (positive) correlation between flux and local detector solid angle (or fraction of detected backscattered projectiles) can be obtained. A positive triple correlation exists when both, the beam and the detector are directed versus the same string of atoms at a small angle with the axis. We show that this yield (giving the 'shoulder' in standard channeling measurements) is extremely sensitive to damage in the near surface region of the crystal.

INTRODUCTION

Information about single crystals can be obtained apart from other methods by RBS/channeling. The angular distributions of the yield of backscattered projectiles give insight into properties of the probe. In the standard single alignment (SA) geometry, a low index axis or plane of a crystal is aligned with a beam of highly collimated swift light ions. In the more sensitive double alignment experiments also the detector has to be aligned. The uni-directional double alignment (UDDA) geometry has among others the advantage that aligning the beam simultaneously means aligning the detector.

We have developed a 180° backscattering facility for UDDA. By this for the first time projectiles with backscattering angles arbitrarily close to 180° can be studied. We will discuss their yield from a statistical point of view. This statistical model has been used already to describe the near surface yield enhancement of amorphous targets (1,2). Now we apply this model to single crystals. We propose an application of the UDDA triple correlation to the investigation of near surface crystal damage.

EXPERIMENTAL SETUP

The facility is extensively described in (3); here we give only a short summary of relevant details. The 180° backscattering assembly is UHV compatible. It is pumped by turbomolecular pumps and by cold fingers filled with liquid nitrogen. The base pressure is in the low 10^{-9} mbar range. The device is connected to the beam transport system of a 700 keV single-ended Van-de-Graaff accelerator, which provides protons and helium ions with energies from 50 keV to 750 keV.

The 180° backscattered projectiles are deflected out of the incoming beam by a vertical magnetic field. To reduce the dispersion of the magnetic field we have added a transversal electric field (Fig. 1) so that the essential part of the spatially dispersed spectrum hits the 300 mm^2 semiconductor detector used for UDDA measurements. The detector and the input stage of the preamplifier are kept at constant temperature of 161 K by a mixture of liquid and solid CS_2. In spite of the large area of the detector, we obtain a resolution of 9.1 keV for 400 keV He projectiles. The detector unit is mounted on a dewar that can be moved in vertical direction by a linear motion feedthrough.

Two baffles (B_1: Ø1.2 mm and B_2: Ø0.8 mm, 2450 mm apart) collimate the incident beam to a maximum half-angle divergence of $\alpha = 0.023°$; the root mean square angle of all trajectories is $\alpha_{rms} = 0.012°$. After being deflected by the electric and magnetic fields the beam passes a vertical slit B_3 of 1.15 mm width. As the slit intercepts the rim of the beam it is made from beryllium to minimize background. The length of the flight path between B_3 and the target is 840mm. In the target chamber, at a scattering angle of 173°, there is a 100 mm^2 semiconductor detector that is used as standard RBS detector and for SA measurements (SA-detector).

The x-component (in the horizontal plane) of the acceptance angle β for backscattered projectiles is given by the width of the vertical slit B_3 ($\beta_x = 0.078°$). The vertical y-component is exclusively determined by the detector baffle. Hence, by moving the detector up or down, we can change this component of the scattering angle and hence the total scattering angle. With the detector positioned in the plane of the incoming beam we measure projectiles scattered at a nominal scattering angle of 180°, and with the detector shifted up by h = 52.5 mm projectiles scattered at 178° are detected. The height of the detector baffle of 3 mm determines the acceptance angle in the vertical plane, which is $\beta_y = 0.078°$. For the root mean square angle we

CP475, *Applications of Accelerators in Research and Industry*,
edited by J. L. Duggan and I. L. Morgan

get β_{rms} =0.039°. Due to the finite emittance of the beam and to the finite acceptance of the detector, with the detector in the 180° position (h = 0) the relevant rms scattering angle is 179.96°.

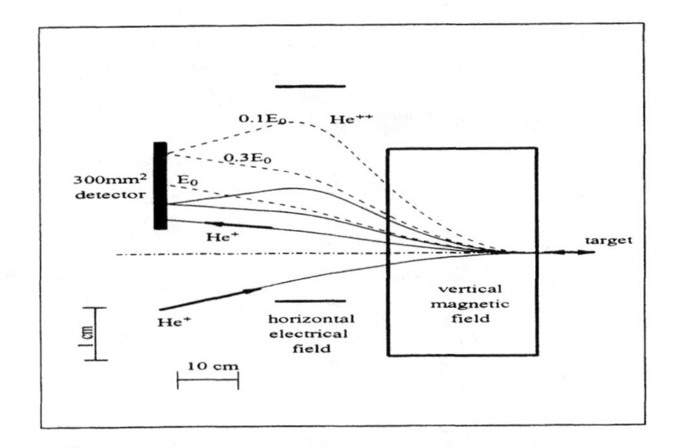

FIGURE 1. Ion optics of the 180° assembly. Shown are the trajectories of He^+ (solid line) and He^{++} (broken lines) backscattered with E_0, 0.3 E_0 and 0.1 E_0, where E_0 is the energy of the incoming He^+ projectiles. Note that the transverse scale is blown up by a factor of 10.

The solid angle $\Delta\Omega_{D0}$ of the double alignment detector is $1.8*10^{-6}$ sr, that is sufficiently small for blocking measurements. The SA-detector has a solid angle of $8.65*10^{-4}$ sr, so that averaging over a great regime of scattering angles avoids blocking effects. This may be compared to the equivalent beam solid angle of $5*10^{-7}$ sr, suitable for channeling with 400 keV He projectiles.

The target is mounted on a computer controlled three-axis goniometer with x-y-z translation. Two of the three axes (tilt and azimuthal angle) are driven by in-vacuum stepping motors. Via a worm drive and an anti-backlash gear-wheel, a single step of the motor results in a rotation of 0.02°. The third axis is rotated by a standard stepping motor outside vacuum, the controller allows ministeps with a width of 0.018° (but with some backlash).

CORRELATION

Theoretical Description

We assume normal incidence (in z-direction) of the incoming beam on the target. The number I(z) of projectiles per second backscattered into the detector from a layer at depth z and of thickness Δz is given by

$$I(z) = \frac{d\sigma}{d\Omega} \int_A f(x,y;z)\, N(x,y;z)\, \Delta z\, \Delta\Omega_D(x,y;z)\, dx\, dy. \quad (1)$$

For scattering around 180° the differential scattering cross section $d\sigma/d\Omega$ is stationary and has been taken out from the integral. All other quantities apply to depth z: f(x,y;z) is the flux distribution function $[m^{-2}s^{-1}]$ and $N(x,y;z)\Delta z$ the areal atomic number density; it is given by

$$N(x,y;z)\Delta z = \Sigma j\ \delta(x-x_j)\ \delta(y-y_j) \quad (2)$$

with j running from 1 to $N_0\Delta z$, where N_0 is the average atomic number density of the target. $(x_j,y_j;z)$ are the atom positions, either regularly ordered or randomly distributed.

$\Delta\Omega_D(x,y;z)$ is the local solid angle of the detector defined by all trajectories starting from position (x,y) and hitting the detector. Obviously, both f(x,y;z) and $\Delta\Omega_D(x,y;z)$, depend on atoms positioned between surface and depth z. If the functions in the integral are not correlated we get the well known equation of the product of their mean values

$$I_0(z) = d\sigma/d\Omega\ I_B\ N_0\Delta z\ \Delta\Omega_{D0}, \quad (3)$$

with

$$I_B = \int f(x,y;z)dxdy \quad (4)$$

the number of incident projectiles per second and with $\Delta\Omega_{D0}$ the standard solid angle of the detector.

Any correlation results in deviations of I(z) from $I_0(z)$, most of them are well known: a positive correlation between f(x,y;z) and $\Delta\Omega_D(x,y;z)$ is called near surface yield enhancement (1), a negative correlation between f(x,y;z) and N(x,y;z) is called channeling in single alignment, whereas a positive correlation gives the channeling shoulder. A negative correlation between $\Delta\Omega_D(x,y;z)$ and N(x,y;z) is called blocking. However, the correlation between all three quantities, f(x,y;z), N(x,y;z), and $\Delta\Omega_D(x,y;z)$, has been investigated to a less extent. It is called uni-axial double alignment, giving either channeling or the channeling shoulder, depending on whether the correlation between f(x,y;z) and N(x,y;z) is negative or positive. This is quite different from bi-directional double alignment geometry e.g., used in measuring reconstruction and relaxation of surfaces (4). In the uni-directional case of double alignment, the correlation between f(x,y;z) and $\Delta\Omega_D(x,y;z)$ is always positive.

Measurements of Correlations

To demonstrate the influence of different correlations on I(z), we bombarded tungsten targets with 400 keV He^+ ions. The random spectra (with and without enhancement) were measured with an amorphous tungsten target to make sure that no accidental alignment of the target can take place. For the other measurements we used a single crystal with (110) oriented surface. The (standard) aligning procedure of the single crystal has been done before. To compare the different measurements, they were normalized to the random spectrum at the low energy part. The single alignment measurements are presented in Fig. 2.

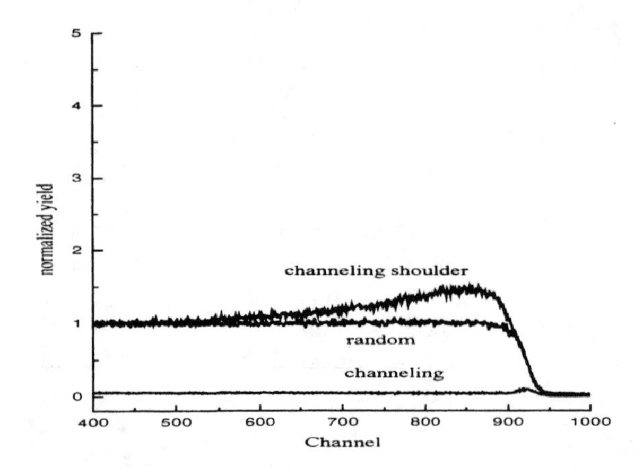

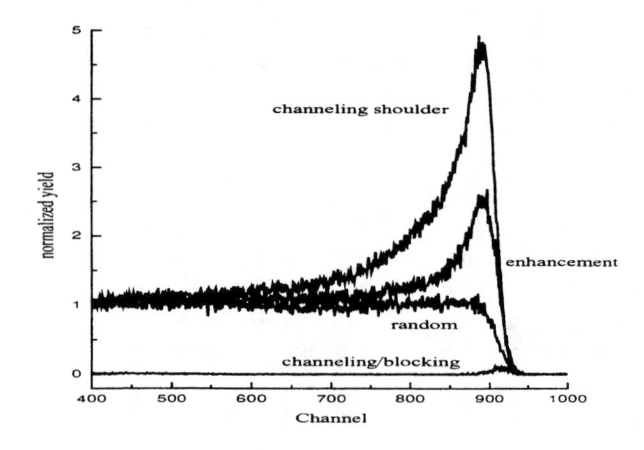

FIGURE 2. Random spectrum and channeling and channeling shoulder spectra of 400 keV He^+ backscattered from W(110) in single alignment geometry. The spectra are normalized to the low energy part of the random spectrum. The dispersion of the detector system is 0.38 keV/channel.

FIGURE 3. The same as Fig. 2 but in double alignment geometry. For better comparison we also keep the scale of Fig. 2.

The random spectrum is used as reference according to Eq. 3. The channeling spectrum is a result of steering the projectiles within the channel by the quasi-continuum potential of the atom strings. In an almost perfect crystal, there are no target atoms in the channel available for head on collisions giving backscattering; so we have a highly negative correlation between f(x,y,z) and N(x,y,z).

A positive correlation between the projectile flux f(x,y,z) and the target atom N(x,y,z) distribution leads to a higher yield called channeling shoulder. This positive correlation originates from focusing the projectiles onto lattice atoms in depth by lattice atoms closer to the surface; to put it differently, at this angle of incidence the border of the shadow cones of the near surface atoms with their high flux density point at lattice positions (5). This increased yield is also called for by the rule of angular compensation formulated by Lindhard (6).

The measurement was done with the beam making an angle of 5° with the <110> axis and of 2° with the {110} plane.

In Fig. 3 we present the same measurements in UDDA geometry. The random spectrum was measured at a backscattering angle of 178° (2), all the others at 180°.

The near surface yield enhancement is a result of a flux peaking effect and is described in detail elsewhere (2). The positive correlation between flux distribution f(x,y,z) and local solid angle $\Delta\Omega_D(x,y,z)$ originates from the fact that a projectile backscattered at an angle close to 180° can follow its inward trajectory on its way back to the surface. Due to focusing by lattice atoms there is even a certain tolerance for the scattering angle to make the backscattered projectile still follow its incoming trajectory.

The most striking effect in Fig. 3 is the exceptionally high channeling shoulder in UDDA. For quantification, we can relate peak areas to numbers of additional atoms per unit area which will give the same yield (excess atoms (7), in units of 10^{15} atoms/cm^2). In the case of the double alignment shoulder we get 804.2, of the near surface yield enhancement 288.6 and for the single alignment shoulder 239.2.

In channeling, the yield dip in the case of UDDA is further lowered by the blocking effect in the same channel.. The minimal yield relative to the random yield amounts to 0.2% in the case of UDDA and to 3.5% in the case of SA.

It follows from the above that the transition from channeling shoulder to the corresponding random spectrum is equivalent to a transition from a crystalline to an amorphous target, and *vice versa*. This will be used for the determination of crystal damage described in the next chapter.

CRYSTAL DAMAGE

As we have seen, the crystalline structure is indispensable for a correlation between flux distribution f and target atom location N. If the target atoms are distributed randomly, the factor N(x,y;z)Δz in Eq. 1 can be replaced by its mean value, $N_0\Delta z$, and taken out from the integral: we get the formula for the „random spectrum“. Evidently, in UDDA the „random spectrum“ corresponds to the spectrum with near surface yield enhancement.

It is common practice in RBS/channeling to examine the crystal quality by determining the minimum yield in single alignment channeling geometry (8). However, most of the damage leads only indirectly to an increased yield by raising the dechanneled part of the beam. These dechanneled projectiles are not immediately backscattered but contribute to the yield in the low energy part of the spectrum, by this obscuring the actual depth of the defect.

Direct evidence is given only when a self-interstitial is positioned in the channel with direct backscattering from this atom.

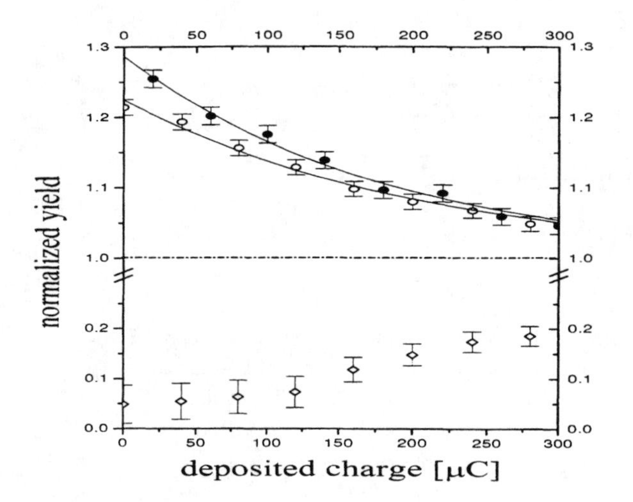

FIGURE 4. Dependence of the relative shoulder areas on the deposited charge in single (open circle) and double (solid circle) alignment geometry and of the channeling minimum yield (rhomb). The line is an exponential fit to the shoulder data.

Let us now look at a crystal adjusted to give the shoulder. As simulations demonstrate, a void is directly disturbing the superposition of projectile flux so that the focusing onto the lattice sites is reduced. In addition, the ejected lattice atom most likely becomes situated at a position with a lower flux density, again reducing the yield. The size of the shoulder therefore seems to be more sensitive to a low defect concentration at shallow depth than normal channeling, as shown in Fig. 4. The sensitivity to the near surface region is due to the fact that the positive correlation between f and N and hence the intensification of the effect is restricted to this region.

To damage the target crystal we used the analysing beam of 400 keV He^+. The target was a Te (0001) single crystal with an appropriate rate of defect production. Measurements were done with the beam at an angle of 5.5° relative to the <0001> axis and 2° from the $\{10\bar{1}0\}$ plane alternating in single alignment channeling, single alignment channeling shoulder and double alignment channeling shoulder geometry. As measure for the (small) defect concentration we use the deposited charge on the target. The area of the channeling shoulder in the spectrum is normalized to the random yield within the same depth range.

At a deposited charge of 300 μC the shoulder has almost disappeared, whereas the minimum yield has only slightly increased. The exponential decay of the SA shoulder areas is not as fast as that for double alignment. The UDDA shoulder is also more sensitive to the crystal quality than the SA shoulder, as one can see at the 0 μC value of the fit curves and the faster decay in the case of UDDA.

SUMMARY

We have given a brief outlook on correlations in ion scattering experiments. The exceptional high channeling shoulder in uni-directional double alignment geometry can be traced to a positive triple correlation of all involved stochasitic quantities, flux districution function f, target atom distribution N, and local detector solid angle $\Delta\Omega_D$. Finally we have found that the magnitude of this channeling shoulder is very sensitive to near surface defects of the lattice and may be used to assess the quality of a single crystal.

ACKNOWLEDGMENTS

This work has been supported by the Österreichischen Fonds zur Förderung der wissenschaftlichen Forschung under Contract No. P12234-NAW.

REFERENCES

1. Pronko P. P., Appleton B. R., Holland O. W., and Wilson S. H., *Phys. Rev. Lett.* 43, 779 (1979)
2. Ellmer H., Fischer W., Klose A., and Semrad D., *Phys. Rev. B.* 55, 2867, (1997)
3. Ellmer H., Fischer W., Klose A., and Semrad D., *Rev. Sci. Instrum.* 67, 1794, (1996)
4. Van Der Veen J.F., *Surf. Sci. Rep.* 5, 199 (1985)
5. Oen O., *Nucl. Instr. and Meth.* 194, 87, (1982)
6. Lindhard J., *Dansk. Vid. Selsk. Mat Fys. Medd.* 34, 14, (1965)
7. Chu W., Mayer J. W., and Nicolet M., *Backscattering Spectrometry*, New York, Academic Press, 1978, ch. 3.5.1, p. 70 and ch. 4.2.2., p. 92
8. Feldman L. C., Mayer J. W., and Picraux S. T., *Materials Analysis by Ion Channeling*, New York, Academic Press, 1982

Development of a Compact High-Resolution RBS System for Monolayer Analysis

K. Kimura,

Department of Engineering Physics and Mechanics, Kyoto University, Kyoto 606-8501, Japan

M. Kimura, Y. Mori, M. Maehara, and H. Fukuyama

Engineering & Machinery Division, Kobe Steel Ltd. 2-3-1 Niihama, Arai-chou, Takasago, Hyogo 676-8670, Japan

A compact high-resolution RBS system consisting of a 90° sector magnetic spectrometer and a 500 kV accelerator is developed. Energy resolution of the spectrometer is designed to be 0.1 % at an acceptance angle of 0.4 msr in an energy range of 27 %. The dimensions of the full system are 3.8 m (L) × 2.8 m (W) × 2.35 m (H) including the accelerator. The system is shown to be capable for monolayer analysis as was designed.

INTRODUCTION

Rutherford backscattering spectroscopy (RBS) is one of the most powerful techniques to measure elemental depth profiles (1). In conventional RBS, a silicon surface barrier detector (SSBD) is used to measure energy spectra of scattered ions. The depth resolution of RBS, which is primarily determined by the energy resolution of SSBD, is about 10 nm. High-resolution RBS (HRBS) systems with either magnetic or electrostatic spectrometers have been developed by several groups to improve the resolution (2 - 8). They used relatively large spectrometers in combination with MV accelerators. For better depth resolution, however, sub-MeV He ions are most suitable (9) and actually the monolayer resolution was demonstrated only with 300 - 400 keV He^+ ions so far (7, 9, 10). For the analysis of these ions, a rather compact spectrometer can be employed. In the present paper, we report on the development of a compact high-resolution RBS system including a 500 kV accelerator and a high-resolution magnetic spectrometer. The magnetic spectrometer is designed to be capable for monolayer analysis with a reasonably wide energy window and a relatively large acceptance angle.

DESIGN OF THE HRBS SYSTEM

The whole HRBS system is schematically shown in Fig. 1. The dimensions of the system are as compact as 3.8 m (L) × 2.8 m (W) × 2.35 m (H). He^+ ions generated by a Penning ion gauge (PIG) ion source can be accelerated up to 500 keV. The ion source and the acceleration tube are enclosed in an isolation tank filled with SF_6 gas. The accelerated ions are analyzed by a Wien filter. The typical beam current is about 30 nA at a beam size of 1 mm. A magnetic quadrupole lens, which has a capability of focussing 1 MeV He^+ beam to less than 5 μm, was installed just before a scattering chamber. The ultimate performance of the lens in this setup has not been examined but a beam of less than 100 μm can be easily prepared at a beam current of 10 nA which is intense enough for the HRBS measurement. Specimens are mounted on a precision goniometer in a scattering chamber. Ions scattered from the specimen are energy analyzed by a magnetic spectrometer and detected by a microchannel-plate position-sensitive-detector (MCP-PSD) placed on the focal plane of the spectrometer. This allows to measure the energy spectrum without sweeping the magnetic field.

The magnetic spectrometer is basically a 90° sector type with 26.6° inclined boundaries for double focussing. The bending radius is 150 mm and the maximum bending power (mE/q^2) is 1.6 MeV. Example of the calculated ion trajectories in the spectrometer are displayed in Fig. 2. The exit boundary of the spectrometer is modified from a straight line so that the focal plane is perpendicular to the central ion trajectory, while the focal plane of the standard double focussing spectrometer inclines about 60° (dashed line in Fig. 2). This improves considerably the detection efficiency of the MCP-PSD because the efficiency drops very rapidly with increasing incidence angle.

Figure 3 depicted the calculated ion position X on the focal plane as a function of the ion energy E. The abscissa shows the energy ratio $\varepsilon = E/E_0$, where E_0 denotes the energy of the central trajectory ion. Using with a MCP-PSD of 80 mm long, the energy range of $0.885E_0$ to

CP475, *Applications of Accelerators in Research and Industry*, edited by J. L. Duggan and I. L. Morgan

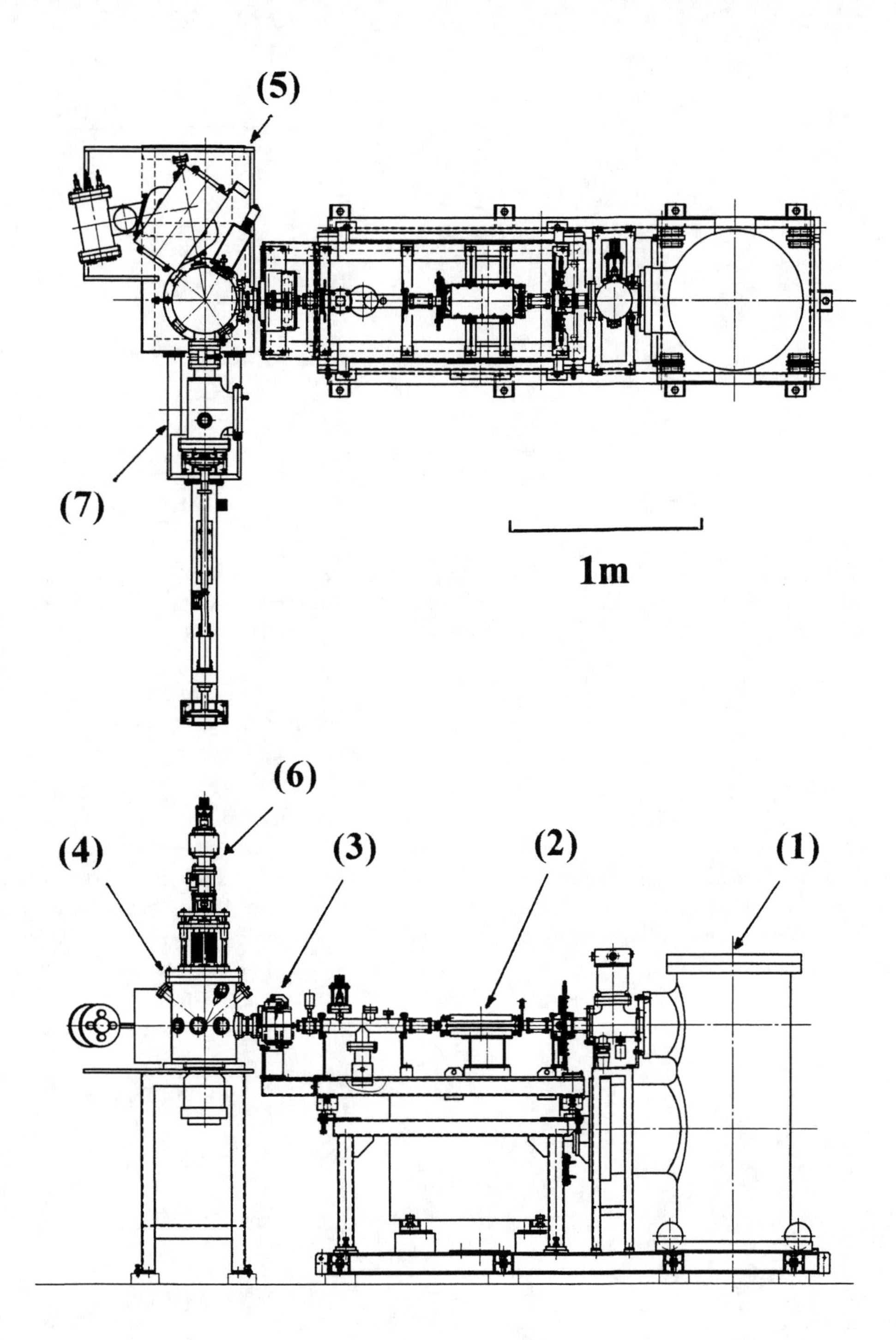

FIGURE 1. Schematic drawing of the HRBS system: (1) isolation tank, a PIG ion source and an acceleration tube are inside, (2) Wien filter, (3) magnetic quadrupole lens, (4) scattering chamber, (5) magnetic spectrometer, (6) high precision goniometer, (7) load lock system for sample exchange.

$1.155E_0$ can be measured. The energy resolution of the spectrometer is designed to be less than 0.1 % in this energy region at an acceptance angle of 0.4 msr if the resolution of the MCP-PSD is better than 0.3 mm. Although the vertical focus is not perfect because of the deviation of the exit boundary from the 26.6° straight line, the vertical spread at the focal plane is less than 2 mm at the acceptance angle of 0.4 msr for the energy range of $0.885E_0$ to $1.155E_0$.

RESULTS AND DISCUSSION

Figure 4 shows an example of the observed position spectrum $F(X)$ for a Au film evaporated on a Si wafer. Using the curve $X(\varepsilon)$ shown in Fig. 3, the energy spectrum

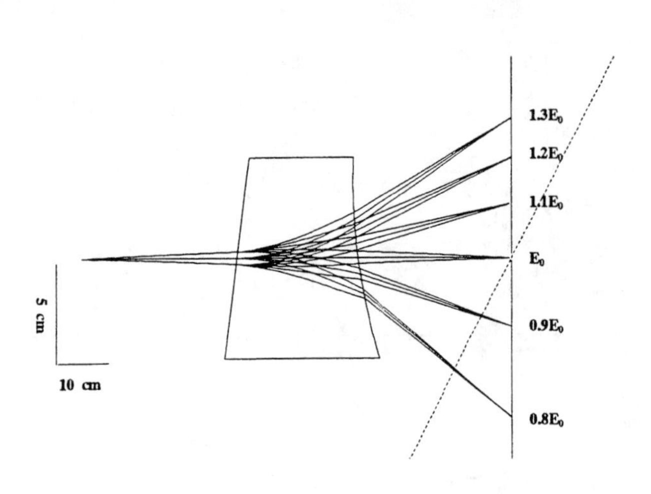

FIGURE 2. Calculated ion trajectories in the spectrometer shown in a horizontal plane. The horizontal axes denotes the distance along the central trajectory.

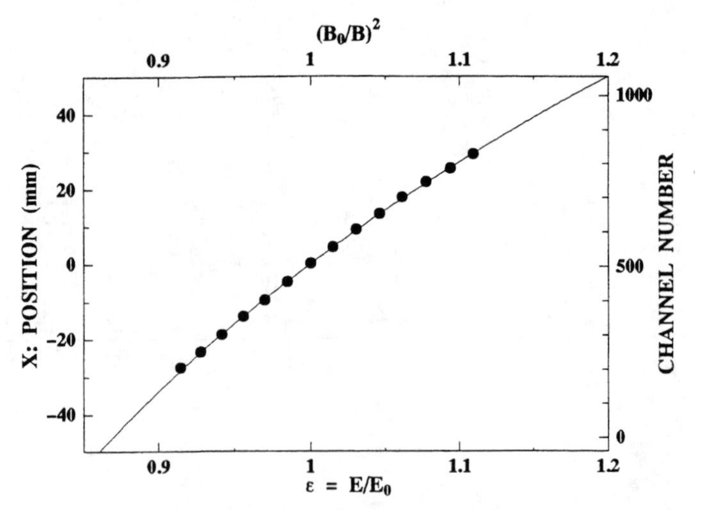

FIGURE 3. Position of the ion in the focal plane as a function of ion energy. The curve shows the calculated result and the circles show the experimental results.

f(E) can be derived from the observed position spectrum *F(X)* as

$$f(E) = F(X(E/E_0))\frac{1}{E_0}\left(\frac{dX}{d\epsilon}\right)_{\epsilon = E/E_0} . \qquad (1)$$

The function $X(\varepsilon)$ can be derived from the measurements of spectra at various ion energies with a fixed magnetic field. Alternatively, it can be obtained from measurements at various magnetic fields B of the spectrometer with a fixed ion energy because the ion trajectory is identical if B^2/E is the same. The latter method has an advantage over the former one, because the precise measurement of B is rather easy as compared with the measurement of ion energy.

The channel number of the Au leading edge is shown as a function of $(B_0/B)^2$ in Fig. 3. The agreement between the calculated $X(\varepsilon)$ and the observed $X([B_0/B]^2)$ is reasonably good, indicating that the spectrometer works as it was designed. Figure 5 shows the energy spectrum obtained from the result of Fig. 4. The energy resolution determined from the Au leading edge is about 1.5 keV. Although the obtained resolution is one order of magnitude better than SSBD, it is worse by a factor five as compared to the calculated resolution of the spectrometer. The most

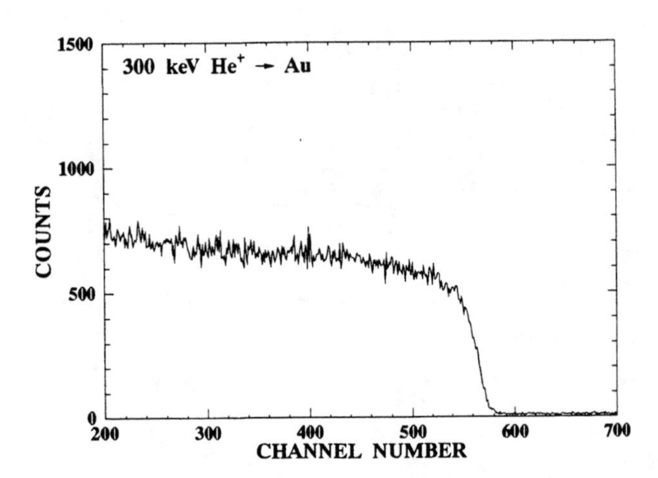

FIGURE 4. Raw spectrum of HRBS for Au/Si. The spectrum is measured using a 300 keV He^+ beam at incidence angle 30° and exit angle 70° (scattering angle 80°).

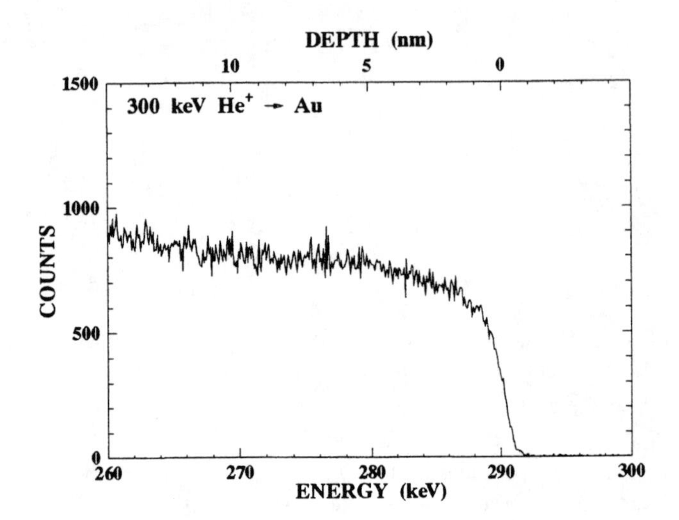

FIGURE 5. Energy spectrum obtained from the raw data shown in Fig. 4. The observed energy resolution is about 1.5 keV, which corresponds to depth resolution of ~ 0.8 nm.

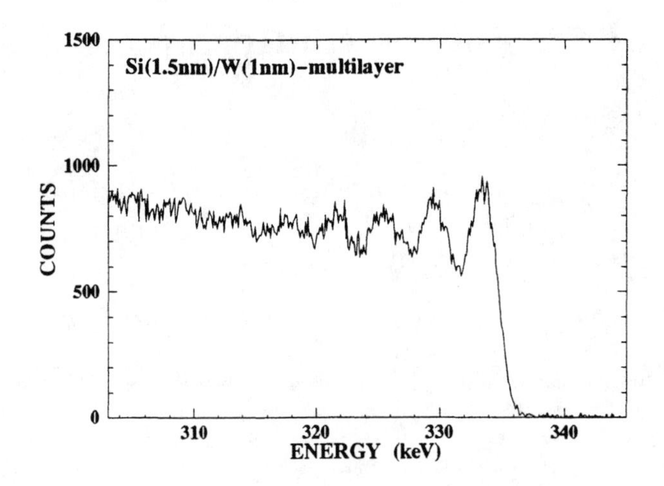

FIGURE 6. HRBS spectrum of Si/W-multilayer. The thickness of each Si (W) layer is 1.5 (1.0) nm. The spectrum is measured using a 350 keV He^+ beam. The W layers can be resolved up to the 8th layer.

probable cause of this discrepancy is the surface contamination. As the pressure of the scattering chamber was ~ 1 × 10^{-8} torr and no surface cleaning procedure was employed beforehand, the surface might be contaminated. Actually, a small energy shift of the leading edge toward lower energy was observed during measurement, indicating that contamination (presumably hydrocarbon) built up on the surface during measurement.

The depth resolution translated from the observed energy resolution is about 0.8 nm in the present case. This can be easily improved up to monolayer resolution [0.24 nm for Au(111)] by using a grazing angle technique, which suppresses the effect of energy loss straggling and is essential for achieving monolayer resolution (2). At an exit angle θ_e of 85°, for example, the energy difference of the 300-keV He^+ ions scattered from adjacent atomic layers is about 2 keV for Au(111). Taking account of the effect of the energy loss straggling, up to the third atomic layer is expected to be resolved as a separated peak (10).

Figure 6 displays another example of the observed HRBS spectrum. A HRBS spectrum of Si/W-multilayer was measured using a 350 keV He^+ beam at a scattering angle 80° and an incidence (exit) angle 38° (62°). The thicknesses of the Si and W layers are 1.5 and 1 nm respectively. The W layers can be resolved up to the 8th layer, indicating that a depth resolution of ~ 2.5 nm can be obtained at a depth ~ 20 nm.

CONCLUSIONS

We have developed a compact HRBS system of dimensions 3.8 m (L) × 2.8 m (W) × 2.35 m (H) including a 500 kV accelerator and a high-resolution magnetic spectrometer. The observed energy resolution is about 1.5 keV for 300 keV He^+ at an acceptance angle of 0.4 msr, which corresponds to a depth resolution of 0.8 nm at the surface of Au. Observation of Si/W-multilayer indicates that a depth resolution of ~ 2.5 nm can be obtained at a depth ~ 20 nm. Using a grazing angle technique, the present system has a capability of monolayer resolution in a surface region.

ACKNOWLEDGEMENTS

We are grateful to Dr. H. Takenaka for providing the Si/W-multilayer. This work was supported in part by a Grant-in-Aid for Scientific Research from the Ministry of Education, Science, Sports and Culture.

REFERENCES

1. See, for example, L.C. Feldman and J.W. Mayer, *Fundamentals of Surface and Thin Film Analysis* (North-Holland, Amsterdam, 1986).
2. E. Bøgh, Phys. Rev. Lett. **19**, 61 (1967).
3. J.K. Hirvonen and G.K. Hulbler, in *Ion Beam Surface Layer Analysis*, eds. O. Meyer, G. Linker and F. Käppeler, p. 457-469, Plenum, New York (1976).
4. A. Feuerstein, H. Grahmann, S. Kalbitzer and H. Oetzmann, in *Ion Beam Surface Layer Analysis*, eds. O. Meyer, G. Linker and F. Käppeler, p. 471-481, Plenum, New York (1976).
5. Th. Enders, M. Rilli, and H.D. Carstanjen, Nucl. Instr. and Meth. **B64**, 817 (1992).
6. W.M. Arnoldbik, W. Wolfswinkel, D.K. Inia, V.G.G. Verleun, S. Lobner, J.A. Reinders, F. Labohm, and D.O. Boerma, Nucl. Instr. and Meth. **B118**, 567 (1996).
7. K. Kimura, K. Ohshima, and M. Mannami, Appl. Phys. Lett. **64**, 2232 (1994).
8. W.A. Lanford, B. Anderberg, H. Enge, and B. Hjorvarsson, Nucl. Instr. and Meth. **B136-138**, 1177 (1998).
9. K. Kimura, K. Nakajima, and M. Mannami, Nucl. Instr. and Meth. **B136-138**, 1196 (1998).
10. K. Kimura, K. Ohshima, K. Nakajima, Y. Fujii, M. Mannami, and H.-J. Gossmann, Nucl. Instr. and Meth. **B99**, 472 (1995).

RBS/Simulated Annealing and FTIR characterisation of BCN films deposited by dual cathode magnetron sputtering

N.P. Barradas and C. Jeynes

School of Electronic Engineering, Information Technology and Mathematics, University of Surrey, Guildford GU2 5XH, UK.

Y. Kusano, J. E. Evetts, and I. M. Hutchings

University of Cambridge, Department of Materials Science and Metallurgy, Pembroke Street, Cambridge, CB2 3QZ, UK.

A dual cathode magnetron sputtering system was used for synthesising carbon-nitride (CN) and boron-carbon-nitride (BCN) films at a nitrogen gas pressure of 2.0 Pa onto Si and NaCl substrates. The stoichiometry and thickness of the films were measured with Rutherford backscattering. The results were analysed with the Simulated Annealing algorithm, which allowed the determination of the B content in samples with as little as 15 at.% B, with a detection limit of 7 at.%. The deposition rate of the CN films increased linearly with graphite target dc power. The deposition rate of the BCN films grown with 100 W rf power to the boron target and between 0 and 100 W dc power to the graphite target was higher than that of the CN films deposited with the same dc power. The absorption bands corresponding to hexagonal-BN (800 and 1350 cm^{-1}), cubic-BN (1050 cm^{-1}), graphite-like carbon bonded with nitrogen (1500 cm^{-1}), nitrile and isocyanate groups (2200 cm^{-1}), and -NH or -OH (3300 cm^{-1}) were detected in Fourier transform infrared (FTIR) spectra. These results show that the B is effectively introduced into the structure of the BCN films.

INTRODUCTION

Intense research activities have been reported on synthesis of novel hard coatings, especially new nitrides and carbides, since they are expected to have some considerable advantages over diamond and diamond-like carbon films, which have high solubility in iron, high etching rate by oxygen at high temperature, and low thermal expansion coefficients [1]. On the other hand, cubic boron-nitride (c-BN) films are thought to be ideal for cutting tools, since their oxidation rate and solubility in iron are substantially low. However, synthesis of c-BN films is difficult and their high internal stress results in poor adhesion to most substrates [2]. Study of carbon-nitride (CN) films has been motivated by the prediction that β-C_3N_4 would exhibit extreme hardness, high elastic modulus and high thermal conductivity, comparable to those of diamond [3]. Numerous attempts to synthesise CN materials with different methods have been reported in the literature[4-13]. Most of them reported amorphous CN phases, except for a few cases with nano-crystallites embedded in amorphous CN [14,15].

The expectation of synthesising a cubic boron-carbon-nitride (c-BCN) compound, which has a similar structure to diamond and c-BN, stimulates the interest in the study of BCN materials [16]. A variety of experimental efforts have been reported to synthesise c-BCN, using chemical vapour deposition [17-19], arc discharge [20], and solid phase pyrolysis [21]. The magnetron sputtering deposition technique has the advantages of coating uniformly on substrates of complex geometry, and of producing electrically insulating materials at low temperature with high deposition rate [22]. However, it has been seldom used to synthesise the BCN compounds and the growth of the films by this method is poorly characterised.

In the present work, we used dual cathode planar balanced magnetron sputtering to synthesise BCN films. The chemical structure was characterised by Fourier transform infrared (FTIR) spectroscopy. The stoichiometry and thickness were measured with Rutherford backscattering (RBS).

EXPERIMENTAL DETAILS

The CN and BCN films were deposited on to sodium chloride (NaCl) crystals for FTIR analysis, and silicon substrates for RBS and thickness measurement with a profilometer. Figure 1 shows the dual cathode planar balanced magnetron sputtering system used for

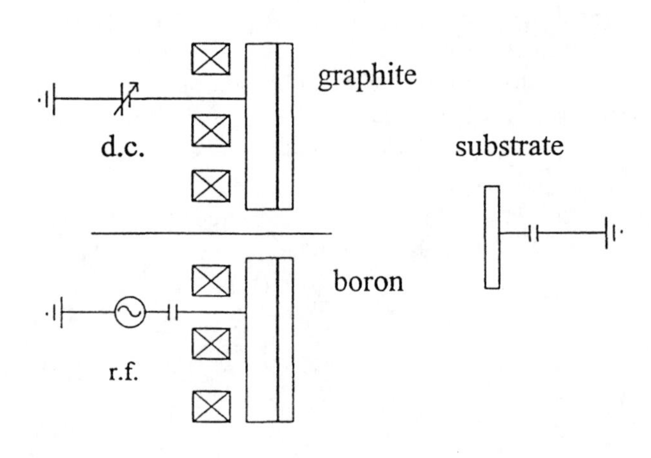

FIGURE 1. Schematic of the magnetron sputtering system.

CP475, *Applications of Accelerators in Research and Industry*, edited by J. L. Duggan and I. L. Morgan

synthesising the films. Graphite and boron targets with dimensions 35x55 mm were mounted on the water cooled cathodes at a distance of 55 mm from the substrates to the centre of each target. The target purity was 99.95 % for graphite and 99.9 % for boron. The racetrack area of the target erosion was about 300 mm^2. The targets were electrically shielded from each other by an aluminium sheet. A stainless steel shutter was placed between the targets and the substrates, which were supported on a platinum holder. The temperature of the substrate holder increased from room temperature to 70°C during deposition, as measured with thermocouples located on the holder just behind the substrates. The vacuum chamber was pumped down and baked for ~13 hours to achieve a pressure of 10^{-5} Pa. The partial pressures of residual gases with mass numbers of 18, 28, 32, and 44 were checked with a quadrupole mass spectrometer, and after baking the chamber their partial pressures normally fell below 2×10^{-7}, 2×10^{-7}, 1×10^{-8}, and 2×10^{-7} Pa, respectively. Mass numbers of 18, 32, and 44 were attributed to H_2O, O_2, and CO_2, respectively. That of 28 indicates the existence of N_2 and/or CO. A glow discharge was generated in pure nitrogen gas at a pressure of 2.0 Pa on the target surfaces by applying between 0 and 100 W dc power to the graphite target, and 100 W rf power on the boron target for 20 minutes for pre-sputtering. The shutter was then opened and thin films approximately 200 nm thick were deposited on to the substrates. After the deposition was completed, the substrates were cooled to room temperature in vacuum, and the chamber then opened to air.

The films were characterised by FTIR and Rutherford backscattering (RBS). Transmission FTIR spectra of the films deposited on the NaCl substrates were collected over a range of 600 to 4500 cm^{-1} with a Perkin Elmer 1720-X spectrometer, with a resolution of 4 cm^{-1}. The stoichiometry and areal density of the samples was determined with RBS using the Surrey 2 MV van de Graaff [23] with 1.5 and 2.0 MeV He$^+$ at normal incidence. The backscattered particles were detected at a 160° scattering angle in the same plane as the beam and the normal to the samples (IBM geometry), and the resolution was 16 keV FWHM. Spectra were collected in random as well as channelled directions, in order to reduce the background due to the Si substrate. The data were analysed using the combinatorial optimisation Simulated Annealing algorithm, which we have previously shown to be able to effectively deconvolute the elements present in different layers [24,25]. The thickness of the samples was determined using a profilometer.

RESULTS AND DISCUSSION

The sensitivity of RBS increases with the square of the atomic number, and is hence low for B. This is aggravated because the small B signal is superimposed on the large signal due to the Si substrate. By measuring in a channelling direction, however, the background signal of the crystalline Si substrate is reduced, while it remains

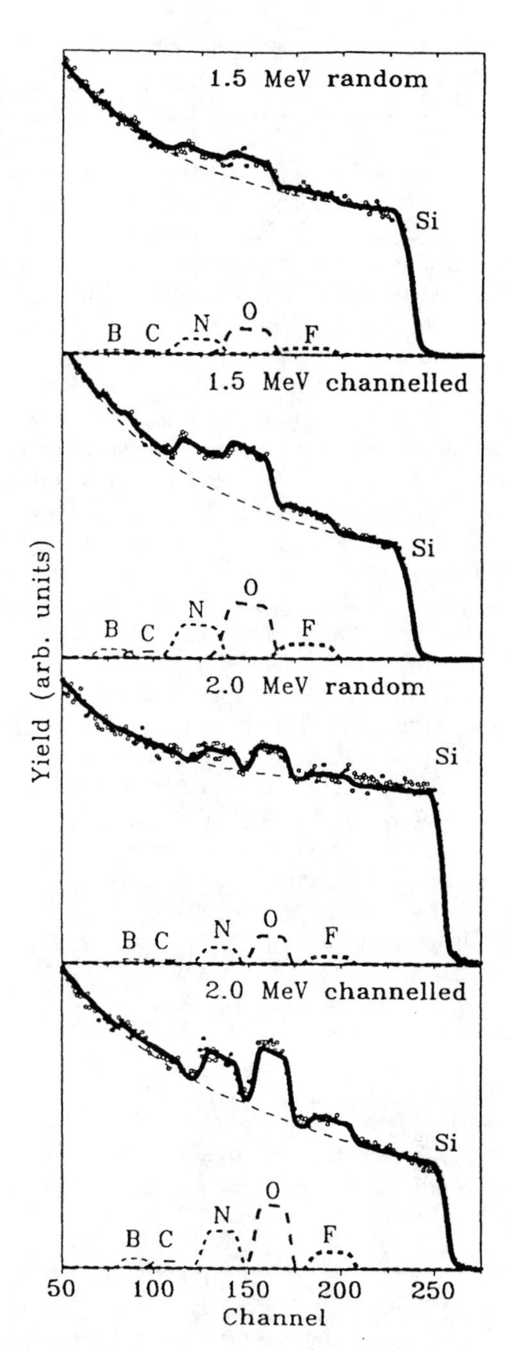

FIGURE 2. RBS data (points) and fit (solid lines) for the sample grown with 0 W on the graphite target.

undisturbed in the BCN film. Further, the thickness of the BCN films leads to a corresponding energy width of the RBS signal depending on the He beam energy. This can lead to superposition of the B and C signals. We therefore measured the samples using two different beam energies. The results are shown in Figure 2 for a sample grown with 100 W rf on the boron target and 0 W on the graphite target. The fits shown were done assuming the same composition for all spectra. This ensures self-consistency and maximizes sensitivity. The B signal, although small, is clearly seen in the channelled spectra. The fit result and statistical error is

TABLE I. C concentration as determined with RBS.

dc C power (W)	0	10	50	100
[C] (at.%)	9.5	33.8	47.5	46.7

$B_{15.6\pm2.1}C_{7.0\pm3.1}N_{30.9\pm0.9}O_{37.7\pm0.6}F_{8.8\pm0.4}$. The detection limit for B is 7 at.% for 99% confidence. A small Fe contamination is also present, probably due to sputtering from the chamber. For higher C content (in at.%) the B content (in at.%) is correspondingly decreased and the film thickness increases leading to superposition of the larger C signal on to the smaller B signal, and the technique is not sensitive to B. The C concentration is shown in Table I.

Figure 3 shows the deposition rates of CN and BCN films as a function of dc power to the graphite target. The rf power to the boron target was 0 W for the CN films and 100 W for the BCN films. As the dc power increased, the deposition rate of CN films increased almost linearly. The gradient of the deposition rate of BCN films was similar to that of CN films, indicating that the difference between these two lines is associated with BN materials in the films, and that the deposition of the BCN films is just the sum of the deposition from each target without significant influence with each other. Therefore, it can be assumed that the BCN films deposited with higher dc power to the graphite target contain a higher amount of CN material, and that the CN content in the BCN films can be controlled independently by the dc power to the graphite target. This is confirmed by the RBS results shown in Table I.

FTIR transmission spectra of the BCN films are shown in Figure 4. The absorption bands detected are: 800 cm^{-1}, corresponding to hexagonal-BN (B-N-B bond bending, out of plane vibration); 1050 cm^{-1}: cubic-BN (reststrahlen band, transverse optical mode); 1350 cm^{-1}: hexagonal-BN (B-N bond stretching, in plane vibration); 1500 cm^{-1} - graphite-like sp^2 hybridised carbon bonded with nitrogen; 2200 cm^{-1}: nitrile (triple bonded CN) stretching and isocyanate groups (N=C=O); and 3300 cm^{-1}: -NH or –OH groups. As the dc power to the graphite target increased, the absorption band at 1500 cm^{-1} associated with graphite-like carbon increased, while those at 1050 cm^{-1} and at 800 and 1350 cm^{-1}, assigned as c-BN and h-BN, respectively, became less well defined. These results indicated that boron was actually introduced into the BCN films and that the BN content decreased as dc power applied to the graphite target increased.

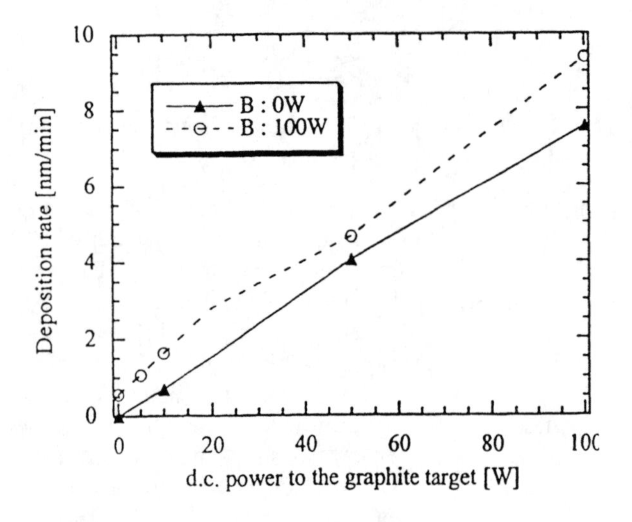

FIGURE 3. Deposition rate of CN and BCN films as function of dc power to the graphite target.

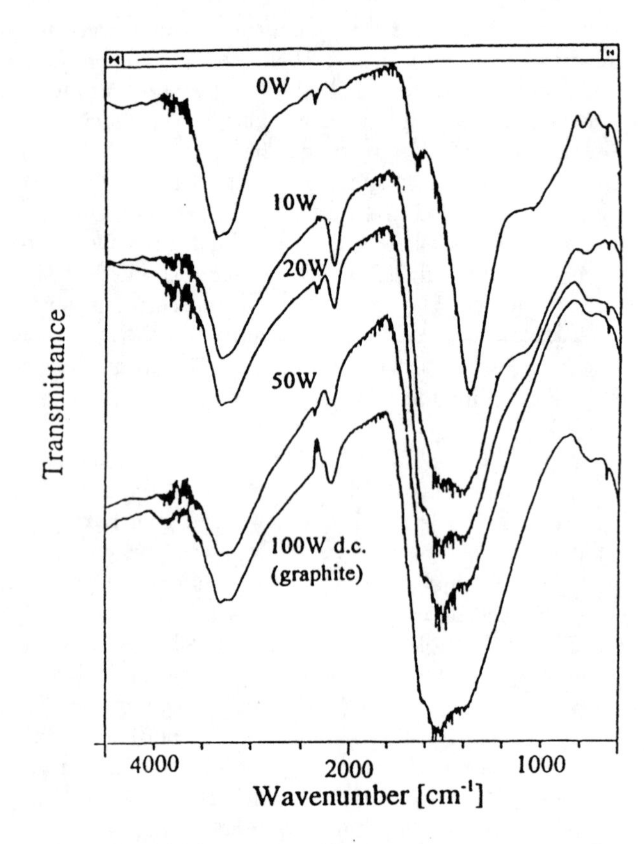

FIGURE 4. FTIR transmission spectra of the BCN films.

CONCLUSIONS

A dual cathode magnetron sputtering system was used to synthesise carbon-nitride (CN) and boron-carbon-nitride (BCN) films on Si and NaCl substrates. The deposition rate of the CN films increased linearly with graphite target dc power. The deposition rate of the BCN films grown with 100 W rf power to the boron target and between 0 and 100 W dc power to the graphite target was higher than that of the CN films deposited with the same dc power. Fourier transform infrared spectra show that the B is effectively introduced into the structure of the BCN films.

ACKNOWLEDGEMENTS

We would like to thank the staff of the D.R. Chick Accelerator Lab. This work was partially funded under EPSRC grant GRL16774.

REFERENCES

1 H. Ehrhardt, Surf. Coat. Technol. **74-75**, 29 (1995).
2 W. D. Sproul, Surf. Coat. Technol. **81**, 1 (1996).
3 A. Y. Liu and M. L. Cohen, Science **245**, 841 (1989).
4 D.C. Nesting and J.V. Badding, Chem. Mater. **8**, 1535 (1996).
5 M. R. Wixom, J. Am. Ceram. Soc. **73**, 1973 (1990).
6 H. Q. Lou, N. Axén, R. E. Somekh, and I. M. Hutchings, Diam. Relat. Mater. **5**, 1303 (1996).
7 N. Laidani, A. Miotello, A. Glisenti, C. Bottani, and J. Perriere, J. Phys. Condens. Matter **9**, 1743 (1997).
8 B. Enders, Y. Horino, N. Tsubouchi, A. Chayahara, A. Kinomura, and K. Fujii, Nucl. Instr. Meth. **B121**, 73 (1997).
9 W. Ensinger, Surf. Coat. Technol. **84** (1996) 363.
10 C. Niu, Y. Z. Lu, and C. M. Lieber, Science **261**, 334 (1993).
11 S. Miyake, S. Watanabe, H. Miyazawa, M. Murakawa, T. Miyamoto, and R. Kaneko, Nucl. Instr. Meth. **B122**, 643 (1997).
12 D. G. McCulloch, and A. R. Merchant, Thin Solid Films **290-291**,99 (1996).
13 F. L. Freire Jr. and D. F. Franceschini, Thin Solid Films **293**, 236 (1997).
14 Z. Zhang, Y. Li, S. Xie, and G. Yang, J. Mater. Sci. Lett. **14**, 1742 (1995).
15 Y.Zhang, Z.Zhou, H.Li, Appl. Phys. Lett. **68**, 634 (1996).
16 T. Lundström and Y.G.Andreev, Mater. Sci. Eng. **A209**, 16 (1996).
17 M. Morita, T. Hanada, H. Tsutsumi, Y. Matsuda and M. Kawaguchi, J. Electrochem. Soc. **139**, 1227 (1992).
18 I. Konyashin, J. Loeffler, J. Bill and F. Aldinger, Thin Solid Films **308**, 101 (1997)
19 K.M. Yu, M.L. Cohen, E.E. Haller, W.L. Hansen, A.Y. Liu and I.C. Wu, Phys. Rev **B49**, 5034 (1994)
20 Z.Weng-Sieh, K.Cherrey, N.G.Chopra, X.Blase, Y.Miyamoto, A.Rubio, M.L.Cohen, S.G.Louie, A.Zettl and R.Gronsky, Phys. Rev. **B51**, 11229 (1995).
21 L. Maya and L.A. Harris, J. Am. Ceram. Soc. **73**, 1912 (1990).
22 M.Z. Karim, D.C. Cameron, M.J. Murphy and M.S.J. Hashmi, Surf. Coat. Tech. **49**, 416 (1991).
23 C. Jeynes, N.P. Barradas, M.J. Blewett, and R.P. Webb, Nucl. Instr. Meth. **B136-138**, 1229 (1998).
24 N.P. Barradas, C. Jeynes, and R. Webb, Appl. Phys. Lett. **71**, 291 (1997).
25 N.P. Barradas, P.K. Marriott, C. Jeynes, and R.P. Webb, Nucl. Instr. Meth. **B136-138**, 1157 (1998).

Rutherford Backscattering and Channeling Studies of Al and Mg Diffusion in Iron Oxide Thin Films

S. Thevuthasan, D.E. McCready, W. Jiang, E.D. McDaniel, S.I. Yi and S.A. Chambers

Environmental Molecular Sciences Laboratory, Pacific Northwest National Laboratory, Richland, WA 99352

Thin films of α-Fe_2O_3(0001) (hematite) and γ-Fe_2O_3 (001) (maghemite) were epitaxially grown on Al_2O_3(0001) and MgO(001) substrates, respectively, using the new molecular beam epitaxy (MBE) system at the Environmental Molecular Sciences Laboratory (EMSL). We have investigated the crystalline quality of these films using Rutherford Backscattering (RBS) and channeling experiments. Minimum yields obtained from aligned and random spectra are 2.7±0.3% for the α-Fe_2O_3(0001) film and 14.5±0.6% for the γ-Fe_2O_3 (001) film. Al and Mg outdiffusion into the hematite and maghemite films were observed at higher temperatures. Indiffusion of Fe atoms from the film into the substrate was observed for the γ-Fe_2O_3(001)/MgO(001) system. In contrast, no Fe indiffusion was observed for the sapphire substrate.

INTRODUCTION

It is important to have high-quality surfaces to understand the surface chemistry on model single crystal oxides. As such, there is a growing interest in the synthesis of epitaxial oxide thin films on various oxide and metal substrates to obtain high-quality surfaces. In particular, the growth of good-crystalline-quality iron oxide thin films is of increasing interest due to their applications in heterogeneous catalysis, magnetic thin films, surface geochemistry, corrosion and integrated microwave devices [1-6]. Recently, several high-quality well oriented single crystal iron oxide films with various stoichiometries have been synthesized using molecular beam epitaxial (MBE) growth and the structural properties have been analyzed by various surface and bulk sensitive techniques[7-12]. Most of these studies are surface related and it has been shown that high-quality well-ordered surfaces can be obtained in these films. On the other hand, the bulk related studies are mostly limited to x-ray diffraction (XRD) studies, and the crystalline quality and impurity concentrations and locations are not known in some cases. We investigated these properties in some of these films using high-energy ion scattering techniques [13,14]. In this paper, we summarize our investigation of the crystalline quality of the epitaxially grown α-Fe_2O_3(0001) (hematite) and γ-Fe_2O_3 (001) (maghemite).thin films and the characterization of the film-substrate interface, using Rutherford backscattering (RBS) and channeling techniques. In addition, we address the issue of interdiffusion as a function of temperature.

EXPERIMENTAL

The films were grown using procedures described elsewhere [15]. After growth the samples were carefully removed from the MBE system and introduced into the channeling end station at the accelerator facility. The details of the accelerator facility and the end stations are described elsewhere [16]. The samples were heated to 200-250°C to desorb hydrocarbons from the surface. For the studies of Al and Mg diffusion into the films, the samples were kept in $1x10^{-6}$ Torr of oxygen during the 20 minute annealing at each temperature. In general the samples were mounted on a molybdenum backing plate using Ta clips and a conventional alumel-chromel thermocouple was attached to the backing plate close to the sample for temperature measurements. These types of temperature measurements reflect the temperature of the back surface of the sample that is in good contact with the backing plate. Temperature ranges of 600°C to 1200°C and 700°C to 900°C were used for Al and Mg diffusion studies respectively. The standard dose of helium ions for one spectrum was $4.4x10^{15}$ ions/cm^2. The backscattering spectrum was collected using a silicon surface barrier detector at a scattering angle of 150°. The primary energy of the ions was 1.04 MeV for the channeling measurements on the α-Fe_2O_3(0001) film and 2.04 MeV for all the other measurements. Also the incident ion beam was directed along the normal to the sample surface.

RESULTS AND DISCUSSION

A. α-Fe_2O_3(0001) Film on Al_2O_3(0001) Substrate

Aligned and random RBS spectra using 1.04 MeV He^+ beam for this film are shown in Fig. 1. A small energy window near the surface region was used to calculate the minimum yield (χ_{min}) for Fe. Another small energy window near the aluminum surface peak region for the substrate was used to calculate the minimum yield for Al. The minimum yield is the ratio of the yield in the channeling geometry to that for a random, non-channeling geometry. For the film χ_{min} is determined to be 2.7±0.3%. In general, the crystalline quality of the film seems to be

CP475, *Applications of Accelerators in Research and Industry*,
edited by J. L. Duggan and I. L. Morgan

very good and the minimum yield is accordingly very low. The minimum yield for the substrate Al is 9.0±0.4%. As reported in Ref. 13, there are five peaks visible in the aligned spectrum (identified by 1, 2, 3, 4, and 5). The first peak at the high energy side is the Fe surface peak at the front of the film. The second peak is attributed to some Fe atoms visible to the ion beam at the interface (back surface of the film). Apparently, the ion beam sees some Al atoms (substrate surface) at the interface as indicated by the third peak. Since the Fe and Al atoms are visible to the ion beam at the interface, there must be some disordering of the iron oxide film at the interface or there must be substrate-film mixing present at the interface. Although mixing of the substrate and the film is possible at the interface, no evidence for mixing has been observed in the random spectrum within experimental resolution and the associated uncertainties. It could be that the lattice mismatch between the substrate and the film is too large [5.4%] to allow a continuous channeling trajectory through the interface between the two crystal structures. A detailed study of the first few layers of α-hematite growth on sapphire has shown that the first monolayer was distorted and considerably expanded with respect to the substrate [17]. Another transmission electron microscopy and selected-area diffraction study [18] reported that there are two distinct types of dislocation arrays at the interface, one of perfect lattice dislocations and the other of partial dislocations of the corundum structure common to the hematite film and substrate. The fourth peak is related to the backscattered ion contribution due to the surface oxygen atoms of the film and the fifth peak is due to the visibility of oxygen to the ion beam at the interface. The visibility of the oxygen atoms to the ion beam again confirms the disordering at the interface. The oxygen atoms in the disordered hematite layers and first few layers in the sapphire substrate contribute to this peak.

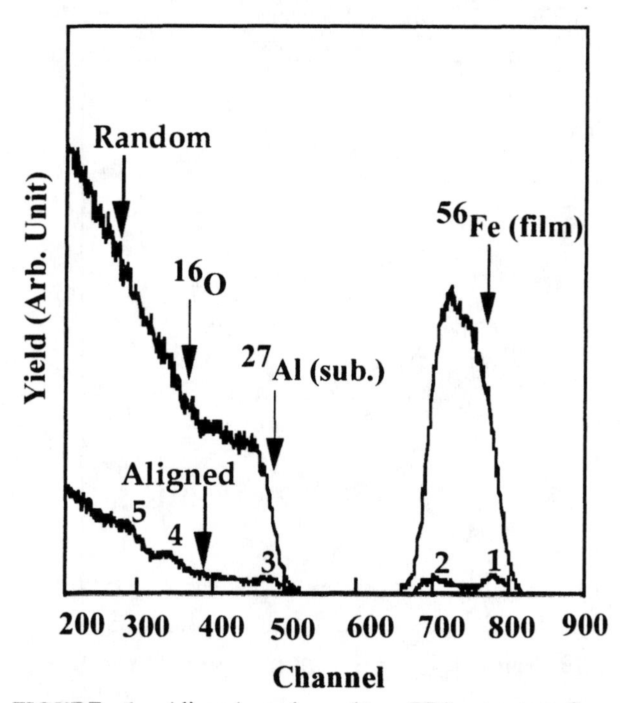

FIGURE 1. Aligned and random RBS spectra for an epitaxially grown α-Fe_2O_3(0001) film on Al_2O_3(0001) substrate. 1.04 MeV incident He^+ ions aligned along the [0001] direction were used.

Another hematite film was grown using higher growth rate [15] to perform the Al diffusion studies. The minimum yield for Fe for this film was about 56% and the crystalline quality of the film was accordingly poor compared to other films. The minimum yield for the Al atoms in the substrate is found to be 46.0±0.5%. In general, this value is significantly higher for a single crystal sapphire substrate and the dechanneling effects in the film due to the poor film crystalline quality can cause slight misalignment of the ion beam in the substrate. As a result, the minimum yield for the substrate Al doesn't reflect the actual crystalline quality of the substrate. The Al diffusion studies were performed at several temperatures in the range from 600°C to 1200°C. The sample was kept in an oxygen environment of 1.0×10^{-6} Torr at each annealing temperature for about 20 minutes. During the annealing time the temperature was controlled within ±5° uncertainty. At the end of each heating the sample was cooled and the channeling and RBS measurements were carried out when the sample was close to room temperature. In addition, since RBS has poor sensitivity to light elements, an Auger spectrum was collected using the cylindrical mirror analyzer (CMA) at the end of each heating cycle. Some indications of an Al Auger signal were observed after the 700°C heating cycle. The Al

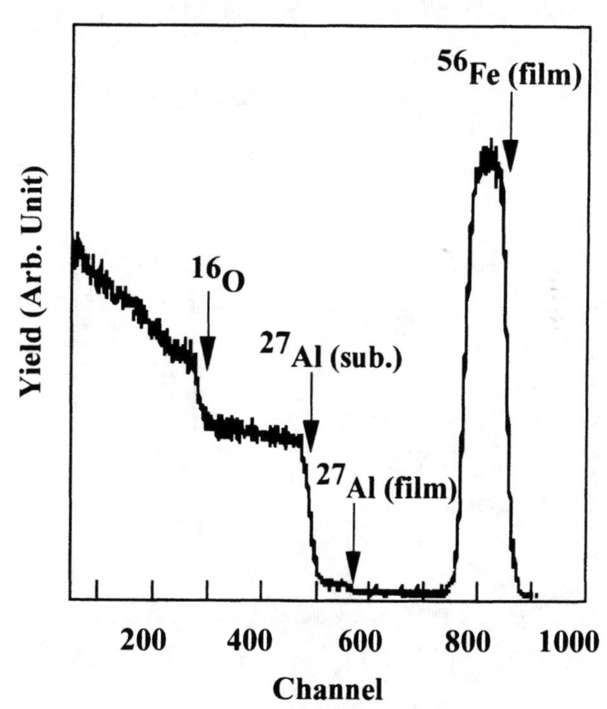

FIGURE 2. Random RBS spectra for an epitaxially grown α-Fe_2O_3(0001) film on Al_2O_3(0001) substrate after 1200°C heating cycle. 2.05 MeV incident He^+ ions aligned along the [0001] direction were used.

signal was very weak and comparable to the noise level in the Auger spectrum. On the other hand, no backscattered helium ions from Al atoms in the film near the surface region were detected above the background in the RBS spectrum. Backscattered He+ ions from Al atoms in the film were detected unambiguously following the anneals at 900°C. After 1200°C heating cycle the Al atoms in the film appear to be increased. In contrast, no Fe indiffusion into the substrate was observed throughout the experiment. The quantity of Al atoms in the films was extracted using RUMP simulations described elsewhere [14]. The experimental random spectrum after the 1200°C heating cycle is shown in Fig. 2.

B. γ-Fe_2O_3 (001) film grown on MgO(001) substrate

Figure 3 shows the aligned and random RBS spectra for the epitaxially grown γ-Fe_2O_3 (001) film on MgO(001) substrate. A small energy window near the surface region was used to calculate the minimum yield for Fe inside the film. Another small energy window near the magnesium surface peak region for the substrate was used to calculate the minimum yield for Mg inside the substrate. For the film χ_{min} is calculated to be 14.5±0.6%. In general, the channeling effect inside the film is poor compared to the α-Fe_2O_3(0001) film. No significant dechanneling peaks due to Fe or Mg at the interface are present in the aligned

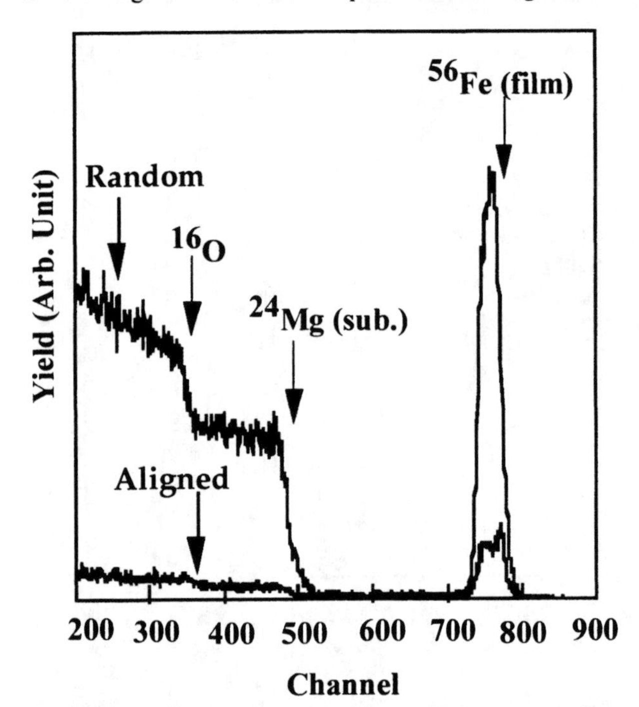

FIGURE 3. Aligned and random RBS spectra for an epitaxially grown γ-Fe_2O_3(001) film on MgO(001) substrate. 2.05 MeV incident He^+ ions aligned along the [001] direction were used.

spectrum. This may be due to the smaller lattice mismatch of 0.89% between the substrate and the film compared to the α-Fe_2O_3(0001)/sapphire system in which the dechanneling peaks due to Fe and Al at the interface are visible because of the larger lattice mismatch of 5.4%. The minimum yield for substrate Mg is 11.6±1.2%.

A more detailed study of Mg and Fe interdiffusion was conducted recently and results are described elsewhere [13]. We show the random spectrum after the 900°C heating cycle in Fig.4. The Mg outdiffusion into the film and Fe

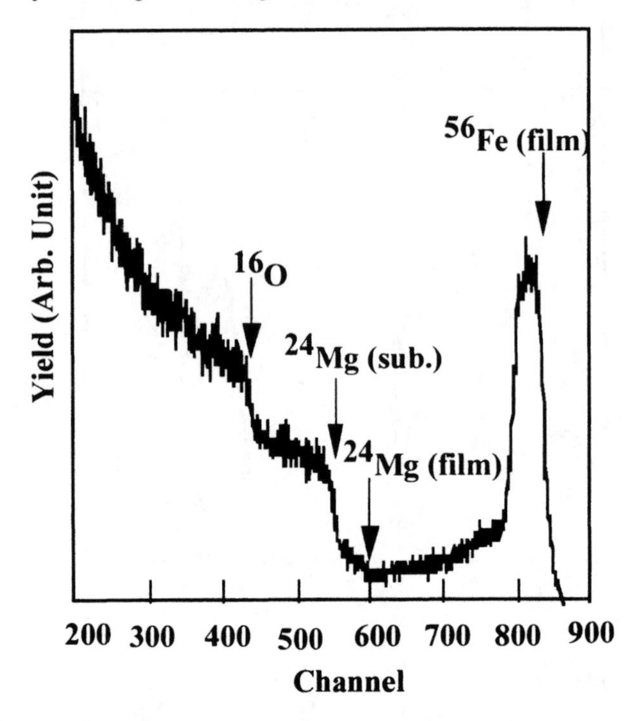

FIGURE 4. Random RBS spectra for an epitaxially grown γ-Fe_2O_3(001) film on MgO(001) substrate after 900°C heating cycle. 2.05 MeV incident He^+ ions aligned along the [001] direction were used.

indiffusion into the substrate are clearly seen in the spectrum. Apparently the Fe indiffusion into the substrate was started around 800°C and the depth of Fe atoms into the MgO substrate was about 560 Å [13]. After the 900°C heating cycle, Fe from the film appears to be diffused much deeper than the 560 Å observed after 800°C heating cycle. Approximately 3-4 atomic percentage of Fe was incorporated into the substrate, more or less uniform with small variations throughout the Fe-diffused region. The quantity of Fe detected in the Fe-diffused region of the substrate was much less than the quantity of Fe required forming the $MgFe_2O_4$ phase within the detection volume. Due to the interference of the diffused Fe and the diffused Mg signals, no attempts were made to extract the atomic concentrations for diffused Mg in the film. The diffused Mg and Fe appears to be mostly substitutional for Fe atoms in the film and Mg atoms in the substrate respectively [13].

CONCLUSIONS

Rutherford backscattering and channeling techniques were used to determine the crystalline quality of epitaxially grown α-Fe_2O_3(0001) film on sapphire and γ-Fe_2O_3 (001) film on MgO(001). The minimum yield obtained from the aligned and random spectra for these films are 2.7±0.3%, and 14.5±0.6%, for α-Fe_2O_3(0001) film, and γ-Fe_2O_3 (001) film respectively. The minimum yields for the alpha phase hematite are quite low and the crystalline quality of the film is correspondingly good. The channeling effect in the γ-Fe_2O_3 (001) film appears to be a little lower compared to the channeling effect in the α-Fe_2O_3 (0001) film. The dechanneling peaks due to Fe, and Al at the interface are visible in the aligned spectrum from α-Fe_2O_3 (0001)/sapphire system due to the larger lattice mismatch between the film and the substrate. On the other hand, the dechanneling peaks due to Fe or Mg are not visible in the aligned spectrum from γ-Fe_2O_3(001)/MgO(001) system due to the smaller lattice mismatch between the film and the substrate. Al and Mg outdiffusion into the hematite and maghemite films were observed at higher temperatures. The indiffusion of Fe atoms from the film into the substrate was observed for γ-Fe_2O_3(001)/MgO(001) system. On the other hand, no Fe indiffusion was observed into the sapphire substrate.

ACKNOWLEDGMENTS

Pacific Northwest National Laboratory is a multi-program national laboratory operated for the U.S. Department of Energy by Battelle Memorial Institute under contract No. DE-AC06-76RLO 1830. The authors gratefully acknowledge partial support from the US Department of Energy, Biological and Environmental Research - Environmental Management Science Program. W. Jiang was partially supported by the United States Department of Energy, Office of Basic Energy Sciences, Materials Science Division.

REFERENCES

1. Geus, J.W., Appl. Catl. 25 (1986) 313.
2. Kung, H.H., *Transmission Metal Oxides: Surface Chemistry and Catalysis* (Elsevier, New York, 1989).
3. Fujii, T., Takano, M., Katano, R., and Bando, Y., J. Appl. Phys. 66 (1989) 3168.
4. Waite, T.D., Rev. Mineral, 23 (1990) 559.
5. Wild, R.K., in *Surface Analysis: Techniques and Applications, Special Publication 84*, edited by D.R. Randell and W. Neagle (Royal Society of Chemistry, London, 1990).
6. Lind, D.M., Berry, S.D., Chern, G., Mathias, H., and Testardi, L.R., Phys. Rev. B 45 (1992) 1838.
7. Anderson, J.F., Kuhn, M., Diebold, U., Shaw, K., Stoyanov, P., and Lind, D., Phys. Rev. B 56 (1997) 1134.
8. Gaines, J.M., Bolemen, P.J.H., Kohlhepp, J.T., Bulle-Lieuwma, C.W.T., Wolf, R.M., Reinders, A., Jungblut, R.M., van der Heijden, P.A.A., van Eemeren, J.T.W.M., aan de Stegge, J., de Jonge, W.J.M., Surf. Sci. 373 (1997) 85.
9. Kim, Y.J., Gao, Y., and Chambers, S.A., Surf. Sci. 371 (1997) 358.
10. Gao, Y., Kim, Y.J., Thevuthasan, S., Lubitz, P., and Chambers, S.A., J. Appl. Phys. 81 (1997) 3253.
11. Gao, Y., Kim, Y.J., Chambers, S.A., and Bai, G., J. Vac. Sci. Technol. A 15 (1997) 332.
12. Gao, Y., and Chambers, S.A., J. Cryst. Growth 174 (1997) 446
13. Thevuthasan, S., Jiang, W., McCready, D.E., and Chambers, S.A., Surf. and Interface Analysis, in press.
14. Thevuthasan, S., McCready, D.E., Jiang, W., Kim, Y.J., Gao, Y., Chambers, S.A., Shivaparan, N.R., and Smith, R.J., submitted to Thin Solid Films.
15. Yi, Y.I., and Chambers, S.A., in preparation.
16. Thevuthasan, S., Peden, C.H.F., Engelhard, M.H., Baer, D.R., Herman, G.S., Jiang, W., Liang, Y., and Weber, W.J., Nucl. Instrum. And Methods in Phys. Res. A, in press.
17. Voogt, F.C., *Ph.D. Dissertation.*
18. Anderson, I.M., Tietz, L.A., and Carter, C.B., Mat. Res. Soc. Symp. Proc. 238 (1992) 807.

Time of Flight Elastic Recoil Detection for Thin Film Analysis

M.S. Rabalais and D.L. Peterson, Jr.

Eric Jonsson School of Engineering, University of Texas at Dallas, Dallas, TX 75080

Y.Q. Wang

Center for Interfacial Engineering, University of Minnesota, Minneapolis, MN 55455

W.J. Sheu and G.A. Glass

Acadiana Research Laboratory, University of Southwestern Louisiana, Lafayette, LA 70504

Time-of-flight elastic recoil detection (TOF-ERD) is a powerful and complimentary technique to Rutherford Backscattering Spectrometry (RBS) for elemental analysis in surfaces and thin films. Its main advantages lie in its capability of not only simultaneously depth profiling light elements (3<Z<9) but also with a superb depth resolution (a few nm). This paper describes the construction and calibration of a TOF-ERD system recently added to the NEC 5SDH-2 1.7 MV Tandem Pelletron[R] Accelerator at the University of Southwestern Louisiana. Initial results on varying-thickness carbon thin foils using MeV gold ion beams yielded a depth resolution of approximately 3.8 nm. TOF-ERD computer software written on site to simulate spectra and to convert time spectra into depth profiles is also presented.

INTRODUCTION

Conventional elastic recoil detection (ERD), first introduced in 1976[1], uses an incident heavy ion beam to kinematically recoil and depth profile low atomic number target atoms (1< Z <9). ERD uses an absorber foil placed in front of a surface barrier detector that separates the scattered particles from the surface recoiled particles. Before ERD, Nuclear Reaction Analysis (NRA) was the primary technique for light element profiling, but the ease and simplicity of ERD has led to its extensive use. However, the range foil used in conventional ERD degrades the energy resolution and hence the depth resolution of the technique. The poor energy resolution of surface barrier detectors restricts this technique to high energies (1MeV/u). Time of flight ERD separates scattered particles from the recoiled particles by their differing flight times and since TOF-ERD does not use an absorber foil or a surface barrier detector, lower energies can be used (0.3 MeV/u). Although more complicated, TOF-ERD is a superior technique for light element depth profiling.

The TOF-ERD technique has been used by several research groups for light element analysis.[2-5] Both microchannel plate (MCP) detectors and half-turn cyclotron electron detectors have been used. While the time resolution of better than 100 ps has been reported for half-turn cyclotron electron detectors,[2] the best time resolution for MCP detectors is reported to be about 200 ps.[3] The values of 500 ps for MCP detectors are more typical. This paper describes the design and calibration of the TOF-ERD system recently implemented at The University of Southwestern Louisiana. A software package that was written on-site to convert a measured time spectrum into a depth profile is also reported. Some initial results on depth profile measurements of several thin carbon foils using 9 MeV gold ion beam are discussed.

FUNDAMENTALS

The fundamental concepts of ERD are derived from classical scattering theory and the technique relies on knowing the energy transferred from the projectile (E_I) to the target (E_R), $E_R = k(\phi)E_I$, for which the kinematic factor is defined as,

$$k(\phi)=\frac{4m_I m_R \cos^2\phi}{(m_I+m_R)^2} \qquad (1)$$

with m_I and m_R representing the masses of the incident and recoiled particles respectively, and ϕ representing the recoil angle in reference to the incident beam. Additionally, a differential scattering cross section giving the probability for the recoil event to occur is usually governed by Coulombic forces that can be modeled as Rutherford, and is expressed in the laboratory reference frame as

$$\sigma_r(E_0,\phi)=\frac{\left[Z_1 Z_2 e^2(m_I+m_R)\right]^2}{\left[2m_R E_0\right]^2\cos^3\phi} \qquad (2)$$

with Z_1 and Z_2 representing the atomic numbers of the incident and target particles, respectively.

In principle, an actual ERD energy spectrum, $N(E_c)$, measured in the experiment is a convolution of an ideal energy spectrum, $N'(E_{c'})$, and the total energy resolution function of the experimental setup, $G(E_{c'} - E_c)$,[6]

$$N(E_c) = \int G(E_{c'} - E_c)\, N'(E_{c'})\, dE_{c'}. \qquad (3)$$

The ideal energy spectrum, $N'(E_{c'})$, is the one to be expected if no energy spread exists in the experiment, and

CP475, *Applications of Accelerators in Research and Industry*,
edited by J. L. Duggan and I. L. Morgan

can be calculated from the concentration vs. depth profile, C(x), based on the following equation,

$$N'(E_{c'}) = Q\varepsilon\Omega\sigma(E)C(x)dx/dE_{c'} \sin\alpha \qquad (4)$$

where $N(E_c)$ is the counts per unit energy at an actual channel energy E_c, $N'(E_{c'})$ is the counts per unit energy at an ideal channel energy $E_{c'}$, $\sigma(E)$ is the recoil crossection at energy E which is directly related to $E_{c'}$ and therefore E_c, as well as the depth X. C(x) is the concentration at a depth of x (depth profile), Q is the collected beam charge for the spectrum, ε is the intrinsic efficiency of TOF spectrometer, and Ω is the solid angle subtended by the spectrometer. The total energy resolution function, $G(E_{c'} - E_c)$, is usually expressed as a Gaussian distribution,

$$G(E_{c'} - E_c) = 1/(2\pi\delta_t^2)^{1/2} Exp[-(E_{c'} - E_c)^2/2\delta_t^2] \qquad (5)$$

where δ_t is the variance of the Gaussian and is related to the full width at half maximum (FWHM) of the total energy spread by δ_t =FWHM/2.356. The energy spread of the system will be discussed in detail in the next section.

For the depth calculation, a slab analysis as shown in Fig. 1 is performed by mathematically partitioning the target into a number of slabs (layers) so that the energy loss of the projectiles and recoiled target atoms are determined as they traverse each slab.

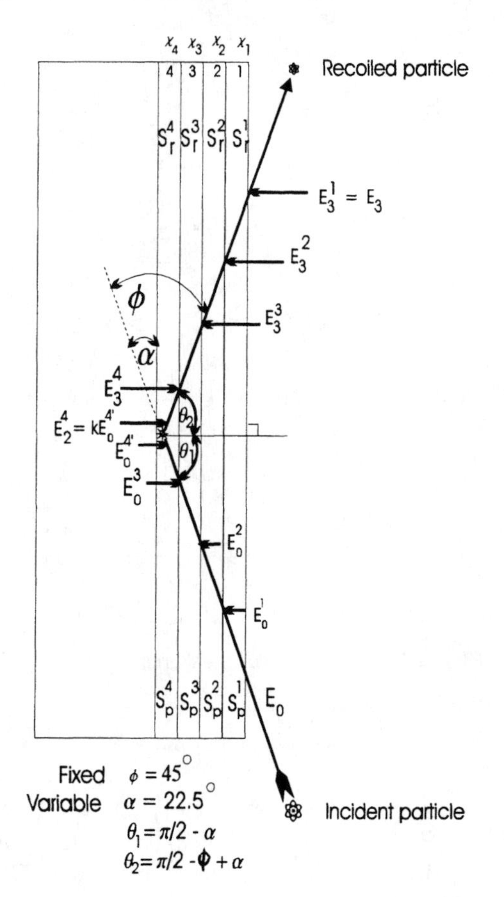

FIGURE 1. Target for slab analysis.

SYSTEM DESCRIPTION

Hardware

Figure 2 shows the TOF-ERD system[7] attached to the NEC 5SDH-2 1.7 MV Tandem Pelletron[R] accelerator at Acadiana Research Laboratory (ARL). Recoiled particles enter into the spectrometer, and, via a 5 mm collimator, pass through a $5\,\mu g/cm^2$ carbon start foil, continue down a drift tube of known length and strike a microchannel plate (MCP) detector. When the recoils pass through the carbon foil, electrons are ejected and detected by a separate MCP detector. These electrons are accelerated so that all or most of the electrons ejected from a single ion are registered as a single pulse in the electron detector. The elapsed time from when the electrons are ejected at the carbon foil and are detected is significantly less than the elapsed time the recoils travel from the foil to the ion detector. The detectors used in our spectrometer are *Comstock* model CP-625C/50F microchannel plate charged particle detectors with a sensitive area of 2.54 cm in diameter.

Figure 3 shows the schematic diagram of the electronic system used with the spectrometer. The signals from the MCPs are first amplified by fast preamplifiers and then fed into fast constant fraction discriminators (CFDs). The outputs of the CFDs are input into a time-to-amplitude converter (TAC) and the converted time spectrum is then registered in a multichannel analyzer (MCA).

Software

In order to obtain the depth profile from the measured spectrum, a program ULTRA[8] was written on-site by one of us (M.S.R.) using FORTRAN-77. ULTRA is a TOF-ERD simulation, data conversion, and analysis program used to process data from the multi-channel analyzer that receives output from the TOF-ERD system. It was intended for use with the system at ARL, however, the ability to modify parameters such as the recoil and target angle via an input file allow ULTRA to be used for other systems.

ULTRA consists of four main parts: (1) inputs of beam and target parameters including crossection data, stopping powers, assumed depth profile, and time-spectrum data; (2) conversion of time-spectrum into energy-spectrum by one of two methods: direct channel to channel conversion via the kinetic energy equation $E=1/2mv^2$, or by the Knapp method [9] that avoids aliasing of data at high count rates; (3) integration subroutines based on Eqs. (1)-(6) and simulated energy-spectrum generation; (4) evaluation of "goodness of fit" between the simulated and measured energy-spectra (Reduced χ^2 test).

ULTRA uses the target slab analysis as shown in Fig.1. The sample was divided into slabs or layers of equal thickness such that each layer was thin enough to respond to the changes of the crossection, the energy loss and the depth profile of the analyzing element. Given a depth profile based on known boundary conditions, the energy spectrum can be calculated. By adjusting the depth profile, a good agreement between the measured and the calculated spectra can be obtained.

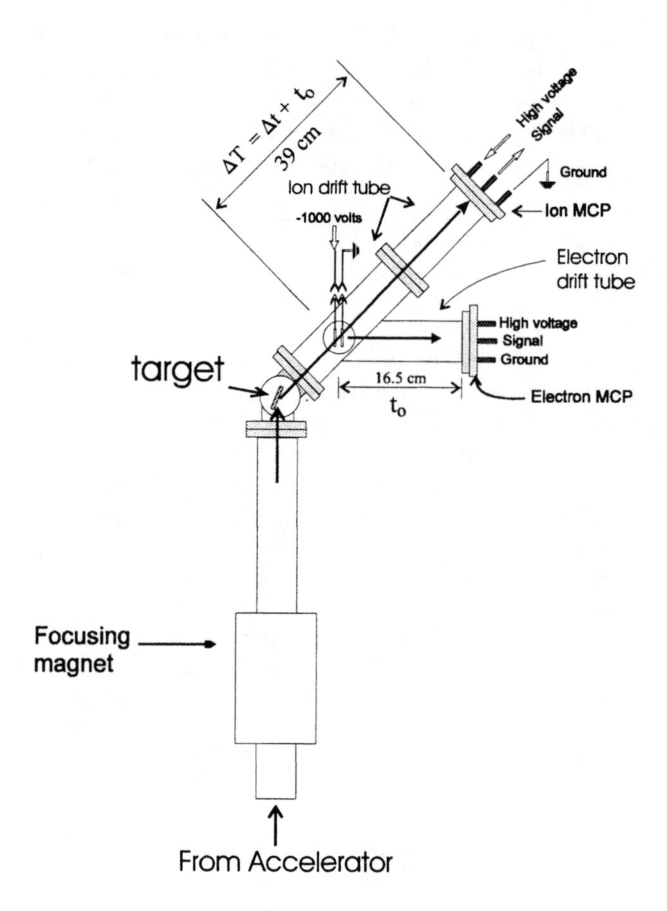

FIGURE 2. TOF-ERD System.

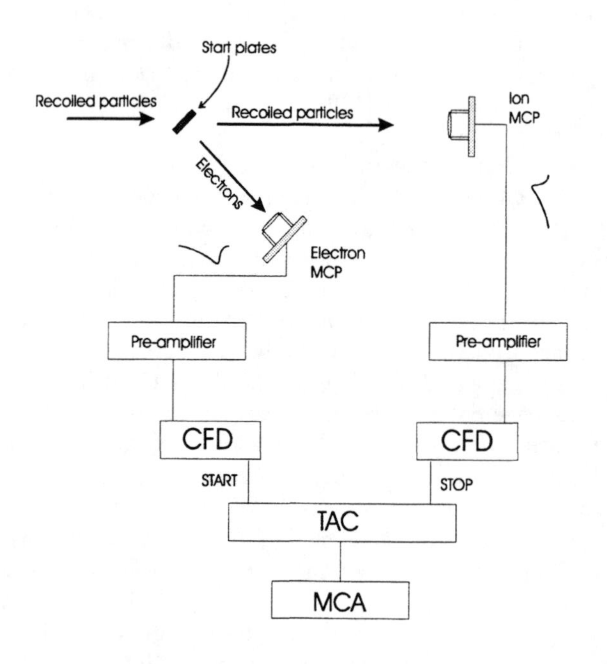

FIGURE 3. Diagram of Electronic System.

System Resolution

Time Resolution

Conventionally, the time resolution of TOF spectrometer is determined by measuring the rise time of the time signal from MCP detectors using a fast oscilloscope. In this work, we used a ^{241}Am alpha source to measure the time resolution of the spectrometer. The TAC-MCA system was first calibrated with a passive nanoseconds delay box and the channel width was determined to be approximately 0.36 ns for a TAC range of 50 ns. The electroplated alpha-source was placed directly in front of the start carbon foil and the collected spectrum is shown in Fig. 4. The spread of the alpha peak is then corresponded to the real time resolution of the TOF spectrometer. The FWHM of the alpha peak indicates a time resolution of approximately 0.49 ns. This time resolution includes not only the intrinsic time resolution of the electronic apparatus (MCPs, preamplifiers, and CFDs) but also the uncertainties introduced by the energy straggling in start foil and the geometrical variation of flight path. The energy resolution (δE_{det}) of the TOF spectrometer resulted from the time resolution (Δt) is determined by the derivative of the kinetic energy equation,

$$\delta E_{\text{det}} = 2\left(\frac{E}{t}\right)\Delta t \tag{6}$$

and therefore changes with depth.

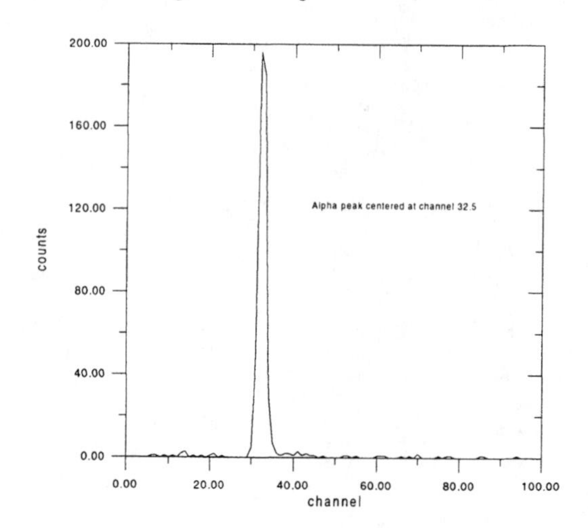

FIGURE 4. ^{241}Am 5.486 MeV alpha spectrum.

Energy Resolution

The accurate determination of the total energy resolution (δE_t) of a TOF-ERD experimental setup is difficult. However, as an approximation, it can be estimated by separately calculating the energy spread due to various contributions and adding them in quadrature to obtain, [6,10,11]

$$\delta E_t^2 = \delta E_{det}^2 + \delta E_g^2 + \delta E_s^2 + \delta E_m^2, \quad (7)$$

where δE_{det} is the energy resolution of the TOF spectrometer, δE_g is the geometric broadening due to the finite detector acceptance angle, δE_s is the energy straggling in the matrix, and δE_m is the contribution due to multiple-scattering in the matrix. The variance in the Gaussian distribution mentioned in equation (5) is defined as $\delta_t = \delta E_t / 2.356$.

The geometric broadening, δE_g, due to the finite detector acceptance angle is represented by [10,11]

$$\delta E_g = \left[2kE_0 \tan\phi - \left[\frac{2kS_1 \tan\phi}{\sin\alpha} + \frac{S_2 \cot(\phi-\alpha)}{\sin(\phi-\alpha)} \right] \delta L \right] \Delta\gamma \quad (8)$$

where k is the kinematic factor and E_0 is the energy of the incident projectile at the surface of the target. S_1 and S_2 represent the incident and recoiled particle's stopping power in the slab while δL is the thickness of the slab as measured normal to the surface (shown in Fig. 1). The effective detector acceptance angle, $\Delta\gamma$, is represented by

$$\Delta\gamma = \frac{1}{D} \left[\frac{s^2 + d^2 \sin^2(\phi-\alpha)}{\sin^2\alpha} \right]^{1/2}, \quad (9)$$

where D is the distance between the target and ion detector, s is the effective width of the detector aperture determined from the solid angle subtended by the start foil aperture and d is the diameter of the incident beam.

The first of the two terms in equation (9) refers to energy spread caused by different recoil angles while the second term quantifies the effect of path-length differences of atoms that, although originating at the same location (and same energy), strike the detector at slightly different locations. The instrumentation registers the two particles as having different flight times thereby giving rise to an uncertainty that is proportional to the path length variation.

The calculation for multiple scattering contribution (angular dispersion and lateral spread) is very complex. Fortunately, for TOF-ERD thin film target analysis, multiple scattering is relatively insignificant in comparison with the geometric broadening.[6,11]

For surface or ultra-thin film analysis, the energy spread caused by energy straggling and multiple scattering can be neglected, so the energy resolution of the TOF-ERD system can be estimated by,

$$\delta E_t = 2kE_0[(\Delta t/t)^2 + \tan^2\phi\, \Delta\gamma^2]^{1/2} \quad (10)$$

where t is the flight time of recoils in the spectrometer and Δt is the time resolution of the spectrometer.

Depth Resolution

Depth resolution is the degree to which the signal coming from recoiled atoms at different depths in a sample can be separated along the energy axis. Utilizing the smallest resolvable detected energy, δE_t, calculated above, an expression for the depth resolution in each slab can be written as

$$\delta x = \delta E_t / \{kS_1/\sin\alpha + S_2/\sin(\phi-\alpha)\}, \quad (11)$$

showing how the *detected* energy of recoils originating in a specific layer changes with a change in depth within *that* layer. By applying this result, a quantitative depth resolution can be found for each of *n* layers in the target.

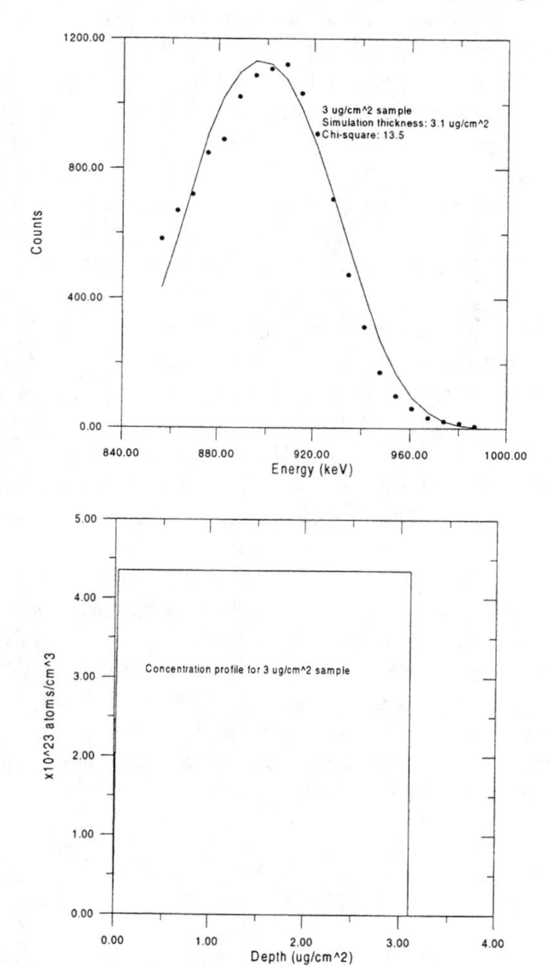

FIGURE 5. (Top) Experimental (dots) and Fitted (line) energy spectra from a 3 μg/cm² carbon foil using 9 MeV Au ions. (Bottom) The carbon depth profile used to obtain the fit above.

APPLICATION

The TOF-ERD system has been used to measure the thickness of several thin carbon foils. A 9 MeV Au^{6+} beam was used to recoil carbon atoms from the films at a tilt angle of $\alpha = \frac{1}{2}\phi = 22.5°$. The particle beam current was kept small (< 1 nA) and the total particle charge was limited to less than 1 μC on each target to reduce the beam irradiation damage on the foil. Four commercially prepared thin-film carbon targets (MicroMatter Co.) were used in the experiment: 3±1, 5±1, 7±3, and 10±3 $\mu g/cm^2$ foils corresponding to a thickness range of 133 - 444 Angstroms for a normal carbon density of 2.25 g/cm^3. The film was first lifted from a microslide substrate in water and then floated onto a fine nickel mesh (80% open area) with the exception that no mesh was used to float 10 $\mu g/cm^2$ film.

Ideally, a well prepared standard sample would have been used, however, at the time of the experiment none was immediately available so the 5 $\mu g/cm^2$ sample was used as a standard for normalization of the spectra from other targets. Using program ULTRA, via the Knapp conversion method[8], the depth profile of the 3 $\mu g/cm^2$ carbon foil was obtained as shown in Fig. 5, where the top shows the measured (dots) and simulated (line) energy spectra, and the bottom shows the corresponding depth profile used in the fit. The fit of the front edge of the recoil carbon peak suggests that the depth resolution of the TOF-ERD system is approximately 0.85 $\mu g/cm^2$ or 3.8 nm for a carbon density of 2.25 g/cm^3. This is comparable to the theoretical value of 3.4 nm as calculated based on Eqs. (10) – (11).

Although the thickness obtained by the TOF-ERD for 3 $\mu g/cm^2$ carbon foil agrees well with that suggested by the Manufacturer, the results for the thicker carbon foils are less consistent. Unlike the uniform profile in 3 $\mu g/cm^2$ foil as shown in Fig. 5, the carbon concentration profile in 7 and 10 $\mu g/cm^2$ foils are found to be not uniform, and the concentration decreases with the depth. While the exact causes for these discrepancies are not well known, there are apparent shortcomings in the measurement and analysis that need to be improved. The preparation of the carbon foil target was difficult and the non-uniform nature or surface roughness caused by wrinkles in preparing the target could increase the path lengths of some ions in the target causing the thickening effect. The charge collection for the foil targets shows little consistency between different targets, which may contribute to the abnormally high carbon concentrations in some of the foils.

CONCLUSION

A TOF-ERD system recently implemented at The University of southwestern Louisiana has been described. The essential part of the system is the TOF spectrometer that utilizes two microchannel plate detectors to generate start and stop time signals. The time resolution of the TOF spectrometer is approximately 490 ps. Computer software (program ULTRA) was written on-site to convert the measured time spectrum into a desired depth profile.

The system has been used to measure the carbon depth profiles of several thin carbon foil targets using 9 MeV Au^{6+} ions as projectiles. The front edge of the carbon spectrum suggests that the depth resolution of the system is approximately 3.8 nm in carbon analysis, which agrees well with the theoretical estimate. While some significant discrepancies remain between the ULTRA-obtained and the manufacturer suggested carbon profiles for some of the carbon foil targets, we believe that to within certain kinematic boundaries, to the extent that the concentration of a standard sample is known, and the consistency in the charge collected on the standard and test samples is maintained, an analysis of thin films can reliably be made using program ULTRA to determine a concentration vs. depth profile of the element of interest.

ACKNOWLEDGMENT

This work is supported by U.S. Department of Energy/Louisiana Educational Quality Support Fund in Cooperative Agreement Number DE-FC02-91ER75669.

REFERENCES

[1] J. L'Ecuyer, C. Brassard, C. Cardinal, J. Chabbal, L. Deschemes, J.P. Labne, B. Terrault, J.G. Martel, and R. St-Jacques, J. Appl. Phys. 47, 881 (1976).

[2] Thomas,J.P., Fallavier,M., and Ziani, A., Nucl. Instrum. Methods, B15, 443 (1986).

[3] Whitlow, H.J., in High Energy and Heavy Ion Beams in Materials Analysis, eds. Tesmer, J.R. et al., Materials Research Society, Pittsburgh, PA, 1990, p. 243.

[4] Knapp,J.A., Barbour,J.C.,and Doyle,B.L., J. Vac. Sci. Technol., A10, 2685 (1992).

[5] Arai, E., Funaki, H., Katayama,M. and Shimizu,K., Nucl. Instrum. Methods, B68, 202 (1992) and B64, 296 (1992).

[6] Wang, Y.Q., Liao, C., Yang, S. and Zheng, Z., Nucl. Instr. and Meth. in Physics Research, B47, 427 (1990).

[7] Peterson, D.L., Jr., Design, Construction and Calibration of a Time of Flight Elastic Recoil Detection System, M.S. thesis, Univ. of S.W. Louisiana, 1996.

[8] Rabalais, M.S., Time of Flight Elastic Recoil Detection for Ultra-thin Film Analysis, M.S. thesis, Univ. of S.W. Louisiana, 1998.

[9] Handbook of Modern Ion Beam Analysis eds. Tesmer, J.R., and Nastasi, M., Materials Research Society, 1995.

[10] Turos, A., Meyer, O., Nucl. Instr. and Meth. in Physics Research, B4, 92 (1984).

[11] Stoquert,J.P., Guillaume,G., Hage-Ali,M., Grob,J.J., Ganter, C., and Siffert,P., Nucl. Instr. and Meth. in Physics Research, B44, 184 (1992).

THE EFFECTS OF MULTIPLE AND PLURAL SCATTERING ON HEAVY ION ELASTIC RECOIL DETECTION ANALYSIS

P.N. Johnston, I.F. Bubb and M. El Bouanani[1]

Department of Applied Physics, Royal Melbourne Institute of Technology, GPO Box 2476V, Melbourne 3001, Australia.

D.D. Cohen and N. Dytlewski

Australian Nuclear Science and Technology Organisation, PMB 1, Menai 2234, Australia.

An increasing number of groups use Heavy Ion Elastic Recoil Detection Analysis (HIERDA) to study a wide range of problems in materials science, however there is no accurate and reliable methodology for the analysis of HIERDA spectra. Major impediments are the effects of multiple and plural scattering which are very significant, even for quite thin (~100 nm) layers of very heavy elements. To examine the effects of multiple scattering, a fast FORTRAN version of TRIM has been adapted to simulate the spectrum of backscattered and recoiled ions reaching the detector. The results of the simulations will be compared with experimental measurements on well characterised samples of thin Au layers on Si performed using ToF-E HIERDA at the Lucas Heights Laboratories of the Australian Nuclear Science and Technology Organisation.

INTRODUCTION

Heavy Ion Elastic Recoil Detection Analysis (HIERDA), with its exceptional and unambiguous multi-elemental depth profiling capabilities, is a powerful analysis tool. The main experimental methods for performing HIERDA involve use of a heavy ion tandem accelerator or cyclotron with one of two basic detector types: (i) Time of Flight and Energy (ToF-E) telescopes which are mass resolving and (ii) Energy Loss and Energy (ΔE-E) telescopes which are nuclear charge resolving. The important capability of the HIERDA method is the quantitative elemental depth profiling of thin films and the near-surface region of solids. The accuracy of the elemental depth profiles is dependent on the understanding of the underlying physical processes governing ion-matter interactions and more importantly their adequate modelling.

Traditionally the analysis of ion scattering data from materials has been done using 'slab' analyses, taking into account the macroscopic features of ion transport, i.e. the scattering process and energy loss. Sometimes additional macroscopic features including energy loss straggling and small angle multiple scattering are included. For light ions, this approach works well in most cases. Small angle multiple scattering often provides the major contribution to depth resolution as the sampled depth gets larger. This exemplified by the broader nature of the low energy edge of spectra features from thin films, e.g. (1). However this technique does not include the contributions to the spectrum from large angle plural scattering, where the exiting ion emerges after two or more large angle scatters in the target. This is a rare event for MeV light ions, but for heavy ions on heavy targets, such as HIERDA, this can be very significant because of the very high scattering cross-sections. Small angle multiple scattering can be considered a continuous or macroscopic process, while plural scattering results from essentially discrete sequences of scattering.

An alternative approach is to use Monte-Carlo modelling of the ion transport to determine the spectra of recoiled and scattered ions reaching the detector. This approach allows some of the 'microscopic' features of ion transport to be included, in particular plural scattering. This approach has traditionally not been considered viable because of the poor computational efficiency of the technique (2), however as computing power increases it becomes more attractive. Monte-Carlo simulation of the spectrum requires some efficiency enhancement

---

[1] Current Address: Department of Physics, University of North Texas, Denton, Texas, USA.

CP475, *Applications of Accelerators in Research and Industry*, edited by J. L. Duggan and I. L. Morgan

approximations especially when used to simulation HIERDA spectra from ToF-E detection systems.

In this paper we compare experimental measurements of scattered and recoiled heavy ions for 60 MeV I ions incident upon on well characterised samples of thin Au layers on Si with spectra simulated by Monte-Carlo modelling using a fast version of TRIM (3). Some considerations to enhance the computational efficiency, needed to make the Monte-Carlo simulation practical, are also discussed.

EXPERIMENT

The experimental data presented in this paper were obtained using 60 MeV $^{127}I^{9+}$ ions from the ANTARES 8 MV FN Tandem accelerator at the Lucas Heights laboratories of the Australian Nuclear Science and Technology Organisation. The HIERDA measurements were made using a ToF-E detector telescope with a flight path of 495 mm and with carbon foils of 25.3 μg/cm^2 as described elsewhere (4). The only significant difference from that earlier experiment being a greater dynamic range for the time of flight system. The time detector further from the target triggers the START in the Time-to-Amplitude Converter (TAC) followed by a STOP signal generated by the nearer time detector delayed by 255 ns. This delay has been increased from previous measurements to include more of the tail of scattered and recoil spectra. The targets were well characterised layers of Au, 60nm, 140 nm and 460 nm on Si substrates.

Calibration

Time calibration is established from the ToF of recoiled and scattered ions from the surfaces of different samples which span a wide range of atomic masses as described by Stannard et al (5).

It was decided to extract time spectra from the raw data as these have better resolution for heavy elements than the energy spectra exhibit (6) as well as being subject to a far simpler and more direct calibration process.

Analysis

The signals from Si, I and Au were extracted from the raw data using the PAW analysis code (7) complemented with a suite of macros called TASS (8). Time spectra were extracted after the individual elemental signals were identified and separated using Mass versus Energy projections, where $M = Ct^2$, E is the pulse height from the Si energy detector, t is the time of flight and C is a constant.

MONTE-CARLO SIMULATION

Monte-Carlo simulations were performed using a FORTRAN version of TRIM (9). The code was modified to allow extraction of the energies and direction cosines of scattered and recoiled ions emerging from the front surface. All emerging scattered and recoiled ions with energies above 1 MeV were recorded and subsequently sorted to generate simulated spectra for ions reaching a region of space identified with the detector.

The Si detector and therefore the entire ToF-E detector telescope subtends 0.02 msr. Choosing only those ions that strike the detector would not allow the Monte-Carlo methods to be sufficiently efficient for spectral simulation.

If we define the scattering and azimuthal angles to be θ and ϕ respectively, then the simulated spectra were chosen on the criteria that (i) $44° < \theta < 46°$ and (ii) ϕ was restricted so that the excess path length in the exiting ion's path is less than 10% in total.

The ion energies were then corrected for energy loss in the first timing foil and converted to time over the known flight length. These flight times were then converted to a time histogram (simulated spectrum) using the experimental time calibration bin width and offset.

Efficiency Issues

The efficiency issues can broadly be separated into geometrical issues of the modelled detection system and approximations employed in modelling .

The acceptance angle used above is three times greater than the actual range of θ which reaches the detector, which slightly exacerbates the geometrical broadening due to variation of the kinematic factor and allows for some ions to travel through up to 8% greater path length on exiting the sample. The effects of kinematic factor and exit path length operate in opposite directions. A much more significant improvement in efficiency is yielded by accepting a wide range of azimuthal angles. Changes in the azimuthal angle affect the energy loss of the exiting ion, but have no impact on the kinematic factor. The chosen range of azimuthal angles only allows for a total increase of exit path length of 10%, but increases the azimuthal acceptance angle by a factor of 50. The net efficiency increase is therefore 150 from geometrical factors.

There are a number of parameters that can be changed in the Monte-Carlo modelling that dramatically affect its efficiency. These are (i) the energy cutoff which stops the transport calculation for a given ion, (ii) the depth cutoff to stop the transport of ions in any irrelevant substrate material and (iii) the minimum kinetic energy transfer allowed for a collision between an ion and a target atom (T_{min}). This minimum kinetic energy transfer corresponds to a maximum impact parameter for nuclear scattering.

In the current modelling the low energy cutoff for ion transport has been set at 1 MeV as scattered and recoiled ions below this energy are not of interest. In fact, I ions below 2.8 MeV and Au ions below 4.5 MeV are too slow to be detected by the experimental apparatus. They fall outside the dynamic range of flight times in this experimental measurement.

The primary interest in these studies is the spectrum of scattered I and recoiled Au. The spectrum of recoiled Si has only been used to assess the adequacy of the thickness estimate of the Au layer. Therefore a significant efficiency can be obtained by only modelling a thin (100 nm) substrate of Si.

The minimum kinetic energy transfer from ion to target atom has a dramatic effect on the speed of calculation, as recoiled atoms do not need to be generated and followed. These recoils have energies below the low energy cutoff mentioned above and therefore it is attractive to ignore them. However it is these collisions that give rise to the classic small angle scattering phenomenon described by Sigmund et al. (10). In Fig.1 the effect of T_{min} of 5 eV and 50 keV are compared.

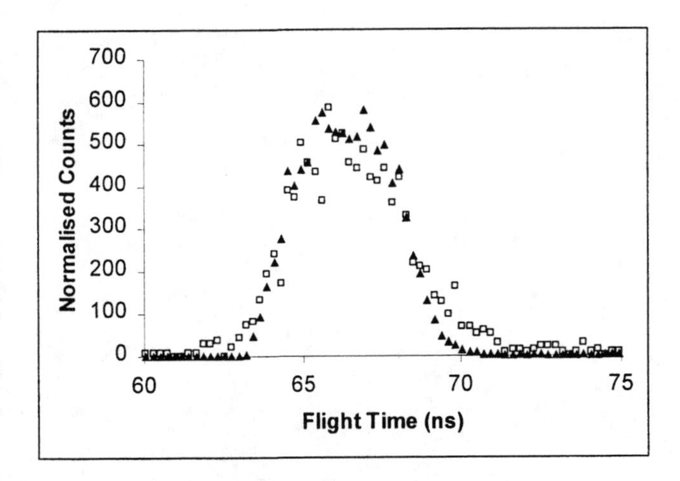

FIGURE 1. The effect of T_{min} of 5 eV (□) and 50 keV (▲) are compared for 60 MeV ^{127}I ions scattered from a structure of 60 nm Au on Si.

RESULTS AND DISCUSSION

Figures 2 and 3 show the scattered I spectrum and recoiled Au spectrum compared with the results of a Monte-Carlo simulation using 75 nm Au thickness and T_{min} of 5 eV.

We see that the simulated spectrum quite accurately follows the low energy side of the main scattering feature. The slope of this low energy edge is determined primarily by small angle multiple scattering. We see that TRIM accurately models this, confirming a previous study (11) which showed that TRIM accurately models the small angle multiple scattering angular distribution predicted by Sigmund et al. (10). As observed by Elliman et al. (1) previously, the stopping power appears greater than predicted by TRIM. The sample of 60 nm nominal thickness is modelled more appropriately in this case using a simulation thickness of 75 nm.

The comparison in Figure 2 also shows a small high energy tail which is observed in the experimental data. This is due to plural, probably double, scattering and can be explained by the fact that the product of kinematic factors for double scattering exceeds the kinematic factor for single scattering.

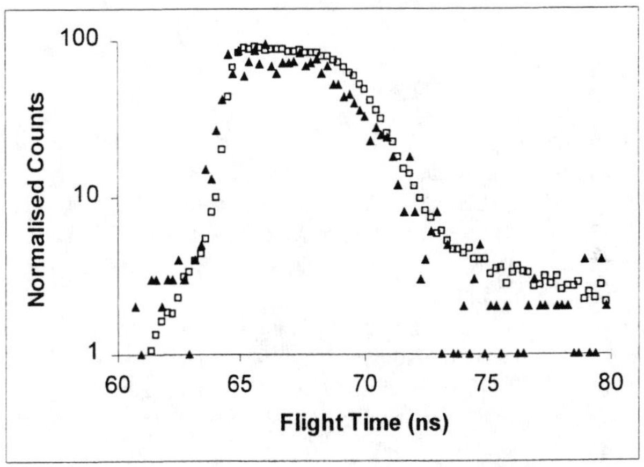

FIGURE 2. Comparison of the scattered I spectrum (□) from a 75 nm Au on Si sample with a Monte-Carlo simulation (▲).

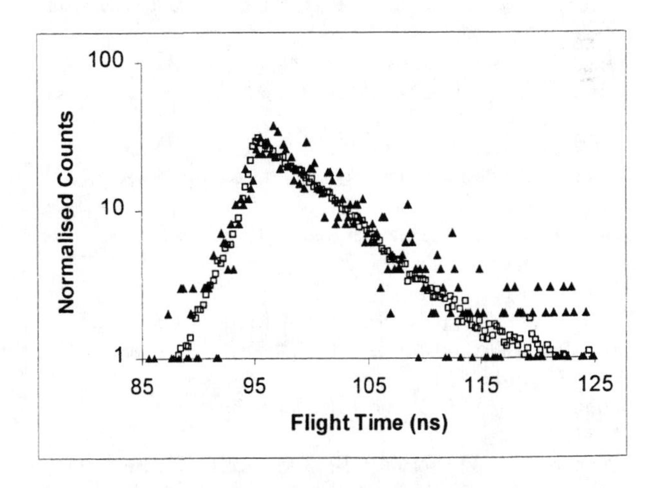

FIGURE 3. Comparison of the recoiled Au spectrum (□) from a 75 nm Au on Si sample with a Monte-Carlo simulation (▲).

The spectra have a totally different shape to that expected for a 'slab' analysis which has very well defined time or energy edges corresponding to the front and back of a simple structure as examined here.

CONCLUSION

The simulations above clearly show that Monte-Carlo simulation reproduces most of the important characteristics of the experimental spectra. As yet, the statistics of this preliminary modelling are insufficient to determine whether the long tails of the spectra are fully accounted for.

This work will be continued through the examination of a range of HIERDA experiments including some where spectra are obtained from targets of pure elements beyond the single scattering critical angle.

ACKNOWLEDGEMENTS

This work is supported by the The Australian Institute of Nuclear Science and Engineering.

REFERENCES

1. Elliman, R.G., Timmers, H., Palmer, G.R. and Ophel, T.R. .. Nucl. Instr. and Meth. B **136-138,** 649 (1998).
2. Davies, J.A., Lennard, W.N. and Mitchell, I.V., Ch. 12, Handbook of Modern Ion Beam Materials Analysis, Eds. J.R. Tesmer and M. Nastasi, Materials Research Society, Pittsburgh, p.352 (1995).
3. Biersack, J.P. and Littmark, U., 'The Stopping and Range of Ions in Solids', Pergamon, New York (1985).
4. Stannard, W.B., Johnston, P.N., Walker, S.R., Bubb, I.F., Scott, J.F., Cohen, D.D., Dytlewski, N., Martin, J.W., Nucl. Instr. and Meth. B **99,** 447 (1995).
5. 'Experimental Studies of Interfacial Phenomena in Barium Strontium Titanate (BST) Devices', Stannard, W.B., Johnston, P.N., Walker, S.R., Bubb, I.F., Scott, J.F., Cohen, D.D., Dytlewski, N. and Martin, J.W., Integrated Ferroelectrics (in press).
6. Johnston, P.N., El Bouanani, M., Stannard, W.B., Bubb, I.F., Cohen, D.D., Dytlewski, N. and Siegele, R.. Nucl. Instr. and Meth. B **136-138,** 669 (1998).
7. CERN, PAW Manual Version 1.14, (Application Software Group, Computing and Networks Division, CERN, Geneva, Switzerland, 1992).
8. Whitlow, H.J., TASS, Internal Report, (Dept. Nuclear Physics, Lund Institute of Technology, Sölvegatan 14 S-223 62 Lund, Sweden, 1993).
9. Hay, H., Department of Electronic Materials Engineering, Australian National University (1995) derived from the original TRIM (11).
10. Sigmund, P. and Winterbon, K.B., Nucl. Instr. and Meth. **119,** 541 (1974).
11. Johnston, P.N., El Bouanani, M., Stannard, W.B., Bubb, I.F., Cohen, D.D., Dytlewski, N. and Siegele, R., Vacuum **48,** 1017 (1997).

Study on Depth Profiles of Hydrogen in Boron-Doped Diamond Films By Elastic Recoil Detection Analysis

Liao Changgeng, Wang Yongqiang*, Yang Shengsheng, and Chen Ximeng

Department of Modern Physics, Lanzhou University, Lanzhou, Gansu Province, 730001, PRC
**Center for Interfacial Engineering, University of Minnesota,Minneapolis, MN 55455, USA*

Depth profiles of hydrogen in a set of boron-doped diamond films were studied by a convolution method to simulate the recoil proton spectra induced by ^{4}He ions of 3 MeV. Results show that the hydrogen depth profiles in these varying-level boron-doped diamond films exhibit a similar three-layer structure: the surface absorption layer, the diffusion region, and the uniform hydrogen-containing matrix. Hydrogen concentrations at all the layers, especially in the surface layer, are found to increase significantly with the boron-doping concentration, implying that more dangling-bonds and/or CH-bonds were introduced by the boron-doping process. While the increased dangling-bonds and/or CH-bonds degrade the microstructure of the diamond films as observed by Raman Shift, the boron-doping significantly reduces the specific resistance and makes semiconducting diamond films possible. Hydrogen mobility (or hydrogen loss) in these films as a result of the ^{4}He beam irradiation was also observed and discussed.

INTRODUCTION

Chemical vapor deposited (CVD) diamond films have some excellent properties, such as good thermal conductivity, wide bandgap, and superior hardness. Therefore, semiconducting diamond films possess great potential applications in high-temperature, high-frequency and high-power electronic devices. In the field of semiconductors, boron is an extensively used dopant. Many papers have studied the performance characteristics of boron-doped diamond films (1-3). It is believed that hydrogen plays a very important role in preparing these films and thus affects the chemical, physical, and electrical properties of the films. Quantitative information about the concentration and depth distribution of hydrogen in boron-doped diamond films is needed to better understand the relationships among the preparation, the microstructure, and the properties of the films. Owing to the difficulty in detecting hydrogen by conventional methods, however, studies on hydrogen depth profiles in boron-doped diamond films are limited. Elastic recoil detection analysis (ERDA) (4-10) is a novel technique in determining hydrogen profiles in thin films and at the surface of bulk materials, with many advantages such as absolute, fast, reliable and nondestructive determination. Furthermore, the use of convolution analysis method (for obtaining the depth profiles of hydrogen concentration from measured recoil proton spectra) substantially simplifies the data calculation and further improves the actual depth resolution of ERDA (6-10). In this paper, the study aims at determining the depth profiles of hydrogen in boron-doped diamond films and also at understanding the relationship between boron/hydrogen concentration and some performance characteristics of the films. Hydrogen mobility (or hydrogen loss) induced by the irradiation of incident ^{4}He ions during the analysis is also studied.

METHODOLOGY

The principles of ERDA have been discussed by many authors (4-10). It employs ions of MeV energies to bombard a sample surface at a glancing angle and to recoil hydrogen atoms out of the sample through elastic collisions. As a result, the energy spectrum of the recoil protons contains information about the hydrogen depth profile in the sample. Although the ERDA experiment is somewhat similar to RBS, simple and straightforward, converting the measured recoil proton energy spectrum into hydrogen depth profile is rather complex. Several ion beam laboratories have developed their own computer codes to do such a conversion (6-10). A brief introduction of the convolution analysis method that we have developed (8) is presented here.

Basic idea of our conversion method is that an actual recoil proton spectrum measured, $N(E_f)$, is considered to be the convoluting result of an ideal proton recoil spectrum, $N'(E_f')$, and an energy spread function, $G(E_f' - E_f)$, of the experimental setup,

$$N(E_f) = \int G(E_f' - E_f)\, N'(E_f')\, dE_f' \qquad (1)$$

where

$$N'(E_f')dE_f' = \varepsilon\Omega Q C(X)\sigma(E_1)(dx/dE_f')dE_f'/\sin\alpha \qquad (2)$$

$$G(E_f'-E_f) = 1/(2\pi\delta^2_f)^{1/2} \exp[-(E_f'-E_f)^2/2\delta^2_f] \qquad (3)$$

Here ε is the intrinsic efficiency of the detector; Ω is the solid angle of the target subtended by the detector; Q is the number of incident ^{4}He ions; C(X) is the hydrogen concentration at depth X; $\sigma(E_1)$ is the recoil crossection at E_1 which is the energy of ^{4}He ions just before the collision at depth X in the specimen; α is the incident tilt angle; and δ_f is the variance of the Gaussian distribution obtained from

CP475, *Applications of Accelerators in Research and Industry*,
edited by J. L. Duggan and I. L. Morgan

the total energy spread. Factors contributing to the energy spread were discussed in detail in References (6-9). For a given experimental condition, the shape of a recoil proton spectrum can be calculated using the above equations by assuming a known depth profile C(X), an optimum fit between experimental and calculated spectra can be found. The assumed depth profile can then be taken as the practical concentration distribution in the sample.

The experiment was done on 2 X 1.7 MV tandem accelerator at Lanzhou University using 3 MeV ^{4}He ions. The tilt angle of sample to incident beam was α=15° and the scattering angle for recoil protons was θ=30°. A range foil of 13.5 μm Mylar was put before the ion detector to avoid the influence of high counting-rate scattered ^{4}He ions. To minimize the beam irradiation-induced hydrogen loss (or hydrogen mobility), the beam current was limited to 3 nA and the total charge for each spectrum was limited to 5 μC. A typical recoil proton spectrum from a 8 μm thick Mylar foil sample is shown in Figure 1, where the circles are experimental data and the solid line is the calculated spectrum using the equations above by assuming a uniform depth profile for hydrogen in the Mylar. The good fit between the experimental and the calculated spectra in Fig. 1 suggests that the conversion software works well. This spectrum is taken as a hydrogen standard in analyzing the boron-doped diamond films.

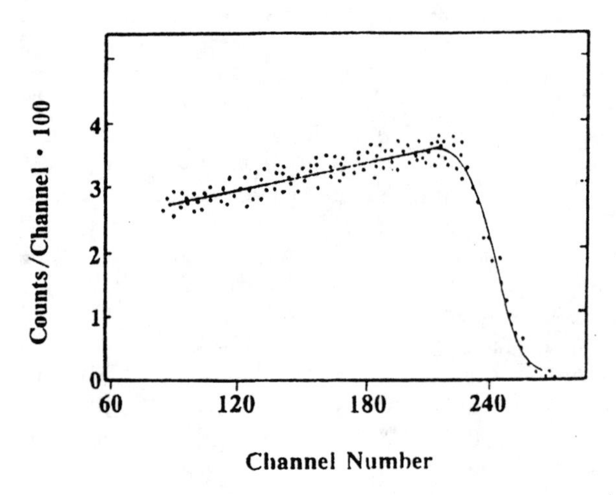

FIGURE 1. Recoil proton spectrum induced by 3 MeV ^{4}He ions for a 8 μm Mylar foil, where the circles and solid line represent the experimental and calculated results, respectively.

Boron-doped polycrystalline diamond films (PDFs) used in this study were grown on (100) silicon single crystal substrates by a hot filament CVD method. A saturated solution of boron trioxide (B_2O_3) powder in methanol (CH_3OH) was thinned with acetone (C_2H_6CO) and the volatilized gas (reactant gas) from this mixture, together with hydrogen gas (H_2), was introduced into the reactant chamber for synthesizing boron-doped PDFs (2). Three films were prepared for this study with different levels of boron-doping concentrations nominally: (a) heavily doped; (b) moderately doped; and (c) lightly doped. The boron-doping concentrations in these films have been analyzed by a strong but broad nuclear resonance reaction $^{11}B(p, \alpha)^8Be$ at E_p= 660 keV (11). The results show that the boron-doping concentrations in these films are uniformly distributed within the films with different concentrations of approximately (a) 2.3x10^{20} at./cm^3, (b) 4.2X10^{18} at./cm^3, and (c) 4.4x10^{17} at./cm^3, respectively.

RESULTS AND DISCUSSIONS

The recoil proton spectra obtained with 3 MeV ^{4}He ions at a beam current of 3 nA and an integral charge of 5 μC for these boron-doped diamond films are shown in Figure 2, where the circles and lines present the experimental and calculated results, respectively. The fits between the experimental and calculated recoil proton spectra are rather good. The corresponding hydrogen depth profiles used to generate the theoretical spectra in Fig. 2 are shown in Figure 3. It clearly shows that the hydrogen depth profiles in these heavily, moderately, and lightly boron-doped diamond films have a similar three-layer structure: the layer of surface absorption, the diffusion region, and the uniform hydrogen-containing matrix. A similar hydrogen profile was also reported for undoped CVD diamond films (12). Figure 3 indicates that the hydrogen concentration in each layer (especially at surface adsorption layer) increases significantly along with the increase in boron-doping concentration. While the thickness of the diffusion layer increases with boron-doping concentration, the thickness of the surface absorption layer (~25 nm) is roughly unchanged in all the films.

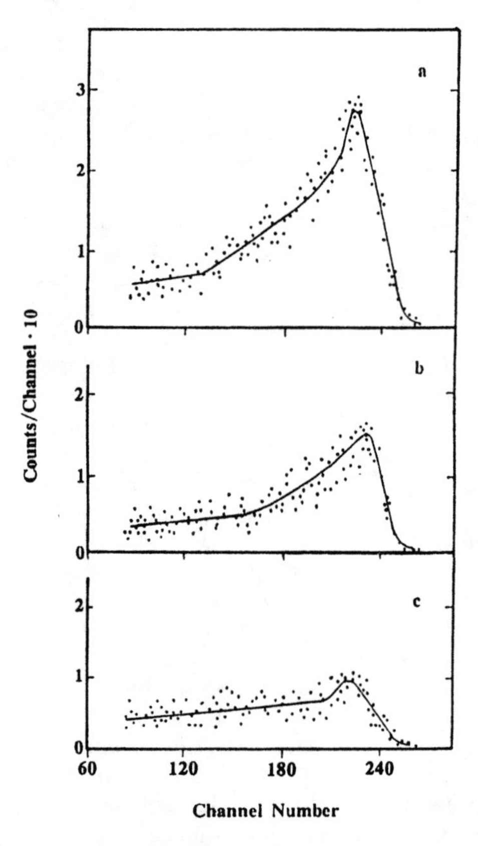

FIGURE 2. Recoil proton spectra induced by 3 MeV ^{4}He ions for CVD diamond films with the boron-doping concentration of: (a) 2.3x10^{20} at./cm^3, (b) 4.2x10^{18} at./cm^3, and (c) 4.4x10^{17} at./cm^3, where circles and solid lines are present experimental and calculated results respectively.

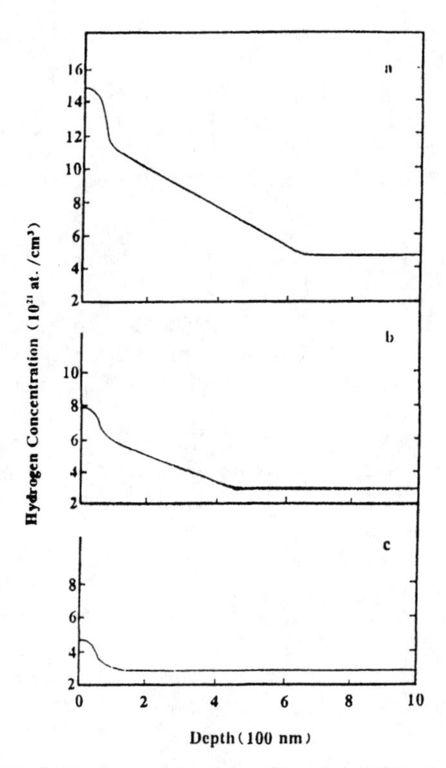

FIGURE 3. Hydrogen depth profiles for CVD diamond films with the boron-doping concentration of: (a) 2.3×10^{20} at./cm^3, (b) 4.2×10^{18} at./cm^3, and (c) 4.4×10^{17} at./ cm^3, respectively.

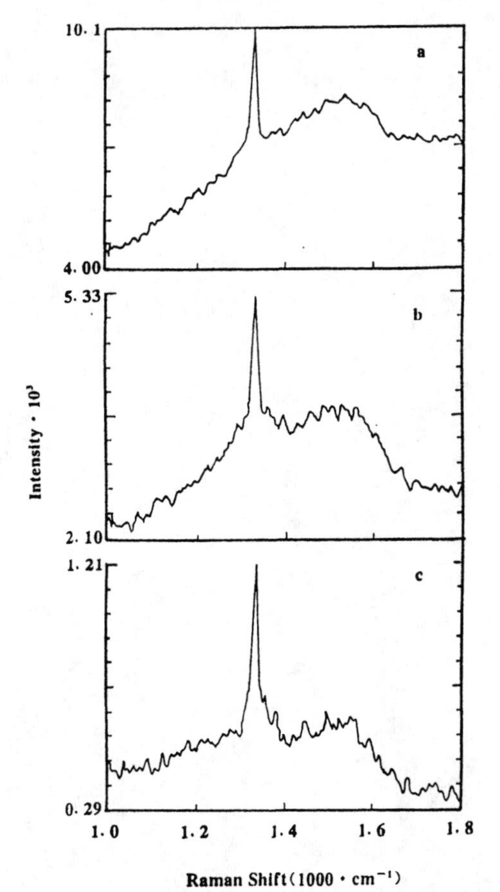

FIGURE 4. Raman Shift spectra from CVD diamond films with boron-doping concentration of: (a) 2.3×10^{20} at./cm^3, (b) 4.2×10^{18} at./cm^3, and (c) 4.4×10^{17} at./ cm^3.

In order to understand how the boron-doping and the hydrogen profile affect the performance characteristics of these diamond films, the Specific Resistance (SR) and Raman Shift (RS) of the films were determined. The Specific Resistance for heavily, moderately, and lightly boron-doped diamond films was measured to be approximately (a) 3×10^5 Ω•cm, (b) 1×10^4 Ω•cm, and (c) 1×10^2 Ω•cm, respectively. The results of Raman Shift spectra are shown in Figure 4, indicating a significant variation of the intrinsic peak at 1332 cm^{-1} for varying boron doping concentrations. The peak-to-base ratio (P/B) of the Raman Shift at 1332 cm^{-1} and the Specific Resistance (SR) are plotted against the hydrogen concentration of the surface adsorption layer as shown in Figure 5.

It is shown that the P/B and SR of the boron-doped diamond films are interrelated and they both decrease as the hydrogen concentration is increased by increasing the boron-doping concentration. On the one hand, the fluorence background scattering in Raman spectra increases with the boron-doping concentration by introducing more dangling bonds and CH-bonds in the diamond films, which in turn decreases the P/B scattering ratio. On the other hand, the increase of boron-doping concentration introduces more electronic energy states between the wide energy gap of the diamond films, and thus decreases the specific resistance of the films.

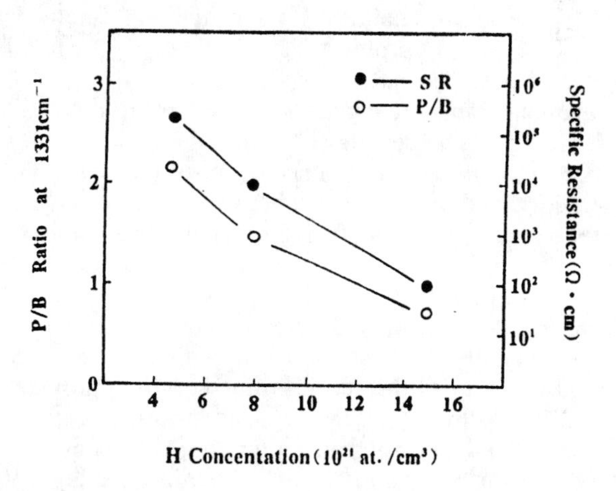

FIIGURE 5. Peak-to-Base ratio (P/B) of Raman Shift at 1332 cm^{-1} and Specific Resistance (SR) of the boron-doped diamond films as a function of hydrogen concentration at the surface.

Hydrogen mobility (or hydrogen loss) in these diamond films induced by the beam irradiation of 3 MeV ^{4}He ions was also observed. The total hydrogen content measured versus the irradiation dose in these films is shown in Figure 6. The fact that hydrogen mobility is more obvious for the heavily boron-doped diamond film than for the others seems to suggest that the heavier the boron-doping, the more free-state hydrogen atoms, and thus more hydrogen dangling bonds may exist in the films. The similar hydrogen mobility behavior was also observed in diamond like carbon films under the irradiation of 6-7 MeV ^{19}F ions (13). Figure 6 also indicates that the hydrogen mobility in these films is fairly small and can be neglected if the total irradiation dose is kept under 5 μC, which is equivalent to the ion dose of 3×10^{14} He/cm^2 in our experiment.

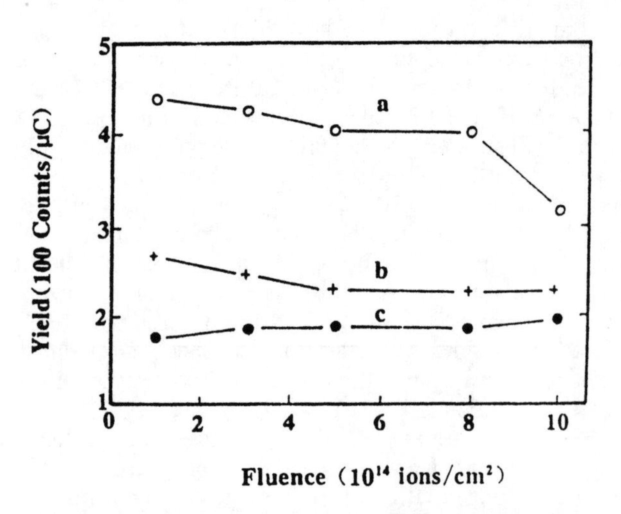

FIGURE 6. Hydrogen mobility induced by 3 MeV ^{4}He ions from the diamond films with boron-doping concentration of: (a) $2.3x10^{20}$ at./cm^3, (b) $4.2x10^{18}$ at./cm^3, and (c) $4.4x10^{17}$ at./ cm^3.

SUMMARY

In this paper the depth profiles of hydrogen in diamond films with different boron-doping concentrations were successfully analyzed by a convolution method to simulate the recoil proton spectra induced by 3 MeV He ions. The possible analysis depth and sensitivity of ERDA for diamond films are estimated to be 1 μm and 0.1 at.%, respectively. The results show that the hydrogen depth profiles for varying boron-doped diamond films are similar in structure and the distinction between the three layers of surface adsorption, diffusion region, and hydrogen-containing matrix is clear. The performance characteristics of the boron-doped diamond films are strongly related to the hydrogen concentration and the boron-doping concentration. In summary, the ERDA combining with the convolution method to simulate the recoil proton spectrum provides a simple, nondestructive, sensitive, and quantitative way for hydrogen depth profiling in boron-doped diamond films, which is valuable for the study of semiconducting CVD diamond films.

ACKNOWLEDGMENT

We would like to thank Professor F.Q. Zhang and her colleagues in Department of Physics at Lanzhou University to provide the diamond films for this study.

REFERENCES

1. Nishimura, K., Das, K. and Glass, J.T., J. Appl. Phys., 1991, **69**, 3142.
2. Zhang, F.Q., Zhang, N.P., Xie, E.Q., Yang, B., Yang, Y.H., Chen, G.H., and Jiang, X.L., Thin Solid Films, 1992, **216**, 279.
3. Zhang, R. J., Lee, S. T., Lam, Y. W., Diamond and Related Materials, 1996, **5**, 1288.
4. L'Ecuyer, J., Brassard, C., Carginal, C., Deschenes, L., Labrie, J.P., Tereault, B., Wartel, J.G., and St. Jacques, R., J. Appl. Phys. , 1988, **47,** 881.
5. Doyle, B. L. and Peercy, P.S., Appl. Phys. Lett., 1979, **34**, 811.
6. Turos, A and Meyer, O, Nucl. Instrum. and Meth., 1984, **B4**, 92.
7. Doyle, B. L. and Brice, D.K., Nucl. Instrum. and Meth. ,1988, **B35**, 301.
8. Wang, Y.Q., Liao, C.G., Yang, S.S., and Zheng, Z.H., Nucl. Instrum. and Meth.,1990, **B47**, 427.
9. Barbour,J.C., SERDAP Computer Program (Sandia National Laboratories, Dept. 1111, Albuquerque, New Mexico (1994).
10. Schiettekatte, F. and Ross, G.G., in *Application of Accelerators in Research and Industry*, edited by J.L. Duggan and I.L. Morgan, 1997, **CP392**, AIP press, New York, p.711.
11. Liao., C.G., Wang, Y.Q., and Yang, S.S., Nucl. Instrum. and Meth., 1998, **A, (**in press).
12. Wagner, W., Rauch, F., Haubner, R., and Lux, B., Thin Solid Films, 1992, **207**, 24.
13. Zheng, Z.H., Wang, Y.Q., Liao, C.G., and Fu, H.B., Vacuum, 1990, **40**, 505.

Use of Rutherford Forward Scattering for the Elemental Analysis of Evaporated Liquid Samples

J. A. Liendo[a,b], A. C. González[b], N. R. Fletcher[c], J. Gómez[a], D. D. Caussyn[c], S. H.Myers[c], C. Castelli[a] and L. Sajo-Bohus[a].

[a]*Departamento de Física, Química y Biología, Universidad Simón Bolívar, Caracas,Venezuela*
[b]*Centro de Física, Instituto Venezolano de Investigaciones Científicas, Caracas,Venezuela*
[c]*Physics Department, The Florida State University, Tallahassee, FL, USA.*

Multielemental analysis of evaporated liquid samples is possible by irradiating the samples with a 16 MeV ^{7}Li beam and detecting the elastically scattered ions at 28°. The method is easily applied when Rutherford scattering dominates. To prepare the targets, the liquid sample is deposited on a formvar backing and dried with vacuum. Preliminary results indicate a possible relationship between sample concentration, uniformity and spectrum energy resolution. Details on the spectrum analysis will be given. The method is mainly useful for multielemental quantification in the low mass region from lithium to fluorine where standard techniques such as PIXE and TXRF are useless.

INTRODUCTION

Elemental analysis of evaporated liquid samples is important for research and industrial purposes. Techniques based on Proton Induced X-Ray Emission [1] and Rutherford Backward Scattering [2] have been useful to analyze different kinds of samples mainly for elements heavier than Sodium. Other methods such as Inductively Coupled Plasma-Atomic Emission Spectrometry and Atomic Absorption Spectrophotometry [3] are adequate to carry out elemental quantification below Sodium but usually they are not multielemental and the sample has to be treated chemically before analysis so that contamination is highly possible and/or the sample is destroyed during the analysis.

Rutherford Forward Scattering offers the alternative of carrying out multielemental analysis of evaporated liquid samples for elements below Sodium without destroying the sample and with minimum manipulation. In principle, this method can be easily applied if Rutherford scattering is predominant. Otherwise, the method is still very useful if the relevant elastic cross sections are determined experimentally.

In this paper, we explain briefly the procedure we have followed to apply elastic scattering at forward angles for the multielemental analysis of any evaporated liquid sample. The selection of beam type, energy and current, detector angle and irradiation time will be addressed as well as sample and target preparation. Some of the results obtained recently will be shown including a discussion concerning the future efforts that we consider important for the improvement of this new method.

EXPERIMENTS AND RESULTS

One of the advantages of this technique is the little manipulation of the sample during its handling. In addition, no chemical treatment is required. Centrifugation (6 minutes at 13,000 G Eppendorf 5412) and dilution with distilled water are only carried out in case an organic sample such as amniotic fluid is analyzed. Centrifugation removes cells and unwanted organic material usually suspended in this kind of sample. Dilution not only makes samples thinner but also more uniform as we have shown recently [4]. Less complicated samples can be analyzed without any treatment. A volume of 10 μl of the liquid sample is deposited on a backing made of a polymer called formvar according to the procedure published in Ref. [5]. The drop is dried with vacuum. A typical drop diameter is approximately 6 mm.

A 16 MeV ^{7}Li beam is a convenient choice for the study of evaporated liquid samples by use of elastic scattering [4,5]. We have proved that detection of elastically scattered ^{7}Li ions at 28° results in spectra such as the one shown in Fig. 1a where an adequate separation between low mass elements such as carbon, nitrogen and oxygen is observed. In Figs. 1b and 1c, we present the spectra collected simultaneously at 20° and 16° respectively to show the convenience of using a detector angle of 28° for elemental separation. These spectra were gathered by bombarding an amniotic fluid sample during 2.5 hours with an average ^{7}Li beam current of 3 nA. Either a higher current and/or a longer irradiation period is not recommended since, under these circumstances, some samples were found to be damaged after irradiation. Beam diameters of approximately 3 mm were observed at the target position. The beam spot was always contained

CP475, *Applications of Accelerators in Research and Industry*, edited by J. L. Duggan and I. L. Morgan

completely inside the sample area. Obviously, if the dried sample residue is not distributed uniformly across the formvar backing, the irradiation of a particular area of the sample may not give reliable results. An appreciable percentage of the detected scattering yield could be due to elements contained in the backing material, not in the sample. This problem needs to be solved in the future.

The use of a lighter beam particle, for example an alpha beam, does not significantly improves the resolution of the spectra [6] and, on the contrary, it may deteriorate its quality due to the lower Coulomb barrier associated with a lower charge beam particle. A comparison carried out between a 16 MeV ^{7}Li and a 24 MeV ^{16}O beams [4] has established that although an oxygen beam seems to be adequate for the analysis of very thin samples, a lithium beam produces much better mass separation in the analysis of evaporated organic liquid samples which, in general, are thick and require to be diluted.

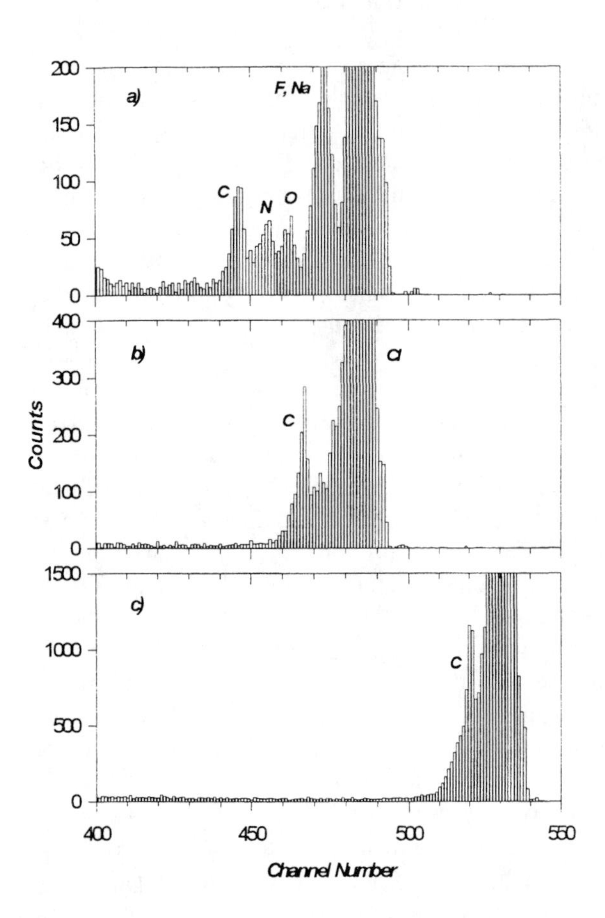

FIGURE 1. Elastically scattered particle spectra collected at a) 28°, b) 20° and c) 16° from the bombardment of an amniotic fluid sample with a 16 MeV ^{7}Li beam during 2.5 hours, beam current ~ 3 nA.

DISCUSSION

A linear energy calibration of the collected spectra was carried out by bombarding a target of gold deposited on a carbon backing with a 16 MeV ^{7}Li beam and the detector set up at 28°. A spectrum gathered during the irradiation

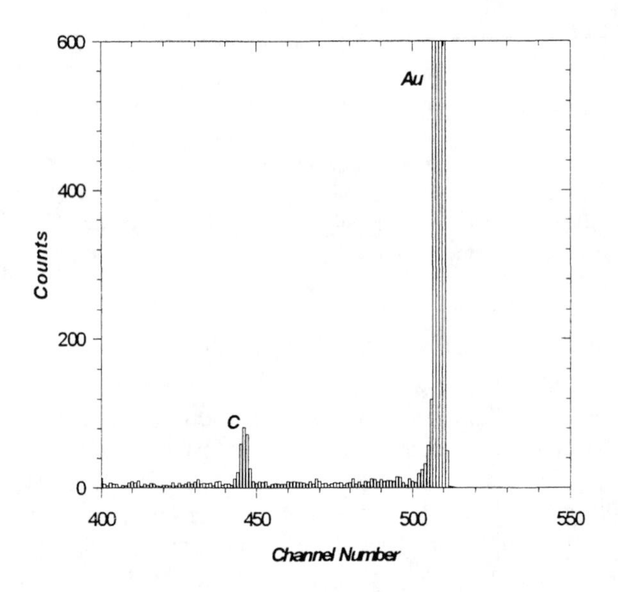

FIGURE 2. Elastically scattered particle spectrum obtained at 28° from the bombardment of a gold on carbon target (~ 20μg/cm^2 Au + 20 μg/cm^2 C) with a 16 MeV ^{7}Li beam.

of this target is displayed in Fig. 2. The centroids of the peaks corresponding to gold and carbon contained in the target are used to determine the channel number vs. energy linear calibration curve necessary to make the elemental assignments shown in Fig. 1.

Once the elemental assignments of a spectrum are made, a chi square Gaussian type minimization fit can be performed. A program called GF2 by D. C. Radford [7] has been used for this purpose. In this program, a gaussian curve is assigned to every element of the spectrum. To start the fitting routine, an initial value for the centroid of every gaussian curve is set by clicking with the computer mouse at the corresponding peak maximum. All the gaussian centroids are allowed to vary during the fitting process. The FWHM values are either kept constant or permitted to change. We have shown [5] that the FWHM of each peak associated to a particular element contained in the target can be expressed as

$$\mathrm{FWHM} = (\Gamma^2 + \Gamma_0{}^2)^{1/2} \quad (1)$$

where

$$\Gamma = (\partial E/\partial\theta)\Delta\theta \quad (2)$$

represents the known kinematic energy dispersion due to the finite dimension of the detector ($\Delta\theta = 0.32°$), E is the kinetic energy of the incident particle after the elastic scattering and Γ_0 is the dispersion introduced in the measurement by the beam, target and electronics characteristics which are common factors to all the peaks observed in a spectrum.

The output of the fitting program provides the area of each gaussian curve , e.g. the number of counts, associated to a particular element contained in the analyzed sample. The expression $T = [0.266\ Z\ A\ Y\ I] / [\Delta\Omega\ Q\ (d\sigma/d\Omega)]$ is then used to calculate the concentration of an element present in the sample in $\mu g/cm^2$ units [5]. In this formula, Z is the equilibrium charge of the beam particle after passing through the target. A is the target element mass number in amu, Y is the area of the gaussian peak usually referred to as the yield or the number of counts, I is a conversion factor between the laboratory and center of mass reference frames, $\Delta\Omega$ is the detector solid angle in msr, Q represents the total accumulated charge in μc and $d\sigma/d\Omega$ is the differential cross section associated to the target element-beam particle interaction in mb/sr. The numerical value of 0.266 results from the units used for the quantities contained in the equation.

From analyzing the spectrum of Fig. 2 with the program GF2, a FWHM of approximately 2.94 channels = 91.8 keV (Γ_{Au}) was obtained for the Au peak. Similarly for the C peak, the width obtained was $\Gamma_C \sim 3.40$ channels = 106.2 keV. The expression given above for FWHM has been used in Table 1 under the assumption that $\Gamma_0 = \Gamma_{Au}$ = 2.94 channels. In this Table, the 4th row contains the expected FWHM values of N, O, F, Na, Cl as if these elements were present in the target used to obtain Fig. 2. Notice that the predicted FWHM for carbon (3.27 channels) is very close to the width value obtained from the spectrum fit (3.40 channels).

For the spectrum presented in Fig. 1a, we found that an adequate fit, shown in Fig. 3, is obtained when the FWHM values of the C, N and O gaussian curves are set to approximately 5 channels = 156 keV and the rest of the fitting parameters are let to vary freely. From the result reported on carbon in the last paragraph, a carbon FWHM value of 3.40 channels would be expected but this is not the case.

The difference between the carbon widths of Fig. 2 and Fig. 3 is due to the fact that the irradiated targets are different: the target used to generate Fig. 2 is made of gold deposited on a carbon support and the one corresponding to Fig. 3 is evaporated amniotic fluid on a formvar substrate. We have evidence [4], which indicates that the deterioration of the spectrum resolution is due to the amniotic fluid target being thicker and less uniform than the gold target. This means that if predicted FWHM values are wanted to be calculated in the amniotic fluid spectrum (Figs. 1a and 3), the value of Γ_0 should be determined first for that particular spectrum. For this purpose, we used the FWHM formula mentioned above with the carbon FWHM set to 5 channels (156 keV) and obtained Γ_0 = 149.6 keV.

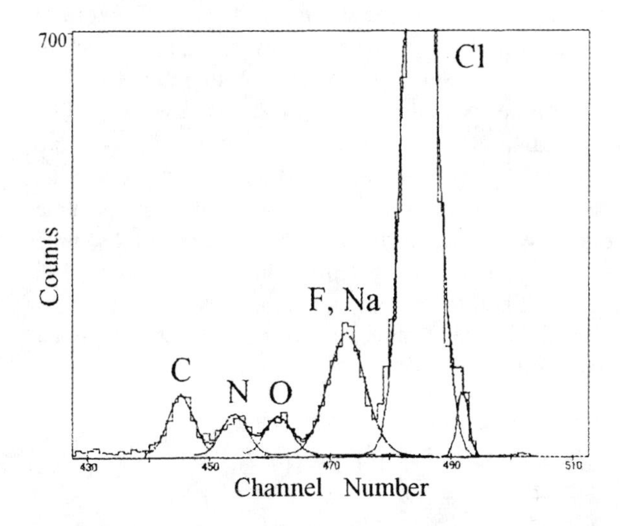

FIGURE 3. Gaussian fit of the spectrum shown in Fig. 1a. The FWHM of C, N and O have been set to 5 channels = 156 keV each one and the rest of the peak widths and centroids have been allowed to vary during the fitting process.

TABLE 1. Predicted FWHM values in channel units (1 channel = 31.23 keV) for different elements assuming they exist in the gold target used for the energy calibration and in an amniotic fluid target. For a 16 MeV ^7Li beam, the energy of the elastically scattered beam particles (E) and the kinematic factor $\partial E/\partial\theta$ at 28° are given. In this table, $FWHM = (\Gamma^2 + \Gamma_0^2)^{1/2}$ where $\Gamma = (\partial E/\partial\theta)\Delta\theta$ and $\Delta\theta = 0.32°$. To generate the results of the 4th row (gold target), $\Gamma_0 = \Gamma_{Au}$ = 2.94 channels = 91.8 keV. For row 5 (amniotic fluid target), Γ_0 = 149.6 keV when a carbon FWHM of 5.0 channels (156 keV) is used.

Element	C	N	O	F	Na	Cl	Au
E(MeV)	13.92	14.21	14.43	14.67	14.89	15.27	15.87
$\partial E/\partial\theta$(keV/°)	138.4	119.8	105.7	89.9	75.1	50.3	9.2
$FWHM_{gold}$ (channels)	3.27	3.19	3.14	3.08	3.04	2.99	2.94
$FWHM_{AF}$ (channels)	5.00	4.95	4.91	4.88	4.85	4.82	4.79

Table 1 (5th row) presents the predicted FWHM values for the amniotic fluid target used in this work. Notice that there is only a difference of less than 4% between the FWHM values of C and Cl which justifies the FWHM values (5 channels) used for C, N and O to obtain the fit shown in Fig. 3. In that figure, the fourth peak from left to right contains Na, F and probably Mg. Any attempt to obtain elemental information above this peak is useless. The number of counts (Y) associated to each Gaussian curve, e.g. the area of every curve, is used to calculate the concentration of the corresponding element by means of the formula given above under the assumption that $d\sigma/d\Omega$ is the Rutherford scattering cross section. In this work, we have considered not necessary to calculate the specific elemental concentrations of the sample shown in Fig. 3. In a previous work [5], we have determined that concentrations as low as 0.1 $\mu g/cm^2$ can be measured with this technique which represents a sensitivity of approximately 3 ppm. Although methods such as Atomic Absorption have better sensitivities, the Rutherford Forward Scattering technique has the advantage of requiring a very small amount of liquid sample for the analysis. Also, no chemical treatment is needed so that contamination is reduced to a minimum and, more important, the possibility of obtaining simultaneously the concentrations of several elements from the same sample permits, for example in the case of a human liquid sample such as amniotic fluid, to carry out correlation studies between different elements of biological and medical interest in order to establish possible connections between physical and medical data.

CONCLUSIONS

Elastic scattering of a 16 MeV ^{7}Li beam at 28° is useful to carry out multielemental analysis of evaporated liquid samples in the low mass region from Li to F. Further work is required in order to improve the energy resolution of the spectra. Sample dilution should be considered for this purpose because it makes targets thinner and more uniform. In addition, the use of a similar beam particle such as ^{6}Li at a energy comparable to the one used in this work may also improve the spectrum quality since ^{7}Li has an excited state at 0.480 MeV that could have been the source of contamination in the spectra presented in this article. A crucial step in this type of multielemental analysis is the understanding of how a liquid drop dries under different conditions. At this moment, we realize this is an issue that needs to be investigated in the future if one wants to have confidence in the results obtained from the method reported here. The degree of uniformity of a dried sample has to be explored in detail because, to the best of our knowledge, there is no previous study on how different elements contained in a liquid sample flow and deposit when drying occurs. We have performed preliminary tests showing very qualitatively that sample uniformity depends on sample dilution [4]. An intensive quantitative research is required on this matter. Finally, the beam-target interaction needs to be studied in more detail in order to investigate the volatilization of certain elements during the sample irradiation.

ACKNOWLEDGMENTS

This research has been funded by project # S1-PN-CB-401, Decanato de Investigación y Desarrollo, Coordinación de Ciencias Básicas, Universidad Simón Bolívar, Caracas, Venezuela, The Florida State University and the National Science Foundation, USA.

REFERENCES

1. Johansson, S. A., and Campbell, J. L., PIXE: A Novel Technique for Elemental Analysis, Chichester, Wiley, 1988.
2. Nicolet, M. A., Mayer, J. W., and Mitchell, I. V., Science **177**, 841-849 (1972).
3. Bussiere, L., Dumont, J., and Hubert, J., Anal. Chem. Acta, **224**, 73-81 (1989); Laitinen, R., Jalanko, H., Kolho, K. L., Koskull, H. Von, and Vuori, E., Am. J. Obstet. Gynecol. **152,** 561-565 (1985).
4. Liendo, J.A., González, A. C., Fletcher, N. R., Gómez, J., Caussyn, D. D., Myers, S. H., Castelli, C., and Sajo-Bohus, L., "Target Preparation and Characterization for Multielemental Analysis of Liquid Samples by Use of Accelerators", presented at the 19th World Conference of the INTDS, October 5-9, 1998, Oak Ridge, Tennessee, USA.
5. Liendo, J.A., González, A. C., Castelli, C., Gómez, J., Jiménez, J., Marcó, L., Sajo-Bohus, L., Greaves, E. D., Fletcher, N. R., Lee, C., Caussyn D. D., Myers, S. H., and P. Barber, Nucl. Inst. and Meth. in Phys. Res. **B140,** 409-414 (1998).
6. Liendo, J.A., González, A. C., Fletcher, N. R., Gómez, J., Lee, C., Caussyn, D. D., Myers, S. H., Castelli, C., Marcó, L., Sajo-Bohus, L., Greaves, E. D., and Barber, P., APH N.S., Heavy Ion Physics **7** (1998) 335-341.
7. D. C. Radford, Computer Code GF2 (Atomic Energy of Canada Limited, Chalk River Nuclear Laboratories, Chalk River, Ontario, KOJ 1P0, Canada), unpublished.

Exchange of Hydrogen Isotopes in Oxide Ceramics and Water-Vapor using ERD Technique

T. Hayashi, T. Horikawa, E. Iizuka, K.Soda and K. Morita

Department of Crystalline Materials Science and Department of Nuclear Engineering, Graduate School of Engineering, Nagoya University, Furo-cho, Chikusa-ku, Nagoya 464-8603, Japan

This paper describes the exchange of deuterium in a tritium breeding ceramics, Li_2ZrO_3, for protium by exposure to air-vapor which has been studied by means of the elastic recoil detection technique with 1.7 MeV He^+ ion beam. The ceramic specimen implanted up to saturation concentration at room temperature with 5 keV D_2^+ ions is exposed to normal air introduced into the vacuum chamber at room temperature. It is found that the retained number of D implanted decreases, while the retain number of H increases as the exposure time to normal air increases. The exchange behavior of D for H in Li_2ZrO_3 is compared with that in $SrCe_{0.95}Yb_{0.05}O_{3-\delta}$ studied previously and the processes for hydrogen isotope exchange observed are discussed.

INTRODUCTION

Dynamic behaviors of hydrogen atoms in oxide ceramics have received intensive attention in applied point of views such as high temperature protonic conductors in electrochemical devices and tritium breeding materials in fusion devices. Recently, dynamic behaviors of hydrogen in the protonic conductor ceramics have been studied by many authors [1], using various experimental techniques such as the conductivity measurement [2], luminescence spectroscopy [3], neutron scattering and diffraction [4] and nuclear reaction analysis with ion beam[5]. The present authors have also studied thermal behaviors of deuterium implanted into a protonic conductor ceramics, $SrCe_{0.95}Yb_{0.05}O_{3-\delta}$ by means of the elastic recoil detection technique with 1.7 MeV He^+ ion beam [6]. It has been found eventually in the present study that D implanted into $SrCe_{0.95}Yb_{0.05}O_{3-\delta}$ is almost completely replaced by H in nomal-air introduced into the vacuum chamber [7].

For tritium breeding ceramics materials, important key issues concerning hydrogen behaviors are recovery of tritium produced during the operation of fusion reactors and reduction of tritium inventory for the maintenance. These issues are closely related to re-emission processes of hydrogen and the experimental results on the isotope exchange of D implanted in $SrCe_{0.95}Yb_{0.05}O_{3-\delta}$ for H in nomal-air is very effective for the tritium breeding materials in point of views from enhancement in the tritium recovery and in the environmental safety.

In this paper, we report the preliminary experimental results on the exchange of deuterium implanted into a tritium breeding ceramics ,Li_2ZrO_3 ,for protium in normal air-vapor at room temperature, which has been measured by means of the elastic recoil detection (ERD) technique. We report the thermal behaviors of hydrogen isotopes in the ceramics. The experimental data are compared with those for $SrCe_{0.95}Yb_{0.05}O_{3-\delta}$ [6,7].

EXPERIMENTAL

The specimen used was a block of Li_2ZrO_3 of 20x50x50 mm^3 in size, which was prepared in Kawasaki Heavy Industries Ltd. The block was sliced into pieces of 2x10x10 mm^3 in size by diamond cutter. The XRD indicated that the specimen was a monoclinic structure of a=5.42660, b=9.03100, c=5.42270, Å and β=112.72°. The atomic composition ratio of the specimen was measured by means of the Rutherford backscattering spectroscopy (RBS) technique with with 1 MeV H^+ ion beam . The RBS spectrum of 1MeV H^+ ion from the specimen heated at 873K for 10min in order to remove out the residuals protium is shown in Fig.1,

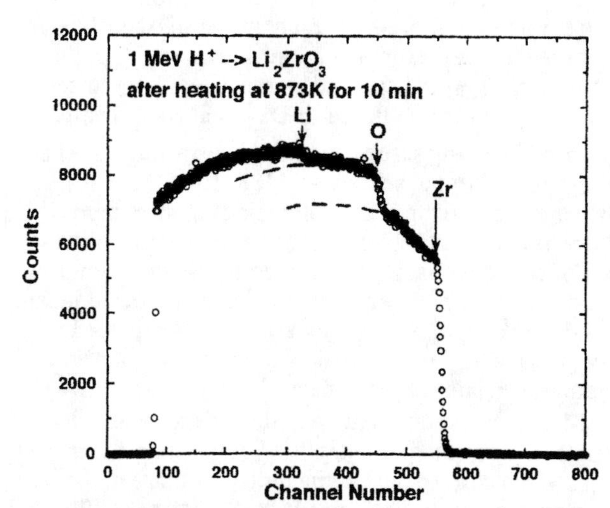

FIGURE 1. RBS spectrum of 1MeV H^+ ion beam from the Li_2ZrO_3 specimen, heated at 873 K for 10 min in order to remove out residual hydrogen (protium). The arrows with elemental sysmbols represents the energies of H^+ ions backscattered from each element.

CP475, *Applications of Accelerators in Research and Industry*, edited by J. L. Duggan and I. L. Morgan

where the energies of H^+ ions scattered from Zr, O and Li elements are shown by the arrows with the symbols. It was confirmed from the spectrum with dashed lines using the standard analysis method that the specimen used kept almost the stoichiometric composition, although the scattering cross-section from Li atoms has some ambiguity due to nuclear reaction.

The Li_2ZrO_3 specimen was placed on a manipulator in contact with a ceramic heater in a conventional UHV chamber, which is schematically shown in Fig.2. The chamber was regularly evacuated to base pressure of 4.0×10^{-7}Pa. Prior to hydrogen ion implantation, the specimen was heated at 873K for 10min in order to remove out residual hydrogen (protium). The temperature of the specimen was measured with a thermocouple of alumel-chromel. The specimen was implanted with 5keV D_2^+ ion at a flux of $6.0 \times 10^{13}/cm^2 \cdot s$ up to saturation concentration at room temperature. The specimen was exposed to normal air at room temperature after the ion implantation. At several stages of the air-exposure, the depth profiles of H and D were measured by means of the ERD technique [7], in which the recoiled $H^+$and D^+ ions were detected through an aluminum foil filter at a forward recoil angle of 20 ° to the incident direction of 1.7 MeV He^+ ion beam which is inclined at 80° to the surface normal of the specimen. The Rutherford backscattering (RBS) spectrum was simultaneously measured at an angle of 150° to the incident direction of 1.7 MeV He^+ ions in order to monitor the fluence.

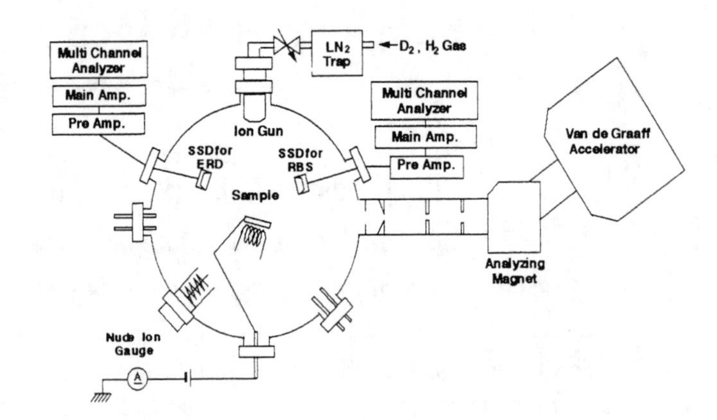

FIGURE 2. A schematic diagram for an experimental arrangement used in the present study.

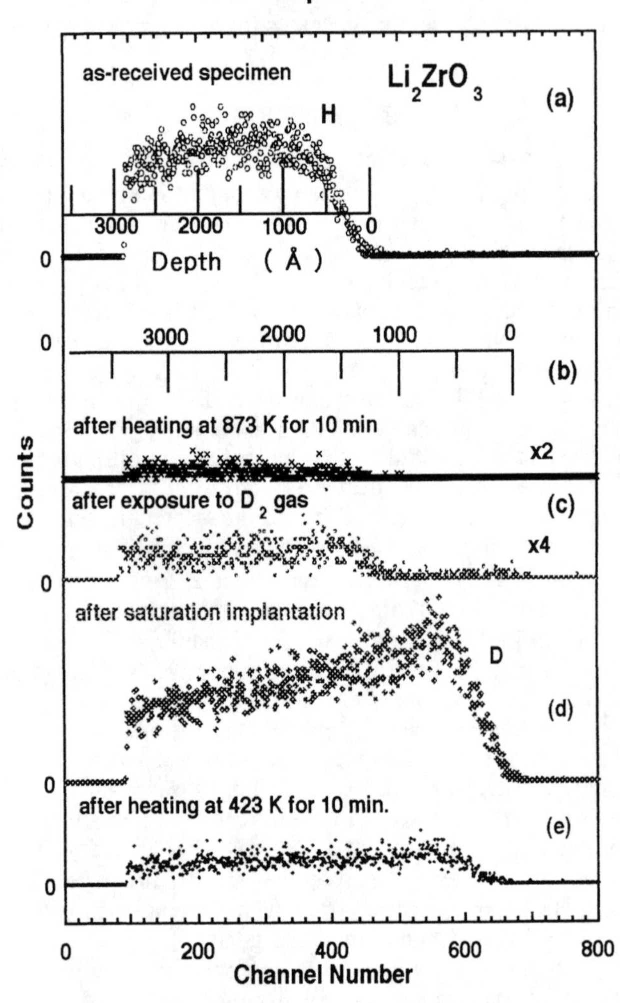

FIGURE 3. ERD spectra of 1.7 MeV He^+ ion beam from the Li_2ZrO_3 specimen as-received (a), heated at 873 Kfor 10 min (b), exposed to D_2 gas (c), implanted up to saturation concentration with 5 keV D_2^+ ions (d), and heated at 423 K for 10 min (e).

EXPERIMENTAL RESULTS AND DISCUSSIONS

Typical ERD spectra of 1.7 MeV He^+ ion beam from the specimen measured at each stage of the standard procedures in the present study are shown in Fig.3, where ERD spectra from the specimen as-received (a), heated at 873K for 10min (b), exposed to D_2 gas at 8.3×10^{-3} Pa (c), implanted up to saturation concentration at room temperature (d) and heated at 423K for 10min (e), are shown and the horizontal axis of channel number represents the kinetic energy of recoiled species. One can estimate the depth of hydrogen isotopes retained in the specimen from the kinetic energy of recoiled species using the standard method as shown in Fig.3 (a) and (b). It is clearly seen from the weak background intensity above 500 channels that deuterium is retained in the specimen by the exposure to D_2 gas. Moreover, it is seen that deuterium implanted is distributed from the peak depth of 500 Å into the deeper layers, which corresponds to the projected range of 2.5 keV D^+ ion in the specimen, and it is almost re-emitted by heating at 423K for 10min. From the standard analysis [8] of the ERD spectrum for the specimen as-implanted, it was found that the ratio of the saturation concentration of D implants to the molar concentration of Li_2ZrO_3 was around 0.22.

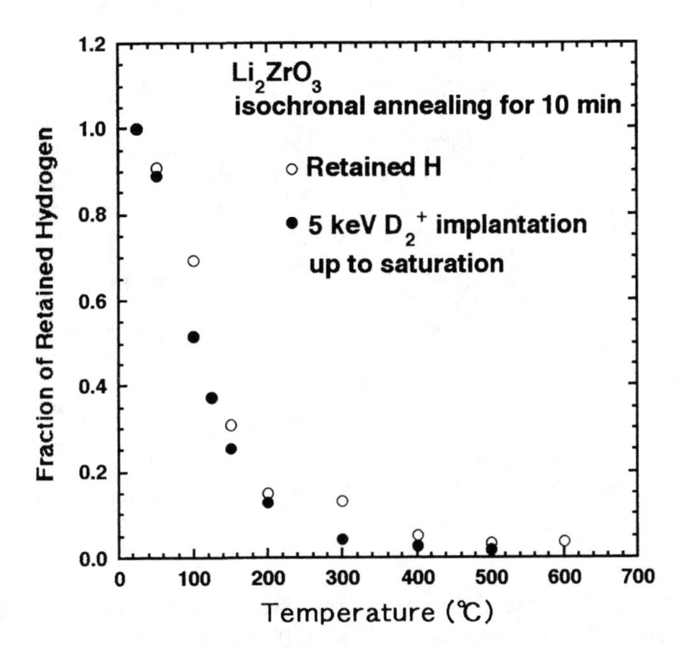

FIGURE 4. Decay curves of H remaing in the specimen and D implanted into the specimen with 5 keV D_2^+ ions up to saturation on isochronal annealing for 10 min. The vertical axis represents the values normalized to the initial values.

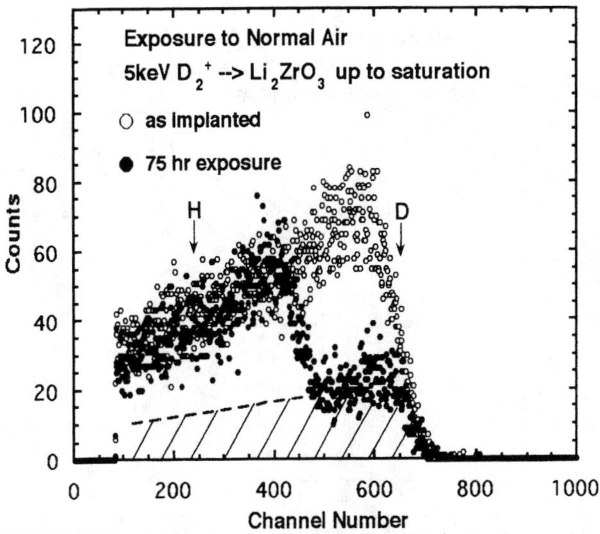

FIGURE 5. ERD spectra of 1.7 MeV He^+ ion beam from the Li_2ZrO_3 specimen implanted up to saturation concentration at room temperature (○) and subseqently exposed to normal air at room temperature for 75 hrs. The arrows with H and D represent the energies of H^+ and D^+ ions recoiled at the surface. The hatched spectrum represenets the ERD spectrum of D after the exposure to air.

The decay curve of the retained number of D implanted into the specimen with 5keV D_2^+ ions up to saturation by isochronal annealing for 10min is shown as s function of temperature in Fig.4, where the decay curve of the retained number of H in the as-received specimen is also shown for comparison and the vertical axis represents the values normalized to the initial values. It is seen from Fig.4 that the retained numbers of H and D are reduced rapidly by heating up to 200℃. The decay rate of D implanted is a little bits faster than that of H remaining in the as-received specimen. When H and D are at the same chemical state, generally, the reaction rate of H is faster than that of D. The result in Fig.4 is ascribed to difference between the chemical states of H and D, or the effects of implantation induced damage on thermal re-emission of hydrogen isotopes. However ,it has not been understand yet.

The specimen of Li_2ZrO_3 implanted with 5 keV D_2^+ ions up to saturation concentration at room temperature was exposed to normal air introduced by ventilation of the vacuum chamber. Typical ERD spectra measured at several stage of the air-exposure are shown in Fig.5, where the spectra for the specimen as-implanted (○) and exposed for 75hrs (●) are shown. The ERD spectrum for the as-implanted specimen in Fig.5 indicates that D atoms implanted into Li_2ZrO_3 are distributed from the surface marked by the arrow (D) into deeper layers. On the other hand, the ERD spectrum for the specimen exposed for 75hrs apparently shows that the intensities between the channel numbers marked by both arrows (H) and (D) are reduced. The latter spectrum, in fact, represents that the ERD spectrum for the as-implanted specimen is reduced to the hatching part by the air exposure and the distribution subtracted by the hatching part represents H atoms which are taken in the specimen by the air exposure. This fact indicates that D atoms implanted into Li_2ZrO_3 are exchanged for H atoms by exposure to normal-air, namely in air-vapor at room temperature, which is very similar to the result observed for $SrCe_{0.95}Yb_{0.05}O_{3-\delta}$.

The peak heights of D and H retained in the specimen measured during the air-exposure process are shown as a function of exposure time in Fig.6. It is seen from Fig.6 that the retained concentration of D decreases and that of H increases as the exposure time increases, while sum of D+H concentrations is kept to almost constant. For comparison, the similar experimental data for $SrCe_{0.95}Yb_{0.05}O_{3-\delta}$. are shown in Fig.7. From comparison between Fig.6 and Fig.7, it is seen that the rate of the D-H exchange in Li_2ZrO_3 is by an order of magnitude smaller than that in $SrCe_{0.95}Yb_{0.05}O_{3-\delta}$.

For $SrCe_{0.95}Yb_{0.05}O_{3-\delta}$, the exchange of H atoms implanted with 5 keV H^+ ions up to saturation concentration for D atoms by exposure to D_2O vapor carried by ventilated dry-air was also measured. However, it was found that almost no exchange of H-D in $SrCe_{0.95}Yb_{0.05}O_{3-\delta}$ [7,9] took place , which indicates an anomalous great isotope effect in the hydrogen isotopes exchange. Based on these results, the reaction processes are proposed to explain the anomalous great isotope effect : dissociative adsorption of OH and H from H_2O molecule at the surface , diffusion of H and re-emission of H due to HD

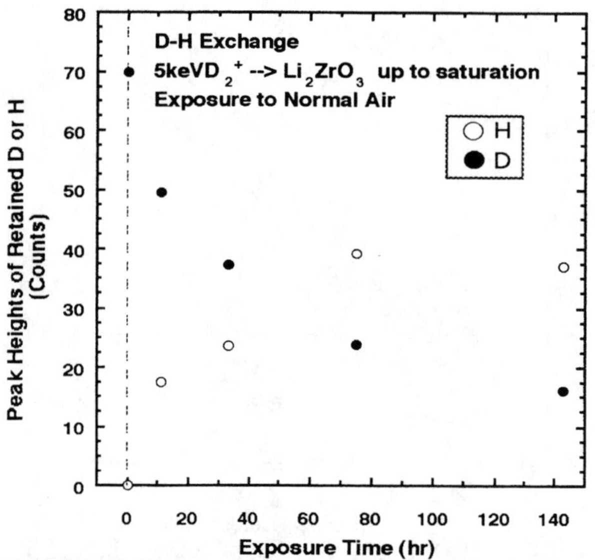

FIGURE 6. The peak heights of D and H retained in the Li_2ZrO_3 specimen, implanted with 5 keV D_2^+ ions up to saturation concentration at room temperature and subsequently exposed to normal air at room temperature, plotted as a function of exposure time.

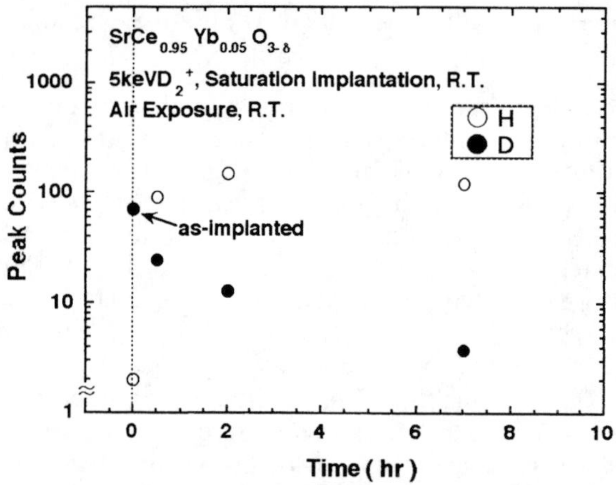

FIGURE 7. The peak counts of D and H retained in the $SrCe_{0.95}Yb_{0.05}O_{3-\delta}$ specimen, implanted with 5 keV D_2^+ ions up to saturation concentration at room temperature and subseqently exposed to normal air at room temperature, plotted as a function of exposure time.

molecular recombination and trapping of H in vacant sites of D as shown schematically in Fig.8. The hydrogen exchange model proposed suggests that the great isotope effect is ascribed to diffusion in the bulk and dissociative adsorption at the surface. It is very interesting to investigate such an isotope effect in Li_2ZrO_3 . The similar experiments with Li_2ZrO_3 are undertaken.

SUMMARY

The exchange of deuterium implanted into Li_2ZrO_3 for protium in air-vapor by the exposure at room temperature has been studied by means of the ERD technique with 1.7 MeV He^+ ion beam. It has been found that the retained number of D implanted into Li_2ZrO_3 decreases, while the number of H taken in by the air-exposure increases as the exposure time increases.

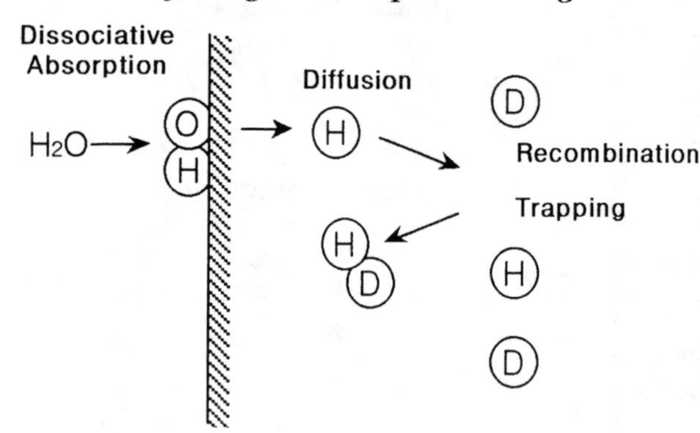

FIGURE 8. A model for exchange of hydrogen isotopes in oxide ceramics and water-vapor.

ACKNOWLEDGEMENTS

The authors are indebted to Professors T. Matsui, T. Miyazaki and H. Iwahara and Dr T. Nagasaki in Nagoya University, for valuable discussion. The authors are grateful to Dr T. Suzuki, in Kawasaki Heavy Industries Ltd for supplying the specimen of Li_2ZrO_3. This research is partly supported by Grant-in-Aid for Scientific Research on Priority Areas (No.260) and (No.271) from The Ministry of Education, Science, Sports and Culture, Japan.

REFERENCES

1. H.Iwahara: Solid State Ionics 77,(1995)289 and 86-88,(1996)9.
2. S. Shin, H. Huange, M. Ishigame and H. Iwahara: Solid State Ionics 40/41(1990)910.
3. Y.Yugami, S. Matsuo and M. Ishigame : Solid State Ionics 77,(1995)195.
4. N. Sata, K. Hiramoto, M. Ishigame, S. Hosoya, N. Nijimura,S. Shin : Phys. Rev. B54, (1996) 15795.
5. N. Matsunami, T. yajima and H. Iwahara : Nucl. Instr. Meth. B65,(1992) 278.
6. B.Tsuchiya, K. Soda, J. Yuhara, K. Morita and H. Iwahara : Solid State Ionics (1998) in press.
7. B.Tsuchiya, E. Iizuka, K. Soda, K. Morita and H. Iwahara : J. Nucl. Mater.258-263 (1998) 555.
8. J. C. Barbour and B. L. Doyle : Handbook of Modern Ion Beam Materials Analysis ed. by J. R. Tesmer and M. Nastasi, MRS. (1995) p. 83.
9. E.Iizuka, T.Horikawa, B.Tsuchiya, K,Morita and H.Iwahara ; Jpn. J. Appl. Phys. to be published.

ADVANTAGES OF A NUCLEAR CHARGE DISPERSIVE (ΔE-E) DETECTION SYSTEM IN HYDROGEN PROFILING BY ERDA

M. El Bouanani[1], P.N. Johnston, I.F. Bubb, P. Jarosch and R.C. Short

Department of Applied Physics, Royal Melbourne Institute of Technology, GPO Box 2476V, Melbourne 3001, Australia

Elastic Recoil Detection Analysis (ERDA) is a well established ion beam method for depth profiling of H in thin films and near surface region of materials. The usual analytical method involves the use of He-4 incident ions with a stopping filter in front of a Si energy detector. The technique has the advantages of (i) less H diffusion and loss than irradiation with heavy incident ions and (ii) unity detection efficiency. However, the depth resolution deteriorates due to ion straggling and thickness variations in the stopping filter. A nuclear charge dispersive detection method based on the use of a ΔE-E silicon detector telescope with ΔE thickness of 4 μm is described. The gains in depth resolution and analyzable depth are demonstrated. Normalization based on the forward-scattered He spectrum is used to eliminate the uncertainties due to the loss of H during irradiation of conventional Kapton and Mylar standards. The quality of the ΔE detector is the critical factor in determining the benefits of using a ΔE-E detector system.

INTRODUCTION

Elastic Recoil Detection Analysis (ERDA) was initiated by L'Ecuyer et al. in 1975 [1] and has become a widely used method for depth profiling light elements in solid materials. ERDA analysis of H using a low energy He beam was first reported by Doyle and Percy [2]. The most important advantage of using He ions is that it causes less H loss during irradiation and analysis compared to the use of heavier incident ions. This is crucial for reliable and accurate depth profiling of H because irradiation-induced diffusion and loss are major sources of error in H-ERDA. In general, the use of Ion Beam Analysis methods including ERDA are subjected to four requirements: (i) good depth resolution, (ii) quantitativity, reliability and accuracy, (iii) sensitivity or low detection limit and (iv) mass resolution and identification.

The optimization of depth resolution has been subject to extensive studies [3-11] of most of the contributing parameters covering the mass of the incident ions, kinematics considerations and different detection approaches [6-9,12]. Considerable effort has been dedicated to make ERDA depth profiling of H quantitatively accurate and reliable. The discrepancies between the published non Rutherford He - H cross sections [13-15] are direct translation of the problem of ion beam irradiation-induced H diffusion and loss. As an alternative, quantitative depth profiles of H are usually calculated by normalizing on the H recoil spectrum from Mylar, Kapton or Polystyrene standards. Any noticeable change in target H content during ion beam irradiation/ or analysis affects the accuracy and reliability of the ERDA measurements. To minimize those effects, light incident ions such as He beams are usually preferred.

The most commonly used detection set-up for He-ERDA measurement of hydrogen is the stopping filter method. A 10 μm Al filter is usually used to stop the forward scattered He ions and to record exclusively the hydrogen energy spectrum in a Si detector. The limitations to the above detection method are (i) the straggling due to the Al filter dominates and deteriorates the surface depth resolution, (ii) the filter detection system is not a mass identification method. Contributions from hydrogen isotopes and electronic noise to H spectrum cannot be resolved. In addition plural scattering in the filter enhances the low energy background affecting the sensitivity/low detection limit. To overcome the above pitfalls, other detection systems performing mass or atomic number identification as well as energy measurement have been developed [6-9,12]. We first initiated and outlined the potential improvements for ERDA hydrogen analysis by using a silicon ΔE-E telescope with 10 μm ΔE detector [16]. Other subsequent studies showed significant improvement in sensitivity [17].

Here we describe the advantages of this new methodology based on a ΔE-E detection telescope with a 4 μm ΔE detector. The fundamental requirements for the optimal thickness of the ΔE detector, the gains in surface depth resolution, the extension of analyzable depth, the elimination of a need for H standards for quantitative depth profiling as well as the improvement of sensitivity are discussed.

---

[1] Current Address: Department of Physics, University of North Texas, Denton, Texas, USA

CP475, *Applications of Accelerators in Research and Industry*,
edited by J. L. Duggan and I. L. Morgan

FUNDAMENTAL REQUIREMENTS

He incident ions of 2.5 MeV with a detection system placed at 30° relative to the direction of the incident He beam is a typical set up for H-ERDA. In the stopping filter method the 10 μm thickness of the Al filter is the minimum thickness required to stop all the forward scattered He ions. In the case of the ΔE-E system, the thickness of the ΔE detector is determined by the fact that the minimum energy loss in the active area of the ΔE detector should be above the electronic noise of the ΔE electronic set-up, the straggling and thickness variations in the ΔE detector and the energy loss in the dead layers.
If one assumes that all electron-holes pairs generated in the active layer of the ΔE detector are collected and the corresponding energy loss is $\Delta E_{active\text{-}layer}$, the minimum thickness t_{min} of the ΔE detector is defined by the lowest acceptable energy loss signal of hydrogen in the active layer of the ΔE detector. In fact t_{min} is optimal when :

$$\Delta E_{active-layer} \geq \Delta E_{res}/2 \qquad (1)$$

for 1.2 MeV hydrogen recoiled ions. ΔE_{res} is the energy loss resolution of the ΔE detector and is defined as the FWHM of the ΔE signal. The thickness t_{min} is then the total thickness of the ΔE detector including the active layer and dead layers. As can be seen in equation (1), the energy loss resolution ΔE_{res} is the most important parameter in determining the minimum usable thickness of the ΔE detector.
In the case of very thin ΔE detector , the major components of the ΔE_{res} are due to the electronic noise/set-up ΔE_{ele} and the ΔE thickness and thickness inhomogeneity-induced straggling ΔE_{str}.

$$(\Delta E_{res})^2 \approx (\Delta E_{ele})^2 + (\Delta E_{str})^2 \qquad (2)$$

The electronic noise ΔE_{ele} of the ΔE detector electronic set-up can be measured using a calibrated pulser. The straggling component ΔE_{str} (straggling due the thickness and the thickness inhomogeneity) of the ΔE detector can be measured during the ΔE energy loss calibration as shown later and described in detail in our earlier work [16]. In this particular study where the thickness of the ΔE detector is about 4 μm, the energy loss resolution ΔE_{res} is dominated by the electronic noise and is about 40 keV. The straggling for 1.2 MeV H ions caused by the ΔE detector is about 33 keV. If one assumes that the topography (thickness inhomogeneity) of both the front and the back surfaces of the ΔE detector are identical , the straggling contribution to the energy loss resolution ΔE_{res} is due mainly to the thickness fluctuations of the front side of the ΔE detector and is approximately (33/2) keV, half the total straggling caused by ΔE detector. This is a good approximation because the straggling is mainly due to thickness inhomogeneity of the ΔE detector.
The high electronic noise is mainly due to the poor impedance matching between the preamplifier and the ΔE detector. The development of special preamplifiers to match the high capacitance of very thin ΔE detectors is required to significantly reduce the electronic noise and improve the energy loss resolution of very thin Silicon ΔE detectors. In fact, the current ΔE energy loss resolution of 40 keV is good enough and the excellent separation between He and H components is clearly seen in the ΔE-E two dimensional spectrum of kapton in Figure 1. However, even if the thickness inhomogeneity-induced straggling for the 4 μm ΔE detector is not a limiting parameter for He-H separation, it remains a dominant component in the optimization of the depth resolution as discussed later.

EXPERIMENTAL PROCEDURE

The measurements have been performed at the RMIT 1 MV Tandem accelerator with 2.5 MeV $^4He^{++}$ beam. A glancing incident beam impinged at 75° to the surface sample normal. The silicon ΔE-E telescope was placed at 30° to the incident beam direction. A 1x3 mm^2 vertical rectangular slit collimated the ΔE-E telescope at 10 cm from the target. The ΔE and E detectors are from ORTEC with thicknesses of respectively 4 μm and 500 μm. The E and ΔE detectors were operated in coincidence mode. The collection of the data use a CAMAC based multi-parameter pulse height analysis system controlled by a MacIntosh Quadra 900 computer and SPARROW KMAX software with event by event data storage to allow for off line analysis.

Energy loss calibration of ΔE

Both detectors E and ΔE were calibrated using 0.8 - 1.8 MeV H beam scattered on a 20 nm thin gold film on silicon substrate in RBS configuration. The thickness of the gold film is chosen so thin that the broadening of the backscattered H is smaller than the energy resolution of the E detector. This allows the use of secondary H beam with well defined energy, acceptable energy dispersion and where the flux of H can be easily controlled.
The ΔE-E telescope was placed at a scattering angle of 150° relative to the incident H direction. The scattered H peak from the thin gold layer is a well defined gaussian with a FWHM of 14 keV. For each incident energy two sets of data collection were performed with and without the ΔE detector in front the E detector. This allows

energy and energy loss calibration for both ΔE and E detectors. A more detailed procedure for the absolute energy loss calibration measurements is described in our earlier work [16]. For 1.0 MeV H, the energy widths (FWHM's) of the signals from E and ΔE detectors in this configuration were 14 keV and 40 keV.

RESULTS AND DISCUSSION

Figure 1 shows a two dimensional spectrum recorded using a 2.5 MeV He-4 beam incident on a Kapton sample. The H and the forward scattered He are well separated. Figure 1 illustrates very well the gain in analyzable depth. The low energy threshold is about 400 keV. This approximately doubles the analyzable depth when compared to the 800 keV low energy threshold in He-ERDA of hydrogen using 10 μm Al filter. Figure 1 shows a very clean spectrum with no other element such as Deuterium or Tritium in the Kapton sample. In addition, the operation of both ΔE and E detectors in coincident mode eliminate any electronic noise contribution to hydrogen energy spectrum which is an important parameter, besides the cross section, in defining the best sensitivity for this method. In fact, it was demonstrated that ΔE-E detection methodology significantly lowers the detection limit and a sensitivity of less than 75 at. ppm H [17] is achieved.

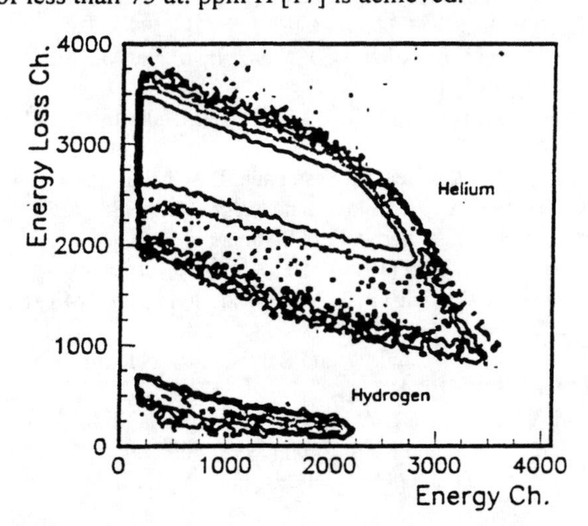

FIGURE 1. ΔE-E ERDA spectrum of Kapton sample.

Figure 2a shows the ΔE-E spectrum of Si_3N_4/Si sample. One can see clearly that the hydrogen contamination is limited to the Si_3N_4 thin film. Figures 2c and 2b show respectively He and H energy projections. As shown in Figure 2c, the total energy projection of the forward scattered He was obtained by adding both E and ΔE signals which gives a much better energy resolution. In Figure 2b the E and ΔE signals for H are not added

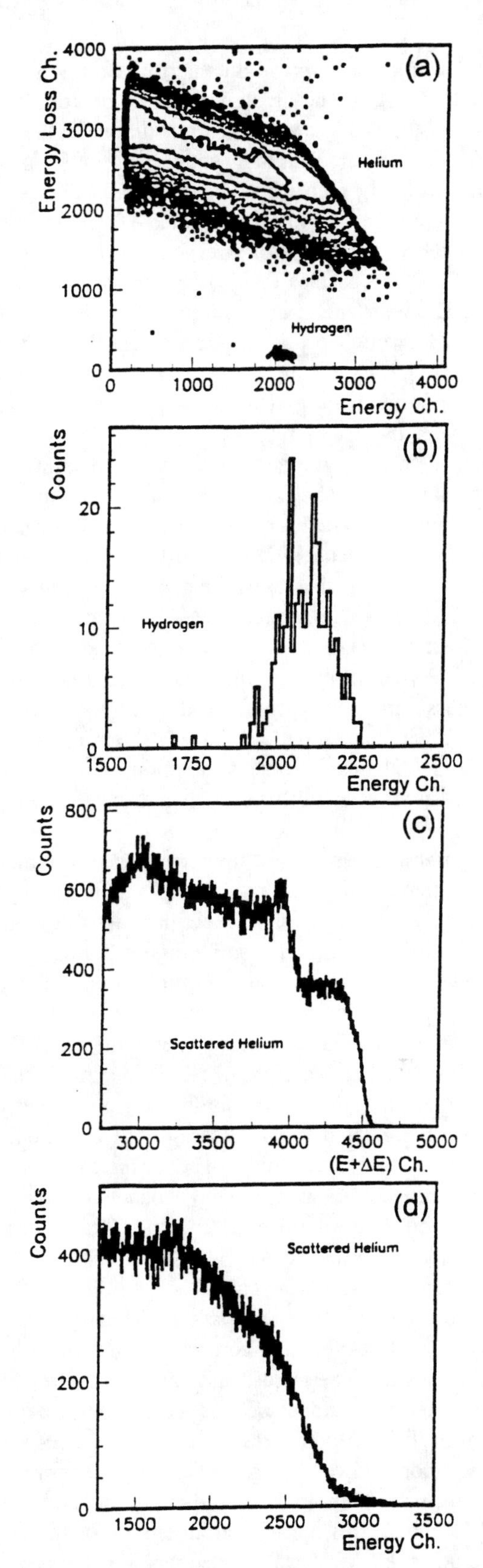

FIGURE 2. (a) ΔE-E ERDA spectrum of Si_3N_4/Si sample. (b) Projected energy spectrum of the H component. (c) Total energy (ΔE+E) projection of the forward scattered He. (d) Energy projection (E) of the forward scattered He.

because it would not improve the energy resolution. The summation of both E and ΔE signals should be done only in the case where the energy loss resolution of the ΔE detectors is much less than their straggling contribution. This can be clearly seen when one compares Figure 2c and 2d. Figure 2c shows well defined (familiar good resolution RBS) front edges corresponding to scattered He from Si at the surface (Si_3N_4) and from the silicon substrate. The poor energy resolution can be clearly seen in Figure 2d through diffusion-like shapes of the front edges.

The gain in surface energy resolution when using ΔE-E system over the stopping filter method was also investigated. To compare the surface energy resolutions, a Kapton sample was analyzed in the same experimental conditions but using ΔE-E system and 10 μm Al filter in front an energy detector. The surface energy resolution was about 69 keV in the case of the Al filter detection method. The gain in surface energy resolution is significant since it is measured to be only 36 keV in the case of ΔE-E system. In fact, this value is dominated by the thickness inhomogeneity-induced straggling of the ΔE detector which is about 33 keV. The fabrication of ΔE detectors with less thickness variation will greatly benefit the surface energy resolution and consequently the surface depth resolution.

The ΔE-E methodology allows the recording of both the forward scattered He ions and the H recoils in one measurement and in the same experimental conditions (total number of incident ions, solid angle). For samples with known composition, auto-normalization of the forward scattered He energy spectrum in a manner similar to Rutherford Backscattering Spectrometry (RBS) allows inter-normalization of the H recoil spectrum. This eliminates the needs for H standards and consequently the uncertainties related to the H loss in the commonly used polymers as H references. The accuracy of the absolute H concentrations are then dependent directly and mainly on both the He-matrix and He-H cross section data. A modified RUMP [18] or other software [19] can be used for such analysis.

The ΔE detector is not a passive absorber and consequently its thickness is a compromise between the minimum acceptable energy loss and the electronic noise of the thin detector and associated electronics (major component of the energy loss resolution) . The use of a 4 μm ΔE detector allowed us to obtain good separation between the electronic noise level and the ΔE signal from 1.8 MeV H beam. This means that an energy of the He beam as high as 3.7 MeV can be used which extends the accessible depth Otherwise, when using 2.5 MeV He beam, a thickness of less than 3 μm can be used for ΔE detector which can improve further both the depth resolution and the accessible depth.

CONCLUSION

The use of ERDA low energy He beams and a ΔE-E silicon detection system with its simultaneous detection of recoiled H and forward scattered He projectiles have been demonstrated to:

- improve significantly the depth resolution
- largely extend the analyzable depth.
- eliminate the need for H standards and their associated uncertainties
- significantly improve the sensitivity by eliminating the electronic noise and separating possible deuterium and tritium contributions from H recoil spectra.

REFERENCES

1. L'Ecuyer, J., Brassard, C., Cardinal, C., Chabbal, J., Deschenes, L., Labrie, J.B., Terreault, B., Mariel, J.G. and St-Jacques, R, J. Appl. Phys. **47** (1976) 881.
2. Doyle, B.L. and Percy, P.S., Appl. Phys. Lett. **34** (1979) 811.
3. Paszti, F., Szilagyi, E and Kotai, E., Nucl. Instr. and Meth. **B 54** (1991) 507.
4. Turos, A. and Meyer, O., Nucl. Instr. and Meth. **B 4** (1984) 92.
5. Nagata S., Yamaguchi, S. and Fujino, Y., Nucl. Instr. and Meth. **B 6** (1985) 533.
6. Ross, G.G., Terreault, B., Gobeil, G., Abel, G., Coucher, C. and Veilleus G., J. Nucl. Mater. **128/129** (1978) 730.
7. Kruse, O. and Carstanjen, H.D.., Nucl. Instr. and Meth. **B 89** (1994) 191.
8. Groleau, G, Gujrathi,S.C. and Martin, J.P., Nucl..Instr. and Meth. **218** (1983) 11.
9. Thomas, J.P., Fallavier, M., Ramdane, D., Chevarier, N. and Chevarier, A., Nucl. Instr. and Meth. **218** (1983) 125.
10. Benka, O., Brandstotter, A. and Steinbauer, E., Nucl. Instr. and Meth. **B 85** (1994) 650.
11. Brice, D.K. and Doyle, B.L., Nucl. Instr. and Meth. **B 45** (1990) 265.
12. Kreissig, U., Grotzschel, R. and Behrisch, R., Nucl. Instr. and Meth. **B 85** (1994) 71.
13. Baglin, J.E.E., Kellock, A.J., Crockett, M.A.and Shih, A.H., Nucl. Instr. and Meth. **B 64** (1992) 469
14. Wang, H. and Zhai, G.Q., Nucl. Instr. and Meth. **B 34** (1988) 145.
15. Benenson, R.E., Wielunski, L.S. and Lanford, W.A., Nucl. Instr. and Meth. **B 15** (1986) 453.
16. El Bouanani, , M., Johnston, P.J., Bubb, I.F. and Whitlow, H.J., in Application of Accelerators in Resesarch and Industry, eds: J.L. Duggan and I.L. Morgan, AIP, New York, 1997, p.647.
17. Sweeny, R.J., Prozesky, V.M., Churms, C.L., Padayachee, J. and Springhorn, K., Nucl. Instr. and Meth. **B 136-138** (1998) 685.
18. Doolittle, L.R., Nucl. Instr. and Meth. **B 15** (1986) 227.
19. Mayer, M., Technical Report IPP 9/113, Max-Plank Institut fur Plasmaphysik, Garching, BRD, 1997.

Status of the Accelerator Facility at the Environmental Molecular Sciences Laboratory

D.E. McCready, S. Thevuthasan, and W. Jiang

Environmental Molecular Sciences Laboratory, Pacific Northwest National Laboratory, Richland, WA 99352

An accelerator facility dedicated to ion beam materials analysis and modification has been completed at the Environmental Molecular Sciences Laboratory (EMSL), a United States Department of Energy collaborative scientific user facility located at the Pacific Northwest National Laboratory in Richland, WA. The EMSL accelerator facility is based on a Model 9SDH-2 3.4 MV tandem ion accelerator manufactured by National Electrostatics Corporation (NEC, Middleton, WI), which includes RF plasma and sputter ion sources. Three beam lines were originally constructed with integral end stations for materials analysis and modification. Recently, the +30° beam line was extended to accommodate an NEC electrostatic microquad assembly, which provides focussed beam spots of 20 μm or less on target. Efforts are currently underway to incorporate particle induced x-ray emission (PIXE), Rutherford backscattering spectrometry (RBS), and nuclear reaction analysis (NRA) capabilities in the microbeam line end station.

INTRODUCTION

Recently, small accelerators have been used extensively for research in materials analysis and modification [1-5]. Common methods of materials analysis using high-energy ion beam analysis include Rutherford backscattering spectrometry (RBS), ion channeling, nuclear reaction analysis (NRA), elastic recoil detection analysis (ERDA), particle-induced x-ray emission (PIXE), and particle-induced gamma ray emission (PIGE). Recent developments, such as nuclear microprobe techniques, the use of radioactive ion beams, and accelerator mass spectrometry (AMS), have spurred additional interest in ion accelerator-based materials research. While traditional methods of ion beam modification of materials (e.g., implantation) have long utilized ion beams with energies on the order of 10 keV to 100 keV, there has been a more recent trend toward the use of MeV ion beams for deep implantation [6-9].

The Environmental Molecular Sciences Laboratory (EMSL) located at the Pacific Northwest National Laboratory (PNNL) in Richland, WA is a new national scientific user facility sponsored by the Office of Biological and Environmental Research of the United States Department of Energy (DOE). The primary general mission of EMSL is the conduct of research aimed at the development of a molecular-level understanding of physical, chemical, and biological processes. In particular, EMSL research is directed toward meeting the scientific challenges associated with the DOE environmental mission. Current topics of study include nuclear and hazardous waste processing and storage, environmental remediation, atmospheric chemistry, and human health effects. EMSL research also supports other national missions, including science and technology, energy, national security, and education.

To achieve these objectives, initial plans for the construction of EMSL called for the incorporation of a wide variety of analytical capabilities, including an ion accelerator facility. The EMSL accelerator facility originally consisted of a 3.4 MV tandem ion accelerator with integral RF plasma and sputtering ion sources, along with three beam lines and end stations. A recent addition to the EMSL accelerator facility has been the extension of one beam line to accommodate a microbeam capability. The microbeam line end station is currently under construction to include PIXE, RBS, and NRA capabilities. The purpose of this paper is to describe the current status of the EMSL accelerator facility and its capabilities. A description of the accelerator, the ion sources, and the end stations, as well as a discussion of the current research in the laboratory are presented.

EMSL ACCELERATOR SYSTEM

Figure 1 is a schematic view of the EMSL accelerator system. Illustrated therein are the integral ion sources, accelerator tank, beam lines, and end stations. The RF plasma ion source is a National Electrostatics Corporation (NEC) Alphatross. The Alphatross source is primarily used for the production of helium ions, but can also be used as a source of H and ^{15}N ions. All other beams are produced by Cs^+ sputtering of solid cathodes using the NEC SNICS ion source. Both the Alphatross and SNICS sources produce singly charged negative ions that are steered into the accelerator by an injector magnet. The NEC Model 9SDH-2 3.4 MV tandem electrostatic accelerator is equipped with twin Pelletron chains which can carry up to 300 μA of charging current to the terminal. Accelerated ions are stripped to positive charge at the terminal by transmission through nitrogen gas or a carbon foil. High-energy beam selection and steering are accomplished using an analyzing (switching) magnet. Three beam lines with four end stations are currently located at +30°, +15°, and -15° in the horizontal plane. Typical base pressure in the low-energy

CP475, *Applications of Accelerators in Research and Industry*,
edited by J. L. Duggan and I. L. Morgan

Schematic View of the Accelerator System

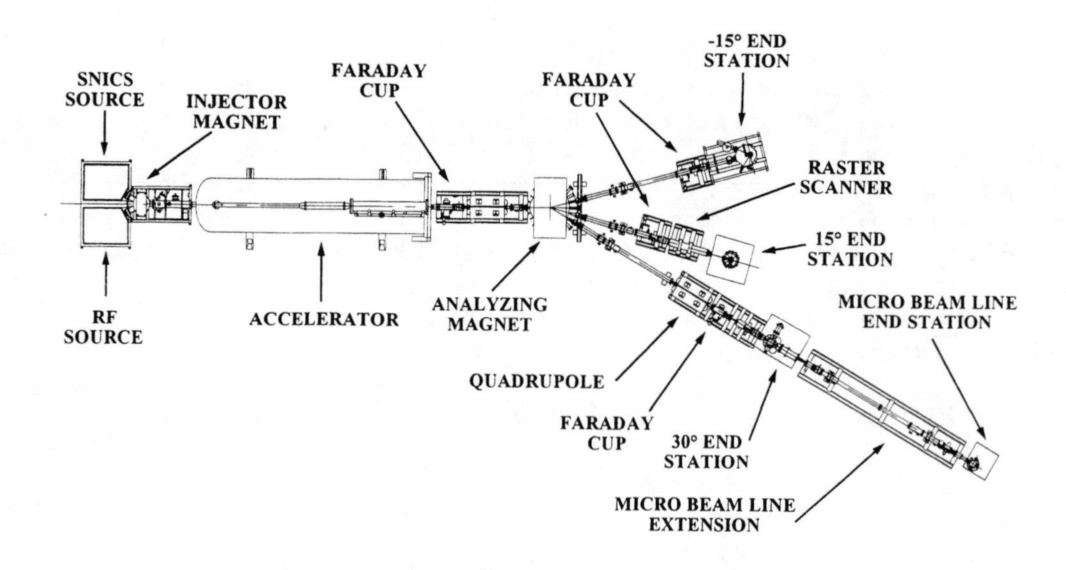

FIGURE 1. Schematic view of the EMSL accelerator system.

portion of the system is on the order of 10^{-6} Pa and 10^{-7} Pa in the high-energy sections. Under normal operation, typical pressures in the high-energy beam lines are in the 10^{-6} Pa range. Thevuthasan, et al provided a more detailed description of the EMSL accelerator facility in a previous study [10].

The UHV end station on the +30° beam line is used for ion channeling, RBS, NRA, and ERDA. A fixed surface barrier detector is located at a scattering angle of 150° and is used for RBS and channeling measurements. A second fixed surface barrier detector is at 30° for ERDA. An additional movable surface barrier detector is available for RBS measurements. There is also a 5-cm x 10-cm bismuth germanate (BGO) detector attached to a linear translator for gamma ray measurement during NRA. The UHV end station has a sample manipulator with non-ambient temperature capability (170K-1300K), three stepper motor-driven axes of rotation (polar, azimuthal, and flip), and three axes of translation (x, y, and z). The angular resolution is 0.01° on the polar axis and 0.001° on the azimuthal and flip axes. The manipulator also utilizes the universal sample transfer platen used throughout EMSL. This technology allows a sample to be synthesized, processed, and characterized without exposure to air [11]. In addition, existing portable vacuum systems for electrochemical testing, optical measurements, specimen processing, or material deposition can be connected to this end station.

A 5-m extension of the +30° beam line accommodates the new microbeam capability. This extension includes beam steerers, a Faraday cup, a selection of apertures (6350 μm to 200 μm), a beam profile monitor, and a microquad assembly. The microquad assembly is an NEC Model EQQ5.5-5 Electrostatic Quadrupole Quadruplet Lens with an entrance aperture of 2.0 mm and an exit opening of 5.5 mm. The focussing elements provide 10-to-1 demagnification of the beam on the target. The microquad assembly is located 4.2 m from the object aperture, and the image position is 152 mm beyond the lens exit [12]. Planned use of the microbeam line end station includes RBS, PIXE and NRA experiments. A surface barrier detector is located at a scattering angle of 135° for RBS measurements. An x-ray detector for PIXE analysis and a BGO detector for NRA measurements have yet to be installed. The integral four-axis manipulator uses the EMSL sample transfer platen and is equipped for non-ambient temperatures (170K-1300K). Initial testing of the microbeam system using 2.04 MeV He^+ and an NEC standard reference target resulted in a beam spot on target of <20 μm in diameter. A CCD zoom camera (100X) provides visual indication of the size and location of the beam spot on target.

The +15° beam line is designed primarily for ion implantation experiments. Although ion implantation is routinely carried out under high vacuum conditions, UHV conditions can be achieved in this end station. In addition, a surface barrier detector is located at a scattering angle of 135° for RBS measurements. A 10-cm x 10-cm NaI detector is mounted at 90° for NRA gamma ray measurements. The end station includes a four-axis manipulator that uses the EMSL sample transfer platen and is equipped for non-ambient temperatures (170K-1300K).

An NEC RC43 end station is attached to the -15° beam line and is configured for several ion beam analytical techniques including RBS, NRA, PIXE, PIGE, and ERDA. A fixed surface barrier detector is located at a scattering angle of 170° for RBS measurements. Another surface barrier detector for ERDA measurements is mounted on a rotating platform. A translatable solid-state detector is used for PIXE analysis, and a 5-cm x 5-cm NaI detector is available for NRA and PIGE measurements. The automated five-axis manipulator uses NEC standard holders capable of supporting multiple small study samples. Equipped with an efficient load lock system, the high vacuum RC43 end station allows for rapid throughput of samples subject to routine analysis.

CURRENT PROGRAMS

Major environmental research presently underway in EMSL includes studies of the surface chemistry of oxides and minerals, radiation effects in nuclear waste forms, remediation of nuclear waste, and long-term storage of spent nuclear fuel. A few of the related programs in the EMSL accelerator facility are described below.

RBS and Channeling Measurements in Single Crystal Oxides

Current EMSL studies of the surface chemistry of oxides and minerals require model single-crystal surfaces. To meet the demand for such study samples, EMSL is equipped with molecular beam epitaxial (MBE) and metal organic chemical vapor deposition (MOCVD) systems which have been used to grow a variety of epitaxial oxide films on a number of oxide and metal surfaces. The RBS and ion channeling tools available in the EMSL accelerator facility are used for the characterization of these films. In particular, ion channeling is used to determine the disordering due to defects and dislocations of the crystal lattice in the film and at the interface. Recent and continuing RBS and channeling studies include investigation of film crystallinity and of the film-substrate interface properties, including interdiffusion [13-15].

Radiation Effects in Oxides and Ceramics

Understanding radiation-induced damage processes, defect formation, defect migration, and crystal recovery in oxides and ceramics is of vital importance to the development of semiconductor device fabrication technology. This information is also of interest in the application of these materials to nuclear waste storage. RBS and ion channeling are used to study damage in these materials as a result of light- and heavy-ion irradiation. Radiation damage is studied as a function of ion species, ion dose, dose rate, and temperature. Damage recovery processes are studied by quantifying residual damage as a function of annealing time and temperature. Recent studies include radiation effects in SiC and GaN, which are important semiconductor materials [16-19]. Other radiation effect studies involve $SrTiO_3$ (perovskite) and $Gd_2Ti_2O_7$ (pyrochlore), which are candidate materials for nuclear waste stabilization and immobilization [20].

Nuclear Waste Remediation

A major environmental issue facing DOE is the stabilization and immobilization of high level radioactive waste (HLW) left over from defense nuclear production during the cold war era. Current plans call for HLW to be vitrified and immobilized in glass or ceramic waste forms prior to permanent disposal [20, 21]. A vital concern for the long-term success of these plans will be the interactions of projected HLW forms with water in the environment. Recent investigations in the EMSL accelerator facility have focussed on hydrogen and oxygen uptake in HLW form glasses leached with ^{18}O-labeled water. These studies were accomplished using $^{19}F(^1H, \alpha\gamma)^{16}O$, $^{15}N(^1H, \alpha\gamma)^{12}C$, d(d,p)T, and $^1H(^{18}O, \alpha)^{15}N$ nuclear reactions. The goal of these experiments is to gain a better understanding of the mechanisms that control leaching of HLW forms by water.

SUMMARY

An accelerator facility for ion beam materials analysis and modification has been commissioned at the Environmental Molecular Sciences Laboratory (EMSL), a national scientific user facility located at the Pacific Northwest National Laboratory (PNNL) in Richland, WA. New additions to the EMSL accelerator facility include a microbeam line and end station. Since its inception in 1997, several major environmental research programs have been established at the EMSL accelerator facility.

ACKNOWLEDGEMENTS

This work was partially supported by the EMSL project, which in turn was funded by the Office of Biological and Environmental Research of the United States Department of Energy. Pacific Northwest National Laboratory is a multi-program laboratory operated for the Department of Energy by Battelle Memorial Institute under Contract No. DE-AC06-76RLO 1830. W. Jiang was partially supported by the United States Department of Energy, Office of Basic Energy Sciences, Materials Science Division.

REFERENCES

1. Hatori, S., Kobayashi, K., Nakano, C., Suohara, Y., and Yamashita, H., in Applications of Accelerators in Research and Industry, Proceedings of the 14th International Conference, ed. J.L. Duggan, I.L. Morgan, Part I and II (1996) p. 1099
2. Oliver, A., in Applications of Accelerators in Research and Industry, Proceedings of the 14th International Conference, ed. J.L. Duggan, I.L. Morgan, Part I and II (1996) p. 1105.
3. Lin, E.K., Wang, C.W., Teng, P.K., and Fou, C.M., Nucl. Instr. and Meth. B **56/57** (1991) p. 996.
4. Oberschachtsiek, P., Weiser, M., and Kalbitzer, S., Nucl. Instr. and Meth. B **56/57** (1991) p. 1010.
5. Menu, M., Calligaro, T., and Salomon, J., Amsel, G., and Moulin, J., Nucl. Instr. And Meth. B **45** (1990) p. 610.
6. Rey, S., Muller, D., Grob, J.J., and Stoquert, J.P., in Applications of Accelerators in Research and Industry, Proceedings of the 14th International Conference, ed. J.L. Duggan, I.L. Morgan, Part I and II (1996) p. 981.
7. Giedd, R.E., Moss, M.G., Kaufman, J., and Wang, Y.Q., in Applications of Accelerators in Research and Industry, Proceedings of the 14th International Conference, ed. J.L. Duggan, I.L. Morgan, Part I and II (1996) p. 993.
8. Stequert, J.P., Grob, J.J., and Muller, D., in Applications of Accelerators in Research and Industry, Proceedings of the 14th International Conference, ed. J.L. Duggan, I.L. Morgan, Part I and II (1996) p. 1011.
9. Polman, A., in Materials Synthesis and Processing Using Ion Beams, Mater. Res. Soc. Proceedings ed. R.J. Culbertson, O.W. Holland, K.S. Jones, and K. Maex.
10. Thevuthasan, S., Peden, C.H.F., Engelhard, M.H., Baer, D.R., Herman, G.S., Jiang, W., Liang, Y., and Weber, W.J., Nucl. Instrum. And Methods in Phys. Res. A, in press.
11. Thevuthasan, S., Baer, D.R., Engelhard, M.H., Liang, Y., Worthington, J.N., Howard, T.R., Munn, J.R., and Rounds, K.S., J. Vac. Sci. Technol. B **13** (1995) 1900.
12. *Instruction Manual for Operation and Service of Electrostatics Quadrupole Quadruplet Lens, Model EQQ5.5-5*, National Electrostatics Corporation.
13. Thevuthasan, S., Shivaparan, N.R., Smith, R.J., Gao, Y., and Chambers, S.A., Appl. Surf. Sci. **115** (1997) 381.
14. Thevuthasan, S., Jiang, W., McCready, D.E., and Chambers, S.A., Surf. and Interface Analysis, in press.
15. Thevuthasan, S., McCready, D.E., Jiang, W., Kim, Y.J., Gao, Y., Chambers, S.A., Shivaparan, N.R., and Smith, R.J., submitted to Thin Solid Films.
16. Jiang, W., Weber, W.J., Thevuthasan, S., and McCready, D.E., J. Nucl. Mater., **257** (1998) 295.
17. Jiang, W., Weber, W.J., Thevuthasan, S., and McCready, D.E.,, Nucl. Instr. and Meth. B **143** (1998) 333.
18. Jiang, W., Weber, W.J., Thevuthasan, S., and McCready, D.E.,Surf. and Interface Analysis, in press.
19. Jiang, W., Weber, W.J., Thevuthasan, S., and McCready, D.E.,Nucl. Instr. and Meth. B, in press.
20. Weber, W.J., Ewing, R.C., Angell, C.A., Arnold, G.W., Cormack, A.N., Delaye, J.M., Griscom, D.L., Hobbs, L.W., Navarotsky, A., Price, D.L., Stoneham, A.M., and Weinberg, M.C., J. Mater. Res., **12** (1997) 1946.
21. Weber, W.J., Ewing, R.C., Catlow, C.R.A., Diaz de la Rubia, T., Hobbs, L.W., Kinoshita, C., Matzke, Hj., Motta, A.T., Nastasi, M., Salje, E.K.H., Vance, E.R., and Zinkle, S.J., J. Mater. Res., **13** (1998) 1434.

SIMNRA, a Simulation Program for the Analysis of NRA, RBS and ERDA

M. Mayer

Max-Planck-Institut für Plasmaphysik, D-85748 Garching, Germany

SIMNRA is a Microsoft Windows 95/Windows NT program with fully graphical user interface for the simulation of non-Rutherford backscattering, nuclear reaction analysis and elastic recoil detection analysis with MeV ions. About 300 different non-Rutherford and nuclear reactions cross-sections are included. SIMNRA can calculate any ion-target combination including incident heavy ions and any geometry including transmission geometry. Arbitrary multi-layered foils in front of the detector can be used. Energy loss straggling includes the corrections by Chu to Bohr's straggling theory, propagation of straggling in thick layers, geometrical straggling and straggling due to multiple small angle scattering. The effects of plural large angle scattering can be calculated approximately. Typical computing times are in the range of several seconds.

INTRODUCTION

Rutherford backscattering spectroscopy (RBS), nuclear reaction analysis (NRA) and elastic recoil detection analysis (ERDA) with MeV ions are powerful tools for the analysis of the near surface layers of solids. During the last decade several computer programs for the simulation and analysis of spectra obtained in MeV ion beam analysis were developed [1], such as the widely used RUMP [2]. The increase in computer power during the last years has made it possible to drop several limitations of previous programs, which were necessary due to computing time limitations. Additionally an error tolerant fully graphical user interface (GUI) for easy use of the program is highly wishful.

This paper describes the physics concepts of the program SIMNRA (version 4.4). More details can be found in [3, 4].

BASIC CONCEPT

The solid is bombarded by ions with energy E_0. For spectrum synthesis the solid is divided into shallow sublayers with thickness Δx. The simulated spectrum is made up of the superimposed contributions of each reaction[1] from each isotope of each element of each sublayer of the solid. This is illustrated schematically in Fig. 1. The energy spectrum of one reaction of a sublayer is called a "brick" [2]. The brick area is determined from the mean reaction cross-section in the sublayer, while its shape (i.e. the heights of the front and back edges) is determined from the cross-sections at the entrance and exit of the sublayer and the change of the stopping power. The brick has to be folded with

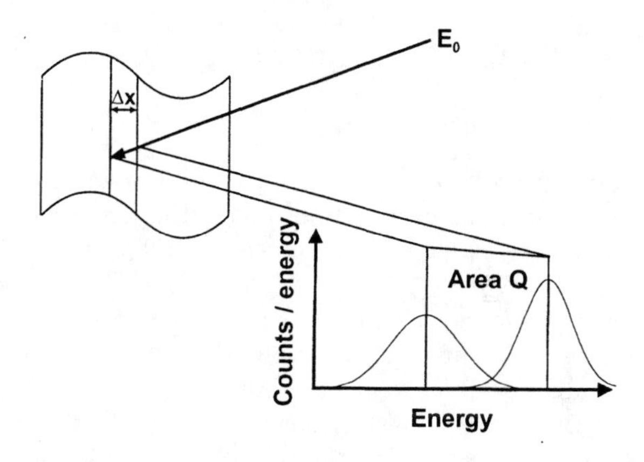

FIGURE 1. Schematic representation of the "brick" concept.

the different energy spread contributions, which vary from the front to the backside of the brick, see Fig. 1. To calculate the content of each channel the folded spectrum has to be integrated over a channel width. This can be combined to a 2-dimensional integral, which is evaluated exactly using Gauss-Legendre integration.

The step width Δx is the most crucial parameter both for the accuracy of the simulation and computing speed. By default SIMNRA uses automatic step width control: The step size Δx is chosen in such a way that the width of the brick is about equal to the full width at half maximum (FWHM) of the energy spread, resulting in small step widths near the surface and larger step widths deep inside the solid, where the depth resolution becomes poor.

CROSS-SECTION DATA

The scattering of two charges is described by the differential Rutherford cross-section, which can be found

[1] A reaction may be, generally, scattering of the projectile ion, creation of a recoil or a nuclear reaction.

CP475, *Applications of Accelerators in Research and Industry*, edited by J. L. Duggan and I. L. Morgan

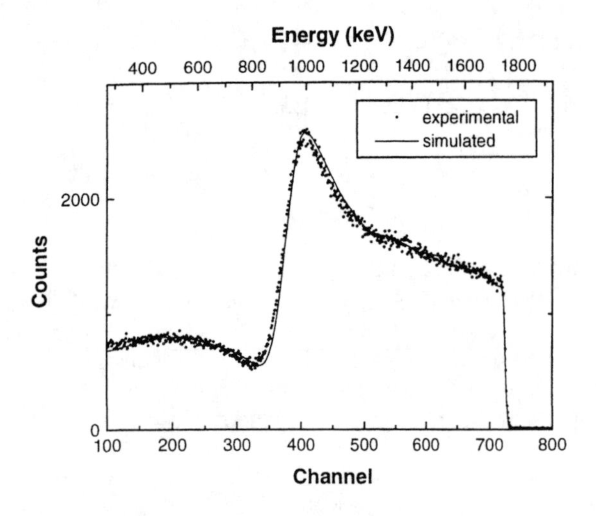

FIGURE 2. Backscattering of 2000 keV protons from silicon, scattering angle $\theta = 165°$. Dots: experimental data; solid line: calculation with SIMNRA, cross-section data from [11].

in many text books, see for example [5]. However, actual cross-sections deviate from the Rutherford cross-section at both high and low energies for all projectile-target pairs.

The low-energy departures are caused by partial screening of the nuclear charges by the electron shells surrounding both nuclei [5, 6]. This screening is taken into account by a correction factor $F(E, \theta)$. SIMNRA uses the angular- and energy dependent correction factor from [6] both for scattered particles and recoils.

At high energies the cross-sections deviate from the Rutherford cross-section due to the influence of the nuclear force [7], and empirical cross-section data have to be used. About 300 different experimental cross-section data sets for non-Rutherford scattering and nuclear reactions for incident protons, deuterons, ^{3}He and ^{4}He ions are included with SIMNRA. Most of the cross-section data were collected by Foster *et al.* [8] and Cox *et al.* [9], they are available at SigmaBase [10]. Some additional cross-section data sets were added by the author. SIMNRA can use differential and total cross-section data. New cross-section data can be added easily by the user.

As an example for the use of non-Rutherford scattering cross-sections Fig. 2 shows the measured and simulated spectra of 2000 keV protons backscattered from silicon.

STOPPING POWER DATA

SIMNRA can use two different sets of electronic stopping power data: The Andersen-Ziegler data for hydrogen and helium ions from [12, 13] and the more recent stopping power data by Ziegler, Biersack and Littmark from [14]. The electronic stopping power of heavy ions in all elements is derived from the stopping power of protons using Brandt-Kitagawa theory [14, 15], the formalism is described in detail in [14].

The nuclear stopping power for helium and heavy ions is calculated with the universal ZBL potential from [14]. The nuclear stopping component for hydrogen isotopes is very small and is neglected.

Stopping in Compounds

SIMNRA uses Bragg's rule [16] for the determination of the stopping power in compounds. However, Bragg's rule assumes that the interaction between the ion and a target atom is independent of the environment. The chemical and physical state of the medium is, however, observed to have an effect on the energy loss, resulting in deviations from Bragg's rule which are most pronounced around the stopping power maximum and for solid compounds such as oxides, nitrides and hydrocarbons. The deviations from Bragg's rule predictions may be of the order of 10–20% [15]. To account for deviations from Bragg's rule SIMNRA offers the option to multiply the stopping power of each layer with a constant correction factor F for each ion species.

ENERGY STRAGGLING

Energy Loss Straggling

When a beam of charged particles penetrates matter, the slowing down is accompanied by a spread of the beam energy which is due to statistical fluctuations of the energy transfer in the collision processes. For thin layers and small energy losses the energy distribution is non-Gaussian and asymmetric [17]. This regime is not implemented in SIMNRA. As the number of collisions becomes large, the distribution of particle energies becomes Gaussian. This regime is described by Bohr's theory [18, 19]. However, Bohr's theory of electronic energy loss straggling is only valid in the limit of high ion velocities. For lower ion energies the deviations caused by the electron binding in the target atoms have to be taken into account. Chu [19, 20] has calculated a correction factor H by using the Hartree-Fock-Slater charge distribution to obtain more realistic values for the electronic energy loss straggling. Graphical representations of H for all elements and various energies can be found in [19, 21]. For low energies and high target Z the Chu straggling theory yields considerably smaller values than the original Bohr theory, for high energies H approaches unity.

For the nuclear energy loss straggling SIMNRA uses Bohr's theory of nuclear straggling [17]. Electronic and nuclear energy loss straggling are independent and are

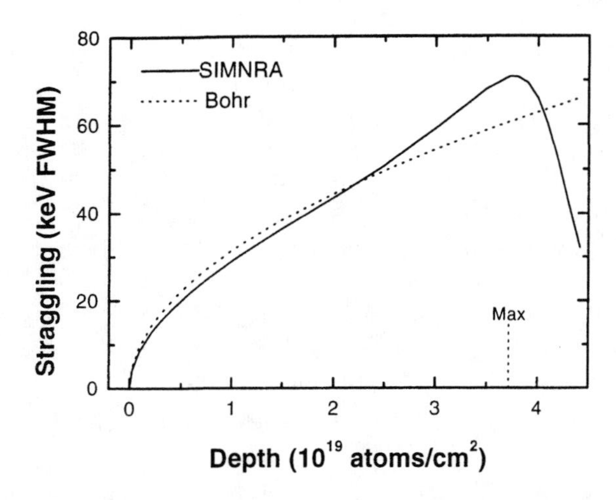

FIGURE 3. Energy loss straggling (FWHM) of 2500 keV ^{4}He ions in Si. Solid line: SIMNRA; dashed line: Bohr's theory. Max: depth at which the particles energy has decreased to the energy of the stopping power maximum.

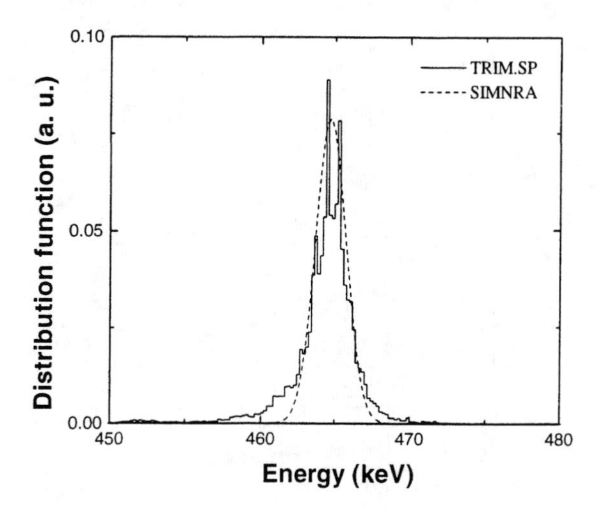

FIGURE 4. Energy distribution due to multiple small angle scattering of 500 keV ^{4}He in Au in a depth of 1.6×10^{17} atoms/cm^2. Incident angle $\alpha = 60°$.

added quadratically.

Propagation of Straggling in Thick Layers

The energy dependence of the stopping power results in a non-stochastic broadening (or squeezing) of an energy-distributed beam. According to [21] the propagation of straggling can be calculated in the following way: If an ion beam with initial mean energy E_i and width ΔE_i penetrates a sublayer with thickness Δx, then the width ΔE_f after penetration of the sublayer is given by

$$\Delta E_f = \frac{S(E_f)}{S(E_i)} \Delta E_i \qquad (1)$$

with E_f the mean energy after the sublayer and $S(E_i)$, $S(E_f)$ the stopping powers at the entrance and exit of the sublayer, respectively. The stochastic effects have to be added to the non-stochastic broadening from Eq. 1.

As an example Fig. 3 shows the energy loss straggling of 2500 keV ^{4}He ions in silicon. The beam broadening and squeezing below the stopping power maximum are clearly visible.

Geometrical Straggling

The finite size of the incident beam and the width of the detector aperture result in a spread $\Delta\beta$ of the exit angle β for outgoing particles [22]. This angular spread leads to energy spread of the particles at the target surface due to a spread $\Delta\theta$ of the scattering angle θ and different path lengths of the outgoing particles in the material. These two contributions to geometrical straggling are not independent of each other and are computed simultaneously.

PLURAL SCATTERING EFFECTS

Usually the trajectories of ingoing and outgoing particles are approximated by straight lines, which are connected at a single point where the reaction took place (single scattering approximation). However, in reality the particle trajectories are determined by a large number of scattering events with small deflection angles, and additional deflections with large deflection angles may occur [23, 24].

Small Angle Scattering

Angular spread due to multiple small angle scattering has been calculated by Sigmund and Winterbon [25] and has been recently reviewed in [21]. SIMNRA uses the same algorithms as presented in [21] for the calculation of multiple small angle scattering, but approximates the energy spread distributions by Gaussian functions. This underestimates the wings of the distributions. The energy distribution of 500 keV ^{4}He in Au is shown in Fig. 4 together with results of the Monte-Carlo code TRIM.SP [26], which takes all collisions into account. The widths of the curves agree well, however, the wings of the distribution are underestimated by SIMNRA.

Large Angle Scattering

Plural scattering with large deflection angles is, for example, responsible for the background below the low energy edge of high Z elements [23, 24]. SIMNRA cal-

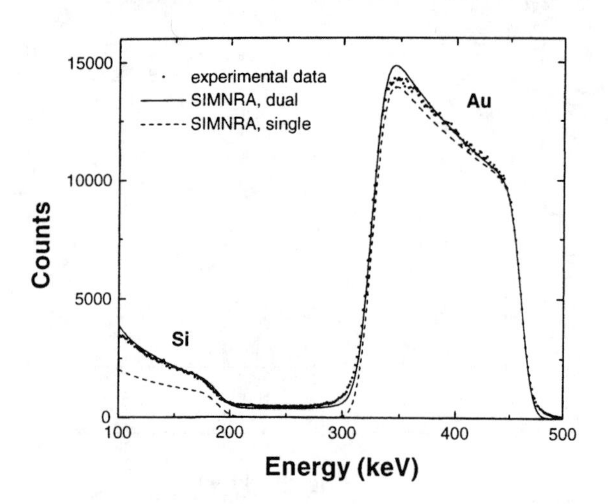

FIGURE 5. 500 keV ^{4}He backscattered from about 100 nm Au on Si, scattering angle $\theta = 165°$. Dots: experimental data; dashed line: SIMNRA with single scattering approximation; solid line: SIMNRA with dual scattering approximation.

culates plural large angle scattering approximately by taking all trajectories with two scattering events into account (dual scattering) [24], which is very similar to the work of Weber *et al.* [27]. As an example the backscattering spectrum of 500 keV ^{4}He from Au on Si is shown in Fig. 5. With dual scattering a better agreement with the experimental data is obtained. The major drawback of dual scattering is the computing time, which increases by a factor of about 200 from several seconds to about 10 min on a Pentium 166 MHz processor.

ACKNOWLEDGMENTS

The TRIM.SP calculations were performed by W. Eckstein, whose help in all questions of plural scattering is gratefully acknowledged.

REFERENCES

[1] E. Kótai, Nucl. Instr. Meth. **B85**, 588 (1994).

[2] R. Doolittle, Nucl. Instr. Meth. **B9**, 344 (1985).

[3] M. Mayer, SIMNRA user's guide, Technical Report IPP 9/113, Max-Planck-Institut für Plasmaphysik, Garching, 1997.

[4] M. Mayer, *SIMNRA User's Guide, Version 4.4*, Part of the program package SIMNRA.

[5] J. Tesmer and M. Nastasi, editors, *Handbook of Modern Ion Beam Materials Analysis*, Materials Research Society, Pittsburgh, Pennsylvania, 1995.

[6] H. Andersen, F. Besenbacher, P. Loftager, and W. Möller, Phys. Rev. **A21**, 1891 (1980).

[7] M. Bozoian, Actual coulomb barriers, In Tesmer and Nastasi [5].

[8] L. Foster, G. Vizkelethy, M. Lee, J. Tesmer, and M. Nastasi, Particle-particle nuclear reaction cross sections, In Tesmer and Nastasi [5].

[9] R. Cox, J. Leavitt, and J. L.C. McIntyre, Non-Rutherford elastic backscattering cross sections, In Tesmer and Nastasi [5].

[10] G. Vizkelethy, Sigmabase: Data base and data server for ion beam analysis, http://ibaserver.physics.isu.edu/sigmabase/.

[11] J. Vorona, J. Olness, W. Haeberli, and H. Lewis, Phys. Rev. **116**, 1563 (1959).

[12] H. Andersen and J. Ziegler, *Hydrogen - Stopping Powers and Ranges in All Elements*, volume 3 of *The Stopping and Ranges of Ions in Matter*, Pergamon Press, New York, 1977.

[13] J. Ziegler, *Helium - Stopping Powers and Ranges in All Elements*, volume 4 of *The Stopping and Ranges of Ions in Matter*, Pergamon Press, New York, 1977.

[14] J. Ziegler, J. Biersack, and U. Littmark, *The Stopping and Range of Ions in Solids*, volume 1 of *The Stopping and Ranges of Ions in Matter*, Pergamon Press, New York, 1985.

[15] J. Ziegler and J. Manoyan, Nucl. Instr. Meth. **B35**, 215 (1988).

[16] W. Bragg and R. Kleeman, Philos. Mag. **10**, 318 (1905).

[17] M. Kumakhov and F. Komarov, *Energy Loss and Ion Ranges in Solids*, Gordon and Breach Science Publishers, New York, London, Paris, 1981.

[18] N. Bohr, Mat. Fys. Medd. Dan. Vid. Selsk. **18** (1948).

[19] J. Mayer and E. Rimini, *Ion Handbook for Material Analysis*, Academic Press, New York, San Francisco, London, 1977.

[20] W. Chu, Phys. Rev. **13**, 2057 (1976).

[21] E. Szilágy, F. Pászti, and G. Amsel, Nucl. Instr. Meth. **B100**, 103 (1995).

[22] D. Dieumegard, D. Dubreuil, and G. Amsel, Nucl. Instr. Meth. **166**, 431 (1979).

[23] P.Bauer, E. Steinbauer, and J. Biersack, Nucl. Instr. Meth. **B79**, 443 (1993).

[24] W. Eckstein and M. Mayer, Rutherford backscattering from layered structures beyond the single scattering model, presented at the COSIRES 98, submitted for publication.

[25] P. Sigmund and K. Winterbon, Nucl. Instr. Meth. **119**, 541 (1974).

[26] W. Eckstein, *Computer Simulation of Ion-Solid Interactions*, volume 10 of *Materials Science*, Springer, Berlin, Heidelberg, New York, 1991.

[27] A. Weber, H. Mommsen, W. Sarter, and A. Weller, Nucl. Instr. Meth. **198**, 527 (1982).

"Magic" Energies for Detecting Light Elements with Resonant Alpha Particle Backscattering

C.J. Wetteland[‡], C.J. Maggiore and J.R. Tesmer

Center for Materials Science

X-M. He and D-H. Lee

Structure/Property Relations

Materials Science and Technology Division, Los Alamos National Laboratory

Resonant backscattering is widely used to improve the detection limit of the light elements such as B, C, N and O. One disadvantage, however, is that several incident energies are normally needed if the sample contains a number of the light elements. There are "magic" energies at which several light elements can be detected simultaneously with suitable sensitivities. When these energies are used along with the elastic recoil detection of hydrogen, multiple elements can be detected without changing the beam energy, and the analysis time is greatly reduced. These reactions along with examples will be discussed.

INTRODUCTION

High-energy alpha-particle resonant backscattering has become a standard tool in ion beam analysis (1). There are several resonances that may be selected (2). Unfortunately there are also detrimental effects that accompany the application of resonances to ion beam analysis. Desirable considerations for choosing the ideal resonance energy for analysis are:

1) multiple elements can be detected simultaneously,
2) the resonances are broad and slowly varying so that depth information can be obtained,
3) the resonance cross sections are large compared to the Rutherford cross sections as well as the substrate cross section,
4) the energy does not produce prompt radiation from the sample or unduly activate the sample being analyzed,
5) there are no unwanted nuclear reactions or resonances from the sample or substrate that complicate the analysis.

The experimenter is lucky if even two of these conditions are met. It is often necessary to change beam energy several times to completely analyze a sample. However, there are several energies where more of the conditions are fulfilled. We refer to these energy regions as "magic." The very first energy region that gained popularity for oxygen detection, 8.6–8.8 MeV, (3)(4)(5) is also useful for carbon and nitrogen. Another example is near 6.6 MeV where boron, carbon and oxygen have enhanced cross sections. Both cases have the disadvantages of prompt radiation and target activation, and a heavy substrate is necessary because of interfering resonances in light substrates such as silicon

In this paper we explore the (α,α) backscattering cross sections for ^{11}B, C, N and O as well as the elastic recoil cross section (α,p) for H in the energy region from 5 to 6 MeV. The energy region near 5.6 MeV allows the simultaneous backscattering measurement of four elements: B, C, N and O, as well as the elastic recoil detection of H (with a change of scattering geometry.) This energy also allows the use of silicon as a substrate—a very common choice.

EXPERIMENTAL METHOD

Sample Preparation

In practice, a sample to determine backscattering cross sections is made with a light substrate on which a layer of a heavy element is deposited. The element of interest is then deposited on this layer. The heavy element, necessary for the cross section determination, is also used to shift the energy of the substrate signal to lower energies so that the signal from the element of interest can be easily observed. The program RUMP (6)(7) was used to aid in the design of the samples before they were deposited.

Boron & Carbon

The boron and carbon samples were deposited on a 5

‡ Now at the Dept. of Ceramics and Materials Engineering, Rutgers University

CP475, *Applications of Accelerators in Research and Industry*, edited by J. L. Duggan and I. L. Morgan

mm thick carbon substrate. The boron in the sample was separated from the substrate by a 4000 Å layer of Mo. This separates, in energy, the boron signal from the carbon substrate signal. The boron layer was a sputter-coated layer of B_4C, 500 Å thick.

For carbon an evaporated 800 Å layer of Au was used followed by a plasma-coated 400 Å layer of carbon.

Oxygen & Nitrogen

The same target was used for both the oxygen and nitrogen measurements. A 2000 Å layer of ZrN_2/ZrO_2 was applied to the carbon substrate. An intermediate layer was not needed because the oxygen and nitrogen signals are naturally separated from the substrate signal.

Hydrogen

A polystyrene layer 8000 Å thick spun onto a silicon substrate was used for the hydrogen elastic recoil cross section measurement.

Backscattering Cross Sections

These measurements were performed in the Ion Beam Materials Laboratory, Center for Materials Science, Los Alamos National Laboratory (8). The He^{++} beam was generated with a National Electrostatics 3MV 9SDH-2 pelletron Data was collected for B, C, N and O over a range of 5–6 MeV at a scattering angle of 167°.

A standard procedure for measuring backscattering cross sections is the ratio method (2). The ratio method can be derived from the equation for the experimental cross section for a given element:

$$\sigma = \frac{Y}{Q(Nt)\Delta\Omega}. \qquad (1)$$

σ is the cross section, Y is the integrated yield, Q is the number of incident particles, (Nt) is the atoms/cm^2 in the target and $\Delta\Omega$ is the solid angle of the detector. While Y can be measured to arbitrarily good statistical uncertainty, the other parameters have higher systematic uncertainties.

In practice, to avoid these uncertainties, a sample is constructed that contains a thin layer of two elements, a light element whose cross section is to be determined, and a heavy element that must be Rutherford at all energies of interest.

The ratio of the yields at an energy where both elements are known to be Rutherford and at the energy of interest allows the cross section ratio to be determined:

$$\left(\sigma/\sigma_R\right)_E \cong \frac{\left(Y_x/Y_m\right)_E}{\left(Y_x/Y_m\right)_{E_R}}. \qquad (2)$$

Y_x/Y_m is the ratio of the yields of the element of interest and the heavy element, E is the energy of interest and E_R is an energy where both elements x and m are Rutherford. The result is an approximation because of the neglect of the mean energy and the screening corrections. These corrections tend to cancel and were not applied in our analysis.

Elastic Recoil Cross Sections

The hydrogen recoil cross sections were measured at scattering angle of 30° with a target tilt of 75°. The cross section in the energy region of interest was obtained by first measuring the yield from the sample at 2.5 MeV, where the cross section is known (9). At both high and low energies the target was thin enough that the yield could be determined by peak integration. This removes the uncertainties associated with stopping cross sections. The yield obtained in the region of 5–6 MeV was then scaled by the 2.5 MeV cross section to determine the cross section.

DATA

The cross section ratios for both backscattering and hydrogen elastic recoil are shown in Fig. 1.

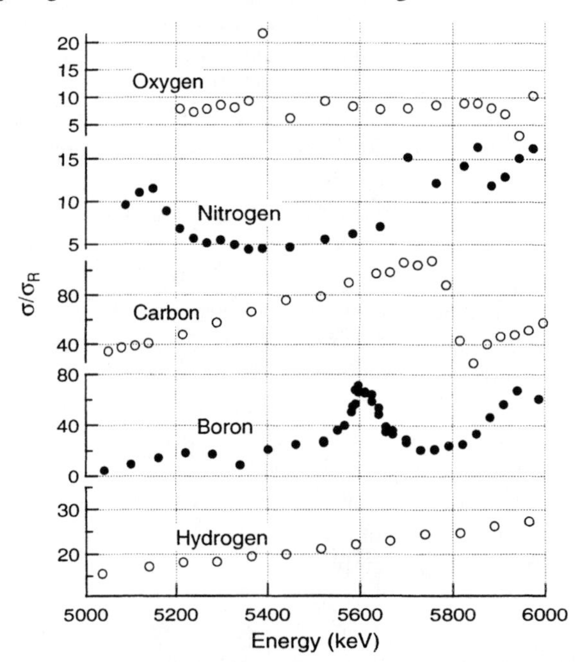

Figure 1. σ/σ_R for alpha resonant backscattering of the elements ^{11}B, C, N and O at a scattering angle of 167° and the elastic recoil cross section of H at a scattering angle of 30°.

The oxygen cross section was not measured below 5.2 MeV because of the interference of the $^{14}N(\alpha,p)$ reaction. The hydrogen cross section was corrected for hydrogen loss and the carbon cross section was corrected for carbon gain during analysis. The statistical uncertainties in the backscattering ratio measurements are ±4%. The uncertainty in the elastic recoil ratios is ≈6% (the uncertainty from (9)).

Choosing the analysis energy in this region depends on the elements expected in the sample. If boron is present in a relatively thin film, the best energy is ≈ 5.625 MeV. At this energy silicon has relatively few resonances in the region of the light element signals and can be used as a substrate material. Table 1 lists the cross section ratios and the cross sections for the elements at this energy.

Element	σ/σ_R	σ (mb/str)
Hydrogen	21.1	538
Boron	65.2	209
Carbon	97.6	468
Nitrogen	6.25	44
Oxygen	8.40	81

Table 1. σ/σ_R and σ for resonant alpha backscattering at 5.625 MeV.

APPLICATION

As an example of the application of this "magic" energy a sample of a plasma ion-deposited sample of ≈1300 Å of a B/C/N/O/H coating on a silicon substrate. Normally, a Mo substrate would be needed and several higher energies would be necessary for analysis.

The backscattering spectra from the sample and pure silicon are shown in Fig. 2a. The subtracted spectrum is shown in Fig. 2b. The Si signal is approximately Rutherford at energies below 1 MeV. The scattering geometry is as described above with the exception that a target tilt of 45° was used. Charge collected was 30 μC and the solid angle of the detector was 2.5 mstr.

The atomic fractions and the 2 σ minimum detection limit of the constituents in the film are given in Table 2.

Element	Atomic Fraction	MDL ($10^{15}/cm^2$)
Hydrogen	.12	<1.2
Boron	.28	7.2
Carbon	.17	2.5
Nitrogen	.29	20
Oxygen	.13	12

Table 2. Measured atomic fractions and minimum detectable amount for elements in a film on a silicon substrate.

The elastic recoil spectrum is not shown. However, the protons from the $^{14}N(\alpha,p)$ reaction produce a small background.

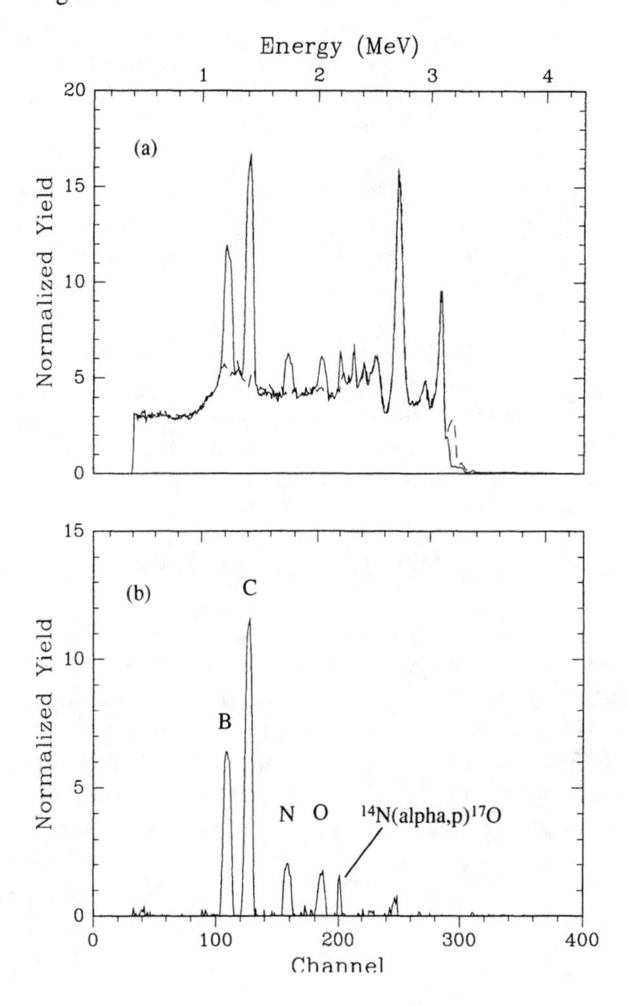

Figure 2. (a)—Spectrum from 1300 Å film of B, C, N, O on Si (solid line) overlaid by the spectrum from pure Si (dashed line). **(b)**—Difference spectrum showing the B, C, N and O peaks.

For this sample the thickness of the film must be less than 4500 Å at normal incidence for the B and C signals to be separated.

DISCUSSION

For samples containing multiple light elements the 5-6 MeV region has many advantages. In particular, the use of the 5.625 MeV "magic" energy for determining hydrogen, boron, carbon, nitrogen, and oxygen concentrations has many practical applications. Examining the wish list from the introduction:

1) four elements, B, C, N and O can be detected simultaneously as well as H (with a change in geometry),
2) the resonances are broad, with the exception of B (≈50 keV) at 5.625 MeV, and slowly varying so

that depth information can be obtained,

3) the resonance cross sections are moderately large compared to their Rutherford cross sections and significantly larger that the Si cross section (there are other useful energies, where the σ/σ_R ratios are larger, that can be used when high sensitivity measurements are needed),
4) this energy does not produce prompt radiation from the sample or significantly activate the sample or substrate,
5) unwanted nuclear reactions or resonances from the sample or substrate is limited and does not significantly complicate the analysis.

While all of the conditions are not totally met, the use of this "magic" energy results in significant timesavings, safety and convenience for researchers.

ACKNOWLEDGEMENTS

We would like to thank Mike Nastasi, Marc Verdier, Brian Gorman and Kevin Hubbard for insightful sample design and construction discussions; and to thank Kevin Walter for assistance in identifying the research samples for measurement. Special thanks to Caleb Evans and Mark Hollander for technical assistance in accelerator operation. This research is supported in part by the U.S. Department of Energy.

REFERENCES

1. Tesmer, J.R., Maggiore, C.J., Nastasi, M., Barbour, J.C., *Materials Chemistry and Physics* **46**, 189-197, 1996.
2. Cox, R.P, Leavitt, J.A. and McIntyre, Jr., L.C., Appendix 7, *Handbook of Modern Ion Beam Analysis*, Tesmer, J.R. and Nastasi, M. Eds., MRS, 481-508, 1995.
3. C. R. Gossett, K. S. Grabowski, and D. Van Vechten, *Proceedings, Thin Film Processing and Characterization of High-Temperature Superconductors,* J. M. E. Harper, R. J. Colton, and L. C. Feldman, Editors, *American Vacuum Society 3*, American Institute of Physics (1988) 443.
4. J. C. Barbour, B. L. Doyle and S. M. Myers, *Phys. Rev. B 38* (1988) 7005.
5. J. A. Martin, M. Nastasi, J. R. Tesmer, and C. J. Maggiore, *Appl. Phys. Lett. 52* (1988) 2177.
6. L. R. Doolittle, *Nucl. Instr. and Meth. B9* (1985) 344.
7. L. R. Doolittle, *Nucl. Instr. and Meth. B15* (1986) 227.
8. Tesmer J.R., Parkin D.M., Maggiore C.J., *"Up Close: The Ion Beam Materials Laboratory at Los Alamos National Laboratory"*, MRS Bulletin, Aug./Sept,1987 page 101-03.
9. Quilliot, V., Abel, F. and Schott, M., *Nucl. Instr. and Meth.* **B83**, 1993, 47-61.

Oxygen Depth Profiling by Nuclear Resonant Scattering

G.T. Gibson, Y.Q. Wang*, W.J. Sheu, and G.A. Glass

Acadiana Research Laboratory, University of Southwestern Louisiana, Lafayette, LA 70504
**Center for Interfacial Engineering, University of Minnesota, Minneapolis, MN 55455*

Nuclear resonance scattering (NRS) $^{16}O(\alpha,\alpha)^{16}O$ at 3.045 MeV (Γ=10 keV) has been used for oxygen depth profiling in various thin oxide films. There are two ways by which the oxygen concentration versus depth profile can be obtained from the experimental data: energy spectrum simulation or yield distribution analysis. Energy spectrum simulation is done using the standard RBS software/Rutherford Universal Manipulation Program (RUMP) where only one spectrum is usually needed from the measurement. Yield distribution analysis is accomplished by using a custom developed software/Resonance Analysis Program (RAP) and involves a series of spectra obtained by stepping up the beam energy above the resonance energy. This article aims at comparing the fundamentals of both methods and also discussing their advantages and disadvantages in terms of the data acquisition and the post data analysis. A thermally grown thick SiO_2 film and a thin titanium oxide film grown by corona point discharge were examined.

INTRODUCTION

Oxide thin films play a critical role in surface science, semiconductor industry, and corrosion engineering industry. Nondestructive analysis of oxide thin films is sometimes preferred, even though oxygen distribution can also be analyzed with such destructive techniques as secondary ion mass spectrometry (SIMS). Because of limitations and/or restrictions imposed by other conventional nondestructive techniques such as Rutherford Backscattering Spectrometry (RBS), Elastic Recoil Detection Analysis (ERDA), and Nuclear reaction Analysis (NRA) in profiling oxygen, the nuclear resonant scattering (NRS) $^{16}O(\alpha,\alpha)^{16}O$ at 3.045 MeV (Γ=10 keV) often becomes an inexpensive and nondestructive choice to study depth profiles of oxygen near surfaces [1-12].

Resonance parameters of NRS $^{16}O(\alpha,\alpha)^{16}O$ near the main resonance 3.04 MeV reported varied from 3.034 to 3.045 MeV, and the resonance width (Γ) varied from 8 to 10 keV. In this work, the resonance parameters (E_R=3.045 MeV and Γ=10 keV) obtained by Cameron [1] were chosen and the NRS crossection data were then generated by a subprogram of the Rutherford Universal Manipulation Program (RUMP) [13]. Figure 1 shows the generated NRS crossection of $^{16}O(\alpha,\alpha)^{16}O$ as a function of incident energy in the range of 2.8 ~ 3.2 MeV at a scattering angle of 165°, where the Rutherford crosection is also included for comparison. The crossection is the same as the Rutherford except near the resonance energy where the cross section is about 18 times of the Rutherford value. The strong and narrow NRS crossection at the resonant energy implies that both good depth resolution and good detection sensitivity for oxygen measurements are possible.

NRS has its uniqueness in elemental depth profiling compared with RBS and NRA. While the kinematics of NRS is identical to RBS, its crossection distribution behaves much like that of a nuclear resonance reaction. So, in principle, data analysis methods used in RBS (energy spectrum simulation) and NRA (yield distribution analysis) should be applicable to NRS. The energy spectrum analysis is to derive the concentration profile from the measured energy spectrum by spectrum synthesis: an assumed depth profile is iteratively modified to obtain a good match of the predicted and the measured energy spectrum. Although this method is more suitable when the energy dependence of the crossection is nearly constant or fairly smooth within the energy range interested, e.g. in RBS analysis, it has been heavily used in oxygen depth profiling by NRS.

The yield distribution analysis involves resonance scanning whereby the incident beam energy is step by step incremented to obtain resonant events at an ever increasing depth, and therefore it is only used when the energy dependence of the crossection exhibits a sharp resonance. The resonance yield (area of the resonance peak) is measured at each incident beam energy to obtain a yield versus beam energy curve. This curve is then converted to a concentration versus depth data of the resonant element. This is a standard procedure for nuclear resonance reaction depth profiling, but it has not yet been popularly used for NRS oxygen depth profiling, partially because of the lack of the commercial software that can conveniently convert yield distributions into concentration depth profiles.

In this article, we compare the fundamentals of both profiling methods and discuss their advantages and disadvantages in terms of the data acquisition and the post data analysis. RUMP software [13] was used to do the energy spectrum simulations, while a Resonance Analysis Program (RAP) [14], developed in this work, was used to convert yield distributions into depth profiles. A thermally grown silicon oxide thick film was first used to examine the effectiveness and accuracy of both analysis methods, then the oxygen depth profile of a titanium oxide thin film grown by a Corona point discharge technique was analyzed.

CP475, *Applications of Accelerators in Research and Industry*,
edited by J. L. Duggan and I. L. Morgan

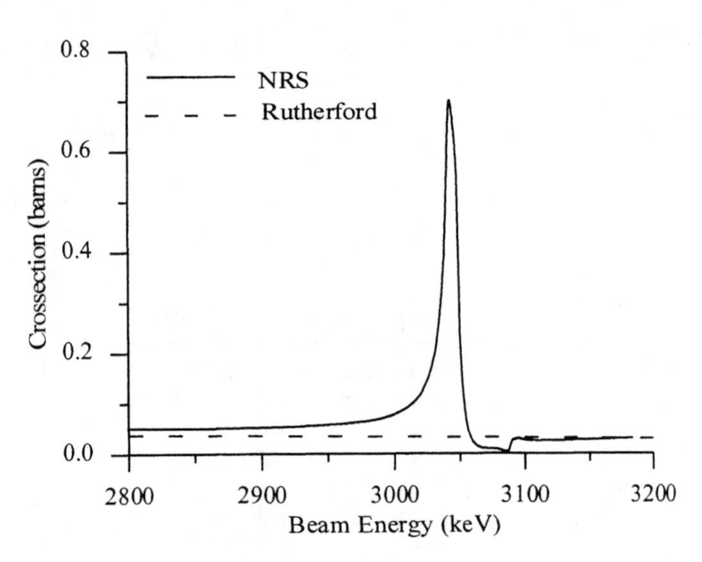

FIGURE 1. Scattering crossection as a function of energy at a scattering angle of 165°, where the Rutherford data are also shown for comparison.

DEPTH PROFILING FUNDAMENTALS

Energy Spectrum Simulation

In principle, an actual energy spectrum, $N(E_c)$, measured in the experiment is a convolution of an ideal energy spectrum, $N'(E_c)$, and the total energy resolution of the experimental setup, $G(E_{c'} - E_c)$,

$$N(E_c) = \int G(E_{c'} - E_c)\, N'(E_{c'})\, dE_{c'} \quad (1)$$

The ideal energy spectrum, $N'(E_c)$, is the one to be expected if no energy spread exists in the experiment, and can be calculated from the depth profile, C(X), based on the following equation for the normal beam incidence,

$$N'(E_{c'}) = Q\varepsilon\Omega\sigma(E)C(X)dX/dE_{c'} \quad (2)$$

where $N(E_c)$ is the counts per unit energy at an actual channel energy E_c, $N'(E_{c'})$ is the counts per unit energy at an ideal channel energy $E_{c'}$, $\sigma(E)$ is the scattering cross section at energy E which is directly related to $E_{c'}$ and therefore E_c, as well as the depth X, $C(X)$ is the oxygen concentration at a depth of X (depth profile), Q is the collected beam charge for the spectrum, ε is the detector intrinsic efficiency, and Ω is the solid angle subtended by the detector. The total energy resolution function, $G(E_{c'} - E_c)$, is usually expressed as a Gaussian distribution,

$$G(E_{c'} - E_c) = 1/(2\pi\delta_t^2)^{1/2} Exp[-(E_{c'} - E_c)^2/2\delta_t^2] \quad (3)$$

where δ_t is the variance of the Gaussian, δ_t = full width at half maximum (FWHM)/2.356, with FWHM being the total energy spread of the experimental setup. The total energy spread includes the detector energy resolution, the energy straggling in the matrix, the geometrical broadening due to the finite detector acceptance angle, and the incident beam energy spread.

Based on these considerations, a computer software RUMP has been compiled and is now widely used in many research laboratories for routine RBS analysis [13]. In the case of NRS, the resonant scattering crossection data must be imported into RUMP. Given a depth profile based on boundary conditions, an actual energy spectrum can be simulated. By adjusting the depth profile, a match between the measured spectrum and the calculated spectrum can be obtained. The imported depth profile is then taken as the practical depth profile.

Yield Distribution Analysis

The depth profile can also be obtained by fitting the yield distribution, i.e. the NRS yield as a function of incident energy. The formalism begins with presenting the functional form of the excitation curve. In principle, the yield is related to the depth profile through the following equation for an incident energy E_0 normal to the surface,

$$Y(E_0) = \iiint Q\varepsilon\Omega\, g(E_0,E_1)\, f(E_1,E_2,X)\sigma(E_2)\, C(X)\, dE_1\, dE_2\, dX \quad (4)$$

where $Y(E_0)$ is the yield of scattered beam for an incident beam energy E_0, $g(E_0,E_1)$ is the energy spread function of incident beam at an energy E_1, $f(E_1,E_2,X)$ is the energy straggling function at a depth X, E_2 is the energy of the incident particle before the scattering, and other parameters are the same as those in Eq. (2).

Finding an analytical solution for depth profile $C(X)$ in Eq. (4) is practically impossible. Finding a numerical solution for Eq. (4) is though possible but very tedious and time-consuming. Fortunately, Eq. (4) can be simplified without introducing a significant error. Since the incident beam energy spread, $g(E_0,E_1)$, and the beam energy straggling, $f(E_1,E_2,X)$, usually follow Gaussian distributions, the integral for these two terms can be closely described by a combined Gaussian function, called the energy response function,

$$F(E_0,E_2,X) = \int g(E_0,E_1)\, f(E_1,E_2,X)\, dE_1$$
$$\approx A/(\delta_b^2+\delta_0^2X)^{1/2} Exp[-(E_2 - E(X))^2/2(\delta_b^2+\delta_0^2X)] \quad (5)$$

where δ_b is the variance of the incident beam energy spread, δ_0 is the Bohr energy straggling prefactor [15], $E(X)$ is the remaining energy of the incident beam with E_0 after penetrating the depth X, and A is the constant for normalization. Combining Eqs. (4) and (5) yields a simplified equation,

$$Y(E_0) = \iint Q\varepsilon\Omega\, F(E_0,E_2,X)\sigma(E_2)\, C(X)\, dE_2\, dX \quad (6)$$

By introducing a depth resolution function, $B(E_0,X)$, Eq. (6) is further simplified as

$$Y(E_0) = \int Q\varepsilon\Omega\, B(E_0,X)\, C(X)\, dX. \quad (7)$$

Here $B(E_0,X)$ can be numerically calculated from Eq. (8),

$$B(E_0,X) \equiv \int F(E_0,E_2,X)\sigma(E_2)\, dE_2. \quad (8).$$

Based on these considerations, a Resonance Analysis Program (RAP) was written in FORTRAN-77 to obtain depth profiles from yield distributions. RAP consists of three main parts [14]: (1) initial inputs of beam and target parameters including crossection data, stopping powers, assumed depth profile, and the experimental yield data; (2) integration subroutines called and yield distribution generated; and (3) evaluation of "goodness of fit" between the calculated and measured yields (Reduced χ^2 test). The double integral in Eq. (6) was performed using an adaptive multidimensional integration subroutine written by A.C. Genz, which was downloaded from the internet GAMS (Guide to Available Mathematical Software) site [16]. The sample was divided into layers of equal thickness such that each layer was thin enough to respond to the changes of the crossections, the energy losses and the oxygen depth profile. Given a depth profile based on known boundary conditions, the yield distribution can be calculated. By adjusting the depth profile, a good agreement between the measured and the calculated yields can be obtained.

APPLICATIONS

Experimental

A 3 μm thick thermally grown SiO_2 film was used as the oxygen standard, and then a titanium oxide thin film (< 100 nm) grown on a well polished titanium substrate by a corona point discharge technique was measured. The experiments were conducted on a NEC 5SDH2 1.7 MV Tandem Pelletron® Accelerator. The beam energy spread of the accelerator is approximately 0.1% of the beam energy (~3 keV for 3 MeV ion beam in our case). The accelerator beam energy was calibrated with a narrow nuclear resonant reaction $^{19}F(p, \alpha\gamma)^{16}O$ at 340 keV and the calibration was also confirmed with nuclear resonant scattering $^{16}O(\alpha, \alpha)^{16}O$ at 3.045 MeV.

In our setup, the $^4He^{++}$ beam was incident perpendicular to the target surface, and the backscattered alpha particles were detected by an ion detector (resolution ~18 keV) at 165°. The solid angle subtended by the detector was ~5 msr. The $^4He^{++}$ beam energy ranged from 2.8 - 3.2 MeV depending on the thickness of the oxide layers. The beam current was kept at about 10-15 nA with a beam size of 1~3 mm in diameter. The charge collection for a single spectrum varied from 3 to 20 μC depending on the incident energy and the specific target.

The stopping powers of alpha particles in pure elements such as O, Si, and Ti were calculated according to the empirical formula by Ziegler et al. [17], then the Bragg's additive rule was used to determine the stopping powers of the ions in compound targets such as TiO_2 and SiO_2. The energy straggling of both incoming and outgoing helium beams was estimated using Bohr's formula [15], and thus was treated as energy independent. For light elemental compound targets such as SiO_2 or even TiO_2, the energy straggling calculated by the simple Bohr model [15] is in a satisfactory agreement [18] with that obtained by more sophisticated models as suggested by Lindhard-Scharff [19] and Chu [20].

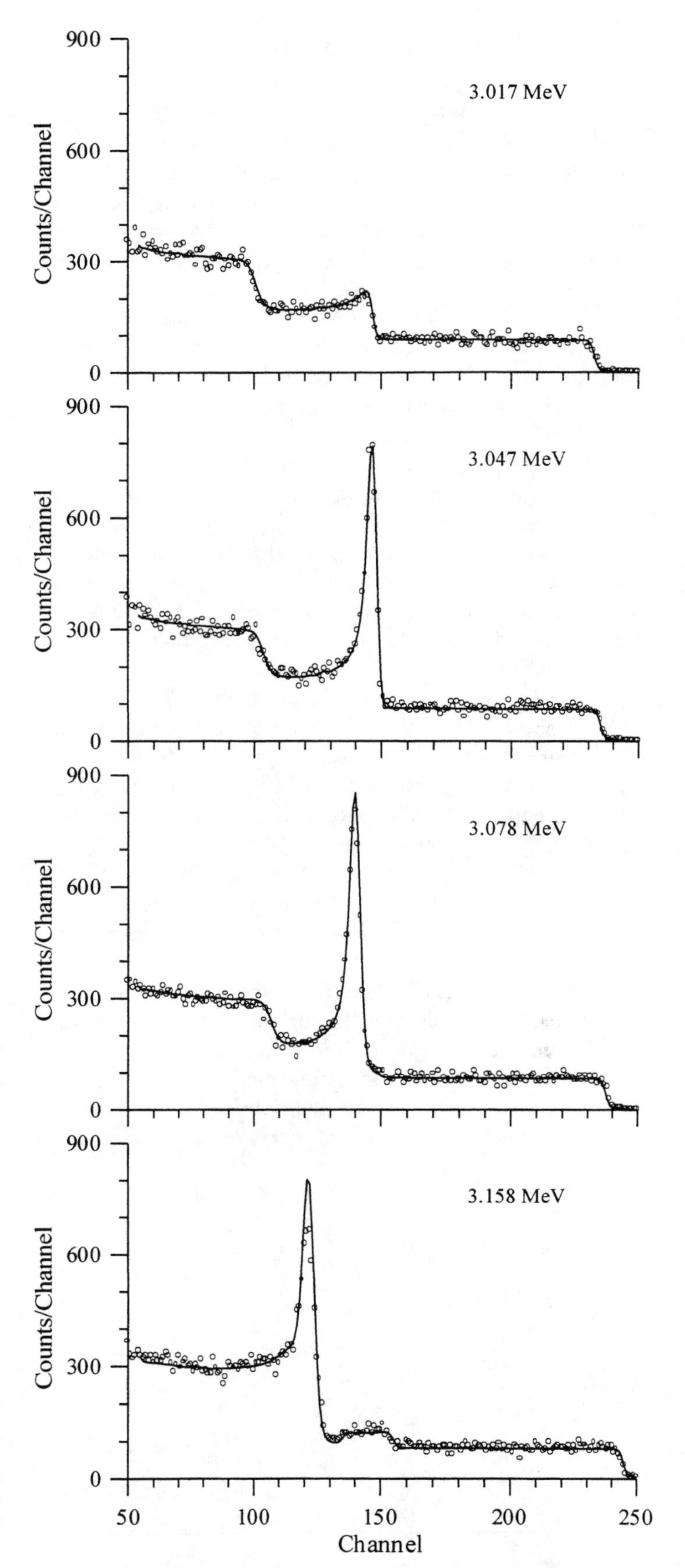

FIGURE 2. Measured (symbols) and RUMP-fitted (solid lines) spectra at different incident energies for 3μm thick SiO_2/Si sample, where the uniform oxygen depth profile was assumed in the simulations.

Thick SiO_2 Film

Figure 2 shows the measured (symbols) and RUMP-fitted (solid lines) NRS spectra for 3μm thick SiO_2/Si sample at different incident beam energies, where the uniform oxygen depth profile was used in the simulations. It clearly indicates that both the intensity and the location of the oxygen peak are dependent on the incident beam energy. The fittings between the experiments and RUMP calculations are quite good except at the highest energy of 3.158 MeV. For the incident beam energy of 3.158 MeV, the oxygen resonant scattering occurs near 0.5μm below the surface as estimated by TRIM code [21]. Figure 2 shows that at such a depth, the Rutherford scattering events from the silicon substrate start to overlap with the nuclear resonant scattering events from oxygen in the film. The Bohr's treatment of the energy straggling (Gaussian distribution), as used in RUMP simulation for both the incoming beam and outgoing beam, may be too simplistic for such a large depth. Another possible explanation for the discrepancy between the fit and the measured data (around the oxygen peak) at 3.158 MeV may be simply due to the fact that our purchased SiO_2 film may not be as uniform as we thought.

Figure 3 shows the yield distribution for the SiO_2/Si sample, where the closed circles and solid line represents the measured and simulated results by RAP, respectively. The uniform oxygen depth profile was assumed in the calculation. The calculated yield distribution by RAP agrees well with the experimental results. It is important to mention that the off-resonance contribution can significantly reduce the depth profiling sensitivity of the resonance peak if not properly subtracted. This is because the yield contributed from the off-resonance scattering as shown in Fig. 1 can be comparable to that only from the resonance peak depending on the energy integration range used in yield production. During the RAP calculation, the integral limits for Eq. (8) needs to correspond to the energy limits used to calculate the yield from the experimental spectrum. In other words, the energy limits in the integral should be translated to the detected energy of the particles.

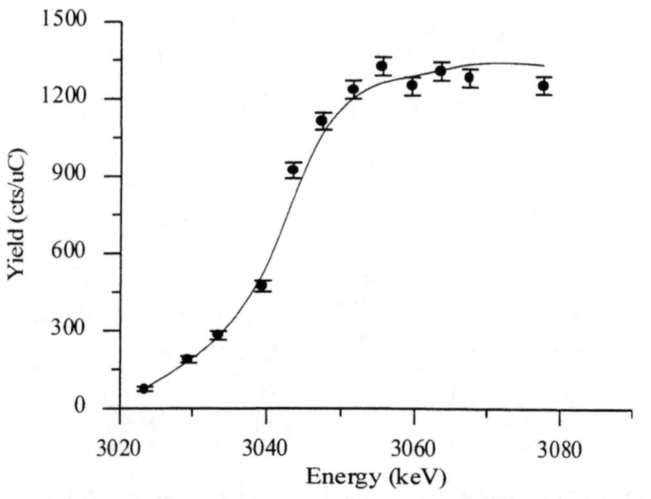

FIGURE 3. Measured (closed circles) and calculated (solid line) yield distributions for 3 μm thick SiO_2/Si sample, where the uniform oxygen depth profile was used in the simulation.

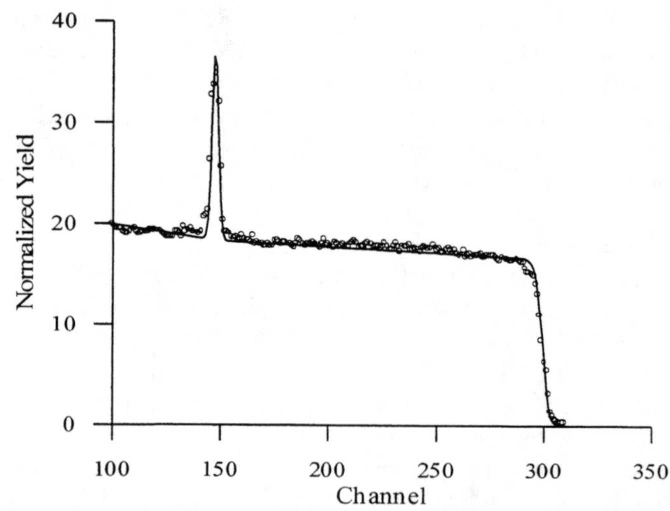

FIGURE 4. Measured (open circles) and RUMP-fitted (solid line) NRS spectra at the incident energy of 3.047 MeV for a thin TiO_2/Ti sample, where the assumed oxygen depth profile is shown in Fig. 5.

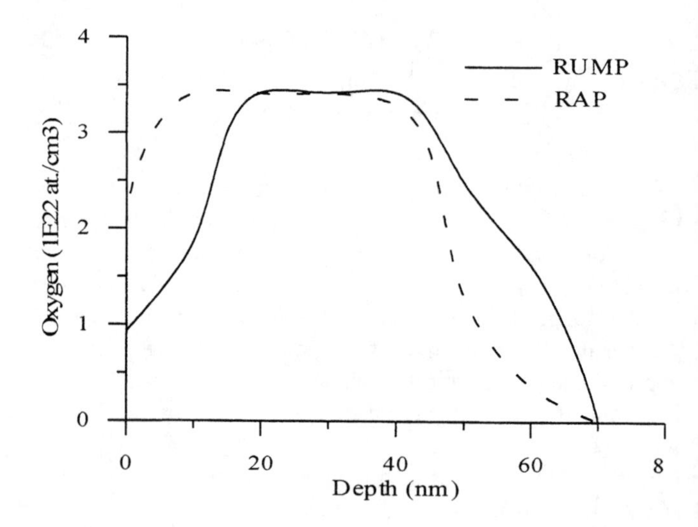

FIGURE 5. Oxygen profiles in TiO_2/Ti sample obtained by the energy spectrum analysis (RUMP) and the yield distribution analysis (RAP).

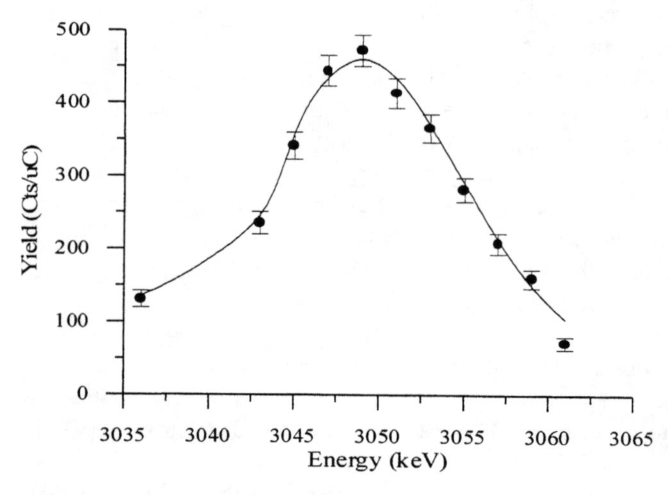

FIGURE 6. Measured (closed circles) and calculated (solid line) yield distributions for a thin TiO_2/Ti sample, where the assumed oxygen depth profile is shown in Fig. 6.

Thin TiO_2 Film

Figure 4 shows the measured (open circles) and RUMP-fitted (solid line) spectra from a thin TiO_2/Ti sample at the incident energy of 3.047 MeV, where the corresponding oxygen depth profile is shown in Fig. 5. Figure 6 shows the yield distributions as a function of the incident beam energy for this sample, where the solid line and solid circles represent the simulated results by RAP and the measured results, respectively. The corresponding oxygen depth profile obtained is also shown in Fig. 5. Figure 6 indicates that the oxygen depth profile in the TiO_2/Ti sample is fairly uniform as expected. It indicates that the results from both energy spectrum and yield distribution methods are in acceptable agreement considering the thinness of the film and the depth resolution of the NRS technique.

DISCUSSIONS

NRS $^{16}O(\alpha,\alpha)^{16}O$ at 3.045 MeV (Γ=10 keV) has been used to measure oxygen depth profiles in oxide films. Both Energy Spectrum Simulation and Yield Distribution Analysis methods were described and discussed. The oxygen depth profiles obtained by two methods for both thick oxide film (3 μm SiO_2/Si) and thin oxide film (~50 nm TiO_2/Ti) are in acceptable agreement considering the depth resolution of the methods.

The energy spectrum simulation method is quick and simple in both measurement and data analysis since usually only one spectrum needs to be collected and analyzed. Beam irradiation effect on the target is usually small, and RBS software such as RUMP can be directly used (NRS crossection data are imported). However, the oxygen depth profile in this method is heavily related to the shape of the spectrum, which is affected by many factors, mainly including detector energy resolution, energy straggling of outgoing beam in the matrix, and geometrical broadening due to the finite angle subtended by the detector. The front edge fit of the silicon signal by RUMP shows a total energy spread of approximately 20 keV in our measurements, which translates into a depth resolution of about 63 nm on SiO_2 surface and 51 nm on TiO_2 surface.

The yield distribution analysis usually involves a smaller energy spread in the calculation, because the energy spectrum is integrated over all the possible scattering energies to generate the scattering yield. In other words, the depth profile is not directly related to the shape of individual energy spectrum. Instead, the depth resolution is determined by the natural width of the NRS crossection distribution (Γ=10 keV), the incident beam energy spread and the energy straggling of the incoming beam in the matrix. Although the total energy spread presented in the yield spectrum method is much smaller (~10 keV at the surface) compared with that presented in the energy spectrum method (20 keV), the surface depth resolution for both methods is actually quite similar since the stopping power of the incoming alpha particles (~3 MeV) is significantly smaller than that of the scattered outgoing alpha particles (1.1 MeV). So, the oxygen depth resolution for the yield distribution method is estimated to be approximately 61 nm on SiO_2 surface and 49 nm on TiO_2 surface.

Yield data usually have smaller statistical uncertainties than counts per channel used in the energy spectrum analysis. However, in the yield distribution analysis, the experiment is more complicated and time consuming, since a series of energy spectra at different incident energies have to be collected to obtain a yield distribution. The collection of many spectra may introduce a significant beam damage on the target, especially when the organic samples are involved. The data analysis is also more complicated since multiple-integrals are involved in the calculation, especially when there is still no commercially available software (like RUMP for RBS) available to use for the yield distribution analysis (RAP is still under development). Therefore the fast and simple energy spectrum simulation method is proved to be more effective in profiling oxygen when the nuclear resonance scattering $^{16}O(\alpha,\alpha)^{16}O$ at 3.045 MeV is used.

ACKNOWLEDGMENT

This work is supported by the U.S. Department of Energy/Louisiana Educational Quality Support Fund in Cooperative Agreement Number DE-FC02-91ER75669.

REFERENCES

1. J.R. Cameron, Phys. Rev. 90 (1953) 839.
2. Z.L. Wang, J.F.M. Westerndorp and F.W. Saris, Nucl. Instrum. Methods, 211 (1983) 193.
3. S.H. Sie, D.R. Mckensie, G.B. Smith and C.G. Ryan, Nucl. Instrum. Methods, B15 (1986) 525.
4. F. Ajzenberg-Selove, Nucl. Phys. A474 (1987) 1.
5. P. Berning and R.E. Benenson, Nucl. Instrum. Methods, B36 (1989)335.
6. J.A. Leavitt, L.C. McIntyre Jr., M.D. Ashbaugh, J.G. Oder, Z. Lin and B. Dezfouly-Arjomandy, Nucl. Instrum. Methods, B44 (1990) 260.
7. D.D. Cohen and E.K. Rose, Nucl. Instrum. Methods, B66 (1992) 158.
8. B. Blanpain, P. Revesz, L.R. Doolittle, K.H. Purser and J.W. Mayer, Nucl. Instrum. Methods, B34 (1988) 459.
9. J. Li, L.J. Matienzo, P. Revesz, Gy. Vizkelethy, S.Q. Wang, J.J. Kaufman, and J.W. Mayer, Nucl. Instrum. Methods, B46 (1990) 287.
10. W. De Coster, B. Brijs, J. Goemans, and W. Vandervorst, Nucl. Instrum. Methods, B66 (1992) 283.
11. W. De Coster, B. Brijs, R. Moous, and W. Vandervorst, Nucl. Instrum. Methods, B79 (1993) 483.
12. R.P. Cox, J.A. Leavitt, and L.C. McIntyre, Jr., Appendix 7, pp. 500, in Handbook of Modern Ion Beam Materials Analysis, eds. J.R. Tesmer and M. Nastasi, MRS Publisher, Pittsburgh, PA, 1995.
13. L.R. Doolittle, Nucl. Instrum. Methods, B9 (1985) 344 and B15 (1986) 227.
14. G.T. Gibson, M.S. Thesis, University of Southwestern Louisiana (1996).
15. N. Bohr, Mat. Fys. Medd. Dan. Vid. Selsk, 18(8) (1948) and 24(19) (1948).
16. A.C. Genz, "Adaptive Multidimensional Integration Subroutine." http:/math.nist.gov/.../AMSUN.html (Oct. 1, 1996).
17. J.F. Ziegler, "Helium Stopping Powers and Ranges in All Elements," Pergoman Press, Inc., Maxwell House, NY, 1977, pp. 26.
18. E. Rauhala, Chapter 2, pp. 14, in Handbook of Modern Ion Beam Materials Analysis, ed. by J.R. Tesmer and M. Nastasi, MRS Publisher, Pittsburgh, PA, 1995.
19. J. Lindhard and M. Scharff, Mat. Fys. Medd. Dan. Vid. Selsk, 27(15), 1953.
20. W.K. Chu, Phys. Rev., A13 (1976) 2057.
21. J.F. Ziegler, "Transport of Ions in Matter," TRIM-96 program package, kindly supplied by Dr. Ziegler.

A Nuclear Reaction Analysis Study of Fluorine Uptake in Flint

Jian-Yue Jin, D. L. Weathers, F. Picton, B.F. Hughes, J. L. Duggan, F. D. McDaniel, and S. Matteson

Ion Beam Modification and Analysis Laboratory (IBMAL), Dept. of Physics, University of North Texas, Denton, TX 76203

Nuclear Reaction Analysis (NRA) using the $^{19}F(p, \alpha\gamma)^{16}O$ resonance reaction is a powerful method of fluorine depth profiling. We have used this method to study the fluorine uptake phenomenon in mineral flint, which could potentially develop into a method of dating archeological flint artifacts. Flint samples cut with a rock saw were immersed in aqueous fluoride solutions for different times for the uptake study. The results suggest that fluorine uptake is not a simple phenomenon, but rather a combination of several simultaneous processes. Fluorine surface adsorption appears to play an important role in developing the fluorine profiles. The surface adsorption was affected by several parameters such as pH value and fluorine concentration in the solution, among others. The problem of surface charging for the insulator materials during ion bombardment is also reported.

INTRODUCTION

As early as 1973, Taylor observed the uptake of fluorine into the surfaces of worked stone material that had been buried in the soil (1). In 1989, Walter et al. noted that there was a strong correlation between the age of a flint artifact and the depth of fluorine penetration into the flint surface after measuring fluorine depth profiles using Nuclear Reaction Analysis (NRA) (2). This suggests a potential method of dating archeological flints. However, the mechanism of how fluorine is incorporated into the flint surface, and how it finally penetrates into the bulk is not well understood.

Walter et al. had studied solid state diffusion of fluorine in silica (3), which is a principle component of flint. Samples were implanted with fluorine, followed by high temperature annealing. An anomalous diffusion behavior was observed. The results argue against the role of solid state diffusion in fluorine transport in flint.

We speculate that the uptake of fluorine in flint artifacts is from ground water and have initiated a systematic study of fluorine uptake in flint samples immersed in hot fluoride solutions. The $^{19}F(p, \alpha\gamma)^{16}O$ resonance nuclear reaction at a proton energy of 872 keV has been used to measure fluorine uptake profiles. A computer program using convolution fitting and deconvolution has been developed to recover the fluorine profiles from the measured raw data (4). Because flint and other forms of silica are insulator materials, surface-charging effects during the measurement have also been studied.

Our preliminary results suggest that fluorine uptake in flint is not a simple phenomenon, but rather a combination of several simultaneous processes, which include surface adsorption and inward diffusion. Surface adsorption appears to play an important role in developing the fluorine profiles. It is affected by several parameters such as fluorine concentration, pH value of the solution, and impurity levels in the flint samples.

EXPERIMENTAL MATERIALS AND METHOD

The primary material used in this study was commercially supplied "Alibates" flint mineral from a single formation in the Amarillo region of Texas, USA. The typical microstructure of flint (or chert) is reported to be granular microquartz cemented with amorphous silica (5,6). Scanning electron microscopy (SEM), transmission electron microscopy (TEM) and x-ray diffraction studies performed in the present work support this microstructural picture. Moreover, x-ray fluorescence (XRF) measurements show that the mineral's major constituents are silicon and oxygen in the stoichiometry of silica (SiO_2). Iron (Fe) (about 0.1 weight%) and calcium (Ca) (about 0.05 weight%) were found, however, to be the major impurities with the exception of water, which is not detected by XRF. Other impurities such as K, Cl, Ge, As, and Br were all found to be at concentrations below the level of 100 ppm.

As a control, the behavior of other forms of silica, was also examined in the experiments. The other forms included were single crystal quartz, poly-crystalline quartz and fused quartz. All sample materials were cut into 2 cm $\times$ 1 cm $\times$ 1 mm pieces with a rock saw. After cutting, the samples were washed with soap, rinsed and finally washed

CP475, *Applications of Accelerators in Research and Industry*, edited by J. L. Duggan and I. L. Morgan

with an ultrasound cleaner, first in methanol and then in distilled water.

Fluorine uptake processing was carried out by immersing samples in a series of sodium fluoride (NaF) solutions with different fluorine concentrations at different temperatures for different times. A buffer solution was added to keep a constant pH value. The constant temperature of the solution was maintained by placing the container in a thermal bath. Nuclear reaction analysis (NRA) measurement of fluorine profiles was carried out on the 3 MV tandem accelerator in the Ion Beam Modification and Analysis Laboratory (IBMAL) of the University of North Texas.

Surface Charging

Because flint and other materials used in this study are insulators, surface charging may become a severe problem in NRA. We used the mesh method to solve the charging problem. In this method, a 98% transparent copper mesh was placed on the surface of the sample; then it was securely wrapped with aluminum foil with a hole to expose the mesh and the sample to the ion beam.

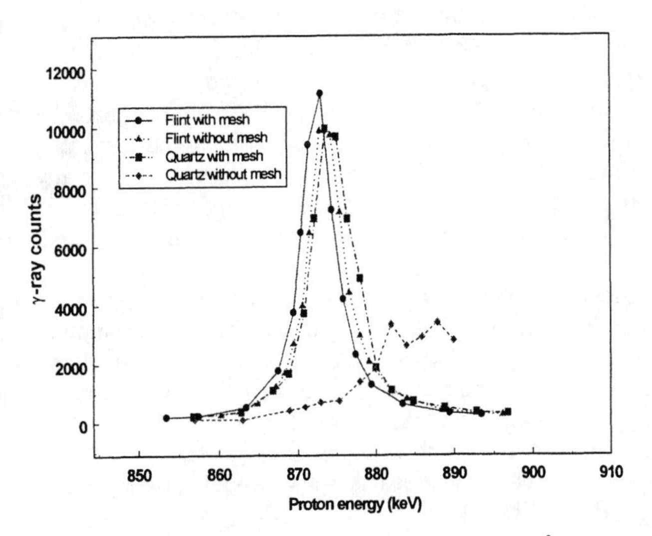

FIGURE 1. Measured NRA resonance peaks of 100 Å CaF_2 on flint and quartz substrates.

The surface charging effect in flint and quartz samples has been checked with and without mesh: A very thin layer of CaF_2 was evaporated on the flint or quartz substrate, and then NRA was carried out to measure the shifts of its resonance peaks. Fig. 1 shows the results. It shows that the surface charging in a quartz sample can be up to 10 kV without the mesh, and the resonance peak is immeasurable due to continuous breakdown of the surface potential. When the mesh method is used, the surface potential fluctuation in quartz is reduced to less than 2 kV, and a well-shaped resonance peak is maintained. Surface charging in flint samples is minor even without a mesh (about 1-2 kV), possibly due to flint's high water content. With the use of a grounded mesh, the surface charging effect in flint is reduced to a negligible value in comparison to the energy stability of the accelerator.

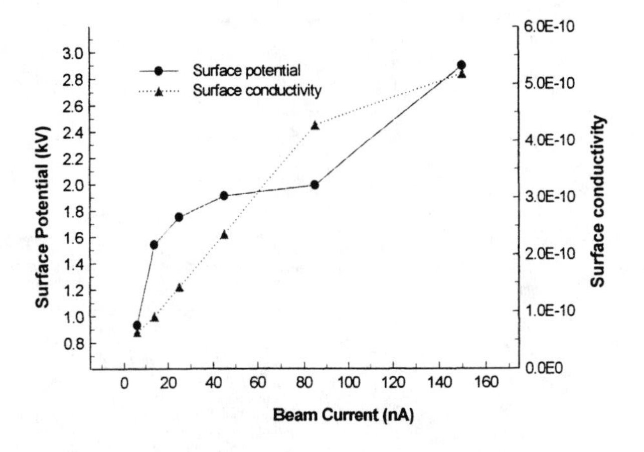

FIGURE 2. Surface potential and surface conductivity versus beam current in NRA measurements for a rock-saw-cut flint sample with 100 Å of evaporation deposited CaF_2.

The beam current effect on the surface charging has also been checked. Fig. 2 shows the relation of peak shift (or surface potential) versus the beam current. The peak shift increases as the beam current increases, but the curve seems to be irregular. However, when the surface conductivity ($\propto$ current/surface potential) is plotted versus current, it is interesting to note that the surface conductivity increases linearly as the beam current increases. This may suggest a radiation-induced conductivity in flint.

RESULTS AND DISCUSSIONS

The differences between a flint sample and other forms of silica were examined first. Fig. 3 shows both the fluorine and hydrogen profiles in different samples. The left figure shows the fluorine profiles of a flint, a mono-crystal quartz, a poly-crystalline quartz, and a fused quartz sample. All the samples were immersed in 300 ppm NaF solution at 80 °C for 6 days. The right figure shows the hydrogen profiles in a flint and a quartz sample. A reversed (p, γ) nuclear reaction, specifically, the $^{1}H(^{19}F, \alpha\gamma)^{16}O$ reaction at a fluorine energy of 6.42 MeV, was employed to measure the hydrogen concentrations. The hydrogen profile should reflect the water concentration in the samples. The differences between the flint sample and the quartz samples are very obvious both in their fluorine and hydrogen profiles. The fluorine uptake in the flint sample is at least 100 times higher than that in other forms of silica. Moreover, the flint sample shows much more

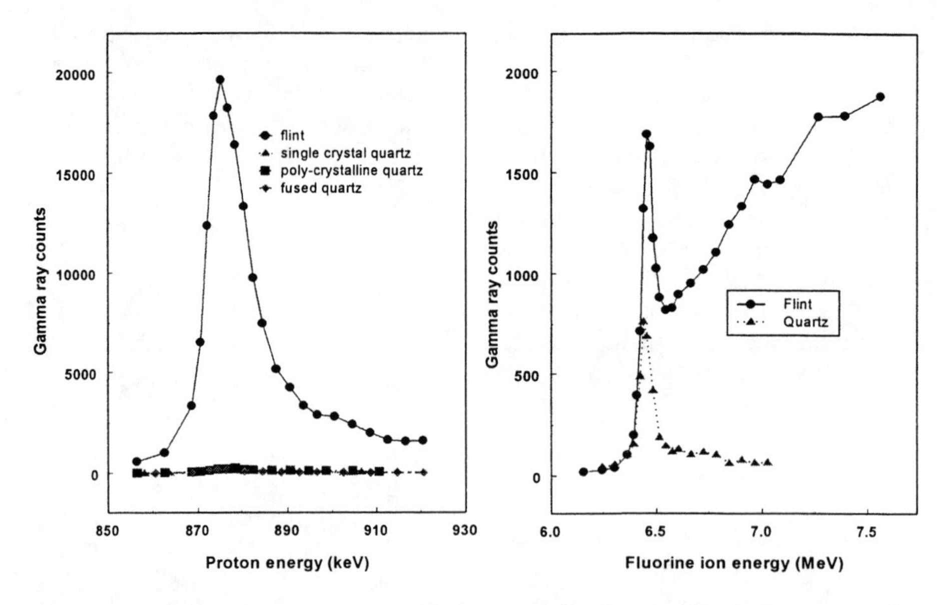

FIGURE 3. Comparison of fluorine profiles and hydrogen profiles between flint and quartz samples.

hydrogen inside the bulk, while the quartz sample shows very little internal hydrogen.

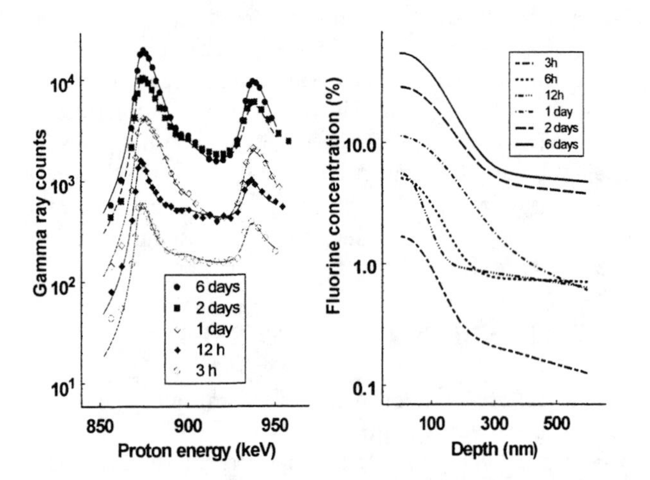

FIGURE 4. NRA measurement of fluorine depth profiles in flint samples immersed for different times in an 80 °C, 300 ppm NaF solution. At left are the experimental data points and their convolution fittings; at right are the fitting results.

It is interesting to note for the flint sample that aside from the hydrogen peak at the surface, hydrogen is less abundant near the surface, and increases with depth into the bulk. This observation is explained as the result of water loss or dehydration from the open voids in the near surface region when the sample was exposed to air or to the vacuum system. This is consistent with the open void structure in flint.

Fig. 4 shows the results of the fluorine uptake as a function of time. Flint samples were immersed in 300 ppm NaF solution at 80 °C for various times from 3 hours to 6 days. The left figure displays the measured raw data, or the so-called γ-ray yield curve, and the right figure shows the fluorine depth profiles generated by our convolution fitting computer program. The results are not as we initially expected: that the depth of the fluorine profile would increase monotonically with the exposure time. The depth of penetration seems not to be very dependent on time, while the fluorine concentration at the surface does increase with time. Moreover, the fluorine concentration at the surface is obviously in excess of its concentration in the solution after a short exposure time. This unexpected result suggests that there is a continuous fluorine surface adsorption when flint samples are exposed to fluoride solution. This continuous adsorption would certainly affect the diffusion profile.

Considering the open void structure in flint, we conjecture that the complicated phenomenon of fluorine uptake into flint is composed of the following separated processes:

1. Fluorine is adsorbed on the flint surface from the solution matrix. Here the flint surface includes the surface areas of those large open volumes (voids) in which the liquid-phase diffusion rate of fluorine ions is similar to that in the solution matrix. The other small open voids are considered to be diffusion channels because the ion diffusion rate in these small voids is substantially slower than that in the matrix solution.

2. Fluorine ions diffuse along the diffusion channels in the liquid phase and are adsorbed by the surfaces (walls) of the channels. This process is a combination of liquid phase diffusion and surface adsorption.
3. Adsorbed fluorine ions migrate along the surfaces of the diffusion channels, exhibiting so-called grain boundary-like diffusion.
4. Adsorbed fluorine into the bulk of micro-quartz or silica spheres, exhibiting so-called solid state diffusion.

The final fluorine uptake profile should be the result of the combination and competition of all the above processes. Therefore, knowledge of all diffusion rates and surface adsorption rates is required to determine the final fluorine profile. In the above four processes, the solid state diffusion seems to be negligible due to its relatively slow diffusion rate. On the other hand, surface adsorption seems to be very important in determining fluorine profiles.

We have investigated fluorine surface adsorption in flint from aqueous solutions with various conditions, such as different fluorine concentrations, pH values, temperatures and impurity levels. The results reveal that the fluorine surface adsorption in flint is strongly influenced by the fluorine concentrations in the solution. Higher fluorine concentrations in the solution dramatically increases fluorine surface adsorption. It also shows that surface adsorption increases as the solution becomes more acidic (low pH value), which is consistent with the general rule of anion adsorption in a charged solid-aqueous interface (7). It is also found that the presence of Ca^{2+} in the solution strongly enhances fluorine adsorption in the silica surface, suggesting that impurities in flint may play important roles in fluorine uptake. The temperature dependent study reveals that fluorine surface adsorption does not depend upon temperature in a simple way. It also shows that fluorine surface adsorption follows the Elovich equation very well (8,9). The Elvich equation is given as:

$$q(t) = b^{-1}\ln(1 + t/k).$$

where $q(t)$ is the amount of adsorption, t is the time, k and b are constants. A more detailed description of fluorine surface adsorption has been reported elsewhere (10).

CONCLUSION

We have studied fluorine uptake in flint samples immersed in aqueous fluoride solutions using the NRA technique. In accordance with the microstructure of the flint samples, the results suggest that the fluorine uptake is not a simple phenomenon, but rather the combination of several simultaneous processes including surface adsorption and inward diffusion. The surface adsorption is affected by several parameters. It is dependent on the pH value, the concentration of fluorine, and Ca^{2+} impurity level in the solution.

ACKNOWLEDGMENT

This work is supported by the Texas Advanced Research Program, NSF, ONR, and the Robert A. Welch Foundation. The authors thank especially Mr. Yangdong Chen and Mr. Robert L. Greeson for their technical assistance.

REFERENCES

1. R.E Taylor, *World Archeology* **1**, 125 (1975).
2. P.Walter, M.Menu, I.C.Vickridge, *Nucl. Instr. And Methods*, **B45**, 119 (1990).
3. P.Walter, M.Menu, J.-C. Dran, *Nucl. Instr. And Methods*, **B64**, 494 (1992).
4. J. Jin, D.L.Weathers, J.P.Biscar, B.F.Hughes, J.L.Duggan, F.D.McDaniel, and S.Matteson, *AIP conference proceedings* **392**, p681, AIP press, New York (1997).
5. H. Graetsch, "Silica: Physical behavior, Geochemistry and Materials Application", Chapter 6, Review in Mineralogy, Volume 29, Editors: P.J.Heaney, C.T.Prewitt, and G.V.Gibbs. Mineralogical Society of America, 1994.
6. W.Shepherd, Flint, Its Origin, Properties and Uses, (Faber and Faber, London, 1972).
7. D.G.Kinniburgh, and M.L.Jackson, "Adsorption of Inorganics at Solid-Liquid Interfaces", Chapter 3, Editors: M.A.Anderson, and A.J.Rubin, Ann Arbor Science Publishers, Inc., 1981.
8. D.O.Hayward, and B.M.W.Trapnell. *Chemisorption* (London: Butterworth's, 1964), p.93.
9. F.J.Higston, "Adsorption of Inorganics at Solid-Liquid Interfaces", Chapter 2, Editors: M.A.Anderson, and A.J.Rubin, Ann Arbor Science Publishers, Inc., 1981.
10. J.-Y. Jin, Ph.D. dissertation, University of North Texas, 1998.

The Recoil Implantation Technique Developed at the U-120 Cyclotron in Bucharest*

C.I. Muntele[1], L. Popa Simil[2], P. M. Racolta[2], D. Voiculescu[2]

[1] *Center for Irradiation of Materials, Alabama A&M Univ., PO Box 1447, Normal, AL 35762*
[2] *Dept. of Applied Physics, IFIN-HH, PO Box MG 6, 76900 Bucharest, Romania*

At the U-120 cyclotron in Bucharest was developed 15 years ago the thin layer activation (TLA) technique for radioactive labeling of metallic components on depths ranging between 100 μm and 300 μm, for wear/corrosion studies. Aiming to extend these kinds of studies on non-metallic components and at sub-micrometric level we were led to the development of the recoil implantation technique for ultra thin layer activation (UTLA) applications. Due to the low energy of the recoils obtained in a sacrificial target from a nuclear reaction, the surface layer of material to be labeled must be as thick as a few hundred nanometers. Also, since the radiotracer is externally created, there are no restrictions for the kind of material to be labeled, except to be a solid. In this paper we present some results of our studies concerning the actual status of this application at our accelerator.

INTRODUCTION

It is more than 25 years since the thin layer activation (TLA) method was developed (1, 2) and almost 15 years since it was implemented in our institute (IFIN-HH), at our U-120 cyclotron. The TLA principle consists of ion beam irradiation on a well-defined area of a machine part subjected to wear, in order to induce a nuclear reaction having as a result a radiotracer. The half time should be at least of the order of a few days, to have enough time for post-irradiation measurement (4) and for transportation to the tribological testing stand. This can be placed in another institute, even in another town, and no special requirements for radioprotection are needed since the activity of the sample is lower than 370 kBq (10 μCi). From the two measuring methods for the wear/corrosion level, we mainly use the circulation one. The lubricating oil takes the debris removed from the sample (irradiated machine part) and passes it through an external flow chamber providing a near 4π detection geometry. The correlation between the measured radioactivity and the thickness of removed layer is established in (5).

This method can be used for systems where a wear/corrosion level of a few tens micrometers is expected to appear for ten or hundred hours of working regime. This is the case of thermal engine piston rings or cylinders, rails, bearings, hydraulic pumps working in salty or sandy environment etc. (6). But for many other mechanisms, where small material loss can appear due to friction or corrosion, the TLA technique becomes inefficient in use. A way to reduce the thickness of the labeled layer is to send the incident beam on the target at an angle smaller than 90 deg. (aiming to obtain the same radioactivity, but in a thinner layer). But even in this case, the radioactive thickness is a few tens of μm.

In 1980 Conlon (3) proposed a method for radioactive labeling of a very thin layer of material by using the recoil implantation technique. A few experiments have been reported by Lacroix (7) on $^{56}Fe(p, n)\,^{56}Co$, giving also an experimental depth profile of the ^{56}Co ($T_{1/2}$=77.1 d) implanted in nickel. The main idea is to produce the radioactive ion in a thin target placed in the front of the surface to be labeled, so that the recoil ions extracted from this "sacrificed" target will implant themselves into the material's surface. Having a low energy (up to 800 keV), their range will be up to 600 nm. The advantage of this technique is the possibility to extend its applicability to any kind of solid material: plastics, ceramics, semiconductors, and metals.

EXPERIMENT

Our laboratory has started a research program to develop the ultra thin layer activation (UTLA) method using the recoil implantation technique for tribological studies at the sub-micrometric level. We made a reaction chamber with a removable modular device as an extension of the existing TLA facility (Fig. 1). The dimensions of the sacrificed target (10 μm Fe foil), sample material to be implanted (15 μm Al foil), and the distance between them (18 mm), were chosen considering the maximum scattering angle of the recoils (51.3 deg. in this case), the spatial resolution needed, and geometric factors. We used a 1 μA, 14 MeV proton beam collimated to a 5 mm diameter.

* Work partially supported by International Atomic Energy Agency, Vienna, Austria

CP475, *Applications of Accelerators in Research and Industry*,
edited by J. L. Duggan and I. L. Morgan

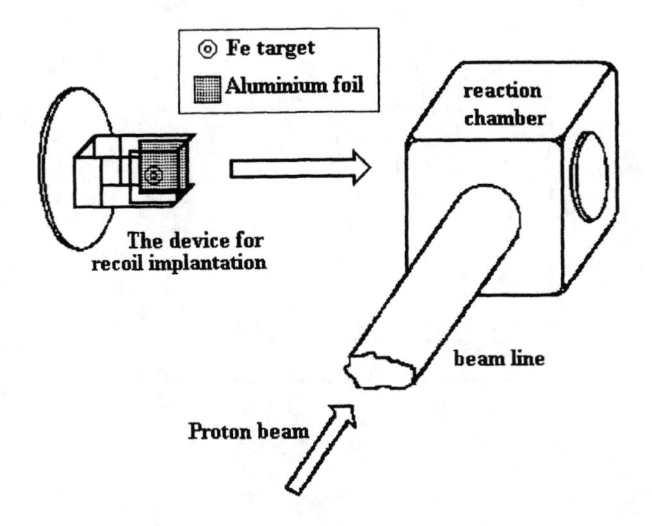

FIGURE 1. Schematic view of the irradiation setup.

RESULTS

At the end of the irradiation time (7 hours) we obtained ~16 kBq (~0.43 μCi) of activity on the entire implanted area. In order to obtain a distribution of recoils vs. scattering angle, the aluminum foil was sectioned in annular concentric parts as shown in Fig. 2. For gamma measurements we used a NaI (Tl) scintillator detector and a multichannel acquisition system. The acquisition time was 1000 s for each part (i.e. 3.3 h for all, compared to $T_{1/2}$=77.1 d), in order to gain a relative error less than 10% (due to the very low activity level of the parts).

The results are presented in Fig. 3, where N_{tot} is the number of recoiled ions on the whole Al foil, N_i is the number of recoiled ions on the part "i", and Ω_i is the corresponding solid angle calculated for the situation presented in Fig. 2. The relative error for the X value is 12%. The maximum value of the implanting dose on each part is shown in table 1, along with other related data.

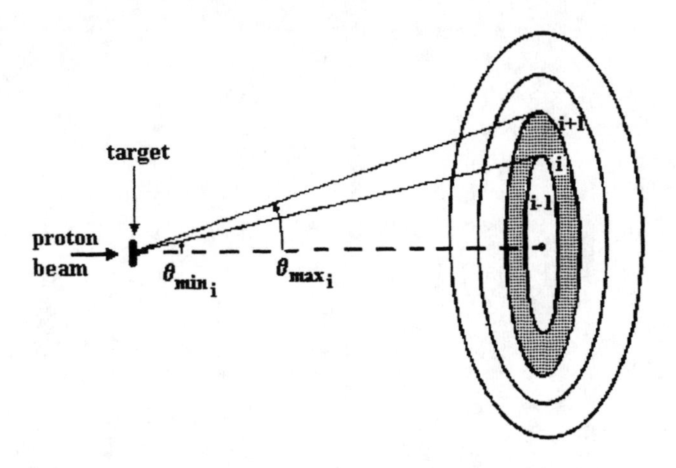

FIGURE 2. The annular sectioning and the measuring mode for the scattering angle.

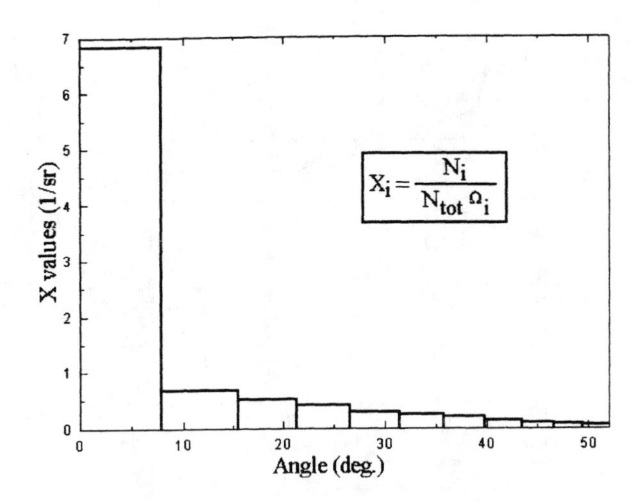

FIGURE 3. The distribution of the recoils vs. scattering angle.

Kinematic calculations were performed for this reaction aiming to obtain the dependence of energy vs. scattering angle (Fig. 4). Ranges and straggles were extracted from TRIM95 (8).

DISCUSSIONS AND CONCLUSIONS

The obtained X values allow us to make a quick evaluation of the needed irradiation data (time and current) for a specific application by using the following formula:

$$I \cdot t = \Lambda_i / (F \cdot X_i \cdot \Omega_i).$$

Here, Λ_i is the desired activity of the sample (depending upon application, but no greater than 370 kBq), F is a constant (=2.2×10^{10} ions/μAxh) derived from the formula N_{tot}/I(μA)xt(h) for sacrificed target thicker than 400 nm, X_i is the value shown in Fig. 3, and Ω_i is the corresponding solid angle which can be easy estimated.

As can be seen in table 1, the irradiation doses have low values for samples positioned at angles over 30 deg. (< 10^{15} cm^{-3}), so that we can assume that no significant changes in the sample's structure will occur. However, the sample must be placed out of outgoing proton beam (~99.997% of incoming current). For simultaneous use of the residual outgoing current, we stopped it into a 100 μm Ti target (5 foils × 20 μm each), for producing ^{48}V positron source via a (p, n) reaction. We use that nuclide for Doppler broadening applications on defects in solids.

The experiment we performed proved that this technique may be a good tool for activating very thin layers of material and can be successfully applied for sub-micrometric tribological phenomena characterization. However, the irradiation time required for that (between 6 hrs. and few days) may be not economically justified.

TABLE 1. Geometrical characteristics and maximum implanted doses for each part.

#	R_{Max}(mm)	θ_{Max}(deg.)	Dose (cm^{-3})
1	2.5	7.9	$2.57*10^{16}$
2	5	15.5	$4.8*10^{15}$
3	7	21.3	$3.3*10^{15}$
4	9	26.6	$2.3*10^{15}$
5	11	31.4	$1.5*10^{15}$
6	13	35.8	$1.06*10^{15}$
7	15	39.8	$7.3*10^{14}$
8	17	43.4	$4.6*10^{14}$
9	19	46.6	$2.7*10^{14}$
10	21	49.4	$1.7*10^{14}$
11	23	51.9	$9.8*10^{13}$

We prepared two samples of different Cu and Cr layers having thicknesses in the range 40 nm - 300 nm (measured by RBS) deposited on Si wafers by using low energy ion beam assisted deposition by magnetron and ion beam sputtering. The samples are to be measured on a gamma detection system, in order to obtain a calibration curve activity vs. material thickness. The results and the results reported in (7) will allow us to adjust a Monte Carlo code able to provide a realistic theoretical curve for all types of material.

The work is still in progress on this research theme, to establish the methodology for calibration of the method.

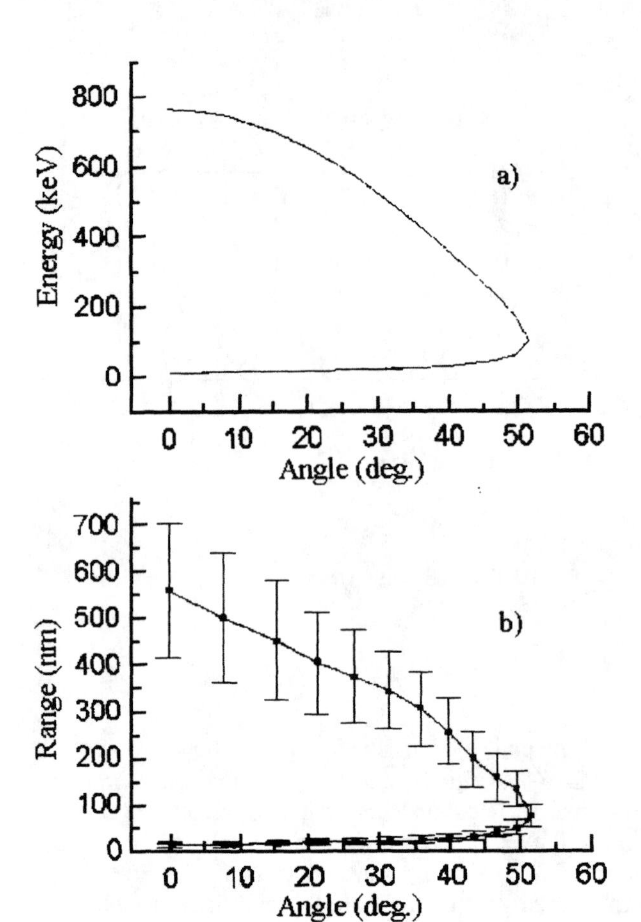

FIGURE 4. a) Energy distribution (kinematics calculation); b) Range distribution (TRIM95 calculation). Straggle is presented as error bar.

REFERENCES

1 Konstantinov, I. O., Krasnov, N. N., *J. Radioanal. Chem.* **8** (1971) 357.

2. Conlon, T. W., *Wear* **29** (1974) 69.

3. T. W. Conlon, Nucl. Instr. and Meth. 171 (1980) 297.

4. Ivanov, E. A., Pascovici, G., Racolta, P. M., *Nucl. Instr. and Meth.* **B 82** (1993) 604.

5. Lacroix, O., Sauvage, T., Blondiaux, G., Racolta, P. M., Popa Simil, L., Alexandreanu, B., *Nucl. Instr. and Meth.* **A 369** (1996), 427.

6. International Atomic Energy Agency, *TECDOC-924*, 1997, pp. 23.

7. Lacroix, O., Sauvage, T., Blondiaux, G., Guinard, L., *Nucl. Instr. and Meth.* **B 122** (1997) 262.

8. Ziegler, J. F., Biersack, J. P., Littmark, U., *The Stopping and Range of Ions in Solids* (Pergamon Press, NY, 1985).

Fragmentation of Macromolecular Ions

I. D. Williams[1], B. Merron[1], R. J. H. Davies[2], I. G. Hughes[3] and V. Morozov[4]

[1]*School of Mathematics and Physics, Queen's University of Belfast, BT7 1NN, UK*
[2]*School of Biology and Biochemistry, Queen's University, Belfast, BT7 1NN, UK*
[3]*School of Science and Technology, Dun Laoghaire Institute of Art Design & Technology, Dun Laoghaire, Ireland*
[4]*Oak Ridge National Laboratory, Oak Ridge, TN 37831-6372 USA*

A novel technique is presented for the study of post-source decay in matrix assisted laser desorption time-of-flight mass spectrometry. By introducing a voltage labelling cell within the field-free region we have been able to distinguish between intact ions, fragmented ions and neutrals in a linear time-of-flight instrument. Furthermore by introducing gas into the cell, collision induced dissociation and charge stripping has been observed. Modelling of the spectra leads to information on the fragment masses, energy release and lifetime of the particular dissociation process. The post source decay and collision induced dissociation of negative ions of bovine insulin is used to illustrate the potential of this technique.

INTRODUCTION

Matrix assisted laser desorption/ionization (MALDI) mass spectrometry has revolutionised the chemical, biological and biotechnological communities. This technique has made it possible to obtain information on large biomolecules which were previously difficult to study by mass spectrometry (1). It is well known that most of the analyte molecular ions are desorbed intact in MALDI (2,3). However, in a linear time-of-flight (TOF) system, fragments formed in the field free region due to post-source decay (PSD) cannot be distinguished from the parent ion since decomposition does not appreciably change the velocity of the products. For the purpose of molecular mass determination, PSD can be considered undesirable because kinetic energy release during fragmentation of the molecular ion leads to peak broadening and loss of mass resolution. In reflectron TOF instruments however it can be exploited to obtain structural information concerning the parent ion (4).

We have developed a novel technique for determining the PSD of macromolecules in a linear MALDI-TOF mass spectrometer without the aid of a reflectron (5). By introducing a voltage labelling cell within the field free region we have been able to distinguish between intact ions, fragmented ions and neutrals on the basis of their differing kinetic energies (and charge for neutral products). This new method, coupled with a computer model for calibration and simulation of macromolecular fragmentation, has the potential for yielding significant information on the fragment masses, energy release and lifetime of the fragmentation process (6).

EXPERIMENTAL APPROACH

Mass spectra are recorded using a Fisons Instruments (Manchester, UK) TOFspec linear time-of-flight mass spectrometer. A nitrogen laser with an output of 337 nm radiation and 4 ns pulse width is used for the desorption and ionization of samples (175 μJ per pulse). Laser irradiance is slightly above the ion formation threshold. Following desorption/ionization, ions are accelerated by biasing the sample with voltages up to 25 kV, and travel through the source region and drift region (60 cm).

A chevron microchannel plate detector operating in current amplification mode is used to detect the ions. The signals from the detector are recorded by a digital oscilloscope (Tektronix TDS 520) coupled to a VAXstation 4000 for data analysis. Saturation effects in the microchannel plate detector caused by high yields of low mass ions from the matrix are avoided by including a quadrupole mass analyzer within the source region to suppress the low mass ion signal.

Collision induced dissociation experiments are carried out by admitting gas into the voltage cell via a needle valve. Nitrogen is used as the target gas for CID experiments. The voltage cell, made from a stainless steel tube 70 mm long, is biased at a negative potential of 5 kV throughout these experiments so that induced collisions occur at an energy of 20 keV. The entrance and exit apertures of the cell are of 3 mm diameter to give a low gas conductance. Tungsten wire mesh grids (92% transmission) cover the entrance and exit of the voltage cell acceleration regions to avoid field penetration. The cell is located in the drift region of the mass spectrometer directly above a 130 ls^{-1} diffusion pump to ensure that the number of collisions outside the cell are kept to a minimum. An ionization gauge positioned near the diffusion pump records the residual gas pressure (triggered on digital oscilloscope) taking into account the varying ionization efficiencies of different gases in the gauge. The

CP475, *Applications of Accelerators in Research and Industry*,
edited by J. L. Duggan and I. L. Morgan

pressure in the cell is approximately 10^3 times higher than the pressure of the surrounding chamber.

The protein bovine insulin (isotopically averaged MW 5733) together with sinapinic acid matrix (MW 224) is used throughout the experiment. About 3 μl of the matrix-protein solution is placed on a stainless steel plate to form a target of around 2 mm diameter. The sample is then allowed to evaporate to dryness at room temperature in air. The molar concentration ratio of matrix to analyte is of the order 5000:1. The vacuum in the ion source chamber and drift chamber is better than 2×10^{-7} mbar in both sections.

RESULTS AND DISCUSSION

Figure 1 shows the PSD of negative $(M - H)^-$ ions of bovine insulin obtained by applying 5 kV to the voltage cell. The effect of the voltage labelling cell is to introduce a velocity dependence of the fragmented ion mass at some part of the ion trajectory. Thus, fragmentation events occurring both before and within the voltage cell show up as distinct peaks on the TOF spectrum. The spectrum in Fig. 1 was obtained by averaging eight separate spectra consisting of 15-20 laser shots each. Two fragment peaks a and b, identified as the A (MW 2335) and B (MW 3397) chains of bovine insulin, can clearly be seen to the left of the intact ion peak i, while the neutral product peak n appears to the right. Peaks corresponding to doubly charged insulin $(M - H)^{2-}$, dimer $(2M - H)^-$ and trimer $(3M - H)^-$ ions, as well as those due to adducts were also observed. However, only the singly charged insulin ion and its fragmentation products are discussed here.

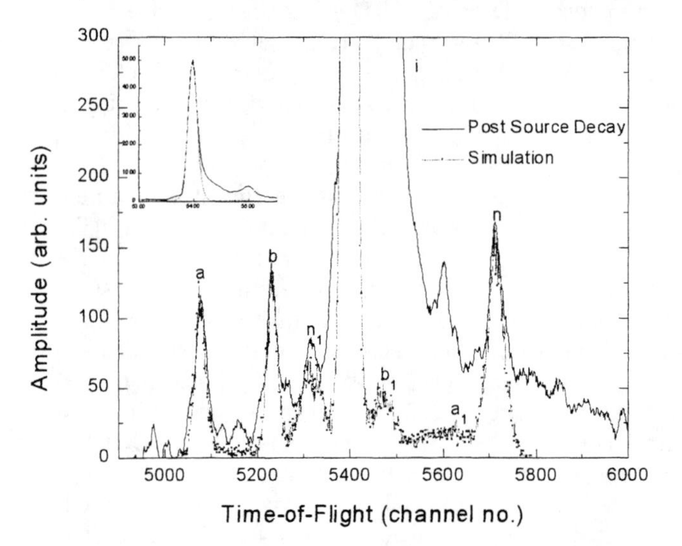

FIGURE 1. The fragmentation pattern of bovine insulin negative ions due to post-source decay together with the computer simulated pattern. The inset shows the fit between experiment and simulation for the intact ion peak i. The main figure shows the comparison between experiment and simulation for the A-chain peak a, the B-chain peak b, and the neutral peak n. Fragmentation events occurring before the voltage cell are labelled a_1, b_1, and n_1.

A model assuming insulin ion fragmentation into intact A and B chains, yielding one charged and one neutral fragment was used to fit the fragment peaks (6). It is assumed that the A and B chain ions are produced with equal probability (7). As a result of the simulation, information on energy release, ionization probability and lifetime before fragmentation is obtained. It is clear that the simulated spectrum is successful in reproducing the characteristics of the experimental spectrum.

A comparison of the PSD and CID fragmentation patterns is shown in Fig. 2. CID experiments were recorded using nitrogen as the neutral collision gas. It is clear that the PSD and CID spectra are quite similar in appearance. However, the CID spectrum shows an increase in the neutral peak contribution and this was recorded for a range of pressures (0.2 – 3.20 × 10^{-6} mbar). A similar model to that used for modelling the PSD spectra was used in simulating the CID spectrum. Whilst good fits to the a and b peak intensities could be achieved, the initial simulation failed to account for the intensity observed in the neutral n peak. This excess of neutrals could not be associated with the fragmentation of the parent ion into other products as no further fragment ion peaks were observed other than the A and B chain ion peaks.

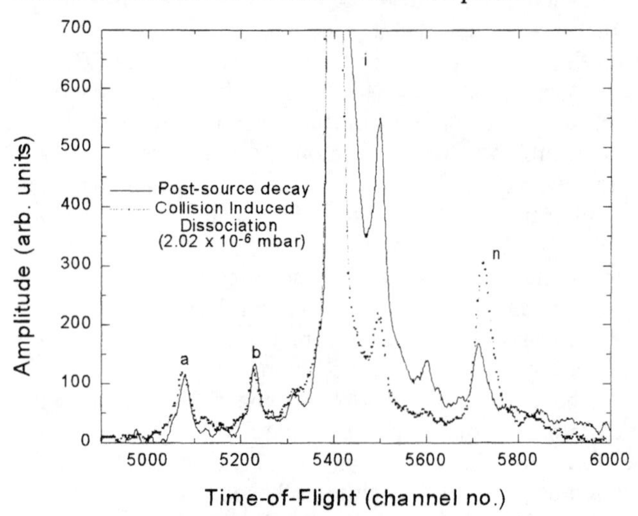

FIGURE 2. The fragmentation patterns of bovine insulin negative ions due to post-source decay and collision induced dissociation (N_2). The fragmented ion peaks, i.e. a and b, are normalised to show the difference between the PSD and CID neutral peaks.

It therefore appears that there is an additional process occurring in conjunction with the PSD and CID of the parent ion in the voltage cell. We suggest that a charge stripping process may be involved, taking place according to

$$(M - H)^- + G \rightarrow (M - H) + G + e^- \quad (1)$$

where G denotes a molecule of the target gas. In order to quantify the effects of the competing processes leading to the spectrum in Fig. 2, let us define C(p) as the fraction of parent ions that decay through PSD and CID. Clearly C(p), a function of pressure, is given by the sum of the areas under the peaks a and b, divided by the area under the parent peak i. Furthermore C(p) can be subdivided into components Y(p), a function of pressure due to CID, and X due to PSD which is independent of target gas density. Hence we have :

$$C(p) = Y(p) + X \quad (2)$$

Similarly, if N(p) is the fraction of neutrals observed, then that fraction of neutrals due to a charge exchange process will be given by :

$$D(p) = N(p) - C(p) \quad (3)$$

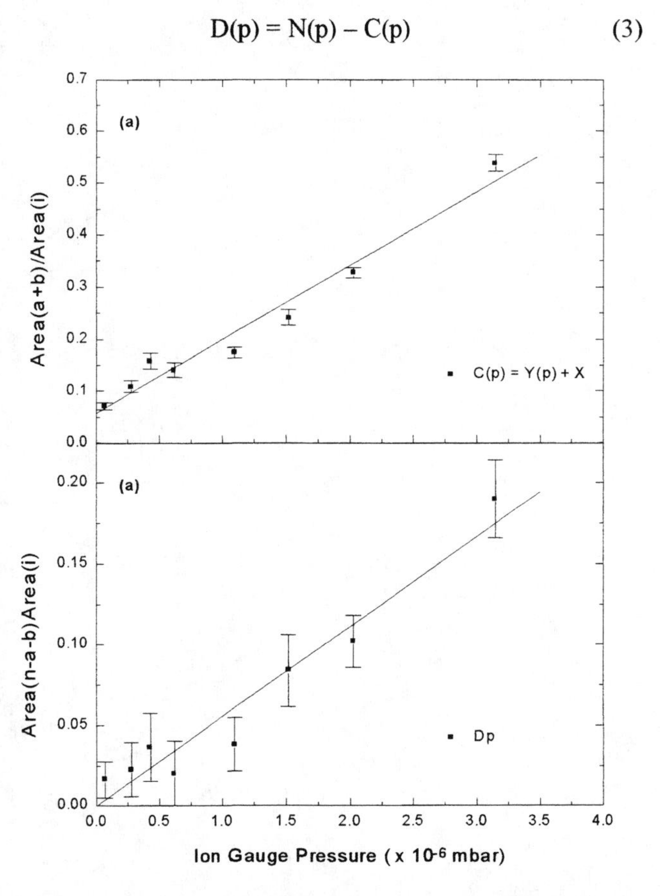

FIGURE 3. **(a)** The total fragmentation C(p) of bovine insulin negative ions into its constituent A and B chain ions due to post-source decay and collision induced dissociation (N_2). The intercept gives the metastable contribution. **(b)** The rate of charge stripping events occurring for CID with nitrogen. Corrections have been made for relative ionization efficiencies.

Figure 3a shows a plot of the product-to-parent peak areas C(p) as a function of the collision gas pressure. The y-intercept, *i.e.* at vanishing residual gas pressure, gives us an indication of the metastable decay contribution X to the CID spectra (8). Thus about 5-6% of the parent ions have undergone fragmentation in the collision cell even though no collisions with the residual gas occurred. Also shown in Fig. 3b is a plot of the excess neutrals D(p) as a function of the collision gas pressure. This gives us an indication of the rate at which charge stripping occurs in the CID of negative ions of bovine insulin with nitrogen.

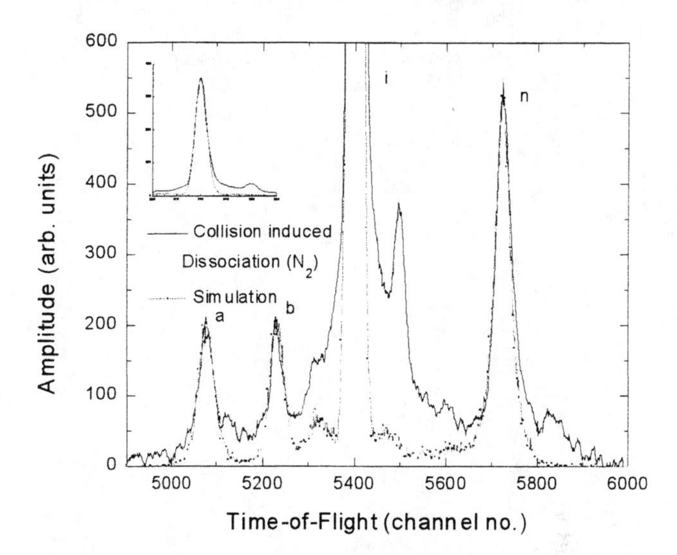

FIGURE 4. The fragmentation patterns of bovine insulin negative ions due to collision induced dissociation with nitrogen together with the computer simulated pattern. The inset shows the fit between experiment and simulation for the intact ion peak i.

Figure 4 shows the simulated pattern of the CID spectrum, taking all of the above fragmentation mechanisms into account.

CONCLUSIONS

The introduction of a voltage labelling cell into the field free region of a linear TOF mass spectrometer has been successful in the separation of fragment ions produced by PSD/CID of large biomolecules in MALDI. Detailed information on the fragmentation mechanisms involved are obtained by modelling the experimental spectra.

ACKNOWLEDGEMENTS

This work has been made possible by funding from ESPRC on grant GR/K72544. Brian Merron is funded by the European Social Fund.

REFERENCES

1. Schriemer, D.C. and Li, L., *Anal. Chem.* **68**, 2721-2725 (1996)
2. Karas, M., Bachmann, D., Bahr, U. and Hillenkamp, F., *Int. J. Mass Spectrom. Ion Processes* **78**, 53-68 (1987)

3. Karas, M. and Hillenkamp, F., *Anal. Chem.* **60**, 2301-2303 (1988)
4. Spengler, B., Kirsch, D., Kaufmann, R. and Jaeger, E., *Rapid Commun. Mass Spectrom* **6**, 105-108 (1992)
5. Hughes, I.G., Morozov, V., Merron, B., Boyle, J., Rainnie, J., Williams, I.D. and Davies, R.J.H., *Rapid. Commun. Mass Spectrom.* **11**, 1509-1514 (1997)
6. Morozov, V., Merron, B., Williams, I.D., Davies, R.J.H. and Hughes, I.G., *Rapid Commun. Mass Spectrom.* **12**, 97-103 (1998)
7. Barber, M., Bordoli, R.N., Elliott, G.J., Tyler, A.N., Bill, J.C., and Green, B.N., *Biomed. Mass Spectrom.* **11**, 182-186 (1984)
8. Bricker, D.L. and Russell, D.H., *J. Am. Chem. Soc.* **108**, 6174-6179 (1986)

Desorption Induced by Atomic and Molecular Ion Collisions on LiF

J. A. M. Pereira and E. F. da Silveira

Depto. de Fisica, Lab. Van de Graaff, PUC-Rio, CP 38071
22952-970 Rio de Janeiro, Brasil

Abstract

Atomic and molecular nitrogen ion beams, produced by the PUC-Rio Van de Graaff accelerator, were used to bombard lithium fluoride thin films. Desorption of secondary ions was measured by means of a time-of-flight mass spectrometer equipped with a double grid acceleration system. The outputs of the experiment are the axial kinetic energy distribution and the desorption yield of the emitted ions. This information allowed determination of the relative contribution to desorption due to collision cascades (nuclear sputtering) and to electronic excitation (electronic sputtering). It was observed that F^- ions are desorbed as a result of collision cascades and that the F^- ion yields depends linearly on the number of constuents in the projectile, i. e., $Y(N_2^+) = 2\,Y(N^+)$. The emission of clusters such as $(LiF)Li^+$ was found to be caused by electronic excitation and the $(LiF)Li^+$ yield revealed a nonlinear dependence: $Y(N_2^+) > 2\,Y(N^+)$. Both processes were found to contribute to Li^+ desorption. These effects are discussed in terms of the density of deposited energy which depends on the projectile velocity and on the electronic stopping power.

INTRODUCTION

This work is devoted to the investigation of sputtering from insulating surfaces induced by heavy projectiles with velocities around the Bohr velocity. This kind of projectile is usually available for Van de Graaff accelerators and is of particular interest for basic research in the field of ion-solid interactions. In this velocity regime, two different mechanisms of ion emission are important: sputtering originated from collision cascades (1) and desorption initiated by electronic transitions (2).

These two contributions can be readily seen in the energy distribution of the emitted ions (3). The description of these phenomena involves concepts of Atomic and Solid state Physics. Lithium fluoride was chosen because of its simple electronic and geometrical properties so the physical processes underlying the phenomenon can be understood more deeply. In addition, this material exhibits a large variety of phenomena when subjected to ionizing radiation (4,5).

Cluster effects have been recognized in the study of secondary ion desorption induced by keV and MeV molecular projectiles (6). It is usually described in terms of the yield enhancement factor:

$$R = Y(n) / nY(1) \qquad (1)$$

where n represents the number of atomic constituents in the projectile and Y represents the yield of a given secondary ion. Only recently the influence of cluster impact on the kinetic energy distribution of emission was investigated (7).

The energy lost in the near surface region is enhanced due to the spatial and temporal correlation achieved with cluster impact. Besides, this energy is deposited in small volumes giving rise to high values of the density of deposited energy and, consequently, nonlinear effects are likely to occur. The volume of energy deposition is usually determined by the projectile velocity (8).

The use of atomic and molecular projectiles, at a given projectile velocity, permits the density of deposited energy to be varied at a constant volume.

EXPERIMENTAL AND RESULTS

The experimental set-up is briefly discussed in this section. A more detailed description can be found in (7). The sample consists of a thin film of LiF which is evaporated onto an aluminum foil.

Beams of N^+ and N_2^+ primary ions with 200 keV / atom impinge the sample at an angle

CP475, *Applications of Accelerators in Research and Industry*,
edited by J. L. Duggan and I. L. Morgan

of 45° (figure 1). The start signal is delivered by electrons emitted from the beam exit target surface are detected in the MCP_1 array. The secondary ions emitted from the beam entrance surface are accelerated in two regions. In the first one, a weak electric field is applied between the sample and the first grid permitting secondary ions with the same mass but different axial energies to spread in time.

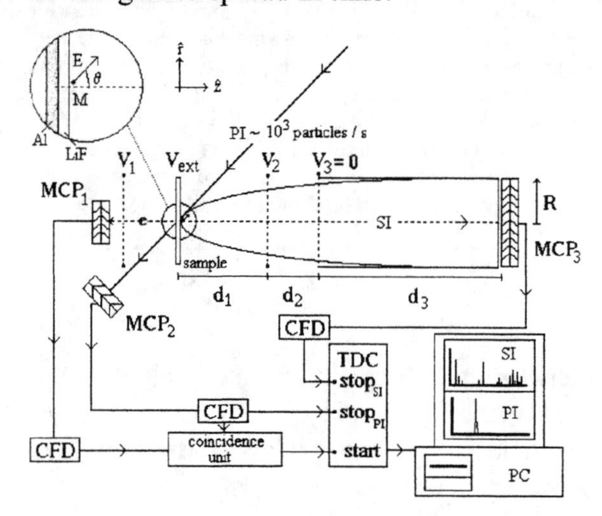

FIGURE 1 - Experimental set-up. The primary ion (PI) beam rate is of 10^3 particles / s. The secondary ions (SI) are accelerated in two steps. The SI time-of-flight is measured by using a TDC and a PC analyzes the data. The chamber pressure is maintained at 10^{-7} mbar by turbo pumps.

In the second acceleration region, the secondary ions receive a kinetic energy appropriated for their detection. This can be done by a suitable choice of the potentials V_{ext} and V_2. The spectrometer can also work with V_2=0. In this case, the secondary ions time-of-flight is weakly dependent on their initial axial energy increasing the mass resolution of the spectrometer. A time to digital converter (TDC) measures the time-of-flight of the primary and secondary ions independently.

Figure 2 shows the changes occurring in the mass spectrum due to the cluster effect. The most affected ion specie is the $(LiF)Li^+$, having mass 33 *u*. The corresponding peak is effectively not observed in the spectra obtained with atomic ions while this secondary ion generates one of the main peaks when a molecular ion beam is used.

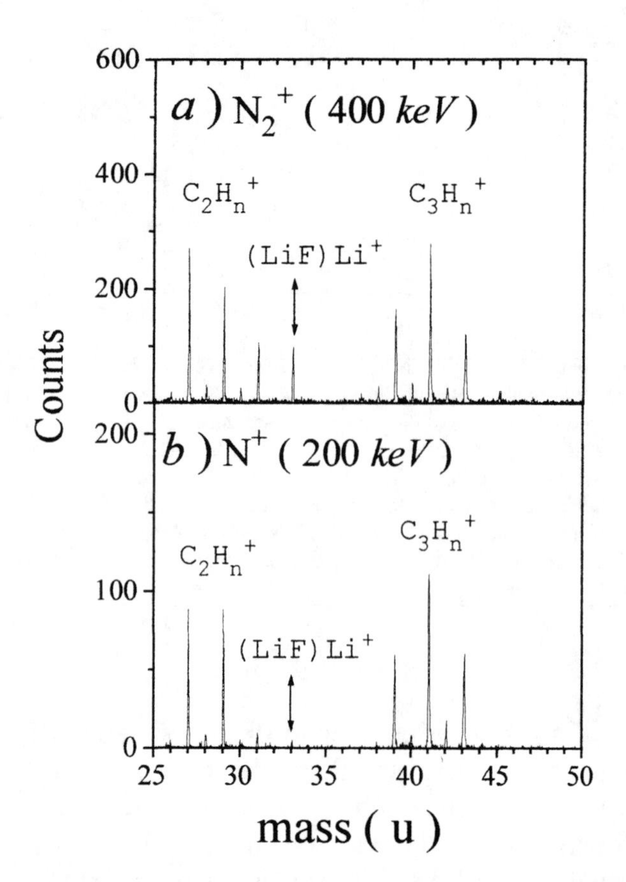

FIGURE 2 - Mass spectra obtained with *a*) molecular and *b*) atomic ion bombardment at E = 200 keV / atom (v = 0.75 v_0). These spectra were obtained with V_2=0 and are not normalized.

Another visible effect concerns to the $C_nH_m^+$ groups. The emission of hydrocarbon clusters with low number of hydrogen is favored by a molecular impact. Table 1 shows the yield enhancement factor for the secondary ions shown in figure 2.

The situation is more complex for Li^+ secondary ion emission. As can be seen in figure 3a, two different mechanisms contribute for Li^+ desorption. The nuclear sputtering contribution is calculated using the collision cascades theory applied to an ionic solid (solid line) (3). This contribution is subtracted from the total energy distribution (data points) in order to

TABLE 1 - Values of R for different secondary ions

Ion	Enhancement factor, $R^{(Y)*}$
$C_2H_3^+$	1.5
$C_2H_5^+$	1.1
$(LiF)Li^+$	5.7
$C_3H_3^+$	1.4
$C_3H_5^+$	1.1
$C_3H_7^+$	1.0

*Calculated according to equation (1)

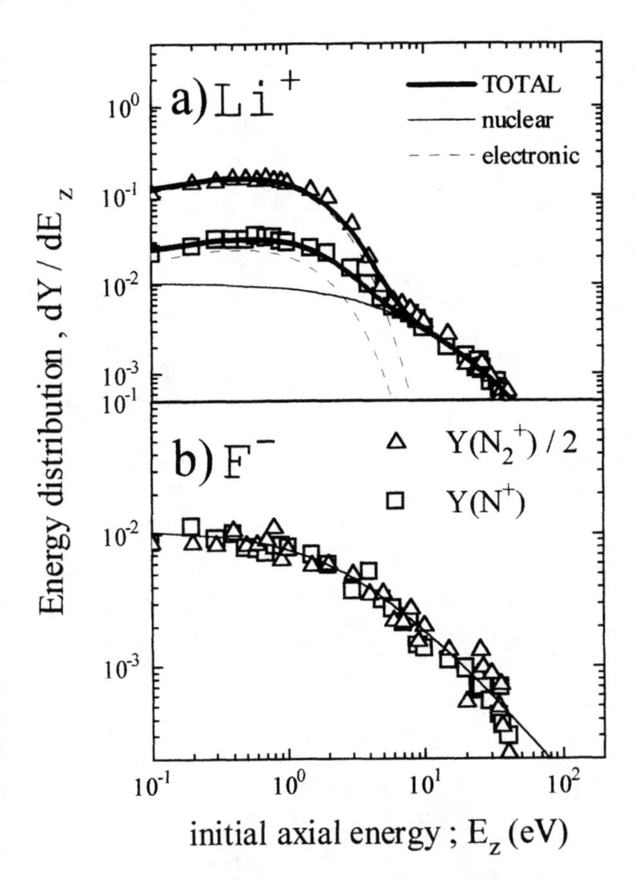

FIGURE 3 - Axial energy distributions for a) Li^+ and b)F^- secondary ions for N^+ and N_2^+ ion bombardment.

obtain the energy distribution due to the electronic sputtering (dashed line). Following this procedure, it is possible to study in separate the electronic and the nuclear sputtering contributions to the desorption yield. An enhancement factor value similar to that of the (LiF)Li^+ is found if only the electronic sputtering contribution of the Li^+ is considered : (R_{Li+}(electronic) = 6).

The collision cascade contribution is linear with the number of projectile atomic constituents in the case of Li^+ and also for F^- as can be seen in figure 3b.

DISCUSSION

The observed non-linear effects on the electronic sputtering can be analyzed in the frame of the density of deposited energy, ε(r), and of the effective energy loss (7). This quantity , S_{eff}, is related to the track characteristic diameters, r_i (infra-track) and r_u (ultra-track), and to the ion electronic stopping power, S_e, according to:

$$S_{eff} = S_e \ln(\varepsilon_0 / \varepsilon_c) / \ln(1+(r_u / r_i)^2) \quad (2)$$

where ε_0 is the energy density at the impact point and ε_c is the energy density threshold for ion emission. This formalism was applied to describe the secondary ion desorption yield dependence on the primary ion velocity (7). The secondary ion yield is nearly proportional to S_{eff}^2 and the ε_c-value for best fit is 0.08 eV / Å^3.

At this point, it is worth to mention that atomic and molecular projectiles produce the same characteristic Li^+ emission energies. In addition, no electronic sputtering was observed for F^- secondary ions. These facts reflect the electronic sputtering mechanism of Li^+ secondary ions (9).

The yield enhancement factor was calculated according to the effective stopping power formalism for 200 keV / atom considering the molecular ion stopping power to present no cluster effect ,i.e., $S_e(N_2^+) = 2\, S_e(N^+)$. The results for this calculation are shown in figure 4 (dashed line) together with experimental values for the electronic sputtering contribution of Li^+ and for (LiF)Li^+.

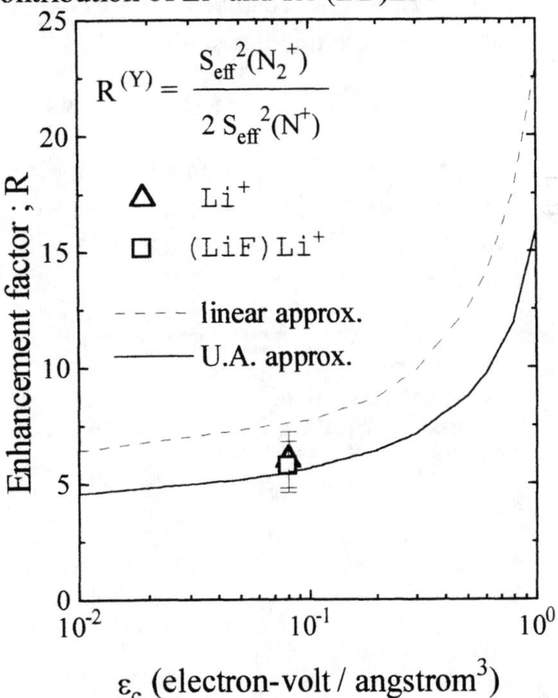

FIGURE 4 – Yield enhancement factor calculated from the effective energy loss.

The deviation is ~25% and can be due to proximity effects on the molecule stopping power. The united atom approximation for the molecular ion stopping power calculation predicts deviations from the stopping power additivity (10). For the present case, the stopping power enhancement is predicted to be $R(S_e) = 0.88$ and one has $S_e(N_2^+) = 1.76\ S_e(N^+)$. If this value is used, one obtain the solid line in figure 4. It can be seen that the sub-linear prediction of the molecular ion stopping power in the united atom approximation produces a better agreement with experimental data.

CONCLUSION

Non-linear effects were found for the electronic sputtering of Li^+ and $(LiF)Li^+$ secondary ions. The effective stopping power formalism was used to describe the experimentally observed enhancement factors for the electronic sputtering. Since at high energies the impact parameters governing nuclear collisions are much smaller than the equilibrium distance between atoms in the solid or in the molecular projectile, proximity effects are of minor importance in the case of nuclear energy loss. This results in the observed linear behavior of the nuclear sputtering of Li^+ and F^-.

ACKNOWLEDGEMENTS

Many thanks are due to Prof. Iosif Bitensky for enlightfull discussions. The Volkswagen Foundation is acknowledged for financial support.

REFERENCES

1. Sigmund,P., *Nucl. Instr. and Meth.* **B27,** 1-20 (1987)
2. Albers, A., Wien, K., Dück, P., Treu, W., Voit, H., *Nucl. Instr. and Meth.* **198**, 69-74 (1982)
3. Pereira, J. A. M and da Silveira, E. F., *Surf. Sci.* **390**, 158-162 (1997)
4. Itoh, N., *Nucl. Instr. and Meth. B* **27**,155-160 (1987)
5. Pereira, J. A. M., da Silveira, E. F., *Nucl. Instr. and Meth. B* **136 – 138**, 779 (1998)
6. Le Beyec, Y., *Int. J. Mass Spectrom. Ion Proc.* **174**, 101-117 (1998)
7. Pereira, J.A.M., Bitensky, I.S., and da Silveira, E.F., *Int. J. Mass Spectrom. Ion Proc.* **174**, 179-192 (1998)
8. Tombrello, T. A., *Nucl. Instr. and Meth. B* **94**, 424, (1994)
9. Pereira, J. A. M and da Silveira, E. F., to be published in *Nucl. Instr. and Meth. B* (1999)
10. Sigmund, P., Bitensky, I.S., Jensen, J., ,*Nucl. Instr. and Meth. B* **112** , 1 (1996)

Sputter-Initiated Resonance Ionization Spectroscopy at the University of North Texas

A.W. Bigelow, S.L. Li, S. Matteson, and D.L. Weathers

Department of Physics, University of North Texas, Denton, Texas 76203

Sputter-initiated resonance ionization spectroscopy (SIRIS), a highly sensitive surface analysis technique that incorporates a primary sputtering ion beam and a tunable laser system to selectively ionize sputtered neutral components, has recently been added to the array of characterization methods available at the University of North Texas (UNT). An overview of the UNT SIRIS system is given, and several of the system components and features developed at UNT are described, including optics of the primary beam and associated diagnostics, target handling, laser beam optics and diagnostics, the detector for the sputtered particles, and the fast pulse triggering scheme for time-of-flight measurements.

INTRODUCTION

Sputtering occurs when an energetic particle impinges on the surface of a target, resulting in the ejection of target components. This erosion process is commonly applied in materials analysis techniques, in thin film deposition processes such as those used in the semiconductor industry, and in the removal of surface contaminants in general. Sputter-initiated resonance ionization spectroscopy (SIRIS) is a technique for analyzing materials through selective resonance ionization of secondary neutral atoms. Coupled with tunable laser technology and position-sensitive particle detection, SIRIS can also be used to investigate nuances of the sputtering process, such as preferential sputtering from compound targets.

APPARATUS

Components of the SIRIS machine at the University of North Texas (UNT) include a low-energy particle accelerator to provide a primary sputtering ion beam, a state-of-the-art target chamber, a tunable laser system for resonance ionization, and a position-sensitive detector. A time-of-flight process, initiated by a short pulse of primary ions on the sample and terminated by a resonance ionization laser pulse at the entrance to the detector, permits velocity measurements of the resonantly ionized particles. Fast timing electronics and data acquisition are handled through a virtual instrumentation program. A schematic view of the SIRIS laboratory at UNT is shown in Fig. 1.

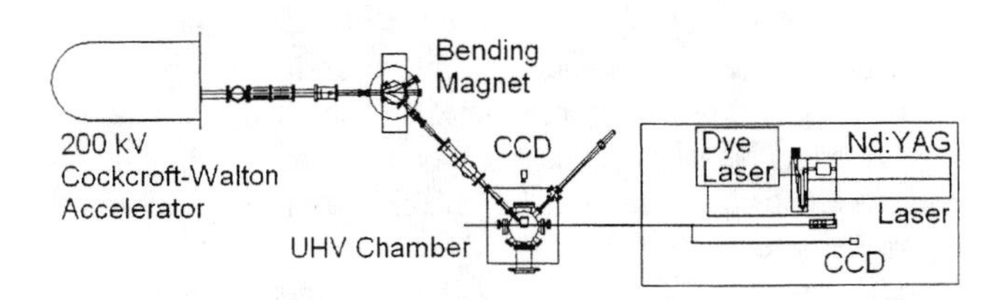

FIGURE 1. Schematic of the SIRIS apparatus. The sample, the position-sensitive detector, and the interaction region are inside the ultra-high vacuum (UHV) chamber

Primary Ion Beam Optics

The source of the primary sputtering beam used in this apparatus is a 200 kV Cockcroft-Walton accelerator. After acceleration, the ion beam successively passes through an electrostatic quadrupole doublet lens, collimating x- and y-slits, a 45-degree singly focussing analyzing magnet, another set of slits, and a final lens before reaching the target. This last lens is a custom-designed electrostatic quadrupole triplet lens (1). It provides point-to-point focussing and a small, round beam on the target to define the point of origin for the sputtered particles. The compression assembly design of the quadrupole triplet lens provides a means of minimizing spherical aberration using a geometrical arrangement proposed by Matteson et al. (2).

Water-cooled collimation slits were built for the system to accommodate the maximum power delivery of the primary ion beam. The slits were designed to dissipate up to 200 W of power, corresponding to a maximum ion current of 1 mA at 200 keV. In addition, the first set of electrically isolated tantalum slits provides a convenient pair of sharp electrodes for fast pulsing of the ion beam by application of short voltage pulses.

CP475, *Applications of Accelerators in Research and Industry*,
edited by J. L. Duggan and I. L. Morgan

For this system, a 45-degree singly-focussing magnet selects primary ions with the appropriate momentum per charge ratio. After an analysis of the magnetic fringe field, shunts were introduced to shorten the effective magnetic field length and to improve the focussing qualities of the magnet (3).

Sample Chamber

Ultra high vacuum (UHV) is attainable in the sample chamber with a combination of pumps. Sorption pumps are used to rough out the system. An ion pump is used to maintain UHV at a base pressure of 5×10^{-10} Torr and a titanium sublimation pump can be activated to further improve the vacuum. Ports on the chamber allow for primary ion beam entrance, sample introduction from a load lock, sample manipulation, viewing, pressure monitoring, laser entrance and exit, reverse view low energy electron diffraction (RVLEED), and the position-sensitive detector.

Of utmost concern in the sample holder design was sample alignment with the ion beam, laser beam, and the detector. An aperture mounted on the sample stage provides a means for sample alignment to the primary ion beam. Sample alignment to the laser is achieved with a shadow imaging method. Also, an electrically conductive indicator incorporated into the sample holder design allows for proper sample positioning with respect to the detector.

Other considerations in the sample holder design were current monitoring and sample heating capabilities. The sample holder is electrically isolated; hence, it can be used as a Faraday cup to measure the primary ion beam current pulses. A radiatively coupled NiCr resistive wire heater and a thermocouple were incorporated into the sample holder for temperature control.

Tunable Laser Beam

To electrically attract sputtered neutral atoms to a detector, the particles must first be ionized. Resonance ionization spectroscopy is a technique for ionizing selected particles through multi-photon excitation schemes. For example, an excitation scheme that works for some elements is a two-photon process: the first photon is for resonant excitation and the second is for ionization.

To this end, a tunable laser system is used to selectively ionize a specific elemental species. This system includes a pulsed 10 Hz neodymium-doped yttrium aluminum garnet (Nd:YAG) pump laser, a harmonic generator, and a tunable dye laser. The laser light enters and exits the sample chamber through fused silica ports. A power meter placed on the exit side of the chamber monitors the incident laser pulses. The pump laser flash lamp signal also initiates the timing sequence for the entire spectrometer.

Optimally, the laser beam should intercept as much of the sputtered flux as possible. At the same time, it should have a narrow waist along the target-detector axis. This last criterion maximizes the velocity resolution of the detector, as will be discussed below. In order to provide consistently this optimal "ribbon focus" at the interaction region between the laser and the sputtered particles, a telescope with a three-meter focal length was developed utilizing a 300 mm diverging cylindrical lens followed by a movable 300 mm converging cylindrical lens. The long focal length minimizes the variation in the beam width throughout the interaction region.

For focussing and profiling the laser beam, a retractable prism is introduced into the beam that guides the laser light to a filtered CCD camera and frame grabber. The diverted path length from the compound lens to the CCD camera is identical to the desired focal length. This arrangement facilitates compensation for chromatic focussing effects when changing the laser frequency.

When properly tuned, the laser light can ionize virtually all the selected neutral atoms within the interaction region. However, as noted by Kobrin (4), at reduced intensities, the ionization probability has a linear dependence on the photon flux. Operating in this regime along the laser beam waist allows the effects from the variations in the photon flux to cancel with those from the variations in the interaction region volume. In the interaction region, a typical laser temporal pulse width is 5 ns, the minimum beam width in the focussed plane is less than 0.5 mm, and the beam height is ~1cm.

Detector

After the particles have been resonantly ionized, they are accelerated into a position-sensitive detector (5). The detector assembly is shown schematically in Fig. 2.

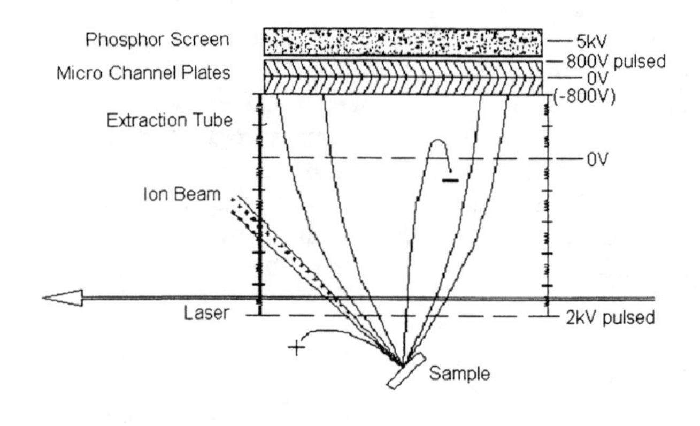

FIGURE 2. Schematic cross section of the interaction region and the position-sensitive detector. Dashed lines represent high-transmission grids.

Velocities of the sputtered neutral particles correlate with the positions that the particles impinge on the detector for a given set of experimental parameters. These parameters are the distance between the sample and the plane of the laser, the time interval between the primary ion pulse and the laser pulse, the distance between the laser plane and the microchannel plate, and the accelerating electric field.

The design of the position-sensitive detector for the SIRIS system was modeled after the detector by Kobrin et al. (4). Sputtered neutral particles leaving the sample's surface sequentially pass through a high-transmission grid and the laser interaction region. In the laser interaction region, only particles of the selected species are ionized. These ionized particles are then accelerated towards the microchannel plate chevron pair by an applied voltage. The microchannel plate output for individual incident ions causes a phosphor screen to fluoresce, producing an illuminated image that is captured by a CCD camera and a frame grabber. Individual detected particles appear as bright spots in the captured image. An added feature to this design is that the direction of the applied electric field in principle reflects all sputtered ions, eliminating this noise component from the experiment.

This detector is retractable to allow clearance for other functions in the sample chamber. It is designed to accept sputtered particles from a 90-degree range of ejection angles with ejection energies up to 120 eV. Note that due to the mass selectivity of SIRIS, there is a direct correlation between the ejection energy and speed of each detected sputtered particle. When the accelerating potential between the grids is turned on, the non-normal trajectories of ionized particles become parabolic with acceleration normal to the detector's surface. Velocity resolution depends greatly on the distance between the target surface and the laser interaction region. If the flight path is increased, the resolutions for both the magnitude and the direction of the velocity improve for the measured particles, but angular acceptance is reduced. The 50 ns duration of the primary ion pulse dominates the uncertainty in the time of flight of the particles and limits the ultimate resolution for the velocity distribution.

The positions for the first grid, the ionization region, and the microchannel plate face were based on optimum spacings determined by Li (5). The first grid is 15.0 mm away from the sputtered surface on the sample, the laser passes 20.0 mm from the surface, and the microchannel plate face is 88.9 mm away from the surface. The diameter of the microchannel plates is 7.62 cm. In order to adjust the velocity resolution, a variable distance for the laser interaction region is needed. The laser path can easily be varied with translational reflective optical elements.

System Control

As mentioned earlier, SIRIS is a pulsed system initiated from a 10 Hz Nd:YAG laser. Precision timing requirements demand automation. Hence, SIRIS control relies on virtual instrumentation. In this case, virtual instrumentation consists of a computer program linked to computer automated measurement and control (CAMAC) modules which, in turn, are linked to the apparatus electronics. Features of the virtual instrument include experimental parameter control and data acquisition.

A user-friendly operator interface constructed using LabVIEW (6) software acts as a graphical virtual instrument base and is shown in fig. 3. LabVIEW is a graphical programming environment for developing electronic instrument applications. The graphical program is symbolic as opposed to the traditional line-by-line text code and is more representative of a flow chart. Data flow occurs along virtual wires that connect the operations of the program. Mathematical operations are represented by graphical signal interconnects, and subroutines are represented by icons. Conventional *for* and *while* loops and case and sequence structures are available. An application program includes both a program diagram and a front panel. The front panel is designated as the user interface.

The SIRIS control program sequentially initializes the instruments, sets the delay times, turns on the detector power supplies, cycles through the data acquisition stage, and turns off the detector power supplies. Other modes of operation are available that keep the power supplies on during a series of experiments.

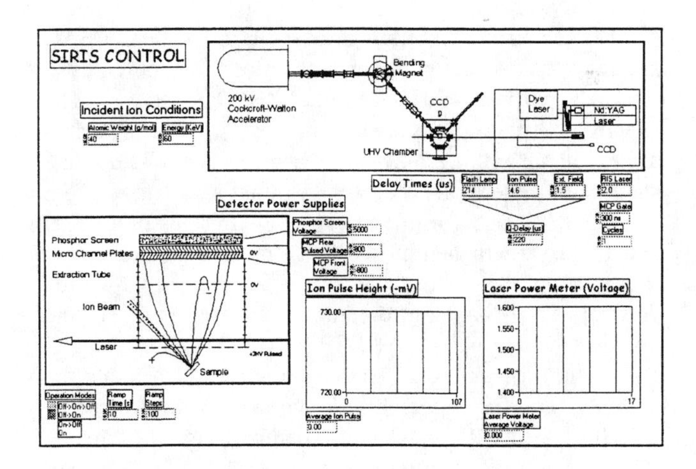

FIGURE 3. Virtual instrument control panel layout for SIRIS.

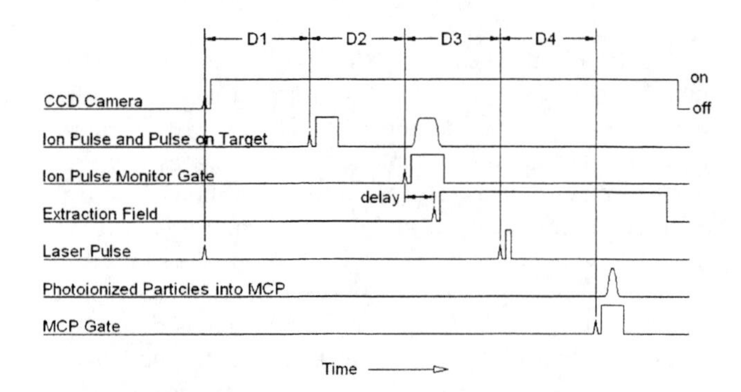

FIGURE 4. Pulse and event timeline for SIRIS. Spikes represent trigger pulses.

Figure 4 is a graphical representation for the SIRIS pulse and event timeline where the delay times are represented by D1 through D4. The first three delay times must always add up to 220 μs, the time interval between the Nd:YAG flashlamp pulse and the Q-switch trigger that delivers greatest laser power. For example, if a 60 keV argon primary ion beam is used, typical values for these delays are: D1=214 μs, D2=4.6 μs, D3=1.5 μs, and D4=2 μs. D2 is the flight time for the primary ion beam pulse from ion beam deflector to sample. The time of flight for the sputtered neutral particles, D3, and the acceleration time to the detector for the resonantly ionized particles, D4, are variable depending on the velocities of interest for the sputtered particles. The additional delay applied before activation of the extraction field is on the order of the ion pulse duration and is provided to protect the primary ion beam from ion optical distortion. The microchannel plate is gated to suppress noise from stray secondary ions.

Fast electronics control is possible through CAMAC modules whose crate is connected to a computer by a general purpose interface bus (GPIB) line. The electronics diagram is shown in fig. 5. The system's timing chain, programmed into two LeCroy 2323A dual delay gate generators, receives an initial start trigger from the Nd:YAG laser. Immediately following this first pulse, the CCD camera behind the detector is triggered and it remains on throughout one cycle of events. A gated LeCroy 2259B analog-to-digital converter registers the incident ion pulses. A DSP 2032 scanning digital voltmeter records the laser power meter voltages and the output from the detector high voltage power supplies. Finally, a DSP 3016 16 channel digital-to-analog converter is used for remote control of the detector power supplies. Other SIRIS system components along the primary ion beam line such as valves, Faraday cups, and viewers can also be controlled through virtual instrumentation.

Data are acquired through a series of CCD images. For a given set of experimental parameters, up to 256 images may be acquired in sequence and averaged. Intensity distributions in the individual images are binary in nature, reflecting the locations of detected particles in the position-sensitive detector. Final averaged images are interpreted using a gray scale that depends on the accumulated intensity from the individual images. Acquired data are normalized with respect to the average number of incident ions per pulse and with respect to the average laser power.

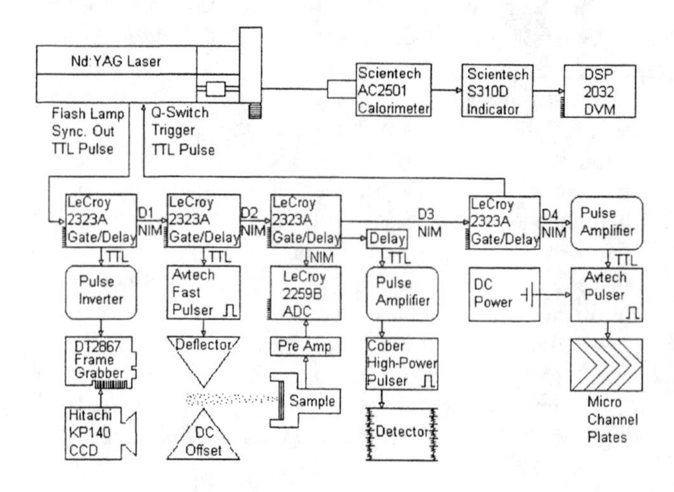

FIGURE 5. Timing electronics diagram for SIRIS. Peripheral hardware is also represented.

CONCLUSION

A SIRIS apparatus has been developed and installed at UNT. The system utilizes a pulsed ion beam, a pulsed tunable laser, and a position-sensitive detector; it is controlled by virtual instrumentation technology. The system is designed to measure energy and angular distributions of sputtered neutral particles, and will be used to study preferential sputtering behavior.

REFERENCES

1. A.W. Bigelow, A.M. Childs, D.L. Weathers, and S. Matteson, presented at the Joint Fall Meeting of the Texas Sections of APS and AAPT, and SPS Zone 13, Arlington, 1996.
2. S. Matteson, D.K. Wilson, D.L. Weathers, F.D. McDaniel, J.L. Duggan, and J.M. Anthony, *Nucl. Inst. and Meth.* **B 56/57**, 1091-1095 (1991).
3. O. Truong, A.M. Childs, A.W. Bigelow, S. Matteson, D.L. Weathers, presented at the Joint Fall Meeting of the Texas Sections of APS and AAPT, and SPS Zone 13, Lubbock, 1995.
4. P.H. Kobrin, G.A. Schick, J.P. Baxter, and N. Winograd, *Rev. Sci. Instrum.* **57 (7)**, 1354-1362 (1986).
5. S.L. Li, A.W. Bigelow, D.L. Weathers, presented at the Joint Spring Meeting of the Texas Sections of APS and AAPT, and SPS Zone 13, Abilene, 1996.
6. National Instruments, 6504 Bridge Point Parkway, Austin, TX, 78730-5039.

Trajectory Effects in Multiply Charged Ion-Surface Interactions

H. Lebius,

Stockholm University, Atomic Physics Department, S-104 05 Stockholm, Sweden and Département de Recherche Fondamentale sur la Matière Condensée, SI2A, CEA Grenoble, 17 rue des Martyrs, 38054 Grenoble Cedex 9, France

W. Huang, R. Schuch

Stockholm University, Atomic Physics Department, S-104 05 Stockholm, Sweden

Ar ions of 4.3 keV q_{in} were scattered at large angles (θ=75°) from a clean oriented surface. By selecting Ar projectiles having a large ionization potential and by using a large scattering angle only ions scattered at the first atomic layer of the surface were detected. Scattered ion energy spectra show peaks of single scattering and double scattering of the Ar projectile ions from one or two surface Au atoms, and the distribution attributed to double collisions splits into two peaks when the scattering plane coincides with a crystallographic plane. Simulations with a MARLOWE code allowed for interpretation of the structure in the double collision peak by in-plane and zig-zag double collisions. Differences in the relative peak heights between the experiment and a MARLOWE simulation were partly explained by different neutralization probabilities with varying trajectories. Yield changes with increasing charge states show interesting possibilities for future experiments with highly charged ions.

INTRODUCTION

The technique of Low Energy Ion Scattering (LEIS) is used to characterize properties of the crystal surface like surface reconstruction and contamination. It was for example applied successfully in the study of the reconstruction of the (110) surface of Au, where a number of models for possible reconstruction patterns were suggested. With LEIS it was possible to discriminate all, but one of them, leading to the now accepted reconstruction model for this surface (1,2).

In a typical LEIS experiment a collimated beam of singly charged ions is directed onto a surface. The scattered ions are then analyzed in an electrostatic analyzer for charge state and kinetic energy. An important issue is the choice of the ion species: in LEIS experiments typically alkali or earth-alkali elements were chosen. These elements have a low first ionization energy, therefore the probability that they survive the interaction with the surface as charged particles is relatively high, and independent on the trajectory. The scattered ions can be easily energy analyzed for determination of effects stemming from different trajectories or surface structures. An example for this is the work by von dem Hagen *et al.* (3), where low-energy K^+ ions were used. We employed ions with high ionization potentials instead, to study the influence of fast neutralization at the surface on the measured energy spectra. This could introduce a method to restrict ion scattering to "above surface" and is of relevance for the scattering of highly charged ions from solid surfaces.

By changing the azimuthal angle of the surface, i.e. the angle between the incident ion beam and a crystal axis on the surface, different scattering trajectories can be suppressed or favored. By using an analyzer with high resolving power it was possible to distinguish between different trajectories in the kinetic energy spectrum of the scattered ions. The high resolving power of the analyzer necessitates a high intensity ion beam. In the present work we therefore were limited to charge states of Ar between 1+ and 3+.

EXPERIMENTAL SETUP

Ar ions with charge states q_{in}=1...3 are produced in a plasmatron ion source and accelerated to a kinetic energy of (4.3 ± 0.1) keV × q_{in}. These ions are charge-state and momentum separated in a magnetic sector field. The ion beam was then collimated by means of two apertures and a skimmer to a beam spot on the crystal of about 2 mm diameter. The incident angle ψ between incoming ion beam and surface was 25°. The energy spectra were normalized to the target current.

At a scattering angle of θ = 75° a 127° electrostatic cylindrical analyzer was mounted to measure the ions. This analyzer has an acceptance angle of 0.5° FWHM and an energy resolution of $\Delta E/E \approx 6 \times 10^{-3}$. The energy width due to the finite acceptance angle is smaller than the energy resolution of the analyzer. The energy scale (spectrometer factor) of the analyzer was determined by analyzing primary ions at different acceleration potentials of the ion source. It is known to within the width of the resolution. Independent measurements with a different setup and the same ion source showed that the systematic error is about (2-3) %. An opening in the outer sector of the analyzer

CP475, *Applications of Accelerators in Research and Industry*, edited by J. L. Duggan and I. L. Morgan

allowed to detect neutral particles. The ions and neutrals were detected by Channeltrons.

A single crystal of Au was used which was cut in the (111) crystal plane. In order to remove contamination from the surface it was sputter cleaned. Due to the used ion currents(about 1 μA of Ar^{1+}) the surface was kept clean throughout the duration of the experiment. The residual vacuum in the experimental chamber was $\approx 5 \times 10^{-10}$ Torr.

The crystal was mounted on a crystal holder which was rotated in two directions. The first axis was situated on the crystal surface and perpendicular to the incoming ion beam, allowing to change the incident angle ψ. The values for ψ chosen in this investigation were 25° and 40°. The second axis of rotation is (within 1.5°) perpendicular to the surface of the crystal. This allows to rotate the crystal azimuthally around the surface normal. By changing the azimuthal angle ϕ the direction of the incident ion beam relative to the crystal axis changes while the incident angle is held constant. In order to determine the azimuthal angle relative to one crystal plane (in this work the (110) plane was chosen as relative zero in azimuth) the intensity of the reflected ions was measured as a function of the azimuthal angle. When the incident beam direction is close to a crystal plane, ions which are not backscattered at the first crystal layer penetrate deeply into the crystal. Therefore the backscattering yield for this channeling direction is smaller than in random directions (a detailed discussion about channeling of high energy ions in single crystals can be found in (4)). By comparing the pattern of yield variations with ϕ to a polar diagram of the crystal structure it was possible to define the azimuthal angle relative to the (110) plane. Typical azimuthal angles used in the experiment are ϕ =25° for a random orientation and ϕ=60°, which is in channeling direction to the (110) plane.

RESULTS AND DISCUSSION

In the case of incoming Ar ions with $q_{in} \leq 3$, only singly charged ions and neutrals were detected within the detection sensitivity of our setup. This is in contrast to measurements with highly charged ions (5), where multiply charged scattered ions were detected. 50% of the backscattered particles are neutral, assuming equal detection efficiencies of the channeltron for atoms and ions. Kinetic energy spectra of the backscattered Ar^+ ions after the interaction of Ar ions with charge state q_{in}+ interacting with the Au (111) surface in channeling and random incidence are shown in Fig. 1.

In the following we will first consider large angle ion-surface scattering by kinematics of binary collisions. Then a 3-D MARLOWE model is used to describe the ion trajectories. These models ignore the neutralization processes in front of the surface, which will be discussed in the final part of this section.

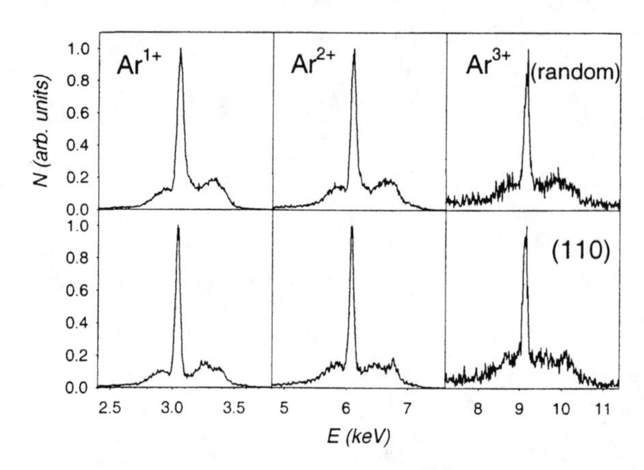

FIGURE 1. Kinetic energy spectra of $Ar^{(1-3)+}$ ions scattered from a Au (111) surface along random (ϕ=26.4°) and channeling (ϕ=62.4°) directions. ψ=25°, θ=75°, E_0=4.3 $q \times$ keV. Only Ar^+ ions were detected in all spectra. The yields are 1 arb. unit ≈ 8 ions/nC, normalized on the target current.

Kinematic considerations

As can be seen in Fig. 1 the main contribution is at a kinetic energy of 3 keV, which means that the Ar ion is transferring about 1.3 keV kinetic energy to the recoil Au atoms. This amount is large compared to the binding energy of a Au atom in the crystal of a few eV. Therefore collisions of Ar ions with Au atoms can be treated as if the Au atoms are free.

It is well known that the 'double collision peak' can be described as two consecutive collisions (3,6). If we furthermore assume that the two collisions are coplanar, then the two scattering angles involved are well defined by $\theta = \theta_1 + \theta_2$. Applying the elastic scattering formula to the two energy positions measured (Fig. 1, Ar^{1+}, (110)), the two scattering angles are θ_1 = 10° and θ_2 = 65° for the peak at lower energy, and θ_1 = 20° and θ_2 = 55° for the peak at higher energy in the 'double collision peak', respectively.

In order to understand why these two distinct trajectories occur (and none in between) a simplified model of the ion-surface interaction was applied. The surface was replaced by two adjacent surface atoms. Only collisions in the plane perpendicular to the surface containing the two surface atoms were studied. The interaction between the incident ion and the surface atoms is described by a screened Coulomb potential (Bohr approximation of the Thomas-Fermi potential). With this potential the scattering angle was calculated as a function of the impact parameter of the incident ion with regard to the first surface atom. In this deflection function two impact parameters can be found, which lead to an overall deflection of 75°: the first at 0.2 Å, leads to the 'single collision peak', and the second at 0.4 Å, describes the 'double collision peak'. The 'double collision' involves two consecutive scattering events with angles of θ_1=22° and θ_2=53°, which is in good agreement with the angles deduced from the measured peak at higher energy in the 'double collision distribution'. This peak therefore can

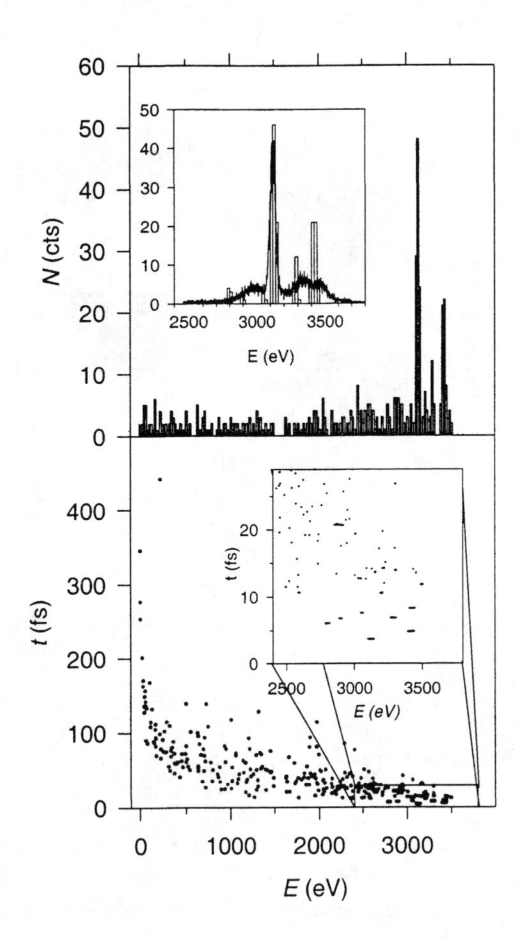

FIGURE 2. Results of a trajectory calculation with the MARLOWE code for Ar particles interacting with a Au(111) surface, ψ=25°, ϕ=60°, E_0=4.3 keV; top graph: kinetic energy spectrum of Ar particles reflected under θ=(75±1)°; bottom graph: interaction time vs. kinetic energy scatter plot of these reflected Ar ions; bottom graph, inset: enlarged view of bottom graph, range corresponds to box in bottom graph; top graph, inset: kinetic energy spectrum of Ar particles with less than 10 fs interaction time, same parameters as in a, the measured spectrum is included for comparison, see text.

be described as 'in-plane double scattering', due to the two-dimensional nature of the model. It should be noticed that the order of scattering angles in the interaction was found to be essential.

Using the formula $\sigma=|b \sin^{-1}\theta \, db/d\theta|$ for the classical, angle-differential cross section and treating the double collision as one process, the ratio of the cross sections of 'in-plane double' to 'single collision' peak is about 0.25, which is larger than the measured intensity ratio of 0.07 ± 0.01. One of the reasons for this overestimation could be, that the model does not take any charge exchange into account, whereas by measuring the charged scattered particles the experiment is sensitive to these processes. This will be discussed further. This two-dimensional model does not reveal the nature of the second structure of the 'double collision peak', therefore the MARLOWE code, treating the trajectories in three dimensions, was used to study this problem further.

A MARLOWE simulation of ion trajectories

The MARLOWE code (7) treats the interaction between the incoming particle and the surface as a series of binary collisions between this particle and the atoms of the crystal. The interaction potential used was the Molière approximation to the Thomas-Fermi potential (8) with Firsov screening length (9). Collision events caused by recoil ions from earlier collisions, the image charge acceleration and thermal vibrations were ignored. Initial positions were taken randomly. A half opening angle of 1° was chosen as the acceptance angle. This is about twice as large as the actual acceptance angle in the experiment due to computing time restraints.

We calculated ≈ 4 × 10^6 trajectories, of which a fraction of 1.2 10^{-4} were scattered into the detection cone. A kinetic energy spectrum of these particles is shown in the upper part of Fig. 2. In contrary to the measured spectrum in Fig. 1 we see a large contribution at the low-energy side of the spectrum. An explanation for this is that the MARLOWE simulation, as discussed above, just calculates the Ar particle trajectories without taking charge exchange processes into account. The analyzer, on the other hand, only selects charged particles.

In order to find criteria for selecting trajectories leading to charged scattered particles, scatter plots of the interaction time vs. the kinetic energy after the interaction are analyzed (shown in the lower part of Fig. 2). Interaction time is the time, during which the Ar particles stay close enough to Au atoms of the crystal to interact. The interaction times obtained in the simulation are between 5 fs and 500 fs. Longer interaction times generally lead to smaller kinetic energies after the collision, as the number of collisions rises with increasing interaction time.

More collisions mean also a higher probability of neutralization, therefore an upper limit of the interaction time was chosen in order to take into account, that only charged particles were detected. In the inset in the lower part of Fig. 2 groups of points are visible with the same interaction time and a finite range in final kinetic energy (the distribution in energy is an effect of the finite acceptance angle in the simulation). The points in one group represent similar trajectories, which will be discussed below. At larger interaction times the points are scattered more randomly. Trajectories with many collisions occur, where the Ar particles are also entering the bulk of the Au crystal and therefore the probability of neutralization is high. A limit of the interaction time of 10 fs was chosen, below which the trajectories stay above or in the first layer of the surface. For this limit a few groups of trajectories appear in the scatter plot. We will show below that the corresponding trajectories are sensitive to the surface orientation and structure.

Comparison of MARLOWE simulation with experimental results

A kinetic energy plot of the scattered Ar particles with these small interaction times is shown in the upper insert in Fig. 2d as the histogram plot. Overlaid is the corresponding measured spectrum from Fig. 1 (Ar^{1+}, (110)). The kinetic energy axis for the experiment was shifted by 2.5% in order to align the energy of the 'single collision peak' in experiment and simulation. This corresponds to the overall uncertainty in the analyzer constant and the primary beam energy of about 3%. As the measured yield was not normalized to the absolute intensity, the experimental yield was normalized to the simulation at the single collision peak.

A double structure of the 'double collision peak' is clearly visible in the simulation, as was seen in the experiment. The relative height (compared to the 'single collision peak') is larger in the simulation than in the experiment, which will be discussed below.

In Fig. 3 a detailed view of the lower insert in Fig. 2 is shown, together with four typical trajectories contributing to the 'single collision peak' and the two 'double collision peaks'. Throughout the following discussion very soft collisions with scattering angles of 1° or less are not considered, as these collisions have only a minor influence on the trajectories.

'Double scattering' at higher final kinetic energies than the 'single collision peak' (group 1) consists of two peaks, the one at lower kinetic energy consists of trajectories belonging to group 2zz (for zig-zag double collision). These trajectories undergo two collision events with scattering angles of $(34.4\pm0.3)^{\circ}$ and $(53.9\pm0.5)^{\circ}$, respectively. The second collision is out of the scattering plane of the first, these trajectories are also called 'zig-zag scattering'. After making a soft collision with one surface atom, the ion makes a harder second collision, but not with the direct neighbor of the atom from the first collision, see the sketch in Fig. 3: it passes between two neighbors of the first surface atom and then makes a hard collision with an atom further away. This is in contrast to Ref. (3), where the 'zig-zag collisions' at lower collision velocities involve two adjacent surface atoms.

Groups 2ip (for in-plane double collision) and 4 (four collisions) are forming the second contribution to the 'double collision peak' at somewhat higher energies than the first contribution from group 2zz. Trajectories from group 2ip have two scattering events (with angles of $(18.3\pm0.1)^{\circ}$ and $(56.7\pm0.8)^{\circ}$, respectively), which are in-plane, as the angles add up to the deflection angle $\theta=75^{\circ}$. Therefore the trajectories are called 'in-plane double scattering'. At the same kinetic energy as the ions from group 2ip those from group 4 can be found, which contribute to 25% to the intensity at that energy. Group d

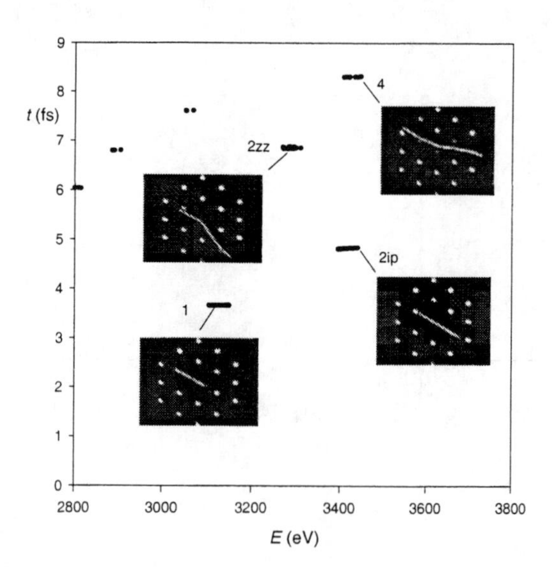

FIGURE 3. Scatter plot of interaction time vs. Kinetic energy for trajectories with interaction times below 10 fs. Trajectories are shown for the single collision group and the two main double collision groups. Group 4 correlates to multiple, out-of-plane collisions. The inlet graphs show a typical trajectory for each group. The view is perpendicular onto the surface. 'Shadows' show the height of the particles above the surface.

has trajectories with four successive scattering events (angles of $(19.96\pm0.06)^{\circ}$, $(15.6\pm0.1)^{\circ}$, $(4.54\pm0.04)^{\circ}$ and $(51.1\pm0.7)^{\circ}$, respectively). All of these trajectories stay above the first crystal layer.

The sum of the two scattering angles for trajectories in group 2zz is close to the sum of the four scattering events for trajectories in group 4, whose final energy is higher than for trajectories of group 2zz. The argumentation is similar to the one given for the energy position of the 'double' relative to the 'single scattering' peak: by using more collisions with smaller individual collisions, the transferred recoil energy is smaller and therefore the final kinetic energy is higher. It is interesting to note that the simulation did not show any trajectories with three collisions resulting in a final energy larger than the energy after double collisions.

However, the calculated energy positions of the double collision peaks in the channeling direction do not exactly match the measured positions. This can not be explained by a misalignment of the azimuthal angle, as calculations with azimuthal angles of $\pm3^{\circ}$ around the channeling direction did not show a sufficient shift in the energy position (typical misalignment is within 1.5°, see experimental setup). At present, no satisfying reason (e.g. approximations in the calculation) for this discrepancy can be given.

Fig. 4 shows kinetic energy spectra for different azimuthal orientations of the crystal with regard to the incident ion beam (the incident angle $\psi=25^{\circ}$ is kept constant). Additionally histograms at $\phi=2.4^{\circ}$ (channeling in the (110) plane) and at $\phi=48^{\circ}$ (off-channeling) show the results of

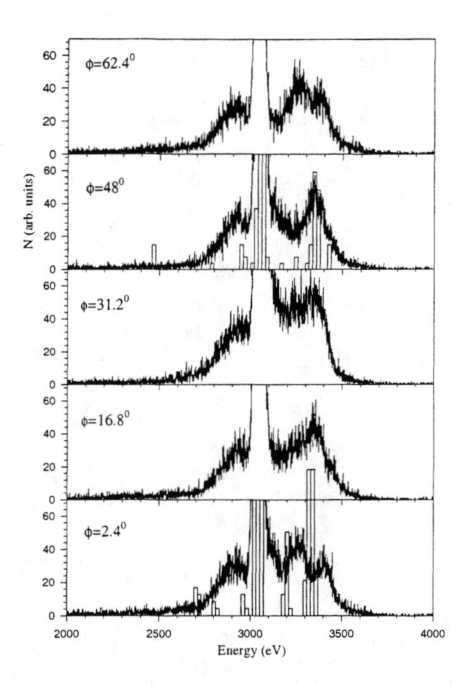

FIGURE 4. Kinetic energy spectra of Ar^+ ions scattered from a Au (111)surface in different orientations. In channeling (ϕ=2.4°) and random (ϕ=48°) orientation spectra obtained with MARLOWE simulations are shown (the maximum yield of the simulation was normalized to half the maximum yield of the experiment).

MARLOWE simulations. To simplify the drawing, the maximum yield of the simulation is normalized to half the maximum yield of the experiment. In channeling the 'double collision peak' in the energy spectrum shows two distinct contributions from 'zig-zag' and 'in-plane double scattering'. As the energetic positions of these two contributions are moving closer to each other in the random orientation, they can not be distinguished in the energy spectrum. This effect was also seen in the experimental spectrum.

Influence of neutralization processes on scattering yields

The upper insert in Fig. 2 shows, that the intensity of the 'double collision peaks' relative to the intensity of the intensity of the 'single collision peak' is much higher in the simulation than in the experiment. One explanation can be that the interaction time for 'double collisions' is higher than for 'single collisions', giving a higher probability for neutralization by Auger decay and recapture during 'double collisions'. A discussion of charge-exchange processes, using rate equations to describe the neutralization process during the collision can be found in reference (5). An increasing neutralization in double scattering as compared to single scattering was observed with increasing ion charge and is reported in reference (10).

In Fig. 4 it is seen that for the experimentally obtained spectra the yield for 'zig-zag double collisions' is higher than for 'in-plane double collisions', in contrast to the

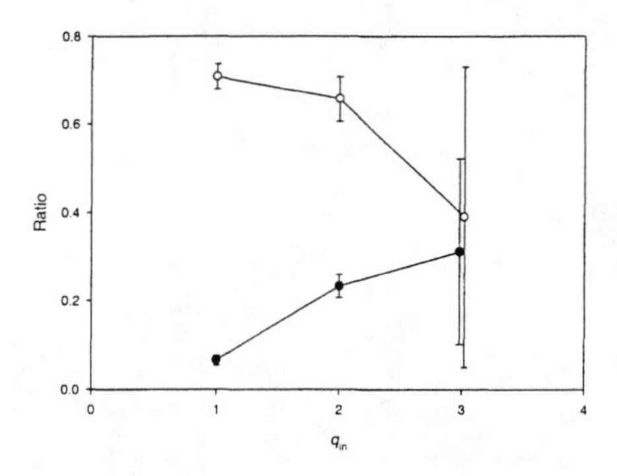

FIGURE 5. Experimentally obtained relative intensities of the 'zig-zag' (open symbols) and 'in-plane' (full symbols) contributions to the 'double collision peak', compared to the 'single collision peak' intensity as a function of q_{in}. The full lines connect the points.

results obtained with the MARLOWE simulation, where the yield for 'in-plane double scattering' is higher. Further MARLOWE simulations were performed with an azimuthal angle around ϕ=60°, in order to see if the relative yields change strongly around the channeling direction. Changes were smaller than the statistical error bar. The reason for this disagreement must therefore be searched elsewhere. It can not be explained with the above mentioned time argument, as the enhanced neutralization probability would decrease the intensity for 'zig-zag double collisions' even further in comparison to 'in-plane double collisions', when charged particles are detected. It could be due to contributions from multiple scattering such as 4 and more collisions which are at higher energy and pile up in this peak.

In the following we discuss the influence of the trajectories on the charge-exchange in front of the surface. Experimentally we obtained reflected ion spectra with q_{in}=1...3, the relative yields of the two 'double collision peaks' are shown in Fig. 5. It should be noted, that the kinetic energy varies with the incident charge state by E_{kin}=4.3 keV × q_{in}. To test, if the observed tendencies are due to the kinetic energy changing with q_{in}, MARLOWE calculations were done for different energies. The obtained relative yields are shown in Fig. 6. The ratios are relative to the 'single collision' yield. For 'in-plane double scattering', the ratio to single scattering predicted by the MARLOWE simulation decreases with increasing q_{in} (because of increasing E), whereas the experimental ratios increase. For 'zig-zag double scattering' the experimentally obtained ratios to single scattering decrease with increasing q_{in}, while the ratios calculated with MARLOWE seem to not depend on q_{in}. The model obviously can not describe the dependence on q_{in} by the rise of the kinetic energy with q_{in} alone. The observed changes in the relative yield as a function of the charge state may therefore origin from charge-exchange processes in front of the surface with cross sections varying with the incident charge state.

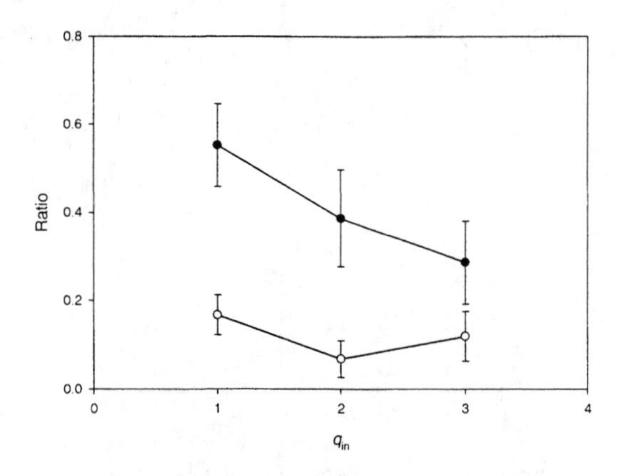

FIGURE 6. Relative intensities of the 'zig-zag' (open symbols) and 'in-plane' (full symbols) contributions to the 'double collision peak', compared to the 'single collision peak' intensity as a function of q_{in}, obtained with the MARLOWE simulation. The incident charge state enters the simulation by the kinetic energy, see text. The full lines connect the points.

According to the simulation, the interaction time of the trajectories leading to 'zig-zag double scattering' is 2.1 fs longer than for 'in-plane double scattering' (in the case of 4.3 keV × q_{in}). As the probability for electron capture rises with the charge state, the additional time in zig-zag scattering could have a larger influence at higher charge states. Therefore the relative yield of scattered ions after zig-zag scattering may decrease with increasing charge state, which in fact was observed in the experiment.

CONCLUSION AND OUTLOOK

Specific double scattering trajectories were examined. It was experimentally possible to distinguish between them by selecting a specific orientation of the crystal (channeling direction) and by using an energy analyzer with a high resolving power. Using a trajectory simulation it was possible to assign the different double collision processes (in-plane and zig-zag) to specific kinetic energies of the scattered ions. In order to explain the experimental results, it was necessary to restrict the simulation to short interaction times, resulting in trajectories above the surface. In detail, the ratio of single to double scattering intensities can be related to neutralization in one additional collision and a longer trajectory at the surface. This revealed insights into electron capture processes close to the surface. The relative intensity of 'in-plane' and 'zig-zag double collisions' could not be explained. In connection with a rate-equation model describing the individual shell populations (5) the charge-exchange processes leading to the decreased yield can be studied in a future investigation. The measured variation of the yields (for the double scattering peaks and the single collision peak) in dependence of the incident charge state is a hint that studying this double structure for even higher incident charge states may give further insight into the atomic processes (i. e. electron capture, Auger ionization) in front of the surface.

ACKNOWLEDGEMENT

We would like to thank L. Hägg for many helpful discussions. This research was supported by the Human Capital and Mobility Program under contract No. CHRT-CT93-0103.

REFERENCES

1. Overbury, S. H., Heiland, W., Zehner, D. M., Datz, S., and Thoe, R. S., *Surface Sci.* **109,** 239 (1981)
2. Hemme, A., and Heiland, W., *Nucl. Instrum. Meth.* **B9**, 41 (1985)
3. von dem Hagen, T., Hou, M., and Bauer, E., *Surf. Science* **117**, 134 (1982)
4. Feldmann, L. C., and Mayer, J. W., *Fundamentals of Surface and Thin Film Analysis*, New York: Elsevier Science Publishing Co., 1986, Chapter 5
5. Huang, W., Lebius, H., Schuch, R., Grether, M., and Stolterfoht, N. *Phys. Rev. A* **56**, 3777 (1997)
6. Datz, S., and Snoek, C., *Phys. Rev.* **134**, A347 (1964)
7. Robinson, M. T., *Phys. Rev.* **B40**, 10717 (1989)
8. Molière, G., *Z. Naturforsch.* **2a**, 133 (1974)
9. Firsov, O. B. *Soviet Phys.-JETP* **6**, 534 (1958)
10. Huang, W., Lebius, H., Schuch, R., Grether, M., and Stolterfoht, N., *Physica Scripta*, accepted for publication (1998)

A New IBA Handbook on ERDA

Y. Serruys, J. Tirira† and P. Trocellier‡

CEA-Saclay, DTA/CEREM/DECM/SRMP, 91191 Gif-sur-Yvette Cedex, France
† CEA-Fontenay-aux-Roses, IPSN, 92265 Fontenay-aux-Roses Cedex, France
‡ CEA-CNRS, Laboratoire Pierre Sue, 91191 Gif-sur-Yvette Cedex, France

Taking as a guideline our recently published handbook "Forward Recoil Spectrometry" , the first one to be dedicated to Elastic Recoil Detection Analysis (ERDA), we give a rapid overview of this IBA technique, that has found more and more applications in the last thirty years. Although most of them are devoted to the profiling of hydrogen isotopes in solids, ERDA offers interesting possibilities in a much broader field. We mainly concentrate onto analysis of multi-elemental targets. In such applications, scattered projectiles and recoil ions of various masses have to be separated or discriminated. This is achieved using such variants of the classical set-up as telescopes, time-of-flight ERDA (TOF-ERDA), discrimination by electromagnetic filters (e.g., ExB-ERDA) or coincidence detection (CERDA). We illustrate the versatility of ERDA by some examples using these methods.

In 1914, Marsden detected recoiling hydrogen nuclei projected in the forward direction by α particles and used a magnetic filter to separate them (1). This was the first Elastic Recoil Detection Analysis (ERDA) experiment. ERDA rapidly developed in the seventies as an analytical technique, particularly after the first multi-elemental profiling experiment by L'Ecuyer and coworkers (2). It is now essentially known as one of the rare tools for quantitative analysis of hydrogen in materials, but its field of applications is indeed much broader. Although chapters were dedicated to ERDA in several IBA handbooks, there was no comprehensive reference about it until we published "Forward Recoil Spectrometry" in 1996 (3). Taking this book as a guideline, we try in this paper to give an overview of the present state of the art and examples of the various applications of this technique with an accent on multi-elemental analysis.

BASIC PHYSICAL PROCESSES

ERDA (also called formerly ERD, ERS, FRD or FRS) consists in detecting target atoms (mass M_2) recoiling after elastic collision with projectile ions (energy E_0 and mass M_1). By many aspects this is similar to RBS, but now we observe the recoil expelled with the energy lost by the scattered ion instead of the scattered ion with its residual energy. Backscattering is possible only for $M_1 > M_2$, hence hydrogen cannot be analysed by RBS. Here we have no such limitation. Energy transferred to the recoil for a fixed scattering angle is characteristic of M_2. It is maximum for equal masses, so we can always find a suitable projectile to have a large enough recoil energy. As in RBS, energy losses of the projectile before collison and of the recoil after collision determine the final energy of the detected particle and allow to compute at which depth the collision occurred.

However, one encounters the same difficulties as for RBS. First, as the stopping power of the target depends on the particle atomic number, on its residual energy and on the initially unknown local target composition, the depth vs. energy relation cannot be directly determined. Second, two recoils with different masses may be detected at the same energy if scattering occurred at different depths such that the difference in energy loss compensates the difference in transferred energy. This is what we call *mass-depth ambiguity*. Finally, all the elements present in the sample contribute to its stopping power, but some of them - let us call them *dummy elements* - may not be detected, like hydrogen in RBS.

The yield of each recoil species produced in a small depth interval is proportional to the local concentration of this species and to its scattering cross-section that depends on the residual energy of projectiles. Here we encounter a major difference with RBS : scattering of heavier particles (e.g., helium) by light particles (hydrogen isotopes) involves non-Rutherford cross-sections. Their dependence versus energy and scattering angle is complex (e.g. the sharp resonance of alpha-deuteron collisions near 2.128 MeV). It is possible to build theoretical models in order to interpolate between available experimental data within limited energy or angle intervals, but available data are not abundant and there is no universal method to give account of experiment in the whole energy range. In our book we discuss in detail recent measurements and interpolation formulae for collisions between helium and hydrogen isotopes, e.g. (4-6). There is still much progress to be expected from both the theoretical and the experimental points of view to acquire a more complete knowledge of these cross-sections, e.g., (7).

CP475, *Applications of Accelerators in Research and Industry*, edited by J. L. Duggan and I. L. Morgan

The numerous minor interactions of particles with electrons and nuclei on their path constitute a stochastic process. As a result, fluctuations around the mean energy loss, known as *energy straggling*, and lateral and angular deflection known as *multiple scattering* are observed. The resulting energy dispersion is a major limitation of analytical depth resolution. For energy straggling, the simple Bohr's formula is no longer applicable when energy loss exceeds some 20% of the initial energy and a more intricate calculation is necessary (8). We have paid particular attention to the intricate problem of multiple scattering, taking into account recent and even unpublished literature. Indeed, owing to the frequent use of grazing incidence and of heavy particles, multiple scattering is much more important in ERDA than in RBS. Despite important theoretical progress in the last years (9-11), there is no general procedure for its evaluation, particularly as concerns compounds (12), and there is not much experimental data to compare with theory. Nevertheless, we proposed a set of fitting formulae for practical evaluation that covers a fairly broad range of sample thickness (3).

CONVENTIONAL ERDA

There are mainly two conventional arrangements for ERDA, called *transmission geometry* and *reflection geometry*. Transmission is limited to thin enough samples (a few microns for a 3 MeV helium beam). One uses normal incidence and the detector is placed in front of the back face of the sample. Detection near 0° allows high sensitivity, up to 10 ppm hydrogen (13), because large detectors can be used with only a weak degradation of depth resolution. In RBS, only backscattered ions can reach the detector. In forward spectrometry, scattered projectiles can reach it as well as recoils and they are undistinguishable from their energy alone. This is called *recoil-projectile ambiguity*. To tackle this difficulty and avoid flooding the detector, an absorber foil is generally placed in front of the detector to stop scattered ions, but this works only if $M_1 > M_2$. With a suitable sample thickness, projectiles are stopped in the sample while recoils are still detectable. When $M_1 > M_2$, there exists a maximum scattering angle and placing the detector at a larger angle eliminates this ambiguity. If $M_2 > M_1$, recoil-projectile ambiguity is always present.

In reflection geometry, both beam incidence and detection are at a glancing angle, typically 70-80°. Such angles improve the depth resolution and there is no limitation of the sample thickness, but the analysable depth is much smaller than in transmission geometry and large errors may be induced by small incertitudes in target tilt or scattering angle. Moreover, sample roughness has a strong influence on depth resolution and can in extreme cases produce large distorsions of the spectrum.

Beside the recoil-projectile ambiguity, multi-elemental analysis exhibits also mass-depth ambiguity. This latter can be resolved up to a limited depth because in some part of the spectrum counts originate from only one element. Although counts at the same energy from different species cannot be distinguished directly, multi-elemental depth profiling is not really untractable, because beside the spectrum we also have the condition that the sum of concentrations is unity at any depth. This condition is useful only if there are no *dummy* elements, which is generally not the case when an absorber is used. Then, one has to resort to concentration standards, difficult to prepare, not always reliable and eventually sensitive to ion beam damaging. Moreover, even if all the elements of the target are detected, poor mass resolution or too large differences in cross-sections may still be obstacles to the interpretation of spectra.

This is why several variants of ERDA have been developed in order to resolve both ambiguities.

OTHER ARRANGEMENTS

Basically, one can imagine two main principles to eliminate ambiguities. By measuring simultaneously the energies of both the recoil and the projectile for given recoil and scattering angles, one can unambiguously distinguish the recoil from the projectile and the determination of depth is possible through the total energy loss of both particles. Another method consists in measuring the mass of the detected particle, or its velocity, which is equivalent once its energy is measured. In the first case, we have coincidence-ERDA, in the second a series of variants : TOF-ERDA, telescope-ERDA, ExB-ERDA, ERDA with mass-spectrometry.

Coincidence-ERDA

Simultaneous detection of both particles requires a transmission geometry, hence it is limited to thin targets. However, coincidence techniques offer several advantages. By properly adjusting both angles, only one species of recoils is detected. Thus, both ambiguities are resolved. This is the principle of coincidence-ERDA or CERDA. In addition, background counts are discarded. For instance, it is possible to discriminate both isotopes of copper using the large difference in recoil angle for a fixed scattering angle obtained with $M_1 \approx M_2$, i.e. with a ^{58}Ni beam (14).

Measuring both energies in coincidence, one can also eliminate the scattering and recoil angles from depth versus energy relations using the relation between these angles. Then detectors with large solid angles can be used to increase considerably the sensitivity or to reduce beam damaging effects by using low currents. This method is called Scattering Recoil Coincidence Spectrometry (SRCS). A sensitivity of 10^{15} at. / cm^2 for deuterium in carbon foils could be reached by this technique (15).

When using SRCS to detect atoms lighter than the projectile, the detector for scattered ions is placed at a small angle, where it can be flooded by heavier recoils. Elastic Recoil Coincidence Spectrometry (ERCS) consists on the opposite to detect simultaneously several light recoil species somewhat heavier than the projectiles, discriminating them through the difference between recoil and scattered energies, e.g., carbon and oxygen in polycarbonate films using a 2 MeV $^4He^+$ beam (16). Annular detectors are particularly suitable for this purpose. Recoils detected at the same angle are discriminated using the energy difference between the scattered ion and the recoil. Position-sensitive detectors have also provided new possibilities for these techniques.

Time-of-flight-ERDA

One possible way to discriminate recoils by measuring their velocity is to use a time-of-flight spectrometer. TOF-ERDA is generally performed with medium mass or heavy ions and a detection angle larger than the maximum scattering angle to avoid recoil-projectile ambiguity. Nevertheless mass discrimination of projectiles resolves this ambiguity when scattered ions can enter the detector. Start and stop signals are generated by secondary electron emission in thin carbon foils, typically distant of 0.5 m. Counts corresponding to different masses form separated tracks in a 3D plot of counts number vs. time-of-flight and energy (corrected for energy loss in carbon foils). Excellent mass resolution can be achieved. For example, complete mass separation of the five nickel isotopes was achieved using a 340 MeV ^{129}Xe beam and 1.34 m flight length (17). Energy loss, straggling and multiple scattering in the carbon foils have to be taken into account, but the main difficulty is to make proper corrections for the decreasing efficiency for light masses (up to boron or carbon) due to unsufficient electron emission.

Telescope-ERDA

A telescope is composed of a thin solid state detector (typically 10-20 µm thick) placed in front of a thick one. From the energy lost in the thin detector, one can deduce the velocity of the recoil and the total energy is the sum of this energy plus the residual energy measured in the second detector. This set-up is mainly used with M1 > M2 at angles larger than the maximum scattering angle because the thin detector would be heavily damaged by abundant scattered ions. Eventually, a gas ionization chamber can replace the thin detector. This technique gives the same type of 3D plots as TOF-ERDA, with the advantage of a mass-independent efficiency. However, it cannot compete with TOF-ERDA for depth resolution because straggling and angular scatter in the thin detector is much larger. It may be used as well for discriminating the three hydrogen isotopes with helium beams, e.g. (18), as for heavy ion ERDA, e.g. (19).

Mass spectrometry and ExB-ERDA

Mass-dependent deflection of ions in electro-magnetic fields allows various mass-spectrometric discrimination set-ups. A magnetic spectrometer separates particles having a different mass/charge ratio, but several optical corrections are necessary to focus particles with different energies. This can be achieved, for example, using a quadrupole followed by three dipoles. Such a device is very large (about 5×5 m) and also very expensive. Its energy resolution may be as low as 4.10^{-4} with an acceptance solid angle as large as 5 msr and an exceptional depth resolution of 0.34 nm at the surface of a sample has been reported (20).

A very simple and inexpensive electro-magnetic mass discriminator is the ExB filter (2, 21). Similarly to a Wien filter, an electric field $\boldsymbol{E}$ and a magnetic field B are applied perpendicularly to each other, but to obtain mass and charge discrimination instead of velocity selection the electric field is chosen as :

$$\boldsymbol{E} = \mathrm{B}\mathrm{v}_0/2 \qquad (1)$$

where v_0 is the ion velocity. The deflection is :

$$\mathrm{y} = \mathrm{q}(\boldsymbol{E} - \mathrm{v}_0\mathrm{B})\mathrm{x}^2/2\mathrm{m}\mathrm{v}_0^2 \qquad (2)$$

almost independent of energy. For each mass it is thus possible to find a value of $\boldsymbol{E}$ such that recoils have an almost constant deflection and can be detected by a movable collimated detector. ExB with 1 to 2.5 MeV He^+ beams or a 2.65 MeV ^{15}N beam has been used for different multi-elemental analyses, particularly for sequential profiling of hydrogen isotopes or of helium and hydrogen. Moreover, ExB-ERDA is the only acceptable set-up with very low energy beams, e.g. 350-400 keV He^+ (21). However, as v_0 is not really independent of energy, there is some divergence of ions with different energies and this limits the filter dimensions, hence the allowable beam energy. Analysable depth is also limited by the excessive divergence of low energy ions. Recently, it has been proposed that reversing the direction of the force somewhere in the filter would contribute to restore convergence. Calculation shows that the energy dependance of the deflection is considerably reduced if ions traverse first a magnetic field, then an ExB field, then again a magnetic field with suitable path lengths in each region. Mass selection is thus achievable for masses up to oxygen in such a B-ExB-B filter when projectiles are 1 MeV He^+ ions (22). However, as selectivity for higher masses requires a B field of 0.5 Tesla, excessive for H and He analysis, one needs a variable B field, hence an electromagnet. A drawback of ExB-ERDA may be that sequential - not simultaneous - multi-elemental analysis may require damaging ion doses delivered to the sample. An important difficulty is also that only one charge state is analysed at a time and that neutrals are not separated. Prior knowledge of charge state ratios is thus necessary and its acquisition may be difficult.

DATA PROCESSING

Once experiment has produced a spectrum, data processing is still needed to retrieve depth profiles. The numerous and complex phenomena involved in ERDA, wherein the sample composition is present as an unknown set of variables, lead to energy versus depth and yield versus composition relations that cannot be inverted directly. In fact, the problem is practically the same as for RBS, except for increased importance of multiple scattering and more frequent difficulties due to *dummy* elements. Despite this resemblance, data processing for ERDA is generally less developed than for RBS. Besides, variants like CERDA, TOF-ERDA, etc., require specific softwares adapted to their particular features.

Facing the impossibility of direct deconvolution, data processing essentially consists of predicting the spectrum that would be obtained given a hypothetical set of depth profiles inside the target, to compare this simulation with experiment and attempt to improve the hypothesis until satisfactory agreement. In the simulation procedure, non-Rutherford cross-sections lead to more intricate and less reliable expressions and including properly multiple scattering effects and their combination with energy straggling is still quite a concern, due to the complexity of theoretical models. Then, a major problem is how to modify the initial hypothesis in order to converge towards the right description of the sample. Most available software, e.g. the well-known RUMP, allow only a multi-parametric minimisation procedure that can deal only with a limited set of parameters. As a consequence, results may have a lower depth resolution than allowed by experimental conditions and the true best fit may well be outside the limited set of hypotheses explored in this way (23). To avoid this drawback, less user-dependent methods with less restrictive hypotheses have been developed for RBS with much success : *profile reconstitution method* (23) and *simulated annealing* (24), which are probably rather complementary than concurrent. In such methods, any change in initial depth profiles is allowed and convergence is driven by an automatic procedure. Their application to ERDA might bring up valuable progress.

CONCLUSION

As can be seen from preceding examples, ERDA has extremely various applications in almost any sort of materials. As a conclusion to this necessarily succinct overview, let us quote W. A. Lanford (foreword to Ref. 3): "While analysis of hydrogen is important, ERDA's future is tied to its ability to provide simultaneous quantitative concentration profiles for all light- and medium-mass elements in any target. It is clear that energy recoil detection has uniquely demonstrated its ability to satisfy this important analytic need".

ACKNOWLEDGEMENTS

We are endebted to Drs. G.G. Ross, H. Hofsäss and N. Dytlewski for their important contributions to chapters of our book on ExB-ERDA, Coincidence-ERDA and TOF-ERDA respectively. We are grateful to Dr. Amsel who gave us access to important unpublished material.

REFERENCES

1. Marsden, E., *Philosophical Mag.* **27**, 824-830 (1914).
2. L'Ecuyer, J., Brassard, C., Cardinal, C., Chabal, J., Deschenes, L., Labrie, J. P., Terrault, B., Martel, J. G. and St. Jacques R., *J. Appl. Phys.* **47**, 381-382 (1976).
3. Tirira, J., Serruys, Y. and Trocellier, P., *Forward Recoil Spectrometry*, New York: Plenum Press, 1996.
4. Tirira, J. and Bodart, F., *Nucl. Instr. Meth. B* **74**, 496-502 (1993).
5. Tirira, J., Trocellier, P., Frontier, J. P. and Trouslard, P., *Nucl. Instr. Meth. B* **45**, 203-207 (1990).
6. Terwagne, G., Ross, G. G. and Leblanc, L., *J. Appl. Phys.* **79**, 8886-8891 (1996).
7. Terwagne, G., *Nucl. Instr. Meth. B* **122**, 1-7 (1997).
8. Cohen, D. D. and Rose, E. K., *Nucl. Instr. Meth. B* **64**, 672-677 (1992).
9. Battistig, G., Amsel, G., d'Artemare, E. and L'Hoir, A., *Nucl. Instr. Meth. B* **85**, 572-578 (1994).
10. Szilágyi, E., Pászti, F. and Amsel, G., *Nucl. Instr. Meth. B* **100**, 103-121 (1995).
11. Schmaus, D. and L'Hoir, A., *Nucl. Instr. Meth.* **194**, 75-85 (1982).
12. Schmaus, D. and L'Hoir, A., *Nucl. Instr. Meth. B* **2**, 187-190 (1984).
13. Wielunski, L. S., Benenson, R., Horn, K. and Lanford, A., *Nucl. Instr. Meth. B* **15**, 469-474 (1986).
14. Rijken, H. A., Klein, S. S. and Devoigt, M. J. A., *Nucl. Instr. Meth. B* **64**, 395-398 (1992).
15. Forster, J. S., Leslie, J. R. and Laursen, T., *Nucl. Instr. Meth. B* **45**, 176-180 (1990) and **66**, 215-220 (1992).
16. Hofsäss, H. C., Parikh, N. R., Swanson, M. L. and Chu, W. K., *Nucl. Instr. Meth. B* **45**, 151-156 (1990).
17. Goppelt, P., Gebauer, B., Fink, D., Wilpert, M., Wilpert, Th. and Bohne, W., *Nucl. Instr. Meth. B* **68**, 235-240 (1992).
18. Prozesky, V. M., Churms, C. L., Pilcher, J. V., Springhorn, K. A. and Behrisch, R., *Nucl. Instr. Meth. B* **84**, 373-379 (1994).
19. Assmann, W., Huber, H., Steinhausen, Ch., Dobler, M., Glückler, H. and Weidinger, A., *Nucl. Instr. Meth. B* **89**, 131-139 (1994).
20. Dollinger, G., *Nucl. Instr. Meth. B* **79**, 513-517 (1993).
21. Ross, G. G., Terreault, B., Gobeil, G., Abel, G. Boucher, C. and Vielleux, G., *J. Nucl. Mater.* **128/129**, 730-733 (1984).
22. Serruys, Y. and Tirira, J., "A modified ExB filter for analysing higher masses with ERDA", presented at the 2nd French-Australian Workshop on Ion Beam Applications, Lucas Heights, Australia, Feb. 1-3, 1995.
23. Serruys, Y., Tirira, J. and Calmon, P., *Nucl. Instr. Meth. B* **74**, 565-572 (1993).
24. Barradas, N.P., Jeynes, C. and Webb, R.P., *Appl. Phys. Lett.* **71**, 291-293 (1997).

Influence of Planar Oscillations on Scattered Ion Energy Distributions in Transmission Ion Channeling

A. A. Bailes III[a] and L. E. Seiberling[b]

[a] *Physics Department, Haverford College, Haverford, PA 19041*
[b] *Department of Physics, University of Florida, P. O. Box 118440, Gainesville, FL 32611*

Utilizing the transmission ion channeling technique and a Monte Carlo simulation of the channeling of He ions in Si, we have been able to determine surface structure by comparing experimental to simulated scattered ion energy distributions. In analyzing data for {110} beam incidence, we have found that planar oscillations persist well past 2000 Å in our Monte Carlo simulations. These oscillations yield no benefit to this method of data analysis but can make analysis more difficult by the requirement for more accurate Si thickness determination.

By comparing experimental to simulated energy distributions of scattered ions, one can make full use of the information obtained in transmission ion channeling experiments to determine surface structure. Whereas the traditional yield versus tilt (angular scan) method uses only the scattering yield from each spectrum, simulated energy distributions allow one to take advantage of the information contained in the position and shape of the peaks in a spectrum. Because of the wide range of energy loss experienced by the ions, spectra taken with beam incidence in the {110} planar direction often exhibit complex shapes, a characteristic that makes this direction particularly interesting to study. We have found, however, that planar oscillations continue well past 2000 Å in our simulations and, combined with our thickness resolution of ~100 Å, make accurate analysis of {110} data unrealizable.

The transmission ion channeling technique (1-3) uses the non-uniform flux of an ion beam as it exits a thin crystal. This non-uniform flux, which changes with the angle of incidence of the ion beam, can then be exploited to locate adatoms at the beam-exit surface by measuring the scattering yield from the adatoms. Scattering yields from adatoms near the middle of the channel (interstitial positions) will be enhanced and yields from adatoms near substitutional positions will be reduced when the angle of incidence of the ion beam is near a low index direction (e.g., the <100> axis). As the incident beam is moved away from the axis in an angular scan, the yield gradually approaches that of the "random" yield, i.e., the scattering yield when the beam enters the crystal in a direction away from all axial and planar influences.

In addition to the scattering yields, we employ another property of channeled ions: the energy loss of the ions as they channel through the crystal. Ions that travel near the center of a channel have lower energy loss than ions that travel though a random direction in the crystal. Thus, the energy distribution of ions scattering from a near-interstitial site will have a different shape and position on the energy axis than one resulting from ions scattering off of a near-substitutional site.

A Monte Carlo simulation of the channeling process generates an ion distribution that gives both the positions and energies of all the ions after they have traversed the crystal (4). These positions and energies are then overlapped with trial adatom sites to produce simulated scattering yields and energy distributions for scattering off of the adatoms. We then can compare the experimental energy distributions to the simulated energy distributions and obtain a χ^2 value to determine goodness of fit. Next we either go through a χ^2 minimization procedure by varying the adatom positions to find the best fit or test the positions corresponding to a given model to see how well it fits the data. We have used both procedures and found this method of analysis to work well, with a typical resolution of 0.1 Å or better for the adsorbate site (2,3).

An interesting feature of simulated energy distributions emerged during an attempt to make a correction for thickness variation in the sample. Simulations for two different sites being compared to the same set of {110} planar data underwent a dramatic and unexpected shift relative to the experimental data and to each other (Fig. 1). In going from a Monte Carlo ion distribution for a thickness of 5125 Å to a distribution for a thickness of 5400 Å, one of the simulations (site A) increased in yield whereas the other (site B) displayed a decreased yield. By increasing the thickness, we expected to see both of the simulated energy distributions shift to lower energy while keeping the same relative yield and shapes. The changes seen here, however, indicate that some other process is playing a significant role. A closer look at other data and simulations for this direction of beam incidence showed similar occurrences.

CP475, *Applications of Accelerators in Research and Industry*,
edited by J. L. Duggan and I. L. Morgan

(~600 Å) creates difficulty in our data analysis because any problems with thickness variation are compounded by the changing shape of the energy distribution. In order to use the {110} data, we would have to know the thickness to within ±50 Å, a resolution that is difficult to attain.

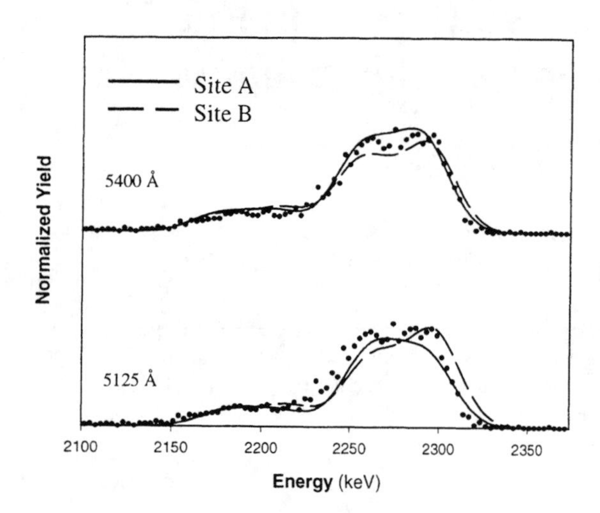

FIGURE 1. Simulated energy distribution flip-flop. The same set of experimental data (dots) is plotted against simulations for two different sites (solid & dashed curves) at two slightly different thicknesses (upper & lower simulations).

Because the only difference between the two simulations was the use of two different thicknesses, a look at the ion distributions should show the cause of the yield flip-flop. Figure 2 shows the ion distributions responsible for the above simulated energy distributions, and indeed the problem is evident. The important feature to notice about these ion distributions is that the band of planar channeled ions in the center is broader and less dense for the 5400 Å distribution shown in Fig. 2(a) than for the 5125 Å distribution in Fig. 2(b).

By examining which ions are responsible for the two simulated energy distributions, we can understand the yield flip-flop. The large black circles in Fig. 2 enclose the ions that are sampled for the site A energy distribution, and the large white circle in the near-interstitial site encloses the ions that contribute to the site B energy distribution. (The size of these circles is two times the 2D vibrational amplitude, as only the ions within that range are considered.) The narrow band in the 5125 Å distribution puts a higher density of ions in the white circle and a lower density in the black circles than does the broader band in the 5400 Å distribution. For a more quantitative comparison, the number of ions in the black circles (site A) changes from 372 to 464 and the number in the white circle (site B) changes from 620 to 556 in going from the lesser to the greater thickness. That 25% increase in the number of ions sampled for site A and 10% decrease in the number sampled for site B results in the reversal in the relative yields for the two sites.

The changing ion distribution is not unique to the two thicknesses discussed in the previous paragraphs. The ion distributions for {110} incidence actually exhibit this oscillatory behavior over the whole range of sample thickness (3000 Å to 6000 Å) in which we are interested. The relatively short wavelength of these oscillations

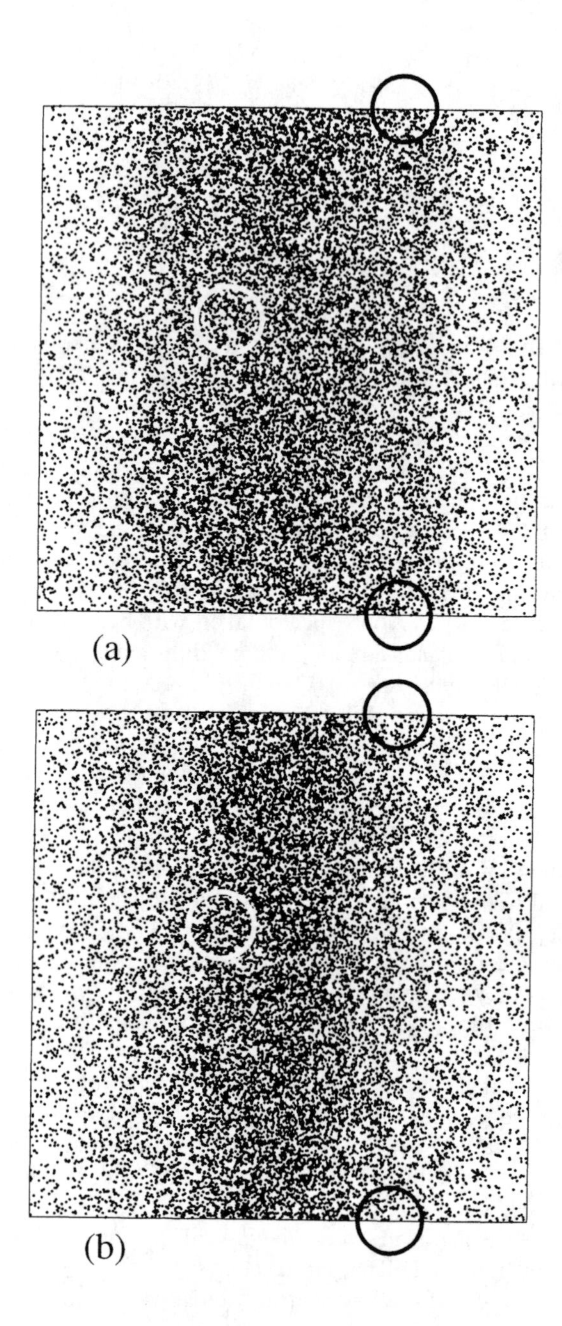

FIGURE 2. Distributions of channeled ions for {110} beam incidence and at Si thicknesses of (a) 5400 Å and (b) 5125 Å. The circles show the ions sampled for the site A (black) and site B (white) energy distributions in Fig. 1. The square region shown is a <100> channel with sides of length 1.92 Å.

This phenomenon of planar oscillations has been documented by Abel *et al.* in a study of oscillations of backscattering yield in planar directions (5) and also by Breese *et al.*, who have imaged these oscillations directly in a transmission channeling experiment using a nuclear microprobe with 3 MeV protons (6). Oscillation of individual ions in a channel is responsible for the oscillation of the band of {110} planar-channeled ions discussed here. This effect demonstrates that the individual, oscillating ion trajectories remain coherent with each other. Furthermore the variation in oscillation wavelength among the different ions is small enough not to damp out the oscillations.

The surprising aspect of our observation of {110} planar oscillations is the thicknesses at which we have seen them. For MeV He ions, it is generally thought that planar oscillations die out after ~2000 Å (5), but we have seen them in the simulated {110} ion distributions at thicknesses greater than 5000 Å. This persistence of the oscillations has recently been observed experimentally as well. Breese has seen them for MeV He ions in Si of comparable thickness to what we have used and also for 3 MeV protons in 1.4 μm Si (7).

The observation of these oscillations for beams incident in the {110} direction raised the question of whether the other directions of beam incidence that we use show the same effect. An appeal to the dimensions involved shows that they are less susceptible. The {110} channel has a width of 1.92 Å, whereas the {100} channel is 1.36 Å across. This smaller channel will cause the oscillations to damp out much more rapidly. For <100> axial channeling, the ions lose coherency even more quickly because of being constrained in two dimensions rather than one. The Monte Carlo ion distributions for the <100> and {100} directions (not shown here) do not show oscillatory behavior, confirming that the oscillations in those directions indeed damp out quickly.

Although they are an interesting feature of ion channeling, planar oscillations only hinder data analysis of the type that we perform. The oscillating ion distributions alter the simulated energy distributions based on the thickness of the sample, whereas we are interested in changes induced by different surface structures. Thus, we gain nothing by their presence. They can affect the analysis adversely, however, because their presence requires a more accurate knowledge of the Si thickness. Since the {110} planar oscillations have a wavelength of ~600 Å, small changes in thickness can have large effects on the simulated ion and energy distributions (Fig. 1 and 2). The normal thickness variation that we see on our thin Si samples, which we produce with a chemical etch, is comparable to the oscillation wavelength. To see the effect of thickness variation on energy distributions, we looked at the spectra obtained for various beam incidences at several spots on the 1 mm diameter sample. In transmission geometry, the high energy edge of the Si peak is created by ions that scatter off of the back side of the sample. The position of this edge is thus indicative of the Si thickness. Figure 3 shows the two extremes of Si thickness, the energy difference between the two corresponding to a 600 Å thickness variation.

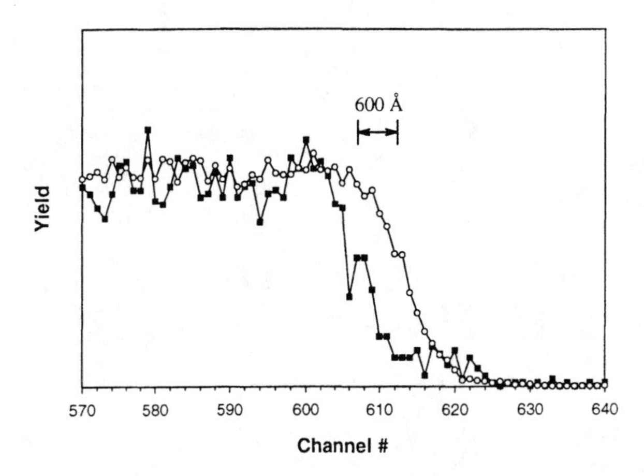

FIGURE 3. Overlapping Si peaks showing a thickness variation of ~600 Å across the sample. Si peak edges at lower energies imply greater thickness and vice versa.

In conclusion, we have found that planar oscillations persist well past 2000 Å in our Monte Carlo simulations of He ions channeling in Si(100) along {110} planes. These oscillations make the comparison of simulated to experimental energy distributions for adsorbate site determination more difficult by requiring a more accurate knowledge of the Si thickness.

ACKNOWLEDGMENTS

This work was supported in part by the NSF (DMR) and the University of Florida Division of Sponsored Research.

REFERENCES

1. Seiberling, L. E., Lyman, P. F., and Grant, M. W., *J. Vac. Sci. Technol.* A **11**, 715-722 (1993).
2. Boshart, M.A., Bailes III, A.A., Dygo, A., and Seiberling, L.E., *J. Vac. Sci. Technol.* A **13**, 2764-2771 (1995).
3. Boshart, M.A., Bailes III, A.A., and Seiberling, L.E., *Surf. Sci. Lett.* **348**, L75-L81 (1996).
4. Dygo, A., Boshart, M. A., Seiberling, L. E., and Kabachnik, N. M., *Phys. Rev.* A **50**, 4979-4992 (1994).
5. Abel, F., Amsel, G., Bruneaux, M., Cohen, C., and L'Hoir, A., *Phys. Rev.* B **12**, 4617-4627 (1975).
6. Breese, M. B. H., King, P. J. C., Grime, G. W., Smulders, P. J. M., Seiberling, L. E., and Boshart, M. A., *Phys. Rev.* B **53**, 8267-8276 (1996).
7. Breese, M. B. H., private communication (1998).

Advanced Capabilities and Applications of a Sputter-RBS System

B. Brijs, J. Deleu, G. Beyer, W. Vandervorst*

IMEC, Kapeldreef 75, B-3001 Leuven, Belgium

**KULeuven, INSYS, Kardinaal Mercierlaan 92, B-3001 LEUVEN*

In previous experiments, sputter-RBS [1] has proven to be an ideal tool to study the interaction of low energy ions. This contribution employs the same methodology to identify surface contamination induced during sputtering and to the determine absolute sputter yields. In the first experiment ERDA analysis was used to study the evolution of Hydrogen contamination during sputter-RBS experiments. Since the determination of Hydrogen concentration in very thin near surface layers is frequently limited by the presence of a strong surface peak of Hydrogen originating from adsorbed contamination of the residual vacuum, removal of this contamination would increase the sensitivity for Hydrogen detection in the near sub surface drastically. Therefore low energy (12 keV) Argon sputtering was used to remove the Hydrogen surface peak. However enhanced Hydrogen adsorption was observed related to the Ar dose. This experiment shows that severe vacuum conditions and the use of high current densities/sputter yields are a prerequisite for an efficient detection of Hydrogen in the near surface layers.

In the second experiment, an attempt was made to determine the sputter yield of Cu during low energy (12 keV) Oxygen bombardment. In order to determine the accumulated dose of the low energy ion beam, a separate Faraday cup in combination with a remote controlled current have been added to the existing sputter-RBS set-up. Alternating sputtering and RBS analysis seem to be an adequate tool for the determination of the absolute sputter yield of Cu and this as well in the as under steady state conditions.

INTRODUCTION

The determination of the composition of thin near-surface layers (often less than 10 nm) becomes more and more important in microelectronics. Whereas this represents a challenging task for studies such as dopant profiling, it is even more complicated in the case of Hydrogen detection (using techniques such as Elastic Recoil Detection Analysis (ERDA)) as a strong surface peak quite often dominates that part of the spectrum related the layer of interest. Since these surface contaminants originate from adsorption from the atmosphere, they are not really linked to the processing scheme of the sample and can be removed without destroying any information related to the process under study. In fact their removal will greatly improve the sensitivity and accuracy for near surface analysis and thus enhance the quality of for instance the ERDA information. In many applications (AES, XPS, ...) a similar problem of surface contamination is faced and quite frequently Ar sputtering is used as an efficient cleaning tool.

The basic objective of the present paper is to study the use of Ar sputtering in combination with ERDA for Hydrogen profiling. Details of the Hydrogen surface peak removal/increase are studied under the influence of the ERDA and the Ar beam. The observed dependencies on vacuum conditions and current density are interpreted in terms of the classical model of adsorption and sputtering, adjusted by enhanced sticking resulting from the ion-beam solid reactions.

CP475, *Applications of Accelerators in Research and Industry*,
edited by J. L. Duggan and I. L. Morgan

In a second experiment, alternating sputtering with in situ RBS analysis was used to monitor the sputter yield changes for Cu bombarded with O_2^+ beams. The choice for an O_2 beam is dictated by the fact that this is one of the most currently used primary beams for SIMS depth profiling which however can cause severe material modifications during the initial phase of the depth profile. The objective here is to study transients in the sputter yield and the stationary state of the altered layer as these determine the final interpretation of a SIMS profile.

EXPERIMENTAL SET-UP

The experiments are carried out in our combined in situ sputter/RBS system described in reference[1,2]. In brief, a mass-filtered Ar-DIDDA ion gun is used to produce craters with the low energy ion (Ar, O_2^+) beam in which center a He beam is focused by a NEC microlens for ion beam analysis. The RBS measurements are performed using a 1 MeV He beam and a detector installed at 15 degrees with respect to the beam; to improve depth resolution, the target is tilted over 65 degrees. Recently, the system has been upgraded with an additional detector for ERDA analysis. The ERDA measurements are performed using a 2 MeV He beam bombarding the target tilted by 75 degrees. The detector is installed at 165 degrees with respect to the beam and 15 degrees with respect to the sample surface. An absorber foil of 9 μm and a collimator of 1 by 4 mm has been installed in front of the detector for optimum Hydrogen detection. In the present experiments, sputtering is carried out with an energy of 12 keV, focused to a diameter of 300 μm and swept over a region of 2 by 2 mm^2. All experiments were performed in an UHV chamber evacuated by a turbomolecular pump and an operating vacuum (beam on target) was better than 10^{-7} mbar. Sputter cycles and ERDA/RBS data collection have been performed automatically in an alternating way under control of the new acquisition code ARIBA[3].

EXPERIMENTAL RESULTS

The present work covers two types of experiments, i.e. ERDA in combination with Ar sputtering on a Hydrogen implanted Si and virgin Si and O_2^+ sputtering of a 50 nm Cu layer on Si.

Evolution of Hydrogen contamination during sputter-RBS

Ar sputtering of a Hydrogen implanted Si sample monitored by ERDA

In line with our objective to remove the surface contamination by Ar sputtering, a Si sample implanted with 1.6 10^{16} Hydrogen at/cm^2 (20 keV) was bombarded with a 12 keV Argon beam under perpendicular incidence with a flux of 3. 10^{13} at/cm^2.sec. Simultaneous ERDA and RBS spectra were collected to study the evolution of the Hydrogen concentration before and after Ar bombardment. The Hydrogen implant serves as a depth marker to elucidate the sample erosion during Ar sputtering.

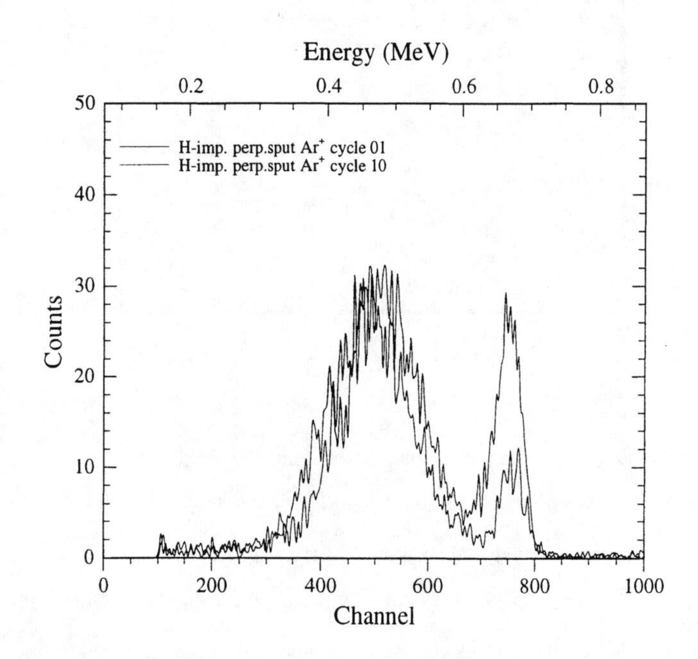

FIGURE 1 2 MeV ERDA spectra of Hydrogen on a Hydrogen implanted Si sample before (cycle 01) and after Argon irradiation (cycle 10).

Figure 1 shows the ERDA spectra of a virgin and bombarded sample. The shift of the H-implant profile to higher energies implies that a significant amount of sputtering (20 nm) has occurred. Despite this sputtering a significant increase of the Hydrogen concentration at the surface is noticed. The Hydrogen accumulated at the surface after removing 10^{17} Si / cm^2 amounts to 10^{15} H/cm^2. A more detailed picture of the evolutions of the surface H as a function of Ar dose is shown in fig. 2. As we perform an ERDA analysis after each sputter cycle we do observe now that already after one sputter cycle, the H dose becomes larger than the adsorption due to two ERDA measurements implying that Ar stimulates the Hydrogen adsorption as well. One must be aware of the H accumulated induced by the He beam as well. As shown in fig.2 (line7), the latter is not negligible either. With larger Ar dose , the H coverage saturates at a level lower as compared to the pure ERDA case.

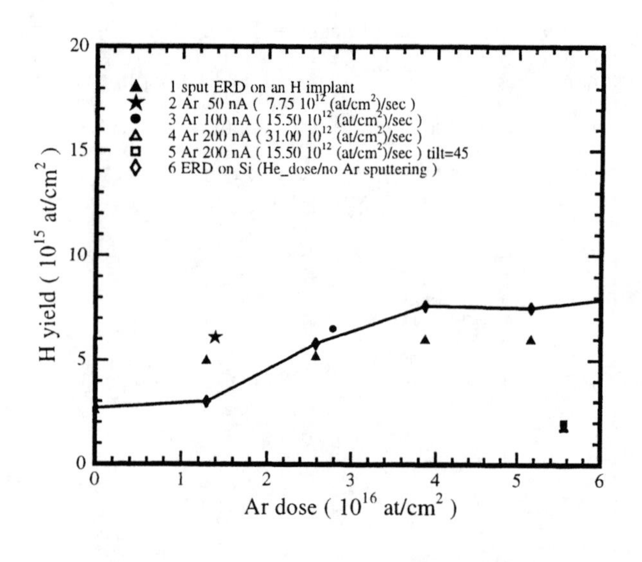

FIGURE 2 This overlay shows the accumulated Hydrogen as a function of Argon (12 keV) and Helium (2 MeV) dose collected by ERDA. Curve 1 shows the accumulated Hydrogen yield before and after incremental Argon doses on a Si sample with an Hydrogen implant of 2.5 10^{16} atoms/cm^2. Curve 6 shows the Hydrogen yield as a function of the Helium dose. Point 2,3,4 and 5 represent the Hydrogen yield after irradiation of a Si sample with different Argon current densities.

The latter is not surprising since besides the enhanced adsorption from the Ar beam also sputtering is induced probably removing some of the ERDA induced H accumulation.

ERDA of a Si sample sputtered by Ar

In order to separate the effect of ERDA and Ar sputtering, a Si sample (without previous ERDA analysis) was bombarded using different current densities : 7.8 10^{12} at/cm^2.sec, 1.5 10^{13} at/cm^2.sec, 3.0 10^{13} at/cm^2.sec. After Ar irradiation for 30 min., ERDA analysis (using 50 μC He) was used to determine the amount of Hydrogen at the surface. The calculated Hydrogen dose values are in fig. 2 and 3. It is now interesting to see that for the Ar current of 200 nA the Hydrogen coverage is much smaller compared to the one used in the combined Ar sputter/ERDA study.

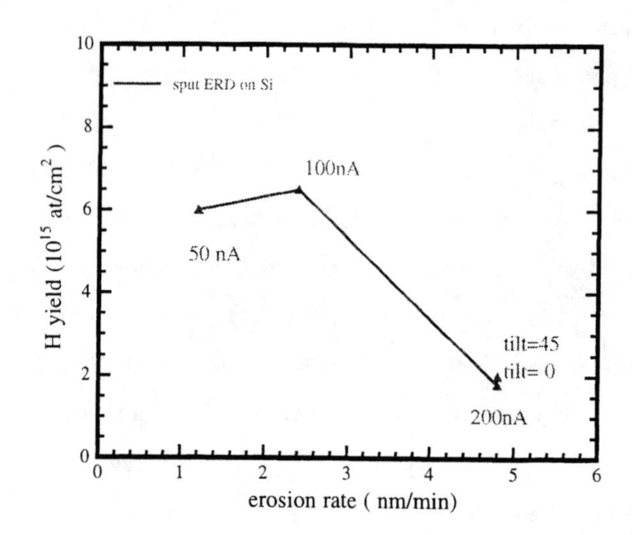

FIGURE 3 . Hydrogen yield on a Si sample as a function of the erosion rate

When plotting the Hydrogen amount as a function of erosion rate (fig. 3) one can observe a very strong reduction for the highest erosion rates. Clearly the stationary coverage is determined by the balance between erosion rate and adsorption rate. The latter can be confirmed by another experiment whereby the current was set again at 200 nA but this time incident under 45 degrees. As the rastered area is increased by a factor 2 because of the tilting, current density will be a factor 50% lower. On the other hand the sputter yield increases roughly by a factor 2 such that the erosion rate remains constant. The Hydrogen coverage in this case coincides with the higher current density at normal incidence.

Discussion

The present experiments clearly show that the adsorption of hydrocarbons during ion beam irradiation is a process which can disturb the correct characterization of the sample significantly. The origin of the problems is of course closely linked to the adsorption of residual gasses (H_2O and hydrocarbons) from the gas phase. This phenomenon is known for some time and not only sets a limit to near surface profiling with ERDA but equal well to the detection limits in SIMS depth profiles. Experiments by Wittmaack [4] and Magee [5] had already elucidated that the surface contamination is set by two processes i.e. the supply of adsorbing species and the probability that a species imping on the sample surface adheres to it (=sticking coefficient). The interpretation of the Ar sputtering results is however more complex as the prime intent of the sputter beam is to remove particles from the surface. The experiments with the H-implant demonstrate that concurrent with the sputtering process sites for enhanced adsorption are created. The dose dependence of this adsorption effect is not straightforward to interpret since the sputtering beam will remove some of the previously adsorbed species as well. However the current density experiments were performed with only one ERDA measurement and since a smaller amount of Hydrogen is observed for the 200 nA case as for the detailed dose studies (where 5 ERDA measurements were included), one must conclude that the sputter process is not very efficient in removing the ERDA contamination.

When analysing the data of fig. 1, one must be able to explain the fact that as a function of dose a constant level gradually is obtained (cfr first data points in fig. 1) whereas the level of this saturation is linked to the current density/erosion rate during the sputter process (fig. 4). The interpretation of these results must again be viewed within the frame of establishing a balance between the adsorbing and sputtered species. If sticking would be zero except when an Ar impact occurs, adsorption proportional with Ar dose follows immediately. Saturation arises from the fact that as the contamination builds up, more "adsorbed" species are exposed to the sputtering process as well leading eventually to a sputtered flux equal to the adsorbing flux.

Within such a scheme it is also possible to explain the observed erosion rate dependence. In essence the number of particles able to adsorb on the surface is independent from the erosion rate and set by the vacuum conditions. If this supply is large enough, the steady state coverage will be set by the balance between the number of sticking sites created relative to the number of already adsorbed species removed by the sputter process. As the number of sticking sites and the erosion rate are directly proportional to the current density a change of the latter one will have no effect (50 nA equal to 100 nA). On the other hand if the sputter efficiency is increased (tilting over 45 degrees), the balance between sticking sites and sputter removal shifts in favor of the latter one and a lower Hydrogen coverage will result for the same current density (100 nA versus 200 nA/tilted). The situation becomes different for high current densities where the supply forms the limiting step. Since in that case the number of "used" sticking sites becomes independent from the current density whereas the sputter removal still increases with it, a lower Hydrogen coverage is to be expected (200 nA case).

12 keV O_2^+ Sputtering of a 50 nm Cu layer on Si.

In a second experiment, a study was made to determine the sputter yield of Cu under a 12 keV O_2^+ bombardment. Alternating sputtering with in situ RBS characterisation gives the opportunity to calculate the number of Cu atoms remaining in the Cu layer before and after sputtering and this to determine absolute sputter yields. The sputter current was monitored in a Faraday cup before and after sputtering and kept constant during the experiment. In order to make a quantitative interpretation of the RBS spectra, detailed simulations using RUMP have been performed. To increase the depth resolution, the target was tilted by 65 degrees. Unfortunately, this resulted in a long tail in the RBS spectra due to plural scattering. During RBS analysis this tail is not taken into account.

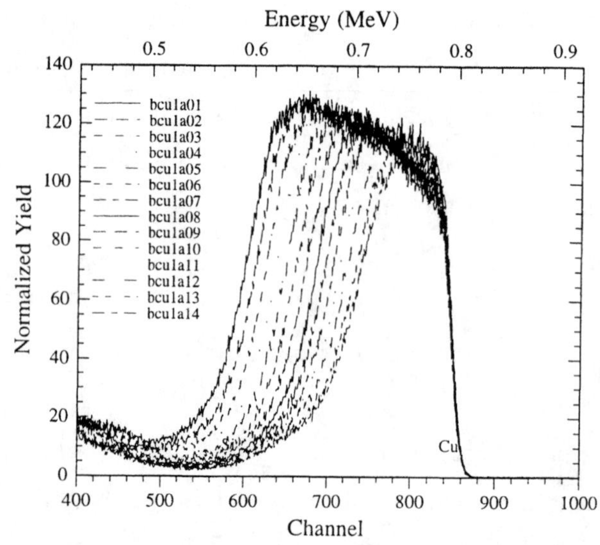

FIGURE 4 Raw RBS data: first spectrum is measured on a virgin Cu sample: other spectra each time after 6.10^{15} O_2^+ /cm^2 bombardment.

Experimental results

A 1 MeV $4He^+$ beam is used to perform the RBS analysis in the sputtered crater. Sputtering is performed using a 12 keV O_2^+ beam of 1.27μA at normal incidence swept over 3.8 mm^2 (i.e. a dose of $4x10^{14}$ at/cm^{2}sec). Each RBS spectrum is accumulated after 30s of sputtering ($12x10^{15}$ at/cm^2). Figure 4 shows the evolution of the raw RBS spectra for increasing oxygen dose. Initially (transient region) the surface Cu intensity is reduced implying that the Cu layer is transformed into a Cu_xO_y layer. Subsequently, the formation of the Cu_xO_y layer is almost complete and no further oxidation occurs. And finally, a severe mixing/segregation phenomena between Si, O and Cu is detected in the interface. Figure 5 shows the calculated Cu and O content in the Cu layer deduced from RUMP simulations of the spectra in figure 4.

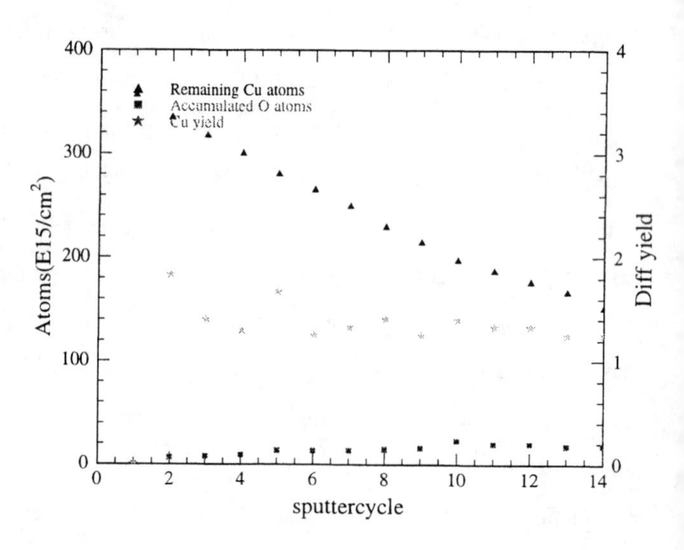

FIGURE 5 Calculated Cu (triangle symbol) and O (square symbol) content in 10^{15} at/cm^2 in the Cu layer in function of the number of sputter cycles. One sputter cycle represents an impingement of 1.2 10^{16} O^+ ions/cm^2. Also the Cu differential yield is plotted (star symbol).

Discussion

The present experiment reveals that the Cu layer bombarded with 12 keV O_2^+ under perpendicular incidence becomes only partly oxidised. The latter is related with the relatively high sputter yield of Cu (=1.5). Note that for instance under similar conditions Si is completely converted to SiO_2 and as a consequence the surface concentration is decreased to 33% of Si.

Notice also that the accumulated O_2 in this experiment (table 1) reaches a maximum of almost 17 10^{15} at/cm^2, which is a factor of 100 lower as in the case of 12 keV O_2^+ perpendicular bombardment on Si.

CONCLUSION

Sputter RBS has proven to be a reliable tool to demonstrate in the first experiment that ERDA analysis for the detection of gaseous species in the near surface region (H, O) is influenced by adsorption induced by the probing beam. Since the effect is directly proportional to the supply from the vacuum and the He dose, high quality ERDA spectra should be collected in very good vacuum with a very limited He dose.

The removal of surface contamination by Ar sputtering does not necessarily lead to better results as the Ar beam induces also some enhanced adsorption. High current densities and high erosion rates are essential to limit the stationary Hydrogen surface contamination.

In the second experiment Sputter RBS in combination with detailed RUMP simulations has been used determine absolute Cu sputter yields.

ACKNOWLEDGEMENTS

This work was supported by the “Fonds voor Wetenschappelijk Onderzoek” . J. Deleu is indebted to the "Flemish Institute for the Promotion of Scientifical and Technological Research in the Industry" (IWT) for financial support.

REFERENCES

[1] Brijs, B., Bender, H., De Coster, W., Moons, R., Vandervorst W., *Nucl. Instr. and Meth.* **B79,** 446 (1993).

[2] De Coster, W., Brijs, B., Bender, H., Allay, J., Vandervorst, W., *Vacuum* **45,** 389 (1993).

[3] Brijs B., De Coster W., Deleu J., Vandervorst W., Wils D. and Vandesteene N., “ARIBA”,presented at the 13 th Int. Conf. on the Application of Accelerators in Research and Industry, Denton, Texas U.S., Nov. 7-10, 1994.

[4] Wittmaack, K., *Radiation Effects* **63** , 205-218 (1982).

[5] Magee C.W. and Botnick E.M., *J. Vac. Sci. Technol.* **19,** 47 (1981).

[6] Deleu, J., Brijs, B., Vandervorst, W., presented at the SIMS workshop, May 10-13, Austin Texas, U.S, (1998).

RAPID ACCURATE AUTOMATED ANALYSIS OF COMPLEX ION BEAM ANALYSIS DATA

P.K. Marriott, M. Jenkin, C. Jeynes, N.P. Barradas, R.P. Webb, and B.J. Sealy

School of Electronic Engineering, Information Technology and Mathematics
University of Surrey, Guildford, Surrey GU2 5XH, UK.

The 'IBA DataFurnace v2.1' code based on a simulated annealing algorithm was released in April 1998. It is capable of handling multiple RBS, EBS and ERD spectra simultaneously and self-consistently.

The DataFurnace automatically solves the general inverse RBS problem: Given the spectrum what is the structure? All practical methods published up to now have depended on an analyst guessing what the spectrum means, with the help of more or less sophisticated tools of various types. There is a large class of relatively simple samples for which this approach works well; however, many samples are not conveniently simple and are either very time consuming to analyse by manual methods or even inaccessible to them in a reasonable time. We give examples of current problems which have generated large numbers of complex spectra which our new code has solved readily.

The algorithm is general, readily extensible to other techniques such as NRA or PIXE, and can easily accommodate stopping power or multiple/plural scattering corrections. We can also take advantage of the Bayesian structure of the formalism to calculate confidence intervals of the solution obtained. We give examples of all of these.

INTRODUCTION

Rutherford backscattering spectrometry (RBS) is a well established and powerful technique for determining the elemental composition of thin films using an energetic light ion beam [1]. Because the data analysis is very simple in principle, RBS can be used to obtain quantitative and traceable information about homogeneous films [2]. However, most interesting real films are not homogeneous. In such cases it is usually very hard to devise a computational method for obtaining depth profiles from the spectra which are sufficiently transparent to give traceable accuracy. Moreover, many spectra obtained from real samples are complex enough to preclude manual extraction of any depth profiles at all in a reasonable time.

In this paper we review the significant progress that has been made in solving the inverse RBS problem, in extending the solution to multiple spectra and multiple techniques, and in putting confidence limits on the solutions obtained. We present new results in the latter case.

The inverse RBS problem - the problem of obtaining a depth profile from a spectrum - has been shown to have a general solution based on the Simulated Annealing algorithm (SA) [3, 4]. We have used the method to analyse complex ion beam synthesised metal silicides [5], ion beam mixed iron silicides [6], implanted waveguides in SiC [7] and ion beam synthesised a-GaN [8].We have extended the method to obtain depth profiles from multiple spectra, which could come from multiple detectors or multiple analysis geometries, and used the extended code to analyse (nominally) BN films [9], ERD/RBS data from SiCx:H films [10] and ToF-ERD data [11].

The quality of solutions obtained cannot be determined from SA but can be determined using Bayesian techniques with Markov Chain Monte Carlo (MCMC) methods [12]. These techniques have already been used to evaluate the confidence limits in the ion beam synthesised a-GaN determination [13] and in high resolution RBS of multilayers [14]. Ambiguity of the solutions in ellipsometry data have also been analysed with a very closely related code [15].

SIMULATED ANNEALING

Simulated Annealing is a global optimisation algorithm designed to find all the global minima (or maxima) for any given function [16]. It is completely general in the sense that in principle it will work on any piecewise continuous function which has a global minimum.

The objective function minimised in our algorithm measures the difference between the observed data Y_{obs} and the spectrum $Y_{the}(\Theta)$ calculated from the proposed layer structure, which we refer to as the "forward model", Θ being all relevant parameters needed. The distance currently used is defined to be

$$f(\Theta,\lambda) = \chi^2 = \sum_i [Y^i_{\exp} - Y^i_{the}(\Theta)]^2 / (1-\lambda n)$$

where n is the number of layers in the proposed layer structure. The parameter λ is a control parameter which

CP475, *Applications of Accelerators in Research and Industry*,
edited by J. L. Duggan and I. L. Morgan

penalises models with large numbers of thin layers. The sum is taken over all channels of the spectrum

The algorithm used is analogous to annealing, hence is called Simulated Annealing (SA). Under the algorithm new candidate solutions are proposed and accepted according to the following probability

$$p = \begin{cases} \exp(-\Delta\chi^2 / T) & \Delta\chi^2 > 0 \\ 1 & \Delta\chi^2 < 0 \end{cases}$$

where $\Delta\chi^2$ is the change in the objective function between the current state and the candidate state. The parameter T is analogous to the temperature of the system. The algorithm starts with a high value of T which is decreased according to a cooling schedule. The algorithm has the property that as T is decreased to zero the accepted solutions will converge to the global minimum of the objective function [16].

BAYESIAN ANALYSIS

The SA algorithm gives a practical way of solving the inverse RBS problem. We turn now to the evaluation of the quality of the proposed solution. Suppose that the result of an RBS analysis of a sample gives a spectrum which we denote by $Y_{obs} = (Y_1, \ldots, Y_k)$ where k is the number of channels in the spectrum. This spectrum can be viewed as a random vector. The randomness in the measurement comes from many sources. These include the fact that the spectrum is collected in finite time, thus is noisy. Other features include detector error and modelling error. However there is an underlying true forward model which depends on a set of quantities Θ. These parameters include some over which we have experimental control, such as the initial energy, incident angle, and detector solid angle. It also contains parameters about which we would like to learn. These include the number of layers and the proportions of elements in each layer. It is convenient to split the parameter vector as $\Theta = (\Theta_s, \Theta_p)$ where Θ_p are the parameters over which we have control, and consider known, and Θ_s those parameters which are unknown and about which we wish to learn. In fact in applying the methodology proposed here we could consider all the parameters as having some variation. Thus for example we could consider the initial energy as having a mean value fixed by design but containing a random variation.

By combining the standard theory of RBS and statistical models of the error mechanisms we can connect Y_{obs} and Θ into a single statistical model, which we denote by $\Pr(Y_{obs} \mid \Theta)$, the probability of seeing the observed spectrum conditional on Θ being the true configuration of the sample. To understand the quality of the SA solution we wish to invert the above probability. This follows since we can write our uncertainty about Θ as $\Pr(\Theta \mid Y_{obs})$. This is the probability that a solution is true given the observed data. This is called the *posterior* distribution since it captures what we know about the sample after the experiment. The inversion is achieved using Bayes' theorem [17] which states

$$\Pr(\Theta \mid Y_{obs}) = \frac{\Pr(Y_{obs} \mid \Theta)\Pr(\Theta)}{\Pr(Y_{obs})} \propto \Pr(Y_{obs} \mid \Theta)\Pr(\Theta).$$

To complete this calculation we therefore need $\Pr(Y_{obs} \mid \Theta)$ the statistical model and $\Pr(\Theta)$ the *prior* probability on the parameters. This prior quantifies the information known about the parameters before the data has been observed.

For the error structure in this paper we concentrate on the most important source of error which is Poisson noise on each channel, although we can easily extend the treatment to other sources of error.

For each channel, *i*, we assume that the actual number of counts, Y_i, is given by a Poisson distribution with mean $\mu_i(\Theta)$. This mean is calculated directly from the forward RBS model. Thus we have

$$\Pr(Y_i = N \mid \Theta) = \exp(-\mu_i(\Theta))\frac{\mu_i^N}{N!}.$$

We shall also assume that the signal in each channel is independent, thus the probability of seeing a complete spectrum will be a product

$$\Pr(S_{obs} \mid \Theta) = \prod_i \exp(-\mu_i(\Theta))\frac{\mu_i^{Y_i}}{Y_i!}.$$

We now consider the prior. This is a distribution on the set of possible Θ values. Recall that the forward model assumes that the sample is made of a set of homogeneous thin layers. Let the number of layers be n say. We decompose the prior as

$$\Pr(\Theta) = \sum \Pr(\Theta \mid n)\Pr(n).$$

The simplest possibility for a prior on n is to have a uniform distribution on { 0, 1, … , Max } where Max is some preset maximum number of layers. There are alternatives though. For example a Poisson distribution on the number of layers acts in the same way as the penalisation term in the SA algorithm.

Finally define the prior on Θ conditional on a fixed n. It is possible to constrain results to ensure some continuity, see [18], but this is not necessary. The simplest prior has all proportions and depths independently uniformly distributed on an appropriate simplex. Note that in general some sensitivity analysis on the form of the prior is advisable.

The Metropolis Hastings algorithm

To implement the above results we need to be able to take a random sample from the posterior distribution. To do this we used the Metropolis-Hastings version of the general MCMC algorithm, [19]. This algorithm is an adaptation of the SA algorithm. The probability of acceptance of Θ_{i+1} as the $i+1$ candidate solution given the current solution is Θ_i is determined by the probability

$$\max\left\{\frac{\Pr(Y_{obs} \mid \Theta_{i+1})\Pr(\Theta_{i+1})q(\Theta_{i+1} \mid \Theta_i)}{\Pr(Y_{obs} \mid \Theta_i)\Pr(\Theta_i)q(\Theta_i \mid \Theta_{i+1})},1\right\},$$

where q is the probability distribution which generates possible new candidate points.

The algorithm generates a sequence of accepted solutions. The distribution of these converges to the required posterior distribution, $\Pr(\Theta \mid Y_{obs})$. We therefore calculate the required error bounds directly from the coverage of this sequence of accepted solutions.

We demonstrate the methodology in the context of a simple example. Figure 1 shows the observed data (bars) and fitted spectrum (solid line) given by the results of the SA algorithm. The data is generated for a Si_7Ge_3 Layer $33.10^{13}/\text{cm}^2$ thick, buried at $333.10^{13}/\text{cm}^2$ deep in Si. Poisson noise has been added to the simulation.

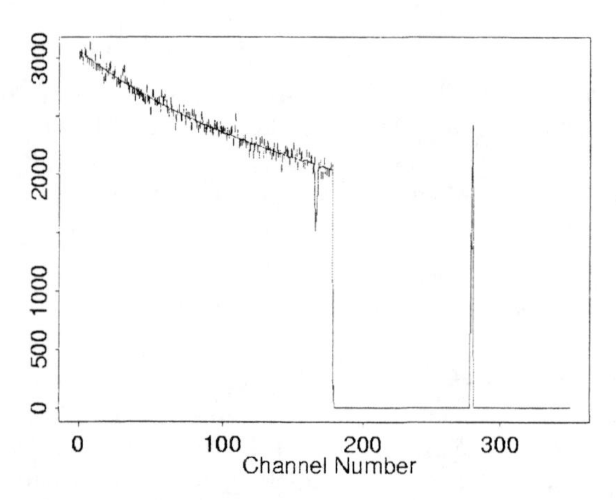

FIGURE 1 Simulated spectrum (bars) for a single Si_7Ge_3 layer in Si. The SA fit is the solid line.

Suppose we concentrate on one particular aspect of this profile, for example the proportion of Ge near the depth of the layer. Figure 2a shows the output of the MCMC analysis. It is a plot of the proportion of Ge against the accepted sample. The general theory of the algorithm shows that the distribution of the accepted values will converge to the equilibrium distribution. This can be seen in the figure. After an initial transient phase, convergence seems to have taken place after about 30000 accepted solutions.

Figure 2b shows a histogram of the accepted solutions, after the 30000 burn-in period. The general theory of MCMC shows that the distribution of this sample will be the posterior distribution of the particular proportion, given the observed data. From the sample we can construct a 95% confidence interval for the proportion as being (0.22, 0.26), shown in the figure as vertical lines.

Various diagnostic tests should be run on the output of any MCMC algorithm. These tests fall into two main types. The first are convergence diagnostics which check if the solution has converged to its equilibrium distribution. The second checks that the algorithm 'mixes' properly, see below.

To test for convergence, plots of moving averages of sample moments against the number of accepted states were used. These plots converge to a constant as equilibrium is reached. Also parallel runs with different initial conditions are used as convergence diagnostics.

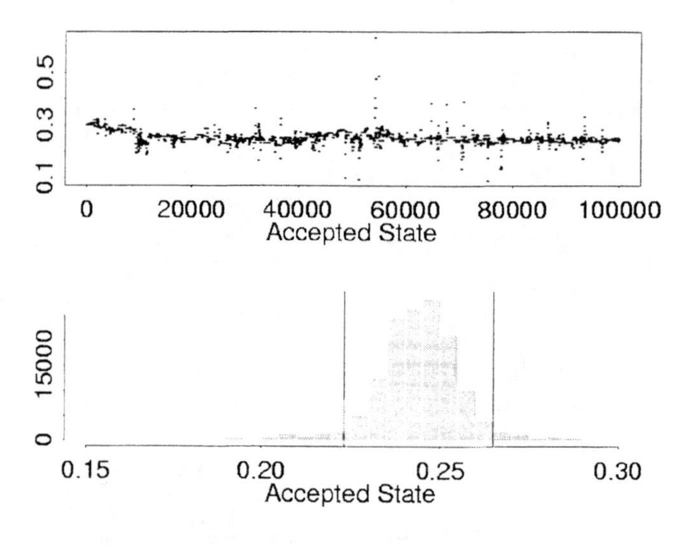

FIGURE 2. (a) Top panel, Ge content as a proportion at a depth of $3.5\times10^{15}/cm^2$ for each accepted solution. (b) Lower panel, Histogram of accepted solutions for second part of the sequence in (a). The 95% confidence interval is shown.

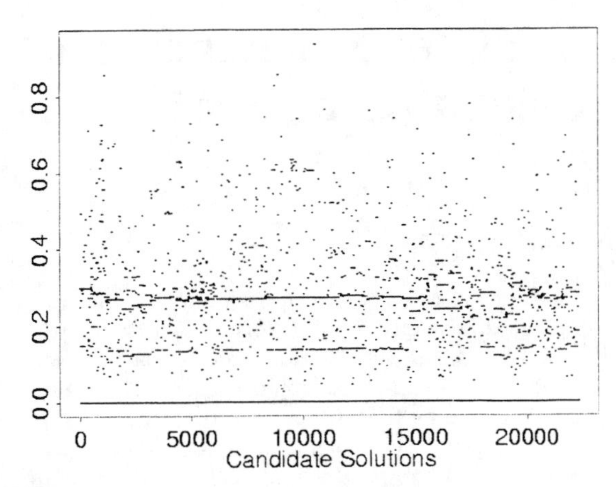

FIGURE 3. Ge content (as proportion) for each candidate solution. Approximately 50% of these solutions were accepted.

Mixing means that the sampling process explores all possible solutions in an efficient manner. Figure 3 shows a portion of the sequence of candidate solutions, including the rejected ones. It can be seen that the attempted solutions reach all parts of the possible space, [0,1]. Thus, at least for this aspect of the solution, all parts of the configuration space have been explored. The efficiency of the mixing is shown by the length of time in which the sequence stays in the same state. Figure 3 indicates that there is a high level of efficiency in our implementation. In general there is a trade-off between making sure that all points in the possible solution space are visited and achieving good levels of efficiency. Good design of the candidate generation function, q, is important in the implementation of the algorithm.

SUMMARY AND CONCLUSIONS

Simulated Annealing recovers the original solution from a simulated RBS spectrum: this is equivalent to solving the inverse RBS problem. Very difficult thin films are readily solved, such as multilayer mixed metal silicides, or ion beam synthesised waveguides in SiC. Samples such as a-SiN:H requiring multiple RBS/ERD spectra are also readily solved self-consistently. The algorithm is very general and ellipsometry data can be included in the self-consistent fit. The computation time using a modern PC is of the order of magnitude of the measurement time, which opens the possibility of on-line automatic data analysis.

We have then gone on to explore the solution space by Monte Carlo Markov Chain computations. We can demonstrate that the whole solution space is explored, and we can also show that our algorithm is remarkably efficient although mathematical details are beyond the scope of this paper. The computation time is about one order of magnitude larger than the measurement time. Confidence intervals for the best solution can readily be extracted from the MCMC calculations, but the structure of the solution space is also explored so that ambiguity (that is, multiple best solutions) will be disclosed if it exists. Thus, the analyst at last has a fully quantitative way to estimate the reliability of the depth profiles obtained from RBS data.

REFERENCES

1. J.R.Tesmer, M.Nastasi, (Eds) Handbook of modern ion beam analysis (Materials Research Society, Pittsburgh, 1995)
2. C.Jeynes, Z.H.Jafri, R.P.Webb, M.J.Ashwin, A.C.Kimber Surf. Interface Anal. 25 (1997) 254-260
3. N.P.Barradas, C.Jeynes, R.P.Webb, Appl.Phys.Lett.71 (1997) 291-3. Also see "IBA DataFurnace v2.1" (April,1998) at http://www.ee.surrey.ac.uk/Research/SCRIBA/ndf/
4. N.P.Barradas, P.K.Marriott, C.Jeynes, R.P.Web, Nucl. Instr. and Methods B136-138 (1998) 1157-1162
5. N.P.Barradas, C.Jeynes, M.A.Harry, Nucl. Instr. and Methods B136-138 (1998) 1163-1167
6. N.P.Barradas, C.Jeynes, K.P.Homewood, B.J.Sealy, M.Milosavljevic, Nucl. Instr. and Methods B139 (1998) 235-238
7. N.P.Barradas, C.Jeynes, S.M.Jackson, Nucl. Instr. and Methods B136-138 (1998) 1168-1171
8. S.A.Almeida, S.R.P.Silva, B.J.Sealy, J.F.Watts, Phil.Mag. 78 (1998) 319 - 324
9. N.P. Barradas, C. Jeynes, Y. Kusano, J.E. Evetts, I.M. Hutchings, These Proceedings.
10. S.J.Toal, H.S.Reehal, S.J.Webb, N.P.Barradas, C.Jeynes, accepted by Thin Solid Films
11. N. P. Barradas, C. Jeynes, R.P. Webb, U. Kreissig, and R. Grötzschel, accepted by Nucl. Instr. and Methods B.
12. N.P.Barradas, C.Jeynes, M.Jenkin, P.K.Marriott, accepted by Thin Solid Films
13. N.P.Barradas, S.A.Almeida, C.Jeynes, A.P.Knights, S.R.P.Silva, B.J.Sealy, accepted by NIMB
14. N.P. Barradas, A.P. Knights, C. Jeynes, O.A. Mironov, T.Grasby, and E.H.C. Parker. Submitted to Phys. Rev. B
15. N.P. Barradas, J.L. Keddie and R. Sackin submitted to Phys. Rev. E.
16. E. Aarts, T. Korst, Simulated Annealing and Boltzman Machines. Wiley, Chichester, 1989.
17. M.J. Schevish, Theory of Statistics, Spinger Berlin 1995.
18. R. Fisher, M. Meyer, W. von der Linden, V Dose. Phys Rev E 55 (1997)
19. W.R. Gilks, S. Richardson, (Eds), Markov Chain Monte Carlo , Chapman and Hall, London 1996.

General Purpose Computational Tools for Simulation and Analysis of Medium-Energy Backscattering Spectra

Robert A. Weller*

Vanderbilt University, Nashville, Tennessee 37235

This paper describes a suite of computational tools for general-purpose ion-solid calculations, which has been implemented in the platform-independent computational environment *Mathematica*®. Although originally developed for medium energy work (beam energies < 300 keV), they are suitable for general, classical, non-relativistic calculations. Routines are available for stopping power, Rutherford and Lenz-Jensen (screened) cross sections, sputtering yields, small-angle multiple scattering, and backscattering-spectrum simulation and analysis. Also included are a full range of supporting functions, as well as easily accessible atomic mass and other data on all the stable isotopes in the periodic table. The functions use common calling protocols, recognize elements and isotopes by symbolic names and, wherever possible, return symbolic results for symbolic inputs, thereby facilitating further computation. A new paradigm for the representation of backscattering spectra is introduced.

INTRODUCTION

In spite of the great advances in the processing power of computers over the last few years most practitioners in the ion-solid interactions community still find it necessary to rely on a collection of books, tables, special-purpose commercial or freeware programs, and locally written software, when doing calculations. Although excellent and very widely used programs such as RUMP[1] exist for special purpose calculations, their output is often cumbersome to combine and use in other environments.[2]

This paper describes a consistent set of computational tools for ion-solid work that has been constructed in order to facilitate prototyping and one-of-a-kind computations. Because it has been implemented in *Mathematica*,[3] this tool set is platform independent and offers the user access to extremely powerful general-purpose mathematical functions both in support of these routines and for combining their results in novel and interesting ways. Of particular note is the availability of symbolic computation. This innovation makes it possible to differentiate, integrate and otherwise manipulate quantities for which closed-form expressions exist. Wherever possible, this tool set returns these closed form symbolic expressions.

The philosophy of the calculations generally follows that of the underlying program *Mathematica* in which computational accuracy, embodied in the selection of algorithms, takes precedence over execution speed. As a result, some functions in the tool set are slow when compared with implementations in a compiled language such as C or Fortran. In a case where greater speed is needed, it is often possible to take advantage of *Mathematica's* own tools, such as those for function approximation, to achieve it. Overall, it has been our experience through several years of development and use that the efficiency of getting final answers to complex problems using this tool set is higher than with any other computational strategy that we have employed.

FUNCTION DESCRIPTIONS

Almost all of the functions defined in the tool set, or the algorithms from which they derive, have been previously documented in the literature. In general, every tool is a function that returns an object as its value. In the case of a simple function like stopping power, the object is an approximate real number. In other cases, several different return values are possible. For example, the backscattering kinematic factor KScatter will return an approximate real number for a backscattering collision when explicit colliding species and an angle are given. However, if these arguments are symbolic, the return value is itself symbolic. In the more unusual forward-scattering case where the kinematic factor is double valued,[4] the return value is a list of approximate real numbers corresponding to the two possibilities. When KScatter is used as a component of other calculations, the user must anticipate the possible inputs and be prepared to deal with the corresponding outputs.

When an element or chemical compound is appropriate as an argument it may be entered symbolically. For example, the element silicon can be represented as "Silicon," without the quotation marks, of course. Specific isotopes are represented as, e.g., Silicon[28]. All the naturally occurring isotopes of the elements in the periodic table are cataloged along with their masses and abundances[5] and several functions are provided for extracting this information. For example, the function IsotopeExpand when applied to a chemical formula returns that formula expanded to include properly weighted isotopes.

Chemical formulae are represented using the *Mathematica* concept of lists. Thus, "SiO_2" becomes "{{Sili-

* Electronic mail: Robert.A.Weller@Vanderbilt.Edu

CP475, *Applications of Accelerators in Research and Industry*,
edited by J. L. Duggan and I. L. Morgan

con,1},{Oxygen,2}}" again without quotations. Multi-layer planar targets are also represented by lists. In this case, each element of the list is itself a list describing a single layer. The list describing a layer has three elements, the chemical formula, the molecular density in molecules/cm^3 and the thickness in cm. The formula for a pure element is, e.g. for elemental silicon, "{{Silicon,1}}." Of course, with fully expanded isotopes, the form for a pure element is identical to that for a compound. A function MolecularDensity is provided for obtaining the number of molecules/cm^3 given a chemical formula and a mass density in g/cm^3.

In order to deal with the common case of a trace element with sub-monolayer coverage, the concept of a zero-thickness layer has been introduced. When the thickness of a layer is given as 0 cm, the density is assumed to be areal density in units of molecules/cm^2. No *a priori* restrictions are placed on the complexity of multi-layer targets or the chemical compounds of which they are composed. Of course, very complex targets, especially with expanded isotopes, may take a long time to process depending upon the operation.

SPECTRUM SIMULATION

Chu, Mayer and Nicolet described the basic physics of backscattering-spectrum simulation,[6] with additional considerations discussed by Doolittle.[1] Doolittle's implementation of the algorithms in the computer program called RUMP is one of the most widely used software tools in the ion-solid community. Our implementation shares many similarities with RUMP but differs in other important details. The most significant of these is the way in which we have chosen to represent spectra.

So far as we know, all previous backscattering simulation implementations, including RUMP, have viewed the backscattering spectrum and the representation of the spectrum as a list of discrete values at equally spaced energies as identical. In other words, the spectrum and the multi-channel analyzer picture of the spectrum have been viewed as synonymous. In our implementation, computation of the spectrum and evaluation of the spectrum are distinct operations.

The computation of a backscattering spectrum is handled by the function SimulateRBS (Fig.1). SimulateRBS requires as arguments (in order) a projectile, e.g. Helium[4], a multi-layer planar target defined as described above, a beam energy in eV, and directions of the beam and the target outward normal specified as {polar, azimuthal} angles in degrees expressed in a consistent spherical coordinate system. SimulateRBS returns a complex object called a ScatteringSpectrum, which is a deterministic rule for obtaining the numerical value of the backscattering spectrum given a numerical value of energy. In other words, SimulateRBS returns a function. You can subsequently evaluate this function at equally spaced channels and obtain the conventional multichannel-analyzer representation of the spectrum. More likely, however, you will use *Mathematica's* Plot function to adaptively select points to provide a high quality graphical representation of the full spectrum or any portion thereof with minimal computation.

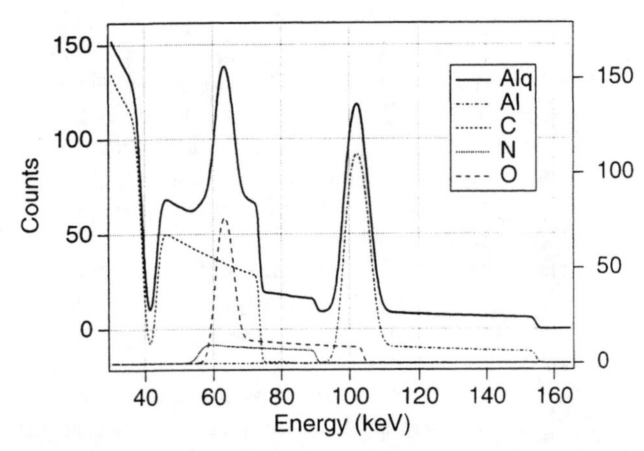

Figure 1. 270 keV He^+ backscattering spectrum (normal incidence, 150° scattering angle) of 120 nm of the organic light emitting diode constituent 8-hydroxyquinoline aluminum ($Al_1C_{27}H_{18}N_3O_3$) on 11 nm Al_2O_3 on a C substrate. Individual elemental constituents are shown as dotted lines offset downward for clarity.

Like RUMP, SimulateRBS produces a "brick," for every distinct constituent of every layer of the target. At any given energy, the full spectrum is the sum of the contribution of all the bricks. However, following the general philosophy of *Mathematica* we do not make numerical approximations. Whenever a value for stopping power or cross section is needed, a call to the appropriate defining function is made. Stopping powers use the procedures originally described of Ziegler, et al.[7] and return numerical values identical to those tabulated by Rauhala,[8] while straggling is computed using the method described by Yang, et al.[9] Lenz-Jensen cross sections are used and computed by a variant on the method described by Mendenhall and Weller.[10] All integrations are handled by *Mathematica's* built in numerical integrator.

By not optimizing the simulation algorithms to fit the capability of current computing, we are, of course, paying a hefty price in execution speed. However, the impact of the penalty continues to drop with time. When the first version of SimulateRBS was written in 1992, it was limited in practicality to simulating at most two or three layer targets without isotope expansion. Presently, it can be used in a closed-loop, non-linear, least-squares fitting procedure for similar targets of typical semiconductor complexity will all isotopes included explicitly, and can return best-fit values in a reasonable time. This functionality has derived from the generality of the design, in spite of the fact that it was never originally intended for closed loop or automated use. Another collateral benefit of this design is that if a user wishes to refine functions such as those for cross sections or stopping power the changes will be automatically reflected in other functions, such as simulation, which use the values.

SimulateRBS represents a brick in a manner chosen by the routine itself. For bricks that are sufficiently narrow, a quadratic polynomial in energy E is used. For thicker bricks, additional basis functions, E^{-1} and E^{-2} are added for a total of five parameters. (This is to be contrasted with RUMP, which uses polynomials.) The use of the inverse powers makes it possible to describe very wide regions of the spectrum without having to manually divide the target into narrow layers. Our experience with medium energy backscattering (at 270 keV primarily) has been that it is never necessary to manually subdivide even the substrate. The choice of which basis to use to represent a brick is made using the concept of singular-value decomposition.[11] If using the full basis set leads to singular values in the solution for the coefficients, the polynomial basis is used.

The most vexing detail of the algorithm used by SimulateRBS was how to deal with the situation that arises if the target is thicker than the range of the projectile, or at least thick enough that some backscattered particles are not energetic enough to emerge from the target. A related issue is how to truncate bricks at the low end of the energy spectrum. This is solved by creating two callable functions in real time. The first gives the energy of the incident particle as a function of depth in the target. The second gives the energy that a backscattered particle must have in order to emerge from a specified depth with a given energy, the minimum energy in the spectrum, taken by default to be 10 keV. These functions make it possible to analyze each brick individually and to truncate it as necessary so that the minimum energy is the preset value. As a result, no special consideration must be given to the thickness of the substrate, although it is desirable from the standpoint of efficiency not to make it unnecessarily thick.

Evaluation of a backscattering spectrum at a specific energy is carried out by rules associated with the term ScatteringSpectrum. These rules assemble the functions describing the individual bricks, evaluate, and sum them. At this stage, the effects of straggling and instrumental resolution are also included. Straggling is intrinsic and is computed by SimulateRBS and included in the ScatteringSpectrum object. Instrumental resolution, which is usually a function of energy for medium energy backscattering spectra, is incorporated when a spectrum is evaluated. Thus, one can explore the effects of various detection strategies (time-of-flight, surface-barrier detectors, etc.) without the time-consuming process of recomputing the intrinsic spectrum itself.

Straggling is handled by convolution assuming that the kernel is a Gaussian with an energy-dependent width. It is assumed that the variance increases linearly from the high-energy to low-energy side of the brick. This energy dependence of the convolution kernel was the most significant mathematical problem encountered in this work. After examining very many approaches to this problem, using built-in *Mathematica* functions, various analytic expansions, and numerical approximations, we concluded that simple Gaussian quadrature used in conjunction with a range-reduction strategy offered the best compromise between generality and efficiency. By dynamically adjusting the integration range for each brick as each point is evaluated, it has been possible to use a fixed-order Gaussian quadrature (order 42) to achieve numerical accuracy exceeding the physical accuracy of the model in a more-or-less definitive way. Surprisingly, this simple procedure is faster than an earlier implementation that used a simple but robust recursive, adaptive quadrature.

The representation of a spectrum by a function instead of by an array of numbers makes it possible to implement functional operators. A functional operator is a mathematical operator that operates on functions to produce other functions. We have, thus far, implemented four such operators, for addition of spectra, multiplication of spectra by a numerical constant, extraction from spectra of individual contributions by location in the target, by element (Fig. 1), or by isotope, and a functional operator which mimics the effects of channeling in the substrate of a target. The first two of these operations occur in response to the use of the ordinary numerical operators for addition and multiplication when the operations involve ScatteringSpectrum objects. Functional operators called ExtractLayer and ExtractElement operate on ScatteringSpectrum objects to produce new ScatteringSpectrum objects that meet the specified criteria. Channeling is approximated by a functional operator called ChannelSubstrate (Fig. 2).

The ordinary arithmetic operators can be used to implement effects that are not easily simulated in an environment that is inherently structured on the concept of uniform layers. For example, in ref. (12) these operators are used to simulate the effect of a non-uniform carbon layer by constructing a Gaussian convolution involving fifty

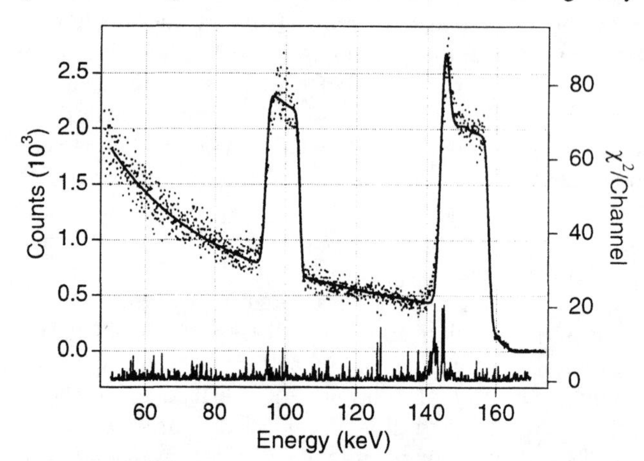

Figure 2. A non-linear least squares fit (solid line) to 270 keV He^+ backscattering data (dots) for an SiO_2 thin film on Si using the function ChannelSubstrate to approximate the effects of channeling in the <110> direction. The film was found to be 9.7 nm thick and to have a surface peak about 50% larger than the theoretical expectation. The channeling χ_{min} was 9%. The overall χ^2 for the fit was about 1.3 per degree of freedom. The curve at the bottom of the figure shows the χ^2 per channel and indicates that the model fits least well in the vicinity of the Si surface peak.

properly weighted individual simulated spectra in which the carbon layer takes on various thicknesses.

Simulating channeled spectra using correct physics is very complex because it involves both a detailed knowledge of the target crystallinity and the use of more advanced potentials.[13] Nevertheless, it is commonplace in the analysis of backscattering data to encounter spectra in which channeling has occurred. Frequently, the channeling is intentional in order to reduce the substrate background beneath spectral features of interest. The backscattering yield from a crystalline target is often described with respect to the yield of a random target of the same substance through a quantity called χ which is the ratio of counts in the channeled spectrum to the number in the random spectrum at the same energy.[13] For well-channeled ions the minimum χ called χ_{min} is often no larger than 5%.

The functional operator ChannelSubstrate approximates the effects of channeling using three user-supplied χ values corresponding to the counts ratio at the left, center and right edges of the bricks representing the substrate constituents. The resulting ScatteringSpectrum is an approximation to the actually observed spectrum that agrees at three points. Because the substrate contribution is ordinarily quite smoothly varying, the overall approximation is quite good. The χ parameters can, of course, be taken to be variables in a fitting operation using SpectrumFit and in this way replace *ad hoc* background subtraction for fits such as SiO_2 on Si. In a case such as this, the physical substrate is best described as two layers one, typically of zero thickness and areal density chosen by theoretical considerations,[13] and the true bulk described in the usual way. By doing this, the simulation accurately captures the surface peak that is observed experimentally (Fig. 2).

The function SpectrumFit is included in the tool kit primarily for experimental prototyping. It uses Marquardt's method[14] with numerical differentiation to perform non-linear, least-squares fits. Any parameter that is adjustable in the simulation or evaluation of a spectrum can, in principle, be fit using this routine. Thus, one may find best-fit values to layer widths, densities, or compositions, or to detector resolution, channeling χ values, etc. Note, however, that as of this writing, such fits can be slow, and may not practical for repetitive use when dedicated tools such as RUMP offer the same functionality.

CONCLUSION

A set of general-purpose computational tools for the simulation and analysis of medium energy backscattering spectra has been implemented in *Mathematica*. Implementations of published algorithms are included for stopping power,[7,8] straggling,[9] cross sections,[10] multiple scattering,[15] and sputtering yield,[16] as well as a comprehensive database of properties of elements.[5,17] The complete tool set is available on-line[18] and is accompanied by a descriptive document[19] which gives examples of the use of each tool.

ACKNOWLEDGMENTS

Thanks to Kyle McDonald for the data in Fig. 2 and to Kyle McDonald and Martha R. Weller for critical comments on this manuscript. Portions of this work were supported by the U. S. Army Research Office under contract number DAAH-04-95-1-0565.

REFERENCES

1. Doolittle, L. R., *Nucl. Instr. and Meth.* **B9**, 344-251, 1985. Doolittle, L. R., *Nucl. Instr. and Meth.*, **B15**, 227-231, 1986.
2. See Leavitt, J. A., McIntyre, L. C., and Weller, M. R., *Handbook of Modern Ion Beam Analysis,* Pittsburgh: Materials Research Society, 1995, ch. 4, pp. 37-81, and references therein, for a comprehensive compilation of backscattering simulation programs.
3. Wolfram, S., *The Mathematica Book, 3rd ed,* Cambridge: Cambridge University Press, 1996. *Mathematica®*, the programming environment for technical computing, is a product of Wolfram Research, Inc. http://www.wri.com.
4. Weller, R. A., *Handbook of Modern Ion Beam Analysis,* Pittsburgh: Materials Research Society, 1995, Appendix 4, pp. 411-416.
5. Audi, G. and Wapstra, A.H., *Nuclear Physics,* **A595,** 409-480, 1995. Data obtained from "The 1995 update to the atomic mass evaluation," National Nuclear Data Center, Brookhaven National Laboratory, http://www.nndc.bnl.gov/nndcscr/masses/RCT1_RMD.MAS95.
6. Chu, W.-K., Mayer, J. W., and Nicolet, M.-A., *Backscattering Spectrometry,* New York: Academic Press, 1978, ch. 4, pp. 89-122.
7. Ziegler, J. F., Biersack, J. P. and Littmark, U., *The Stopping and Ranges of Ions in Matter,* New York: Pergamon, 1985, pp. 218-222.
8. Rauhala, E., *Handbook of Modern Ion Beam Analysis,* Pittsburgh: Materials Research Society, 1995, ch. 2, pp. 3-19 and pp. 385-410.
9. Yang, Q., O'Connor, D. J., and Wang, Z., *Nucl. Instr. and Meth.* **B61**, 149-155, 1991.
10. Mendenhall, M. H., and Weller, R. A., *Nucl. Instr. and Meth.* **B58**, 11-17, 1991.
11. Press, W. H., Teukolksy, S. A., Vetterling, W. T., and Flannery, B. P., *Numerical Recipes in C,* Cambridge: Cambridge University Press, 1992, ch. 2, pp. 59-65.
12. McDonald, K. and Weller, R. A., *Nucl. Instr. and Meth.* **B**, submitted, 1998.
13. Feldman, L. C., Mayer, J. W., and Picraux, S. T., *Materials Analysis by Ion Channeling,* New York: Academic Press, 1982, pp. 20-26, and pp. 160-166.
14. Press, W. H., Teukolksy, S. A., Vetterling, W. T., and Flannery, B. P., *Numerical Recipes in C,* Cambridge: Cambridge University Press, 1992, ch. 15, pp. 681-688.
15. Mendenhall, M. H., and Weller, R. A., *Nucl. Instr. and Meth.* **B93**, 5-10, 1994.
16. Yamamura, Y., and Tawara, H., *Atomic Data and Nuclear Data Tables* **62**, 149-253, 1996.
17. Emsley, J., *The Elements,* New York: Oxford University Press, 1989.
18. http://particlesolid.vuse.vanderbilt.edu
19. Weller, R. A., *Particle-Solid Tools,* http://particlesolid.vuse.vanderbilt.edu

CHARACTERIZATION OF THIN FILM CCD FILTERS ON BOARD THE GERMAN ASTRONOMY SATELLITE ABRIXAS BY SOFT X-RAY TRANSMISSION MEASUREMENTS

K.-H. Stephan[a], H. Bräuninger[a], F. Haberl[a], P. Predehl[a]
H.J. Maier[b], J. Friedrich[c], D. Schmitz[d], F. Scholze[d], G. Ulm[d]

[a]*Max-Planck-Institut für extraterrestrische Physik, 85748 Garching, Germany.*
[b]*Sektion Physik, Ludwig-Maximilians-Universität München, 85748 Garching, Germany.*
[c]*Institut für Angewandte Physik, Heinrich Heine Universität Düsseldorf, 40255 Düsseldorf, Germany.*
[d]*Physikalisch-Technische Bundesanstalt, 10587 Berlin, Germany.*

We have developed optical filters for the German X-ray astronomy satellite ABRIXAS (A BRoadband Imaging X-ray All Sky Survey)(1). Specific CCD's (2) will be used as detectors in the focal plane on board the observatory. Since these detectors are sensitive from the X-ray to the near infrared spectral range, X-ray observations require optical filters, which combine high transmittance for photon energies in the soft X-ray region and a high absorptance for ultraviolet and visible radiation. With respect to the mission goal in orbit a spectral transmission function is required attenuating radiation below photon energies of 10 eV by more than 7 orders of magnitude and transmitting soft X-ray photon energies above 1000 eV by more than 90 percent. This was realized by a 0.80 μm thick polypropylene foil, which is coated with approximately 60 nm aluminum on both sides. The filters have an effective diameter of 73 mm without any support structure. Environmental tests have been performed and proved the filters to be resistant against sound pressure and vibrational load stresses during the launch of the spacecraft. Synchroton radiation was used to characterize the properties of the filters in the soft X-ray photon energy range 60 eV < E < 2000 eV. We describe the measurements determining the spectral transmittance function in the center of the filters, and the transmission topography at discrete photon energies across the effective area, and present the resulting performance data.

INTRODUCTION

ABRIXAS is a small satellite project in the German national space program managed by the Deutsches Raumfahrtzentrum (DLR). The scientific institutions involved are the Max-Planck-Institut für Extraterrestrische Physik, the Astrophysikalisches Institut Potsdam and the Institut für Astronomie und Astrophysik Tübingen. ABRIXAS will scan the whole sky in an energy band between 0.5 keV and 10 keV starting in spring 1999. The imaging system consists of seven identical Wolter I telescopes with a focal length of 1.6 m, each consisting of 27 nested mirror shells. The focal planes of the telescopes are arranged side by side on a 60 mm x 60 mm large pn CCD camera of the type used for XMM (3). Four identical optical filters are mounted on a filter wheel in front of the focal plane camera. Figure 1 presents the schematic drawing of the optical system of the observatory.

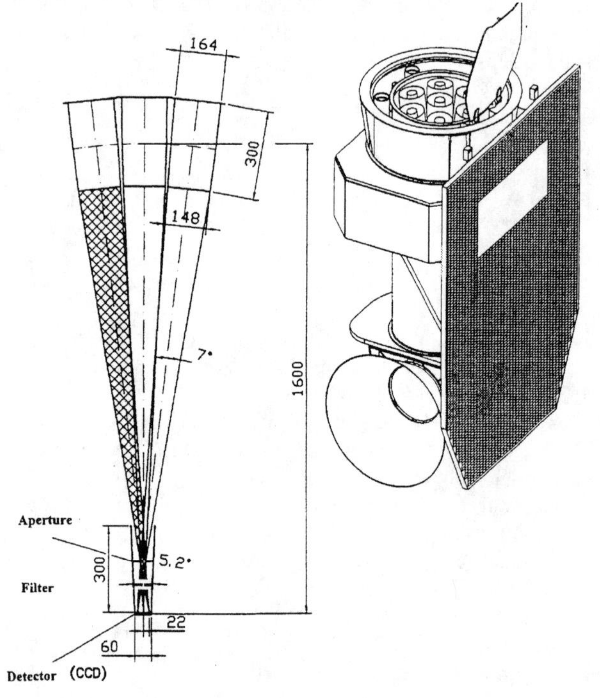

FIGURE 1. Optical system of the ABRIXAS observatory.

THE FILTER

We have calculated some realistic combinations of filter efficiency versus star type and astronomical hydrogen

TABLE 1. Design for the filter layer composition.

Material	Mass density in μg/cm^2
Aluminum	16
Polypropylene	73
Aluminum	16

CP475, *Applications of Accelerators in Research and Industry*, edited by J. L. Duggan and I. L. Morgan

column density values. As a consequence the filter was designed to suppress visible and ultraviolet radiation by more than 7 orders of magnitude. The corresponding spectral transmittance function was estimated from optical data given in the literature (4). The design parameters are given in table 1.

Polypropylene, which has the stoichiometrical composition (C_3H_6), is most suitable as filter carrier foil because of its low absorption coefficient for soft X-rays combined with an excellent mechanical strength. The filters were manufactured cooperatively by the Max-Planck-Institut für extraterrestrische Physik (MPE) and the Technology Group of the Ludwig-Maximilians-Universität (TL-LMU) at Garching. The polypropylene carrier foils of approximately 0.8 μm thickness were fabricated by using a stretching facility (5) at the MPE, vacuum coating was performed at the TL-LMU (6). A low oxygen content of 5-10 atomic % is expected for the Al-layers, resulting from the residual gas present in the vacuum chamber during the Al deposition. This promisses good opacity in the visible range. Finally the filter foils were glued on circular frames having an effective diameter of 73 mm by a space-qualified glue (trademark Solitan).

OPTICAL TRANSMITTANCES

Energy dispersive measurements

The test-facilities given in table 2 were used to measure the spectral transmittance in the soft X-ray region.

TABLE 2. Test facilities and used photon energy range.

Energy Range	Facility	Institute
60 eV to 1900 eV	SX-700	PTB/BESSY
60 eV to 1500 eV	BUMBLE BEE	HASYLAB/DESY

The spectral transmittance function was measured in the center of the filters with a reflectometer in the Physikalisch-Technische Bundesanstalt (PTB) radiometry laboratory at the electron storage ring BESSY I (7). The measurements were performed with synchroton radiation monochromatized by an SX-700 (8) monochromator. Behind the monochromator the radiation is refocused by a toroidal mirror. The measured transmittance is shown in figure 2. The absorption edges of Al at 72.5 eV (L_{III}) and 1559 eV (K) are clearly visible. Around the C-K edge at 248.2 eV no data points are shown because the photon flux was not sufficient due to carbon contamination of the optical elements of the beamline. The contribution of higher order radiation to the photon flux is suppressed effectively in the PTB beamline by operating the monochromator in a non-standard mode (8) and by using an appropriate set of filters. Around the Al-K edge which is used to determine

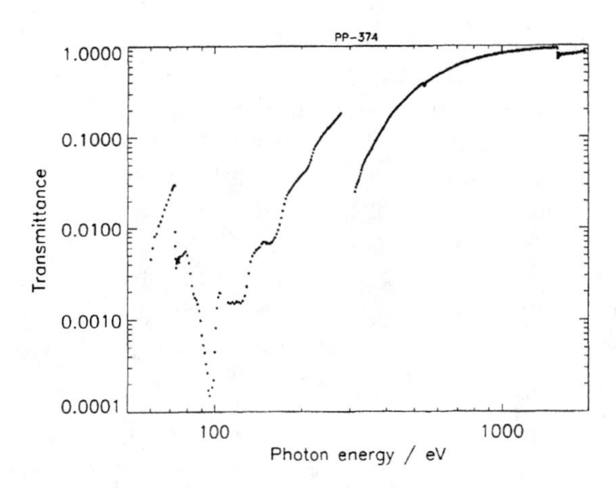

FIGURE 2. Measured spectral transmittance of a filter.

the surface mass density of Al, the uncertainty in the transmittance is ±1%.

The structure between 529 eV and 540 eV is related to the O-K edge, caused by some oxygen content present in the filter. From the observed transmittance ratios at the Al-K jump we arrive at typical mass densities of the total aluminum deposit on the filters of 35 $\mu g/cm^2$ using the law of the attenuation of electromagnetic radiation by matter. As a worst case estimation the oxygen content bound in the layers due to the deposition procedure at a pressure of 10^{-6} mbar was found to be less than 10 atomic percent.

Local transmission topography

The measurements to determine the transmission topography across the filters were carried out at the synchrotron radiation laboratory HASYLAB at the reflectometer described in (9). Computer controlled rotation and translation allow to scan the filters in 2 dimensions with a spatial resolution of less than 1mm. Schottky type diodes (Hamamatsu G1127) served as detectors. Their suitability has been investigated and is reported elsewhere (10,11). Since the sample can be removed from the beam, the normalization was done just by detecting the direct beam. The corrections for changes of the incoming photon flux are made by monitoring the total electron yield from the last focusing mirror of the beamline. The monochromatized radiation in the soft X-ray region 30 eV to 1000 eV is supplied by the monochromator BUMBLE BEE, whose principles and characteristic data are described in detail elsewhere (12,13). Figure 3 gives an example for a scanned transmission profile at 91 eV.

Profile of the local mass density distribution

The topographic distributions of the mass densities corresponding to the Al layers and the carbon content in the polypropylene carrier were calculated from the local transmittance profiles below and above the corresponding absorption edges. Figure 4 shows the mass density distribution of the aluminum deposit across the filter derived from the transmittances below and above the jump edge at 70 eV and 91 eV, respectively. The average mass density of the total aluminum layer was found to be 35 μg/cm^2 ±5 %. According to our experience the variations of the mass densities are caused by the differing sticking coefficient across the foil. The instrumental settings of the aluminum evaporation condensation process are highly reproducible.

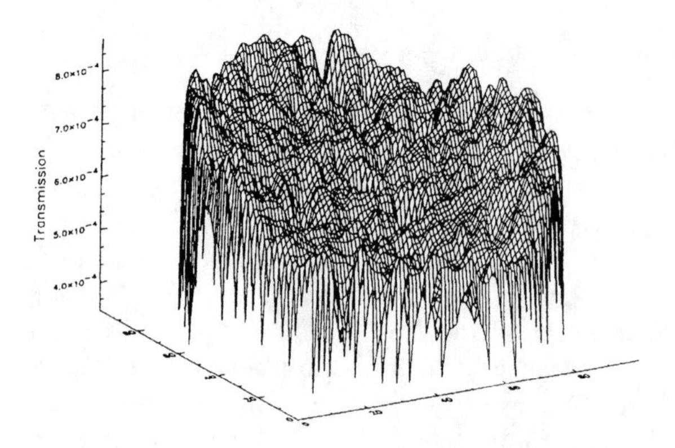

FIGURE 3. Measured topographic transmittance distribution across the filter above the Al-K edge at 91 eV.

Figure 5 shows the the mass density distribution of the carbon content in the polypropylene foil across the filter derived from the transmittances below and above the jump edge at 284 eV and 292 eV, respectively.

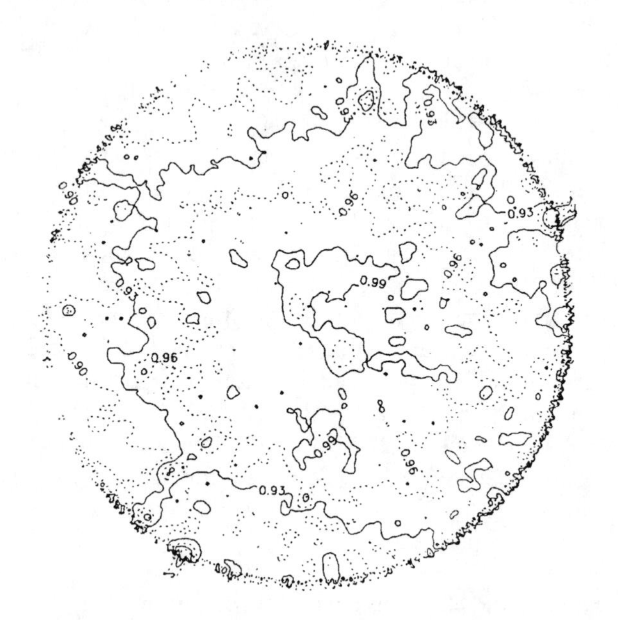

Figure 4. Contour plot of the mass density distribution of the aluminum deposit across the filter. Normalized to center.

Moreover, we measured the topographic transmittance across the filters at the photon energy of 500 eV, which is the lower limit of the photon energy range of interest during the mission of the observatory in orbit. The result is shown in the figure 6. We derive from it that the variations of the transmittance are less than ±10 %, which meets the requirements.

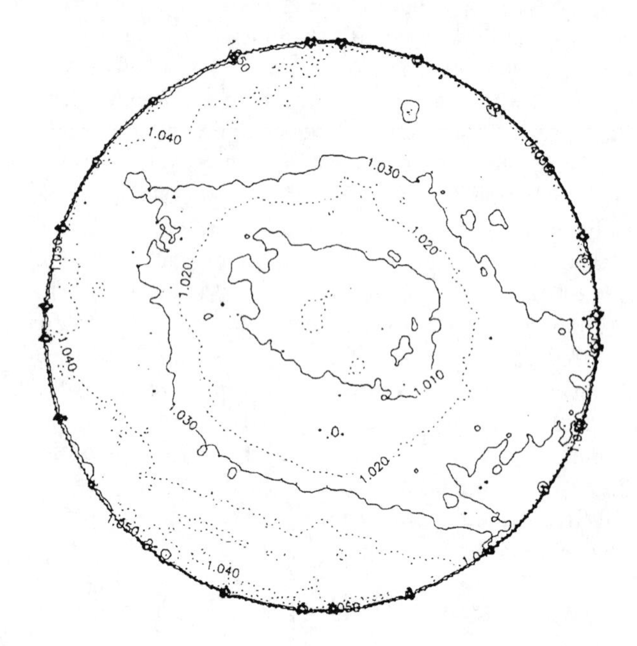

Figure 5. Contour plot of the mass density distribution of the carbon content in the polypropylene carrier across the filter. Normalized to center.

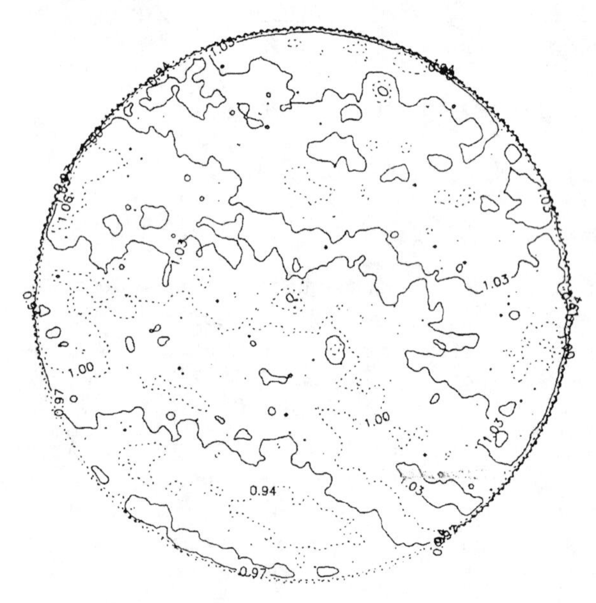

Figure 6. Contour plot of the local transmittance distribution across the filter at a photon energy of 500 eV. Normalized to center.

AGEING EFFECTS

Long term deteriorations of the filter opacity to visible radiation may occur due to the oxidation of the Al in ambient atmosphere. However, we do not expect the filters to significantly increase the transmittance during storage and mission time. Assuming the formation of an Al_2O_3-layer of 3 $\mu g/cm^2$ on top of an Al film of 30 $\mu g/cm^2$, the oxidation effect causes an increase of T_{vis} from 10^{-8} to $5*10^{-8}$. Al_2O_3 is a "dense" oxide, which is characterized by the fact, that a thin layer prevents further oxidation of the lower lying Al film. This is in accordance with our experience with self-supporting Al foils produced in the TL-LMU, which did not show significant oxidation during storage in ambient atmosphere over several years. We plan to make an oxygen depth profiling by soft x-ray direct reflection measurements at HASYLAB at time intervals of half a year and will control the opacity in the visible range.

SUMMARY AND CONCLUSIONS

We have manufactured and tested a total of 15 filters and have provided four items to the ABRIXAS flight model PN CCD-camera. Simulations performed on test foils showed that the filters are stable against the environmental stresses during the launch of the spacecraft. The transmittance of the selected flight model filters in the visible photon range varies in the limits between $1*10^{-8}$ and $5*10^{-10}$. The method to measure the global suppression of visible light by the filters is described in (14). The detection limit of the transmittance obtained with this method is found to be as low as 10^{-10}. The selected flight model filters met the required transmittances in the soft X-ray spectral range. We found that one order of magnitude of the transmittance in the visible causes an increase of transmittance at 500 eV of only approximately 0.8 %. We will control the ageing effects of the stand-by filters on a longterm scale by measuring oxydation criteria and the opacity in the visible photon energy range.

ACKNOWLEDGEMENT

We wish to express our thanks to Marieluise Hirschinger and Dagmar Frischke, who carefully prepared the polypropylene carrier foils, performed the mechanical strength qualification tests , made the aluminum vacuum evaporation condensation and determined the transmittance in the visible photon energy range.

REFERENCES

1. J. Trümper, G. Hasinger, and R. Staubert, "ABRIXAS-ABRoad-band Imaging X-ray All-sky Survey", X-Ray Surveys Workshop, Potsdam, June 1997, Astron. Nachrichten 319, 113-116(1998).
2. L. Strüder et al., "The MPI/AIT X-ray-imager-high-speed-pn CCDs for X-ray detection", Nucl. Instrum. and Meth. A288(1), 227(1990).
3. K.O Mason, G.F. Bignami, A.C. Brinkman, and A. Peacock, "The ESA high throughput X-ray spectroscopy mission XMM", Milano Preprint Series in Astrophysics 138 (1994).
4. B.L. Henke, E.M. Gullikson, and J.C. Davis, "Atomic Data and Nuclear Data tables 54", 181-342 (1993).
5. D.M. Barrus, R.L. Blake, "Technique for producing ultrathin polypropylene films", Rev, Sci. Instrum., Vol. 48, No. 2.
6. K.-H. Stephan, C. Reppin, F. Haberl, M. Hirschinger, H.J. Maier, D. Frischke, M. Wedowski, P. Bulicke, G. Ulm, J. Friedrich, P. Gürtler, "Optical filters for the EPIC CCD-camera on board the XMM astronomy satellite", EUV, X-Ray, and Gamma-Ray Instrumentation for Astronomy VIII, SPIE 3114, 166-173 (1997).
7. D. Fuchs, M. Krumrey, P. Müller, F. Scholze, and G. Ulm, "Soft x-ray reflectometer for large and complex samples using synchrotron radiation", Rev. Sci. Instrum. 66, 2248(1995).
8. F. Scholze et al., "Plane grating monochromator beamline for VUV", Rev. Sci. Instrum. 65, 3229 (1994).
9. H. Hofgreve, D. Giesenberg, R.-P. Haelbich and C. Kunz, "A new VUV-refectometer for UHV-applications", Nucl. Instr. and Meth. 208, 415 (1983).
10. J. Barth, E. Tegeler, M. Krisch and R. Wolf, "Characteristics and applications of semiconductor photodiodes from the visible to the x-ray region", Proc. SPIE 733, 481 (1987).
11. M. Krumrey, E. Tegeler, J. Barth, M. Krisch, F. Schäfers and R. Wolf, "Schottky type photodiodes as detectors in the VUV and soft x-ray range", App. Opt. 27, 4336 (1988).
12. W. Jark, R.-P. Haelbich, H. Hofgreve and C. Kunz, "A new monochromator for the energy range 5 eV - 1000 eV", Nucl. Instr. and Meth. 208, 315 (1983).
13. W. Jark,and C. Kunz, "Output diagnostics of the grazing incidence plane grating monochromator BUMBLE BEE (15 - 1500 eV)", Nucl. Instr. and Meth. A 246, 320(1986).
14. K.-H. Stephan, C. Reppin, M. Hirschinger, H.J. Maier, D Frischke, D. Fuchs, P. Müller, P. Gürtler, "On the Performance of an Optical Filter for the XMM Focal Plane CCD-Camera EPIC", EUV, X-Ray, and Gamma-Ray Instrumentation , SPIE 2808, 421-437(1996).

Robust Fitting of Spectra to Splines with Variable Knots

R. L. Coldwell

Department of Physics, University of Florida, Gainesville, FL 32611 and
Constellation Technology Corporation, 7887 Brian Dairy Road, Largo, FL 33777

Spectra consist of continuum features that vary over many channels, and peaks that vary over few channels. In a fit to the continuum the peaks appear as outliers. Robust methods in which the weights associated with the peaks are reduced allow the continuum to be fitted almost independently of the peaks. This requires a very smooth background. Cubic splines, which are continuous with continuous first and second derivatives, are a good choice for this task when the locations of the discontinuities in the third derivatives, the knots, are included in the parameters optimized in the fitting process. An extension to the Marquardt method allows a Newton-Raphson method to be used in this optimization.

INTRODUCTION*

Spectra consist of continuum and peaks. It seems intuitive that the continuum should be smooth and as close to the bottom of the peaks as possible. This paper shows that this can be accomplished by fitting the continuum as the exponential of a cubic spline with adjustable knots. Most of the ability to fit the continuum comes simply from the stiffness of the splines. The addition of outlier regression, however, enables this stiffness to fit the continuum with no reference to the peaks.

The methods described here are embodied in the code RobWin being developed at Constellation Technology with windows programming by George P. Lasche, and analysis programming by the author.

The first few sections give a sketch of the mathematics of the non-linear regression methods that allow the continuum to be fitted with respect to the knot locations.

Curve fitting is tested only when the error estimates on the data points are very small. The data shown here was counted for 128 hours with distant sources and for 16 hours with close sources. In the first, an example, which easily separates into continuum and peaks, a nuclear spectrum taken with an intrinsic germanium detector, is shown. Then in the second, which tests the method presented here more severely, an ^{152}Eu source was placed relatively close to a thin HgI_2 Constellation Technology spectral-grade detector.

* This work supported in part by the Department of Defense Nuclear Treaty Programs Office through the United States Army Space and Missile Defense Command agent for the Defense Advanced Research Projects Agency and performed at the Pinellas Science, Technology, and Research Center under Grant DASG-609-610-007.

NON-LINEAR MINIMIZATION

Define

$$\chi^2 = \sum_i \left(\frac{f_A(\vec{c}, x_i) - f_i}{\varepsilon_i} \right)^2 \qquad (1)$$

In equation 1 $\vec{c}$ represents the set of constants in the fit and ε_i is the error estimate for each data point f_i. The object is to minimize χ^2 with respect to $\vec{c}$. The $\vec{c}$ such that the derivatives of χ^2 with respect to its components are all zero, starting from a $\vec{c}_0$ sufficiently close to $\vec{c}$ has components given by equation 2.

$$c_K = c_{K,0} - \sum_M \left. \frac{\partial^2 \chi^2}{\partial c_M \partial c_K} \right)_{\vec{c}_0}^{-1} \left. \frac{\partial \chi^2}{\partial c_M} \right)_{\vec{c}_0} \qquad (2)$$

To find a minimum with the Newton-Raphson method of equation 2, it is necessary that $|\vec{c} - \vec{c}_0|$ be less than the radius of convergence of a quadratic expansion of equation 1. When this is not true the quantity minimized should be $|\vec{\lambda} \bullet (\vec{c} - \vec{c}_0)|^2 + \chi^2$, where $\vec{\lambda}$ is a multidimensional version of the Marquardt parameter. For sufficiently large values of $\vec{\lambda}$, this added term dominates and the minimum is found at $\vec{c} = \vec{c}_0$, while in the limit, the Newton Raphson method is recovered. Owing to the ease with which the problem of linear dependence of the constants is solved with an extremely small value of $\vec{\lambda}$ some work with this gives the false impression in references (1) and (2) that the actual value of $\vec{\lambda}$ is unimportant. The Robmin method used in this work solves iteratively for the decrease in χ^2 that can be

CP475, *Applications of Accelerators in Research and Industry*,
edited by J. L. Duggan and I. L. Morgan

achieved in each step. It uses the difference between the vector of first derivatives and the array of second derivatives for the predicted changes in $\vec{c}$ to find the ratio of the λ_K's to each other. Finally it uses a one dimensional Newton's method to find the final $\bar{\lambda}$ for which the predicted χ^2 is as much lower than the present one as can be achieved. The method is working properly far from the minimum when every third prediction results in a lower χ^2. In the event that convergence appears to be established, but is very slow, Aitkin's extrapolation (3) is used in the early steps. At the extrema in χ^2, $|\lambda| \to 0$. The best description of the current Robmin is in reference (4), while descriptions of earlier versions are in (5) and (6). In practice, the above simply gives $\vec{c}$ such that the derivatives of χ^2 are zero. If the weighted least squares fit is carried out as given with data that varies by many orders of magnitude, there are numerous extrema in addition to the desired global minimum. The solution is to begin with large error estimates in equation 1, slowly decreasing them to the maximum of those given by the actual statistics and smaller and smaller percentage error estimates.

SPLINES

The spline is the function representing the data with the smoothest possible second derivatives. (7) (8). A spline can be defined as

$$s(x) = \sum_{i=0}^{3} a_i x^i + \sum_{i=1}^{M} c_i (\xi_i - x)_+^3$$

$$\frac{ds(x)}{dx} = \sum_{i=1}^{3} i a_i x^{i-1} - 3\sum_{i=1}^{M} c_i (\xi_i - x)_+^2 \qquad (3)$$

$$\frac{d^2 s(x)}{dx^2} = \sum_{i=2}^{3} i(i-1) a_i x^{i-2} + 6\sum_{i=1}^{M} c_i (\xi_i - x)_+$$

$$\frac{d^3 s(x)}{dx^3} = 6a_3 - 6\sum_{i=1}^{M} c_i \theta(\xi_i - x)$$

Where

$$(x)_+ = \begin{cases} x & x \ge 0 \\ 0 & x < 0 \end{cases}; \quad \theta(x) = \begin{cases} 1 & x \ge 0 \\ 0 & x < 0 \end{cases} \qquad (4)$$

With c_1=1 and ξ_1 and all others 0, this is

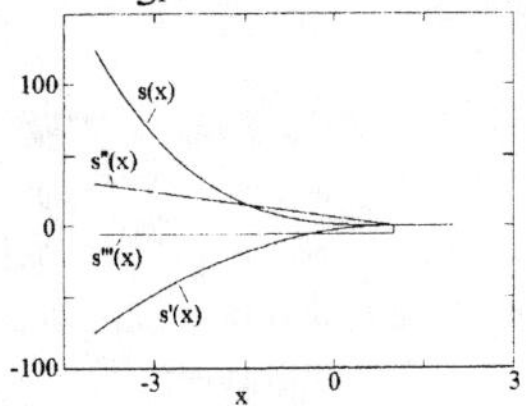

FIGURE 1 A single knot spline and its derivatives

Consider a peak like bump which is 1 at x=0, extending from -1 to 1 and is zero elsewhere, which we want to fit to a spline. This b-spline is

$$S_B(x) = \frac{3}{2}(-1-x)_+^3 - \frac{27}{2}\left(-\frac{1}{3}-x\right)_+^3 + 24(-x)_+^3 - \frac{27}{2}\left(\frac{1}{3}-x\right)_+^3 + \frac{3}{2}(1-x)_+^3 \qquad (5)$$

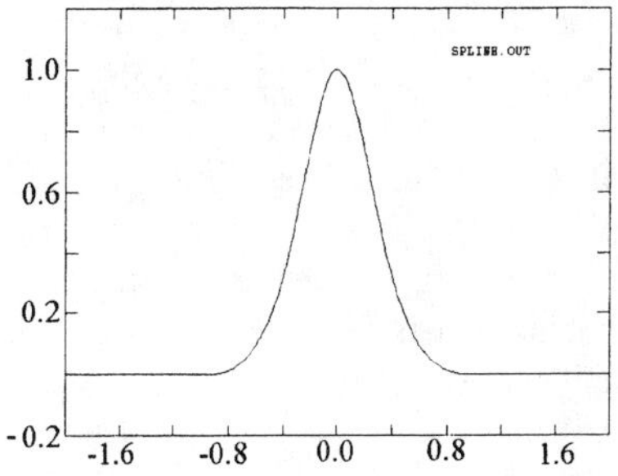

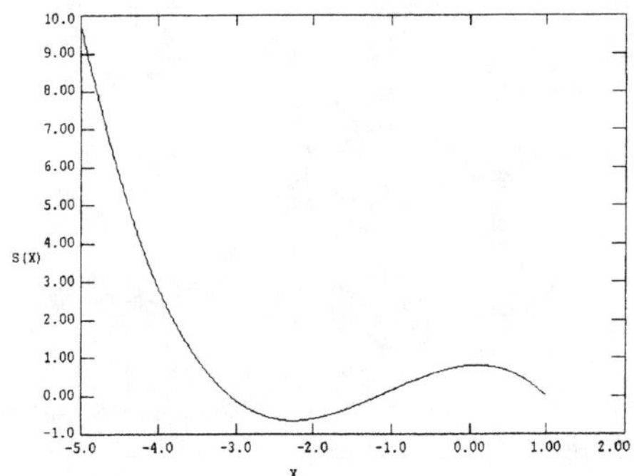

FIGURE 2 The bspline of equation 5

Five knots from -1 to +1, spanning the range of twice the fwhm of the peak are required to produce this. It is interesting to attempt to produce this result with only three knots.

FIGURE 3. Attempt to produce a bspline with 3 knots

The practical value of this is that a cubic spline will be able to reproduce a peak only if it has approximately 5 knots within a few half widths of the peak. In addition, the natural tendency of the splines defined here is to become large for small values of x. This is also the natural tendency for spectral data. The only coding needed to take advantage of this reluctance of a spline to produce a peak is the requirement that new knots be introduced at a distance from the old knots.

A positive-definite continuum is best fitted to the exponential of this spline. The hard part of fitting splines to data is the fitting of the knots. Allowing these to vary, however, significantly decreases the number of constants required to fit the continuum and more importantly eliminates the usual curve fit problem of oscillations in the fitting function.

The final detail needed to keep the background below the peaks is the outlier regression, which increases the error estimates for data above the continuum and decreases it for those below. This is described in detail in reference (6) pp. 50-54.

EXAMPLE WITH NARROW PEAKS

The following is an almost ideal example. It is the longest count time spectra from a test devised by George P. Lasche, of Constellation Technology. Kenneth W. Neufeld and Jan A. Nobel took the spectrum at Constellation Technology's Pinellas Plant just north of St. Petersburg Florida. It is composed of 16,382 channels over a nominal energy range from 0 to 3 MeV, and was counted for 128 hours. It was taken with a set of 10 certified laboratory calibration radioactive sources, each of about 10 microcuries, mounted together at a distance 20.48 meters (67.2 feet) with a 100%-efficient high-purity germanium detector.

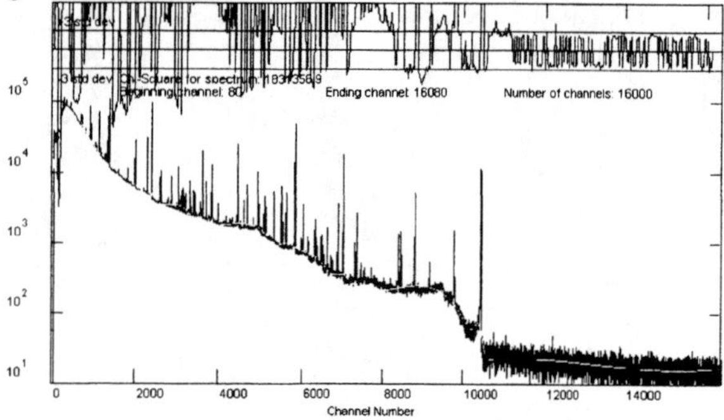

FIGURE 4. White line is a 60-constant approximation to the continuum, fitted as the exponential of a spline.

The results of simply fitting the spectra in terms of the exponential of a spline are shown in figure 4. It appears fair, but the residuals contain negative regions, and a close up of the low energy region indicates that the numerous peaks in that region have confused the spline, which fits too high. The full featured fit, whose low energy region is shown in figure 5 and a high energy region in figure 6, adds outlier suppression after half the requested number of knots have been inserted and is biased to put the first group of knots at low channel numbers. This is a fit to the spectrum with no peaks, only a continuum. After this fit, the difference between the background and the peaks is ready to be fitted as either peaks or as a nuclide response function.

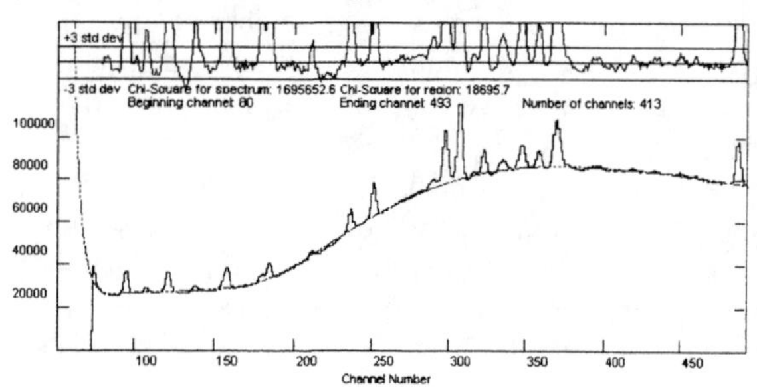

FIGURE 5. Low energy region with outlier regression.

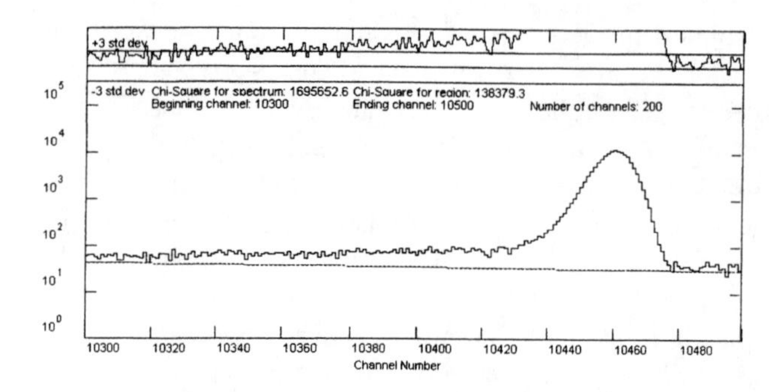

FIGURE 6. Continuum near the 2614 KeV ^{208}Tl line.

Note that the residuals in this fit are almost entirely positive owing to the utilization in the outlier regression equations of the fact that the continuum is lower than the peaks.

EXAMPLE WITH WIDE PEAKS

The source was 462 ±0.5% KBq of ^{152}Eu held at 16.8 ± 0.5 cm from crystal. The detector, an HgI_2 crystal about 1 in^2 x 2mm, was set for 0 to 3000 ± 4 KeV. Amanda M. Gerrish of Constellation Technology made the settings and took the 16-hour spectrum in March 1998.

Somewhat fewer constants need to be used on the first fit to avoid fitting peaks as part of the continuum. The result of fitting with 48 constants is shown in figure 8.

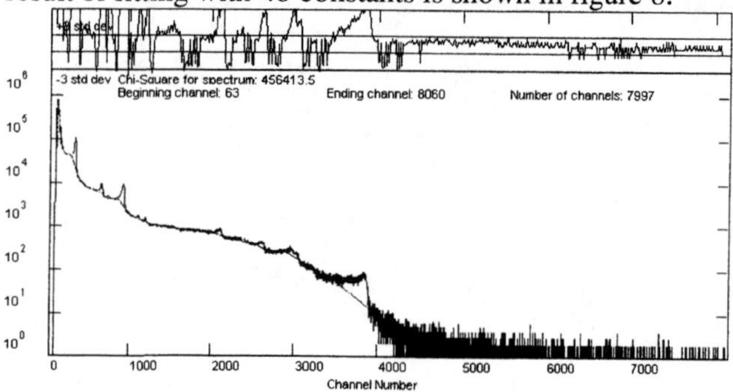

FIGURE 7. 48 constants fitted from channel 63 to 8060

It is not immediately obvious that this fit to the background allows the underlying ^{152}Eu to properly be fitted. That this is so is best illustrated by completing the fit.

The Nuclide Response Function

The code RobWin allows the user to select the isotopes to be fitted to a nuclide reference function, which is then generated from the Brookhaven Nuclear Data Library (9). The conversion from a set of lines and intensities to a response function requires an energy calibration, a full width at half-maximum fit as a function of energy, and detector efficiency as a function of energy. The code was set to generate these as fits to the data. In addition templates representing the response of the detector to single isolated peaks are needed. The code was started with Gaussian templates, then a standard template was constructed from the 1408 KeV line by fitting the difference between the data and the background, the results of which are shown in figure 8.

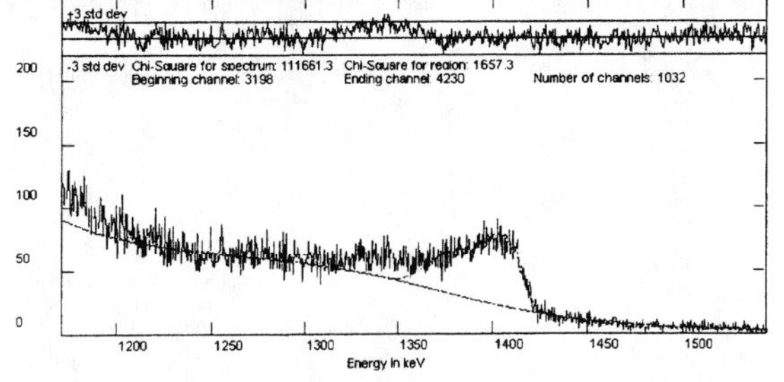

FIGURE 8. High-energy template as fitted

The 121 KeV peak was also made into a standard and a small set of unspecified peaks was allowed. The fitting proceeded by making small changes, and allowing the rest of the fit to adapt to them. At one point the continuum and the efficiency function conspired to make the continuum go low. The continuum fit was then restarted with the unspecified peaks present to help at low energies, but without the nuclides. Then the nuclide fit was re-introduced, but the continuum not allowed to vary. The final fit was at a local minimum and all constants were allowed to vary. The results near the origin and in an intermediate region are shown in figures 9 and 10.

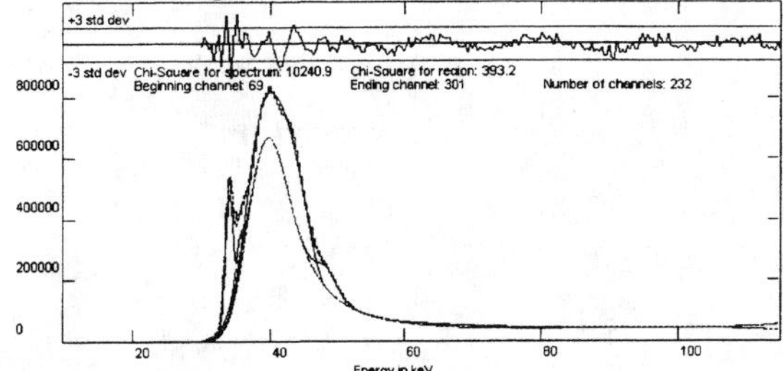

FIGURE 9. Final fit to HgI_2 spectrum at low energies.

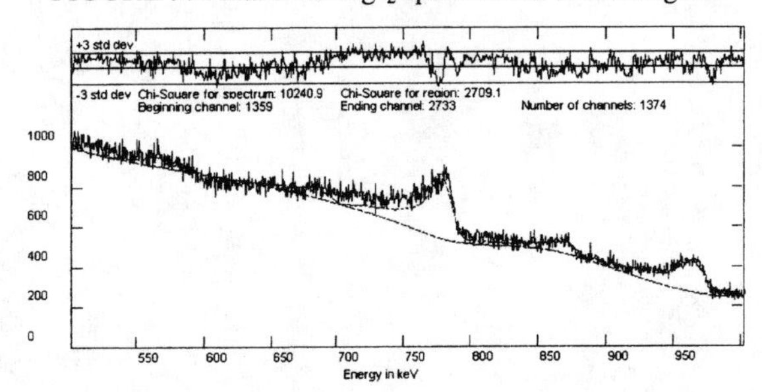

FIGURE 10. Fit in the intermediate energy region

The efficiency of the detector is in the nuclide response function as an exponential of a linear term and a 2-knot spline. The final rather innocuous looking efficiency is shown in figure 11. If it had been known in the beginning, as would be the case in future fits to this detector, the fit would have been much easier.

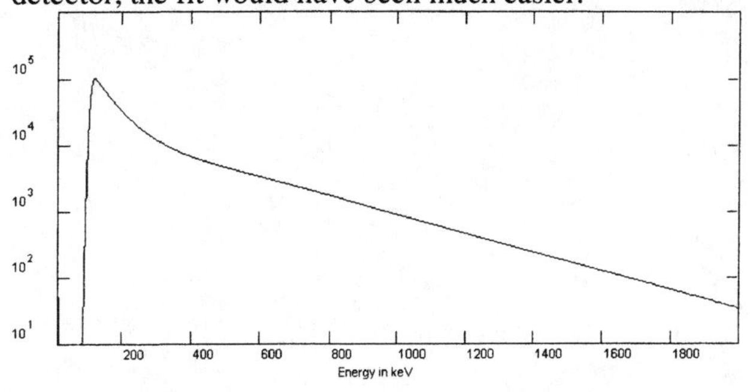

FIGURE 11. Efficiency of HgI_2 detector.

CONCLUSION

The final χ^2 of 10,240 for 7998 channels indicates that the fit is good, but not perfect. The ability to fit the continuum as the exponential of a cubic spline, however, has been demonstrated.

REFERENCES

1. Bevington, P. R., *Data Reduction and Error Analysis for the Physical Sciences,* New York, McGraw Hill (1969)
2. Press, W. H., Teukolsky, S. A., Vetterling, W. T. , Flannery, B. P., *Numerical Recipes : The Art of Scientific Computing,* Cambridge University Press, Boston 1986 p. 523
3. Hildebrand, F. B., *Introduction to Numerical Analysis,* McGraw-Hill (1956) pp. 443-445.
4. Alexander, S. A. and Coldwell, R. L. "Atomic Calculations Using Variational Monte Carlo", in *Recent Advances in Quantum Monte Carlo Methods,* ed W. A. Lester Jr., World Scientific (1977) pp. 56-58
5. Coldwell, R.L., in *Radiative Properties of Hot-Dense Matter,* Singapore, World Scientific (1983), pp. 315-349
6. Coldwell, R. L. and Bamford, G. J., *The Theory and Operation of Spectral Analysis Using Robfit,* AIP, New York (1991) pp. 50-54.
7. Hollady, J. C.,"A Smoothest Curve Approximation", Math. Tables and Aids to Computation, **11,** 233-243 (1957)
8. Ahlberg, J. H., Nilson E. N., and Walsh, J.E., *The Theory of Splines and Their Applications.* New York, Academic Press (1967), p. 3.
9. PCNUDAT Nuclear Data master database used by permission of NNDC at B.N.L.

Author Index

A

B

C

D

E

F

G

K

L

M

N

O

P

Q

R

S

T

U

V

W

X

Y

Z